5th International Congress on Energy Fluxes and Radiation Effects (EFRE 2016)

Journal of Physics: Conference Series Volume 830

Tomsk, Russia
2 – 7 October 2016

ISBN: 978-1-5108-4041-6
ISSN: 1742-6588

Printed from e-media with permission by:

Curran Associates, Inc.
57 Morehouse Lane
Red Hook, NY 12571

Some format issues inherent in the e-media version may also appear in this print version.

Printed by Curran Associates, Inc. (2017)

For permission requests, please contact the Institute of Physics
at the address below.

Institute of Physics
Dirac House, Temple Back
Bristol BS1 6BE UK

Phone: 44 1 17 929 7481
Fax: 44 1 17 920 0979

techtracking@iop.org

Additional copies of this publication are available from:

Curran Associates, Inc.
57 Morehouse Lane
Red Hook, NY 12571 USA
Phone: 845-758-0400
Fax: 845-758-2633
Email: curran@proceedings.com
Web: www.proceedings.com

Table of contents

Volume 830

5th International Congress on Energy Fluxes and Radiation Effects 2016

2–7 October 2016, Tomsk, Russian Federation

Accepted papers received: 21 March 2017
Published online: 4 May 2017

Preface

Papers

HIGH-CURRENT ELECTRONICS

MODIFICATION OF MATERIALS WITH PARTICLE BEAMS AND PLASMA FLOWS

RADIATION PHYSICS AND CHEMISTRY OF CONDENSED MATTER

IOP Conf. Series: Journal of Physics: Conf. Series **830** (2017) 011001 doi:10.1088/1742-6596/830/1/011001

Special Issue of the Materials of the V International Congress on Energy Fluxes and Radiation Effects (Tomsk, Russia, 2016)

Maksim Trigub, Georgiy Osokin, Alexander Konovod
National Research Tomsk Polytechnic University, 30, Lenin Avenue, Tomsk, Russia
email: geosokin@tpu.ru

The present issue of the journal is based on the materials of the V International Congress on Energy Fluxes and Radiation Effects 2016 (EFRE 2016) that was held on October 2 to 7, 2016 in Tomsk (Russia). This large scientific forum gathers together scientists, developers and representatives of knowledge-intensive enterprises that have relevance to physics and technology. This year, the Congress was dedicated to the 120th anniversary of the National Research Tomsk Polytechnic University.

The history of the Congress as a joint scientific event dates back to 2000; it traditionally includes three conferences: International Symposium on High-Current Electronics (SHCE), International Conference on Radiation Physics and Chemistry of Condensed Matter (RPC) and International Conference on Modification of Materials with Particle Beams and Plasma Flows (CMM). However, each of these large conferences has its own lasting history.

In 2016, the International Symposium on High-Current Electronics was arranged for the 19th time. The participants have presented the results of fundamental studies and applied outcomes in the fields of high-power pulsed energy engineering and electronics (Pulsed Power), physics and application of high-power electron and ion beams, high- and low-temperature gas discharge plasma, physics of high-energy treatment and extreme states of matter, electric pulsed technologies. Noteworthy, the symposium took place in the year of the 40th anniversary of USSR scientific discovery of explosive electron emission. This physical phenomenon—being one of the basics of high-current electronics—determines the processes in pulsed vacuum discharge, enables the operation of high-current electron accelerators, finds application in beam pumping of high-power pulsed gas lasers, switches and other electric pulsed devices.

The 17th International Conference on Radiation Physics and Chemistry of Condensed Matter is traditionally organised starting from 1969. This conference is the largest both on national and international scale in the fields of radiation physics and chemistry of inorganic substances, investigation of fundamental problems of elementary, nonlinear and catastrophic processes in condensed matter under fluxes of hard radiation.

The International Conference on Modification of Materials with Particle Beams and Plasma Flows was held for the 13th time beginning from 1988. The high level of this event is due to the participation of the representatives of outstanding scientific schools that stand at the origins of this scientific and technological field.

Regular organisation of EFRE Congress strengthens the interdisciplinary bonds between three grand fields of physics and technology playing increasingly important role in the technological development of society. Pulsed energy generation and high-current electronics provide equipment for electric discharge, plasma beam and radiation technologies that intensely fill and update the technological niches in various branches of industry, defence, chemistry, biology and medicine. The participants of the forums get comprehensive picture of scientific and applied problems and have an opportunity to look at them from the perspective of theoretical and experiment studies and engineering.

The place of the event has been chosen for a reason: Tomsk is a city having strong scientific and educational network including outstanding physical and technological schools. It was Tomsk where many of the scientific topics of the Congress germinated. Each of the forums gathers the representatives of abundant student pool of Tomsk.

The main organisers of the Congress—Tomsk Polytechnic University and Institute of High-Current Electronics of SB RAS, which soon will celebrate its 40th anniversary—have strong bonds in the field of scientific research, educational activities, management, and represent a splendid example of real integration of academic and university-level science.

In 2016, the Congress gathered leading national and international scientists, manufacturers and consumers of vacuum and electron beam equipment, accelerators and radio-frequency devices, representatives of field-specific organisations and business-structures. The 500 participants of the Congress included specialists from Russia, Republic of Belarus, Brazil, Germany, Israel, Italy, Kazakhstan, People's Republic of China, Republic of Korea, USA, Ukraine, France and Czech Republic.

The conferences were conducted with the financial support of the Russian Foundation for Basic Research (projects nos. 16–02–20530, 16–08–20408, 16–02–20464).

Peer review statement

All papers published in this volume of *Journal of Physics: Conference Series* have been peer reviewed through processes administered by the proceedings Editors. Reviews were conducted by expert referees to the professional and scientific standards expected of a proceedings journal published by IOP Publishing.

IOP Conf. Series: Journal of Physics: Conf. Series **830** (2017) 012001 doi:10.1088/1742-6596/830/1/012001

Run-away electron preionized diffuse discharge as a source of efficient laser emission in the IR, UV, VUV

A N Panchenko, N A Panchenko and V F Tarasenko

Institute of High Current Electronics SB RAS, Academichesky Av., 2/3, Tomsk, 634055

Email: alexei@loi.hcei.tsc.ru

Abstract. REP DD was suggested as an excitation source of various gas lasers. The efficient lasing was obtained in the IR, UV, and VUV spectral ranges. The ultimate intrinsic efficiency of non-chain chemical lasers on HF(DF) molecules was achieved. REP DD pumped N_2 laser with an ultimate electrical efficiency of 0.2% was developed. Lasing on N_2 molecules with 2 or 3 peaks in successive REP DD current oscillations was obtained for the first time. The laser action on F_2* at 157 nm and rare gas fluorides (KrF*, XeF*) under REP DD pumping was obtained for the first time, as well. It has been shown that the volume stage of REP DD in mixtures with fluorine can last over 50 ns during several current half-cycles. Therewith, the efficiency and the pulse duration of lasers on rare gas fluorides and VUV F_2* laser parameters under REP DD excitation are comparable with those obtained in suitable transverse discharges. The results allow the conclusion that the REP DD homogeneity in mixtures with F_2 and SF_6 is high enough for attaining high laser efficiency.

1. Introduction

It is known that in active mixtures, the main parameters of gas discharge lasers such as pulse duration, output, and efficiency are essentially dependent on the uniformity and stability of self-sustained volume discharge. Currently, it is generally recognized that the necessary conditions for volume discharge development are: the use of properly shaped electrodes having no strong edge effects, preionization from different sources (VUV radiation, x-rays, electron beam and so on) producing certain electron number density, over-voltage pulse applied to a laser gap [1–3].

Nevertheless, the formation of quite uniform discharges in various gases at high pressure and efficient lasing are possible even without any preionization, if voltage pulses with an amplitude over 100 kV and sub-ns rise-time are applied to needle or blade electrodes [4]. In this case, the volume discharge is formed due to preionization by the beam of run-away electrons. This discharge form is called the run-away electron preionized diffuse discharge (REP DD) [5].

The present paper reports the parameters of gas lasers pumped by REP DD.

2. Experimental set-up and measurement procedure

The discharge chamber described in detail in [4] was used in the experiments. The REP DD was formed by the nanosecond pulses of RADAN-220 generator between the blade-shaped electrodes of 30 cm length, which were placed at a distance of 1.8 cm [6]. In the impedance matched excitation mode, the discharge current duration was 10 ns, while in the oscillating mode, it could reach 50 ns within the half-cycle of 10 ns. The active volume was usually $V \approx 20$ cm^3.

The circuitry of RADAN generator includes a high voltage pulsed forming line (PFL) with the capacitance C=50 pF and the oil insulation, which is charged from a Tesla transformer (TT) and than is switched on a load with commercial two-electrode high-pressure nitrogen spark gap R-49.

One of the basic parameters of a gas laser is its efficiency, which in turn, is dependent on the self-breakdown voltage of P-49 spark gap. Two spark gaps with breakdown voltages U_1=240±10 kV and U_2=280±10 kV were used in the experiments. Therewith, the energy stored in the generator transmission line was not higher than E_1=1.56 J or E_2=2.1 J.

Optical cavity mirrors were fixed at the chamber ends; there was an additional window in the sidewall, to shoot the discharge and record the spontaneous emission spectra and pulses.

A flat cavity used in the experiment consisted of a mirror with the reflection coefficient R≈100%, at the laser wavelength, or an aluminum coated mirror and MgF_2 plate. Mirrors with R=20–90% were mounted behind the exit plate.

The radiation energy was measured using an OPHIR calorimeter with a PE-50BB sensor head. The discharge current was measured using a shunt. The time variations in radiation pulses at λ=193 nm were recorded using a high-speed Photek PD025 Spec.No. SPD025/S01-01 photodiode, and at λ=200–800 nm, with a FEK-22 SPU photodiode. The amplitude–time characteristics of IR radiation were detected using a FSG-23 photoresistor. Photodetectors were mounted close to the output mirrors.

A standard photocathode was replaced by a copper plate in the FEK-22 photodiode, to measure pulses in the VUV region. The modified photodetector was connected to the output plate and evacuated. This detector was tested using pulses of N_2, XeF*, and KrF* lasers. The measured KrF* laser pulse shape coincided with the pulse shape measured using the FEK-22 SPU photodiode. The radiation at λ = 337 and 351 nm was not recorded by the modified photodetector.

The visible and the UV discharge radiation spectra were measured using a HR4000 spectrometer (Ocean Optics B.V.). Spectra in the IR range were recorded using an MDR-12 monochromator equipped with a 300 lines mm^{-1} grating and FSG-23 photoresistor.

The integral glow of discharge was shot using a Sony A100 digital photo camera. A TDS3054B digital oscillograph (0.5 GHZ, 2.5 × 10^9 samples/s) was used for recording the electric signals.

3. Efficient lasing in REP DD

3.1. Non-chain HF(DF) lasers

The volume discharge in SF_6 mixtures with hydro- and deuteron-carbons can be formed in the transverse geometry without preionization [7]. However, the discharge arising in SF_6 mixtures with H_2 or D_2, under conventional excitation regimes, is unstable, and hence, the electrodes providing the uniform electrical field in the discharge gap and the intense preionization are required to improve the discharge stability [8]. Therewith, the maximal electrical efficiency of HF(DF) lasers in the transverse geometry in the mixtures with $H_2(D_2)$ was obtained using short excitation pulses (~20 ns) [9].

Figure 1 illustrates the laser parameters of non-chain chemical lasers under PEP DD excitation. In our experiments, the duration of REP DD current pulse was about 10 ns without any oscillations. The input electric power was as high as 15 MW·cm^{-3} providing a short delay time of laser pulse appearance. Laser pulses have one peak with about 75 ns duration at half-maximum followed by an exponentially decaying tail. The maximal output of HF (110 mJ) and DF lasers (75 mJ), with the peak radiation power over 1.2 MW, was obtained, similarly to [8], in the mixtures with H_2 and D_2.

The energy deposed into the laser active medium was estimated from the current and voltage waveforms, and was equal to E_{dep}=1 J. This energy deposition corresponds to the specific pumping energy of 50–65 J l^{-1}, which is optimal for a non-chain discharge HF(DF)-laser [8]. It is easily to see that the internal HF laser efficiency (relative to E_{dep}) is as high as η_{int}=10%, which corresponds to the limiting value for the discharge HF laser [8]. The obtained DF laser generation efficiency (η_{int}=7.5%)

is also close to the limiting value for the DF discharge laser (η_{int} = 8%).

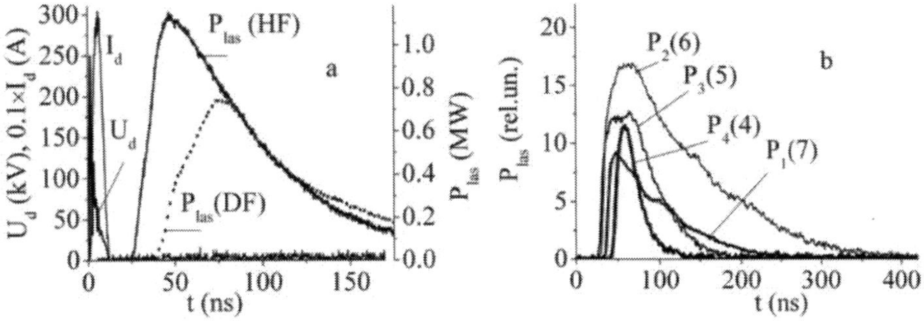

Figure 1. Waveforms of the voltage across the laser gap U_d, REP DD current I_d, and integral laser pulses on HF (P_{las} (HF) and DF P_{las} (DF) molecules (a). the power distribution P_{las} over laser lines in cascade transitions of the DF laser (b). Gas mixtures SF_6 : $H_2(D_2)$ = 8 : 1 at p=300 Torr are used, E_2=2.1 J.

The number of lasing lines attained 16 in mixtures of SF_6 with hydrogen. The radiation energy distribution over HF bands was $Q(P_1):Q(P_2):Q(P_3)$=1:0.62:0.1 and intensive cascade transitions were observed in the generation spectrum. The maximal energy was emitted at the P_1 band of HF molecules, while up to 85% of laser output was emitted in the cascades. A number of DF lasing lines attained 26, and lines of the P_2 band had the maximal intensity.

Intense cascade transitions prove the high uniformity of REP DD, which provides the high uniformity of energy deposition into an active medium. It is known that the cascade transitions do not occur in a non-uniform discharge, while the integral radiation pulse in this case exhibits the well-pronounced spike-mode character [8, 10].

The cascade transitions increase the efficiency of energy extraction from the active medium of non-chain chemical lasers, because a single excited molecule HF* (v=3) or DF* (v=4), where *v* is the vibrational quantum number, may emit up to 3–4 photons. The results allow the conclusion that the high homogeneity of a REP DD satisfies the main conditions for volume discharge development in the mixtures with SF_6. In this case, due to a high power pumping pulse, the lasing occurring on separate lines started 15–20 ns later than the discharge gap breakdown, with a jitter of 5 ns, which decreases the energy loss for the attainment of lasing threshold. These factors provide the high efficiency of REP DD pumped non-chain laser.

3.2. Nitrogen laser

For obtaining the generation on the $C^3\Pi_u \rightarrow B^3\Pi_g$ transition of nitrogen molecule (λ=337.1 nm), the value of parameter E/p in the discharge gap should be at least 100 V cm^{-1} Torr^{-1}. Under pumping by a self-sustained discharge, the voltage across the gap falls in several ns. Thus, to obtain high efficiency under such conditions, one has to design the supply circuits providing fast ($\approx$10 ns) injection of high energy ($\approx$0.1 J cm^{-3}) into the working medium, in order to reach rapidly the lasing threshold. Hence, the pulse generator used in the experiments is optimal for pumping a nitrogen laser. However, the high impedance of generator requires the high discharge resistance, which was achieved using SF_6 additions into nitrogen. The high voltage pulse amplitude allows us to form a diffuse discharge without additional preionization in N_2-SF_6 mixtures at pressures up to several atmospheres.

Two operation modes of N_2 laser under REP DD excitation were determined from the calculations and in experiments. The ultimate electric efficiency of nitrogen laser is characteristic of the first operating mode. In this case, the main part of energy, which is stored in the PFL, is deposited into the discharge plasma during 10 ns. Under REP DD pumping, the generation domain width was 0.5 cm with the uniform distribution of laser radiation power over the discharge aperture. The maximal energy of UV radiation was Q=4.2 mJ, while the peak radiation power reached 1.4 MW with E_2=2.1 J. The electrical efficiency (with respect to E_2) was as high as $\eta_0 \approx 0.2$ %. Such efficiency is close to the ultimate theoretical value [11] and to maximal efficiencies obtained experimentally for the nitrogen laser at λ=337.1 nm, pumped by the transverse discharge with UV preionization [11, 12].

The characteristic feature of the second oscillation mode of N_2 laser is two or three radiation peaks during one excitation pulse (see figure 2). This operation mode can be achieved in N_2-SF_6 mixtures at pressure below ≈100 Torr. In the case of REP DD excitation, the second and the third laser peaks appear during discharge current oscillation. It was found that the addition of He into N_2-SF_6 mixture allows to increase intensities of the second and the third laser peaks.

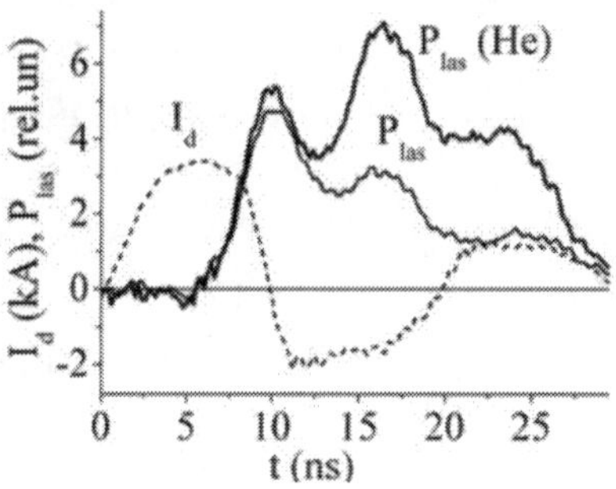

Figure 2. Waveforms of REP DD current I_d, laser pulses at 337.1 nm P_{las} in mixture N_2:SF_6 =30:30 Torr, the waveform P_{las}(He) is obtained in the He:N_2:SF_6=100:30:30 Torr mixture, E_2=2.1 J.

The REP DD in this operation mode comprises the wide diffuse streams starting for both blade electrodes, its emission spectra being typical for the volume discharge in nitrogen and consisting of the intense lines of nitrogen second positive system. It is known, that in the contracted discharge in nitrogen, the emission spectrum is dominated by broad continua, bands, and lines of the cathode-electrode material, as well as by the lines of nitrogen and oxygen atoms, and ions [13]. This fact shows the sufficient stability of REP DD in mixtures of nitrogen with SF_6, which volume stage duration in our experiments lasts longer than 50 ns. The radiation energy in the operation mode with several radiation peaks is about 1 mJ.

3.3. Excimer lasers

Gas lasers on XeF*, KrF*, ArF*, F_2* molecules are usually pumped by the transverse volume discharge with preliminary ionization of a working volume. As a rule, duration of laser radiation pulses in mixtures of rare gases with fluorine is substantially limited by a discharge arcing or its filamentation. The discharge non-uniformities usually develop in a short time (10–30 ns) after the breakdown of laser gap [14, 15] As in case of the nitrogen and non-chain chemical lasers, it leads to a need to use pumping generators forming short excitation pulses, and the set-up, on the base of the RADAN pulser, can be optimal for this laser application.

In our experiments, the maximal radiation energy was reached in mixtures with helium as a buffer gas. In mixtures with neon, the laser output decreases dramatically due to low discharge resistance and as a result, the input electric power was insufficient for attaining the efficient lasing on rare gas fluorides. Additives of argon in working gas led to REP DD transformation into a channel form.

Figure 3 shows the integrated photo of REP DD and laser spot in a typical mixture of XeF* laser. The discharge has the form of channels, which begin at the electrodes and then quickly extended to the center of the discharge gap, forming a uniform volume glow without signs of contraction. Similar discharge view was observed in the KrF* laser mixtures. The laser spot width of XeF* and KrF* lasers was 4–5 mm.

IOP Conf. Series: Journal of Physics: Conf. Series **830** (2017) 012001 doi:10.1088/1742-6596/830/1/012001

Figure 3. Integral image of REP DD and XeF* laser spot in He:Xe:F_2=3 atm:15:5 Torr mixture.

Experimental results describing XeF* laser operation under REP DD excitation are presented in figure 4. Mixtures with two fluorine donors, F_2 and NF_3, were used. In both cases, an oscillatory character of REP DD, with duration exceeding 50 ns, was evident. The current decay in mixtures with fluorine was faster, which meant a higher resistance of the volume discharge plasma in F_2 based mixtures.

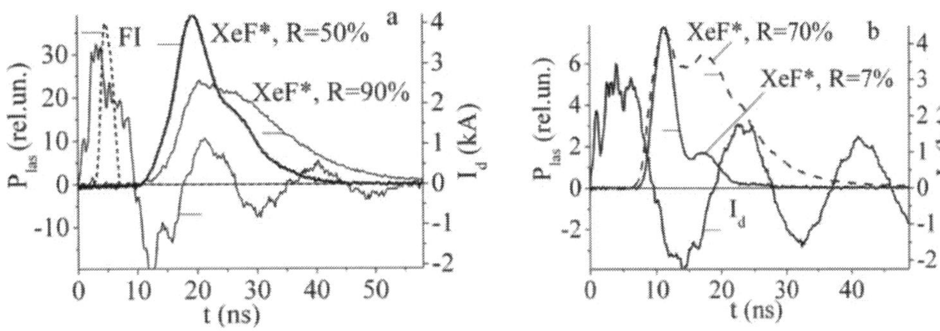

Figure 4. Waveforms of REP DD current and integral laser pulses on FI lines and XeF* molecules for different reflectance R of the output mirror, mixture He:Xe=3.5 atm:15 Torr with 5 Torr F_2 (a) and NF_3 (b) is used, E_l=1.56 J.

In mixtures with fluorine, a short generation peak on red transitions of atomic fluorine was observed within 2–3 ns after the discharge formation. The delay time of XeF* laser pulse appearance in mixtures with F_2 increases, and the energy of the radiation was 3 times lower than in mixtures with NF_3. This is due to the fluorine absorption at 351–353 nm and quenching of XeF*molecules by F_2.

Maximal output of XeF* laser was as high as 10 mJ, which corresponded to the electrical efficiency η_0 = 0.65%. The efficiency value is close to the parameters of XeF* lasers pumped by the transverse volume discharge [16]. Laser emission was continued during 2–3 current periods, and full length of laser pulse on XeF* molecules in mixtures with fluorine reached 45 ns. In mixtures with NF_3 the radiation pulse was 10 ns shorter. This means maintaining of the REP DD diffuse phase burning even after the repeated change of current direction in the laser gap.

Laser parameters obtained in He-Kr-F_2 working mixtures are presented in figure 5. As in the case of XeF* laser, a short radiation pulse on atomic fluorine transitions was observed in few ns after the REP DD formation. Lasing on KrF* molecules started within 10 ns and lasted for two current half-cycles, while total duration of the laser pulse was as long as 30 ns. This indicates the high homogeneity and stability of REP DD in the mixtures of helium and krypton with fluoride. The radiation energy increased linearly with He pressure and reached Q=20 mJ, which corresponded to the electrical efficiency η_0=1.25%.

The input energy in our experiments was estimated to be E_{in}=0.64 J. Hence, the internal efficiency of the KrF* laser under REP DD excitation can reach η_{int} > 3%, which is comparable with the parameters of the KrF* laser pumped by a transverse discharge with UV preionization [17]. At the same time, the input electric power into the REP DD plasma reaches 2 MW cm^{-3}, which is also close to the conventional modes of KrF* lasers pumping by the electric discharge [17].

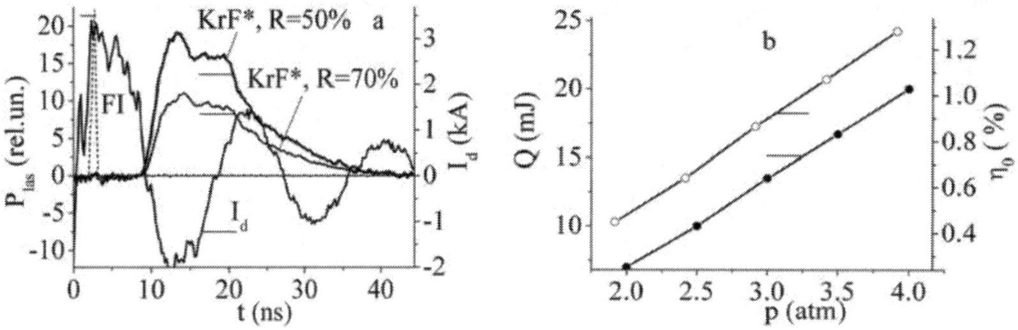

Figure 5. Characteristic waveforms of the discharge current I_d, and laser pulses P_{las} on atomic fluorine lines (FI) and at on KrF* molecules for the output mirrors with R = 50% and 70%, He Kr : F_2 = 3.5 atm: 100: 5 Torr gas mixture is used (a) and KrF* laser output Q and electric; the efficiency □$_0$, as a function of gas mixture He: Kr: F_2 = X: atm: 100: 5 Torr pressure, R = 50% (b), E_1=1.56 J.

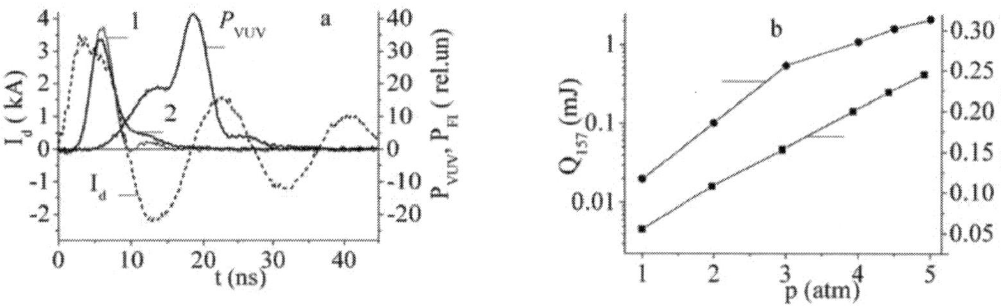

Figure 6. Waveforms of REP DD current I_d and laser radiation on FI (curves 1, 2) and F_2 (P_{VUV}) in He:F_2=3 atm:5 Torr mixture for R=7% (1, P_{VUV}) and 90% (2), (a) and laser output on red lines (Q_{FI}) and 157 nm (Q_{157}) as functions of He pressure in mixture with 5 Torr F_2, E_2=2.1 J.

A laser action on the VUV transition of F2 molecules at 157 nm was obtained in He-F2 mixtures for the first time, under REP DD excitation (see figure 6). The VUV laser pulse on F2 molecules had two distinctpeaks, and one weak peak during three half-periods of the REP DD current, lasing duration being as long as 25 ns. The VUV radiation power in the peaks increased exponentially with He pressure. The maximal VUV energy of 2 mJ was achieved at He pressure of 5 atm. Therewith, the electrical efficiency of F_2 laser was as high as 0.1%, which, similarly to XeF* and KrF* lasers, is comparable with the parameters of F_2 lasers pumped by the transverse discharge with preionization [18–19]. Simultaneously with VUV emission, a laser action on atomic fluorine with the maximal output of 0.25 mJ was observed.

4. Conclusion

Laser radiation parameters in mixtures of rare gases, H_2 (D_2) and nitrogen, with fluorine containing molecules, pumped by the REP DD are studied. It is shown that the stable volume discharge of long duration can be formed in high-pressure mixtures with F_2, NF_3 and SF_6, in compact systems formed by the blade electrodes without additional illumination source.

Promising prospects of REP DD application for exciting series of gas lasers are demonstrated. A laser action on molecules N_2, HF, and DF, with ultimate efficiency, was obtained.

Novel operation modes of N_2 laser with two and three radiation peaks was demonstrated under pumping by the oscillating run-away electron preionized discharge.

A laser action on F_2*, KrF* and XeF* was obtained in REP DD for the first time. It was shown that volume stage of REP DD in mixtures with fluorine could last >50 ns during several current half-cycles. Therewith, the efficiency and pulse duration of lasers on rare gas fluorides and VUV F_2* laser parameters under REP DD excitation are comparable with the laser characteristics obtained in a convenient transverse volume discharges with preionization.

Acknowledgement
This work was made under the support of the Russian Foundation for Basic Research, project No. 14-08-00074.

References

[1] Palmer A J 1974 *Appl. Phys. Lett.* **25** 138
[2] Levatter J I., Lin S C 1980 *J.Appl. Phys.* **51** 210
[3] Stappaerts E A 1982 *Appl. Phys. Lett.* **40** 1018
[4] Vil'tovskii P O, Lomaev M I, Panchenko A N, Panchenko N A, Rybka D V, and Tarasenko V F 2013 *Quant. Electron.* **43**, 60
[5] Baksht E H, Burachenko A G, Kostyrya I D Lomaev M I, Rybka D V, Shulepov M A and Tarasenko V F 2009 *J. Phys. D.* **42**, 185201
[6] Mesyats G A, Korovin S D, Rostov V V, Shpak V G and Yalandin M I 2004 *Proceedings of the IEEE* **92** 1166
[7] Apollonov, V V, Belevtsev A A, Kazantsev S Yu, Saifulin A V Firsov and K N 2000 *Quantum Electronics* **30** 207.
[8] Panchenko A N, Orlovsky V M and Tarasenko V F 2004. *Quantum Electronics* **34** 320
[9] Midorikawa K, Sumida S, Sato Y, Obara M, Fujioka T 1979 *IEEE Journal of Quanum Electronics*. **15** 190
[10] Barano, V Yu, Vysikailo F I, Dem'yanov A V, Malyuta D D, Tolstov V F 1984 *Soviet Journal of Quantum Electronics*. **14** 791
[11] Tarasenko V F. 2001 *Quantum Electronics* **31** 489
[12] Iwasaki C, and Jitsuno T 1982 *IEEE Journal of Quantum Electronics* **18** 423
[13] Tao S, Tarasenko V F, Zhang C, Lomaev M I, Sorokin D A, Yan P, Kozyrev A V and Baksht E Kh 2012 *J. Appl. Phys.* **111** 023304
[14] Kuscner M J 1991 *IEEE Transactions on Plasma Science* **19** 387
[15] Mathew D, Bastiaens H M J, Boller K J and Peters P J M 2006 *Applied Physics Letters* **88** 101502
[16] Kumagai H and Obara M 1989 *IEEE Journal of Quantum Electronics* **25** 1874
[17] Borisov V M, Bragin I E, Vinokhodov A Yu, Vodchits V A 1995 *Quantum Electronics* **25** 507
[18] Ishchenko V N, Kochubei S A, Razhev A M 1986 *Soviet Journal of Quantum Electronics* **16** 707
[19] Kitamura T, Arita Y, Maeda K, Takasaki M, Nakamura K, Fujiwara Y, and Horiguchi S 1997 *Journal of Applied Physics* **81** 2523

The effect of angular divergence and space charge on transmission of an electron beam through a magnetic mirror

V T Astrelin[1,2], I V Kandaurov[1], V V Kurkuchekov [1,2], V M Sveshnikov [2,3] and Yu A Trunev [1]

[1] Budker Institute of Nuclear Physics SB RAS, Novosibirsk, Russia
[2] Novosibirsk State University, Novosibirsk, Russia
[3] Institute of Computational Mathematics and Mathematical Geophysics SB RAS, Novosibirsk, Russia

E-mail: V.T.Astrelin@inp.nsk.su

Abstract. An experiment on generation and transmission of an intensive electron beam through a magnetic mirror has been carried out at the GOL-3 magnetic trap. An adapted theory and numerical modelling were used to find a maximal beam current for which no beam electrons are reflected. The effect of space charge and magnetic mirror of the reflection were taken into account. A comparison of the computed limiting current with experimental value allowed us to validate a hypothesis that the appearance of reflected electrons is responsible for electric breakdown in the electron gun.

1. Introduction

Early experiments performed before 2014 on the open magnetic trap GOL-3 at the Budker Institute of Nuclear Physics SB RAS had been carried out using electron beam generation in a multi-aperture diode with a plasma emitter (Fig.1). The beam with small angular spread is formed in the moderate magnetic field B_0~10 mT, transmits through a liner and is injected into the GOL-3 trap along the magnetic field which increases up to 1 T inside the trap.

The experiments show that the beam duration (0.1-0.5 ms) is limited by the electric breakdown in the diode. The breakdown can be caused by the electrons reflected by the enhanced magnetic field at the entrance mirror of the trap. These electrons, falling onto the grounded anode, can generate a dense plasma on its surface. This plasma then penetrates into the diode and breaks its operation. The paper presents the analytical and the numerical investigation of the transmission and reflection of the electrons in a convergent magnetic field. The study takes into account a space charge of the beam, its azimuthal magnetic field, an initial angular spread of the beam electrons, and a radial distribution of the beam current density.

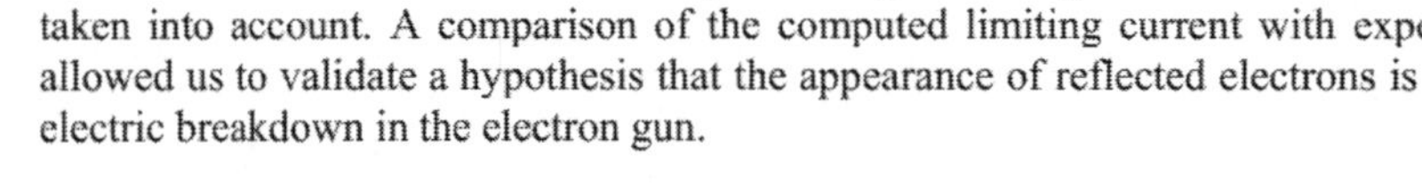

Figure 1. Scheme of the GOL-3 experiment

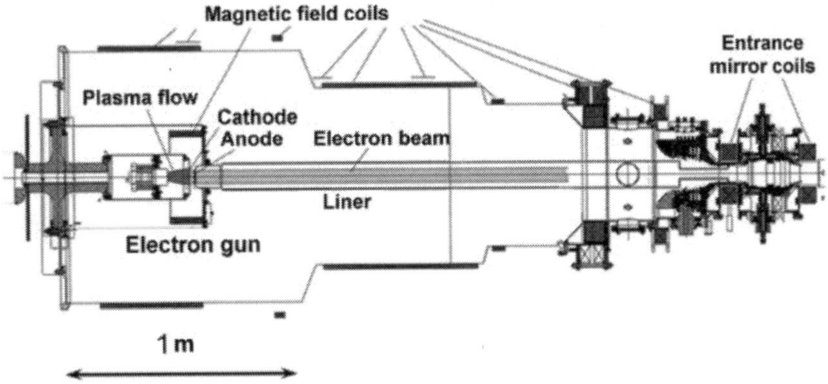

Figure 2. Layout of the electron gun in the tank of the GOL-3 trap

2. Experiment

The experiment is performed on the multi-mirror trap GOL-3 [1]. The trap assembles a 12 m length stainless steel pipe, installed inside a solenoid with the corrugated magnetic field (B_{max}/B_{min}~ 1.7 T/1.2 T). The pipe is attached to a vacuum tank with a lower magnetic field. The source of the electron beam [2] is placed (see Fig.1 and 2) inside the tank in a region with almost homogeneous magnetic field $B_0 \sim 0.01$ T. The electron gun can generate a multi-jet beam with the energy up to 100 keV, the current up to 100 A, the initial diameter of ~ 5 cm, and the calculated divergence of electrons velocity of not higher than 0.05 rad. The beam passes through the apertures in a flat grounded anode and then moves through a liner of ~ 10 cm diameter. Beyond the liner, the beam enters the trap through a rising magnetic field at the entrance mirror with a maximal field $B \sim 1$ T.

Reduction of the beam pulse down to ~80-100 μs is observed in the experiments. It occurs as the beam current increases up to ~85 A with the beam energy of ~ 90 keV. To prove the effect of the electron reflection in the transmitting channel on the breakdown of the diode, we need to find theoretically the conditions for the occurrence of reflection and compare these conditions with the experimental ones. Below, we use analytical estimates and a numerical simulation to figure out these conditions.

3. Estimates of angles and beam current limit

Increase of the beam current injected into the trap is limited by the increase of the angular divergence in a rising magnetic field and a possible increase of the negative potential on the beam axis as the electron space charge increases. We evaluate the initial pitch angles of electrons at the exit of the source. The azimuthal magnetic field is $B_{\varphi 0} \sim 8\cdot10^{-4}$ T on the periphery of a homogeneous beam with the total current $I_b \sim 100$ A and $R_{b0} \sim 2.3$ cm radius. This field determines the maximal pitch angle $\theta_B \leq B_{\varphi 0}/B_0 \sim 0.08$ rad for electrons pushed along the axis through the anode apertures. Inside the beam, the pitch angles are proportional to radius. In addition, the angles acquired by the electrons inside the diode ($\theta_D \leq 0.05$ rad) due to an imperfect optic, may be added.

At the beginning of the beam path inside the liner, the beam angle spread can be affected by the electric field of the beam. Assuming that the radial electric field $E \leq 2I_b/(V_z R_{b0})$ accelerates the electrons during the time interval $\tau \sim \pi\gamma mc/2eB$, we obtain $V_\perp/V_z \sim \pi I_b c/(2\gamma B R_{b0}\cdot eU/m) \sim 0.3$ for the peripheral electrons. Corresponding pitch angles inside the beam ($\theta_E \leq 0.3$ rad) are approximately proportional to the radius. Therefore, the electrons at the beam periphery are reflected by the magnetic field first prior to the electrons inside the beam.

The magnetic force lines have a helical shape with the step $h_0 \sim 2\pi R_{b0} \cdot B_0/B_{\varphi 0} \sim 200$ cm along the transmission channel. A step of the Larmour orbit helix for an electron with the energy eU ~100 keV is $z_h \sim 67$ cm, i.e., $h_0 \sim 3\ z_h$. Furthermore, due to irregular distribution of the magnet coils along the channel (see Fig.2), the magnetic field is significantly heterogeneous at the distance comparable with the step of the helix. It means that the motion of the electrons is not strictly regular and adiabatic, so that the changes of the pitch angles along the length of the system can be estimated only approximately.

To estimate the space-charge-limited current of the beam, we use a formula derived for a beam with a radially uniform current density and the pitch angle distribution of the electrons [3]. When deriving this formula, the authors of [3] assumed that the electron gyroradius is significantly smaller than the diameter of the beam, and the magnetic field is uniform. As the result, the derived beam-current limit is

$$I_{\max} = -\frac{\mathrm{mc}^3\left(\gamma_0^{2/3} - \gamma_{0\perp}^{2/3}\right)^{3/2}}{|e|\left(1 + 2\ln\left(R_c / R_b\right)\right)}, \tag{1}$$

where $\gamma_{0\perp} = \left(1 - v_{0\perp}^2 / c^2\right)^{-1/2} = \left(1 - \sin^2\theta_0 \cdot \left(1 - 1/\gamma_0^2\right)\right)^{-1/2}$ is the relativistic factor evaluated on the initial transverse velocity $v_{0\perp}$, $\gamma_0 = \gamma_{0\perp} + |eU_0| / \mathrm{mc}^2$, U_0 the potential of the cathode, θ_0 the pitch angle of electrons at the exit of the source, and $\sin\theta_0 = v_{0\perp} / v_0$, $v_0 = c\left(1 - 1/\gamma_0^2\right)^{-1/2}$. The notations R_c and R_b are, respectively, the radii of the transmission channel and the beam at the magnetic mirror, where the maximal value of a geometry factor is $\ln(R_c/R_b) \sim 2.5$.

To find a limiting value of the beam current in the convergent magnetic field, we take into account a dependence of the transversal relativistic factor $\gamma_\perp$ for the beam electrons on an arbitrary magnetic field B. Assuming a conservation of the adiabatic invariant, $\mu = p_\perp^2 / 2B = \text{constant}$ (and consequently, $v_\perp \gamma / \sqrt{B} = v_{0\perp}\gamma_{0\perp} / \sqrt{B_0}$), we can write

$$\gamma^{-2} \equiv 1 - \left(v_\parallel^2 + v_\perp^2\right)/c^2 = 1 - \left(v_\parallel^2 + v_{0\perp}^2 \cdot B / B_0 \cdot \gamma_{0\perp}^2 / \gamma^2\right)/c^2, \tag{2}$$

where γ, $v_\perp$, $v_\parallel$ θ and B are the beam characteristics for any point of the transmitting path. Hence, it follows that the longitudinal velocity depends on the ratio B/B_0, $\gamma_{0\perp}$ and total energy γ as

$$v_\parallel^2 = c^2\left[1 - \frac{\gamma_{0\perp}^2}{\gamma^2}\left(\frac{1}{\gamma_{0\perp}^2}\left(1 - \frac{B}{B_0}\right) + \frac{B}{B_0}\right)\right]. \tag{3}$$

In case of the homogeneous magnetic field $B = B_0$, Eq. (3) becomes $v_\parallel^2 = c^2\left(1 - \left(\gamma_{0\perp}/\gamma\right)^2\right)$, as in Ref. [3]. Hence, following [3], we obtain a modified expression for the limiting beam current by making a substitution of $\gamma_{0\perp}^2$ in Eq. (1) with the combination from the Eq. (3): $\left\{(1 - B/B_0) + \gamma_{0\perp}^2 B/B_0\right\}$, which is valid for the case of barely convergent magnetic field:

$$I_{\max} = -\frac{\mathrm{mc}^3\left(\gamma_0^{2/3} - \left\{(1 - B/B_0) + \gamma_{0\perp}^2 B/B_0\right\}^{1/3}\right)^{3/2}}{|e|\left(1 + 2\ln\left(R_c / R_b\right)\right)}. \tag{4}$$

For the comparison of the limiting current with the experimental value, we express the current through the initial pitch angle of electrons at the anode, where $B \approx B_0$:

$$\sin\theta_0 = p_{0\perp} / p = \sqrt{\left(\gamma_{0\perp}^2 - 1\right)/\left(\gamma_0^2 - 1\right)}. \tag{5}$$

Here $p_{0\perp}$ is the initial transversal component of the total electron momentum p. The function $I_{max}(\theta_0)$ is drawn in Fig.3 for three values of the ratio $K = B/B_0$. As is seen from Fig.3, the limiting current strongly depends on the value of the initial pitch angle for a large K. If the beam current exceeds I_{max},

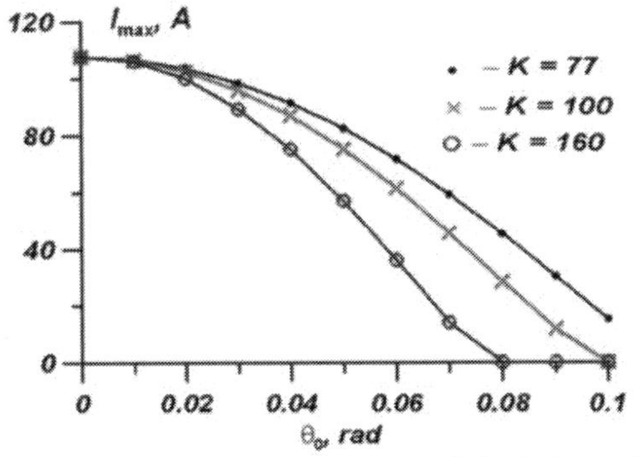

Figure 3. Dependence of current limit on initial pitch angle of electrons

the beam electrons are reflected by the enhanced magnetic field at the mirror, as well as by the space charge. As $\theta_0 \to K^{-1/2}$, the current drops to zero. Evidently, this case corresponds to a "pure" reflection of all the beam electrons by the converging magnetic field, even if the effect of space charge is neglected.

The case described by Eq. (4) can occur in an experiment with injection of the beam into the gas at the beginning of the beam pulse. Later, due to ionization of the gas by the beam electrons in the channel of transmission, the plasma is produced; it can neutralize the space charge of the beam.

The above rough estimate shows that the dynamics of the beam and its current limit significantly depend on the initial angular spread of electrons and the degree of the charge neutralization. To obtain a more accurate picture, numerical simulation is needed.

4. Numerical model

Numerical simulations are performed using the code ERA [4]. It implements a second-order finite-difference scheme for computing trajectories of relativistic electrons. It also computes intrinsic magnetic field of the beam using the algorithm described in [5].

We assume that the electrons are injected along the axis into a complex computational domain from the side of a flat anode with an initial velocity determined by the diode voltage. The anode and the walls of the transmission channel are assumed to be grounded. The geometry of the system is shown in Fig.2. A multi-mirror trap, which is not shown here, is located at the end of the transmission channel behind the second mirror coil. The boundary condition under electric field $E_z = 0$ parallel to the system axis is put on the output butt-end of the computational domain.

The electron beam is emitted from an area with the diameter of 4.6 cm at the input end of the computational domain. At the first run of simulation, a uniform beam with the initial energy of 90 keV is assumed. An initial angular velocity spread associated with the imperfection of diode optics is neglected since we intend to investigate the effect of intrinsic fields on the beam transmission. At the next runs, the variants of the beam transmission with a full and separate influence of its electric and magnetic fields are considered.

5. Results of numerical simulation

A series of calculations with increasing beam currents are done with the aim of finding the current limit of the beam. Calculations are performed for the left part of the channel (see Fig. 4) with the length 263 cm where the magnetic field is compressed in $K \sim 100$ times. The result of the simulation is shown in Fig. 4 for the beam with the current 50 A, the space charge being taken into account. The figure shows the profile of the magnetic field on the axis of the beam, pitch angles, and electron trajectories of the beam. Under these conditions, all the electrons penetrate the trap without reflection. Note that the scales of the coordinate axes in the bottom drawing are different.

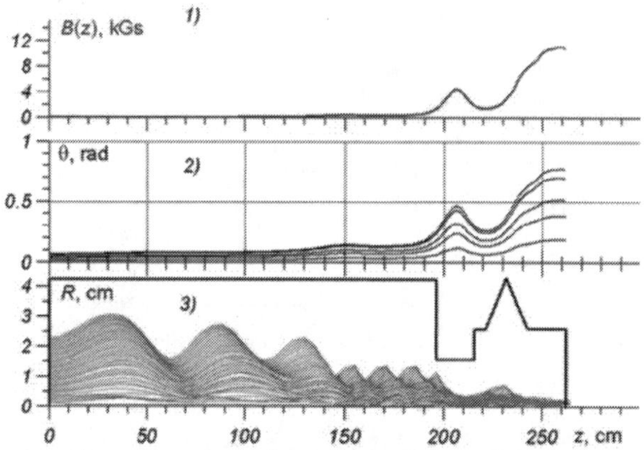

Figure 4. Beam current $I = 50$ A; Electric and magnetic fields of the beam were taken into account; 1) - Magnetic field, 2) - Pitch angles, 3) - Electron trajectories

As the current increases up to 60A and above, the pitch angles rise and at $I_b \sim 70$A, the reflection of the electrons from the second mirror of the trap at Z ≈ 260 cm occurs (see Fig. 5). The axial distribution of the pitch angles for few trajectories (every 10th of 50) shows that they rise with radius and are maximal on the periphery of the beam (Fig.5(1)). Accordingly, the reflection of the electrons starts at the periphery of the beam. In Fig. 5(2), only 5 external trajectories are reflected, they correspond to ~ 20% of the beam current. Presumably, the amount of energy released on the anode ~ 30 kW/cm^2 is sufficient for plasma generation leading to the breakdown of the diode. We find, however, that the beam current in the experiment is larger as compared to the computed one. This means that the beam charge is partially neutralized by the plasma. The neutralization reduces the space charge and radial electric field, leading to the rise of pitch angles. Thus, the neutralization increases the critical current, at which the electron reflection begins.

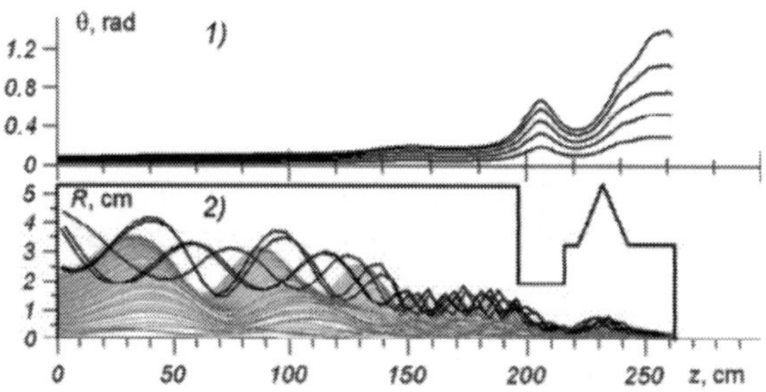

Figure 5. Beam current $I = 70$ A; Electric and magnetic fields of the beam were taken into account; 1) Pitch angles, 2) Electron trajectories; dark trajectories are reflected

To test the effect of the beam space charge, the following calculations are carried out with a completely compensated space charge. The magnetic field of the beam current is accounted for in our numerical model. For these conditions, the beam with the current up to 70 A passes into the trap without reflection. With the growth of the current, the pitch angles increase almost proportionally to the current. In the beam with the current of 90 A, the electrons on the periphery of the beam containing ~ 30% or more of the total current, are reflected by the magnetic field. In the experiment, the appearance of reflected electrons presumably leads to producing the plasma on the surface of the anode and inside the transmission channel, which can bring to the breakdown of the diode.

In addition, numerical simulation of the beam with Gaussian radial distribution of the current

$$j_0(r) = \frac{I_b}{\pi r_{b0}^2 \left(1 - \exp\left(-R_{b0}^2 / r_{b0}^2\right)\right)} \cdot \exp\left(-r^2 / r_{b0}^2\right), \tag{6}$$

where $R_{b0} = r_{b0} = 2.3$ cm, with initial pitch angles $\theta_0 = 0, \pm 0.025, \pm 0.05$ rad, and the charge neutralization is made. In this case, the reflection of the electrons begins at the beam current 80A. Returned current comprised, approximately, 6% of the beam current (8 trajectories from 250). The result of the simulation is drawn in Fig.6.

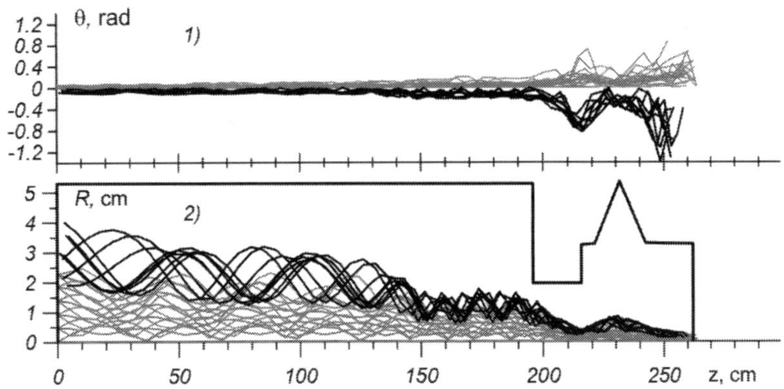

Figure 6. Beam current $I = 80$ A; Electric field of the beam **was not taken** into account; 1) Pitch angles, 2) Electron trajectories (only 15 of 250 are drawn in brown), all reflected trajectories are drawn in black.

6. Conclusion

A comparison of the numerical simulation results and the theory of the limiting beam current with the experimental data has revealed a few statements.

The magnitude of the beam current, at which the reflection of the beam electron from the magnetic mirror begins, in the numerical simulation approximately meets the experimentally measured current, at which the breakdown in the diode occurs. Some difference between these currents indicates a partial or a complete compensation of the beam space charge inside the plasma produced by the ionization of the residual gas by the beam electrons.

Acknowledgements

The theoretical part and computer simulation of this work were supported by the Russian Science Foundation (project N 14-50-00080). The experimental part was supported by RFBR Grant No. 16-08-00785. Authors are thankful to Dr. S. Sinitsky for helpful discussions and to Prof. I. Kotelnikov for the help in translation of the text.

References

[1] Burdakov A, Arzhannikov A, Astrelin V, Batkin V, Beklemishev A, Burmasov V, Derevjankin G, Ivanenko V, Ivanov I, Ivantsivskiy M, et al. 2009 *Fusion Science and Technology* **55** 2 63

[2] Kurkuchekov V V, Astrelin V T, Avrorov A P, Burdakov A V,. Bykov P V, Davydenko V I, Derevyankin G E, Ivanov A A, Kandaurov I V, Rovenskikh A F, et al. 2012 *Abstracts of 9th Int. Conf. on Open Magnetic Systems for Plasma Confinement* Novel Injector of Intense Long Pulse Electron Beam for Linear Plasma Devices (Tsukuba) http://www.prc.tsukuba.ac.jp/OS2012/abstract-download/

[3] Sinitsky S L and Arzhannikov A V 2012 *Pulse Power Beams* ed E P Voytenko (Novosibirsk: Novosibirsk State University) p 140 (in Russian)

[4] Sveshnikov V M 2006 *Numerical simulation of intense beams of charged particles: Thesis for*

the degree of doctor of physical and mathematical sciences (Novosibirsk) p 333 (in Russian)
[5] Astrelin V T and Sveshnikov V M 1979 *Rus. J. of Appl. Mekhanics and Technical Physics* V **20**, Issue 3, pp 261–265 (in Russian)

Amplitude-time characteristics of runaway electron beams during the breakdown phase in high-pressure gases

V F Tarasenko

Institute of High Current Electronics SB RAS, 2/3 Akademichesky, Tomsk, 634055, Russia

E-mail: VFT@loi.hcei.tsc.ru

Abstract. Results of experimental studies of the amplitude-temporal characteristics of a supershort avalanche electron beam (SAEB) with a picosecond time resolution are presented. It is shown that the maximum SAEB current and the voltage drop in the gap are timed to tens of picoseconds and that the use of sharp-ended cathode improves the stability of SAEB generation.

1. Introduction

The generation of runaway electrons in high-pressure gases is a fundamental physical phenomenon. Now, many papers dealing with the runaway electron beams and X-rays, which are produced at 1 atm pressure and higher are available. In the last ten years, the greatest progress was reached in this field due to the development of high-voltage pulses and optimum gas diode designs, subnanosecond and picosecond current sensors, and high-resolution real-time oscilloscopes. There are also theoretical models of the processes occurring in gas diodes within several and split picoseconds. The latest results of experimental and theoretical research in runaway electrons and in diffuse discharges are summarized elsewhere [1, 2]. Analysis of these results shows that the parameters of runaway electron beams obtained by researchers differ greatly. It seems to be a consequence of different experimental and measurement conditions, as well as due to the limited number of studies, in which the oscilloscopes operating at ~30 GHz and 100 GS/s are used. We know about only several studies, in which the time resolution was close to the maximum one of those available today. They are applied by the Institute of High Current Electronics, Tomsk [3, 4], and the Institute of Electrophysics, Ekaterinburg [5].

Here, we report the experimental study of a picosecond resolution allowing further researches in the field of runaway electron beams generated in subnanosecond breakdowns in the air of an atmospheric pressure. In the study, we measured the amplitude-time characteristics of runaway electron beams and the breakdown voltage, as well as followed the time correlation between the maximum supershort avalanche electron beam current and the voltage drop across the gap.

2. Experimental setup

The setup shown in figure 1 was used in our experiments. It assembled a SLEP-150 pulser, a gas diode, and a measuring system. The collector receiving part was 3 mm in diameter. One of the electrodes of high-voltage line *1* was the peaking switch *2*, allowing us to decrease the line length and to form a voltage pulse of about ≈ 1 ns FWHM at a matched load. The voltage rise time was determined by the peaking switch and was about ≈ 250 ps at a level of 0.1–0.9 of the switch optimum operation. The gas diode *5* was connected to a high-voltage line *1* via a short transmission line *3* with a wave impedance of 100 Ohm. The voltage amplitude in the transmission line depended on the breakdown voltage of peaking switch *2* and could range from 120 to 200 kV. In our experiments, the breakdown voltage of the switch was ~180 kV.

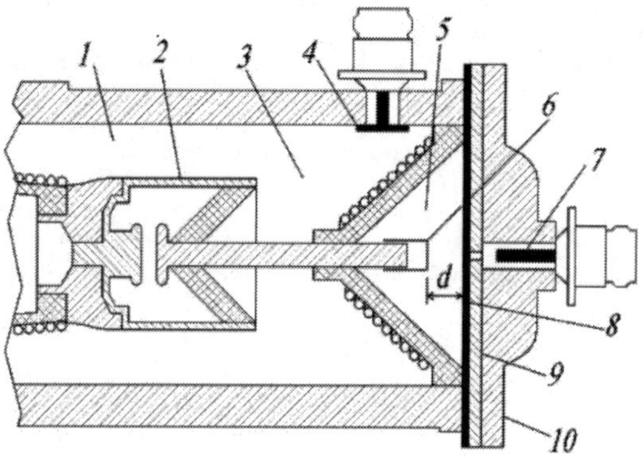

Figure 1. Schematic of the output of the SLEP-150 pulser, gas diode, and collector with the 3-mm-diameter receiving area: (1) high-voltage line of the pulser, (2) peaking spark gap, (3) transmission line, (4) capacitive voltage divider, (5) gas diode, (6) tubular cathode, (7) receiver part of the collector, (8) foil, (9) 5-mm-thick collimator with a 1-mm-diameter hole, and (10) collector case.

The parameters of a SAEB were measured with two cathodes, one of which was a stainless steel foil tube of 6 mm diameter and 100 μm thickness (cathode 1). Another one was a stainless steel ball of 9.5 mm diameter (cathode 2). The anode of gas diode was an aluminum foil *8* of 10 μm thickness, which was reinforced with a grid or a collimator *9* from the collector side. In a series of experiments, the foil was removed, and the collimator served as the anode. Thus, the SAEB was measured through 1-mm collimator hole without any attenuation by aluminum foil. The time resolution of collector reached 20 ps [4].

The measuring equipment also included a capacitive voltage divider and a current shunt. The signals from the divider, shunt, and collector were transmitted to a LeCroy WaveMaster 830Zi-A real-time digital oscilloscope (the bandwidth 30 GHz, the sampling increment 12.5 ps) via RG58-A/U high-frequency cables (Radiolab) of 1 m length with N-type (Suhner 11 N-50-3-28/133 NE) and SMA-type connectors (Radiall R125.075.000). To register the signal of voltage attenuation, we used 142-NM high-frequency attenuators (Barth Electronics) with a bandwidth of up to 30 GHz. The signal from the collector was transmitted to the oscilloscope without attenuators, either with or without anode foil. The voltage and the SAEB current were measured simultaneously in each pulse. The timing accuracy of SAEB and voltage pulses were not worse than 10 ps. The SAEB generation time with respect to the voltage pulse was determined from the capacitive current fed to the collector. For this purpose, the collimator was removed, and the foil was replaced by a grid of 64 % transparency. This procedure is described in details elsewhere [6]; the voltage timing accuracy and SAEB pulses were not worse than 50 ps. The SAEB and X-rays were detected using the blackening of RF-3 film, in a black paper envelope of 120 μm thickness placed downstream to Al foil anode. The discharge plasma glow in the gap was photographed using Sony A100 camera.

3. Results and discussion

The nanosecond voltage pulse applied to the gap with the sharp-ended cathode, in an inhomogeneous electric field, gives rise to a discharge, in which runaway electrons are generated [1, 2]. The discharge form depends on the gap width *d*. At large gap widths *d* or small voltage amplitudes of the sharp-ended cathode, a corona discharge is formed. As far as the gap width d decreases, the corona discharge during the voltage pulse life transforms to a diffuse discharge, the plasma of which fills the entire gap. This discharge mode is termed a runaway electron preionized diffuse discharge or REP DD [7] and is described in detail elsewhere [1]. The glow of diffuse discharge is much brighter than that of the corona. As far as d is further decreasing, we observed the sequence of three discharge forms in the gap: a corona discharge (detected by high-resolution CCD camera), a REP DD, and a spark, into which the REP DD was transformed during the voltage pulse. The duration of diffuse and spark discharges as well as the energy deposited in the gas by every discharge depends on the gap width, cathode design, air pressure, and voltage pulse parameters. In the course of the REP DD formation, an ionization wave starts from the sharp-ended electrode. When the wave front reaches the opposite electrode, the runaway electron beam of the highest amplitude is detected downstream of the thin foil anode.

In our experiments, the SAEB parameters were studied when the cathode 1 was at an interelectrode gap of 4–35 mm and when the cathode 2 was at the gap of 4. 6 and 8.5 mm. The imprint left on the film by the electron beam downstream to the anode foil without the collimator was 54 mm in diameter, i.e. covering the entire foil surface. The maximum film exposure was observed when we pplied the tubular cathode 1 at a gap of 12 mm or less, near the sharp end of the cathode. Near the central axis the exposure decreased. The decrease of film exposure was more noticeable at small interelectrode gaps and when the distance from the central region of the foil increased.

Waveforms of the voltage and SAEB current for the cathode 1 at d = 12 mm are presented in figure 2.

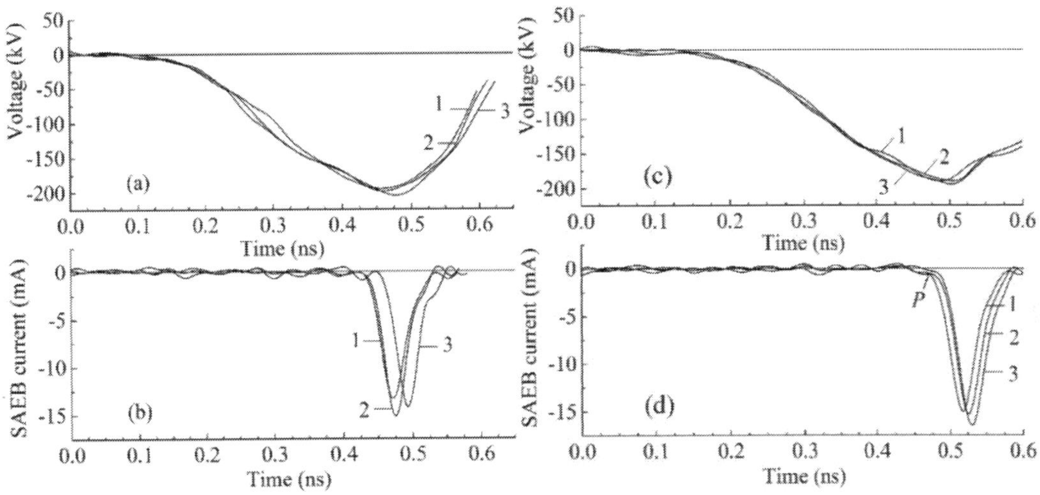

Figure 2. Waveforms of the voltage (a,c) and the SAEB current (b,d) pulses. The SAEB was registered by a collector with the 3-mm-diameter receiving area, which was located behind a collimator with 1-mm-diameter hole and 10-μm-thick aluminium foil (a,b) and behind a collimator without a foil (c,d). The gap width is d = 12 mm. The cathode is a tube of 6-mm-diameter.

The runaway electron current was measured through the collimator hole (1 mm diameter) with and without anode foil of 10 μm thickness. The FWHM and the amplitude of the SAEB without foil changed insignificantly, which indicated that the collector did not detect electrons of an energy higher than 32 keV. However, the SAEB pulse rise times in these two cases differed. With no anode foil, there was a prepulse (figure 2(d), symbol P), whereas the beam current measured with the anode foil rose steeper (figure 2(b)). Hence, the energy of runaway electrons that contribute to the prepulse is no greater than 32 keV. It is seen from the figures for the timed SAEB pulse and the voltage drop.

Waveforms of the voltage and SAEB current for the cathode 1 at d = 4 mm and d = 18 mm are presented in figure 3.

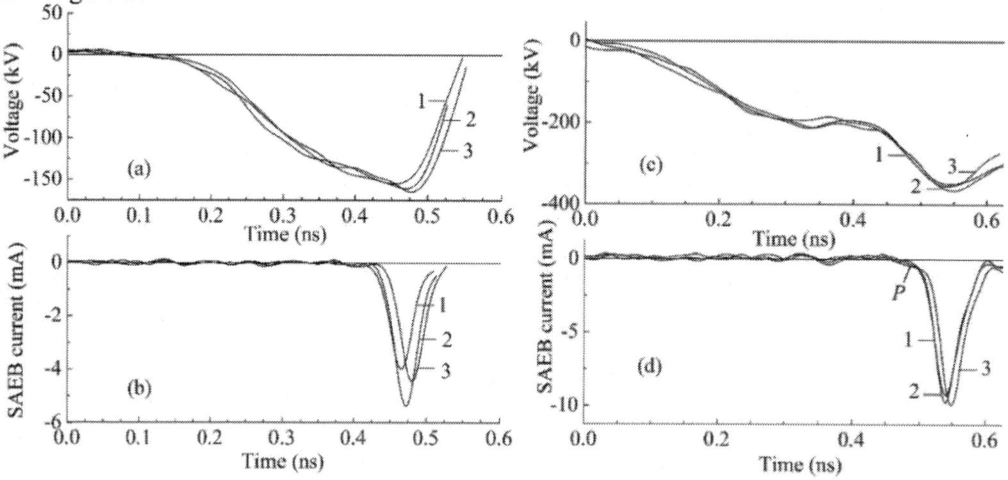

Figure 3. Waveforms of the voltage (a,c) and the SAEB current (b,d) pulses. The SAEB was registered by a collector with the 3-mm-diameter receiving area, located behind a collimator with 1-mm-diameter hole and 10-μm-thick aluminum foil (a,b) and behind a collimator without foil (c,d). The gap width was $d = 4$ (a,b) and 18 (c,d) mm. The cathode is a tube of 6-mm-diameter.

The runaway electron current was measured through the collimator hole (1 mm diameter). A decrease of the gap width d decreased the voltage amplitude, as well as the SAEB amplitude and FWHM. The interelectrode gap $d = 12$ mm was optimal for the cathode 1. At $d = 18$ mm, the voltage amplitude increased but the SAEB amplitude decreased. An increase of the interelectrode gap increased the FWHM of SAEB, which can be explained by the longer time of ionization wave transition through the gap at large d. As the distance to the anode foil increased, the prepulse was detected (figure 3(d), symbol P). Figure 3 shows that the SAEB pulses and the voltage drop are timed, similar to the case when the gap was equal to 12 mm.

Figure 4 shows waveforms of the voltage and SAEB for the cathode 2 at $d = 6$ mm.

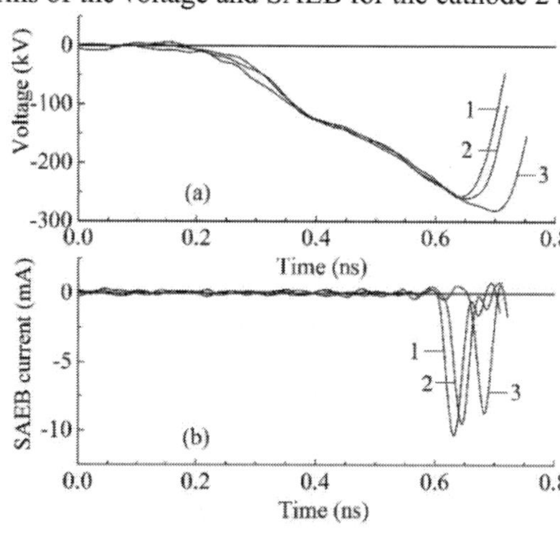

Figure 4. Waveforms of the voltage (a) and the SAEB current (b) pulses. The SAEB was registered by a collector with the 3-mm-diameter receiving area, located behind a collimator with a 1-mm-diameter hole and a 10-μm-thick aluminium foil. The gap width is $d = 6$ mm. The cathode is a stainless steel ball of 9.5 mm diameter.

The average electric field strength at the surface of cathode 2 was lower than that at the sharp end of the cathode 1, and the breakdown voltage increased. The spread of breakdown delay times and breakdown voltage amplitudes became larger. The beam current measured through the collimator hole decreased. However, the SAEB generation and the voltage drop in the gap were timed as before (figures 2 and 3). The increase of breakdown delay time proportionally increased with delay time of the SAEB generation.

Analyzing the oscilloscope traces obtained with high time resolution, we can distinguish the following features. First, the maximum of the SAEB current and the voltage drop in the gap are timed to high accuracy for both cathodes. At the moment when the collector records the maximum SAEB current, the voltage across the small and optimal gaps decreases steeply. In air of an atmospheric pressure, when the voltage amplitude is of hundred kilovolts, the FWHM of SAEB current measured through the collimator holes of small diameters, with the cathode 2 and d = 4 mm can be ≈25 ps. Downstream to the entire anode foil surface, the FWHM of SAEB is ~100 ps.

4. Conclusion

Thus, the experimental research of the amplitude-time characteristics of runaway electron beams with picosecond resolution and the discharge characteristics in air of an atmospheric pressure allows the following conclusions.

1. The maximum SAEB current and the voltage drop in the gap are timed to an accuracy of tens of picoseconds.
2. As far as the gap with cathode foil tube of 6 mm diameter and 100 μm thickness increases, we detect electrons, the energy of which was less than 32 keV before the main beam current pulse.
3. The use of sharp-ended cathode 1 stabilizes the breakdown delay time and voltage, ensuring better stability of SAEB generation.

The amplitude-time characteristics of runaway electron can be caclulated using new hybrid mathematical model [8]. Detailed results will be given in the nearest future.

Acknowledgments

I am grateful to I.D. Kostyrya and D.V. Rybka for their help in performing these experiments. The work is performed in the framework of the Russian Science Foundation (the project #14-29-00052).

References

[1] Edited by Tarasenko V F 2014 *Runaway Electrons Preionized Diffuse Discharges* (New York, USA: Nova Science Publishers, Inc)
[2] Edited by Tarasenko V F 2015 *Generation of Runaway Electrons and X-rays in the Discharges of High Pressure* (Tomsk: STS) in Russian
[3] Tarasenko V F, Rybka D V, Burachenko A G, Lomaev M I, Balzovsky E V. 2012 *Rev. Sci. Instrum.* **83** 086106
[4] Balzovsky E V, Rybka D V, Tarasenko V F 2015 *Instruments and Experimental Techniques*. **58** (5) 640-645
[5] Sharypov K A, Shpak V G, Shunailov S A, Ul'masculov M R, Yalandin M I 2013 *Rev. Sci. Instrum.* **84** 055110
[6] Burachenko A G, Tarasenko V F 2010 *Technical Physics Letters* **35** 1185-1194
[7] Baksht E Kh, Burachenko A G, Kostyrya I D, Lomaev M I, Rybka D V, Shulepov M A, Tarasenko V F 2009 *J. Phys. D: Appl. Phys.* **42** 185201
[8] Kozyrev A V, Kozhevnikov V Yu, Lomaev M I, Sorokin D A, Semeniuk N S, Tarasenko V F 2016 *EPL* **114**, 45001

The influence of the resonance effects on the radiative characteristics of helium plasma

E V Koryukina[1] and V I Koryukin[2]
[1] Plasma Physics Department, National Research Tomsk State University, 36 Lenin Avenue, Tomsk 634050, Russia
[2] Department of Physics with a Course of Higher Mathematics, Siberian State Medical University, 2 Moskowski Trakt, Tomsk 634055, Russia

E-mail: evk@phys.tsu.ru

Abstract. In the present work, the influence of an alternating circularly polarized electric field on the energy spectrum of the He atom is studied. The calculations are performed by the method of the energy matrix diagonalization of an atom in the electric field. This method has allowed us to study the behaviour of the helium energy spectrum from the same numerical procedure under resonant and non-resonant excitations by the electric field. Based on the calculation results, we have found that the resonance effects take place not only in the vicinity of resonance, but they influence the shift directions of the Stark states even under non-resonant excitation. Additionally, we have established that the helium energy spectrum behaves consistently in the electric field. The results obtained have allowed us to clarify mechanisms of the influence of the resonance effects on the radiative characteristics of helium plasma.

1. Introduction

The study of the influence of an alternating electric field on atomic spectra is a topical problem of modern physics. After first experiments, where the dynamic Stark effect was detected, general theoretical methods were developed for calculating the interaction of laser fields with atoms [1, 2 and the references therein]. In the framework of these methods based on perturbation theory, formulas for calculating the Stark state shifts in electric fields of different polarization were derived [1, 3 and the references therein]. All formulas in these papers were derived for the case of non-resonant excitation by electric fields. In the case of resonant excitation, it is necessary to use the resonance approximation [4, 5]. Both for resonant and non-resonant excitations, limitations of non-stationary perturbation theory (perturbation by the electric field is not very great, atomic levels are isolated, and atoms under consideration are Rydberg ones) must be satisfied [1 – 3, 5]. In so doing, by virtue of computational complexity in the framework of perturbation theory, one has to restrict oneself to the two-, three-, and maximum four-level approximations in calculating the AC Stark effect [6, 7].

By the present time, a lot of other excitation sources (in particular, high-frequency electrodeless discharge lamps, light-emitting diodes, superpower lasers, and so on) have appeared. These sources generate electric fields with parameters essentially different from those of optical lasers. Atomic energy spectra induced by these electric fields do not satisfy above-mentioned conditions of validity of perturbation theory. Thus, a new theoretical approach is necessary for calculating atomic spectra excited by alternating electric fields.

In the present work, a theoretical method free from limitations of perturbation theory is used for calculating the helium emission spectrum in an alternating circularly polarized electric field. This numerical method suggested and developed by us in [8, 9] is based on diagonalization of the energy matrix of an atom in the electric field. The suggested approach is valid both under resonant and non-resonant excitations by the electric field, and allows us to take into account the Stark state interactions.

2. Calculation method

In a circularly polarized electric field, the non-stationary Schrödinger equation is written as

$$\mathrm{i}\frac{\partial \psi_n(\mathbf{r},t)}{\partial t} = (H_0(\mathbf{r}) - F(x\cos\omega t \pm y\sin\omega t))\psi_n(\mathbf{r},t), \quad (1)$$

where ψ_n is the wave function of the n-th state of the system, H_0 is the unperturbed Hamiltonian, and the operator $-F(x\cos\omega t \pm y\sin\omega t)$ describes perturbation induced by the interaction of an atom with a circularly polarized electric field of frequency ω and strength F. The "+" and "–" signs correspond to the right and left polarization of the field, respectively. For the case of a circularly polarized field, the non-stationary Schrödinger equation can be reduced to the stationary one due to separation of spatial and time variables within the framework of the rotating-wave approximation [10].

Within this approximation, the wave function in the coordinate system rotating about the Z-axis with the frequency ω, has the form

$$\varphi(\mathbf{r},t) = \exp(i\omega t J_z)\psi(\mathbf{r},t), \quad (2)$$

where J_z is the z-component of the total angular momentum operator. Substituting the equation (2) in the equation (1), we get

$$\mathrm{i}\frac{\partial \varphi(\mathbf{r},t)}{\partial t} = Q\varphi(\mathbf{r},t), \qquad Q = (H_0 - \omega J_z \pm FD_x)\,. \quad (3)$$

As seen from the equation (3), the operator Q is time-independent. Hence, in the rotating-wave approximation, it is possible to go from the non-stationary Schrödinger equation (1) to the stationary one, and we have

$$Q\varphi(\mathbf{r}) = \varepsilon\varphi(\mathbf{r})\,, \qquad \varphi(\mathbf{r},t) = \exp(-i\varepsilon t)\varphi(\mathbf{r}), \quad (4)$$

where Q is the energy operator of an atom in the electric field, and ε and $\varphi(\mathbf{r},t)$ are the energy and wave function of an atom in the electric field in the rotating coordinate system. It is obvious that the Schrödinger equation (4) can be solved within stationary perturbation theory, but this way of solution has been found too complicated [1, 2, 11]. We have found that instead of solving the Schrödinger equation (4) within a perturbation theory, it is much more convenient to solve it using a diagonalization procedure. It was shown that the wave functions and energies of an atom, being solutions to the Schrödinger equation (4), can be found from diagonalization of the energy matrix of an atom in the electric field with the following elements

$$Q_{mn} = E_n^{(0)}\delta_{mn} - \omega < \varphi_m^{(0)}(\mathbf{r})|J_z|\varphi_n^{(0)}(\mathbf{r}) > \pm F < \varphi_m^{(0)}(\mathbf{r})|D_x|\varphi_n^{(0)}(\mathbf{r}) >. \quad (5)$$

In equation (5), $\varphi_n^{(0)}$ and $E_n^{(0)}$ are the wave function and energy of the n-th state of an atom in the absence of the electric field; J_z is the z-component of the total angular momentum operator; and D_x is the x-component of the dipole moment operator. The matrix elements of the D_x operator are determined as

$$< \varphi_m^{(0)}|D_x|\varphi_n^{(0)} > = < \gamma JM|D_x|\gamma' J'M' > = \frac{(-1)^{J-M}}{\sqrt{2}}\left[\begin{pmatrix} J & 1 & J' \\ -M & -1 & M' \end{pmatrix} - \begin{pmatrix} J & 1 & J' \\ -M & 1 & M' \end{pmatrix}\right] < \gamma J\|D\|\gamma' J' >, \quad (6)$$

Details of calculating the matrix elements (6) are reported in [9, 12, 13].

The energy matrix with the elements (5) is obtained in the coordinate system rotating about the *Z*-axis with frequency ω. Upon diagonalization of this matrix, we get the energies ε_n and wave functions φ_n as

$$\varphi_n(\mathbf{r},t) = \exp(-i\varepsilon_n t)\sum_k C_{nk}(F,\omega)\varphi_k^{(0)}(\mathbf{r}) \tag{7}$$

for the n atomic states in the electric field in the rotating coordinate system. The coefficients C_{nk} in the wave function (7) are the expansion coefficients of the wave functions of an atom under the action of the electric field in terms of unperturbed wave functions. These coefficients depend on the electric field strength and frequency.

To go to the initial coordinate system and find the average atomic energies in the electric field, it is necessary to perform averaging over the oscillation period of the electric field. After this averaging, the wave functions and average energies of an atom in the electric field in the initial coordinate system are written as

$$\psi_n(\mathbf{r},t) = \exp(-i(\omega J_z + \varepsilon_n)t)\sum_k C_{nk}(F,\omega)\varphi_k^{(0)}(\mathbf{r}), \tag{8}$$

$$\overline{E}_n = <\psi_n(\mathbf{r},t)|H(\mathbf{r},t)|\psi_n\psi_n(\mathbf{r},t)> = \varepsilon_n + \omega < \varphi_n(\mathbf{r})|J_z|\varphi_n(\mathbf{r})>. \tag{9}$$

It follows from the above reasoning that the proposed theoretical approach is free from limitations of perturbation theory and can be used for calculating the dynamic Stark effect in a circularly polarized electric field with the strength and frequency changing in wide ranges. Since the energies and wave functions in the electric field are determined by the diagonalization procedure, it allows us to take into account the Stark state interactions. The algorithm of our theoretical approach is implemented in a special software package StarkD [14].

In this work, the described above method is applied to the study of the dependence of the helium energy spectrum on the electric field frequency.

3. Results and discussion

It is well-known from the perturbation theory that the Stark shifts of atomic states induced by an alternating circularly polarized electric field are calculated as

$$\Delta E = -\frac{1}{2}\alpha F^2, \tag{10}$$

where α is the the polarizability of an atom, and F is the electric field strength. If the electric field frequency equals the transition frequency between the Stark states, the curve characterizing the dependence of the polarizability on the electric field strength has resonances. In these resonances and near-resonance ranges, the polarizability calculated using the formulas obtained in [1 – 3] is infinite, therefore, in the case of resonant excitation by the electric field, the polarizability α must be calculated in the resonance approximation [4, 5, 15]. After the transition through resonance, the polarizability changes in the sign [16]. Hence, the Stark state shifts have big values in resonances, and, by virtue of equation (10), the Stark state shifts change their directions after the transition through resonances.

Since the influence of the electric field leads to mixing the Stark states, it is desirable to take into account the Stark state interactions in calculating atomic spectra induced by electric fields. In the framework of our numerical method, taking into account of these interactions occurs automatically, because the wave functions and energies of atoms in the electric field are determined by the diagonalization procedure.

Let us consider the results of calculating the ω-dependence of the shift directions of the Stark states obtained in the framework of our theoretical method for the He atom.

First of all, we have carried out the calculations of the helium spectrum under resonant excitation by the electric field. In the framework of our theoretical approach, there are no difficulties in calculating the Stark state shifts just in resonances. The calculation results for the *S* state shifts of the He atom in resonances are listed in table 1. In this table, the Stark state shift is defined as $\Delta E = E - E_0$, where E is the energy of an atom in the electric field calculated using equation (9), and E_0 is the energy of this atom in the absence of the electric field. The values of the electric field frequency in the second column are the resonance frequencies of the corresponding transition from the first column. The Stark state shifts in resonances have been computed at the electric field strength F=1.2 kV/cm. It should be noted that the 6 $^3S_1$ energy level as well as the $^1S_0$ levels does not split in the electric field, because this level weakly interacts with neighbouring energy levels when the electric field strength is so small.

Table 1. Shifts of the S states of the He atom in resonances (F=1.2 kV/cm).

Resonance	ω (MHz)	ΔE (cm^{-1})
$1s^2\,{}^1S_0 - 2\,{}^1P_1$	$5.1305\cdot10^9$	5906.058
$1s^2\,{}^1S_0 - 3\,{}^1P_1$	$5.5820\cdot10^9$	27456.021
$1s^2\,{}^1S_0 - 4\,{}^1P_1$	$5.7408\cdot10^9$	11022.708
$6\,{}^1S_0 - 6\,{}^1P_1$	$0.47979\cdot10^7$	78.5978
$6\,{}^1S_0 - 7\,{}^1P_1$	$2.89066\cdot10^7$	480.4672
$6\,{}^3S_1 - 6\,{}^3P_2$	$7.69337\cdot10^6$	126.8043
$6\,{}^3S_1 - 6\,{}^3P_1$	$7.69345\cdot10^6$	126.9727
$6\,{}^3S_1 - 6\,{}^3P_0$	$7.69440\cdot10^6$	127.6617

As it follows from table 1, the Stark state shifts in resonances have finite values, and there is no necessity to use the resonance approximation.

In order to compare the ω-dependence of the Stark state shifts obtained by our numerical method with those calculated in the framework of non-stationary perturbation theory, we have plotted the curves characterizing dependences of the ground and excited S states of the He atom on the electric field frequency. These curves calculated at the electric field strength F=1.2 kV/cm are shown in figures 1 – 3. As it is seen from these figures, when the electric field frequency equals the resonance frequency, we get a drastic increase in the Stark state shifts in resonances and near-resonance ranges.

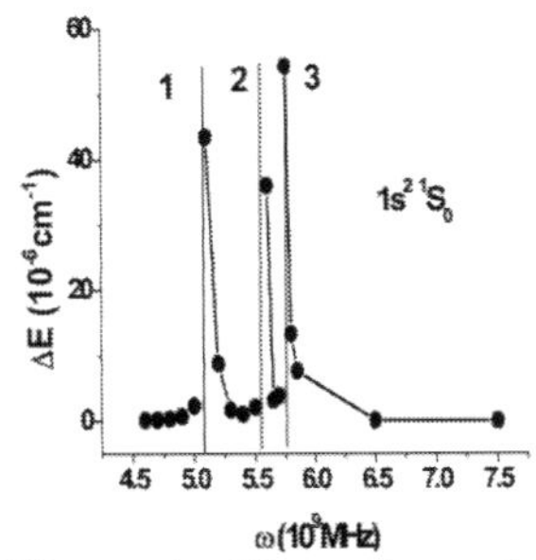

Figure 1. The ω-dependence of the ground state 1s^2 ^{1}S$_0$: 1 – $1s^2\,{}^1S_0 - 2\,{}^1P_1$ resonance, 2 – $1s^2\,{}^1S_0 - 3\,{}^1P_1$ resonance, 3 – $1s^2\,{}^1S_0 - 4\,{}^1P_1$ resonance.

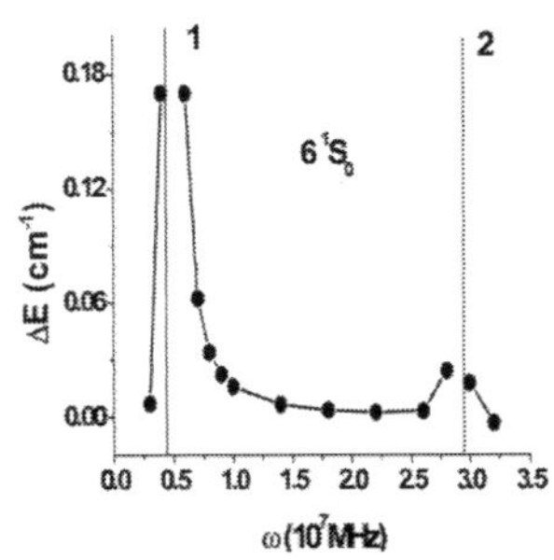

Figure 2. The ω-dependence of excited the 6 ^{1}S$_0$ state: 1 – $6\,{}^1S_0 - 6\,{}^1P_1$ resonance, 2 – $6\,{}^1S_0 - 7\,{}^1P_1$ resonance.

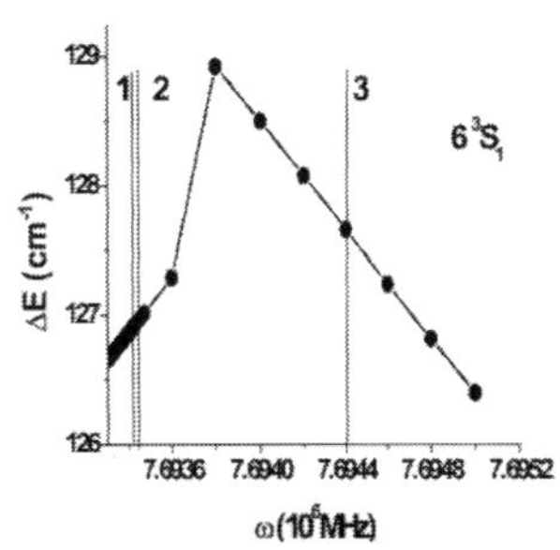

Figure 3. The ω-dependence of excited the 6 ^{3}S$_1$ state: 1 – $6\,{}^3S_1 - 6\,{}^3P_2$ resonance, 2 – $6\,{}^3S_1 - 6\,{}^3P_0$ resonance, 3 – $6\,{}^3S_1 - 6\,{}^3P_0$ resonance.

This result agrees with the perturbation theory that confirms a validity of our theoretical approach. It is interesting to note that peaks characterizing the Stark state shifts in resonances are very sharp for singlet states (see figures 1 and 2, and table 1), whereas these peaks are not very high for triplet states of the He atom (see figure 3).

Now, let us consider the case of non-resonant excitation of the helium spectrum. The calculations of the helium state shifts are carried out for different frequencies of the electric fields generated by real excitation sources, namely, ω=100 MHz (a high-frequency discharge HFD), ω=1.5191·10^5 MHz (a

NH_3-laser), ω=2.4352·10^6 MHz (a HCN-laser) and ω=2.4181·10^7 MHz (a CO_2-laser). The calculation results for singlet and triplet S states of the He atom are shown in figures 4 and 5. These figures demonstrate that an increase in the electric field frequency leads to the resonance effects (that is, to the changes of the shift directions of the Stark states) even under non-resonant excitation by the electric field. Based on the calculation results, we can conclude that the presence of the resonance effects results from two reasons.

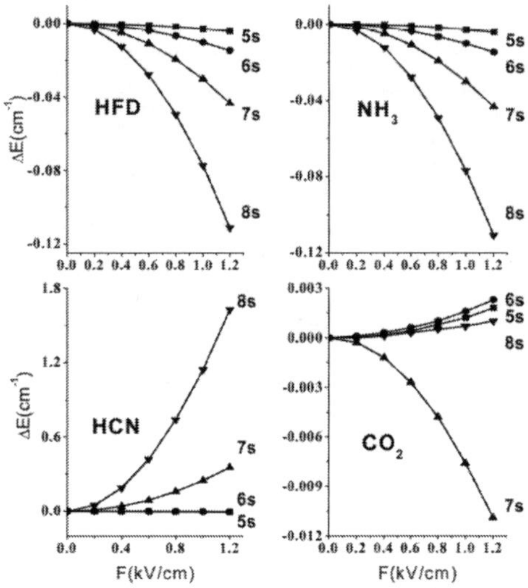

Figure 4. Dependence of the Stark state shifts on the electric field frequency for the 1sns $^1S_0$ energy levels.

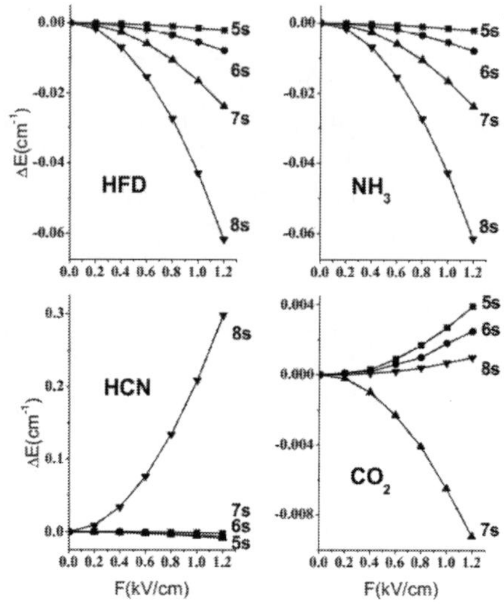

Figure 5. Dependence of the Stark state shifts on the electric field frequency for the 1sns $^3S_1$ energy levels.

Firstly, an increase in the electric field strength leads to restructuring of the energy spectrum, that is, to changing the positions of the Stark states relative to each other. To illustrate restructuring of the helium spectrum in the electric field, we list energy state numbers for the $^1S_0$ and $^3P_0$ states of the He atom for different ω (see table 2). It follows from this table that owing to restructuring of the energy spectrum the transition frequencies change with an increase in the electric field frequency and it leads to appearance of resonances.

Table 2. Dependence of the Stark state numbers on the electric field frequency in the helium energy spectrum (F=1.2 kV/cm).

Level		ω (HFD)	ω (NH_3)	ω (HCN)	ω (CO_2)
$^1S_0$	5s	121	121	123	158
	6s	185	185	200	232
	7s	249	249	264	294
	8s	313	313	336	347
$^3P_0$	5p	130	127	139	162
	6p	192	191	211	235
	7p	258	255	276	299
	8p	322	319	332	350

Secondly, since the wave functions and energies of atoms in the electric field are determined by the diagonalization procedure, all atomic states included in the energy matrix behaves consistently, that is, if the action of the electric field leads to changing the shift directions of the *np*-states, then the *n's*- and *n"d* -states also change their shift directions. The reason for such consistent behavior is the fact that, according to equation (5), all Stark states in the helium spectrum are linked by the matrix elements of the *x*-component of the dipole moment operator D_x.

As an illustration of a consistent behavior of the helium spectrum in the electric field, in addition to figures 4 and 5, we present figures 6 and 7. These figures show the dependence of the *np* state shifts on the electric field frequency for the He atom. In figure 7, the values of the magnetic quantum number M for the 7 $^1P_1$ level are presented in a black frame. Comparison of figures 5 and 6 illustrates a consistent change in the shift directions of the 8 $^3S_1$ and 7 $^3P_0$ levels of the He atom.

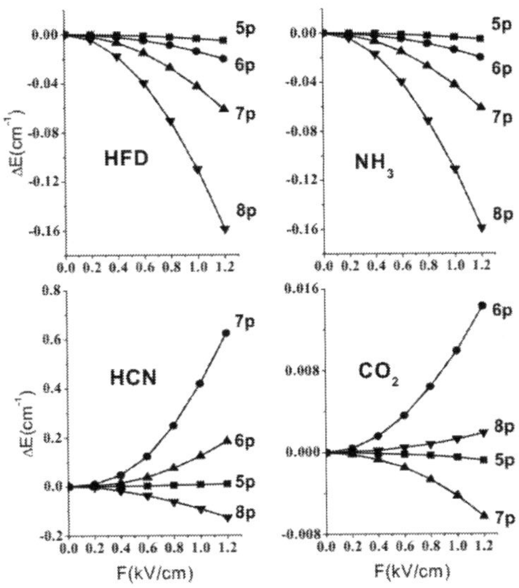

Figure 6. Dependence of the Stark state shifts on the electric field frequency for the 1*snp* $^3P_0$ energy levels

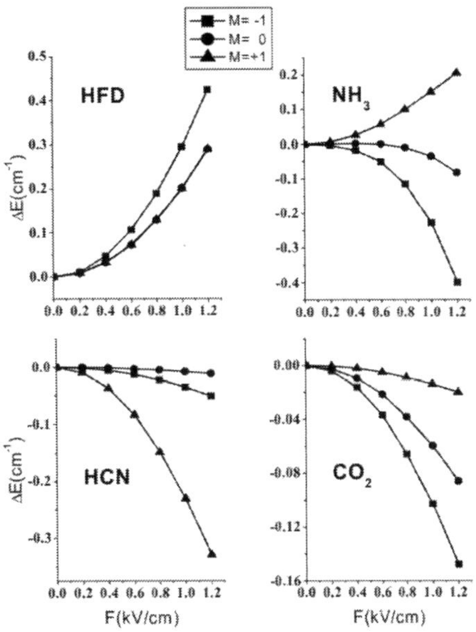

Figure 7. Dependence of the Stark state shifts on the electric field frequency for the 1*s*7*p* $^1P_1$ energy level

As it is seen from figures 4 – 7, for the *np* states as well as for the *ns* states, non-resonant excitation by the electric field leads to drastic changes in the shift directions of the Stark states. This phenomenon is obtained for the energy levels both non-splitting (see figures 4 – 6) and splitting (see figure 7) in the electric field. For the energy levels, non-splitting in the electric field, the resonance effects result in the breakdown of a monotonous dependence of the Stark state shifts on the principal quantum number *n* of the outer-shell electron (see figures 4 – 6). For the energy levels, splitting in the electric field, the presence of the resonance effects also leads to changing the shift directions of the Stark states (see figure 7).

Thus, the calculations carried out in the framework of our theoretical method allow us to reveal the reasons for the changes in the shift directions of the Stark states of the He atom under non-resonant excitation by the electric field. It is obvious that these changes influence the radiative characteristics of helium plasma, in particular, they lead to changing the colour characteristics of plasma emission. Changing the colour characteristics of lamps with an increase in the electric field frequency was observed in experiments [17 and references therein].

4. Conclusions

In this paper, it is shown that the method of the energy matrix diagonalization of an atom in the electric field enables to carry out calculating atomic emission spectra by the diagonalization procedure both under resonant and non-resonant excitations by the electric field. In addition, it has been found that the resonance effects take place under non-resonant excitation of the He atom, and restructuring of the energy spectrum is the reason for these effects. In so doing, the emission spectrum of the He atom behaves consistently. Finally, it has been shown that the resonance effects are the reasons for changing the shift directions of the Stark states under non-resonant excitation by the electric field.

These theoretical results obtained for the first time allow us to clarify mechanisms of the formation of the helium spectrum in the electric field and have practical applications in a plasma spectroscopy, a gas discharge physics and a laser physics.

Acknowledgements

This work was supported by the Programme of Improving TSU's International Competitiveness.

References

[1] Rapoport L B, Zon B A, and Manakov N L 1978 *Theory of Many-Photon Processes in Atoms* (Moscow: Atompress)
[2] Delone N B and Krainov V P 1984 *Atoms in Strong Light Fields* (Moscow: Atompress)
[3] Delone N B, Zon B A, Krainov V P and Hodovoj V A 1976 *Progr. of Phys. Sciences* **120** 1
[4] Bayfield J E 1979 *Phys. Reports* **51** 317
[5] Rautian S G, Smirnov G I and Shalagin A M 1979 *Non-linear resonances in atomic and molecular spectra* (Novosibirsk: Nauka)
[6] Averbuch I Sh and Perelman N F 1985 *Zh. Eksp. Teor. Fiz.* **61** 1091
[7] Changjiang W, Suter D, Andrew S M, Windsor A S M and Manson N B 1998 *Phys.Rev* A **58** 2310
[8] Koryukina E V 2003 *Rus. Phys. J.* **46** 1069
[9] Koryukina E V 2005 *J. Phys. D: Appl. Phys.* **38** 3296
[10] Bunkin F V and Prokhorov A M 1964 *Zh. Eksp. Teor. Fiz.* **46** 1091
[11] Sambe H 1973 *Phys. Rev.* A **7** 2203
[12] Jucys A P and Savukynas A J 1973 *Mathematical Foundations of the Atomic Theory* (Vilnius: Mokslas)
[13] Koryukina E V 2006 *Atmospheric and Oceanic Optics* **19** 519
[14] Koryukina E V 1998 *Vestnik TGPU, Series: Natural and exact sciences* Iss. **5** 41
[15] Delone N B and Krainov V P 1999 *Progr. of Phys. Sciences* **169** 753
[16] Chernov V E, Dorofeev D I, Kretinin I Yu and Zon B A 2005 *Phys. Rev.* A **71** 022505
[17] Konomi M http://hdl.handle.net/10228/3729

4-kW multi-phase battery powered power supply

A V Ponomarev[1], S R Korzhenevskiy[1], A A Komarskiy[1], A S Chepusov[1], Y I Mamontov[2] and S V Ponomarev[3]

[1] Institute of Electrophysics, 106 Amundsena Street, Ekaterinburg, 620016, Russia
[2] Ural Federal University, 19 Mira Street, Ekaterinburg, 620002, Russia
[3] Tomsk Polytechnic University, 30 Lenina Street, Tomsk, 634034, Russia

E-Mail: apon@iep.uran.ru

Abstract. A method was developed for building the powerful battery power supplies. Using the method, the battery power supply with a 4 kW max power and up to 93% efficiency was developed to supply the "Yasen" X-ray apparatus. Two 60 A·h series-connected starter lead-acid batteries were used as a primary power supply. A DC output voltage of the source is stable over the entire power range and equals to 310 V. The power supply is based on a 5-phase HF-inverter. There is no difficulty in designing such power supplies with different power outputs. It can be done by the increasing or the decreasing number of phases (of inverter channels). This approach is not limited by the increased number of the inverter channels. The maximum output power will be determined by the battery characteristics only. The power supply is mounted on a mobile trolley, to increase the mobility of the entire set of equipment. The unit dimensions are 410x320x440, the weight is about 40 kg. The unit is forced air-cooled. A power operating mode is short and periodic.

1. Introduction

The self-contained power supplies are commonly used to power various electrical installations and devices in case the connection to the mains is not available. These are divided into two main types – the electromechanical and the battery powered power supplies. The electromechanical power supply is an electric power generator driven by a combustion engine. A release of harmful substances from the engine into the atmosphere reduces the scope of its application. Furthermore, this type of devices does not provide even a short-time increase in the power taking off that exceeds an average value. In contrast to the electromechanical power sources, the battery powered power supplies are more environmentally friendly and allow a short-time high power output. However, a limited capacity of the battery, and hence the energy stored, does not allow supply of the high power equipment within an extended period

This paper describes a design, an operating principle, and the characteristics of the 4 kW battery powered supply intended for the "Yasen" X-ray apparatus [1].

It should be noted that the market offers a large variety of the battery powered power sources of various capacities produced by Russian and foreign manufacturers. Initially, attempts were made to use these units to provide feeding of the "Yasen" X-ray apparatus. However, these attempts were not successful due to the requirements specified to the X-ray apparatus power supply. All the employed units are fed with the 220-volt AC household electrical equipment or the electro-driven tool, which is not sufficiently sensitive to a decrease of the supply voltage. Thus, the power supplies =12/~220 V made both by Russian and foreign manufacturers demonstrated a significant output voltage reduction at the maximum rated power. As a result, those were not appropriate to power the "Yasen" X-ray apparatus, which requires a stable supply voltage over the entire range of the rated input, since the magnetic compressor included in the circuit is sensitive to the supply voltage.

2. Problems occurring during development of powerful low-voltage power supplies

Lead-acid car starter batteries are most often used as a primary power source in the powerful self-contained power supplies. The output voltage of these batteries is 12 V. The study revealed that the output power of the device with a single 60 A·h battery is limited by about 2–3 kW. This is due to the fact that under loading the battery output voltage reduces. The reducing depends on the state of charge, capacitance, the charge level of the battery, and other factors. A high-ampere current flowing through a primary electric circuit causes a voltage drop across its components, which results in a further decrease of the voltage supplied to the high-frequency transistor inverter, which is an integral part of these devices. In addition, we should consider the voltage drop across the inverter transistors: as the current flowing through the circuit grows, it increases. Due to low supply voltage of the inverter (several volts), the filtering capacitance becomes ineffective. In this case, the pulsed power taking off occurs mainly from the battery. Additional losses in the circuit are caused by the skin effect occurring in conductors. Thus, as the output power grows, the supply voltage of the transistor inverter decreases, and the voltage drop across the transistors increases. This leads to a vicious circle, when the supply voltage drop requires an increased current input. The increased current input leads to the further reduction in the circuit voltage supply.

The power of the supply can be increased up to 4 kW or higher by increasing the supply voltage via two 12 V series-connected batteries. However, even in this case the primary circuit current reaches 200 A at maximum output power. The current flow of this magnitude poses special requirements for the high-frequency transistor inverters. Though, there are semiconductor devices with a similar operating current level, however, the single transistor inverter, in this case, was not effective. We propose to use a multi-phase inverter transistor. It should be noted that the idea is not innovative, and it is widely used to provide a computer processor power supply. Microchips to control this type of high-current low-voltage inverters are produced by a number of leading companies. Also, the multiphase power supplies with a battery backup exist. A 1 kW unit developed by Orchid Technologies is presented in Orchid Technologies leaflet (http://orchid-tech.com/wp-content/uploads/2014/12/ OTEC_Design_Note_90_web.pdf). The multi-phase two-kilowatt high-voltage regulated power supply for space applications is described in [2].

3. Design description

The block diagram of the device is shown in Figure 1. The main component of the system is a 5-phase inverter. It consists of 5 similar transistor inverters made in accordance with a push-pull circuit. The maximum output power of every inverter is up to 800 W. The development and design costs of the inverter of this power level are significantly lower than those in the case of a single 4 kW inverter.

The control system generates on/off signals for the inverter transistors. In addition, the control system measures the output voltage and the inverter current input during operation and ensures the device protection. The control circuit also provides a number of service functions and indication of the operating modes. A 32-bit STM32F030C8 controller is used to perform measurement, control, and protection.

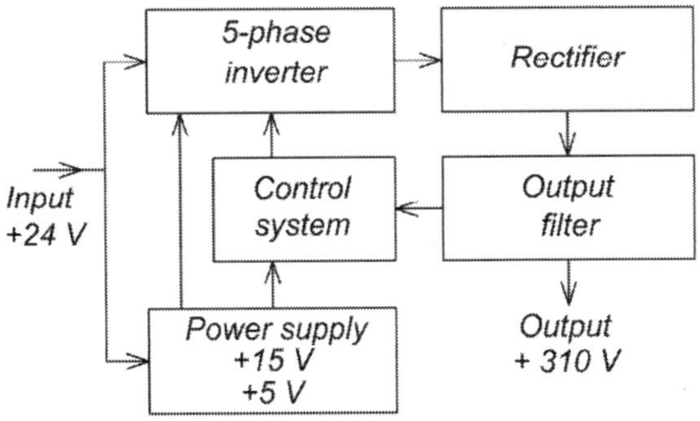

Figure 1. The block diagram of the device.

The CPLD EPM240T100C5N is used to control the inverter. This microchip receives a 7 bit binary control signal from the microcontroller and generates 5 pairs of output pwm-signal sequences. The output signals are displaced in phase by 3600/5=720. The diagrams of the control and output signals are shown in Figure 2. The data were obtained through CPLD operation simulation using Quartus 9 software. During simulation, the input values from the microcontroller were being increased and the duration of the transistor control pulses was growing.

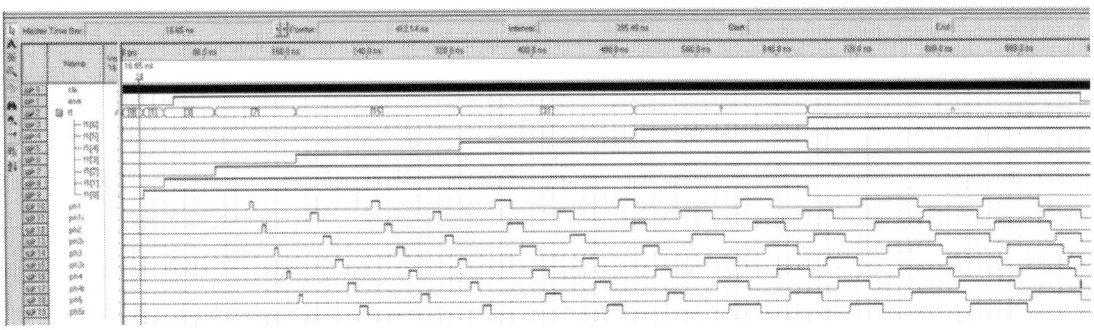

Figure 2. Results of EPM240T100C5N simulation.

Input signals fl[0:6] are fed by the microcontroller and carry a binary code information on the value of the inverter transistors turn-on angle. The ena signal is an enabling signal for five pairs of the ph1 ph5 control signals to travel to the inverter transistors. The diagram shows when the control pulse width increases, at a certain point, one of the inverter transistors keeps open. This means that the current from the battery is continuously fed by switching between the inverters. The output frequency of the device is 5 times greater than the inversion frequency (5*33 kHz=165 kHz). The increased frequency of the output voltage reduces the requirements to the output filter.

An auxiliary power unit is used to provide +5 V for the control system and +15 V to power the inverter drivers. The power supply is simple in its structure; it is made of standard components forming a step-down circuit.

4. Experimental results

The results of the power supply testing are shown in Figure 3. The values of the power consumed by the "Yasen" apparatus at different operating frequencies are plotted along the X axis. The Y axis indicates the values of the device efficiency, voltage U_b and battery current I_b. The currents were measured by standard shunts with 0.5 accuracy class. M381 pointer-type ammeters of 1.5 accuracy class were used as indicators in current measurement. The voltage was measured with digital devices MY-62. The output voltage of the unit remained stable and was equal to 310 V over the entire operating range. The output voltage of the power supply started to decrease at an operation output power above 4.3 kW. At this power the operation frequency of "Yasen" apparatus was 3.5 kHz. The power supply attained its maximum power efficiency of 93% at 4.1 kW. A further increase in the power input caused the efficiency drop, and the output voltage was observed to decrease. The device efficiency at a low output power level was found to be below the maximum, as well. One of the causes of this phenomenon is that the losses, occurring during the reversal magnetization of the transformer cores and the switching losses in semiconductors, are virtually stable over the entire range of the output power. However, in the low output power range, their impact on the efficiency is more significant. Furthermore, in the high-power range, the battery current is not fed by pulses, but continuously due to the phase shift. This cancels out the skin effect, which is equivalent to the reduction of the ohmic losses in the primary high-current circuit. The duration of the "Yasen" apparatus operating during all the measurements was equal to 2 s.

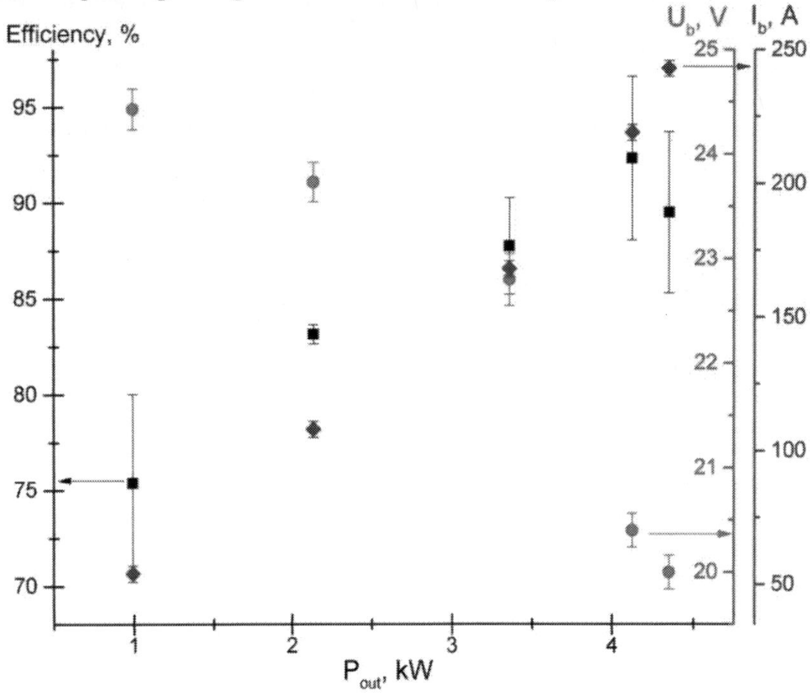

Figure 3. The results of the power supply testing.

The power supply is mounted at a mobile trolley to increase the mobility of the entire equipment set. The unit dimensions are 410x320x440, the weight is about 40 kg. Lithium-ion batteries can be used instead of lead-acid ones, to reduce the power supply dimensions and weight. The unit is forced air-cooled. The power operating mode is short and periodic.

5. Discussion and conclusion

Findings of the research demonstrate the possibility of designing a battery pack in accordance with the proposed approach. It should be noted that the constructed and tested prototype is only an experimental setup. The authors suggest that an optimization of the device can improve its efficiency. The optimization is possible through both the changed number of the inverter channels (phases) and an additional circuitry and design solutions. The reduced number of channels decreases the switching losses since the total number of transistors and transformers in the inverter reduces. However, the increased current loading per every channel increases the conduction losses. To design an effective device, a study is to be conducted to find the optimum number of the inverter channels in every case. The decreased operating frequency of the inverter will also reduce the switching losses.

Further enhancement of the device efficiency is possible by optimizing a technique of the residual energy recuperation. The residual energy is the energy stored in the transformer cores of the inverter and conductors at the time of the current flowing through the inverter transistors. In our circuit, this energy is stored in an intermediate capacitor and then transferred back to the primary power source

IOP Conf. Series: Journal of Physics: Conf. Series **830** (2017) 012005 doi:10.1088/1742-6596/830/1/012005

through the power resistor. The Joule's losses in the resistor can be significantly reduced by its replacing with an active circuit of the residual energy recuperation. An active recuperation circuit can be a separate high-frequency inverter to transform the energy from the intermediate capacitor to the primary power supply.

The design and parameters of the inverter transformers are also important targets for a device optimization. The topology of power conductors should also be considered when designing this type of device.

The proposed approach can be readily used as a basis for creating the powerful battery power supplies in a wide range of capacities for various applications.

References

[1] Bessonova V A, Gavrilov P V, Komarskiy A A, Korzhenevskiy S R and Chepusov A S 2016 *Medical Radiology and Radiation Safety* **2** pp 53-57

[2] Garth D R, Muldoon W J, Benson G C and Costague E N 1971 *Power Electronics Specialists Conf. (Pasadena, CA)* vol 1 (New Jersey: IEEE) pp 110-16

IOP Conf. Series: Journal of Physics: Conf. Series **830** (2017) 012006 doi:10.1088/1742-6596/830/1/012006

Novel control system of the high-voltage IGBT-switch

A V Ponomarev[1,3], Y I Mamontov[2], A I Gusev[1] and M S Pedos[1]

[1] Institute of Electrophysics, 106 Amundsena Street, Ekaterinburg, 620016, Russia
[2] Ural Federal University, 19 Mira Street, Ekaterinburg, 620002, Russia
[3] Author to whom any correspondence should be addressed

E-Mail: apon@iep.uran.ru

Abstract. HV solid-state switch control circuit was developed and tested. The switch was made with series connection IGBT-transistors. The distinctive feature of the circuit is an ability to fine-tune the switching time of every transistor. Simultaneous switching provides balancing of the dynamic voltage at all switch elements. A separate control board switches on and off every transistor. On and off signals from the main conductor are sent to the board by current pulses of different polarity. A positive pulse provides the transistor switch-on, while a negative pulse provides their switch-off. The time interval between pulses defines the time when the switch is turned on. The minimum time when the switch is turned on equals to a few microseconds, while the maximum time is not limited. This paper shows the test results of 4 kV switch prototype. The switch was used to produce rectangular pulses of a microsecond range under resistive load. The possibility to generate the damped harmonic oscillations was also tested. On the basis of this approach, positive testing results open up a possibility to design switches under an operating voltage of tens kilovolts.

1. Introduction

Nowadays, the power semiconductor switching units are widely used in a Pulsed Power technology. The high-power switching requires a high-current equipment with a high working-voltage. However, the development of a single solid-state high-voltage device has physical limitations. A common method to produce a high-voltage solid-state switch uses a series connection of single devices on a stack [1]. Full control switch can be developed using MOSFETs or IGBTs. Today, such transistors are delivered by a number of companies. "Behlke" (www.behlke.com) offers the most extensive range of MOSFET and IGBT-based high-voltage switches. Unfortunately, a high cost and a limited variety of models promote the development and manufacturing of similar devices.

One of the most important aspects of the series operation of devices is to equalize the static and dynamic balancing of the voltage. The static voltage balancing can be simply achieved by connecting small balancing resistors in parallel with each device. The dynamic voltage balancing during the switching transient is much more difficult to achieve. Two dynamic voltage balancing techniques are available: the load-side balancing and the gate-side balancing. The load-side balancing employs a snubber circuit and/or a clamp circuit. The snubber circuit and/or clamp circuit provide the dynamic voltage balancing by limiting the device voltage rising rate and/or clamping the peak voltage. This technique, however, cannot be used for high-power applications since much loss is involved in the snubber and clamp circuits, which is proportional to the switching frequency. To solve this problem, an active gate control technique has been presented in the last few years, which does not degrade the switching transient characteristic. Each gate drive circuit should be actively controlled, so that all device voltages increased or decreased at the same rate [2].

This paper introduces a new method of designing a control circuit using the IGBT-based high-voltage switch with an active gate control technique.

2. Methods of designing control circuits using high-voltage switch units

There are several standard methods to control high-voltage switches by transistors. The most common control method uses an insulating transformer [3]. A series of isolated windings allows us synchronously to apply and remove a control voltage from switch transistors. Despite a seeming simplicity of the transformer control circuit, it has a number of disadvantages. The major drawback is its limited duration of a control pulse. This stems from the finite voltage-time integral of a control transformer core. Thus, a control circuit pulse width cannot exceed several to tens of microseconds. Moreover, the control circuit transformer design should be closely observed to reduce its leakage inductance. This will enable to reduce a control signals edge time. Another disadvantage of the control circuit transformer is a control signal's edge dependence on the number of controlled transistors and their gates capacity. Thus, the charging and discharging time of transistor gates will be proportional to the root of L and C product, where L stands for the secondary transformer circuit inductance and C stands for the total capacity of transistor gates.

Along with the transformer control circuits, there are opto-isolator circuits [4]. In this case, every transistor is controlled by a separate driver. The driver is triggered by a single fiber optic cable. This method is significantly more complicated than the previous one. To function, every isolated driver must be provided with a power supply for the circuit. To implement a parallel connection of light with the same intensity for each control fiber optic cable is a challenging task. Besides, the large dimensions of the device are caused by the numerous fiber optic control cables. The advantage of this circuit is that it allows us to deliver the control signals of any duration. In addition the length of transistors control signals edges does not depend on their number and their gates' capacity.

This paper proposes an original circuit design that encompasses all the benefits of the two methods mentioned above. To control the transistors, it uses a separate driver board with an isolated power supply. The control signals are supplied by a single wire loop (current loop) with the current pulses of different polarities. A positive current pulse determines the turn-on time of a transistors' switch signal. The negative current pulse turns off the transistors. Thus, the minimum pulse generation time equals to several microseconds, while the on-state switch duration time is not limited. Rogowski coils are installed on the driver boards and serve as current transducers. The advantage of these coils is their ability to differentiate the current input waveforms. As a result, Rogowski coil transmits a short-edge trigger signal to the driver's circuit. Shortening the edge of the control signal allows us to increase a precision of the transistors' switching time.

3. High-voltage switch control circuit using current loop

A simplified control circuit of the high-voltage switch using IGBTs is shown on Figure 1. A bipolar current control signal is generated by a bridge converter using VT1-VT4 transistors. The signal is generated in the form of pulses only at the moment of switching of the converter's transistors. The length of this pulse edge is determined by the C1 capacity and the inductance of the current loop, which, in its turn, is measured by the sum of the current loop inductance and the inductance of primary windings of T1-Tn Rogowski coils. Rogowski coils are made without any magnetic cores and have a low value of the magnetic coupling between windings. Thus, the inductance of the T1-Tn primary windings slightly exceeds the inductance of a usual wire. It can be concluded that the current loop inductance is determined only by its geometry. Considering the circuit characteristics, the half-period length of a current pulse was about 500 ns. When the signal was differentiated by Rogowski coils, the length of the signal's front edge from its secondary windings was shortened to a value of about 70 ns.

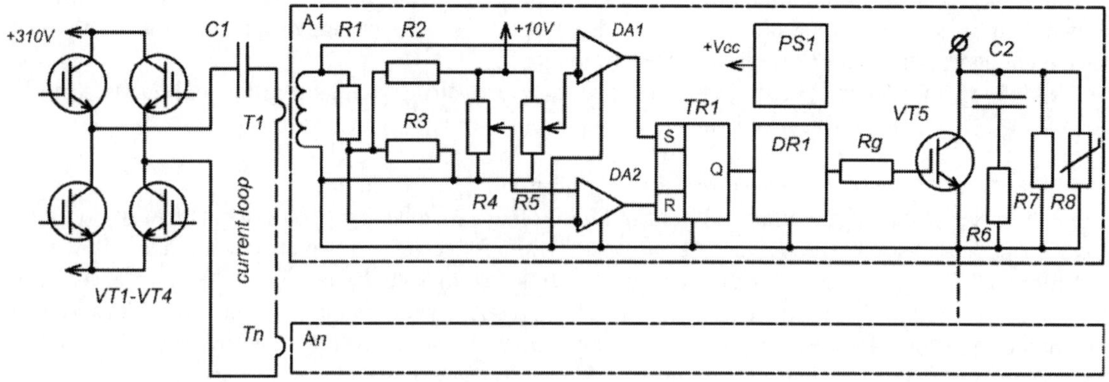

Figure 1. Control unit – circuit diagram.

The signals from Rogowski coils were sent out to the input of DA1 and DA2 high-speed comparators. While a positive signal switched on the DA1 comparator, the negative one switched on the DA2 comparator. The signals were sent out from the comparator outputs to TR1 RS-trigger, making it to toggle. The signals from the trigger were sent out to the DR1 driver. DR1 driver, in its turn, controlled the IGBT of VT5 switch.

The novelty and merit of the offered circuit configuration is the use of DR1 and DR2 comparators in the circuit. The comparators allow us to set independently the on/off time of every transistor switch with high precision. R4 and R5 trimmer potentiometers are used to change a reference voltage for the comparator response. The comparator checks this voltage with the input signal from the Rogowski coil (Figure 2b).

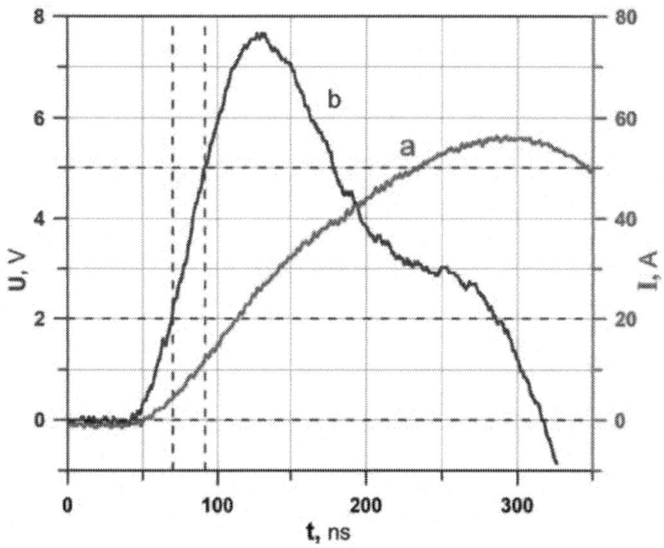

Figure 2. Control signals: a – current loop signal; b – output signal from the Rogowski coil.

Changing of the reference voltage changes the comparators switch on time. Figure 2 shows that when the comparator reference voltage to 3V (from 2 V to 5 V) changes, the response time will change to about 20 ns. Thus, changing the reference voltage, it becomes possible to change the on/off time of the transistor switch. The stabilized PS1 power supply allows us to change the reference voltage accurate to tenths and hundredths mV. This change of the reference voltage allows us to set the circuit response time accurate to split nanoseconds. The precise setting of the response time of every transistor provides a dynamic voltage balancing of its elements.

4. Experimental setup and methods

The efficiency and the operational capacity of the offered high-voltage switch control circuit method was tested and underwent an experimental verification. The switch prototype with a working-voltage of 4 kV was assembled for these experiments. The switch was made on two high-speed power IGBT-modules X2G75FD12P1 by Hivron. Each module included two series connected transistors with a maximum voltage of 1200 V and an average current of 75 A. The turn on and off of the transistors was carried out by four identical control plates.

The electric measurement facilities included a P6015A high-voltage probe and current shunts. The shunts had a proper signal rise time not exceeding 1 ns. A TDS 684B oscilloscope with 1 GHz bandwidth and Barth Electronics wide band signal attenuators were used for recording the output pulses.

5. Experimental results

The developed switch was tested in 2 modes. The first mode demonstrated the switch ability to generate the rectangular pulses of a microseconds range on a resistive load. The parallel-connected TVO resistors with a low inductance and a total resistance of 48 Ohms served as a load. Figure 3a shows the current and voltage signals on the load during rectangular pulse generation with the length of about 7 μs. Figure 3b and 3c shows the voltage and current curves at the time of the device switching on and off.

The second mode demonstrated the switch ability to generate the pulses of a damping harmonic oscillation. The received oscillation frequency slightly exceeded 1 MHz. In this test, the switch was installed into the gap of oscillating LC-circuit. Figure 3d shows the circuit current without an outfeed of the energy. The damping decrement, in this case, was determined by the losses in semiconductor elements and the resistive losses in conductors and instrument shunt. Considering a relatively low working-voltage of the switch, the maximum amplitude of the current did not exceed 20 A.

6. Discussion and conclusion

This paper introduces a high voltage solid state switch control circuit. On its basis, a switch prototype using 4 IGBTs at an applied voltage of up to 4 kV was designed. The switch underwent the tests, during which its characteristics were studied. Based on the conducted research, the following conclusions were made:

- A full control high voltage solid-state switch is able to generate the pulses of the length from several microseconds to the unlimited maximum length. These characteristics were obtained due to an offered control circuit method, using a current loop.
- The inductance of the controlling current loop has an insignificant effect on the values of generating pulses. Thus, its geometric dimensions are not limited, which, in its turn, will allow us to design the switches with a large number of steps and a high working-voltage.
- The tests of the designed switch revealed that on its basis, it is possible to build the square-wave generators, as well as the damped harmonic oscillation pulse generators.
- It has been observed that there is no overvoltage on the switch at its turn-off. It is probably connected to the existence of tail currents that are typical for the IGBTs. The accumulated charges in the collector of a transistor do not allow us to break the running current abruptly, preventing, in this way, the voltage spike on the stray circuit inductances.

The device demonstrated a high reliability. To set up the switching turn-on and turn-off times of every transistor is simple and takes up a minimum time. Another advantage of the device is its reduced size compared to the circuits using a fiber-optics technology.

The disadvantage of the device is a lack of the feedback in the control circuit reported in [5]. As a result, an electrical circuit drift and a dynamic voltage imbalance on the switch elements can occur.

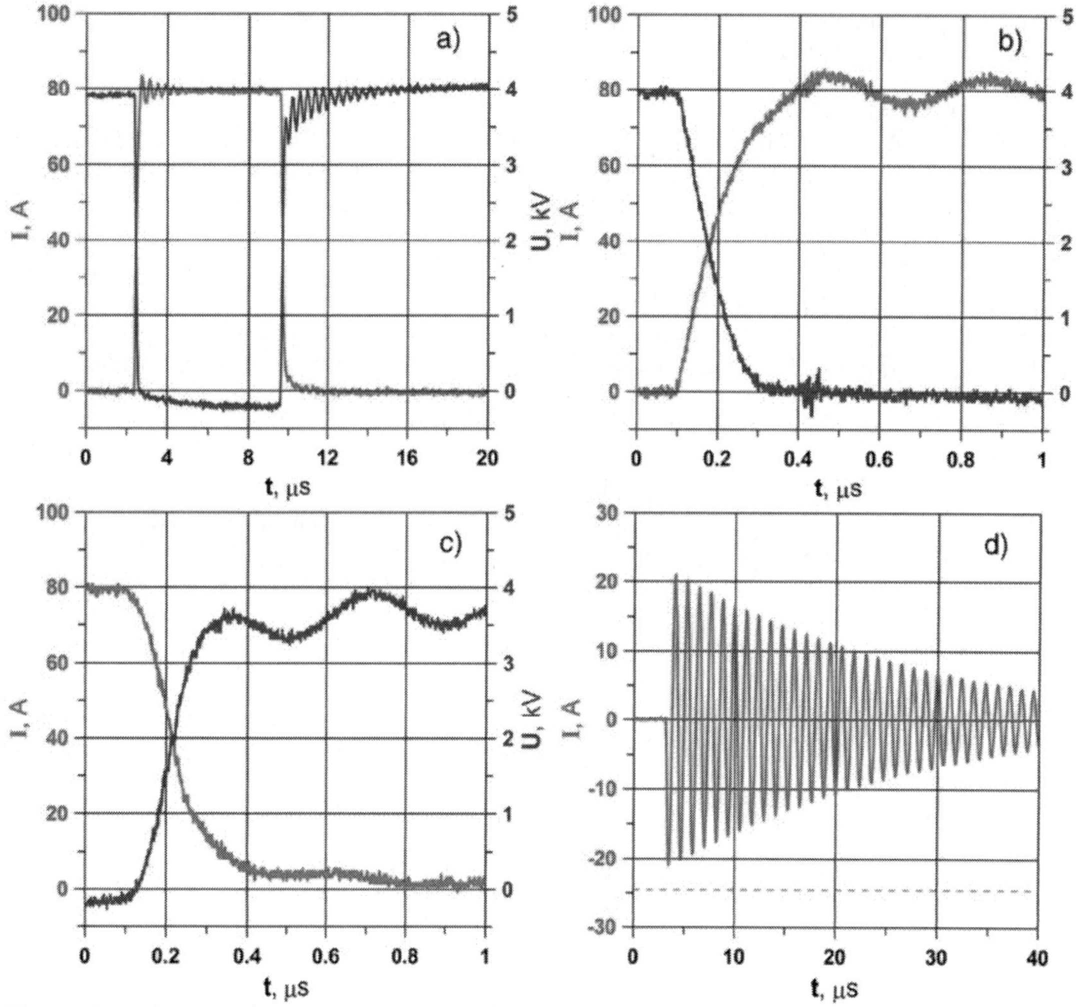

Figure 3. Voltage and current curves of switch operation in two modes: a-c – the formation of a rectangular pulse; d – the formation of damped harmonic oscillations.

Acknowledgements
The reported study was supported by Russian Foundation for Basic Research (RFBR), research project No. 15-08-01707-a.

References

[1] Jiang J, Ma G M, Luo D P, Li C R, Li Q M and Wang W 2014 *Review of Scientific Instruments* **85** 024706
[2] Baek J W, Yoo D W and Kim H G 2001 *IEEE Trans. on Industry Appl.* **37** 1832-39
[3] Moshkunov S I, Rebrov I E and Homich V U 2013 Russia? *Uspekhi prikladnoi fiziki* **1** pp 630-635
[4] Ponomarev A V, Gusev A I, Pedos M S and Mamontov U I 2014 *J. Izvestiya Vischih Uchebnih Zavedeniy: Phizika* **57** pp 250-254
[5] John V, Bum-Seok Suh and Lipo T A 1999 *IEEE Trans. on Industry Appl.* **35** pp 1108-1117

Parameters of runaway electron beam generated during excitation by nanosecond voltage pulses in short gaps filled with nitrogen

D A Sorokin[1,2], **V F Tarasenko**[1,2] **and M I Lomaev**[1,2]

[1] Institute of High Current Electronics SB RAS, 2/3 Akademichesky, Tomsk, 634055, Russia
[2] National Research Tomsk State University, 36 Lenina Ave., Tomsk, 634050, Russia

Email: SDmA-70@loi.hcei.tsc.ru

Abstract. Parameters of a supershort avalanche electron beam (SAEB) in nitrogen excited by the triangular-shaped voltage pulses with 45kV amplitude in incident wave and full width at half-maximum (FWHM) of ~ 1 ns were investigated. In experiments, cylindrical-shaped cathodes made of aluminum and stainless steel were used. An interelectrode distance in the gas-filled diode was 3, 5, and 8 mm. It was established that the highest values of SAEB's current were registered with an aluminum cathode. It was shown that, in contrast to the case of 8-mm gap length, when the interelectrode distances were 3 and 5 mm, the amplitude of SAEB current pulse in nitrogen began to decrease with high values of the pressure – 100 and 50 kPa, respectively.

1. Introduction

To date, a runaway electron beam (REB) and X-rays which take place during the nanosecond discharges in various gases at atmospheric pressure is registered by many research teams (see the monograph [1] and references therein). Nevertheless, there are relatively few detailed studies on the REB parameters behind the anode foil in a wide range of experimental conditions. Only three research teams have registered the REBs in atmospheric-pressure gases with collectors at the temporal resolution of a registration system not worse than 100 ps [2-4]. As well, there are few data on the REB parameters in gases at pressures ranging from 100 to 1 kPa. Although, still in [5] it had been shown that when the gas pressure was reduced from 100 to ~ 6 kPa, the number of electrons in a beam increases. However, due to the lack of equipment with subnanosecond temporal resolution, duration of the beam current pulse was not measured by the authors of [5]. The result of [5] was later confirmed in [6]. In addition, according to [6], in helium, hydrogen, neon, nitrogen, argon, methane, sulfur hexafluoride, krypton and xenon, the pressure of which ranged from 30 to 100 kPa, the REB had the full width at half-maximum (FWHM) of about 100 ps. More detailed studies of the REB current pulse duration when the pressure of nitrogen, hydrogen, and helium was decreased to those below atmospheric showed that the FWHM values of ~ 100 ps remained up to very low pressures of these gases (1.3, 4 and 8 kPa, respectively) [7].

The aim of this study was to investigate the amplitude and duration of the REB generated in short interelectrode gaps pressured with nitrogen in the range of 1-100 kPa, which was excited by

1-ns-duration triangular-shaped voltage pulses having incident wave's amplitude of 45 kV. According to [8], an electron beam generated in the gas-filled diode and registered behind the anode foil will be called the supershort avalanche electron beam – SAEB.

2. Experimental setup

The experimental setup was assembled of a gas-filled diode, a pulser FPG-100-01 (FID Technology), and a registration system. The pulser formed the triangular-shaped voltage pulses having the FWHM of ≅ 1 ns, the risetime of ≅ 0.7 ns, the falltime of ~ 5 ns, and the amplitude on high-resistance load of 90 kV which was applied to a cathode in the gas-filled diode. The cathode was 6-mm-diameter cylinder with the wall thickness of 1 mm and 100-μm-thickness edge at the end facing towards the anode (i.e., the cathode had a small radius of curvature). The cathodes were made of aluminum and stainless steel. The gas pressure in the diode varied from 1 to 100 kPa. A flat anode of the gas diode was an aluminum foil having thickness of 10 μm. The distance between electrodes in experiments was 3, 5, and 8 mm. The diode was evacuated to ~ $1.3 \cdot 10^{-4}$ Pa, using the forevacuum pump, and filled with high purity nitrogen containing not more than 0.001 % of impurities. The amplitude and the FWHM of the REB's current pulse were registered behind the anode foil using the collector of 20-mm-diameter receiving part. The signal from the capacitive voltage divider located near the gap was a triggering signal. The pulse repetition rate was 1 Hz. Waveforms of pulses from the capacitive voltage divider and collector were recorded using the digital real-time oscilloscope Agilent Technologies DSO-X6004A (bandwidth was 6 GHz, sampling rate was 20 GS/s). Waveforms were recorded in both the single pulse mode and averaging mode (over 16 pulses).

3. Results and discussion

Dependencies of the SAEB current pulse amplitudes from nitrogen pressure for the values of interelectrode distance d = 3, 5 and 8 mm are presented in Figs. 1, 2, and 3.

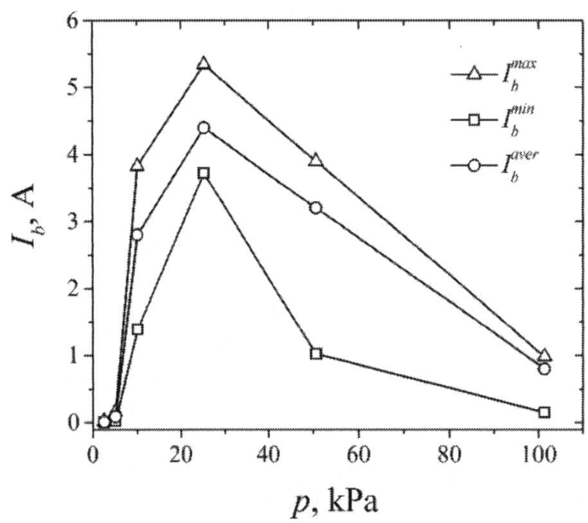

Figure 1. The amplitude I_b of a SAEB's current pulse versus nitrogen pressure p. I_b^{max} – a maximal value of the amplitude; I_b^{min} – a minimal value of the amplitude; I_b^{aver} – a value of the amplitude averaged over 16 pulses. The cathode is a 6-mm-diameter aluminum cylinder. The interelectrode distance is 8 mm.

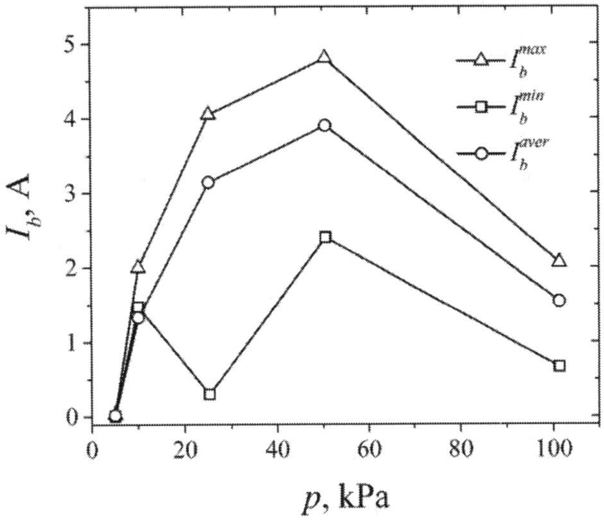

Figure 2. The amplitude I_b of a SAEB's current pulse versus nitrogen pressure p. I_b^{max} – a maximal value of the amplitude; I_b^{min} – a minimal value of the amplitude; I_b^{aver} – a value of the amplitude averaged over 16 pulses. The cathode is a 6-mm-diameter aluminum cylinder. The interelectrode distance is 5 mm.

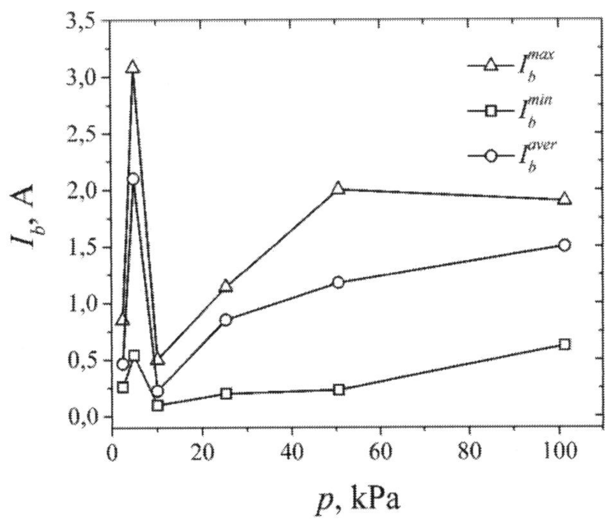

Figure 3. The amplitude I_b of a SAEB's current pulse versus nitrogen pressure p. I_b^{max} – a maximal value of the amplitude; I_b^{min} – a minimal value of the amplitude; I_b^{aver} – a value of the amplitude averaged over 16 pulses. The cathode is a 6-mm-diameter aluminum cylinder. The interelectrode distance is 3 mm.

There are maximal, minimal, and average (over 16 pulses) values of the SAEB current pulse amplitude in the figures. The presented data were obtained using aluminum cathode. When the interelectrode distance d decreased below 8 mm and gas pressure - below 100 kPa, it was found that, the SAEB current pulse amplitude started to decrease at relatively high pressure values. So, when d was 8, 5, and 3 mm, the decrease of the SAEB current pulse amplitude was observed at nitrogen pressures of 25, 50 and 100 kPa, respectively. Thus, the highest amplitude values of SAEB current pulses at the interelectrode distance of 3 mm take place at high pressure. The dependencies shown in figures above differ from those obtained earlier in [2], where the interelectrode distance between the cylindrical cathode and the anode was $d \geq 12$ mm. The duration of the current pulse of an electron beam generated in the SAEB mode did not change with gas pressure and interelectrode distance and had the value of FWHM of ~ 110 ps. At low pressures, lower than 10 kPa, as in [7], the transition to the vacuum-diode mode of electron beam generation occurred. In this mode, the FWHM of an electron beam current pulse increased. The gas pressure corresponding to change of REB generation mode was depended on the distance between the electrodes and had the maximal value at d = 3 mm. The amplitude of an electron beam current pulse in the transitional mode usually increased. In the vacuum-diode mode (at pressure lower than 0.1 kPa) the amplitude of an electron beam current pulse was lower than that in the transitional mode.

The results obtained for small d can be explained by the delay of SAEB generation increasing with a decrease of the interelectrode distance. The voltage pulse having a triangular shape was used in these experiments for initiation of a breakdown in the gas-filled diode. Therefore, the decrease of pressure at short interelectrode distances led to the breakdown delay, and the SAEB was generated during the voltage pulse falltime. Accordingly, the voltage reduced during SAEB generation. The delay of breakdown and the SAEB generation process had maximal values at the minimal interelectrode distance (d = 3 mm). Therefore, at this interelectrode distance, the amplitude of a beam current pulse began to decrease under 100kPa nitrogen pressure.

Dependencies of the amplitude of SAEB current pulse on the nitrogen pressure at d = 5 mm for the stainless steel cylindrical cathode are demonstrated in Fig. 4.

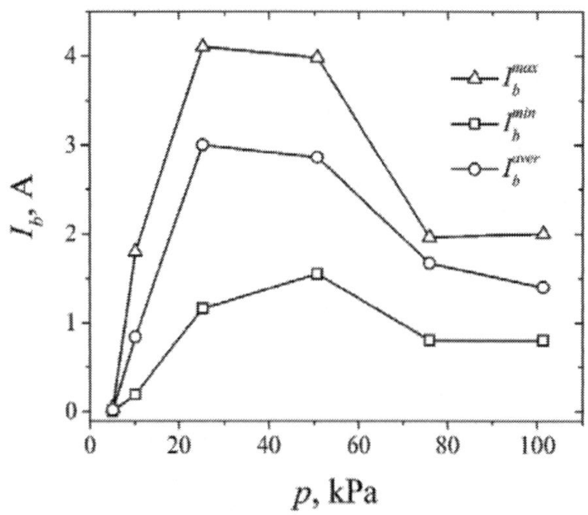

Figure 4. The amplitude I_b of a SAEB current pulse versus nitrogen pressure p. I_b^{max} – a maximal value of the amplitude; I_b^{min} – a minimal value of the amplitude; I_b^{aver} – a value of the amplitude averaged over 16 pulses. The cathode is a 6-mm-diameter stainless steel cylinder. The interelectrode distance is 5 mm.

As in the previous figures, there are maximal, minimal, and average (over 16 pulses) values of the SAEB current pulse amplitude in Fig. 4. Comparison of the results obtained for cathodes made of different materials indicates that under conditions of our experiments, the cathode material has a different effect than those reported in [8]. Both our experiments and a number of experiments reported in [8] demonstrated that the voltage pulse risetime did not exceed 1 ns. However, when we used the stainless steel cylindrical cathode, the SAEB current pulse amplitude did not only demonstrate any increase, but even decreased compared to the aluminum cathode. This fact can be explained by differences in the waveform and the amplitude of voltage pulses, which were used in our experiments and reported in [8]. At lower voltage, the stainless steel cathode applied in our experiments did not increase these values during SAEB generation process. It is known that the work function of aluminum electrons is lower than that of the stainless steel. Accordingly, when the voltage applied across the gap is approximately equal, the amplitude of an electron beam current should be higher when we use an aluminum cathode. We observed this in our experiments, see Fig. 2 and 4.

The parameters of runaway electron beams in the gaps with an interelectrode distance of some millimeters can be caclulated using new hybrid mathematical model [9], which allows simulation of the SAEB generation mode. A comparison of calculation results and the experimental data will be reported in the nearest future.

4. Conclusion

The investigation have shown that the parameters of SAEB, which was generated in the diode with an interelectrode distances 3, 5, and 8 mm, filled with nitrogen, and was registered behind the anode foil, depend on the interelectrode distance, parameters of the voltage pulse, and the material of potential cathode with a small radius of curvature (cylindrical cathode in this case). At interelectrode distance of 5 mm and excitation by triangular-shaped voltage pulses having the amplitude in the incident wave of 45 kV, the risetime of $\cong 0.7$ ns and the FWHM of $\cong 1$ ns the highest amplitudes of the SAEB current pulses is registered when the cathode is made of aluminum. A decrease in the SAEB current pulse amplitude at interelectrode distances 3 and 5 mm with aluminum cathode begins at essentially higher nitrogen pressures (100 and 50 kPa, respectively), than it had been reported in [7]. The obtained dependencies can be explained by a substantial influence of the waveform, duration, and the amplitude of voltage pulse on the value of breakdown voltage for a gap used. It follows from the experimental data that, comparing the SAEB parameters obtained using different experimental equipment, it is necessary to take into account all pulsers parameters, a gas diode design, as well as a design, sizes, and a material of potential cathode with a small radius of curvature.

Acknowledgments

The work is performed in the framework of the Russian Science Foundation (the project #14-29-00052).

References

[1] Edited by Tarasenko V F 2015 *Generation of Runaway Electrons and X-rays in the Discharges of High Pressure* (Tomsk, Russia: STS) in Russian

[2] Tarasenko V F, Baksht E Kh, Burachenko A G, Kostyrya I D, Lomaev M I and Rybka D V 2008 *Plasma Dev. Oper.* **16** (4) 267–298

[3] Mesyats G A, Yalandin M I, Reutova A G, Sharypov K A, Shpak V G. and Shunailov S A 2012 *Plasma physics reports* **38** (1) 29-45

[4] Alekseev S B., Lomaev M I, Rybka D V, Tarasenko V F, Shao T, Zhang Ch and Yan P 2013 *High Voltage Engineering* **39** (9) 2112-2118

[5] Tarasova L V, Khudyakova, L N, Loiko T V and Tsukerman V A 1974 *J. Tech. Phys.* **44** (3) 564–568 [in Russian]

[6] Baksht E Kh, Lomaev V I, Rybka D V and Tarasenko V F 2006 *Tech. Phys. Lett.* **32** (11) 948–950

[7] Tarasenko V F, Baksht E Kh, Burachenko A G, Lomaev M I, Sorokin D A and Shut'ko Yu V 2010, *Tech. Phys. Lett.* **36** (4) 375-378

[8] Zhang Ch, Tarasenko V F, Shao T, Baksht E Kh, Burachenko A G, Yan P and Kostyrya I D 2013 *Laser and Particle Beams* **31** 353- 364

[9] Kozyrev A V, Kozhevnikov V Yu, Lomaev M I, Sorokin D A, Semeniuk N S and Tarasenko V F 2016 *EPL* **114**, 45001

Composition of pyrolysis gas from oil shale at various stages of heating

S M Martemyanov, A A Bukharkin, I A Koryashov and A A Ivanov

National research Tomsk polytechnic university, Russia, Tomsk, Lenin str., 30, 634050

Email: martemyanov@tpu.ru

Abstract. Underground, the pyrolytic conversion of an oil shale in the nearest future may become an alternative source of a fuel gas and a synthetic oil. The main scientific problem in designing this technology is to provide a methodology for determination of the optimal mode of heating the subterranean formation. Such a methodology must allow predicting the composition of the pyrolysis products and the energy consumption at a given heating rate of the subterranean formation. The paper describes the results of heating of the oil shale fragments in conditions similar to the underground. The dynamics of composition of the gaseous products of pyrolysis are presented and analyzed.

1. Introduction

Retorting of an oil shale has long been regarded as a source of alternative energy and chemical technology raw materials. The main its products are a synthetic oil and a combustible gas. Before retorting the extracted shale, the rock is ground to a desired fractional composition and charged to a specific reactor. There the oil shale is exposed to heat and thermal degradation.

The composition and yield of the pyrolysis products is dependent on the process parameters of retorting – the fractional composition of raw rock, the heating rate, the slag removal rate, the technological scheme of movement of the rock in the retort. Many technological regimes have been developed for processing of the oil shale in retorts - a number of designs and processing parameters to achieve the maximum yield of products are already known [1-4].

At the same time, the necessity of extraction of the rocks on the surface is an inevitable limitation of the retorting. This operation determines the cost of obtained products. In this regard, the underground pyrolytic conversion, which consists in a direct heating of the rocks in situ, looks promising. However, in this case, obtaining the gas with constant composition is practically impossible. Heating the continuous medium, which takes place in a subterranean formation, a temperature gradient and the movement of liquid and gaseous products of pyrolysis are unavoidable. Calculation of such heating requires taking into account many factors [5].

Heating of the subterranean formation by an electric current can be used for the conversion of the underground reservoirs of the oil shale. In this case, electrodes connected to the surface generator are placed into two wells. The source of heat is the cylindrical area between the electrodes through which the current flows. The heat spreads from the source due to the thermal conductivity and the motion of pyrolysis products. The gaseous pyrolysis products are selected directly in the heating process through the wells, and the shale oil may be selected from the reservoir after extraction of the electrodes.

At the initial stage of heating of the underground formation, the electric current forms a thin conductive channel, which has a high temperature because of the high energy density. Then, the conductive channel expands and the heat is being distributed in the surrounding space. The paper describes an experimental simulation of the heating of the subterranean formation on a laboratory scale. The purpose of the study is a definition of the dynamics of the composition of pyrolytic gas during the heating of the oil shale at different stages of heating.

2. Experiment conditions

Experimental studies were carried out for the oil shale of Huadian deposit (China). A sample of the rock was heated under conditions, similar to the underground.

The sample was placed in a concrete shell, which simulated the effect of enclosing rocks. For creating the shell, the sample with the rebar cage was placed into a pouring form and poured by a concrete. The sample was covered with a thin layer of polyethylene film for waterproofing from the concrete. The concrete was reinforced by a fiberglass rebar of 5 mm thick. The fiberglass is an insulator and does not affect the distribution of the field and the current flow. The poured sample was dried to a constant weight under room conditions (about 30 days).

The temperature of the sample was measured by tungsten-rhenium thermocouples, which were inserted into the sample in the ceramic covers. The covers were used as dielectric insulators to prevent the damage of a measuring unit and thermocouples.

The electrodes were made as graphite rods of 10 mm diameter, spaced 250 mm apart and inserted into the sample till a depth ~120 mm. The temperature was measured at 4 points. The first point was located in the center of the inter-electrode distance. The second, third, and fourth point were removed from the axis between the electrodes at 30 mm, 60 mm, and 90 mm, respectively (Figure 1).

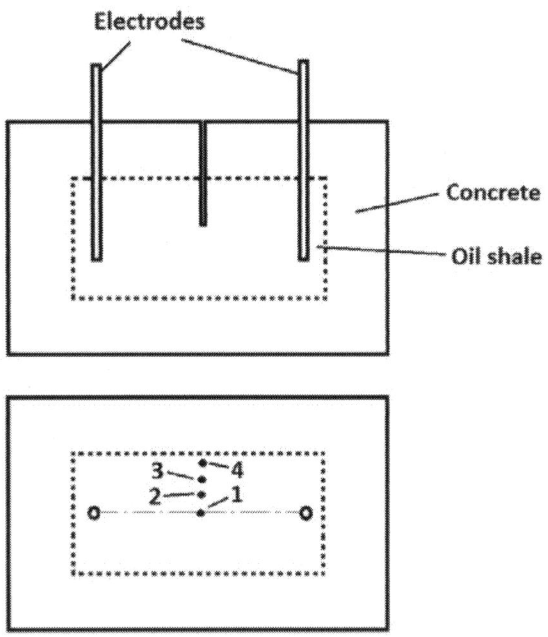

Figure 1. A diagram of thermocouples placing (a side view and a top view): 1, 2, 3, 4 – thermocouples

The graphite conductive paste was used to improve an electric contact between the electrode surface and the rock surface. The sample was placed inside a chamber, which was evacuated to remove oxygen and then, filled with nitrogen to a pressure of 5 kgf/cm^2. The increased pressure inside the chamber simulated the pressure of fluids in the underground reservoir.

During heating, the gas was sampled from the chamber without stopping the process. The gas was filtered from an aerosol and disintegrated by the filter-reducer LFR-1/4-D-5M-MINI (Festo, Germany) with a pore size of 5 microns. The gas chromatograph was supplied, under a slight positive pressure (about 0.01 kgf/cm^2). A gas was cooled directly during movement of the sample through the tubes, thus, the gas got the chromatograph at room temperature. We used Festo valves and tubes.

Gas analysis was performed using a gas chromatograph analyser Agilent 7890A, with a thermal conductivity detector using a column "ShinCarbon ST 100/120 mesh" 2 m long and of 1 mm diameter, according to the standard method for an absolute calibration. The chromatograph was calibrated using a standard gas mixture.

3. Results

In a given experiment, the heating power was about 500 W during all heating time. If we assume a cylindrical heating channel, it is reasonable to act with the power density, divided by the length of the inter-electrode gap. In the described experiment, this value was 2 kW/m. The resulting heating characteristics are shown in Figure 2. The heating was carried out continuously for 19 hours.

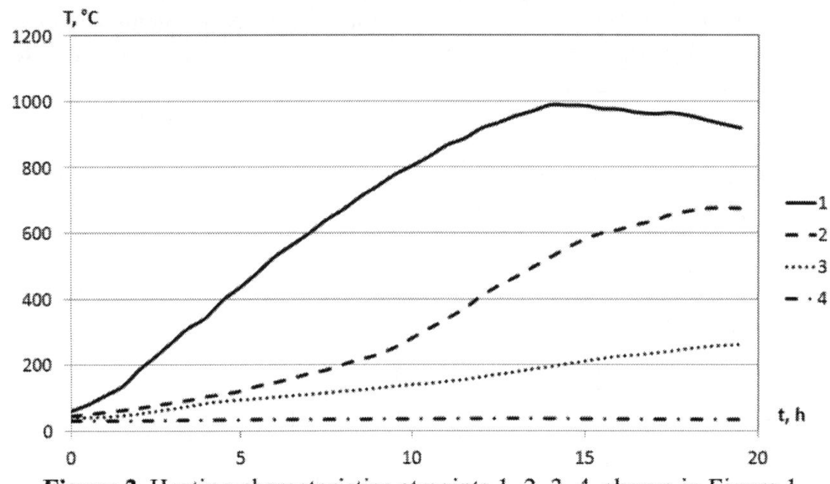

Figure 2. Heating characteristics at points 1, 2, 3, 4, shown in Figure 1

The gas was sampled from the chamber at intervals of about 1 hour. The dynamics of the gas composition is shown in Figure 3.

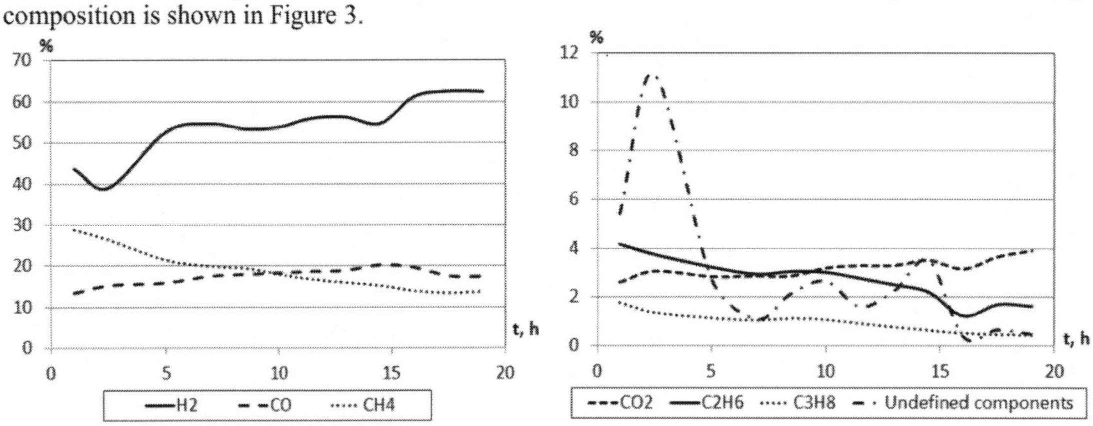

Figure 3. Pyrolysis gas composition dynamics

With the change of the gas composition, its calorific value also changes. The dynamics of the calorific value during heating is shown in Figure 4.

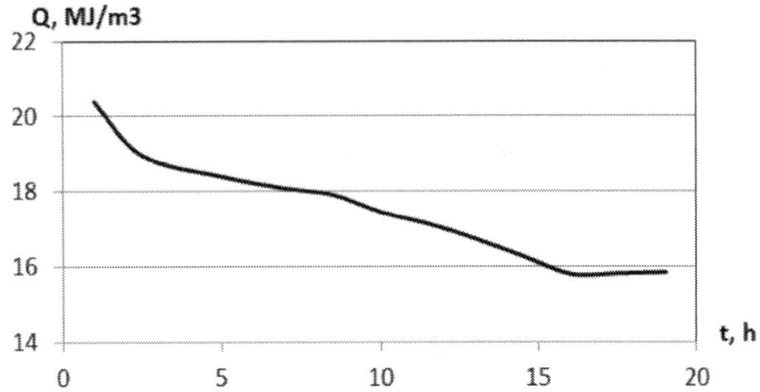

Figure 4. A changing calorific value of the pyrolysis gas

4. Discussion

The experimental data confirmed the estimates of the low rate of thermal energy distribution in an oil shale [5, 6]. In this experiment, the temperature in the center of a heated zone at the time, when a temperature at 90 mm apart from the channel center did not reach 100°C, exceeded 900°C. In such a situation, the increasing input power will reduce an efficiency for the following reasons:

- an unproductive temperature rise in the center of a channel, which has been already depleted;
- a consumption of thermal energy in the power-consuming reactions of decomposition of rock mineral components in the channel center, which occurs at temperatures above 900°C;
- an oxygen release due to mineral decomposition and oxidation of the fuel conversion products.

Thus, a viable option for heating and pyrolysis of large volumes of the rock under underground conditions is to increase the heating time at a fixed power. Calculations based on a mathematical model and experimental studies have shown that to achieve a temperature of 400°C at a point distanced 1 m apart from the center of the channel takes about 40 days.

The pyrolysis gas consists mainly of such combustible components as hydrogen, carbon monoxide, and methane. At the beginning of heating, only a small area in the vicinity of the electro breakdown channel undergoes the pyrolysis. The temperature in this channel is high, then, as far as the larger rock volume heats, it begins to decrease. According to the results, the number of undetectable by chromatograph components decrease during the heating. Most likely, these are the sulfur compounds (sulfur dioxide, hydrogen sulfide), the output of which is associated with a decay of the weak bond C=S, and also a water steam.

At the first stage of the heating, the gas contains an abnormally high volume fraction of methane, which is not typical gas for the pyrolysis from solid fuels. This gas is not a product of thermal decomposition of the organic part of the rock, but it is mainly a gas desorbed from the oil shale by high temperature. This is confirmed by an approximately constant fraction of methane in the final stage of the experiment, when the heating is close to a thermodynamic balance.

The component composition diagram allows concluding that the composition of the gaseous products is appropriate for the production of liquid synthetic fuels due to the Fischer-Tropsch process. The most suitable ratio H_2/CO has a value, which is equal to or greater than 2. In this experiment, the H_2/CO ratio was about 3.5.

During heating, the balance of carbon monoxide and dioxide is shifted towards CO. At the same time, the content of alkanes is reduced, resulting in a reduction of the calorific value, since they are the most calorific components of the gas. The increase of the ratio of hydrogen/alkanes can be explained by the decay of alkanes by an increasing temperature.

Acknowledgement
The reported study was funded by RFBR, according to the research project No. 16-38-60006 mol_a_dk.

References

[1] Brandt A R 2009 *Energy & Fuels* **23** 6253
[2] Kann J et al 2004 *Oil Shale* **21** 195
[3] Jiang X M, Han X X and Cui Z G 2007 *Energy* **32** 772
[4] Soone J and Doilov S 2003 *Oil Shale* **20** 311
[5] Knyazeva A G, Martemiyanov S M and Maslov A L 2015 *EPJ Web of Conferences* **82** 01063
[6] Martemyanov S M and Maslov A L 2014 *Advanced Materials Research* **1040** 620

IOP Conf. Series: Journal of Physics: Conf. Series **830** (2017) 012009 doi:10.1088/1742-6596/830/1/012009

Mason's equation application for prediction of voltage of oil shale treeing breakdown

S M Martemyanov

National research Tomsk polytechnic university, Russia, Tomsk, Lenin str., 30, 634050

Email: martemyanov@tpu.ru

Abstract. The application of the formula, which is used to calculate the maximum field at the tip of the pin-plane electrode system was proposed to describe the process of electrical treeing and treeing breakdown in an oil shale. An analytical expression for the calculation of the treeing breakdown voltage in the oil shale, as a function of the inter-electrode distance, was taken. A high accuracy of the correspondence of the model to the experimental data in the range of inter-electrode distances from 0.03 to 0.5 m was taken.

1. Introduction

Electrical treeing, which is the formation of tree-like channel structures, is observed in the low conductive materials under high voltage. In [1], the treeing phenomenology for polymeric insulation materials is widely described. It is ascertained that the growth of channel structures called dendrites occurs due to degradation of the material by partial discharges (PD). Many treeing diagnostic methods are based on measurement of the partial discharge characteristics [2, 3].

Depending on the dielectric material and its molecular structure, the dendrites can be conductive and nonconductive [4, 5]. In the case of non-conductive dendrites, the treeing is going on due to an increased voltage within the dendrite. The main part of the PDs is concentrated in the area of the dendrite base. The PDs have a relatively large apparent charge, which increases by lengthening of the dendrite. The structure is lengthened when the discharge streamers reach the tips of the dendrite. This forms a hollow channel structure with dielectric walls. In this case, the intensity of partial discharge weakens by the growth and branching of the structure.

In the case of conductive dendrites the structure lengthening occurs due to the increased voltage, in the material near the tips of the dendrites. The PDs act mainly on the tips of the dendrite branches and extend it by a thermal effect of the partial discharge. The PDs have a relatively low apparent charge, which is almost independent of the size of the dendrite. The structure lengthens more rapidly than in the case of non-conductive dendrites. By the growth and branching of the tree, the intensity of the PDs increases. The channel structure is coated with a low-ohmic layer of products of a dielectric thermal destruction [4].

The partial discharges and treeing can occur in the carbonaceous rocks, such as the brown and black coal and the oil shale. [6], leading to the breakdown of large interelectrode distances at relatively low voltage. This effect can be used for breakdown and underground heating of solid fuels formation in the technologies of in situ processing [7]. However, we need to know the voltage required for the breakdown of a given inter-electrode distance.

In dielectrics with the non-conductive dendrites, the breakdown moment can be predicted with the help of the apparent charge value [2, 3]. In dielectrics with the conductive dendrites, the PD characteristics do not correlate with a breakdown time. The reason can be that in the structures with the conductive dendrite, the PDs arise mainly on the tips of the tree, so the discharge characteristics depend a little on the size of the tree.

We observed treeing in solid fossil fuels (the black coal, the brown coal, the oil shale) and its correlation with the characteristics of the partial discharges. Observations of treeing in the solid fuels showed the following features. At a constant value of the applied voltage, the dendrite growth is going on for some time, after which the intensity of partial discharge is reduced. As far as the voltage is increasing, the intensity of the PDs rises again and continues for a while, then decreases again. This behavior is related to the fact that the voltage change on the electrodes will affect the value of the field concentrated at the tips of the dendrite. The Mason's article [8] presented a formula that describes the field at the tip of the electrode in the tip-plane electrode system. Use of this equation can help to determine the field on the tips of the dendrites.

The article describes a model, which can be used to predict the treeing breakdown voltage in the solid fossil fuels, as a function of the inter-electrode distance, using the Mason's formula.

2. Experimental setup and results

In the experimental setup, the dependence of the treeing breakdown voltage on the distance between the electrodes was investigated (Figure 1).

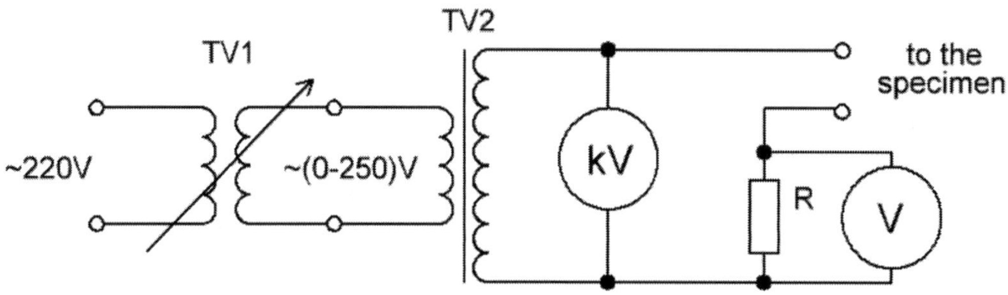

Figure 1. Experimental setup circuit

In the setup, the voltage ~220 V is regulated by a transformer with a variable magnetic coupling over the voltage range from 0 to 250 V. This voltage supplies a step-up transformer with an output voltage up to 100 kV. The measured parameters in the experiments were the voltage at the sample and the current flowing through this sample, which were detected using a resistive shunt. The electrodes were made of a steel rod and inserted into the rock in 5-10 cm, to provide an electrical contact of the electrodes with an interlayer cut for an efficient dendrite growth.

In the experiments, the voltage was gradually raising till the moment of breakdown. During the experiment, the growth of dendrites was diagnosed by the presence of current spikes caused by a large number of short pulses in the circuit, which is typical for partial discharges. It was noticed that at a constant value of the applied voltage, the treeing stopped after some time, as if reaching a potential barrier. When the voltage increased, the treeing was detected again, but after some time, stoped. The voltage increasing rate was 100 V/min. This rate was enough for treeing processes to complete and reduce the activity, waiting for a further increase of the voltage.

The voltage of the treeing breakdown, as a function of the inter-electrode distance, for the oil shale of Huadian deposit (China) is shown in Figure 2.

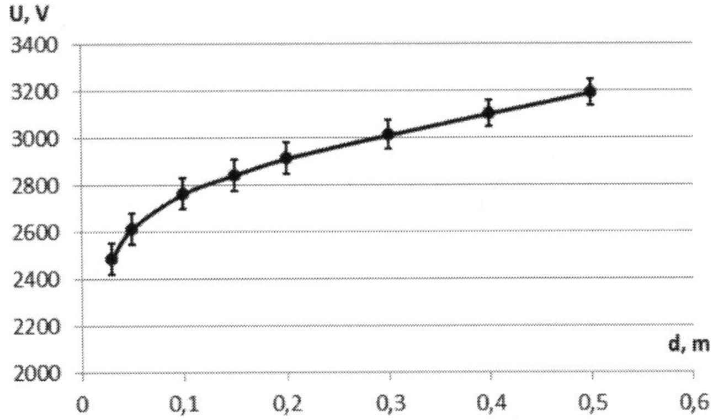

Figure 2. Voltage of treeing breakdown of oil shale as a function of inter-electrode distance

The voltage is dependent on the distance nonlinearly, as would be in the case of a pure electric breakdown. The shape of curve induces a supposition that the breakdown at large distances between the electrodes requires a relatively low voltage. This effect can be used in the technologies of in situ heating and the pyrolysis of solid fuels, where the first stage is a breakdown of the inter-electrode distance, and the second one is the heating of the formation by a joule heat in a breakdown channel [7]. In this case, the distance between the electrodes can reach tens to hundreds of meters, and the voltage, at the same time, is technically achievable.

This application of the treeing breakdown needs a way, to predict the voltage required for the breakdown of the given inter-electrode distance. If we assume that each rock can be tested in a laboratory, at small distances, the obtained characteristic can be extrapolated to larger distances. In this case, a low of approximation of the experimental data should be reasonably chosen. For the purpose of such an approximation, we have considered the use of Mason's formula for the treeing breakdown modeling. The calculations were performed for the oil shale of Huadian deposit (China).

3. Mason's equation application for breakdown modelling

The following model of treeing and treeing breakdown process was assumed for phenomena modelling. The partial discharges occur at a tip of the dendrite, as a breakdown of a small part of the material in the area, where the field is extremely high and exceeds the electrical strength of the rocks (Figure 3).

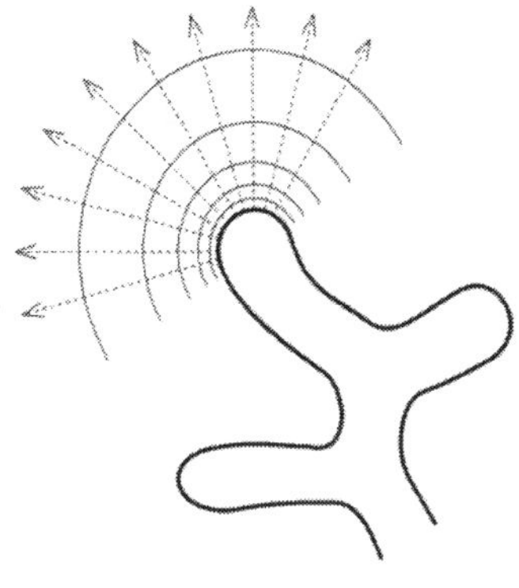

Figure 3. Field concentration near the tip of the dendrite

The maximum field near the tip can be calculated by the Mason's formula [8]:

$$E = \frac{2V}{r\ln(1+\frac{4d}{r})} \quad (1)$$

where E is the maximum field near the tip of the dendrite, V is the voltage between the electrodes, r is the radius of the tip of the dendrite, d is the distance between the electrodes.

When the field near the tip exceeds the electrical strength of the rock, the local electrical breakdown (i.e. a partial discharge) occurs in this fragment of the rock. Suppose that the electric strength of the rock is approximately constant at all points, and the partial discharge occurs when the maximum field at the tip E exceeds a certain amount, called a critical field E_{cr} in similar models [9]. Thus, on the basis of (1), the PDs will occur, when the voltage between the electrodes is:

$$V = \frac{E_{кр}}{2} r\ln(1+\frac{4d}{r}) \quad (2)$$

Thus, using the formula (2), we obtain the dependence of the voltage between the electrodes, which will enough for a steady occurrence of partial discharges and treeing development, at a distance d, with a given value of the critical field E_{cr} and the tip radius r. If we assume that under this condition, the treeing development and breakdown will occur, then the (2) may be considered as a dependence of the breakdown voltage on the inter-electrode distance.

The radius of the tip of the dendrite depends on the material structure and composition. Suppose that for one rock, it is an approximately constant value. For the considered oil shale, the average radius was determined by an electron microscopy scanning and amounted to about 5 microns (Figure 4).

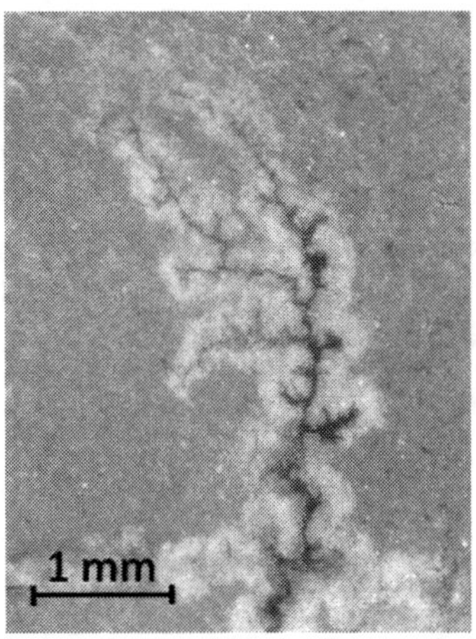

Figure 4. Dendrite in the oil shale

The value of the critical field E_{cr} can be determined using the approximation of the experimental data by obtained analytical characteristic. The approximation by least-squares method showed that the value of the critical field is $E_{cr} = 98\times10^6$ V/m (Figure 5a).

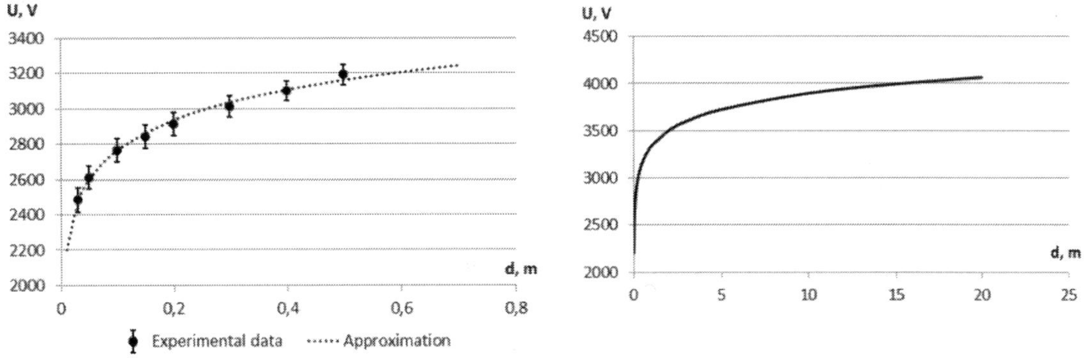

a b

Figure 5. Curves of the experiment and the model: a) the approximation of experimental data, b) extrapolation of obtained characteristic to larger inter-electrode distances

The average error of the approximation, relative to the average value of the experimental data, is less than 1%, suggesting that there is an exact correspondence between the model and the real processes.

An extrapolation of the obtained analytical characteristic to larger inter-electrode distances (Figure 5b) showed that the breakdown of the oil shale of 20 m requires the voltage of 4060 V.

4. Summary

The proposed approximation method can be used to predict of the voltage required for the treeing breakdown of solid fuels at a given distance between the electrodes. The advantage of this approximation method are a true account of a physical principle of the electrical treeing and physical parameters of the medium. The disadvantage of the proposed model is a possible spread in the values of the critical field and the radius of the tip of the dendrite relating to an inhomogeneity of the material of fossil rocks. It was noted that the real effect of the partial discharge starts at the voltage, which is significantly lower than this estimate.

For the given oil shale, the model has correspondence to the experimental data with high accuracy at small inter-electrode distances. At large inter-electrode distances, the compliance of the model should be checked under field conditions.

Acknowledgement

The reported study was funded by RFBR, according to the research project No. 16-38-60006 mol_a_dk.

References

[1] Dissado L A and Fothergill J C 1992 *Electrical degradation and breakdown in polymers* (London: Peter Peregrinus)
[2] Champion J V, Dodd S J and Alison J M 1996 *J. Phys. D: Appl. Phys.* **29** 2689
[3] Vogelsang R et al 2002 *Proc. 15th International Conference on Electrical Machines (Bruges)* vol 2002 (New-York: John Wiley & Sons ltd) 15
[4] Champion J V and Dodd S J 2001 *J. Phys. D: Appl. Phys.* **34** 1235
[5] Dodd S J, Chalashkanov N M and Fothergill J C Proc. 2010 International Conference on Solid Dielectrics (Potsdam)
[6] Bukharkin A A at al 2016 *Key Engineering Materials* **685** 672
[7] Martemyanov S M and Maslov A L 2014 *Advanced Materials Research* **1040** 620
[8] Mason J H 1955 *Proc. IEEE C* **102** 254
[9] Champion J V, Dodd S J and Stevens G C 1994 *J. Phys. D: Appl. Phys.* **27** 1020

IOP Conf. Series: Journal of Physics: Conf. Series **830** (2017) 012010 doi:10.1088/1742-6596/830/1/012010

Mitochondrial respiration inhibition after exposure to UWB pulses as a possible mechanism of antitumor action

L Zharkova[1,2], I Romanchenko[2], M Bolshakov[1,2] and V Rostov[1,2]

[1]National research Tomsk state university, Tomsk, Russia
[2]Institute of high current electronics SB RAS, Tomsk, Russia

Email: Zharkova_lubov@mail.ru

Abstract. The respiration of isolated mice liver mitochondria after exposure to nanosecond UWB pulses (0.15 – 36 kV/cm, 0.6 – 1.0 GHz centre frequency, 3 – 20 ns pulse duration) has been investigated. The respiratory control (RC, the ratio of oxygen consumption) was estimated. The possibility of mitochondrial membrane electroporation was detected as the decrease in the electrical resistance, according to the β-dispersion of the electric current. The monotonous decrease of RC after 1000 UWB pulses from 0.15 kV/cm was observed, the ohmic resistance of mitochondria suspension was reduced. The obtained data indicate the inhibitory effect of UWB pulses on a state of irradiated mitochondria and its membrane.

1. Introduction

It is known now that a pulsed ultra-wideband (UWB) radiation efficiently influences the biological objects. The UWB pulses of a microsecond duration have been used already for a tumor therapy in a combination with a chemotherapy, for an efficient targeted delivery of the chemotherapy, according to the mechanism of tumor cell membrane electroporation [1]. Even one nanosecond pulse of an electric field (20 kV/cm or more) can cause abnormalities in the development of irradiated fish eggs [2]. The nanosecond and ultrashort pulses of the UWB radiation (tens of kV/cm) can damage the membranes of intracellular organelles and temporary (reversible) violate the integrity of the plasma membrane, as well essentially increase the level of free intracellular calcium. All of them affect the level of cellular metabolism [3]. It is also known that microsecond electric pulses may reduce an ATP level in the irradiated cells [4]. The reducing amount of ATP, first, must be associated with a decrease in the production of ATP (by uncoupling of the oxidation and phosphorylation, and disruption of the electron transport chain (ETC) in the inner mitochondrial membrane). The UWB pulses are characterized by short pulse durations (tens of nanoseconds or shorter) and high pulse powers, and they should provide even more efficient action on the biological structures, such as mitochondria, and exercise their influence in a similar way. Mitochondria are the major cell organelles, in which the free chemical energy of food substrates is transferred and stored into the energy ATP molecules and further is expended for metabolic needs of a cell. From this point of view, it would be reasonable to study the effect of nanosecond UWB pulses on the coupling of oxidation and phosphorylation of isolated mitochondria (via assessing the intensity of mitochondrial respiration), as well as to identify a dependence of the effects on the field parameters and the number of affecting pulses.

2. Materials and methods

The mitochondria of an outbred white mice liver, which was isolated by the differential centrifugation [5], were used for the experiments. The suspensions of mitochondria were once exposed to 16 – 1000 nanosecond UWB pulses with a frequency of 0.6 – 1.0 GHz, a duration of 4 – 25 nanosecond, an amplitude of 0.1 – 36 kV/cm, and a pulse repetition rate of 13 pules per second (pps). Generation of high-frequency pulses was carried out by exciting the gyromagnetic precession in a nonlinear transmission line with a saturated ferrite. The RF pulse from a nonlinear transmission line was converted from a coaxial TEM mode into a TE_{01} mode of a rectangular waveguide by a specially designed converter [6-7]. The Eppendorf test tube with a suspension was placed in the center of the waveguide, where the amplitude of the electric field had a maximum value. The RF pulse was absorbed in the alcohol load after interacting with the sample. The mitochondria respiration was measured by an oxygen meter AKPM-02 (Russia). The respiratory control (the ratio of oxygen consumption) was calculated. The possibility of mitochondrial membrane electroporation was detected as a decrease in the electrical resistance, according to the β-dispersion of the electric current (measured by the LCR meter, Belarus). The irradiated and the sham mitochondrial suspensions were used. The sham suspensions were subjected to the same manipulations, as irradiated one, but without activating the radiation source. Six replicates were performed for each exposure experiment and sham. Statistical comparisons were made using the nonparametric Mann-Whitney test in Statistica 6.0 software.

3. Results

The respiratory control by Chance [8] is the most informative indicator for the analysis of a mitochondria functioning. The RC is equal to the ratio of the mitochondrial oxygen absorption rate in the presence of the ADP to the oxygen absorption rate after the exhaustion of the ADP in suspension. The RC equal to 5 shows a good and efficient conjugation of the oxidation and the phosphorylation [5] for our isolated mitochondria (Figure 1, sham).

It was found that the nanosecond UWB pulses affect the rate of mitochondrial respiration. The effect was shown in a decrease of the oxygen consumption by mitochondria in the presence of ADP. As a result, the value of the RC was decreased. The effect depends on the electric field. The RC decreased monotonically with increasing a field strength to a value of 0.15 kV/cm or more (Figure 1). This indicates the inhibitory effect of the UWB pulses on mitochondria.

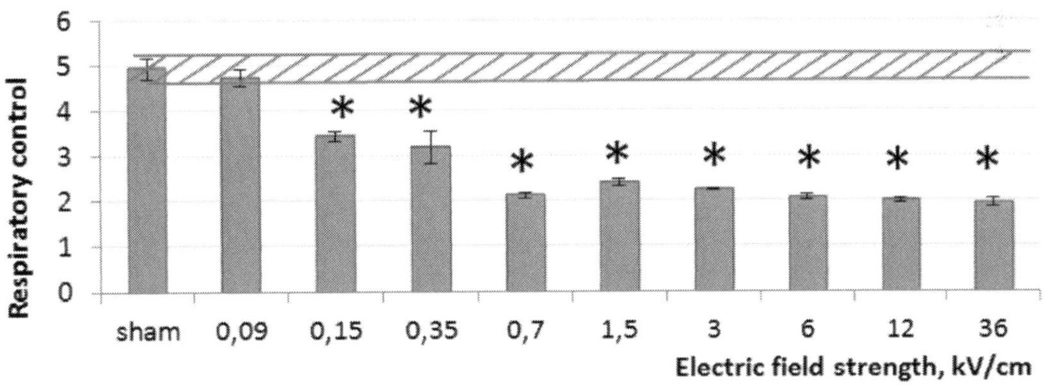

Figure 1. Changes of the RC value after exposure to 1000 nanosecond UWB pulses with field of 0.09–36 kV/cm per pulse. Shaded space is ± standard error of the mean of RC for sham mitochondria. * – differences of indicators are significant as compared with sham ($p \leq 0.05$).

The obtained effect of the respiration inhibition could be caused either by an electric field or by a total absorbed UWB radiation energy. Clarifying experiments were conducted in order to determine the contribution of the energy impact on the effect. In these experiments, the effect was measured after an exposure to the pulses of different intensity and different amounts, provided that the energy of RF pulses in the bunch (the RF power * pulse number = const) was the same. The RF energy was measured according to the waveforms from the E-sensor in the waveguide. It was found that the effect of the electromagnetic pulse with the same total energy from the UWB source provided the formation of an equal biological effect. In particular, there was a 30% reduction of the RC after an exposure to the UWB pulses with an incident total energy of 2.5 mJ per session (Figure 2A) and a 61% reduction after an exposure to the UWB pulses with a higher incident total energy of 10 mJ per session (Figure 2B). Thus, the increasing of energy leads to an increase in the biological effects.

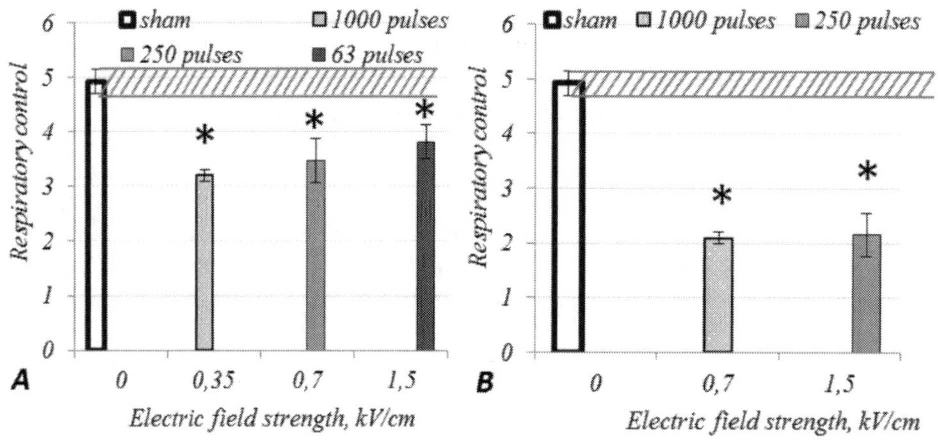

Figure 2. RC changes after exposure to 1000 nanosecond UWB pulses with incident total energy of 2.5 mJ (***A***) and 10 mJ (***B***). (Field intensity from 0.35 to 1.5 kV/cm per pulse, the number of pulses from 1000 to 63, center frequency 0.6 GHz, pulse duration 25 ns). Symbols are similar to Figure 1

The drop of the RC indicates a decreased conjugation of an oxidation and phosphorylation, as well as a broken integrity of the mitochondrial membrane. Under such conditions, it should be a disturbed ATP production by the irradiated mitochondria. A confirmation of this assumption validity could be the data of Frandsen S.K. [4], in which the decrease in the ATP level after an exposure to microsecond pulses was described. The authors of the work linked this effect with an electroporation of the cell membranes and the intracellular structures. Our additional experiments were conducted to check the possibility of an electroporation after the UWB pulses. In those experiments, the resistance of a mitochondrial suspension to the alternating current (on AC frequencies 10^4-10^6 Hz, β-dispersion) was measured. The resistance to a current of such frequencies characterized the physicochemical state of the biological membranes. It has been found that the UWB reduces a resistance of mitochondria, measured at frequencies of β-dispersion. In particular, an exposure to 1000 pulses of the UWB with a field of 3 kV/cm reduced an ohmic resistance by 23%. The observed decrease in the mitochondrial membrane resistance may be due to the formation of additional structures of a nonspecific conductivity (electropores) induced by the exposure to the UWB radiation.

4. Conclusions

The obtained data indicate that the UWB pulses of a nanosecond duration reduce the level of the mitochondrial respiration and break the oxidative phosphorylation coupling. The observed changes could be due to a break of a membrane structure in some irradiated mitochondria.

The reduced mitochondrial suspensions resistance after the exposure to UWB pulses is a result of the electroporation of a lipid bilayer of an inner mitochondrial membrane. This should lead to a decrease in a proton potential on the inner mitochondrial membrane, and consequently to the oxidation and the phosphorylation uncoupling, and therefore, reduce the RC. As a result, it can affect either changes in a cellular metabolism or a deficiency of the ATP, or the generation of reactive oxygen species (as a result of a disruption of the respiratory chain), or the oxidative modification of the lipid and protein components of mitochondrial membranes [9, 10]. Moreover, the RC reduction may be due to the loss of mitochondrial cytochrome *c* ratio or a shift of the ATP synthetase /ATPase activity relations (caused by a decrease of the proton potential) [10, 11]. All these processes can explain the observed decrease in the oxygen consumption of mitochondria, reducing the RC and the ohmic resistance of mitochondrial suspensions.

Nevertheless, the effects depended on the electric field and the number of pulses, as well as on the total energy per pulse. This circumstance is consistent with the opinion of T. Schunc [12] that it is the amount of energy, which determines the efficiency of the electroporation. The temperature of irradiated samples increases due to the energy growth, and then the increase in the effect, upon more energy absorption, can be explained by the addition in a mobility of fatty acid residues of the mitochondrial membrane lipids, which will provide the increase of the electroporation probability. Changes of intracellular processes are possible in cases when the UWB radiation affects the functioning of mitochondria. This makes it possible to control the metabolism of cells, including tumor cells, by the UWB radiation with optimal parameters. The observed dependence of the effect on the number of pulses is fundamentally important because it will minimize the effect on any cell (including the cells around the tumor) while retaining the required result. In addition, a research in this area could contribute to the development of the perspective methods of a directed transport of biologically active substances into cells using the electroporation induced by the UWB exposure.

Acknowledgments
The work was supported by the Russian Foundation for Basic Research (grant No. 16-38-60040).

References

[1] Jordan DW, Uhler MD, Gilgenbach RM, Lau YY 2006 *Journal of applied physics* **99**, 094701
[2] Dong Koo Kang, S Hamid R Hosseini, Eri Shiraishi, Masato Yamanaka and Hidenori Akiyama 2012 *IEEE Transactions on Plasma Science* **40** (10) pp 2379– 2387
[3] Ravi P Joshi, K H Schoenbach 2010 *Critical Review in Biomedical Engineering* vol **38** (3) pp 255–304
[4] Frandsen SK, Gibot L, Madi M, Gehl G and Rols MP 2015 *Plos ONE* **10** (12): e0144028
[5] Knjazeva IR, Ivanov VV, Zharkova LP, Kutenkov OP, Rostov VV and Bolshakov MA 2011 *Tomsk state university journal of biology* vol **4** p 113–123
[6] Gubanov VP, Gunin AV, Kovalchuk OB, Kutenkov VO, Romanchenko I V and Rostov VV 2009 *Tech. Phys. Lett.* vol **35** (7), pp 626–628
[7] Romanchenko IV, Rostov VV, Gubanov VP, Stepchenko AS, Gunin AV and Kurkan IK 2012 *Rev. Sci. Instr.* **83** (7) 074705–4
[8] Chance B 1977 *Annu. Rev. Biochem.* vol **46** pp 967–980
[9] Zharkova LP, Ivanov VV, Knjazeva IR, et al. 2011 *Tomsk State University Journal. Biology* **№3** (15) pp 161-170
[10] Schwan HP and Foster KR 1980 *Proc. of the IEEE* vol **68** pp 104–113
[11] Zorov DB, Juhaszova M, Sollott SJ 2014 *Physiological reviews* **94**(3) pp 909-950
[12] Schunc T, Bieth F, Pinguet S, Delmote Ph 2016 *Electomagnetic biology and medicine* vol **35** (1) pp 84-101

About the selection of transverse modes in the X-band oversized oscillator with 2.5 GW output power

R V Tsygankov, V V Rostov, A V Gunin, A A Elchaninov, A B Markov, and G E Ozur

Institute of High Current Electronics, Siberian Branch, Russian Academy of Sciences, 2/3 Akademichesky Avenue, Tomsk, 634055, Russia

Email: tsygankov@lfe.hcei.tsc.ru

Abstract. The paper describes the numerical and experimental results of the microwave O- type oscillator based on an oversized slow wave structure (SWS). The feedback is applied to the design scheme, which provides intense modulation of the electron beam in the cathode-anode region and two special cavities before SWS. The selectivity of TM_{02} operating mode occurs due to increased diffraction loss of parasitic modes in the cathode part. The slow wave structure consists of two identical sections with the phase-shifting region in between. The use of this configuration leads to the formation of a locked TM_{01} wave, having good conditions for the transformation into the working mode TM_{02}. In the experiments, a stable generation regime with pure TM_{02} mode at a frequency of 10 GHz with an efficiency of about 30% and the output power of 2.5 GW in the magnetic field below the cyclotron resonance was obtained.

1. Introduction

The use of the high-current electron beams demonstrates that the most advanced systems, considering the power augmentation and the energy conversion efficiency of the electron beam into the energy of the electromagnetic-microwave pulse are the relativistic Cherenkov-type oscillators, and in particular, the backward wave oscillator (BWO). The generation in such systems is usually performed at lowest modes due to the lower starting current against the higher modes. In X-band, raising of the generated power up to the level of ≈ 1 GW results in the necessity for enlarging the diameter D of the slow wave structure (SWS) in comparison with the wavelength λ to prevent the microwave breakdown [1]. The reduction of maximal electric fields at the SWS has been achieved in the BWO with resonant reflector at $D \approx 1.5\ \lambda$ [2] which allowed us to increase the output microwave power up to 3 GW at the pulse width of 20 ns [1, 3]. The further grow of D leads to the mode competition. The special methods of mode selection should be used to discriminate parasitic modes.

In this paper, the new approach for development of the low Q-factor oversized resonant SWS ($D \approx 2.5\ \lambda$) is considered.

The slow-wave system under study (figure 1) consists of two similar sections arranged in the way to meet the conditions for the radiation of the slow wave TM_{01} only into the TM_{02} wave. The low-Q cavity includes a hollow cylindrical cathode, which functions as a reflector as well. The output signal is partially transmitted to the cathode and modulates the electron beam in the SWS.

The fundamental research and simulation of steady-state processes of such systems (BWO with modulation of emission, [4]) reveals the potential to increase the performance up to 50%.

IOP Conf. Series: Journal of Physics: Conf. Series **830** (2017) 012011 doi:10.1088/1742-6596/830/1/012011

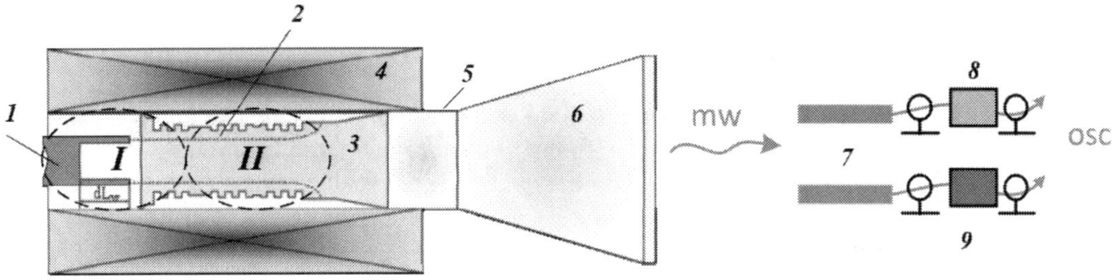

Figure 1. Experimental setup: 1- cathode; 2- electron beam; 3- SWS; 4- solenoid; 5- collector; 6- horn antenna; 7- receiving antenna; 8- detector on hot media; 9- heterodyne; I- reflection region backward wave; II- region of oversized SWS.

At the cathode side of the SWS, a matching element in the form of several grooves is placed. Here, the wave TM_{02} counter to the electron beam is transformed into the mode mixture causing the minimal diffraction losses into the region of the hollow cathode. The cathode-anode gap can be considered as a region of the combined quasi-optical system where the parasitic modes provide essential diffraction losses. The losses of the operating wave are of 1 – 2%. To improve the selective properties of the hollow cathode, a resonant reflector was used instead of the plane [5]. The operating wave is reflected from the reflector inserted into the cathode, so, the efficient wave selection is provided.

2. Computational modelling

The final sizing and the optimization of the mode for the stationary generation were performed with the KARAT 2.5D code [6]. The average radius of the oversized slow wave system was 3.7 cm. The corrugation amplitude was about 0.6 cm. The diode was fed by TEM wave to obtain the cathode voltage of 680 kV. The electron beam was emitted from the cathode side-face with the outer radius of 3.2 cm. The electron beam current was about 12 kA. At the output crosscut behind the collector, a pure enough mode of the circular waveguide TM_{02} was registered. The guiding magnetic field of the solenoid selected above the cyclotron resonance was about 3 T [7]. As figure 2(a) shows, the transient process lasts several nanoseconds (the voltage rise time is 4 ns). The optimal generation mode at the frequency of 10 GHz corresponded to the power of 3.8 GW (figure 2(a, b)) in the numerical experiment.

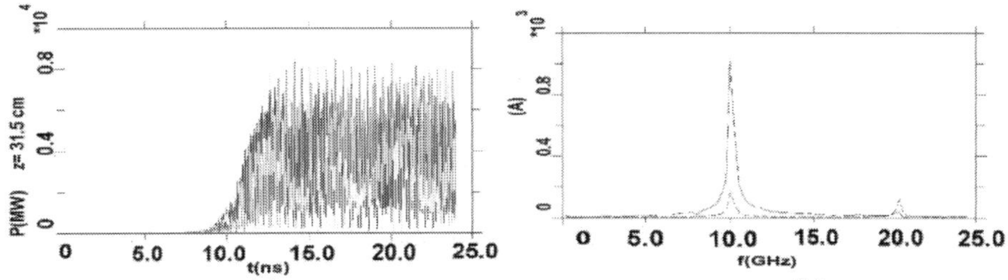

(a) (b)

Figure 2. (a)- changing the output power of the microwave oscillations in time; (b)- the emission spectrum

Here, the maximal voltage of the electrical field at the SWS rounding is estimated to be about 0.7 MV/cm. Such a field level provides the operation without microwave breakdown with the pulse durations up to 10^{-8} ns and requires a certain surface treatment when increasing the radiation duration. The maximal calculated performance efficiency was 47%. The calculation results reveal the dependence of the generator efficiency on the diode accelerating voltage as well as on the position of the reflector inside the cathode hollow. When other parameters are fixed, the geometric variations of corrugation near the adapter significantly affect the generator efficiency. The optimal geometry deviation of 5% results in the efficiency fall up to 30%.

3. Experimental research at SINUS-7

Before the experiment, the special treatment of the SWS surface was provided. The internal part of the electrodynamic structure (figure 3) was treated with the low-energy electron beam of 35 keV [8] using the OCTAGON-2. This technique provided the higher quality of the surface finish and the reduction of the regions where locally the electric field strengthening occurred. The surface finish as well as the structural composition was monitored with the electron-beam microscope.

In preliminary studies, with untreated SWS surface, we observed an unstable operation of the generator [9].

It was previously demonstrated [10] that the limitation of the microwave pulse duration in a nanosecond pulsed RBWO operating at a gigawatt power level could be related to the development of an explosive emission from the SWS surface under the action of intense microwave fields. A key role in suppression of the microwave generation is played by positive ions emitted from the near-surface plasma, which remove the space-charge limitation on the electron emission current. Thus, the microwave pulse duration is limited by the total time of the development of explosive emission and the accumulation of ions in the SWS volume [11, 12].

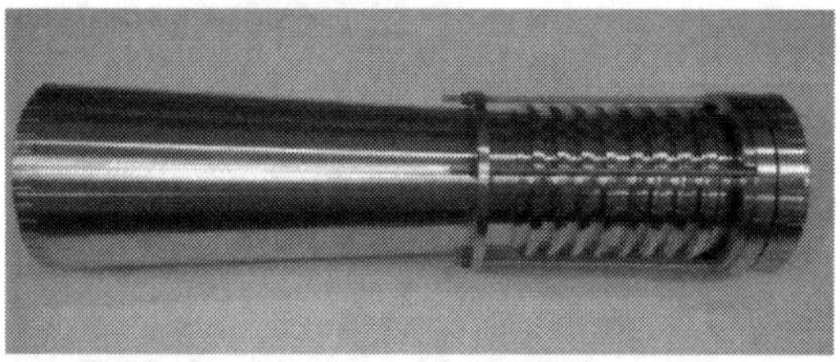

Figure 3. Common view of electrodynamic structure

After preparation, the electrodynamic system was studied at the SINUS-7 accelerator. The operating diode voltage and the electron beam current of 670 kV and 13 kA, respectively, were close to the simulations. The microwave radiation pulses were recorded by the magnetic loop antenna placed in the far field zone. The microwave pulse energy was measured by the aperture liquid calorimeter with the enhanced recording system. The calorimeter was placed near the output horn antenna. The stable microwave signals at the defined frequency of 10 GHz (figure 4(a)), which were repeated from pulse to pulse, were generated under the magnetic field of about 1 T. To measure a frequency, a heterodyne measuring technique was used. In the experiment, the heterodyne reference frequency was 9 GHz, so the frequency values indicated in the waveform were about 1 GHz that corresponded to 10.0 GHz of the output microwave pulse.

In the magnetic fields below the cyclotron resonance, the microwave radiation power measured by the calorimeter agrees well with the calculated power and is 2.7 GW. The directivity diagram obtained in the far field zone of the horn antenna indicates the generation of the symmetric mode TM_{02}. The numerical calculations with KARAT demonstrate that the optimal magnetic field for this generator is 3 T. Here, the calculated power should be about a third higher than in the fields below the cyclotron absorption. However, providing this mode experimentally, the amplitude of the microwave signal decreased, and another lower frequency of 8.5 GHz appeared which corresponded to the non-symmetrical wave TE_{11} (figure 4(b)).

A resonant reflector was installed into the cathode instead of the former reflector in the shape of a plane wall. The calculation results obtained with the ANSYS HFSS [13] software reveal that this resonant reflector at 10 GHz provides almost full reflection of the operating modeTM_{02}-TM_{01} while the non-symmetrical parasitic modes pass into the cathode at most.

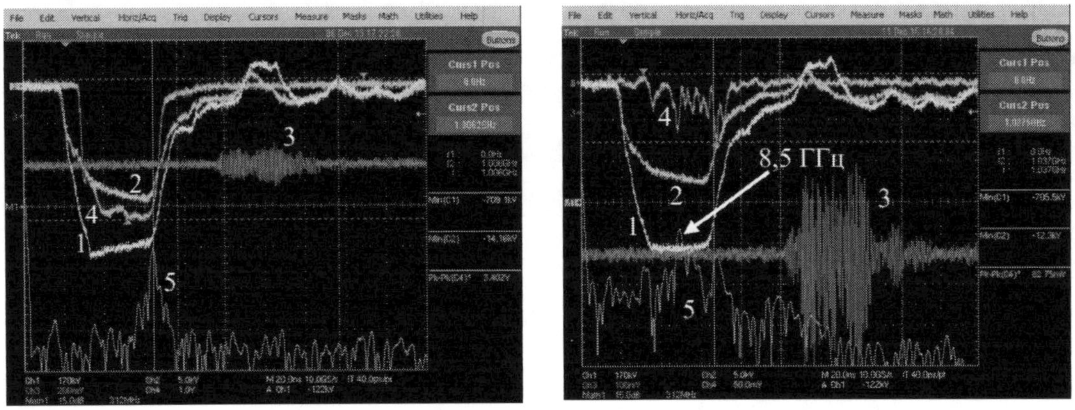

(a) (b)

Figure 4. Waveforms (TDS6404): (a)- low magnetic field, (b)- strong magnetic field.
1- diode voltage, 2- diode current, 3- microwave signal with a heterodyne circuit (reference frequency 9 GHz) 4- detected microwave signal, 5- spectrum of microwave signal

In the experiment, a microwave absorber was placed behind the reflector. So, the mode selection was provided due to the diffraction losses of the parasitic oscillations in the coaxial region as well as inside the cathode.

The dependence of the microwave radiation normalized power on the diode voltage at different positions of the reflector inside the cathode (figure 5) confirms a good agreement between the experimental data and the numerical simulations.

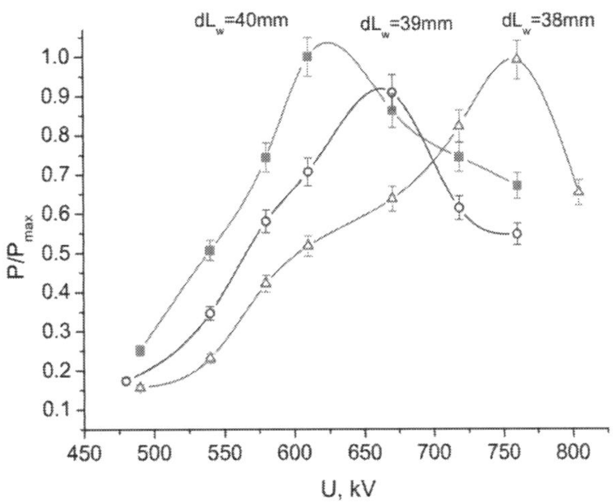

Figure 5. The experimental dependence of the microwave radiation normalized power on the diode voltage at different positions of the reflector inside the cathode dL_w

The curves in figure 5 indicate that within a certain voltage range, there is an optimal reflector position dL_w, which corresponds to the reflection phase of the counter wave, resulting in the maximal output microwave power.

In the numerical simulations, the output microwave power reached 3.8 GW in a strong magnetic field above the cyclotron resonance. In experiments, however, the output microwave power was unstable (figures 4 (b), 6) when the guiding magnetic field was 3 T.

The research demonstrated that the electron beam modulation in the cathode-anode region decreases the starting current for the operating mode and provides the increase of the generator efficiency. The unstable mode in the strong magnetic fields was revealed even under the slight beam axis misalignment.

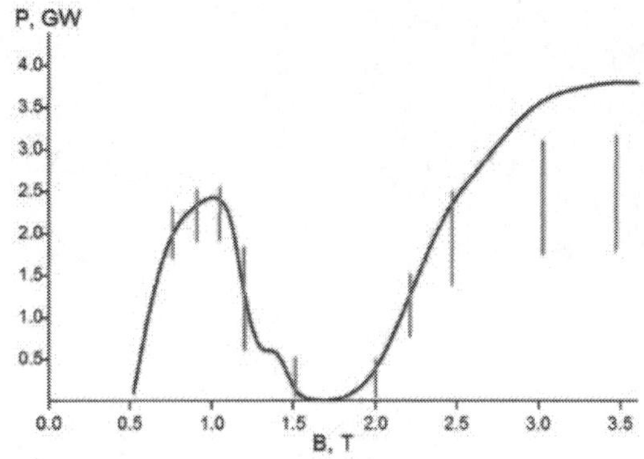

Figure 6. Output power P on the magnetic field B. Solid curve – numerical calculation on PIC-code KARAT at 670 kV, 12 kA, vertical lines – range of experimental results.

This can indicate the modest suppression of the competitive non-symmetrical modes.

4.Conclusion

So, the microwave radiation power in the stable generation mode reached about 2.5 GW in the experiment as well as in the simulations. It showed the 30% efficiency at the frequency of 10.0 GHz just in the magnetic fields of about 1 T. The numerical simulations demonstrate the perspective of the efficiency up to 48% when generator operating in a strong magnetic field of 3 T. It can be provided by improving the selectivity of the generator.

Acknowledgments

The authors are grateful for the help in carrying out the work to Kisanov S A, Sukhov M Y, Petrov V I.
This work was supported by the Russian Foundation for Basic Research grant № 16-02-01088 A.

References

[1] Gunin A V, Klimov A I, Korovin S D, Kurkan I K, Pegel' I V, Polevin S D, Roitman A M, Rostov V V, Stepchenko A S and Totmeninov E M 1998 *IEEE Transactions on Plasma Science*, vol **26**, no 3, pp 326–331
[2] Kurkan I K, Rostov V V, Tot'meninov E M 1998 *Techn. Phys. Lett.*, vol **24**, no 5, p 388-389
[3] Korovin S D, Rostov V V, Tot'meninov E M 2005 *Techn. Phys. Lett.*, vol **31**, no 5, p 411-413
[4] Nazarova M V, Solntsev V A, and Melikhov V O *2011 Journal of Communications Technology and Electronics* vol **56**, no 4, p 511
[5] Ju J C, Li W, Yang H W, Zhang J, Elchaninov A A, Tsygankov R V, Gunin A V, Rostov V V *Izvestiya VUZov Phizika*, vol **57**, no12/2, pp 42-46
[6] Tarakanov V P 1992 *User's Manual for Code KARAT* (Springfield: BRA)
[7] Rostov V V, Tsygankov R V, Gunin A V, Elchaninov A A, Ju J C, Li W, Yang H W, Zhang J 2015 ICOPS-2015 *IEEE Proc. Int. Conf. on Plasma Science. (Belek, Turkui)* p 435
[8] Markov A B, Yakovlev E V, Petrov V I 2013 *IEEE Trans. Plasma Sci.*, vol **41**, Issue 8, p 2177
[9] Tsygankov R V, Rostov V V, Elchaninov A A, 2014 *Izvestiya VUZov Phizika*, vol **57**, no12/2, pp 304-309
[10] Korovin S D, Mesyats G A, Pegel' I V, et al. 2000 *IEEE Trans. Plasma Sci.* vol **28**, p 485

[11] Batrakov A V, Karlik K V, Kitsanov S A, Klimov A I, Konovalov I N, Korovin S D, Mesyats G A, Ozur G E, Pegel' I V, Polevin S D, Proskurovsky D I, Sukhov M Y 2001 *Techn. Phys. Lett.*, vol **27**, no 2, pp 150–152
[12] Volkov S N, Karlik K V, Kovalchuk B M, Kurkan I K, Ozur G E, Pegel I V, Polevin S D, Proskurovskiy D I, and Sukhov M Yu 2008 *Techn. Phys. Lett.*, vol **34**, no 7, pp 581–583
[13] Ansys High Frequency Structure Simulator (HFSS) 2016 (Ansys Corporation, Pittsburgh, PA, 15317)

IOP Conf. Series: Journal of Physics: Conf. Series **830** (2017) 012012 doi:10.1088/1742-6596/830/1/012012

Synthesis of ultrawideband radiation of combined antenna arrays excited by nanosecond bipolar voltage pulses

V I Koshelev, V V Plisko and E A Sevostyanov

Institute of High Current Electronics SB RAS, Tomsk, Russia

E-mail: koshelev@lhfe.hcei.tsc.ru

Abstract. To broaden the spectrum of high-power ultrawideband radiation, it is suggested to synthesize an electromagnetic pulse summing the pulses of different length in free space. On the example of model pulses corresponding to radiation of combined antennas excited by bipolar voltage pulses of the length of 2 and 3 ns, the possibility of twofold broadening of the radiation spectrum was demonstrated. Radiation pulses with the spectrum width exceeding three octaves were obtained. Pattern formation by the arrays of different geometry excited by the pulses having different time shifts was considered. Optimum array structure with the pattern maximum in the main direction was demonstrated on the example of a 2×2 array.

1. Introduction

The spectrum width of high-power ultrawideband (UWB) electromagnetic pulses is determined by characteristics of radiator and exciting voltage pulse. To broaden the radiation spectrum, it was suggested in [1] to synthesize an electromagnetic pulse through the summation in free space of the pulses of different length (frequency band) radiated by an array of combined antennas. The possibility of synchronous operation of four independent bipolar pulse formers (BPF) exciting a 4-element array of combined antennas has been presented in the investigation [2]. The amplitude of voltage pulses, the length, and the time instability σ were equal to 50 – 60 kV, 3 ns, and 50-70 ps, respectively. The summation of fields from separate antennas in the array was realized in the far-field zone. Using the independent BPF, it is possible to create high-power UWB sources with combined antennas excited by bipolar voltage pulses of different length and, as a result, to increase the frequency band of the electromagnetic pulse. It will allow increasing the possibility of UWB radar as well as the study of susceptibility of electronic systems to the influence of strong electromagnetic fields.

At the first stage of the study, an UWB source based on a 4-element array of combined antennas excited by bipolar pulses of the length 2 and 3 ns has been developed. Below, the results of the numerical simulation of the antenna array for this type of radiation source are presented.

2. UWB pulses and their spectra

We will consider the increase of the spectrum width of a synthesized pulse using an example of pulses radiated by the previously developed antennas excited by the bipolar voltage pulses of the length 2 [3] and 3 ns [4]. To simulate the pulse radiated by the combined antenna, we will use the derivative of the sum of two Gaussian functions:

$$E(t,\tau)=\frac{8}{\tau}\left(\frac{4t}{\tau}-2\right)e^{-\left(\frac{4t}{\tau}-2\right)^2}-\frac{8}{\tau}\left(\frac{4t}{\tau}-4\right)e^{-\left(\frac{4t}{\tau}-4\right)^2}, \quad (1)$$

where τ is the length of the bipolar voltage pulse exciting the antenna by the level of 0.1 of the amplitude. In [5], it is shown that formula (1) allows a reasonably good simulation of the real pulses E(t) radiated by the combined antennas as well as their spectra S(f). Fig. 1a presents the pulses calculated by the formulae (1), while Fig. 1b presents their spectra.

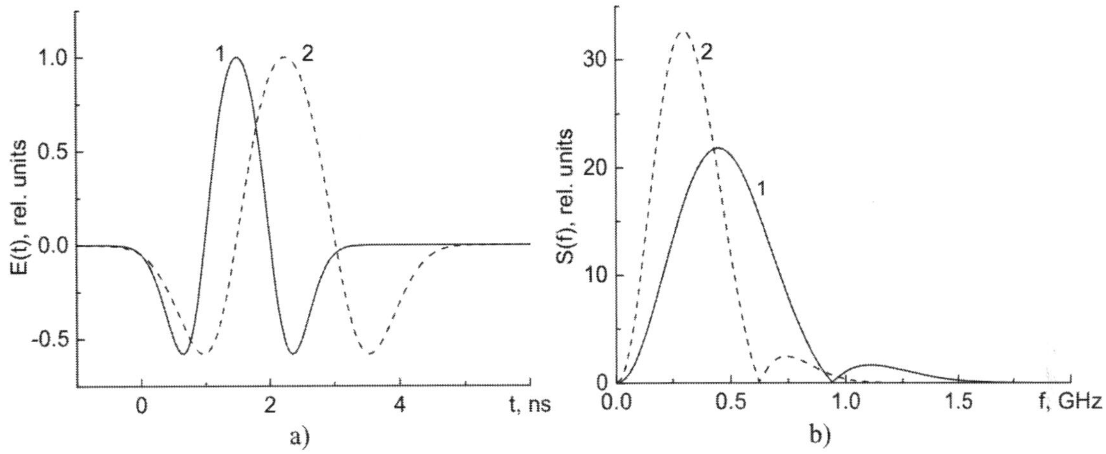

Figure 1. Radiation pulses (a) at the antenna excitation by bipolar voltage pulses of the length 2 ns (1) and 3 ns (2) calculated by the formula (1) and the spectra corresponding to these pulses (b).

3. Summation of radiation pulses

The result of summation of radiation pulses is determined by both the pulse characteristics and their time delays relative to each other. Thus, changing the delay between the radiation pulses of different length (spectrum width), we can obtain pulses with different characteristics: the amplitude and the spectrum width.

We will optimize the synthesized pulse by two parameters: peak amplitude of the field and the maximum spectrum width. Note that the pulse spectrum can be characterized by the following parameters:

1) the spectrum width $\Delta f=f_H-f_L$, where f_L is the lower and f_H is the upper boundary frequency of radiation spectrum by the level of -10 dB;

2) the relative spectrum width $\Delta f/f_0$, where $f_0=(f_H+f_L)/2$ is the central frequency of radiation spectrum;

3) the ratio of the upper and the lower boundary frequencies f_H/f_L.

To analyze the characteristics of a synthesized pulse, we will plot the peak strength of the field E_p and the spectrum width Δf by the level of -10 dB versus the time delay Δt between the pulses (Fig. 2).

Maximum peak field strength is realized at the delay Δt=0.76 ns which corresponds to the synchronization of the pulse amplitudes (Fig. 3). Herewith, pulse amplitudes are summed but the ratio of the upper and the lower boundary frequencies of the total pulse increases insignificantly, approximately by 10%, in comparison with the pulses radiated at the antenna excitation by bipolar pulses of the length 2 and 3 ns.

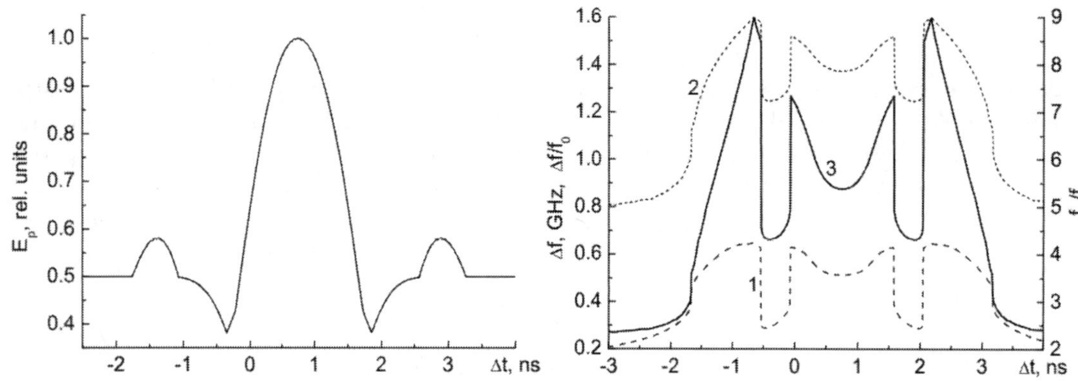

a) b)

Figure 2. Peak field strength (a) and total pulse spectrum by the level of -10 dB (b) versus the delay between the pulses. 1 – spectrum width $\Delta f = f_H - f_L$, 2 – relative spectrum width $\Delta f/f_0$, 3 – ratio of the upper and the lower boundary frequencies f_H/f_L.

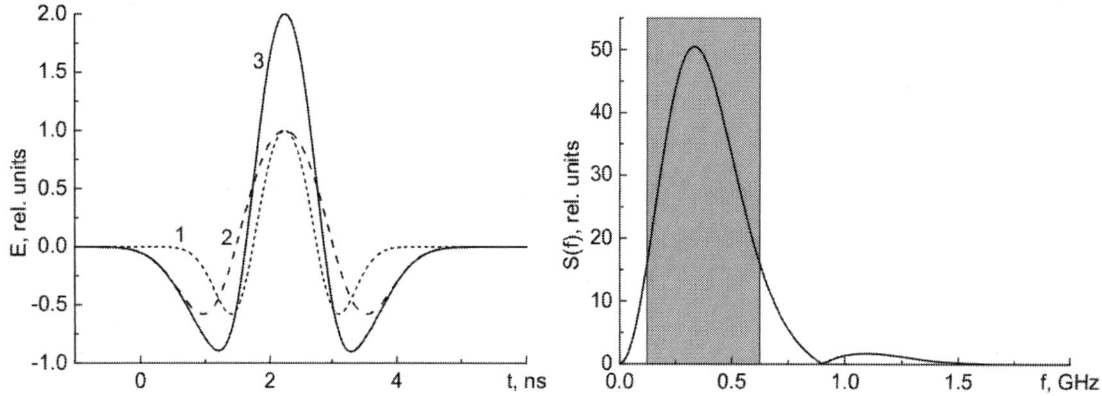

a) b)

Figure 3. Pulse synchronization by the amplitude maximum. a) 1 – radiation pulse at the antenna excitation by a bipolar pulse of the length 2 ns, 2 – radiation pulse at the antenna excitation by a bipolar pulse of the length 3 ns, 3 – total radiation pulse. b) total radiation pulse spectrum.

Fig. 2 (b) demonstrates that the maximum values for all three parameters characterizing the pulse spectrum are accounted for the delays of -0.66 and 2.16 ns. Fig. 4 presents the total pulse for the delay Δt=-0.66 ns and it spectrum. In comparison with a single pulse, the ratio of the upper and the lower boundary frequencies of the total pulse increased by a factor of 1.85 amounting to more than three octaves. Herewith, the field amplitude of the total pulse decreased by 7% relative to the amplitude of the initial pulses.

Table 1 presents the frequency characteristics for the singular (Fig. 1) and total pulses for the delays of 0.76 and -0.66 ns (Fig. 3 and 4). The Table demonstrates that the shift of the lower boundary frequency f_L up to 83.5 MHz. makes the substantial contribution into the bandwidth widening of the total pulse with the delay of Δt=-0.66 ns.

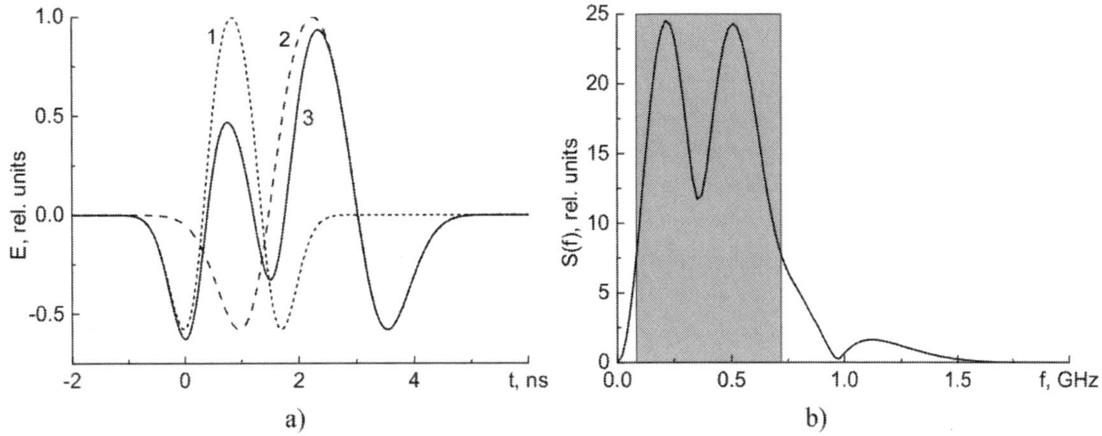

Figure 4. Summation of the pulses with the delay of -0.66 ns. a) 1 – radiation pulse at the antenna excitation by a bipolar pulse of the length 2 ns, 2 – radiation pulse at the antenna excitation by a bipolar pulse of the length 3 ns, 3 – total radiation pulse. b) total radiation pulse spectrum.

Table 1. Parameters of radiation pulses

UWB radiation pulse	f_L (GHz)	f_H (GHz)	f_0 (GHz)	Δf (GHz)	$\Delta f/f_0$	f_H/f_L
2 ns	0.1658	0.7686	0.4672	0.6028	1.290	4.636
3 ns	0.1104	0.5125	0.3115	0.4021	1.291	4.643
2+3 ns, max amplitude	0.1209	0.6232	0.3721	0.5023	1.35	5.153
2+3 ns, max spectrum	0.0835	0.72	0.4017	0.6364	1.584	8.618

4. Patterns

Similar to [3, 4], we will use the pattern by the peak power E_p^2 for antennas and arrays radiating ultrawideband pulses. When the pulses of the length 2 and 3 ns are synchronized by the field maximum (Δt=0.76 ns), their pattern will not differ from the array pattern that is excited synchronously (Δt=0) by the pulses with equal length (Fig. 5, curve 1). However, for the total pulse with maximum spectrum width (Δt=-0.66 ns), the direction of the pattern maximum depends on the array configuration. Here, we consider the array of 2×2 combined antennas, in which two elements are excited by a bipolar voltage pulse of the length 2 ns, while another two elements are excited by a bipolar voltage pulse of the length 3 ns. Figure 6 demonstrates two possible configurations. Numbers 2 and 3 mark the array elements excited by the bipolar pulses of the length 2 and 3 ns, respectively. We will calculate the array patterns by the peak power of radiation in the H-plane by means of direct summation of the pulses radiated by the array elements assuming that the element has a cardiod pattern and the radiation pulse waveform is independent of the angle. For the array configuration depicted in Fig. 6a, the pattern maximum is shifted by the angle of 50° (Fig. 5, curve 2). To return the pattern maximum to the initial direction (perpendicularly to the array plane), it is necessary to use the array configuration presented in Fig. 6b. In this case (Fig. 5, curve 3), the pattern does not practically differ from the pattern for the radiation pulses synchronized by the field maximum.

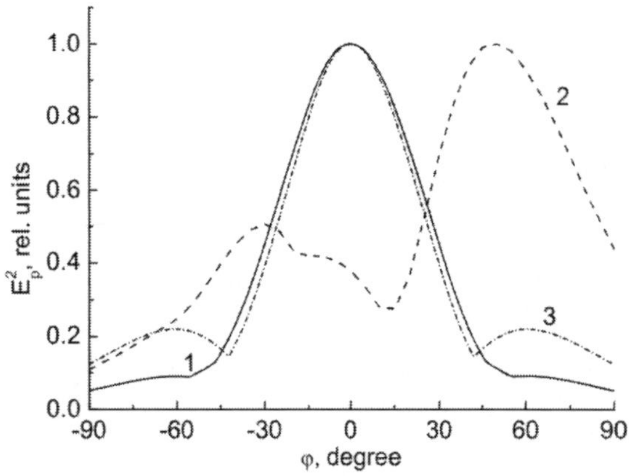

Figure 5. Patterns of the 2x2 arrays in the H-plane. 1- at synchronous excitation (Δt=0 ns) by the pulses of equal length; 2 – at excitation by the pulses of 2 and 3 ns with Δt = -0.66 ns (variant 6a); 3- at excitation by the pulses of 2 and 3 ns with Δt = -0.66 ns (variant 6b).

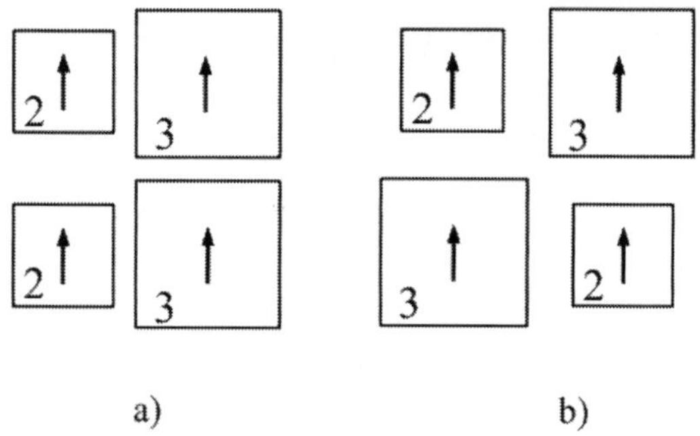

Figure 6. Variants of configuration of a 2x2 array. The arrows indicate the polarization plane of vector E.

5. Conclusion

By means of numerical simulation, we have shown a possibility to broaden the spectrum of ultrawideband radiation using the summation of the pulses of different length (frequency band) in a free space. In case the antennas are excited by the bipolar voltage pulses of the length 2 and 3 ns, it becomes possible to obtain radiation pulses with the spectrum bandwidth up to three octaves. To obtain the pattern with the maximum in the main direction, the array elements should be located diagonally.

Acknowledgment

The work was supported by the Russian Science Foundation, the project #16-19-100081.

References

[1] Andreev Yu.A., Buyanov Yu.I., Koshelev V.I., Plisko V.V., Sukhushin K.N. // *Proc. Ultra-Wideband, Short-Pulse Electromagnetics* **4** – 1999 – pp 181-186

[2] Efremov A.M., Koshelev V.I., Kovalchuk B.M., Plisko V.V. // *Instruments & Experimental Techniques* – 2013 – **56** – 302

[3] Gubanov V P, Efremov A M, Koshelev V I, Koval'chuk B M, Korovin S D, Plisko V V, Stepchenko A S and Sukhushin K N 2005 *Instruments & Experimental Techniques* **48** 312

[4] Andreev Yu A, Efremov A M, Koshelev V I, Koval'chuk B M, Plisko V V and Sukhushin K N 2011 *Instruments & Experimental Techniques* **54** 794

[5] Koshelev V I, Plisko V V and Sevostyanov E A 2015 *Izv. Vyssh. Uchebn. Zaved., Fiz.*, vol 58 No 8/3 pp 54–58

IOP Conf. Series: Journal of Physics: Conf. Series **830** (2017) 012013 doi:10.1088/1742-6596/830/1/012013

The Initial Stage of Neck Formation in an X-Pinch

V I Oreshkin[1,2], A P Artyomov[1], S A Chaikovsky[3,1,4], E V Oreshkin[4] and M M Tsventoukh[4]

[1] Institute of High Current Electronics of the Siberian Branch of the Russian Academy of Sciences, Akademichesky Avenue 2/3, Tomsk 634055, Russia
[2] National Research Tomsk Polytechnical University, Lenin Avenue 30, Tomsk 634050, Russia
[3] Institute of Electrophysics of the Ural Branch of the Russian Academy of Sciences, Amundsen 106, Ekaterinburg 620016, Russia
[4] Lebedev Physical Institute of the Russian Academy of Sciences, Leninsky Avenue 53, Moscow 119991, Russia

E-mail: oreshkin@ovpe.hcei.tsc.ru

Abstract. A model is proposed to describe the initial stage of neck formation in an X-pinch that proceeds in three stages: the electrical explosion of metal wires that generates the X-pinch; the expansion of the wire material that occurs due to an excess of the gas-kinetic pressure over the pressure of the magnetic field. The model allows one to predict the minimum rate of current rise at which the formation of a "hot spot" in an X-pinch is possible. The minimum current rise rate is determined by the thermodynamic parameters of the wires at a critical point; it is of the order of 1 kA/ns.

1. Introduction

In the 80s of the last century, the method of production of high-temperature dense plasmas by electrically exploding crossed wires was proposed at P. N. Lebedev Physical Institute (Russia) [1]. As this version of wire load consists of wires crossed as if they form of the letter "X", it has received the name X-pinch. In X-pinch experiments, two or more metal wires 10–50 μm in diameter are used, through which a current pulse of amplitude from 100 kA to several megaamperes with a rise time of 100–500 ns is passed [2–7]. In the wire crossing region, a plasma hot spot of several micrometers in size forms, which lifetime is about 1 ns. The parameters of this type of X-ray sources make them irreplaceable for use in projection radiography of various short-lived physical objects. In particular, we used X-pinches in studying the thermal instabilities [8–10] that develop in electrically exploded conductors.

Notwithstanding that X-pinches and their applications have been the topic of a rather large body of research, not nearly all features of X-pinches are now well studied. In particular, the criterion for the rise rate of the current passing through an X-pinch that was derived from experimental data [4],

$$\frac{dI}{dt} > (1-1{,}5)\frac{kA}{ns}, \qquad (1)$$

has not yet been substantiated. This criterion is very important, as in the experiments where the current rise rate was lower than that obeying (3), no hot spot formed irrespective of the current amplitude and, hence, a bright radiation source was not produced.

2. The Initial stage of Neck Formation

Let us consider the processes that occur in an X-pinch produced by the explosion of two or more crossed wires. In this case, the mass per unit length of the X-pinch is given by the formula $m_0 = \frac{Nm_w}{\cos\varphi}$, where N is the number of wires, φ is the angle of inclination of the wires to the X-pinch axis, m_w is the mass per unit length of an individual wire, r_0 is the initial radius of a wire, ρ_0 is the density of the wire metal under normal conditions, and φ is the angle of inclination of a wire to the z-axis. For analysis, it is convenient to use the effective initial radius of a wire $R_0 = \sqrt{\frac{m_0}{\pi\rho_0}}$.

The process of formation of a neck in an X-pinch is assumed to proceed in three stages.

Stage I is the electrical explosion of the wires. At this stage, the wires melt and further heat up to the point where they cease to exhibit metallic conduction [11,12]. Denote the duration of this stage by t_1.

Stage II is the expansion of the wire material caused by the excess of the gas-kinetic pressure over the magnetic field pressure immediately after the explosion. The expansion proceeds until the magnetic pressure becomes higher than the gas-kinetic pressure due to the current rise [13]. Denote the duration of this stage, from the explosion to the cessation of expansion, by t_2.

Stage III is the compression of the wire material which is accompanied by the outflow of matter from the neck region. At the final stage of the neck formation, a hot spot forms whose dimensions are less than the length of the neck [3]. The radiation pulse arises at the hot spot when the mass is almost completely lost. Denote the duration of this stage, from the onset of compression to the occurrence of a radiation pulse, by t_3.

The time of the overall process, from the onset of current passage to the occurrence of an X-ray pulse, is the sum of the times of these stages: $t_f = t_1 + t_2 + t_3$.

Let us consider details of the first two stages.

Stage I, the electrical explosion of the wires

During the electrical explosion the wire material passes through all phases from a condensed state to plasma. The explosion of the metal occurs near a critical point [11] at which the magnetic pressure fails to restrain the expanding material. The critical point corresponds to the point of the phase diagram at which the lines separating the liquid, the gas, and the two-phase region meet. The parameters of the material at the critical point (the temperature T_{cr}, the pressure p_{cr}, and the density ρ_{cr}) for some metals are given in Table I.

The explosion time can be defined in terms of a quantity, called the integral of the specific action of current, as [14, 15]

$$h = \int_0^{t_1} j^2(t)dt , \tag{2}$$

where $j = \frac{I}{S}$ is the current density in a wire, I is the current flowing through the pinch, and $S = N\pi r_0^2 = \frac{m_0}{\rho_0}\cos\varphi$ is the total cross-sectional area of the wires. Values of the integral of specific current action for some metals are given in Table I.

For a linear rate of current rise, the time dependence of the current density can be described as $j(t) = \frac{\rho_0}{m_0 \cos\varphi} \frac{dI}{dt} t$. Substituting this expression into the integral (2), we obtain an estimate for the explosion time:

$$t_1 = \left[\frac{3hm_0^2 \cos^2\varphi}{\rho_0^2} \left(\frac{dI}{dt} \right)^{-2} \right]^{1/3} . \quad (3)$$

As the explosion of a wire occurs near the critical point and the density at the critical point, ρ_{cr} , is a factor of 3–4 less than the normal density of the metal, ρ_0 , the radial dimension of the wire increases about twofold by the time of its explosion.

According to (3), for the typical parameters of X-pinches (the effective wire radius $R_0 = 2\times10^{-3}$ cm, $\cos\varphi = 35^0$ and the current rise rate $\frac{dI}{dt} = 1\frac{kA}{ns}$) the explosion time is estimated as $t_1 \approx$ 5.3 ns for Fe, 5.7 ns for Au and Al, 5.8 ns for W, and 7.5 ns for Cu.

Stage II, the expansion of the wires

Let us estimate the wire expansion time. This quantity can be estimated only roughly because of the complexity of the processes occurring at this stage. The expansion starts after the explosion of the wire; that is, at the time t_1, and stops at the time $t_1 + t_2$ when the magnetic field pressure becomes higher than the gas-kinetic pressure. As the explosion occurs near the critical point, the parameters of the wire material during expansion can be characterized by its thermodynamic parameters at the critical point. Therefore, the condition for the wire material to stop expanding can be written as

$$\frac{B^2(t)}{8\pi} \approx ap_{cr} , \quad (4)$$

where $B(t) = \frac{2I(t)}{cR(t)}$ is the magnetic field at the boundary of the wire, R (t) is the effective wire radius, c is the velocity of light in vacuum, a > 1 is the dimensionless coefficient (hereinafter, a = 2). As mentioned, the pinch radius at the wire crossing by the time of explosion is about $2R_0$; therefore, the time dependence of the wire radius can be approximated by the relation $R(t) \approx 2R_0 + vt$, where v is the velocity of expansion of the wire. The characteristic expansion velocity can be estimated as the thermal velocity of atoms at a temperature equal to the temperature of the metal at the critical point; that is, as $v \approx \sqrt{\frac{2kT_{cr}}{m_i}}$, where k is Boltzmann's constant, m_i is the atomic mass, and T_{cr} is the temperature at the critical point.

Substituting the expressions for the wire radius and for the magnetic field at the wire boundary into (4), in view of , we obtain an estimate for the wire expansion time:

$$t_2 \approx \frac{4cR_0\sqrt{\pi p_{cr}} - t_1 \frac{dI}{dt}}{\frac{dI}{dt} - \left(\frac{dI}{dt} \right)_{\min}} , \quad (5)$$

where $\left(\frac{dI}{dt} \right)_{\min} = \sqrt{\frac{8\pi c^2 p_{cr} kT_{cr}}{m_i}}$.

IOP Conf. Series: Journal of Physics: Conf. Series **830** (2017) 012013 doi:10.1088/1742-6596/830/1/012013

As it follows from the expression (5), for the current rise rate equal to $\left(\frac{dI}{dt}\right)_{min}$ the wire expansion time tends to infinity; hence, this value can be used as an estimate of the minimum current rise rate at which the stage of expansion can go to the stage of compression. The value of $\left(\frac{dI}{dt}\right)_{min}$ is determined only by the parameters of the metal; the values of $\left(\frac{dI}{dt}\right)_{min}$ for some metals are given in Table I. The values of this quantity for all metals appear to be close to each other because of the closeness of the critical pressures and temperatures. As it can be seen from Table I, to satisfy the criterion (1); that is, to create conditions for the formation of a neck in an X-pinch, the rate of current rise should be greater than $\left(\frac{dI}{dt}\right)_{min}$, and the criterion (1) can be written as

$$\frac{dI}{dt} > (2-3)\cdot\left(\frac{dI}{dt}\right)_{min}. \tag{1a}$$

As the rate of current rise in X-pinches is always greater than $\left(\frac{dI}{dt}\right)_{min}$, expression (5) can be rewritten as

$$t_1 + t_2 \approx 4cR_0\sqrt{\pi p_{cr}}\left(\frac{dI}{dt}\right)^{-1}. \tag{6}$$

Table I. Parameters of metals [14].

Article I. Metal	Article Iu	Article Iu	Article 1	Ag	Ni	Fe	W	Ti	Mo	Zn
h, 10^9 $A^2 \cdot s/cm^4$	4.1	1.8	1.8	2.8	1.9	1.4	1.85[14]	0.8	–	–
ρ_{cr}, g/cm^3	2.39	5.68	0.64	2.93	2.19	2.03	4.85[16]	1.13 [16]	3.18	2.29
T_{cr}, eV	0.72	0.77	0.69	0.61	0.89	0.83	1.38[16]	0.75[16]	1.39	0.275
p_{cr}, kbar	7.46	6.1	4.47	4.5	9.12	8.25	11.8 [16]	4.78[16]	12.63	2.63
$\left(\frac{dI}{dt}\right)_{min}$, kA/ns	0.45	0.24	0.54	0.26	0.58	0.55	0.47	0.43	0.66	0.17

The expression (6) allows one to estimate the net time of the wire explosion and expansion. For the typical X-pinch parameters (effective wire radius $R_0 = 2\times10^{-3}$ cm and current rise rate $\frac{dI}{dt} = 1\frac{kA}{ns}$) the net time of explosion and expansion is estimated as $t_1 + t_2 \approx 12.9$ ns for Fe, 11.1 ns for Au, 9.7 ns for Al, 15.4 ns for W, and 12.2 ns for Cu. Comparing these values with the explosion time estimated above by (3), we see that for the given X-pinch parameters the expansion times t_2 are comparable to the explosion times t_1 and $t_1 + t_2 << t_3 < t_f$.

3. Conclusion

The model allows one to find the minimum current rise rate at which X-pinch necking is possible. The minimum current rise rate is determined by the dynamics of expansion of the wire material and depends on the wire parameters (pressure and temperature) at a critical point, the point in the phase diagram where the lines separating the liquid, the gas, and the two-phase (liquid plus vapour) region meet. As for all metals the critical pressures and temperatures are close to each other, the minimum current rise rates are almost the same for them (about 1 kA/ns).

Acknowledgements

The work was supported in part by the Russian Foundation for Basic Research (grants No. 15-38-20617, 16-08-00969 and 15-08-03845) and by a scholarship of the President of the Russian Federation (grant No. SP-951.2016.1).

References

[1] Zakharov S M, Ivanenkov G V, Kolomenskij A A, Pikuz S A, Samokhin A I and Ulshmid I 1982 *Sov. Tech. Phys.* Lett. **8** 456
[2] Mesyats G A, Shelkovenko T A, Ivanenkov G V, et al. 2010 *Journal of Experimental and Theoretical Physics* **111** 363–370
[3] Ivanenkov G V, Pikuz S A, Sinars D B, Stepnievski V, Hammer D A and Shelkovenko T A 2000 *Plasma Physics Reports* **26** 868–874
[4] Shelkovenko T A, Pikuz S A, Douglass J D, McBride R D, Greenly J B and Hammer D A 2006 *IEEE Transactions on Plasma Science* **34** 2336–2341
[5] Artyomov A P, Fedyunin A V, Chaikovsky S A, Oreshkin V I, Lavrinovich I V and Ratakhin N A 2013 *Technical Physics Letters* **39** 12–15
[6] Oreshkin V I, Chaikovsky S A, Artyomov A P, Labetskaya N A, Fedunin A V, Rousskikh A G and Zhigalin A S 2014 *Physics of Plasmas* **21** 102711
[7] Oreshkin V.I, Oreshkin E.V., Chaikovsky S.A, Artyomov A.P 2016 *Physics of Plasmas* **23** 092701
[8] Oreshkin V I 2008 *Physics of Plasmas* **15** 092103
[9] Rousskikh A G, Oreshkin V I, Chaikovsky S A, Labetskaya N A, Shishlov A V, Beilis I I and Baksht R B 2008 *Physics of Plasmas* **15** 102706
[10] Oreshkin V I, Rousskikh A G, Chaikovsky S A and Oreshkin E V 2010 *Physics of Plasmas* **17** 072703
[11] Oreshkin V I, Chaykovsky S A, Ratakhin N A, Grinenko A, and Krasik Y E *Physics of Plasmas* **14** 102703
[12] Krasik Y E, Grinenko A, Sayapin A, et al. 2008 *IEEE Transactions on Plasma Science* **36** 423–434
[13] Sedoi V S, Mesyats G A, Oreshkin V I, Valevich V V and Chemezova L I 1999 *IEEE Transactions on Plasma Science* **27** 845–850
[14] Mesyats G A 2007 *Pulsed power* (Springer Science & Business Media)
[15] Oreshkin V I, Barengolts S A and Chaikovsky S A 2007 *Technical Physics* **52** 642–650
[16] Fortov V E, Khishchenko K V, Levashov P R and Lomonosov I V 1998 *Nuclear Instruments and Methods in Physics Research Section A: Accelerators, Spectrometers, Detectors and Associated Equipment* **415** 604–608

Effect of emission on subnanosecond breakdown in a gas diode at low pressure

E Kh Baksht[1], S Ya Belomyttsev[1], A G Burachenko[1,2], A A Grishkov[1] and V A Shklyaev[1,3]

[1]Institute of High Current Electronics, Russian Academy of Sciences, 2/3 Akademichesky Ave., 634055 Tomsk, Russia
[2]Tomsk State University, 36 Lenin Ave., 634050 Tomsk, Russia
[3]Institute of High Technology Physics, National Research Tomsk Polytechnic University, 30 Lenin Avenue, 634050 Tomsk, Russia

Email: beh@loi.hcei.tsc.ru

Abstract. The paper presents experimental and numerical research results on the operation of gas diode at low pressure. A high dispersion in the runaway electron beam current (from 20 to 100 A) with respect to the average one (~50 A) is observed for a tubular cathode with a working edge radius of 30 μm, nitrogen pressure of 30 Torr, and an interelectrode gap of 6 mm. Numerical simulation data show that the low beam current (~20 A) is due to the early electron emission from the cathode (at the stage of low-voltage prepulse), in which the runaway electron beam is formed from the boundary of plasma layer developing early in the breakdown. The high beam current (~100 A) is due to the delayed electron emission from the cathode, which increases the diode voltage and the runaway electron beam current. In the latter case, the runaway electron beam is formed directly at the cathode.

1. Introduction

At present, many research papers dealing with generation of runaway electron beams in gas diodes with highly inhomogeneous electric field, considering the possibility to obtain such beams in various gases, at various pressures, the mechanisms of their generation and their parameters, depending on the interelectrode gap width as well as reporting about the cathode materials and their design, about the range of their possible applications, etc. have been published [1–4]. However, the generation stability of runaway electron beams, which is also very important, is reported only in [5, 6], in which general attention is paid to the onset of cathode operation in the steady-state mode (a conditioned cathode) without considering the dependence of beam current scattering on the gas diode parameters.

In experiments with the gas diodes, the current of runaway electron beam can differ greatly from pulse to pulse, under the same operating conditions. The stability of beam current is influenced by different factors. Our previous studies have shown that in nitrogen, at 30 Torr pressure, the duration of beam current is limited by the anode foil, which precludes the passage of electrons with the energies below threshold depending on the foil thickness [7, 8]. In this case, the beam current depends on the maximum voltage across the discharge gap (the breakdown voltage), and at the same voltage rise time, it depends, in fact, only on the emission characteristics determined by the state of the cathode surface.

Thus, the only factor that can be responsible for the beam current scattering under such conditions is the state of the cathode surface varying from pulse to pulse.

The objective of our experimental and numerical study was to investigate the effect of cathode emission on the generation stability and parameters of the runaway electron beam at the onset of the cathode to a steady-state mode. Clearly, the stability of beam parameters depends strongly on the voltage rise time, discharge chamber geometry, cathode design, gas pressure, and the state of cathode emitting surface. Whereas most of these parameters can be fixed or recorded by the diagnostic equipment, the state of cathode surface changes from shot to shot, and this can lead to an uncontrollable deviation of the runaway electron beam parameters. To study the influence of cathode emission on the runaway electron beam parameters, we considered the pulses, in which the rise time of incident voltage wave and the voltage amplitude were identical. It has been shown in [8], that the duration of beam current in nitrogen at 30 Torr pressure is limited by the foil, which separates the diode and the beam-measuring unit precluding the passage of electrons with the energies below 40 keV. That is, the beam current amplitude at such pressures depends only on the time of the onset of emission processes at the cathode.

2. Experiment

In our experiments, we used a SLEP-150 nanosecond generator connected to a transmission line filled with a transformer oil and to a gas diode (Figure 1). The anode was an Al foil of thickness 15 μm (3) and a metal diaphragm of diameter 10 mm (2). The foil at its back was reinforced by grids of the total transparency 6.6 %, which in addition to reinforcement, provided attenuation of the beam current. The cathode (1) was a stainless steel tube of diameter 7.3 mm with a working edge 30 μm thick. The amplitude of incident voltage wave in the transmission line was ~120 kV with a rise time of ~250 ps at a level of 0.1–0.9, and its FWHM with a matched load was ~1 ns.

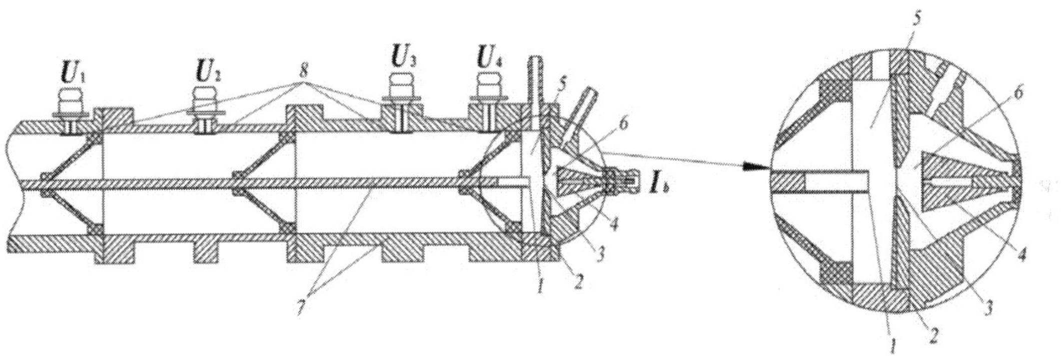

Figure 1. Transmission line and gas diode: 1 – cathode; 2 – diaphragm; 3 – Al foil; 4 – collector; 5 – gas diode; 6 – drift space; 7 – transmission line; 8 – capacitive dividers.

Figure 2 presents experimental data on the stability of the runaway electron beam current in a sample of 300 pulses. It is seen that the beam current ranges from 20 A to 100 A. Reasoning that all pulses were produced in a single series at a pressure of 30 Torr, it can be stated with confidence that it is the state of the emission surface, which influences the beam current under these conditions.

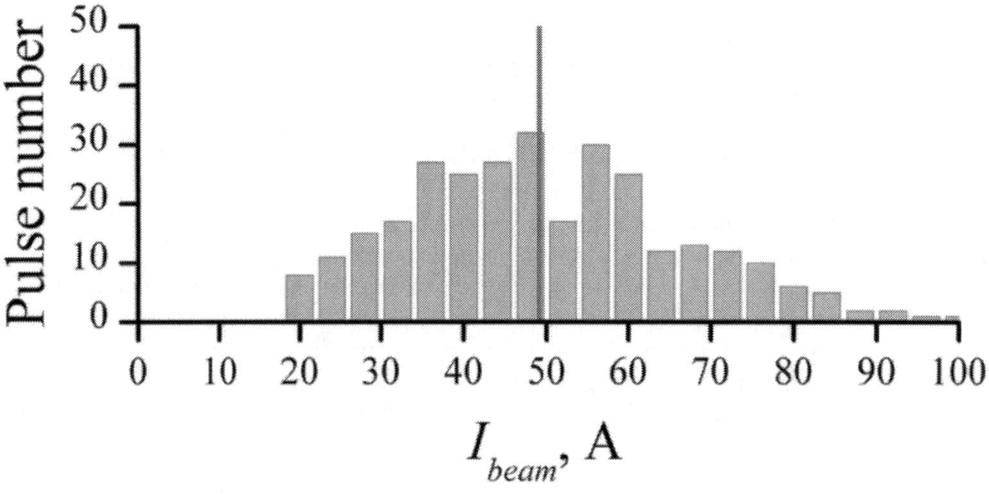

Figure 2. Histogram of the runaway electron beam current (gap width 6 mm, cathode rounding-off radius 30 μm, nitrogen pressure 30 Torr).

3. Numerical simulation

Elucidating the effect of mission surface on the beam current amplitude, we simulated the breakdown developing under experimental conditions using the 2.5D (2D3V) axisymmetric PIC-code XOOPIC [9].

The computational domain included the diode with part of the transmission line and beam current measuring unit.

The emission was described using a model [8] covering both the field emission (the early breakdown stage) and the transition to explosive emission (the unlimited emission ability of cathode). The field emission was described by a modified Fowler–Nordheim formula for a cathode covered with microprotrusions having different amplification factors β. The above-mentioned formula includes an average amplification factor $<\beta>$ characterizing the state of its emission surface:

$$\langle j_{FN} \rangle (E) = \sqrt{\pi} AB^2 \varphi^2 \cdot \alpha^{3/2} \cdot \exp\left(-\frac{2}{\alpha}\right) \cdot \left\{1 + 2\sqrt{\frac{\alpha}{\pi}} + \frac{\alpha}{2}\right\}, \quad (1)$$

where $\alpha = \sqrt{\frac{\langle\beta\rangle \cdot E_{macro}}{B\varphi^{3/2}}}$; $\langle\beta\rangle$ is the average electric field amplification at the microprotrusions: $\langle\beta\rangle = 4-8$ [10].

The average current density estimated by formula (1) was multiplied by the total surface area $I_{FN} = \langle j_{FN} \rangle \cdot S_C$ with calculation of the coefficient K for field-to-explosive emission transition:

$$\eta_t = \sum_t I_{FN}(t)^2 \cdot \Delta t, \quad (2)$$

$$K = \frac{\eta_t}{\eta_0}, \quad (3)$$

where η_0 is equivalent to h being the specific action for explosion [11] but, unlike h calculated for the current density, η_0 is selected for the total current from the cathode. This is because in the axisymmetric approximation used, we cannot describe individual microprotrusions, and experiments give only a general idea about emission properties of the cathode. The emission current was calculated by the expression

$$I_{em} = (1 - K) \cdot I_{FN} + K \cdot I_{ex} \tag{4}$$

where I_{ex} is the explosive emission current calculated using the Gauss theorem [9]. The electron emission form the cathode was specified by two parameters: $\langle\beta\rangle$ for the onset of emission early in the breakdown and η_0 for the time of field-to-explosive emission transition.

The emission parameters $\langle\beta\rangle$ and η_0 were varied so, that the beam current measured by the collector and the voltage at the third divider would coincide with their experimental values corresponding to the highest and lowest currents on the histogram in Figure 2, because the generation of runaway electrons and the breakdown dynamics at these values could differ greatly.

4. Simulation results

Figure 3 presents experimental waveforms of the voltage from the divider U_3 in the transmission line (grey solid curves) and the respective waveforms calculated by integrating the r-component of the electric field for the point of the divider position along the z axis (black solid curves). It is seen that the experimental data corresponding to the left (Figure 3a) and right (Figure 3b) branches of the beam current histogram in Figure 2 agree well with the calculations allowing, respectively, for the explosive emission by the Gauss law (Figure 3a) and for the field-to-explosive emission transition (Figure 3b). It is also seen from Figure 3 that according to integration of the electric field z-component from cathode to plane anode, the accelerating voltage across the diode (thin black curves) in two models of emission differs greatly.

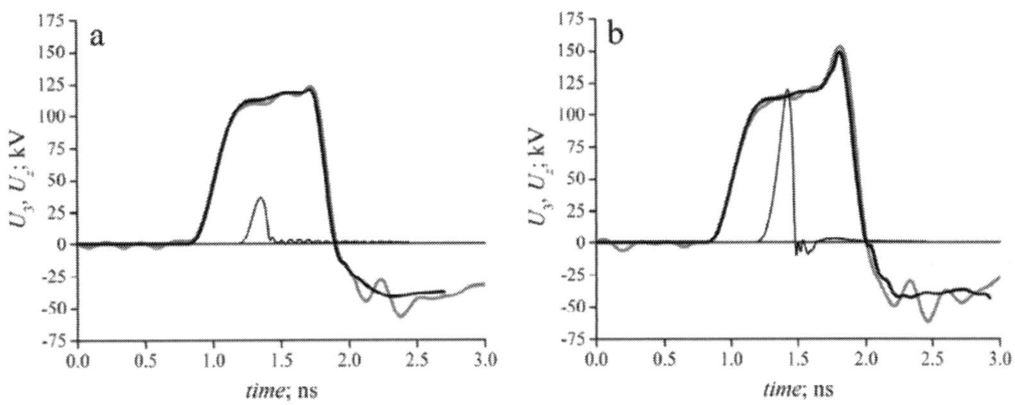

Figure 3. Experimental waveforms of the voltage from divider U_3 (grey solid curves) for the left (a) and right (b) branches of the histogram in Figure 2 and respective calculated waveforms (black solid curves) for explosive emission (a) and for field-to-explosive emission transition (b); U_z – accelerating voltage across the diode (thin black curves).

The difference is due to different dynamics of breakdown and runaway electron generation in two emission models. In the model of explosive emission, the cathode becomes emitting almost immediately after arrival of the voltage when U_z is no greater than several kilovolts. Near the cathode, a plasma layer is formed (Figure 4a, 1.325 ns), giving rise to a runaway electron beam from its boundary and to a breakdown. Due to the early electron emission and pulsed breakdown, the maximum voltage U_z is not higher than 30 kV. In the model allowing for the field emission, we have a delay of an electron emission from the cathode and an increase in U_z to 125 kV, and the beam is formed directly at the cathode, in the region of a strong electric field. The runaway electron beam carries away a negative space charge, leaving only a positive space charge of ions resulting from the impact ionization of a neutral gas by electrons. Near the cathode, the cathode layer is formed (Figure 4b, 1.425 ns) with a potential fall reaching, by the estimates, tens of kilovolts (in contrast to the model of explosive emission). During transition from the field to the explosive (unlimited) emission, the cathode layer disappears (Figure 4b, 1.45 ns).

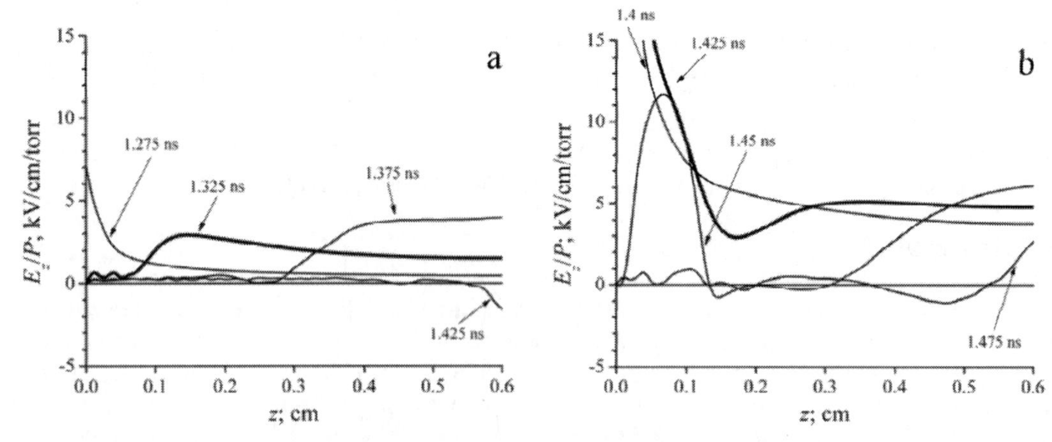

Figure 4. Distribution dynamics of the electric field z-component in the diode for explosive emission (a) and for field-to explosive emission transition (b).

It is seen from Figure 4 that the electric field distributions in the diode for two emission models differ greatly, and hence, the beam currents in two cases differ greatly as well (Figure 5). In the model of explosive emission, the beam current at the collector is not higher than 20 A (Figure 5a), which corresponds to the left branch of the experimental histogram (Figure 2). In the model allowing for the field emission, the beam current reaches 110 A (Figure 5b), which is demonstrated by the right branch of the histogram (Figure 2).

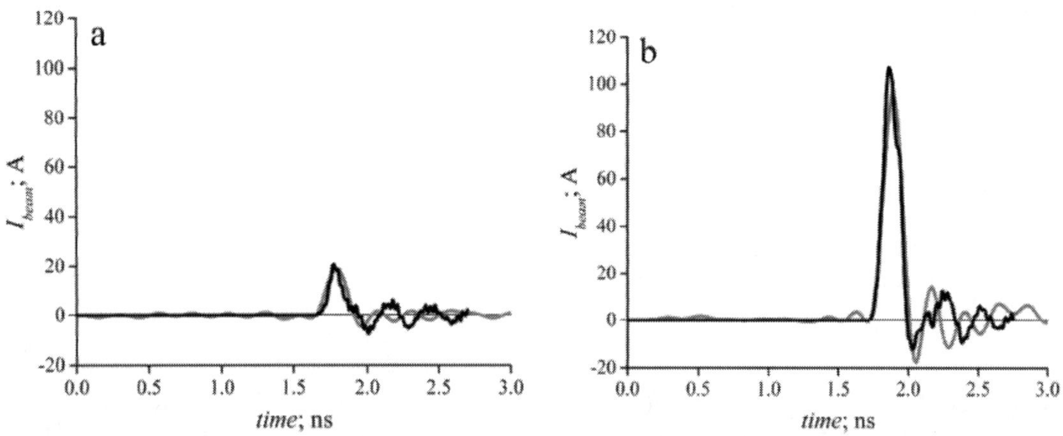

Figure 5. Experimental waveforms of the beam current (grey curves) and respective calculated waveforms (black curves) for explosive emission (a) and for field-to-explosive emission transition (b).

It should be noted that the calculated waveforms are timed. Our numerical simulation demonstrates that the two emission models give different points in time, at which runaway electrons are detected at the collector. In the model of explosive emission, the beam current at the collector peaks at 1.79 ns, and in the model allowing for the field emission, the beam current is delayed by almost 100 ps and its maximum falls on 1.88 ns.

5. Conclusion

Thus, we analyzed the scattering of experimental data in a large sample of pulses produced in a single series of experiments under the same conditions. The runaway electron beam current in the sample ranges from 20 to 100 A, and its deviation is associated mostly with the state of the emission surface of the cathode.

Our numerical simulation of the formation of a runaway electron beam and breakdown under conditions close to the experimental ones show that the modes of the processes and their dynamics differ in different models of emission.

Acknowledgments

The work was supported by RFBR grants No. 14-02-00136 and No. 15-08-03983.

References

[1] *Runaway Electrons Preionized Diffuse Discharges* 2014 ed Tarasenko V F (New York: Nova Science Publishers, Inc.)

[2] *Generation of runaway electrons and X-rays in high pressure discharges* 2015 ed Tarasenko V F (Tomsk, STT) In Russian

[3] Babich L P, Loiko T V and Tsukerman V A 1990 *Phys. Usp.* **33** 521

[4] Tarasenko V F and Yakovlenko S I 2004 *Phy. Usp.* **47** 887

[5] Yalandin M I, Reutova A G, Ul'maskulov M R, Sharypov K A, Shpak V G, Shunailov S A, Klimov A I, Rostov V V and Mesyats G A 2009 *Tech. Phys. Lett.* **35** 804

[6] Mesyats G A, Yalandin M I, Reutova A G, Sharypov K A, Shpak V G and Shunailov S A 2012 *Plasma Phys. Rep.* **38** 29

[7] Baksht E Kh, Belomyttsev S Y, Burachenko A G, Grishkov A A, Shklyaev V A and Tarasenko V F 2014 *Izv. Vyssh. Uchebn. Zaved. Fiz.* **57** 121 In Russian.

[8] Shklyaev V A, Baksht E Kh, Belomytsev S Ya, Burachenko A G, Grishkov A A and Tarasenko V F 2015 *J. Appl. Phys.* **118** 213301

[9] Verboncoeur J P, Langdon A B and Gladd N T 1995 *Comp. Phys. Commun.* **87** 199

[10] Kozyrev A V, Korolev Y D, Mesyats G A 1987 *Sov. Phys. Tech. Phys.* **32** 34

[11] Mesyats G A and Yalandin M I 2009 *Dokl. Phys.* **54** 63

25 keV electron beam formation based on a three-electrode system of the obstructed glow discharge in H2 and D2

Yu S Akishev[1,2], V B Karalnik[1], M A Medvedev[1], A V Petryakov[1], N I Trushkin[1] and A G Shafikov[1]

[1]SC SRC RF TRINITI, 108840, Moscow, Troitsk, Pushkovykh Str., 12, Russia,
[2]NRN University MEPhI, 115409, Moscow, Kashirskoe shosse 32, Russia

E-mail: akishev@triniti.ru

Abstract. The results of experimental study on the formation of high-current electron beams using three-electrode system are presented. This system enables to create a diffusive obstructed glow discharge in a pulsed regime with an amplitude of applied voltage up to 25 kV. The influence of such experimental parameters as the pressure of plasma forming gas and the amplitude of applied voltage on the obstructed glow discharge duration was investigated. Both the transverse and the longitudinal structure of formed electron beam was determined.

1. Introduction

The electron beams (e-beams) of moderate energy ($\varepsilon \leq 25$ keV) and the non-thermal plasma created by these beams are widely used in various fields of science and practice: gas lasers excitation [1], surface modification of various materials [2,3], creation of new materials including nano-powders [4], etc. Some information about the formation of such electron beams can be found in [5,6]. The e-beams of moderate energy are perspective for the creation of small-sized neutron source based on the super-strong charging of dust particles [7]. In this source, the e-beam provides both the plasma around dust particles and their super-strong charging up to the potential energy equal to e-beam energy. Positive ions accelerate in the strong electric field of charged particle and increase their energy to high magnitude. The impact of high-energy ions with a particle leads to nuclear fusion reactions accompanied by neutron release. The numerical modeling of dust particles charging and nuclear fusion reactions [7] enabled us to formulate the requirements to the discharge and e-beam parameters providing the optimum experimental conditions for generation of neutrons. It was shown also that the most suitable plasma forming gas was deuterium. In this paper, the experimental results on the generation of pulsed high-current e-beam with the energy up to 25 keV are presented. The experiments were performed in H2 and D2 at pressure P = 0.5 - 2 Torr. The e-beam was generated by the obstructed glow discharge using a triple electrode system of an original construction. This e-beam source will be used for generation of neutrons due to super-high charging of dust particles by the e-beam.

2. Experimental setup

The e-beam with the energy up to 25 keV was generated by the obstructed glow discharge using a three-electrode system of an original construction (see figure 1a). The solid cathode of 24 mm diameter made of stainless steel and two anodes made of metallic grids with geometrical transparency of 60% and 70% were used. The first and second grids were placed at 2 mm and 9 mm distance from the cathode. The electrode system was placed into a quartz tube of 110 mm inner diameter and 300 mm length. The main discharge was excited in a narrow gap between the cathode and the first anode. This pulsed discharge was powered by the applied voltage up to 25 kV and characterized as a high-current obstructed glow discharge with the amplitude of total current up to 150 A. Namely this main discharge formed the e-beam. The low-current discharge between the cathode and the second anode worked in the regime of normal glow discharge. The low-current discharge played complimentary but very important role - this discharge created the pre-ionization of gas gap and increased the stability of main discharge. Each discharge has been fed by its own electrical power. A three-electrode system enabled us to generate the pulsed and steady-state e-beams with the energy up to 25 and 10 keV, respectively.

The e-beams were formed in H2 and D2 of high purity (99.99%) at pressure P = 0.5 - 2 Torr. The experiments were performed using weak gas blowing through the reactor in order to keep the initial high purity of plasma forming gas. The applied voltage was measured by HV divider PINTEK HVP-39 (1000:1, 40 kV, 200 MHz). The total discharge current was measured by low-inductive shunt with the resistance of 0.024 Ohm. The Faraday cylinder was used to measure the current of e-beam. The scheme of optical measurements of the dynamics of optical emission, which was excited in gas by the e-beam, is shown in figure 1b. The emission of different parts of reactor was collected by quartz lens and transferred to the line of quartz light guides, which were connected with appropriate photomultiplier tubes (PMT) FEU-100. All electrical signals were recorded by oscilloscopes such as Tektronix TDS 520, Tektronix TDS 2012 и Tektronix DPO2024. Spatial distribution of optical emission from the quartz reactor was registered by digital camera Canon EOS 550, which formed the image integrated in time. The transverse structure of e-beam was studied using flat phosphor, which was placed at various distances from the outlet of electron gun (e-gun). The plane of phosphor was oriented perpendicularly to the e-beam.

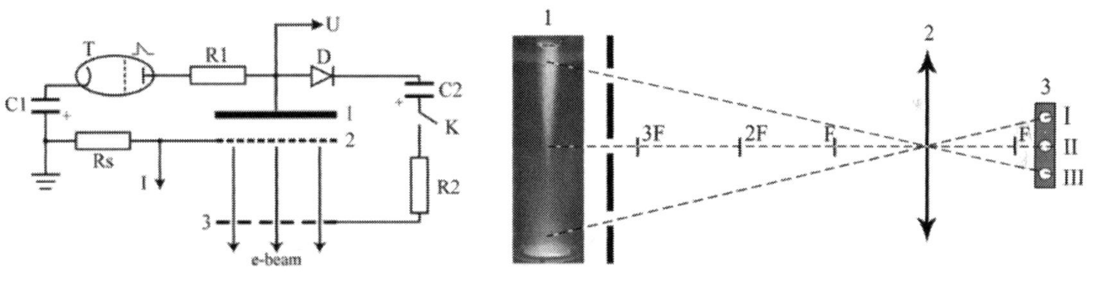

a) b)

Figure 1. Total scheme of the experimental setup. a) Electrical scheme: 1 - cathode, 2,3 – grid anodes of main and auxiliary discharges, R1=56 Ohm, R2=1 MOhm – ballast resistors, Rs =0.024 Ohm - current shunt, D – high-voltage diode KC 201E, C1=12.5 nF, C2 =100 μF – pulsed capacitors, K – switch for charging the capacitor of the auxiliary low-current discharge, T – thyratron TGI-1000/25. b) Optical scheme: 1 – the integrated in time image of optical emission of gas excited by the e-beam (the outlet of e-gun is at bottom), 2 - quartz lens with focus length F=11.2 см, 3 – the line of quartz light guides connected with PMT I, II, III.

3. Experimental data

The typical waveforms of the current and voltage of main discharge in a narrow gap as well as the synchronized with them signals of PMT I and PMT III are presented in figure 2a. In this figure, one can see that the high-voltage regime in D2 lasts approximately 200 ns, during which the current falls down from 150 A to 5 A. After that, the discharge abruptly transfers from a high-voltage regime to a high-current regime with a low voltage. In fact, depending on experimental parameters (gas pressure, the voltage applied to an auxiliary and main discharge), duration of high-voltage regime can vary from 0.15 μs to 5 μs. The e-beam is generated by the discharge under high-voltage obstructed regime. As a rule, the current of e-beam is equal to 95% of the total discharge current. An optical gas emission initiated by e-beam appears with some delay regarding to the appearance of e-beam at this place. The image of pulsed discharge is presented in figure 2b. The photo was taken from the side of anode grid. This photo gives the discharge image integrated over a long time period. It means that the images of diffusive discharge under high-voltage regime and the constricted discharge under low-voltage regime with high-current cathode spots are imposed.

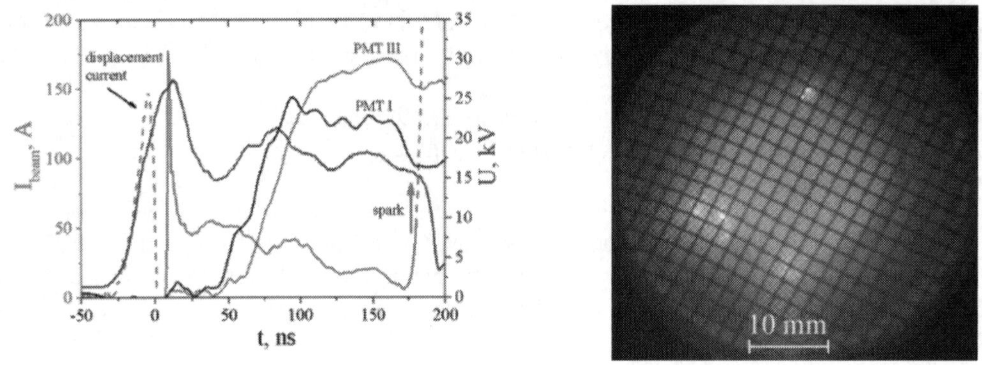

a) b)

Figure 2. a) The current and voltage waveforms of the obstructed discharge and signals of PMT I and PMT III, which collect the light from the places located at x=0 and x=22 cm (x=0 corresponds to the e-gun outlet). b) The image of pulsed discharge taken from the side of anode grid. U_d = 24 kV. In both cases the plasma forming gas is D2 at pressure P=1 Torr; the voltage magnitude U_{aux} applied to the auxiliary discharge is equal to 2 kV.

The dependence of maximum current in the high-voltage obstructed discharge on gas pressure P at different amplitudes U_d of the applied voltage is presented in figure 3 (curves 1 and 2) and figure 4. In these figures, one can see that the higher P and U_d, the higher current in obstructed discharge. Contrariwise, the duration of high-voltage obstructed discharge decreases with increasing gas pressure (curves 3 and 4 in figure 3).

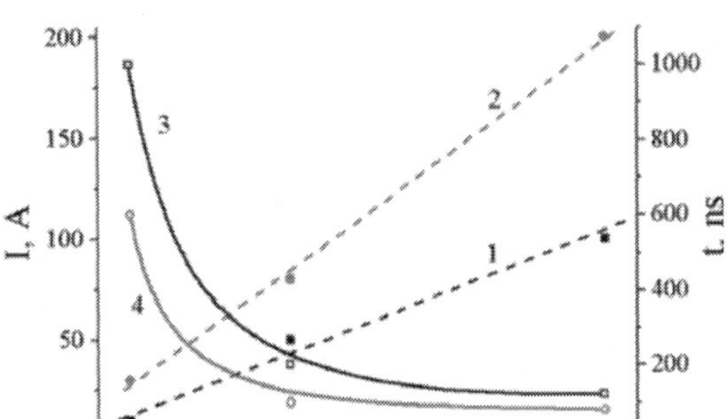

Figure 3. The maximum current in high-voltage obstructed discharge vs deuterium pressure. U_d = 20 kV (curve 1); U_d = 24 kV (curve 2). The obstructed discharge duration vs pressure of D2. U_d = 20 kV (curve 3); U_d = 24 kV (curve 4).

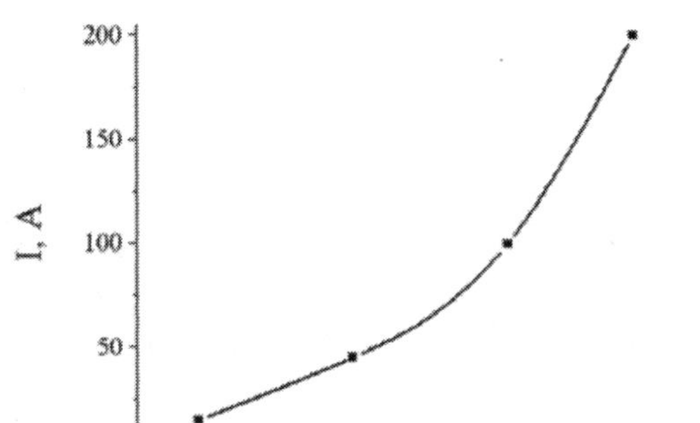

Figure 4. The maximum current in high-voltage obstructed discharge vs the applied voltage amplitude. Gas - deuterium at pressure P = 2 Torr.

The side view of reactor showing the distribution of optical emission of H2 excited by the pulsed e-beam with the energy ε = 20 keV is presented in figure 5. Each photo in figure 5 is taken with a long exposure time and therefore, gives the integrated image though in fact, the optical emission varies in space and time. The gas pressure is equal to P=0.5, 1.0 and 2.0 Torr in figures 5a, 5b, and 5c, respectively.

a)

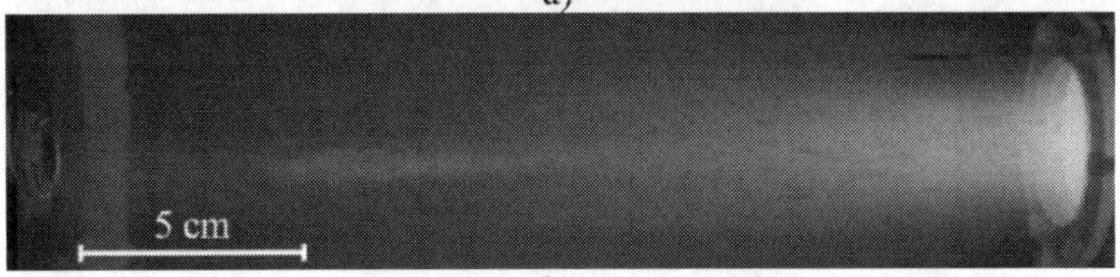

b)

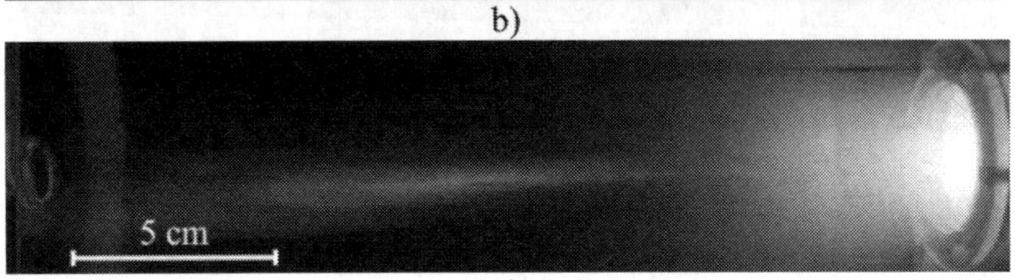

c)

Figure 5. The integrated in space and time distribution of optical emission of H2 excited by the pulsed e-beam with the energy ε = 20 keV. The voltage magnitude U_{aux} applied to the auxiliary discharge is equal to 2 kV. The gas pressure was P=0.5 Torr (a), P=1 Torr (b) и P=2 Torr (c).

The spatial distribution of optical emission in the reactor is determined by the distribution of current density in the e-beam. In such a case, the set of images in figure 5 reflects the evolution of spatial structure of the e-beam with increasing gas pressure. One can see that the e-beam enters the reactor as a broad diffusive beam. However, in the course of its propagation through the gas, the e-beam can constrict itself and be concentrated approximately around the tube axis. At gas pressure P=1 Torr, the e-beam has 36 mm diameter at the e-gun outlet, however, at the distance of 8 cm from the outlet it constricts itself to 8 mm diameter. Further, the increase of e-beam diameter up to 15 mm is observed. The intensity of optical emission from the constricted e-beam is higher, compared to that of diffusive e-beam.

The length of e-beam diffusive mode diminishes with increasing gas pressure but the length of its constricted mode grows. For instance, at gas pressure P=2 Torr, the e-beam constricts itself to the diameter 5 mm already at 4 cm away from the outlet, and further, the e-beam diameter is increasing insignificantly. The appearance of e-beam constriction with increasing gas pressure correlates with increasing e-beam current under growing gas pressure (figures 3 and 4). We assume that the constriction of e-beam happens due to the magnetic field, which is formed by its own current (pinch-effect [8]).

The e-beam transverse structure correlates with the spatial distribution of the emission of flat phosphor placed perpendicularly to e-beam. Figure 6 presents the set of images of the luminescent phosphor. These images show the evolution of transverse e-beam structure, depending on the distance from the e-gun outlet. The initial e-beam energy ε is equal to 20 keV. The variable experimental parameters are: the gas pressure P and the voltage U_{aux} of auxiliary discharge.

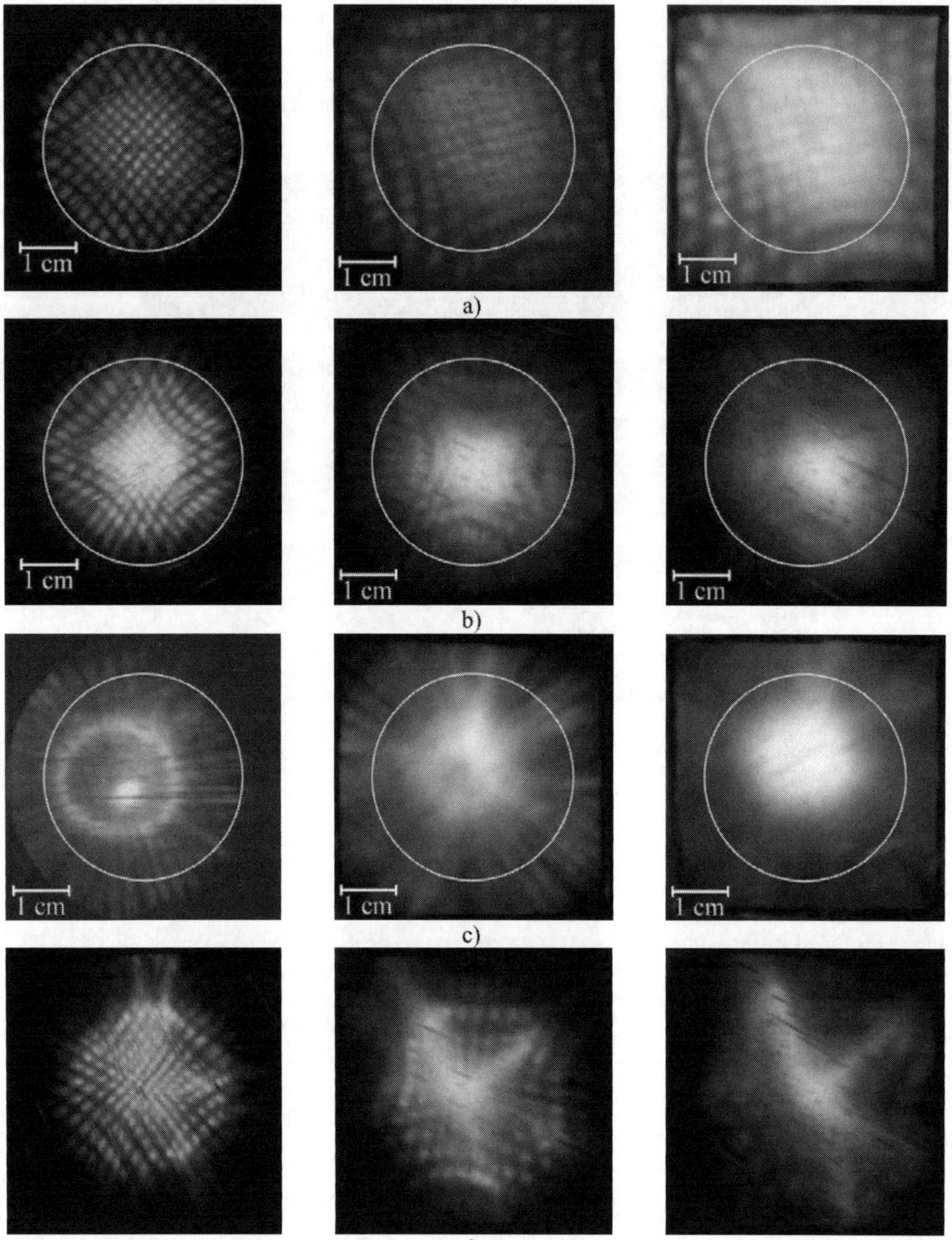

4cm 8cm 12cm

Figure 6. The set of photos of the luminescent phosphor emitted by the e-beam with the initial energy ε=20 keV. The distance of phosphor from the e-gun outlet is shown at the bottom of each column. The circle in each photo marks the boundary of e-gun outlet (36 mm diameter). a) P=0.5 Torr, U_{aux} =2 kV; b) P=1 Torr, U_{aux} =2 kV; c) P=2 Torr, U_{aux} =1 kV; d) P=1 Torr, U_{aux} = 0 kV.

In figure 6, one can see that the distribution of current density in the e-beam is modulated by the grid at the e-gun outlet. In other words, the e-beam formed by the e-gun is split into a set of thin e-beams be the grid. At low pressure, the modulation is kept over a long distance despite of the beam scattering and its increasing transverse sizes. The modulation disappears in the constricted mode of e-beam.

The absence of auxiliary discharge (U_{aux} = 0 kV) leads to the instability of main discharge ignition. For instance, at P = 0.5 Torr, applied voltage amplitudes U_d = 20 kV, and U_{aux} = 0 kV, the ignition start of obstructed high-voltage discharge in a narrow gap happens, on the average, one time of four attempts. The absence of the auxiliary discharge leads also to the strong inhomogeneity of e-beam. The transverse inhomogeneity of e-beam exhibits itself in the appearance of a star-like structure (see figure 6d). The drawbacks pointed above do not allow the practical usage of high-voltage obstructed discharge without auxiliary low-current discharge. So, the developed three-electrode system has a good perspective in practical using for stable generation of the high-current and transversely homogeneous pulsed e-beam with the energy up to 25 keV.

Acknowledgments
The work was wholly supported by Russian Scientific Foundation with the grant # 16-12-10458.

References

[1] Wernsman B, Ranea-Sandoval HF, Rocca JJ and Mancini H 1986 *IEEE Trans. Plasma Sci.* **14** 518-22
[2] Ozur GE, Proskurovsky DI, Rotshtein VP and Markov AB 2003 *Laser & Particle Beams* **21** 157–74
[3] Burdovitsin VA, Klimov AS, Medovnik AV and Oks EM 2010 *Plasma Sources Science and Technology* **19** 20–26
[4] Muller G, Schulteiss C 1994 *Proc. 10th Int. Conf. on High Power Particle Beams* (San Diego, USA) pp 833–836
[5] Kozhevnikov VYu, Kozyrev AV, Korolev YuD, 2006 *Plasma Physics Reports* **32** 1027
[6] Klimenko KA, Korolev YuD, 1992 JTPh **60** (9) 138.
[7] Akishev YuS, Karal'nik VB, Petryakov AV, Starostin AN, Trushkin NI and Filippov AV 2016 *Plasma Physics Reports* **42** 14–24
[8] Lowson JD 1958 J.Electron Control. **5** 146-51

IOP Conf. Series: Journal of Physics: Conf. Series **830** (2017) 012016 doi:10.1088/1742-6596/830/1/012016

Investigation of thyristor-based switches triggered in impact-ionization wave mode

A I Gusev, S K Lyubutin, S N Rukin, B G Slovikovsky and S N Tsyranov
Institute of Electrophysics, 106 Amundsen Street, Yekaterinburg, 620016, Russia

E-Mail: gusev@iep.uran.ru

Abstract. An operation of the thyristor-based switches triggered in impact-ionization wave mode has been investigated. The thyristor switch contained two series connected tablet thyristors having a silicon wafer of 56 mm diameter. Applying across the switch a triggering pulse with a voltage rise rate *dU/dt* of over 1 kV/ns, the thyristors transition time to a conductive state was reduced to shorter than 1 ns. It is shown that the maximum amplitude of a no-failure current is increased with increasing *dU/dt* at the triggering stage. A possible mechanism of the *dU/dt* value effect on the thyristors breakdown current is discussed. Under a safety operation regime at *dU/dt* = 6 kV/ns (3 kV/ns per a single thyristor), the switch discharged 1-mF capacitor, which was charged to a voltage of 5 kV, to a resistive load of 18 mΩ. The following results were obtained: a peak current was 200 kA, an initial *dI/dt* was 58 kA/µs, a FWHM was 25 µs, and a switching efficiency was 0.97. It is shown also that a temperature of the silicon wafer is one of the main factors that affects on the thyristor switching process.

1. Introduction

In recent years, semiconductor switches on the basis of dynistor structures with a time of transition into a conducting state of <1 ns were developed at the Ioffe Physical Technical Institute [1-2]. Such switches – deep-level dynistors (DLDs) – are able to switch a current of several kiloamperes of a capacitive energy storage at a current rise rate of up to 200 kA/µs; at present, this cannot be attained using any other semiconductor switches [2]. The DLD is switched into a conducting state upon an application of an overvoltage pulse with a rise rate of >1 kV/ns. This allows the electric field in the plane of the reversely biased collector junction of the dynistor to be increased to values of 300–350 kV/cm, at which an intense ionization of deep levels in silicon is initiated, within a few nanoseconds.

The injection of electrons from these centers into the structure region with a high field initiates an impact-ionization wave, which passes through the structure of a device and uniformly fills its entire area with a dense electron-hole plasma. The wave motion velocity is considerably higher than the saturated velocity of carriers in silicon; in connection with this, the structure of a device with a typical wafer thickness of several hundred microns and a carrier transit time of a few nanoseconds is filled with a plasma in a time of several hundred picoseconds. The further current flow through the structure is maintained by the injection of carriers into the base regions through the injector junctions. Detailed descriptions of the physical principles of the DLD operation can be found, e.g., in [3] and [4].

Recently, it was shown in [5] and [6] that ordinary thyristors, which, as dynistors, have a four-layer semiconductor structure, can operate in such a mode. A thyristor switch, which contained several series-connected commercial thyristors of a tablet design with a diameter of a silicon wafer of 40 mm, was triggered by an external overvoltage pulse with a few nanosecond rise time [5]. The switch operated at a current pulse amplitude of over 20 kA and a current rise rate *dI/dt* of over 100 kA/μs. Compared with the traditional mechanism for triggering the thyristor by a current pulse through the gate electrode, the *dI/dt* value was increased by about 250 times.

In [6], the impact-ionization switching of power thyristors, implemented by an overvoltage pulse with a nanosecond rise time, applied to main thyristor electrodes, was studied experimentally and by numerical simulation methods. It was shown that the voltage rise rate at the triggering stage was the main factor affecting the time of switching the thyristor from the blocking state to the conducting state.

This work continues the research started in [5] and [6]. The results of experiments on switching current pulses with an amplitude of up to 200 kA during discharge capacitors with over 10 kJ of stored energy are presented. An effect of the voltage rise rate at the triggering stage on the thyristors main switching characteristics is shown and discussed. Finally, we also show how a temperature of the silicon wafer affects on the thyristor switching process.

2. Experiments with thyristors operating at 200-kA peak current

In the experiments, the thyristor switch was assembled of two series-connected thyristors of T453-800-24 type with a DC operating voltage of 2.4 kV, a surge current amplitude of 20 kA, a critical current rise rate of 400 A/μs when triggered through the gate electrode, and diameter of silicon wafer of 56 mm.

The electrical circuit is shown in Figure 1. The capacitive store contained 5 identical modules connected in parallel. The measured values of every module elements were as follows: C was ~220 μF, L was ~0.33 μH, and R was ~72 mΩ. In the experiments, the charging voltage U_0 was equal to 5 kV. The value of the peak current through the thyristor switch was changed by changing the number of parallel modules included in the discharge circuit and by changing the total value of the load resistance. The maximum amplitude of the discharge current for 5 modules connected in parallel was ~230 kA. A pulse duration (the FWHM) was ~25 μs. Equivalent values of the capacitance and resistance of the total discharge circuit were 1.1 mF and 14.5 mΩ, respectively.

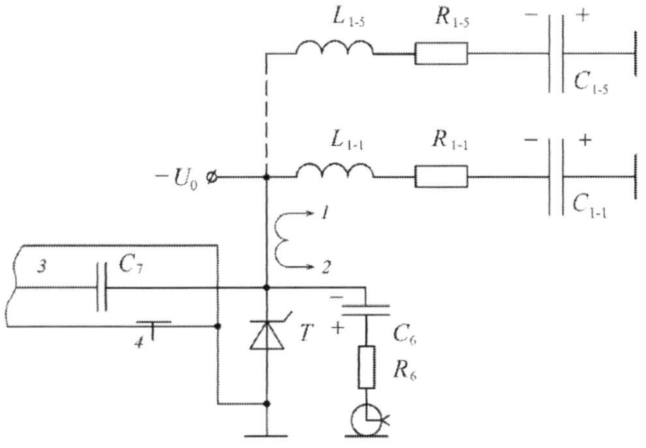

Figure 1. Experimental discharge circuit.

For the thyristor switch T triggering, a compact solid state generator is used. An output unit of the generator contains a coaxial oil-filled 50-Ω line (an item 3 in Figure 1). The triggering pulse of a negative polarity is delivered to the thyristor T via an isolating ceramic capacitor C7 (4.7 nF), and is registered by means of a capacitive probe 4. The output pulse in the line has a rise time of ~1 ns. The output pulse amplitude can be adjusted within 50 to 100 kV. The current was measured by Rogowski coil (outputs *1* and *2* in Figure 1). The resistive voltage divider (elements C6, R6, and 50-Ω measuring cable) was connected across the whole switch of two thyristors.

It was found that the voltage rise rate *dU/dt* at the triggering stage was the main factor affecting the thyristors switching characteristics. In the experiments, we varied *dU/dt* value in the range of 0.5 to 3.0 kV/ns. Typical waveforms across the thyristor switch at different *dU/dt* are shown in Figure 2.

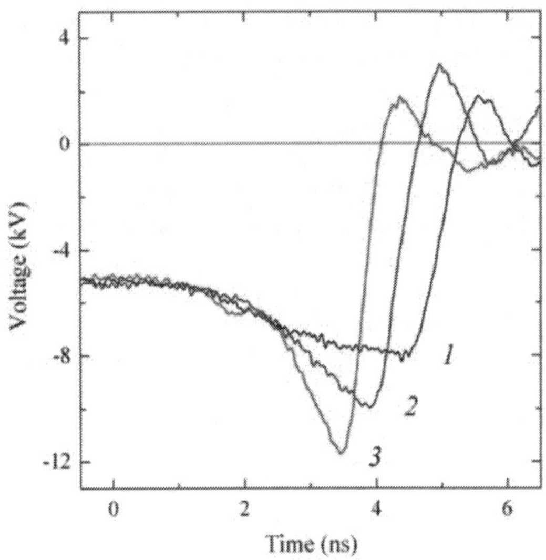

Figure 2. Waveforms of the voltage across 56-mm thyristors during switching process: *dU/dt* is ~0.5 (curve *1*), ~1.3 (curve *2*), and ~3.0 kV/ns (curve *3*).

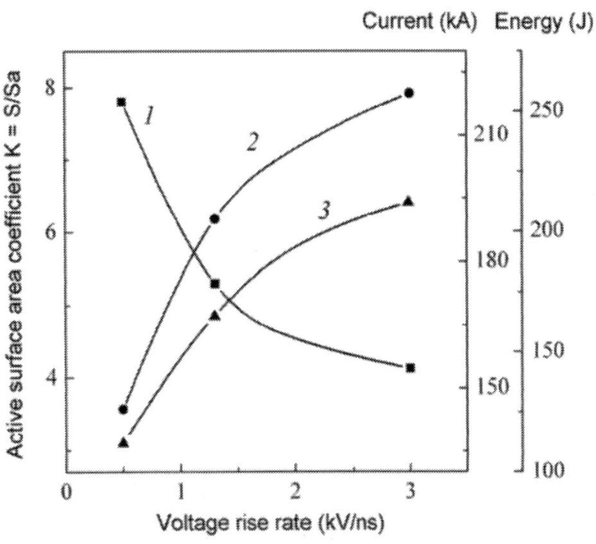

Figure 3. Coefficient of the structure active surface area K = S/Sa (curve *1*), amplitude of the failure current (curve *2*) and thyristor dissipated energy (curve *3*) as a function of the voltage rise rate at the triggering stage.

At each triggering mode shown in Figure 2, we increased the discharge current through the switch until the failure of the tested thyristors occurred. The results of these tests are shown in Figure 3. It is seen that increasing the *dU/dt* value from ~0.5 to ~3.0 kV/ns allowed us to increase the failure current from ~140 up to ~220 kA (the curve *2* in Figure 3).

To explain the experimental results obtained, we calculated a temperature increase of the silicon wafer during a current pulse through the waveforms of the dissipated power for the failure current values shown in Figure 3. Assuming that the surface area of wafer was heated uniformly, the following calculated results for the structure temperature increase were obtained: 61, 90, and 115 ^{0}C for *dU/dt* values of 0.5, 1.3, and 3.0 kV/ns, respectively. It is well known that the failure of the power silicon devices due to overheating occurs when the temperature of the structure reaches 400–600 ^{0}C. At this temperature range, due to a process of electron-hole plasma thermal generation, the current flowing through the structure pinches to the overheated spots of the structure that leads to the device failure.

The various temperature values of the structure obtained in our experiments could mean that the current flowed through a part of the total surface area of the structure. In this case, an increase of the *dU/dt* values at the triggering stage led to an increase of an active surface area of the structure through which the current flowed. This nonuniformity of the current distribution could be defined by a coefficient $K = S/S_a$, where S is the total surface area of the structure, and S_a is its active part. Assuming that in all the experiments, the failure temperature was equal to 500 ^{0}C we calculated the values of the coefficient K, as a function of *dU/dt* (curve *1* in Figure 3). It shows that the active area of the structure increases from ~13 to ~25% of the total area when *dU/dt* is increased from ~0.5 to ~3.0 kV/ns. This question needs to be considered in an independent study.

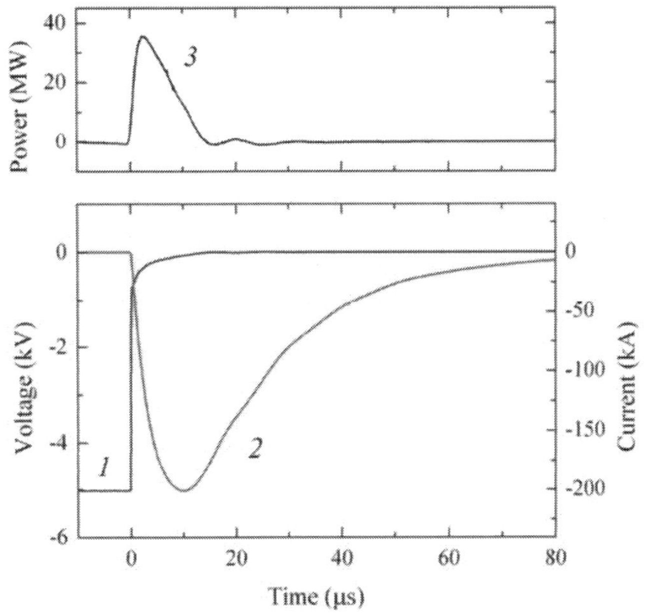

Figure 4. Waveforms of the voltage (*1*), current (*2*), and power losses (*3*) for the thyristor switch triggered by *dU/dt* of 3.0 kV/ns (curve *3* in Figure 2) and operated at peak current of 200 kA.

Finally, we tested the thyristor switch under the no-failure operation regime. The waveforms are shown in Figure 4. In the experiment, the switch was triggered by an overvoltage pulse, in accordance with curve *3* in Figure 2. A 1.1-mF capacitor, which was charged to a voltage of 5 kV, was switched to a resistive load of 18 mΩ. The following discharge parameters were realized: the discharge current amplitude of 200 kA, the maximum current-rise rate of 58 kA/μs, the current rise time (0.1-0.9 level) of 5.5 μs, the pulse duration (FWHM) of ~25 μs, and the switching efficiency of 0.97. During the performed tests, thyristors were enabled over 100 times. After each 5 shots, the leakage current through the thyristors was measured. At a nominal DC voltage of 2.4 kV per thyristor, it was around 4 μA and did not change during the tests.

3. Effect of temperature on the thyristor switching process

An experimental electrical circuit for studying the process of switching a thyristor from the blocking state to the conducting state at different temperatures is shown in Figure 5. In the experiments, we used a commercial low-frequency tablet thyristor of T133-320-22 type, made in Russia (http://www.proton-electrotex.com). The thyristor had the DC operating voltage of 2.2 kV and a silicon wafer of 32 mm diameter. The circuit contains a capacitor C1 = 0.1 μF charged to a bias voltage U_0 = 2.2 kV of a negative polarity. The bias voltage is applied under DC or pulse modes to the tested thyristor T through a resistor R1. For thyristor triggering, the same triggering generator was used. A resistor R2 served to change the triggering voltage rise rate *dU/dt* across the thyristor T.

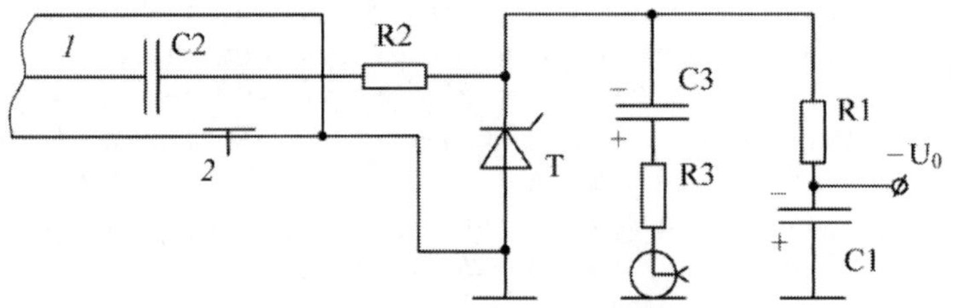

Figure 5. Circuit diagram of the experimental setup.

The voltage across the thyristor T was measured by means of a resistive divider having the elements C3 and R3. To avoid the influence of the inductance of the thyristor contacts located inside the casing, a semiconductor wafer was extracted from the casing, and the divider was connected directly across the wafer. In the experiments, the tested thyristor T was heated up from a room temperature of 25 °C up to 175 °C using an external heater. The temperature of the semiconductor wafer was measured by means of a thermocouple. The experimental results are shown in Figures 6 and 7.

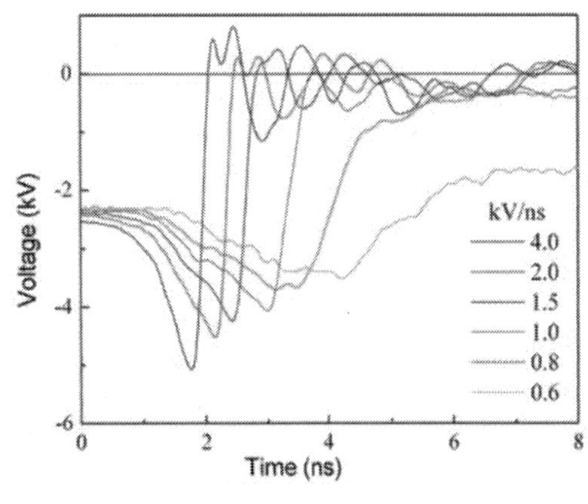

Figure 6. Waveforms of the voltage across the thyristor during switching process at room temperature (T = 25 °C) and different values of the voltage rise rate *dU/dt*.

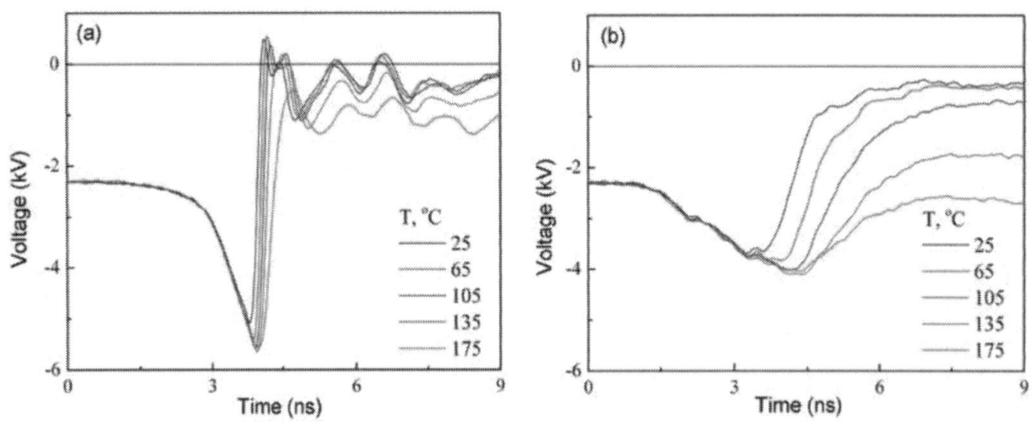

Figure 7. Waveforms of the voltage across the thyristor during switching process at different temperatures: $dU/dt = 4$ kV/ns (a) and 0.8 kV/ns (b).

The obtained results show the existence of two main factors that affect the thyristors switching process – the voltage rise rate dU/dt at the triggering stage and the temperature of the semiconductor wafer. These two factors affect the switching process in an opposite manner. An increase of the dU/dt values leads to an increase of the voltage across the thyristor before its switching into the conductive state (Figure 6). For this reason, the impact-ionization front in the structure is formed at a higher electric field and has a higher velocity. Finally, the higher velocity of the impact-ionization front decreases the thyristor transition time to the conductive state (Figure 6).

And vice versa, the increase of the structure temperature has a negative effect on the switching process, since the temperature growth reduces the intensity of carrier avalanche multiplication processes. This factor leads to the decrease of the impact-ionization front velocity, the concentration of the electron-hole plasma, and as a result, to the increase of the residual voltage across the thyristor after the switching process (Figure 7).

At high values of dU/dt, the switching process depends weakly on the temperature up to T ~105 °C (Figure 7a, 4 kV/ns). At T > 105 °C, the increase of a transition time to the conductive state and the residual voltage is observed. At low values of dU/dt, the temperature growth makes the switching process dramatically worse (Figure 7b, 0.8 kV/ns), and at T > 105 °C, the switching process disappears.

Acknowledgements
The study was supported by Russian Foundation for Basic Research (RFBR), research projects Nos. 14-08-00625 and 15-08-07821.

References

[1] Aristov Yu V, Voronkov V B, Grekhov I V, Kozlov A K, Korotkov S V and Lyublinskii A G 2007 *Instrum. Exp. Tech.* **50** pp 224–227.
[2] Korotkov S V, Aristov Yu V, Voronkov V B, Zhmodikov A L, Korotkov D A and Lyublinskii A G 2009 *Instrum. Exp. Tech.* **52** p 695.
[3] Grekhov I V, Korotkov S V and Rodin P B 2008 *IEEE Trans. on Plasma Science* **36** pp 378-382.
[4] Grekhov I V 2010 *IEEE Trans. on Plasma Science* **38** pp 1118-1123.
[5] Gusev A I, Lyubutin S K, Rukin S N and Tsyranov S N 2015 *Instrum. Exp. Tech.* **58** pp 376-380.
[6] Gusev A I, Lyubutin S K, Rukin S N and Tsyranov S N 2016 *Semiconductors* **50** pp 394-403.

Experimental research of neutron yield and spectrum from deuterium gas-puff z-pinch on the GIT-12 generator at current above 2 MA

R K Cherdizov[1], F I Fursov[1], V A Kokshenev[1], N E Kurmaev[1], A Yu Labetsky[1], N A Ratakhin[1,3], A V Shishlov[1,3], J Cikhardt[2], B Cikhardtova[2], D Klir[2], J Kravarik[2], P Kubes[2], K Rezac[2], G N Dudkin[3], A A Garapatsky[3], V N Padalko[3] and V A Varlachev[3]

[1]Institute of High Current Electronics SB RAS, 2/3 Akademichesky Ave., Tomsk, 634055, Russia
[2]Czech Technical University in Prague, Faculty of Electrical Engineering, Technicka, 16627 Prague 6, Czech Republic
[3]National Research Tomsk Polytechnic University, 30 Lenina Ave., Tomsk, 634050, Russia

E-mail: rustam.k.cherdizov@gmail.com

Abstract. The Z-pinch experiments with deuterium gas-puff surrounded by an outer plasma shell were carried out on the GIT-12 generator (Tomsk, Russia) at currents of 2 MA. The plasma shell consisting of hydrogen and carbon ions was formed by 48 plasma guns. The deuterium gas-puff was created by a fast electromagnetic valve. This configuration provides an efficient mode of the neutron production in DD reaction, and the neutron yield reaches a value above 10^{12} neutrons per shot. Neutron diagnostics included scintillation TOF detectors for determination of the neutron energy spectrum, bubble detectors BD-PND, a silver activation detector, and several activation samples for determination of the neutron yield analysed by a Sodium Iodide (NaI) and a high-purity Germanium (HPGe) detectors. Using this neutron diagnostic complex, we measured the total neutron yield and amount of high-energy neutrons.

1. Introduction

A Z-pinch is a type of plasma confinement system that uses the electrical current to compress plasma magnetically to the axis. For half a century, Z-pinches have been studied as a source of neutron radiation [1].

We conducted experiments with different types of deuterium gas-puff loads [2]. Since 2013, we used a single-shell deuterium gas-puff surrounded by an outer plasma shell consisting of carbon and hydrogen ions. The outer plasma shell allows us to increase the efficiency of energy transfer from the current generator to the load and improve the stability of a pinch. This type of load was successfully used for neutron production in DD-reaction in our experiments. Neutron yield was increased by an order of magnitude (from 10^{11} to 10^{12}) at the peak current of 2.7 MA. Careful measurements of the neutron yield and neutron spectra are an essential and important part of the research. Since 2014, we

IOP Conf. Series: Journal of Physics: Conf. Series **830** (2017) 012017 doi:10.1088/1742-6596/830/1/012017

have been started to use neutron activation analysis for the neutron measurements in addition to other diagnostic techniques.

This paper summarizes the experimental results on measurements of the neutron yield and the amount of high-energy neutrons with the help of neutron activation method.

2. Apparatus and diagnostic setup

The experimental sessions with deuterium gas-puff were carried on the GIT-12 generator at the Institute of High Current Electronics in Tomsk, Russia. The GIT-12 is a pulse current generator, which consists of 12 primary capacitive energy storage modules, an intermediate inductive storage, and a microsecond plasma opening switch (POS). At a 50 kV charging voltage, the generator stores an energy of 2.6 MJ. When the POS is applied, a load current reaches a value of about 2.5 MA with a 200 ns risetime and a rate of up to 20 kA/ns. Without the POS, the generator provides a load current of 4.7 MA with a rate of 3 kA/ns and a current rise time of 1.7 μs. In these experiments, the generator operation mode without the POS was used.

In the 2014-2015experimental sessions, the load was composed of a single shell deuterium gas-puff with an outer plasma shell of carbon and hydrogen ions. A fast electromagnetic valve was used to form the gas-puff. The diameter of the gas-puff was 80 mm, and the injected linear mass was from 60 μg/cm to 100 μg/cm. The outer plasma shell with a linear mass of about 5 μg/cm was generated by 48 plasma guns located at the diameter of 350 mm [2].

To determine the neutron yield and other parameters of neutron radiation, we used a neutron diagnostic complex containing various detectors. Time-of-flight detectors measured neutron energy. These detectors are composed of Bicron BC-408 scintillators and Hamamatsu H1949-51 photomultipliers. To prevent effect of hard x-rays, detectors were shielded by lead. In our experiments, four nToF detectors were located in the radial direction at distances of 1.5 or 2 m, 5.7 m, 10.1 m, and 25.8 m, and one detector was located in axial direction below cathode at a distance of 4.8 m. Using signals from these detectors, one can reconstruct neutron spectrum [3]. The total neutron yield was measured by a set of BTI bubble detectors, which were calibrated in situ. Five BTI detectors were placed at different angles and distance from the pinch, and the yield was determined as the average of the obtained values. A silver activation counter (SAC) was another diagnostics for the neutron yield measurements. SAC was located radially at a distance of 4.55 m from the pinch. SAC is assembled from polyethylene block (as moderator) and SBM-19 Geiger counter, which is covered by 1 mm silver foil. This technique is based on the activation of silver foil by neutrons from the pinch and subsequent registration of the radioactive decay intensity of silver isotopes. Initially, SAC was calibrated for distances, which are significantly less, than that used in our experiments. Therefore, it was cross-calibrated with the bubble-detectors.

A new independent technique, which was employed for neutron measurements in our experiments, was the neutron activation analysis. The method is based on activation of a sample by a neutron flux [4]. The sample is bombarded by neutrons, causing the elements to form radioactive isotopes. The radioactive emissions and radioactive decay paths for each element are known. Using special diagnostic equipment, it is possible to study spectra of the emissions from the activated sample and determine the concentrations of the radioactive isotope within it. Knowing the number of target and produced nuclei, one can calculate the amount of neutrons passing through the sample. In our experimental sessions, we used Canberra GC 1020 high-purity Germanium (HpGe) detector and Sodium-Iodide (NaI) gamma-spectrometer with special software for registration of gamma rays. Both detectors were calibrated by radioactive samples before every experimental campaign.

We can calculate the amount of activated particles as

$$N_p = N_t \cdot \sigma_{aver} \cdot \Phi \qquad (1)$$

where N_p is a number of the activated (produced) nuclei in the sample, N_t is number of the target particles in the detector, σ_{aver} is the average cross section of the reaction, Φ is the neutron flux through the sample.

Thereby, we can calculate neutron fluence as

$$\Phi = \frac{N_p}{N_t \cdot \sigma_{aver}} = \frac{Se^{\lambda t}A}{\varepsilon I N_A mk(1 - e^{-\lambda \Delta t})\sigma_{aver}} \quad (2)$$

where S is the area of the gamma rays peak, λ is the decay constant of radioactive material, t is the time between the beginning of activation and start of measurement, A is the atomic mass of the target isotope, ε is the detection efficiency of the gamma-spectrometer for a certain energy of gamma rays, I is the gamma ray emission probability, N_A is the Avogadro constant, m is the mass of the sample, k is the abundance of target isotope in the sample, Δt is the time of measurement. For neutron yield calculation, we should multiple the neutron fluence on the geometry factor $4\pi R^2$, where R is the distance from the pinch (neutron source) to the activation sample in radial direction. All data about nuclei and properties of nuclear reactions we can found in the IAEA databases [5][6]. The average cross-section of the reaction was calculated, using neutron spectra, reconstructed from nToF-detectors signals.

Selecting samples for activation, we were guided by the following considerations: availability, purity (lower impurity content), high concentrations of the desired isotope in the element, high cross-section of reaction (> 100 mbarn). An isotope that appears as a result of the activation should have a half-life in the interval from 2 minutes to 5 hours (otherwise gamma rays intensity will be very low for the detectors used).

Activation measurements were conducted with indium and copper samples. Determination of the total neutron yield was carried out with the help of indium samples. In 2014, the measurements were based on ^{115}In(n,γ)^{116m}In neutron radiative capture reaction, but this reaction is sensitive to scattered neutrons. Therefore, ^{115}In(n,n')^{115m}In nuclear excitation reaction with 2 MeV energy threshold was used in 2015. The number of high-energy neutrons (with energy > 12 MeV) was measured by ^{63}Cu(n,2n)^{62}Cu reaction.

3. Experimental results

Conducting neutron activation measurements during the experimental sessions in 2014 and 2015, we pursued the following goals: first, to master the activation technique and analysis routine; second, to obtain the data on the total neutron yield with the help of an independent diagnostic method; third, to determine the part of high-energy neutrons in the full spectrum. In subsequent experiments, it is planned to expand the application of the activation method for the study of neutron production from deuterium gas-puff z-pinch, for instance, to characterize the anisotropy of neutron radiation and to obtain quantitative data on the neutron yield in different ranges of energy spectrum.

For the total neutron yield measurement, indium samples were placed for activation outside the experimental chamber at the distance of 37 cm from the pinch in the radial direction. The neutron yield determined with the help of two different neutron activation reactions is shown in Figure 1. The data obtained by the SAC and the bubble detectors are shown for each shot for comparison. In principle, the neutron yield determined by both reactions correlates well with the data obtained by other diagnostics. However, the results of measurements carried out with the help of neutron radiative capture reaction ^{115}In(n,γ)^{116m}In is systematically higher, than the total neutron yield measured by the mean of the SAC and the bubble detectors. The most likely reason for this discrepancy is a high reaction cross section in the region of neutron energies below 2 MeV and, therefore, sensitivity of this reaction to the scattered neutrons. From our point of view, the measurements performed with the help of ^{115}In(n,n')^{115m}In nuclear excitation reaction with 2 MeV energy threshold provide more accurate data on the total neutron yield.

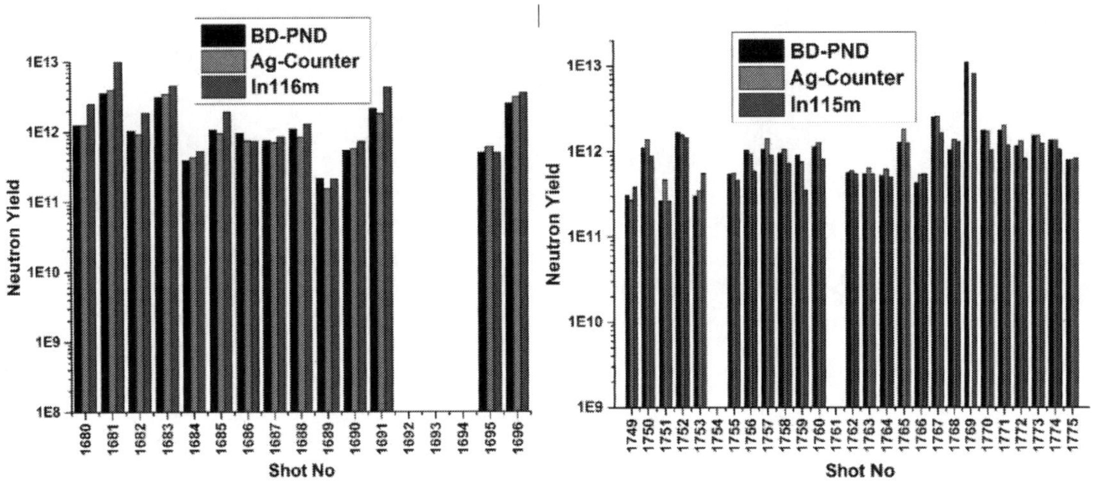

Figure 1. Neutron yields measured by activation of indium samples in comparison with the data obtained by means of the bubble detectors and silver activation counter

In the early experiments with triple shell deuterium gas-puff, the high energy tail of the neutron spectrum reached the energies up to 6 MeV. In the experiments with a single-shell deuterium gas-puff surrounded by an outer plasma shell, the neutron spectrum was significantly broader, and the energy of neutrons exceeded 12 MeV. In order to determine the amount of high-energy neutrons, the activation of the copper samples was employed. The reaction used in activation analysis has a high energy threshold (11.9 MeV), and that makes it possible to cut off the low-energy neutrons. The measured amount of high-energy neutrons was about 10^9 neutrons/shot in a typical shot with the total neutron yield above 10^{12} neutrons/shot. The fraction of the high-energy particles in the total neutron yield (≤0.5 %) is consistent with the neutron spectrum data obtained with the help of nToF diagnostics.

4. Conclusion

The neutron production from the single-shell deuterium gas puff with the outer plasma shell was studied in the experiments on the GIT-12 generator. The neutron activation method has been employed during two experimental campaigns. The neutron yields determined with the help of activation method and by other diagnostics are in good agreement. The measured amount of the high-energy neutrons is 10^9 neutron/shot. The part of high-energy neutrons in the neutron spectra determined by the activation method is consistent with the results obtained by means of the nToF diagnostics.

Acknowledgments

This research has been supported by the Grant Agency of the Czech Republic (Grant No. 16-07036S), the Czech Ministry of Education (Grant Nos. LD14089, LG13029 and LH13283), the Ministry of Education and Science of the Russian Federation, Project № 3.1170.2017, the IAEA (Grant No. RC17088), the CTU (Grant No. SGS 10/266/OHK3/3T/13).

References

[1] Ruytov D D, Derzon M S, Matzen M K 2000 *Reviews of Modern Physics* Vol. **72** No. 1
[2] Klir D *et al* 2015 *Plasma Phys. Control. Fusion* 0741-3335, **57**, 4
[3] Rezac K, Klir D, Kubes P and Kravarik J 2012 *Plasma Phys. Contr. Fusion* **54**, 105011
[4] Kuznetsov R A 1974 *Activation analysis* (Moscow: Atomizdat) p 343
[5] Experimental nuclear reaction data IAEA – Nuclear Data Section, Vienna International Centre,

available on-line: https://www-nds.iaea.org/exfor/exfor.htm
[6] Evaluated nuclear reaction libraries IAEA – Nuclear Data Section, Vienna International Centre, available on-line: https://www-nds.iaea.org/exfor/endf.htm

Hybrid MHD/PIC simulation of a deuterium gas puff z pinch

D L Shmelev[1], V I Oreshkin[2] and S A Chaikovsky[1]

[1] Institute of Electrophysics, RAS, 106 Amundsen St., Ekaterinburg, 620016, Russia
[2] Institute of High Current Electronics SB RAS, 2/3 Akademichesky Ave., Tomsk 634055, Russia

E-mail: shmelev@iep.uran.ru

Abstract. We present the hybrid MHD/PIC simulations of the time evolution of a deuterium gas puff z pinch. Recent experiments with 3-MA current pinches [6] made in a novel configuration have shown that the neutron yields can reach 3.6×10^{12}. It was shown that the observed neutron spectra could be explained by a suprathermal distribution of deuterons with a power law fall off in the ion energy distribution function at large energy. In order to perform the numerical simulation of gas puff z pinch a new hybrid model was developed. The described hybrid model treats the electrons as a massless fluid and ions as macroparticles. The macroparticle dynamic is calculated with the use of PIC method. Ion-ion Coulomb collision is considered with the use of MC method. In the model simulation, in the configuration close to described in [6], it was obtained the neutron yields up to 1.2×10^{12}. Most neutrons are not thermonuclear. This level of the neutron yield is reached only when a strongly nonuniform neck-like constriction of z-pinch plasma occurs. In this case, the obtained deuteron spectra (with energy up to several tens MeV) have suprathermal high energy tail. These simulations demonstrate the utility of the developed hybrid model for the z-pinch simulation.

1. Introduction

Z-pinches are known as efficient sources of x-ray radiation, fast electrons, ions and fast neutrons [1]. It is believed that as a neutron source the Z-pinch could be useful in radiation material science, radiobiology, nuclear medicine, recycling of nuclear waste and for a fusion-fission hybrid reactor. Significant efforts have been made to obtain neutron numbers sufficient for these applications.

Traditionally, the study of plasma liners is in line with the problems of inertial fusion [2-4]. Plasma liners along with the classic pinches and plasma foci can serve as sources of intense neutron radiation with a neutron yield in the range 10^9 - 4×10^{13} neutrons per pulse [5]. The highest neutron yield for pinches ($\sim 4\times10^{13}$) was received at the Z-machine (SNL, USA) in experiments on the compression of deuterium liners [5].

Recent experiments [6, 7] made in a novel configuration have shown that the neutron yields can reach 3.6×10^{12} at current about of 3 MA. Earlier, this neutron yield could be surpassed only on the 15-MA pinch. Energy neutron spectrum obtained in experiments [6, 7] has a high energy non-Maxwellian tail. Possible reasons of this high energy tail origin are discussed, for example, in [8]. It was shown that the observed neutron spectra could be explained by a suprathermal distribution of deuterons with a power law fall off in the ion energy distribution function at large energy. These high energy deuterons with energy up to 40 MeV (highest achievement for Z-pinches experiments) were detected in [6, 7]. The mechanism of generation of such deuterons is still not clear [8]. Computer simulation of

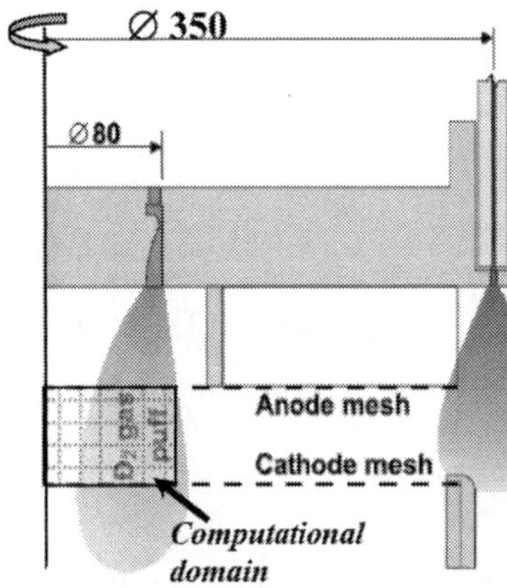

Figure 1. Sketch of experimental configuration [2, 3] and computational domain used in present work.

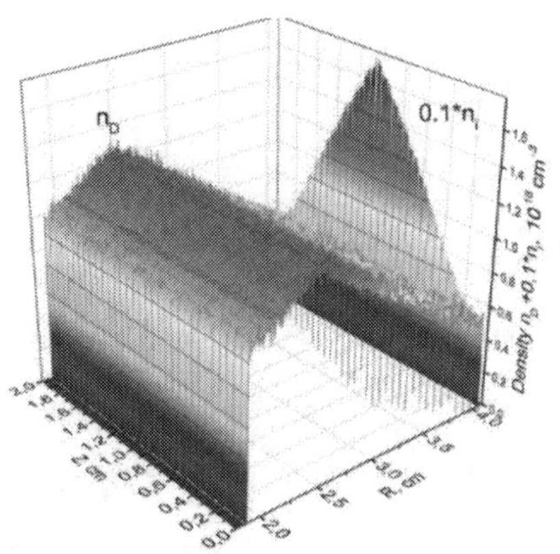

Figure 2. Initial density distribution.

Z-pinch implosion is a method to research the mechanism. MHD approach in this case is hardly applicable because the essential part of the deuterons has a suprathermal distribution. Various hybrid [9] and kinetic [10] models were developed to simulate different stages of Z-pinch dynamics. Here we describe a relatively new hybrid model, which has certain advantage in comparison with previously developed hybrid models, but is still simpler than the kinetic model [10].

2. Model description

A two-dimensional hybrid quasineutral model was developed to study the dynamic of the deuterium gas puff z pinch. According to the hybrid approach the electron subsystem is described as an inertialess fluid, but ions are described as macroparticles. The developed model is analogous to the model [11, 12], but the described model is modified to account for the deuterium ionization and neutron production due to DD fusion.

The governing equations in case of the 2D axial symmetrical geometry are the following:

$$\frac{d\vec{r}_i}{dt}=\vec{V}_i; \qquad m_i\frac{d\vec{V}_i}{dt}=e_iE+\frac{e_i}{c}\vec{V}_i\times\vec{B}+\frac{m_e n_e}{n_i}\frac{\vec{u}_e-\vec{u}_i}{\tau_{ei}}, \tag{1}$$

$$n_i=\frac{1}{H_c}\sum_{k=1}^{N_c}S(\Delta x_k); \quad \vec{u}_i=\frac{1}{H_c}\sum_{k=1}^{N_c}\vec{V}_k S(\Delta x_k); \quad T_i=\frac{1}{H_c}\sum_{k=1}^{N_c}\frac{m_i}{3}\left(\vec{V}_k-\vec{u}_i\right)^2 S(\Delta x_k) \tag{2}$$

$$n_e=\frac{1}{e}\sum_i e_i n_i; \qquad \vec{u}_e=\frac{1}{en_e}\left(\sum_i e_i n_i\vec{u}_i-\vec{J}\right), \tag{3}$$

$$\frac{\partial B_\theta}{\partial t}+\frac{\partial u_{ez}B_\theta}{\partial z}+\frac{\partial u_{er}B_\theta}{\partial r}=\frac{c^2}{4\pi}\left(\frac{\partial}{\partial r}\left(\frac{1}{r\sigma}\frac{\partial rB_\theta}{\partial r}\right)+\frac{\partial}{\partial z}\left(\frac{1}{\sigma}\frac{\partial B_\theta}{\partial z}\right)\right), \tag{4}$$

$$\vec{J}=\frac{c}{4\pi}\nabla\times\vec{B}, \qquad \vec{E}=-\frac{\vec{\nabla}P_e}{en_e}-\frac{1}{c}\vec{u}_e\times\vec{B}-\frac{m_e}{e}\sum_i\frac{\vec{u}_e-\vec{u}_i}{\tau_{ei}}, \tag{5}$$

$$\frac{3}{2}n_e\left(\frac{\partial T_e}{\partial t}+\vec{u}_e\nabla T_e\right)+P_e\vec{\nabla}\vec{u}_e+\vec{\nabla}\chi\vec{\nabla}T_e=\frac{\vec{J}^2}{\sigma}-\frac{3m_e}{\tau_{ei}}(T_e-T_i)-W_r \tag{6}$$

where V_i—velocity of ion (of type i) macroparticles, r_i—ion macroparticles radius vector, S—shape function of macroparticle (bilinear interpolation was used), H_c—form factor of cell, N_c—number of macroparticles in cell, n_i—ion density, n_e—electron density, u_i—ion drift velocity, u_e—the electron

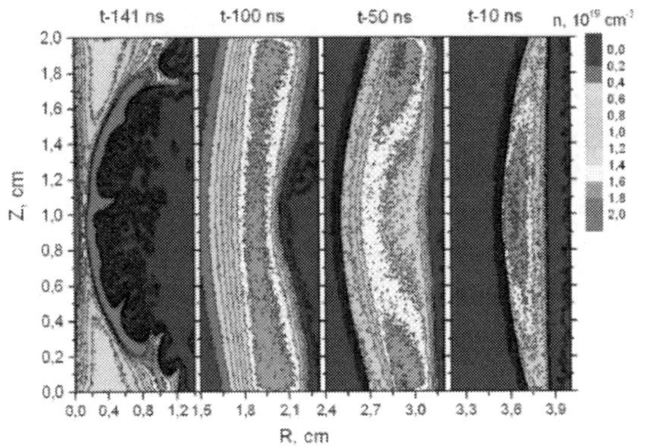

Figure 3. Time sequence of plasma density distributions.

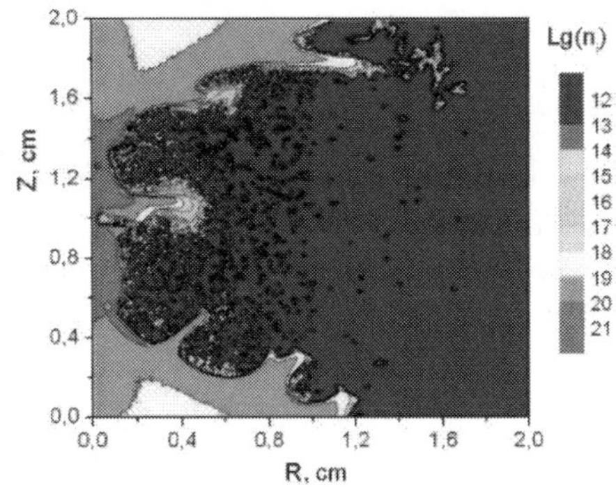

Figure 4. Plasma density distribution at t=144 ns, appearance of neck-like constriction.

drift velocity, *J*—total arc current density, τ_{ei}— electron-ion collision time, σ—plasma (Spitzer) conductivity, *B*—magnetic field (only θ component exists in a given geometry), *E*—electric field, P_e—electron pressure, χ—thermal conductivity, W_r—energy loss due to bremsstrahlung radiation. The change of the ionic chaotic energy as a result of collisions with electrons is taken into account using the method described in [13]. The macroparticles undergo ion-ion Coulomb collisions. These collisions have been considered with the help of DSMC-like method developed in [14]. Processes of deuterium ionisation and DD fusion have been considered with the help of direct simulation Monte Carlo (DSMC) method.

The geometric configuration of the computational domain has been chosen in order to match the geometry of the experiment [6, 7]. Figure 1 shows sketch of the experimental setup and the computational domain. It is seen that the computational domain contains only the inner gas shell. The outer plasma shell is far away from the domain. The model does not consider the motion of the plasma shell. Our calculation starts at the instant when the plasma shell reaches the inner gas shell. The shell parameters at this point are the additional parameters of the problem. We varied velocity and linear mass distribution of the outer plasma shell in order to get the neutron yield as close to the results [6, 7] as possible. Figure 2 shows an example of the initial density distribution. Deuterium density (n_D) in the inner shell is of about 10^{18} cm^{-3} (linear mass 120 μg/cm). Density of hydrogen plasma (n_i) in the outer shell is of about 10^{19} cm^{-3} (linear mass 5 μg/cm). Velocity of the incident plasma shell $6 \cdot 10^7$ cm/s.

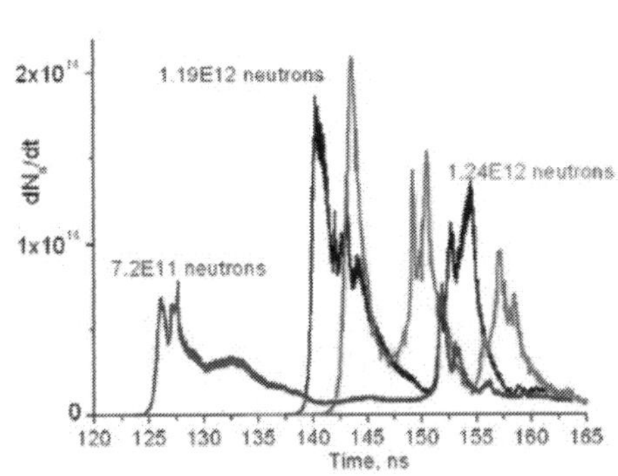

Figure 5. Time dependence of neutron production intensity for the different positions of inner gas shell.

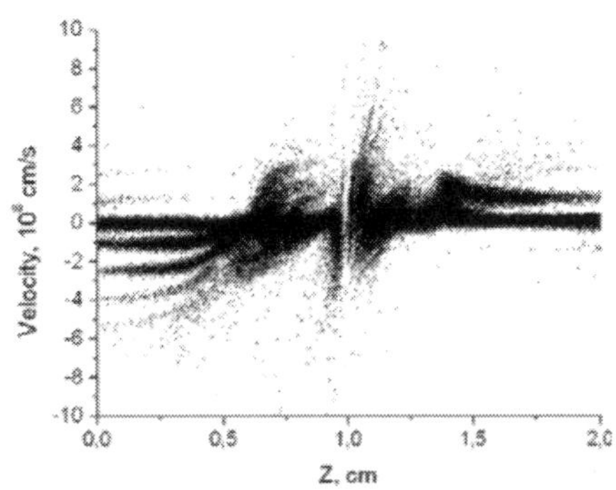

Figure 6. Ion phase portrait along the axis for t=144 ns.

IOP Conf. Series: Journal of Physics: Conf. Series **830** (2017) 012018 doi:10.1088/1742-6596/830/1/012018

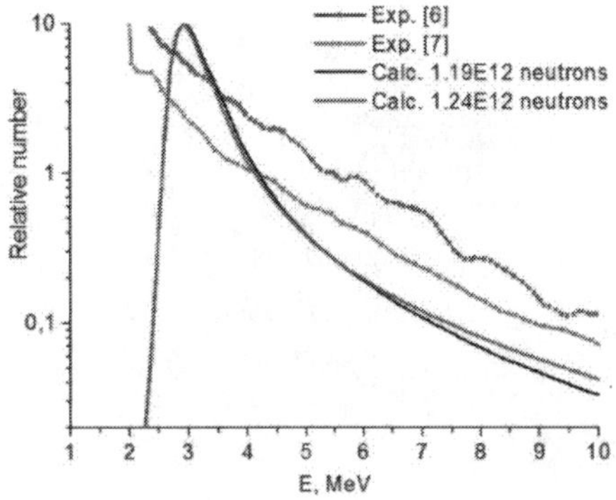

Figure 7. Neutron energy distributions obtained in the calculation in comparison with experimental ones.

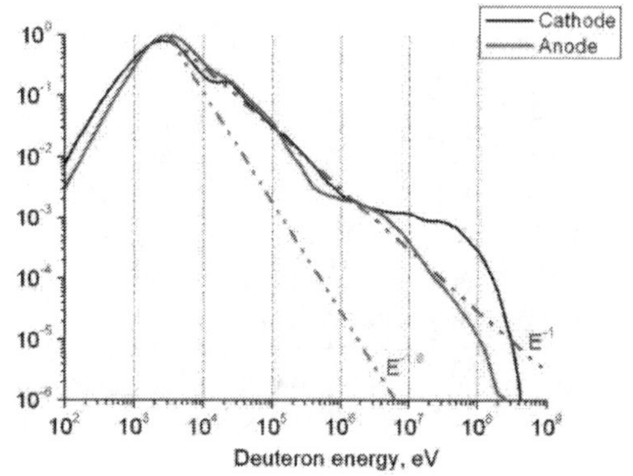

Figure 8. Deuterons energy distributions obtained in the calculations for the deuterons come to cathode and anode within the circle of radius 0.5 cm.

3. Results and discussion

Possible mechanisms of neutron generation in DD fusion are the following [8]: thermonuclear mechanism (i.e. neutron production due to high ion temperature), ion acceleration due to the axial plasma outflow from a constriction, generation of fast ion in the low dense plasma surrounding the pinch column, target mechanism [15], coulomb explosion [16]. The last two mechanisms are not considered in our model because of assumption of plasma quasineutrality. However, we believe that the contribution of these mechanisms in the neutron generation is insignificant because in the case of their implementation the neutron pulse should be very short (less than 1 ns), while in the experiments it can be tens of nanoseconds.

The highest neutron yield reached in our simulation at the mentioned liner masses of the shells and current 2.8 MA is $1.24 \cdot 10^{12}$ neutrons that is in two time less than the yield obtained in the experiment. This result was obtained only at a certain density distribution of the incident shell. The shell density distribution has a pronounced maximum at the center, as shown in Figure 2. If the density distribution of incident shell has this form then the further compression of the inner shell remains stable almost up to the axis, as it shown in Figure 3. Two plasma streams appeared in the shell in this case, one directed to cathode and other directed to anode. Evidently, these streams stabilise the compression. However, the inner shell loses up to 25 percent in mass due to these streams. If the outer shell has for example a flat density distribution then instabilities quickly develop and the compression effectiveness decreases. Neutron yield in this case is reduced by an order.

Due to the particular form of the shell (Figure 3) the compression at the axis starts approximately at the center (along Z) producing a neck-like constriction (Figure 4). Following the first constriction the next one or two constrictions appears during of about 20 ns (Figure 5). Due to this no uniform constriction a plasma outflow from the neck appears (Figure 6). As it seen from the figure, the ions escaping from the neck looks like beams propagating through relatively quiet plasma. The source of the most energetic deuterons in this case is the cave (hot spot) at Z of about 1 cm (compare Figures 4 and 6). This cave can be treated as the beginning of the neck disruption. Low density plasma in this cave can be accelerated up to several MeV with the help of Laurence force.

It is clear that this picture cannot be obtained with the help of MHD approach. Appearance of the high energy ion beams can in principle explain the suprathermal high energy tail of neutron energy distribution (Figure 7).

Energy distributions of deuterons come to the electrodes within the circle with radius 0.5 cm are shown in Figure 8. It is seen that deuterons can reach energy up to 100 MeV. Note that deuterons with energy up 50 MeV were detected in experiments [6, 7]. Authors of paper [6] have shown that the measured suprathermal neutron energy distribution can be explained if one assume that deuterons have an isotropic distribution function with power law fall $E^{-1.8}$. As it seen in Figure 8 the power low of high energy deuterons obtained in our calculation is rather E^{-1}. However, this contradiction with [6] can be easily explained by strongly anisotropic energy distributions of deuterons in plasma in our case (Figure 6).

Let us discuss briefly a possible mechanism of the production of suprathermal ions and neutrons. Average ion temperature in our simulation is of about 2 keV. Thus, the thermonuclear mechanism makes a certain contribution in total neutron yield but it is not a source of suprathermal neutrons. Plasma acceleration due to the axial plasma outflow from the constriction can accelerate ion up to 100 keV. As it shown in [8] the ion distribution function in this case should be proportional to E^{-2}. The main input in the production of suprathermal ion in our simulation is contributed by the ions accelerated in low dense plasma surrounding the pinch column. In our case the low dense plasma appears because the ion Larmor radius becomes more than the size of hot spot, and ions due to the Larmor rotation can escape from the hot spot area. In contrast to MHD models ([8] in example) where the described mechanism is hardly possible, our hybrid model is able to consider this mechanism.

Acknowledgments
The work was supported by Russian Science Foundation; Project No. 16-19-10142.

References

[1] Giuliani J L, Commisso R J 2015 *IEEE Trans. Plasma Sci.* **43** 2385-2453
[2] Smirnov V P 1991 *Plasma Phys. Controlled Fusion.* **33** 1697-1714.
[3] Ryutov D D, Derzon M S, Matzen M K 2000 *Rev. Modern Phys.* **72** 167-223.
[4] Haines M G. 2011 *Plasma Phys. Control. Fus.* **53** 093001.
[5] Coverdale C A, Deeney C, Velikovich A L *et al* 2007 *Phys. Plasmas* **14** 056309
[6] Klir D, Shishlov A V, Kokshenev V A *et al* 2015 *Plasma Phys. Control. Fusion* **57** 044005
[7] Klir D, Shishlov A V, Kokshenev V A *et al* 2016 *Phys. Plasmas* **23**, 032702
[8] Vikhrev V V, Korolev V D 2007 *Plasma Phys. Rep.* **33** 356-380
[9] Sotnikov V I, Leboeuf J N, Deeney C *et al* 2005 *Phys. Plasmas* **12** 092701
[10] Welch D R, Rose D V, Thoma C *et al* 2010 *Phys. Plasmas* **17**, 072702
[11] Shmelev D L, Uimanov I V 2015 *IEEE Trans. Plasma Sci.* **43** 2261-2266
[12] Shmelev D L, Barengolts S A *et al* 2015 *J. Phys.: Conference Series* **652** 012041
[13] Rambo P W, Procassini R J 1995 *Phys. Plasmas* **2** 3130-3145
[14] Bobylev A V, Potapenko I F 2013 *J. Comp. Phys.* **246** 123-144
[15] Trubnikov B, Plasma Physics and the Problem of Controlled Thermonuclear Reactions [in Russian], Academy of Sciences of the USSR Press, Moscow **4**, 87-97 (1958).
[16] Oreshkin V I, Oreshkin E V, Chaikovsky S A *et al* 2016 *Phys. Plasmas* **23** 092701

IOP Conf. Series: Journal of Physics: Conf. Series **830** (2017) 012019 doi:10.1088/1742-6596/830/1/012019

Rail-type gas switch with preionization by an additional corona discharge

O S Belozerov and E G Krastelev

Joint Institute for High Temperatures, Russian Academy of Sciences, st. Izhorskaya 13/19, Moscow, 125412 Russia

E-mail: ekrastelev@yandex.ru

Abstract. Results of an experimental research of a rail-type gas switch with preionization by an additional negative corona discharge are presented. The most of measurements were performed for an air insulated two-electrode switch assembled of cylindrical electrodes of 22 mm diameter and 100 mm length, arranged parallel to each other, with a spark gap between them varying from 6 to 15 mm. A set of 1 to 5 needles connected to a negative cylindrical electrode and located aside of them were used for corona discharges. The needle positions, allowing an effecient stabilization of the pulsed breakdown voltage and preventing the a transition of the corona discharge in a spark form, were found. It was shown that the gas preionization by the UV-radiation of the parallel corona discharge provides a stable operation of the switch with low variations of the pulsed breakdown voltage, not exceeding 1% for a given voltage rise-time tested within the range from 40 ns to 5 μs.

1. Introduction

A long-term stable operation of high-voltage pulsed power generators is mainly determined by a lifetime and stability of the parameters of switching devices involved in their operation. This primarily concerns the spark gas switches, which are used most often in designing the high-voltage generators, in particular, in almost all types of Marx generators and others [1].

Traditionally, an increase of the lifetime of switches is achieved by increasing a working surface of the electrodes and by excluding the sharp edge electrodes with high electric field gradients and a short time delay of discharge, which quickly fail due to erosion. The deterioration of the dynamic characteristics of gas switches with thick and smooth electrodes may be compensated by pre-ionization of the insulating gas. The method, which is used most frequently in practice, is based on the photoionization by the ultraviolet (UV) radiation of spark sources [1,2].

Below, there are the results of a study of characteristics of an experimental rail-type switch with long linear electrodes, in which an additional corona discharge is used as a source of the UV-radiation for pre-ionization. The corona discharge is ignited by a set of needle electrodes, electrically connected to the negative main electrode. Together with the second positive rail electrode, the needles form a corona discharge gap, which works in parallel to the main spark gap of the switch. Two separate discharge gaps - the main spark gap between the rail electrodes and the corona discharge gap - provide an opportunity to optimize the electrodes for a long-term high-current operation with a stable dynamic performance. This method was successfully tested in a design of coaxial switches [3].

2. The design of the experimental switch

The main spark gap is formed by two parallel cylindrical electrodes of 22 mm diameter and 100 mm length of the working cylindrical part ended by semi spherical edges. The electrodes are bolted to the insulating plates made of Plexiglas, which are adjusted parallel to each other with the help of 4 long bolts at the corners of the plates. During the experiments, the spark gap between the electrodes was changed within the range from 6 mm to 15 mm, using the sets of spacers of an appropriate length installed at corner bolts.

To produce a highly divergent field, the corona discharge needles are located aside from the negative cylindrical electrode. They are fixed to a movable metal plate, which is connected directly or through a current viewing resistor to the negative cylindrical electrode. Steel sewing needles and segments of wires of 0.5 mm diameter, made of W, Cu and Fe, were used as the corona needles.

The photo of the assembled switch and a sketch of the cross section of its electrode system are shown in Figure 1. The designation of dimensions used in the next sections of this paper and the calculated plot of the equipotential lines characterizing the electric field distribution in the main and the corona gaps are shown on the sketch in Figure 1(b).

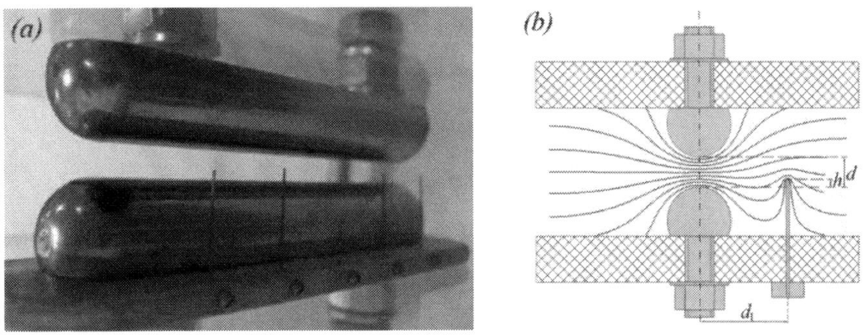

Figure 1. Photo (a) and calculated pattern of equipotential lines in a cross section (b) switch. Step equipotential lines is 0.1 U, where U voltage between the electrodes.

The equipotential lines in the gap between the main electrodes are distributed nearly uniformly. They are concentrated near the top of the corona needle indicating a highly divergent electric field in this region. According to tree-dimensional calculations, the electric field at the top of the needle is approximately an order of magnitude higher than the field in the gap.

Three-dimensional calculations of the electric field distribution were carried out by a finite element method. The results were used for the calculations of a static breakdown voltage U_{br} and for the estimations of the field at corona needles, depending on their position – the distance from the plane of symmetry of the switch d_1 and the elevation above the top of located near cylindrical electrode h.

3. The switch characteristics in the static mode

3.1. Breakdown of the main spark gap.

Being measured without corona needles, the static breakdown voltage U_{br} increases almost linearly when the gap between the electrodes increases from ≈19 kV at d=6 mm to ≈44 kV at d=15 mm. A difference between the measured values U_{br} and those obtained by calculation does not exceed 5% for the entire range of the spark gap variations.

3.2. Characteristics of corona discharge

A high electric field at the corona needles provides an ignition of the corona discharge at the voltage levels, which are lower than those of the spark gap breakdown. Throughout the range of changes, the needle height h was from -3 mm to +3 mm (relative the top of a cylindrical electrode, see Fig.1) and the distance d_1 was up to 35 mm. Measured in this ranges values of the threshold voltage U_{cr}, which is necessary for an ignition of the corona discharge for different needles, varied from 6-8 kV to 12-13 kV. A comparison of the experimental and the calculated data shows that the variation of the threshold voltage correlates with the corresponding variations of the calculated electric field strength near the tip of needles.

Figure 2 shows the volt-ampere characteristics of the corona discharge for two positions of a single corona needle: d_1=26 mm and d_1=21 mm for the fixed gap d=10 mm and the needle elevation h=+1 mm.

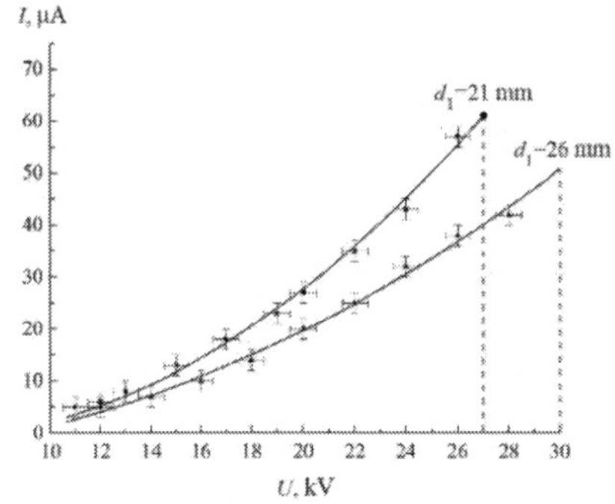

Figure 2. The volt-ampere characteristics of corona discharge for two positions of the corona needle d_1 for the main gap d=10 mm and needle height h=+1 mm.

Measured characteristics are in a good agreement with the those shown in the same figure approximations by a quadratic function: $I \sim kU(U - U_{cr})$, where U_{cr} – a critical voltage.

The principal difference between the two shown volt-ampere characteristics is as follows. The first of them, corresponding to the distance d_1=26 mm, is completed by the spark breakdown of the main gap at U=30 kV. In the second case, (d_1=21 mm) at U≈27 kV voltage, the corona transfers to a spark form prior to the breakdown of the main gap.

Figure 3 shows the dependences of the main gap breakdown voltage and the voltage for a transition of the corona discharge to the spark form for the distance d_1 of the needle location. Experimental data are shown for two main gaps: d=10 mm and d=13 mm, and for a fixed elevation of the corona needle h=1 mm. In this figure, light symbols correspond to the main gap breakdown, and black symbols – to the transition of the corona discharge to the spark form.

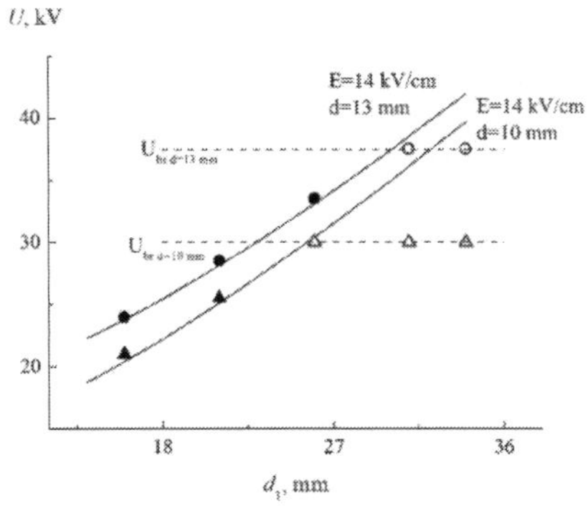

Figure 3. Dependence of the sparking voltage on the corona needle location d_l for two main gaps: d=10 mm and d=13 mm. Light symbols – main gap sparking, blacked symbols – corona gap sparking

It is seen that for a far-located needle, the gap between the main cylindrical electrodes is sparking first and the breakdown voltage does not depend on the distance of needle location d_l remaining the same as the static breakdown voltage U_{br}, shown in the picture by the dotted lines. For a closer located needle, starting from some critical position, the spark breakdown takes place in the corona gap. This breakdown voltage depends on the position of the corona needle and it is limited by a critical value of an average field in the corona gap, which determines the transition of the corona discharge to the spark [1, 4]. An analysis of the data obtained for the conditions of the experiments show that for an air of an atmospheric pressure, it corresponds to the average field $E \approx 14$ kV/cm in the corona gap. The calculated voltages for E=14 kV/cm, for two given spark gaps are shown in Figure 3 by solid lines.

Similar dependencies obtained for various values of the spark gap, needles number, and their positions showed that the average field in the gap between the tip of the corona needles and the positive cylindrical electrode is a crucial condition for the transition of the negative corona discharge to the spark.

Taking E=14 kV/cm as the average field limit for an air insulated gap, the positions of the needles, excluding a possible transition of the corona discharge to the spark form, may be calculated. For example, for the main gap d=10 mm, the condition E <14 kV/cm is satisfied by moving the corona needles to the distance $d_l \geq 26$ mm. These requirements are important for a normal switch operation in a static mode, or for the DC pre-stressed, or the slowly-varied voltage switches. For the pulsed mode of the switch operation, significantly higher fields are allowed due to a qualitatively different dynamics of the corona discharge.

4. The switch characteristics in the pulsed mode

The dynamic characteristics of the switch were studied by applying single pulses with a voltage rise time from 40 ns to 5 μs and a voltage amplitude up to 60 kV.

Without corona needles, the variations of the pulsed breakdown voltage exceeds 20%. An installation of one or several corona needles qualitatively changes the picture by a strong stabilization of the breakdown voltage. Simultaneous measurements of the applied to the switch pulsed voltage and the corona discharge current show that a short pulse of the corona current precedes to the breakdown of the main gap. This corona current pulse, so called Trichel pulse, appears when the rising voltage exceeds the critical voltage U_{cr} defined as a threshold voltage for a corona ignition under the static mode.

Figure 4 shows 10 superimposed traces of the switch voltage and a signal from a current viewing resistor of 75 Ohm in series with the corona needle. Traces are recorded using the probe Tektronix 6015 (bandwidth 60 MHz) for the switch with one needle installed at d_l=23 mm and h=+1 mm. The dotted line shows the static breakdown voltage U_{br} for the given gap of d=10 mm.

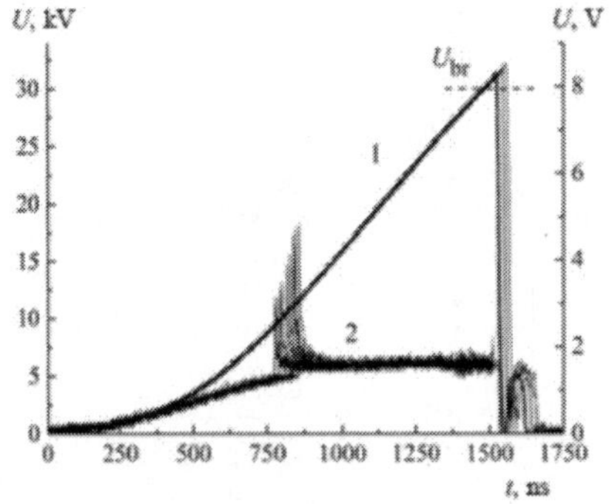

Figure 4. Traces of switch voltage (1) and the signal from current viewing resistor of 75 Ohm in series with the corona needle (2).

As we can see, the switch breakdown occurs in a short time after reaching the level of the static breakdown voltage U_{br} and is preceded by a short pulse of the corona discharge at the voltages of 10-11 kV (U_{cr}, for the given switch parameters, is about 8.5 kV).

The the breakdown is reproduced with very high stability, both of a time and an amplitude of the pulsed breakdown voltage. The scatter of the breakdown voltage is about 1%.

The traces of the corona discharge shown in figure 4 don't reflect a real shape and a real amplitude of the corona current since a long oscilloscope sweeping is needed for an observation of the entire voltage pulse. In this case, they serve only as a time indicator of a corona discharge. A trace podium is due to the presence of a capacitive coupling between the corona needle and the high-voltage positive electrode. Together with the current viewing resistor, the stray capacitance forms a differentiating circuit. As a result, the recorded voltage drop across the resistor is proportional to the derivative of the applied pulsed voltage. Using the recorded traces, the stray capacitance for a single needle and a given geometry of the switch was defined as ≈1.5 pF.

Real traces of the pulsed corona current recorded at a faster oscilloscope sweeping show that the duration of the corona current pulse at the half maximum is about 5 ns. The current amplitude reaches 0.35 A, and it is two orders of a magnitude above the measured for Trichel pulses in a static mode.

The corona current traces recorded for two needles connected to the common current viewing resistor looks like one half of an amplitude of one needle current pulse or a pair of separated in time pulses. A similar picture was observed for three or more corona needles. The synchronous flash of the corona discharge of a few needles, which are different in their position and sizes, may be explained in terms of a mutual UV-radiation resulting in the generation of free electrons initiating the discharge.

A consequence of several Trichel pulses was observed as a longer voltage pulses applied to switch and longer gaps between the main electrodes before the breakdown. Note, that the amplitude of the first pulse is greater than that of the subsequent pulses. It is consistent with the results of model calculations, which assume that the first current pulse in the corona is generated in a free space charge gap, and its appearance after the first flash weakens the electric field in the ionization region and reduces the current of the subsequent pulses.

4.1. The stability of pulsed breakdown voltage

The measurements of the breakdown voltage, as the functions of the needles parameters, show that the needle elevation h has the main influence on a stability of the breakdown voltage and its deviation. It was found, that the breakdown voltage variation was minimal for the corona needle, slightly elevated over the top of the main negative electrode. In particular, for the fixed parameters of measurements, the variation of the corona needle elevation from h=-3 mm to h=+1 mm resulted in a decrease of the breakdown voltage deviation from ≈1 % to ≈0.5%. Further elevation of the corona needle had no effect. This dependence was observed for other experimental conditions.

This dependence may be explained by the UV-radiation of the working surface of the cylindrical electrode resulting in an appearance of additional photoelectrons emitted from the electrode. Since the work function of the metal is lower than the ionization potential of the gas, the photoelectric effect at the surface can have a greater value for the appearance of, the so-called, "effective" electrons, responsible for the beginning of the switch closing. Moreover, the appearance of free electrons at the surface of the negative electrode is most favorable for the development of anode directed avalanches.

4.2. Distribution of discharges along the length of the electrodes

Visual measurements of the spark discharge distribution along the cylindrical electrodes show that for single corona needle, more than 50% of the total number of discharges were localized within ±1 cm from the axial position of the needle. Increasing the number of corona needles extended the length of the discharge area. The discharge distribution, which was close to be uniform throughout the total electrode length of 100 mm, was achieved for five corona needles set with axial step of 2 cm. Note that the delay time and the breakdown voltage does not depend on the number of needles and a position of the discharge zone.

5. Conclusion

The results of the research presented in this paper show that the gas pre-ionization of parallels corona discharge by the UV-radiation provides the stable operation of the rail-type switch with low variations of the pulsed breakdown voltage. The measured variations did not exceed 1% for the given voltage rise-time within the range from 40 ns to 5 μs. The requirements for the corona needles geometry were found, for preventing the transition of the corona discharge to the spark form and for an effective stabilization of the pulsed breakdown voltage.

References

[1] Mesyats G A 2004 *Pulse Energy and Electronics* (Moscow: Nauka) p 704.
[2] Schaefer G, Kristiansen M and Guenther A 1990 *Advances in Pulsed Power Technology* vol **2** (New York: Plenum).
[3] Krastelev E G, Sedin A A and Tugushev V I 2015 *Physics of Atomic Nuclei* V **78** № 14 p 1635
[4] Kozlov B A and Solovygo V I 2006 *Technical Physics* V **51** № 7 p 821.

High power radiators of ultra-short electromagnetic quasi-unipolar pulses

V M Fedorov[1], V E Ostashev[2], V P Tarakanov[3] and A V Ul'yanov

Joint Institute for High Temperatures of Russian Academy of Sciences (JIHT of RAS) Izhorskaya 13/2, Moscow, 125412, Russia

E-mail: [1] vmfedorov@ihed.ras.ru, [2] ostashev@ihed.ras.ru, [3] karat8tarak@gmail.com

Abstract. Results of creation, operation, and diagnostics of the high power radiators for ultra-short length electromagnetic pulses (USEMPs) with a quasi-unipolar profile, which have been developed in our laboratory, are presented. The radiating module contains: the ultra-wideband (UWB) antenna array, the exciting high voltage pulse semiconductor generator (a pulser), the power source and the control unit. The principles of antenna array with a high efficiency aperture about 0.9 were developed using joint four TEM-horns with shielding electrodes in every TEM-horn. Sizes of the antenna apertures were (16-60) cm. The pulsers produced by "FID Technology" company had the following parameters: 50 Ohm connector impedance, unipolar pulses voltages (10-100) kV, the rise-time (0.04-0.15) ns, and the width (0.2-1) ns. The modules radiate the USEMPs of (0.1-10) GHz spectrum, their repetition rate is (1-100) kHz, and the effective potential is $E*R = (20\text{-}400)$ kV, producing the peak E-field into the far-zone of R -distance. Parameters of the USEMP waves were measured by a calibrated sensor with the following characteristics: the sensitivity 0.32V/(kV/m), the rise-time 0.03 ns, the duration up to 7 ns. The measurements were in agreement with the simulation results, which were obtained using the 3-D code "KARAT". The USEMP waves with amplitudes (1-10) kV/m and the pulse repetition rate (0.5-100) kHz were successfully used to examine various electronic devices for an electromagnetic immunity.

1. Introduction

The high power impulse radiators of the ultra-short duration electromagnetic pulses (USEMPs) are successfully used in a number of applications such as checking of the electromagnetic immunity of electronic systems and in the experiments devoted to stop the transport facilities [1, 2]. Some advantages of the USEMP waves in comparison with the pulsed radiation of sinusoidal filling (the type of radar pulses) – they are lower. The USEMP waves generate the electromagnetic (EM) fields with the ultra-wideband (UWB) frequency spectrum of $f_h/f_l > 10$ –(ratio of the upper and the lower frequencies) that allows reduce of the number of tests. These tests often require both high amplitudes of E(t,R) –the strength of electric field (typical levels are (1-20) kV/m – peak values at the distance R placing the objects), and relatively high repetition rate pulses. The USEMP radiation allows also fabrication of the radiators with non-high average power, because it employs a high duty cycle as a division of the long period of pulse repetition $T_{p,p}$ at a pulse width of radiation impulses $T_{w,p}$, typically $T_{p,p}/T_{w,p} > 10^4$. This gives also lower dangers as an ether soiling or physical damages in the examined electronic systems.

IOP Conf. Series: Journal of Physics: Conf. Series **830** (2017) 012020 doi:10.1088/1742-6596/830/1/012020

1.1. Radiators. Construction of the pulsed high power radiators met the complex problems like: the EMC of systems with high-voltage pulsed sources, the triggering and the control devices with low-voltage pulses; and the realization of coherent radiation from few radiators at various conditions; providing both the reliable wideband frequencies at radiation and the electrical insulation inside the antenna units with high-voltage pulses. We had built a series of the autonomous module radiators with a highly efficient antenna array using the UWB TEM-horns [3-6]. The modules can radiate the USEMP waves with (1-100) kHz repetition rate, UWB frequency spectrum (0.1-10) GHz, and with the effective potential of $U_{rad} = E_p(R)*R = (20\text{-}400)$ kV – (producing of the peak E-field into a far-zone of the distance R).

In our works, we solved a problem of increasing the radiation aperture efficiency $\eta_A = A_{eff}/A_a$ for antennas using the TEM-horns. This η_A is a ratio of the antenna effective aperture A_{eff} to the physical aperture $A_a = (h{\cdot}b)$, where h is the gap and b –the width of antenna aperture. Note, that the ratio A_{eff}/A_a is widely used to analyze the microwave radars, and it is useful also for the impulse radiators if the analysis will use the reciprocity theorem [7]. It predicts that the effective apertures A_{eff} will be equal in the receiver and the transmitter modes (see details in section 2). Basic TEM-horns with square cross-sections and long electrodes (triangular plates with a separation angle $< 30^o$ and a length $L_H > c{\cdot}T_{w,p}$) are known as good UWB antennas but with an average efficiency of $\eta_{A,H} = A_{eff,H}/A_{a,H} \approx 0.6$ [8]. Using both the antenna array with joining four TEM-horns and with shielding electrodes in the every TEM-horn, we obtained the increasing efficiency up to $\eta_A = 0.9$ [3-6, 9] for antenna apertures 16...60 cm.

The autonomous module assembled: the UWB antenna array with four TEM-horns, the exciting high voltage pulse semiconductor generator (G_pV), the power source and the control unit. The G_pV generators were manufactured by "FID Technology" company [10] with the following parameters: 50 Ohm impedance of the output coaxial feeder, mono-polar pulses (10-100) kV, the rise-time (0.04-0.15) ns, the FWHM of (0.2-1) ns, and (1-100) kHz repetition rate, fed by DC power ~0.3 kW. The time deviation of G_pV pulse sync was ~ 30 ps, which allows the use of multi-module radiators with an efficient potential of a multi-megavolt level.

1.2. Measurements and analysis. A creation of the effective pulsed high power radiators requires both the reliable measurements of radiator parameters, and the computer electromagnetic modeling. We used the numerical full electromagnetic 3-D code "KARAT" [11] for simulation of the non-stationary processes occurring in real electro-dynamic principal sites. The pulse electric fields **E**(t) in the transverse electromagnetic (TEM) traveling waves were measured using various antenna sensors like those described in [3-6, 9, 12]. These antennas had the receiving tracts with an ultra-wideband range with the upper boundary of $F_h > 10$ GHz and a rise time $T_{0.1\text{-}0.9} \leq 0.03$ ns for pulsed EM waves.

2. UWB strip-line sensor with small cross-section. Principle of Reciprocity for Pulsed Antennas

The design of the last version of strip-line sensor having the long dielectric stalk with high permittivity $\varepsilon = 16$ and small sizes of cross-section (1.5x1.5 mm^2) [6, 12] allowed us to solve two main problems of the pulsed UWB electromagnetic probes: 1) to realize the direct calibrations of sensors; 2) to enlarge the permissible time interval up to 7ns in measurements of the electric fields E(t) in the traveling waves.

1.3 Results of simulation by code "KARAT". The sensor model in the simulations had the length of 8 cm (5 times lower than the real sensor of StF4) and a mall cross-section about $h_{CS}{\times}b_{CS} = 1.5{\times}1.5$ mm^2 (like a real sample). The dielectric stalk (the dark tape in figure 1) had a high permittivity $\varepsilon = 16$. At its face edge (z = 1 cm), a thin low conductive region was placed as a load Z_{CS} = 52 Ohm. This resister is connected with a signal foil tape placed over a dielectric stalk. Note, in the real sensor, a connecting coaxial signal cable with the impedance 50 Ohm played the role of this input load. At the end of strip-line (the point z = 9 cm in figure 1), a short circuit for the signal foil was done. The sensor was placed into the TEM-line-cell with 1cm gap (figures 1, 2) between two plane electrodes.

The plane TEM-wave with field components $E_y(t)$, $H_x(t)$ (step-like pulses with a rise-time $T_{0.1\text{-}0.9} \approx 0.02$ ns) is transmitting through an entry window (z = 0 cm in figure 1). This wave was coming into the gap along the axis z between two plane electrodes ($y_G = 0$; $y_V = 1$ cm), and it went away through an end window (z = 10 cm) without visible reflections. These results show that the strip-line sensor with a small aperture $A_{a,CS} = (1.5\text{x}1.5\ \text{mm}^2)$ will produce low disturbances in the TEM-wave in the gauge line-cell with a gap of more than 1 cm. The calculated by KARAT electric field $E_{CS}(t)$ at a load Z_{CS} equals $E_0(t)/5$, where $E_0(t)$ –the electric field in the line-cell wave. Using this ratio and the h_{CS} = 0.15 cm, we calculate $V_{CS}(t)$ –a "measured voltage" at the load Z_{CS} = 52 Ohm and the sensor sensitivity K_{CS} in (1):

$$V_{CS}(t) = E_S(t)\cdot h_{CS} = E_0(t)\cdot K_{CS}\ ;\quad E_{CS}(t) = E_0(t)/5\ ,\quad K_{CS} = V_{CS}(t)/E_0(t) \approx 0.3\ \text{V/(kV/m)} = 0.3\ \text{mm};\quad (1)$$

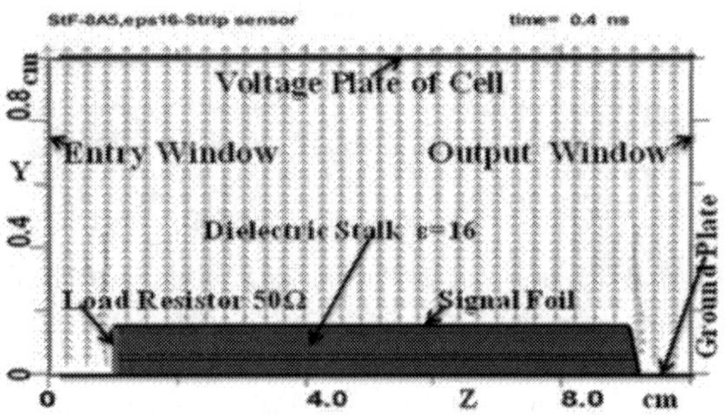

Figure 1. Sensor model (YZ-section, x=0.8 cm) in the gauge cell with 1cm gap. Vertical arrows are $E_0(t)$- electric field vectors at the moment t=0.4 ns.

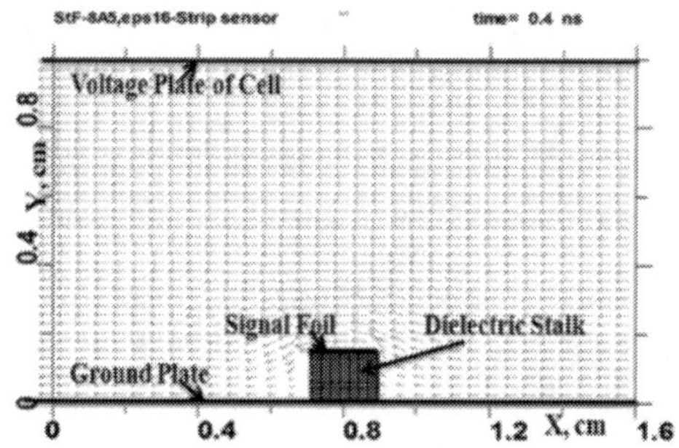

Figure 2. XY-cross-section (z=5.1 cm) inside the gauge cell. Horizontal arrows are magnetic field vectors at the moment t=0.4 ns.

We estimate the effective aperture $A_{eff,CS}$ in the sensor by the equality of P_{AL} –the electrical power in a load Z_{CS} and $P_{E,CS} = A_{eff,CS}\cdot(E_0(t))^2/Z_0$ -the coming wave power in this effective aperture, Z_0 =377 Ohm, as:

$$A_{eff,CS}\cdot(E_0(t))^2/Z_0 = (V_{CS}(t))^2/Z_{CS} = P_{AL},\quad A_{eff,CS} = (K_{CS})^2\cdot Z_0/Z_{CS} \approx 0.3\cdot(h_{CS})^2,\quad \eta_{A,CS} = A_{eff,CS}/A_{a,CS} \approx 0.3 \quad (2)$$

Note, the symbols η_A or A_{eff} in (2) and others we introduced in the section 1.1 as the characteristics of TEM-horns, but they will be used further for any antennas. The specifics of antenna types like those presented in figure 1 is a matched load in a face aperture (in a real sensor, instead, it is a connected coaxial signal cable, a so-called antenna feeder). In this case, the full absorbed power from the flow of wave power in the face aperture will equal to the double power in the antenna load $2\cdot P_{AL}$. May be, it will be correct to use $A_{eff} = 2\cdot A_{eff,CS}$. But this is a wrong idea, because the only value of the sensor effective aperture $A_{eff,CS}$ from (2) is an identical value of radiating aperture for the sensor used in the transmitting antenna mode. That is a demand in agreement with the reciprocity theorem (see section 2.3).

2.2 Experiments. The E-field sensor antenna of StF4 is a non-symmetrical two-wire line with a high dielectric permittivity ε = 16, 40 cm length, and the cross-section $A_{a,StF4} \approx 1.5\times1.5\ \text{mm}^2$ (a face real aperture). This line was placed on a brass plate of 0.5 mm thickness, 8 cm width, 50 cm length. The coaxial cable with the impedance Z_S = 50 Ohm and the length 1.5 m is the input load. The sensor was contained into a foam plastic box of 1.9 cm thickness, 8 cm width, and 70 cm length. The StF4 data were directly measured by means of placing its inside compact the TEM-line-cell. The line-cell data were: 2 cm gap, 8.3 cm width of signal foil, 70 cm work length, and wave impedance of 58 Ohm. Each end of the cell was connected with the TEM-half-horn of 17 cm length, one half-horn was loaded by 58 Ohm. Another one was connected to 50 Ohm coaxial feeder for test generator voltage $V_G(t)$. An absence of the reflections in the line-cell was tested by the Tektronix DSA8200 with 80E08 TDR. A rise-time in the line-cell input was ≈ 0.03 ns. The electric field $E_{gl\text{-}c}(t)$ in the gauge line-cell was calculated as:

$$E_{gl\text{-}c}(t) = V_G(t)\cdot(116/108)/2 = 0.54\cdot V_G(t)\ (V/cm), \quad \text{or} \quad E_{gl\text{-}c}(t) = 54\cdot V_G(t)\ (V/m) \qquad (3)$$

Experimental data for the signals of the StF4 strip-line sensor in the TEM wave traveling along the gauge TEM line-cell with 2 cm gap and the $E_C(t)$ electric field in the cell are shown in figures 3 and 4.

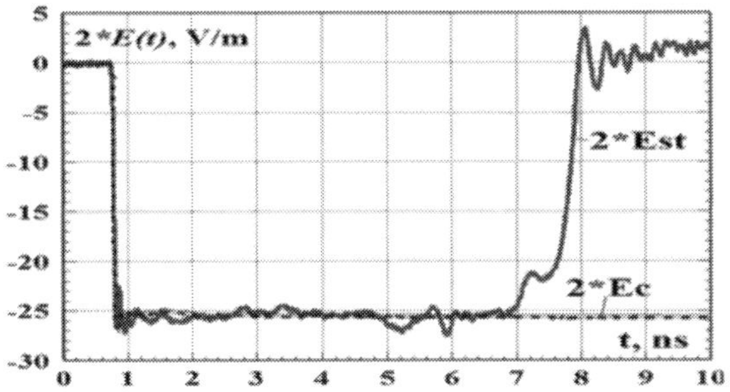

Figure 3. $E_{st}(t)=V_{StF4}(t)/K_{StF4}$- signal from StF4-sensor. $E_C(t)=V_{gl\text{-}c}(t)/0.02m$- in the gauge line-cell which was excited by step-like pulsed voltage

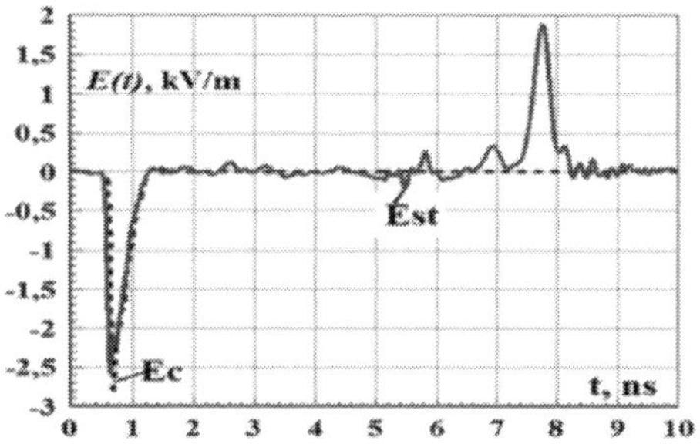

Figure 4. $E_{st}(t)=V_{StF4}(t)/K_{StF4}$- signal from StF4-sensor. $E_C(t)=V_{gl\text{-}c}(t)/0.02m$- in the line-cell it was excited by bell-type voltage

In figure 3, the gauge line-cell was excited by 80E08 module (E_C –a dotted line) and in figure 4, it was excited by TMG60 test generator [13] (E_C –a dotted line). We used the test results to calibrate the StF4 sensor parameters for any measurements realized in the TEM traveling-waves. The parameters are the following: the sensitivity $K_{StF4} = 0.32$ V/(kV/m), $T_{0.1\text{-}0.9} < 0.03$ ns- the rise-time, and the work time gap of $T_{max} = 7$ ns. We calculate the effective aperture $A_{eff,StF4}$ and the efficiency $\eta_{A,StF4}$ for the StF4 sensor antenna (4) in the same way as (2) :

$$A_{eff,St}\cdot(E_0(t))^2/Z_0 =(V_{St}(t))^2 /Z_S\ , \quad A_{eff,St}=(K_{StF4})^2\cdot Z_0/Z_S = 7.7\cdot10^{-3}\ cm^2, \quad \eta_{A,StF4} = A_{eff,St} /A_{a,St} \approx 0.34 \quad (4)$$

Note, the computer simulation of the sensor model gives the values (1, 2), which are close to the experimental ones. Note, the design of StF4 sensor is a type of directional wave coupler with a finite length. The interval finite time work gap $T_{max} \approx 7$ ns is clearly displayed in figures 3, 4 (the signal duration is lower than T_{max}).

We measured the K_{StF4} -sensitivity for various directions of the coming waves [12]. In the first, the StF4 sensor is an antenna with a linear polarization (along the normal to the base plate). In the second, the K_{StF4} -sensor sensitivity does not change within the broad angles in the H-plane, and vice versa in the E-plane (the normal to the base plate), it is increasing by a factor of 1.3 (it is maximum) for 30° angle deflection. Measurements with an accuracy 5% using the StF4 sensor demonstrated a deflection of the aim angle towards the radiator within the interval of (0-6) degrees in the plane normal to the base plate and ±20 degrees in the plane of the base plate (see results in figure 5). Note, for other types of sensor as the TEM-half-horn antenna we measured [12] relative more effects both for increasing sensitivity and for distortions of pulse shapes. Applications of the TEM-half-horn antenna sensors to measure the EMP waves from high power radiators are in [14, 15].

2.3 Principle of Reciprocity for Pulsed Antennas. It is known, the so-called the reciprocity theorem is used often to analyze the linear electrical systems, radio antennas, and sometimes, for the antennas with pulsed non-sinusoidal signals [7, 16]. This theorem predicts that the received pulsed signals will be identical when we change places of the receiving and radiating antennas. In this cases, it is necessary to provide some conditions as reported in [7]: 1) the cables connecting the antennas with a generator and with a measuring device should be of a sufficient length and have to be equal to the impedances 50 Ohm; 2) there should be a sufficiently large "free space" around antennas for decreasing parasitic reflections.

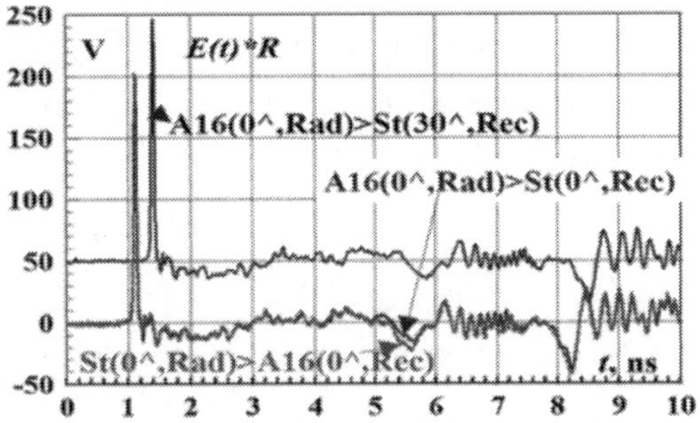

Figure 5. E(t)*R –waveforms. (TMG40R+A16) irradiated the StF4-receiver at (0°; 30°) angles in H- plane. Inferior curve is (TMG40R +StF4)-radiator irradiated A16-receiver

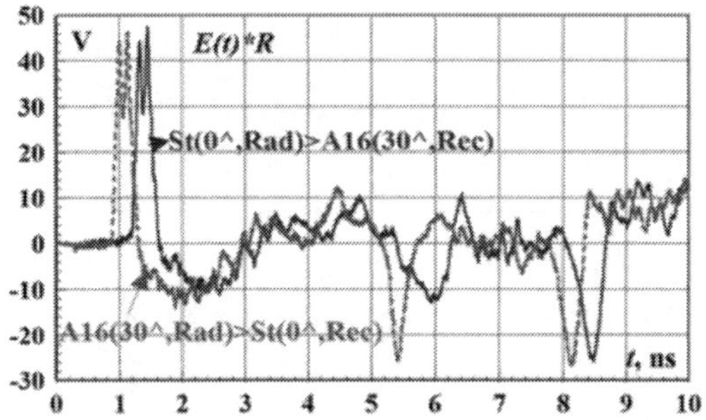

Figure 6. E(t)*R –waveforms. StF4 -(receiver/radiator)+3.5m+A16-(rad/rec) (turned by angle 30° at H -plane) +TMG40R as excitation generator

Figures 5, 6 show the experimental results for the "Rad+3.5m distance +Rec" set-up, which have been obtained using the devices: 1) antennas of the StF4-(a small aperture) and the A16-(16×16cm^2 =$A_{a,A16}$, in [3, 6]); 2) TMG40R ([13], $T_{f,p} \approx 25$ ps, $T_{w,p} > 1$ ns, $V_G \approx 43$ V, 100 kHz) as an exciting source of step-like pulses; 3) the digital sampling oscilloscope DSA8200 (0-30 GHz bandwidth, 12 bit resolution) with 80E07-09 remote modules. Figure 5 shows the curves (top and bottom) of the StF4 sensitivity without changes for various set-ups. This figure shows lower two similar superposition curves for the set-ups of StF4 as (Receiver/Radiator) and the A16 as (Radiator/ Receiver), that demonstrates the efficacy of the reciprocity principle. A better test of the reciprocity principle is demonstrated in figure 6 for the following conditions: 1) the step pulsed voltage generator with a rise-time $T_f \approx 25$ ps ; 2) the A16 antenna has the L_a =16 cm>>$c \cdot T_{f,p}$ -a "big" aperture and the StF4 antenna has a small aperture, a distance between them is 3.5 m ; 3) the A16 opening surface is turned by 30° angle towards the beam between two antennas. The curves in figure 6 were moved apart in 0.3 ns at the time axis for the best comparison. These signals are equally excepting a jump occurring at the time moment $t \approx 5.4$ ns (a parasitic reflection by surrounding objects). Note, the inner structure of A16 is detected by the data in figure 6: two peaks in the radiated pulse are separated by a delay ≈ 0.14 ns, i.e. the spatial gap $\Delta L \approx 0.14\times30 = 4.2$ cm. This is the distance between the centers of two "small horns", which are turned at 30° angle: $\Delta z_{hh} = 8 \cdot \sin(30^o) = 4$ cm. The A16 was made by four "small horns" of 8×8 cm^2 size.

2.4 Summaries. The test results confirm that we can change the places for any receiving and transmitting antennas, and that the received pulsed signals will have an identical shape and magnitude. It can be realized: if the A_{eff} - antennas effective apertures would be equal in the receiver and the transmitter modes. In (2), it is given that the effective antenna aperture $A_{eff} = P_{AL}/\Pi_{EA}$ equals to the power in the matched receiver antenna feeder load $P_{AL} = (V_{AF})^2/Z_{AF}$ divided by the flow density power in the TEM-wave coming in the normal to the antenna face $\Pi_{EA} = (E_{TE})^2/Z_0$. They are: $Z_{AF} = 50$ Ohm – the feeder impedance; $Z_0 = 377$ Ohm – the free space impedance; $V_{AF}(t)$ –the receiving voltage in the feeder; $E_{TE}(t)$ –the electric field of the incoming wave in the antenna face. According to these equations, A_{eff} is a function of (t) – the time, though, for the antenna, there may be some pulse signals received with small distortions. For example, the StF4 sensor antenna has the constant A_{eff} for the USEMPs with the rise-time $T_{f,p} > 30$ps and the pulse duration $T_{w,p} < 7$ns, (see (4), and the data of figures 3, 4). The antenna parameters can be measured by the experiments similar to figure 5. The transmitter with the StF4 –as an antenna, which was excited by a step-like pulses would radiate the

USEMPs waves with a shape like δ(t) –a delta function. Any antenna, receiving such wave impulse, will produce in its feeder $V_{\delta,r}(t)$ –the voltage response, according its pulsed characteristic.

In the analysis of antennas of high power impulses we ordinarily use simple approximations [7, 12, 15] (see equation (5) below): $A_{a,A}$ –the antenna face aperture; η_A –the antenna efficiency for pulses with the "average time" parameters: $T_{f,p} > T_{f,A}$ and $T_{w,p} < T_{0.9,A}$. Here: $T_{f,A}$ –the antenna transient rise-time and $T_{0.9,A}$ –the antenna "work time gap", for signal impulse data of $T_{f,p}$ –the rise-time, and $T_{w,p}$ –the pulse width. The antenna impulse characteristics are received as written above. Often $T_{0.5,A}$ –the full width at half-maximum (FWHM) for the $V_{\delta,r}(t)$ is used instead of $T_{f,A}$ –the rise-time ($T_{f,A} \approx 1.05 \cdot T_{0.5,A}$).

We can note that the reciprocity principle is very useful in numerical simulation (an example reported in [12]) of the radiation pattern for pulsed antennas. It is well known, to solve this task, we needed calculations for EM-fields at a free space in the far-field zone of the radiating antenna. A size of this far-field zone should be much greater than that of the antenna. Simulations of the pulsed antenna as a receiver of the EMPs waves allowed us to decrease spatial sizes around the antenna. Although, one variant of calculations giving the antenna characteristics for the radiation/receiver is only for one spatial direction.

3. High Power Radiators for USEMP waves. Calculations, Experiments

Many our radiating antennas have been built as the antenna array joining four "small TEM-horns" like the A16. Each "small TEM-horn" was connected to its two-wire line and then to the HV-HF coaxial antenna feeder. This feeder had the coaxial connector with the wave impedance 50 Ohm. The connectors were of enlarging sizes for higher voltages at the range 10–100 kV for various pulsed generators as reported in [3-6]. For instance, A16 was supplied by connectors of "HN" male type for sub-nanosecond pulses up to 30 kV, and "N" male type for the amplitudes about 10 kV. The A16 in 10 kV -version had a transit rise time of about 0.04 ns, according to our results (see figure 5). Each "small TEM-horn" in the antenna array contained the dielectric inset between its two electrodes to condense the wave power and to make the even transit times for waves on the way to a face of antenna. The over electrodes of these "small horns" were mounted for return currents shielding rectangular plates. One edge of this plate was connected with the electrode on the face of antenna, and another far edge was connected with absorption resistors. An assembly of the absorption (for damping) resistors was equal to 50 Ohm.

For pulsed radiators, the equations estimating the effective potential $U_{rad} = E_p(R) \cdot R$ – (the product of the peak E-field in the far-zone by R- distance, see above in 1.1) exist in various approximations. In [9], it was manufactured on the basis of the well-known Maxwell's extended equations and the Kirchhoff-Huygens principle: secondary radiating sources residing at the antenna face surface, the so-called antenna aperture A_a or the antenna opening surface (more detailed in [17]), that is:

$$U_{rad}\,(A+G) = E_{p,AG}(R) \cdot R \approx (\eta_A \cdot A_{a,A})^{0.5} \cdot (30/(Z_G \cdot \pi))^{0.5} \cdot V_G/(c \cdot T_{f,ef}) \approx 0.44 \cdot V_G \cdot (\eta_A \cdot A_{a,A})^{0.5} /(30 \cdot T_{f,ef}) \quad (5)$$

In (5), we used symbols like those employed in equation (2) and the additionals: 1) U_{rad} – the effective potential for the radiator with the antenna A, which excited by the generator G (V or kV); 2) $A_{a,A}$ – the face antenna aperture and $A_{eff,A} = \eta_A \cdot A_{a,A}$ – the effective antenna aperture (cm^2) ; 3) η_A – the antenna efficiency (for our antennas $\eta_A \approx 0.9$) ; 4) $Z_G = 50$ Ohm –the generator/ feeder wave impedance, $(30/(Z_G \cdot \pi))^{0.5} = 0.437$; 5) V_G – the amplitude of a rapid mono-polar pulse for exciting the generator of the G_pV (V or kV); 6) c = 30 cm/ns – the ight velocity in a free space; 7) $T_{f,ef} \approx ((T_{f,G})^2 + (T_{f,A})^2)^{0.5}$ –the rise-time (ns) for a pulsed wave in the antenna plane aperture, $T_{f,G}$ –the rise-time of the generator pulse and $T_{f,A}$ –the rise time for the antenna feeder-aperture transient time (or antenna impulse characteristic: $1.05 \cdot T_{0.5,A} \approx T_{f,A}$, see above in section 2.4).

In figure 5, the experimental result demonstrated for the radiator (A16+TMG40R) is U_{rad} (A16+TMG40R)$_{exp} \approx 200$ V and calculated by (5) is U_{rad} (A16+TMG40R)$_{cal} \approx 210$ V ≈ 200 V , for $V_G = 43$ V , $T_{f,ef} \approx 0.045$ ns. This $T_{f,ef}$ equals to $1.05 \cdot T_{0.5,AG}$ for U_{rad} (t; A16+TMG40R) – the transmitter impulse in the experiments of figure 5. Using these data, we estimate the transient time of the antenna A16 by the value $T_{f,A16} \approx 40$ ps.

Figure 7 shows $E(t;R;A30)\cdot R$ at $R = 8$ m in the experiments with A30 antenna (30×30 cm^2–aperture) and the generator G_p100 (70-100 kV, 0.1-0.15 ns, 1 kHz [10]). Figure 8 shows the set-up of a similar autonomous radiator ($A36+G_p100$): A36 antenna is placed in the left-side; G_p100 and DC source (the accumulator assembly) with control devices are placed in right-side.

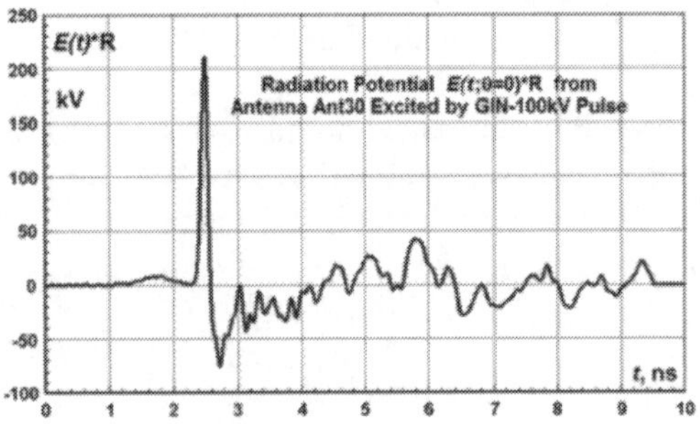

Figure 7. Radiating potential for *E(t;* R*;*A30*)·R* from the (Ant30+ G_p100) radiator

Figure 8. The radiator ($A36+G_p100$), into right-side box are placed G_p100, DC source (accumulator assembly) and control devices

The amplitude of $E(t)\cdot R$ curve shown in figure 7 is about 210 kV. Using this value and the equation (5), we can apply it for an approximate definition of the parameters of GIN100 by (6):

$$U_{rad}(A30+G_p100) \approx 0.44\cdot V_{G100}\cdot(\eta_A\cdot A_{a,A36})^{0.5}/(30\cdot T_{f,ef}) \approx 4.15\cdot(V_{G100},kV)\cdot(0.1ns/T_{f,ef}) = 210\text{ kV} \quad (6)$$

Using (6) and the FWHM $1.05\cdot T_{0.5,AG} = T_{f,ef} \approx 0.14$ ns in figure 7, we do the estimation: $V_{G100} \approx 70$ kV. These values are realistic data for the ($A30+G_p100$)-radiator. In experiments with the (A30 + TMG40R)-radiator, similar to figure 5, we measured the rise-time for the antenna A30: $T_{f,A30} \approx 0.09$ ns. Using this data and $T_{f,ef} \approx 0.14$ ns for the ($A30+G_p100$)-radiator, we estimated the rise-time $T_{f,G100} \approx 0.11$ ns for the generator G_p100. The magnitude of $V_{G100} \approx 70$ kv is a rapid part of the impulse of the generator G_p100. We estimate a full amplitude of the voltage $V_{max}(G_p100) \approx 80$ kV by adding the voltage of pre-pulse base (for G_p100 it is about 15%). It is close to the measured voltage. Figure 8 shows a typical design of the high power USEMP source of the autonomous high power radiator ($A36+G_p100$).

4. Conclusion

There was considered a new electromagnetic field sensor of type StF4, which is a non-symmetrical two-wire line small cross-section of about 1.5×1.5 mm^2 and 40 cm length. The pulsed characteristics of the StF4 sensor antenna were calibrated using the gauge TEM line-cell. The StF4 parameters were: the sensitivity $K_{StF4} = 0.32$ V/(kV/m), the rise-time $T_{0.1\text{-}0.9} < 0.03$ ns, and the work time gap of the $T_{max} = 7$ ns. The transmitter with StF4 as an antenna, which was excited by the step-like pulses, would radiate the "calibrated" USEMPs waves with shape like the delta-function. Such source is a good device for the measurements of pulsed antennas. Non-stationary processes in the receiver/radiator were studied analytically and by numerical 3-D code "KARAT". At our laboratory, we had constructed the models of the autonomous compact high power radiators, using highly efficient, up to $\eta_A = 0.9$ antenna apertures (the size of one side of 16...60 cm) of the antenna array joining four TEM-horns and shielding the electrodes in every TEM-horn [3-6, 9]. The UWB radiators were built with a highly effecient antenna array, with the generators of pulsed voltage (10-100) kV, and the repetition rate (1-100) kHz with a control synchronization. These radiators generated the radiating effective potentials $U_{Rad} = E_{max}*R = (20\text{-}400)$ kV. The radiators have been successfully tested for the electromagnetic immunity of various electronic systems.

IOP Conf. Series: Journal of Physics: Conf. Series **830** (2017) 012020 doi:10.1088/1742-6596/830/1/012020

Acknowledgments

The work was supported by RAS scientific program "Elemental base and power semiconducting electronics devices on the basis of silicon and silicon carbide" (coordinator of this program is academician of RAS I.V. Grekhov).

References

[1] Camp M and Garbe H 2006 *IEEE Trans. EMC* vol **48** no 4 pp 829-833

[2] Parfenov Y V, Titov B A, Zdoukhov L N, Radasky W A 2015 *Conf. ASIAEM-2015* (Korea), *Book of Abstracts, ID #78* http://www.asiaem.org

[3] Fedorov V M, Lebedev E F, Ostashev V Ye, Ul'yanov A V 2010 *Proc. 16-th SHCE.* (Tomsk, Russia) pp. 437-440, (http://www.hcei.tsc.ru/conf Proceedigs 2010)

[4] Fedorov V M, Lebedev E F, Ostashev V E, Tarakanov V P, Ul'yanov A V 2011 *IEEE EMC Proc. 9th in Sump.on EMC and EM Ecology* (St. Petersburg, Russia) pp. 394-397

[5] Fedorov V M, Lebedev E F, Ostashev V E, Tarakanov V P, Ul'yanov A V 2012 *Proc. Prog. in Elec. Res.* (Moscow, Russia) pp 1461-1466, (http://www.piers.org/piers2012moscow/)

[6] Fedorov V M, Ostashev V E, Tarakanov V P, Ul'yanov A V 2015 *Sympos. EMC2015-Dresden, Proc. IEEE: CPF15EMC-USB* pp 1127-1132

[7] Fedorov V M 2011 *Proc. All-Russia 5th Conf. on Radiolocation and Wireless Communication* (Moscow, IRE of RAS, November 21-25, 2011) pp 196-200 (in rus)

[8] Podosenov S A, Sokolov A A 2000 *Radiation and Mesurement of Pulsed Electromagnetic Fields* (Moscow: "Sputnic", 2000, 250 pp, in rus.) chapter 5 pp 142-167

[9] Fedorov V M, Ostashev V E, Ul'yanov A V 2012 *17-th SHCE.* (Tomsk, Russia), *Russian Physics Journal* vol **55** no 10/3 pp 79-83

[10] "FID Technology", http://www.fidtechnology.com

[11] Tarakanov V P 2003 *Mathematical Simulation. Problems and Results* (Moscow: Nauka, 2003, in rus) pp 456-476

[12] Fedorov V M, Ostashev V E, Tarakanov V P, Ul'yanov A V 2014 *EFRE-2014* (Tomsk, Russia) *Russian Physics Journal* **57** No 12/2 pp 21-24

[13] "TRIM Ltd.", http://www.trimcom.ru

[14] Andreev Yu A, Efremov A M, Koshelev V I, Kovalchuk B M, Plisko V V, Sukhushin K N 2011 *Instruments and Experimental Techniques* No 6 pp 51-60 (in rus)

[15] Belichenko V P, Buyanov V I, Koshelev V I 2015 *Ultrawideband pulsed radiosystems* Editor by Koshelev V.I., (Russia, Novosibirsk. Nauka, in russia), ISBN: 978-5-02-019197-6. 481 p.

[16] Feld Ya N 1943 *Dokl. Akademii Nauk USSR* **41**, No 7 pp 294-297

[17] Ostashev V E, Ul'yanov A V, Fedorov V M 2016 Estimation of efficiency of energy conversion in UWB radiator: from electric energy to energy of directed radiation *Report on this Conference of EFRE-2016* (Tomsk, October 2-7)

Linear transformer and primary low-inductance switch and capacitor modules for fast charging of PFL

Yu A Bykov, E G Krastelev, G V Popov, A A Sedin and V F Feduschak

Joint Institute for High Temperatures, Russian Academy of Sciences, st. Izhorskaya 13/19, Moscow, 125412 Russia

E-mail: ekrastelev@yandex.ru

Abstract. A step-up linear pulse transformer and a modular primary powering system were developed for fast (≈350 ns) charging of a pulse forming line (PFL) of a high-current electron accelerator. The linear transformer is assembled of a set of 20 inductors with circular ferromagnetic cores and one–turn primary windings. The secondary turn is formed by housing tube walls and a voltage adder with a film-glycerol insulation installed inside of the inductors. The primary powering system assembles 10 modules, each of them is a low-inductance site of two capacitors of 0,35 μF and one gas switch mounted at the same enclosure. The total stored energy is 5.5 kJ at the charging voltage of 40 kV. According to test results, the equivalent parameters at the output of the transformer are the next: a capacity – 17.5 nF, an inductance – 2 μH, a resistance – 3.2 Ohms.

1. Introduction

Linear transformers take a special place among the variety of existing high voltage pulse transformers. They assemble a series set of one-turn transformers with primary windings fed separately and the common one-turn secondary winding. The coaxial version with the volume primary winding and the secondary winding formed by a common housing and a central stalk-voltage adder have the lowest possible induction and a minimal resistance of windings. It reduces both the energy transfer time and losses [1-3].

The following are the design and the basic characteristics of linear pulse transformer with an output voltage of 800 kV and the modular capacitive energy storage in the form of low-inductance capacitor-switch modules, designed for fast (≈350 ns) charging of a pulse forming line of a nanosecond high-current accelerator.

2. Linear pulse transformer

2.1. Linear pulse transformer design

The design concept of the linear pulse transformer is shown in Figure 1.

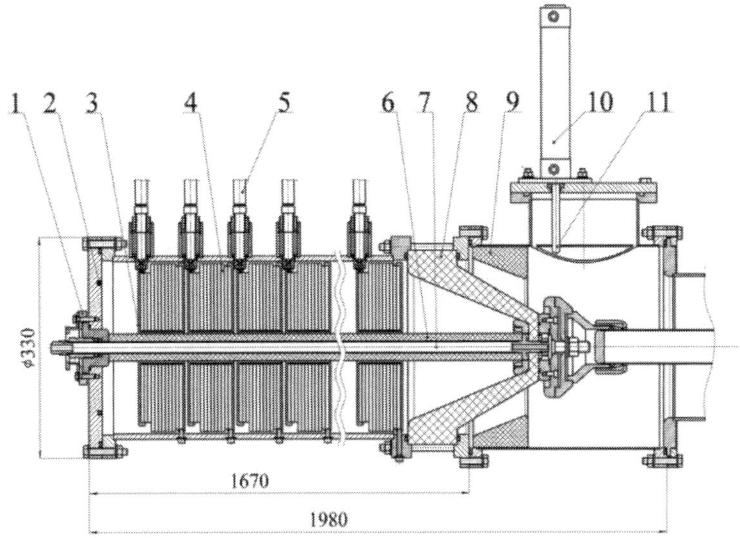

Figure 1. The design of the linear pulse transformer. 1 – small annular gap air switch, 2 - Rogowski coil, 3 – primary winding, 4 – ferromagnetic core, 5 – cable feedthrough, 6 – multi-layer film-glycerol insulation of stock-adder, 7 - stock-adder, 8 – conical insulator, 9 - dielectric insert, 10 - pneumatic actuator, 11 - grounding rod.

The linear pulse transformer assembles 20 series inductors (ferromagnetic sections) placed inside a metal housing with an inner diameter of 275 mm. Each of the inductors assembles an annular ferromagnetic core inside of the primary winding of the rectangular cross-section. The one-turn winding is formed by the housing part, a cylinder segment inside of the annular core and two discs, one of which connects one end of the cylinder segment to the housing, and the second one with a smaller outer diameter is connected to the four high-voltage cable feedthrough. The feedthrough design provides a tight connection with the KVIM cables that are used for the transmission the energy of storage capacitors to the primary windings. The inner diameter of the primary winding is 40 mm, its length - 70 mm.

The outer housing and a central stalk-voltage adder of 18 mm diameter with a multilayer film-glycerol insulation inside all of the 20 ferromagnetic sections form a secondary winding. At the low-voltage side of the transformer, the stock insulated from the housing by an annular air switch is mounted for feeding the magnetization current. A small (1-1.5 mm) air insulated gap of the annular switch is quickly closed at the beginning of the high-voltage pulse generation, grounding this end of the stock during the working cycle. The second high-voltage end of the stock is connected to the load through the output conical insulator. The internal volume of the transformer including the ferromagnetic system and the central stock is filled with glycerol. The conical insulator separates it from an intermediate section between the transformer and the pulse forming line with water insulation.

The central stock is the most stressed element of the pulse transformer. The electric field in the last section of the transformer exceeds 1.1 MV/cm at the maximum output voltage of 800 kV for the selected diameter of the stock 18 mm and the internal diameter of the primary winding of 40 mm. A multi-layer insulation made of a polyethylene film and impregnated with glycerol provides the required dielectric strength [4]. Based on the available data, the resource of the insulation is estimated for 1000 pulses, which is comparable with the lifetime of cables and capacitors.

2.2. Ferromagnetic system

A ferromagnetic core of the inductors is made of 2NSR alloy with the full-scale induction of 3 T in the magnetic field of 800 A/m [5].

Each of the cores is assembled of 12 rings wound by a tape of 5 mm width and of 30 microns thickness. The inner diameter of rings is 50 mm, an outer diameter of the core is 250 mm in the middle section (7 rings) while those located near the ends of the primary winding have a diameter 240 mm (2 rings) and 220 mm (3 rings). With a given fill factor of 0.7, an efficient cross section of the core is S=42 cm^2. The polymer films are used for the insulation of ferromagnetic rings from each other and from the primary winding. The total weight of the ferromagnetic system of the transformer is about 300 kg.

The measured volt-second integral of all twenty inductors is 0.24 volt-second. Energy losses for the full scale of magnetization are about 30 J, which correlates well with the estimates based on the known values of specific losses for the core material ≈0.1 J/kg.

Prior to the generation of a high-voltage pulse, the transformer cores are magnetized by a pulsed current in the secondary winding. The magnetization current generator is connected to the central stock at the low voltage side of the transformer where the stock is insulated from the housing by the annular air switch. During the magnetization, a pneumatically operated grounding rod in the transition section with water insulation grounds the high-voltage end of the central stock. The unipolar magnetization current is generated by discharging a 100 μF capacitor charged to a voltage of 1 kV, through a thyristor switch and a connected in series inductor of 35 μH. This inductor reduces the current rate providing a reliable switching off of the thyristor at zero current and protects the generator circuit from surge prior to the closing of the ring switch.

2.3. High-voltage feedthrough to the intermediate section with water insulation

A conical insulator made of polyethylene is used as a barrier between the transformer filled by glycerol and an intermediate section with water insulation. Its length is determined by the voltage distribution along the film-glycerol insulation at the output part of the stock outside of the last inductor. In addition, a shaped ring insert made of polyethylene is placed immediately after the insulator in the section with water insulation. It acts as a diverging lens for an electric field in water and equalizes the voltage distribution along the insulator surface.

The geometry of the transition region together with the calculated pattern of the equipotential lines is shown in Figure 2.

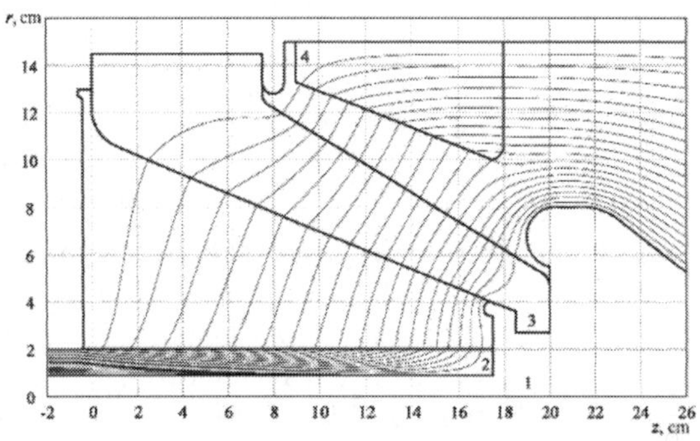

Figure 2. Geometry of transition section with plot of equipotential lines. 1 – high-voltage electrode, 2 – film-glycerol insulation of the stock, 3 – insulator, 4 – plastic insert. Step of the lines is 0.05U.

3. Primary energy storage and switching system

The primary driving system assembles 10 energy storage and switching modules, each of which is a set of two capacitors and one gas switch. The modules are mounted on both sides of the cylindrical housing of the transformer all along its length. The outputs of four capacitors of two modules, which are located at the opposite sides of the transformer, are connected to the primary windings of the four inductors (sections). High-voltage cables KVIM with the length of 0.5 m are used for energy transmission from the modules to the transformer inductors.

A photo of the linear transformer assembled together with the primary energy storage and switching modules is shown in Figure 3.

Figure 3. 800 kV linear transformer together with energy storage and switching modules. View from the low-voltage side of the transformer.

3.1. The design of the capacitive storage and switching module

The capacitive energy storage and the switching module is a compact capacitors-switch set, each of which assembles two capacitors of 0.35 μF, one common triggered gas switch and 8 output cables – 2 groups of 4 from each of the capacitors. The design of the module is shown in Figure 4.

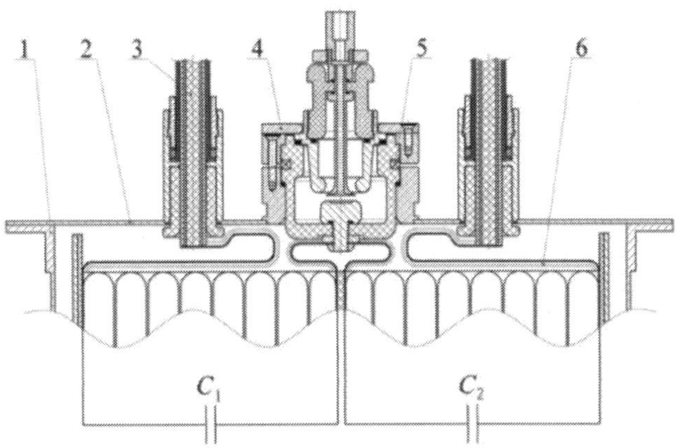

Figure 4. Cross-section of the energy storage and switching module. 1 - housing, 2 – cap, 3 - cable output, 4 - gas switch, 5 - Rogowski coil, 6 oil-barrier insulation.

The design of the module is based on the use of capacitor packages of the industrial capacitor IK50-3. One of four sections of IK50-3 capacitor is used in one module. The every section consists of two connected in series groups of 8 capacitor packs, tightly fastened into a single structure by metal shrouds between the end metal plates. Every group of the packs is a high-current capacitor of 0.35 μF for the voltage of 50 kV. The section is placed into a metal housing of the module with the dimensions 314x314x160 mm^3 and isolated by a paper-oil insulation for the full voltage of 50 kV.

The output from the midpoint of the two capacitors is connected to the high voltage electrode of the triggered gas switch installed on the cover in front of the midpoint. Outer packages of the capacitors are connected to the central conductors of the output cables. Wide buses isolated by the paper-oil insulation are used for connections inside the module. Very close locations of all elements of the module together with a low-inductance design of the connecting buses and the use of 4 output cables KVIM in each arm provide a low total inductance of the discharge circuit (≤40 nH).

A three-electrode gas switch with a "field distortion" is used for fast switching of the capacitors. The sketch of the switch is shown in Figure 4.

The switch consists of an insulating housing made of polyamide plastic "caprolon", which is installed in a metal base welded to the module cover, two main electrodes and a disk trigger electrode. Dry air with the pressure up to 4 bars is used as an insulating gas. One of the main electrodes, installed in the bottom of the insulator, is connected to the midpoint of the capacitors and a high-voltage charging cable (not shown in Figure 4). The second electrode is grounded. The position of the disk edge of the trigger electrode corresponds to the equipotential surface with a potential of 1/3 of the voltage U applied to the switch. The same voltage of $U/3$ is applied to the trigger electrode via a resistive divider in the trigger pulse generator unit.

Closing of the switches of all 10 modules is initiated by supplying pulses of an opposite polarity with an amplitude of the voltage up to 40 kV from one common generator to the trigger electrodes using 3 m long coaxial cables. The rise time of the trigger pulse is ≈15ns. Measurements have shown that the average time delay for the switch closing, relative the beginning of the trigger pulse, is about 25 ns, and the its deviation does not exceed ± 5 ns for system of all 10 switches.

4. Test results

The assembled system was tested in several operation modes which differed by the load at the output of the transformer.

4.1. Short-circuit mode

The tests in a short circuit mode were performed at a reduced to 20 kV capacitor charging voltage, corresponding to the total stored energy of 1400 J.

Figure 5 shows the current waveform measured by Rogowski coils in the secondary circuit of the transformer and the switch current of one module.

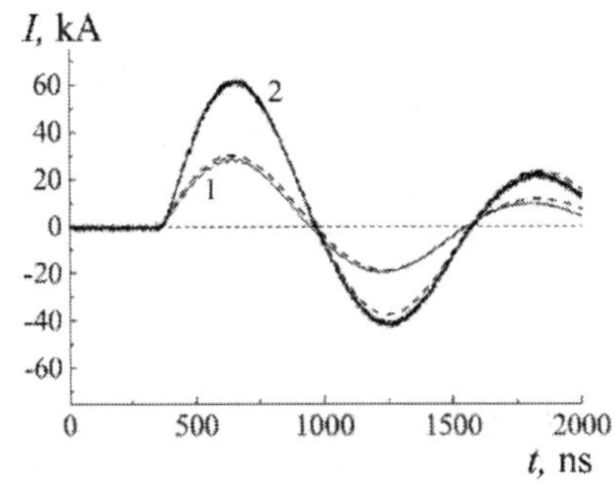

Figure 5. Waveforms of currents at the output of the transformer (1) and through the switch of one of the modules (2) for short circuit mode. The dashed lines show the calculated currents.

The amplitude of the short-circuit current at the transformer output for the charging voltage of 20 kV is ≈30 kA, while measured in the circuit of the switch discharge current of two capacitors, was two times higher and reached ≈60 kA.

As a result of measurement, the parameters of the simplified equivalent circuit in the form of the RLC-circuit have been determined as the transformer output: a capacity - 17.5 nF, an inductance - 2.03

μH, and the total series resistance 3.2 Ohms. The results of the test calculation using the developed equivalent circuit are shown in Figure 5 as a dotted line.

4.2. Resistive load mode

Two loads with a resistance of 20 and 40 Ohms were used to test the dielectric strength of the insulation for long aperiodic discharge mode. High-voltage liquid resistors were formed by filling the volume of the transition section, which was closed by a flange with a solution KCl.

Figure 6 shows the waveforms of the load current and the module switch current obtained for the load resistance of 20 Ohms and the primary charging voltage of 30 kV. The dashed lines show the currents calculated using the equivalent circuit.

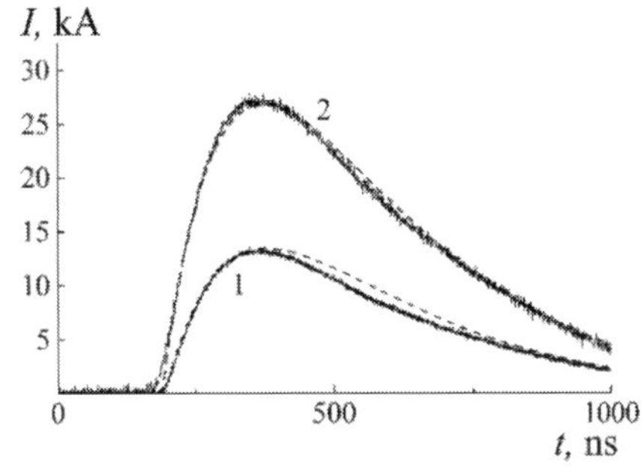

Figure 6. Waveforms of the load current at the output of transformer (1) and module switch current (2) for resistive load of 20 Ohm and charging voltage of 30 kV. The dashed lines show the calculated currents.

The results of measurements in this mode indicate the absence of breakdowns and parasitic leakages due to an insulation failure under increasing exposure to an electrically stressed state.

5. Conclusion

Now, the designed linear pulse transformer together with 10 module primary driving system is successfully used for the fast (≈350 ns) charging of a forming line of a nanosecond high current accelerator. The experience shows a high repeatability of the output pulses due to a stable synchronous operation of all 10 modules of the primary energy storage and a switching system. The modular design of the system, the limited voltage on its elements, low levels of an electromagnetic interference due to a closed design of the transformer and the modules with connecting coaxial cables, as well as the absence of a transformer oil and a related peripheral equipment ease up an installation operation.

The work was supported by the Russian Science Foundation (project 16-15-10355).

References

[1] Mesyats G A 2004 *Pulse Energy and Electronics* (Moscow: Nauka).
[2] Vizir V A, Elchaninov A S, Zagulov F Ya, Kovsharov N F, Sorokin S A and Feduschak V F 1986 *Pribori i technika experimenta (Instruments and Experimental Techniques)* V**5** p 95 [in Russian].
[3] Lavrinovich I V, Zharova N V, Petin V K, Ratakhin N A, Fedushchak V F, Shlyakhtun S V and Erfort A A 2013 *Instruments and Experimental Techniques* V**56** №3 p 329.
[4] Kovsharov N F and Feduschak V F 1984 *Pribori i technika experimenta (Instruments and Experimental Techniques)* V**3** p 217 [in Russian].
[5] http://www.amet.ru

Corona pre-ionized gas switches with an increased lifetime for Marx generator of the lightning test complex

Yu A Bykov and E G Krastelev

Joint Institute for High Temperatures, Russian Academy of Sciences, st. Izhorskaya 13/19, Moscow, 125412 Russia

E-mail: ekrastelev@yandex.ru

Abstract. A new design of triggered spark gas switches with an increased operation life and stable dynamic performances for 2.4 MV, 4 MJ Marx generator of the lightning test complex is developed. An operation mode of switches in the test complex is the following: the total voltage – up to 80 kV, the discharge current – up to 50 kA, the charge flowing – up to 3.5 C/pulse, the working gas – dry air at an atmospheric pressure. An increased operating life is achieved by using the torus-shaped electrodes with an increased working area and by the application of a thick disk with a hole, as a trigger electrode, installed between the two torus-shaped electrodes. A short breakdown time delay and a high stability of the breakdown voltage under dynamic conditions are provided by gas pre-ionization in the spark gap by UV-radiation of an additional corona discharge in the axial region of the switch.

1. Introduction

A mobile lightning test complex constructed at Joint Institute for High Temperatures RAS (Moscow) in collaboration with TRINITI (Troitsk) is intended for tests of lightning protection systems, measurements of the conductivity of soil and grounding elements at current levels of 50 kA and pulse duration up to 150 μsec [1].

A pulsed power generator of the test complex consists of two discharged in series Marx generators providing the total peak voltage up to 2.4 MV. The full stored in Marx capacitors energy is as high as 4.2 MJ.

Each Marx generator includes 15 stages of bipolar charging up to ±40 kV capacitors switched by 16 air insulated gas switches. Seven of them are three-electrode triggered switches and others are two-electrode switches closed by overvoltage. All switches have the same semi-spherical (R=46 mm) main electrodes of 78 mm diameter. They are installed radially in sections of a plastic (polyethylene) tube of 250mm inner diameter assembled together, to form columns of UV-coupled switches. The triggered switches of a "field distortion" type have the third electrodes in the form of a thin disk with a central hole of 50 mm diameter installed at the mid-plane between the main electrodes. The chosen spark gap of 50 mm between the electrodes of all switches corresponds to the self-breakdown voltage of about 115 kV, which is ≈1.5 times higher than the maximal voltage of 80 kV appeared by the end of the capacitor charging to ±40 kV.

The experience of the full-scale lightning tests showed that a stable synchronous operation of two Marx generators is limited by a degradation of characteristics of triggered switches. High flowing currents and the high charge – up to 3.5 C/pulse lead to melting and local thermal deformation of sharp edges of the thin disc electrodes, which in its turn results in increasing the switching time delay and its deviation. To solve this problem, a new design of triggered switch with stable dynamic characteristics during a long time was developed. It is based on the design of the developed earlier long life time switches with toroidal main electrodes and a thick trigger disc in a combination with parallel corona discharge in the axial region of the electrode system [2]. Gas pre-ionization in the spark gap by UV-radiation of the corona discharge provides a short time delay of the switch closing and a high stability of the pulsed breakdown voltage. The design of the new electrode system is compatible with the construction of the original switches

2. Design and main characteristics of new switch

2.1. Switch electrode system

The electrode system chosen for a new triggered switch combines two discharge gaps working in parallel. First of them, the main high-current spark gap, is formed by two torus shaped main electrodes and a thick trigger disc electrode installed between them. The second is an additional corona discharge gap. It is formed by a corona needle installed in the cavity of one (negative) main electrode and the other electrodes positively charged relative to a negative corona needle. A correct design of this electrode system is able to provide the following consequence of the discharges. The first is that the corona discharge sets up at some critical voltage U_{cr}, which is lower than the breakdown voltage U_{br} of the main high-current gap. Then, when the applied voltage reaches the level of U_{br}, a fast breakdown of the main gap takes place. The short time delay of the main gap closing is provided by the appearance of free electrons, as a result of gas photoionization by the ultraviolet (UV) radiation from the corona discharge localized in an axis region. At these conditions, the breakdown time delay is determined mainly by the time of plasma channel formation [3].

The new electrode system of the triggered switch was developed on the basis of the outlined requirements,. The size of the electrodes was chosen for a maximal compatibility with an original switch construction.

Figure 1 shows the cross-section of the new electrode system installed in the plastic tube section of the original switch. The electric field distribution is illustrated by the calculated plot of equipotential lines shown in figure 1. This pattern of equipotential lines corresponds to the initial state of the switch before closing with a zero potential at the middle trigger electrode and the equal in magnitude voltages U of opposite polarity applied to the main electrodes. The step of equipotential lines is 0.1U.

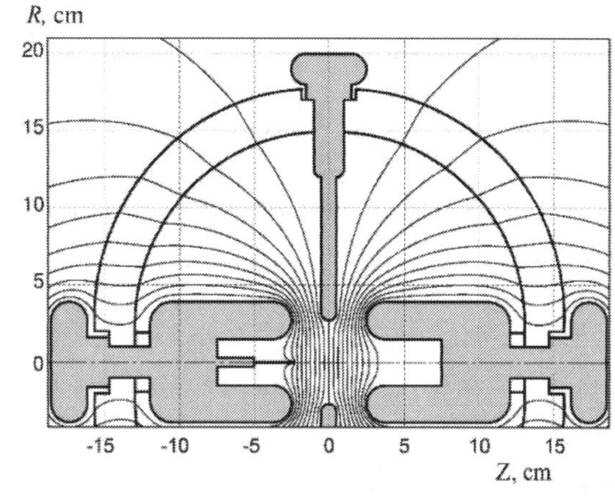

Figure 1. Cross-section sketch of the new triggered switch electrode system with calculated plot of the equipotential lines. Electrodes and their mounting parts are colored.

2.2. The main spark gaps

The main spark gaps are formed by two electrodes of the external diameter of 78 mm with the toroidal working surface of r=12 mm and the trigger electrode in form of disc of 10 mm thick with the central hole of 54 mm diameter with rounded edges located midway between them. The outer diameter of electrodes is equal to the outer diameter of the electrode holders of the original switches. It provides the possibility of reusing the existing holders after little mechanical improvement. The minor radius of the tor r=12 mm was found as optimal for achieving approximately the same voltage, as for original switch breakdown, for the same gap d=25 mm between the top of the toroidal electrode and the trigger disc. This result of the breakdown voltage calculations was supported by direct measurements of the breakdown voltage for two sets of electrodes – an original semispherical and a new toroidal one.

Figure 2 shows the plots of measured self-breakdown voltages U_{br} vs the gap distance d for the semispherical (1) and for the toroidal (2) electrodes with the optimized minor radius r=12 mm.

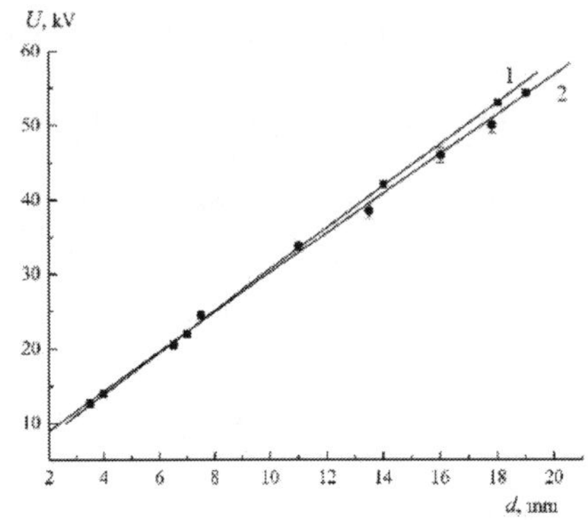

Figure 2. Breakdown voltage vs the gap distance: 1 – for original semispherical electrodes, 2 – for toroidal electrodes with minor radius r=12 mm

As you can see, the new toroidal electrodes provide approximately the same static breakdown voltages as measured for the original electrodes for the same gap distances. The difference of the measured and calculated breakdown voltages is lower than 5%.

2.3. Characteristics of corona discharge

The corona needle electrode is made of a tungsten wire with a diameter of 0.5 mm. The needle is installed in the cavity of the main toroidal electrode (negative during the work). The tip of the needle is located at the same axial position as the top of the tor or slightly outside. As it was shown during the dynamic test of the switch, the outside position of the needle tip of $h\approx1$ mm, relative to the top of the tor, increases the efficiency of the pre-ionization and provides a lower spread of the pulsed breakdown voltage. It is supposed, that the UV-illumination of the working surface of the main electrode may be more efficient, because of the lower work function for metals, compared to the ionization potential of the gas.

The volt-ampere characteristics of the stationary corona discharge for two spark gaps d=20 and d=25 mm are shown in Figure 3. The discharge current was measured by an analog (integrating the Trichel pulses) microampere meter connected in series with the negative high voltage electrode with the corona needle. The positive voltage of the same magnitude was applied to the second main electrode, and the trigger disc electrode was grounded via the resistor of 1 kOhm.

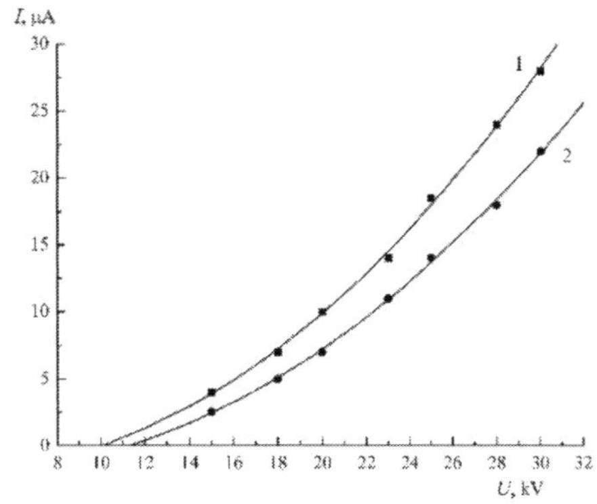

Figure 3. The volt-ampere characteristics of corona discharge for two gap distances: 1 – d=20 mm and 2 – d=25 mm. The needle tip is outside of the tor for h≈1 mm.

The presented characteristics show that the corona set up critical voltage U_{cr} is in a range 10÷12 kV and much lower than the breakdown voltage for the both gaps. Measured characteristics are in a good agreement with the shown by solid lines approximations by a quadratic function: $I \sim kU(U - U_{cr})$, where U_{cr} – the critical voltage.

It is important to note that the second microampere meter connected in series with the positive main electrode showed approximately the same currents and the readings of the microampere meter in the circuit of the trigger electrode were close to zero up to 25÷30 kV, when they reached several microamperes. It means that the most of the electrons from the corona discharge and formed in air negative oxygen ions move in the axis region and do not fall in the gaps. The main reason for the appearance of free electrons in the spark gaps is the gas photoionization by the corona UV-radiation.

2.4. *Switch dynamic test*

Dynamic characteristics of the switch were measured at a test stand by applying the static voltage of charged to ±U capacitors to the main electrodes and the fast rising pulse to the disc trigger electrode. The measurements were performed for the static voltages within ± 30 kV limited by the capability of the test stand. To satisfy these conditions, the spark gap distance of the experimental switch was reduced to 13÷15 mm. The pulsed voltage at the trigger electrode was measured by a high voltage probe Tektronix 6015, and the capacitors charging voltage was measured by the electrostatic kilovolt meter C1-35. The resulting voltage at the gaps was defined as a sum of the readings of the kilovolt meter and the recorded voltage pulse at the trigger electrode.

Figure 4 shows the traces of the voltage at the trigger electrode recorded for the applied trigger pulse of a positive polarity for two levels of a static voltage at the main electrodes (the capacitors charging voltage). The gap distance for these measurements is d = 13 mm, which corresponds to the self-breakdown voltage of 38 ± 0,5 kV, the critical voltage for the corona discharge set up is U_{cr}≈9 kV. The trace marked as 1 was recorded for the charging voltage U = ± 20.5 kV and the trace 2 – for U = ± 22 kV. These charging voltages correspond to 0.54 and 0.6 of the levels of self-breakdown voltage, respectively. The dotted lines in figure 4, marked as U_1 and U_2 , show the levels of the total voltage equal to the sum of charging and pulsed trigger voltages corresponding to the static breakdown for the given gap d=13 mm.

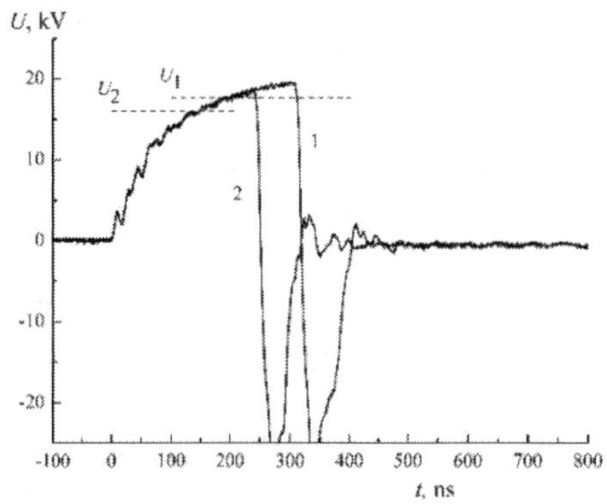

Figure 4. Voltage traces at trigger electrode for static charging voltages: 1 – U=±20.5 kV and 2 – U=±22 kV, dotted lines – levels of total voltage corresponding to static breakdown for the given gap d=13 mm.

As you can see, the closing of the first gap takes place shortly after the total voltage, charging plus triggering, reaches the level of the breakdown voltage shown by a dotted line.

For the charging voltage of U=±20.5 kV (the trace 1), the breakdown of the first gap occurs at the peak trigger voltage of 19.4 kV, when the total voltage at the gap reaches 39.9 kV which exceeds the static self-breakdown voltage by 1.9 kV or lower than by 5%. The time delay, measured from moment of reaching the level of the breakdown voltage until the gap closing, is about 108 ns. After that, the doubled charging voltage of 41 kV is applied to the second gap. For this voltage, the breakdown of the second gap takes about 85 ns.

An increase of the charging voltage to U = ± 22 kV (trace 2) leads to the lowering of the trigger voltage level, at the moment of breakdown in the first gap, and to the shortening of the closing time of both gaps. So, the time delay of the second gap breakdown for the increased to 2U≈44 kV total voltage is reduced to ≈70 ns. All measured time delays are consistent with the estimations of a time needed for the plasma channel formation [4].

A long time tests show that the scattering of the pulsed breakdown voltage amplitude for the first gap does not exceed 0.5 kV or is not higher than 1.5% of the self-breakdown voltage. For comparison, the dynamic tests of previously used original switch under the same conditions, with the decreased to 11 mm gaps, show a much longer time delay, up to 1 ms, and the breakdown voltage spread is higher than 5 kV.

2.5. Switch design

The developed design of the new electrode system is compatible with the construction and all electrical connections of the original switches. The chosen sizes of the electrodes provide re-using of the existing elements in the new switches: the unchanged tube sections of switch housing, the mounting elements, the outer connectors and the holders of the trigger electrode. Newly manufactured are the toroidal and trigger electrodes, the mounting elements for the corona needle. All mentioned parts for the new switch are made of stainless steel.

Figure 5 shows the photos the new (a) and the original (b) switches installed in the same plastic tube sections.

IOP Conf. Series: Journal of Physics: Conf. Series **830** (2017) 012022 doi:10.1088/1742-6596/830/1/012022

(a)

(b)

Figure 5. Photos of the new (a) and the original (b) triggered switches.

3. Conclusion

The new high-current, high-coulomb triggered air insulated switch was developed for the Marx generator of the lightning test complex. The increasing of the life time is achieved by using the torus-shaped main electrodes with an increased working area and by the application of the thick disc trigger electrode installed at a mid-plane between the main electrodes. The low breakdown time delay and the high stability of the breakdown voltage under dynamic conditions are provided by the gas pre-ionization by UV-radiation of the additional corona discharge in the axial region of the switch. The design of the new electrode system is compatible with the construction of the used now original switches.

References

[1] Fortov V E 2015 2015, Russia, *Teplofizika Vysokikh Temperatur*, 2015, Vol. **53**, No. 6, pp. 819–823.
[2] Krastelev E G, Sedin A A and Tugushev V I 2015 *Physics of Atomic Nuclei* V **78** № 14 p 1635
[3] Mik J M and Craggs J D 1958 *Electrical breakdown of gases* Oxford at the Clarendon Press.
[4] Martin T H 1989 *In proceed. of 7-th IEEE International Pulsed Power Conference* Monterey, California p 73.

IOP Conf. Series: Journal of Physics: Conf. Series **830** (2017) 012023 doi:10.1088/1742-6596/830/1/012023

Pseudospark switches commutation depending on trigger characteristics

A V Akimov[1,2], O V Anchugov[1], A F Baidak[1], P A Bak[1] and D A Shvedov[1]

[1]Budker Institute of Nuclear Physics, 630090, 11 Lavrentieva, Novosibirsk, Russia
[2]Novosibirsk State University, 630090, 2 Pirogova, Novosibirsk, Russia

Email: avakimov@inp.nsk.su

Abstract. The commutation characteristics of TPI-type pseudospark switches, depending on the parameters of trigger circuits, are studied. An influence of the trigger pulse rise time on the pseudosparks time jitter is explored. It has been shown that the pseudospark switches can be triggered on a pre-ionization electrode with a cathode and a grid, simultaneously grounded. The double and the triple gapped deuterium and hydrogen switches were explored. The time jitter values of 1-2 ns on the deuterium switches triggered on the pre-ionization electrode have been achieved.

1. Introduction

The 20 MeV, 2 kA induction linear accelerator LIA-20 is developed in BINP. It can operate in a triple-pulse mode. To supply the induction cells of the linac, the pulse modulators are used, which provide 21 kV pulses of 60 or 360 ns duration [1]. The modulator's switches are the TPI-type cold cathode thyratrons or the pseudo spark switches ranked at 50 – 75 kV anode voltage [2]. One of the main questions of the thyratron operation in the LIA's pulse system is the time jitter.

2. Cold cathode thyratrons's jitter statistics at LIA-2 facility

Earlier in BINP, the 2 MeV, 2 kA induction linac LIA-2 was designed that can be considered as a future LIA-20's injector [3, 4]. The LIA-2 has been used in VNIITF (Snezhinsk) for more than 6 years. For the LIA-2, about 100 thyratrons TPI1-10k/50 were tested. To control the thyratrons, a trigger board was designed in BINP, in accordance with manufacturer recommendations [5], figure 1.

The trigger board provides a complex thyratron's power supply that includes a hydrogen reservoir DC power supply Ures with current stabilization, a pre-ionization arc discharge power supply UDCpretrig that is turned on for 50-100 ms, a pulse pre-ionization 3 kV power supply Upretrig that is turned on 500 – 800 ns earlier than a main grid 3 kV pulse power supply Utrig. Such complex control allowed to achieve a quite stable thyratrons commutation. 100 thyratrons's jitter statistics are presented in figure 2. As you can see, the most thyratrons operate with the jitter of about 5 ns, although there are a lot of devices with a lower and bigger jitter.

3. Single pre-ionization pulse thyratron control

In the LIA-20 facility, it is proposed to use few hundreds of thyratrons, so it is desired to simplify a thyratron control and keep the jitter at the same level or better. In collaboration with IHCE (Tomsk), it was found that it is possible to control the thyratrons with a grounded grid and a cathode with a single trigger pulse applied to a pre-ionization grid and keeping a pre-ionization auxiliary arc discharge (figure 3).

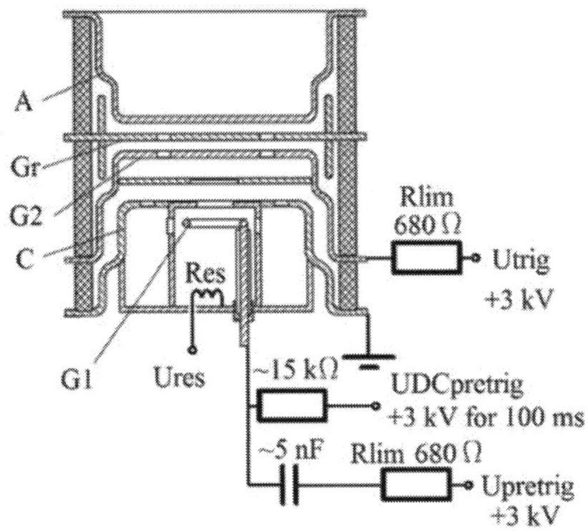

Figure 1. Principal schematic of double-pulse thyratron control with electrodes indicated: A – anode, C –cathode, Gr – gradient, G1 – grid of a pre-ionization, G2 – main grid.

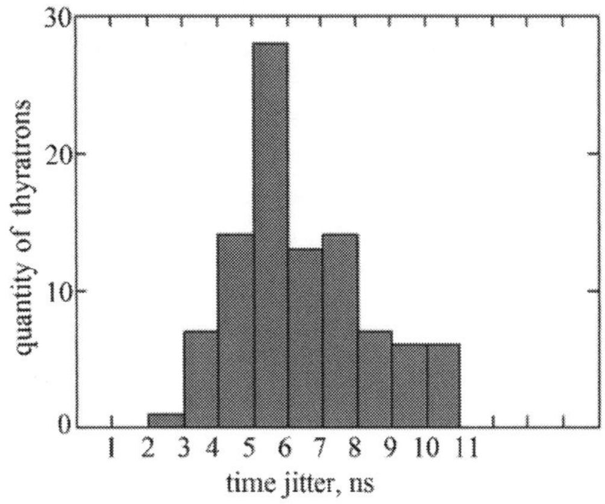

Figure 2. Statistics of LIA-2 thyratrons's jitter.

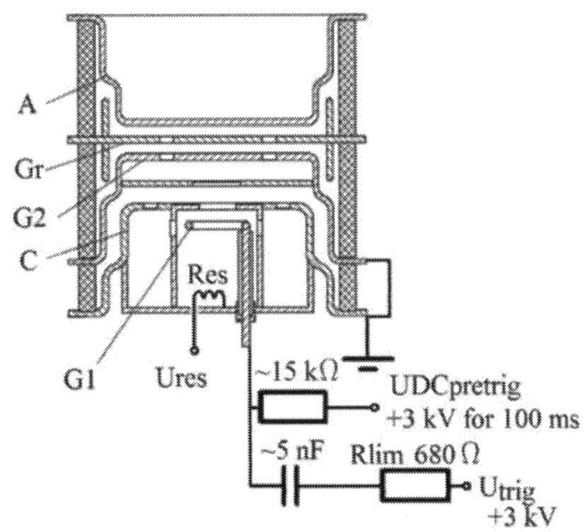

Figure 3. Principal schematic of single pre-ionization pulse thyratron control.

An advantage of such control is a much lower voltage spike applied to the trigger board circuits due to thyratron high current conduction. It can be explained due to a smaller part of an anode current that flows to a pre-ionization grid which is well shielded. For example with a main grid triggering (figure 1), a voltage spike amplitude can overstep 10 kV value during self-triggering of the thyratron that may lead to a trigger board failure. From the other hand, there is a question about a reliability of the thyratron switching-on process using a pre-ionization grid triggering. In that case, a grounded main grid is a potential barrier for the charge carriers which are supposed to penetrate from cathode part of the thyratron to an anode one, to provide a stable switching-on. Nevertheless, using the pre-ionization grid triggering with the grounded main grid and the cathode, a quite stable thyratrons switching-on process with an average 5 ns jitter has been achieved.

4. Thyratron switching-on stability depending on trigger parameters

The thyratron switching-on stability, using a double-pulse control (figure 1), was studied depending on the trigger and pre-trigger pulses voltage amplitude measured at the trigger board output. The data taken at a 3-gapped hydrogen thyratron test are presented in table 1. It is possible to lower the thyratron jitter from 10 to 1 ns rising the trigger pulses amplitude. A rise time of the trigger and pre-trigger pulses was about 35 ns. A main grid current Itrig was measured using a limiting resistor Rlim = 680 Ohm.

Table 1. 3-gapped hydrogen thyratron jitter depending on trigger pulses voltage amplitude with double-pulse control.

Upretrig (kV)	Utrig (kV)	Itrig (A)	jitter (ns)
1	1	1.3	10
1	2	2.6	2.5
1	3	3.9	1.5
2	3	3.9	1.3
3	3	3.9	1

We also have studied the influence of the trigger parameters on the thyratron switching-on process in a single-pulse trigger schematic shown in figure 3. It was discovered that rising a trigger pulse slope value from 30 to 100 kV/μs, the thyratron's jitter was getting much shorter - from 10 to 2 ns.

Table 2. 3-gapped deuterium thyratron jitter depending on trigger pulse voltage amplitude with single-pulse control.

dUtrig/dt (kV/μs)	Utrig (kV)	jitter (ns)
30	3	10
100	3	2

For both considered approaches of the thyratron control, the following conclusion can be made. The thyratron switching-on process stability depends on the value of trigger grid current in a time interval. A higher current corresponds to a larger quantity of the charge carriers extracted from a cathode zone of the thyratron to an anode one that makes the switching-on process more stable.

Also, it can be mentioned that, using a new trigger approach, the different thyratron modifications were tested: such as the 2-gapped and the 3-gapped hydrogen and deuterium devices. The pre-ionization grid trigger parameters are: Utrig = 3.8 kV, dUtrig/dt ~ 30 kV/μs. An absolute jitter value depends on a thyratron type and can vary from 2 – 3 ns to 10 – 20 ns for various species in a batch. In the course of testing of over 40 thyratrons, it was observed that the 2-gapped devices have a lower jitter than the 3-gapped ones. On the other hand, the deuterium devices have a lower jitter than the hydrogen ones with the same design. It can be explained by that the deuterium thyratrons can be operated at higher levels of a gas pressure where a better switching-on stability is provided. Hereinafter, it is proposed to develop a more preferable thyratron's design to provide the jitter insignificantly depending on different species in a batch.

5. Summary

The cold cathode thyratrons are quite simple to control. With proper trigger parameters a good switching-on stability with jitter of 1 – 2 ns can be provided.

References

[1] Akimov A, Bak P, Logachev P and Nikitin O 2015 *Proc. IEEE Pulsed Power Conf. (Austin)*

pp 541-543
[2] Bochkov V D, Dyagilev V M, Ushich V G, Frants O B, Korolev Yu D, Sheirlyakin I A and Frank K 2001 *IEEE Trans. on Plasma Science* vol **29** no 5 pp 802–808.
[3] Logachev P *et al.* 2011 *Proc. Int. Particle Accelerator Conf. (San Sebastian)* pp 1906-1908
[4] Akimov A *et al.* 2015 *Proc. IEEE Pulsed Power Conf. (Austin)* pp 538-540
[5] Akimov A V, Bak P A, Kazarezov I V and Korepanov A A 2006 *Problems of Atomic Science and Technology. Series: Nuclear Physics Investigations* (Kharkov) no **3(47)** pp 92–94

Generation of accelerated electrons in nanosecond electrical discharges using extensive slot cathodes limited by dielectric walls

N A Ashurbekov, K O Iminov and A R Ramazanov

Dagestan State University, 43a M. Gadjieva St., Makhachkala, 367002, Russia

Email: nashurb@mail.ru

Abstract. The spatio-temporal formation dynamics and basic electrical and optical characteristics of nanosecond discharges with an extended slot cathode in helium are investigated. It is established that electrons ribbon beams are formed in such discharges with an energy of several hundred eV. It is demonstrated that the restriction of the discharge gap in dielectric walls leads to the trapping of electrons in the gap and an increase in the efficiency of excitation and ionization of atoms of beam electrons, which leads to a substantial increase in the current density and intensity of optical radiation of the limited discharge. Experimental data obtained on the relaxation of charge density on the deposited surface of the charge limiter and the value estimates of the electric potential of the dielectric surfaces, were analysed. It is demonstrated that the variation in the discharge structure restricted by the dielectric walls corresponds to the variation in the distribution of the electric field in the gap, as well as under the influence of the surface potential.

1. Introduction

One of the significant features of high-voltage nanosecond discharges is that they are developing in gas pressure gaps, which form when high-energy electrons are generated in the discharge gap [1-5]. This phenomenon is of a fundamental importance since the high-energy electrons and X-ray radiation, which are generated during deceleration of high-energy electrons on the anode and in the gap, influence the ionization process in the gas; this consequently determines the pulse discharge characteristics. The discharge properties of the generated high-energy electrons are largely determined by conditions during the discharge development phase, electrode geometry and, to some extent, by the processes on the wall of the discharge chamber. In this regard, the nanosecond discharges by means of extending the cathode slits, where almost all the applicable voltage to the discharge gap is concentrated in a narrow cathode layer interspacing between the cathode plasma discharge chamber and the cathodic surface, are of particular interest [6-9]. The main function in the systematic establishment of these categories plays a developing role, which results in secondary electron emission on the cathode surface, the electrons acquired in the cathode layer being used up as an energy for excitation and ionization of the gas resulting from subsequent oscillations in the cathode chamber. The acceleration of electrons in the cathode layer, in order to redistribute the field in the discharge with a slot cathode, enables the reception of beams of high-energy electrons at much lower values of applied voltage. Such plasma-beam discharge (PBD) enables the generation of currents up to hundreds of amperes, as a result of medium and high gas pressures required to produce, for example, optical generation in gas lasers. In comparison with similar discharges, the advantages of PBD with a slot cathode are manifested in a significantly greater volume of discharge, low operating voltages, and the ability to control the beam properties.

The objective of the present work is an experimental study of the effect of limiting the discharge area of the dielectric walls on the generation of accelerated electrons and the main characteristics of nanosecond PBD with an extended slot cathode.

2. Experimental methods and technique

An automated experimental setup was used, the detailed description of which is given in [7]. The setup consists of a high-voltage nanosecond pulse generator (VPG), a discharge chamber (figure 1 (a)), pumping and gas pressure control systems, a system for inferring the electrical and optical characteristics of the discharge with PC registration of measurements. A Blumlein type VPG, enabling the voltage doubling, which generated the voltage pulses with a leading edge of 10–15 ns and had an amplitude regulator up to 10 kV and a pulse repetition rate of up to 100 Hz, was manufactured.

The discharge chamber consisted of a quartz tube of 5cm diameter and a length of 50 cm, into which an electrode system, consisting of two aluminium electrodes spaced by 0.6 cm, was placed. The cathode consisted of a cylindrical rod having a length of 40 cm and a diameter of 1.2 cm, along which a rectangular cavity (slit) of 0.2 cm width and a depth of 0.6 cm had been cut. The anode consisted of a flat plate with a length of 40 cm, a width of 2 cm, and a thickness of 0.5 cm (figure 1 (a)). In order to limit the discharge field, flat plates of a polished glass-bonded dielectric material were used. Limiters were installed along the electrode at a distance of 2 mm (the width of rectangular cavity in the cathode) from each other (see figure 1 (b)). The design of the discharge chamber enabled the spatial structure of the discharge to be observed, and the spatial distribution of optical radiation in the discharge gap and in the cathode chamber to be recorded.

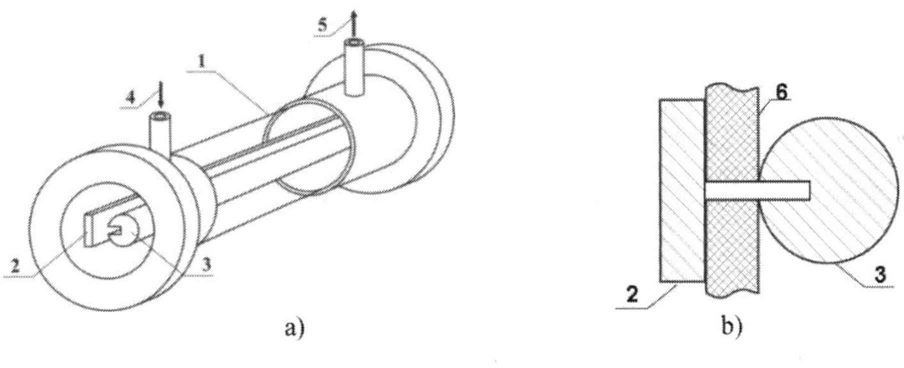

Figure 1. a – discharge chamber type (1 – quartz tube; 2 – anode; 3 – cathode; 4, 5 – gas inlet and outlet); b – discharge boundary schema (2 – anode, 3 – cathode; 6 – dielectric loading).

The discharge formation and its development dynamic were analyzed by means of the images produced by the high-speed time-lapse photography over time, using a Princeton Instruments PI-MAX3 ICCD Camera. Following the detection on the CCD, the image matrix was read with a controller, digitized, and transmitted to a computer for processing via a broadband connection.

In order to study the changes in the dielectric properties of the discharge limiters following their interaction with nanosecond discharge, the investigated dielectric samples were placed between two plates of a parallel-plate capacitor, as a dielectric. The facings of this capacitor are connected to the digital (WK6500P) capacitance and tgδ-dielectric loss meter. Measurements were carried out both at a fixed frequency and when scanning this frequency within the range from 1 kHz to 15 MHz. In the

IOP Conf. Series: Journal of Physics: Conf. Series **830** (2017) 012024 doi:10.1088/1742-6596/830/1/012024

work, the dielectric permittivity and tgδ of the dielectric loss of the discharge limiter were measured before and after the interaction with the nanosecond PBD.

3. Results and discussions

Systematic experiments were performed to study both the electrical and optical characteristics, and the dynamics of formation of the spatial structure of the discharge within the gap and the slot of the cathode, using helium, to measure the gas pressure variability in the chamber (p = 1-100 Torr), and the amplitude of the voltage applied to the electrodes (U_0 = 0.5 - 5 kV). In similar conditions, the properties of the dielectric material of the wall limiting the discharge were also examined before and after the interaction with the PBD.

Experimental investigations have shown that the oscillogram of the discharge current (I_{br}) and the burning voltage (U_{br}) differ greatly in the form, the magnitude, and the duration for unlimited and limited discharge with identical end values U_0 and p. Figure 2 shows an example of the oscillogram U_{br}, I_{br} and the corresponding image of the spatial distribution of optical radiation filming 70 ns after the beginning of the current pulse.

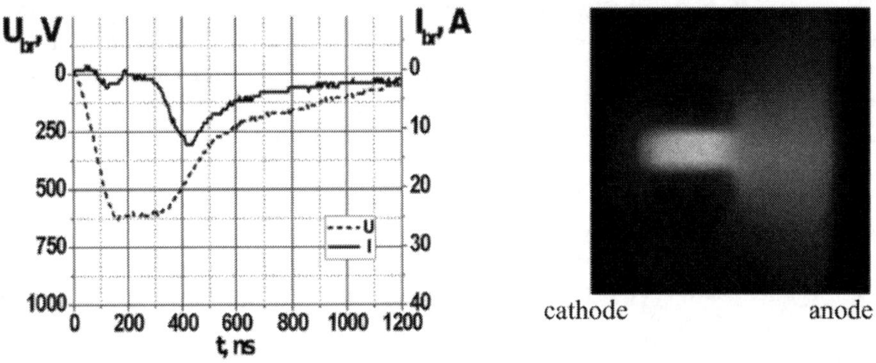

a)

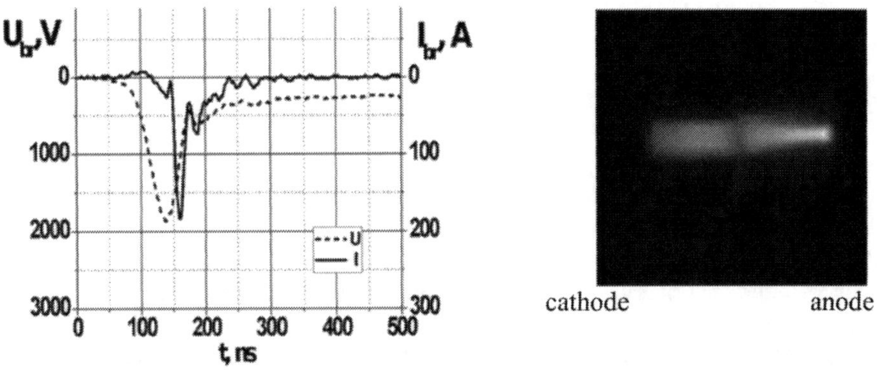

b)

Figure 2. Oscillograms of voltage and burning of the discharge current and the images of spatial distribution of optical radiation: a-unlimited discharge; b-limited discharge in helium (p = 10 Torr; U_0 = 1 kV)

In the figure, it is shown that the peak value of the pulse U_{br} is three times higher, and I_{br} is more than one order higher than a limited discharge. Limitation of the discharge gap leads to a substantial excitation of pulses U_{br} and I_{br}. The rate of increase of current rises, when the discharge limit reaches up to 5 · 10 10 A / s.

The research of the spatio-temporal formation dynamics of the optical discharge radiation has enabled the characteristics of the formation and development of unlimited and limited discharges in helium gas at moderate pressures to be established. In the unlimited discharge, a glow first appears on the cathode surface at the exit of the slot. Approximately 30 ns after the start of the current pulse, a diffused glow uniformly fills the cathode chamber and the gap between the electrodes. Subsequently, there is an increase in discharge emission intensity, which reaches a maximum at 65–70 ns. Moreover, the intensity of radiation in the cathode chamber grows faster and much higher in the active phase of the nanosecond discharge than in the gap between the electrodes (figure 3 (I)). When the gas pressure is raised above 40 Torr, the structure of the unlimited discharge changes significantly. Within the cathode gap, the discharge is pressed against the walls of the chamber and the dark space is formed towards the center, which comes to the surface of the anode. The limited discharge first glow appears simultaneously on the cathode surface and in the gap between the electrodes. Further, the discharge permeates the cathode chamber and increases the intensity of the discharge radiation. In the limited discharge, the opposite holds true: the radiation intensity grows faster and earlier in the limited space between the electrodes. Under certain external conditions, a focusing of the optical radiation and discharge is observed towards the center of the anode surface (figure 3 (II)). Under gradual increase $U_0 \geq 2.5$ kV at the focal site along the surface of the anode, an extensive bright spot appears, and the glow is filling the limited gap with maximum intensity in the center. The results of the study of the formation of optical radiation dynamics of the discharge show that the limiting of the discharge gap leads to a significant change in the spatial structure of the discharge and a redistribution of the intensity of the optical radiation in the discharge gap.

I

II

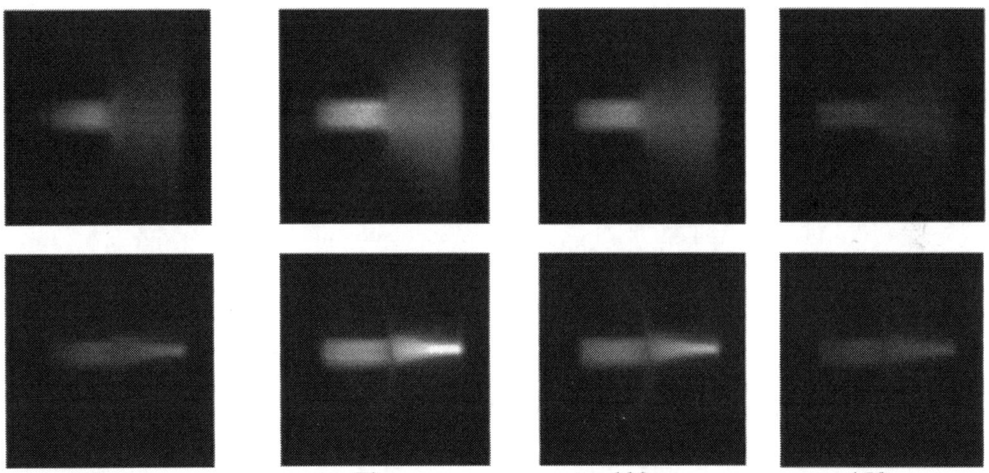

t = 40 ns t = 70 ns t = 100 ns t = 150 ns

Figure 3. Optical image: I - unrestricted discharge; II - restricted discharge in helium (p = 20 Torr; U_0 = 1 kV)

A comparative analysis of the primary parameters of unlimited and limited discharges was conducted under the same external conditions. The density of the discharge current $j_D = I_{br}/ S$ was estimated from experimental values I_{br} at the maximum cross-sectional area of the discharge S. At the moment when the gas pressure reached p = 10 Torr, the value, which shows that the length of the cathode potential drop (CPD) area $d_c \approx 0.01$ cm, and the length of the mean free path of electron, is $\lambda = 1 / (N\sigma_0) \approx 0.08$ cm [8]. Since $\lambda > d_c$, electrons emitted from the cathode surface pass a check point area without collisions and gain the energy $W = eU_c$, where $U_c = 3 \cdot U_{br}/ 5$ [9].

Oscillograms $I_{br}(t)$ and $U_{br}(t)$ can be used to estimate the electron beam current density ($j_e(t)$). It is known, the anode current density is expressed by the formula: $j_D = en_e v_{dr}$, where n_e is the number of electrons formed in the plasma beams and by means of secondary electrons when colliding with gas atoms. As it is shown in the table, the beam's electrons are accelerated and gain the energy of hundreds eV. These accelerated beam electrons lose their energy ionizing the gas atoms during collisions with them in the gap, and the total number of created secondary electrons can be evaluated by means of the ratio: $z_i \approx W / E_I$, where E_I is the energy created by a pair of electron - ion (for helium $E_I = 46$ eV). Assuming that the anode current is exclusively conditioned by secondary electrons, and proceeding from the condition of continuity of current in an electric circuit, we can then approximately evaluate $j_D \approx z_i j_e$. The evaluations show that the discharge limiting leads to an elevated efficacy of ionization of atoms by accelerated electrons; this, in turn, results in the increase of discharge current density. The most favorable conditions for the formation of accelerated electron beams are developed in an unrestricted discharge, where the magnitude of the beam current at the anode surface reaches up to 20% of the discharge current (see table 1).

Table 1. The main parameters of the discharge in helium are p = 10 Torr.

		unlimited discharge		
U_0, V	1000	2000	3000	4000
U_{br}, V	390	750	960	1120
J_e, A / cm^2	2.0	13.3	25.0	50.0
W, eV	234	450	576	672
J_e, A / cm^2	0.4	1.3	3.0	3.4
J_e/j_D	0.20	0.10	0.08	0.07
		limited discharge		
U_0, V	1000	2000	3000	4000
U_{br}, V	750	1000	1200	1300
J_e, A / cm^2	45.0	107.5	120.4	125.0
W, eV	450	600	720	780
J_e, A / cm^2	4.6	8.3	7.7	7.4
J_e/j_D	0.10	0.08	0.06	0.06

Such considerable changes in the electrical, optical characteristics and spatial structure of the unrestricted and restricted discharge patterns are caused by a change of the electric field spatial distribution in the discharge gap under discharge limitations associated with any number of surface phenomena at the dielectric-plasma boundary.

The dielectric properties of the wall material limiting the discharge dependent on the time function and the degree of interaction with the PBD were studied, in order to establish this connection. Through the interaction of PBD on the dielectric wall limiter, some quantity of dielectric charge is captured. The sign of the developing charge coincides with the sign of the particles charge bombarding the surface. The quantity of the captured charge depends on the energy current density of charged particles settling on the surface, and the duration of pulse discharge. To determine the surface charge density at a low frequency range, the ratio: $\sigma \approx \omega\varepsilon_0\varepsilon \mathrm{tg}\delta$ can be used, correlating $\omega \mathrm{tg}\delta$ with its conductivity [10].

Estimates of the captured charge density using measurements of the values tgδ and ε enabled the construction of a relaxation curve for the captured charge density material following the cessation of interaction with the PBD at a determined frequency (figure 4):

$$n \approx \frac{\omega \varepsilon \varepsilon_0 \mathrm{tg}\delta}{e\mu} \quad , \tag{1}$$

where μ - the mobility of electrons in glass-bonded dielectric material.

The estimates show that the charge density of captured dielectric material in the discharge reaches 10^9 cm^{-3}, after the cessation of interaction with PBD a relaxation surface charge density occurs and the process duration lasts tens of minutes (figure 4).

The change in the electrical characteristics, while limiting the discharge, can be explained by means of the following: when limiting the discharge by dielectric walls, a portion of the line of electric force becomes isolated on the negatively charged surface of the limiter; in order to penetrate into the cathode field, a higher field value is required, which leads to an increased burning voltage value. With more efficient ionization of emitted atoms by the electron limiter surface, free electrons, and free electrons captured in the discharge gap in the case of limited discharge, the density of plasma between the electrodes is much higher. Before this dense plasma attains its respective field values, it penetrates quickly into the cathode chamber, which leads to a faster growth and higher values of the discharge current than in the case of unrestricted discharge (figure 2).

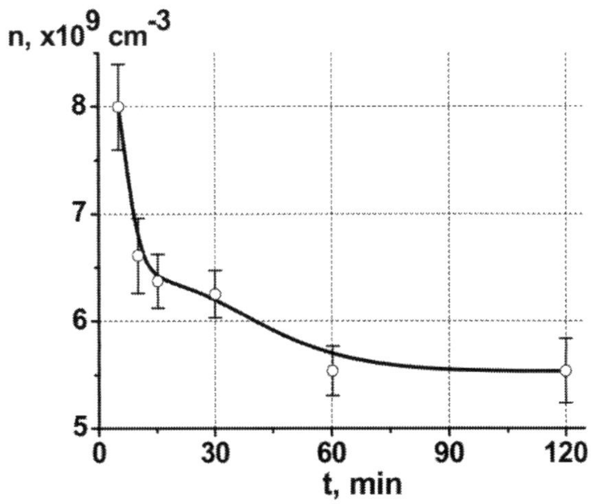

Figure 4. Density relaxation of captured charge material after cessation of interaction with the PBD when pulses surpass a frequency of 100 Hz

Changes of the discharge spatial structure located along the center of the gap with a focus on the anode surface, and similarly the redistribution of radiation while limiting the discharge are primarily due to a change in the electric field distribution in the gap. The charge is obtained due to irradiation of the dielectric surface by an electron beam. The charging time of the dielectric surface depends on the conditions of its irradiation and ends at such a time when deep traps are completely filled [11-14]. The total electric charge Q, resulting from capture of the accelerated electrons by traps of the dielectric, during the process of their interaction with the PBD on the dielectric surface, gives rise to the emergence of a surface potential which exhibits a significant influence on the distribution of the electric field in the limited discharge gap. The surface potential of the limiter can be evaluated by means of the following relation [11]:

$$V_s = \frac{Q}{\pi\varepsilon_0(\varepsilon+1)R} \quad , \tag{2}$$

where ε_0 and ε is the dielectric penetrability of the vacuum and dielectric, and R is the path depth of primary electrons. The depth of penetration of primary electrons depends on the accelerating voltage and on the element composition of the glass-bonded dielectric material, which, in our experiment, was a few microns thick. Under the experimental conditions influenced by PBD, the limiter surface charges negatively with a value of the surface potential within several hundred volts. Consequently, electrons are reflected from the negatively charged limiter wall and under the influence of the changed field distribution in the gap, are focused towards the center of the gap. This leads to a change in the spatial structure of the discharge due to its limitations. The discharge gradually contracts toward the center and is focused in a line along the anode surface (figure 2 (b)).

4. Conclusions

The experimental research and evaluations made thereon show that electron ribbon beams with an energy of several hundred eV were formed in the nanosecond discharge with the help of an extended slot cathode. It is demonstrated that the restriction of the discharge gap in the dielectric plates leads to electrons capturing in the gap and the increase of excitation and ionization efficiency of beam electron atoms, which also leads to an increase of the current density and intensity of optical radiation. The relaxation of the captured dielectric surface charge density takes tens of minutes, which affects the electric field distribution between the electrodes and causes a change in the spatial structure of the discharge.

Acknowledgments

This work was financially supported by the project part of the state assignment of the Ministry of Education and Science of Russia in scientific activities, project 3.1262.2014K.

References

[1] *Generation of runaway electrons and X-rays in the discharges of high pressure* 2015 ed V F Tarasenko (Tomsk: STT) p 568
[2] Mesyats G A 2005 *Pulsed Power (Springer US)* 568
[3] Naidis G V and Walsh J L 2013 *J. Phys. D: Appl. Phys.* **46** 095203
[4] Tarasenko V F and Yakovlenko S I 2004 *Phys. Usp.* **47** 887
[5] Babich L P, Bochkov E I and Kutsyk I 2014 *JETP Letters* **99**(7) 386–90
[6] Ashurbekov N A and Iminov K O 2015 *Technical Physics* **60**(10) 1456-63
[7] Ashurbekov N A, Iminov K O, Ramazanov A R and Shakhsinov G Sh 2014 *Technical Physics Letters* **40**(8) 685-88
[8] Ashurbekov N A, Iminov K O, Kobzev O V and Kobzeva V S 2010 *Technical Physics. The Russian Journal of Applied Physics* **55**(8) 1138-44
[9] Ashurbekov N A, Iminov K O, Kobzev O V and Kobzeva V S 2007 *High Temperature* **45**(4) 433 – 38
[10] *The electrical properties of polymers* 1986 ed B Sajin (Leningrad:Chemistry) p 224
[11] Cazaux J 1999 *J. Appl. Phys.* **85** 1137
[12] Cazaux J 2006 *Nucl. Instrum. Meth. Phys. Res. B* 244 307
[13] Song Z G, Ong C K and Gong H 1996 *J. Appl. Phys.* **79**(9) 7123-28
[14] Melchinger A and Hofrnann S 1995 *J. Appl. Phys.* **78** 6224-32

Activation of the explosive-emission-cathode in conditions of artificial initiation of field emission

S A Shunailov[1], M I Yalandin[1], K A Sharypov[1], M D Kolomiets[1], M R Ulmasculov[1], V G Shpak[1], V V Rostov[2] and G A Mesyats[1,2,3]

[1]Institute of Electrophysics UB RAS, 620016, Ekaterinburg, Russian Federation
[2]Institute of High Current Electronics SB RAS, 634055, Tomsk, Russian Federation
[3]P.N. Lebedev Physical Institute RAS, 119991, Moscow, Russian Federation

E-mail: ssh@iep.uran.ru

Abstract. Low inertial transition of the field emission into explosive electrons emission at the cathode and related fast-rise-time electron beam formation are of interest in a phase-stable excitation of relativistic high-power microwave oscillators. We present experiments on the cathode activation using emission initiation during specially prepared advance voltage pulse (prepulse) possessing variable amplitude of tens-hundreds kilovolts and the width adjustable from tenths to units of nanosecond. Such a prepulse can be separated in time with a main accelerating pulse as well as adjoined to the fast-rise-time voltage front.

1. Introduction

For fixing the phase of relativistic microwave generator towards the front of the voltage at the cathode [1,2] a high-current electron beam with a subnanosecond rise time and a high current rate (dI_e/dt) up to 10^4 A/ns is required [3-6]. This justifies the relevance of studying the dynamics of explosive electron emission (EEE) cathodes [7] in the interval <1 ns and conditions of its initiation, which affect the front beam stability and the magnitude of dI_e/dt. Modern technique of forming high-voltage pulses [8-11], methods of checking the characteristics of the vacuum diode (VD) [12], beam current broadband sensors [13], and modern digital oscilloscopes allow picosecond resolution to study the dynamics of the field emission (FE) into EEE stage for microscopic cathodes with different shape and amplitude of applied voltage pulses [14-16]. This work is devoted to the research of the effects of double pulses [17] - activation of EEE on the cylindrical edge of the cathode with the leading voltage pulses. These effects were studied for the voltage pulses of not less than one hundred kilovolts, with subnanosecond fronts, comparable pulse durations and for different intervals between them. Cathode material, VD geometry and the parameters of the external magnetic field were typical for microwave experiments.

2. Experimental arrangement

Figure 1 presents the layout of experimental set-up, where we have used the special formers of a short high voltage pulses (3,8,10) [8-10]. The installation was operated in three variants, but the gaps of RADAN-303 drivers were always stabilized by triggering pulses from the generator 2. In option #1 (components 8, 9-11 were not used), the cathode was fed by pulse with tunable amplitude up to -330 kV and a duration of $\geq$ 1 ns. Switching the nitrogen spark gap under the pressure of up to 70 atm in the inductive-capacitive former 3 provided the front voltage shorter than 200 ps, but there was a nanosecond prepulse (tens of kilovolts) because of the interelectrode capacitance. The duration of the main pulse could be shortened to hundreds of picoseconds (FWHM) with a cut-off discharger at the output of block 3. The transmission line (44 Ohm) between VD 5 and the driver 3 had such a delay that the first unipolar reflected pulse was supplied to the cathode 6 in ~8 ns.

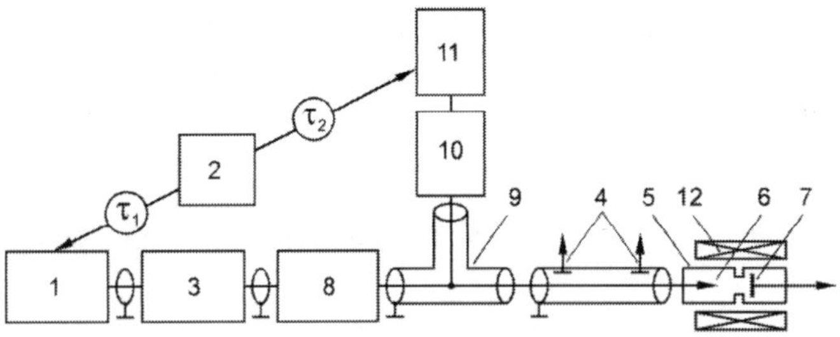

Figure 1. Layout of experimental set-up. *1,11* – nanosecond driver RADAN-303; *2* – triggering pulse generator; *3* – pulse forming L-C network; *4* – capacitive dividers of DTDR section; *5* – vacuum diode; *6* – cathode; *7* – beam collector; *8* – NLTL; *9* – high-voltage T-branch; *10* – pulse slicer; *12* – pulsed solenoid.

Option #2 amended by block 8, representing the gyromagnetic nonlinear transmission line with saturated ferrite (NLTL, [10]). NLTL transformed the pulse from generator 3 in the sequence of decreasing amplitude oscillations with a frequency of ~2 GHz, that is, the interval between the first peaks was ~500 ps. Increasing the delay before initiating pulse and restructuring its shape/length (Option #3) was provided with inclusion of units (9-11). Meanwhile, the delay from the central connection point of the high-voltage T-branch 9 to blocks 8 and 10 determined the valid ahead of the leading pulse from the generator 10 without the influence of reflections on the main pulse delivered to the cathode from the block 8. Fixed change in the delay between pulses was carried out by changing the lengths of the transmission lines from the generator 2, that is, the time τ_1 and τ_2.

Pulse solenoid 12 provided the magnetic field of ~ 2 T in the VD. In the experiments the sequences of the voltage pulses were recorded with identical capacitive dividers 4. For these signals by the method of dynamic time-domain reflectometry (DTDR the current and VD impedance was determined [12]. In part of the experiments, the electron beam current was recorded by the collector sensor 7 [13] with a resolution of ~20 ps.

3. Cathode emissivity variation under action of sequential voltage pulses

For the dynamics of the cathode emission properties the polishing effect [18] is typical. With the increase in the number of HV pulses (N) the morphology changes of the emitter affect at the time and amplitude characteristics of the VD current [14,15,19]. The effect is most pronounced for the cathodes made from flexible materials, such as steel [19,20], and at times less than a nanosecond, when the mechanism of regeneration of the emitters is not valid. At the stage of transition of FE into EEE for individual microemitters the condition $j^2t_d \approx const$ known in the EEE theory is in force. Due to the loss of mass and fusion (smoothing) of microemitter the field amplification is reduced, the density of emission current (j) decreased nonlinearly, and as a result, the EEE delay (t_d) may gradually exceed the duration of the voltage and current are generally not recorded [21].

The situation changes if a second voltage pulse with nanosecond delay is applied to the cathode, and this refers to the effects of a "double pulse" [17]. During the first voltage pulse FE occurs at the cathode and maybe EEE, but the field on the emitter in the interval between pulses is missing. Therefore, an important factor is the residual temperature of the emitter, dynamics of dissipation, as well as the presence and expansion of plasma formation [17], occurring on a leading pulse. That is, the level of cathode activation in the first pulse and the time interval to the second are important.

From the expanding plasma, the second pulse pulls the electron current. As a result, plasma formation near the center of the leader charges positively and increases the field in the area of adjacent emitters facilitating their electrical blast. Thus, both factors (the expansion of the plasma and field amplification) are provided a larger emission of current at the second pulse even when its amplitude is smaller than the first one. This mode is illustrated in figure 2, which used the installation option #1 with cut-off discharger block 3 and the steel tubular cathode (flange 100 μm). The ratio of the amplitudes of two pulses at the cathode was almost the same in all cases and is close to (U_1/U_2), since the maximum electron beam current (55 A) is not great. Since $U_2<U_1$, then before training (N=1) $I_{21}\approx 0.7I_{11}$. One hundred pulses were enough to obtain $(I_{12}/I_{11}) < 2$ (polishing!), $I_{22}\approx 1.2I_{21}$, and, most importantly, $(I_{22}/I_{12}) > 2$. With further training (up to N=300) the polishing effect reduced the current of the first pulse more than 20 times as compared to I_{11}. The polishing effect is beginning to emerge at the second pulse also. The current here is decreased, but it exceeded by 5 times the first one.

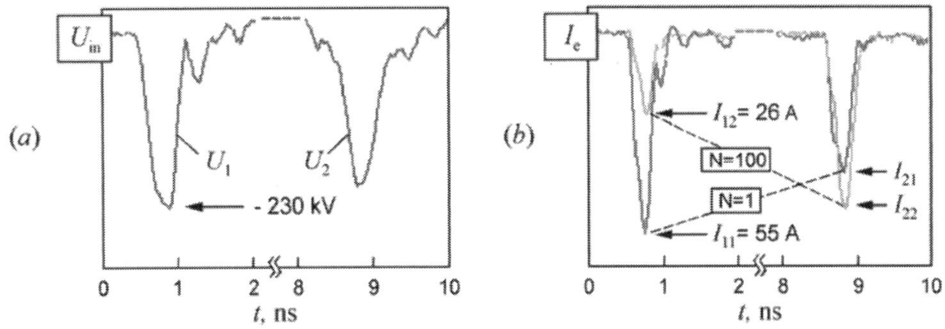

Figure 2. (a) High-voltage pulses U_1 and U_2 applied to stainless steel tubular cathode. (b) Beam current measured by collector for a virgin cathode (I_{11} and I_{21}) and after N=100 shots (I_{12} and I_{22}).

The effect of "polishing" the surface is also observed for the cathodes made of graphite, although it manifests itself with a much larger number of shots [14,19]. When exposed to sequential pulses of voltage, as in figure 2,a, the tendency of changes in current amplitudes is preserved, although the differences ($I_{11} \rightarrow I_{12}$) and ($I_{12} \rightarrow I_{22}$) were significantly less. For a single voltage pulse of comparable duration (~300 ps FWHM), the ratio of $(I_{12}/I_{11}) \sim 0.5$ is reached after N~10^4 shots [22]. With the same N the delay growth of the front current at ~200 ps was observed when the duration of the voltage (FWHM) was ~1.5 ns [16]. However, the decline in the amplitude of the current in the latter case does not exceed 10%. We believe that in contrast to steel, the change in the morphology of the edge of a virgin graphite cathode has been associated primarily with entrainment of material in electro explosions [15], where the most small-scale microemitters disappear and a new inhomogeneity with a small structure reveal.

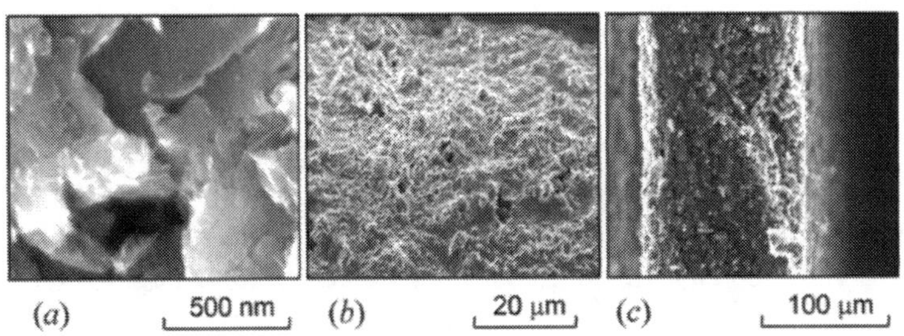

Figure 3. Microphotography of the graphite cathode edge.

4. Graphite cathode emissivity with affect of advance voltage pulse

In the VD with tubular graphite cathode, there are several characteristic spatial scales: the size of microemitters (l_1<100 nm) representing the edge, the edges of the plates, the faces of the crystallites, as shown in figure 3,a; the distance between the protruding agglomerate emitters ($l_2 \leq 10$ μm, figure 3,b); emissive edge of the cathode ($l_3\approx$100 μm, figure 3,c); the gap between the cathode and the anode (l_4~1 cm). The source of electrons is the plasma after the electric explosion of microemitters expanding with a typical speed ~ 2x10^6 cm/s. Estimating the time of plasma expansion on these scales l_i, it is possible to get times τ_i, in which the cathode emissivity will be different. The set of these times includes the intervals t_1~5 ps, t_2~500 ps, t_3~5 ns, and t_4~500 ns.

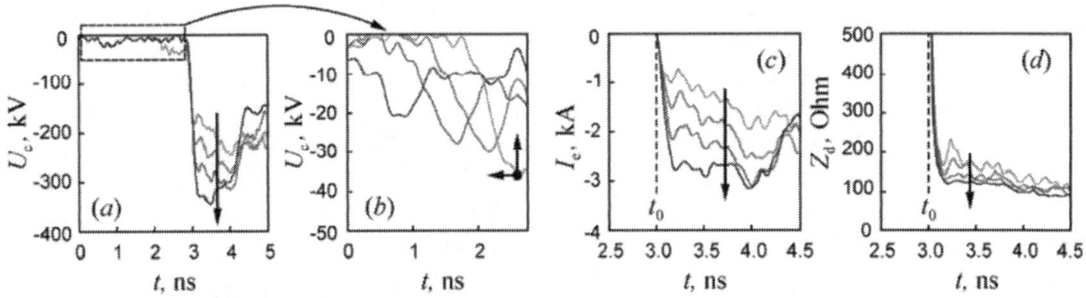

Figure 4. Voltage pulses (a) and prepulses (b) at the cathode. (c) Emitted current. (d) VD impedance. All DTDR-restored waveforms have absolute time reference.

Figure 4 presents an experiment with a graphite tubular cathode on the installation for version #1 (without the chopper spark gap). With the increase of the peaking discharger gap, the voltage breakdown and the amplitude of the generated pulse are increasing (figure 4,a). This reduces the amplitude of the prepulse and increases its duration (figure 4,b). From the abovementioned rule $j^2t_d \approx const$ it follows that in a certain time interval the relatively low prepulse (figure 4,b) does not provide transition of FE into EEE stage. The emitted current during the time of their influence reaches not more than a dozen(s) amperes and does not exceed the oscillographic registration noise. However, the current waveforms in figure 4,c show that the activation of the cathode occurs during the prepulse. With increasing of the prepulse duration (figure 4,b) dI_e/dt at the front of the current pulse at the level of 500 A varied from 15 kA/ns to 22 kA/ns. In addition, for all the voltage amplitudes of the main pulse with the presence of the prepulse leads to a low-inertia emission of kiloamper current at the time of the order of ~100 ps (the beginning of this interval is linked to $t_0 = 3$ ns). Apparently, it is connected with the current selection from micron plasma formations close to the individual emitters or the most protruding agglomerates of such emitters in the interval shorter than t_2. The low inertia of pre-activated emitters leads to the fact that in ~200 ps the envelope of the current (figure 4,c) repeats the burst and the local downturn in the voltage front (figure 4,a). These characteristics are correct for variations of $Z_d(t)$ (figure 4,d) as well. Previously, we observed the effect of activation of the graphite cathode with different prepulses for close amplitudes and durations of sub-nanosecond high-voltage pulses [23]. The effect was to increase the amplitude of the current and the charge of generated beam by a longer activation of the cathode on the prepulse. However, it is clear that, ceteris paribus, the voltage difference during activation of the cathode is a more fundamental factor than the time of exposure. This is due to the nonlinear dependence of emission current from the electric field at the emitter. As a result, with the smallest voltage amplitude (figure 4,a) after ~ 1 ns the VD impedance has not been established. If the largest voltage the impedance at t > 4 ns (figure 4,d) is almost stabilized. Apparently, the influence of plasma expansion is already expressed, which in a few nanoseconds ($t \rightarrow \tau_3$) including the duration of the prepulse can cover not only the area between the agglomerates' emitters (figure 3,b) but the emissive edge of the cathode as a whole (figure 3,c).

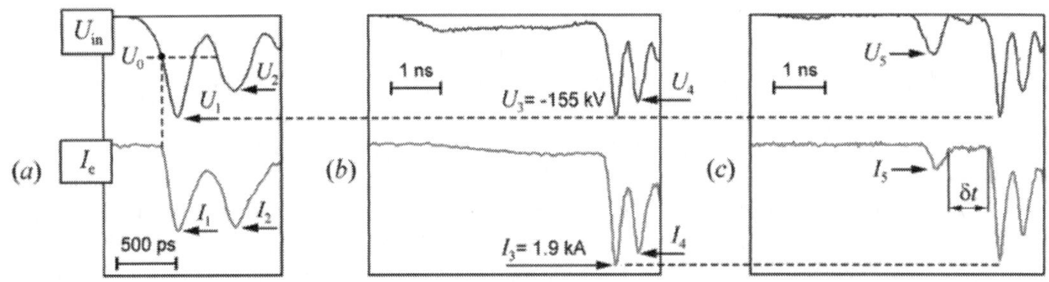

Figure 5. The emission of the beam when applying pulses with a delay of 500 ps (a), for activation of the cathode by a prepulse (b) and with a short pulse (c). The current registration – the collector sensor.

Complementary effect on the VD current of electric field exposure, its magnitude and EEE-plasma dispersion after the advancing voltage pulse is observed in the experiments shown in figure 5. When there is no prepulse (set-up option #2), its role is play the first surge delivered from NLTL (figure 5,a). If $(U_2/U_1)\approx 0.7$ at the second pulse the increased current $(I_2/I_1)\approx 1$ is emitted. There is an obvious role of plasma expansion in the interval of t_2~500 ps, and the activation of a larger number of EEE centers in the field, strengthened by the expanding plasma. Note the fact of the inertia of the emission of the cathode in the presence of prepared plasma: between the peaks the voltage falls below the voltage level of the initial activation of the cathode (U_0), but the current does not stop. The second important point is that the current pulse I_1, that occurs when a significant voltage on the cathode (-60 kV) has a fairly steep front: $dI_e/dt \approx$ 14.5 kA/ns at the level of I_1=500 A.

When all installation components are in operation (option #3) it is possible to form the elongated prepulse (figure 5,b). EEE-current occurs during the prepulse when the front passes at the “shelf”. This occurs when the voltage U_0 is by half less than in the case in figure 5,a, but the time of voltage rise until the current appears (~1 ns) is by 5 times more. For 2.5 ns the current during the prepulse increases and then there is a tendency for its decrease because of the decline of the prepulse voltage. In this mode of the cathode activation in the first voltage peak the maximum current $I_3\approx$1.9 kA ($dI_e/dt \approx$ 6.5 kA/ns at the level of I_e=500 A) is emitted and the VD impedance is stabilized. Indeed, $(U_4/U_3)\approx(I_4/I_3)$ is achieved. Almost the same ratio of the peaks of current and voltage, and the maximum amplitude of the current is maintained when the cathode is fed by a short lead pulse (figure 5) with the amplitude U_5 increased in comparison with a prepulse in the previous case. Large enough EEE - current $I_5\approx$400 A corresponds to U_5 pulse. On the δt interval this current practically disappears, but the plasma expansion and a greater number of activated "hot" centers lead to the effective emission of the beam on subsequent pulses.

5. Conclusions

In the conclusion, advance high-voltage pulses demonstrate pronounced influence on the cathode emission initiation. This effect includes preliminary transition of FE into EEE stage and depends on the advance voltage value (i.e., E-field) applied, a spread of the EEE plasmas nearby emitters, and amplification of the microscopic E-field strength by charged plasma clouds resulted in activation of even more EEE-centres. As a result, electron beam pulse may achieve the rate of current rise exceeding 20 kA/ns which is attractive in the tasks of phase-stable high-power microwave generation.

Acknowledgements

The work was supported by the research program of IEF UB RAS project №0389-2014-0005, and the grant of RFBR №14-08-00111. The authors note a contribution of Ural Federal University’s electrophysics department to the possibility of performing test measurements using the Tektronix DPO73304D ultra-wideband oscilloscope.

References

[1] Afanas'ev K V, Bykov N M, Gubanov V P, El'chaninov A A, Klimov A I, Korovin S D, Rostov V V and Stepchenko A S 2006 *Tech. Phys. Lett.* **32**(11) 925-27

[2] Grishin D M, Lyubutin S K, Mesyats G A, Rostov V V, Rukin S N, Slovikovsky B G, Timoshenkov S P, Ul'maskulov M R, Sharypov K A, Shpak V G, Shunailov S A and Yalandin M I 2008 *Tech. Phys. Lett.* **34**(10) 822-24

[3] Rostov V V, Elchaninov A A, Romanchenko I V and Yalandin M I 2012 *Appl. Phys. Lett.* **100** 224102

[4] Rostov V V, Elchaninov A A, Romanchenko I V, Shunailov S A, Ul'maskulov M R, Sharypov K A, Shpak V G, Rukin S N and Yalandin M I 2014 *Radiophysics and Quantum Electronics* **56**(8-9) 475-91

[5] Sharypov K A, El'chaninov A A, Mesyats G A, Pedos M S, Romancheko I V, Rostov V V, Rukin S N, Shpak V G, Shunailov S A, Ul'masculov M R and Yalandin M I 2013 *Appl. Phys. Lett.* **103** 134103

[6] Ginzburg N S, Cross A W, Golovanov A A, Mesyats G A, Pedos M S, Phelps A D R, Romanchenko I V, Rostov V V, Rukin S N, Sharypov K A, Shpak V G, Shunailov S A, Ulmaskulov M R, Yalandin M I and Zotova I V 2015 *Phys. Rev. Lett.* **115** 114802

[7] Mesyats G A, Bugaev S P and Proskurovskii D I 1972 *Sov. Phys. Usp.* **14**(4) 536-37

[8] Mesyats G A, Shpak V G, Shunailov S A and Yalandin M I 1994 *Proc. SPIE Int. Symp.: Intense Microwave Pulses* vol **2154** 262–8

[9] Yalandin M I, Lyubutin S K, Oulmascoulov M R, Rukin S N, Shpak V G, Shunailov S A and Slovikovsky B G 2002 *IEEE Trans. Plasma Sci.* **30** 1700-4

[10] Romanchenko I V, Rostov V V, Gubanov V P, Stepchenko A S, Gunin A V and Kurkan I K 2012 *Rev. of Sci. Instr.* **83** 074705

[11] Gusev A I, Pedos M S, Rukin S N, Timoshenkov S P and Tsyranov S N 2015 *Rev. of Sci. Instr.* **86** 114706

[12] Sharypov K A, Ul'masculov M R, Shpak V G, Shunailov S A, Yalandin M I, Mesyats G A, Rostov V V and Kolomiets M D 2014 *Rev. Sci. Instr.* **85** 125104

[13] Yalandin M I, Reutova A G, Sharypov K A, Shpak V G, Shunailov S A, Ul'maskulov M R, Rostov V V and Mesyats G A 2010 *IEEE Trans. On Plasma Sci.* **38** 2559–64

[14] Korovin S D, Litvinov E A, Mesyats G A, Murzakaev A M, Rostov V V, Shpak V G, Shunailov S A and Yalandin M I 2004 *Tech. Phys. Lett.* **30** 813-16

[15] Korovin S D, Litvinov E A, Mesyats G A, Rostov V V, Rukin S N, Shpak V G and Yalandin M I 2006 *IEEE Trans. On Plasma Sci.* **34** 1771-76

[16] Belomyttsev S Y, Rostov V V, Romanchenko I V, Shunailov S A, Kolomiets M D, Mesyats G A, Sharypov K A, Shpak V G, Ulmaskulov M R and Yalandin M I 2016 *J. Appl. Phys.* **119** 023304

[17] Bazhenov G P and Rotshtein 1974 *Powerful pulsed sources of accelerated electrons* ed G A Mesyats (Novosibirsk: Nauka) pp 67-71

[18] Mesyats G A, Proskurovskii D I, Yankelevich E B and Tregubov V F 1976 *Sov. Phys. Dokl.* **21** 228

[19] Gunin A V, Landl' V F, Korovin S D, Mesyats G A and Rostov V V 1999 *Tech. Phys. Lett.* **25** 922-26

[20] Yalandin M I, Mesyats G A, Rostov V V, Sharypov K A, Shpak V G, Shunailov S A and Ulmaskulov M R 2015 *Appl. Phys. Lett.* **106** 233504

[21] Mesyats G A, Yalandin M I, Reutova A G, Sharypov K A, Shpak V G, Shunailov S A 2012 *Plasma Phys. Reports* **38**(1) 29-45

[22] Yalandin M I, Reutova A G, Ul'maskulov M R, Sharypov K A, Shpak V G, Shunailov S A, Klimov A I, Rostov V V and Mesyats G A 2009 *Tech. Phys. Lett.* **35** 804-7

[23] Mesyats G A and Yalandin M I 2005 *Phys. Usp.* **48**(3), pp. 211-229

IOP Conf. Series: Journal of Physics: Conf. Series **830** (2017) 012026 doi:10.1088/1742-6596/830/1/012026

The role of high-energy electrons in the formation of the transverse profile of high-speed ionization wave fronts in gases

N A Ashurbekov, K O Iminov and G S Shakhsinov

Dagestan State University, 43a M. Gadjieva St., Makhachkala, 367002, Russia

Email: nashurb@mail.ru

Abstract. A comprehensive experimental study of the spatio-temporal structure of the fronts of discrete, high-speed ionization waves (HSIW) in shielded discharge tubes filled with inert gases is carried out. It is shown that the ionization wave front at low gas pressures (below 1 Torr) has a volumetric structure, but at gas pressures above 10 Torr assumes a cylindrical form, compressed into the dielectric boundary of the discharge tube. The high-energy electrons generated at the ionization wave front have a significant impact on its structure and their energy relaxation modes. Evaluations of the energies of electrons accelerated at the HSIW front are provided along with an analysis of the influence of the energy relaxation regimes of the high-energy electrons on the structure and dynamics of the development of the ionization wave front in the shielded dielectric discharge tubes.

1. Introduction

High voltage nanosecond discharges, in which internal electric fields exceed the critical value required to transfer electrons in continuous acceleration mode, are characterised by intensive generation of high-energy electrons in the electrical breakdown of the gas. There exist various discharge systems, in which electron beams or a group of high-energy electrons are formed during electrical breakdown in a gaseous environment [1-9]. One such type of discharge is the nanosecond electrical discharge in long shielded dielectric tubes, which is developing under conditions of the formation and propagation of high-speed ionization waves [7–11]. Such discharges have a number of features related to the mechanism of gas breakdown by high-speed ionization waves. This primarily relates to the spatio-temporal structure of the discharge, the complexity of which is due to non-simultaneity and heterogeneity of gas ionization along the discharge tube. It is known, the application of high voltage pulses with a nanosecond rise time leads to the formation of the ionizing potential gradient waves in the discharge tube propagating from the high voltage to the grounding electrode with a velocity of 10^8–10^{10}cm/s. If the high-speed ionization wave front duration is significantly lower than the ionization wave propagation time from one electrode to another, the gas excitation along the discharge tube takes place in the direction from the high-voltage to the grounded electrode with a delay determined by the HSIW velocity. The localisation of a strong electric field at the HSIW produces the final source of high-energy electrons.

The aim of this study is to investigate the role and energy relaxation modes of high-energy electrons in the formation of the spatio-temporal structure of the ionization wave front in shielded dielectric discharge tubes filled with inert gases.

2. Methods and experimental techniques

In order to study the formation dynamics of the high-speed ionization wave front structure, the authors of this paper studied nanosecond discharges in shielded glass tubes fitted with internal electrodes having a length of 50 cm and an internal diameter of 8 mm. A schematic view of the gas-discharge system is represented in figure 1. Aluminium electrodes were constructed in the form of hollow cylinders, through which the optical discharge radiation was recorded along the tube. Discharge tubes were placed inside a metal screen having a diameter of 2 cm. In order to record the electrical characteristics of the HSIW, several capacitive sensors were placed along the discharge tube at predefined distances from each other. In order to eliminate the influence of edge effects, edge sensors were installed on the borders of the screen at a distance greater than double the diameter of the metal screen. A light sensor in the form of a length of optical fibre was installed next to each capacitive sensor in order to study movement of the optical radiation of the high-speed ionization wave front along the tube. The signal from a 50-Ohm low-induction resistance shunt, included between the screen and the grounded electrode, was used for measuring the rate of discharge. A wideband digital recording device with a frequency band of 350 MHz was used as a recording system.

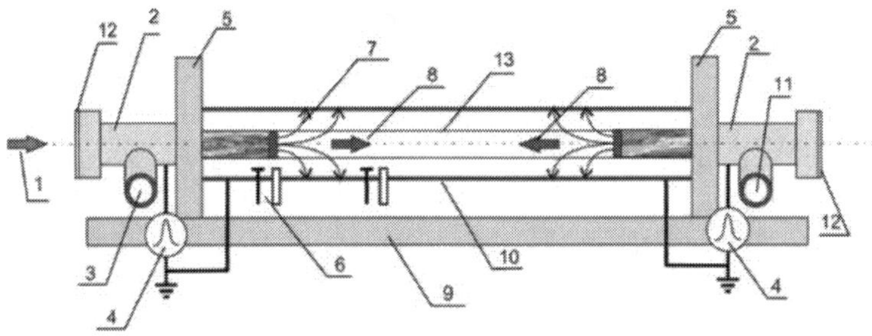

Figure 1. Flowchart of the discharge tube and the formation of HSIW. 1 – laser beam; 2 – hollow electrodes; 3 – input gas valve; 4 – nanosecond pulse generator; 5 – discharge limiters (isolator); 6 – capacitive and optical sensors; 7 – electric field lines; 8 – direction of HSIW propagation; 9 – stand (isolator); 10 – metallic screen; 11 – output valve of vacuum pump; 12 – quartz windows; 13 – quartz tube.

A block diagram of the experimental setup is shown in figure 2. A high-voltage pulse generator (VPG) for the formation of HSIWs was assembled on the transformer circuit according to a coaxial design in which the primary winding was composed of four turns and the secondary of two coils with 12 turns each. A TGI-130/10 ceramic thyratron was used as a switching device in the VPG. This generator allows two synchronised high-voltage pulses to be simultaneously produced with adjustable amplitude up to 40 kV, a repetition rate of 100 Hz, and a voltage pulse duration at half maximum of about 100 ns.

The system of capacitive and optical sensors allowed the velocity of HSIW propagation to be determined by measuring the delay time of signals from any two sensors. In addition, an analysis of the amplitude of the signals from the capacitive sensors allowed the attenuation coefficients of the HSIW to be determined in the process of propagation of ionization wave along the tube.

The density of the electrons behind the HSIW front was evaluated relative to the magnitude of the reverse shunt current and HSIW damping coefficient. The voltage across the discharge gap was measured using a calibrated voltage divider.

The transverse spatiotemporal structure of the HSIW front was investigated using a high-speed photographic recorder based on the ICCD of a Princeton Instruments PI-MAX3 camera model, allowing the capture of images with exposure times of 2 ns showing the optical emission discharge structure of a cross-section of the discharge tube.

Since metastable atoms play an important role in the process of gas ionization in the inert gases, the occupation dynamics of the metastable states of neon and helium atoms was investigated separately, with a nanosecond time resolution for the HSIW front, using the laser absorption spectroscopy method. The methodology of such studies is described in [8, 9]. In the present work, a tunable Nd:YAG model LQ629-100 laser with a LP601 parametric light generator generating the wavelengths between 410 to 680 nm was used as a source for probing the optical radiation plasma.

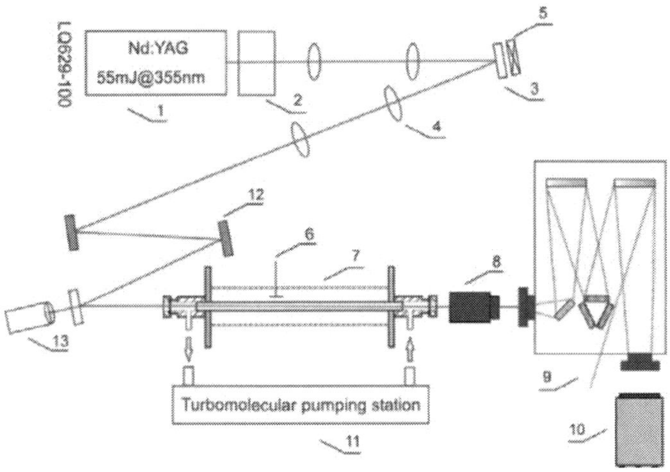

Figure 2. Block diagram of the experimental setup. 1 – Nd: YAG LQ629-100 laser; 2 – LP601 parametric light generator with a range of 410 to 680 nm wavelength tuning generation; 3 – glass plate; 4 – lenses; 5 – beam lock; 6 – capacitive sensors mounted along the discharge tube; 7 – discharge tube; 8 – photographic objective; 9 – MS7504i spectrograph; 10 – CCD camera; 11 – turbomolecular pumping station; 12 – mirrors; 13 – photocathode for controlling the laser pulse characteristics.

3. Experimental study

The transverse structure of the ionization wave front is highly dependent on the distribution of the electric field over the front, i.e. the modes and mechanisms of high-energy electron formation and relaxation. In order to obtain information about the transverse structure of the HSIW front, a series of experimental studies was carried out on the HSIW transverse profile by means of high-speed photographic recording of the transverse distribution of the optical radiation from the front in time-lapse mode. The time resolution was about 2 ns and the exposure of each frame was between 5 and 20 ns, depending on the development stage of the ionization wave. These studies were carried out in helium, neon, and argon gases at pressures ranging from 1 to 100 Torr and voltage pulse amplitudes of 40 kV, using a Princeton Instruments PI-MAX3 ICCD camera model. Synchronization was performed from the start of the HSIW near the high-voltage electrode and the opening of the electro-optic shutter of the PI-MAX3 high-speed camera. In this way, a set of optical cross-sectional pictures of the high-speed ionization wavefront was obtained in helium, neon, and argon gases at different pressures and voltage pulse amplitudes.

The studies showed that there are certain regularities in the formation of the transverse structure of a high-speed ionization wave.

Figure 3 presents typical photographs of the cross-section of the discharge for four time moments following the start of the HSIW in neon near the high voltage electrode and the corresponding contour

IOP Conf. Series: Journal of Physics: Conf. Series **830** (2017) 012026 doi:10.1088/1742-6596/830/1/012026

lines of equal intensity. These photos and the corresponding optical radiation intensity distributions from different regions of the HSIW front demonstrate that at gas pressures above 10 Torr near the boundary of the discharge tube, there is a region of intensified gas ionization. Its thickness is less than 1 mm but optical radiation from the central regions of the plasma waveguide is almost absent. Furthermore, it should be noted that the thickness of the reinforced area of ionization is virtually unchanged in proportion to the degree of excitation and ionization, from which it can be concluded that the ionization shock waves are generated towards the centre of the discharge tube during propagation.

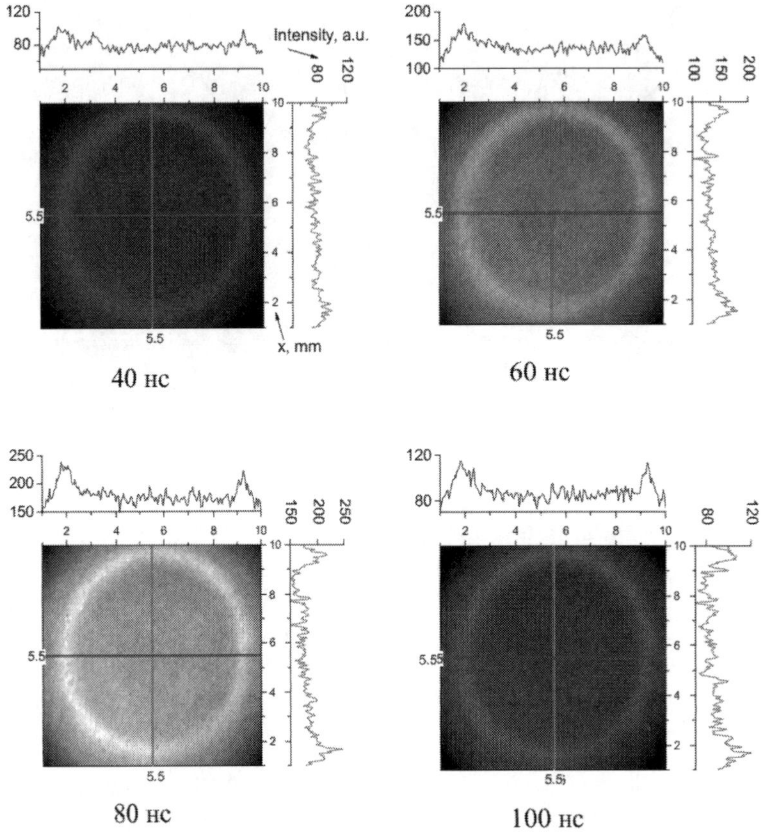

Figure 3. Synchronised time-lapse high-speed photography of the high-speed ionization wavefront in neon in the cross section of the plasma waveguide under the formation conditions of two adverse ionization waves. Pressure of neon – 25 Torr; amplitude of voltage pulses – 28 kV; exposure time – 5 ns.

The electrodes in the discharge tube are supplied with voltage pulses of an opposite polarity. The points in time from the beginning of the glow are given below the photos.

Figure 4 shows the analogous synchronised time-lapse photographs of the high-speed ionization wavefront in neon at different times at a gas pressure of 1 Torr. As it is seen from these pictures, at low pressures, the degree of ionization and excitation of the plasma gas in the centre of the plasma waveguide is higher than at the boundary of the discharge tube and the HSIW is formed and distributed throughout the volume of the discharge tube.

The electrodes in the discharge tube are supplied with voltage pulses of opposite polarity. The points in time from the beginning of the glow are given below the photos.

The analysis and systematisation of the experimental data allows us to establish the general laws of formation of the transverse structure of the high-speed ionization wavefront in plasma waveguides, provided with hollow cylindrical electrodes. In particular, the following patterns are noted:

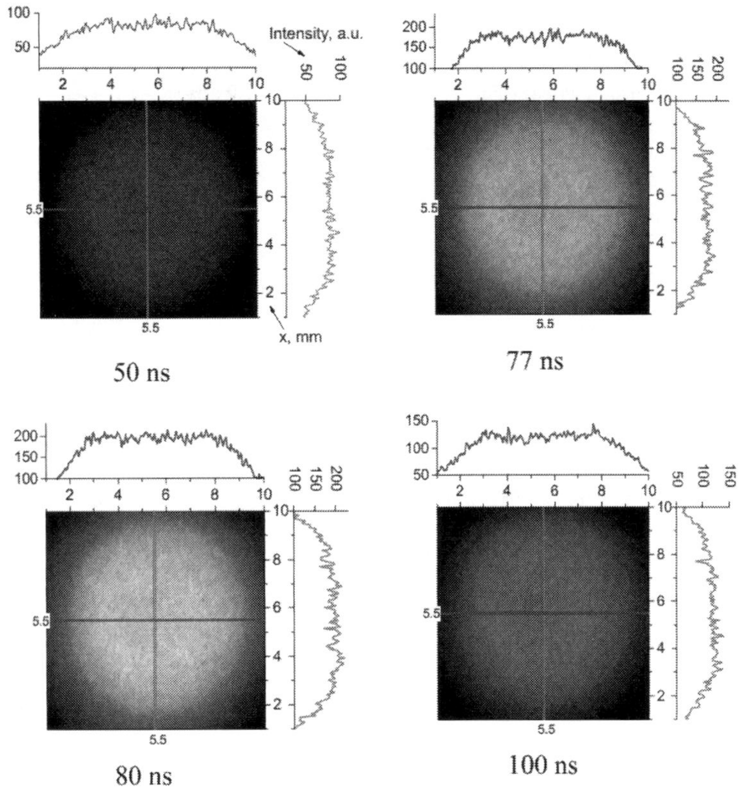

Figure 4. Synchronised time-lapse high-speed photography of the high-speed ionization wavefront in neon in the cross section of the plasma waveguide under the formation conditions of two adverse ionization waves. Pressure of neon – 1 Torr; amplitude of voltage pulses – 28 kV; exposure time – 20 ns.

1. For all three gases at gas pressures below 1 Torr, a HSIW is formed and distributed throughout the volume of the discharge tube across the whole range of investigated voltage amplitudes;
2. At gas pressures below 0.1 Torr, a significantly uniform distribution of the integrated intensity of the optical radiation becomes established in the cross section of the discharge tube;
3. With increasing gas pressures, the homogeneous structure of the high-speed ionization wavefront transfers into a grazing form via the dielectric surface with the localisation of the highly ionized region near the wall in the cylindrical layer having a thickness of less than 2 mm.
4. In proportion to the operating time of the degree of ionization of the gas, the thickness of the ionized gas region hardly changes and ranges within a value of 1 mm. With reduction of gas pressure to levels below 1 Torr, the thickness of ionized gas increases to a few millimetres.
5. The pressure transition boundary of the volumetric structure of the HSIW front to the grazing structure grows from argon to neon and further to helium. In argon, the HSIW front transitions from volumetric to grazing form along the surface of the dielectric gas at pressures above 0.5 Torr; in neon, this takes place at gas pressures above 3 Torr; in helium – at gas pressures above 10 Torr. With

an increase in the amplitude of the voltage pulses, the pressure boundary at the transition of the volumetric form of the HSIW reduces to grazing form.

With grazing form of the HSIW front, the maximum intensity of the optical radiation near the walls of the discharge tube can exceed the analogous quantity from the central regions of the discharge tube by up to 5-6 times.

The dynamics of the operating time of the metastable atoms of inert gases was investigated under similar conditions. The plasma density was measured by forming a narrow laser beam to probe metastable atoms in the central portion of the discharge tube and in the vicinity of its borders. Figure 5 shows the time dependence of the measured density of metastable neon atoms. The maximum value of the density of metastable atoms in the HSIW front appears to be comparable with the maximum value of the electron density, which is a characteristic of stepwise ionization of inert gases in nanosecond discharges.

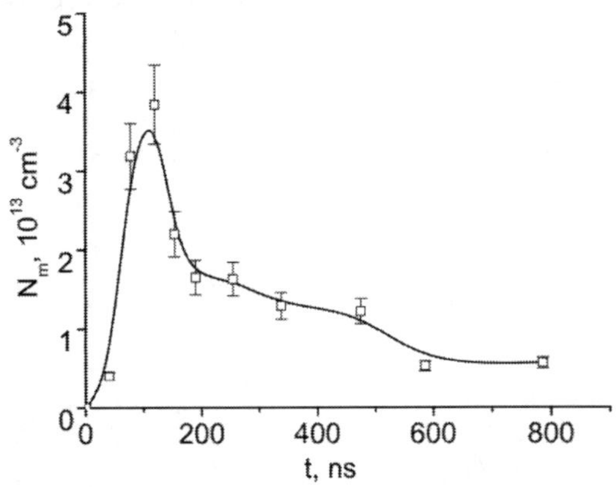

Figure 5. Time dependency of the density of excited neon atoms in the $3s^3P_1$ state, measured along the centre of the tube. Pressure of neon – 2 Torr; amplitude of voltage pulses – 36 kV.

4. Discussion of experimental results

With the wave function of electrical breakdown of the gas, almost all the electric field is concentrated in the ionization wavefront. In this area of intensified electric field, free electrons are accelerated and gain the energy ε, which can be estimated from the following equation [1,10]: $\frac{d\varepsilon}{dx} = eE - F(\varepsilon)$, where $F(\varepsilon) = \frac{2\pi e^4 ZN}{\varepsilon} \ln(\frac{2\varepsilon}{I})$ is the so-called effective ionization stopping power, I is the average energy of inelastic losses of electrons, and E is the electric field strength. Under the condition $eE > \max(F)$ of a proportion of the electrons being able to pass into the regime of continuous acceleration, high-energy electrons are formed, according to the known "runaway electron" effect.

The characteristic size of the region of intensified field (the geometric length of the HSIW front) is of the order $x_0 \sim \tau\upsilon$, where τ is the duration of the leading edge of the HSIW and υ is the velocity of the HSIW. With negative polarity, high-voltage pulses applied to the electrodes of the plasma waveguide, and the ionization wave characteristics are mainly determined by the current high-energy electrons. The processes in the HSIW front are similar to the processes at the "head" of the electron beam propagating through the gas. In this case, the ionization wavefront acts as a moving virtual cathode, emitting high-energy electrons with energies of $\varepsilon_b \sim m\upsilon^2/2$. In particular, in [9, 10] it is shown that the critical electron velocity value near $\upsilon_0 = \sqrt{\frac{2.72I}{m}}$ the electrons undergoes relaxation oscillations under the condition:

$$\frac{E_0}{N} > \frac{\beta\gamma}{2.72\alpha} = \frac{2\pi e^3 z}{2.72I} \quad (1)$$

It means that the energy of high-energy electrons in the HSIW front fluctuates near the value $\varepsilon_b \sim m\upsilon^2/2$.

Table 1 shows the values of the electric field at the wave front ionization, the HSIW parameters, and energy of electrons accelerated at the HSIW front in neon when the ionization waves have a negative polarity. These values are calculated from the waveform of the electrical potential gradient measured with calibrated capacitive sensors for the VPG, with voltage pulses having a duration of about 100 ns.

Comparing of the calculated values of the reduced electric field strength with the critical values for continuous acceleration [1, 4, 10, 12], we can conclude that the conditions for continuous and efficient acceleration of electrons forming high-energy electrons on the HSIW front with a voltage pulse duration of about 100 ns for the VPG exist only at gas pressures below 1 Torr. At the same time, the energy of the high-energy group of electrons is of the order of several tens of eV; at these energies of electron excitation, the ionization rates of helium and neon atoms also have maximum values.

High-energy electrons generated in the HSIW front, carry out effective electronic excitation and ionization of the gas atoms, thereby gaining free electrons and excited atoms at the ionization wave front. In addition, a proportion of the high-energy electrons also produce preliminary ionization of the gas at the negative polarity HSIW front. The spatial dimensions of the relaxation region of the high-speed electron energy affect the structure and the uniformity of the HSIW front. At the front of the ionization wave, the function of the electron energy distribution function (EEDF) consists of two parts – fast, with the energy $\varepsilon_b \sim m\upsilon^2/2$, and slow, consisting mainly of a plasma formed by secondary electrons. It is known that a local or non-local EEDF formation mode can be analyzed by introducing a relaxation parameter [13, 14]:

$$k(\varepsilon) = (\nu_e + \delta\nu_a + \nu^*)\tau_r \quad (2)$$

where ν_e is the frequency of inter-electron collisions; $\delta = 2m/M$; ν_a is the frequency of elastic collisions between electrons and atoms; ν^* is the frequency of inelastic collisions of electrons with atoms; τ_r is the time of withdrawal of electrons from the ionization region as a result of transport phenomena. If $k(\varepsilon) \gg 1$, the energy of the high-energy electrons is relaxed in the ionized volume, without ranging beyond the HSIW front. A significant part of the energy of such electrons is expended on the excitation and ionization of the gas atoms in the specified area.

Table 1.

P, Torr	U, kV	E, V/cm	E/P, V/(cm · Torr)	τ, 10^{-9}s	υ, 10^8cm/s	x_0, cm	ε, eV
1	28	1481	1481	54	3.5	18.9	35
1	40	1894	1894	48	4.4	21.1	55
10	28	1221	122	62	3.7	22.9	39
10	40	1867	187	51	4.2	21.4	50
20	28	1275	64	61	3.6	22.0	37
20	40	1429	71	56	5.0	28.0	71
60	28	1176	20	70	3.4	23.8	33
60	40	1587	26	60	4.2	25.2	50

Such a regime can be formed at moderate gas pressures, when the region of enhanced ionization is localized in a small layer near the discharge tube wall, where the high-energy electrons have been initially formed. In the case of the volumetric form of the high-speed ionization wave front (for the considered conditions at gas pressures below 1 Torr and low amplitudes of voltage pulses), the flux of fast electrons in the discharge tube wall is small, and the value of the wall potential in the particular case of the Maxwell distribution of the main groups of electrons is determined by the average energy of the plasma electrons [13]:

$$\Delta V = \frac{kT_e}{e} \ln\left[\frac{\sqrt{M}}{3\sqrt{\pi m}}\right] \sim \frac{kT_e}{e} \tag{3}$$

When $k(\varepsilon) \ll 1$, if $e\Delta V < \varepsilon_b$, the faster electrons depart to the wall in free diffusion mode, and the high-speed EEDF part the energy peak. The withdrawal of these electrons to the walls can increase the value of the near-wall potential to the value of the order $\frac{\bar{\varepsilon}_b}{e}$, since for the fulfillment of conditions of the quasi-neutrality of plasma, in this case, it is necessary to withhold a proportion of the fast electrons in the volume. As a result, fast electrons being locked in volume leave this energy region, which is also determined by collisional processes in the volume and forms the volumetric form of the high-speed ionization wave front structure. In this case, as in $k(\varepsilon) \gg 1$, a continuous electron spectrum is formed, and the energy of fast electrons is expended in the excitation and ionization of the gas atoms in the entire volume of the discharge tube.

5. Conclusions

Thus, the results of the pilot study and our analysis show that under the considered conditions, the HSIW front has two formation modes – namely the volumetric form of the HSIW structure (when gas pressures are below 1 Torr) and a cylindrical shape pressed towards the walls of the discharge tube (at gas pressures above 10 Torr). At the HSIW front, the high-energy electrons are generated, their energy relaxation has a significant impact on the structure of the HSIW front. In the formation and propagation of the cylindrical HSIW front, the spatial distribution of plasma parameters attains cylindrical symmetry with a maximum density of excited atoms near the walls and the minimum in the centre of the tube. The size of the wall thickness of the plasma ionization is determined by the maximum length of the energy relaxation of a group of high-energy electrons when, in the process of ionization growth, the shock waves towards the sides of the discharge tube do not form.

Acknowledgments

This work was financially supported by the project part of the state assignment of the Ministry of Education and Science of Russia in scientific activities, project 3.1262.2014K.

References

[1] Mesyats G A 2005 *Pulsed Power* (Springer US) p 568
[2] *Generation of runaway electrons and X-rays in the discharges of high pressure* 2015 ed V F Tarasenko (Tomsk: STT) p 568
[3] Tarasenko V F, Yakovlenko S I 2004 *Phys. Usp.* **47** 887
[4] Babich L P 2005 *Phys. Usp.* **48** 1015–37
[5] Ashurbekov N A, Iminov K O, Taibov K T and Yusupova G M 2015 *Modern Physics Letters B* **29** 1550102
[6] Ashurbekov N A, Iminov K O, Kobzev O V and Kobzeva V S 2010 *Technical Physics* **55** 1138–44
[7] Vasilyak L M, Kostyuchenko S V, Kudryavtsev N N and Filyugin I V 1994 *Phys. Usp.* **37** 247-68
[8] Ashurbekov N A, Kurbanismailov V S, Omarov O A and Omarova N O 2000 *High Temperature* **38** 795-810
[9] Ashurbekov N A, Iminov K O, Shakhsinov G S and Ramazanov A R 2015 *High Temperature* **53** 627-37
[10] Ashurbekov N A, Borisov V B, Egorov V S and Kardashov V R 1995 *Optics and Spectroscopy* (in Russian) **78** 999-1003
[11] Starikovskaia S M, Anikin N B, Pancheshnyi S V, Zatsepin D V and Starikovskii AY 2001 *Plasma Sources Science and Technology* **10** 344–55

IOP Conf. Series: Journal of Physics: Conf. Series **830** (2017) 012026 doi:10.1088/1742-6596/830/1/012026

[12] Tkachev A N, Fedenev A A and Yakovlenko S I 2005 *Technical Physics* **50** 447-54
[13] Tsendin L D 2010 *Phys. Usp.* **53** 133-57
[14] Tsendin L D and Golubovskii Yu B 1977 *Sov. Phys. Tech. Phys.* **22** 1066

IOP Conf. Series: Journal of Physics: Conf. Series **830** (2017) 012027 doi:10.1088/1742-6596/830/1/012027

Gyromagnetic nonlinear transmission line generator of high voltage pulses modulated at 4 GHz frequency with 1000 Hz pulse repetition rate

M R Ulmasculov, K A Sharypov, S A Shunailov, V G Shpak, M I Yalandin, M S Pedos and S N Rukin,

Institute of Electrophysics UB RAS, 620016, Ekaterinburg, Russian Federation

E-mail: marat@iep.uran.ru

Abstract. Results of testing of a generator based on a solid-state drive and the parallel gyromagnetic nonlinear transmission lines with external bias are presented. Stable rf-modulated high-voltage nanosecond pulses were shaped in each of the four channels in 1 s packets with 1000 Hz repetition frequencies. Pulse amplitude reaches -175 kV, at a modulation depth of rf-oscillations to 50 % and the effective frequency ~4 GHz.

1. Introduction

Multichannel microwave sources [1-7] considerably increase a power flux density of rf-emission in the far-field region of the major maximum of the directional pattern (DP) and sharpen a beam [1-3]. Maximum increase in the flux power density is quadratic to a quantity of sources. So, it is clear that there is an interest in the systems when a power of every channel reaches hundreds of megawatts and higher [1-5]. The on-line control of delays of the output pulses allows an adjustment of the mutual phasing [5-7] permitting the integral DP scanning in space. Therefore, the accuracy of a time-lock of the microwave sources is required.

The four-channel rf-source is presented below. The external axial magnetic field biased (H_z) coaxial nonlinear transmission lines (NLTL) partially filled with ferrite rings were used as the sources of high-power microwave pulses [3-10]. The source rf modulation frequency was increased to 4 GHz, a fairly high output peak power of hundreds of megawatts, and operated at the maximum pulse repetition frequency (PRF) of 1000 Hz.

The microwave oscillations come due to the gyromagnetic phenomena. The gyromagnetic precession excitement takes place under the fast consequent action of a couple of the orthogonal magnetic fields on the magnetic moment of the fully magnetized ferrites. An external bias H_z-field forms the ferrite saturation in an axial direction, while an incident high-voltage pulse creates an azimuthally magnetic reversal field. As an incident pulse with a sufficient amplitude and rise time propagates through the NLTL, and the sufficient external bias is applied, the ferrite magnetic moments rotate around. The damped gyromagnetic precession excites microwave oscillations.

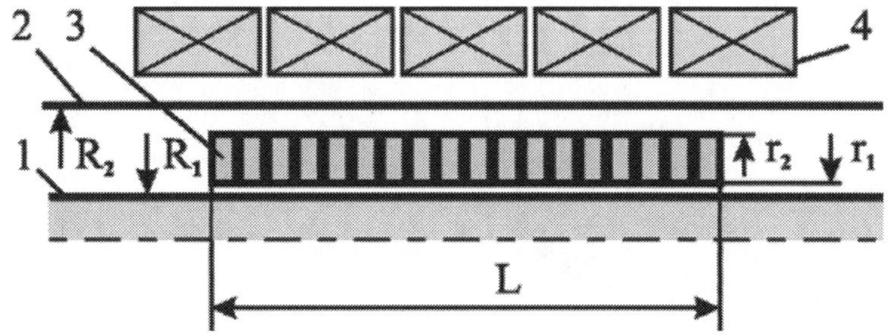

Figure 1. Schematic of coaxial NLTL.1,2- outer and inner electrode of coaxial line, 3-ferrite rings, 4-solenoid

2. Experimental setup

At this work, the two-section NLTLs were used [3, 5-7, 9]. The input section operated as a pulse sharpening device and sharpened a front, sufficient for the second NLTL section operation in the oscillator's mode. Both sections had integrated a construction depicted in figure 1. The ferrite line implements a continuous unit of NiZn-rings of a type M200VNP ($2R_1$=25 mm; $2R_2$=8 mm; $2r_1$=10 mm; $2r_2$=16 mm) of the total length 700 mm. A direct current solenoid was assembled of five sections, symmetrically placed over a case around ferrite. The NLTL had a partition in the sharpening and generating parts by a self-contained power supply. Experimentally, it has been found that it is optimal to utilize the two first coils for sharpening and delay line tuning of the pulse. The bias field H_z of the remaining NLTL part was optimized in the range of 35-55 kA/m to form the rf oscillations. The design of NLTL and the biasing field data gave the acceptable delay tuning range (more than the rf half-period), the maximum rf modulation deepness, and the minimized distortion of excited oscillations envelop for a full scale of the delay control.

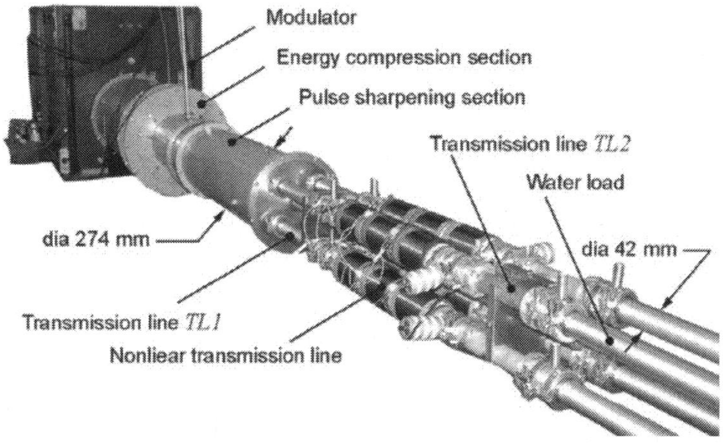

Figure 2. A view of the 4-channel rf oscillator on NLTL

The installation view and the basic components of the rf-oscillator are illustrated in figure 2. The full solid state high-voltage driver S500 [11, 12] served as a supply source. The driver formed a pulse with a rise time of ~1 ns, the duration (FWHM) of ~5 ns, and the amplitude V_d≈-500 kV in a single pulse mode across a 50 Ohm load. When the system operated at 1000 Hz PRF, the voltage amplitude reduced to V_{d1000}≈-430 kV.

An output driver pulse was split into four parallel channels by a mismatched cross-type power divider (4x48 Ohm) presented in figure 2. The instability of the pulse amplitude (a total spread 10%, a standard deviation σ~2%) was related to the main voltage ripple across a capacitive filter of the primary charging device (PCD).

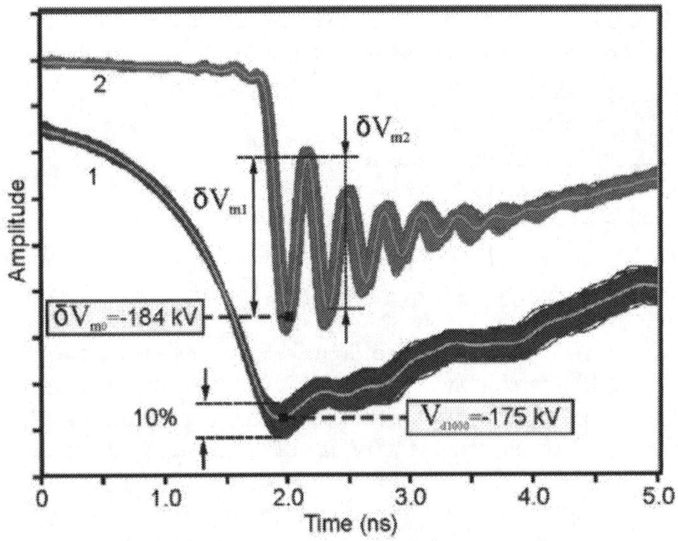

Figure 3. Waveforms for a single channel system at repetition rate of 1000 Hz. Pulses of the driver (1) and NLTL (2)

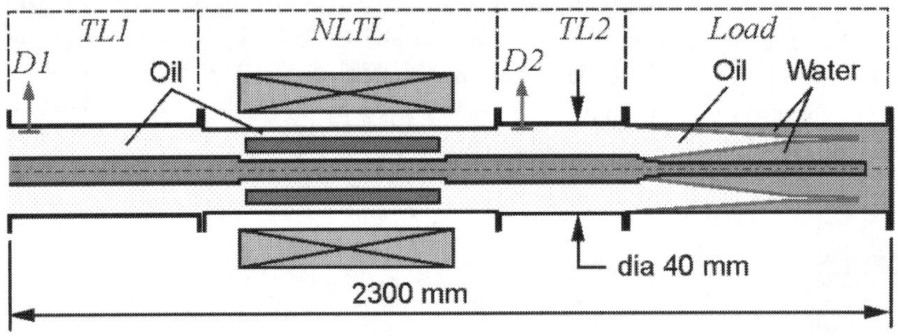

Figure 4. Schematic of the NLTL: terminated by a coaxial water line

Every channel was terminated by an absorbing water load (figure 4). The coaxial transition to the water load had a small-angle long conical insulator to minimize the reflections. The voltage pulses were recorded in the oil-filled transmission lines (TL1 and TL2), using the capacitive dividers capable to reproduce ~70 ps rise time. The accuracy of amplitude calibration of the dividers D1, D2 was of ~5%. The location of the dividers in the lines provided about 5 ns delay of the reflected signals.

3. Examination of individual NLTL

Testing of the NLTL electric strength and a stability of the output parameters of exited oscillations were held by a single channel terminated by a water line at the repetition frequency 1000 Hz during 1 second. An average input pulse amplitude was $V_0 \approx -175$ kV, a standard deviation of $\sigma = 3.5$ kV. An electric strength of the NLTL was enough to operate without breakdown. Waveforms of the output pulse burst are presented in figure 3. The mean value of the first peak of the rf-modulated pulse was $|V_{m0}| = 184$ kV. An electric field strength on the surface of the NLTL inner electrode of 8 mm achieved ~425 kV/cm, and ~255 kV/cm in ferrite. An insignificant increase of amplitude due to the shock wave has been formed at the incident pulse front. The maximum increase was obtained by utilizing ferrite rings of the greater diameter [4].

The NLTL excitation mode was chosen so that the modulation depth of the oscillations was maximum after the leading peak (V_{m0}). This was achieved by varying the bias field. Varying the solenoid bias field H_z, the NLTL excitation mode have been adjusted to maximize the modulation depth (δV_{m1}), which followed the leading peak. The average amplitude and the standard deviation (σ_0) of the peak (V_{m0}), the second peak amplitude (δV_{m2}), and their ratio to V_0 were measured. The effective frequencies (f_1-f_3) of the three initial peaks were defined as the inversely proportional double time within points at the half-high rise and fall times of the corresponding peaks. The obtained values are listed in table 1.

Table 1. Output parameters

Parameters	Value	Parameters	Value	Parameters	Value	Parameters	Value
$\lvert V_{m0}\rvert$, kV	184	V_{m0}/V_0, %	105	f_1, GHz	2.7	H_z, kA/m	55
$\lvert \delta V_{m1}\rvert$, kV	84	$\delta V_{m1}/V_0$, %	51	f_2, GHz	3		
$\lvert \delta V_{m2}\rvert$, kV	83	$\delta V_{m2}/V_0$, %	47	f_3, GHz	3.5		

4. High-PRF operation of multi-channel oscillators

The four-channel system was tested to investigate the mutual stability of operation of individual NLTLs under the excitation of rf oscillations. Analysis of obtained results was carried out for a packet of 100 pulses at the pulse repetition frequency of 1000 Hz. The NLTL identity was provided by an equal quantity of the rings and symmetric location of the solenoids over the rings. All channels were terminated by the coaxial water load.

The external bias field of three output coils was identical and was lowered to Hz1=40 kA/m. This field gave a maximum delay time turning range and a more sharpened leading peak rise time up to tf=120 ps. The effective frequency increased to ~ 4 GHz, but the mean value of the output pulse peak amplitude decreased down to -165 kV, figure 5. The output pulse waveforms coincided within a reproduction accuracy of the divider D2 of each channel. The spectrum (the Fast Fourier transformation) of a typical output pulse from the packet is presented in figure 6. The spectrum maximum values (the effective frequency) correspond to ~4 GHz.

The channel delay times were various, because the ferrite ring parameters under high voltage, short pulse durations, and the frequency of ~4 GHz had not been controlled. The time-lock of a pulse waveform was achieved by current turning in two beginning sections of the solenoid at the output of NLTLs. The identical delays corresponding to calculated fields Hz2 generated by the coils were 4; 40; 48, and 57 kA/m.

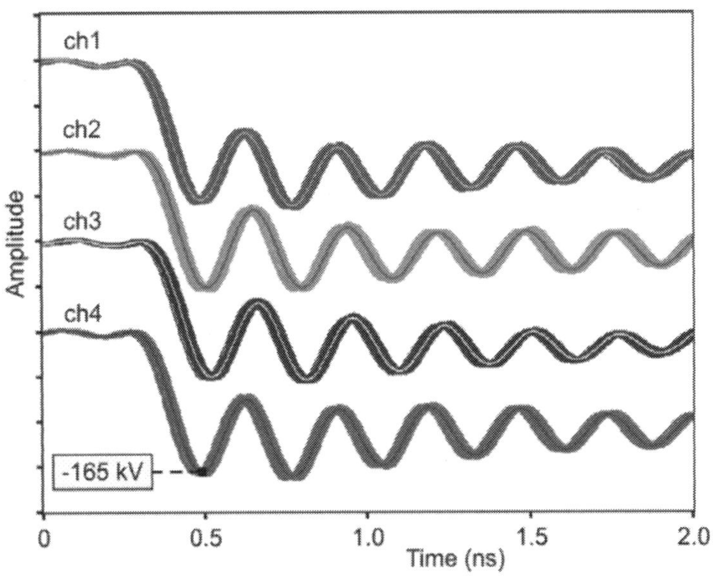

Figure 5. Waveforms for a 4-channel systems at repetition rate of 1000 Hz

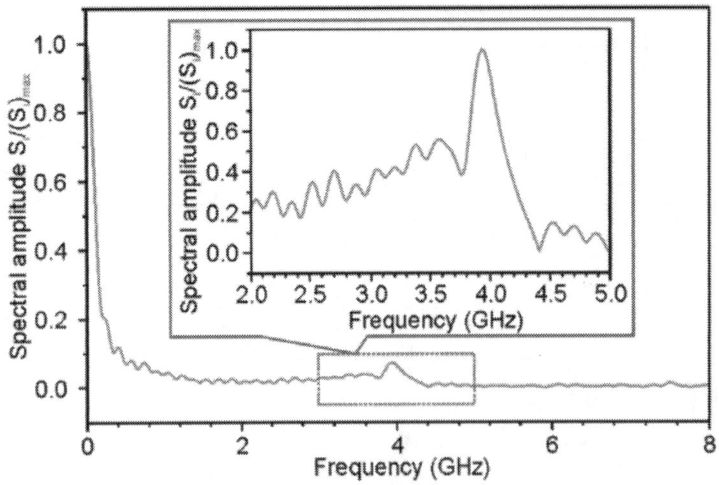

Figure 6. FFT spectrums of the output pulse of NLTL

The time instability of generation of the rf-modulated pulses depended on H_{z2} in the channels. A cross-coupling comparison of the channels showed the maximum relative spread in time-locking of δt=55 ps (15 ps on average) with a standard deviation of σ_δ=20 ps and had only one channel with a lower minimum H_{z2}=4 kA/m. The minimum δt=35 ps (8 ps on average) and σ_δ=1.4 ps were between two channels with sub-equal and high values H_{z2} 40 kA/m и 48 kA/m, which were practically identical to H_z of the NLTL oscillator section. The decrease in instability with an increasing bias field is predictable, because the deeper ferrite saturation by an external field reduces the ferrite medium state uncertainty in an action of a previous pulse. It is necessary to note, that in the system with the unseparated sharpening and main (gyromagnetic) NLTLs, the channel with H_{z2}=55 kA/m (in the sharpening NLTL), which is higher than H_{z1}=40 kA/m (in the main NLTL), had a high σ_δ=9 ps.

5. Conclusions

The designed four-channel rf-source based on a solid-state driver and gyromagnetic NLTLs produced high voltage pulses with a maximum modulation frequency of 4 GHz, with the peak voltage at the NLTL output of 175 kV. The electric strength provided an operation at a pulse repetition rate of 1000 Hz during 1 second. The integral packing of the delay and the modulating sections of NLTLs allowed the electronic time locking. The electronic timing demonstrates the capability of irradiated RF beam steering.

Acknowledgements

This work was supported by research program of IEP UB RAS 0389-2014-0005 and by RFBR grant No. 16-08-00058.

References

[1] Yalandin M I, Shunailov S A, Ul'maskulov M R, Sharypov K A, Shpak V G, Rostov V V, Romanchenko I V, El'chaninov A A, and Klimov A I 2012 *Tech. Phys. Lett.* **38**(10) 917-20

[2] Sharypov K A, El'chaninov A A, Mesyats G A, Pedos M S, Romancheko I V, Rostov V V, Rukin S N, Shpak V G, Shunailov S A, Ul'masculov M R, and Yalandin M I, 2012 *J. Appl. Phys.* **103** 134103

[3] Rostov V V, El'chaninov A A, Klimov A I, Konev V Y, Romanchenko I V, Sharypov K A, Shunailov S A, Ul'maskulov M R, and Yalandin M I, 2013 *IEEE Trans. On Plasma Sci.* **41** 2735–41

[4] Ulmaskulov M R, Pedos M S, Rukin S N, Sharypov K A, Shpak V G, Shunailov S A, Yalandin M I, Romanchenko I V, and Rostov V V, 2015 *Rev. of Sci. Instr.* **86** 074702

[5] Romanchenko I V, Rostov V V, Gunin A V, and Konev V Y, 2015 *J. Appl. Phys.* **117** 214907

[6] Johnson J M, Reale D V, Krile J T, Garcia R S, Cravey W H, Neuber A A, Dickens J C, Mankowski J J, 2016 *Rev. of Sci. Instr.* **87** 054704

[7] Johnson J M, Reale D V, Cravey W H, Garcia R S, Barnett D H, Neuber A A, Dickens J C, and Mankowski J J, 2015 *Rev. of Sci. Instr.* **86** 084702

[8] Bragg J W, Dickens J, and Neuber A, 2013 *J. Appl. Phys.* **113** 064904

[9] Reale D, Bragg J W, Gonsalves N, Johnson J, Neuber A, Dickens J, and Mankowski J, 2014 *Rev. of Sci. Instr.* **85** 054706

[10] Romanchenko I V, Rostov V V, Gubanov V P, Stepchenko A S, Gunin A V, and Kurkan I K, 2012 *Rev. of Sci. Instr.* **83** 074705

[11] Rukin S, Bushlyakov A, Lyubutin S, Ponomarev A, Slovikovsky B, Timoshenkov S, Tsyranov S, 2007 *Pulsed Power Conference. Digest of Technical Papers. IEEE International* 691-701

[12] Gusev A I, Pedos M S, Rukin S N, Timoshenkov S P, and Tsyranov S N, 2015 *Rev. of Sci. Instr.* **86** 114706

Oversized interference microwave switches with distributed power of a switching wave

S N Artemenko, S A Gorev and V S Igumnov

Institute of Physics and Technology, Tomsk Polytechnic University, Tomsk, 634050, Russia

Email: snartemenko@mail.ru

Abstract. These are the results of research of the oversized (multimode) interference microwave switch at operating mode TE_{01}. The switch was produced from two H-plane T-junctions being connected to each other through straight arms (series connection). Two types of oversized tees were considered. One was based on an oversized waveguide with 25x58 mm^2 in cross-section and another was a package of regular H-tees operating at TE_{10} mode with a mutual switched arm. The operating frequency was 9.15 GHz. The conducted simulation showed the conditions of the "proper" TE_{01} mode at "open" and "close" states of the switch. Moreover, the relations between the arm' lengths and field intensity distribution were compared with the similar relations of a regular cascade microwave plasma switch. In additional, experiments were carried out at low power level.

1. Introduction

The microwave compressor is a passive amplifier of microwave pulses. It accumulates the input power into its own cavity and then rapidly radiates the power into the load. There are two types of compressors' switching, active and passive, differing by method of their switching.

Passive switches do not change resonance characteristics of cavities. On the contrary, an active switch changes the Q-factor of the compressor's cavity from high to low value. This feature provides higher amplification. The speed of commutation impacts on the gain and the envelope of output pulses. The switch should have the fast speed of switching and the high transmission attenuation in order to reach high amplification. One of implementation of such switches is a microwave interference switch.

The interference microwave switch is based on H-plane T-junction. One straight arm of the T-junction is coupled to the cavity and one of the others arms is shorted at a half wavelength. The last one is the output.

The shorted arm has the discharge gap at a quarter wavelength from the short. At Off state the waves from the cavity and the shorted arm radiate with the same amplitudes in anti-phase. At On state the waves' phases coincide.

The cross section of the shorted arm limits the electrical strength of the switch. The quartz tube, the blown contribute into the decrease of electrical strength.

One way for decreasing the switched power is distribution switched power among few switching units[1,2]. A construction of the cascade interference switch is studied in the paper[3].

Moreover, the oversized H-tee (figure 1) can provide greater electrical strength than in regular H-tee. Two implementations of the oversized H-tee were considered in [4,5,6]. The first implementation is an H-tee produced from an oversized waveguide and the second one is a package of the regular H-tees with the mutual switched side arm.

Geometrical similarity between the regular cascade switch and the oversized one gives a reason to suppose their similarity in operating.

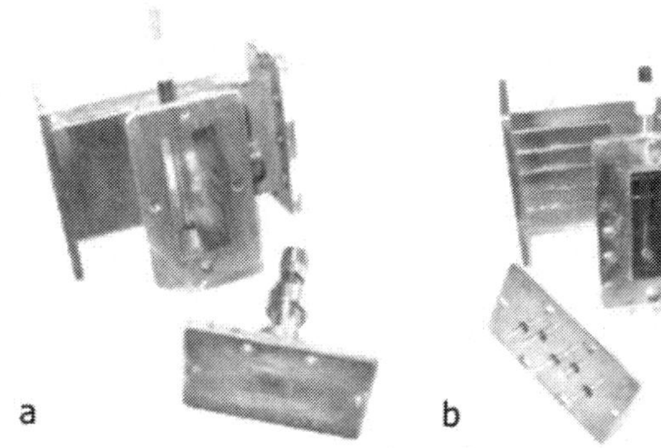

Figure 1. The interference microwave switches produced from an oversized waveguide (a) and a package of regular waveguides with the mutual side arm (b).

2. Condition of the "clear" operating mode

This paragraph concerns technical details appears in oversized waveguides.

The transmission attenuation depends on the correctness of the operating mode TE_{01} that is not primary mode for such waveguide. For exciting "clear" TE_{01} mode the waveguide taper turning TE_{10} mode of the regular waveguide was considered. The simulation demonstrated one of such tapers (figure2) that was performed for further experiments.

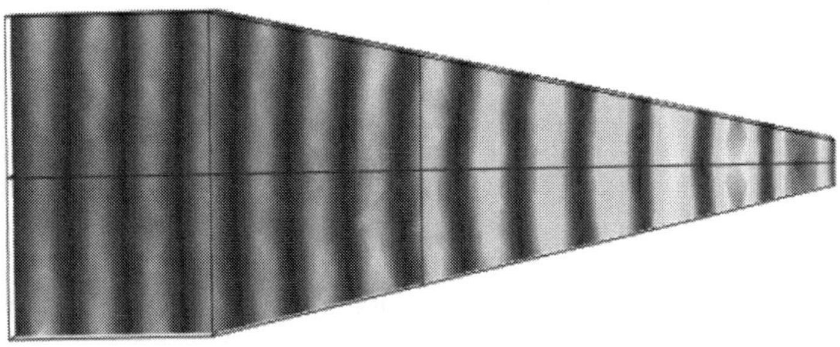

Figure 2. Figure shows the field structure of operating mode formed by a horn taper.

The problem of oversized units is their own geometrical distortions influence on field structure and transmission attenuation such as distortion of the short or the side arms. According to estimations the transmission attenuation is steadily at slope of distortions no greater than 0.01. The attenuation does not decrease less than -55 dB. The small geometrical distortions effect the field structure distortions insignificantly. Assess was carried out with simulation.

3. Experiment

Experiment was carried out with a model of the switch including two oversized H-tees with cross-section of 58x25 mm^2 at X-band. Each oversized H-tee was produced from five identical regular H-tees with cross-section of 25x10 mm^2. The switched arm was mutual for each regular H-tee (figure 3). The induction loops embedded in the short pick up signals from side arms in the view of control of switched power level. The detectors and loops were calibrated. The On state was imitated by copper wire injected along electric field lines at low power level. The operating frequency was 9150 MHz Excitation was through an input iris and the taper simulated in the previous section.

Figure 3. The oversized cascade interference switch based on two packages of regular H-tees with mutual switched arm.

A figure 4 demonstrates changes of wave amplitudes when one of the switches is commutated. As one can see a wave amplitude in the switch that is turned off increases significantly and electromagnetic field in the commutated switch almost disappears. This disappearance is caused by a node of a standing wave. Such behavior is identical to the regular cascade switch.

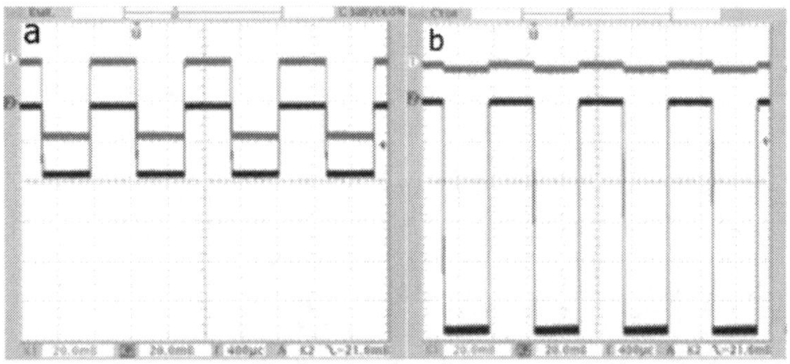

Figure 4. Oscillograms of wave amplitudes in the side arms of the switches. Amplitudes before commutation of any switch (a) and amplitudes after commutation of all switches (b).

4. Conclusion

The oversized cascade switch allows increasing efficiency of a compressor. Since, switched power is decreased. The relations of the cascade switch remain for oversized implementation.

Acknowledgments

This research was performed with partial financial support from the National Research Tomsk Polytechnic University Competitiveness Programme and with support of the Russian Foundation for Basic Research (Grant No.15-08-01853a).

References

[1] Guo J and Tantawi S 2007 *IEEE Particle Accelerator Conference* Albuquerque pp 4189-4191 (New York: IEEE)

[2] Tantawi S. 1998. *8th Workshop on Advanced Accelerator Concepts* vol **472** pp 959-966 (Md: Amer Inst Physics)

[3] Artemenko S N, Avgustinovich V A, Gorev S A and Igumnov V S 2015 *12th International Conference on Gas Discharge Plasmas and Their Applications (GDP)* v **652** 012017, 6 p. (Russia: Iop Publishing Ltd)

[4] Avgustinovich V A, Artemenko S N and Zhukov A A 2013 *Technical Physics Letters* v **39** pp 492-494.

[5] Artemenko S N, Avgustinovich V A and Arteev M S *Technical Physics Letters* v **39** pp 26-32

[6] Artemenko S N, Avgustinovich V A, Gorev S A, Igumnov V S, Kaminsky V L, Novikov S A and Yushkov Y G 2014 *XXIV Russian Particle Accelerator Conference* pp 183-185 (Novosibirsk: BINP)

Numerical simulation of electrical explosions in megagauss magnetic fields

V I Oreshkin[1,2], S A Chaikovsky[3,1,5], K V Khishchenko[4] and E V Oreshkin[5]

[1] Institute of High Current Electronics of the Siberian Branch of the Russian Academy of Sciences, Akademichesky Avenue 2/3, Tomsk 634055, Russia
[2] National Research Tomsk Polytechnical University, Lenin Avenue 30, Tomsk 634050, Russia
[3] Institute of Electrophysics of the Ural Branch of the Russian Academy of Sciences, Amundsen 106, Ekaterinburg 620016, Russia
[4] Joint Institute for High Temperatures of the Russian Academy of Sciences, Izhorskaya 13 Bldg 2, Moscow 125412, Russia
[5] Lebedev Physical Institute of the Russian Academy of Sciences, Leninsky Avenue 53, Moscow 119991, Russia

E-mail: oreshkin@ovpe.hcei.tsc.ru

Abstract. The paper reports on a magnetohydrodynamic simulation of electrical explosions of conductors in megagauss magnetic fields. It is shown that in a plane geometry, the time of plasma formation at the surface of a metal conductor does not depend on the rate of rise of the magnetic field and is determined by the properties of the metal; the absolute values of the magnetic field at which plasma is formed are 5±0.25 MGs for copper, 4.25±0.2 MGs for tungsten, 3.85±0.15 MGs for aluminum, and 3.6±0.25 MGs for titanium. In cylindrical geometry, the time of plasma formation does depend on the rate of field rise.

1. Introduction

The electrical explosion of conductors (EEC) has been studied for a long time and has a number of practical applications. An EEC mode of interest in the generation of superstrong magnetic fields and in the energy transfer, using vacuum transmission lines, is the current skinning mode [1-4]. For this mode, the time of energy delivery to the conductor is shorter than, or comparable to the time of magnetic diffusion into it. The basic processes inherent in the current skinning mode are the propagation of a nonlinear magnetic diffusion wave [5], the formation of a low-temperature plasma at the conductor surface, and the development of thermal instabilities [6,7].

Nonlinear magnetic diffusion features, a speed of field penetration into a conductor are anomalously high compared to a conventional magnetic diffusion. The high diffusion speed is related to a decrease in conductivity of the metal due to its heating by an electric current. A nonlinear diffusion can occur only in a rather strong magnetic field whose induction, for the fairly often used metals, should be not lower than 250-450 kGs [1,2].

2. Magnetohydrodynamic model

The electrical explosions of the conductors are generally simulated in the magnetohydrodynamic (MHD) approximation, requiring a knowledge of the equations of a state for a wide range of thermodynamic parameters and the transport coefficients, among which the electrical conductivity is the most important one. For describing the thermodynamic properties of metals, various semiempirical models and databases are available, whereas the issues related to the transport coefficients, in a metal–dielectric transition region and near the critical point, are not explicit.

The electrical explosion of conductors was modeled in a single-temperature MHD approximation. In a cylindrical geometry, the MHD equations have the form [8]

$$\frac{d\rho}{dt}+\frac{\rho}{r}\frac{\partial v}{\partial r}=0\ ; \qquad (1)$$

$$\rho\frac{dv}{dt}=-\frac{\partial p}{\partial r}-j_z B_\varphi\ ; \qquad (2)$$

$$\rho\frac{d\varepsilon}{dt}=-\frac{p}{r}\frac{\partial v}{\partial r}+\frac{j_z^2}{\sigma}+\frac{1}{r}\frac{\partial}{\partial r}(\kappa\frac{\partial T}{\partial r})\ ; \qquad (3)$$

$$\frac{1}{c}\frac{\partial B_\varphi}{\partial t}=\frac{\partial E_z}{\partial r};\ j_z=\frac{c}{4\pi r}\frac{\partial (rB_\varphi)}{\partial r}\ ; \qquad (4)$$

$$j_z == \sigma E_z; \qquad (5)$$

$$\varepsilon=f(\rho,T);\ p=f(\rho,T) \qquad (6)$$

where $\frac{d}{dt}=\frac{\partial}{\partial t}+v\frac{\partial}{\partial r}$ is the substantial derivative; ρ, T are the density and the temperature of a matter; v is the radial velocity component; p, ε are the pressure and the internal energy; B_φ is the azimuthal component of the magnetic field strength; E_z is the axial component of the electric field strength; j_z is the axial component of the current density; κ; σ are the heat and electrical conductivities.

Equations (1–6) were solved numerically using a one-dimensional MHD code written in Lagrangian coordinates [9]. Hydrodynamic equations (1–3) in the code were solved using a cross scheme [8], which is an explicit difference scheme with a combined (linear and quadratic) pseudo-viscosity for shock wave calculations. Maxwell equations (4) supplemented with Ohm's law (5) and the heat conduction equation were solved using implicit difference schemes based on the data-flow sweep method.

The boundary conditions imposed on the Maxwell equations have the form

$$B_\varphi(R)=\frac{2I_n}{cR}\ , \qquad (7)$$

where R is the outer radius of a wire; I_n is the current through the wire.

The system of MHD equations is completed by the equations of state (6) derived for metal wires at the Joint Institute for High Temperatures RAS [10]; these are the wide-range semiempirical equations based on a model [11] that takes into account the high-temperature melting and the evaporation, as well as the metastable liquid and gaseous states.

The electrical conductivity is determined using the conductivity tables based on the calculation and experimental data [12,13], considering that this parameter represents a certain empirical function of the density and the specific energy deposited in a matter and depends parametrically on the form of equations of state of the matter.

The initial data for constructing the conductivity tables are the following: (1) the dependence $\sigma_1(T,1)$ at the normal density of a matter (its table value) and (2) the dependence of conductivity calculated for the gas flame region by classical formulae [14]. In the transition region near the critical point, the conductivity is given in a parametric form in which, compared to [12], the parameter γ is not identically equal to unity:

$$\lg\frac{\sigma(T,\delta)}{\sigma_1(T,1)} = \Phi(T,\delta)\lg\frac{\sigma_{кр}}{\sigma_1}\left(\frac{\lg\delta}{\lg\delta_{кр}}\right)^{\gamma}; \tag{8}$$

where σ_{cr} is the conductivity at the critical point; $\delta=\rho/\rho_0$ is the relative density of a matter; ρ_0 is its normal density; δ_{cr} is the relative density at the critical point; $\Phi(T,\delta)$ is a function of order 1 dependent on the phase boundary position. Constructing the conductivity tables, the quantity σ_{cr} is assumed to be a variable parameter, independent of temperature. The conductivity at the critical point is chosen to provide the best agreement between the results of MHD simulation and experimental data.

3. Simulation results and comparison with experimental data

The objective of our simulation was to predict how the magnetic induction, responsible for the formation of plasma at the conductor surface, depends on the rate of rise of the magnetic field in the mode of current skinning.

In the simulation, the rate of rise of the magnetic field was assumed to increase linearly; that is, the magnetic field at the conductor surface increases according to the law

$$B(R) = \frac{dB}{dt}t. \tag{9}$$

The rate of rise $\frac{dB}{dt}$ was varied in the range from 3×10^{12} to 3×10^{14} Gs/s.

First, we performed the computations in a plane geometry for the penetration of a linearly increasing magnetic field in a metal conductor occupying an infinite half-space, reasoning that when the skin effect is strong, the skin layer thickness is much lower than the conductor radius, and hence, the geometry is plane.

The instant of the plasma formation at the conductor surface was the point at which the surface reached 2 eV, because the temperature at the critical point (the sublimation temperature) for the majority of metals is close to 1 eV (0.8 eV for copper, 1.3 eV for tungsten), and hence, the metal at 2 eV is knowingly transformed to the state of the weakly ionized gas.

Figure 1 presents simulation data for copper, tungsten, aluminum, and titanium.

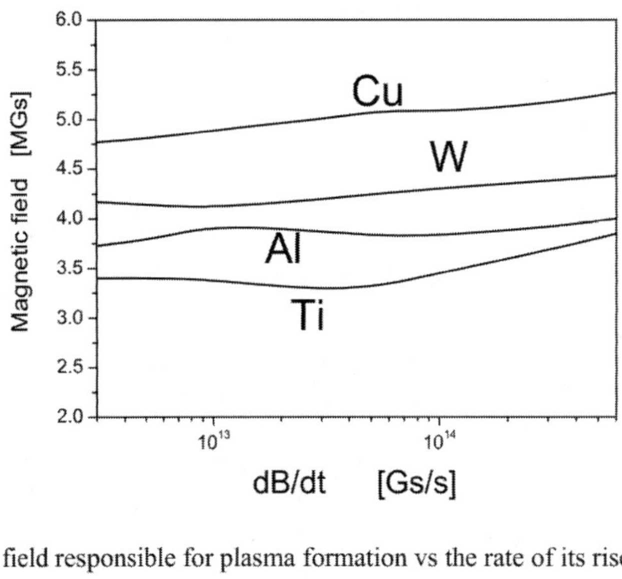

Figure 1. Magnetic field responsible for plasma formation vs the rate of its rise in plane geometry.

It is seen from the figure that in the plane geometry, the magnetic field at which plasma is formed depends very weakly on the rate of a field rise. The absolute values of the magnetic field are 5±0.25 MGs for copper, 4.25±0.2 MGs for tungsten, 3.85±0.15 MGs for aluminum, and 3.6±0.25 MGs for titanium.

Figures 2 and 3 present the simulation data in a cylindrical geometry for copper and titanium, respectively. The curves 1, 2, and 3 correspond to various conductor diameters: 2 mm (1), 1.5 mm (2), and 1 mm (3). It is seen that in the cylindrical geometry, the magnetic field, at which plasma is formed at the conductor surface, depends on the rate of the field rise. As the rate of rise of the magnetic field is increased, the magnetic induction, responsible for the formation of plasma, increases and tends to the value found in the plane geometry.

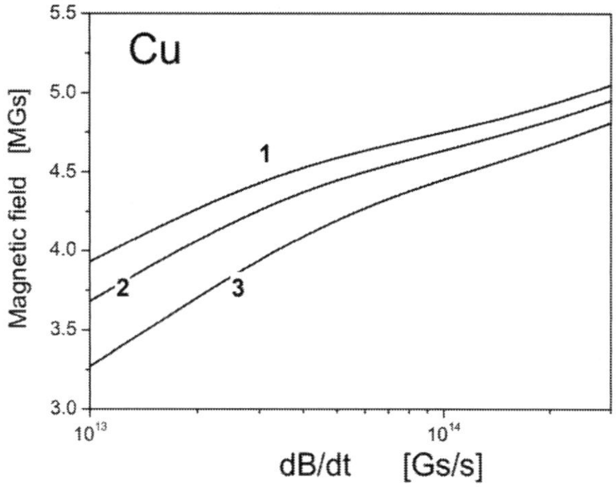

Figure 2. Magnetic field responsible for plasma formation vs the rate of its rise in cylindrical geometry for copper conductors of different diameters: 1 – 2 mm, 2 – 1.5 mm, 3 – 1 mm.

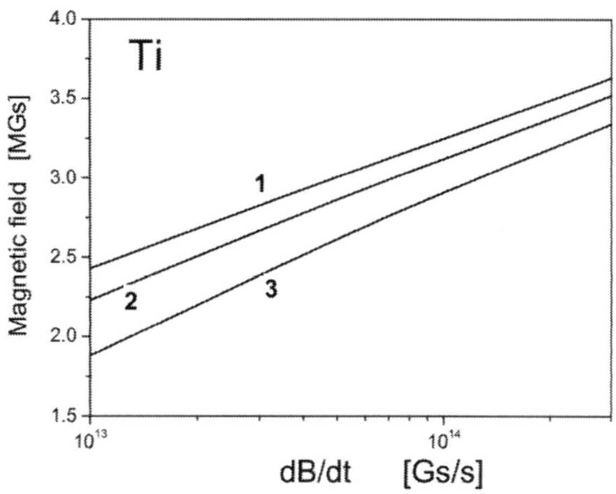

Figure 3. Magnetic field responsible for plasma formation vs the rate of its rise in cylindrical geometry for titanium conductors of different diameters: 1 – 2 mm, 2 – 1.5 mm, 3 – 1 mm.

The simulation results agree well with the experimental data [4,15,16], suggesting an adequacy of the used model.

4. Conclusion

The electrical explosions of conductors are generally simulated in the magnetohydrodynamic (MHD) approximation, requiring a knowledge of the equations of state for a wide range of thermodynamic parameters and the transport coefficients, one of the most important of which is the electrical conductivity. For describing the thermodynamic properties of metals, various semiempirical models and databases are available, whereas the issues related to the transport coefficients in the metal–dielectric transition region and near the critical point are implicit.

Acknowledgements

The work was supported in part by the Russian Foundation for Basic Research (grants No. 15-38-20617, 16-08-00969 and 16-32-00631).

References

[1] Knoepfel H 1972 Pulsed high magnetic fields [in Russian] Mir Moscow
[2] Shneerson G A 1973 *Zh. Tekh. Fiz* **18** (2) 419
[3] Awe T J, Peterson K J, Yu E P, McBride R D, Sinars D B, Gomez M R, Jennings C A, Martin M R, Rosenthal S E, Schroen D. 2016. *Physical Review Letters* **116(6)** 065001
[4] Chaikovsky S A, Oreshkin V I, Mesyats G A, Ratakhin N A, Datsko I M and Kablambaev B A 2009 *Physics of Plasmas* **16** 042701
[5] Oreshkin V I and Chaikovsky S A 2012 *Physics of Plasmas* **19** 022706
[6] Oreshkin V I 2008 *Physics of Plasmas* **15** 092103
[7] Baksht R, Tkachenko S, Romanova V, Mingaleev A, Oreshkin VI, Ter-Oganes'yan A, Khattatov T, Shelkovenko T, Pikuz S 2013 *Technical Physics.* **58(8)** 1129.
[8] Samarsky A, Popov Y P 1992 Difference Solution Methods of Gas-Dynamic Problems Nauka Moscow
[9] Oreshkin V I, Chaykovsky S A, Ratakhin N A, Grinenko A and Krasik Y E 2007 *Physics of Plasmas* **14** 102703
[10] Lomonosov I, Fortov V, Frolova A, Khishchenko K, Charakhchyan A, Shurshalov L. 2003 *Technical Physics* **48(6)** 727-735
[11] Bushman A V and Fortov V E 1987 *Sov. Tech. Rev. B: Therm. Phys* **1** 219
[12] Bakulin Yu D, Kuropatenko V F and Luchinsky A V 1976 *Zh. Tekh. Fiz. (in Russian)* **20** 1963
[13] Oreshkin V I, Khishchenko K V, Levashov P R, Rousskikh A, Chaikovskii S A 2012 *High Temperature* **50** 584
[14] Kalitkin N.N. 1978 Properties of matter and the MHD programs [in Russian] Preprint No 85 of Keldysh Institute Applied Mathematics Moscow
[15] Chaikovsky S A, Oreshkin VI, Datsko I M, Labetskaya N, Ratakhin N A 2014 *Physics of Plasma* **21** 042706
[16] Chaikovsky S A, Oreshkin V I, Datsko I M, Labetskaya N, Rybka D, Ratakhin N A 2015 *Physics of Plasmas* **22** 112704

IOP Conf. Series: Journal of Physics: Conf. Series **830** (2017) 012030 doi:10.1088/1742-6596/830/1/012030

Formation of long nanosecond rectangular pulses in the active RF pulse compression system with a compact resonant cavity

S N Artemenko[1], S A Gorev, V S Igumnov[2] and Yu G Yushkov

National Research Tomsk Polytechnic University, Tomsk, Russia

Email: [1]snartemenko@mail.ru, [2]vladislavigumnov@yahoo.com

Abstract. This work presents the results of the study of an active microwave pulse compression system capable of forming the rectangular pulses with duration ~10-100 ns, while its dimensions are several times smaller than the radiated wave train. Such compression system is based on the compact planar-voluminal resonant cavity constructed in the shape of a meander from waveguide sections and H-plane tees. The resonant cavity sections are parallel and are in the same plane with tees. The energy input element is located in the input end of the first section. The output device designed as an H-plane tee interference switch is connected to the output end of the last section. Each end of remaining sections is connected through a straight arm to an H-tee with a short-circuited quarter-wave second straight arm. The side arm of each tee is connected with the side arm of the tee in the next section, thus coupling the sections. The short-circuited arm provides the "open" mode and transmission of the wave from tee to tee. We determined the expressions for wave amplitudes in the components of the meander resonant cavity made of three sections and analyzed the expressions as the functions of the parameters determining the oscillation range and energy distribution in the resonant cavity. Experiments demonstrated that under certain conditions the compressors with such resonant cavity could generate nearly rectangular pulses with duration equal to the time of wave two-way traveling along the resonant cavity, and with the power compatible with that of the wave in the resonant cavity, and the length of the radiated wave train several-fold exceeding the size of the compressor. At pulse duration equal to 25 ns, the gain coefficient was 13 dB and pulse power was 40 MW. The work demonstrates the possibility to change the geometry of the resonant cavity by rearranging its components without changing the output pulse parameters.

1. Introduction

Work [1] describes the concept of active microwave pulse compressions systems which, being of a relatively small size, are capable of forming rectangular nanosecond RF pulses with the radiated wave train length several times exceeding the compression system dimensions. The concept is based on replacing a traditional energy-storing resonant cavity with a structured resonant system with variable geometry. The structural element of the system is replaced with a one- or moderately multimode waveguide section with H-plane tees or their analogues on both ends and a resonant length. Such elements with limiting tees with a shorted straight arm and a free side arm can transmit a working wave with very little reflection. This feature of structure elements and the capability of the tees to change the wave direction allow us to build compact storage systems with a relatively long time of working wave travel. When energy is extracted via an interference RF switch on the system's output port, this time determines the duration of the output rectangular pulse.

Paper [1] presents the first results of the experimental study of a microwave pulse compression system operating on the basis of the proposed concept and equipped with a resonant system made of two waveguide sections assembled as a meander half-period. The work demonstrates the compression system performance: practically rectangular pulses with duration 15 ns, gain 15 dB and power 60 MW were formed at 2800 MHz. The length of the radiated wave train four times exceeded the compression system dimensions.

This work contributes to the development of the concept by studying the RF compression system similar to one studied in the work [1] and by designing the form of a meander using three rather than two waveguide sections connected through limiting H-tees, as shown in Figure 1. The study focuses on the specific details of the work of such compression systems.

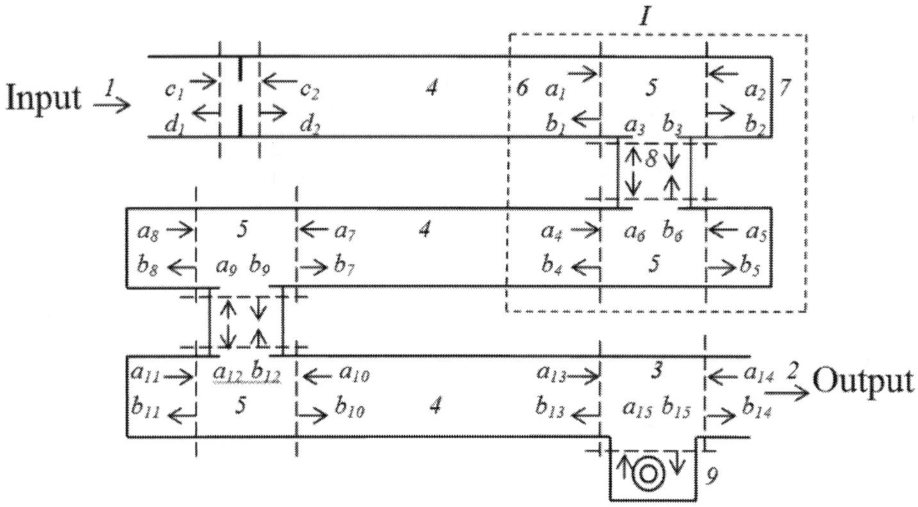

Figure 1. Schematic of the active microwave compression system with a compact storage cavity in the form of a meander. 1 – the input port with the energy input element; 2 – the output port with the energy extraction device 3; 4 – the waveguide sections; 5 – limiting H-tees with an input straight arm 6, a short-circuited second straight arm 7 and a shared side arm 8; 9 – the microwave switch.

2. Analysis

Particular interest represent those characteristics of the system, which determine the parameters of compression system input pulses. Thus, we need to determine the dependencies of wave amplitudes in system elements from the electrical length of the component parts of each element. These amplitudes show the energy distribution in the system and substantially influence the pulse power and the shape. Also, we need to determine the range of oscillations in the system and the energy distribution by oscillations depending on the section and tee arms lengths.

The analysis will be performed using the dispersion matrix method [2], where we can write down the following equations for the necessary amplitudes:

$$\begin{pmatrix} d_1 \\ d_2 \end{pmatrix} = S_1 \begin{pmatrix} c_1 \\ c_2 \end{pmatrix}, \begin{pmatrix} b_1 \\ b_2 \\ b_3 \end{pmatrix} = S_2 \begin{pmatrix} a_1 \\ a_2 \\ a_3 \end{pmatrix}, \begin{pmatrix} b_4 \\ b_5 \\ b_6 \end{pmatrix} = S_2 \begin{pmatrix} a_4 \\ a_5 \\ a_6 \end{pmatrix}, \begin{pmatrix} b_7 \\ b_8 \\ b_9 \end{pmatrix} = S_2 \begin{pmatrix} a_7 \\ a_8 \\ a_9 \end{pmatrix}, \begin{pmatrix} b_{10} \\ b_{11} \\ b_{12} \end{pmatrix} = S_2 \begin{pmatrix} a_{10} \\ a_{11} \\ a_{12} \end{pmatrix} \quad (1)$$

where a_i, c_i and d_i, b_i are amplitudes of the incident and reflected waves in the elements, S1 and S2 are the dispersion matrix of the energy feed device and H-plane tee with transmission coefficient h on the side of the side arm and

$$S_1 = \begin{pmatrix} -\sqrt{1-k^2} & jk \\ jk & -\sqrt{1-k^2} \end{pmatrix}, S = \begin{pmatrix} \frac{-1+\sqrt{1-h^2}}{2} & \frac{1+\sqrt{1-h^2}}{2} & \frac{h}{\sqrt{2}} \\ \frac{1+\sqrt{1-h^2}}{2} & \frac{-1+\sqrt{1-h^2}}{2} & \frac{h}{\sqrt{2}} \\ \frac{h}{\sqrt{2}} & \frac{h}{\sqrt{2}} & -\sqrt{1-h^2} \end{pmatrix} = \begin{vmatrix} s_{11} & s_{12} & s_{13} \\ s_{21} & s_{22} & s_{23} \\ s_{31} & s_{32} & s_{33} \end{vmatrix} \quad (2)$$

In addition, wave amplitudes in the system are connected with the following evident correlations:

$$a_1 = d_2 e^{-0.5\alpha - 0.5j\varphi}, c_2 = b_1 e^{-0.5\alpha - 0.5j\varphi}, a_2 = -b_2 e^{-\beta - j\psi}, a_3 = b_6 e^{-0.5\gamma - j0.5\xi}, a_6 = b_3 e^{-0.5\gamma - j0.5\xi}, a_5 = -b_5 e^{-\beta - j\psi},$$
$$a_4 = b_7 e^{-0.5\alpha - 0.5j\theta}, a_7 = b_4 e^{-0.5\alpha - 0.5j\theta}, a_8 = -b_8 e^{-\beta - j\psi}, a_9 = b_{12} e^{-0.5\gamma - j0.5\kappa}, a_{10} = -b_{10} e^{-0.5\alpha - j0.5\chi}, a_{11} = -b_{11} e^{-\beta - j\psi},$$
$$a_{12} = b_9 e^{-0.5\gamma - j0.5\kappa} \quad (3)$$

where α,β, γ, and φ, ψ,ξ,χ are, respectively, attenuation constants and phases of waves at the wave two-way traveling along the waveguide section between limiting tees and shorted straight and shared side arms of tees. Using (1)-(3), we can obtain expressions for wave amplitudes in all components of the system as the functions of transmission coefficient h and phasing constant in components:

$$b_6 = \frac{US_2}{1 - VS_2}, b_2 = A + Bb_6, b_1 = s_{11} e^{-0.5\alpha - 0.5j\varphi} - s_{12} e^{-\beta - j\psi} b_2 + s_{13} e^{-0.5\gamma - 0.5j\xi} b_6, b_{12} = Rb_9$$

$$b_3 = Ud_2 + Vb_6, b_5 = Cb_7 + Db_3, b_8 = Eb_4 + Fb_{12}, b_9 = Jb_4 + Kb_{12}, b_{10} = \frac{(M - LP)b_9}{1 - NL}, b_{11} = \frac{(P - MN)b_9}{1 - NL}$$

$$b_7 = \frac{XS_1Ud_2}{(1 - VS_2)(1 - WS_1)}, d_2 = \frac{jk}{1 + (1 - k^2)^{0.5} e^{-0.5\alpha - j\varphi} b_1}, b_4 = \frac{Ud_2}{(1 - VS_2)(1 - WS_1)}, b_{10} = \frac{(M - LP)b_4}{(1 - KR)(1 - NL)}$$

c

$$A = \frac{s_{21} e^{-0.5\alpha - 0.5j\varphi}}{1 + s_{22} e^{-\beta - j\psi}}, B = \frac{s_{23} e^{-\beta - j\psi}}{1 + s_{22} e^{-\beta - j\psi}}, C = \frac{s_{21} e^{-0.5\alpha - 0.5j\theta}}{1 + s_{22} e^{-\beta - j\psi}}, D = \frac{s_{23} e^{-0.5\alpha - 0.5j\psi}}{1 + s_{22} e^{-\beta - j\psi}}, F = \frac{s_{23} e^{-0.5\gamma - j0.5\kappa}}{1 + s_{22} e^{-\beta - j\psi}}, E = C$$

$$G = s_{11} e^{-0.5\alpha - 0.5j\theta} - s_{12} e^{-\beta - j\psi} E, H = s_{13} e^{-0.5\gamma - 0.5j\xi} - s_{12} e^{-\beta - j\psi} F$$

$$U = s_{31} e^{-0.5\alpha - 0.5j\varphi} - \frac{s_{32} s_{21} e^{-\beta - j\psi} e^{-0.5\alpha - 0.5j\varphi}}{1 + s_{22} e^{-\beta - j\psi}}, V = s_{33} e^{-0.5\gamma - 0.5j\xi} - \frac{s_{32} s_{23} e^{-\beta - j\psi} e^{-0.5\gamma - 0.5j\xi}}{1 + s_{22} e^{-\beta - j\psi}}$$

$$W = s_{11}e^{-0.5\alpha-0.5j\theta} - \frac{s_{12}s_{21}e^{-\beta-j\psi}e^{-0.5\alpha-0.5j\theta}}{1+s_{22}e^{-\beta-j\psi}}, X = s_{13}e^{-0.5\gamma-0.5j\xi} - \frac{s_{12}s_{21}e^{-\beta-j\psi}e^{-0.5\gamma-0.5j\xi}}{1+s_{22}e^{-\beta-j\psi}}$$

$$J = s_{31}e^{-0.5\alpha-0.5j\theta} - s_{32}e^{-\beta-j\psi}E, K = s_{33}e^{-0.5\gamma-0.5j\kappa} - s_{32}e^{-\beta-j\psi}F$$

$$L = \frac{s_{12}e^{-\beta-j\psi}}{1+s_{11}e^{-0.5\alpha-j0.5j\chi}}, M = \frac{s_{13}e^{-0.5\gamma-j\kappa}}{1+s_{11}e^{-0.5\alpha-j0.5j\chi}}, N = \frac{s_{21}e^{-0.5\alpha-j0.5j\chi}}{1+s_{22}e^{-\beta-j\psi}}, P = \frac{s_{123}e^{-0.5\gamma-j\kappa}}{1+s_{22}e^{-\beta-j\psi}}$$

$$R = -s_{31}e^{-0.5\alpha-0.5j\chi}\frac{M-PL}{1-LN} - s_{32}e^{-\beta-j\psi}\frac{P-MN}{1-LN} + s_{33}e^{-0.5\gamma-0.5j\kappa}$$

$$S_1 = s_{11}e^{-0.5\alpha-0.5j\theta} - \frac{s_{32}s_{21}e^{-\beta-j\psi}}{1+s_{22}e^{-\beta-j\psi}} + \left[s_{13}e^{-0.5\alpha-0.5j\theta} - s_{12}s_{23}e^{-\beta-j\psi}\frac{e^{-0.5-0.5j\theta}}{1+s_{22}e^{-\beta-j\psi}}\right]\frac{JR}{1-KR}$$

$$S_2 = s_{31}e^{-0.5\alpha-0.5j\theta} - \frac{s_{32}s_{21}e^{-0.5\alpha-0.5j\theta}e^{-\beta-j\psi}}{\left(1+s_{22}e^{-\beta-j\psi}\right)}\frac{XS_1}{1-WS_1} + \left[s_{33}e^{-0.5\gamma-0.5j\xi} - s_{23}s_{32}e^{-\beta-j\psi}\frac{e^{-0.5\gamma-0.5j\xi}}{1+s_{22}e^{-\beta-j\psi}}\right]$$

The obtained expressions will help us to find the behaviors of the system parameters dependencies in question. Thus, Figure 2 shows the dependencies of the wave power gain in the sections from the electrical length of the first section at different values of transmission coefficient h of the tee on the side of the side arm and two different lengths (quarter-wave or half-wave) of shorted straight arms. As is seen from the dependencies, the gain in sections is equal for all h (lines 1-3 и 4-6) at quarter-wave arms. However, this gain is reduced as h values become lower. For example, when h falls from 1 (the tee is matched on the side of the side arm) to 0.8 (the tee is not matched on the side of the side arm), the gain in sections drops by 50%, i.e. falls by 2 dB. The reason for this reduction becomes clear if we determine the gain distribution in all components of the system. The character of the distribution suggests that the gain reduction in the sections is connected with growing losses in the side arms due to the growth of the accumulated energy density. As h becomes lower, the low-Q cavity represented by the shared side arm increases the load on the main storing capacity of the system, i.e. the capacity of sections. When h falls from 1 to 0.8, the wave power in side arms grows from the half-power in the sections to the level two times higher. When h = 0.945, the wave powers are equal in the sections and side arms. In the given case all sections of the system and side arms of tees are half-wave ones. The coupling aperture transmission factor exceeded the optimal value for one section but was less than this value for three sections.

Changing the connection with the side arm results in changing the working frequency of the system. The same result is observed with different lengths of the sections. However, this difference does not influence the distribution of energy among sections. At the same time, when h is changed, the energy in the system is redistributed, which is manifested in the changes of energy correlations in side arms and sections.

After we change the length of a shorted quarter-wave arm to half-wave, the energy in the first section grows, while the stored energy in the section and third sections falls close to the zero level. This situation is shown by lines 7,8 and 9, respectively (Figure 2). This redistribution of energy can be obviously used for the formation of pulses of various shape, duration and gain, which is supported by the specific dependency of the gain in the parts of the resonant system element on the length of the shorted straight arm. As is seen from the dependencies in Figure 3, the change in the arm length exerts strong influence on gain only near the half-wave length of the arm. This factor allows us to implement control of pulse parameter changes.

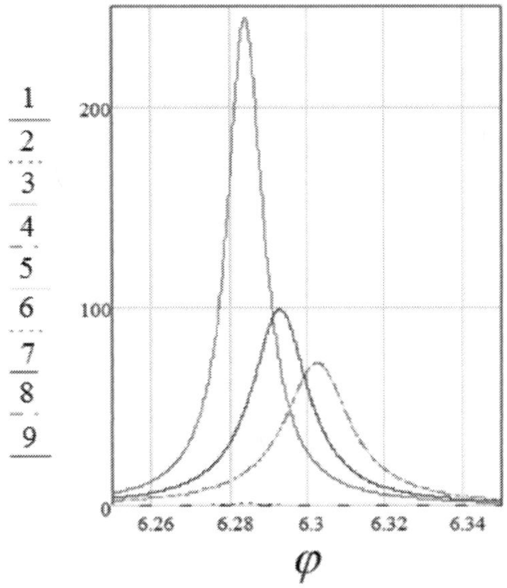

Figure 2. Dependencies of the wave power gain in the sections from the electrical length of the first section at different values of transmission coefficient of the tees on the side of the side tee and two different values of length ψ of shorted straight arms. Lines 1-3 and 4-6 – h =1, 0.8 respectively, ψ = π/2; line 7 – h = 1, ψ = π, gain in the first section; lines 8,9 - h = 1, ψ = π, gain in the second and third sections.

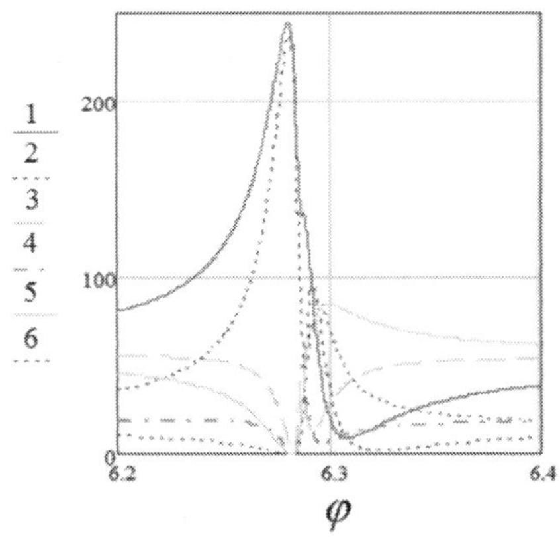

Figure 3. Dependencies of wave power gain in sections from the electrical length of a short-circuited arm near its half-wave length for a tee matched (lines 1-3) and unmatched on the side of the side arm.

After we change the length of a shorted quarter-wave arm to half-wave, the energy in the first section grows, while the stored energy in the section and third sections falls close to the zero level. This situation is shown by lines 7,8 and 9, respectively (Figure 2). This redistribution of energy can be obviously used for the formation of pulses of various shape, duration and gain, which is supported by the specific dependency of the gain in the parts of the resonant system element on the length of the shorted straight arm. As is seen from the dependencies in Figure 3, the change in the arm length exerts strong influence on gain only near the half-wave length of the arm. This factor allows us to implement control of pulse parameter changes.

3. Experiments

As in work [1], the experimental studies were carried out using the S-band compression system with the planar-volumetric storage cavity designed as a meander but made of two structure elements (instead of three). Structure elements, as in work [1], were also designed as identical sections of a round waveguide with 90 mm diameter limited with H-tees made of the same waveguide (Figure 4). The coupling aperture was located on the input end of the first section. The second end was connected to the straight arm of the first limiting tee. The second straight arm of the tee was made with a quarter-wave length and was short-circuited. The side half-wave arm was connected to the corresponding arm of the limiting tee of the second section. The tee in the second section was connected with the tee in the third section in the same manner. Thus, the side arms of the section were connected in an integrated flat resonant system where each tee could be transformed into the energy extraction device. This could be simply performed by replacing the short-circuiter of the straight arm with an interference RF switch.

The main goal of the experiment was to demonstrate that this compression system could generate nanosecond rectangular pulses which would be longer than those described in work [1] and thus to prove the efficiency of the concept. For this purpose, the energy was extracted from the interference RF switch connected to the output end of the last (third) section. The switch tee was coplanar to the limiting tees, while the compression system output port was the open straight arm of the switch, and the closed half-wave side arm housed the MW switch. The switch tee on the side arm side was not matching and had the 0.8 transmission coefficient on the side of this arm. As in work [1], the switch in the switch unit was a trigatron RF discharge switch a vented gas discharge tube. The tube was in the electric field peak and oriented by the waveguide diameter in the plane of polarization of the transmitting wave. The electrical discharge switch of the lighting of the switch discharge gap was in one end of the tube.

Figure 4. Active microwave compression system with a compact storage cavity in the form of a meander.

At working frequency ~2800 MHz, the cavity length was 650 mm from the axis of the one limiting tee side arm to the axis of another tee. The length of the shorted straight arm was ~110mm from the short circuitor to the axis of the side arm, i.e. was close to the quarter-wave geometrical length of the arm. The total length of side arms between the axes of the section was ~178 mm. With this length of the sections and the shared arm of limiting tees, H11(31) oscillations were excited in the resonant cavity. Twenty seven half waves were in the sections and four half waves were in the shared arms.

The measured intrinsic Q-factor of the resonant cavity was ~2.1×104. The estimated two-way travel time of the transmitted wave along all system components was ~20ns, while the maximum estimated gain coefficient of the compression system equal to the gain of the resonant cavity excited with the infinitely long pulse was equal to ~18 dB. With account for the accumulation time limited with 3 μs and the corresponding duration of the input pulse in the experiments with high power level and considering ~2-3dB loss at the extraction, the anticipated gain coefficient was ~14-15dB.

The magnetron oscillator with pulse power up to ~2MW was used as a source of input pulses. The insulating gas in the cavity was gaseous nitrogen at 3-4 ATMG with 10-12% insulating gas. Switching to output mode was performed in the argon medium in the discharger tube under pressure 2-3 ATMG or in the mixture of argon with ~5-10% insulating gas at 0-1 ATMG.

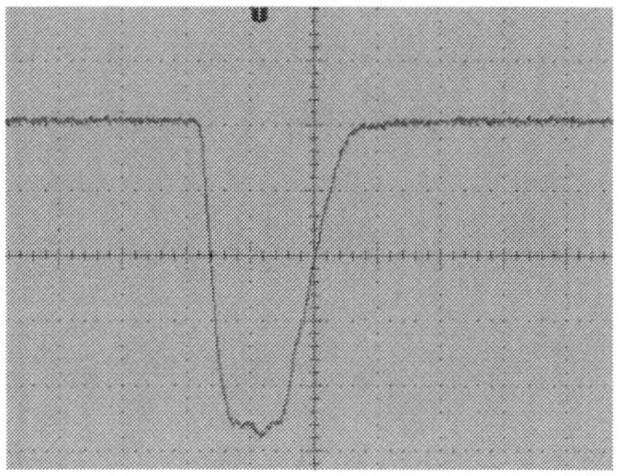

Figure 5. Oscillogram of the output pulse of the RF compression system with switching in argon with 7-10% insulating gas: gain 13 dB, power 40 MW, duration 25 ns.

Figure 5 shows the oscillogram of a RF pulse with a nearly rectangular form generated by the compression system. The pulse reached 13 dB gain, 24-25 ns duration and approximately 40 MW power. The pulses were formed with switching in argon with a relatively significant share of insulating gas. When the share of insulating gas was decreased up to its complete removal from the mixture, the pulse shape was deformed, as shown in the oscillograms in Figure 6. An adequately supported explanation of such deformation has not been found yet. The reason for the distortion of the pulse shape will be the subject of further research.

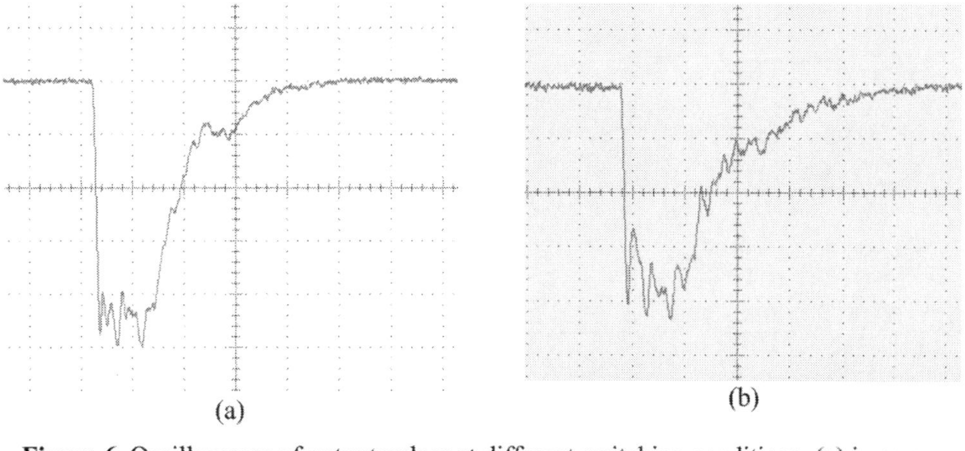

Figure 6. Oscillograms of output pulses at different switching conditions, (a) in argon; (b) in argon with 5-7% insulating gas.

4. Summary and Conclusions

The paper summarizes the study of the active microwave pulse compression system with energy storage in the resonant system made in the form of a meander. Due to such form, the system has relatively small dimensions and the wave travel time in the system is much larger than the travel time in the system of maximum dimensions. The system is designed on the basis of the structure element made as a waveguide section whose ends are limited with two H-tees. The connection of the sections through tees allows building the resonant cavity as a one-, two- or three-dimensional structures. It was proved that the compression system of this kind is capable of generating pulses with the pulse shape very close to rectangular and the length of the radiated wave train several times exceeding the compression system dimensions. In fact, these configurations are a set of traditional RF compression systems based on H-tees aligned along a straight line, either in one plane or in three dimensions. Due to this approach, we can build compact active RF compression systems that can generate long rectangular nanosecond pulses and can be fitted effectively into the RF unit.

Acknowledgments

This work was supported in part in the scope of the Program on National Research Tomsk Polytechnic University Competitiveness Enhancement Program and partially supported by grant RFBR № 15-08-01853.

References

[1] Artemenko S, Beverly III R, Gorev S, Igumnov V, *Proceedings of 2016 IEEE International Power Modulator and High Voltage Conference*, 2016
[2] Altman J L, *Microwave Circuits*, Princeton 1964, N.J.: Van Nostrand

IOP Conf. Series: Journal of Physics: Conf. Series **830** (2017) 012031 doi:10.1088/1742-6596/830/1/012031

Angular distribution of beam electrons in a source with arc plasma emitter

V Kurkuchekov, V Astrelin, I Kandaurov and Yu Trunev

BINP of SB RAS, Novosibirsk, Russia

kurkuchekov.victor@gmail.com

Abstract. Results on studying the angular characteristics of an electron beam, generated in a multi-aperture diode with an arc-discharge plasma emitter are reported. The main beam parameters were as follows: the electron energy up to 120 keV, the emission current up to 100 A, the pulse duration 0.1 - 0.3 ms, and the initial diameter ca. 8 cm. The beam was formed and transported to a metal target in an adiabatically converging magnetic field. The diagnostic technique based on an X-ray imaging of the profiles of individual beamlets passed through the pepperpot-like mask was developed and used to investigate an angular distribution of the beam electrons. The spatial resolution of the diagnostic was evaluated in a special test experiment and found to be not worse than 4 lp/cm at a 10 % contrast level. It was demonstrated that an angular distribution of the beam electrons fits well by the Gaussian function with the RMS width ~ 0.067 rad. The data on the angular distribution measured with pepperpot diagnostic are in a good agreement with those obtained in the experiments on the beam passage through a magnetic mirror.

1. Introduction

The electron beam source has been developed in Budker Institute, Novosibirsk. It is based on an arc-discharge plasma cathode and a multi-aperture electron optical system (EOS) [1]. The diode-type EOS assembles two plain molybdenum electrodes with 241 round apertures drilled in a hexagonal grid. An efficient diameter of EOS was ca. 8 cm. The main beam parameters were as follows: the electron energy up to 120 keV, the emission current up to 100 A, and the pulse duration of a sub-millisecond range. The beam was transported and compressed adiabatically in the converging magnetic field. A metal target was exposed to a power density of 1-10 GW/m^2 that is necessary for some material science and engineering applications. To achieve such power density, the 30-fold or more compression of the initial beam current density is required. Assuming the adiabatic invariance of the magnetic moment, a high degree of compression needs some requirements to the initial pitch angles of electrons. Thus, the information about the beam angular characteristics is of a paramount importance.

In the presented work, we study the angular distribution of electrons in a beam generated in a multi-aperture, plasma-emission diode. For this purpose, the diagnostic system based on pepperpot technique was developed. It allows us to measure the angular distribution of electrons over the beam cross-section in a single shot. The spatial resolution of the diagnostic system was evaluated in a special test experiment and was found to be not worse than 4 line pairs per centimeter at the 10 % contrast level. The algorithm of experimental data processing is also described. The experimental results are compared with those obtained in the experiments on the beam transportation through a magnetic mirror.

2. Experimental layout

The experiments were carried out using the electron beam test facility designed for the simulation of high thermal loading on the construction materials [2]. A layout of the experiments is shown in **Figure 1**. The hollow high-voltage electrode (*1*) was mounted inside a vacuum tank on a PMMA bushing (*3*). The inner surface of the high-voltage electrode serves as the distributed anode of an arc discharge that creates an emission plasma (*2*). The molybdenum plate at the face end of the hollow electrode with 241 round openings of 3 mm diameter is a cathode of an accelerator diode. The grounded anode electrode of the diode (*4*) employed 4.4 mm openings aligned precisely with their cathode counterparts.

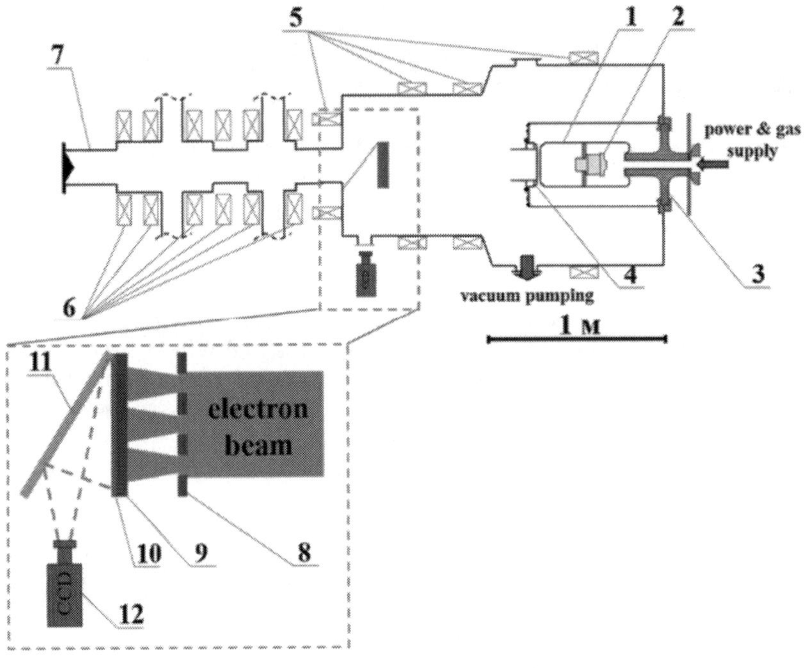

Figure 1. Experiment layout: 1 - high voltage electrode (hollow anode of the arc discharge) with emission electrode; 2 - arc plasma generator; 3 - high voltage isolator; 4 - extraction electrode (diode anode); 5 - vacuum tank magnetic field coils;6 – solenoid section coils; 7 – Faraday cup; 8 – pepperpot mask; 9 - beam collector (x-ray converter); 10 - phosphor screen; 11 - 45°-angled mirror; 12 - fast CCD-camera;

An accelerated beam was transported in a guiding magnetic field to the Faraday cup located at 2 m from the beam source. The field of mirror configuration was produced using coils placed at the vacuum tank exterior (*5*) and at the solenoid section of the facility (*6*). The pepperpot system was installed at 0.85 m from the beam source on a movable holder. The beam was received onto a stainless steel mask with an array of pinholes of 2mm diameter, arranged with a pitch of 14 mm horizontally and vertically, covering the square of 120x120 mm. As a result, the number of beamlets were cut out from the initial beam. These beamlets were absorbed in a thin metal plate placed downstream of the pepperpot mask and served as the X-ray converter. The phosphor screen (Gd_2O_2S:Tb phosphor on the plastic substrate) was attached directly to the back side of the converter plate for imaging the beam X-ray footprint. The luminous spots on the screen were viewed via a 45° mirror using a fast CCD camera. The luminescence decay time for Gd_2O_2S:Tb was about 0.6 ms, whereas a typical beam duration in the experiments was 0.1 – 0.3 ms. Thus, the time integrated measurements were realized.

The increase of the distance between the pepperpot mask and X-ray converter allows us to obtain a better resolution of the diagnostic. However, in this case, an overlapping of the luminous spots may occur, that complicates the results of interpretation. The electron trajectories are also affected by the guiding magnetic field, if the following condition is not met:

$$L \ll \lambda = V_e \frac{2\pi}{\Omega}, \quad (1)$$

where L – the distance between the mask and the x-ray converter, λ - a step of Larmor spiral. In our case ($B \sim 0.01$ T, $E_e \sim 80$ keV), the Larmor step is $\lambda \approx 587$ mm. Thus, the distance between the mask and the X-ray converter was chosen as 2 cm.

The material and thickness of the beam converter are crucial for the amount of light emitted by the phosphor screen. The number of emitted photons S can be calculated as:

$$S = k \cdot \int_E N(E) \cdot T(E) \cdot A(E) \cdot E dE, \quad (2)$$

where E – the X-ray photon energy, $N(E)$ – the spectral response of the X-ray converter, $T(E)$ – the converter transmittance, $A(E)$ – the absorption coefficient of the phosphor layer and k is the light yield of the phosphor (k =35 photon/keV). The spectral response of the collector was determined using the model reported in [3, 4]. The phosphor absorption coefficient was calculated, taking into account the surface density of the active substance – 55 mg/cm2. The number of photons was calculated for various materials: W, Ta, and a stainless steel with the beam parameters as follows: $E_b = 100$ keV, $I_b = 100$ A, $\tau_b = 0.15$ ms. All calculations use the material properties specified in the NIST database [5].

The number of photons vs. the collector thickness is shown in the Fig. 3. Over the entire range of the converter thickness, tantalum produced the highest light yield. Thus, tantalum was chosen for the X-ray converter as the most optimal and the easiest material for processing. The thickness of the converter was taken as 0.01 cm.

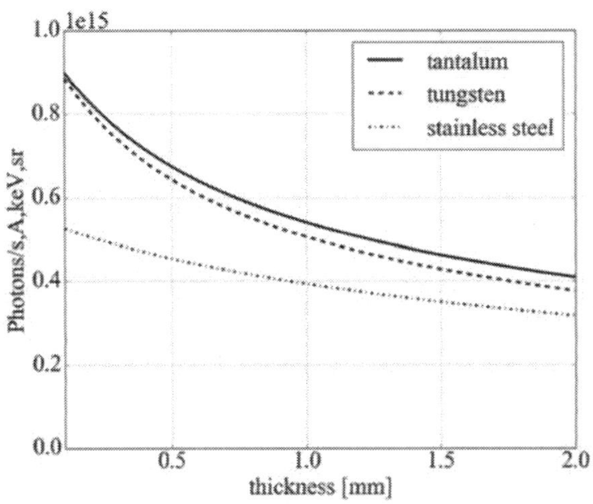

Figure 2. Dependence of light yield on x-ray converter thickness for different materials.

3. Experiments and results

A typical pepperpot image is shown in figure 3 for the beam with the following parameters: 80 keV, 70 A, 0.1 ms. In the upper part of the figure, an image intensity profile is depicted. It was measured along the horizontal dotted line. As it is seen in the figure, the X-ray footprints of separate beamlets are clearly distinguishable. Due to non-uniformity of the beam current profile, the brightness of the footprints has a maximum in the center of the beam and decreases to the periphery. A visible dark strip in the image is a result of the converter design, which comprises two overlapping pieces of the tantalum foil. The pepperpot image reveals a substantial non-uniform background glow that complicates the recovery of angular beam characteristics. This background may be caused both by the electrons with large angles and the X-rays, that were generated at the pepperpot mask and passed onto the phosphor screen directly through the mask material.

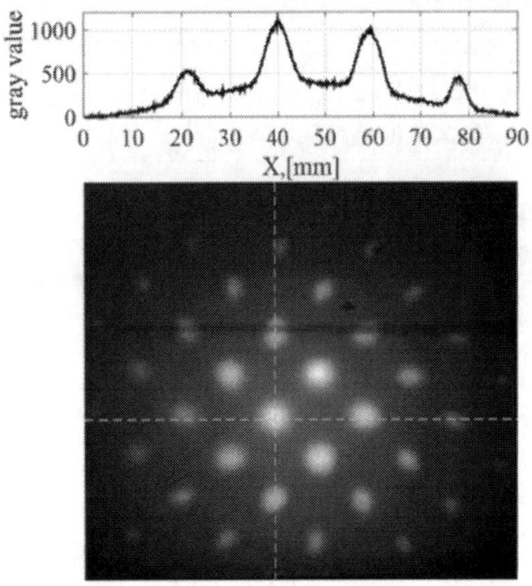

Figure 3. Typical X-ray image of the beamlets, and intensity profile measured along horizontal dashed line crossing the center of the beamlet.

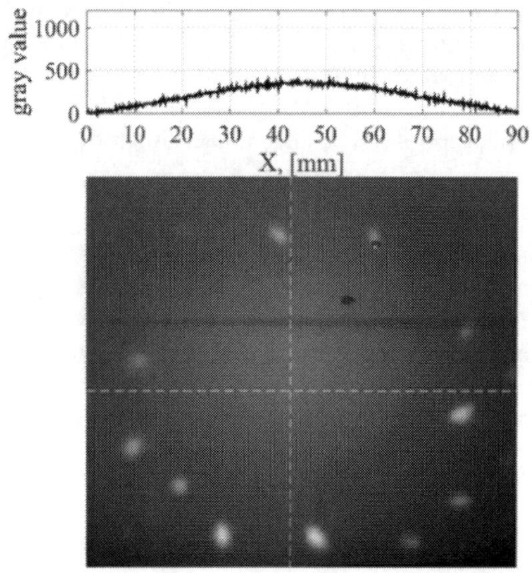

Figure 4. Image of background, and intensity profile measured along horizontal dashed line.

To evaluate the background origin, a special experiment was carried out, where the major part of pepperpot pinholes was covered with a stainless steel foil of 100 μm thick. An example of the resulting image is shown in figure 4. The luminescence intensity profile was measured along the horizontal line crossing the center of image. It can be seen, the profile of the background light did not changed indicating that it was caused only by the X-rays passing through the pepperpot plate material. Under further image processing, this background should be subtracted from the signal.

An additional experiment on measuring the sharp edge spread function has been performed to determine a spatial resolution of the imaging technique. In this experiment, the pepperpot mask was dismantled, and the beam was stopped directly by the X-ray converter. The edge was formed with a tungsten plate of 0.1 cm thickness, which covered a part of the converter surface and completely shadowed the X-rays. On a basis of the measured edge profile, the line transfer function and the modulation transfer function were calculated [6]. The diagnostic's spatial resolution was found not worse than 4 line pairs per cm at the 10% contrast level.

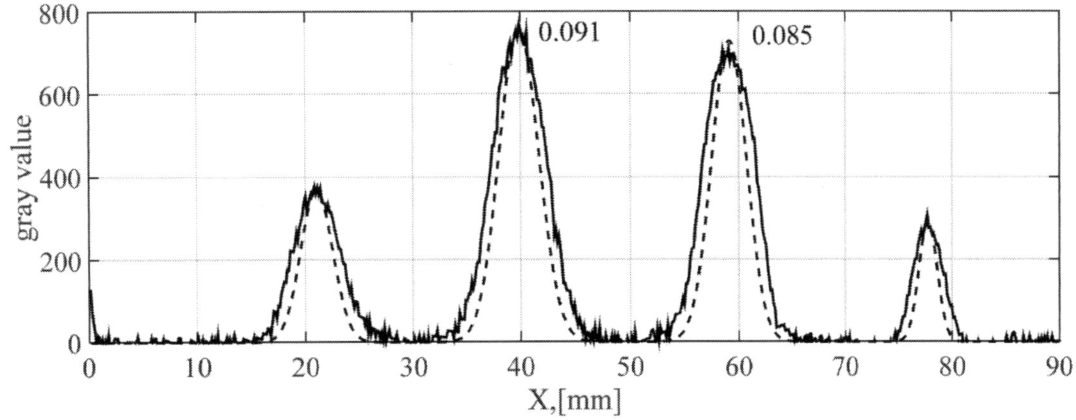

Figure 5. Beamlets intensity profile processing: solid curve - intensity profile after background subtraction; dashed curve – resulting beamlets profile.

The image intensity profile after processing is shown in figure 5. The solid curve corresponds to the intensity profile after background subtraction. For the convenience of further processing, the profile of every beamlet was interpolated with a Gaussian function:

$$F(x) = a_1 \cdot e^{\frac{-(X-a_2)^2}{2a_3^2}},$$

where a_1, a_2, a_3 – interpolation parameters. The experimentally observed profiles are the result of the convolution of the electron angular distribution function with the step function, which describes the pepperpot orifice and the line spread function. Thus, the deconvolution procedure was carried out, to find the angular distribution function of the beam electrons. The result is shown in figure 5 by a dashed curve. For two most bright beamlets, the RMS width of the Gaussian approximation was found (in radian terms) as 0.091 and 0.085. It should be noted that the magnetic field value at the diode and at the diagnostic location was 6.5 mT and 12 mT, respectively. Thus, assuming the conservation of the magnetic moment, the RMS width of the angular distribution function at the diode exit should be found as ca. 0.067 rad.

To verify the adiabaticity of the beam magnetic compression, a number of experiments, where the guiding magnetic field at the pepperpot location varied from 12 mT to 32 mT, while the field at the diode region was constant, was carried out. The resulting dependence of the beam angular spread vs. the magnetic field value is shown in figure 6.

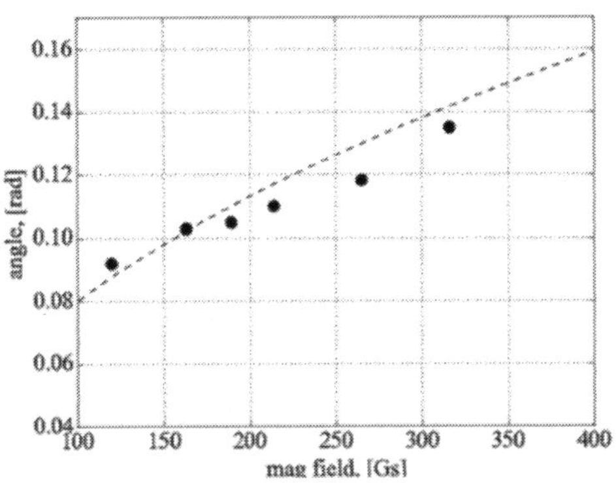

Figure 6. Dependence of the beam angular spread vs. magnetic field value.

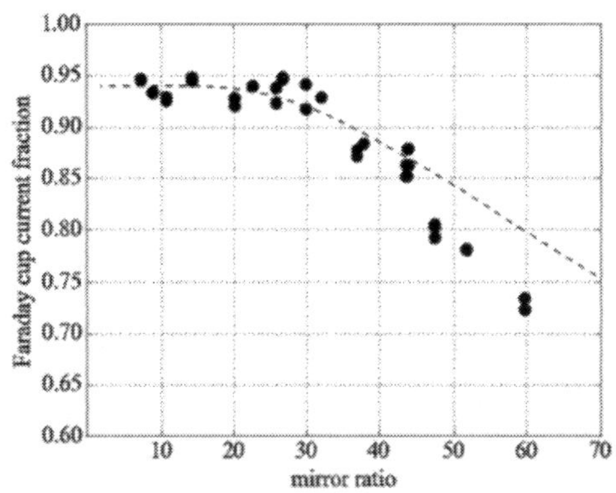

Figure 7. dependence of the beam transmission vs. mirror ratio

The points corresponds to the Gaussian RMS width measured in the experiments. The dashed curve is derived from the invariance of the magnetic moment and is given by the formula:

$$\theta(B) = arctg\left(\mathrm{tg}(\theta_0) \cdot \sqrt{\frac{B}{B_0}}\right),$$

where θ_0, B_0 – the RMS angle and the magnetic field value at the diode location, θ and B – the angle and the field at the diagnostic location. The previously obtained values, 0.067 rad and 6.5 mT, were used as θ_0 and B_0 , respectively. As it can be seen from the figure, the experimental dependence correlates satisfactory with theoretical predictions. That proves the adiabaticity of the beam compression under our experimental conditions.

This fact allows us to compare the measured beam angular distribution with the beam transition through the magnetic mirror. In these experiments, the beam parameters were constant, while the magnetic field mirror ratio varied within the range from 7 to 60. The beam current through the magnetic mirror was measured with a Faraday cup (FC). The resulting dependence of the beam transmission vs. mirror ratio is shown in figure 7. As it can be seen, about 95 % of the beam emission current passed to the FC for the mirror ratio value from 7 to 30. With a further increase of the mirror ratio, the current on the FC decreased gradually. For the mirror ratio of 60 only about 72 % of the initial current passed to the FC. For higher mirror ratios, the measurements were not carried out because of the beam pulse shortening due to the electrons reflected from the mirror. The dashed curve is derived within an assumption that the beam angular distribution is described by the Gaussian function with the RMS width of 0.067 rad. Overall, the theoretical curve agrees well with the experimental data, except the highest mirror ratio values when the FC current declines faster. This difference, apparently, can be explained due to the effect of the beam space charge [7].

4. Conclusion

The angular characteristics of the electron beam generated in the source with plasma cathode and the multi-aperture diode have been studied. For this purpose, the imaging X-ray diagnostic tool based on the pepperpot technique was developed. Using this diagnostic, it was demonstrated that the beam angular distribution could be fairly good described with the Gaussian function with the RMS width of ca. 0.067 rad. The results are in a good agreement with those obtained in the experiments on the beam pass through the magnetic mirror.

Acknowledgements

The work at the electron beam facility was supported by Russian Science Foundation (project N 14-50-00080). The development of beam angular distribution diagnostic supported by RFBR, research project No. 16-38-00739 mol_a.

References

[1] V. V. Kurkuchekov, et al. 2013 *Fusion Science and Technology,* **63** № 1T 292-294

[2] Leonid V, Arakcheev A, Burdakov A, Kandaurov I, Kasatov A, Kurkuchekov V, Mekler K, Popov V, Shoshin A, Skovorodin D, Trunev Yu and Vasilyev A 2016 *AIP Conf. Proc.* **1771**

[3] Horst E 1999 *X-ray Spectrometry* **28** 255–266

[4] Horst E 2006 *Advances in X-Ray Analysis* **49** 267 – 273

[5] Hubbell J and Seltzer S 2004 *Available: http://physics.nist.gov/xaamdi. National Institute of Standards and Technology, Gaithersburg*

[6] Smith S, 1999 *California Technical Publishing* 423

[7] Astrelin V, Kandaurov I, Kurkuchekov V and Trunev Yu 2016 *AIP Conf. Proc.* **1771**

IOP Conf. Series: Journal of Physics: Conf. Series **830** (2017) 012032 doi:10.1088/1742-6596/830/1/012032

Study of electron beam uniformity in large-area multi-aperture diode with arc plasma cathode

I V Kandaurov[1], V V Kurkuchekov[1,2] and Yu A Trunev[1,2]

[1] Budker Institute of Nuclear Physics of SB RAS, Novosibirsk, 630090, Russia
[2] Novosibirsk State University, Novosibirsk, 630090, Russia

E-mail: I.V.Kandaurov@inp.nsk.su

Abstract. The use of plasma emission cathode in the conjunction with a multiple apertured electron optical system (EOS) is promising for the multi-MW class electron beams of a large cross-sectional area. In a multi-aperture source, the beam parameters could be raised simply due to increase of the number of apertures (i.e. an effective emission area), if a uniformity of the electron emission over a large-area plasma cathode is ensured. In the presented paper, the cross-sectional distribution of the emission current density was investigated using the X-ray diagnostic technique for two versions of the diode-type EOS, with electrodes performed as flat molybdenum "grids". The first one had 241 apertures arranged hexagonally inside a circle with a diameter of 8.3 cm and the second had 499 apertures within a circle of 11.8cm diameter. The emission plasma is produced using a single arc-discharge plasma generator placed on the axis at 20 cm from the EOS. It was demonstrated that multi-aperture systems with a single on-axis plasma generator can be effectively employed to obtain large-area beams, even in the presence of the guiding magnetic field. All apertures are emitting in the 499-apertured EOS. The beam current density is quite uniform up to the radius 2.5cm and gradually decreases to the periphery.

1. Introduction

The high-power beams of subrelativistic electrons with a cross-sectional area ~ 10^2 cm^2 are used in a number of scientific and technical applications, particularly, in a material science and engineering, for surface modification of solids, modeling of high thermal fluxes on plasma facing components in fusion reactors, etc. The use of a plasma electron emitter is very promising for the large area beam application due to a lack of engineering issues inherent to the large area thermionic cathodes.

An electron beam source based on the plasma cathode and the multi-aperture electron optical system (EOS) has been developed at Budker Institute, Novosibirsk. The beam parameters (non-simultaneous) are as follows: the energy up to 120 keV, the total current up to 250 A, the pulse length up to 1 ms. The beam extraction/acceleration needs a multi-aperture diode optics. The cathode and the anode are as flat molybdenum “grids” with multiple round openings, which are drilled in a hexagonal order. The emission plasma is produced with a single arc discharge plasma generator placed at an axis. Details of the beam source design are given elsewhere [1].

The concept of the multi-aperture diode is well suited for generation of the beam with a large cross-sectional area. An increase of the emission current here can be achieved, in general, simply by increasing the number of grid openings (i.e., an effective emission area). However, in the scheme with a single on-axis emission plasma generator, this approach faces the challenge of radial uniformity of the emission plasma, in other words, the efficiency of use of the peripheral apertures in the EOS with a large cross-section. The problem becomes exacerbated, as the guiding magnetic field is applied to transport and to compress the electron beam, producing a required current density on a target. The experiments performed earlier have shown that the presence of the magnetic field leads to the contraction of an arc discharge in the near-axis region.

This paper presents an experimental study of the emission current density distribution over the multi-aperture EOS for two versions of various sizes. The first one had 241 round aperture arranged in a hexagonal order inside a circle with a diameter of 8.3 cm and the second had 499 apertures within a circle with a diameter of 11.8 cm. A diameter of the emission openings was 0.3 cm and 0.26 cm, and an effective emission (open) area was ca. 17 cm^2 and ca. 26 cm^2, respectively. The diode gap was 1 cm in both cases.

An X-ray imaging diagnostic technique was developed to measure the beam profile on a flat metal target. The diagnostic system allows obtaining a 2D-picture of the cross-sectional distribution of the beam current density in every single shot. The spatial resolution of this diagnostic was found using special tests: 4 pairs of lines per 1 cm at a 10% contrast level.

2. Experimental setup

The experiments have been carried out employing an electron beam test facility designed for the simulation of pulsed high heat loads effect on the construction materials [2]. The schematic of the experimental setup is shown in figure 1. The beam source is installed inside the vacuum barrel. It is immersed in an axial magnetic field produced by external magnetic windings on the barrel. The field typical value is 7 - 10 mT at the diode region and increased by 1.7 times at the beam collector located at 0.85 m downstream of the diode. The accelerating voltage is applied to the diode cathode which is performed as a hollow cylinder, made of a stainless steel. An emission grid, through which electrons are extracted, is mounted at the end face of the cylinder. The emission plasma is generated by an arc discharge source of a washer-gun type [3, 4] placed at a distance of 20 cm from the emission grid.

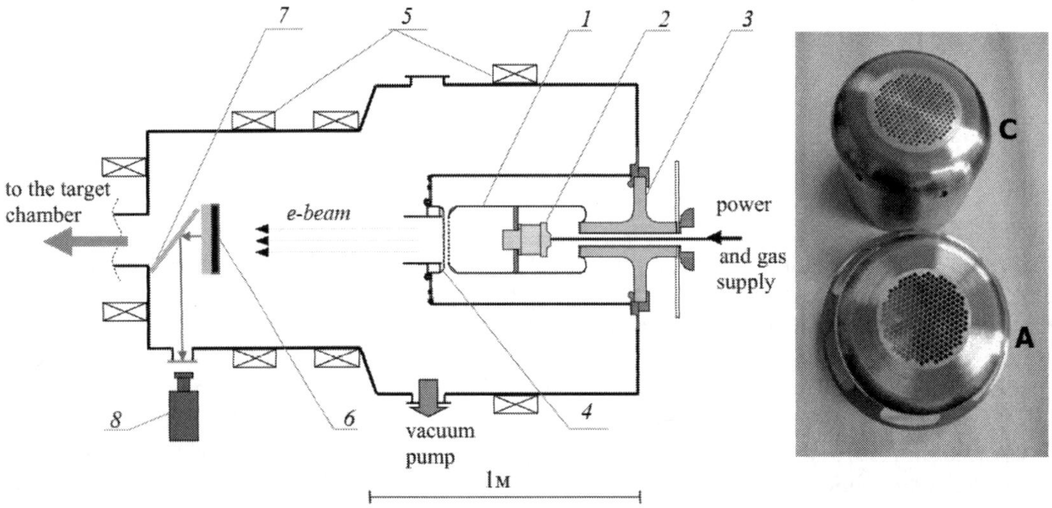

Figure 1. Schematic layout of the experimental setup: *1*– hollow high-voltage electrode with emission grid (cathode); *2*– washer gun plasma source; *3*– feedthrough insulator; *4*– extracting grid (anode); *5*– magnetic coils; *6*– beam collector with attached phosphor screen; *7*– 45°-mirror; *8*–fast CCD-camera.

An emission plasma generator is equipped with a small magnetic coil, which provides a magnetic field up to 800 mT in the arc channel (the initial segment of the arc discharge path). Thus, the magnetic field lines are strongly divergent in a space between the arc generator and the emission grid. In this way, in principle, the conditions for spreading the discharge plasma over the emission grid, providing a better emission uniformity across the wide-area EOS, are created.

2.1. X-ray diagnostic system

The beam collector is designed as a thin metal plate and serves as an X-ray converter. The phosphor screen (Gd_2O_2S:Tb phosphor on a thin plastic film) is attached directly to the back side of the collector plate for imaging the beam X-ray footprint. The phosphor glow is registered with the use of a 45° mirror and a fast CCD camera. It should be noted that the luminescence decay time for Gd_2O_2S:Tb is of ca. 0.6 ms, whereas a typical beam duration in the presented experiments is of ca. 0.2 ms. Thus, the image current density distribution on the collector integrated over the beam period could be imaged.

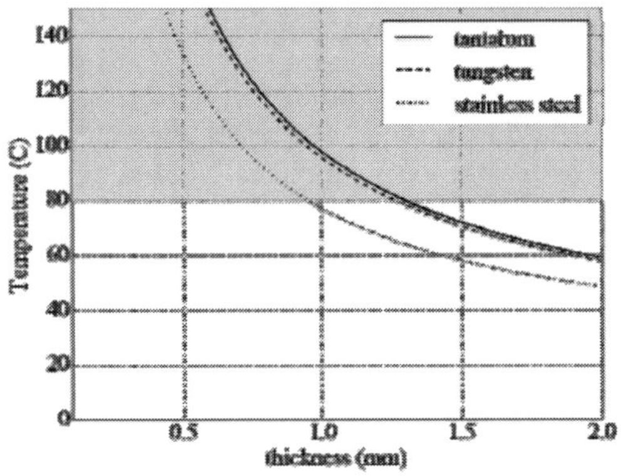

Figure 2. Collector back side temperature vs. collector plate thickness for different materials.

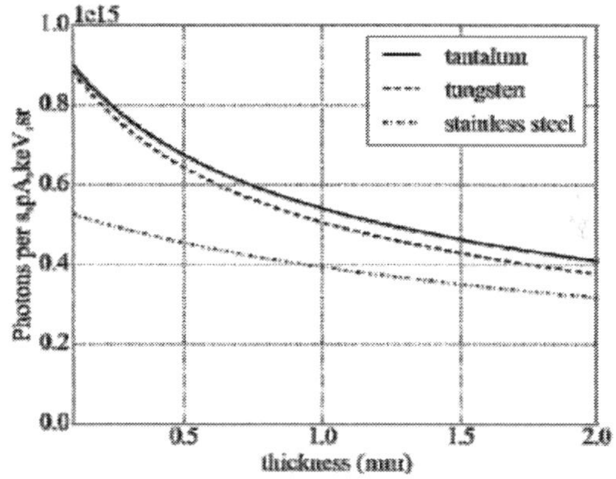

Figure 3. Phosphor screen yield vs. collector thickness for different materials.

For ~ 100 keV beam electrons, an angular distribution of X-rays produced in the target is nearly isotropic. Therefore, to achieve a better spatial resolution of the image, the phosphor screen should be placed as close to the collector surface as possible, which in its turn requires a minimum thickness of the collector plate. A small thickness is needed also to reduce the attenuation of the X-rays passing through the collector material onto the phosphor screen. However, the use of a thin collector is complicated by its high heating under the electron beam load. The phosphor screen will be damaged if a temperature of the backside of the beam collector exceeds 80° C. The heating of the collector plate has been calculated for various materials (W, Ta, and a stainless steel) for the beam parameters as follows: $E_b = 100$ keV, $I_b = 100$ A, $t_b = 0.15$ ms and the beam diameter of 7 cm. It is assumed that the absorbed heat is uniformly distributed over the beam footprint. The electron backscattering factor for all materials was taken into account in the calculations. The temperature curves versus the collector thickness are shown in figure 2. As it can be seen, an acceptable backside collector temperature is attained for the material thickness of 0.1 cm or more.

The material and the thickness of the beam collector also affect the amount of the light emitted by the phosphor screen. The number of emitted photons S can be calculated as:

$$S = k \cdot \int_E N(E) \cdot T(E) \cdot A(E) \cdot E \, dE$$

where E stands for the X-ray photons energy, $N(E)$ is a spectral response of the collector, $T(E)$ is the collector transmittance, $A(E)$ is the absorption coefficient of the phosphor layer, and k is the light yield of the Gd_2O_2S:Tb phosphor (k = 35 photon/keV). The spectral response of the collector is calculated using the model outlined in [5, 6]. The absorption coefficient calculations based on the surface density of the phosphor active substance: 55 mg/cm^2. All calculations use the material properties specified in the NIST database [7], the electron beam parameters and the collector materials were taken the same as for the heating calculations.

The number of light photons versus the collector thickness are shown in figure 3. With the collector thickness of over 0.1 cm, a difference in the emitted light is inessential for all considered materials. Thus, the stainless steel was taken as an inexpensive and easy-to-process material. Ultimately, the 0.1 cm thickness of the collector plate was chosen.

3. Experimental results

A typical X-ray image of the beam (80 keV, 70 A, 0.1 ms) produced with the 241 aperture EOS is shown in figure 4. The beam cross-section is very close to a circle. The image shape and the center of mass are stable from shot to shot. In the upper part of figure 4, the beam current density profile, which is measured along the dotted line crossing the center of the image, is shown. The profile is quite flat, with a steep fall at the edge. The beam current density reaches 3.2 A/cm^2 at the center of the beam. An effective beam diameter on the target, defined as a diameter of a circle, which encloses 90% of the beam current, is found to be ca. 7 cm.

To obtain the spatial resolution of the imaging diagnostics, special experiments on measuring the sharp edge spread function have been performed. The edge is formed with a tungsten plate of 0.1 cm thickness imposed upon the converter, the overlapping part of the beam footprint, and completely shadowing the X-rays. The resulting image is shown in figure 5. The normalized intensity profile is measured along the line perpendicular to the edge and crossing the center of the beam footprint. On the basis of the edge profile, the modulation transfer function of the diagnostic was calculated (e. g. [8]). The limiting spatial resolution was found as 4 line pairs/cm at 10% level.

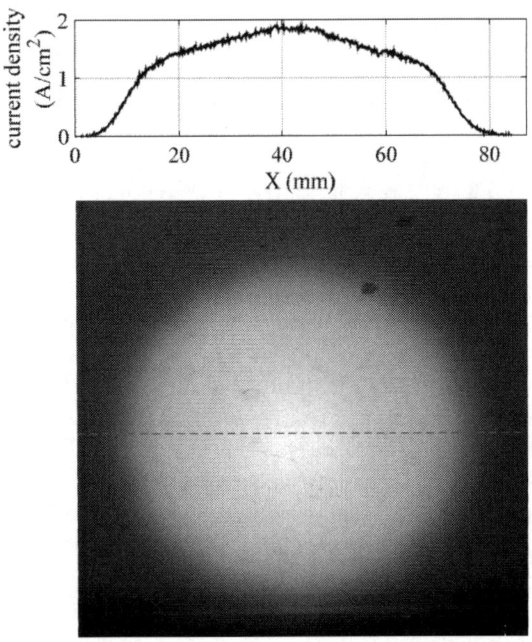

Figure 4. Typical X-ray image of the beam and current density profile measured along dotted line crossing the center of the beam.

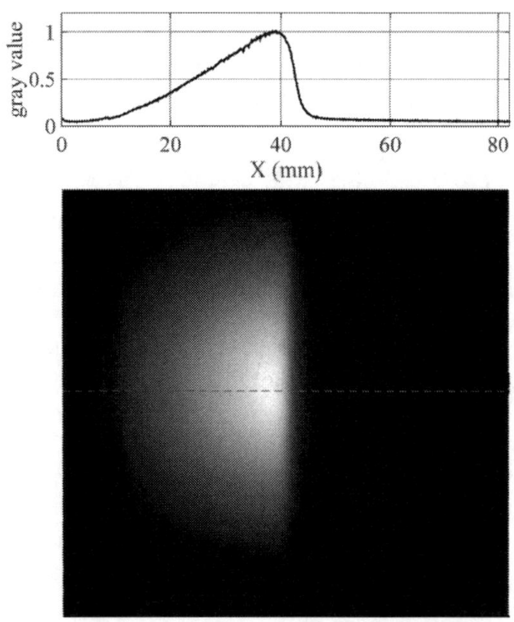

Figure 5. Image of the edge response and intensity profile measured along dotted line crossing the center of the beam.

Since the diagnostic measures the current density distribution on the collector at a distance of 0.85 m downstream of the beam source, it is possible to evaluate the current density distribution directly in the multi-aperture diode.

Originally, the beam consists of many separate jets (beamlets). In a guiding magnetic field, the beamlet envelope oscillates radially due to transversal velocities of the electrons. That leads to a periodic focusing of the beamlet, ultimately, to its initial radius. Thus, varying the magnitude of the accelerating voltage and the magnetic field it is possible to reproduce the initial beam structure in the collector footprint and evaluate the emission yield of every single aperture. An example of such image is shown in figure 6. As we can see from the picture, the beam structure completely resembles the EOS structure, all of the 241 apertures are utilized in the beam generation. An emission is quite uniform all over the EOS, the current density decreases by 20% of the maximum at a radius of 2.5 cm and falls almost twice at the EOS periphery (at the radius of 4 cm). It should be noted that the discrete structure of the beam on the target disappears with a change of the magnetic field and the energy of electrons, which is also confirmed by the melting imprint at the collector.

At the next experimental step, the number of apertures was nearly doubled (499 vs. 241) and the EOS diameter increased ca. 1.4 times (11.8 cm vs 8.3 cm). This step allowed to increase the maximum beam current and the emission current of 250 A at the 0.25 ms pulse was achieved. The beam structure in the 499-apertured EOS is shown in figure 7. All of 499 apertures are employed, as with the previous EOS, the current density decreases by 20% of the maximum at the radius of ca. 2.5 cm and falls to ~ 0.25 of the maximum in the peripheral apertures (at the radius of 5.5 cm).

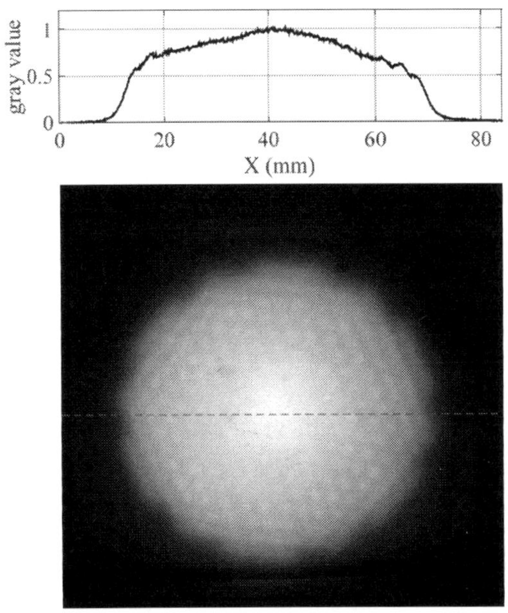

Figure 6. Image of the beam structure *for 241-apertured EOS* and intensity profile measured along dotted line crossing the center of the beam.

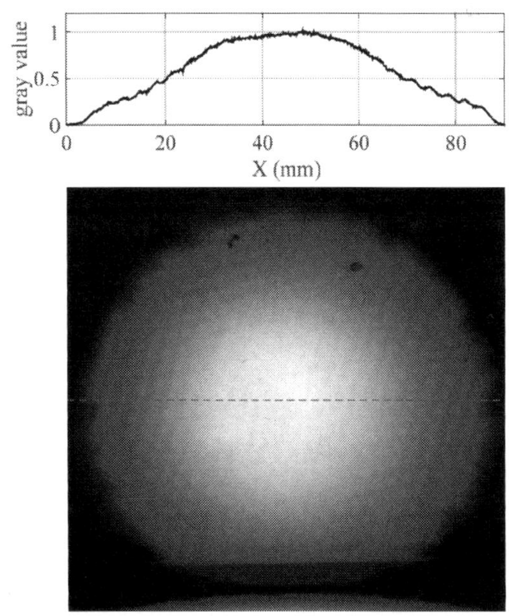

Figure 7. Image of the beam structure for *499-apertured EOS* and intensity profile measured along dotted line crossing the center of the beam.

4. Conclusions

An electron beam current density distribution in the multi-aperture diode with the plasma cathode have been studied with the help of the imaging X-ray diagnostic tool. It was demonstrated that the large-area electron beams could be effectively formed in the multi-aperture system with the single on-axis arc-discharge plasma generator even in the presence of the guiding magnetic field. All apertures are emitting in the 499-apertured EOS with the diameter ca. 12 cm. The beam current density is quite uniform up to the radius 2.5cm and gradually decreases to the periphery.

It should be noted that the beam emission profile is weakly dependent on the magnetic field strength in the arc channel of the washer gun. The effective diameter of the beam imprint practically does not change when the field varies in the range from 100 to 800 mT. On this basis, one can assume that the geometry of the magnetic flux in the hollow anode of the arc is not the key factor determining the cross-sectional size of the discharge plasma near the emission grid.

Acknowledgments
The authors are grateful to V. T. Astrelin for numerous helpful discussions, and also thank the staff of the GOL-3 complex for technical support of the work.

The work at the electron beam test facility was financially supported by Russian Science Foundation (project N 14-50-00080). The development of the X-ray diagnostic technique was partially supported by RFBR project 16-38-00739.

References
[1] Kurkuchekov V V et al. 2013 *Fusion Sci. Technol.* **63** 1T 292-4
[2] Vyacheslavov L et al. 2016 Novel electron beam based test facility for observation of dynamics of tungsten erosion under intense ELM-like heat loads *AIP Conferences Series* to be published
[3] Davydenko V I, Morozov I I, Roslyakov G V and Savkin V Y 1987 *Instrum. Exp. Tech.* **29** 6
[4] Fiksel G, Almagri A F, Craig D, Iida M, Prager S C and Sarff JS 1996 *Plasma Sources Sci. Technol.* **5** 78-83
[5] Ebel H 2006 *Advances in X-Ray Analysis* **49** 267 – 73
[6] Ebel H 1999 X-Ray Tube Spectra *X-Ray Spectrometry* **28** 255-66
[7] Hubbell J H and Seltzer S M 2004 *Tables of X-Ray Mass Attenuation Coefficients and Mass Energy-Absorption Coefficients (version 1.4).* (National Institute of Standards and Technology http://physics.nist.gov/xaamdi)
[8] Smith S W 1997 The *Scientist and Engineer's Guide to Digital signal Processing* (California Technical Publishing http://www.dspguide.com/) chapter 25

IOP Conf. Series: Journal of Physics: Conf. Series **830** (2017) 012033 doi:10.1088/1742-6596/830/1/012033

Study of thin metallic film explosion in vacuum

A S Zhigalin[1], A G Rousskikh[1], V I Oreshkin[1,2] and R B Baksht[1]

[1]Institute of High Current Electronics SB RAS, 2/3 Academichesky ave., Tomsk, 634055, Russia

[2]National Research Tomsk Polytechnic University, pr. Lenina 30, Tomsk, 634050 Russia

Email: Zhigalin@ovpe2.hcei.tsc.ru

Abstract. An experiment with Al, Cu and Ni exploding foils was carried out at a current density of (0.5-1) $\times 10^8$ A/cm^2 through the 6-μm foil with a current density rate of about (0.5-1) $\times 10^8$ A/cm^2·s. To record the metal foil effervescence during the foil explosions, a two-frame radiographic system was used. It was shown that the duration of the explosion resistive phase was considerably lower than the metal boiling time. The foil energy deposition is equal to 30-70% of the sublimation energy.

1. Introduction

In recent years, the wire explosion in vacuum has been the subject of quite many studies [1-5] due to the wide application of wire arrays for the production of high-density hot plasmas. The foil explosion in vacuum still remains the least understood phenomenon. However, the electrical explosion of thin beryllium foils is a key point in the MagLIF (Magnetized Liner Inertial Fusion) project [6] aimed at producing high-density, high-temperature plasmas by compressing a magnetic field with a light metal liner. In this connection, it is of interest to study the main features of thin metal foil explosions in vacuum, such as the energy deposition in the exploding foil material and the material boiling. This paper presents the results of an experiment on the explosion of thin metal (Al, Cu, and Ni) foils at a current density of (0.5-1)$\times 10^8$ A/cm^2 in the foil and a current density rise rate of (0.3-0.8)$\times 10^{15}$ A/cm^2·s. Measurement of the standard characteristics of the exploding foils was performed simultaneously with their x-ray imaging. The paper is organized as follows. Section 2 describes the experimental setup and procedures. The experimental results are presented in Section 3 and discussed in Section 4, which is followed by Conclusion.

2. Experimental setup

The foil explosion experiment was carried out using a setup comprising three current generators. The WEG-1 generator [7] was operated as a driver to initiate foil explosions, the XPG-1 and XPG-2 generators (radiographs) [8] were used for diagnostics. The spatial image of the exploding wire was taken using 2-ns x-ray flashes generated in the load unit of the XPG-1 and XPG-2 generators. The three generators were started from an external power supply by means of switches S1, S2, and S3, respectively. The delays between the operations of the XPG-1, XPG-2, and WEG-1 generators were set with the help of a DPG trigger pulse generator; the operation jitter was 50 ns for each of the three generators.

A resistive voltage divider was used for voltage measurements, and the current through the exploding foil was measured with a B-dot probe. The voltage divider measured only the voltage across the circuit section where the foil was connected.

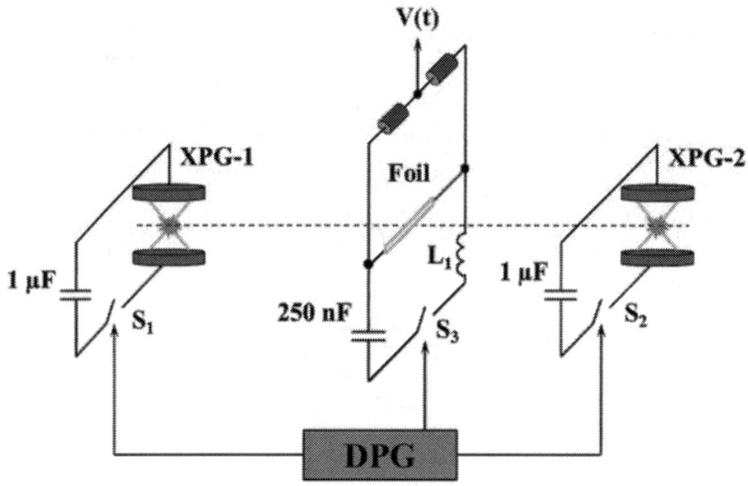

Figure 1. Schematic of the electric circuit of the experimental setup.

Constructing the voltage waveforms, we subtracted the inductive component from the readings of the divider. The B-dot probe was placed near the ground electrode. The probe was calibrated in reference to the readings of a calibrated shunt. The parameters of the entire circuit of the WEG-1 generator were determined from the current and voltage waveforms obtained in the short-circuit mode. The test foil was fastened in a special holder with its contacts soldered to the electrodes [3]. The chamber was evacuated to a pressure less than $6 \cdot 10^{-5}$ Torr with an oil-vapor pump. Bare x-ray diodes were used to detect the x-ray flashes that were generated by the XPG generators. The backlighting system for x-ray imaging of the exploding foils is described in detail elsewhere [9].

3. Experimental results

The experiment was performed with Al, Ni, and Cu foils of 20mm length, 1mm width, and 6 μm thickness. For exploding Al and Ni foils, the charge voltage of the WEG-2 generator capacitor was 20 kV, providing a current rise rate of $\sim 0.7\times10^{15}$ A/cm^2·s. The authors of Ref.2 demonstrated for a W wire explosion that the increase of the current rise rate results in the energy input increase. To estimate the effect of the current rise rate on the deposited energy for exploding Cu foils, the WEG-2 generator capacitor was charged to 20 and 10 kV, at which the current rise rate was $\sim 0.3\times10^{15}$ and $\sim 0.7\times10^{15}$ A/cm^2·s, respectively.

3.1. Electrophysical measurements

Figure 2 presents the typical current and inductively corrected voltage (V_{corr} = V - LdI/dt) waveforms recorded during a foil explosion in vacuum. Herein, the energy ε(t) deposited in the discharge is also shown.

The energy ε(t) was determined as

$$\varepsilon(t) = \int_0^t I\left(V - L\frac{dI}{dt}\right) dt .$$

As it follows from the waveforms shown in Fig. 2, V_{corr} = V - LdI/dt reaches a maximum within 200-300 ns after the onset of current passage through the foil, which is followed by its sharp decrease.

This voltage behavior is quite similar to that featured by wire explosions. Following a usual nomenclature [1, 3], we treat the time, at which V_{corr} reaches a maximum, as the collapse time t_{coll}. At $t = t_{coll}$, the foil resistance also reaches its maximum value $R_{foilmax}$; for the shot presented in Fig. 2, we have $R_{foilmax} = 0.87\ \Omega$. Assume that the energy deposited in the foil material, ε_{dep}, is equal to the energy deposited in the discharge during the period from $t = 0$ to $t = t_{res}$ at which $R_{foil} = 0.1R_{foilmax}$. According to the nomenclature of Ref. [1], the time t_{res} corresponds to the period of the resistive (Joule) heating. The energy deposited in the discharge increases further after $t = t_{res}$.

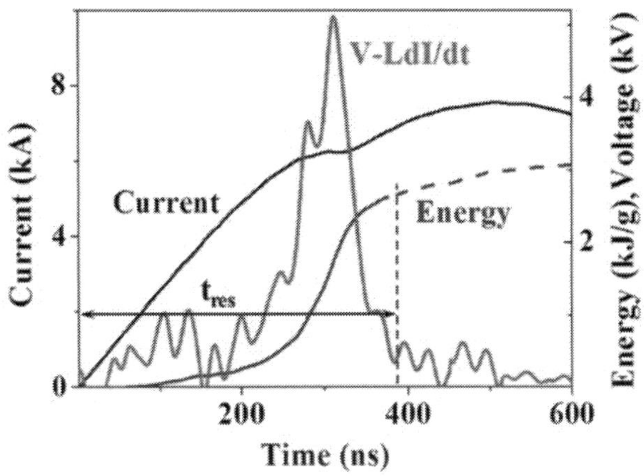

Figure 2. Waveforms of inductively corrected voltage V_{corr}=V - LdI/dt and current I(t), and the time behavior of the energy ε(t) deposited in the discharge for a Cu foil exploded at V_0 = 10 kV (0.3×10^{15} A/cm^2·s). The time t_{res} corresponds to the period of Joule heating of the foil material.

In Fig. 2, the portion of the curve ε(t) at $t > t_{res}$ is indicated by a dashed line. Note that a similar behavior of ε(t) at $t > t_{res}$ was observed for Ag wires of 2-3 cm lengths (see Fig. 8 in Ref. 1). It seems that at $t > t_{res}$, the energy is deposited in the plasma corona that shunts the foil core. Table I summarizes the deposited energy averaged over seven shots. The data were derived from the voltage and current measurements for all tested foils.

Table 1. The measured specific deposited energy averaged over seven shots. As discussed above, the energy is deposited in a foil during the initial resistive heating stage.

Foil material	Current density rise rate $dj/dt \times10^{15}$, A/cm^{2}s	Deposited energy ε_{dep}, kJ/g	Melting energy ε_{melt}, kJ/g	Evaporation energy ε_{evap}, kJ/g	Percent of evaporation energy deposited in foil, %
Al	0.7 ± 0.1	5.3 ± 0.5	0.4	10.85	49
Ni	0.6 ± 0.1	1.3 ± 0.1	0.3	6.3	21
Cu	0.71 ± 0.1	3.3 ± 0.3	0.205	4.75	69
Cu	0.33 ± 0.06	2.4 ± 0.2	0.205	4.75	50

The tabulated data show that, similar to exploding wires, the energy deposited in a foil exploding in vacuum is significantly greater than the melting energy, making 20% for materials with high boiling point (Ni) and 50-70% for materials with low boiling point (Al, Cu). These data almost coincide with those given in Ref. 1. Thus, according to our experimental data, the material of an exploding foil turns into a liquid not later than 100 ns after the onset of current passage through the foil. Based on these data, we may suppose that, as with exploding wires, during a foil explosion, a shunting discharge starts developing over the surface of the molten metal at $t = t_{coll}$. Note, that the Cu foil data present the experimental proof of an increase of energy deposition into a foil core with the current rate for exploding foils in a vacuum just it has been shown before in Ref.2 for wire explosion.

3.2. X-ray backlighting

When backlighted at $t < t_{res}$, a foil is opaque to x rays, and during this period, the magnetic field of the current flowing through the foil hinders its expansion. The x-ray patterns taken at a time after $t = t_{res}$ show small irregularities (bubbles) transparent to x rays. Figure 3a shows the x-ray pattern of an Al foil taken 365 ns after the onset of current passage through the foil; for this shot, the time t_{res} was 190 ns. In the magnified x-ray fragment, the bubbles of a characteristic size of several micrometers are clearly seen. Once bubbles have occurred, they rapidly grow in number. The time dependence of the number of bubbles per 1 mm^2 obtained for foils of various materials, $N_{bubble}(t)$, is shown in Figure 3b. We denote the time corresponding to the onset of a sharp rise of the function $N_{bubble}(t)$ as t_{boil}, the time at which the foil material starts boiling.

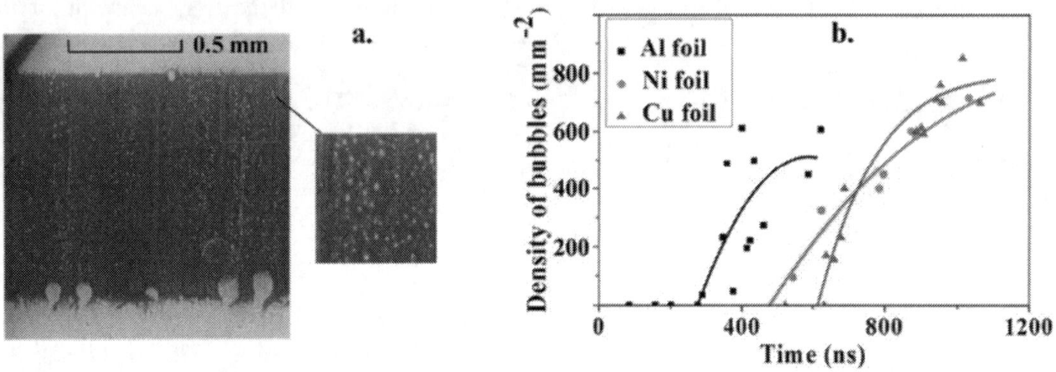

Figure 3. X-ray pattern of an Al foil taken 365 ns after the onset of current passage through the foil; the time t_{res} for this shot was 190 ns (*a*); the number of bubbles per 1 mm^2 versus time for exploded Al, Ni, and Cu foils (*b*).

Table 2. The resistive heating time and the boiling time for different foil materials.

Foil material	Current density rise rate $dj/dt \times 10^{15}$, A/cm^{2}s	Resistive heating phase duration t_{res}, ns	Time of the foil boiling t_{boil}, ns
Al	0.7 ± 0.1	175 ± 5	275 ± 20
Ni	0.6 ± 0.1	104 ± 5	483 ± 20
Cu	0.71 ± 0.1	209 ± 5	n/a
Cu	0.33 ± 0.06	286 ± 5	600 ± 20

As it follows from the data presented in Table 2, the period between $t = t_{res}$ and the onset of boiling (t_{boil}) is some hundreds of nanoseconds, being significantly greater than the statistical measurement error. The observed delay between the onset of boiling of the liquid foil material and the cessation of its heating can be accounted for by metastable states arising in the liquid metal of the exploded foil. The decay of the metastable states in our experimental conditions is described in detail elsewhere [10].

4. Discussion

The occurrence of a shunting corona during an electrical explosion of wires is well known for a long time [1-3, 11], but the medium in which the corona develops remains the subject of discussion. In principle, the medium in which the shunting occurs may be the low-density plasma resulting from the expansion of the conductor material [2, 12, 13], or the shunting is a consequence of the breakdown

IOP Conf. Series: Journal of Physics: Conf. Series **830** (2017) 012033 doi:10.1088/1742-6596/830/1/012033

developing in the cloud of the gas desorbed by the hot metal surface [4, 10, 11]. In our foil explosion experiments, the delay of the metal boiling relative to t_{res} on its own may indicate that the shunting discharge develops in the desorbed gas. Below, we shall discuss how the "desorption model" aids to interpret our experimental data.

The shunting of conductors electrically exploding in vacuum is discussed in detail elsewhere [4]. The scenario of the shunting of an exploding foil can be described in brief as follows. The molten surface of the foil intensely desorbs gas; the gas expansion velocity is determined by the foil temperature. Until the gaseous shell is broken down, the pressure of the magnetic field induced by the current flowing through the foil hinders the expansion of the foil material, the foil temperature increases and can reach $T_{foil} = 0.2$-0.6 eV [10]. For hydrogen, which is the main component of the gas absorbed on the metal surface, the gaseous shell expansion velocity (implying $T_{gas} \approx T_{foil}$) is in the range $(0.6\text{-}1.1)\times10^6$ cm/s. As the gaseous shell expands further, the gas density decreases, and so does the electric strength of the shell. Simultaneously, the resistance of the foil increases with temperature. Thus, we have two factors that determine the collapse time:

i. the increase of the voltage across the electrode gap due to the increase of the foil resistance,
ii. the decrease of the breakdown field strength of the gaseous shell, $\mathbf{E}_{gas}$, due to the decrease of the gas density that it is correct for the left part of the Paschen's curve.

As a result, the gaseous shell is broken down at $t = t_{coll}$, and a high-conductivity corona is formed during the period between t_{coll} and t_{res}. According to the above reasoning, $\mathbf{E}_{gas}$ should not depend on the foil material. Actually, at $\mathbf{dj/dt} = 0.71\times10^{15}$ A/cm^{2}s, according to our data, $\mathbf{E}_{gas}$ is equal to 4-6 kV/cm and does not depend on the foil material.

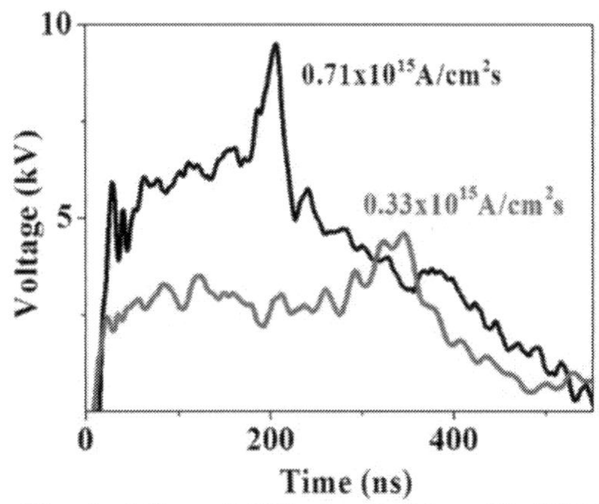

Figure 4. Waveforms of the voltage across an exploding Cu foil for the rise rate of the current through the foil $\mathbf{dj/dt} = 0.71\times10^{15}$ and 0.33×10^{15} A/ cm^{2}s.

The doubling of **dj/dt** resulted in a 30-40% increase of the deposited energy (see Table 1) mainly due to the increase of the foil flashover voltage. It follows from the plots showing the time behaviour of the voltage across the electrode gap, $\mathbf{V}(t) = R_{foil}I + LdI/dt$ (Fig. 4), that were obtained for exploding Cu foils at different **dj/dt**. The waveforms clearly indicate that the increase of the hold-off voltage and the energy deposited in the foil material is accompanied by the decrease of pulse duration of the voltage across the gap.

5. Conclusion

The experimental study of the electrical explosion of Al, Cu, and Ni foils in vacuum has shown that the specific energy deposited in the foil material, ε_{dep}, makes several tens of percents of the evaporation energy ε_{evap}. The obtained $\varepsilon_{dep}/\varepsilon_{evap}$ ratio is close to that measured previously for exploding wires [1, 4]. It has also been revealed that the foil material starts boiling 300-600 ns after the cessation of Joule heating. Based on the experimental observations, it has been concluded that the shunting of a

IOP Conf. Series: Journal of Physics: Conf. Series **830** (2017) 012033 doi:10.1088/1742-6596/830/1/012033

foil exploding in vacuum is due to the discharge developing in the cloud of gas desorbed from the foil surface.

Acknowledgements
This work is supported by the Russian Science Foundation (grant No. 17-08-00459).

References
[1] Sinars D B, Min Hu, Chandler K M, Shelkovenko T A, Pikuz S A, Greenly J B, Hammer D A and Kusse B R 2001 *Physics of Plasmas*. **8**, No 1, p 216.
[2] Struve K W and McDaniel D H 2004 *Phys. Plasmas* **11**, 4573.
[3] Beilis I I, Baksht R B, Oreshkin V I, Russkikh A G, Chaikovskii S A, Labetskii A Yu, Ratakhin N A and Shishlov A V 2008 *Physics of Plasmas* **15**, 013501.
[4] Rousskikh A G, Baksht R B, Labetskii A Yu and Oreshkin V I 2004 *Plasma Physics Reports*. **30**, No 11, p. 944.
[5] Shi Z, Wang K, Shi Y, Wu J and Han R 2015 *Journal of Applied Physics*, **118**, 243302.
[6] Gomez M R, Slutz S A, Sefkow A B, Hahn K D, Hansen S B, Knapp P F, Schmit P F, Ruiz C L, Sinars D B, Harding E C *et al* 2015 *Phys. Plasmas* **22**, 056306.
[7] Ratakhin N A, Fedushchak V F, Erfort A A, Zharova N V, Zhidkova N A, Chaikovsky S A and Oreshkin V I 2007 *Russ. Phys. J.*, **50,** No 193.
[8] Artyomov A P, Fedyunin A V, Chaikovsky S A, Oreshkin V I, Lavrinovich I V and Ratakhin N A 2013 *Tech. Phys. Lett.* **39,** No 1, p. 12.
[9] Baksht R B, Rousskikh A G, Zhigalin A S, Oreshkin V I and Artyomov A P 2015 *Physics of Plasmas* **22**, 103521.
[10] Oreshkin V I, Zhigalin A S, Rousskikh A G and Kuznetsov V V 2013 *Journal of Engineering Thermophysics* **22**, No 4, p. 288.
[11] Karakhanov S M 1978 *Sov. Phys. Tech. Phys*. **23**, p. 832.
[12] Bobrova N A, Razinkova T L and Sasorov P V, 1992 Sov. J. *Plasma Phys*. **18**, p. 269.
[13] Chittenden J P, Lebedev S V, Ruiz-Camacho J, Beg F N, Bland S N, Jennings C A, Bell A R, Haines M G, Pikuz S A, Shelkovenko T A and Hammer D A 2000 *Phys.Rev.* **61**, 4371.

IOP Conf. Series: Journal of Physics: Conf. Series **830** (2017) 012034 doi:10.1088/1742-6596/830/1/012034

RF pulse formation dynamics in gyromagnetic nonlinear transmission lines

I V Romanchenko, P V Priputnev and V V Rostov

Physical Electronics Department, Institute of High Current Electronics, SB RAS, 2/3 Akademichesky Ave., Tomsk 634055, Russia

Email: dr.romanchenko@gmail.com

Abstract. We present the results of rf oscillations growth dynamics as the high voltage pulse propagates through the gyromagnetic nonlinear transmission line (NLTL). Several equidistant electrical probes were placed inside the NLTL in order to examine the oscillatory wave formation and the role of higher order modes in this process. RF pulses are generated at 2 GHz frequency with nonaveraged peak power of 100 MW. Azimuthal asymmetry of the wave propagating through NLTL is detected.

1. Introduction

Generation of high power microwave (HPM) pulses by gyromagnetic nonlinear transmission lines (NLTLs) as a possible alternative to traditional HPM tubes attends intensified attention of researchers for last ten years [1-6]. Optimal conditions are found for the generation of rf pulses with frequencies from 0.6 GHz [1] to 5 GHz [5], rf power from tens MW [4] to several hundred MW [2], pulse repetition rate from 800 Hz [6] to 65 MHz [7]. Based on these studies, multichannel rf sources with electronically controlled beam steering are designed and investigated [8-9]. In all experiments with gyromagnetic NLTLs, the traditional experimental method includes registration of the high voltage pulse incident on NLTL and the pulse coming out of the NLTL. At that, the intrinsic dynamics of rf pulse formation inside NLTL remains unclear. In this paper we present results of direct measurement of the wave propagating through NLTL by sensors placed inside NLTL.

2. Shock wave formation dynamics

2.1. Experimental set-up

The general scheme of experimental set-up is presented in figure 1. SINUS-150 high voltage pulse generator was used to produce pulses with 5 ns duration and amplitude of 120 kV. Two D-dot sensors were place in transmission lines before and after NLTL. The length of ferrite filling was 700 mm. Three pairs of identical D-dot sensors were placed inside NLTL with 220 mm axial shift and 90° azimuthal shift within the pair. The high voltage pulse was absorbed in an RF resistive load (HVR). The waveforms were recorded by TDS6604 Tektronix Digital Oscilloscope with 6 GHz bandwidth.

The rf cables between oscilloscope and three pairs of sensors inside NLTL were identical and calibrated by Network Analyzer Agilent N5227A.

2.2. Results

At first, the situation without a bias magnetic field was examined. The input and output waveforms as far as waveforms at three cross-sections inside NLTL are presented in figure 2. One can see the shock wave formation with a stationary rise time and velocity. The pulse sharpening is accompanied by pulse shortening because of magnetic losses at the shock front [10].

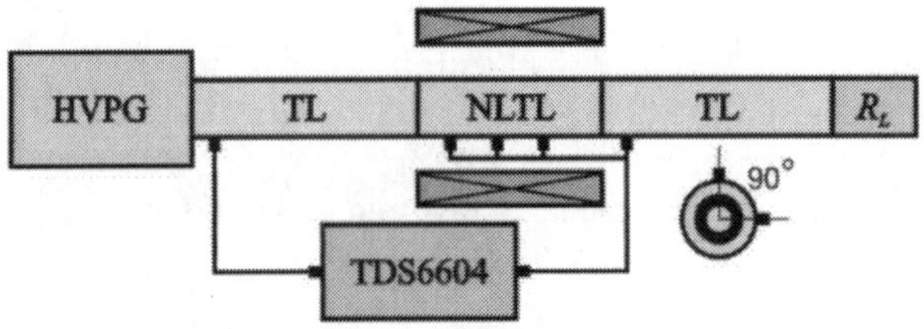

Figure 1. General scheme of experimental set-up. HVPG – high voltage pulse generator, TL – transmission line, RL – rf resistive load, TDS6604 – Tektronix Digital Oscilloscope.

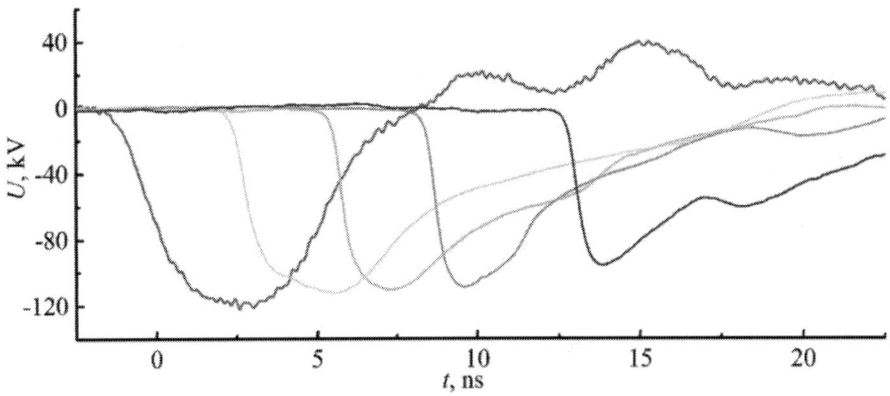

Figure 2. Input and output waveforms together with waveforms at three equidistant cross-sections inside NLTL. No bias.

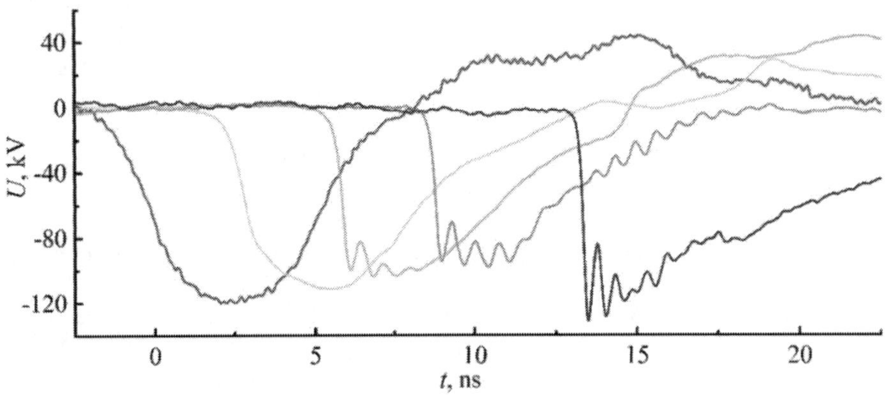

Figure 3. Input and output waveforms together with waveforms at three equidistant cross-sections inside NLTL. Bias magnetic field of 32 kA/m.

If the bias magnetic field is applied, rf oscillations with a frequency close to 2 GHz appear, figure 3. One can see similar dynamics with not so distinct pulse shortening because of different sharpening mechanism. The shock wave velocity is changed with the bias magnetic field which is a basis for multichannel NLTL systems with a bias control of phase in each channel [7-8].

3. Azimuthal wave asymmetry

The azimuthal wave symmetry was checked in three cross-sections by D-dots with 90° azimuthal shift. The results are presented in figure 4.

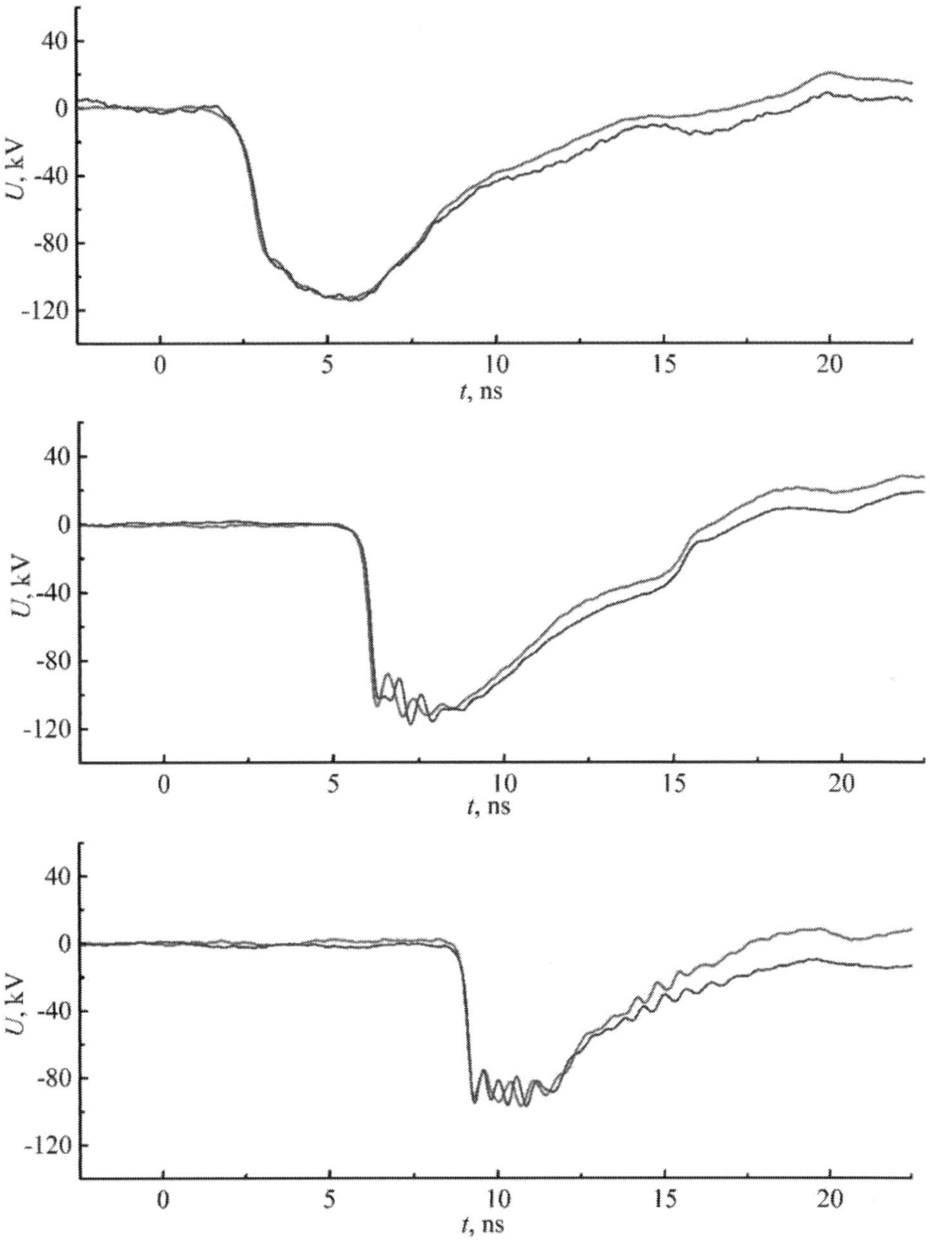

Figure 4. Waveforms at three equidistant (220 mm) cross-sections recorded by two sensors with 90° azimuthal shift. Bias magnetic field of 32 kA/m.

At the first cross-section (at 130 mm from NLTL beginning), the waveforms seem to be very similar to incipient oscillations at the shock front. At the second cross-section while the rise time is the same for both sensors the oscillations are quite different. One sensor shows decaying oscillations while at another the oscillations start is decayed. That could be because of a difference in the group velocities of rf oscillations and a shock front. At the third cross-section there is a significant difference in frequencies of oscillations from both sensors – at the same interval, one sensor shows five oscillations while another - four. These results indicate that the modes of electromagnetic wave inside NLTL are more sophisticated than TEM-mode and additional analysis is needed.

4. Conclusions

The presented results show that traditional view on gyromagnetic NLTLs as systems with single TEM wave should be revised. Different waveforms at the same cross-section of NLTL indicate that azimuthal asymmetry takes place as a result of excitation of at least one higher order mode. That should be checked for other geometries and ferrite materials. Maxwell equations are more preferable for description of wave dynamics in NLTL than traditional for NLTLs Telegraph equations.

Acknowledgments

The work was supported by IHCE SB RAS Project No. SP-3354.2015.1. Authors acknowledge assistance of Tomsk Center of Collective Use SB RAS for providing Network Analyzer Agilent N5227A.

References

[1] Gubanov V P, Gunin A V, Kovalchuk O B, Kutenkov V O, Romanchenko I V and Rostov V V 2009 *Tech. Phys. Lett.* **35**
[2] Romanchenko I V and Rostov V V 2010 *Tech. Phys.* **55**, 1024
[3] Rostov V V, Bykov N M, Bykov D N, Klimov A I, Kovalchuk O B and Romanchenko I V 2010 *IEEE Trans. Plasma Sci.* **38**, 2681
[4] Bragg J W B, Dickens J C and Neuber A A 2013 *IEEE Trans. Plasma Sci.* **41**, 232
[5] Chadwick S J F, Seddon N and Rukin S 2011 *IEEE Pulsed Power Conference, Chicago, IL* 178
[6] Ulmaskulov M R, Pedos M S, Rukin S N, Sharypov K A, Shpak V G, Shunailov S A, Yalandin M I, Romanchenko I V and Rostov V V 2015 *Rev. Sci. Instrum.* **86**, 074702
[7] Bragg J W B, Sullivan III W W, Mauch D, Neuber A A and Dickens J C 2013 *Rev. Sci. Instrum.* **84**, 054703
[8] Romanchenko I V, Rostov V V, Gunin A V and Konev V Yu 2015 *J. Appl. Phys.* **117**, 214907
[9] Johnson J M, Reale D V, Krile J T, Garcia R S, Cravey W H, Neuber A A, Dickens J C and Mankowski J J 2016 *Rev. Sci. Instrum.* **87**, 054704
[10] Afanasyev K V, Kovalchuk O B, Kutenkov O V, Romanchenko I V and Rostov V V 2008 *Instrum. Experim. Techn.* **51**, 406

IOP Conf. Series: Journal of Physics: Conf. Series **830** (2017) 012035 doi:10.1088/1742-6596/830/1/012035

Pulsed plasma chemical synthesis of $Si_xC_yO_z$ composite nanopowder

G Kholodnaya, R Sazonov, D Ponomarev and G Remnev

Institute of High Technology Physics, Tomsk Polytechnic University, 30 Lenin Avenue, 634050, Tomsk, Russia

E-mail: galina_holodnaya@mail.ru

Abstract. $Si_xC_yO_z$ composite nanopowder with an average size of particles about 10-50 nm was produced using the pulsed plasma chemical method. The experiments on the synthesis of nanosized composite were carried out using a TEA-500 pulsed electron accelerator. To produce a composite, $SiCl_4$, O_2, and CH_4 were used. The major part of experiments was conducted using a plasma chemical reactor (quartz, 140 mm diameter, 6 l volume). The initial reagents were injected into the reactor, then a pulsed electron beam was injected which initiated the chemical reactions whose products were the $Si_xC_yO_z$ composite nanopowder. To define the morphology of the particles, the JEOL-II-100 transmission electron microscope (TEM) with an accelerating voltage of 100 kV was used. The substances in the composition of the composite nanopowder were identified using the infrared absorption optical spectrum. To conduct this analysis, the Nicolet 5700 FT-IR spectrometer was used.

1. Introduction

In modern solid-state physics, the properties of nanocomposite structures are actively studied, which are applied for obtaining materials of pre given properties. The components of such combinations can be both metals and their oxides, as well as semiconductors and dielectrics. Regarding the increasing industrial production of hybrid automobiles and Li-ion battery (LIB) electromobiles [1-2], in recent years, a demand for manufacturing a new electrode material for a negative electrode with a higher specific capacity than synthetic graphites (theoretical quantity 372 mAh/g) and improved LIBs has increased. A great attention was paid to alloys of lithium with tin, aluminum, and silicon. A special interest was paid to silicon [3], which showed the highest theoretical specific capacity during electrochemical alloying with lithium (3579 mAh/g) for Li15Si4 [4], hardly yielding to a lithium metal - 4235 mAh/g [5].

The extraction of high amount of lithium is associated with great changes in a volume and structural transformations inducing significant inner mechanical stresses and particle cracking which result in a contact fault with a shunt. As a result, the electrodes based on lithium alloys quickly lose the reversible capacity during cycling. A possible solution of the problem of silicon electrode cyclability is the location of silicon particles inside a conducting matrix. The material for such matrix can be carbon with its high electron conductivity (as high as 10^3 S/cm), low specific weight, and insignificant increase in volume after lithium introduction. The silicon-carbon composites are obtained using grinding of the silicon and graphite mixture in a ball grinder [6-8], coating of silicon particles with carbon, using the methods of chemical and thermal deposition from the gas phase [9-11], or high-energy plasma [12], the pyrolysis of a precursor mixture [13-14].

The purpose of the given work is the experimental production of $Si_xC_yO_z$ composite nanopowders using a pulsed plasma chemical method.

2. Methodology

The $Si_xC_yO_z$ composite nanopowder has been obtained using the pulsed plasma chemical method. The given method of synthesis is based on the bulk excitation of the reaction gas by a pulsed electron beam and organization of the process of reaction over the whole region of excitation.

The beam energy inputs are significantly lower than the energy of the chemical endothermic reactions of the synthesis. In this way, the potentially high performance in practical realization of the method was attained. The use of the plasma, which was generated by a pulsed electron beam, in the synthesis enabled to decrease a nonproductive energy loss, to vary the properties (the size of particles, crystal lattice) of the final product, to make the process more technological and easy-to-control. In Fig. 1, the scheme of the $Si_xC_yO_z$ composite production is presented.

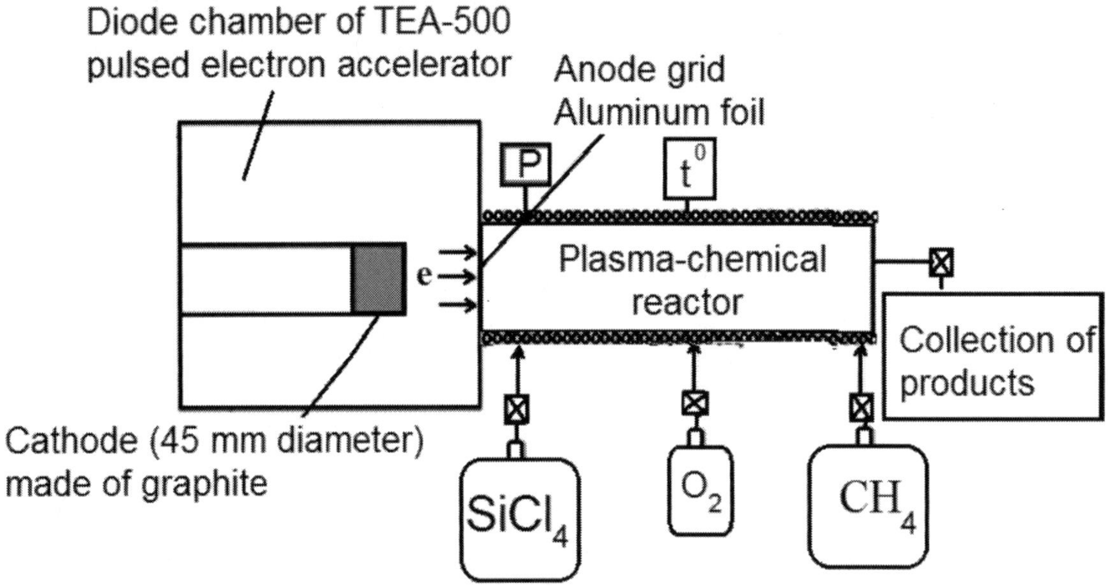

Figure 1. Scheme of an experimental facility

The experiments on the synthesis of the nanosized composite were carried out using a TEA-500 pulsed electron accelerator [15-17]. To obtain the $Si_xC_yO_z$ composite powder, $SiCl_4$, O_2, and CH_4 were used. Most experiments were done using the plasma chemical reactor (quartz, 140 mm diameter, 6 l volume). The reactor was equipped with a manometer, a vacuum meter, a pressure sensor, a shut-off and control valve of the initial reactant mixture inlet and the gas pumping out. The plasma chemical reactor was pumped out up to a pressure of ~ 1-5 Torr before puffing the gas. An electron beam was injected from the end of the reactor. A series of samples was manufactured, the inital concentration of CH_4 varying from sample to sample (the sample 1: SiCl4 170 Torr, O_2 152 Torr, CH_4 57 Torr; the sample 2: SiCl4 170 Torr, O_2 152 Torr, CH_4 76 Torr; the sample 3 SiCl4 170 Torr, O_2 152 Torr, CH_4 152 Torr).

3. Results and Discussion

A wide-spread method of analysis of a chemical composition and a solid substance strucute is the measurement of the optical reflection spectrum in the infrared region (400-4000) cm^{-1}. The use of this method to study the nanoparticles enables to conduct the bulk analysis of the matter, since the penetration depth of IR-radiation (several microns) exceeds the geometrical size of the particles. To carry out this analysis, the Nicolet 5700 FT-IR Spectrometer was used. The technical characteristics of the Nicolet 5700 FT-IR Spectrometer are: the spectral range is as high as (7400 – 350) cm^{-1}; the resolution is 0.09 cm^{-1}; a signal-noise ratio is 40 000:1 (peak-to-peak for 1 minute of scanning); the wavenumber precision is 0.01 cm^{-1} [18-22].

Fig. 2 demonstrates the typical infrared radiation absorption spectra of the $Si_xC_yO_z$ composite powders (samples 1 and 3). The powder under study was preliminary mixed with KBr and was pressed in the tablet. The reflection spectrum of pure KBr was subtracted from the mixture reflection spectrum.

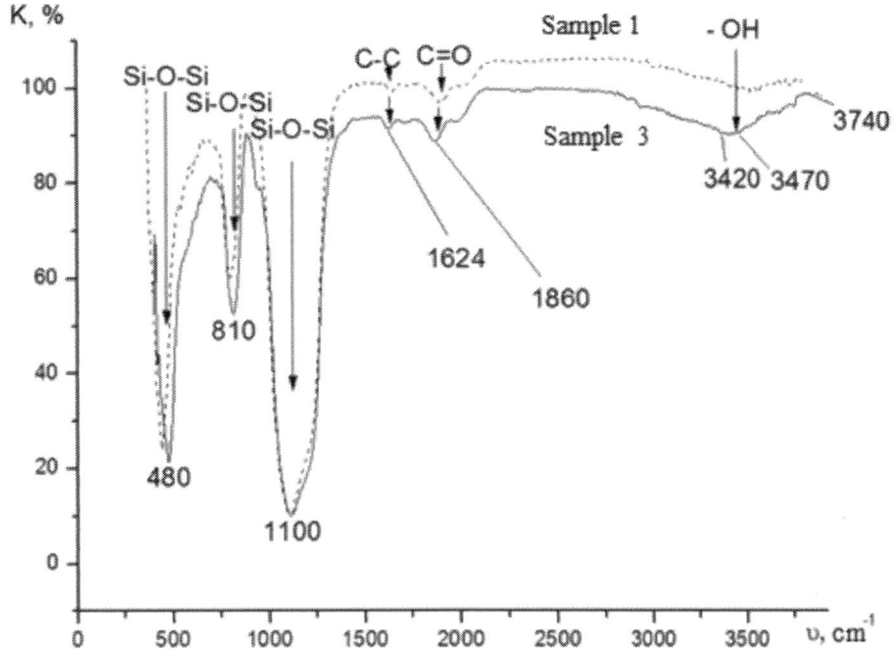

Figure 2. Infrared reflection spectrum of the synthesized $Si_xC_yO_z$ nanocomposites

For the studied samples, the peaks of 1100 and 480 cm^{-1} are typical, they are responsible for Si–O–Si and O–Si–O bond fluctuations. The peak of the C – C bond is recorded. A peak of 810 cm^{-1} can be referred both to the Si – C bond fluctuation (from 750 to 805 cm^{-1}) and to the Si – O – Si bond fluctuation (from 770 to 860 cm^{-1}).

The size of the powder particles was determined using the JEM-100CX II transmission electron microscope (TEM). Fig. 3 presents the histograms from the scanned slides obtained using TEM. The histograms were built with over 1000 particles.

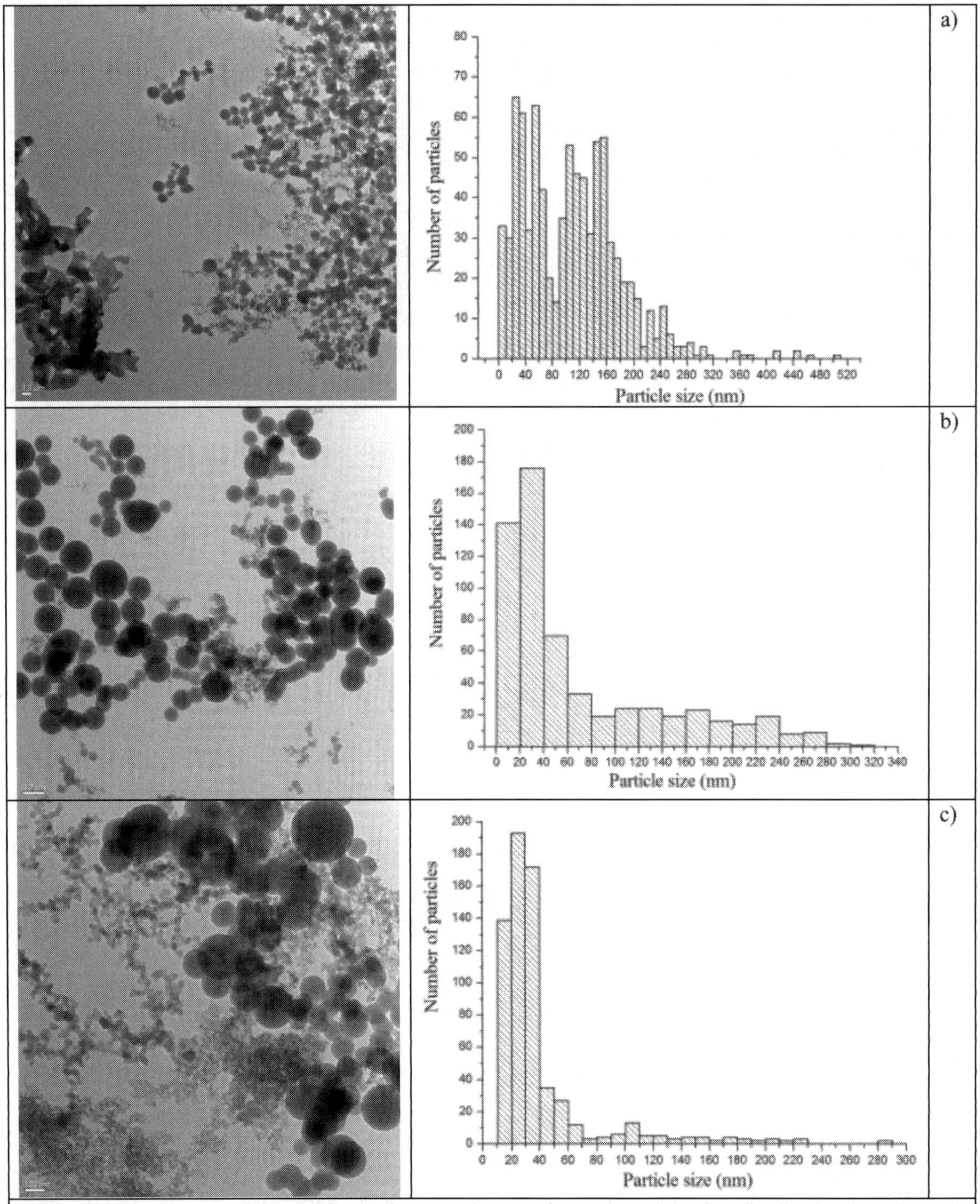

Figure 3. The images of the synthesized composite powder and particle size distribution bar chart: a) sample 1; b) sample 2; c) sample 3

Fig. 3 shows that the particles are spherical; the built bar charts make it possible to conclude that the size of particles varied from 20 to 200 nm. For sample 1, the presence of the irregularly-shaped particle agglomerations is typical. There are the particles of various sizes – large and very small. The particles of sample 2 are uniform and spherical in morphology. There is a large amount of tiny particles in sample 3, larger particles are rare, compared to samples 1 and 2.

For every $Si_xC_yO_z$ sample, the micro-diffraction patterns were taken. In all studied samples, the weak reflexes are observed. But the brightest reflexes were observed in sample 3 - $Si_xC_yO_z$ (Fig.4) that enabled to conclude about the presence of the crystalline structure in the synthesized nanocomposite. The X-ray phase analysis also justified the presence of the crystal structure in the synthesized nanocomposite. Nevertheless, to specify the obtained data, additional investigations should be carried out.

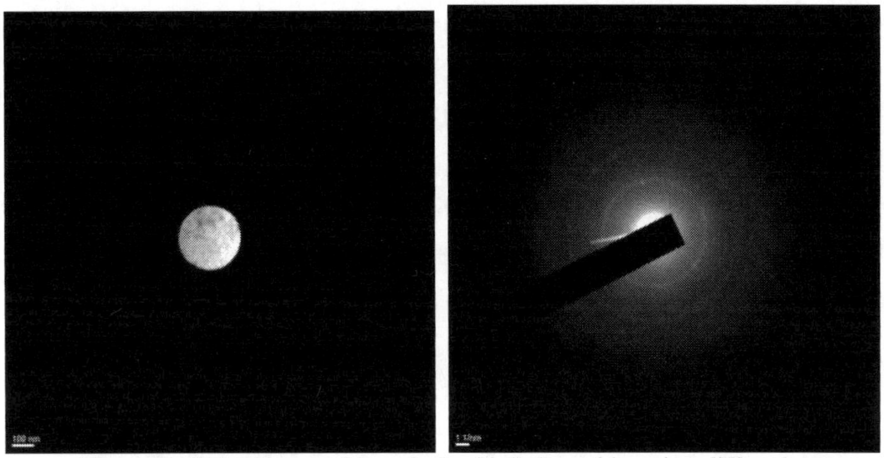

Figure 4. TEM-image of sample 3 - $Si_xO_yC_z$ and its microdiffraction pattern

The synthesized composite $Si_xC_yO_z$ nanopowders were studied using EDX method, which enabled to make a conclusion about the consistency of the chemical composition for the nanoparticles of the different size produced by the method of the plasma chemical synthesis (Fig. 5, sample 3).

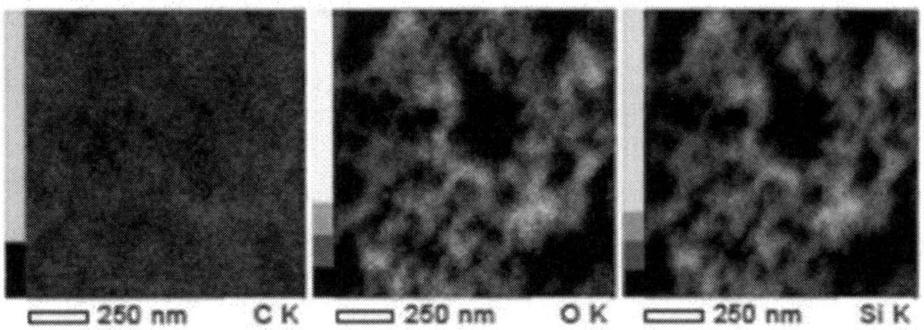

Figure 5. Color windows reflecting the elements of sample 3

EDX-analysis shows that the carbon distribution in powder is uniform.

4. Conclusion

Thus, in the plasma chemical process initiated by the pulsed electron beam, the $Si_xC_yO_z$ powders are synthesized, which are composed of the spherical particles with a diameter of 20-200 nm and with various carbon concentration. According to the results of the X-ray phase analysis and micro-diffraction patterns, we can conclude that the synthesized samples have a crystal lattice. For all composite powders obtained using pulsed plasma chemical synthesis, the presence of the Si–O–Si and C–C bonds is typical.

IOP Conf. Series: Journal of Physics: Conf. Series **830** (2017) 012035 doi:10.1088/1742-6596/830/1/012035

Acknowledgements
The reported study was partially supported by RFBR, research project No. 16-38-00055 mol_a. The physicochemical properties of nanosized composites are studied in TPU Nano-Centre and Department "Nanomaterials and Nanotechnologies" of TPU.

References

[1] S.P. Kuksenko, Yu.A. Tarasenko, I.O. Kovalenko, N.T. Kartel. *J. Chemistry, Physics and Technology of Surface.* **15** (2009) pp 144-153.
[2] T. Kojima, T. Ishizu, T. Horiba, M. Yoshikawa, *J. Power Sources.* **189.** (2009) pp 859-863.
[3] U. Kasavajjula, C. Wang, A. Appleby, *J. Power Sources.* **163** (2007) pp 1003–1039.
[4] M. Obrovac, L. Christensen, *J. Electrochem. Solid-State Lett.* **7** (2004) A93–A96.
[5] M. Holzapfel, H. Buqa, F. Krumeich, P. Novak, F.-M. Petrat, C. Veit, *J. Electrochem. Solid-State Lett.* **8** (2005) A516–A520.
[6] Wang G.X., Yao J., Liu H.K., *J.Electrochem. Solid-State Lett.* **7** (2004) A250-A253.
[7] B. C.Kim, H.Uono, T.Sato, T.Fuse, T.Ishihara, M.Senna, *J. Solid State Ionics.* **172** (2004) pp 33-37.
[8] S.P. Kuksenko, *J. Chemistry, Physics and Technology of Surface.* **14** (2004) pp 123–128.
[9] N. Dimov, S. Kugino, M. Yoshio, *J. Electrochim. Acta.* **48** (2003) pp 1579–1587.
[10] S.-H.Ng, J.Wang, D.Wexler, K.Konstantinov, Z.-P.Guo, H.-K.Liu, *J. Angew. Chem. Int. Ed.* **45** (2006) pp 6896–6899.
[11] S.P. Kuksenko, V.G. Lutsenko, *T.Grotthuss Electrochemistry Conference.* Vilnius, June 5-8, (2005) p 95.
[12] A.A. Arie, J.O. Song, J.K. Lee, *J. Mat. Chem. Phys.* **113** (2009) pp 249–254.
[13] M.K. Datta, P.N. Kumta, *J. Power Sources.* **165** (2007) pp 368–378.
[14] J.Saint, M.Morcrette, D.Larcher, L.Laffont, S.Beattie, P.Pérès, D.Talaga, J.-M.Tarascon, *J. Adv. Funct. Mater.* **17**(2007) pp 1765-1774.
[15] G.E. Remnev, E.G. Furman, A.I. Pushkarev, S.B. Karpuzov, N.A.Kondrat'ev, D.V. Goncharov, *J. Instruments and Experimental Techniques.* **47** (2004) pp 394-398.
[16] Y.I. Isakova, A.I. Pushkarev, G.E. Kholodnaya, *J. Instruments and Experimental Techniques.* **54** (2011) 183-186.
[17] D. Ponomarev, G. Kholodnaya, G. Remnev, M. Kaikanov, R. Sazonov, *Journal of Physics: Conference Series.* **552** (2014) Article number 012024.
[18] M. Machida, K. Norimoto, T. Watanabe, et al., *J. of Materials Science.* **34** (1999) pp 2569-2574.
[19] Y.-G. Kwon, Se-Y. Choi, E.-S.Kang, et al., *J. of Materials Science.* **35** (2000) pp 6075-6079.
[20] G. Takahiro, K. Takayuki, A. Yoshimoto, *J.of Sol-Gel Science and Technology.* **13** (1998) pp 975–979.
[21] G. Kholodnaya, D. Ponomarev, R. Sazonov, G. Remnev, *J. Radiation Physics and Chemistry.* **103** (2014) pp 114-118.
[22] D.V. Ponomarev, G.E. Remnev, R.V. Sazonov, G.E. Kholodnaya, *J. IEEE Trans. on Plasma Sc.***41** (2013) pp 2908-2912.

IOP Conf. Series: Journal of Physics: Conf. Series **830** (2017) 012036 doi:10.1088/1742-6596/830/1/012036

Electron and phonon spectra dynamics and features of phase transitions in sodium at P=0-100 GPa

N B Volkov and E A Chingina

Institute of Electrophysics, Russian Academy of Science, Ural Branch, Amundsen Street 106, Yekaterinburg 620016, Russia

E-mail: zhenya@iep.uran.ru

Abstract. An electron and phonon spectra dynamics, features of structural phase transitions, and melting of sodium under the pressure ranging from 0 to 100 GPa are investigated. The electron and the phonon spectra of crystal sodium are calculated ab initio within the density functional theory by means of the software package LmtART-7 (see [1-3] and references herein), allowing the fully potential method of linear muffin-tin orbitals (FP-LMTO method). Earlier, this method was used in papers [4-6] for the calculation of a metals band structure within an atomic-spheres approximation (LMTO-ASA method). Using the Lindemann measure and the calculated phonon spectra, the theoretical values of melting points corresponding to the experimental data are obtained. Features of the electron and the phonon spectra dynamics in the melting curve maximum vicinity and within the structural transition range are discussed cI2 $\rightarrow$ cF4.

1. Introduction

Modern methods of research of a structure of substances with high pressures [7, 8], allow us to reveal not simple behavior of simple metals (e.g. see reviews [9-12] and references therein). Firstly, a low - symmetric and incommensurate host-guest of crystal structures arose from structural phase transitions. Secondly, it is the abnormal character of melting curves $T_m(P)$, showing an existence of the maxima and the minima, and also – the sites in the curve with $dT_m/dP<0$ [13, 14]. Thirdly, it is the superconductivity [15, 16] and the transitions of metal-to-semiconductor [17] and semiconductor-to-metal [18] in Li, and the transition of metal-insulator in Na [19].

In experiments [13, 14] the anomalous behavior of the melting curve $T_m(P)$ of sodium was revealed: the maximum of $T_m = 1000\,\mathrm{K}$ with P=30 GPa and its subsequent falling to room temperatures at $P = 118 \div 120$ GPa (see curves 1 and 2 designated in Fig. 1a by star symbols at Fig. 1). The negative melting slope of sodium is within the pressure range, in which the stable solid structures are known to be a body-centered cubic (b.c.c. or cI2 in Pearson notation [20]) and to be a more compact face-centered cubic (f.c.c. or cF4 in Pearson notation [20]) above 65 GPa [21]. Under higher pressure of $P = 115$ GPa, after transition from cF4 to a lower symmetric cubic structure with a 16-atom cell (cI16) [13, 21], the melting temperature is $T_m = 300$ K, which is lower than that at a normal pressure 371 K. See also Fig. 1a, where the sodium structure in the melting curve minimum vicinity, according to the crystallographic researches of the authors of [14], is shown in a boxed insert.

It can be seen, that the transition range from crystal structures cI16 and oP8 at $P = 118$ GPa and $T = 296.6$ K to the liquid with the diffuse nature of X-ray radiation scattering equals to $\Delta T = 9.4$ K temperature interval. Herein, within the range of $296.6 \le T \le 301.4$ K, the sodium structure represents a mixture of crystal structures - a tetragonal tI50 and a rhombic oC120 containing 50 and 120 atoms in unit cells, respectively, and within the range $301.4 \le T \le 306$ K – a monoclinic structure mP512 containing 512 atoms in the cell. Therefore, one may assume that the transition range, bounded from the bottom by the melting curve passing through (306 K, 115 GPa), (296.6 K, 118 GPa), (306 K, 119.4 GPa) points, and from the top by a liquid melt, represents rather a polycluster amorphous body (see [24]), than a mono - or a polycrystalline body. Along the line $T_m = 306$ K, both a density and a volume of the liquid are non-uniform. Thus, the derivatives $\partial V_{Lm} / \partial P|_{T_m = \mathrm{const}}, \partial V_{Sm} / \partial P|_{T_m = \mathrm{const}}$ are negative and cross each other at $P = 118$ GPa, where $V_{Sm} = V_{Lm}$.

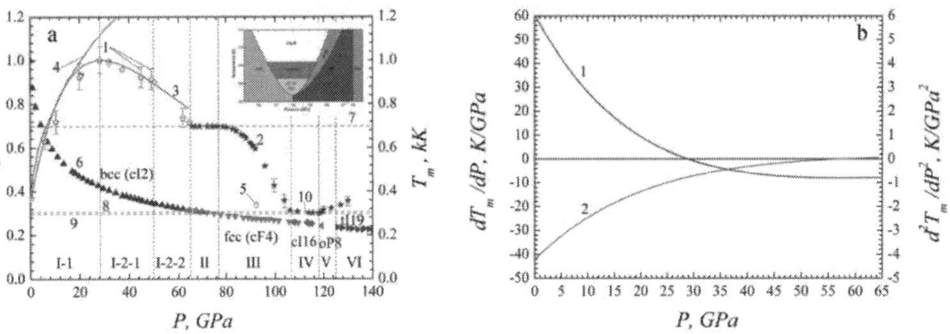

Figure 1. Left (a): Points (1, 2 - stars) - Na melting temperature experimental values according to [13]; curve 3 - our approximation of [13] data within the range $0 \le P \le 64.5$ GPa; curve 4 - melting temperature calculated values, obtained in previous paper [22] where genuine two-phase equation of state was used; points (5, diamonds) - melting temperature calculated values, obtained in current paper (onward introduced); triangles (6) represent crystalline sodium relative atomic volume and its structure variations at pressure rise [13, 14, 23]. On a boxed insert part of the melting curve minimum is shown with transition range structures being indicated according to [14]. Horizontal dash line (7) corresponds to T_m within transition range cI2 → cF4, where $V_{Sm} \simeq V_{Lm}$; line (8) corresponds to T_m of Na having cI16 structure (see boxed insert); line (9) corresponds to the temperature, below which sodium stays in crystalline state with structure, marked on figure with vertical dashed lines; (10) - range, corresponding to the melting curve minimum, where Na liquates from being solid. Right (b): First (1) and second (2) approximation (curve 3 on Fig. 1a) derivatives.

It should be mentioned, that since a cesium melting curve anomalous behavior has been experimentally discovered in 60's of the 20th century [25, 26] (see also reviews [27, 28] and the references therein), numerous attempts of its theoretical explanation have been performed: e.g. see [29-34]. In papers [29-31] the solid-liquid transition is considered as an order-disorder transition within a Lennard-Jones and Devonshire model of liquids [35, 36] under Bragg-Williams approximation which is technically applicable to alloys (solutions) only. Despite the faint applicability of Lennard-Jones and Devonshire theory and Bragg-Williams approximation for simple monoatomic liquids, such as melts of alkali metals (see some reasonable criticism of the theory in Ya.I. Frenkel's book [37]), in the noted papers, a qualitative confirmation of a chance of a maximum emergence resulting from a long-range order deviation in solids has been obtained. For the cause of maximum formation on melting curve papers [32-34] calls repulsive potential and its softness to attention and also point at possible dependence of the vicinity. I. e. the repulsive potential probable relation to the near-order local structure of a solid metal and its melt. The paper [38] should also be noted, since it reports that the substance model used in [29-34] cannot provide any clear statement for the melting curve abnormal behavior. The authors of [38] rather convincingly showed the possibility of a model fluid, with only one type of a local structure, to have a wide features range of melting, structural transitions in solids under condition of interparticle repulsion extreme softening, leading to the nearest neighbors division reduction. In other words, the extreme softening of the interparticle repulsion promotes the fluid clustering (a polycluster amorphous body in terms of A.S. Bakai [24]). For the extra proof, let's consider a liquid phenomenological model suggested in our papers [22, 39]. According to the model, a liquid consists of the clusters having a quasicrystal short-range ordering and an intercluster space occupied with a monoatomic gas (the monoatomic clusters). A sodium melting curve (see curve 4 on Fig. 1a) and a curve of the pressure relative change in the course of melting (see curve 2 on Fig.2a) were obtained within a two-phase approximation. A comparison with experimental data shows, that up to 15 GPa, a good coincidence between theoretical and experimental melting curves is observed. The melting temperature growth deceleration and the volume jumped reduction acceleration is observed under increasing pressure. However, this fact is not sufficient for gaining a zero-value volume jump at $P = 28.3$ GPa. See for example curve 3 in Fig. 3a (marked with stars), we obtained the extrapolation to zero for the experimental data represented in [28] (the curve 1 marked with triangles in Fig. 2a). In our estimation, it was concerned both the actual two-level model and the identical number of atoms in clusters, thus not allowing to obtain a denser, in comparison with a solid body, packing of atoms in its melt, which was observed on the sites of the melting curve with a negative slope.

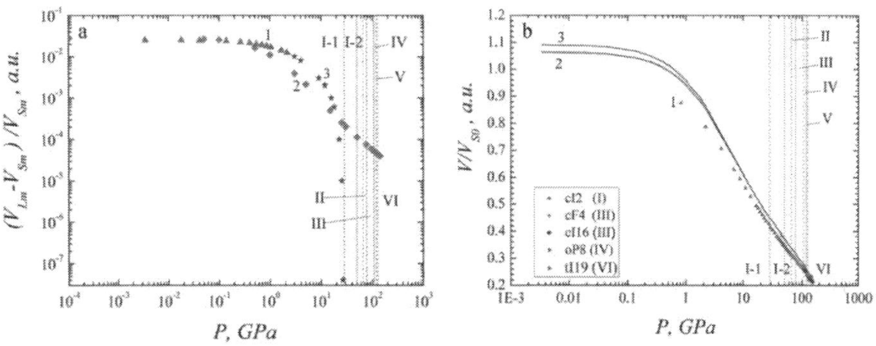

Figure 2. Left (a): Relative volume variation of Na while melting within the range I-1 according to experimental date from [28] - triangles (1), our calculations from [22] - diamonds (2) and our extrapolation to zero of data provided by [28] - stars (3). Right (b): Relative volume of solid and liquid sodium qualitative behavior versus pressure according to the experimental data and our interpolation (see Fig. 2 (a)).

At present time, in order to describe the melting curves behavior, the so called ab initio methods are used, in which forces acting on atoms enforcing them to gain their instantaneous arrangement (well-ordered or not) are computed within quantum dynamical methods [40-47]. For example, the ab initio molecular dynamics simulation of Na solid and liquid phases has been carried out in papers [40, 41] reporting about the coherence between a structure and a density and determining the reasons of the melting curve negative inclination. In papers [42, 43, 45], the explanation of sodium melting is suggested via ab initio calculations of the phonon spectrum and the mean square deviation of atoms, where only the matter solid phase properties are used. The cited papers show the rebuilding of sodium with an essential role of the phonon spectrum in a melting curve abnormal character within a single-phase approximation, using the well-known Lindemann measure as the melting criterion. In our opinion, little attention is payed to the analysis of the lattice atomic zero fluctuations contribution, to the features of a melting curve behavior, because under pressure, the lattice atomic zero fluctuation energy increases. Thus, close to zero temperatures, the equilibrium locations of atoms may not coincide with the ideal lattice nodes. In favor of this assumption, it is possible to adduce the cI16 equilibrium crystal structure, which can be a superposition of eight cI2 cells, however atoms of the latter would not reside in nodes of it.

The analysis is also necessary because the anomalous behavior of melting curves under increasing pressure may occur not only with alkali metals, but also with other metals, including noble and transitional, but under significantly greater pressure values. For instance in [48], the structural transition in Au from a cubic centered (fcc) to a hexagonal-close-packed (hcp) structure at $P = 240$ GPa has been found experimentally; the same transition for Al at $P = 217 \pm 10$ GPa has been observed in [49]. The theoretical paper [50] suggested possibility of existence for Al of an incommensurate host-guest (h-g) structure Al16-I4/mcm at $P = 3.2 - 8.8$ TPa being similar to the incommensurate host-guest structure of phase IVa of barium.

The objective of the current paper is a careful analysis of the electron and phonon spectra dynamics and the lattice atomic zero fluctuation contribution to the cI2→cF4 structural transition features and the melting curve behavior for sodium within the pressures range from 0 to 100 GPa.

2. Calculation method

We used the LMTART-7 software package [2, 3] for the ab initio phonon and electron spectra determination, as well as of the total interaction energy between the electrons and the sodium nuclei. The dynamics of a lattice was computed using a linear response method within a well-known density functional theory (see details of the method in the reviews [51-53]) by LMTART-7. The exchange-correlation potential (Vosko-Wilk-Nussair) [54] was considered in a generalized gradient approximation [55]. We used a pseudo-potential method in the plane wave expansion (PLW). The PLW expansion is a full potential approximation which uses non-overlapping muffin-tin spheres, where the potential is represented via the spherical harmonics, the expansion, and the interstitial region where the potential is expanded in plane waves. The full potential regime provides the best accuracy by an increasing computational time [2, 3]. The number of plane waves 14146 was used, the plane-wave energy cutoff was 138.1 Ry (up to 250 Ry). The integration on Brillouin zone was carried out on grid 32*32*32 special (Monkhorst-Pack) k-points. The phonon spectra were calculated by means of the interatomic power constants, on the basis of 35 basic q-vectors, in the irreducible Brillouin zone (a grid 8*8*8). The calculated via LMTART-7 cI2 Na phonon states densities for various pressure values correspond to normalization requirements within the accuracy, which is substantially lower than one percent. The highest deviation, 1.6%, is obtained for $P = 0.263$ GPa.

To obtain Na melting temperature, similar to [42, 43, 45], we use a well-known Lindemann measure

$$\sqrt{\ll u^2(T_m) \gg}\ L d_{NN} \tag{1}$$

where $\ll u^2(T_m) \gg$ - is an atom displacement standard deviation at a melting temperature T_m; L, d_{NN} - is a Lindemann measure and a mean nearest-neighbor distance, correspondingly.

The atomic displacement standard deviation depends on a normalized phonon states density $g_{ph}(\omega)$ using the relation [56]:

$$\ll u^2(T) \gg = \frac{\hbar}{2M} \int \frac{g_{ph}(\omega)}{\omega} \coth\frac{\hbar\omega}{2T} d\omega \tag{2}$$

Normalization requirement $g_{ph}(\omega)$ is:

$$\int_{\omega_{min}}^{\omega_{max}} g_{ph}(\omega) d\omega = 1, \tag{3}$$

where $g_{ph}(\omega_{min}) = g_{ph}(\omega_{max}) \equiv 0$. It should be noted that ω_{min} values, obtained using the LMTART-7 software package, might become negative (results of negative frequencies minimal values emergencies will be further analyzed). Herein, according to our calculations, the normalization requirement (3) is fulfilled with high accuracy.

At T≫$\hbar\omega$, the equation (2) transforms to:

$$\ll u^2(T) \gg = \frac{T}{M}\int \frac{g_{ph}(\omega)}{\omega^2} d\omega \tag{4}$$

The resultant expressions, we used for T_m calculations in both a quantum case and a classical limit, are written as:

$$(Ld_{NN})^2 - \frac{\hbar}{2M}\int_{\omega_{cut}}^{\omega_{max}} \frac{g_{ph}(\omega)}{\omega} \coth\frac{\hbar\omega}{2T} d\omega = 0; \tag{5}$$

$$T_m = (Ld_{NN})^2 M \left(\int_{\omega_{cut}}^{\omega_{max}} \frac{g_{ph}(\omega)}{\omega^2} d\omega \right)^{-1}. \tag{6}$$

The replacement of ω_{min} in (5) and (6) by $\omega_{cut} > \omega_{min}$ is caused by an infeasibility to obtain the physically sensible data. The curve $T_m(P)$ calculations within full frequency range $[\omega_{min}, \omega_{max}]$ ($\omega_{cut}(P), \omega_{max}(P)$ behavior will be analyzed later). We should note that our ω_{max} values calculated using the LMTART-7 are close to those obtained in [43] using the Quantum ESPRESSO software package [57].

Calculations of T_m using (5) and (6) were performed in MATHEMATICA 7 software package. For Na taken, the Lindemann constant value was L=0.14, which was similar to [43].

3. Results. Discussion

An important characteristic of the FP-LMTO method is a non-touching muffin-tin sphere radius, R_{MT}. It separates the electrons belonging to an atom core from the electrons of external orbitals. Fig. 3 shows the pressure dependent relation R_{MT}/R_c (R_c - an average nearest neighbors distance determining the macroscopic density (i.e. a volume)). It can be seen that for the given lattice type, the value is constant (curve 2), while the relative volume per atom (curve 1) smoothly decreases. Within the region II, where the structural transition cI2→cF4 takes place, it decreases abruptly. Inside this region, at least within the range P≃65÷75 GPa, the simultaneous existence of the phases cI2 and cF4 is observed.

Let's return to Fig.1. We obtained the analytical approximation of the experimental results in the region I (cI2 lattice, the curve 3 on Fig.1a) as follows:

$$T_m = T_{m0} + A\left(\exp(-\alpha_1(P-P_0)) - \exp(-\alpha_2(P-P_0))\right), \tag{7}$$

where T_{m0}371 K; $A = 1.70981 \cdot 10^8$ K, $\alpha_1 = 3.49197 \cdot 10^{-2}$ GPa^{-1}; $\alpha_2 = 3.492 \cdot 10^{-2}$ GPa^{-1}. The first and the second derivatives of the melting curve are shown in Fig.1b. It is obvious that the T_m maximum point does not belong to a simple second-order phase transition, if the existence of the implicit local order parameter is not assumed, suggested like a tensor irreducible part in paper [58]

$$T_{\alpha\beta\gamma\delta}(\mathbf{x}) = \sum_{i=1}^{3} n_\alpha^i(\mathbf{x}) n_\beta^i(\mathbf{x}) n_\gamma^i(\mathbf{x}) n_\delta^i(\mathbf{x}), \tag{8}$$

where the orthogonal unit vectors $\{\mathbf{n}^i\}(\mathbf{x})$ $(i = 1,2,3)$ define the local cubic order direction at a point $\mathbf{x}$. The real situation is apparently much more complicated. Firstly, because the order parameter (8) can be applied to a simple cubic lattice only. Secondly, the melting curve and a dashed vertical line at

$P = 28.3$ GPa divide the phase plane into four areas with various local densities and symmetries. Within the region I-1, a metal density below $T_m(P)$ is greater, than the melt density above $T_m(P)$. Within the region I-2 - situation is vice versa: above $T_m(P)$, the melt density exceeds the metal density below $T_m(P)$. Unfortunately, the single-phase Lindemann measure along with *ab initio* molecular dynamics methods are not capable of solving the problem uniquely. The calculated values (diamonds, Fig. 1a), we obtained over the relations (5) and (6), are well consistent with the experimental data. However, this agreement is achieved by cutting off the initial frequencies of the phonon states density (see curve 2 on Fig. 4a). Moreover, the behavior of curve $\omega_{cut}(P)$ qualitatively repeats the melting curve behavior, its more flat maximum being shifted to the side of larger pressures. It indicates that the melting curves data are to be accepted cautiously when we use the Lindemann measure or similar molecular dynamics methods. In our opinion, investigating the melting curves abnormal behavior at high pressures, it is necessary to pay more attention to the research of a condensed matter structure and a near order symmetry.

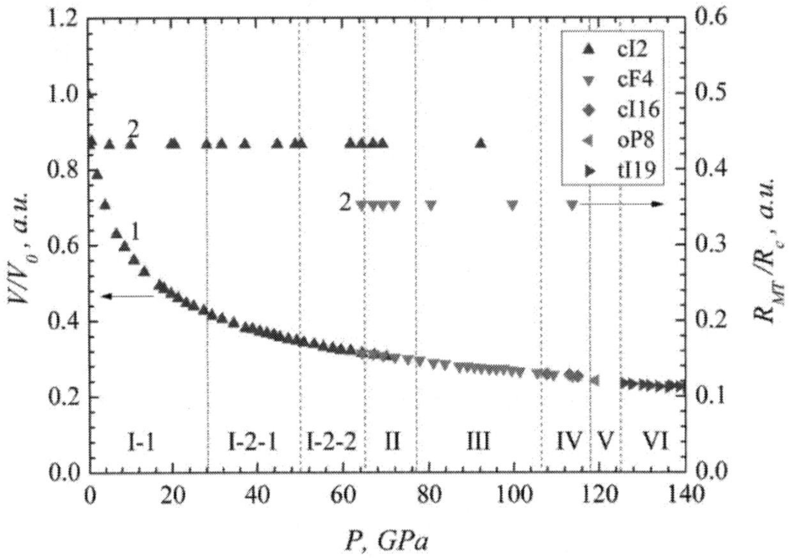

Figure 3. Relative volume [23] (1) and R_{MT} / R_c ratio calculated via LmtART-7 [2] dependences versus pressure comparison.

Fig. 4b shows the dependence of the zero fluctuation energy relation to the latent melting heat at normal conditions, obtained within the Debye model of a metal with a boundary frequency $\omega_{max}(P)$ (curve 1) and

$$\varepsilon_{vac} = \frac{\hbar}{2}\int_{\omega_{min}}^{\omega_{max}} \omega g_{ph}(\omega) d\omega$$

(curve 2). It is obvious that in the region I, the vacuum fluctuation energy can reach up to 20% of the latent melting heat, thus, encouraging the disorder growth when the pressure rises. The suggested conclusion assigns a necessity of considering a zero fluctuations contribution to both cases of a lattice long range order destruction during melting, and a crystalline structure reconstruction at high pressures. The additional pro argument can be found in paper [59]. It claims that the growth of the pressure promotes the anharmonicity increase of lattice fluctuations, which suppresses a superconductivity and leads to a superconductor-to-semiconductor transition of lithium at $P = 80$ GPa

(See [18] and references therein).

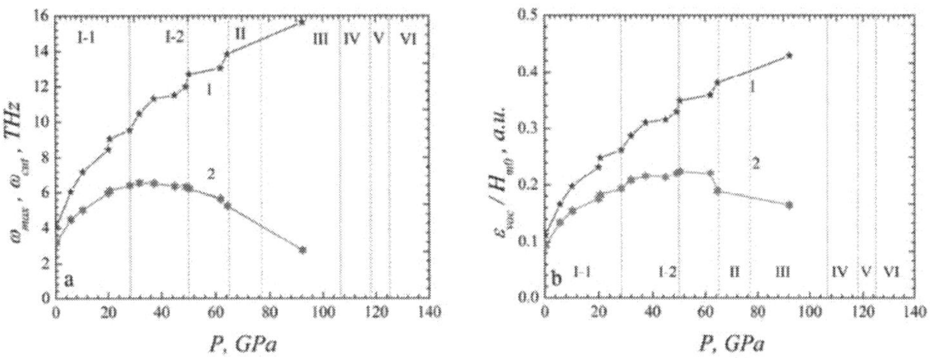

Figure 4. Values of boundary (1) and cutoff (2) frequencies of phonons (a) and vacuum fluctuations energy to latent melting heat ratio at normal conditions(b). Curve 1 is calculated within Debye model approximation taking into account the boundary frequency value; curve 2 is obtained according to phonon states density computed via LMTART-7 [2].

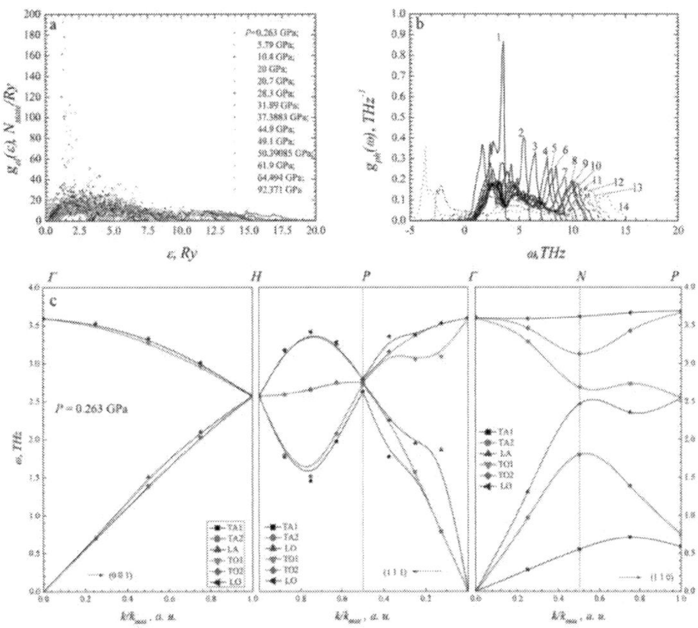

Figure 5. Electron (a) and phonon (b) states density computed via LMTART-7 [2], and also phonon spectrum (c) for P = 0.263 GPa.

Let's consider a dynamics of a phonon spectrum and a density of electron and phonon states near the melting curve maximum and at the beginning of a phase transition cI2→cF4. Fig. 5 shows the electron (a) and the phonon (b) states densities, and the phonon spectrum for $P = 0.263\,\text{GPa}$ (c). The curves numbers in Fig. 5b correspond to the pressure values designed in Fig. 5a. Since all calculations were made for the unit cell containing two atoms, three acoustic branches and three optical branches do exist. Our calculated acoustic spectrum qualitatively coincides with the sodium spectrum given in textbooks, e.g. [60].

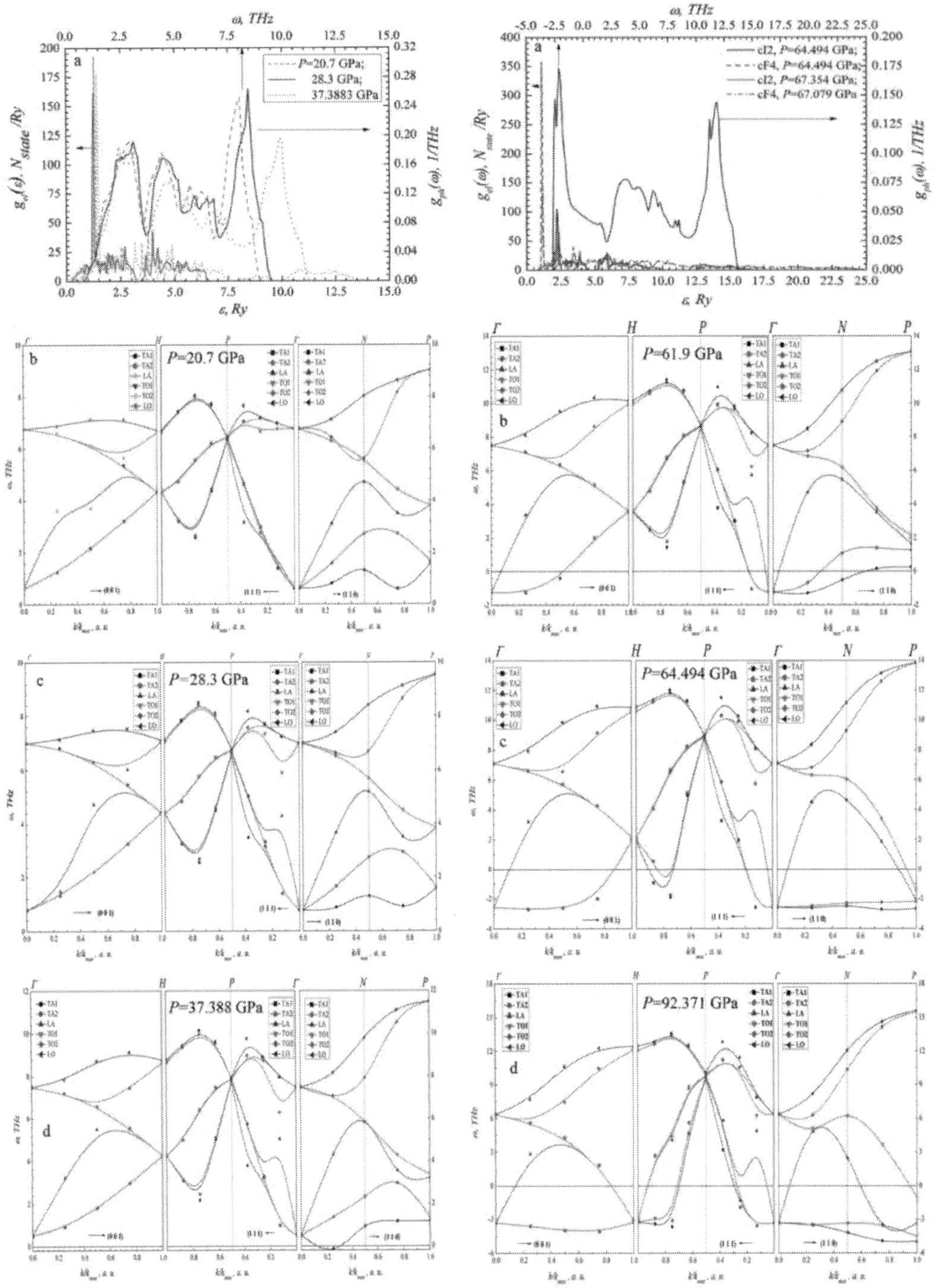

Figure 6. Left: Electron and phonon densities of states (a) and a phonon spectra of Na b.c.c. lattice in the vicinity of melting curve maximum:(b, c, d); Right: Density of electron b.c.c. and f.c.c. lattices states and phonon b.c.c. lattices states for sodium in the range of b.c.c.-f.c.c. transition (a), phonon spectrum of sodium b.c.c. lattice in the vicinity of transition b.c.c.-f.c.c. (b, c) and in the range of only f.c.c. lattice existence (see also Fig. 1).

Fig. 6 shows the electron and the phonon densities of states (a) and the phonon spectra (b, c, d), in the vicinity of T_m maximum (left); the electron (for b.c.c. and f.c.c. lattices) and the phonon densities of states for the b.c.c. lattice at $P = 64.494$ GPa (a), the phonon spectra of the b.c.c. lattice for three pressure values: (b) 61.9 GPa, (c) 64.494 GPa, and (d) 92.371 Gpa (right). The features of the phonon spectra, in the vicinity of the melting curve maximum are the existence of a non-zero positive frequency value at $k = 0$ and the small negative transverse acoustic branch frequency at $k\,/k_{max} \simeq$ 0.2. It demonstrates the sodium local structure transformation. The existence of larger segments at the acoustic and the optical branches lying in the region of frequencies negative values testifies the instability of these branches and the lattice itself.

4. Conclusion

Thus, in present paper we obtained the energy distribution of the electrons and the phonons as well as the density of states for all experimental points within the range of Na b.c.c. lattice existence. Using the one-phase Lindemann measure and the calculated phonon spectra, the theoretical values of the melting points matching the experimental data has been obtained. An energy of zero fluctuations was defined, and the necessity to account its contribution to the dynamics of the electron and the phonon spectra has been shown. Features of the volume jump behavior during melting under the increasing pressure are discussed.

Acknowledgments

The authors appreciate S.Y. Savrasov for providing the LMTArt-7 software package for electron and phonon spectra of sodium calculations at various pressures as well as K.A. Nagayev for help in preparing of English version of the article. The work is carried out within the state order No.~0389-2014-0006 and under the partial financial support of the RFBR (project No.~16-08-00466) and the Ural Branch of RAS within the UB RAS fundamental research program "Matter at high energy densities" (project No. 15-1-2-8).

References

[1] Savrasov S Y and Savrasov D Y 1992 *Phys. Rev. B* **46**(19) 12181-12195
[2] Savrasov S Y 1996 *Phys. Rev. B* **54**(23) 16470-16486
[3] Savrasov S Y 2004 Program LMTART for electronic structure calculations (*Preprint* cond-mat/0409705)
[4] Andersen O K 1975 *Phys. Rev. B* **13**(8) 3060-3083
[5] Andersen O K and Jepsen O 1984 *Phys. Rev. Lett.* **53**(27) 2571-2574
[6] Andersen O K, Jepsen O and Gloetzel D 1985 *in "Highlights of Condense- MatterTheory"* (Amsterdam, New York: North Holland) pp 59-176
[7] Jayaraman A 1983 *Rev. Mod. Phys.* **55**(1) 65-108
[8] McMahon M I 2015 *Phil. Trans. R. Soc.A* **26**(6) 20130158
[9] Kolobyanina T N 2002 *Phys.Usp.* **45**(12) 1203-1211
[10] Maximov E G, Magnitskaya M V and Fortov V E 2005 *Phys. Usp.* **48**(8) 761-780
[11] McMahon M I and Nelmes R J 2006 *Chem. Soc. Rev.* **35** 943-963
[12] Degtyareva O 2010 *High Pressure Research* **30**(3) 343-371
[13] Gregoryanz E and et al 2005 *Phys. Rev. Lett.* **94**(18) 185502
[14] Gregoryanz E and et al 2008 *Science* **320** 1054-1057
[15] Shimizu K and et al 2002 *Nature* **419** 597-599
[16] Matsuoka T and et al 2008 *J. Phys.: Conf. Series* **121** 052003
[17] Matsuoka T and Shimizu K 2009 *Nature* **458** 186-189
[18] Matsuoka T and et al 2014 *Phys. Rev. B* **89** 144103
[19] Ma Y and et al 2008 *Nature* **458** 182-186
[20] Pettifor D G 1995 *Bonding and Structure of Molecules and Solids* (Oxford: Clarendon Press)
[21] Hanfland M, Loa I and Syassen K 2002 *Phys. Rev. B* **65**(18) 184109

[22] Volkov N B and Chingina E A 2015 *J. Phys.: Conf. Series* **653** 012084
[23] Guillaume C L and et al 2011 *Nature Physics* **7** 211-214
[24] Bakai A S 1987 *Polycluster Amorphous Solids* (Moscow: Energoatomizdat)
[25] Kennedy G C, Jayaraman A and Newton R C 1962 *Phys. Rev.* **126**(4) 1363-1366
[26] Stishov S M and Makarenko I N 1968 *Sov. Phys. JETP* **27**(3) 378-380
[27] Stishov S M 1969 *Sov. Phys. Usp.* **11**(6) 816
[28] Stishov S M 1975 *Sov. Phys. Usp.* **17**(5) 625
[29] Rapoport E 1967 *J. Chem. Phys.* **46**(8) 2891-2895
[30] Rapoport E 1968 *J. Chem. Phys.* **48**(4) 1433-1437
[31] Yoshida T and Okamoto H 1971 *Prog. Theor. Phys.* **45**(3) 663-681
[32] Yoshida T and Kamakura S 1971 *Prog. Theor. Phys.* **47**(6) 1801-1815
[33] Yoshida T and Kamakura S 1974 *Prog. Theor. Phys.* **52**(3) 822-839
[34] Ogura H and et al 1977 *Prog. Theor. Phys.* **58**(2) 419-433
[35] Lennard-Jones J E and Devonshire A F 1939 *Proc. Roy. Soc. A* **169** 317
[36] Lennard-Jones J E and Devonshire A F 1939 *Proc. Roy. Soc. A* **170** 464
[37] Frenkel J I 1946 *Kinetic Theory of Liquids* (London: Oxford University Press)
[38] Prestipino S, Saiya F and Malescio G 2010 *J. Phys. Chem. Phys.* **133** 144504
[39] Volkov N B and Chingina E A 2014 *Russian J. Physics* **57**(12/3) 146-151
[40] Raty J Y, Schwegle E and Bonev S A. 2007 *Nature* **449** 448-451
[41] Ko`ci L and et al 2008 *Phys. Rev. B* **77**(13) 132101
[42] Martinez-Canales M and Bergara A 2008 *J. Phys. Chem. Solids* **69**(9) 2151-2154
[43] Lepeshkin S V, Magnitskaya M V and Maksimov E G 2009 *JETP Lett.* **89**(11) 586-591
[44] Grabovski B and et al 2009 *Phys. Rev. B* **79** 134106
[45] Lepeshkin S V and et al 2012 *JETP* **115**(1) 105-111
[46] Glensk A and et al 2015 *Phys.Rev. Lett.* **114** 195901
[47] Minakov D V and Levashov P R 2015 *Phys. Rev. B* **92** 224102
[48] Dubrovinsky L and et al 2007 *Phys.Rev. Lett.* **98** 045503
[49] Akahama Y and et al 2006 *Phys.Rev. Lett.* **96** 045505
[50] Pickard C J and Needs R J 2010 *Nature Materials* **9** 624-627
[51] Savrasov S Y and Maximov E G 1995 *Phys. Usp.* **38**(7) 737-760
[52] Savrasov S Y, Savrasov D Y and Maximov E G 1997 *Phys. Usp.* **40**(4) 337-358
[53] Baroni S and et al 2001 *Rev. Mod. Phys.* **73**(2) 515-562
[54] Vosko S H, Wilk L and Nusair M 1980 *Can. J. Phys.* **58** 1200-1211
[55] Perdew J P, Burke K and Ernzerhof M 1996 *Phys. Rev. Lett.* **77**(18) 3865-3868
]56] Kosevich A M 1981 *The Physical Mechanics of Real Crystals* (Kiev: Naukova Dumka)
[57] Giannozzi P and et al 2009 *J. Phys.: Condens. Matter* **21** 395502
[58] Mitus' A S and Patashinskii A Z 1981 *Sov. Phys. JETP* **53**(4) 798-803
[59] Errea I, Rousseau B and Bergara A 2012 *J. Phys.: Conf. Series* **377** 012060
[60] Katsnelson M I and Trefilov A V 2002 *Crystal Lattice Dynamics and Thermodynamics* (Moscow: Atomizdat)

Wideband TEM-TE11 mode convertor for HPM applications.

D N Bykov[1], N M Bykov[1] and I K Kurkan[1, 2]

[1] Institute of High Current Electronics, 2/3 Akademichesky Ave., Tomsk, 634055, Russia

[2] National Research Tomsk Polytechnic University, 30 Lenin Ave., 634050 Tomsk, Russia

E-mail: ikk@lfe.hcei.tsc.ru

Abstract. The mode convertor design of fundamental coaxial TEM to the lowest asymmetric TE11-mode of a circular waveguide was proposed and optimized with ANSYS HFSS software. It includes axially aligned parts: the input coaxial line with the high voltage insulator, conical coaxial matching line, wave-coax transition section and output circular waveguide. The most losses in this type of convertor caused by the wave of coaxial TE11-mode running back to the microwave source. To minimize these losses, there is the matching conical coaxial line with the cut-off insertion for coaxial TE11-mode. Characteristics of the convertor are as follows: the maximum input peak power – 3GW, the input impedance – 28Ohm, the central operating frequency – 1.14GHz. The power conversion efficiency to the output mode is from 90% upto 100% in the frequency band of 20%.

1. Introduction

The interest in wide-band transformers of the principal coaxial wave (TEM) to a low asymmetric mode (TE11) of the circular waveguide is kept up by the development of generators of the high-power microwave pulses with the subnanosecond and nanosecond duration, where the microwave pulse is formed in coaxial lines: the generators of the ultra wide band (UWB) pulses, sources based on Non-linear transmission line (NLTL), coaxial vircators and etc.[1-3]. Other devices that needed the wideband output section are generators with wideband frequency tuning [4].

The short pulse duration as well as the basic principles of operation result in HPM pulses with the frequency band of about 10% or more. There is a number of patents and papers devoted to the coax-to-waveguide transitions where axes of the waveguide and the coaxial are orthogonally placed. Among them the "door knob" type is the only that meets HPM requirements.

The closest structural analogue of the transformer proposed had been developed more than fifty years ago [5]. There are the coaxial and the rectangular waveguide sections that placed axially. They are electromagnetically coupled by the coaxial inner conductor extended into the waveguide and angled to the axis. Sometimes, the axial coaxial-to-waveguide transition is preferable to develop a compact design of an assembly. This paper stated the high quality of the matching the transmission lines with the power reflection level of about 1% in the frequency band of 2.6-3.6GHz. That corresponds to the operating band width of about 35%. However, the device is not intended for the high power operation because of the step-like coax-to-waveguide junction and the sharp edges of the matching elements.

As opposed to the converter described [5], the paper proposed the structure with a circular waveguide instead of the rectangular one. The axial geometry of the elements simplifies their production with the increased mechanical and electric strength. Another big difference is the matching coaxial – conical line used to eliminate the step-like coaxial-waveguide junction. These distinguishing features are the reasons of the significant increase in the converter's breakdown level.

The aimed maximal peak power of 3GW was calculated under condition of the 300kV/cm upper limit for electric field strength at regular sections. This value matches the value of the static breakdown for most dielectrics including the insulating oil used in a high voltage equipment. The structure requires the output section filling with SF6 up to 5atm of overpressure. The vacuum insulation is allowed as well.

ANSYS HFSS software [6] was used to analyze the wave modes presented in the structure. It was revealed that the key factor which impacted the efficiency and the band width of the converter was the excitation level of the lowest coaxial non-symmetrical wave TE_{11coax}. To minimize the power fraction transformed into the TE_{11coax} mode, a short rejecting line was provided in the converter. That enabled us to essentially increase the efficiency of the frequency wideband conversion into the operating mode TE_{11}. In numerical simulation, the power conversion into the TE_{11} mode from 0.9 upto 1.0 was obtained within the 20% band with the central frequency of 1.14GHz.

The structure, which is similar to the proposed one excluding the rejecting line, was used with the horn antenna for emission of the microwave radiation in the work [2]. As the simulation proved, the use of the structure without the rejection line decreased the conversion efficiency of the TEM wave power into the operating mode for about 20%. The experiments demonstrated the breakdown in the gas filled section of the converter. Using the SF_6 under the overpressure of 2-3atm helped eliminate the breakdown at the output power level of 260MW.

2. The Converter Structure

The structure of the proposed converter presented in figure 1 includes the following sections.

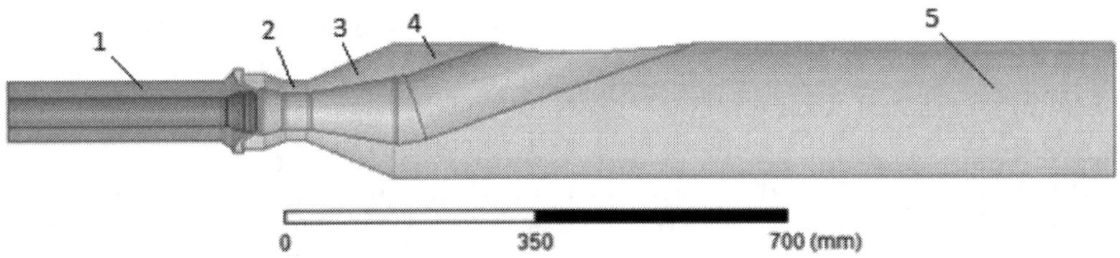

Figure 1. The view of the mode convertor: 1 - oil filled input line, 2 – TEM_{11coax} rejecting line, 3 – conical coaxial matching line, 4 – coax-to-waveguide transition, 5 – output waveguide.

Input section (1) is the coaxial line with the impedance of 28Ohm filled with insulation oil. The inner diameter of the outer conductor is 8 cm. The input line is separated from the gas filled sections by a high-voltage Nylon6 insulator. The rejecting coaxial line (2) has the impedance of 28Ohm. The inner diameter of its outer conductor is 8cm. The conical coaxial matching line (3) has the length of 12cm. The angular aperture is of 45^0. The impedance at the larger base crossection is about 41.5Ohm. The coax-to-waveguide transition (4) was designed as the prolongation of the inner conductor of the matching line angled at 20 degrees from the axis. The output section (5) is the circular waveguide with the inner diameter of 18cm.

3. Numerical Simulation

The proposed structure (figure 1) is the result of modifying the axis coaxial waveguide transition, where the coaxial outer diameter is equal to the diameter of the circular waveguide (see figure 2). The coaxial and waveguide are coupled by the junction, designed as the prolongation of the inner coaxial conductor angled to the axis and electrically linked to the common outer conductor. This configuration was numerically simulated with the discrete variations of the inner conductor diameter and of the junction slope angle. The junction slope angle was varied within the range of 10-90 degrees, the coaxial impedance was within the range of 20-60Ohm. The research was aimed at the achievement of the maximal band width with the highest conversion coefficient of the TEM wave to the circular waveguide mode TE_{11}.

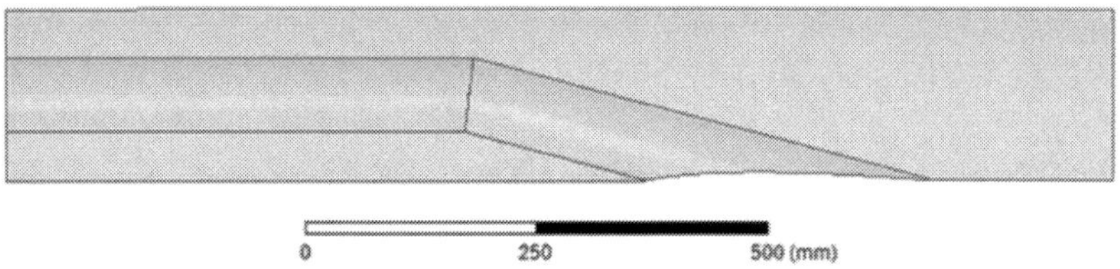

Figure 2. The coaxial-to-waveguide transition with the common outer conductor.

The most successful results in this configuration were obtained when the coaxial impedance ρ was within 30-50 Ohm and the junction slope angle β was 15 – 20 degree. It was revealed that the conversion coefficient K_{conv} approximate to 1.0 was hardly possible to be obtained under various ρ and β combinations even at the narrow frequency band. The typical value of K_{conv} was 0.8.

The simulation results for the instance with the $\rho = 50$Ohm and $\beta = 15^0$ are presented in figure 3. The power conversion coefficient K_{conv} to the output TE_{11} mode is about 0.8 in the band of 1-1.4GHz (figure 3a). The rest power fraction of 0.2 is transformed to the reflected TEM and TE_{11coax} modes (figure 3b). In this case, the most losses are attributed to the TE_{11coax} mode, while the fraction of reflected TEM mode is of few percent near the central frequency. Any changes in the coaxial impedance and the junction slope angle result in the reproportioning between the reflected modes and irregularity for the output power dependence on frequency.

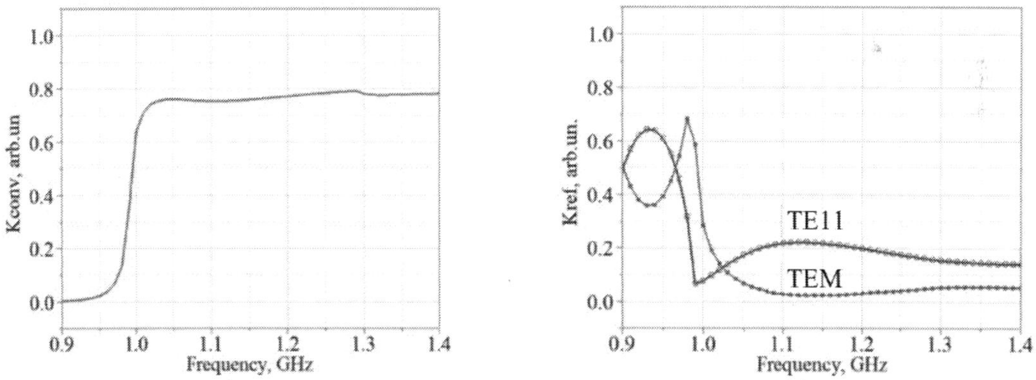

a) b)

Figure 3. Simulation results for the coaxial-to-waveguide transition with the common outer conductor: a – conversion power coefficient to the TE_{11} output mode, b – reflected power coefficients for TEM and TE_{11coax} modes.

The numerical simulation results of the optimized construction of the mode convertor are presented in figure 4. The dependence of power conversion coefficient Kconv to the output mode on frequency is shown in figure 4a. The operating bandwidth determined by the level Kconv = 0.90 is of 20% with the central frequency of 1.14GHz. The reflected power coefficients for coaxial TEM and TE_{11} modes are presented in figure 4b.

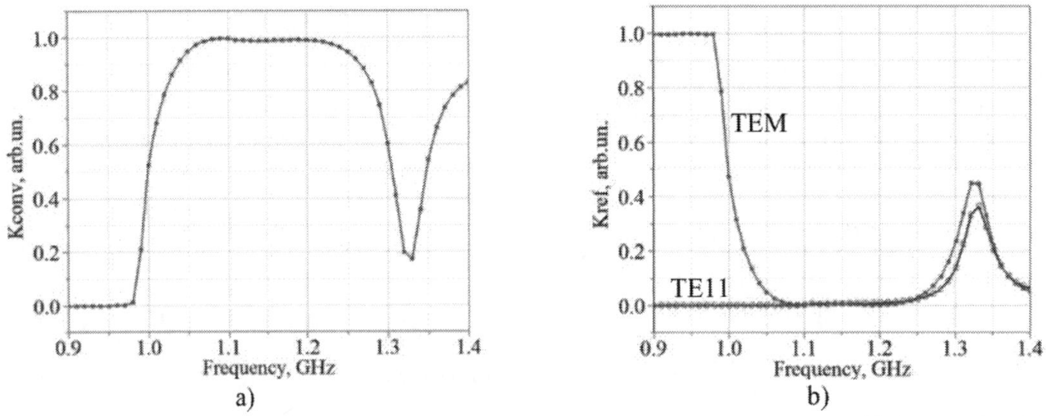

Figure 4. Simulation results for the proposed convertor: a – conversion power coefficient to the TE_{11} output mode, b – reflected power coefficients for TEM and TE_{11coax} modes.

4. Conclusion

The mode convertor from the coaxial TEM mode to the fundamental TE_{11} mode of circular waveguide with axially aligned input and output was proposed and numerically simulated. The power conversion efficiency obtained was 0.90 – 1.00 within the 20% band and the central frequency was 1.14 GHz. The maximum operating power is supposed to be about 3GW using SF_6 gas with about 5atm overpressure in the output section. The main effect constraining the operating band is the exciting of the coaxial TE_{11} mode. The recent experiments [2] proved that such type of mode convertor is capable to operate with the microwave pulse of a few nanosecond duration with GW power level in the L band.

References

[1] Totmeninov E M, Klimov A I, Konev V Yu, Pegel I V, Rostov V V, Tsygankov R V and Tarakanov V P 2014 *Tech. Phys. Lett.* **40(2)** 152-156

[2] Romanchenko I V, Rostov V V, Gubanov V P, Stepchenko A S, Gunin A V and Kurkan I K 2012 *Rev. of Sci. Instr.* **83** 074705

[3] Kovalchuk B M, Polevin S D, Tsygankov R V and Zherlitsyn A A 2010 *IEEE Trans. on Plasma Sci.* **38(10)** 2819-2824

[4] Bogdankevich I L, Ivanov I E, Loza O T, Strelkov P S, Ul'yanov D K and Garate E 2007 *Tech. Phys. Lett.* **33(6)** 480-482

[5] Wheeler G J 1957 *Nat. Conv. Rec. I.R.E.* **5** pt I 182-185

[6] ANSYS High Frequency Structure Simulator (HFSS) 2016 Ansys Corporation Pittsburgh PA 15317 URL: http://www.ansys.com/Products/Electronics/ANSYS-HFSS

IOP Conf. Series: Journal of Physics: Conf. Series **830** (2017) 012038 doi:10.1088/1742-6596/830/1/012038

Radiographic research of the Bi plasma jet formed by the vacuum arc discharge

A P Artyomov[1], A G Rousskikh[1], A V Fedunin[1], S A Chaikovsky[1,2], A S Zhigalin[1] and V I Oreshkin[1]

[1]Institute of High Current Electronics SB RAS, 2/3 Akademichesky Ave., Tomsk, 634033, Russia
[2]Institute of Electrophysics UD RAS, 106 Amundsen St., Ekaterinburg, 620016, Russia

E-mail: aap545@gmail.com

Abstract. The results of experiments on a soft x-ray radiography (≈ 1-2 keV) of a bismuth plasma formed by the high-current vacuum arc discharge are represented. The plasma gun with the arc current ≈ 60 kA and the current rise time ≈ 7 μs was used to produce the Bi plasma jet. The compact pulsed radiograph XPG-1 (250 kA, 220 ns) with an X-pinch load consisting of four Mo wires with a diameter 25 μm was used as a source of the soft X-ray radiation. The X-ray backlighting images of the researched plasma jet and the Bi step-wedge with a step thickness of ≈ 100 nm were recorded simultaneously in the course of the experiment. A comparison of the plasma jet x-ray image with the current trace has enabled to estimate dependencies of the linear mass on the arc current. The experiments have shown that when the arc current density reaches $\approx 3 \cdot 10^5$ A/cm^2, the evaporation rate of the electrode material reaches ≈ 100 μg/μs, that under the plasma velocity ≈ 0.5 cm/μs, provides a plasma jet linear mass ≈ 200 μg/cm. At a distance of ≈ 1-2 mm from the arc cathode surface, the sharp increase of the jet linear mass (up to ≈ 500 μg/cm) occurred.

1. Introduction

Currently, along with the Z-pinches based on the use of a wire array [1, 2] and a gas-puff [3, 4], Z-pinches based on the use of the vacuum arc discharge systems are employed [5 - 10]. The advantage of plasma liners based on arcs is that such a shell has not a problem with a "cold start", because their initial conductivity reaches significant values of $\sim 10^4$ Ohm$^{-1}\cdot$m^{-1} [11]. One of the most important parameters of the liner shell at the formation stage is the substance distribution (both radially and along its length). The most informative method for investigation of the dense plasma stream distribution is a soft X-ray radiography. The X-pinch high-temperature dense plasma was used as a radiation source for the X-ray radiography [12, 13, 14]. The X-ray radiography based on the X-pinch has a spatial resolution of 2÷3 μm, while a temporal resolution is about a nanosecond. The diagnostics, in a combination with the step-wedge manufactured from the same material as the investigated plasma, allows us to carry out not only a qualitative but also a quantitative analysis of the plasma flow structure. Knowledge of a distribution of the plasma flow linear mass is extremely important both for the mathematical description of the Z-pinch implosion and a study of the fundamental laws of the electrodes material evaporation under the high current arc electrical discharge.

2. Experimental setup

The experiment was carried out on a special stand consisting of a plasma gun [5 - 10] (the arc current ≈ 60 kA and the current rise time ≈ 7 μs) and a compact pulsed radiograph XPG-1 [12, 13, 14] (the current amplitude of 250 kA and a rise time of 220 ns). The plasma gun was used to form a bismuth plasma jet and the generator XPG-1 with the X-pinch load was used as a source of the soft X-ray radiation. The X-pinch load consisted of four Mo wires with a diameter 25 μm. A drawing of the plasma gun is presented in figure 1. The plasma gun was fed by a capacitor bank with $C = 20$ μF which has been charged up to 20 kV. The bismuth cathode with a diameter of 5 mm was placed in a center of the plasma gun.

The experimental scheme with the soft x-ray radiography of a bismuth plasma jet is presented in figure 2. The scheme magnification is 1.48. In the experiment, a bismuth plasma image was registered using a Micrat-ORTO film. The photographic film was filtered by Kimfol 4μm +Al 0.4 μm+ Polypropylene 6 μm, to protect it from the exposure of arc self-radiation. Bismuth step-wedge with a step thickness of ≈ 100 nm was deposited on a polypropylene filter surface. Bismuth step-wedge was used to define a linear mass of Bi plasma produced by the vacuum arc.

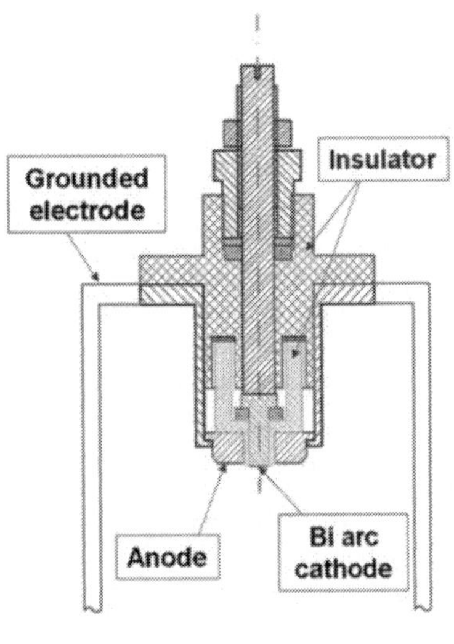

Figure 1. The drawing of plasma gun.

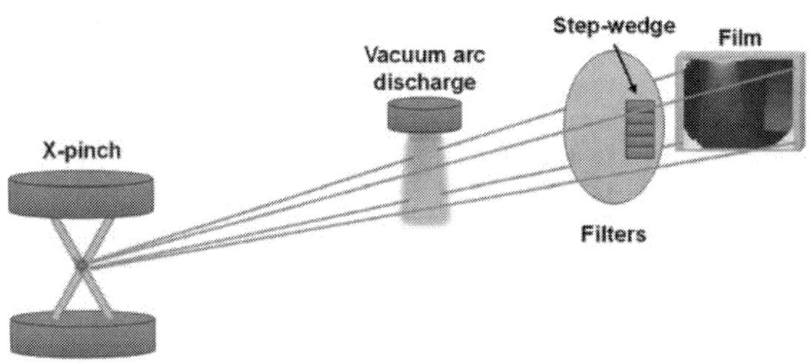

Figure 2. Sketch of soft x-ray radiography (≈ 1 ÷ 2 keV) of a bismuth plasma jet formed by the high-current vacuum arc discharge.

The X-pinch and the plasma gun currents were measured with a Rogowski coil and a B-dot probe, correspondingly. The X-pinch x-ray emission was detected with a copper-cathode x-ray diode (XRD) filtered in the spectral region $h\nu > 0.8$ keV.

Both generators were synchronized. We expect that the time of the maximum intensity of the substance evaporation must correspond to the maximum arc current. The X-pinch x-ray probing pulse occurred approximately at 1.5 - 2.0 μs after the plasma gun current maximum. Such delay was chosen because of a substance with a maximum density needs some time to evaporate and fly a distance of about 1 cm, so that we were able to register it. Therefore, it was expected that the most contrast x-ray backlighting image would be registered at this time interval.

3. Experimental results

The main aim of the work was to determine the distribution of the bismuth plasma jet mass per unit length along the direction of jet propagation away from the cathode surface. For this purpose, in the experiments we recorded simultaneously two shadow images: 1 – bismuth plasma jet and 2 - bismuth step wedge. These images are presented in figure 3.

The step wedge image with an indicated thickness of steps is shown in figure 4. It is known that the optical density D is proportional to the radiation intensity I, transmitted through the test substance with the layer thickness h.

$$D \propto I \propto I_0 e^{-\mu\rho h}, \tag{1}$$

where ρ and μ – the density and the absorption coefficient of the substance, through which the light passed. At the points, at which the optical density of the plasma D_{pl} and the optical density of one of the step wedge layer D_{sw} are the same, we can write the equation:

$$\rho_{pl} h_{pl} = \rho_{sw} h_{sw}, \tag{2}$$

where h_{pl} and ρ_{pl} - the thickness of the absorption layer and the averaged plasma layer density, h_{sw} and ρ_{sw} - the thickness and the density of the step wedge layer.

The registration of a shadow image of the bismuth step wedge allowed mathematically describe a direct relation between the value $\rho_{sw} \cdot h_{sw}$ and the optical density D:

$$\rho h(D) = aD^2 + bD + c, \tag{3}$$

where the coefficients a, b, and c depend on the intensity and the spectrum of a radiation source. Therefore, these coefficients should be determined for every image.

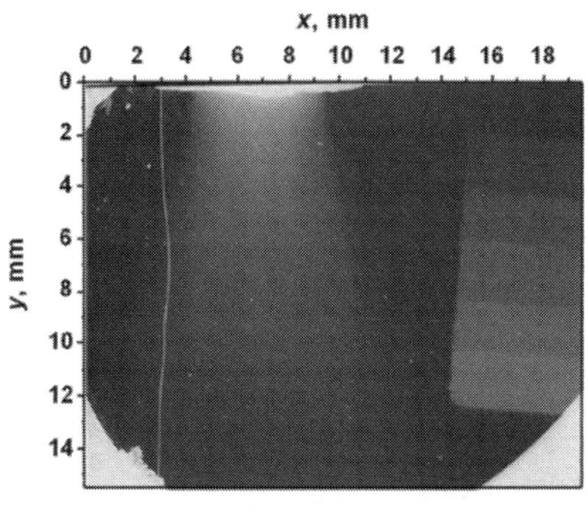

Figure 3. The Bi plasma and step-wedge backlight images.

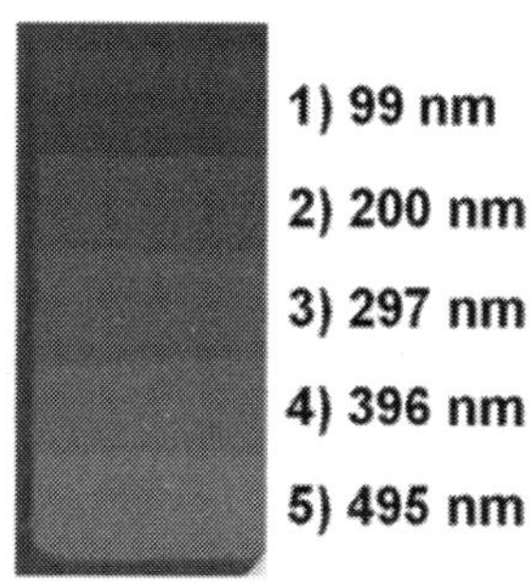

Figure 4. Step-wedge image with indicated thickness of steps.

The optical density distribution $D(r)$ across the plasma jet image (the white lines in figure 5) was registered along the axis X (figure 3) in several cross-sections. Every cross-section was situated at some distance from the arc cathode. This distance to the cross-section corresponds to the time interval from the moment of evaporation of the cathode substance to the moment of image registration. Optical density distributions for several distances from the arc cathode are shown in figure 5. The thickness of every cross-section equals to 1 pixel of image. Therefore, dependences $D(r)$ on the various distances from the arc cathode were used to study the dynamics of the arc plasma linear mass changes. In the experiments, the distance from the X-rays source to the investigated plasma jet is much greater than the jet diameter. Therefore, the incident radiation to the object under study can be approximately regarded as a flat, and the radiation absorbing layer depth at every point of the plasma jet radius is equal to the chord length at this point.

Using (3), we can plot the distribution of the $h_{pl}{\cdot}\rho_{pl}$ value as a function of the cross section radius. The examples of $h_{pl}{\cdot}\rho_{pl}$ distributions, as a function of r , that are built for the cross sections located at various distances from the cathode are shown in Figure 5.

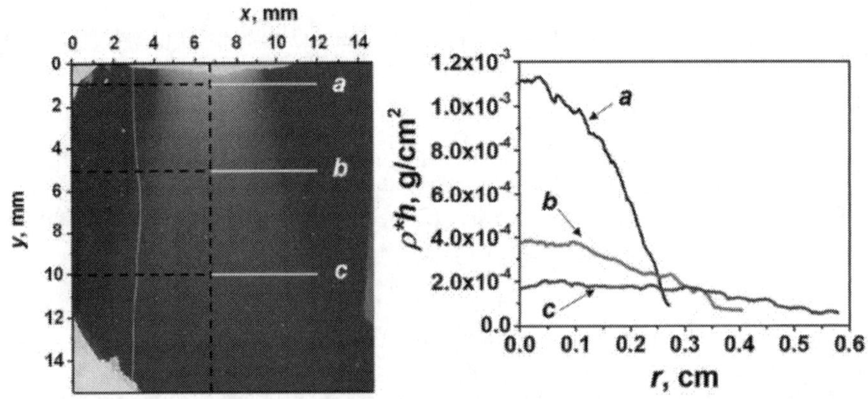

Figure 5. The multiplication of the plasma density and the radiation absorbing layer depth distribution $h_{pl}{\cdot}\rho_{pl}$ as a function of radius r for several cross-sections located in various distances from the cathode.

The linear mass of an every plasma layer was calculated by integrating the multiplication of the density and the radiation absorbing layer depth along the radius:

$$m_p = 2\int_0^R \rho \cdot h(r)dr . \tag{4}$$

The resulting set of the linear mass values for the cross sections located at various distances from the cathode surface allowed to estimate the dynamics of the linear mass evolution in time (see green points in figure 6). In order to determine the relationship between the distance from the arc cathode and the timeline, in this work we used the bismuth ions velocity equal to 0.5 cm/μs (for the arc current of 500 A), obtained in [15]. Traces of the vacuum arc current (black trace), the XPG-1 current (red trace), and the XRD signal (blue trace) are presented in figure 6 for comparison.

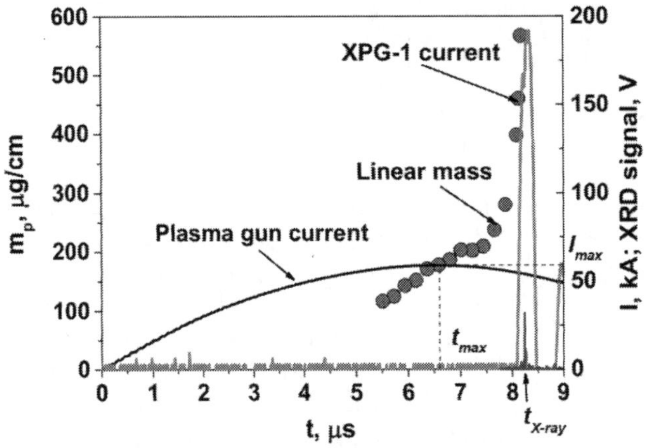

Figure 6. Dependence of the linear mass on time (green points) with traces of arc current (black), XPG-1 current (red) and XRD signal (blue).

4. Conclusion

The experiments have shown that when the arc current density reaches $\approx 3\cdot10^5$ A/cm^2 , the evaporation rate of the electrode material reaches $\approx$ 100 μg/μs, that under the plasma velocity $\approx$ 0.5 cm/μs provides a plasma jet linear mass $\approx$ 200 μg/cm. At a distance of $\approx$ 1 - 2 mm from the arc cathode surface, the sharp increase of the jet linear mass (up to $\approx$ 500 μg/cm) occurred.

This experimental fact can be explained in two different ways. The first explanation is that when the arc current density exceeds the some threshold value, the evaporation process begins to acquire an explosive character. Another explanation is the presence of molten metal droplets near the surface of the arc cathode. To clarify this experimental fact, it is necessary to carry out further studies of this physical phenomenon.

Acknowledgments

This work was supported in part by the Russian Foundation for Basic Research (grant No.15-08-03845-a and No. 14-02-00382-a)

References

[1] Mazarakis M G *et al* 2009 *Phys. Rev. E.* **79** 016412

[2] Aleksandrov V V, Barsuk V A, Grabovski E V, Gritsuk A N, Zukakishvili G G, Medovshchikov S F, Mitrofanov K N, Oleinik G M and Sasorov P V 2009 *Plasma Phys. Rep.* **35** 200

[3] Coverdale C A *et al* 2010 *High Energy Density Physics* **6** 143

[4] Coverdale C A *et al* 2007 *Phys. Plasmas* **14** 022706

[5] Russkikh A G, Shishlov A V, Zhigalin A S, Oreshkin V I, Chaikovaskii S A and Baksht R B 2011 *Plasma Sources Sci. Technol.* **20** 035011

[6] Rousskikh A G, Zhigalin A S, Oreshkin V I, Chaikovsky S A, Labetskaya N A and Baksht R B 2011 *Phys. Plasmas* **18** 092707

[7] Baksht R B, Rousskikh A G, Zhigalin A S and Oreshkin V I 2013 *IEEE Trans. on Plasma Sci.* **41** 1

[8] Rousskikh A G, Zhigalin A S, Oreshkin V I, Labetskaya N A, Chaikovsky S A, Batrakov A V, Yushkov G Yu and Baksht R B 2014 *Phys. Plasmas* **21** 052701

[9] Rousskikh A G, Oreshkin V I, Zhigalin A S, Chaikovsky S A, Baksht R B and Mitrofanov K N 2015 *Journ. of Phys.: Conf. Series* **652** 012065

[10] Rousskikh A G, Zhigalin A S, Oreshkin V I, Frolova V, Velikovich A L,Yushkov G Yu and Baksht R B 2016 *Phys. Plasmas* **23** 063502

[11] Rousskikh A G, Oreshkin V I, Zhigalin A S and Yushkov G Yu 2016 *Tech. Phys. Lett.* **42** 223

[12] Ratakhin N A, Fedushchak V F, Erfort A A, Zharova N V, Zhidkova N A, Chaikovskii S A and Oreshkin V I 2007 *Rus. Phys. J.* **50** 193

[13] Artyomov A P, Fedyunin A V, Chaikovsky S A, Zhigalin A S, Oreshkin V I, Ratakhin N A and Rousskikh A G 2013 *Instr. Exp. Tech.* **56** 66

[14] Baksht R B, Rousskikh A G, Zhigalin A S, Oreshkin V I and Artyomov A P 2015 *Phys. Plasmas* **22** 103521

[15] Bugaev A S, Gushenets V I, Nikolaev A G, Oks E M and Yushkov G Y 2000 *Technical Physics*. **45** 1135

IOP Conf. Series: Journal of Physics: Conf. Series **830** (2017) 012039 doi:10.1088/1742-6596/830/1/012039

High-intensity pulsed beam source with tunable operation mode

A V Nashilevskiy, G G Kanaev, V V Ezhov, V I Shamanin

National Research Tomsk Polytechnic University, Lenina avenue 30, 634050, Tomsk, Russia

E-mail: naw@tpu.ru

Abstract. The report presents the design of an electron and an ion pulsed accelerator. The powerful high-voltage pulse generator of the accelerator and the vacuum bushing insulator is able to change the polarity of the output voltage. The low-inductance matching transformer provides an increase in the DFL output impedance by 4 times. The generator based on a high voltage pulse transformer and a pseudo spark switch is applied for DFL charging. The high-impedance magnetically insulated focusing diode with Br magnetic field and the "passive" anode was used to realize the ion beam generation mode. The plasma is formed on the surface of the anode caused by an electrical breakdown at the voltage edge pulse; as a result, the carbon ion and proton beam is generated. This beam has the following parameters: the current density is about 400 A/cm2 (in focus): the applied voltage is up to 450 kV. The accelerator is designed for the research on the interaction of the charged particle pulsed beams with materials and for the development of technological processes of a material modification.

1. Introduction

Pulsed nanosecond electron and ion beams are widely used in applied researches of radiations and a plasma chemistry, a materials science, a medicine, a biology [1-2].

The most widespread and well-known electron and ion beam sources are the nanosecond pulsed high-voltage linear accelerators based on the generators with pulse-forming lines with distributed parameters [2- 5].

The main goal of the pulse-forming line is storing the energy rather slowly in order to transfer it rapidly to the load. A high-pressure gas spark gap is used as a switch in such devices. It is preferred to use the pulse-forming line with deionized water in the accelerators with pulse duration lower than 1 μs and the stored energy over 100 J. It allows designing the generators of a relatively small weight and size. To provide matching of the generator and the diode impedance, the high-voltage transformer is used after the pulse-forming line [5]. To charge the pulse-forming line, Marx generators are mainly used [2, 4-5]. A disadvantage of the Marx scheme is that a large number of spark gaps is required with synchronous triggering. In a frequency mode, it could lead to an instability of the output parameters of the accelerator.

IOP Conf. Series: Journal of Physics: Conf. Series **830** (2017) 012039 doi:10.1088/1742-6596/830/1/012039

In this paper, we consider the high-voltage pulsed accelerator of electrons and ions based on the high-voltage generator with a double pulse-forming line (DFL). To improve the DFL charging stability, the generator with a pulsed transformer and the thyratron (a pseudo spark switch) were used. The design of the accelerator is able to change the polarity of the output voltage without changes in the design.

2. The high-voltage pulsed accelerator

Figure 1 shows the electrical circuit of the accelerator.

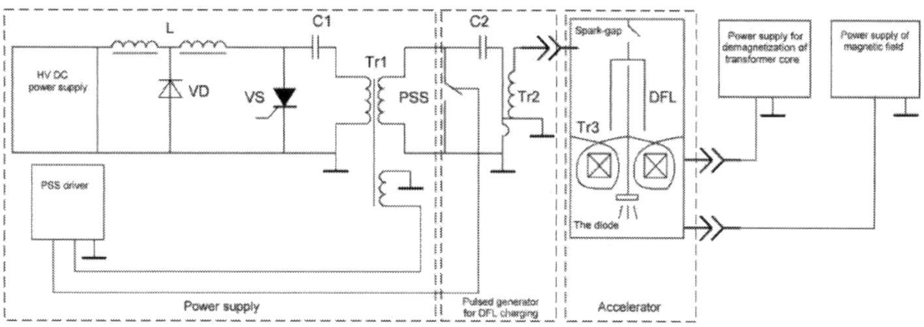

Figure 1. The electrical circuit of the accelerator. L,VD – diode and throttle of protective circuit, C1-power supply capacitance storage, VS- thyristor, Tr1 –pulsed transformer (2kV/40kV), PSS – pseudo spark switch (thyratron), C2 – pulsed generator capacitance storage, Tr2 - pulsed transformer (40kV/250kV), DFL – double forming line, Tr3 -matching pulsed transformer (250kV/500kV)

The HV DC is used as a charging unit. The capacitive storage C1 is charged to the required voltage through the primary winding of the pulsed transformer Tr1. During this charging, the ferromagnetic core of Tr1 is demagnetized by the charging current. Then, a control unit of the HV DC power supply triggers the thyristor VS, and the storage C1 discharges to the primary winding of the transformer Tr1.

The voltage induced across the secondary winding of the transformer Tr1 charges the capacitance storage C2. When the voltage is maximized at the storage C2, the PSS driver triggers the thyratron PSS, and the storage C2 discharges to the primary winding of the transformer Tr2. The induced voltage charges the DFL capacitance to the voltage when the gas spark gap is triggered. The DFL is discharged to the matching transformer Tr3. The core of Tr3 was preliminary demagnetized using the pulse from the special power supply. The high-voltage pulse from the secondary winding Tr3 is applied to the high-voltage output where an ion or electron diode is installed.

The two-winding throttle L and the diode VD provide a protection of the HV DC power supply output circuit against a reverse current. The power supply of magnetic field supports the required current in ion diode coils.

Figure 2 shows the design of the accelerator. The accelerator facility consists of four units (Fig.1): a power supply, a pulsed generator for DFL charging, an accelerator, and a diode system. In Figure 2, the pulsed generator for DFL charging and the accelerator are presented.

The generator based on the pulsed transformer [6] is applied as a pulsed generator for DFL charging. The thyratron PSS, the capacitance storage C2, and the pulsed transformer are located in an oil-filled tank. The thyratron TDI4-100k /45PD is produced by Pulsed Technologies Ltd. The pulsed generator can provide a charging voltage from 150 to 250 kV at 10 pps at the DFL.

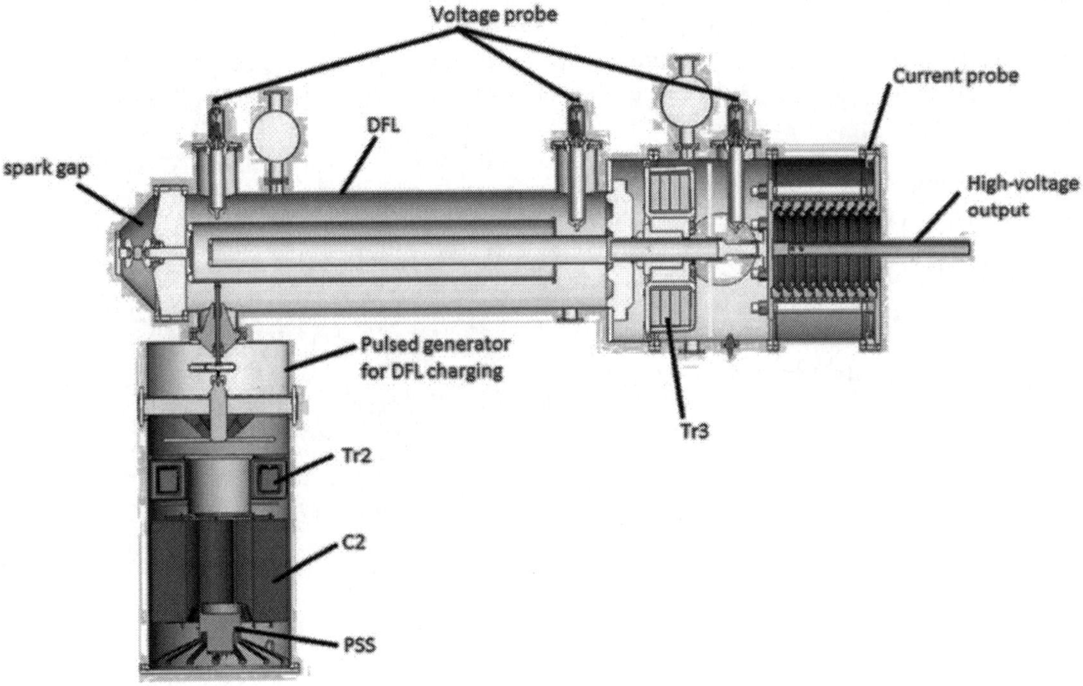

Figure 2. Design of the accelerator. PSS – pseudo spark switch (thyratron), C2 – pulsed generator capacitance storage, Tr2 - pulsed transformer (40kV/250kV), DFL – double forming line, Tr3 - matching pulsed transformer (250kV/500kV)

DFL electrodes are made of stainless steel. Deionized water is used as the dielectric. Two organic glass insulators limit the DFL volume. The accelerator is equipped with a water preparation system based on demineralization of water by an ion exchange method and a water tank. Two voltage probes are located on the DFL to measure the charging voltage and the voltage before the pulsed transformer Tr3.

The two-electrode gas spark gap switch is used. The spark gap is equipped with a gas circulation system to operate in the frequency mode.

The matching pulsed transformer Tr3 is located at the end of the DFL in an oil-filled volume. The magnetic core of the matching pulsed transformer is made of 50NP permalloy. The primary winding is the end of the DFL inner electrode. The secondary winding is realized in a form of twelve parallel-connected cables (RC-50-13-17) without armor [5]. The current and voltage probes are located after matching the transformer on the high-voltage output to measure the output accelerating parameters. The vacuum gradient insulator divides the oil and the vacuum volume.

The vacuum chamber for the ion mode is equipped with a turntable for samples.

3. Experimental results

Experiment was carried out in the electron mode. A material of the cathode is graphite. A diameter of the cathode is 45 mm. An anode-cathode distance is 20 mm. The 30-μm titanium foil was used as the anode. Figure 3 shows a typical waveform of the beam current and the accelerating voltage.

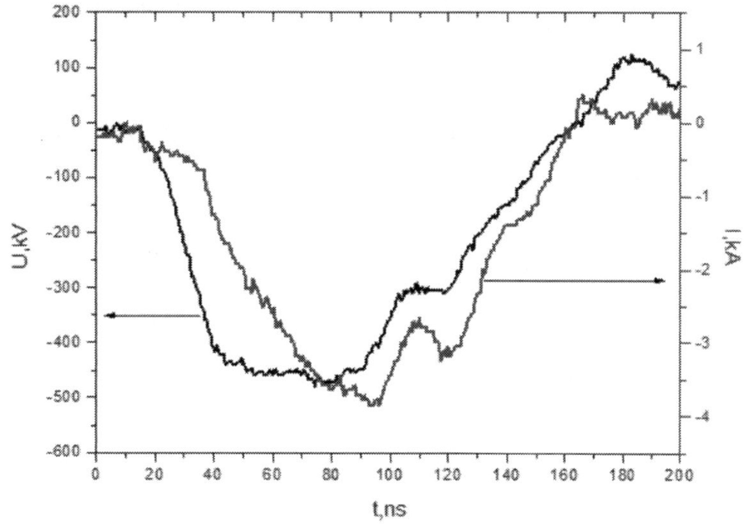

Figure 3. Typical waveforms of beam current and accelerating voltage

The accelerating voltage is 450kV. The output current is 3.7kA. The diode described in [7] was used as ion diode. The anode-cathode distance is 5mm. An emission surface of the anode was polyethylene. The magnetic field at the surface of the anode is provided by a special power supply and amounted to 0.8T. The power supply for the magnetic field is synchronized with the accelerator. When a high-voltage pulse is applied to the anode, the current in diode coils is maximum.

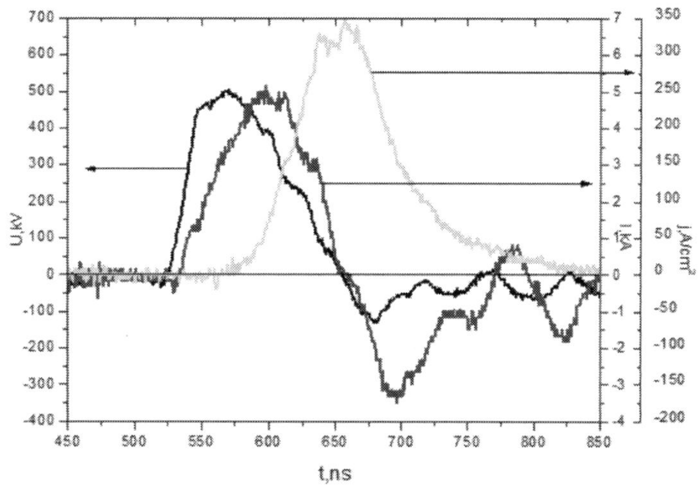

Figure 4. Typical waveforms of accelerating voltage, diode total current, and current density.

The current density of the ion diode was measured using a Faraday cup with magnetic insulation of electrons (0.4T). Figure 4 shows the typical waveforms of the accelerating voltage, the diode total current, and the current density.

4. Summary

The high-intensity pulsed beam source with a tunable operation mode was developed as a research facility for investigation of the interaction of charged particle pulsed beams with materials and for development of the technological processes of a material modification. The replacement of the Marx generator to the pulsed generator with pulsed transformer for DFL charging improved the stability of the DFL charging voltage. A stability of the output parameters in the frequency mode is determined by the spark gap.

Acknowledgments
The authors thanks Stepanov A.V. for the help on development and adjustment of ion diode.

References

[1] Bluhm H 2006 *Pulsed power systems: principles and applications* (Germany Berlin - Springer-Verlag)
[2] Mesyats G A 2004 *Pulsed Energetics and Electronics* (Moscow- Nauka)
[3] Bykov N M, Gunin A V, Korovin S D and Korolev Yu D 2008 Operating features of a high-voltage spark gap switch with gas blow normal to the breakdown path in a repetitively pulsed mode *Instruments and Experimental Techniques* **6** 838-45
[4] Remnev G E et al. 1997 High Power Ion Beam Sources for Industrial Application *Surface and Coatings Technology* **1** 103-9
[5] Remnev G E, Furman E G, Pushkarev A I , Kondratiev N A, Goncharov D V 2004 High-current pulsed accelerator with matched transformer: construction and exploitation characteristics *IEEJ Transactions on fundamentals and materials* **6** 491–5
[6] Kanaev G G, Kukhta V R, Lopatin V V, Nashilevskii A V, Remnev G E, Uemura K and Furman E G 2010 A High-Voltage Pulse Generator for Electric-Discharge Technologies *Instruments and Experimental Techniques* **1** 95–9
[7] Stepanov A V, Remnev G E 2009 Influence of the Configuration of the Magnetic Field of an Ion Diode on the Parameters of an Ion Beam *Instruments and Experimental Techniques* **4** 565–8

Peculiarities of the ionized fronts formation and development in pre-ionized gas

V S Kurbanismailov[1], O A Omarov[1], G B Ragimkhanov[1] and D V Tereshonok[2]

[1]Dagestan State University, M Gadjieva st. 43a, Makhachkala, Russia 367003
[2]Joint Institute for High Temperatures of RAS, Izhorskaya st. 13 Bd.2, Moscow, Russia 125412

Email: gb-r@mail.ru

Abstract We present the results of the investigation of the initial stage of volume discharge formation in helium of atmospheric pressure. The investigation of the non-locality of energy distribution influence on the ionization wave parameters in the front area is carried out on the basis of a two-dimensional axial-symmetry drift-diffusion model. The results of numerical calculations are in a satisfactory agreement with the experimental data.

1. Introduction

The impulse volume discharge can be used for a wide range of technological applications as the power laser, plasma shields and screens, plasma-chemical reactors due to the unique proceeding processes. Physics of these processes is quite complicated and now is under consideration [1-5].Crucial improvement of performance data of plasma devices demands reliable quantitative data about the processes of discharge formation. One of the main characteristics of the discharge are the electric field dynamics, the distribution of the charged particles concentration in a discharge gap, the elementary processes frequency, the discharge current, and the gap-drop voltage.

This paper studies the detailed analysis of the impulse volume discharge formation process in the high pressure helium [6, 7] on the basis of numerical simulation.

Now, there are three approaches for the numerical simulation of plasma: the kinetic, in which Boltzmann equations are numerically solved either directly, or using the microparticle method with regard to collisions, according to Monte Carlo method [8, 9], the hydrodynamical one, in which the equations are solved numerically for several first moments of Boltzmann equation [9, 10], and the hybrid method [11]. The choice of the model is defined by a free path length for every sort of the particles. For discharge under the atmospheric-pressure, the hydrodynamical and the hybrid methods are usually used for numerical simulation.

In recent works [1-2], we have presented the experimental results of electric, spectral, and optical properties of initial formation stages of the impulse volume discharge in the atmospheric pressure helium. In particular, the research results with the use of high-speed photoelectronic subnanosecond time-resolved register (PER2-1) have allowed us to establish the successive steps of the formation and development of the impulse volume discharge in the atmospheric pressure helium.

This research work is the continuation of the cycle of the works performed earlier.

2. Physical Model and Numerical Approach

For discharge description, we use a drift-diffusion model, which includes equations for electron and ions, Poisson equation for the electric field [12]. The energy equation for the neutral gas taken into account does not change dramatically the results. That is why in our present numerical simulation we do not consider the energy equation for the neutral gas.

$$\frac{\partial n_e}{\partial t} + \nabla \cdot \vec{\Gamma}_e = \alpha_{nl} \left| \vec{\Gamma}_e \right| - \beta n_e n_i,$$

$$\frac{\partial n_i}{\partial t} + \nabla \cdot \vec{\Gamma}_i = \alpha_{nl} \left| \vec{\Gamma}_e \right| - \beta n_e n_i, \tag{1}$$

$$\nabla \cdot \vec{E} = \frac{e}{\varepsilon_0}(n_+ - n_e)$$

Here, the electron and ion fluxes, which can be expressed like that $\vec{\Gamma}_e = -n_e \mu_e \vec{E} - D_e \nabla n_e$, $\vec{\Gamma}_+ = n_+ \mu_+ \vec{E} - D_+ \nabla n_+$, is the current density, n_e and n_+ are the density of electrons and ions, α_{nl} is a nonlocal first Townsend coefficient, and β recombination coefficient, $\mu_e(E/N)$ и $\mu_+(E/N)$ electron and ion mobility, $D_e = \frac{kT_e \mu_e}{e}$ and $D_+ = \frac{kT_+ \mu_+}{e}$ electron and ion diffusion coefficients, ε_0 is the dielectric constant; $\vec{E}$ is the electric field strength, φ is the electric potential, N - concentration of neutral particles, $T_e(E/N)$ - electron temperature, $T_+ = T_0$ - ion temperature, k - Boltzmann constant, e - elementary charge.

The geometry of calculated domain is a gap between two plate electrodes with radius $r = 2$ cm. The distance between these electrodes is 1 cm. This gap is connected to the charged capacitor $C = 15$ nF with voltage $U_0 = 7$ kV through the ballast resistance $R_b = 2$ Ohm.

Transport coefficients, ionization, and deionization parameters are given by dependences from [13, 14].

At the initial time moment, the everywhere temperature is $T_0 = 293$K, the initial pressure is $p = 760$ $_{\text{S}}$Top. Initial distribution of the electric field is found in the solution to Poisson equation. The electron and ion density at the initial time inside the computational domain is: $n_e(x, y, 0) = n_+(x, y, 0) = n_0$. ($x$ - a dimension in radial direction).

Boundary conditions for the cathode are $y = 1$ cm, $\Gamma_e = \gamma \Gamma_+$, $\partial n_+ / \partial y = 0$, $\varphi_k = 0$, where $\gamma = 0{,}1$ is a secondary emission ratio, for the anode they are $y = 0$ cm,

$$\partial n_e / \partial y = 0,\ n_+ = 0,\quad \varphi_k = U(t) - I(t)R_b,\ I(t) = 2\pi \int_0^r j_a(x,t) x dx,\ \partial \varphi / \partial y = 0.$$

$I(t)$ is a total current in electrical circuit, $j_a(x,t)$ is a distribution of the current density on the anode. Calculations were performed on a nonuniform grid with an increased mesh resolution near the cathode and the anode. The number of the cells between the electrodes is 300, and a radially uniform grid has 200 cells.

The calculation method is explicit with the first-order accurace in space and time. Poisson equation for an electric field was solved by the iterative method. The calculation was made in a two-dimensional axisymmetric position.

A non-local ionization coefficient was taken from [15]:

$$k_{i\_nl}Nn_e = k_i\left(1+\frac{\vec{E}D_e\nabla n_e}{\mu_e n_e E^2}\right)Nn_e = \alpha\left|\vec{\Gamma}_e\right|\left(1+\frac{\vec{E}D_e\nabla n_e}{\mu_e n_e E^2}\right) = \alpha_{nl}\left|\vec{\Gamma}_e\right| \,, \tag{2}$$

Here α is a local first Townsend coefficient.

The non-locality for electron mobility is calculated in the same way:

$$\mu_{e\_nl} = \frac{1}{n_e m_e v_m E_x}\left(n_e eE_x + \nabla(n_e kT_e)_x\right) = \mu_e\left(1+\frac{\nabla(n_e kT_e)}{en_e E_x}\right). \tag{3}$$

Here m_e is an electron mass, $p_e = n_e kT_e$ - the electron gas pressure, v_m - the frequency of the electron collisions with neutral gas particles, k_{i_nl} is the non-local ionization rate constant, μ_{e_nl} is the non-local electron mobility coefficient.

Let's examine the results of discharge numerical study.

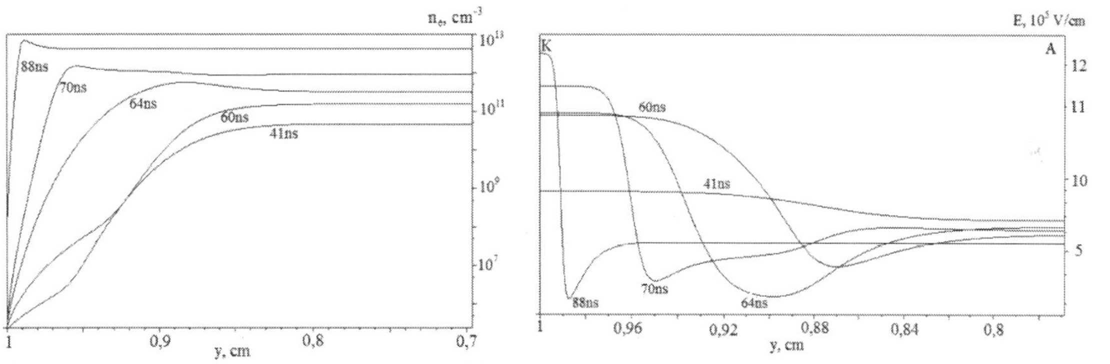

a) b)

Figure 1. Electron concentration distribution (a) and electric field strength (b) in discharge gap: U0=7 kV, p=760 torr, d=1 cm.

3. Result and Discussion

The electron concentration distribution in a discharge gap at various time moments is presented in figure1a. According to the figure 1a, from the beginning plasma polarization occurs near cathode. Since the drift electrons leave the cathode, a region and the area with a small electron density appears and electric field strength increases (fig.1b), which leads to the high direct ionization. The electron density increases in a polarized layer with increasing the distance from the cathode due to a decrease of their drift speed in a non-uniform field. The positive charge accumulation in the cathode region leads to the field enhancement and to the most intensive collision ionization. This means a transfer of the electron and ion maximum density to the cathode. Because of opposite drift directions of the charged particles maximum values, the density of electrons and ions is displaced relative to each other, and the electric field profile has an inflection and is minimum. It should be mentioned that the main process in the cathode layer is a collision ionization. When the ionization wave approaches the cathode, the plasma parameters and electric field strength become established in the cathode layer. A settling time is defined by the ion drift velocity in a strong electric field.

The results of calculations show that electric field strength in the cathode layer is up to ~10^4-10^5V/cm at the stage of volume discharge. This is especially important at the initial stage of discharge formation, when the development of the cathode directed ionization wave occurs.

IOP Conf. Series: Journal of Physics: Conf. Series **830** (2017) 012040 doi:10.1088/1742-6596/830/1/012040

Figure 2a, b shows the electron concentration distribution and the electric field strength without (1) and with taken into account (2) non-local effect. We can see, that taking into account the non-local effect leads to an increase of the wave-ionization velocity.

Figure 3 presents temporal variation of the discharge current. The given dependences show the importance of the taken into the account non-local effect on the stage of the current growth.

Figure4 presents maximum amplitude values of the discharge current for the different applied voltage. Taking into account the non-local effect for the ionization constant rate and for the electron mobility doesn't change dramatically the result. The discharge current under U_0=10 kV approaches to its maximum at 90 ns, and experimentally measured time [1, 2] is ≈95 ns.

This means that, the results of the numerical simulation are in a good agreement with the results of the experiment. The characteristics of time dependency for the voltage after the swift drop show a step, which conforms to the stage of the overall combustion. At this stage, the basic discharge parameters remain guise-steady.

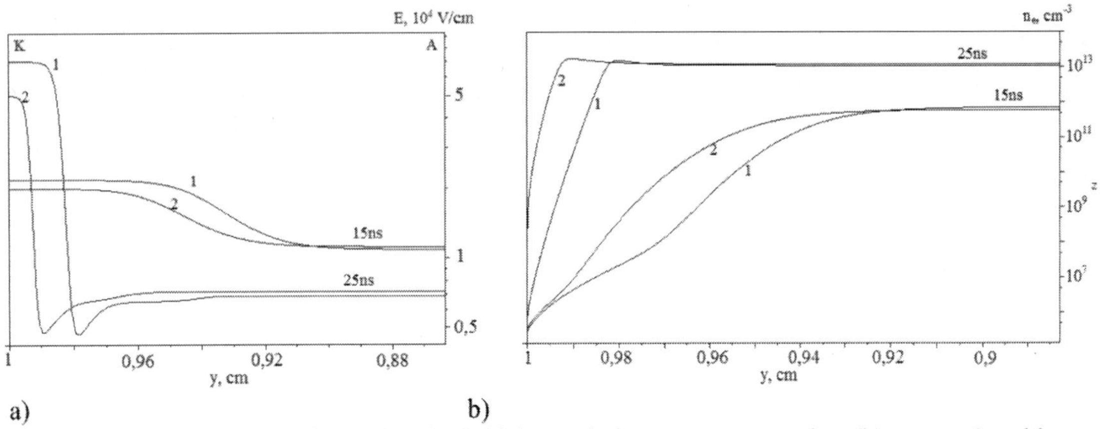

Figure 2. Distribution of the electric field (a) and electron concentration (b): curve 1- without taking into the account the non-locality; curve 2- taking into the account the non-locality (U_0=12кВ).

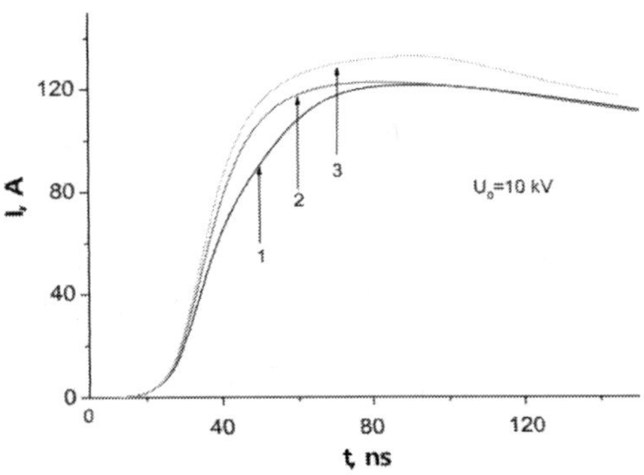

Figure 3. Temporal variation of discharge current: 1- without taking into the account the non-locality; 2- taking into the account the non-locality of ionization constant rate; 3- taking into the account the non-locality effect for the ionization constant rate and electron mobility.

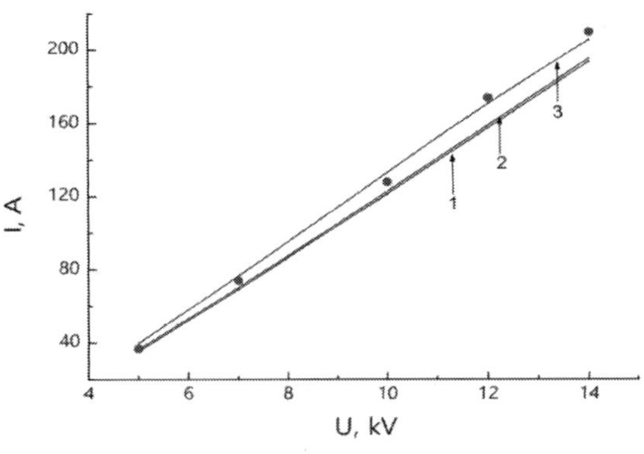

Figure 4. Maximum amplitude values of discharge current for different applied voltage: 1- without taking into the account the non-locality; 2- taking into the account the non-locality of ionization constant rate; 3-taking into the account the non-locality effect for the ionization constant rate and electron mobility.

4. Conclusion

So, the use of non-local corrections leads to an increase of the ionization wave velocity, but the main discharge characteristics do not change dramatically and may be calculated in the assumption of the local approximation.

Acknowledgments

The experimental part was supported by the Ministry of Education and Science of the Russian Federation for scientific research works.

Project № 2644 of the basic part of state task of Russian Ministry of Education and Science in scientific activities.

The theoretical part was supported by the President of Russia Scholarship for young scientists (no 3812.2016.1).

References

[1] Tarasenko V F, Baksht E and Burachenko A G 2010 *Technical Physics Letters.* **36**(4), pp.375-78.

[2] Naidis G V and Walsh J L 2013 *J. Phys. D: Appl. Phys.* **46** (095203), p.13.

[3] Tarasenko V F and Yakovlenko S I 2004. *Phys. Usp.* **47**(887).

[4] Babich L P, Bochkov E I and Kutsyk I 2014 *JETP Letters.***99**(7), pp. 386–90.

[5] Osipov V V 2000 Physics-Uspekhi. **170**(3), pp. 225-245.

[6] Kurbanismailov V S, Omarov OA, Ragimhanov G B, and Aliverdiev A A 2014 *Plasma Physics and Technology.* **1**(1),pp. 28-29.

[7] Kurbanismailov V S, Omarov OA, Ragimhanov G B, Ali Rafid Abas Ali. 2015 *Advanced in Applied Physics.***3**(2),pp. 154-161.

[8] Birdsall C K 1991 *IEEE Trans. Plasma Sci.*.**19**(2), pp. 65-85.

[9] Kim H C, Yang S S, etal. 2005 *J.Phys. D: Appl. Phys.* **38**, pp. 283-301.

[10] Georghiou G E, Papadakis A P, Morrow R A, and Metaxas A C 2005 *J.Phys. D: Appl. Phys.*. **38,** pp. 303–328.

[11] Kushner M J 2005 *J. Phys. D: Appl. Phys.* **38**, pp. 1633-1643.

[12] Tereshonok D V 2014 *Technical Physics Letters.* **40**(3), pp. 83-89

[13] Raiser U P. *Physics of gas discarge* /MFTI. Dolgoprudny: Intellect. 2009. 736p.

[14] Surzhikov S T *Physical mechanics of the gas discharges.* M.: Publishing house MSTU under the name of N.E. Bauman, 2006. 640 p.

[15] V R Soloviev and V M Krivtsov 2009 *J. Phys. D: Appl. Phys.* **42**(125208) p.13 pp.

Computer simulation of energy release modes in discharge channel and its influence on stress-strained state formation in solid material under electro-blasting technology

A S Yudin, N V Voitenko and N S Kuznetsova

Tomsk Polytechnic University, Lenina av. 30, Tomsk, 634028, Russia

E-mail: tevn@hvd.tpu.ru

Abstract. In an electro-blasting technology for the solid destruction, the pulse power generators with different types of switches is used. One of them is a triggered vacuum switch, that for an easy operation, has a suitable lifetime of 10^4–10^5 commutations in average and can pass about 100 coulombs of charge. However, in most cases it passes only a half-cycle of the current in the ringing mode operation. In this paper, the influence of the ringing current pulse duration on the stress-strained state formation is investigated. Simulation results of the energy release modes in a discharge channel are given. The crowbar mode of operation is also investigated and a comparison with the ringing mode are presented. It is shown that in a dumping oscillations mode with a damping factor of about 7, the second and the subsequent oscillations contribute only about 5 % of a total energy into the shock-wave.

1. Introduction

An electro-blasting technology and its pros and cons, in comparison with traditional approaches for the rock and the concrete destruction and demolition, have been extensively described in the literature [1-4]. To produce a blast effect, the underwater electro-discharge wave dynamics principles [5, 7] and the high current generators are employed. A common high current generator is a capacitor bank electrical energy storage with a charger, a commutator, and a triggering unit. But under closer examination, there is a certain difference in the topology of circuit diagrams. As we know from the network analysis, the R-C-L circuit could release its energy in three different modes: the discharge with damping oscillations, the discharge without oscillations, and the critical mode. There are various types of capacitors and switches to consume all the stored energy in a most efficient way. It should be noted, however, that the formation of a steep current front, which can be obtained only in the damping oscillation mode, is preferable in the electro-blasting technology, which needs the high pressure pulse generation. The higher the initial pressure pulse, the better the blasting efficiency [8]. In the damping oscillation mode, it is better to use the switches that can pass all current oscillations, for example, the high pressure gas switches or the implement crowbar mode.

In this paper, an attempt to estimate the influence of the second and the subsequent current oscillations on the shock-wave dynamics and the tensile stress state is made. A comparison with the crowbar mode is also given.

2. Mathematical model description

In this study, the 1D model of the electro-burst is implemented. A complete description of the model is given in [7]. The model involves the transient process equations for the R-C-L circuit (figure 1), the discharge channel energy balance equation, the hydrodynamic equations of wave disturbances description, the plasma channel, and the surrounding medium state equations.

The processes of electrodynamic oscillation in the R-C-L circuit were characterized by the voltage balance equation (taken from the Kirchhoff's laws):

$$L\frac{di(t)}{dt} + (R_{ch} + r)i(t) + \frac{1}{C}\int_0^t idt = U_0\ , \quad (1)$$

where: L is the circuit inductance, H; R_{ch} is the channel resistance, Ω; $i(t)$ is the circuit current, A; U_0 is the initial voltage of the capacitor, V; r is the active resistance of the generator, Ω.

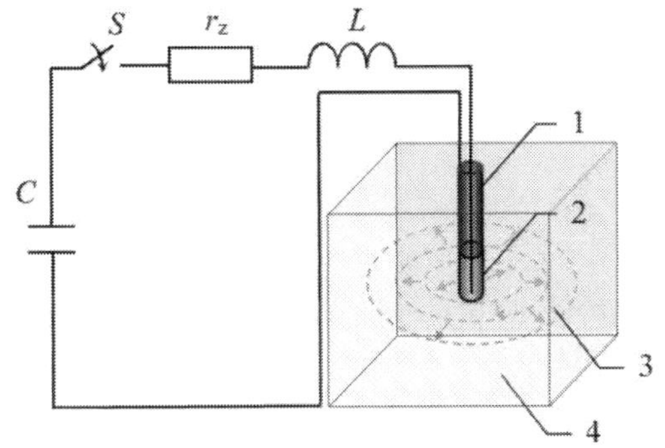

Figure 1. The scheme of shock-wave action on the system of water–concrete under electro-blast: 1 – coaxial cable, 2 – electro-discharge system, 3 – shock-wave disturbances, generated due to expansion of the discharge channel, 4 – destructible material

The current and the voltage waveforms vs time have been calculated with the help of equation (1) along with the empirical equations for the plasma channel resistance $R_{ch}(t)$ which is defined via the integral of a current action in the form of Weizel–Rompe [9]:

$$R_{ch}(t) = K_{ch} \cdot l_{ch}\left(\int_0^t i^2(t)dt\right)^{-\frac{1}{2}}, \quad (2)$$

where: K_{ch} is a spark constant [10].

Electrical plasma channel energy conversion into the plasma energy and mechanical work, performed by the expanding plasma channel:

$$\frac{1}{\gamma-1}\frac{d}{dt}(P_{ch}V_{ch}) + P_{ch}\frac{dV_{ch}}{dt} = i(t)^2R_{ch}\ , \quad (3)$$

where: γ is the effective ratio of specific heats; P_{ch} is the channel pressure, Pa; $V_{ch} = \pi r_{ch}(t)^2 l_{ch}$ – channel volume, m^3; r_{ch} and l_{ch} are the channel radius and length respectively, m.

The wave propagation is based on a system of the equations including the differential equations of laws of the mass conservation, the momentum, and the energy in the adiabatic approximation, the equations of the media and the plasma state, the correlation of the stress tensor, the spherical tensor and the deformation deviator, equations of the wave dynamics [7]. Kirchhoff's equations and the power balance equation of the discharge channel were jointly solved with the state equations of the surrounding medium, the equations of the continuity, the internal energy, the rations for the components of the stress deviator. The system of equations describing the evolution and the dissipation of shock-wave disturbances was solved self-consistently and approximated by the difference scheme, according to Wilkins method [11].

On the basis of the described equations, the numerical algorithm and the software were developed that allowed us to carry out the electro-blast computer experiments in the water–concrete condensed media.

3. Calculation results of tensile stress state at different modes of current oscillations

The computer simulations were made under the conditions listed in tables 1–3.

Table 1. Electric parameters of simulation.

U_0, kV	L, μH	r, mΩ	C, μF	l_{ch}, cm	γ	K_{ch}, V·s$^{0.5}$·m^{-1}
13	1.5	0.022	1120	9	1.26	210

Table 2. Parameters of concrete.

ρ, g/cm^3	μ, GPa	Y_0, MPa	α, m/sec	b
2200	17	77	2330	1.51

Table 3. Parameters of borehole filling media (polyethylene).

ρ, g/cm^3	α, m/sec	b
940	2900	1.48

The four operating modes were compared with each other: 1) – the half period mode; 2) – the one period mode; 3) – the one and a half period mode; 4) – the crowbar mode. These are the typical modes realized in our experiments, except the crowbar mode.

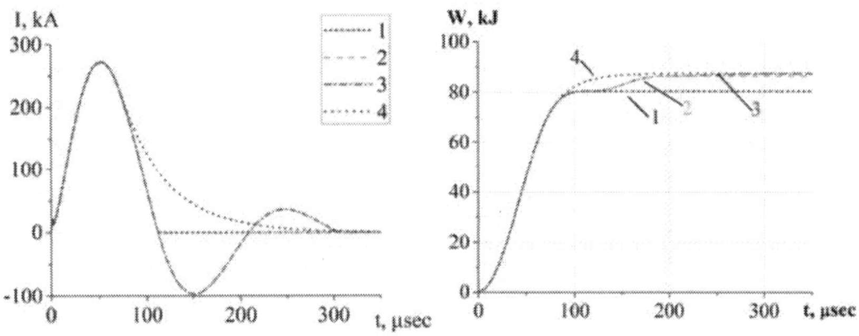

Figure 2. The current and energy curves: 1) – the half period mode; 2) – the one period mode; 3) – the one and a half period mode; 4) – the crowbar mode.

The analysis of the shock-wave dynamics have shown that the discharge channel emits the intense wave disturbances only at an initial phase of its expansion, and the main fraction of the energy is released in the first oscillation of the current. The second and the subsequent oscillations have a little effect on the energy of the shock-wave. The graphs show that in the modes with two and three current oscillations, 9 % and 10.2 % more energy is released, respectively, in relation with the mode 1). Before the time point of 100 μsec, when there is no difference in the energy input (figure 2), there is no difference in the tangential tensile stresses as well (figure 3, 100 μsec). After 200 μsec, when the shock-wave comes to the free border, in the modes 2) and 3), the tangential tensile stress become higher, up to 5% in average (figure 3, 200-900 μsec), that is evidently caused by an addition of the energy from the buttery storage. One can see that a tensile stress amplitude is greater and acting longer in time, if the energy from the generator continue to release in the discharge channel (figure 3, the modes 2), 3) and 4)). A comparison of the ringing mode with the crowbar one shows that in the crowbar mode, there is no significant differences in the tangential tensile stresses for a corresponding time interval, except the time points of 400 and 900 μsec, when the stress reaches up to 5% over the magnitude in the relation of the ringing mode with two and three half-waves of the oscillation (figure 3, 400, 900 μsec). So, this mode could be preferable in a pulse power technology for the rocks and concrete destruction, but its implementation involves either an extra spark gap switch or a semiconductor rectifier, that actually is more complicated.

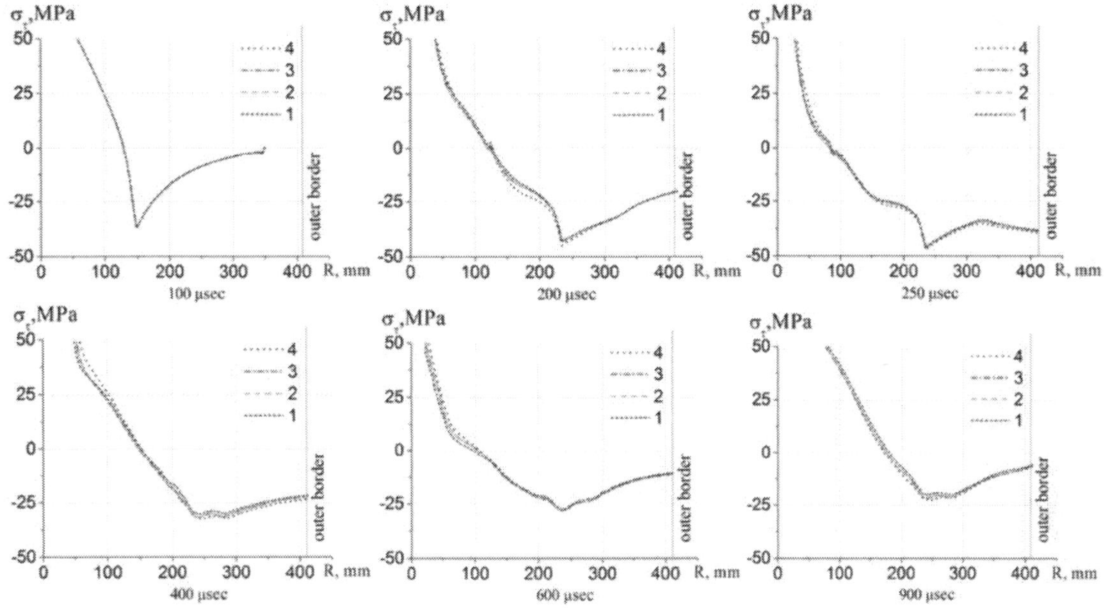

Figure 3. Distribution diagrams of tangential stresses in the pressure wave in a concrete at time of 100, 200, 250, 400, 600 and 900 μsec, R – a radial distance from the discharge channel.

4. Conclusion

The data obtained in this research show that the crowbar mode is preferable in the pulse power technology for the rocks and concrete destruction. It should, however, be admitted that a system with one commutator, for example the trigatron or the TVS, is more simple in technical implementation. Moreover, the crowbar mode gives only about 10% profit in the tensile stresses in comparison with the half wave ringing mode with a damping factor of about 7 that can be easily realized with the TVS. In some cases, the loss of 10% of the energy is not so high price for a simplicity of the entire pulse power system.

Despite the ability of the TVS to recover, in some cases, after 0,5T of current oscillation, it has a low inductance geometry, wide operating voltage ranges, a high current and charge carrying capability. Also, they are extremely robust devices that can survive after an over-voltage, an over-current, and a reverse-current fault which would destroy other types of switching devices. With 10^4–10^5 pulses lifetime these devises have a good application potential for the low pulse repetition rate generators like the pulse power systems designed for the rock fracturing.

Acknowledgements

This work was supported by the Federal Target Program, project No. 14.575.21.0059 (RFMEFI57514X0059).

References

[1] Yutkin L 1986 *The Electrohydraulic Effect and its Application in Industry*, (St. Petersburg, Engineering) p 253 (in Russian).

[2] Pronko S, Schofield G, Hamelin M and Kitzinger F 1993 *Ninth IEEE International Pulsed Power Conference Digest of Technical Papers (21-23 June 1993)* pp 15–18.

[3] Bluhm H, Frey W, Giese H, Hoppé P, Schultheiß C and Sträßner R 2000 *IEEE Trans. on Diel. and El. Insul* **7** 5, pp. 625-636.

[4] Silva C.M.M, Stellin A, Hennies W.T. and Costa E.G. 2002 *Int. J. Min. Reclamat. Environ* **16** 4 pp 261–269.

[5] Krivitskij E.V, Shamko V.V.1979 *Transient Processes at the High-Voltage Discharge in Water* (Kiev, Naukova dumka) p 207 (in Russian).

[6] Otsuka M, Okamoto N, and Itoh S 2008 *Materials Science Forum.* **566** pp 225-230.

[7] Burkin V.V., Kuznetsova N.S. and Lopatin V.V. 2009 *J. Phys. D: Appl. Phys* **43** pp 185-204.

[8] Stefanov Yu. P., Bakeev R.A., Yudin A.S., and Kuznetsova N.S. 2015 *AIP Conference Proceedings* **1683**.

[9] Rompe R and Weizel W 1944 *Zs. Physik B* **122** pp 9–12.

[10] Sjomkin B, Usov A and Ziniviev N 2000 *Transient Processes in Electro-Discharge Installations* (St. Petersburg: Nauka) p 276 (in Russian).

[11] Wilkins M 1964 *Calculation of Elastic-Plastic Flow, Methods in Computational Physics* (New York: Academic) pp. 211–264.

IOP Conf. Series: Journal of Physics: Conf. Series **830** (2017) 012042 doi:10.1088/1742-6596/830/1/012042

Low-inductance switch and capacitor energy storage modules made of packages of industrial condensers IK50-3

Yu A Bykov, E G Krastelev, A A Sedin and V F Feduschak

Joint Institute for High Temperatures, Russian Academy of Sciences, st. Izhorskaya 13/2, Moscow, 125412 Russia

E-mail: ekrastelev@yandex.ru

Abstract. A low-inductance module of a high-current capacitive energy storage with an operating voltage of 40 kV is developed. The design of the module is based on the application of capacitive sections of the industrial condenser IK50-3. The module includes two capacitors of 0.35 μF each, one common low-jitter triggered gas switch and 2 groups of output cables of 4 from each capacitor. A bus bars topology developed for the switch and cables connections provides a small total inductance of the discharge circuit, for the module with the output cables KVIM of 0.5 m long, it is lower than 40 nH. The set of 10 modules is now used for driving the 20 stages linear transformer for a fast charging of the pulse forming line of the high-current nanosecond accelerator. A design of the module and the results of tests of a single module and a set of 10 are presented.

1. Introduction

The increased rate of energy transfer from the primary capacitive storage to the load requires a reduction of the total discharge circuit inductance, including the inductance of the capacitor, the switch, and the connecting bars. One of the implementable ways to reduce the inductance is to reduce a length of the connecting bus with an optimization of their topology by combining into a single structure of the storage capacitor and the switch, and reducing the inductance of the outputs, which are not under a constant high voltage. Such devices, called a capacitor-switch module, produce pulse currents with the amplitude of more than 100 kA at a pulse duration ≈200-300 ns [1, 2]. These parameters are achieved using specially developed high-voltage high-current foil capacitors, a design of which together with a gas switch and output electrodes forms a low-impedance coaxial system [3]. Due to their uniqueness and the lack of mass production, they are expensive and difficult to obtain.

Below, it is given a description and the basic characteristics of the capacitor-switch module based on the use of packages of the industrial condensers IK 50-3, which significantly simplifies and reduces the cost of its manufacture. The assembly is a single module of the 10-module primary power supply of the 20-stage linear pulse transformer for a voltage of 800 kV, used for the fast (300-400 ns) charging of a forming line of a high-current electron accelerator.

IOP Conf. Series: Journal of Physics: Conf. Series **830** (2017) 012042 doi:10.1088/1742-6596/830/1/012042

2. Module design

The compact capacitor-switch module combines in a single package two series-connected capacitors of 0.35 μF each, a gas switch, 8 output cables jointed in two group of 4 from each capacitor, and one charging cable. The module design is shown in Figure 1.

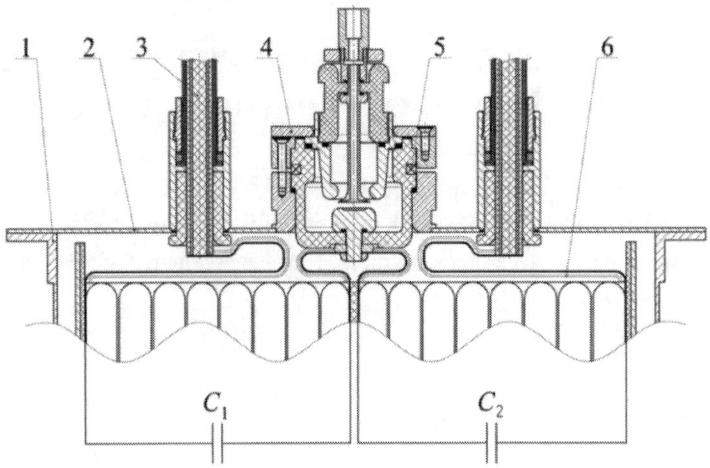

Figure 1. The cross-section of capacitor-switch module: 1 – housing, 2 – cover, 3 – output cables, 4 – gas switch, 5 – Rogowsky coil, 6 – oil-barrier insulation.

The design of the module is based on the using of capacitor packages of the industrial capacitor IK50-3. One of four sections of IK50-3 capacitor is used in one module. The section consists of two connected in series groups of 8 capacitor packs each, tightly fastened together into a single structure by metal shrouds between the end metal plates. Each of the packs group is a high-current capacitor of 0.35 μF for voltage of 50 kV. The section is placed into a metal housing of the module, with dimensions 314x314x160 mm3, and isolated from it by a paper-oil insulation, for the full voltage of 50 kV.

The midpoint of the two capacitors of the section is connected to the high voltage electrode of a triggered switch, and the outlets from the outer packages of capacitors are connected to the central conductors of the output cables.

2.1. Gas switch

The three-electrode gas switch of a "field distortion" type is used for switching the both capacitors. Dry air with a pressure up to 4 bars is used as an insulating gas.

The design of the switch is shown in Figure 1. It assembles an insulating housing made of polyamide plastic "caprolon", installed in a metal base welded to the module cover, two main electrodes with a toroidal working surface, and a disk trigger electrode. The main electrode, mounted at the base of the insulating casing, is connected to the midpoint of the capacitors and to the central wire of a charging cable (not shown in Figure 1). The second main electrode is connected to the grounded base welded to the cover.

The trigger electrode is placed into the main gap through the cavity in the main grounded electrode. It is a hollow rod, gradually widening and turning into a sharp-edged disk with a diameter of 16 mm. The outer edge of the disc is placed at a distance of ≈0.7 mm from the top of the grounded electrode. Its position corresponds to the position of the equipotential surface with a potential of 1/3 of the voltage U applied to the switch. The same voltage of $U/3$ is applied to the trigger electrode from the outside located resistive voltage divider (installed in a trigger generator unit)

The plot of equipotential lines in the spark gap calculated for an initial state, for the trigger electrode voltage $U_{tr}=U/3$, is shown in Figure 2a.

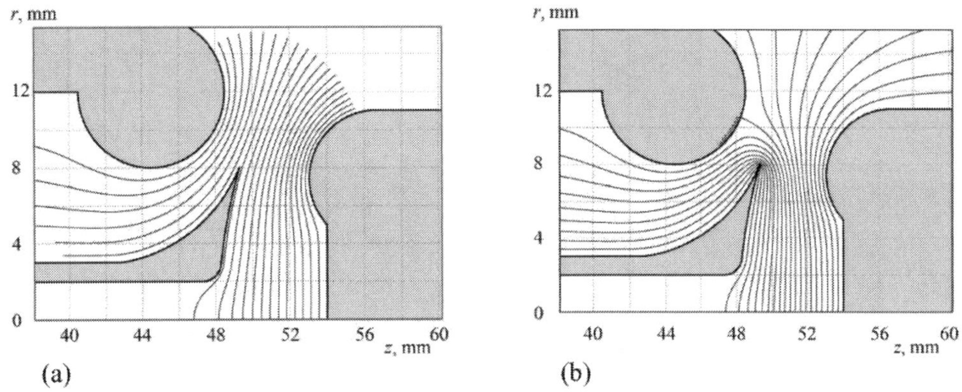

Figure 2. The calculated pattern of the equipotential lines in the gap of the switch:
a – $U_{tr} = U/3$, b – $U_{tr} = -U$.

The closing of the switch is initiated by supplying a trigger pulsed voltage of the opposite polarity to the trigger electrode. Figure 2b shows the calculated plot of the equipotential lines corresponding to the trigger voltage $U_{tr} = -U$, where U is the charging voltage of the capacitor. The high field enhancement near the edge of the trigger electrode provides conditions for the fast development of the discharges on both sides with a low time delay of a gas breakdown, which under these conditions is determined by the time of the development of spark channels.

2.2. The connecting buses

Connections inside of the module are made of wide buses isolated by a paper-oil insulation to reduce the inductance of the discharge circuit. A cross-section of the buses is shown in Figure 1. It was designed to satisfy the requirements of a minimal stray inductance at an acceptable electric field distribution, providing the required electric strength of the isolation. The gaps between the buses, connecting the outer capacitor packages with the output cables and connecting the midpoint of the capacitors with the switch high voltage electrode, were determined mainly by electric field distribution over the switch insulator.

A calculated plot of the equipotential lines in the area of the switch, before and after switching, is shown in Figure 3. The pattern of the lines in Figure 3a shows that for the selected configuration of the connecting buses and the gap between them, the voltage distribution on the switch insulator is close to uniform. For the buses topology shown in the picture, there are no zones of the local field enhancements.

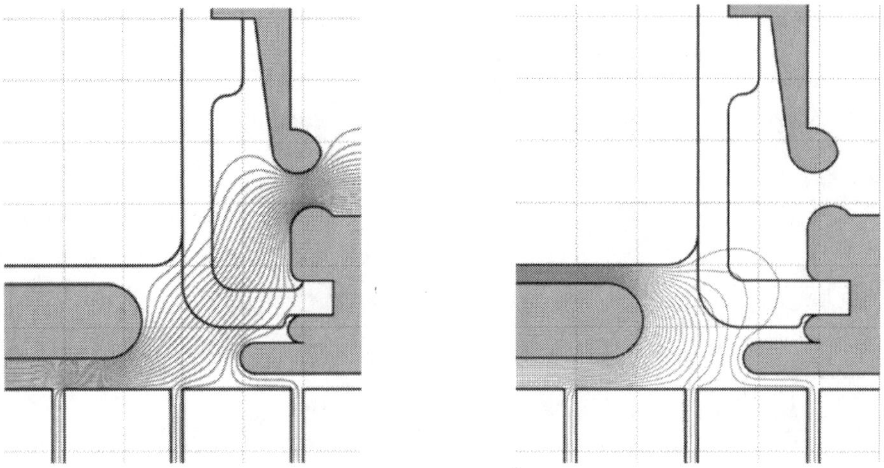

Figure 3. The calculated plot of the equipotential lines: a – before switching and b – after the switch closing. The trigger electrode is not shown.

After switch closing, as shown in Figure 3b, a high field is observed only in the areas between the output buses and the module cover. On the basis of the calculated field values, the parameters of the oil-barrier insulation have been determined.

The estimated inductance of the designed connecting buses is in the range from 20 to 25 nH.

3. Test results

The main characteristics of the manufactured modules were determined experimentally during the several stages of testing.

3.1. Module testing in the static mode

Tests included the slow charging (30 seconds) of the capacitor module to the operating voltage of 40 kV, the exposure for 60 sec, and the subsequent discharge on to the high-resistance load. The tests were aimed to identify possible defects in the insulation and contacts. Additionally, the values of the electric capacity and the dielectric loss (tgδ) were measured before and after the implementation of a charge-discharge cycle. The absence of breakdowns during the static measurements and differences in the measured values of the capacitance and the dielectric loss showed that there are no defects in the module.

3.2. Module testing in the pulsed mode

The tests were aimed at a determination of the parameters of an equivalent circuit of the module and obtaining the amplitude and the time characteristics of the discharge circuit as a whole. Rogowski coil mounted into the housing of the switch was used for measurements of the discharge current, the pulse voltage was measured by a HV probe (divider 1:1000) Tektronix 6015. The measurements were carried out within the range of charging voltages from 20 to 40 kV. The switch was triggered by a fast rising pulse with the voltage amplitude up to 60 kV from the external trigger generator.

As a result of measurements in various modes, the parameters of all elements of the discharge circuit were determined and the equivalent circuit of the module was constructed. The equivalent circuit of the module in the mode of the short circuit at the ends of the output cables is shown in Figure 4.

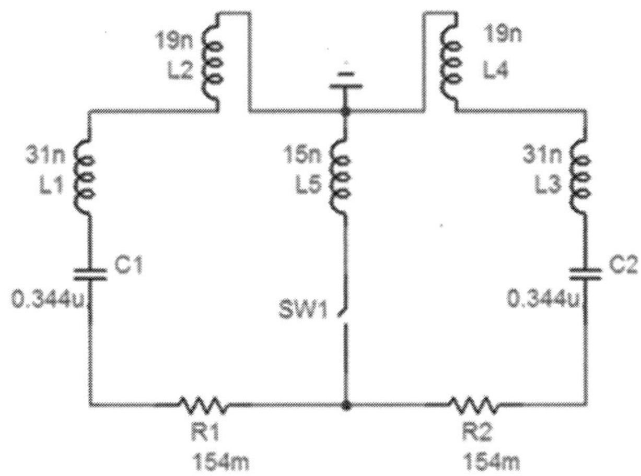

Figure 4. The equivalent circuit of the discharge circuit

L1 and L3 – the total inductances of the capacitor and the connecting bus inside the module, L2 and L4 – the total inductances of 4 KVIM cables connected in parallel, L5 is the inductance of the switch, R1 and R2 are the series resistances of the two branches of the circuit, SW1 – switch, C1 and C2 – measured capacitance of the module capacitors.

The pulses of the discharge current in the short circuit mode for the charging voltage of 20 kV are shown in Figure 5, the dotted line is the calculated current for the given equivalent circuit.

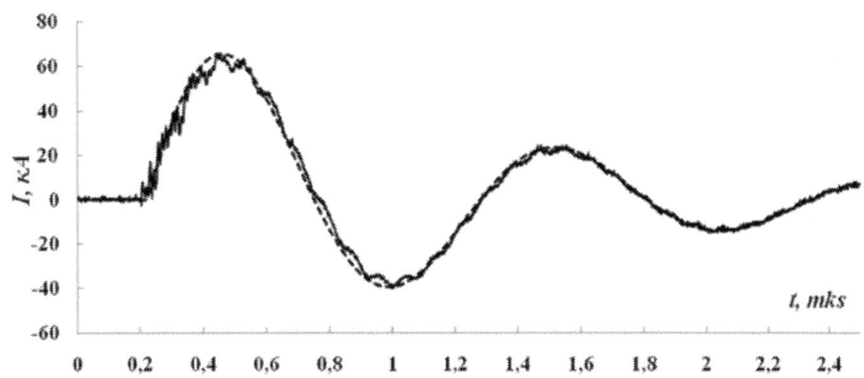

Figure 5. Discharge current pulses in the short circuit mode at the cable outlets. Charging voltage 20 kV.

The measured and the alculated currents practically coincide during the entire discharge cycle. The amplitude of the first peak of the current ≈ 60 kA, the duration of the first half-wave of the current is close to 450 ns. At the charging voltage of 40 kV, the measured amplitude of the total current flowing through the switch was as high as 120 kA.

3.3. Synchronous working of 10 modules in the power supply system of the linear pulse transformer
The power system of the linear pulse transformer includes 10 modules that are installed on both sides of the transformer casing, opposite to the corresponding inputs of the primary windings of the inductors.

The scheme of the connections of modules to the primary windings of the transformer is shown in Figure 6.

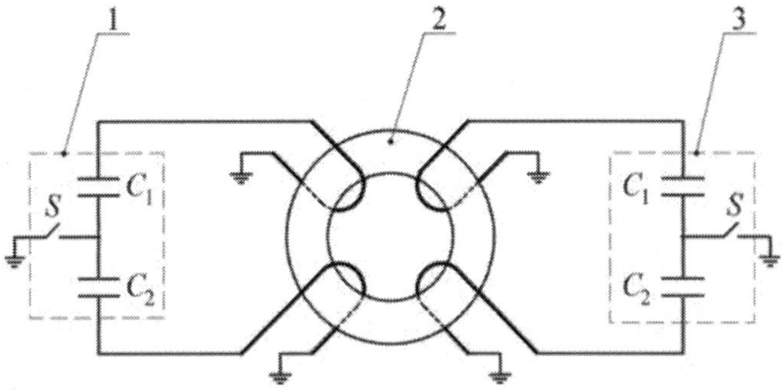

Figure 6. The scheme of the connections of modules to the primary windings of the transformer. 1, 3 – modules, 2 – inductor with ferromagnetic core and one-turn primary winding.

Each one-turn primary winding of the transformer has 4 entry points which are connected to one of the four output cables from both capacitors from the two modules located on both sides of the transformer. Simultaneously, the same voltage with the other cables is supplied to the primary windings of three neighboring inductors. Thus, 4 capacitors of two modules are discharged to the primary turns of the four inductors.

The strong coupling between modules for this transformer powering system, chosen for reduction of the lengths of the connecting cables, imposes strong requirements on the dispersion of the switching time of the modules. A single, common trigger pulse generator is used to provide a mall temporary variation of the closing time for all 10 switches of all 10 modules.

During the test of the 10 modules power supply system, the following data were obtained. For a trigger pulse with a rise time ≈15 ns, an average time delay of the switch closing, relative to the beginning of the pulse, is ≈25 ns. It provides a stable simultaneous operation of all 10 switches with the maximal deviation of the closing time within ±5 ns.

4. Conclusion

The parameters of the developed module meet the requirements of its application in the system of a primary powering of the pulsed linear transformer for fast charging of a pulse forming line of a high-current electron accelerator. The use of the packages of industrial condensers IK50-3 simplifies the module production and reduces its cost. The designed topology of the connecting buss bars, the use of 4 output cables in each arm, and the merge into a single structure with the switch provide the low total inductance of the module (≤40 nH).

The work was supported by the Russian Science Foundation (project 16-15-10355).

References

[1] Feduschak V F et al 2007 *Proceedings of the All-Russian Conference "Electrophysics materials and facilities"*, Novosibirsk, p 223-232.

[2] Ratakhin N I et al 2006 *Russian Physics Journal*, No.**11**. Ad. C, p 511 – 513.

[3] http://hctcap.com

IOP Conf. Series: Journal of Physics: Conf. Series **830** (2017) 012043 doi:10.1088/1742-6596/830/1/012043

Characteristics of capillary discharge channel and its effect on concrete splitting-off by electro-blasting method

N S Kuznetsova, A S Yudin and N V Voitenko

Institute of High Technology Physics, Tomsk Polytechnic University, 30 Lenin avenue, Tomsk, 634050, Russia

Email: Natalia_Kuznetsova@tpu.ru

Abstract. The numerical simulation results on fracture of a concrete block due to dynamic explosive loads applied to the walls of a blast hole are presented. The influence of the pulse shape on the shock-wave dynamics is considered. A comparison of mechanical stresses in direct and reflected pressure waves induced in the concrete block by explosion pulses of various durations and amplitudes shows that the shorter pulses with higher amplitudes and steeper rise times provide a higher blasting efficiency. The wire application for the discharge initiation enables the operating voltage of the generator to decrease, the discharge gap to increase, and hence, the channel energy to lead to the demolition build-up at electro burst. The significant dependence of the stress-wave profile on the pressure pulse wave shape at the borehole wall, which is determined by the rate of electrical energy release in the plasma channel, has been shown. An analysis of the stress-wave dynamics has shown that the rapid power deposition into a plasma channel tends to shift an amplitude of the tangential stresses in a reflected wave to the higher values and to extend the region of tensile tangential stresses initiating the main crack propagation from the borehole walls to a free material surface.

1. Introduction

Research of the electro-blasting destruction and the splitting-off of solid materials, due to intensive development of the electro-discharge technologies, is of great theoretical and practical interest. Practical interest is provided with the development of technologies of solid processing and destruction by the use of electrical discharge and its advantage over the mechanical methods. The main advantages are as follows: the impulse loading of the material, providing the brittle material fracture with a predominance of tensile and shear stresses and, hence, the destruction at lower mechanical stresses compared to the destruction using compression, like in the mechanical methods.

Thus, the electro-discharge technologies of material processing and the destruction are energy efficient, since the solid destruction occurs mainly due to the tensile and shear stresses, rather than compressive ones, as in mechanical methods. Accordingly, the specific energy consumption is reduced proportionally to the ratio of material strength, by 2-5 times. The pulse loading provides a brittle solid fracture with minimal energy loss for the plastic deformation.

High energy density of the plasma channel (of about 25-30 kJ/cm^3 [1]) allows creation of the conditions for the crack initiation in a solid at relatively low values of a single pulse energy, which is equivalent to grams of trotyl. The possibility of regulation over a wide range of the mode of channel energy release allows the creation of the optimal conditions for solid loading, crack propagation, and material destruction depending on the kind and size of the destructible solid.

The efficiency of electro-discharge method, in comparison with the traditional mechanical ones, strongly depends on the electrical properties of the destructible material and almost does not depend on its mechanical properties, and with increasing of the rock strength, it increases. Therefore, although the method can be applied to materials of a various mechanical strength, the greatest technical and economic effect of its application is reached for hard rocks and materials, which is a priority area of its use.

The expanding conducing plasma channel of a capillary discharge is a promising physical tool for the electro discharge technologies of a solid fragmentation, boring, cutting, a surface cleaning, and others [2–5]. Most of such experiments normally involve an electrical discharge in a water-filled cavity drilled in the rock. One of the ways of the plasma channel initiation in the condensed materials is the wire electro-explosion. The plasma channel, initially only 10 to 50 μm wide, expands and launches a pressure wave into the surrounding solid material causing the tensely-deformed material state and forming the mechanical compressive, tensile, and shear stresses. The mechanical stresses are in the range of hundreds of MPa and cause the solid material destruction. This effect is comparable to that of a chemical explosion in a preformed borehole.

A fracture pattern of the surrounding solid material depends on a pressure profile Pg(t), produced at the borehole wall. This pressure, in turn, depends on the spatiotemporal distribution of electrical power deposition in the discharge channel, and on the dynamics of shock-wave propagation through the material. The main goal in raising the efficiency of electro-discharge technology is therefore matching of the rate of electrical energy deposition with the profiles of the generated shock waves.

The absence of experimental information about the general laws determining the amplitude and profile of the pressure waves launched by the plasma channel expansion in the condensed media does not allow estimation of the generated pressure in the surrounding media according to the measured electrical discharge characteristics (the current and the voltage). Some attempts have been made for the discharge in liquids by the authors of monographs [6, 7], but their formulation of practical problems did not assume recording of the shock wave profile. The analysis of the shock-wave processes, the understanding of mechanisms of the shock wave propagation are, thus, essential for an improvement of the electro-discharge technologies, optimizing the pulse generator parameters and the discharge modes coordinated with the wave dynamics in the destructible material. The eventual aim of this research is to develop the physical insight into the shock wave dynamics in condensed media under the capillary electrical discharge. To progress towards it, the present work has focused on the simulation of computer blast-hole experiments with copper wire electro-explosion in the combination of materials "polyethylene – concrete" and an analysis of the shock and pressure wave dynamics, depending on the rate of electrical energy release in the channel.

2. Numerical electro-blast model

For revealing the basic laws of the energy realize to the discharge gap and transformation of a part of this energy to a shock-wave resulting in solid material destruction, the mathematical model of electro-blast in solid has been developed. The model consistently describes the operation of the discharge circuit of pulse generator, the plasma channel expansion, the generation and propagation of the shock and pressure waves in the polyethylene and concrete.

The electric circuit diagram of the pulse generator that was used in the experiments is shown in figure 1.

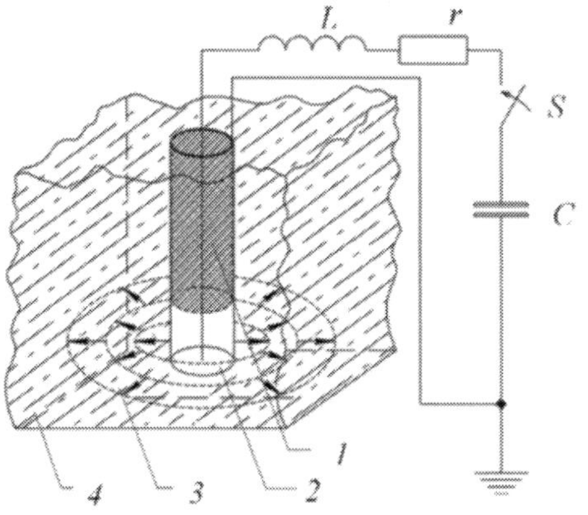

Figure 1. Scheme of plasma-blasting fracture of concrete: 1 – coaxial cable; 2 – electro-blast cartridge; 3 – shock-wave disturbances, generated by expanding plasma channel; 4 – concrete test specimen.

To increase the efficiency of electrical energy transformation into the pressure wave energy, the amplitude of shock wave, and efficiency of its transfer to the destructible material, the polyethylene cartridge with copper wire in its center line was used [8]. The cartridge was placed in a drilled hole in the concrete sample. During discharge, the wire explosion and the plasma channel formation occur. After the electrode gap bridging by a plasma channel, the great amount of electrical energy is transferred from the capacitor bank *C* into the plasma within a very short time. Joule heating of the plasma channel results in the extremely rapid channel pressure buildup and the subsequent plasma expansion. A steep pressure rise takes place on the walls of the hole, generating a shock wave in the surrounding material. The propagation of the shock and pressure wave disturbances forms a tensely-deformed material state and results to the formation of mechanical compressive, tensile, and shear stresses, causing the solid destruction. An enhancement of the mechanical fracture work (the pressure wave energy) produced by the plasma channel is the main goal of energy optimization.

The physical and mathematical model of the investigated system consist of equations describing the wave conversion of the electrical energy, shock-wave effects, and the dynamics of deformed solid material.

Electrodynamic oscillating processes in the discharge circuit obeys the Kirchhoff's law:

$$L\frac{di(t)}{dt} + (R_{ch} + r)i(t) + \frac{1}{C}\int_0^t i dt = U_0\,, \tag{1}$$

where L is the circuit inductance, μH; R_{ch} is the channel resistance, Ω; *i(t)* is the current in the circuit, A; U_0 is the charge voltage of a capacitor bank, V; *r* – an active resistance of the generator, Ω.

The electrical channel energy transformation into the plasma and shock wave energy is determined as:

$$\frac{1}{\gamma-1}\frac{d}{dt}(P_{ch}V_{ch}) + P_{ch}\frac{dV_{ch}}{dt} = i(t)^2 R_{ch}\,, \tag{2}$$

where γ is the effective ratio of specific heats; P_{ch} is the plasma channel pressure, Pa; $V_{ch} = \pi r_{ch}(t)^2 l_{ch}$ – the channel volume, m^3; r_{ch} and l_{ch} are the channel radius and length accordingly, m.

The dynamics of shock and pressure waves in the polyethylene and concrete was analyzed due to the hydro-dynamical equations, in the form of equations of the conservation of momentum, mass, and energy in the Lagrangean coordinates for a cylindrical symmetry [9, 10].

The empirical equation for the discharge channel resistance before the wire explosion was determined as [6]:

$$R_{ch}(t) = R_{ch0}(1 + 68.065\frac{w}{W_s} - 233.188\left(\frac{w}{W_z}\right)^2 + +1526.428\left(\frac{w}{W_z}\right)^3 - 57.867\left(\frac{w}{W_z}\right)^4, \tag{3}$$

where R_{ch0} is the initial resistance, *w* is specific energy of the copper wire. The wire initial resistance is defined as

$$R_{ch0} = \frac{l_{ch}}{s}\rho_{cop}\,. \tag{4}$$

Here, s is a wire cross section area, and ρ_{cop} is a specific resistance of copper. A specific energy of the copper wire is defined as follows:

$$w = \frac{1}{m}\int_0^t i^2 \cdot R_{ch} \cdot dt, \quad (5)$$

where m is the mass of wire.

The capillary plasma channel resistance $R_{ch}(t)$ after the wire explosion is determined through the integral of a current action in the form of Vaicel-Rompe [11]:

$$R_{ch}(t) = A_{ch} \cdot l_{ch} \left(\int_0^t i^2(t)dt\right)^{-\frac{1}{2}}, \quad (6)$$

where A_{ch} is a spark constant (in case of polyethylene A_{ch}=210 V·s$^{0.5}$·m^{-1} [3]).

Polythene compressibility has been described by the state equation. Since under electro-discharge technologies, the maximum level of pressure achieved in the solid is lower than 1000 MPa, it is possible to use the barotropic dependence in the form of $P=f(\rho)$ or a linear dependence of the shock wave speed D on the mass speed U in front of the wave:

$$D = a + bU. \quad (7)$$

By means of Rankine-Hugoniot equations [9], this equation was transformed into:

$$P = \rho_0 a^2\left(1-\frac{\rho_0}{\rho}\right)\bigg/\left[1-b\left(1-\frac{\rho_0}{\rho}\right)\right]^2, \quad (8)$$

where P – the pressure in a material covered with a wave, ρ_0, ρ – the initial and current material density. For polyethylene, $\rho_0 = 0.94$ g/cm^3, $a = 2901$ m/s, $b = 1.481$ [9].

3. Blast-hole simulation results

The research and analysis of the energy conversion laws and the shock-wave dynamics were carried out for the electric circuit shown in figure 1. Several plasma blasting computer experiments were undertaken, using the concrete specimens of about 300×300×600 mm size. The diameter of the drilled hole is 26 mm, and its depth is about 300 mm. The hole was drilled at the distance of 300 mm from the concrete block surface. The length and diameter of a copper wire initiating the discharge were 100 mm and 0.2 mm, respectively. The generator circuit parameters varied within the ranges: U_0=(5–15) kV, L=1.22 μH, C=(224–1120) μF.

The simulation begins at the moment of switching on the generator commutator S (figure 1). The Joule heating of the plasma channel results to an increase of its conductivity. The internal energy, leads to the current amplitude growth (figure 2, a, b). The time period of current oscillations is about T=(0.3–0.5) ms for the chosen generator parameters. The positive feedback between the energy release and channel conductivity causes a build-up of internal energy of the plasma channel (figure 2, c). The electrical power deposition reaches (4–5)·10^8 W.

The explosive electrical energy of the channel results in a sharp increase of the plasma pressure up to (2–2.5) GPa, leading its expansion. The channel pressure is not stable; an increase in the channel volume causes the pressure reduction and consequently, the slowdown of its expansion. The speed of pressure reduction depends on the dynamics of electrical power deposition in the plasma channel. The channel expansion leads to the pressure pulse formation on the borehole walls (figure 2, d). An increase in the rate of energy deposition into the plasma channel tends to shift the pressure wave amplitude at the borehole walls to the higher values.

The generator capacitance and inductance leads to a damped oscillation in the circuit, thereby the discharge gap acts as a nonlinear resistance. During the oscillations, concerted changes of the discharge current, the voltage, and the channel resistance occur. The current and voltage oscillations lead to fluctuations of a power of the energy release in the plasma channel. Thus, during the first half period of the current oscillation, (50–80) % of the total stored energy is allocated in the channel. The ohmic losses in the circuit do not exceed (15-25) %. Further fluctuations of the energy have little effect on the plasma channel pressure and the radius. Thus, the main shock action on a destructible material is generated during the first current oscillation.

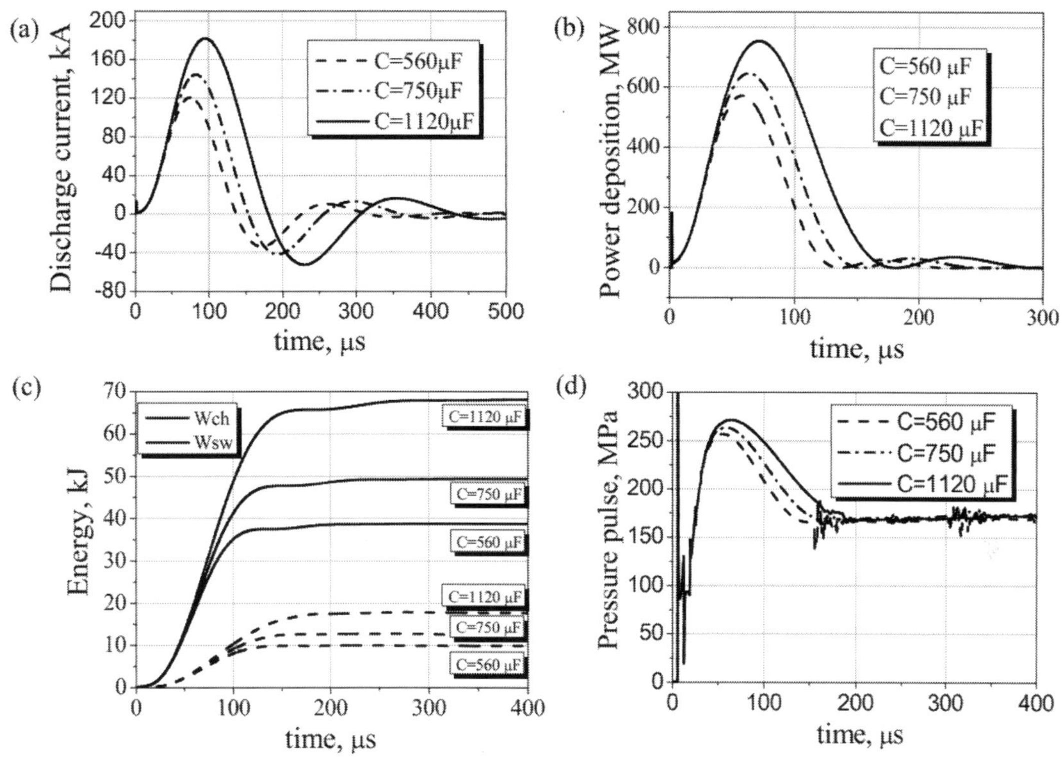

Figure 2. The time dependencies of the discharge channel current (a), electrical power deposition (b), channel (W_{ch}) and shock wave (W_{sw}) energy (c) and pressure pulse on the borehole walls for different rates of electrical energy deposition.

The amplitude and profile of the mechanical stresses causing the concrete destruction strongly depend on the pressure waveform produced at the borehole walls which, in its turn, is determined by the spatiotemporal distribution of the electrical power deposition in a plasma channel. The small channel radius leads to a sharp decrease in the stress-wave amplitude in the vicinity of the channel caused by the wave-energy divergence in the radial direction. The shock-wave disturbances in a concrete are damped along the way of their propagation, due to the wave energy dissipation. The figure 3 illustrates the radial and tangential stress distribution before the wave reflection from a free concrete specimen surface for various rates of the electrical energy deposition in a channel.

The obtained results showed that the rapid power deposition results in a higher amplitude of compressive stresses of the direct wave both in radial and tangential directions. However, a decrease of the rate of energy deposition into the discharge channel results to the formation of tensile tangential stresses causing the crack initiation at the smaller radial distance from the channel (figure 3, b).

The necessary condition for the main crack generation in a concrete is the formation of tensile mechanical stresses. The crack is initiated when one of the stress components exceeds the concrete tensile strength of about (5–10) MPa.

The wave dynamics shown in figure 3, demonstrates that the lower energy deposition rate leads to the higher amplitude of tensile stresses in tangential directions. In the case of the rapid deposition, the region of tensile tangential stresses of the direct wave is formed later. Figure 3 (a) shows, that mechanical stresses in radial direction are compressive along the distance from the channel till the free concrete surface.

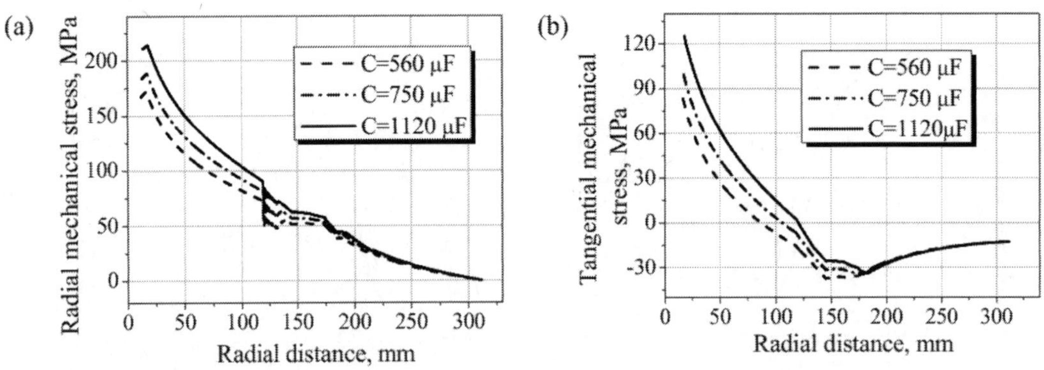

Figure 3. Radial distribution diagrams of radial (a) and tangential (b) mechanical stresses in the direct pressure wave in concrete (t=130 μs).

The fracture pattern of concrete specimen is defined by the resultant tensely-deformed solid state, formed by an interaction of the direct and reflected waves. A mechanical stress distribution, with an intricate profile generated by a superposition of the direct and reflected waves, is shown in figure 4.

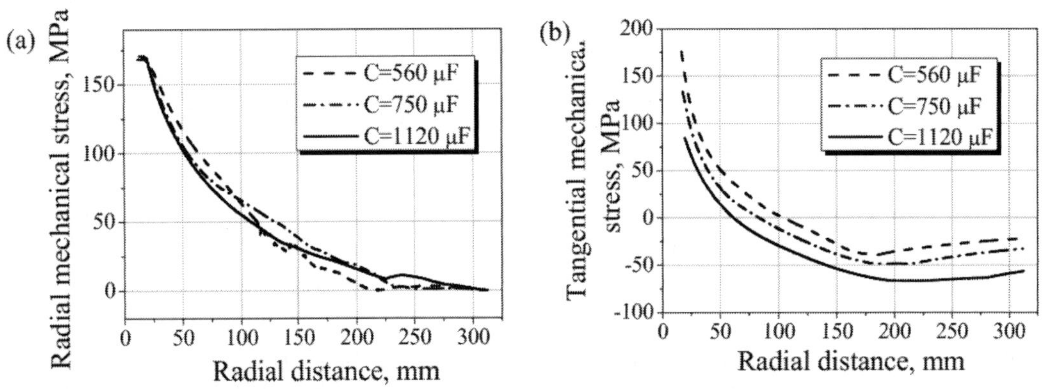

Figure 4. Radial distribution diagrams of radial (a) and tangential (b) mechanical stresses of the reflected pressure wave from the free concrete surface (t=280 μs).

The amplitude of compressive mechanical stresses of the wave is relatively law and lies in the ranges of material elastic deformation. It is shown that the region of tangential stresses is formed in the direct wave, but in the reflected wave, its amplitude is doubled. Formation of this area leads to the formation and development of radial cracks, oriented from the borehole wall to the free concrete surface. The increase of the rate of electrical energy deposition into the plasma channel tends to shift the amplitude of tangential stresses to higher values and extend the region of tensile tangential stresses from the borehole walls to the free material surface (figure 4, a). The mechanical stress level for the crack growth is rather low, in comparison with that of the crack initiation. That is why, the stresses shown in figure 4 are redundant for the crack propagation in a concrete.

Thereby, the results of electro-blast simulation have shown that the rapid power deposition produces an increase in the peak pressure at the borehole wall. However, because it is produced by a shock, the pressure pulse to the borehole walls falls rapidly, whereas slow power deposition produces a sufficiently sustained pressure pulse

4. Conclusion

At this stage of research and development of the blast-hole electro fracture of concrete, an essential gain of an efficiency of the solid disintegration has been achieved, and the possibility of material splitting-off at the free surface with a radial distance 0.3 m has been demonstrated. The results of simulation have shown a significant dependence of the stress-wave profile on the pressure pulse wave shape at the borehole wall which, in its turn, was determined by the rate of electrical energy deposition into the plasma channel. The analysis of stress-wave dynamics has shown that the rapid power deposition results in the higher amplitude of compressive stresses of the direct wave both in radial and tangential directions. The increase of the rate of electrical energy deposition into the plasma channel tends to shift the amplitude of tangential stresses of the reflected wave to higher values and extend the region of tensile tangential stresses from the borehole walls to the free material surface.

Acknowledgements
This work was supported by grants of RFBR № 14-08-01088 and RFBR № 16-48-700278.

References

[1] Burkin V Burkin V, Kuznetsova N and Lopatin V 2009 *J. Phys. D: Appl. Phys.* **43** 185204
[2] Vorobjev A and Vorobjev G 1971 *Pulse breakdown and destruction of dielectrics and rocks* (Tomsk: Publishing house TGU) p 223.
[3] Sjomkin B, Usov A and Kurets V 1995 *The principles of electric impulse destruction of materials* (St. Petersburg: Nauka) p 276
[4] Bluhm H, Frey W, Giese H, Hoppe P, Schultheis C and Sträbner R 2000 IEEE Trans. On Dielecrics and Elecr. v **7** pp 625–636
[5] Timoshkin V, Mackersie J and MacGregor S 2004 *IEEE Trans. Plasma Sci.* **5** 2055
[6] Naugolnyh K and Roy N 1971 Electric discharges in water (Moscow: Nauka) p 238
[7] Krivitskij E Dynamics of electro explosion in a liquid (Kiev: Naukova Dumka) p 208
[8] Kuznetsova N, Lopatin V, Burkin V, Golovanevskiy V, Zhgun D and Ivanov N 2011 *Pros. 17th IEEE Int. Pulsed power Conf.* (Chicago) pp 267–273
[9] Kinslow R 1970 *High-velocity impact phenomena* (New York: Academic Press) p 533\
[10] Rihtmyer R and Morton K 1967 *Difference methods for initial-value problems* (New York: Interscience)
[11] Wilkins M 1964 *Calculation of elastic-plastic flow, methods in computational physics* (New York: Academic) pp 211–264

PIN-diode diagnostics of pulsed electron beam for high repetition rate mode

Ivan Egorov[1], Yu Xiao[2] and Artem Poloskov[1]

[1] Institute of high technology physics, Tomsk polytechnic university, 2a Lenin Avenue, Tomsk, 634028, Russia
[2] School of physics and nuclear energy engineering, Beihang University, No.37 Xueyuan Road, Beijing, 100191, China

E-mail: egoris@tpu.ru

Abstract. This work describes the operating principle and test results of the diagnostics for measuring the pulsed electron beam parameters under repetitive operation mode. The diagnostics is based on a PIN-diode, which is used as a bremsstrahlung detector. The signal from a PIN-diode was converted to a pseudo constant voltage signal which can be measured by a conventional voltmeter. Then the signal acquired by the voltmeter was compared with a reference signal indicating the normal operating regime of the accelerator, thus information about the shot-to-shot reproducibility of the electron beam parameters was given. The system was developed and tested for the ASTRA-M accelerator with the following operating parameters: 470 kV accelerating voltage, 120 ns beam duration and up to 50 pulses per second repetition rate.

1. Introduction

The field of electron beam applications is growing up rapidly [1, 2]. For applications, which require a high level of an average electron beam power [2, 3], the continuous and the linear electron accelerators are usually used. The pulsed accelerators are used for relatively lower average power applications and, meanwhile, allow for a compact facility size [4, 5]. As a rule, for pulsed systems, a higher pulse repetition rate means a higher productivity. The ASTRA-M pulsed electron accelerator [6] uses the 40 pulses per second (pps) continuous operation mode for the purpose of wastewater treatment, as well as single pulses for research applications [7 - 9]. The ASTRA-M accelerator generates electron beams with the kinetic energy of electrons of up to 470 keV and the pulse width of 120 ns at half maximum [6]. During the use of accelerator for some applications, the electron beam parameters cannot be measured directly, thus, the accelerating voltage and the total diode current are used for controlling the electron beam parameters. However, such indirect information about the beam parameters is not enough to ensure an accurate operation of the accelerator, especially when the electron beam is ejected into the atmosphere. An alternative approach to control the electron beam parameters is measuring the bremsstrahlung radiation, which occurs when the electron beam interacts with a matter. The bremsstrahlung intensity for the same measurement geometry depends on the electron beam parameters. One of the most convenient ways to measure the bremsstrahlung with a high time resolution is to use PIN-diodes [10, 11]. However, when a great number of pulses, e.g. 10^6 shots, need to be done every day, it is impossible for the operator of the accelerator to check every pulse. To control the bremsstrahlung pulses in the automatic mode, a measurement system with a nanosecond resolution and the complicated algorithms for signal processing are required. An alternative simplified method with a compact schematic, high-reliability, and a satisfactory performance is described in this paper.

2. Experimental setup

The PIN-diode based system was built to control parameters of the electron beam inside the reaction chamber for a wastewater treatment [6]. Figure 1 shows the schematic of the reaction chamber with the PIN-diode detector installed. The exit window of the ASTRA-M accelerator is located at the top of the reaction chamber (pos. 7 in Fig. 1). The PIN-diode detector was installed in an area where the most intense radiation occurs (pos. 6 in Fig.1).

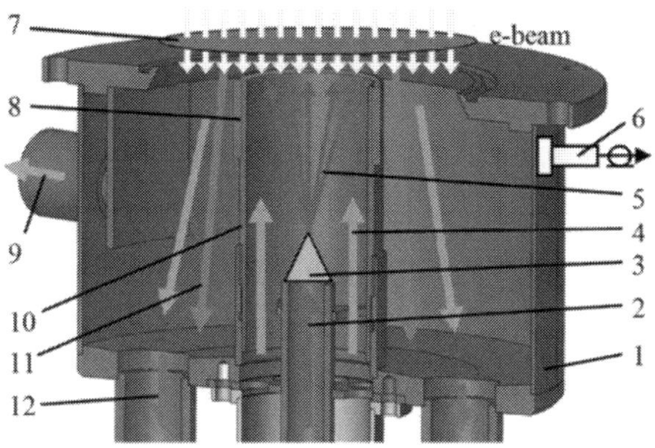

Figure 1. Schematic of the PIN-diode installation: 1 – stainless steel housing of the reaction chamber; 2 – water supply; 3 – sprayer; 4 – atmospheric air flow; 5 – sprayed water; 6 – isolated PIN-diode detector; 7 – exit window of the accelerator; 8 – area of an intensive bremsstrahlung generation; 9 – output air flow; 10 – divider; 11 – processed water; 12 – output water flow.

The electrical circuit for the PIN-diode diagnostics is shown in Figure 2. The measuring scheme consists of two main blocks. The detecting module converts a current signal of the PIN-diode detector to a voltage signal, which is further measured by an oscilloscope. The integrating circuit is used for a repetitive operation mode only and will be described below.

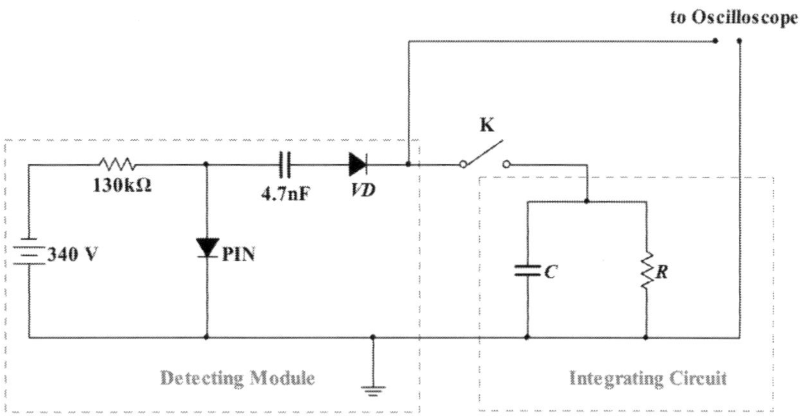

Figure 2 Electrical circuit of the PIN-diode diagnostics.

The basic principle of the operation of the PIN-diode bremsstrahlung detector can be described as follows. When the intrinsic zone of the PIN-diode is irradiated by the bremsstrahlung photons, the electron-hole pairs are excited. The generated charges are driven by the external electrical field and then a pulsed signal is generated. The described design (Fig. 2) was assembled in a single case and fixed in the accelerator base. The signals from the detecting module were transmitted to an oscilloscope in the operator room using 8-m long cables.

3. Results and discussion

Figure 3 shows the waveforms of the accelerating voltage, the electron beam current measured by a capacitive voltage divider and Faraday cup, respectively. The low-load operation regime of the accelerator has been chosen for tests to monitor the lowest value of the electron beam current. Such regime is accompanied by a delay of 150 ns between the emission current and the post pulses of voltage due to the mismatching of the generator and the diode impedances. Figure 3 also shows the calculated waveform of the electron beam power and a signal obtained from a PIN diode detector. The output voltage signal of the detecting module is proportional to the calculated electron beam power. The time resolution of the PIN-diode measurement system, typically of about 10 ns under the biased mode, satisfies the control function for the pulses of the hundreds nanosecond duration with a reasonable accuracy.

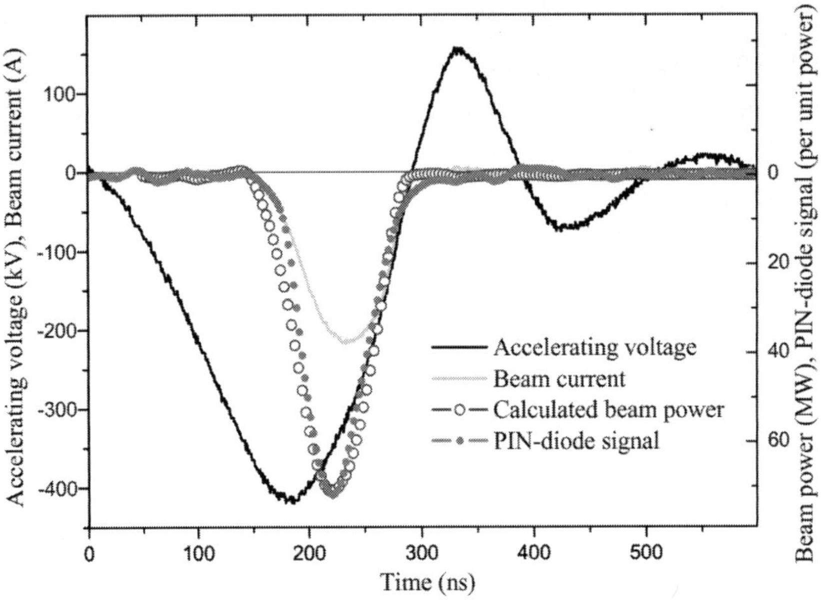

Figure 3 Typical waveforms of the accelerating voltage, electron beam current and a signal from the PIN-diode detector compared with the calculated beam power.

When the switch K is closing, the signal at the output of the detecting module charges the capacitance C of the integrating circuit (see Fig. 2). It is necessary to use the diode VD to avoid the influence of the detecting module on the discharge parameters of the integrating circuit. A discharge time constant of the integrating circuit can be adjusted by changing the capacitance C and the resistance R (Fig. 2). The parameters of the cables and the oscilloscope should also be taken into account.

The optimal value of the discharge time constant was chosen in accordance with the pulse repetition rate of the electron beam during repetitive operation of the accelerator. Figure 4 explains the principle of the signal formation at the output of the integrating circuit (U_{out}) for a chosen discharge time constant. The fluctuations of the output voltage (U_{out}) depend on the ratio of the discharge time constant and the time interval between two pulses. The range for a normal operation of the accelerator was chosen to fit in between the maximum and the minimum values of U_{out} (Fig. 4c).

The voltage pulses of a milliseconds scale on every accelerator shot (Fig. 4) were converted into a digital form. Then, they were transmitted to a master controller of the accelerator for comparison with the value of a chosen operation range. This procedure has been automated for the continuous operation of the accelerator.

There are two undesirable incidents which should be controlled during accelerator running in the high repetition mode. These incidents can be easily detected using the developed diagnostics. The first one is the missing of one electron beam pulse, in the case of which U_{out} will drop down significantly, as it is shown in Figure 4 (c, 1). The reasons of the missing of the beam pulse can be critical for a further operation, for example, it can indicate the shortcut of the electron diode, so the accelerator operation should be stopped immediately. The second is when U_{out} slightly exceeds the operation range as shown in Figure 4 (c, 2). If it happens periodically, it means that the electron beam parameters are gradually changing. The shot-to-shot variation of the electron beam parameters during the repetitive operation occurs due to, for example, the running-in of the cathode at the beginning of the operation and the degradation of the emission properties at its end. The operation range can be varied for different operation modes.

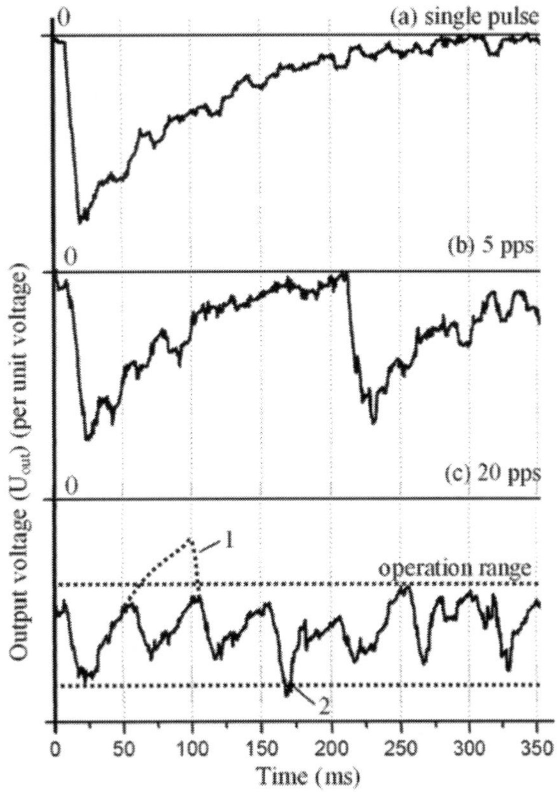

Figure 4 Typical waveforms of the signal at the output of integrating circuit for different pulse repetition rate of the accelerator.

4. Conclusion

The high-frequency electron beam diagnostics for the repetitive pulse electron accelerator has been developed and tested. The diagnostics allows for an automated control of the electron beam parameters by means of measurement and analysis of the bremsstrahlung pulses (with a high repetition rate). The diagnostics can be used to detect both the events of missing pulses in a pulse train and slowly changing parameters of the generated electron beam in the continuous mode.

Acknowledgments
The authors would like to thank Tomsk Polytechnic University for technical support in performing experiments. This work was partially supported by RFBR grant No 16-32-00287mol_a. Xiao Yu would like to thank China Scholarship Council (Grant No.201206020099) for support to visit Tomsk Polytechnic University in Russia and contribute to this work.

References

[1] Parejo Calvo W A, Duarte C L et al 2012 *Radiation Physics and Chemistry* vol **81** iss 8 (Amsterdam: Elsevier) p 1276
[2] Pikaev A K 2002 *Radiation Physics and Chemistry* vol **65** iss 4-5 (Amsterdam: Elsevier) p 515
[3] Kim Y, Kim J, Han B, Choi J-S, Lee M-J 2011 *Journal of the Korean Physical Society* vol. **59** No 6 (Seoul: Korean Physical Society) pp 3489-93
[4] Sokovnin S Yu, Balezin M E, Shcherbinin S V 2013 *Instruments and Experimental Techniques* vol. **56** iss 4 (Moscow: Pleiades Publishing) pp 411-3
[5] Engelko V, Yatsenko B, Mueller G, Bluhm H 2001 *Vacuum* vol **62** iss 2-3, (Amsterdam: Elsevier) pp 211-6
[6] Egorov I, Esipov V, Remnev G, Kaikanov M, Lukonin E, Poloskov A 2012 *Proc. of the 2012 IEEE Int. Power Modulator and High Voltage Conf.*(San Diego: IEEE Publishing) art. no. 6518845, pp. 716-9.
[7] Mostovshchikov, A.V., Ilyin, A.P., Egorov, I.S., Ismailov, D.V. 2016 *Key Engineering Materials*, **712**, pp. 60-64.
[8] Mostovshchikov, A.V., Ilyin, A.P., Azanov, A.A., Egorov, I.S. 2016 *Key Engineering Materials*, **685**, pp. 639-642.
[9] Poloskov, A., Egorov, I., Ezhov, V., Remnev, G. 2016 *Key Engineering Materials* vol **685** (Pfaffikon: Trans Tech Publications) pp. 667-71.
[10] Hussain S, Ahmad S, Sadiq Mehboob, Lee P, Zakaullah M, Waheed A 2009 *Journal of Fusion Energy* vol **28** iss 1 (Springer US) pp 124–9
[11] Owen R L, Holton J M, Schulze-Briese and Garman E F 2009 *Journal of synchrotron radiation* vol **16** (Chester: International Union of Crystallography) pp 143-51

Macrophage and tumor cell responses to repetitive pulsed X-ray radiation

M A Buldakov[1,2], M S Tretyakova[2], V B Ryabov[3], I A Klimov[1], O P Kutenkov[4], J Kzhyshkowska[2,3], M A Bol'shakov[2,4], V V Rostov[4] and N V Cherdyntseva[1,2]

[1] Department of Molecular Oncology and Immunology, Cancer Research Institute, Tomsk NRMC, Kooperativnii str. 5, 634050 Tomsk, Russia
[2] Laboratory for Translational Cellular and Molecular Biomedicine, Tomsk State University, Pr. Lenina, 36, 634050 Tomsk, Russia
[3] Department of Innate Immunity and Tolerance, Institute of Transfusion Medicine and Immunology, Medical Faculty Mannheim, University of Heidelberg, Theodor-Kutzer-Ufer 1-3, 68167 Mannheim, Germanye
[4] Physical Electronic Department, Institute of high-current electronic SB RAS, 2/3 Akademichesky Avenue, 634055 Tomsk, Russia

E-mail: buldakov@oncology.tomsk.ru

Abstract. To study a response of tumor cells and macrophages to the repetitive pulsed low-dose X-ray radiation. Methods. Tumor growth and lung metastasis of mice with an injected Lewis lung carcinoma were analysed, using C57Bl6. Monocytes were isolated from a human blood, using CD14+ magnetic beads. IL6, IL1-betta, and TNF-alpha were determined by ELISA. For macrophage phenotyping, a confocal microscopy was applied. "Sinus-150" was used for the generation of pulsed X-ray radiation (the absorbed dose was below 0.1 Gy, the pulse repetition frequency was 10 pulse/sec). The irradiation of mice by 0.1 Gy pulsed X-rays significantly inhibited the growth of primary tumor and reduced the number of metastatic colonies in the lung. Furthermore, the changes in macrophage phenotype and cytokine secretion were observed after repetitive pulsed X-ray radiation. Conclusion. Macrophages and tumor cells had a different response to a low-dose pulsed X-ray radiation. An activation of the immune system through changes of a macrophage phenotype can result in a significant antitumor effect of the low-dose repetitive pulsed X-ray radiation.

1. Introduction

It has previously been demonstrated by several studies that a pulse-modulated non-ionizing radiation with a frequency range of 6–25 pulse/sec can induce different types of biological responses [1-3]. However, short pulses of the ionizing radiation failed to produce a significant increase of biological responses [4]. The exposure of hamster cell lines to a few pulses of X-ray radiation with 5 min interval between pulses and an absorbed dose of 0.1 Gy did not affect a cell survival [5]. A low efficiency of the used regimens can be attributed to the suboptimal pulse repetition frequency, the pulse duration, the total dose, and the number of fractions. Previously, we have demonstrated that the low-dose nanosecond pulsed X-ray radiation inhibited the tumor cell proliferation in vitro up to 95% [6].

However, tumor cell reactions in vivo and possible mechanisms of antitumor effects of the low-dose nanosecond pulsed X-ray are insufficiently studied.

It is known that an immune system plays an important role in antitumor effects of the low dose ionizing radiation. Several studies demonstrated a reduced lung metastasis in animals irradiated with the low dose radiation (<1 Gy) [7-9]. This effect can be associated with an activation of natural killer cells and macrophages, as well as an increased production of cytokines that play an important role in anti-metastatic effects of the ionizing radiation. The ability of tumor associated macrophages to respond to a plethora of environmental factors (including the ionizing radiation) makes them perspective targets for a cancer therapy.

The application of pulsed X-ray radiation as a perspective therapeutic tool is based on the fact that biological responses are strongly modified when the pulsed radiation regime is used (i.e. every pulse is divided into a series of short pulses of a nanosecond duration and a certain repetition frequency). Consequently, this can result in the alteration of macrophage function, and the stimulation of immune responses, which influence the key stages of tumor progression, thereby enhancing the effects of anticancer therapy.

2. Materials and methods

2.1. Irradiation

Irradiation of the animals was performed using a generator of a pulsed periodic X-ray ("Sinus-150", Institute of High-Current Electronic SB RAS, Tomsk, Russia). A high-voltage pulse had a half-height duration of 4 ns. The calculated photon energy spectrum had a maximum at 90 keV, and most of the quantum flux was within 60-200 keV range. The pulse amplitude stability of the diode voltage in the operating mode was constantly controlled by a capacitance sensor. The absorbed doses were verified, using thermoluminescent dosimeters LiF. Direct-reading dosimeters ("Arrow-Tech", USA) in the plastic casing were also used for a continuous dose control. The pulse repetition rate was 10 pulses per second.

2.2. Analysis of tumor growth and metastasis

Female C57Bl6 mice (6-8 weeks old) were subcutaneously injected with 2×10^6 LLC cells. The irradiation started when the tumor volume reached 430±25 mm^3. The animals were exposed daily, for 5 days, to the local irradiation at 1.8 mGy/min so that a daily absorbed dose was equal to 0.02 Gy, and the total absorbed dose was 0.1 Gy. Control mice were sham-exposed (a generator at the off-mode) under identical conditions. The solid tumor weight and the size was measured on the 19th day after the tumor cells injection. For measurement of spontaneous metastases, lungs were fixed with Bowen fixative and amounts of colonies in every mouse were calculated under a dissecting microscope [10]. All animal studies were carried out in accordance with the regulations and with permission of the European Convention for the Protection of Vertebrate Animals used for Experimental and Other Scientific Purposes (Strasbourg, 1986).

2.3. Monocyte isolation and cultivation of macrophages

The isolation and cultivation of human monocytes/macrophages was done, using a magnet separation. Briefly, the cells were purified from individual buffy coats of healthy donors. Peripheral blood mononuclear cells (PBMC) were separated, using a ficoll, and monocytes were collected using a percoll gradient followed by CD14+ magnetic cell sorting, using a monocyte isolation kit (MiltenyiBiotec, Germany). Macrophages were cultured at 1 × 106 cell/mL in X-vivo 10 serum free medium supplemented with human M-CSF (Peprotech, USA) at a final concentration of 10 ng/ml. Monocytes were irradiated with absorbed doses 0.001, 0.01, and 0.1 Gy.

2.4. Cell viability assay
The viability of cells was measured as has been previously described in [11]. Briefly, 24 h after irradiation, the cells were incubated in 20 μL MTT solution (5 mg/mL), and formazan crystals were solved with 100 μL DMSO. The absorbance was determined at 490 nm, using a microplate reader Multiscan (Thermo Scientific, USA). The cell survival rate was measured as an absorbance compared to that of untreated controls. The results represent the average of 3 independent experiments.

2.5. ELISA analysis
IL1-betta, TNF-alpha and IL6 content in supernatants from monocytes cultures was determined by ELISA (R&D Systems, USA).

2.6. Immunofluorescence
The macrophage phenotype was analyzed using the confocal microscopy. Macrophages were stained using CD68 antibody (1:100) (PG-M1, Dako, Germany) and a custom-made rabbit anti-mouse/human stabilin-1 antibody (RS1), as a marker of M2 phenotype (1:1000) (PSL, Germany). As a fluorescent secondary antibodies, the goat anti-mouse Alexa488 (1:400) (Dianova, Germany) and the goat anti-rabbit Cy3 (1:400) (Dianova, Germany) were used. The samples were analyzed by the confocal microscope LSM 780NLO (Carl Zeiss, Germany).

2.7. Statistical analysis
The Mann-Whitney U test for non-parametric trials was used for the statistical analysis of the results, and p values lower than 0.05 were regarded as significant.

3. Results and discussion

The pulsed X-ray local irradiation of mice led to a significant antitumor activity against LLC tumors. A delay of tumor growth measured both by taking into account a tumor size and a tumor weight reached 42 % (figure 1A). To analyze an anti-metastatic activity of pulsed X-ray radiation, a number of lung colonies was used. The fractioned local irradiation, using pulsed X-ray, significantly decreased the number of metastatic colonies (figure 1B). The average number of colonies in a control group was 55.9±4.41. The irradiation of mice with a total absorbed dose of 0.1 Gy inhibited the colony formation by 60 % (22.2±1.2). The results of the present study demonstrate that the local irradiation of mice with the doses of pulsed X-ray below 1 Gy leads to a significant suppression of the tumor growth and tumor colonies in a lung. Some authors showed that the traditional (non-pulsed) X-ray irradiation inhibits the growth of the primary Lewis lung carcinoma by 50 %, if we use 2 fractions of 20 Gy for each or 5 fractions of 8 Gy for each (40 Gy total absorbed dose) [12]. In the present study, the tumor growth delay was observed when we applied the 200-times lower total dose of the repetitive pulsed X-ray.

Presently, there is a lack of data about the in vivo antitumor efficacy of X-ray doses lower than 0.2 Gy. However, under certain experimental conditions, the delayed tumor growth in mice exposed to the low doses radiation was observed. Kojima et al. observed a significant inhibition of a tumor progression in mice only when the irradiation was applied before the tumor inoculation [8]. Ito M., et al. found that the total-body irradiation of mice with 0.2 Gy before the tumor cell inoculation significantly prolonged the mean time of the tumor progression [13]. The irradiation of mice 2 h before the tumor cell inoculation (the absorbed dose 0.1 – 0.2 Gy) resulted in 30-50% decrease of pulmonary tumor colonies [14]. Authors noted that the mechanism of such antitumor efficacy might be related to the activation of an immune system, i.e. the stimulation of cytotoxic activities of peritoneal macrophages and natural killer cells, which resulted in a decrease of pulmonary tumor colonies [8, 15-17].

We demonstrated that the irradiation of human monocytes with the total absorbed doses 0.1 and 0.5 Gy of the pulsed X-ray increased the secretion of IL6 and IL1-betta but not TNF-alpha by macrophages 3 days after irradiation (figure 2).

IOP Conf. Series: Journal of Physics: Conf. Series **830** (2017) 012045 doi:10.1088/1742-6596/830/1/012045

The macrophage phenotype was analyzed using the 6-day culture (5 days after irradiation). As it is shown in figure 3, the non-irradiated macrophages had CD68+/RS1- phenotype (M1 phenotype). After irradiation, an increase of the stabilin-1 expression was observed, indicating a transition to M2-skewed CD68+/RS1+phenotype. The tumor associated macrophages which make up to 50% of total cells in certain tumors, are the potential targets for an anticancer therapy (including the ionizing radiation). In this case, the application of pulsed X-ray radiation, as a perspective therapeutic tool, is based on the fact that the biological responses are strongly modified when the pulsed radiation regime is used. Wunderlich R., et al. suggest that the X-ray radiation of low and moderate doses may alter the cytokine expression, the chemotaxis, the migration, and the phenotype of macrophages [18]. El-Saghire H., et al. showed that in contrast to the high doses, the low doses of radiation can be more efficient for the monocyte activation. Monocytes exposed to the low doses of ionizing radiation induced the immune-stimulatory and the pro-survival responses, while those exposed to the high doses induced the immuno-suppressive responses [19]. In addition, an exposure of macrophages to the low doses of X-ray radiation can lead to an increase of the phagocytic and the proliferative activity [20, 21].

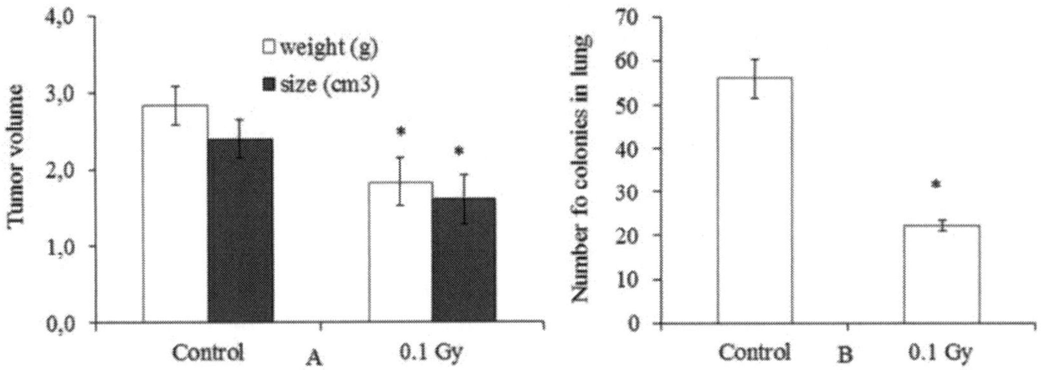

Figure 1. Volume of a solid tumor (A) and the number of metastatic colonies in the lung (B) of C57Bl6 mice after the exposure to pulsed X-ray at an absorbed dose 0.1 Gy. Error bars indicate the standard error of the mean (SEM) for n=3 independent experiments; * - indicates statistical significance ($p \leq 0.05$).

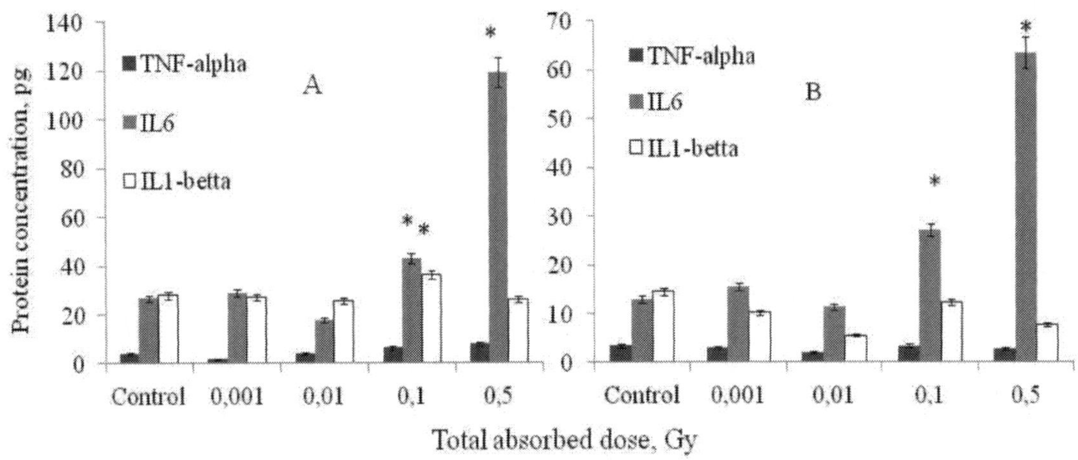

Figure 2. The cytokine concentration 3 days (A) and 6 days (B) after the monocytes exposure to the pulsed X-ray at the absorbed dose 0.001 – 0.5 Gy. Error bars indicate the standard error of the mean (SEM) for n=5 independent experiments; * - indicates statistical significance ($p \leq 0.05$).

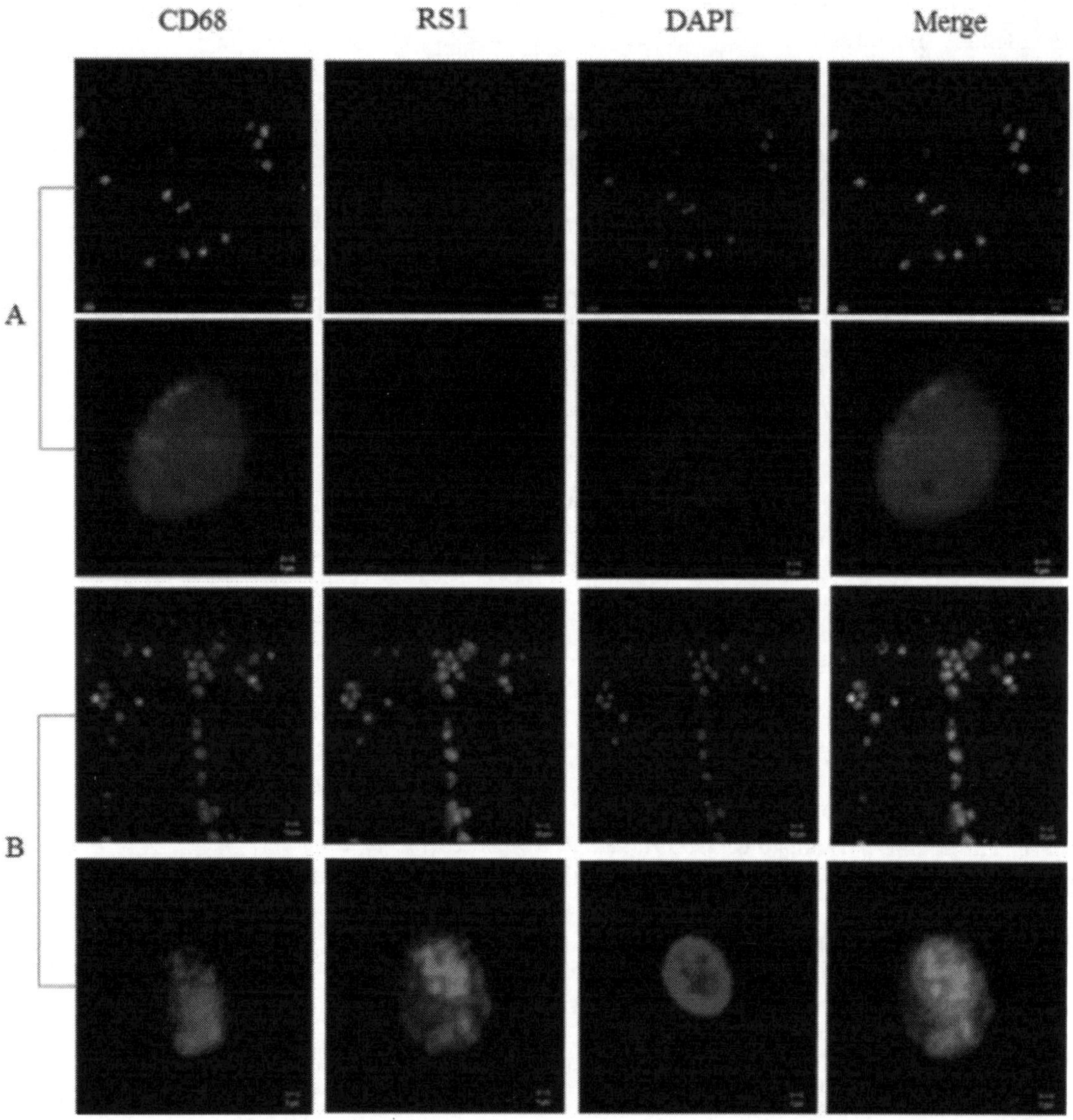

Figure 3. The volume of solid tumor (A) and the number of metastatic colonies in the lung (B) of C57Bl6 mice after exposure to the pulsed X-ray at the absorbed dose 0.1 Gy. Error bars indicate the standard error of the mean (SEM) for n=3 independent experiments; * - indicates statistical significance ($p \leq 0.05$). x630/6300

4. Conclusion

The described experimental data provide an explanation of the inhibitory effects of the low-dose repetitive pulsed X-ray radiation on the development and progress of tumors. The induction of the immune system through the macrophage activation can result in the tumor growth inhibition and the decrease of lung metastatic colonies.

Funding

Work was supported by "The Tomsk State University Academic D.I. Mendeleev Found Program" under grant № 8.1.62.2015. The study was performed on equipment of Tomsk regional common use center, with the support of the Russian Ministry of the Agreement No.14.594.21.0001 (RFMEFI59414X0001).

References

[1] Darenskaya N G, Nasonova T A and Aleshin S N 1997 *Radiats. Biol. Radioecol.* **37(3)** 336-42
[2] Dromi S, Frenkel V, Luk A, Traughber B, Angstadt M, Bur M, Poff J, Xie J, Libutti S K, Li K C and Wood B J 2007 *Clin. Cancer. Res.* **13(9)** 2722-27
[3] Fang H Y, Tsai K C, Cheng W H, Shieh M J, Lou P J, Lin W L and Chen WS 2007 *Int. J. Hyperthermia.* **23** 371-80
[4] Shinohara K, Nakano H, Miyazaki N, Tago M and Kodama R 2007 *J. Radiat. Res.* **45** 509-14
[5] Kreder N C, ten Cate R, Rodermond H M, van Bree C, Franken N A, Zdzienicka M Z and Haveman J 2004 *J. Radiat. Res.* **45** 385-91
[6] Buldakov M A, Litviakov N V, Klimov I A, Kutenkov O P. Bol'shakov M A, Rostov V V and Cherdyntseva N V 2011 *Siberian Journal of Oncology* **6** 47-51
[7] Nowosielska E M, Wrembel-Wargocka J, Cheda A, Lisiak E and Janiak M K 2006 *J. Radiat. Res.* **47** L229-36
[8] Kojima S, Nakayama K and Ishida H. 2004 *J. Radiat. Res.* **45** 33-39
[9] Nowosielska E M, Cheda A, Wrembel-Wargocka J and Janiak M K 2010 *Dose Response* **8(2)** 209-6
[10] Gorski D H, Mauceri H J, Salloum R M, Halpern A, Seetharam S and Weichselbaum R R 2003 *Cancer. Res.* **63** 308-11
[11] Pirayesh Islamian J, Mohammadi M, Baradaran B, Farajollahi A, Aghamiri S M, Asghari Jafarabadi M, Karami H, Monfaredan A and Shanehbandi D 2016 *Bioimpacts* **6(2)** 93-8
[12] Gorski D H, Mauceri H J, Salloum R M, Gately S, Hellman S, Beckett MA, Sukhatme V P, Soff G A, Kufe D W and Weichselbaum R R 1998 *Cancer Res.* **58** 5686-89
[13] Ito M, Shibamoto Y, Ayakawa S, Tomita N, Sugie Ch and Ogino H 2008 *J. Radiat. Res.* **49** 197-201
[14] Hosoi Y and Sakamoto K. 1993 *Radiother. Oncol.* **26** 177-9
[15] Novosielska EM, Cheda A, Wremblel-Wargocka J and Janiak M K 2011 *Int. J. Radiat. Biol.* **87(2)** 202-12
[16] Hashimoto S, Shirato H, Hosokawa M, Nishioka T, Kuramitsu Y, Matushita K, Kobayashi M and Miyasaka K 1999 *Radiat. Res.* **151** 717-24
[17] Cheda A, Wrembel-Wargocka J, Lisiak E, Nowosielska E M, Marciniak M and Janiak MK 2004 *Radiat. Res.* **161(3)** 335-40
[18] Wunderlich R, Ernst A, Rödel F, Fietkau R, Ott O, Lauber K, Frey B and Gaipl U S 2015 *Clin. Exp. Immunol.* **179(1)** 50-61
[19] El-Saghire H, Michaux A, Thierens H and Baatout S 2013 *Int. J. Mol. Med.* **32(6)** 1407-14
[20] Pandey R, Shankar B S, Sharma D and Sainis K 2005 *Int. J. Radiat. Biol.* **81(11)** 801-12
[21] Liu S Z 2006 *Dose Response* **5(1)** 39-47

IOP Conf. Series: Journal of Physics: Conf. Series **830** (2017) 012046 doi:10.1088/1742-6596/830/1/012046

Simulation of pre-breakdown phase of electrical discharge in reinforced concrete

N S Kuznetsova, A S Yudin and N V Voitenko

Institute of High Technology Physics, Tomsk Polytechnic University, 30 Lenin avenue, Tomsk, 634050, Russia

Email: Natalia_Kuznetsova@tpu.ru

Abstract. The application of an electrical discharge technology for destructive recycling of the reinforced concrete is considered. Its main advantages, in comparison with the mechanical methods, are that the electrical discharge channel acting as a rock-breaking tool has an unlimited service life, and a lifetime of the electrode systems is much higher. The physical and mathematical model of the discharge development is described. The simulation results have shown that the discharge channel propagation velocity and the trajectory depend on the reinforcement locality and the voltage amplitude. The voltage affects the average speed of the discharge structure development which can reach the value of up to $\upsilon=5\cdot 10^3$ m/s. It is also shown that the reinforcing elements located between the electrodes attract the growing discharge structure. The lower the distance between the vertical axis of the high voltage electrode and the metal reinforcement position, the more probability of the discharge channel orientation towards this element.

1. Introduction

A big amount of apartments and industrial buildings were built using a precast reinforced concrete in the 50-s and 70-s of the last century in Russia. Taking into account that a service life of such buildings is approximately 50–100 years, they need a massive demolition and recycling in the near future. Under low demolition, the fairly conditioned structural elements could be used in low-rise and private buildings or transported to the landfills, or finally, employed in new construction projects. An increased the volume of the construction and demolition work over the next decades does not allow us to employ such schemes for economic and environmental reasons. It raises the question of recycling of the building refuse and overage concrete elements.

Undoubtedly, the easiest method in terms of the utilization is a mechanical one, nevertheless it has some disadvantages. The limiting factors are: firstly, the large weight and an overall size of the equipment; secondly, the above-mentioned characteristics and operating costs rise nonlinearly with an increase of the destroyed products strength, and impose a limit on the maximum strength of products that could be destroyed. These problems are absent in the technologies of electro-impulse and electro-hydraulic fracture of concrete products [1, 2]. Moreover, with the increasing material strength and therefore, it brittleness, an efficiency of the method increases.

The method is based on the following processes: the shock waves generated by the penetrated discharge channel expansion, as well as the tensile and compression waves generated at further stages of the discharge lead to the progressive crack growing that results to the splitting-off of the concrete fragments with a subsequent release of the concrete reinforcement elements. The advantage of such technology, in comparison with the mechanical ones, is that the electrical discharge channel acting as a rock-breaking tool has an unlimited service life, and the life of electrode systems is much higher.

Nevertheless, the proposed technology has some disadvantages. As noted in [3], the destruction process is working more or less successfully only with the relatively deep bedding of the reinforcement cage and widely spaced reinforcement elements. Otherwise, there is a possibility of direct contact of the high voltage electrode with the reinforcement cage, leading to the almost complete cessation of the concrete item fracture process due to the lack of breakdown in the bulk of the item when a high voltage pulse is applied.

The use of mathematical modeling has a great potential for the quantitative description of the discharge development and prediction of the discharge trajectory.

In this paper, the results of computer simulation of the discharge channel formation under various conditions of the reinforcement position in a concrete and the voltage wave parameters are presented. The simulation is focused on a pre-breakdown phase of the electrical discharge in the reinforced concrete which starts from the channel initiation at the high-voltage electrode tip to the gap bridging by the discharge structure. The computer simulation is carried out with the use of numerical modeling on the base of a software developed in Moscow Engineering and Physical Institute.

2. Model of the discharge development

The discharge development is simulated with the help of a numerical discrete model created on the basis of the stochastic-deterministic approach [4–7]. The discrete algorithm involves a stochastic description of the discharge channel growth and a deterministic calculation of the electric field by means of Poisson equation, a charge dynamics on the basis of Ohm law and a charge conservation law, and the channel conduction variation in the form of a modified Rompe-Wiezel equation.

The equivalent scheme describing the operation of the pulse-voltage generator and the discharge structure growth is shown in figure 1. This scheme is determined by the type of generators which are used in the high voltage laboratories and pulse electrical discharge technology.

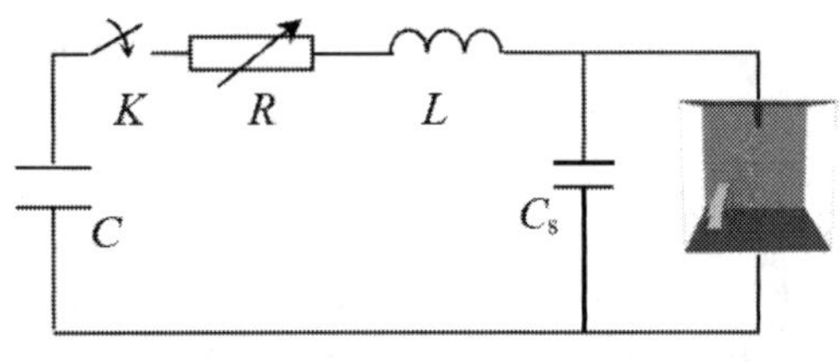

Figure 1. The equivalent scheme of reinforced concrete breakdown: C is the generator capacitance, K is the commutator, L is the circuit inductance, R is the variable resistance of the circuit, C_s is the stray capacitance.

The main equations of this model are presented in [7–9]. The circuit resistance R exponentially decreases:

$$R = R_1 + (R_0 - R_1)e^{-t/\theta_R}, \quad (1)$$

where R_0 is the initial resistance at the time $t=0$, R_1 is the minimal value of the resistance, which is equal to (1–3) Ohm for the 6-10 stepped Marks multiplying circuit; θ_R is the resistance decrease characteristic time. The stray capacitance C_s is the sum of the stray capacitances of the circuit elements.

The dynamics of the current and voltage in the discharge generator circuit is calculated on the basis of the finite-difference approximation of Kirchhoff equations [9].

The probability density ω_n of the discharge channel growth in a direction $\vec{n}$ is assumed to be proportional to the square of the local field projection E_n along this direction, if the projection value exceeds the critical strength:

$$w_n = \alpha\theta(E_n - E_c)E_n^2, \tag{2}$$

where α is the growth rate coefficient, E_c is the critical field for the discharge channel growth, $\theta(x)$ is the step function ($\theta(x)$=1, for $x > 0$ and $\theta(x)$=0 for $x \leq 0$).

The channel conduction during the discharge development is estimated on the basis of the modified Rompe-Wiezel equation [5]:

$$\frac{\partial\gamma}{\partial t} = \chi\gamma E_l^2 - \xi\gamma, \tag{3}$$

where χ and ξ are parameters of the rate of increase and decrease of the conduction, respectively. The first term in the right part of equation (3) is related to the conduction increasing due to Joule energy release within the discharge channels. The second term describes the conduction decrease due to energy dissipation into the surrounding material.

The presented model describes the discharge channels propagation, charges motion in the channels, dynamics of the electric field potential and the change of channels conductivity. Discharge development is considered in the homogenous dielectric for the needle-plane electrode geometry (figure 2).

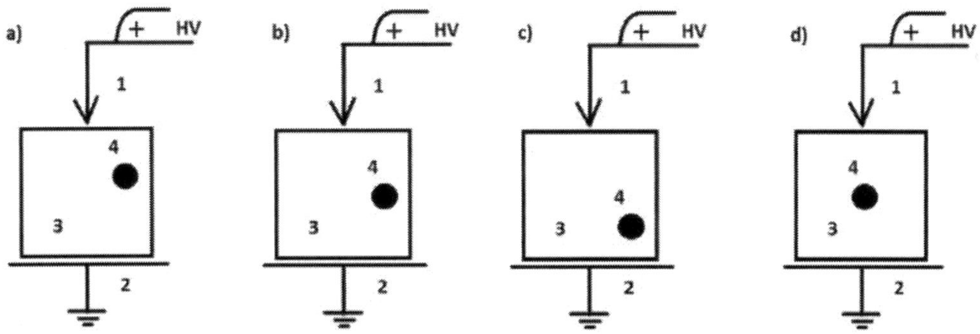

Figure 2. The schemes of the simulation region: 1 – HV electrode; 2 – ground electrode; 3 – concrete; 4 – reinforcement element.

3. Simulation results of the discharge development in the reinforced concrete

Simulation of the discharge development has been carried out in the needle-plain electrode geometry for the concrete. The computer experiments were made for the following parameters of the discharge circuit: U_0=450 kV, C=14,8 μF, L=10 μH, Cs=0,2 nF, R_0= 10 MOhm, R_l=2 Ohm, θ_R=10 ns. The results of computer investigations of the electrical discharge in concrete by means of the presented model under the different reinforcement positions are presented in figure 3.

The electrical discharge initiates with the formation of one or several channels at the initiating needle, when the electric field strength exceeds the critical value (t = 216 ns). The rate of channel growth increases with the voltage build-up and discharge structure eventually approaches the ground electrode. The average speed of the discharge development equals to υ=3,97·10^3 m/s. As a result of the discharge structure propagation, the electrode gap bridging by the main discharge channel occurs (the main discharge channel is shown in figure 3 by the thick line). As the most part of the current flows through the main discharge channel, the conduction of the remaining channels quickly decreases and they attenuate. As a result, only the main discharge monochannel remains.

The discharge structure orientation in the concrete strongly depends on the reinforcement location. The reinforcing elements located between the electrodes attract the discharge structure. The less the distance between the vertical axis of the high voltage electrode and the metal reinforcement position, the more the probability of the discharge channel orientation towards this element.

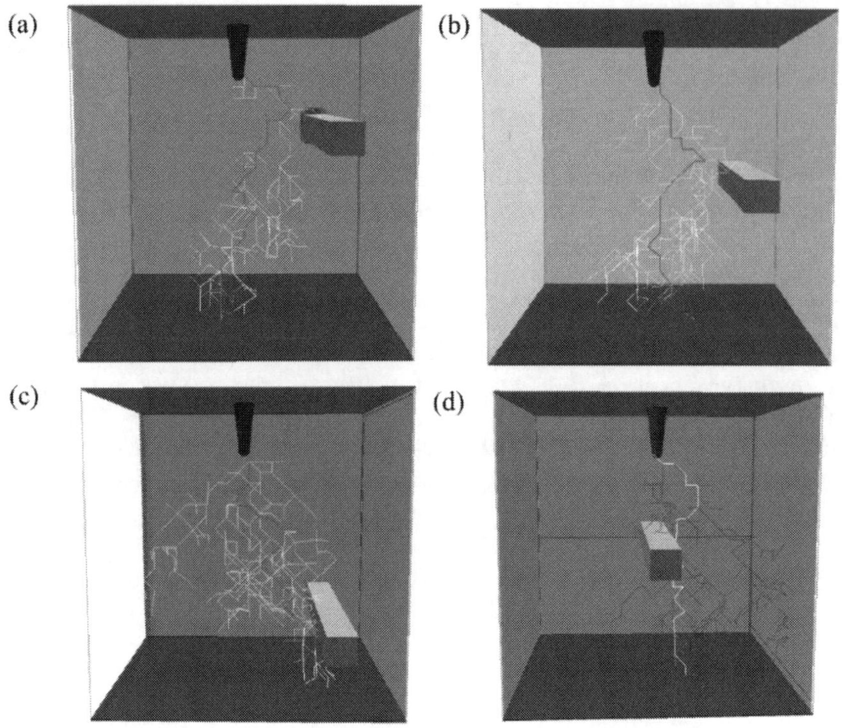

Figure 3. Discharge channels structure at the various positions of the concrete reinforcement

After the bridging of the electrodes gap by the main discharge channel, the abrupt increase in the discharge current and the decrease in the discharge voltage occur, whereas the inductive voltage rise. After the concrete breakdown, the damped oscillations begin in the generator circuit (figure 4). During this process, the main discharge channel operates as the variable nonlinear resistance.

The period of the oscillations is determined by the generator capacitance and inductance. At the given simulation parameters, the oscillations stop occurs in a twenty microseconds. The oscillation of the discharge voltage has a trapezoidal form. It is caused by the variation of the mono-channel conduction during the electrical discharge.

The dynamics of the discharge channel energy release is caused by the current and voltage oscillations (figure 5).

The energy stored in the pulse generator capacitor is equal to 14,98 kJ. The whole energy which released within the main discharge mono-channel amounts to 20 % of the generator energy. Almost half of the plasma channel energy is released during the first period of the current oscillation. Other part of the energy is released within the active resistance R of the generator.

Oscillation of the plasma channel conduction occurs due to the power oscillation (figure 6). The minimal values of the channel conduction correspond to the time moments when the discharge current takes on a zero value.

The voltage increase leads to the growth of the discharge structure branching, a speed of the channel propagation and the discharge current (figure 7). The voltage build-up affects the average speed of the discharge structure development which can reach the value of up to $\upsilon=5\cdot 103$ m/s. The channel trajectory is placed mainly close to the reinforcement cage (the main mono-channel is shown in figure 7 by the thick line).

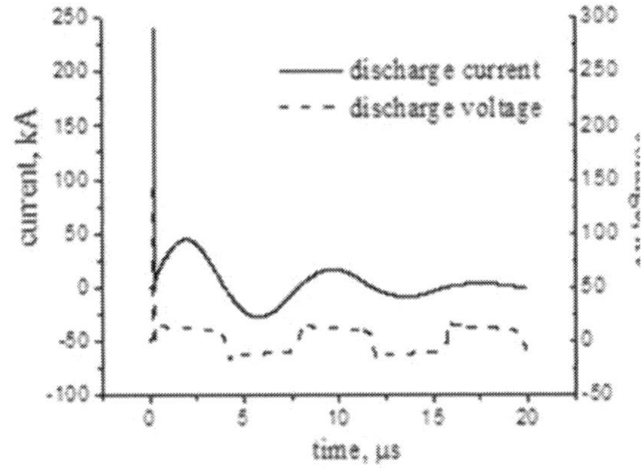

Figure 4. Time dependencies of the discharge voltage and current.

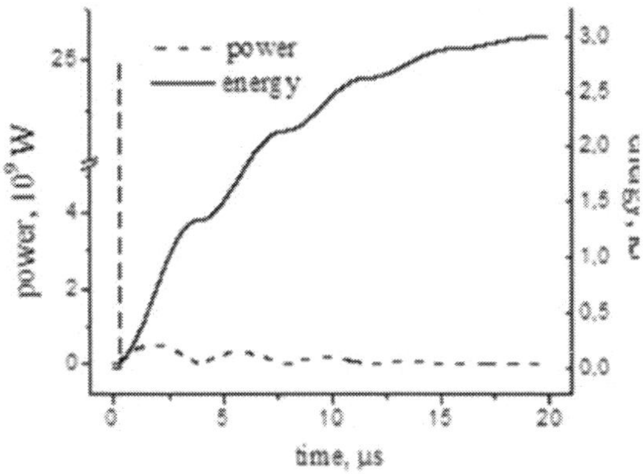

Figure 5. Time dependencies of the total energy and Joule power release within the discharge channel.

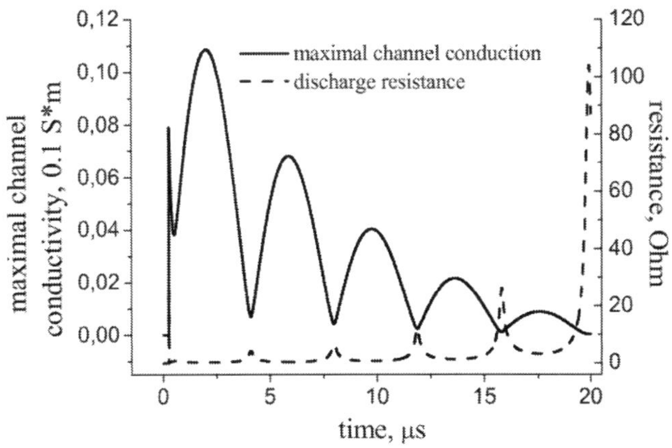

Figure 6. Time dependencies of the maximal conduction of the discharge channels and electrode gap resistance

(a) (b)

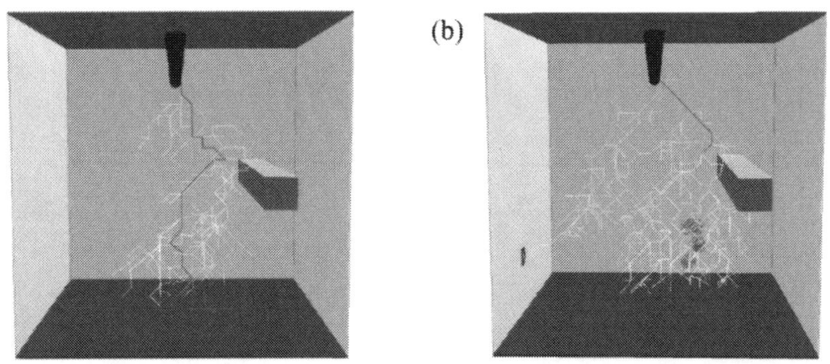

Figure 7. Simulated discharge structure (a) U_0=450 kV, D=2.25, t=260 ns; (b) U_0=600 kV, D=2.67, t=252 ns.

The heterogeneity of spatial distribution of the concrete conductivity and presence of the reinforcing steel in the concrete affect the discharge development. The discharge channel locality is controlled by an existence of the concrete reinforcement with a high conductivity. The reinforcing steel of the high conductivity located between the electrodes attracts the growing structure. The shorter the distance between the high voltage electrode vertical axis and the metal reinforcement position, the higher efficiency of the concrete distraction nearby it.

4. Conclusion

The presented mathematical model takes into account the main physical processes determining the reinforced concrete breakdown, consistently with the high-voltage generator operation. The obtained simulation results are helpful in understanding of a breakage behavior of the reinforced concrete in the electro-discharge technology of solid destruction.

5. Acknowledgements

This work was supported by grants of RFBR № 14-08-01088 and RFBR № 16-48-700278.

References

[1] Sjomkin B, Usov A and Kurets V 1995 *The principles of electric impulse destruction of materials* (St. Petersburg: Science) p 276
[2] Kapishnikov N, Lebedev V and Lipov G 2013 *Izvestija Vuzov. Physica* **12**
[3] Kapishnikov N, Lebedev V and Lipov G 1996 *Izvestija Vuzov. Physica* **4** 109–14
[4] Lopatin V, Noskov M, Badent R, Kist K and Schwab A1998 *IEEE Trans. Diel.* EI **5**, 250
[5] Noskov M, Lopatin V and Kurets V 2001 *Proc. Int. Conf. on Pulse Power Applications* (Gelsenkirchen) pp.F.06/1– F.06/6.
[6] Noskov M, Malinovski A, Cooke C, Wright K and Schwab A 2002 *J. Appl. Phys.* **92** 4926
[7] Lopatin V, Noskov M, Usmanov G and Cheglokov A 2006 *Izvestija Vuzov. Physica* **3** 11
[8] Noskov M, Lopatin V, Usmanov G and Cheglokov A 2006 *Izvestija Vuzov. Physica* **49** 11–18
[9] Lopatin V, Noskov M, Usmanov G and Cheglokov A 2006 *Russian Physics Journal* **49** 243–250

IOP Conf. Series: Journal of Physics: Conf. Series **830** (2017) 012047 doi:10.1088/1742-6596/830/1/012047

Design and testing of fluid resistor for repetitive high-voltage pulse generator

A V Poloskov, I S Egorov and M A Serebrennikov

National Research Tomsk Polytechnic University, Tomsk, Russia, (Lenin ave. 30, Tomsk 634050 Russia)

E-mail: poloskow@tpu.ru

Abstract. The paper presents a design and the results of testing of the liquid resistive load for a repetitive high-voltage generator (200 kV, 0.5 ms). The load uses a sealed dielectric case, which is to be placed into a vacuum volume ($5x10^{-4}$ Torr) for electrical strength ensuring. Repetitive testing of the generator with the load (10 pps) caused the electrolyte heating, the load resistance decreasing, and the changing of a generator mode. An expansion tank is used to compensate thermal expansion of the electrolyte, which makes it possible to absorb up to 1 MJ of energy in the load without seals breaking. A generator load curve can be obtained for one experiment with a help of the fluid load without any additional depressurization of the vacuum volume.

1. Introduction

To test the facilities and devices manufactured for research and application purposes, due to their unique characteristics in various research centers, often needs a complicated diagnostic equipment, parameters of which essentially differ from its industrial prototypes. The current paper examines the results of testing of the fluid resistive load, which was designed for an operation in a vacuum chamber of the high-voltage generator of the ASTRA repetitive electron accelerator [1], under residual atmospheric pressure of $5x10^{-4}$ Torr. Such pressure remained in the course of accelerator operation, to provide dielectric strength of the high-voltage vacuum insulator of the generator and to provide the conditions of pulsed electron beam generation. The accelerator has found its application in a wide range of research and application problems [2-5]. The parameters of an electron beam are very significant for the defined applications. These parameters are determined by the characteristics of an accelerating voltage pulse formed by the generator. The generator was designed as a capacitive storage discharge through the high-voltage pulse transformer [1]. According to [6], the pulse parameters depend on matching of the generator impedance ($Z_g \approx 80\ \Omega$) with an abrupt non-linear characteristic of the impedance of the vacuum electron diode [7], which transforms the stored electrical energy into a kinetic energy of beam electrons. Thus, there is a possibility to examine the operation parameters of the high-voltage pulse generator under different values of the load resistance, i.e its load characteristic.

To meet the application conditions, the generator resistive load should provide the pulse dielectric strength over 200 kV [1, 7], and the change in *Zn* load resistance should range from *Zg* to 2*Zg*. The load should provide the energy absorption, using the pulsed power, as high as 500 MW [1] under an average power of 400-500 W. The graphite resistors with a ceramic case (TVO), which are traditionally used to build the low-inductive loads for the generators with a high pulse power, can be used in a series connection, to provide the dielectric strength of a resistor surface. However, the power dispersed by the resistors abruptly decreased when they operate in vacuum making impossible to work with a high pulse repetition rate. It is also not very convenient to replace the resistors to adjust the load resistance that requires resealing of the vacuum chamber and significantly complicating a revival of the load characteristic of the generator.

It seems to be more convenient to use the fluid resistors described in [8]. There, the electrolyte is utilized as a volume resistance, whose initial conductivity can be easily adjusted by a salt concentration. In the current work, the solution KCl (2%) was used as an electrolyte of the fluid load.

2. Experimental setup

Basing on the data described in [8], in terms of the conditions defined before, an original construction of the fluid resistor was developed. The body of the fluid load was made as a dielectric tube, hermetically plugged from the ends by the metal flanges operating as electrodes (Fig. 1). The structure of the fluid load was vertically installed, with a filler neck being directed upward. The length of the body over the insulator was 400 mm. The volume of water was 3.38 l.

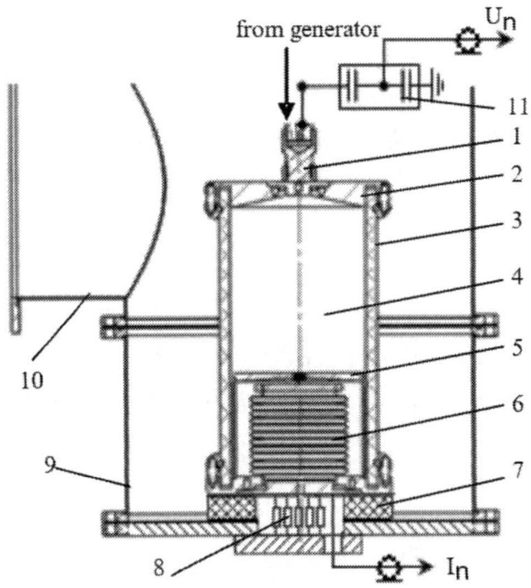

Figure 1. A structure of the fluid load with pressure damper and a scheme of testing of the generator with a load. 1 – a high-voltage output of the generator; 2 – a potential electrode, a filler neck; 3 – a dielectric tube (a glass fiber plastic); 4 – a volume with an electrolyte; 5 – a grounded electrode; 6 –a pressure damper; 7 – a base insulator; 8 – a low-inductive shunt (0.05 Ω); 9 – a body of the vacuum chamber; 10 – a pumping branch; 11 – a capacitive voltage divider.

The load was supplied with a pressure damper of the electrolyte. The grounded electrode was made in such way that when changing the volume of the damper, the geometry of the gap between electrodes remained unchanged. The estimated calculation of the heat expansion of the electrolyte showed that about 1 MJ could be released in the load without a considerable increase in pressure inside the body, due to the volume compensation. The structure of seals enabled to use the fluid load under residual atmosphere pressure of 5x10-4 Torr.

To carry out the tests, the load was installed into the vacuum chamber and connected to the high-voltage output of the generator. The grounded electrode was installed on the insulator base and connected to the generator body, using a low-inductive shunt allowing us to control the current, passing through the load (Fig. 1). The voltage applied to the load was controlled using a capacitor voltage divider (Fig. 1).

3. Results and discussion

At the moment of assembling, the load resistance *Zn*=226.5 Ω was determined using the immittance device (E7-21). During the experiment, we assumed that the load resistance over pulse duration remained unchanged. In the course of experiment, the load voltage *Un* and the current *In* oscillograms were recorded (Fig. 2). They were employed to determine the *Zn* load resistance at the moment of maximum current using Ohm's law.

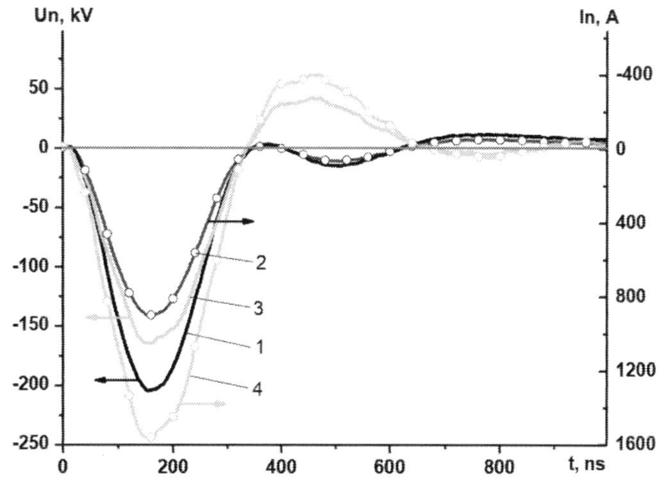

Figure 2. Waveforms of the voltage (1, 3) and the current (2, 4) of generator under the load resistance 226.5 Ω (1, 2) and 105 Ω (3, 4).

The curves given in Fig. 2 were also used to estimate the energy released in the load *En* per pulse duration:

$$Pn\ (t) = Un \cdot In;$$

$$En = \int Pn\ dt,$$

where *Pn* is a pulse power developed in the diode;

To determine the load characteristic, the high-voltage generator loaded on the high-voltage fluid resistor was tested at a constant value of the energy stored in the storage under pulse repetition rate 10 pps for 20 minutes. The voltage and current curves were recorded for 2400 pulses. The energy absorption by the electrolyte increased its temperature and, as a result, decreased the load resistance.

The load current and voltage curves recorded for 12000 pulses enabled to build a variation diagram of the resistance load (the load characteristic) and the energy released in it (Fig. 3). The diagram in Fig 3 is confirmed by a computational model described in [9].

The change from 226.5 Ω to 105 Ω in the generator resistance caused a mismatching of the generator and load impedances. It was exhibited in a change of the energy amount released in the load (Fig. 2-2). It should be noted that reasoning from the purpose of the generator, the energy released for duration of the first voltage pulse can be treated as useful (Fig. 2-3). The energy of electrons generated during an afterpulse is not enough for its practical use. Besides, when the load resistance is lower than $2Zg \approx 160$ Ω, the discharge process of the generator is periodical, that negatively affects the lifetime of the high-voltage vacuum insulator of the generator [10].

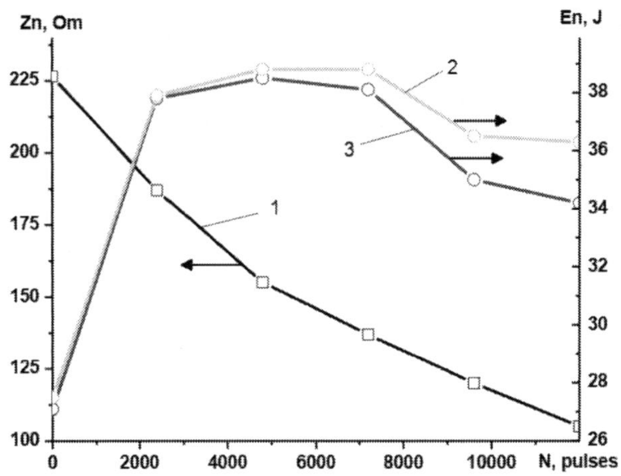

Figure 3. Diagrams of the load resistance (1) and the energy released in the load per pulse (2, 3) with an increasing number of pulses, at a pulse repetition rate of 10 pps.

The temperature of the fluid load before and after the experiment was recorded using a thermal imager (Fluke Ti10); it was 20°C and 53°C, respectively.

The energy, released in the load for the whole time of the experiment procedure E_Σ, was estimated as:

$$E_\Sigma = c \cdot m \cdot \Delta T = 460.7 \text{ kJ},$$

where c was the specific heat capacity of the electrolyte (2% KCl concentration), $c \approx 4046$ kJ/(kg·°C);

ΔT was an electrolyte temperature variation, $\Delta T = 33°$;

m was the mass of electrolyte, m=3.45 kg.

Using the data of the diagram in Fig. 3, we obtained the calculated value of energy $E_{\Sigma P}$ released in the load for the whole time of the experiment procedure:

$$E_{\Sigma P} = \int En\, dN = 441.4 \text{ kJ}.$$

So, the measured and calculated values of the energy released in the load during the experiment differ by 5% that can be explained by the inaccuracy of measuring instruments.

4. Conclusion

The suggested structure of the fluid resistive load and the described method of its application enabled to design the load for the generator operating under changing load resistance resulting from its heating. We used high-voltage pulses (over 200 kV) with 10pps pulse repetition rate. In the course of the experiment, the load remained leak-proofed under a residual atmosphere pressure of $5x10^{-4}$ Torr. Due to this load, 450 kJ ± 5% of energy was released that caused its heating by 33°.

Acknowledgements

This work was partially supported by RFBR grant #16-32-00287mol_a.

References

[1] Egorov I., Esipov V., Remnev G., Kaikanov M., Lukonin E., Poloskov A. 2012 *Proceedings of the IEEE International Power Modulator and High Voltage Conference*, art. no. 6518845, pp. 716-719.2.

[2] Mostovshchikov A.V., Ilyin A.P., Azanov A.A., Egorov I.S. 2016 *Key Engineering Materials*, **685**, pp. 639-642.

[3] Egorov I.S., Esipov V.S., Lukonin E.I., Poloskov A.V. 2015 *Instruments and Experimental Techniques*, **58** (1), pp. 64-66.

[4] Mostovshchikov A.V., Ilyin A.P., Egorov I.S., Ismailov D.V. 2016 *Key Engineering Materials*, **712**, pp. 60-64.

[5] Poloskov A., Egorov I., Ezhov V., Remnev G. 2016 *Key Engineering Materials*, **685**, pp. 667-671.

[6] Remnev G.E., Furman E.G., Pushkarev A.I., Karpuzov S.B., Kondrat'ev N.A., Goncharov D.V. 2004 *Instruments and Experimental Techniques*. V. **47**. № 3. C. 394-398.

[7] Egorov I.S., Ezhov V.V. Poloskov A.V. *Fundamental research*. – 2014. – № **9-9**. – C. 1964-1968 (in Russian).

[8] Gerasimov A.I. *Instruments and Experimental Techniques*. 2006. V. **49**. № 1. C. 1-26.

[9] Egorov, I. 2014 *Review of Scientific Instruments*, **85** (6), art. no. 066112.

[10] Egorov, I. 2016 *IEEE Transactions on Dielectrics and Electrical Insulation*, **23** (4), art. no. 7556492, pp. 2174-2180.

High-current electron gun with a planar magnetron integrated with an explosive-emission cathode

P P Kiziridi and G E Ozur

Institute of High-Current Electronics, SB RAS, Tomsk, 634055, Russia

E-mail: ozur@lve.hcei.tsc.ru

Abstract. A new high-current electron gun with plasma anode and explosive-emission cathode integrated with planar pulsed powered magnetron is described. Five hundred twelve copper wires 1 mm in diameter and 15 mm in height serve as emitters. These emitters are installed on stainless steel disc (substrate) with 3-mm distance between them. Magnetron discharge plasma provides increased ion density on the periphery of plasma anode formed by high-current Penning discharge ignited within several milliseconds after starting of the magnetron discharge. The increased on the periphery ion density improves the uniformity of high-current electron beam produced in such an electron gun.

1. Introduction

Low-energy (10-30 keV), high-current (10-25 kA) electron beams (LEHCEBs) are of great interest for material surface treatment and found already rather wide application in practice: increasing the corrosion resistance of metals and alloys, smoothing (polishing) of the surface, improvement the electric strength of vacuum insulation and so on [1, 2].

To produce LEHCEBs, the axis-symmetric electron guns with an explosive-emission cathode and plasma anode based on high-current Penning discharge are generally used [1-3]. One of the key parameter of LEHCEBs is their uniformity of energy density distribution via cross section and the requirements to the beam uniformity grow constantly. Multiple experiments have shown that to improve LEHCEB it is needed to increase ion density on the periphery of plasma anode in comparison with its near-axis region.

We suggest a new method for producing of plasma anode with such profile of ion density. For this purpose, we install a planar magnetron inside the cathode assembly. Planar magnetron produces an annular plasma column as well as a flux of neutrals due to cathode sputtering. At further ignition of high-current Penning discharge, the ion density at the periphery of plasma anode becomes higher than in the central region. Besides, we expected that magnetron discharge plasma will improve the stability of cathode emission. The present paper describes the developed high-current electron gun with explosive-emission cathode integrated with planar pulsed powered magnetron.

2. Description of electron gun

Principal design of electron gun is given in Fig. 1. Planar magnetron consists of two permanent magnets (disc 1 and ring 2) and magnetic circuit 3 and is placed just behind the cathode substrate 5. Magnets produce an arched magnetic field 4 over the cathode surface. Such magnetic field localize plasma of magnetron discharge mainly in the vicinity of magnetic field maximum. Thus, plasma cloud acquires a ring shape occupying the periphery region over the cathode surface. The cathode disc substrate 5 is manufactured from stainless steel, on which the emitters 3 made of copper wires with 1 mm diameter and 15 mm height are regularly installed with the step of 3 mm.

The principal problem of the operability of the suggested scheme is the difference in working gas (argon) pressure required for magnetron discharge operation and for production of LEHCEB. For ignition of the magnetron discharge, the pressure of 1.5×10^{-3} Torr or higher is required (at the 1-kV of triggering pulse and 500–600 Gs of magnetic field induction on the cathode surface). But for LEHCEB production, the pressure should be about three times lower. The are, at least, two ways how to solve this problem:

1. It is possible to decrease the working gas pressure threshold needed for magnetron discharge ignition if increase the magnetic field induction near the cathode surface. In our case, it was provided by decreasing of the cathode substrate to 1–2 mm instead of standard 4–5 mm. The pulsed mode of magnetron discharge excludes water cooling of the cathode substrate and allows one to decrease its thickness.

2. The amplitude of the triggering voltage pulse was increased from standard 800–1000 V to 3 kV.

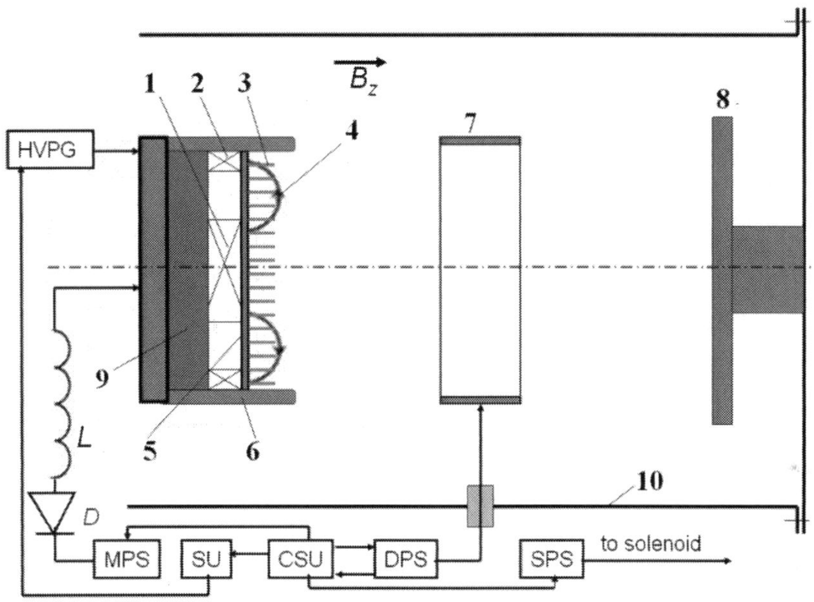

Figure 1. The block-diagram of LEHCEB source. 1 and 2 – permanent magnets; 3 – copper emitters; 4 – magnetic field lines B_r; 5 – cathode substrate; 6 – body of the cathode assembly; 7 – anode of reflective discharge; 8 – collector of electron beam; 9 – magnetic circuit; 10 - body of electron gun; HVPG - high-voltage pulsed generator; MPS – magnetron power supply; SU – starting unit of HVPG; CSU – control and synchronization unit; DPS – discharge (reflective) power supply; SPS - solenoid power supply; D - diode assembly; L – protective inductance.

After several milliseconds from the starting of magnetron discharge with the current of 1–5 A, the processes in electron gun will develop in traditional way [1]. Pulsed powered solenoid produces longitudinal guide (for further produced LEHCEB) magnetic field B_z, and then with the delay of several milliseconds else (when it reaches the value of $B_z \approx 500$–1500 Gs), 5-kV positive voltage pulse is applied to the anode 7 and Penning (reflective) discharge starts. The grounded collector 8 and explosive-emissive cathode serve as the cathodes of reflective discharge (RD). After RD transits to high-current stage (discharge current up to 200 A), a cylindrical column with characteristic density of $(2–5)\times10^{12}$ cm^{-3} is formed between the cathode and collector. After formation of this column (plasma anode), an acceleration voltage pulse with an amplitude of 15–30 kV and 10–50 ns rise time is applied to the cathode. Electric field is concentrated in a narrow near-cathode layer and achieves significant values (up to 300–400 kV/cm). The electric field is amplified on the tips of emitters, and explosive-emission centers (dense plasma clouds or cathode spots) appear quickly (within several ns or tens of ns). Plasma anode ions bombardment of the emitters stimulate their operation. After the cathode spots appeared, acceleration voltage is concentrated in a double layer between the cathode and anode plasmas. Electrons emitted by cathode plasma are accelerated in double layer and transported through anode plasma onto the collector in a guide magnetic field B_z. Due to charge neutralization of the beam, it is transported practically without the lost of current. The lost of kinetic energy of the beam electrons is also negligible [4].

The time of RD transition into high-current stage makes up usually tens of microseconds, but at the presence of magnetron discharge plasma this time decreases several times because of quick inclusion of the electrons of this plasma into the process of ionization of the working gas. Since these electrons are located in the peripheral part of plasma column, so the degree of ionization and ion density are expected to be higher just in this region. Finally, just such an increase of the ion density is desirable for improvement the LEHCEB uniformity [4].

In order to transverse magnetic field of the permanent magnets do not hinder RD operation as well as LEHCEB formation, the emitters have the height of 15 mm. Since the induction B_r falls dramatically with the distance from the cathode surface, so at the tips of emitters B_r does not exceed 50–70 Gs and does not distort the guide magnetic field B_z, therefore RD and beam formation are running without problem.

To prevent MPS from over voltage, the inductance coil $L \approx 500$ μH is installed (Fig. 1). An additional and very important protection is provided by the diode assembly *D*, consisting of 50 diodes HER308 (1000 V, 3 A, time of recovery 100 ns) connected in series.

Acceleration voltage pulse is measured by resistive divider, cathode current and beam current onto collector are measured by wide-band Rogowsky coils.

LEHCEB's energy density is measured with the use of thermal imager TESTO 875-1. At these measurements, 200-μm stainless steel foil serves as the beam collector. At energy density of 3–10 J/cm^2, such a foil is heated by tens of degree Centigrade per pulse and this heating which reflects beam energy density distribution is panoramically measured through an infrared window. Characteristic spatial resolution of such a method is defined by the pause between the beam pulse and the moment of thermogram fixing. This time makes up about 1 s which corresponds to spatial resolution about 2–3 mm.

Fig. 2 presents typical waveforms of the pulses, and Fig. 3 presents a temporal diagram of the processes. The consequence of the processes in time and time delays t_1, t_2, t_3 are regulated by control and synchronization unit.

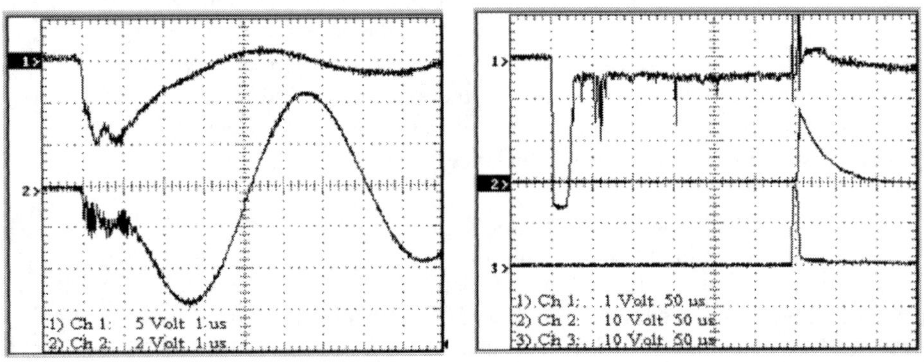

Figure 2. Typical waveforms. (a): acceleration voltage (Ch1, 10 kV/div); cathode current (Ch2, 12 kA/div); (b): magnetron discharge burning voltage (Ch1, 320 V/div), RD current (Ch2, 80 A/div); RD burning voltage (Ch3, 2 kV/div).

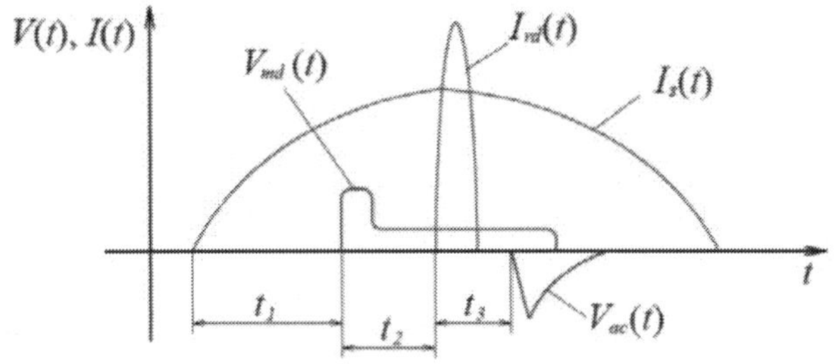

Figure 3. Temporal diagram of the processes in electron gun. $I_s(t)$ is solenoid current, $I_{rd}(t)$ is RD current, $V_{md}(t)$ is magnetron discharge burning voltage, $V_{ac}(t)$ is acceleration voltage of electron gun.

Fig.4 presents a thermogram and corresponding energy density distribution via beam cross section. It can be seen that non-uniformity in the central part of 4.5 cm in diameter does not exceed ±10 %.

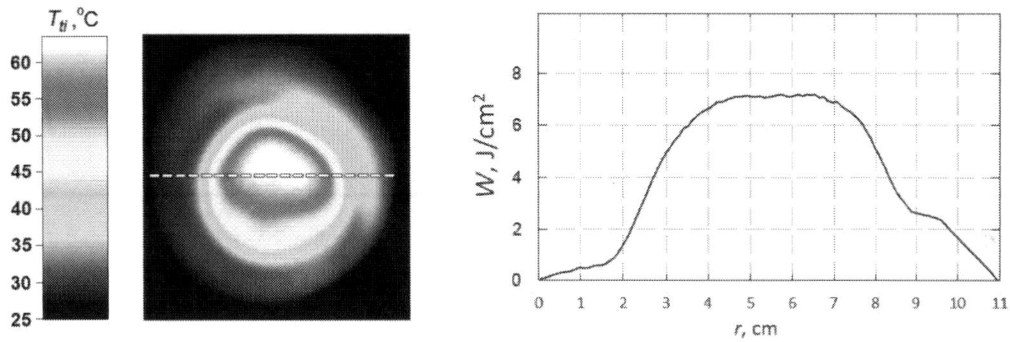

Figure 4. Thermogram and corresponding energy density distribution via beam cross section (along the dashed line). Charge voltage of HVPG is 25 kV, the distance between the cathode and the collector is 20 cm, argon pressure is 0.5 mTorr.

3. Summary

The new high-current electron gun, in which plasma anode is formed with the use of hybrid discharge matching pulsed magnetron and reflective (Penning) discharge, has been developed and tested. The suggested scheme of high-current electron beam production looks promising from the point of view of the beam uniformity and stability.

Acknowledgment
The work was supported by Russian Foundation of Basic Research (grant # 16-48-700239-r_sibir-a).

References

[1] Ozur G.E., Proskurovsky D.I., Rotshtein V.P., Markov A.B. *Laser and Particle Beams*, 2003, v. **21**, No. 2, pp. 157–174.
[2] Uno Y., Okada A., Uemura K. et al, *Precision Engineering*, 2005, v. **29**, No. 4, pp. 449–455.
[3] Batrakov A.V., Markov A.B., Ozur G.E., Proskurovsky D.I., Rotshtein V.P.. *The European Physical Journal Applied Physics*, v. **43**, issue 3, 2008, pp. 283–288.
[4] Nazarov D.S., Ozur G.E., Proskurovsky D.I. *Instruments and Experimental Techniques (Soviet)*, 1996, No. **4**, pp. 83–88.
[5] Kiziridi P.P., Ozur G.E., Zjulkova L.A., Batrakov A.V. *Russian Physics Journal*, 2012, vol. **55**, No. 6/2, pp. 42–48.
[6] Kiziridi P.P., Markov A.B., Ozur G.E., Frolova V.P., *Russian Physics Journal*, 2014 г., vol. **57**, No. 3/2, pp. 114–117.

Numerical simulation of the coaxial magneto-plasma accelerator and non-axisymmetric radio frequency discharge

V V Kuzenov [1,2], S V Ryzhkov[1] and P A Frolko[1]

[1] Department of Thermal Physics, Bauman Moscow State Technical University, 2 Baumanskaya Street 5, 1, Moscow 105005, Russia
[2] N.L. Dukhov All-Russian Research Institute of Automatics (State Corporation «ROSATOM» company), Suschevskaya Street 22, Moscow 127055, Russia

E-mail: vik.kuzenov@gmail.com

Abstract. The paper presents the results of mathematical modeling of physical processes in electronic devices such as helicon discharge and coaxial pulsed plasma thruster. A mathematical model of coaxial magneto-plasma accelerator (with a preionization helicon discharge), which allows estimating the transformation of one form of energy to another, as well as to evaluate the level of the contribution of different types of energy, the increase in mass of the accelerated plasmoid in the process of changing the speed. Main plasma parameters with experimental data were compared.

1. Introduction

The theoretical work presented in this contribution applies a radio-frequency (RF) wave-heated (helicon) plasma as an alternative concept for ion sources in various applications [1–7]. An approximate mathematical model of a helicon type plasma source (coaxial pulsed plasma thruster) was developed and calculation based on the developed multiparameter computational model was previously done.

The approximated mathematical model of radio-frequency ion source of helicon-type plasma and coaxial multi-plasma accelerator was created in [9–12]. Such devices can be used for high temperature plasma applications involving materials science. The initial verification of computing results on the basis of a group of experimental data was done. These computations were made with the use of self-approved approximated mathematical model of RF plasma source. The estimate of various power types contributions into the process of acceleration and transformation of one type of energy into the other was performed. Analyzed RF devices can be used as the plasma sources to solve problems in materials science [13].

The aim of this work is to compute main electrophysical characteristics of a coaxial magneto-plasma accelerator with the voluminous energy source. The coaxial magneto-plasma accelerator (CMPA) with a preionization helicon discharge [8–12] is considered.

2. Calculation of the main characteristics of RF ion source

The RF source extraction system has the following sizes: the length of the cathode channel $l = 3$ mm, the diameter of the channel $d = 0.6$ mm. The gas discharge bulb is made of quartz and has the outer diameter of 0.03 m and length of 0.2 m.

RF power system consists of a main oscillator with an operating frequency $f = 27.12$ MHz, the power amplifier "Acom-1000" and matching unit, providing power up to 400 W in continuous mode [14]. The sources used a spiral antenna that contains 4 coils of copper wire with a diameter of 4 mm.

Computational and theoretical modeling is performed based on the developed numerical method. The experiment has been conducted using an ion beam accelerating device [14], which is suitable for injectors and other applications. Figures 1–4 present theoretical and experimental curves for the following discharge parameters: antenna frequency $f = 27.12$ MHz, magnetic field $B = 3\cdot10^{-2}$ T, gas pressure $p = 1.33$ Pa, antenna radius $R = 0.015$ m, and antenna length $L = 0.13$ m. The electron density n_e^{Σ} and ion current density j versus the RF power P_{hel} for discharge in helium are given in Figures 1 and 2. The electron density n_e^{Σ} as a function of RF power P_{hel} in argon discharge is presented in Figure 3, and the power W_{hel} embedded in RF plasma ion source with hydrogen plasma is shown in Figure 4 for Ar.

In Figures 1 and 3 we have plotted the change in electron density n_e^{Σ} versus power P_{hel}, supplied to the helicon antenna. Curve 1 is obtained from the mathematical model described in this work, while the experimental curve 2 is taken from Ref. 14.

The ion current density j increases with the RF power P_{hel} as shown in Figure 2: the calculated values (curve 1) and experimental results (curve 2). When comparing experimental data with the theoretical values, the mathematical model developed by the authors is used.

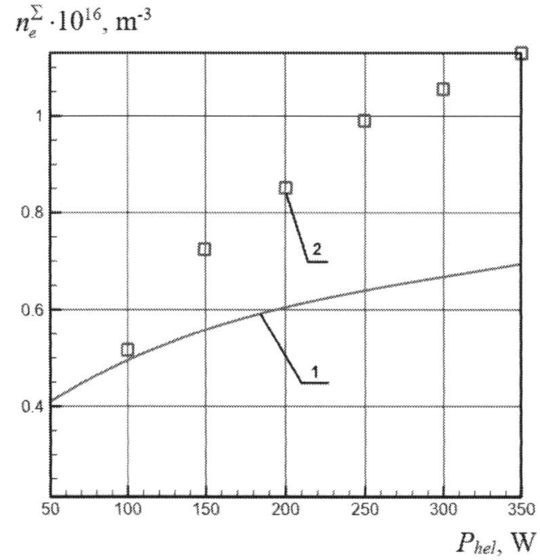

Figure 1. The variation of the electron density n_e^{Σ} on the RF power P_{hel} for He discharge: 1) calculation, 2) experiment.

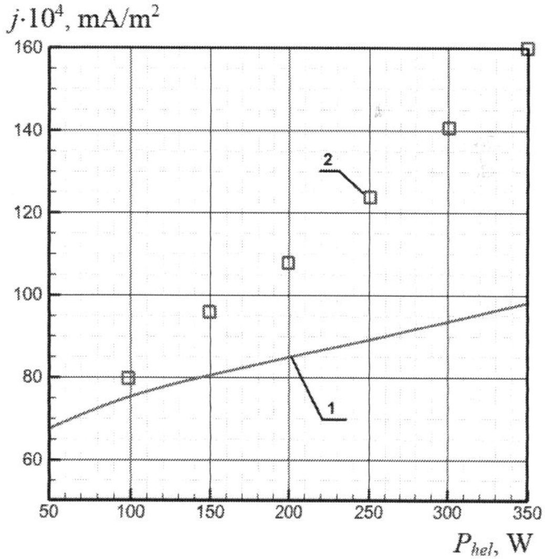

Figure 2. Comparison of calculated (1) and experimental (2) dependences of the ion current density j on helicon power P_{hel} in He as working gas.

The power W_{hel} embedded in the RF plasma source versus the electron density n_e^{Σ} is shown in Figure 4. The experimental and calculated dependencies of practically important characteristics, as the ion current density j (Figure 2) and the electron density in the RF source (Figure 3), show that the error lies in the range of 15–20% for power $P_{hel} \leq 150$ W, injected into plasma.

When the power level is imbedded in the discharge plasma $P_{hel} > 200$ W, there is a maximum in error value >35%. However, in this case, the ion current density j is strongly depending on the electrical parameters of the extraction system which is not taken into account in the mathematical modeling.

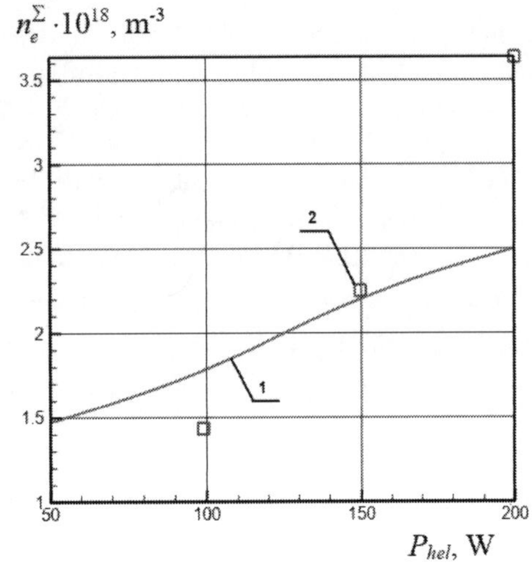

Figure 3. The calculated (1) and experimental (2) electron density n_e^{Σ} as a function of RF power P_{hel} in Ar discharge.

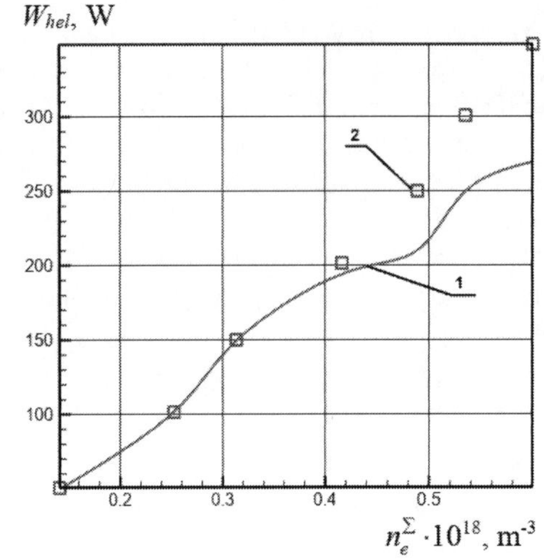

Figure 4. Comparison of calculated (1) and experimental (2) dependencies of the power W_{hel}, embedded in RF plasma ion source in H as working gas.

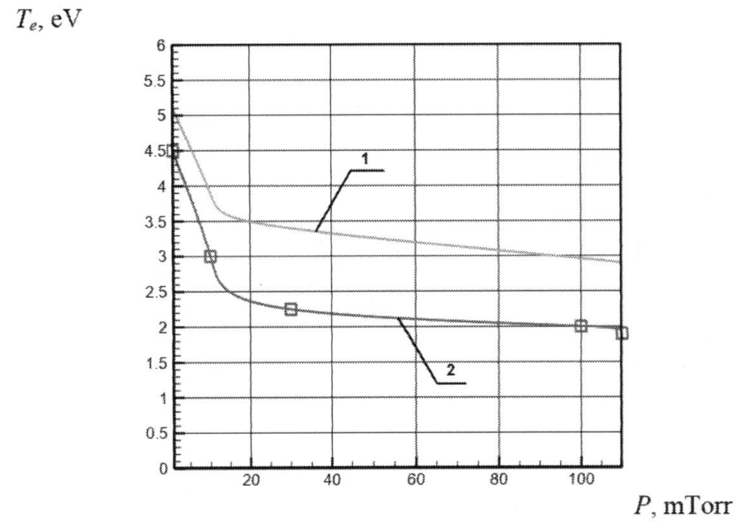

Figure 5. Comparison of 1) calculation with 2) experimental data on electron temperature T_e vs. Ar pressure P. f = 13.56 MHz, P_{hel} = 0.8 kW, $B = 1.2\cdot10^{-2}$ T, R = 0.23 m, L = 0.3 m.

IOP Conf. Series: Journal of Physics: Conf. Series **830** (2017) 012049 doi:10.1088/1742-6596/830/1/012049

Let us consider the following results of power W_{hel} in the RF ion source: experimental and calculated data (Figure 4) agree satisfactory. Figure 5 shows the pressure P dependence of the electron temperature T_e and one can see that the behavior of the curve 1 is qualitatively similar to that of curve 2. At the same time, it is difficult to judge quantitatively because the verification of calculations is not presented in [15].

In general, the comparison gives that the precision of the calculation we have carried out for ion current density j depends on the accuracy of mathematical model describing the process of ion extraction from the end surface of a RF source, especially at $P_{hel} > 200$ W. Increase of calculation accuracy of the energy fraction lost by radiation, ionization and recombination processes may increase the accuracy of the determination of electron density n_e^Σ and temperature T_e.

At the same time, it is clear that the most precise and accurate results can be received by considering the inductive and helicon discharge as non-linear, self-consistent physical system.

Therefore, joining of computer codes of two basic elements of a mathematical model into one code, which allows us to solve the self-consistent problem (to perform an initial assessment) for the calculation of RF electric and magnetic fields in a cylindrical plasma source, is performed for finding the excitation conditions and the analysis of wave absorption mechanism in specific models of plasma source. RF fields were excited by antenna located on the side surface of the plasma cylinder.

Calculations carried out taking into account the thickness of collision δ_c and collisionless δ_a skin layer under experimental conditions [16] shows that the specified values are $\delta_c \approx 0.09$ m and $\delta_a \approx 0.02$ m. This difference between collision thickness and anomalous skin depth allows speaking about a possible non-uniformity of the radial distribution of the electron density n_e^Σ.

Calculated dependence of density, electron and ion temperature in argon plasma on time are given in Figures 6–8 with the following values of electrotechnical and geometrical characteristics of RF source: $P_{hel} = 0.1$ kW, $B_0 = 0.28 \cdot 10^{-2}$ T, $R = 0.075$ m, $L = 0.2$ m. Systems operate at 13.56 MHz.

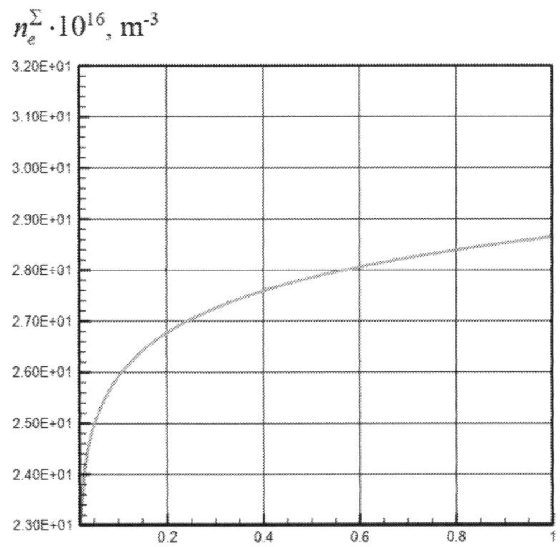

Figure 6. The time dependence of plasma electron density.

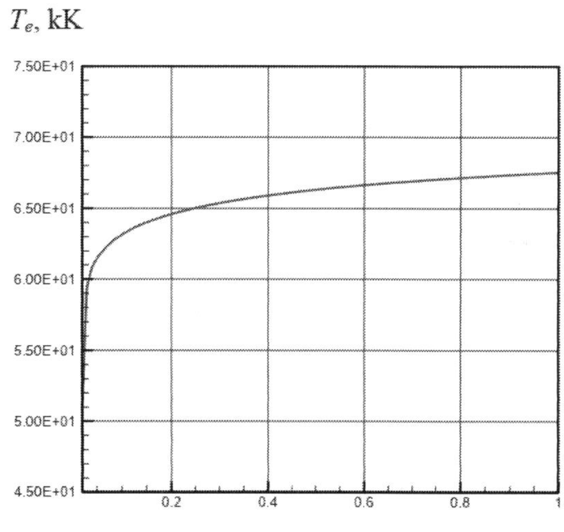

Figure 7. The calculated dependence of electron temperature T_e on time.

An evaluation by the formula from [17] shows in this case ($10^{-3} \geq B_0 \leq 5 \cdot 10^{-3}$ T, $P = 0.67$ Pa, $T_e \approx 7$ eV) that helicon waves excited in plasma are poorly absorbed.

Figures 6–8 show that the most significant changes in heat characteristics are observed at the initial stage ($t \leq 0.05$ s) of energy supply to the low-temperature rarefied plasma source.

In the plasma for the above values of electrical and geometrical characteristics of the RF discharge, a strong discontinuity is observed in the ion T_i and electron T_e temperature: $T_e \approx 70$ kK, and

$T_i \approx 0.7$ kK, i.e. the energy supplied to plasma from an external power generator is mainly accumulated in the internal energy of the electrons and only partially changes the internal energy of ions.

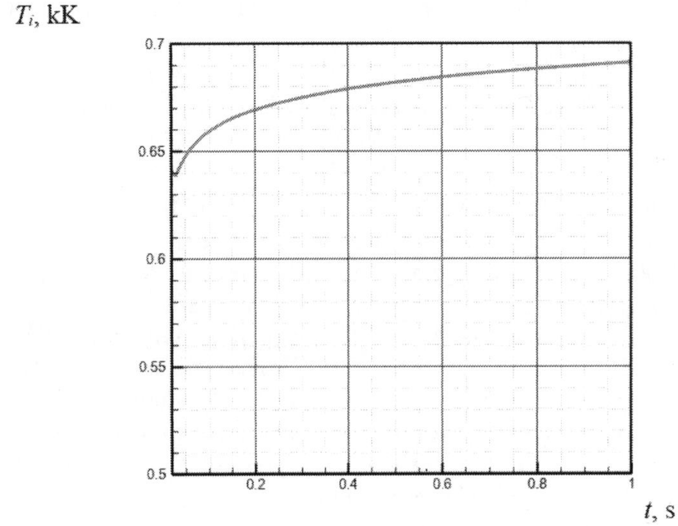

Figure 8. The time dependence of ion temperature.

3. Conclusion

The simplified kinetic scheme, developed by authors, was used to calculate the component composition of molecular hydrogen plasma in the helicon discharge chamber. The kinetic scheme includes the processes of ionization of molecules and hydrogen atoms by electron impact recombination processes, various mechanisms dissociation of hydrogen molecules by electron impact. The results of numerical modeling of the kinetics of ionization and dissociation of molecular hydrogen in a partially ionized plasma, helicon discharge are presented.

The calculations revealed that the most significant factor (along with the formation of a shock wave), limiting the amount of plasmoid velocity in the CMPA channel, is attached mass increasing over time.

References

[1] Melazzi D, Cardinali A, Manente M and Pavarin D 2012 *J. Phys.: Conf. Ser.* **401** 012015
[2] Chen F F 2015 Helicon discharges and sources: a review, *Plasma Sources Science and Technology* **24** 014001
[3] Vavilin K V, Kralkina E A, Nekludova P A, Petrov A K, Nikonov A M, Pavlov V B, Airapetov A A, Odinokov V V, Pavlov G Ya and Sologub V A 2016 *J. Phys.: Conf. Ser.* **669** 012033
[4] Fredriksen Å, Mishra L N, Gulbrandsen N and Miloch W J 2010 *J. Phys.: Conf. Ser.* **257** 012019
[5] Ryzhkov S V, Chirkov A Yu and Ivanov A A 2013 *Fusion Science and Technology* **63** (1T) 135
[6] Fortov V E, Khrapak A G, Khrapak S A, Molotkov V I and Petrov O F 2004 *Phys. Usp.* **47** 447
[7] Ryzhkov S V 2008 *Proc. 35th EPS Conf. on Plasma Physics and Contr. Fusion* ECA **32D** P1.114
[8] Chabert P and Braithwaite N 2011 *Physics of radio-frequency plasmas* (New York: Cambridge University Press)
[9] Kuzenov V V and Ryzhkov S V 2015 *Applied Physics* **2** 37
[10] Kuzenov V V, Polozova T N and Ryzhkov S V 2015 *Problems of Atomic Science and Technology* **4** (98) 49
[11] Kuzenov V V and Frolko P A 2015 *International Workshop on Computer Science and*

Engineering: Information Processing and Control Engineering-2015, WCSE 2015-IPCE 48
[12] Frolko P A 2016 *AIP Conference Proceedings* **1771** 070013
[13] Sivkov A, Shanenkov I, Pak A, Gerasimov D and Shanenkova Y 2016 *Surface and Coatings Technology* **291** 1
[14] Alexenko O V, Miroshnichenko V I and Mordik S N 2014 *Problems of Atomic Science and Technology* **5** (93) 153
[15] Aleksandrov A F, Vavilin K V, Kral'kina E A, Nekludova P A and Pavlov V B 2014 *Applied Physics* **1** 9
[16] Kral'kina E A 2008 *Phys. Usp.* **51** 493
[17] Raizer Yu P 2001 *Gas Discharge Physics* (Berlin: Springer)

Acknowledgments
This work was supported by the Russian Science Foundation (Project No. 16-11-10275).

Modernization of high-power (5 kW) broad ion beam source

D R Emlin[1], N V Gavrilov[1], P V Tretnikov[1], V F Nasyrov[2], A Z Timerbaev[2]

[1] *IEP UB RAS, 106 Amundsen St., Yekaterinburg, 620016, Russia*

[2] *Scientific and Production Association "Technopark of Aviation Technologies", 5/1 Tramvaynaya St., Ufa, RB, 450027, Russia*

E-mail: erd@iep.uran.ru

Abstract. In the course of the long-term performance (during 5 years) of a high-power source of gas ions (25 keV, 0.2 A, 600 cm^2) with a plasma emitter based on cold cathode discharge, the character and rate of key constructive elements faults were determined, which allowed to calculate the inter-repair time, complexity and cost of the repair. The peculiarities of the gas-discharge system and the ion beam forming system limiting the effectiveness of ion beam treatment were revealed as well. Conditions favorable for the decrease in the discharge voltage by 50–200 V and igniting voltage up to 1.5-2 times are determined. The possibilities of lowering the minimal flow of working gas are demonstrated. The design of the discharge system with reduced sputtering rate of local areas of the hollow cathode is offered. The changes added to ion source design aimed to enhance the lifetime of the plasma chamber that is exposed to cyclic heating by the back electron beam leading to the development of through cracks, and to enlarge the rupture life of glow discharge hollow cathode by optimizing its configuration and the conditions of discharge ignition and burning, are described. The upgraded design of a multislit ion-optical system with enhanced performance ensures uniform surface distribution of ion fluence.

1. Introduction

The gas ion source with a large beam cross section was elaborated for modifying large surfaces of materials by ion implantation and for using it in coating deposition processes assisted by ion beam (IBAD) [1]. To generate ion-emitting plasma, the injection of fast electrons into extended cathode cavity (plasma chamber) containing low-area anode was used. The electrons are emitted from plasma of self-sustained glow discharge with hollow cathode through grid electrode and accelerated in double layer of space charge while crossing the grid electrode [2]. As a result of electron oscillations in the cathode cavity, volume plasma is formed whose space inhomogeneity depends on gas pressure across the cathode cavity and method used for feed gas input. It was stated that under one-side injection of electrons into the plasma chamber the acceptable (±10%) inhomogeneity of plasma density on 600 mm length are reached by regulated gas fed from opposite ends of plasma chamber [3]. In order to reduce the rate of ion sputtering of hollow cathode its area was enlarged up to ~ 1300 cm^2. To rise the grid electrode resource, a space between the grid and the cathode aperture was enlarged till 100 mm, as a result, the electron flow after passing cathode aperture of 0.5-1 cm^2 area broadened up to 100 cm^2. 100-fold reduction of emitting plasma density allowed to use a perforated electrode with 4 mm apertures instead of fine-grained grid to solve the problem of the grid resource limited by ion sputtering [4]. The choice of multislit ion-optical system for ion beam formation was specified both by quasi-linear geometry of the beam cross-section and by the need to adjust many apertures at a long distance and to keep the adjustment while the temperature of electrodes varies in a wide range. One-side rigid fixing of electrodes and single rods forming the slits as well as their floating arrangement from the opposite side provided reliable functioning of ion-optical system. The set of required parameters of ion beam (25 – 40 kV, 0.2 A) also impeded its formation as demanded a long (6 – 8 cm) accelerating gap.

All these not only complicated the aperture adjustment, but also led to the increase of the rate of gas ionizing in the gap at plasma-forming gas pressure 0.05 Pa and caused the back flow of accelerated electrons. As a result, heating of the source and exceptional consumption of power of high-voltage supply take place [1]. Substantial extra energy release required changes in design and productivity of the cooling system.

Therefore, in the course of designing and testing of the ion source the main problems hampering its operation were determined and measures for their overcoming taken. In the present work, the lifetime of key elements of the source under conditions of real pilot use in the installation for modifying surfaces of turbine blades was estimated. Results of experiments aimed at the improvement of technical and operational characteristics of the ion source realized in the new source design are described.

2. Service conditions and technical maintenance of the ion source

The ion source is used in vacuum coater «Victoria-2M» for ion-plasma modifying of the details of aviation technique in Scientific-Production Association «Technopark of Aviation Technologies» working in cooperation with Public Joint Stock Company "Ufa Engine Industrial Association" and Ufa State Aviation Technical University during 5 years. The ion source is used for treatment of low and high-pressure compressor blades and allows to provide ion cleaning (etching) and ion implantation of 200 oversized details by one cycle. Modifying the surfaces of gas-turbine motor components by ion implantation leads to enhancement of fatigue strength of titanium alloys by 6 – 10%; nickel alloys by 12 – 14%. Besides, ion beam treatment increases the lifetime of components, their microhardness and adhesion strength after successive coating depositions.

Source parameters are as follows: accelerating voltage is 25 kV, beam current of nitrogen or argon ions is 0.2 A, the initial cross section of the beam is 600 x 100 mm2, the operation gas pressure in treatment chamber equals 0.05 Pa. The ion source is exploited in cyclic mode: 1 – 3 cycles a day, one cycle taking 150 min. Total running hours of the source in continuous mode of beam generation made ~ 3500 hours, on the average, 5 hours a day. The ion source design is shown in Fig. 1. At both ends of water-cooled hermetic case 1 ceramic high-voltage bushing insulators 2 are arranged; at one of them hollow cathode 3 is mounted, at the opposite one – plasma chamber 4 and rod tungsten anode 5 are fixed. Plasma chamber and cathode are placed inside metallic screen 6, which equalizes electric field strength in a high voltage gap and blocks plasma feeding from the gap between grid electrode 7 and cathode 3. The large length of the gap (30 mm) is caused by the significant thermal expansion of the plasma chamber in axial direction when ion source is used for a long time. The ion beam is formed by the ion-optical system consisting of emitter electrode 8 and accelerating electrode 9 and propagated into the treatment chamber through the rectangular aperture in output flange 10.

Routine technical maintenance of the ion source was accomplished by working staff and consisted in periodic (1 time per 6 months, after 600-700 hours of operation) replacement of diaphragm 11 made of stainless steel 12X18H10T in hollow cathode (3, 13, 14) because owing to ion sputtering a diameter of the aperture increased from 8 to ~ 15 mm (Fig. 2). This aperture increase led to the growth of the discharge voltage and required rise of gas flow. Tungsten igniting electrode 16 (Fig. 1) was replaced simultaneously because of its shortcutting on 10 – 15 mm. The reduction of electrode length worsened conditions of the discharge ignition, which also led to the necessity to increase the minimal gas flow. Tungsten rods (Ø2 mm) (12, Fig. 1) of accelerating electrode 9 of ion-optical system (IOS) (8, 9) sputtered by ion beam were periodically (after 1300-1500 hours) replaced. More complicated prophylactic repair was made once after 2500 hours of operation. It included general maintenance or

IOP Conf. Series: Journal of Physics: Conf. Series **830** (2017) 012050 doi:10.1088/1742-6596/830/1/012050

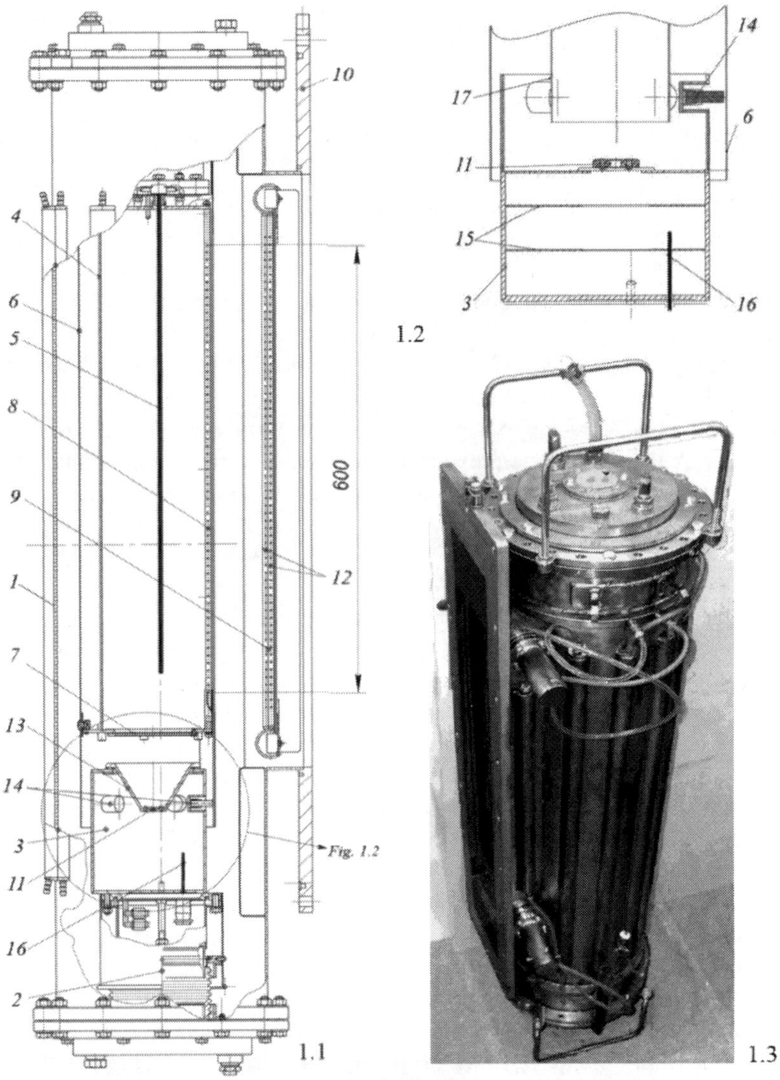

Figure 1. Ion source. 1.1 and 1.3 – general drawing and image of ion source, 1.2 – draft of the cathode.

1 – case; 2 – high-voltage insulating assembly; 3 – hollow cathode; 4 – plasma chamber; 5 – rod anode; 6 –high-voltage screen; 7 – plasma cathode grid; 8, 9 – screen and accelerating electrodes of beam forming system; 10 – output flange; 11 – diaphragm with constricting aperture; 12 – tungsten rods; 13 – conic cathode insertion; 14 – centering insulators; 15 – cathode insertions, 16 – ignition electrode; 17 – anode of constricted discharge.

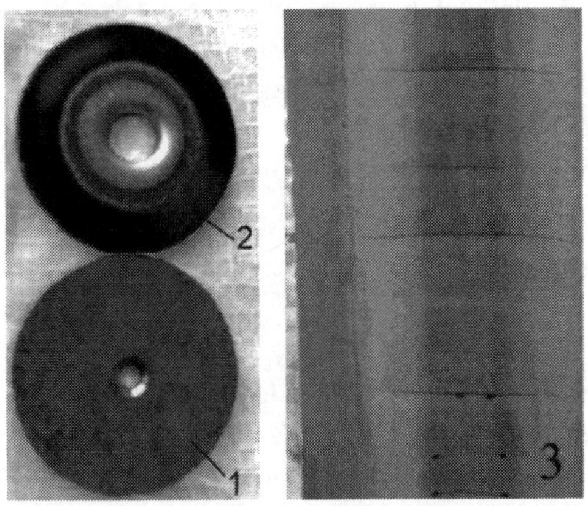

Figure 2. Diaphragm in initial state (1) and after 400 hours of operation (2), segment of plasma chamber, with through cracks (3).

plasma chamber 4 (Fig. 1) and hollow cathode 3 replacement.

Ionization of gas in accelerating gap of the IOS by accelerated ions and formation of back electron flow lead to heating of uncooled case of plasma chamber 4 up to temperatures more than 500°. Multiple cyclic heating and cooling lead to loss of stainless steel plasticity and through cracks development (Fig. 2), hence plasma chamber construction loses its rigidity. The simplest solution of this problem is placing a cylindrical replaceable insert into plasma chamber. Gradual degradation of this insert does not result in emergency effects.

Rather a complicated configuration of the hollow cathode including elements 3, 11, 13, 14 is specified by the necessity of special space for electron flux expansion towards the grid 7; for disposal of fixing

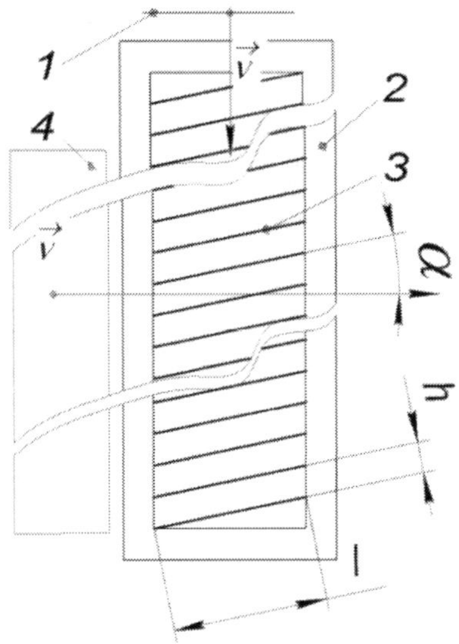

Figure 3. Scheme of ion-optical system with slant disposition of rods. 1– moving collector, 2 –electrodes of IOS, 3- rod, 4 - moving article.

elements of changeable diaphragm 11 with constricting aperture and fixing of insulators 14 to centre position of the plasma chamber. Intensive ion etching of constructive elements of cathode 14 protruding into the cavity resulted in the loss of leak-tightness and final destruction of hollow cathode. Optimized design of the hollow cathode, with a reduced number of intensive sputtering zones, is shown in Fig. 1.2. Arrangement of two cathode insertions 15 with central apertures Ø50 mm inside the cavity provided reduction of discharge voltage from 750 – 800 V up to 600 V; hence, the rate of cathode case sputtering decreased. Sputtering of aperture edges in the insertions does not disturb the source operation.

3. Optimizing the conditions of beam formation and discharge operating

The known disadvantage of the mode of broad ion beam formation by multi-slit ion-optical system (IOS) is that even at high homogeneity of plasma ion emitter such IOS provides almost regular distribution of beam current density along the large axis of its cross section only at definite distance from the beam formation system and definite modes of its operation, conditioned by the combination of beam current and accelerating voltage.

Regularity of distribution is provided as a result of angular divergence of elementary beams formed in singular apertures and their overlapping in the space of beam drift. At less than optimal distances from ion-optical system, irregularity of beam current density distribution is observed, characterized by interchange of maximum and minimum values of ion current density along large axis of the beam. Therefore, even at insignificant change of beam generation modes or at treatment of complicated surfaces fluence of ion irradiation of the surface of articles moving crosswise the long axis of beam section also will be irregularly distributed across the surface.

The proposed way to reduce inhomogeneity of ion beam treatment implies the change of the angle of slope of the slits in ion-optical system electrodes: the angle between slit and short axis of beam's cross section has to differ from 0°. The minimal angle of deviation $\alpha_{\min} = arctg(h/l)$ is conditioned by length l and width h of single slit aperture. As a result, all parts of surface of the article moving across the long axis of the beam receive approximately equal fluence of ion irradiation. Design and operation principle of IOS are shown in Fig. 3.

In Fig. 4 distributions of ion current measured by collector (length l and width $h/4$) moving lengthwise the long axis of the beam for optic with rectangular slit orientation and for optic with angle of slits rotation $\alpha = 9°$ are shown. Measurements were made at 3 cm distance from ion-optical system, at which the level of inhomogeneity of the beam section is high. Even in such conditions the inhomogeneity of fluence made <±15%.

The research on optimizing conditions of ignition and burning of self-sustained glow discharge with hollow cathode was purposed at reduction of minimum value of gas flow, ignition voltage of pulse discharge in cathode cavity and DC voltage, applied between the cathode and anode, providing stable ignition and burning of constricted discharge. The research showed that the change in the potential of diaphragm 11 with outlet aperture imposes significant effect on the conditions of the discharge ignition between the cathode and anode. At cathode potential of the diaphragm disruption of cathode sheath in constricting aperture is required for the discharge development; this is achieved by the increase in the cathode-anode gap's voltage and growth of pulse discharge current. At the anode potential of the diaphragm, conditions of pulse ignition are practically stable; the ignition voltage of

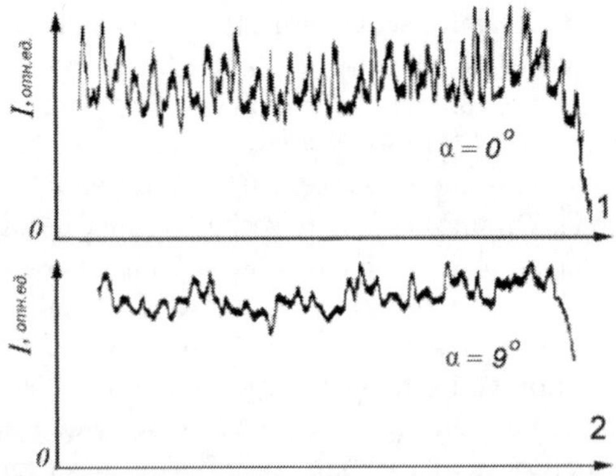

Figure 4. Distributions of ion current to collector moving lengthwise the long axis of IOS. 1 – system with rectangular orientation of slits, 2 – system with $\alpha = 9°$.

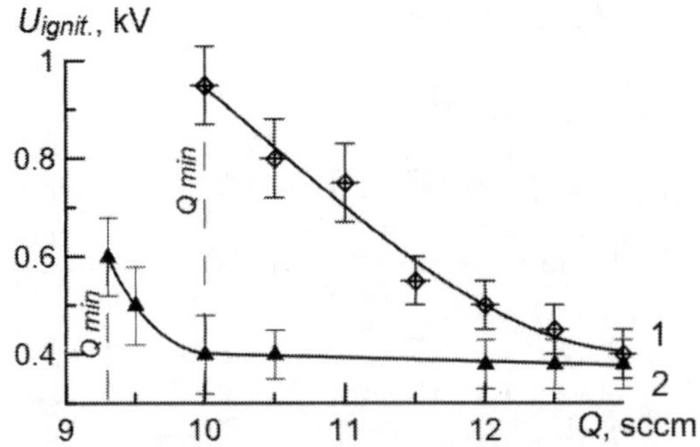

Figure 5. Ignition voltage of constricted discharge vs gas (argon) flow. Potential of diaphragm: 1 – of cathode, 2 – of igniting electrode. Initiating pulse discharge current is 12 A.

the constricted discharge decreases by 300 – 500 V (Fig. 5); and the discharge voltage falls by 50 V (Fig. 6.1). Besides, the diaphragm with the potential close to the anode one (40 – 50 V) is not subjected to ion etching. The effect of cathode insertions (15, Fig. 1) on burning voltage of the constricted discharge is shown in Fig. 6.2. The growth of the amplitude of ignition pulse current from 12 to 50 A leads to the reduction of the minimum flow of feeding gas, at which the constricted discharge ignition is provided, from 11 to 7 sccm.

4. Conclusion

The experience of the long-term (during 5 years) pilot use of the high power source of the broad gas ion beam (25 kV, 0.2 A, 600 cm^2) with a plasma emitter based on the discharge with a cold cathode was summarized. The source is easy and reliable in exploitation and does not require frequent repair. Current prophylactic repair of the ion source held twice a year consisted in replacement of the cathode diaphragm because of the significant increase of outlet aperture dimensions as a result of ion sputtering, and also the change of tungsten rods in accelerating electrode of ion-optical system, sputtering by ion beam.

Regular (biyearly) prophylactic maintenance consisted in repair or total change of plasma chamber and hollow cathode. The loss of plasticity of metal and development of through cracks took place owing to cyclic heating of plasma chamber wall by back electron flow. The integrity of hollow

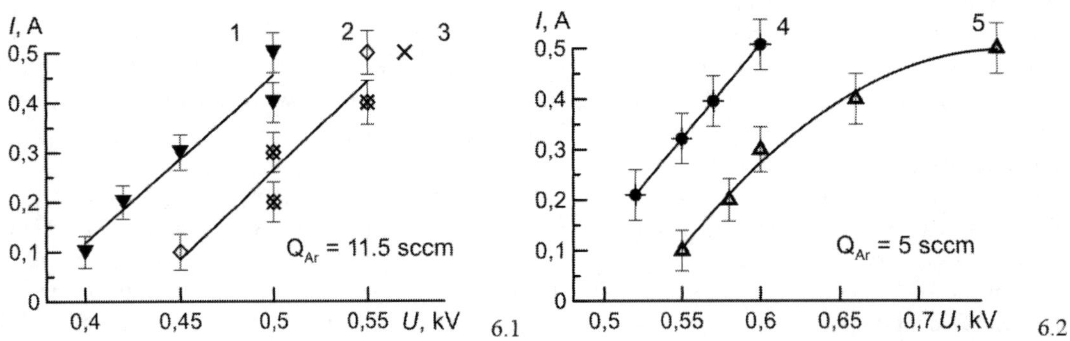

Figure 6. Volt-ampere characteristics of constricted discharge. Diaphragm with constricting aperture has potential: 1, 4, 5 – of igniting electrode, 2 – of cathode, 3 – floating; 4 - cathode with insertions, 5 - without insertions.

cathode design is destroyed because of intensive scattering of protruding parts. In order to increase the lifetime of these components a new design of hollow cathode was elaborated and use of changeable insertion inside plasma chamber was recommended.

The results of experiments allowed to increase the lifetime of the diaphragm with constricting aperture owing to altering its potential from the cathode to anode one and to reduce the inhomogeneity of ion beam treatment using the multislit ion-optical system with the angle between slits and long axes of beam section different from 90°.

References

[1] Emlin D R and Menshakov A I 2012 *Bulletin of the SUSU. Ser. Mechanical Engineering Industry* **20**(33) 131
[2] Vizir' A V, Oks E M, Shchanin P M and Yushkov G Yu 1997 *Technical physics* **42(6)** 611
[3] Gavrilov N V, Kamenetskikh A S, Emlin D R, Bureyev O A and Menshakov A I 2008 *Proc. 9th Int. Conference on Modification of Materials with Particle Beams and Plasma Flows*, Tomsk, Russia. Publishing house IAO SB RAS, pp 7-10
[4] Gavrilov N V and Kamenetskikh A S 2007 *Technical physics* **52(3)** 301

Acknowledgments

The work was supported in terms of the item of State task No. 0389-2014-0006 and partially supported by the Program of fundamental research of Ural Branch of RAS (Project No. 15-17-2-28).

Characteristics of cold atmospheric plasma source based on low-current pulsed discharge with coaxial electrodes

O A Bureyev, Yu S Surkov and A V Spirina

Institute of Electrophysics, Ural Branch of the Russian Academy of Science, 106 Amundsen St., Ekaterinburg, 620016, Russia

E-mail: bureyev@iep.uran.ru

Abstract. This work investigates the characteristics of the gas discharge system used to create an atmospheric pressure plasma flow. The plasma jet design with a cylindrical graphite cathode and an anode rod located on the axis of the system allows to realize regularly reproducible spark breakdowns mode with a frequency ~ 5 kHz and a duration ~ 40 µs. The device generates a cold atmospheric plasma flame with 1 cm in diameter in the flow of various plasma forming gases including nitrogen and air at about 100 mA average discharge current. In the described construction the cathode spots of individual spark channels randomly move along the inner surface of the graphite electrode creating the secondary plasma stream time-average distributed throughout the whole exit aperture area after the decay of numerous filamentary discharge channels. The results of the spectral diagnostics of plasma in the discharge gap and in the stream coming out of the source are presented. Despite the low temperature of atoms and molecules in plasma stream the cathode spots operation with temperature of ~ 4000 °C at a graphite electrode inside a discharge system enables to saturate the plasma by CN-radicals and atomic carbon in the case of using nitrogen as the working gas.

1. Introduction

The source under discussion belongs to the class of devices named in literature «Cold Plasma Jet». This term is used for sources of non-equilibrium atmospheric pressure plasma flow with temperature of heavy particles in the range of 300-1000 K that is at least one order of magnitude lower than the electron temperature in such plasma [1]. Application of these devices allows treatment of thermo-sensitive materials with simultaneous reduction of cost as compared with low-pressure plasma technologies as there is no need to use vacuum installations.

The results of works on modification of materials in cold atmospheric plasma [2,3] show that such treatment most often leads to simultaneous development of surface morphology (increase of roughness after plasma-chemical etching) and increase of hydrophilicity because of doping of polar functional atomic groups containing oxygen and nitrogen onto the surface. One of important advantages of cold atmospheric plasma treatment as compared with thermal treatment or electrolytic processes is modification of the only surface layer without mechanical or thermal damage of the material bulk. Energy consumption and environmental hazard of production process are reduced as well.

Characteristics of the source of a new type creating a jet of cold atmospheric plasma near 1 cm in diameter for different working gases including nitrogen and atmospheric air are studied in this work. The main feature of the source operation is the chaotic movement of cathode spots of low-current pulse discharge inside a cylindrical graphite cathode with the creation of secondary plasma flow after decay of numerous spark current channels.

IOP Conf. Series: Journal of Physics: Conf. Series **830** (2017) 012051 doi:10.1088/1742-6596/830/1/012051

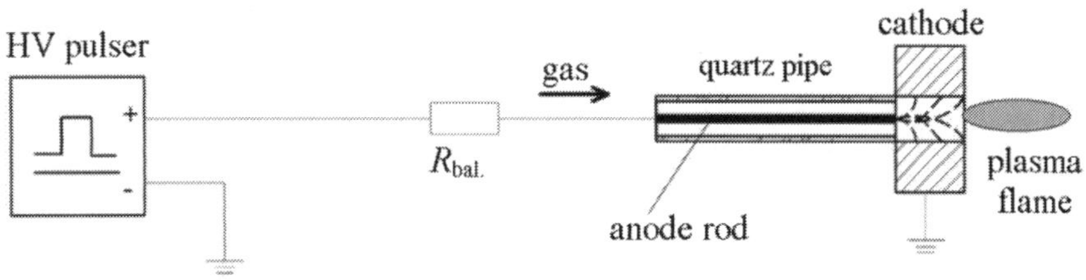

Figure 1. Plasma source diagram.

2.Methods of experiment

Plasma source diagram is presented in figure 1. Discharge system with coaxial geometry contains a steel rod anode on the axis and a hollow tubular cathode made of isotropic graphite mounted at the end of quartz pipe through which the working gas is fed into the discharge gap under excess pressure. In the process of source operation, decaying plasma is forced out of the discharge gap by the flow of the working gas through the cathode aperture ⌀ 8 mm, forming a plasma flame. Technically pure argon and nitrogen were used as working gases as well as atmospheric air of natural humidity pumped by a compressor.

Electric power for discharge system was supplied by a high voltage pulse power source with a pulse rate regulated in the range of 10 Hz – 7.5 kHz. To prevent the discharge from transition into the arc mode, ballast resistor 4 kΩ was installed into the discharge circuit. The amplitude of high voltage pulses of positive polarity was 5 kV. Pulses of open circuit voltage have rise time ~10 μs, the time of descent being approximately 100 μs. The duration of discharge current pulses was 40 μs.

The measurement of average values of discharge current and discharge voltage was fulfilled by means of magneto-electric system devices. Pulse discharge current was measured by 1.3 Ω shunt. Low discharge currents were measured by the potential drop on a ballast resistor. Pulse voltage on discharge gap was measured with high-voltage probe with galvanic decoupling of input and output signals. Current and voltage pulses were registered using a digital oscilloscope.

Spectrograph ISP-30 equipped with photo-electronic cassette MORS-6 with a spectral range 196 - 930 nm was used for spectral diagnostics of the plasma. Exposure could be altered from 240 ms to 25 s. This provided a possibility of spectrum registration in linear regime of variation of the intensity of studied spectrum lines. The measurement was performed from 20 cm distance to the spectrograph slot: in the line of the hollow cathode axis and against the axis of plasma flame expansion.

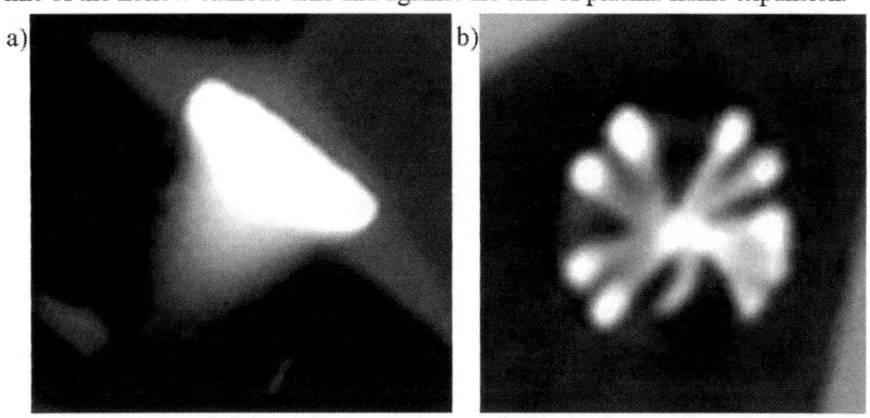

Figure 2. Photos taken along the normal (a) and along the axis of propagation of the plasma flame (b).

3.Results and discussion

Figure 2a presents a photograph of the glow generated by plasma jet of the source when compressed air is used as a working gas. Maximum length of plasma jet as estimated by the zone of visible glow reaches 0.5-2 cm and varies depending on the features of the pulse discharge burning: frequency, amplitude, duration of discharge current pulses, the kind of the working gas and speed of its flow. Operation range of the average discharge current (pulse frequency) in which plasma jet is formed is 50-150 mA (2.5-7.5 kHz). The lower limit is connected with substantial weakening of plasma jet when the area of discharge working zone is decreased at the inner surface of a graphite cathode at low average currents (at low frequency). The upper limit of average current was restricted by maximum power of the used pulse power supply. The gas flow providing stability and cooling (nonequilibrium) of plasma jet varies in the range 1-10 l/min.

The photo of the sequence of radially distributing spark current channels (from the end of discharge system) is presented at figure 2b. The parameters of gas discharge system and gas flow are selected in such a way that chaotic movement of cathode spots of discharge inside the aperture in the graphite cathode generates a flow of secondary plasma distributed upon the total area of cathode outlet after the decay of numerous filamentary current channels. According to [4], low-current cathode spots on isotropic graphite are characterized by chaotic movement and lead to uniform erosion of all the working surface of the electrode.

The increase in the area of the inner surface of the cylindrical cathode at which cathode spots are generated and chaotically move, can be visually observed by the increase in the pulse frequency of the power supply (along with growth of average discharge current). When working at a minimum frequency of power supply ~10 Hz the spark channels are situated only at the shortest distance between the cathode and anode. The point of binding of current channels on the end of rod anode remains at that almost immovable from pulse to pulse.

In figure 3 the comparison of two oscillograms – of discharge current and voltage drop at ballast resistor when working with nitrogen is given for two magnitudes of pulse frequency of power supply – 70 Hz and 5 kHz. At high working frequency providing average discharge current at the level of 100 mA, weak current of glow discharge between pulses of magnitude 2-3 mA is observed. When the high voltage pulse is applied, spark breakdown of the gap happens with the monotonous current increase up to ~ 460 mA. At low frequency (70 Hz) the absence of current between pulses is characteristic and the growth of the main spark current takes place after passing of the pre-breakdown pulse with amplitude 18 mA with delay of the main pulse for 10 μs.

In figure 4 the waveforms of discharge current and voltage on the gap at average current 100 mA and frequency 5 kHz, working on nitrogen, are presented. The voltage of low-current glow discharge between pulses remains practically equal to discharge voltage (~ 500 V) during the main current pulse

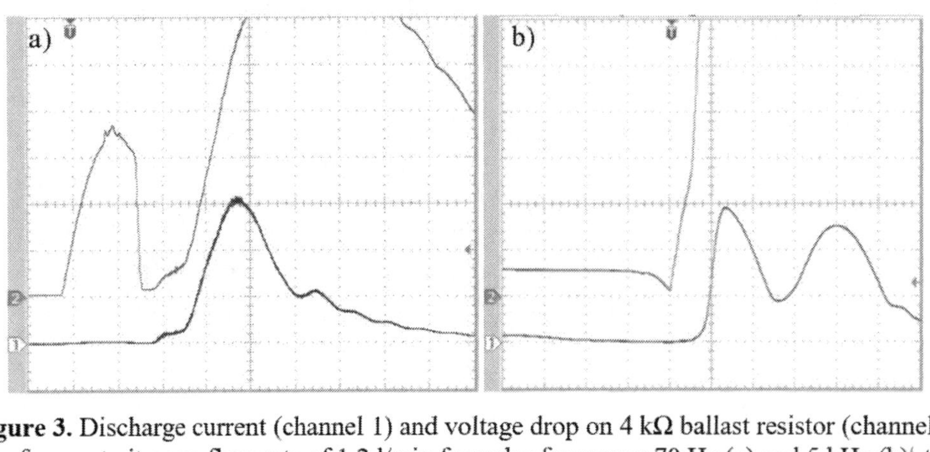

Figure 3. Discharge current (channel 1) and voltage drop on 4 kΩ ballast resistor (channel 2) waveforms at nitrogen flow rate of 1.2 l/min for pulse frequency 70 Hz (a) and 5 kHz (b); time scale: 10 μs/div.; current scale: 150 mA/div.; voltage scale: 20 V/div.

passing. The amplitude and duration of current pulses along with the change of the kind of gas and speed of flow are not substantially altered, being at the level of 500 mA and 40 microseconds (for values of average discharge current 100 mA and frequency 5 kHz).

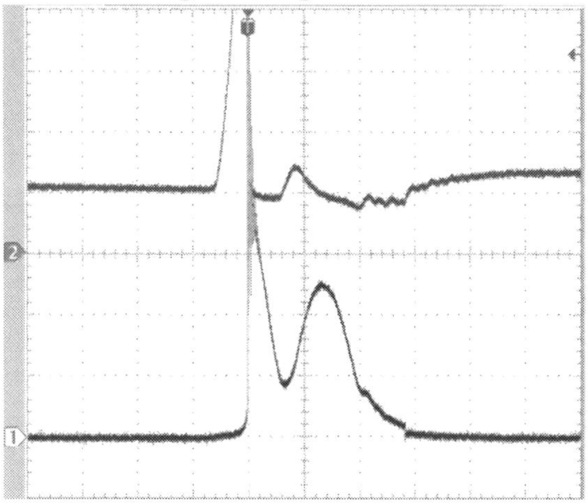

Figure 4. Current (channel 1; 150 mA/div.) and voltage (channel 2; 500 V/div.) waveforms for discharge in nitrogen flow; time scale: 20 μs/div.

Gas temperature in plasma jet was evaluated according to the data of spectral diagnostics basing on rotational temperature of N_2 molecules. Such estimate of temperature is allowed for nitrogen at atmospheric pressure because the condition of equilibrium between rotational and translational temperatures of molecules [5] is fulfilled. The rotational temperature was estimated by 337.13 nm band of the second positive system $C^3\Pi_u \rightarrow B^3\Pi_g$ of nitrogen. The determined temperature of heavy plasma particles is 370 ± 19 K at discharge current 150 mA. The temperature of the gas in the jet may be decreased to 50 °C along with the reduction of the average discharge current and increase in the distance from the cathode outlet.

The data of spectral diagnostics also permitted to evaluate the temperature of substance in cathode spots on the graphite cathode; its maximum value at the conditions of study is ~ 4000 °C. The process of carbon sublimation during the functioning of cathode spots at isotropic graphite allowed to get plasma containing a substantial part of CN-radicals as well as ions of atomic carbon, according to obtained optical emission spectra of plasma jet in nitrogen flow (figure 5). In addition to CN molecules, the highest peaks on the derived spectra of plasma jet were generated by N_2, NO and N_2^+ molecules. For plasma jet in air flow, beside radiation bands of excited molecules CN, N_2, NO and N_2^+, lines of atomic oxygen have sufficient intensity as well. The presence of these components in a plasma jet can be used for modification of properties and for plasma-chemical etching of the surfaces of materials,

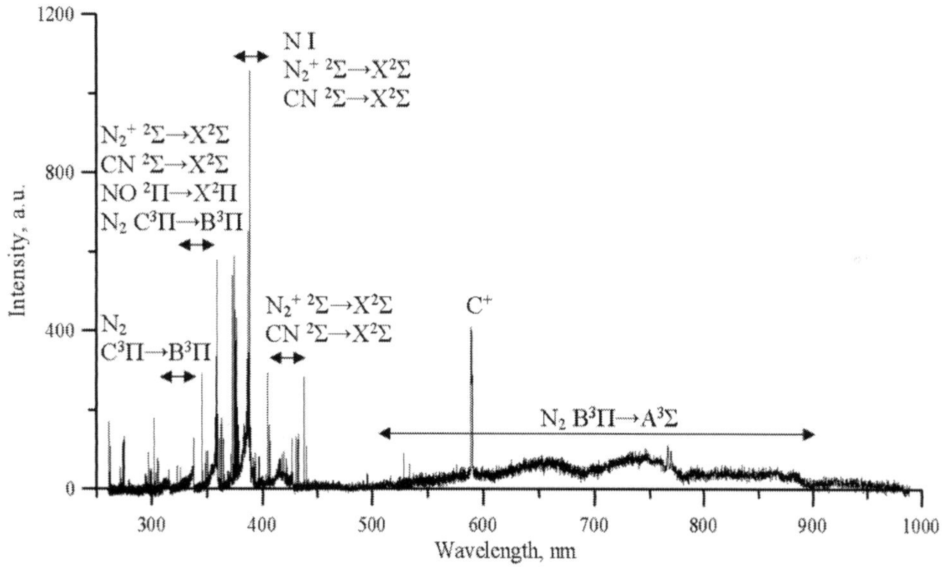

Figure 5. Overview emission spectrum of the plasma jet for nitrogen at 150 mA average current.

including polymers because the developed plasma source permits selection of moderate temperature regime of treatment.

To study the effect of a plasma jet on carbon materials, a series of experiments on treatment of carbon fiber and pyrolytic graphite samples in air and nitrogen plasma was held. In figure 6 microphotos of carbon fiber treated in plasma made with the electronic microscope are shown.

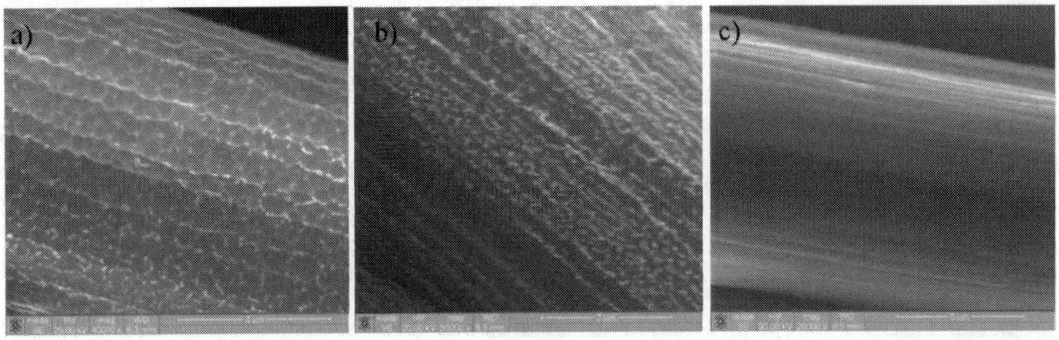

Figure 6. Microphotos of the surface of the carbon fiber after treatment in nitrogen (a) and air (b) plasma; (c) – the untreated carbon fiber surface.

Because of plasma treatment of the samples, development of regular surface relief (figure 6) is observed that leads to the increase in roughness. The results of measurement of roughness parameter R_a, obtained with a probe microscope when scanning 1x1 μm^2 sites on samples of treated and untreated pyrolytic graphite, show approximately threefold increase in roughness after processing by air plasma of the studied source.

4.Conclusion

The characteristics of discharge and plasma of the atmospheric plasma jet source based on pulsed spark discharge with a hollow graphite cathode with coaxial geometry have been studied. Conditions of stable formation of plasma jet on the outlet of the source were determined for different plasma generating gases. It was shown that source plasma on base of air or nitrogen is non-equilibrium and temperature of heavy particles does not exceed ~100 °C at average discharge current up to 150 mA.

In the composition of plasma jet in air or nitrogen flow, beside the main excited molecules specific for this class of devices, a substantial part of CN-radicals was observed. Spectroscopic temperature measurements on cathode surface indicate the presence of cathode spots on graphite having temperature sufficient for carbon sublimation. Functioning of such spots leads to saturation of plasma with CN molecules in nitrogen or air flow.

The hybrid composition of cold plasma jet including besides plasma forming gas, atoms and molecules formed during graphite sublimation allows to enhance the possibility of usage of cold atmospheric plasma sources for materials treatment and coating deposition.

References

[1] Winter J, Brandenburg R and Weltmann K-D 2015 *Plasma Sources Sci. Technol.* **24** 064001
[2] Kusano Y, Andersen T L and Michelsen P K 2008 *J. Phys.: Conf. Series* **100** 012002
[3] Penkov O V, Khadem M, Lim W-S and Kim D-E 2015 *J. Coat. Technol. Res.* **12** 225
[4] Batenin V M, Glina V, Klimovskii I I and Selezneva L A 1991 *High Temp.* **29** 1204
[5] Belmonte T *et al* 2015 *Plasma Sources Sci. Technol.* **24** 064003

Acknowledgments

The work was carried out in the framework of theme of State task No. 0389-2014-0006 and was partially supported by the Program of fundamental studies of UB RAS (project No. 15-17-2-28).

Generation of plasma flows by arc plasmatrons

A S Anshakov[1], P V Domarov[2], V A Faleev[1], A A Danilenko[2]

[1]Kutateladze institute of thermophysics SB RAS, Lavrentiev av. 1, Novosibirsk, 630090, Russia
[2]Novosibisrsk State Technical University, Karl Marx av. 20, Novosibirsk, 630073, Russia

E-mail: anshakov@itp.nsc.ru

Abstract. Plasma technologies of material modification and deposition of coatings with different functional properties can be implemented if the efficient sources of plasma flows are available. These sources are the electric-arc generators of thermal plasma (plasmatrons). The effect of different plasmatron schemes on distribution of temperature and velocity of plasma flows is considered in the current study. For example, it is shown that a partitioned interelectrode insert and gas injection between the sections ensure a plateau in temperature distribution T(r). A similar result was obtained for the plasma flow at the exit of the plasmatron with a stepped output electrode.

1.Introduction

Despite the variety of designs of electric-arc plasma generators, they are divided into two types. The first type includes the jet plasmatrons, and the second one joins the melting plasmatrons. In this paper, we will focus on the jet plasmatrons, designed for surface treatment of products. The latter includes many processes, such as deposition of powder materials of various functional purposes, hardening the surfaces of machine mechanisms, modification of surfaces of the metal and non-metal products, etc.

Each of these and other processes requires high-performance long-life electric-arc plasmatrons, meeting the requirements of a particular technology. To implement any process, a set of certain plasmatron parameters and characteristics, making it possible to achieve the required quality, performance, repeatability, and cost-effectiveness, is required. For this reason, various design schemes of plasmatrons, which can ensure full implementation of the specified process parameters through the use of plasma flows with their individual characteristics on temperature and velocity distribution, are being developed.

2.Features of plasma flows for different plasmatrons

2.1.Plasmatron with partitioned interelectrode insert (IEI).

The plasmatron scheme is shown in Fig. 1 [1]. The interelectrode insert (IEI) is designed to obtain a higher voltage on the arc than that in other plasmatron constructions. To reduce heat losses in discharge chamber walls, cold gas injection g_i between the sections is used.

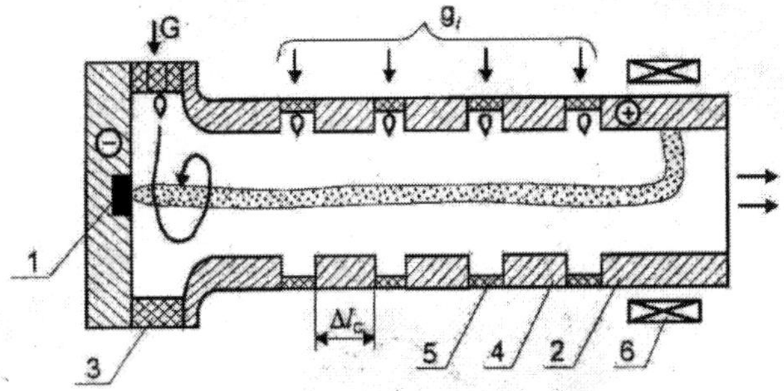

Figure 1. Plasmatron with interelectrode insert (IEI): 1 – cathode, 2 – anode, 3 – unit of main gas flow G input, 4 – IEI section, 5 – unit of intersectional gas flow g_i input, 6 – solenoid.

We will present the experimental data on the study of the plasma flow temperature field for the plasmatron with IEI. The plasmatron power versus arc current is 150-350 kW. The diameter of sections is $10 \cdot 10^{-3}$ m, the diameter of output electrode is $d_A = 14 \cdot 10^{-3}$ m, and IEI length is $a = 0.3$ and 0.4 m. The total air flow rate is $G = 10 \cdot 10^{-3}$ kg/s, $g_i = 0.3 \cdot 10^{-3}$ kg/s. The jet temperature was measured by the method of relative intensities of the spectral lines of copper CuI 5105.5 nm and CuI 5153.2 nm by a quartz spectrograph ISP-28. Local thermal equilibrium in the plasma jet was assumed. Radial temperature distributions in the plasma air jet at different arc currents, measured at distance $z = 1 \cdot 10^{-2}$ m from the nozzle edge, indicate constant temperature along the jet radius with the coefficient of profile flattening from 40% at I = 70 A to 95% at I = 180 A.

The plateaus in the radial distribution of jet temperature are caused by intense turbulent mixing of cold gas, supplied uniformly along the channel length, with gas heated in the arc chamber. In addition, a step in the channel, formed by a sharp transition from IEI diameter to the diameter of the output electrode, also facilitates the alignment of the profile of heated gas temperature at the plasmatron outlet.

The temperature field of the studied plasma jet, plotted by the results of 2-4 temperature measurements in different cross-sections is shown in Fig. 2. The type of radial temperature distributions with plateau T (r) is kept up to high values of z.

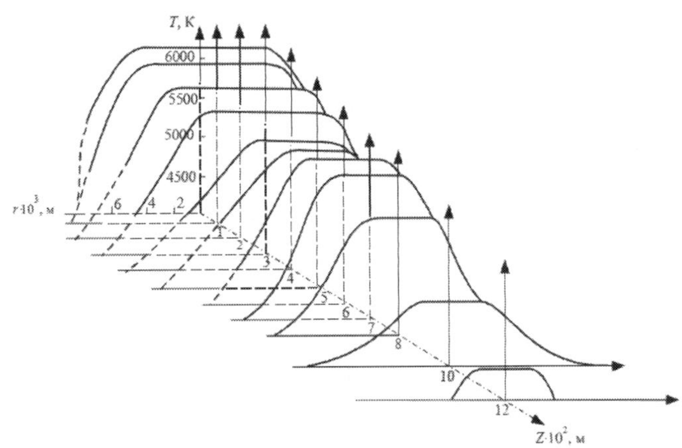

Figure 2. Temperature field of a plasma jet from the plasmatron with a partitioned IEI: $d = 1 \cdot 10^{-2}$ m; $\bar{a} = 30$; $G = 10 \cdot 10^{-3}$ kg/s; $I = 120$ A; $d_A = 1{,}4 \cdot 10^{-2}$ m.

The curves in Fig. 2 indicate that up to $z = (5\text{-}6) \cdot 10^{-2}$ m the temperature profile leveling decreases and then increases again. Moreover, in the zone of the mentioned cross-sections of the jet, the temperature at the axis increases slightly, and the temperature profile in the boundary layer changes (it becomes flatter). The experimental data allowed us to determine the border of a potential core (initial region of the jet) and zone of transition. In our case, the jet core extends to $z = (7\text{-}8) \cdot 10^{-2}$ m. In the initial region, the temperature along the jet decreases slowly (the axial temperature gradient is about 50 C/cm) and its drop is determined by energy removal through radiation. Starting from cross-section $z = 8\cdot10^{-2}$ m, the temperature along the jet axis decreases significantly faster (the axial temperature gradient is 300 C/cm).

To explain the features of radial temperature distributions along the jet, the studied jet was recorded by a high-speed camera SKS-1M. The sequences of frames demonstrate that the end of the high-temperature core is subjected to spiral rotation, caused by residual swirl of stabilizing gas. It leads to more intense mixing of the boundary layer with the high-speed potential core, intensifying boundary layer removal in this area, which, in turn, causes cold gas inflow from the periphery and, as a sequence, temperature reduction in the considered region ($z = (4\text{-}6)\cdot10^{-2}$ m in Fig. 2).

2.2.Plasmatron for deposition of powder coatings.

Plasma deposition of powder materials on the components and products is a common electro-technological process in mechanical engineering, power engineering, electronics, aviation and space technologies. The technology is very effective for deposition of many types of protective coatings. The sources of high-speed and high-temperature flows for powder spraying are the compact electric-arc plasmatrons with the power from 10 to 50 kW. The plasmatrons with partitioned IEI and fixed average arc length are the most common. The plasma-forming gases are: argon, nitrogen, natural gas, air, and mixtures thereof.

The principle scheme of a plasmatron with IEI is shown in Fig. 3. The arc discharge length is determined by the length of IEI. Nitrogen flow rate is $G_1 = 0.8\cdot10^{-3}$ kg/s, argon flow rate is $G_2 = 0.2\cdot10^{-3}$ kg/s.

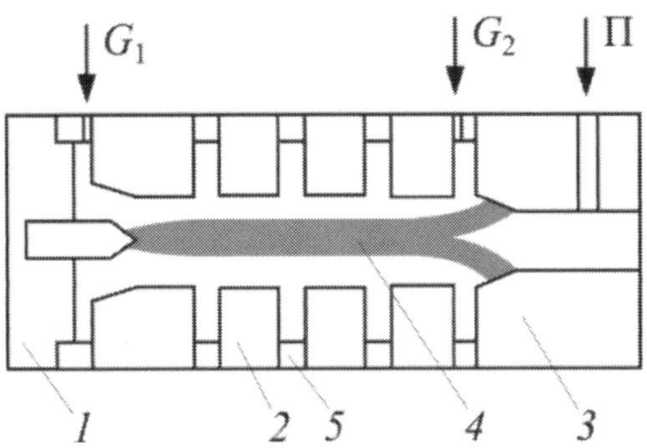

Figure 3. Plasmatron with IEI for spraying: 1 – cathode, 2 – IEI section, 3 – anode, 4 – electric arc, 5 – intersectional insulators, П – point of powder input.

The linear circuit plasmatrons for spraying differ from those with gas-vortex arc stabilization by the absence of swirl gas flow in the discharge chamber. This is caused by transverse dispersion of powder, introduced into the jet at the plasmatron outlet due to exposure of particles to the centrifugal forces. Therefore, axial supply of the working gas is mainly used in the spraying plasmatrons or specific measures are taken to eliminate the circumferential component of the jet velocity.

Comparative data on the plasma flow temperatures in the plasmatron with self-sustained arc length GN-5R (curves 1-3) and plasmatron with IEI (curve 4) are shown in Fig. 4 [2]. The efficiency of plasma-forming gas heating in the plasmatron with a fixed length is obvious.

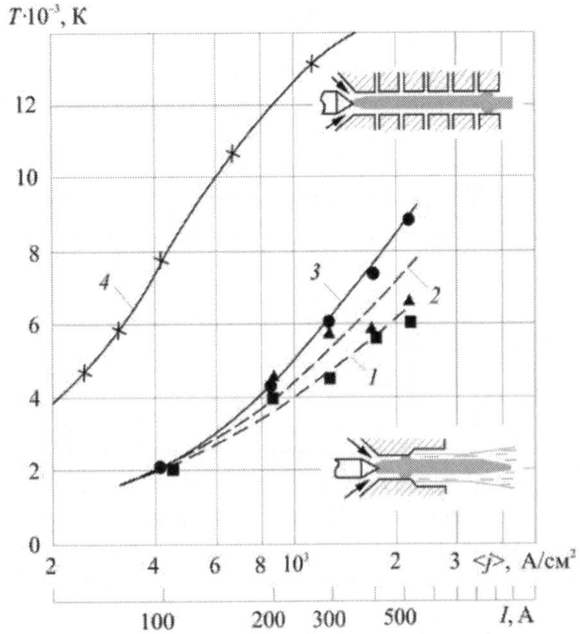

Figure 4. Mean mass temperature of the argon plasma jet for the plasmatron with IEI and GN-5R at the following gas flow rates, x10^{-3}kg/s: 1 – 0.39, 2 – 0.67, 3 – 1.23, 4 – 0.67.

Differences in the parameters and characteristics of discharge glow in plasmatrons with different arc lengths affect the characteristics of the effluent plasma jets, responsible for the properties of coatings: adhesion strength, density, structure, etc. The studies have shown that the shape and characteristics of the jet depend on location of the reference arc spot in the output electrode and have an asymmetrical form. When the anode spot goes to a new location due to the processes of arc bridging, gas flow pulsations or other reasons, the jet parameters at the point of powder supply change and it affects the quality and repeatability of the process. In a partitioned plasmatron, the jet is symmetrical and more compact. Consequently, the angle of jet opening with the powder is smaller than that of the plasmatron with the tubular or stepped output electrode. Accordingly, the use factor of powder and coatings quality increase.

The plasma jet parameters were studied experimentally on the plasmatron with the interelectrode insert EDP-167, developed at the Institute of Thermophysics SB RAS (diameter of IEI sections is $8{\cdot}10^{-3}$ m, IEI length is $60{\cdot}10^{-3}$ m, and diameter of the anode nozzle is $6{\cdot}10^{-3}$ m) using a water-cooled Gray probe. At that, time-averaged dynamic pressure Δp and enthalpy increment Δh were determined. The probe was moved along plasma jet axis z and its radius r using a coordinate device. Radial distributions of enthalpy and dynamic pressure near the nozzle have a sharp peak at the jet axis and with an increase in coordinate z these profiles flatten out noticeably. Distributions of Δp and Δh along the jet axis for argon and nitrogen-argon mixture are shown in Figs. 5 and 6.

It is evident that enthalpy at the argon jet within the acceleration area and powder particles heating is considerably lower than that of the nitrogen-argon jet. This result is a consequence of an increase in the plasmatron power with increasing percentage of nitrogen in the plasma-forming gas (at the constant arc current and gas flow rate, the transition from argon to nitrogen leads to an increase in discharge voltage more than twice).

2.3.Plasmatron with the stepped output electrode.

The simplest scheme of plasmatron is shown in Fig. 7 [1]. An abrupt channel expansion (step) generates the gas-dynamic conditions for predominant arc bridging just after the zone of flow detachment and ensures the constancy of its average length in a wide range of current strength, gas flow rate and pressure for constant l and d.

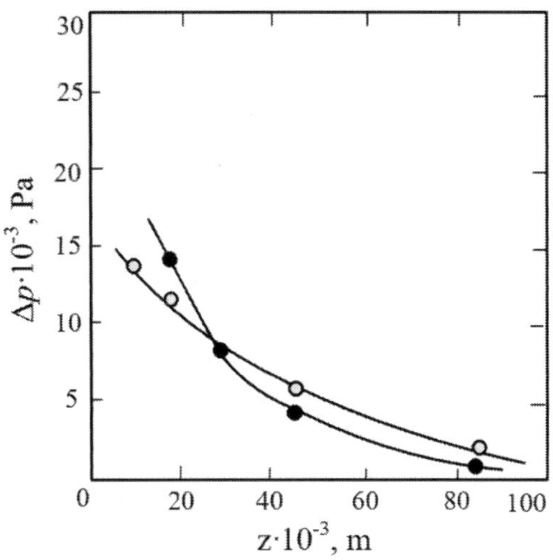

Figure 5. Distributions of dynamic pressure along the plasma jet. I = 100 A, $G = 1{\cdot}10^{-3}$ kg/s; o – Ar, • – 80% N_2 + 20%Ar.

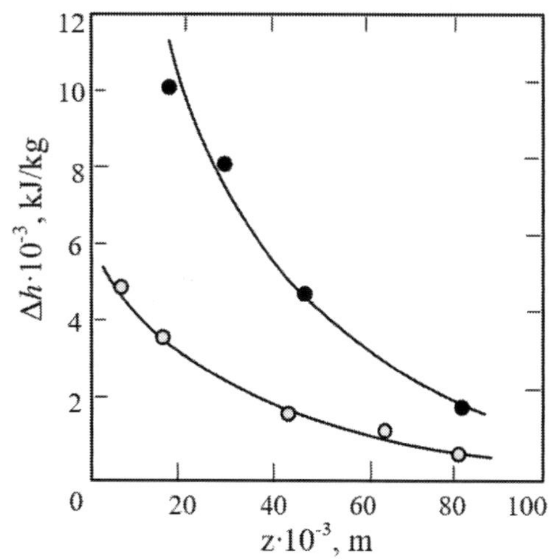

Figure 6. Distribution of enthalpy along the plasma jet.I = 100 A, $G = 1{\cdot}10^{-3}$ kg/s ; o – Ar, • – 80% N_2 + 20%Ar.

According to spectral and pneumatic measurements [4], plateau T(r) in temperature distribution along the radius is observed with 10-% accuracy in the plasma jet core at the outlet of nozzle with diameter D. The profiles of plasma flow are uniform (Fig. 8), and this proves intensive mixing of the hot and cold gases at flow separation on a step in the output electrode.

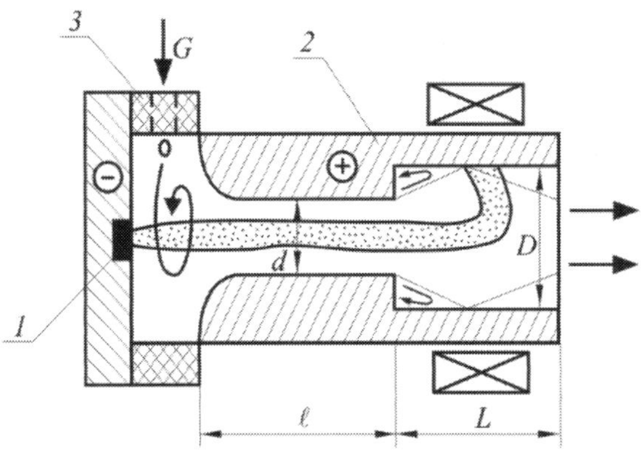

Figure 7. Scheme of plasmatron with fixation of the average arc length by a step: 1 – inner electrode; 2 – outer electrode of stepped geometry; 3 – vortex chamber for the gas input with flow rate G.

The plasma flows of the two-jet plasmatron are of a particular scientific and practical interest [1]. The arc discharge beyond the electrodes increases significantly the flow temperature of different plasma-forming media and opens a broad perspective of their practical application in the field of modifying the product surfaces.

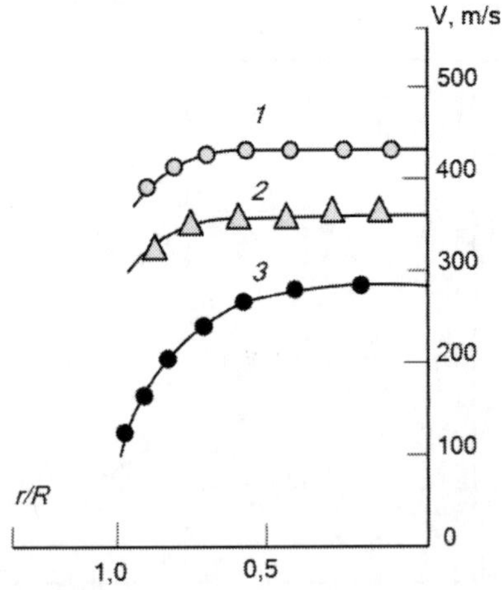

Figure 8. Radial distribution of plasma flow velocity. $D = 40\cdot10^{-3}$ m, $G = 40\cdot10^{-3}$ kg/s, $I = 500$ A. Distance from the nozzle edge: *1* – $z = 0.5D$, *2* – $z = D$, *3* – $z = 2D$.

3.Conclusion

The interconnection of a plasmatron structural scheme with plasma flow parameters is shown. It was determined for different designs of jet plasmatrons that the plasma flow structure is determined by the flow of plasma-forming gas in the discharge chamber.

The investigation results are required not only for application to a specific technology, but also for understanding the physical processes of arc glow at forced heat exchange with the plasma-forming gas.

References

[1] Cherednichenko V S, Anshakov A S, Kuzmin M G 2011 *Plasma Electrotechnological Installations* (Novosibirsk: NSTU)

[2] Donskoy A V, Klubnikin V S 1979 *Electroplasma Processes and Installations in Mechanical Engineering* (Leningrad: Mashinostroenie)

[3] Timoshevskiy A N, Vasilkovskaya A S, Ginal S V 1977 *VII Proc. All-Russian Conf. on Generators of Low-Temperature Plasma* vol 1 (Alma-Ata: Energy Institute) pp 242–245

Acknowledgments

This work was financially supported by the Russian Ministry of Education and Science, according to Subsidy Agreement No.14.607.21.0118 (unique project identifier RFMEF160715X0118).

Operational testing of self-heated cathode made from compacted TiN powder

N V Gavrilov, A S Kamenetskikh, D S Koleukh, S N Paranin, V S Pasmurov, A V Spirin

Institute of Electrophysics, 106 Amundsen str., Yekaterinburg, 620016, Russia

E-mail: alx@iep.uran.ru

Abstract. The results of tests of a self-heated hollow cathode made by magnet-pulse pressing of the mixture of TiN (90 %) and Ti (10 %) powders with further high-temperature annealing and fusing during operation of the compact as a cathode in high-current (10 - 45 A) discharge are presented. It was found that the rate of the cathode mass loss during operation in Ar/N_2 mixture made $2.3*10^{-7}$ g/C. The possibility of the cathode use for oxygen-argon plasma generation at separated gas feeding (argon – through cathode cavity, and O_2 – to anode area of discharge) was shown. Testing of massive tubular cathodes with the increased thickness of the wall (up to 2.5 mm) and large inner diameter (up to 12 mm) possessing an enhanced resource (300 - 500 h) was carried out.

1. Introduction

High-current (10 - 100 A) discharge with a self-heated hollow cathode (SHHC) is used in physical and chemical vapor deposition for ion sputtering, activation of plasma-chemical processes in a working chamber and for ion assistance [1]. SHHC are most often made of high-melting metals and lanthanum hexaboride. The lifetime of such cathodes is restricted by the intensive erosion of the high-temperature active zone (this area is usually ten times smaller than area of the cathode). The active zone is shifted in the direction of the outlet aperture of the hollow cathode with the growth of the discharge current, and its area is reduced. The rate of cathode erosion increases as a result. The erosion rate of niobium cathode makes ~ $3 \cdot 10^{-7}$ g/C under discharge currents near 10 A, which restricts the lifetime of a tubular cathode Ø 8 mm with the wall 1 mm thick to 250 h [2]. Lifetime of a self-heated cathode can be enhanced by increasing its diameter and wall thickness. However the high cost of cathode materials traditionally used for SHHC (tantalum, niobium) makes hollow cathodes noncompetitive in comparison with rod thermo-emission cathodes even at attaining acceptable lifetime. The authors' work experience with SHHC of titanium nitride (TiN) showed high effectiveness and the stability of emission of such cathodes in nitrogen-argon mixtures and comparable with metallic cathodes rate of erosion ~ $1 \cdot 10^{-7}$ g/C [3, 4]. However, the method of TiN cathodes preparation by high-temperature nitriding of titanium tube does not permit to significantly increase the cathode wall thickness because of the reduction of the nitriding rate. In the present work, the method of production of tubular cathodes with a wall 2.5 mm thick and inner diameter up to 12 mm by magnet-pulse pressing of TiN powder with further annealing in nitrogen flow and agglomeration in high-current discharge plasma is described. Phase composition of compacts in the initial state and after 70 h tests as SHHC was studied. The results of tests of such a cathode at discharge currents up to 45 A in DC mode and up to 100 A in AC mode are shown.

2. Experimental Technique

2.1. Preparation of hollow cathode from titanium nitride

Cathodes were made in the form of tubes of two type-sizes, one having external diameter 11.1 mm and thickness 2.4 mm, and the other – 16.9 mm and 2.6 mm, correspondingly. The mixture of powders, containing TiN (Technical Conditions 6-09-112-75, produced by JSC «Vecton») with grain no more than 100 μm, 10 wt.% of Ti with grain no more than 50 μm (State Standard 17746-79) and 2 wt.% of polyvinylbutyral (PVB), was used as precursor. Polymer and metallic components were introduced as binding and plastifying agents. The mixture was prepared in a form of suspension on the base of isopropanol by mixing of components in a low-speed device of «drunken barrel» type with steel balls. Dried obtained agglomerate was milled in a ball grinder by steel balls and then sifted. Only the fraction smaller than 40 μm was used. Optic microphotograph of composite powder particles are shown in figure 1. They have an irregular form of splints with clear edges not changing under the treatment of powder in blender and ball grinder.

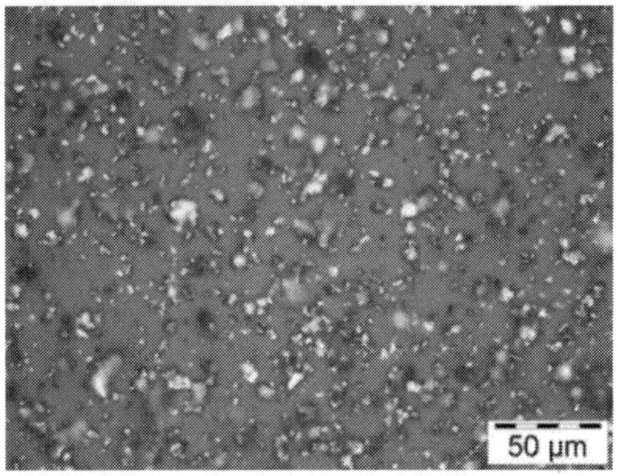

Figure 1. The composite powder particles.

Tubular blanks from composite powder were formed by a magnet-pulse pressing method [5]. A steel spiral inductor with a working channel Ø 22.5 mm was connected to a pulse current generator on the base of capacitive energy accumulator PCG-135 (425 μF, 25 kV). Press-stem with a layer of composite powder placed between external copper casing and coaxially located form-generating rod of tempered steel was installed into the inductor channel. The press-stem was encapsulated and pumped till residual pressure reached ~ 5 Pa, which level was maintained during all the process of pressing. Powder was pressed due to the compression of the copper casing under the action of pulse magnetic field of inductor. Magnetic field with amplitude up to 25 T provided pressure up to 0.25 GPa on the conductor. After pressing tubular compacts were extracted from press-stem and annealed in a vacuum furnace in nitrogen flow during 1 h at pressure 2.7 Pa and temperature 600 °C for thermolysis of PVB and removal of its products from the compact.

2.2.Methods of tests of a self-heated hollow cathode

The tests of TiN compacts as SHHC were accomplished in a facility shown in figure 2. Tubular compact *1* with inner, external diameters and height: 6, 11 и 70 mm, correspondingly, was placed inside of water-cooled case *2*, thermally isolated from other elements of the system and used as a hollow cathode. A hollow anode *3* Ø of 80 mm and 100 mm height was installed opposite the outlet aperture of hollow cathode *1*. A mesh with 1.2 x 1.2 mm cells was placed at the flat end of the hollow anode. The gas mixture with the ratio of Ar/N_2 flows rate 28/5 (sccm) was leaked through the cathode cavity. Ar (30 sccm) was fed through the cathode cavity and O_2 (7 до 30 sccm) directly into vacuum chamber *4* in tests with Ar/O_2 working gas. Gas mixture pressure in the vacuum chamber made 0.2-0.5 Pa. The glow discharge with current up to 4 A and voltage ~ 500 V was ignited to provide hollow cathode heating to temperatures at which

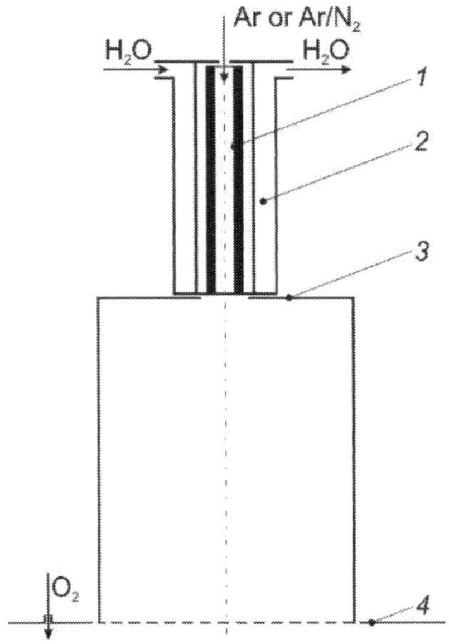

Figure 2. The facility for tests of SHHC. *1* – SHHC, *2* - water-cooled case, *3* – hollow anode, *4* - vacuum chamber.

thermionic emission becomes the main mechanism of the discharge maintenance. The discharge voltage reduced to values less than 100 V in this regime. The value of the discharge current was set in 5 – 45 A range. SHHC was tested in AC discharge which provided pulse heating of the near surface layer of the cathode (several tens of μm thick) by $10\text{-}10^2$ ℃ and the increase in the discharge current by several hundreds of A [4].

The duration of cyclic tests of SHHC was from 0.5 to 6.0 h. Mass changes for the compacts were measured after every cycle by means of analytic scales VL-210 (Gosmeter) with 0.0001 g discretion.

Phase composition of SHHC material was studied by means of XRD using diffractometer XPert PRO (PANalitical) in copper radiation with beta-filter on diffracted beam. X-ray patterns were exposed to total-profile analysis using XPert High Score Plus software. The coherent-scattering region (CSR) was determined by Scherer's method based on reflexes in narrow angles of dispersion (form factor 0.9).

Compacts density and porosity were measured by the attested method of hydrostatic weighing. The density was determined based on the results of weighing a sample fragment in air and in water. Compacts were covered with a thin waterproof layer of nitrocellulose for weighing in water. A fractional error of density measurement did not exceed 1 %. Open porosity was determined by water-consuming. Compacts were saturated by water in a vacuum chamber with water feeding and boiling. A fractional error of open porosity measurement did not exceed 4 %.

The structure of compacts in the initial state and after the tests was studied by means of optic microscope Olympus BX51TRF-5 with maximum magnification 100. Fragments of compacts were grinded in the plane of the cross section of the tube by means of grinding-polishing mill Phoenix Beta 1c (Buehler, Germany) using diamond suspension with dispersion not more than 1 μm in the finish phase.

3. Results and discussion

The appearance of compacts before tests and after operation in discharge during 70 h is shown in figure 3. Density and porosity of initial compacts made 3.56 g/cm^3 и 29.6 %, correspondingly. During test procedure, the outer diameter of the cathode was reduced to 9.1 mm in the medial part, corresponding to the active zone, to 10.1 mm on the upper flat-end and to 10.9 mm on the lower flat-end (initial values being: 11.1, 10.1, 11.2 mm, correspondingly). The compact density grew up to 3.93 g/cm^3, and porosity was reduced to 25.8 % after the tests.

The dependence of SHHC mass on the electric charge of DC discharge in Ar/N_2 mixture is presented in figure 4. At the initial period (shown by dashed line at the figure 4) when the charge is less ~ $3.6 \cdot 10^5$ C,
The

Figure 3. The compacts before (at the left) tests and after (at the right) operation in discharge during 70 h.

cathode mass increased due to the conversion Ti →TiN. The process of nitrogen diffusion through the nitride layer is described by relation: $(\Delta m/S)^2 \sim Kt$, where Δm is the mass of consumed nitrogen, S is the area of diffusion, t is the period of nitrogen consuming [6]. The constant K of nitrogen consuming rate by titanium, as estimated by test results, made $1.8 \cdot 10^{-8}$ $g^2/cm^4 \cdot s$, that is several times less than cited values [7]; that can be connected with the restricted access of nitrogen to the compact bulk.

The increase in weight corresponds to total transition of 10 % Ti addition into TiN of stoichiometric composition, which is confirmed by the results of XRD. XRD patterns of the compact after vacuum annealing at 600 °C and of the sample after cyclic tests as SHHC for 70 h are shown in figure 5. The position and intensity of peaks of TiN phase with FCC lattice and of α-Ti correspond to a standard sample [8, 9]. The portion of TiN and Ti phases makes 92.4 and 7.6 wt. %, and the average size of the CSR – 85 and 25 nm, correspondingly. Peaks of α-Ti phase are absent on the XRD pattern of SHHC sample passed cyclic tests. Crystallites of TiN have lattice constant 0.424 nm that corresponds to defect-free samples synthesized in equilibrium conditions [8]. The average size of crystallites grew up to 115 nm due to sample exposure at high temperatures.

SHHC operation after conversion of Ti into TiN is accompanied by mass reduction (see figure 4) caused by the erosion of the cathode inner surface in the gas discharge. The rate of erosion made $2.3 \cdot 10^{-7}$ g/C independently of the discharge current. The measured value is comparable with the magnitude of the erosion rate of SHHC produced by the method of nitriding of Ti tube in a discharge [3]. At comparable values of erosion rate, the cathode made from TiN powder possesses higher lifetime owing to increasing the wall thickness from 1 to 2.5 mm. The SHHC from TiN compact, unlike the Ti cathode, does not require long-continued training in N_2 flow which provides the formation of heat-resistant TiN across the overall thickness of the cathode wall.

The calculated time of increasing the erosion depth to 1mm makes ~500 h under discharge current 10 A for SHHC with inner diameter 6 mm. Estimation was done under the hypothesis that cathode erosion takes place within the limits of 3.5 cm^2 active zone. The cathode lifetime could be significantly enhanced in the case of the increase in the cathode diameter or acceptable erosion depth. It must be noted that the erosion rate of hollow cathodes made of pure high-melting materials, i.e. tantalum [10] in Ar atmosphere is close to that obtained in this work (~ 10^{-7} g/C).

The cross-section of samples are shown in figure 6. A dense layer of material near 0.1 mm thick had been formed in the active zone from the inner side of the tube surface while samples were tested as SHHC. Closed pores with linear sizes from units till tens of μm are present in SHHC bulk. The porous structure with crystallites of unequal size is observed in an untreated compact sample. Thus, during cathode operation at characteristic temperatures of 2000 - 2200 °C agglomeration of powder compact takes place, which allows to make massive cathodes of TiN with enhanced wall thickness and increased lifetime.

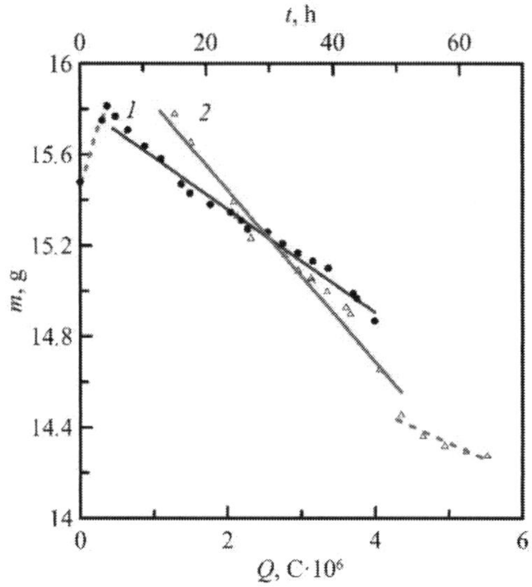

Figure 4. The dependence of SHHC mass on electric charge of DC discharge in Ar/N_2 (*1*, dashed segment of *2*) and Ar/O_2 (2) mixture.

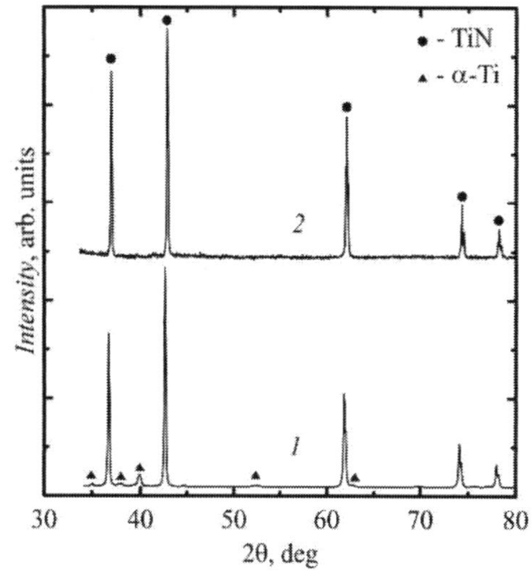

Figure 5. XRD patterns of compact after vacuum annealing at 600 °C (*1*) and of sample after cyclic tests as SHHC for 70 h (*2*).

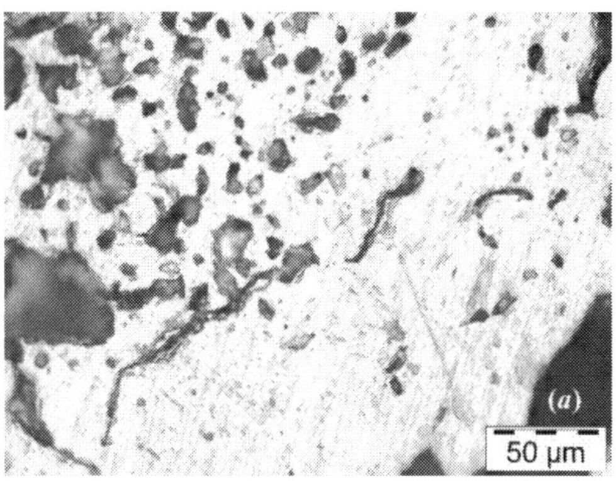

Figure 6. The cross-section of the sample tested as SHHC (*a*) and untreated sample (*b*).

The SHHC from compacted TiN powder was tested in an electron source generating the assisting plasma in a magnetron sputtering system for reactive deposition of oxide coatings. SHHC operates steadily under partial O_2 pressure in a working chamber up to 0.5 Pa. The duration of working cycles made 2 - 6 h. It is known that the lifetime of metallic thermo-emission cathodes is significantly reduced at O_2 partial pressure more than 10^{-5} Pa. [11]. Poisoning of cathodes based on metal borides (i.e., LaB_6) in the presence of both O_2 or N_2 leads to multiply reduction of the cathode emission current [12]. The result of tests of SHHC from compacted TiN powder during operation in Ar/O_2 atmosphere is presented in figure 4. It was determined that the erosion rate of SHHC grew by 1.6, as compared with the regime of work in Ar/N_2 mixture, and reached $3.6 \cdot 10^{-7}$ g/C. It was shown in the work [13] that intensive oxidation of TiN begins at temperatures more than 700 - 800 °C. The compounds of Ti and O_2 transform into a liquid phase when temperature exceeds 1900 °C (the characteristic of the working range of tested SHHC) [14]. Obviously, the erosion rate of SHHC of TiN increased during the operation in Ar/O_2 gas mixture because of the formation and sublimation of titanium oxides. However, the rate of this process in experimental conditions is relatively low. SHHC had operated for 70 h in Ar/O_2 atmosphere (carried charge $4*10^6$ C) and further tests of the cathode were held in Ar/N_2 mixture. In figure 4 the segment of dependency *2* marked with dashed line corresponds to the additional stage of tests. As can be seen the rate of cathode erosion was reduced to values close to value for SHHC that operated in Ar/N_2 mixture only. Thus, the tested SHHC is able to restore its resource characteristics after a long-term work in Ar/O_2 atmosphere that points at the absence of total poisoning of the cathode bulk.

In figure 7 the current-voltage characteristic of the discharge with SHHC operating in Ar/N_2 mixture is shown. Transition into the regime of thermo-electronic emission accompanied with the reduction of the discharge voltage to 100 V takes place after cathode heating in glow discharge plasma (4 A, 500 V).

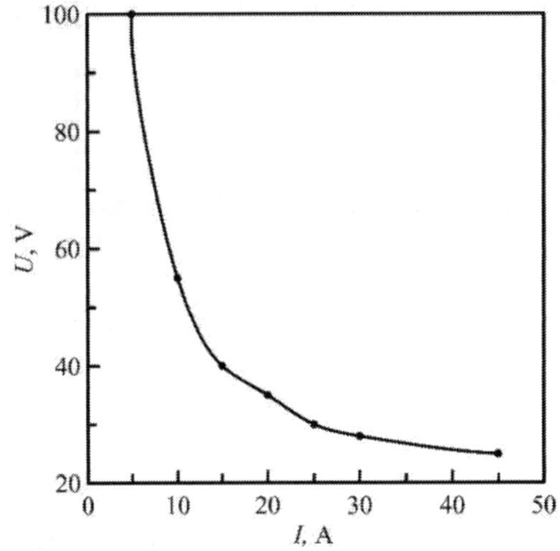

Figure 7. The current-voltage characteristic of discharge with SHHC operating in Ar/N_2 mixture.

Discharge voltage monotonously decreased to ~25 V as discharge current grew up to 45 A. The oscillograms of current and voltage of discharge in high-current pulsed mode are shown in figure 8. The discharge is stable in a wide range of frequency (1-1000 Hz) and pulse duration (10-1000 μs) if the surface temperature of the cathode during a pause does not fall below minimum level providing initial current of thermo-electronic emission when voltage pulses are applied across the discharge gap. The value of initial current depends on conditions of discharge transition into low-voltage burning regime supported by thermo-electronic emission of the cathode. The initial current made ~ 4 A in experiments and minimum necessary temperature of cathode surface was ~ 1700 °C. The pulse current of the discharge reaches 100 A and the discharge voltage makes 240 V at frequency 200 Hz and pulse duration 200 μs.

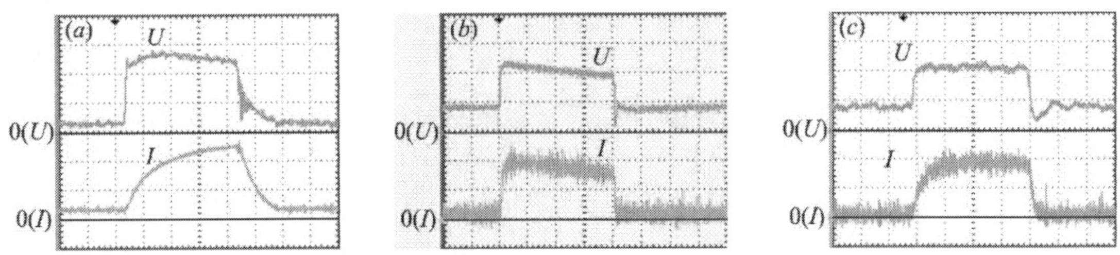

Figure 8. Oscillograms of current (*I*) and voltage (*U*) of discharge with SHHC made from TiN. Frequency: *a*, *b* – 200 Hz; *c* – 1 kHz. Pulse duration: *a*, *c* – 200 μs; *b* – 1 ms. Scale division of *I*: *a* - 50 A/div; *b*, *c* – 10 A/div; and of *U* - 100 V/div.

IOP Conf. Series: Journal of Physics: Conf. Series **830** (2017) 012053 doi:10.1088/1742-6596/830/1/012053

4. Conclusion

Testing of SHHC made of titanium nitride by the method of magnet-pulse pressing of powder material in the form of tubes with walls 2.5 mm thick, inner diameter 6 and 12 mm was accomplished. The work of tubular compacts as a SHHC provides their agglomeration across all the wall thickness that allows forming of mechanically strong massive cathodes of TiN with the enhanced wall thickness and increased operational resource.

Cathodes show stable work in a DC mode under the following conditions: discharge currents ~ several tens of A, discharge voltage less than 100 V and in pulse mode in a wide range of frequencies (10-1000 Hz) and pulse duration (10-1000 μs) at average current ~ 4 A, current amplitude up to 200 A and discharge voltage up to 300 V. It was shown that SHHC of titanium nitride is suitable for stable operation in Ar flow at high (up to 0.5 Pa) partial O_2 pressure in the anode area of discharge. The rate of cathode erosion makes $2.3 \cdot 10^{-7}$ and $3.7 \cdot 10^{-7}$ g/C at plasma generation in Ar/N_2 and Ar/O_2 gas mixtures, correspondingly, which provides the lifetime not less than 500 and 300 h (supposed that erosion depth does not exceed 1 mm) under discharge current 10 A.

References

[1] Muhl S., Pérez A. 2015 *Thin Solid Films* **579** 174
[2] Gavrilov N V, Men'shakov A I 2011 *Instruments and Experimental Techniques* **54(5)** 732
[3] Gavrilov N V, Men'shakov A 2012 *Technical Physics Letters* **38(11)** 1031
[4] Gavrilov N V, Men'shakov A 2016 *Technical Physics* **61(5)** 669
[5] Lee J G et al. 2015 *Metallurgical and Materials Transactions A* **46A** 3139
[6] Gulbransen E.A., Andrew K.F. 1949 *Trans. Am. Inst. Min. (Metal) Eng.* **185** 741
[7] Samsonov G V et al. 1972 Metodi polucheniya, svoistva i primenenie nitridov. (Kiev: Izdatel`svo IPM AN USSR, ONTI) (in russian)
[8] Wong-Ng W et al. 1987 (NBS, Gaithersburg, MD, USA., ICDD Grant-in-Aid) CAS Number: 12033-66-8
[9] Sailer R, McCarthy G 1993 (North Dakota State University, Fargo, North Dakota, USA., ICDD Grant-in-Aid) CAS Number: 7440-32-6
[10] Grishin S D et al. 1977 TVT **15(4)** 906 (in russian)
[11] Kresanov V S at al. 1987 Visokoefectivniy emitter electronov na osnove gesaborida lantana (Moscow, Energoatomizdat) (in russian)
[12] Gallacher H.E. 1969 *J. Appl. Phys.* **40** 44
[13] Samsonov G V 1969 Nitridi (Kiev, Naukova dumka) (in russian)
[14] Hazin L G 1970 Dvuokis` titana (Leningrad, Himiya) (in russian)

Acknowledgement

The study was carried out within a state program task No.0389-2014-0006 and was partially supported by the Program of fundamental research of Ural Branch of RAS project № 15-17-2-28.

Low-pressure low-frequency inductive discharge with ferrite cores for large-scale plasma processing

M V Isupov[1], A V Fedoseev[1], G I Sukhinin[1,2], V A Pinaev[1]

[1]Kutateladze institute of thermophysics SB RAS, Lavrentiev av. 1, Novosibisrsk, 630090, Russia
[2]Novosibisrsk State University, Pirogova 2, Novosibirsk, 630090, Russia

E-mail: isupovmikhail@yandex.ru

Abstract. Electrophysical characteristics of a low–frequency (100 kHz) low–pressure (10–100 Pa) argon inductive discharge with ferrite cores have been investigated for discharge currents of 1–50 A, discharge chamber diameter of 230 mm. The dependencies of electric field strength on the argon pressure and discharge current were measured. Radial profiles of electron density and temperature were determined with double electric probes. Experimental results were compared with numerical results obtained by a self-consistent kinetic model of the low-frequency ICP, based on the simultaneous solution of a non-local Boltzmann equation for electron energy distribution function and balance equations for the metastable argon atoms density and gas temperature. Comparison of numerical and experimental results is presented.

1.Introduction

Low-pressure, high-density radio-frequency inductively coupled plasma (RF ICP) is widely used in many areas of semiconductor industry for plasma etching, plasma enhanced chemical vapour deposition, and plasma-immersion ion implantation. However, the principle of ICP generation has a few limitations that significantly complicate the scaling up of ICP sources for large-scale plasma processing systems (for example, for the future 450 mm wafers technology) [1]. ICP has a weak magnetic coupling between the ICP coil and plasma ($k\approx0.2$–0.7), a low power factor of the coil ($\cos\varphi<1$) respectively, therefore the use of a resonant-matching network is necessary to improve the ICP power transfer efficiency, leading to high current and high voltage in the coil. The high coil current leads to high power losses in the coil and matching network, while the high coil voltage affects plasma through a capacitive coupling between the coil and plasma, leading to an intense ion bombardment of the discharge chamber. As the coil voltage is increasing with the coil size, this undesirable effect becomes a growing problem for large-scale plasma systems. Another limitation of ICP is a high driving frequency of its generation (typically, 13.56 MHz). As a result, a standing wave effect may arise with the increase in the ICP coil length, disturbing plasma uniformity (which is very important for plasma processing).

These limitations of RF ICP can be overcome with the enhancement of the magnetic coupling between the ICP coil and plasma by a closed ferromagnetic core [1]. In this case, the magnetic flux is concentrated almost in the ferromagnetic core, therefore magnetic coupling and power factor of the coil are high, leading to the increase in the ICP power transfer efficiency. It also allows to separate the ICP coil and discharge chamber in space, thereby the capacitive coupling between the ICP coil and plasma is eliminated. As the magnetic permeability of the core is high, the driving frequency can be significantly decreased, to a few kilohertz, thereby the standing wave effect is also eliminated. As a result, new devices for large-scale plasma processing can be developed on the basis of the low-pressure low-frequency inductively coupled plasma sources enhanced with ferromagnetic cores [1].

It is necessary to underline that existing models of RF ICP cannot be used for the low-frequency ICP with a ferromagnetic core due to a principal difference in discharge configuration: in the low-frequency ICP with a ferromagnetic core, the discharge current is directed axially to the discharge chamber while in RF ICP – azimuthally. That is why the development of a model of the low-pressure low-frequency ICP enhanced with ferromagnetic cores and its experimental verification is an important task.

2.Experimental setup

A principal scheme of the experimental setup is shown in figure 1. The discharge chamber of the setup consists of a wide (230 mm internal diameter, 1 meter length) part 1 and a narrow (40 mm internal diameter) *U*-shaped part 2 forming together with the part 1 a closed path for discharge current. The gas discharge chamber is made of stainless steel water cooled sections insulated and sealed with silicon rubber gaskets. On the *U*-shaped part 2, eight ferrite cores 3 with six turns of primary winding 4 are mounted. All primary windings are connected in parallel to a power supply 6 (50–100 kHz, 500 V) through a matching network 5 (variable *LC* circuit). To pump out the chamber a fore-vacuum pump 7 is used. Gas pressure is controlled with a leak valve (not shown) and measured with a MKS baratron 626A PR4000B 8. Discharge current is measured with a Rogowski coil 9. To determine an average electric field strength *E*, inter-section voltage *U* is measured with a voltmeter 10 ($E=U/L$, where *L* is the discharge path). To determine electron density and temperature, a double Langmuir probe 11 made of tungsten wire with a diameter of 200 μm is used. The probe can be moved both along the discharge chamber axis and radius to determine a spatial distribution of plasma parameters in the discharge chamber.

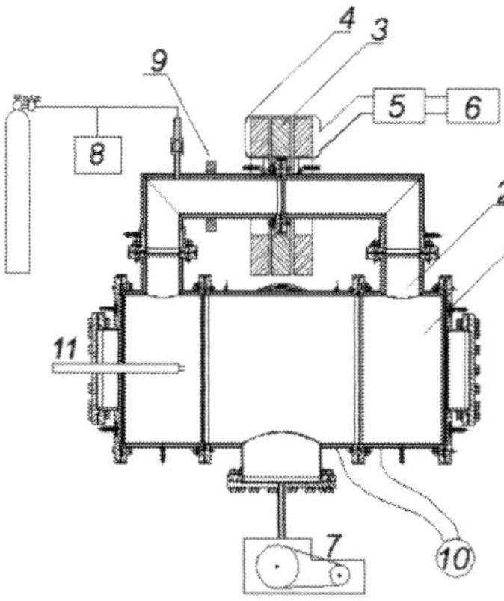

Figure 1. Experimental setup: 1 – wide part of gas discharge chamber (ID of 230 mm), 2 – narrow U-shaped part of the chamber (40 mm ID), 3 – ferrite cores, 4 – primary winding, 5 – matching network (variable LC circuit), 6 – power supply 100 kHz, 500V, 7 – fore-vacuum pump, 8 – MKS baratron, 9 – Rogowski coil, 10 – voltmeter to measure inter-section voltage, 11 – Langmuir probe.

3.Numerical model

A one-dimensional radial model of the low-frequency argon ICP has been developed based on the assumption of plasma stationarity and uniformity in axial direction. To calculate a radial distribution of the electron component of plasma, equation (1) for isotropic part of electron energy distribution function (EEDF) $f_0(\varepsilon, r)$ is used [2], where $\varepsilon=U-e_0\cdot W(r)$ is the total electron energy, U is a kinetic electron energy, $W(r)$ is a radial distribution of plasma potential, $N_g(r)$ is a radial distribution of gas density, $T_g(r)$ is a radial distribution of gas temperature, $Q_d(U)$ and $Q_k^{in}(U)$ are the scattering cross sections of electrons in elastic and inelastic collisions with gas atoms, with the total momentum losses $H(U)=N_gQ^d(U)+\Sigma_kN_gQ_k^{in}(U)$:

$$\frac{1}{r}\frac{\partial}{\partial r}\left[\frac{rU}{3H(U)}\frac{\partial}{\partial r}f_0(\varepsilon,r)\right]+\frac{\partial}{\partial\varepsilon}\left[2\frac{m_e}{M}U^2N_gQ^d(U)\left(f_0(\varepsilon,r)+\frac{3}{2}\frac{\partial f_0(\varepsilon,r)}{\partial\varepsilon}T_g(r)\right)\right] \quad (1)$$

$$-\sum_k UN_g Q_k^{in}(U) f_0(\varepsilon, r) + \sum_k (U + U_k^{in}) N_g Q_k^{in}(U + U_k^{in}) f_0(\varepsilon + U_k^{in}, r) + S = 0$$

The last term S in equation (1) takes into account the processes of direct, stepwise and Penning ionisations. The inelastic collisions in argon are divided into four groups (k = 1–4), namely, the allowed transitions are grouped and have a cross section with an 11.623 eV threshold, two forbidden transitions are with a 12.9 eV threshold, for 13 forbidden transitions the threshold is 11.273 eV, and the ionization potential is 15.6 eV. For simplicity, the radial distribution of plasma potential $W(r)$ is taken as a parabolic function of radius, with a wall potential of 12 V, near the threshold for metastables excitation. This assumption is used to describe the radial distribution of plasma potential in DC glow discharges [3].

Anisotropic parts of the EEDF are related to the isotropic part with equations (2) and (3):

$$\frac{\partial f_0}{\partial r} - e_0 E_r(r)\frac{\partial f_0}{\partial U} + H(U) f_r = 0 \qquad (2)$$

$$-e_0 E_z \frac{\partial f_0}{\partial U} + H(U) f_z = 0 \qquad (3)$$

where E_r and E_z are radial and axial components of electric field strength.

Solving the Boltzmann equation (1) for isotropic part of EEDF and equations (2), (3) for anisotropic parts, macroscopic plasma parameters can be determined as the corresponding integrals of EEDF: a radial distribution of electron density $n_e(r)$, a radial distribution of electron temperature $T_e(r)$, radial and axial components of electron flux density $j_r(r)$ and $j_z(r)$.

Metastable density $N_m(r)$ is described by a balance equation (4):

$$\frac{\partial N_m}{\partial t} - \frac{1}{r}\frac{\partial}{\partial r}\left(r D_m \frac{\partial N_m}{\partial r}\right) = S_{im}(r) + k_{rec} n_e(r) n_i(r) - k_{em} n_e(r) N_m(r) - 2k_{mm} N_m^2(r) \qquad (4)$$

where S_{im} is a rate of metastable atoms excitation, k_{rec} is a rate constant of radiation recombination (n_i is an ion density), k_{em} is a rate constant of stepwise ionisation from metastable atoms, D_m is the metastables diffusion coefficient, k_{mm} is a coefficient of Penning ionization. Here we assume the condition of electrical neutrality, n_e=n_i.

Also we take into account gas heating by the electron current and heat conduction to the discharge tube wall. For this purpose we solve the following thermal balance equation:

$$\frac{\partial T_g}{\partial t} - \frac{1}{r}\frac{\partial}{\partial r}\left(r\kappa_g(T_g)\frac{\partial T_g(r)}{\partial r}\right) = P^{el}(r) \qquad (5)$$

where k_g is the heat conductivity coefficient, P^{el} is the gas heating due to elastic collisions of electrons and argon atoms. Gas heating may result in changing of the radial profile of the gas density $N_g(r)$ thus affecting the isotropic and anisotropic parts of the EEDF. Rate constants S_{im}, k_{em} and gas heating P^{el} are calculated by the integration of EEDF with the corresponding cross sections.

3.Results and discussion

In figure 2, the dependencies of electric field strength on argon pressure are shown, for various discharge currents. Also, literature data [4] for a low-frequency ICP enhanced with ferrite cores are shown for comparison. It is seen that $E(p)$ dependence has a minimum at the pressure of about 100 Pa (except for the lowest discharge current of 1 A). The same pressure dependence was observed for the low-frequency ICP in a 50 mm discharge tube [4]. For argon pressure of 100 Pa, calculations give nearly the same values of the electric field strength of about 0.5 V/cm.

In figure 3, calculated EEDFs are shown on the discharge chamber axis, for argon pressure of 100 Pa and discharge currents of 1 and 10 A. It is seen that calculated EEDFs are non-Maxwellian: the tails of EEDFs are depleted for electron energies higher than the threshold of argon excited states (11.3 eV). The number of electrons with energies higher than argon ionization potential is negligible, the main ionization mechanisms are multistep and Penning ionizations (that is why the wall potential is chosen near the threshold for argon metastables excitation).

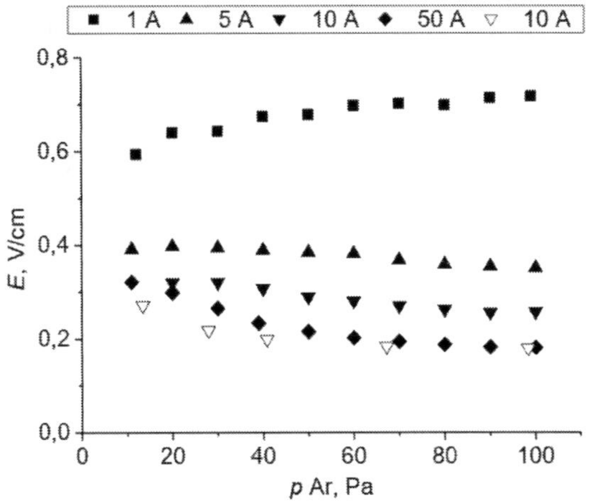

Figure 2. Electric field strength vs. argon pressure. Black symbols – experimental results, hollow symbols – literature data [4].

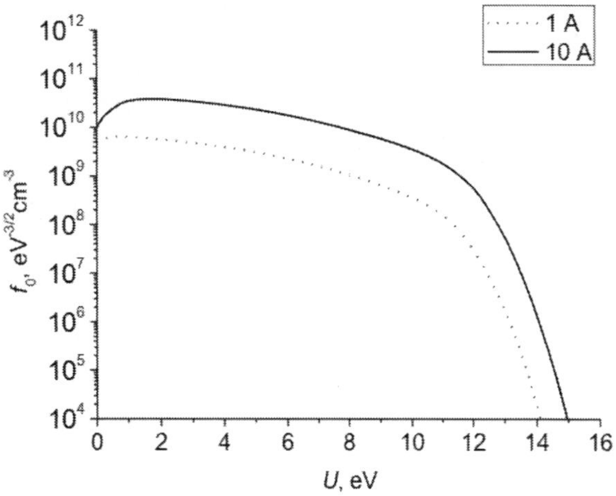

Figure 3. Calculated EEDF on the discharge chamber axis, argon pressure of 100 Pa.

In figure 4, calculated and experimentally measured radial distributions of electron density are shown, for argon pressures of 100 and 30 Pa, discharge currents of 1 and 10 A. The asymmetry of experimentally measured radial profiles of electron density is caused by a branch pipe placed on the front side of the gas discharge chamber (not shown in figure 1). The branch pipe was used to insert a double Langmuir probe into the discharge chamber radially, thus plasma was spread into the branch pipe resulting in a higher value of electron density in the near wall region. Calculated radial profile of the electron density is more contracted than experimentally measured due to a higher gas pressure, which leads to the decrease in the electron flux to the tube walls and increase in the gas temperature and current density on the discharge chamber axis.

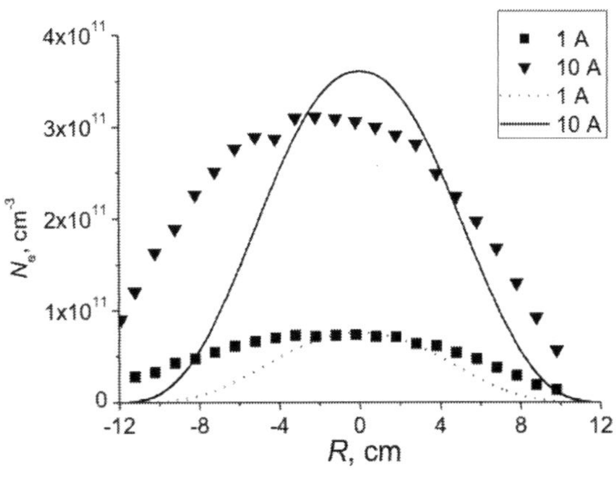

Figure 4. Radial distributions of electron density. Symbols – experimental results (Ar pressure of 30 Pa). Lines – numerical results (Ar pressure of 100 Pa).

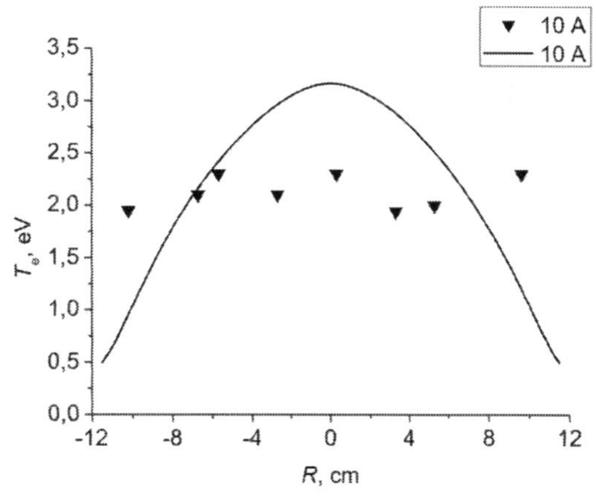

Figure 5. Radial distribution of electron temperature. Symbols – experimental results (Ar pressure of 30 Pa). Line – numerical results (Ar pressure of 100 Pa).

In figure 5, calculated and experimentally measured radial distributions of the electron temperature are shown, for argon pressures of 100 and 30 Pa, and the discharge current of 10 A. It is seen that experimentally measured values of the electron temperature are almost constant along the radius of the discharge chamber, while calculated radial profile has a parabolic-like distribution, due to the abovementioned assumption of plasma potential as a parabolic function of radius. Parabolic profile of electron temperature also leads to the contraction of gas discharge, while "flat" electron temperature means that ionization occurs even in the near wall region (figure 4). The observed discrepancy between experimental and numerical radial profiles of electron temperature indicates that the assumption of parabolic distribution of plasma potential may be incorrect for the case of the low-pressure low-frequency ICP with a ferrite core. It is necessary to underline that the assumption was made for DC glow discharges with a tube diameter of a few centimeters [3], while the low-frequency ICP discharge chamber diameter is 23 centimeters.

In figures 6, 7 the calculated radial distributions of gas temperature and density are shown, for argon pressure of 100 Pa, discharge currents of 1 and 10 A. It is seen that gas heating does strongly affect the radial profile of gas density at the pressure of 100 Pa, which in turn affects all equations (1)–(5). For example, decreasing the gas density N_g leads to an increase in the current density on the discharge chamber axis due to increasing the electron drift velocity $v_e(E/N_g)$, which gives a contribution to the calculated effect of contraction of the electron density profile.

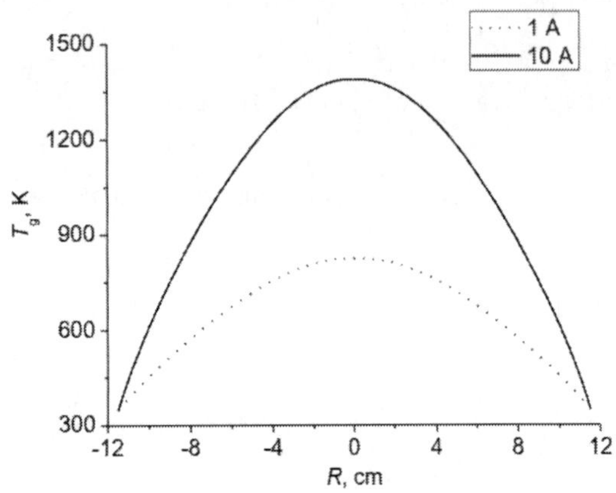

Figure 6. Calculated radial distribution of gas temperature, argon pressure of 100 Pa.

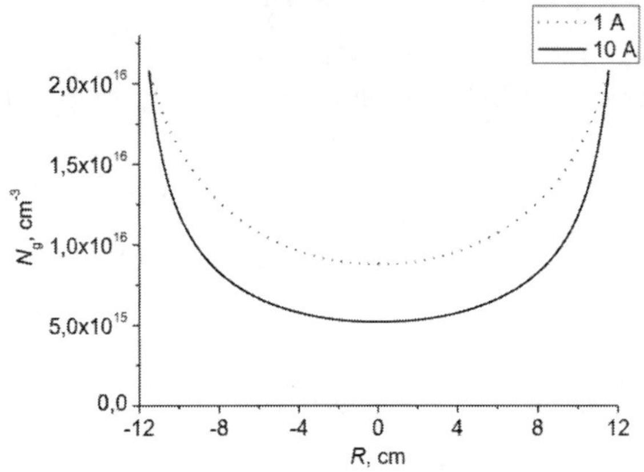

Figure 7. Calculated radial distribution of gas density, argon pressure of 100 Pa.

4.Conclusion

A radial model of the low-pressure low-frequency ICP in argon, based on the simultaneous solution of a non-local Boltzmann equation for EEDF and balance equations for metastable argon atoms density and gas temperature has been developed. EEDF is found to be non-Maxwellian, with a depletion of EEDF for energies higher than a threshold for metastable excitation. Numerical results are compared with the results of electric probe measurements. A model assumption of parabolic plasma potential taken from the case of DC glow discharges seem to be incorrect for our case, the radial distribution of plasma potential in the low-pressure low-frequency ICP enhanced with ferrite cores should be investigated more carefully.

References

[1] Godyak V 2013 *J. Phys. D: Appl. Phys.* **46** 283001
[2] Fedoseev A V and Sukhinin G I 2004 *Plasma Physics Reports* **30** 1061–70
[3] Uhrlandt D 2002 *J. Phys. D: Appl. Phys.* **35** 2159–68
[4] Piejak R, Godyak V and Alexandrovich B 2001 *Journal of Applied Physics* **89** 3590–3

Acknowledgments

This work was supported by the Russian Foundation for Basic Research (grant No. 15–08–02770).

Effect of electron temperature on the distribution of plasma parameters on the phase planes for thermionic diode with low temperature emitter

V P Zimin and T A Piskunova

National Research Tomsk Polytechnic University, 30 Lenin Avenue, 634050 Tomsk, Russia

E-mail: zimin@tpu.ru

Abstract. Influence of temperature distribution of thermal and fast electrons on the distribution of the ion current density in the electrode gap diode has been studied. Studies on the phase planes, especially in the plane of the plasma density – ion current density have been carried out. It is shown that the generation of ions in the electrode gap is specified by the fast electron temperature.

1. Introduction

To effectively control the surface plasma processing technology products it is necessary to know the distributions of plasma parameters and mechanisms of their formation. One critical parameter of the plasma is the distribution ion current density in the interelectrode gap which occurs due to the mechanism of generation of ions. For plasma processing of the surfaces of products, as a rule, glow discharge is used, rather complex structure on the phenomenon. Probe characteristics this discharge have two extended linear section [1, p. 316]. This is interpreted as the temperature distribution of the two group electrons. The impact of these distribution on ionization processes remains unclear. On the other hand, cesium low-temperature plasma of thermionic diode has similar probe characteristics [2]. But cesium discharge thermionic diode is simpler in structure as compared with the glow discharge and therefore it is easier to study the mechanism of generation of ions.

At a certain combination of parameters thermionic diode, in particular at low temperatures emitter probe semilog cesium plasma arc characteristics have two linear areas. Processing of these characteristics gives two electrons temperature distribution in the interelectrode gap (IEG) [2, 3]: in the low (thermal) energy of the electrons and the electrons in the high energy range. The latter group is associated [2] with the electrons which have major influence on the processes of generation of excited atoms and cesium ions.

With a low temperature plasma model [2] we can obtain the distribution of plasma parameters which are not measured in the experiment: the ion current density, electron energy density, function generating ions in the plasma volume; for spectroscopic measurements – the potential of the space occupied by the plasma. It is interesting to study the influence of two electron temperature distribution in the distribution of unmeasured plasma parameters, especially the function of the ion generation electrode in the diode gap.

As a method of research we used the analysis of experimental distributions and unmeasured plasma parameters on the phase planes [4–8]. Such a representation of distributions allows, among other things, designing qualitative and quantitative criteria for the comparison of distributions of plasma parameters.

In this paper we investigate the effect of the distributions of two electron temperature distribution in the experimental and unmeasured (calculated) cesium plasma parameters for a thermionic diode with a low temperature emitter. On the experimental material checked the results of theoretical research model distributions of plasma parameters on the phase planes [8].

2. The model of calculating the distribution of the ion current density

The equation for the ion current density in the IEG thermionic diode (taking into account the friction forces on the ions, electrons $R_{ie} = J_e/\mu_{ei}$) can be written as [2]:

$$J_i = \left(\frac{1}{\mu_{ea}} + \frac{1}{\mu_{ia}}\right)^{-1} \left[-\frac{J}{\mu_{ea}} - (\beta_e - 3/2)kn\frac{dT_e}{dx} - (\beta_i - 3/2)kn\frac{dT}{dx} - k(T_e + T)\frac{dn}{dx} \right]; \quad (1)$$

continuity equation for the ion current density can be written as:

$$\frac{dJ_i}{dx} = -\Gamma(n, n_a, T_e), \quad (2)$$

$$\Gamma(n,n_a,T_e) = en_a\bar{v}_e(T_e)\sigma(T_e)\left[1 - n^2/n_{sh}^2(n_a,T_e)\right], \ \sigma(T_e) = 1{,}44\cdot 10^{-12}\exp\left[-3{,}21/(kT_e)\right],$$

where e – the electron charge; k – the Boltzmann constant; n – plasma density; T_e, T – electron temperature plasma and heavy components (atoms and ions) temperature; $\bar{v}_e$ – the thermal velocity of the electrons; μ_{ea}, μ_{ei}, μ_e, μ_{ia} – coefficients of diffusion and mobility of electrons and ions; β_e, β_i – coefficients of proportionality, taking into account the scattering of electrons and ions on atoms; Γ – function of generating ions in the IEG; $\sigma(T_e)$ – step ionization cross section; $n_{sh}(n_a,T_e)$ – Saha density plasma.

The good approximation for the temperature of the heavy component of the plasma in a weakly ionized plasma diode is of a linear variation along the gap from the emitter temperature to the collector temperature. The distribution density of the atoms $n_a = n_a(x)$ in the IEG is determined from the equation of state for the pressure of the saturated cesium vapor:

$$p_{Cs} = n_a kT. \quad (3)$$

Using characteristics probes, we can obtain distribution $n = n(x)$, $T_e = T_e(x)$, and the potential of the space occupied by the plasma $V = V(x)$, and then evaluated $J_i = J_i(x)$. Spectroscopic measurements make it possible to obtain only the first two distributions. In this case, the distribution $V = V(x)$ also relate to unmeasured plasma parameters.

Thus, the expressions (1)–(3) and for the known distribution $T_e = T_e(x)$ and $n = n(x)$ allow calculating the distribution of ion current density and its derivative.

3. Results and discussion

Experimental distribution of the plasma density and two temperatures of the electrons were obtained [2, 3] for two current density J_1=0.26 A/cm^2, J_2=0.33 A/cm^2 and the following parameters of the thermionic diode: temperature emitter T_E = 915 K, pressure of saturated cesium vapors of the reservoir p_{Cs}=0.44 torr, interelectrode spacing d =0.1 cm. The temperature collector is assumed to be T_C = 600 K.

To construct the distribution of plasma parameters on the phase planes we must calculate their spatial derivatives, so the experimental distribution approximated. We used the following approximating functions: for $T_{et} = T_{et}(x)$ – exponential; to $n = n(x)$ – sine; for $T_{et} = T_{et}(x)$, $T_{eg} = T_{eg}(x)$, $n = n(x)$ – polynomial.

Figure 1 shows the experimental distribution of the plasma parameters and their best approximation in the phase plane. Red lines denote the temperature distribution of fast electrons $T_{eg} = T_{eg}(x)$, and black lines – the temperature distribution of slow electrons $T_{et} = T_{et}(x)$. The solid lines meet current density J_1=0.26 A/cm^2, the dotted lines – J_2=0.33 A/cm^2.

Experimental data were obtained [3] in the range $0.15 \leq x \leq 0.85$ cm, x – distance measured from the emitter. The boundaries of the extrapolation from the emitter marked cross (x), while the collector – a circle (o).

In the following figures notations are similar (if not stated otherwise): the red color is used to indicate the distribution of the plasma parameters obtained with the use $T_{eg} = T_{eg}(x)$, the black – with $T_{et} = T_{et}(x)$; solid lines – J_1, dashed lines – J_2.

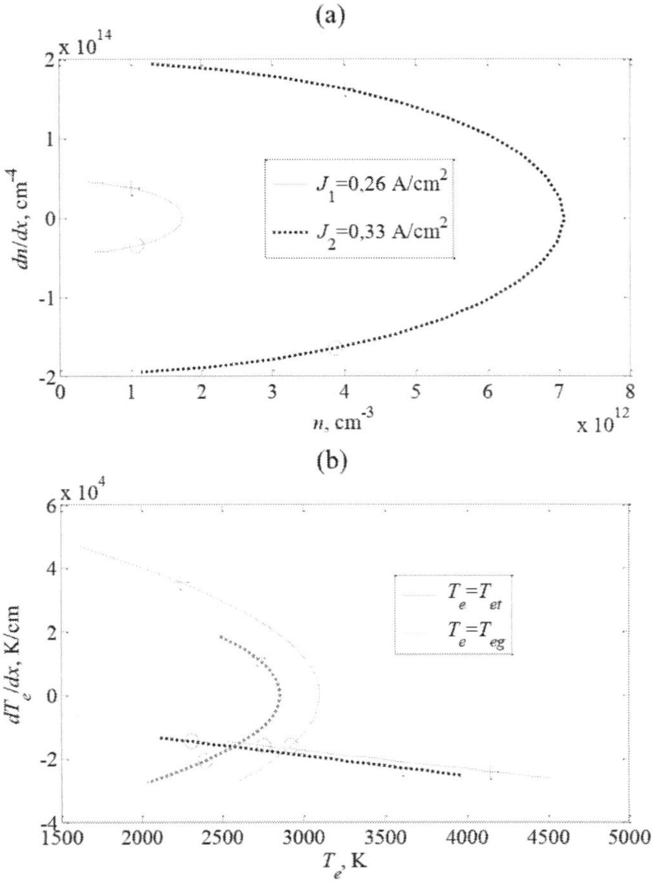

Figure 1. Experimental distribution of the plasma parameters and their best approximation represented in the phase plane. (a) (n, dn/dx). (b) (T_e, dT_e/dx).

The final choice of type of approximation was based on four criteria of fitness: the sum of squared errors (SSE), the square of determination (RS), adjusted R-squared (RSA) and the root of the mean squared error for (RSMSE).

The behavior of all four criteria the fitness was the same for selected approximating functions of distributions $T_{eg} = T_{eg}(x)$ and $n = n(x)$. Therefore, it is a quadratic polynomial for the distribution $T_{eg} = T_{eg}(x)$ the best approximation (figure 1 (b)), and for $n = n(x)$ – sinusoidal dependence (figure 1 (a)). When fitting the experimental data $T_{et} = T_{et}(x)$ the best approach to fitness criteria (exponential dependence) was SSE and RSMSE and a linear relationship was for the remaining two criteria of fitness. Perhaps this is a reflection of the fact that on the one hand the fundamental solution of the heat equation is the exponential function and on the other hand because of the smallness of the IEG solution is actually linearized. In further calculations we used the exponential approximation (see, figure 1 (b)).

Figure 1 (a) shows that both the density distribution of the plasma current densities are symmetric, their maxima differ by approximately 4 times and they are at the point $x \approx 0.5d$. By increasing the current density the temperature distribution $T_{et} = T_{et}(x)$ decreases and the distribution of temperature $T_{eg} = T_{eg}(x)$ increases and these distributions are converging. If the temperature difference is maintained for J_1 (figure 1 (b)) in the range $0.15 \leq x \leq 0.75$ cm then the range is considerably reduced for J_2: $0.15 \leq x \leq 0.35$ cm. Approximation based on both temperatures vary throughout the measuring range (see, figure 1 (b)). By increasing the current density, the plasma density is increased by 3–4 times, respectively, the thermal conductivity of low-energy electron gas increases and the temperature $T_{et} = T_{et}(x)$ decreases.

The size of the electrode sheaths with non-equilibrium plasma is ~ 10^{-3} cm, which it is much less IEG. Relaxation length of the electron energy distribution function for the Maxwell distribution for the electrode of the plasma parameters L_{M1}=0.06 cm is compared with IEG, and it decreases with increasing current density – L_{M2}=0.03 cm. The area of influence of the electron beam from the emitter on the electron energy distribution function takes for J_1 approximately half of the gap, and it decreases almost doubled for J_2. The temperature distribution $T_{et} = T_{et}(x)$ and $T_{eg} = T_{eg}(x)$ differ from each other at the same distances from the emitter (see, figure 1 (a)) for different current densities.

The mean free path of electrons in the atoms Cs l_{ea} varies in the range 0.002 ... 0.006 cm ($l_{ea} \ll d$) and the diffuse arc mode is realized in the gap. The length of the Coulomb scattering of electrons in the Cs ions l_{ei} is to order more l_{ea}, so it does not affect the transport processes in plasma.

The presence of two electron temperature distributions raises the question of their impact on the unmeasured plasma parameters. To answer this question we can use the simulation, but the current models of plasma does not exist. It is possible to evaluate the influence of a distribution on the plasma parameters calculated using experimental data. Using the expression (1), the distribution $J_i = J_i(x)$ is calculated, is constructed and is analyzed the phase curves for the two distributions the temperature electrons $T_{et} = T_{et}(x)$ and $T_{eg} = T_{eg}(x)$.

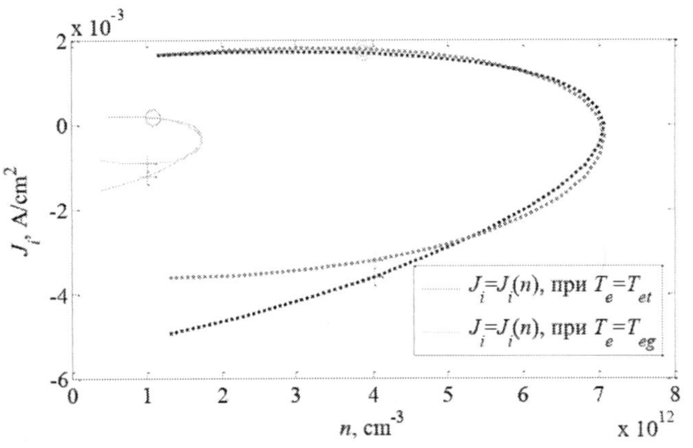

Figure 2. The distribution of plasma parameters are in the phase plane (n, J_i) represented for different densities of the diode current.

Figure 2 shows the distribution of the plasma parameters in the phase plane in the coordinates of the plasma density – the ion current density. The phase curves are of regular and similar for the different densities current of the diode. Constant displacement phase curves $J_i = J_i(n)$ has due to a member of $J\,\mu_{ia}/\mu_{ea} \approx -0.4 \cdot 10^{-3}$ A/cm^2 in the expression (1) for J_i. Thermal diffusion terms in the expression (1) for J_i have little effect on the behavior of the phase curves. The biggest differences for different electron temperature distributions are expressed in the asymmetry of the phase curves $J_i = J_i(n)$ near the electrode sheaths for temperature distribution $T_{et} = T_{et}(x)$.

Analysis of the experimental distributions of plasma parameters showed that the volume recombination is proportional to the square $n/n_{sh}(T_{eg})$, around the electrode gap it is small, excepting the area by emitter for J_1.

There is maximum ratio $n/n_{sh}(T_{eg})\approx 0.5$ at $x=0$. This behavior is explained by extrapolating $T_{eg}=T_{eg}(x)$ in this region, where the electron temperature decreases monotonically decreases rapidly and n_{sh}, according to the expression (2) $n_{sh} \sim \exp(-1/T_{eg})$.

Figure 3 (a) shows the behavior of ion generation functions $\Gamma=\Gamma(n,n_a,T_e,x)$ on the plane configuration for the two electron temperature distributions. Functions $\Gamma=\Gamma(n,n_a,T_{et},x)$ in region by emitter are to order more similar functions $\Gamma=\Gamma(n,n_a,T_{eg},x)$.

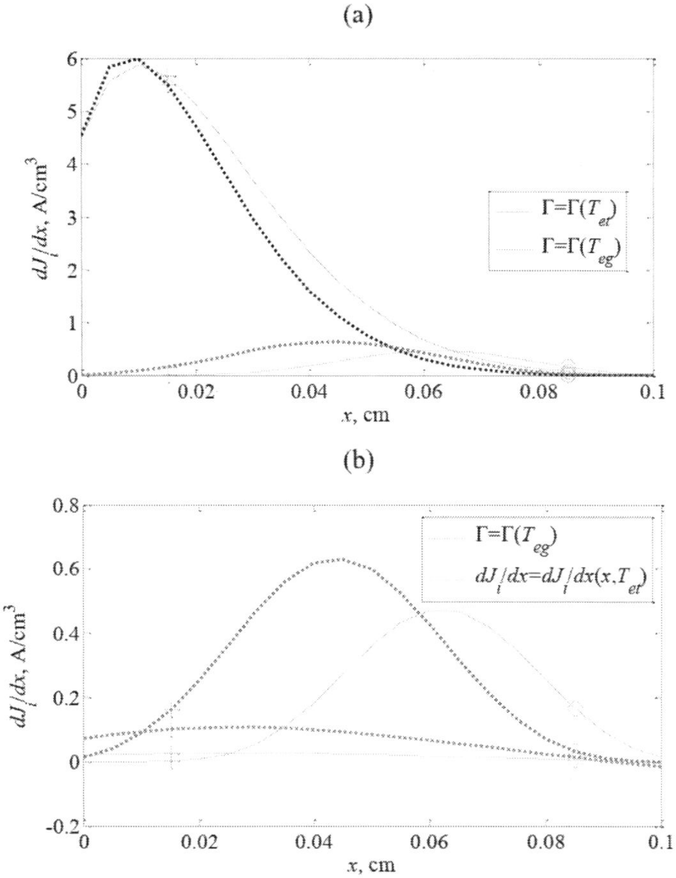

Figure 3. Distribution of ion generation represented in the IEG. (a) for different electron temperature distributions. (b) for the distribution of electron temperature $T_{eg}=T_{eg}(x)$.

Furthermore, the function of ion generation for distribution $T_{et}=T_{et}(x)$ decreases with increasing current density through the diode, but for distribution $T_{eg}=T_{eg}(x)$ is increasing against. Differentiate depending $J_i=J_i(x)$ on x (purple curves in figure 3 (b)) also have an increase with the increase J.

Moreover, the functions $\Gamma=\Gamma(n,n_a,T_{eg},x)$ and dJ_i/dx are comparable in the value. This difference becomes even less expressed with decreasing cross section of ionization $\sigma=\sigma(T_e)$, which is typical for the values of the plasma density $n\le 10^{13}$ cm^{-3} [2]. These results confirm the fact that the processes of the volume ionization and recombination are determined by the fast electron temperature $T_{eg}=T_{eg}(x)$.

4. Conclusion

The results based on the analysis on the phase planes of the influence the two electron temperature distribution types (slow – $T_{et} = T_{et}(x)$ and fast – $T_{eg} = T_{eg}(x)$) on the distribution of unmeasured parameters of the low-temperature plasma cesium thermionic diode allow to draw the following conclusions:

1. Experimental plasma density distributions $n = n(x)$ are almost symmetrical. The maximum $T_{et} = T_{et}(x)$ decreases with increasing current density (increasing the plasma density) distribution and the maximum of the distribution $T_{eg} = T_{eg}(x)$ increases. Thus, the distribution of two types of temperature distributions of electrons converge together.
2. There is significant difference in dependencies $J_i = J_i(n)$ calculated for two types of temperature distributions of electrons in the region close to the emitter.
3. To show the decisive role of the fast electron temperature distributions $T_{eg} = T_{eg}(x)$ by the analysis of the impact on the processes of ion generation in the electrode gap for different types of electron temperature distributions.

References

[1] Granovsky V L, Musin A K, Golubev V S, Solntsev G S, Sena L A and Golant V E 1971 Elektricheskiy tok v gaze. Ustanovivshiysya tok (The electrical current in a gas. Current) (*in Russian*) (Moscow: Nauka)
[2] Baksht F G, Dyuzhev G A, Martsinkovsky A M, I Mojzes B Y, Pikus G E, Sonin E B and Yuryev V G 1973 Termoemissionnyie preobrazovateli i nizkotemperaturnaya plazma (Thermionic converter and low-temperature plasma) (*in Russian*) (Moscow: Nauka)
[3] Baksht F G, Dyuzhev G A, Korobova I L, Kaplan V B, Martsinkovsky A M, Mojzes B Y Shakhnazarov G A and Yuryev V G 1968 *Journal of Technical Physics* (*in Russian*) **38** 1075
[4] Zimin V P 2011 *Bulletin of the Tomsk Polytechnic University* (*in Russian*) **318** 29
[5] Zimin V P 2012 *Bulletin of the Tomsk Polytechnic University* (*in Russian*) **321** 17
[6] Zimin V P 2013 *Bulletin of the Tomsk Polytechnic University* (*in Russian*) **322** 11
[7] Zimin V P 2013 *Bulletin of the Tomsk Polytechnic University* (*in Russian*) **323** 158
[8] Zimin V P 2014 *Bulletin of the Tomsk Polytechnic University* (*in Russian*) **325** 7

IOP Conf. Series: Journal of Physics: Conf. Series **830** (2017) 012056 doi:10.1088/1742-6596/830/1/012056

The one-particle approximation in the reflecting discharge simulation

N N Schitov and D I Yurkov

Federal State Unitary Enterprise "All-Russian Research Institute of Automatics n.a. N.L. Dukhov", Sushchevskaya st. 22, Moscow, 127055, Russia

E-mail: nschitov@mail.ru

Abstract. The method of some reflecting discharge (Penning discharge) characteristics computation, based on the one-particle approximation is proposed. This discharge is widely used in ion sources aimed at surface modification. However, only the steady state of this discharge is sufficiently described, whereas pulsed modes are preferable in many cases. In fact, the proposed method is similar to the approach used in the early times of first glow discharge investigations and crossed fields ion sources. It may be applied for the early discharge stages (the Townsend regime) description. It is somehow simpler than the diffusion-drift approximation used as a rule for the stationary state description, because plasma does not exist yet. On the other hand, one need not use most of usual diffusion-drift simplifications e.g. 1 or 2D models, uniform magnetic field etc. So the process of discharge formation may be described exactly for different kinds of Penning cells geometries and fields configurations. The discharge ignition condition for the Penning cell, analogous to the Townsend law is evaluated. It allows one to appreciate the discharge formation time as a function of cell geometric parameters, field configurations, anode voltage and Townsend's coefficients α and γ. This time, or exactly the trajectory length during this time, plays the role of the Townsend parameter d – the distance between electrodes. The calculated values of such times show good agreement with experimental data.

1.Introduction

The diffusion-drift approximation used as a rule in Penning discharge simulation [1-3] allows one to evaluate key dependencies of the discharge characteristics upon the process parameters in a steady state. However, the initiation or discharge formation regime is quite interesting for some applications. Its adequate description is important from the fundamental point of view too. On the other hand, such description is somehow simpler than the one for the steady state as plasma does not exist yet and fields are not disturbed by free charges and currents. Charge trajectories may be easily calculated exactly even including "friction" due to charge-neutral collisions. Moreover, in the Townsend mode, when plasma does exist, charges do not disturb fields significantly. That is why the so called "one-particle approximation" may be used allowing one to calculate the exact trajectory. It makes possible to give the proper formulation of the discharge ignition condition analogous to Townsend's law because in a Penning cell one need to employ the trajectory length instead of the inter-electrode distance and consequently to determine the time needed to strike the discharge.

2.The calculation of non-uniform fields and electron trajectories in a Penning cell

The calculation of the electrostatic field ***E*** distribution implies the Laplace equation solution with the constant scalar potential $\varphi(\vec{r})$ boundary conditions on metallic surfaces and calculated on the basis of the "image method (mirror-image method)" on dielectric ones. The solution for the simple Penning cell geometry in series expansion form is given in [4, pp 490-491]. The magnetic fields ***B*** as well as vector potential are also calculated by the instrumentality of series expansion with known (measured) ***B*** axial component [5]. The three terms approximation of the vector potential is quite approved. It is not evident *apriori* which field is better (from the discharge ignition time viewpoint) – uniform or nonuniform creating magnetic traps, "plugs" etc.

The electron trajectories calculation in the residual atmosphere is performed according to nonrelativistic equation of motion: $m\dot{\vec{v}} = e\vec{E} + e[\vec{v},\vec{B}] - m\nu_m\vec{v}$ (by inserting the transport cross-section ν_m – [6], system SI, cylinder coordinates, electron starting from cathodes). The deceleration due to the ionization is not separated in this equation from other processes but the ionization frequency is much smaller as a rule than the transport one [6]. But the transport frequency itself in the case of the Penning cell is much smaller than the electron oscillation frequency. For example, the hydrogen transport frequency is ν_m=4.8*10^9*p [6], that is around 5 MHz at 10 mTorr which is significantly smaller (three orders of magnitude) than the oscillation one and even more the cyclotron frequency.

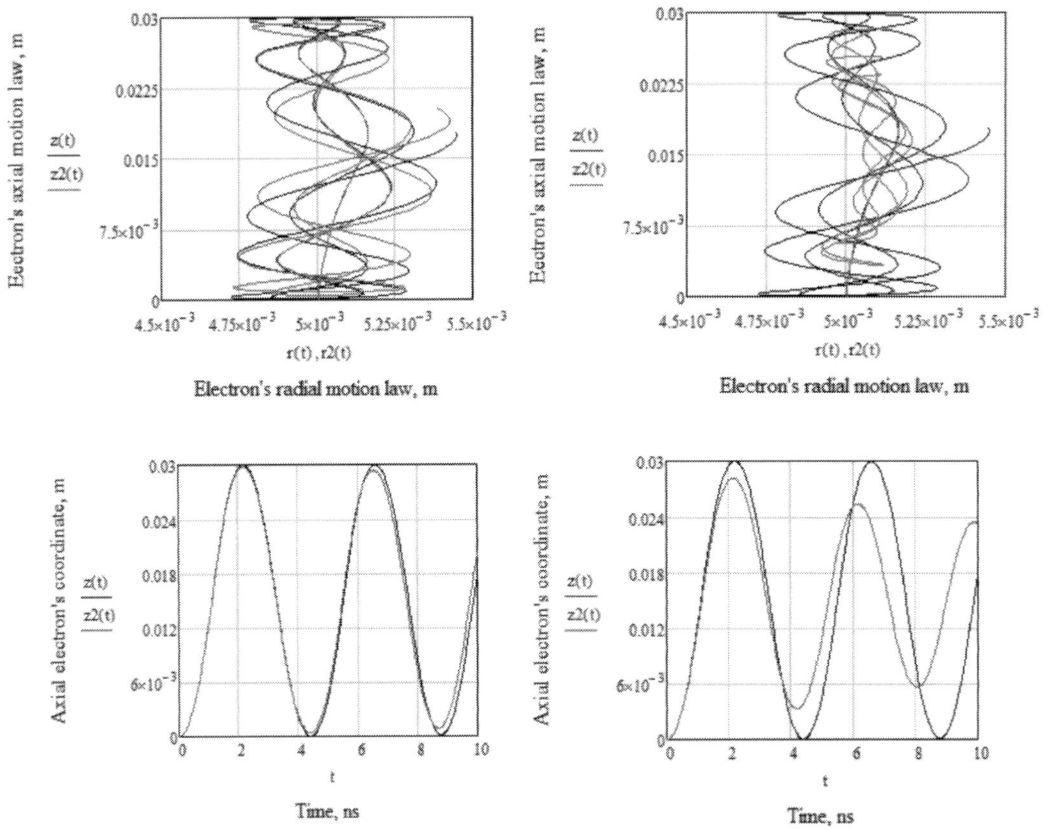

Figure 1. The electron trajectories and motion laws along the Z axe in the Penning cell taking into account the friction (red lines) and without friction (p=2 mTorr – left, 20 mTorr – right).

Figure 1 shows the projections of the electron trajectories z(r) and z2(r2) on the {r,0,z} plane as well as their motion laws z(t), z2(t) calculated according to the solution mentioned above [4, pp 490-

491]. The geometrical parameters of the cell are as follows: radius – 1 cm, length – 3 cm, anode height – 5 mm. The anode potential – 2 kV. The initial conditions are as follows: an electron starts from the cathode point 5 mm aside from its axis with zero initial velocity in the uniform and constant magnetic field 50 mT. The motion laws are given taking account the friction (index 2) and without it with the transport frequency 10^7 Hz (left plots) which approximately corresponds to 2 mTorr pressure. It is seen that the oscillation period is of 4 ns order whereas the cyclotron period for such an induction is only 0.715 ns. So during the oscillation period an electron makes approximately 6 turns around its trajectory. On the right plots, the trajectories and motion laws are given for the case when the transport frequency is equal to 10^8 Hz, corresponding to 20 mTorr pressure – an order of magnitude greater. The difference is well seen but as in the former case, it influences only the oscillation damping decrement but practically does not change its period. Consequently, one may not take into account the friction due to different collisions in the electron trajectories calculation in the before-discharge stage.

The magnetic field nonuniformity introduces naturally big changes in the trajectory forms. The examples of the electron trajectory at hydrogen pressure 1.5 mTorr calculated in uniform and nonuniform fields are shown in figure 2. The electron trajectories starting from the point (r=12 mm, z=3 mm) calculated during 40 ns are shown: in the 29 mT uniform (equal to the averaged across the cell nonuniform field) – at the right,and nonuniform magnetic field – at the left. It is well seen that in the nonuniform field the electron moves practically across the whole cell volume, its trajectory is much longer than in the case of the uniform field. In the uniform magnetic field electron oscillates practically along the straight line declining from it towards anode only by few mm tenths.

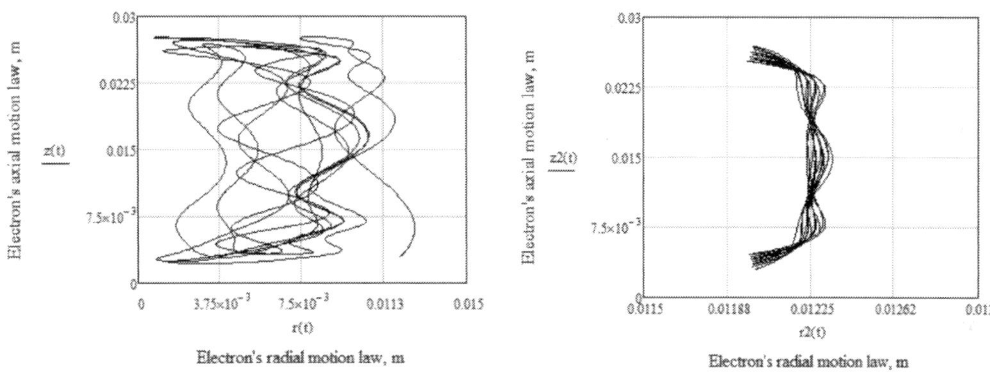

Figure 2. The trajectories of electrons starting from the same cathode point in the uniform (at the right) and nonuniform magnetic fields, U_{an}=800 V.

3.The Penning discharge ignition condition

The discharge ignition condition (the Townsend rule) [5]: $\gamma(\exp(\alpha d)-1)=1 \Rightarrow \alpha d=\ln(1/\gamma+1)$ in which Townsend's coefficients are implied – α (ionization) and γ (or generally Γ), characterizing the secondary ion-electron emission as well as other near electrodes electron generation processes and the inter-electrode distance – d, cannot be used in the genuine form and must be modified. The Penning cell differs significantly from the two parallel flat infinite electrodes investigated by Townsend. That is why the inter-electrode distance – d in the Townsend's law must be substituted by the trajectory length L(t) along which an electron generated on the cathode (or anticathode) moves oscillating between cathodes until it recombines or falls on the anode. However, the whole current is not determined by these "falls", as in the parallel flat electrodes system.

While determining the ignition time one has to calculate the trajectory length during the given time interval. Besides, as the 1st Townsend's coefficient strongly depends upon the electron energy which is the function of the cell point (or the time if trajectories are time parameterized) the integral must be in the exponent instead of the simple product *αl*. The 2nd Townsend's coefficient – γ – depends upon the electron energy too, but because of the weak (logarithmic) dependence of the Townsend's law may be taken constant. Finally, one has to take into account the fact that the trajectory type and length may largely depend on the start point (distance from the cathode axis) and the whole current is the integral upon the cathodes surfaces.

All these circumstances urge one to modify the ignition condition this way:

- It is not the current density that must become infinite in the arbitrary point of the infinite cathode, but the whole electron current from the cathode and anti-cathode, in some points the density may be zero;
- The ions generated in the cell's part (half for symmetric) adjacent to the cathode may fall only on the cathode, not anti-cathode because of their small initial energy and large anode potential, the same for the cell's part adjacent to anti-cathode;
- In the last case one has to account the fact that part of them quit the cell through the exit orifice, which will not generate secondary electrons from anti-cathode.

Assuming the knowledge of the α dependence upon the energy let us repeat the Townsend law evaluation taking into account the proposed obvious amendments. However, the Townsend's 1st coefficient explicit form $\alpha=A\cdot p\cdot\exp(-Bp/E)$ does not suit. One has to approximate experimental dependencies by some appropriate functions. For example, the H_2 α dependency [6] upon the electron energy may be approximated by the sum of the parabola and exponent.

The electron motion law knowledge allows one to write the explicit form of such an approximation and the whole ionization probability as time functions:

$$\alpha_{apr}(t)=\alpha_{apr}(T_{kin}(t))=\alpha_{apr}((\dot{r}(t))^2+(\dot{z}(t))^2+(r\dot{\psi}(t))^2),\ P_{ion}(\tau)=\int_0^\tau \alpha(t)\cdot p\cdot dl(t)$$

The electrons trajectories and consequently the ionization coefficients are the start point functions or the distance from the axis – $r0$ (for cathode) and $\rho 0$ (for anti-cathode). Accounting for the point dependence of α and different contributions of ions born in different cell parts in the secondary electrons generation on the cathode and anode let us introduce the short denominations for different integrals that will appear in the next evaluation:

$$In_{k,1}(r0,\tau)=\begin{cases}\int_0^\tau \alpha(r0,t)dl(r0,t), & z\le Lc/2\\ 0, & z>Lc/2\end{cases}\quad In_{k,2}(r0,\tau)=\begin{cases}\int_0^\tau \alpha(r0,t)dl(t), & Lc/2<z\le Lc\\ 0, & z<Lc/2\end{cases}$$

$$In_{ak,1}(\rho 0,\tau)=\begin{cases}\int_0^\tau \alpha(\rho 0,t)dl(\rho 0,t), & Lc/2<\zeta<Lc\\ 0, & \zeta<Lc/2\end{cases}\quad In_{ak,2}(\rho 0,\tau)=\begin{cases}\int_0^\tau \alpha(\rho 0,t)dl(\rho 0,t), & \zeta<Lc/2\\ 0, & Lc/2<\zeta<Lc\end{cases}$$

Let us account now the possible difference in generalized 2nd Townsend coefficients for the cathode and anti-cathode introducing coefficients Γ_κ and $\kappa\Gamma_{a\kappa}$. The κ coefficient describes the part of ions born in the anticathode cell half-part (they will be extracted in the acceleration gap). In the first approximation this coefficient may be taken as the cathodes surface area ratio. Repeating Townsend let us introduce the following functions to reduce the notations:

$$f_1(r0,\tau)=\exp(In_{k,1}(r0,\tau))-1,\ f_2(r0,\tau)=\exp(In_{k,2}(r0,\tau))-1,$$

$$f_3(\rho 0,\tau)=\exp(In_{ak,1}(\rho 0,\tau))-1,\ f_4(\rho 0,\tau)=\exp(Ina_{k,2}(\rho 0,\tau))-1.$$

Then introducing initial currents from the cathode and anti-cathode – I_{k0}, I_{ak0}, let us write the current expressions taking into account the electrons contributions on the cathode and anticathode (R_{ap} is the exit aperture radius):

$$2\pi j_k \int_0^{R_k} r0(1-\Gamma_k f1(r0,\tau))dr0 = I_{k0} + \Gamma_{ak} 2\pi j_{ak} \int_{R_{ap}}^{R_{ak}} \rho 0 f4(\rho 0,\tau)d\rho 0$$

$$2\pi j_{ak} \int_{R_{ap}}^{R_{ak}} \rho 0(1-\kappa\Gamma_{ak} f3(\rho 0,\tau)d\rho 0 = I_{ak0} + \kappa\Gamma_{ak} 2\pi j_k \int_0^{R_k} r0 f2(r0,\tau)dr0.$$

Solving this system of equations relatively to the current density on the cathode j_k, for instance, we will get the ignition condition as the coefficient at this density vanishing:

$$\int_0^{R_k}(1-\Gamma_k f1(r0,\tau))r0dr0 \int_{R_{ap}}^{R_{ak}}(1-\kappa\Gamma_{ak} f3(\rho 0,\tau))\rho 0 d\rho 0 = \kappa\Gamma_k\Gamma_{ak}\int_0^{R_k} f2(r0,\tau)r0dr0 \int_{R_{ap}}^{R_{ak}} f4(\rho 0,\tau)\rho 0 d\rho 0$$

This is the Penning cell ignition condition during τ according to all assumptions made above. It is not so easily interpreted as the classical Townsend law and even hardly calculated if more simplifying assumptions due to the electrons and ions trajectories properties are not made. Fortunately, the calculations show that the ionization probability as a point function is almost constant (in contrast to trajectory length). It permits one to significantly simplify the ignition condition by retrieval of the averaged current density from the integral. Introducing the electrodes areas – S_k and S_{ak} and assuming $\kappa=S_{ak}/S_k$ (for the 1st approximation) we will obtain this condition in such a form:

$$(1-\Gamma_k f1(\tau))(1-\frac{S_{ak}}{S_k}\Gamma_{ak} f3(\tau)) = \kappa\cdot\Gamma_k\Gamma_{ak} f2(\tau)f4(\tau) \qquad (1)$$

As it can be easily seen besides the absence of the integration the condition does not contain the start point dependencies of the f_i but only the discharge ignition time τ (because of the dependencies absence one can take the only one trajectory, e.g. with the start point in the middle of the distance from the axis). If the anti-cathode area tends to zero (for example if it is made from the high transparency mesh) the ignition condition becomes practically Townsend's:

$$(1-\Gamma_k f1(\tau)) = 0 \Rightarrow \int_0^{\tau}\alpha(l)dl\Big|_{z\le 0,5Lc} = \ln(1+\Gamma_k^{-1})$$

with a simple modification consisting in the interchange of the inter-electrodes gap d by the ignition time τ and the integral is calculated in the cathode part of the cell. To solve graphically the equation (1) is to find out the crossover of its left side – LS(t) and right side – RS(t) (dimensionless) time dependencies – figure 3 (argument Θ_i varies discretely – 0.5 ns per step – to truncate the calculation time, p=15 mTorr). For example, the ignition condition is achieved during 200 ns in the symmetric cell (2 cm long, 2 cm diameter, 1 cm anode height) in hydrogen residual atmosphere (1.5 mTorr), magnetic induction 50 mT and anode potential 1 kV if cathodes areas are practically equal and Γ coefficients are equal to 0.514. If only the cathode secondary emission is accounted its value must be raised to a factor 3, that is approximately 1.5 to conserve the ignition time. So this time strongly depends upon the "integer" coefficients Γ. For the stainless steel electrodes and hydrogen pressure around 1 mTorr the calculated ignition time is around 400 ns which agrees well with the experimental data.

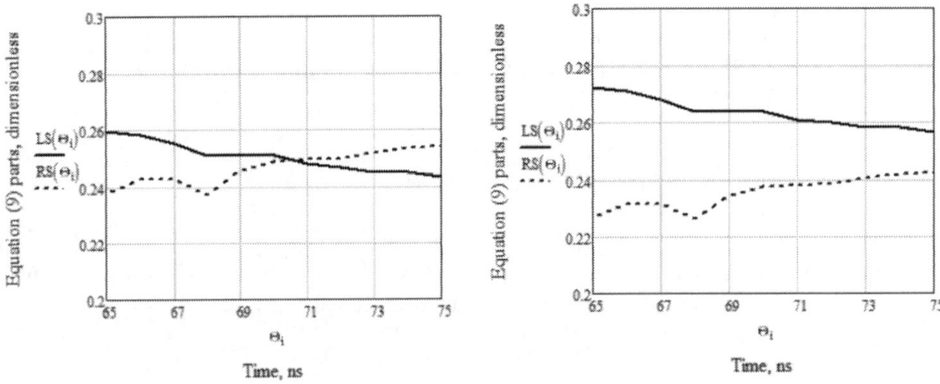

Figure 3. Two examples of calculated plots with slightly different parameters (Γ_k=1.1 – at the left, Γ_k=1.05 – at the right).

Varying different cell parameters, residual pressure and anode potential one can see the ignition time dependencies upon these parameters. One need not every time find out the crossover of the (1) left and right hand sides plots – it takes too much time. To determine the trends, it is sufficient to investigate the dependence of the e.g. the right part upon these parameters during a rather short integration time. It is possible because the left side is always greater in the beginning. That is why the right side rise during given time at some parameter variation means the ignition time reduction and vice versa. One can also use the "two point linear approximation" of the equation (1) left and right sides, thus radically shortening the calculation, but it leads to greater ignition times. For instance, this approximation gives 88 ns ignition time instead of 70 ns in the figure 3 (left) example.

We need to note that with the one-particle approximation usage the calculation of the ignition and extinction potentials in the Townsend mode may also be done with the introduction of explicit dependence of the ionization coefficient on the electric field strength similar to the Townsend's way with all modifications made above.

4.Conclusion

The use of the one-particle approximation at the early Penning discharge stages simulation allows one to avoid numerous simplifications used in diffusion-drift models. Particularly the influence of nonuniform fields in asymmetric cells on the discharge formation rate and ignition/extinction potentials may be investigated numerically. The characteristic times and potentials may be calculated using the ignition condition introduced above analogous to the Townsend law for flat infinite electrodes system. This approach allows one to optimize different Penning ion sources parameters according to specific surface modification tasks.

References

[1] Shuurman W 1967 Investigation of a low pressure Penning discharge *Physica* **36** 136
[2] Nikulin S P 1998 *Russian J. Tech. Phys.* (in russian) **68(7)** 56
[3] Zjulkova L A, Kozyrev A V and Proskyrovskiy D I 2004 *Russian J. Tech. Phys. letters* **30(16)** 19
[4] Vladimirov V S 1976 *The Mathematical Physics Equations* (in russian) (Moscow: Nauka) p 527
[5] Kelman V M and Yavor S A 1968 *Electronic Optics* (in russian) (Leningrad: Nauka) p 487
[6] Raizer Yu P 1987 *Gas Discharge Physics* (in russian) (Moscow: Nauka) p 591

Plasma source based on an unbalanced magnetron sputtering system

D P Borisov, V A Slabodchikov and V M Kuznetsov

Tomsk State University, Tomsk 634050, Russia

E-mail: borengin@mail.ru

Abstract. The paper presents research results on the capabilities of an unbalanced magnetron sputtering (UMS) system with a coefficient of geometrical unbalance K_G=0.3 to produce gas discharge plasma far from its target. Using argon as the working gas and silicon as the target material, it is shown that the proposed UMS system provides the generation of plasma with an ion current density of ≈ 0.2 mA/cm^2 in the region of treated material at 440 mm from the Si target. The research data on the maximum power at which the UMS system produces high-density plasma without melting the Si target are also presented.

1.Introduction

Magnetron sputtering systems are an efficient tool for deposition of high-adhesion coatings [1] and plasma-assisted ion implantation [2] using an argon plasma generator [3] which provides preliminary surface cleaning of metal substrates and better penetration of atoms into their surface layers. In technological terms, it is, however, desirable to use as fewer plasma devices as possible: for example, only magnetron sputtering systems to provide surface cleaning as well as plasma-immersion doping and plasma-assisted coating deposition. To do this the ion current density from the plasma produced by a magnetron sputtering system must be higher than 0.2 mA/cm^2 at an ion energy to a substrate from several to hundred electron-volts [4–7]. Here we study the possibility of producing gas discharge plasma of required density with an unbalanced magnetron sputtering system.

2.Experimental equipment and research technique

The unbalanced magnetron sputtering system (UMS) under study had a planar circular configuration; a simplified schematic of its cross-section is shown in figure 1 *(a)*. The UMS system was mounted on one of the side flanges of a cylindrical vacuum chamber made of stainless steel with a diameter of 1200 mm and height of 600 mm. The cathode-target of the UMS system was a silicon disk of diameter 80 mm and thickness 8 mm mounted at the face of the UMS system casing (not shown in figure 1) by tightly pressing the target to a water-cooled copper base of thickness 2 mm which, being as such the face of the UMS system, separated the water-cooled volume of the magnetic system and the vacuum volume. The axially symmetric circular magnetic system was formed by an annular permanent magnet (NdFeB) and a cylindrical magnetic core (steel) with an axial channel for cooling water supply to the copper base-face.

Figure 1 *(a)* also shows a field pattern calculated for the magnetic system. It is seen that on the system axis, there is a point at which the magnetic field lines reverse their direction, and hence, the system is unbalanced. The calculated behavior of the magnetic field is confirmed by figure 1 *(b)*, which shows an image of the discharge and plasma produced in the UMS system. It is seen that the unbalanced field lines limit the electron scattering region because the unbalance distorts the arch shape of the field in the magnetic trap region (region of bright plasma glow), decrease the field in the radial direction, and allow electrons to leave the magnetic trap [8, 9]. The electrons escaped from the trap move to a substrate, rotating about the divergent unbalanced lines and ionizing the working gas (argon).

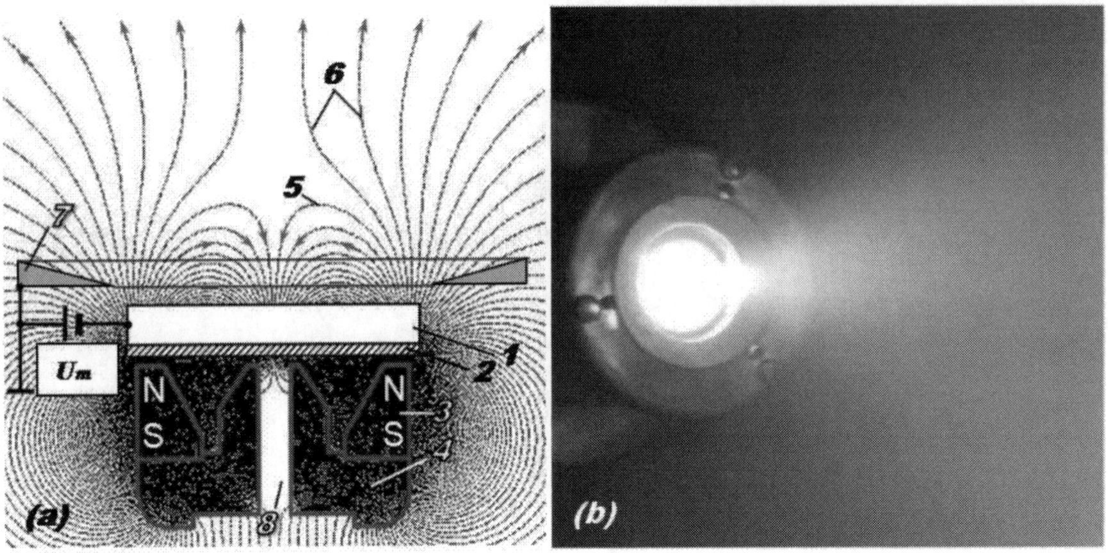

Figure 1. Simplified cross-sectional schematic of the UMS system with calculated magnetic field lines *(a)* and image of the discharge glow *(b)*: *1* – Si target; *2* – water-cooled Cu base; *3* – annular permanent magnet; *4* – cylindrical magnetic core; *5* – balanced field lines; *6* – unbalanced field lines; *7* – grounded anode; *8* – cooling water channel; U_m – discharge power supply.

The magnetic field on the axis and at the cathode surface was measured with a meter based on Hall sensors. At the surface of the Si target, the axial (normal) component of the magnetic field was 0.095 T. At a distance of 11.4 mm along the axis, the magnetic field reversed its direction, reaching its maximum equal to 0.022 T at 28 mm from the target. Then, the axial component decreased gradually and vanished at 180 mm from the target. The measurements confirmed the calculation. The coefficient of geometrical unbalance for the UMS system was K_G=0.3, being defined as the ratio between the distance along the system axis from the target surface to the point of field reversal and the average diameter of the target sputtering zone [10].

The anode of the UMS system was anode *7* (figure 1 *(a)*) as well as the entire grounded vacuum chamber. The argon pressure in the vacuum chamber was controlled by varying the gas flow rate to the chamber. The ion current density at the center of the chamber was measured using a one-sided plane probe with a conducting area of 200 cm^2 located perpendicular to the UMS system axis at a distance of 440 mm from the target. The composition of the plasma produced in the UMS system, in particular the content of the ionized component of one or another element, was analyzed using a COLIBRI-2 multichannel atomic emission spectrometer in the region where treated objects were located. The test materials were NiTi substrates with a Ni content of 50.9 at %. The elemental composition of surface layers after treatment in different modes was analyzed using a Shkhuna-2 Auger spectrometer (Tomsk Polytechnic University).

3.Results and discussion

Figure 2 shows current-voltage characteristics of the discharge in the UMS system, demonstrating a huge effect of the Ar pressure on the discharge parameters.

IOP Conf. Series: Journal of Physics: Conf. Series **830** (2017) 012057 doi:10.1088/1742-6596/830/1/012057

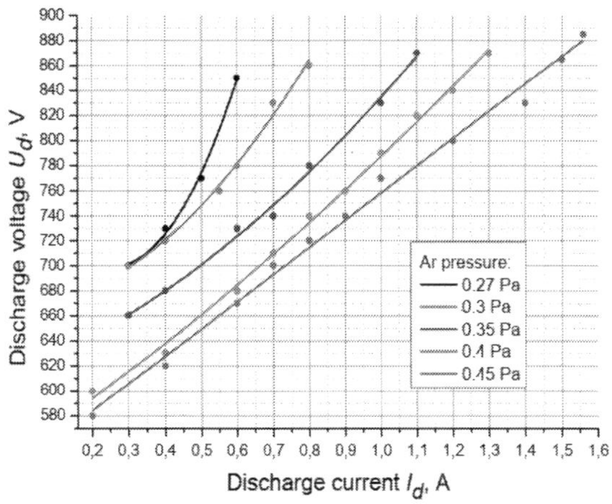

Figure 2. Current-voltage characteristics of the discharge in the UMS system at different operating pressures.

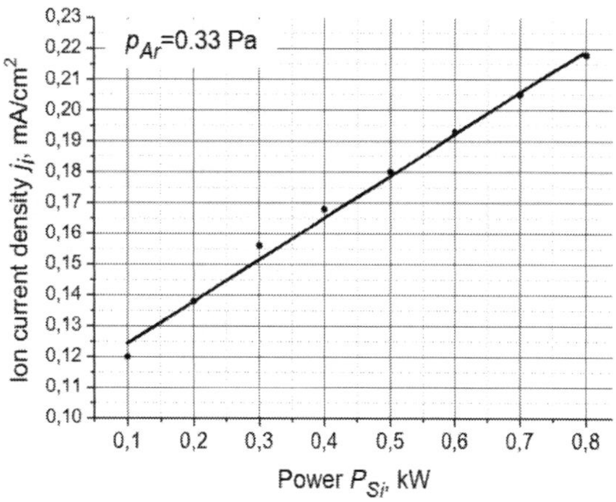

Figure 3. Ion current density at the center of the vacuum chamber vs the power of the UMS system.

The current-voltage characteristics are ascending; that is, for increasing the discharge current the discharge voltage should be increased with the increase in the power of the UMS system. The most important characteristic of the UMS system is its capability of producing high-density plasma in the region of treated objects for their efficient heating, cleaning, doping of their surface, and plasma-assisted coating deposition. The plasma density was estimated from the saturation ion current density of the one-sided plane probe at its negative bias of 200 V with respect to the vacuum chamber. Figure 3 shows the dependence of the ion current density on the power of the UMS system at an operating pressure of 0.33 Pa.

As can be seen from figure 3, the designed UMS system is capable of providing ion current densities from the plasma which are sufficient to greatly influence the energy and structural phase states of the substrate and its modified surface layers even at the so large distance from the substrate as 440 mm. Our experiments show that in the UMS system, the use of a Si target tightly pressed to the Cu base and cooled through it does not allow a considerable increase in the power compared to that in figure 3 due to overheating and melting of the target.

Figure 4 shows an image of the Si target of the UMS system after its operation under the conditions indicated in figure 3 but at a power higher than 0.8 kW. It is seen that the target surface is melted and that the formed liquid droplet does not fall but flows down the target and becomes pointed at its end as if it were lying on the magnetic field lines near the target. Such a situation was observed in all experiments: the droplet of melted material was always held at the bottom of the target and the surface of its pointed end was similar to the configuration of magnetic field lines near the target.

We determined the maximum power at which the Si target of the UMS system with indirect cooling through the Cu base was not melted. Figure 5 shows the maximum power of the UMS system as a function of operating pressure.

Atomic emission spectrometry of the plasma produced by the UMS system in an operating mode with low Si target consumption at a dissipated power of 0.2 kW (the mode most used by the authors) suggests a rather low ionization degree of the plasma. This judgment is supported by experimental data on the ion current density from the plasma at respective Ar pressures. According to atomic emission spectrometry, almost the whole ion current is the current of Ar ions. The fraction of Si ions is low, lying on the verge of the spectrometer sensitivity ($\leq$ 1 % of the Ar ions). The flow of the sputtered material in the region of the treated substrates is dominated by Si atoms. Thus, the high density of Ar ions in the plasma and the high content of Si ions in the substrate region make it possible to vary the elemental composition of modified surface layers by varying the bias applied to the substrates.

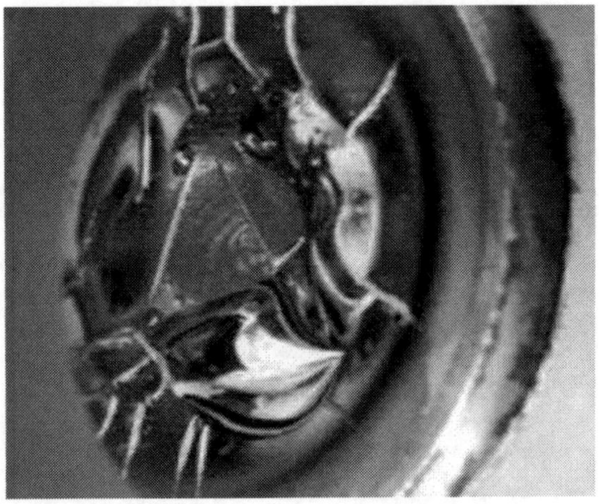

Figure 4. Image of the melted Si target of the UMS system.

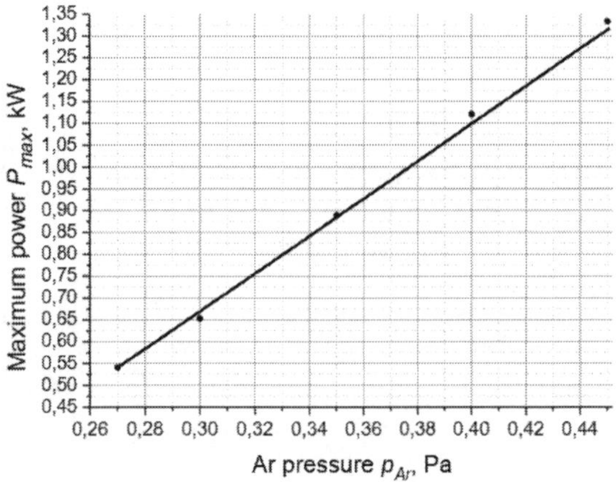

Figure 5. Maximum power of the UMS system vs the operating pressure.

Figure 6 *(a–d)* shows Auger profiles of the chemical elements in NiTi surface layers treated in the plasma of the UMS system at a negative bias of 200 *(a)*, 400 *(b)*, 600 *(c)*, and 1000 V *(d)* to the specimens with respect to the anode (vacuum chamber).

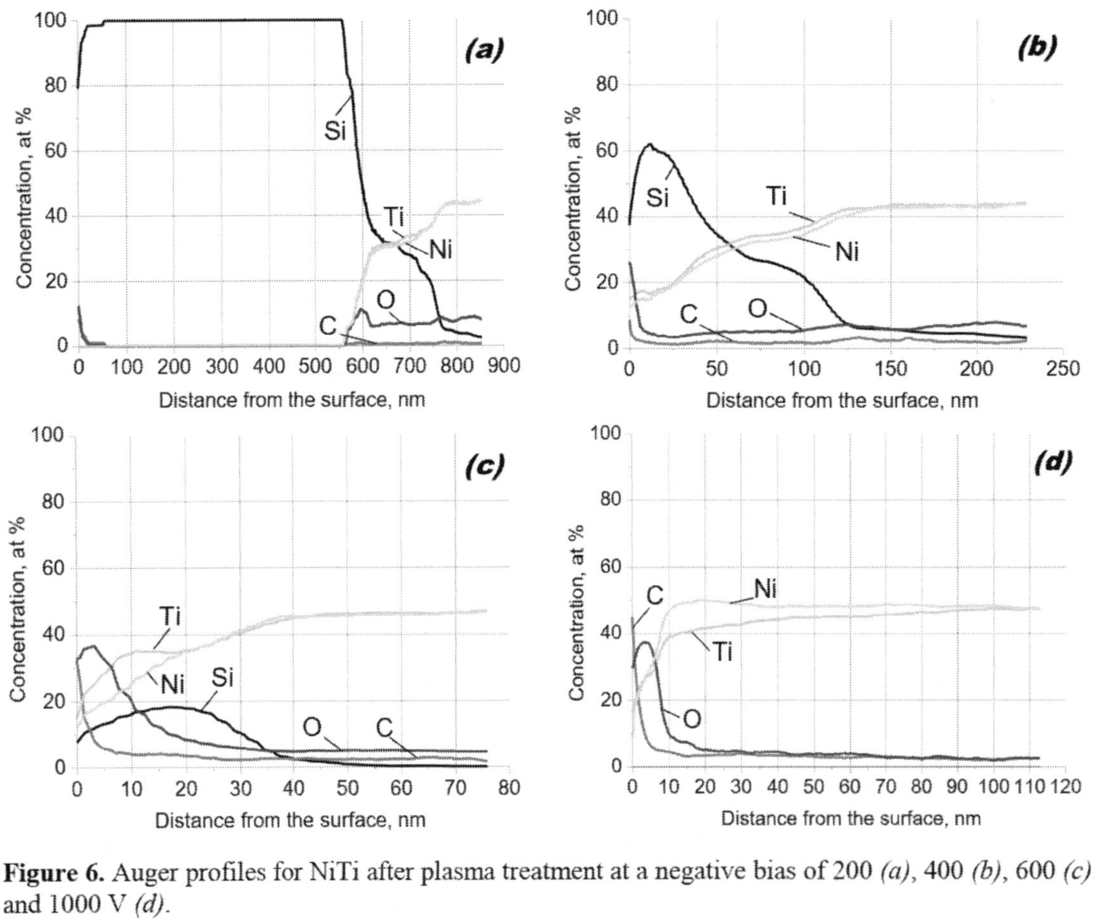

Figure 6. Auger profiles for NiTi after plasma treatment at a negative bias of 200 *(a)*, 400 *(b)*, 600 *(c)*, and 1000 V *(d)*.

It is seen from figure 6 that increasing the bias provides different modes of plasma immersion surface treatment of the material: deposition of a Si coating, plasma immersion ion doping and etching (surface cleaning) without coating deposition, and surface doping with Si. At a negative bias of 200 V (figure 6 *(a)*), a coating of thickness 600 nm is deposited on the surface with a considerable depth of Si penetration into the material (300 nm) and almost without any gas impurity (oxygen, carbon, etc.) contamination of the coating, suggesting that the plasma of the UMS system provides high-quality surface cleaning of the substrates [2]. At a negative of bias 400 V (figure 6 *(b)*), the coating is etched, but the depth of Si penetration into the material remains large (200–250 nm) with a maximum Si content of 60 at% [2]. Increasing the bias to 600 V (figure 6 *(c)*) decreases the depth of Si penetration into the NiTi substrate to 50–60 nm and the maximum Si content to less than 20 at%, suggesting that etching in this mode becomes more and more prevailing. Finally, at a negative bias of 1000 V (figure 6 *(d)*), the surface layer is fully etched such that the treatment represents surface etching and cleaning. The oxygen- and carbon-saturated thin (< 10 nm) surface layer formed in this mode is likely due to the interaction of the elements with etching-activated titanium in air on removal of the specimens from the vacuum chamber.

4.Conclusion

Thus, our experiments demonstrate high efficiency of the proposed unbalanced magnetron sputtering system as a plasma source for realization of different types of surface treatment: coating deposition, doping, etching, and cleaning. This opens up new possibilities for simplifying the design of vacuum plasma setups and for increasing the technological efficiency of vacuum plasma treatment of metal articles.

References

[1] Borisov D P, Kuznetsov V M, Slabodchikov V A 2015 *Journal of Physics: Conference Series* **652** 012020
[2] Slabodchikov V A, Borisov D P, Kuznetsov V M 2015 *AIP Conference Proceedings* **1683** 4932905
[3] Borisov D P, Koval N N, Kuznetsov V M et al. 2013 *IEEE Trans. Plasma Sci.* **41** 2183
[4] Levchuk D 2007 *Surf. and Coat. Technol.* **201** 6071
[5] Ribeiro E, Malczyk A, Carvalho S et al. 2002 *Surf. and Coat. Technol.* **151-152** 515
[6] Zhang X B, Xiao J Q, Pei Z L et al. 2007 *J. Vac. Sci. Technol.* A. **25** 209
[7] Mukherjee S, Prokert E, Richter E, Moller W 2005 *Surf. and Coat. Technol.* **200** 2459
[8] Window B, Savvides N 1986 *J. Vac. Sci. Technol. A.* **4** 453
[9] Rossnagel S M, Kaufman H R 1986 *J. Vac. Sci. Technol. A.* **4** 1822
[10] Solov'ev A A, Sochugov N S, Oskomov K V et al. 2009 *Plasma Physics Reports* **5** 399

Acknowledgments

This study (state task No.11.1655.2014/K) was supported by the Ministry of Education and Science of the Russian Federation.

IOP Conf. Series: Journal of Physics: Conf. Series **830** (2017) 012058 doi:10.1088/1742-6596/830/1/012058

Analysis of work and efficiency increase of of the ILUR-03 installation magnetron system for tubular specimens outer surface modification

B A Kalin[1], N V Volkov[1], R A Valikov[1], A S Yashin[1], V P Krivobokov[2], S N Yanin[2] and Yu N Yuriev[2]

[1] National Research Nuclear University MEPhI (Moscow Engineering Physics Institute), 115409, Russia, Moscow, Kashirskoe highway, 31
[2] National Research Tomsk Polytechnic University, 634050, Russia, Tomsk, Lenin Avenue, 30

E-mail: nvvolkov@mail.ru, bakalin@mephi.ru, yashin_itf@mail.ru

Abstract. The method of material near-surface layers doping by mixing of alloying elements films with ion beam is widely used in science and technology. Three magnetrons with independent power systems, integrated in installation for ion-beam treatment of long-range products ILUR-03, were used as deposition systems. Targets for magnetrons were in the form of disks 60 mm diameter and 5 mm thickness and consisted of the following elements: Al, Fe, Mo, Zr, Cr of purity better than 99.99 at.%. Deposition was performed in argon atmosphere at 1-5 Pa pressure and room temperature in stable current mode at 30-100 mA. Analysis of the obtained films on the surface of cylindrical specimens from zirconium alloys with the outer diameter of 9.15 mm showed high uniformity of coating on length of 300 mm, good adhesion and absence of discontinuities in the films body.

1.Introduction

The method of magnetron coating is the deposition of the target atoms, which are sprayed in the magnetron plasma, on the substrate. This method allows us to put films with accuracy up to tens of nanometers on a wide class of substrates, and therefore it is widely used in various fields of science and technology. For example, the effectiveness of the surface modification of metal materials by mixing of the surface layer by ion beam depends on the quality and uniformity of previously deposited films of alloying elements. The magnetron sputtering method enables multi-component substrate doping as film deposition is not a thermally-activated process [1].

2.Materials and tools

Installation ILUR-03, which is a system for the complex ion-beam processing of long cylindrical products, is used in this paper [2]. As can be seen from figure 1, the installation vacuum system consists of a preliminary evacuation chamber, an ion source chamber and a chamber for deposition, in which the three independent magnetrons are installed. The specimens are delivered through the gateway system using a manipulator operating in automatic mode.

IOP Conf. Series: Journal of Physics: Conf. Series **830** (2017) 012058 doi:10.1088/1742-6596/830/1/012058

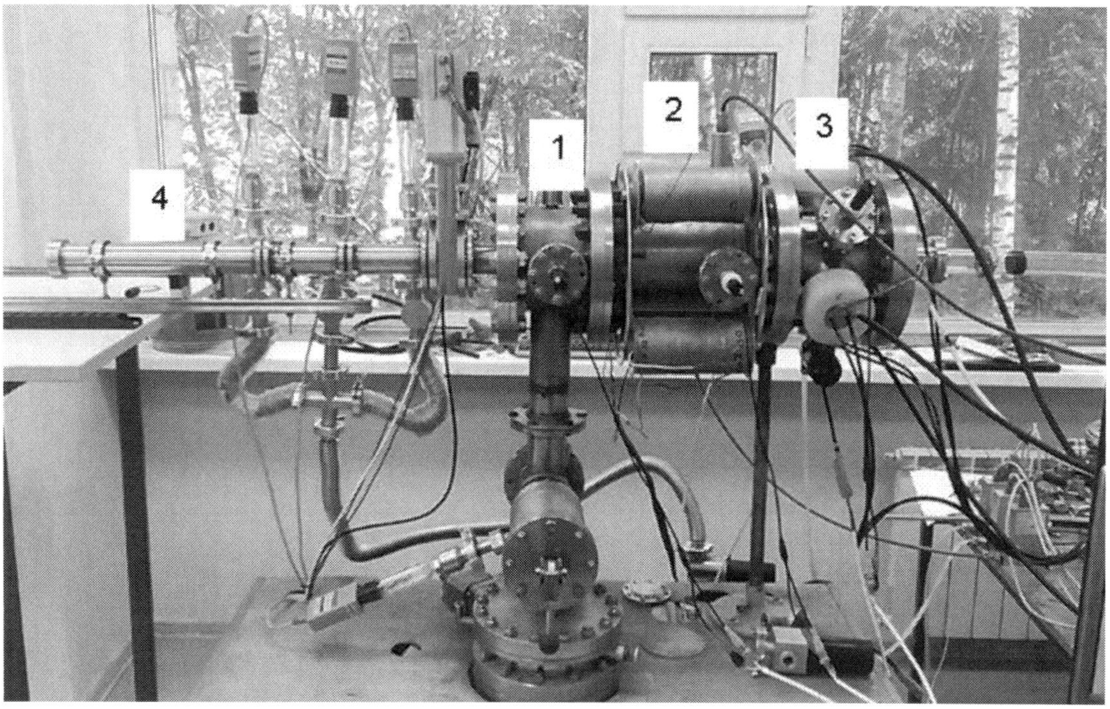

Figure 1. Installation ILUR-03: preliminary evacuation chamber (1), ion source chamber (2), chamber for deposition (3) and the gateway system (4).

Voltage is applied to the magnetron through specialized power supplies that can operate in a mode both DC and impulse voltages up to 800 V. The targets are in the form of disks 60 mm in diameter and 5 mm thick. Metal plates from Al, Fe, Mo, Zr, Mg, Cr with purity better than 99.99 at.% were used as materials for sputtering. The substrates were pieces of cladding tubes from E110 alloy (Zr-1%Nb) with the outer diameter of 9.15 mm and a length of 300 mm. The uniformity of coating was achieved by the continuous rotation of the tube and its movement in the longitudinal direction. Rotation and movement speed was between 1-10 rpm and 5-50 mm per min respectively.

The processing modes were selected during the preliminary experiments. Deposition was performed in an argon atmosphere at a pressure of 1-5 Pa in the discharge current stabilization mode at 30-150 mA. Residual gas pressure was maintained at 10^{-3} Pa. The appearance of the magnetron discharge at different current values is shown in figure 2.

To improve the adhesion surface the substrates were pre-treated by a radial beam of Ar^+ ions with the mean energy of 3-4 keV and total dose irradiation 10^{17}-10^{18} ion cm^{-2}. As a result of this treatment the contaminants and particles adsorbed with the surface of the tubes are removed, and the relief becomes smoother [3]. After the coating doping of the substrate surface layer was carried out by ion mixing of material by Ar^+ beam with the total radiation dose of 10^{18}-10^{19} ion cm^{-2}.

The current-voltage characteristics of the discharge, taken in the operating pressure range for a variety of targets are shown in figure 3. As can be seen on the example of Al, Mo and Cr, functions are close to linear in this voltage range, and the voltage discharge initiation is about 250-350 V.

These samples were examined using optical and scanning electron microscopes. The thickness of the films was determined by the gravimetric method on the GH-252 A & D analytical balances with a sensitivity of 0.01 mg. The elemental composition of the surface layer of the samples was monitored by X-ray microanalysis.

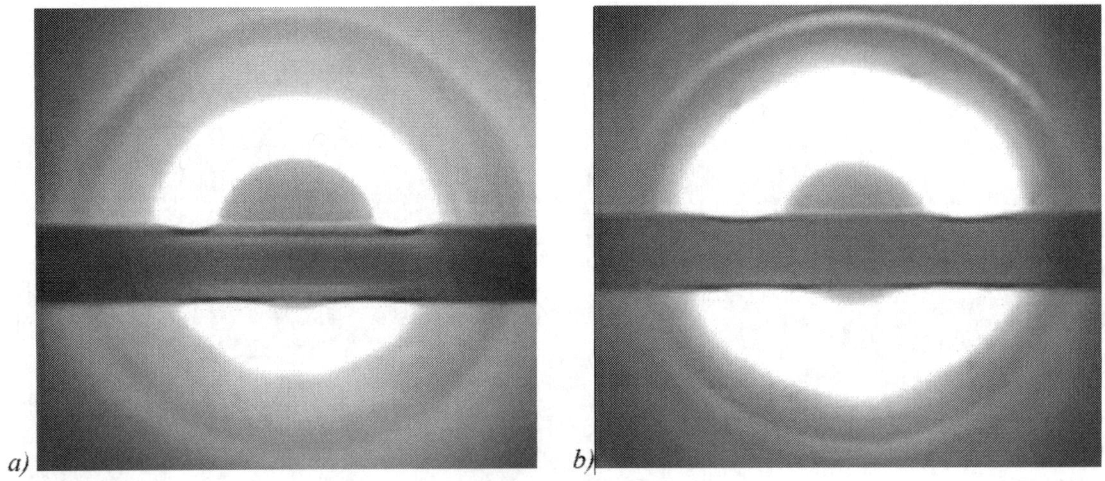

Figure 2. Appearance of magnetron discharge on a target of Al at a current of 50 mA (a) and 100 mA (b), a dark line in the foreground - a cylindrical substrate.

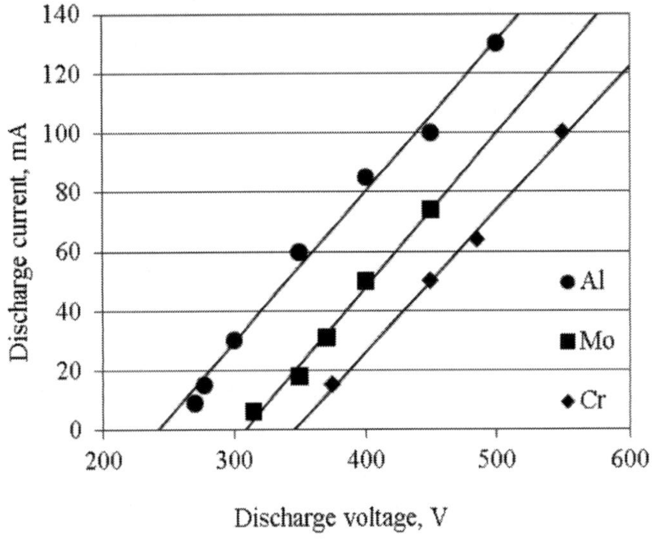

Figure 3. The current-voltage characteristics of the magnetron discharges on targets of Al, Mo and Cr, solid lines indicate a linear approximation.

3.Results and discussion

Figure 4 presents images of the sample with the iron sprayed layer obtained in the mode of the secondary and back-scattered electrons as an example. No significant irregularities of the deposited film are visually observed, and the film itself follows the surface topography of the original sample. Thus, the technological scratches formed during the finish machining of the tube and the inclusions of less than 1 μm size are clearly distinguished.

Table 1 summarizes the results of the elemental composition of the surface layer of doped samples obtained by X-ray microanalysis. As it can be seen, the atomic percentage of material in element films whose thickness was from 60 to 150 nm after the doping ranges from 0.3 to 3.5.

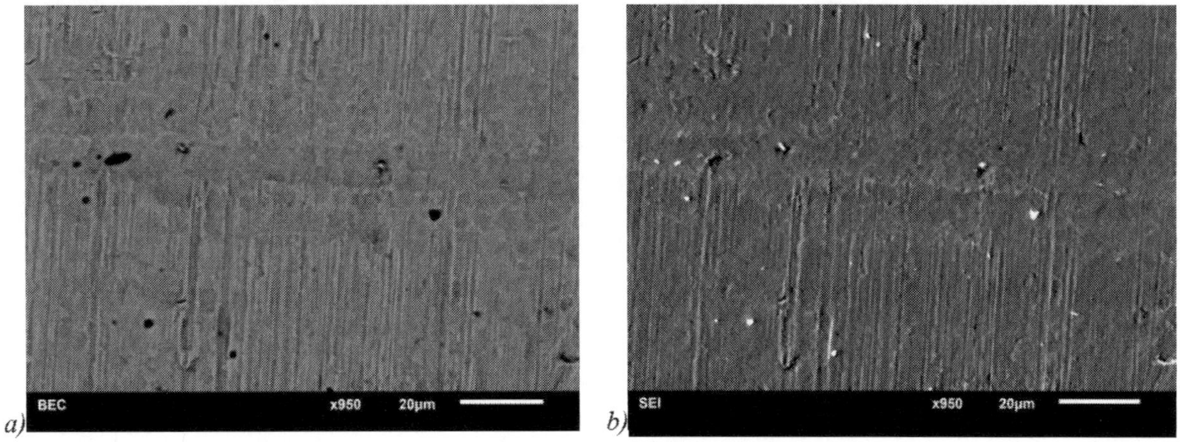

Figure 4. Images of the sample surface with iron sprayed layer obtained in the mode of secondary (a) and back-scattered electrons (b).

Table 1. The results of X-ray microanalysis of the sample surface after doping by ion mixing of deposited films.

	№1 Al-alloying	№2 Mg-alloying	№3 Fe-alloying	№4 Cr-alloying	№5 Mo-alloying
Zr (at %)	98.4	98.7	94.9	97.9	98.3
Nb (at %)	0.5	0.8	0.9	0.9	0.8
Fe (at %)	-	-	3.5	-	-
Cr (at %)	-	-	-	0.6	-
Mo (at %)	-	-	-	-	0.5
Al (at %)	0.8	-	-	-	-
Mg (at %)	-	0.3	-	-	-
Film thickness (nm)	80	90	150	60	70

4.Conclusion

The operation of the magnetron system of the ILUR-03 installation for ion beam processing of long cylindrical parts is analyzed. The range of operating parameters to ensure the effectiveness deposition of the films of several alloying elements Al, Fe, Mo, Mg, Cr on samples for subsequent surface doping in ion mixing mode is determined.

References

[1] Kalin B A, Volkov N V, Valikov R A, Yashin A S and Yakutkina T V 2016 *IOP Conf. Series: Materials Science and Engineering* **130** 012039

[2] Kalin B A, Volkov N V, Valikov R A, Yashin A S, Krivobokov V P, Yanin S N, Asainov O Kh and Yuriev Yu N 2016 *IOP Conf. Series: Materials Science and Engineering* **130** 012051

[3] Kalin B A, Volkov N V, Yashin A S, Valikov R A and Yakutkina T V 2016 *Bulletin of the Russian Academy of Sciences. Physics* **80** 165-167

IOP Conf. Series: Journal of Physics: Conf. Series **830** (2017) 012059 doi:10.1088/1742-6596/830/1/012059

Plasma jet characteristics in vacuum arc with diffused cathode spot

R Kh Amirov, N A Vorona, A V Gavrikov, G D Liziakin, V P Polistchook, I S Samoylov, V P Smirnov, R A Usmanov and I M Yartsev

Joint Institute for High Temperatures of the Russian Academy of Sciences, Izhorskaya 13 Bldg 2, Moscow 125412, Russia

E-mail: ravus46@yandex.ru

Abstract. Diffused vacuum arc, which is characterized by the absence of microparticles in cathode erosion products and by the irregular voltage oscillations, is considered to be a perspective plasma source for plasma reprocessing technology of spent nuclear fuel (SNF). The development of this technology requires data on ions energy in plasma jet. In this work parameters of plasma jet in diffused vacuum arc with a gadolinium cathode were studied by a retarding field analyzer, Langmuir and condensation probes. Gadolinium is regarded as a substance simulating SNF plasma. Ion energy spectrum was studied at arc currents of 30-75 A and voltages of 4-15 V at the distance of 20 cm above the arc anode. Dependencies of spectrum widths and most possible ion energies on arc voltages were obtained. The measured electron temperature was 2 eV, the maximum ion energy reached 70 eV. Experimental data were used to calculate adiabatic plasma expansion through the anode outlet.

1.Introduction

For the development of any plasma technology it is extremely important to have full information about parameters of used plasma. Nowadays a plasma reprocessing technology of spent nuclear fuel (SNF) is one of dynamically developing technologies. Conversion of condensed SNF in a plasma state with required parameters is one of the tasks that must be solved during the development of this technology. Vacuum arc with stationary diffuse cathode spot (DCS) is considered to be a perspective plasma source for plasma separation technology. This discharge can produce plasma with parameters required for plasma separation, such as high ionization degree (close to 100 %) and average charge of the ions close to unit [1]. The main distinction between a vacuum arc with DCS and contracted vacuum arcs on cold cathodes is a relatively low cathode current density (10-100 A/cm^2) that is achieved by a high cathode temperature (up to 2 kK). In these conditions voltage oscillations and technically harmful cathode droplets are absent in the vacuum arc with DCS [2].

Experimental studies of the vacuum arc with DCS on the thermionic gadolinium cathode have demonstrated that plasma generated by this discharge generally satisfied the mentioned requirements for the plasma separation technology [3]. However, the lack of values of ion energies prevents further development of this technology. Calculations [4, 5] show that separation of heavy and light ion groups becomes possible with ion energy about one electron-Volt.

Ion kinetic energies in cathode jets of contracted vacuum arcs were studied for most of the conductive materials from lithium to uranium [6]; these energies varied in the range of 20–150 eV. In this discharge ion energy slightly depends on the arc current and geometry of the electrodes unit. In

IOP Conf. Series: Journal of Physics: Conf. Series **830** (2017) 012059 doi:10.1088/1742-6596/830/1/012059

the vacuum arc with DCS on a thermionic cathode there is a possibility of arc voltage control by changing of the cathode temperature and correspondingly saturated vapor pressure. Thus, ion energies behavior in vacuum arc with DCS is very attractive. The main purpose of our work was to study an after anode plasma characteristics on the gadolinium cathode, including the dependence of ion energies on arc current and voltage. Also, it should be noted that properties of vacuum arcs on gadolinium and uranium cathode are quite similar [3].

2.Experimental setup

The scheme of the experiment is shown in figure 1. The discharge was initiated in the vacuum chamber with residual gas pressure below 10 mPa. A rectifier with an output voltage of 380 V was used as an arc power supply. Gadolinium (arc cathode) was placed in a heat-insulated molybdenum crucible with the height of 14 mm and external diameter of 25 mm. The diameter of the crucible outlet was 14 mm. Under the crucible there was an electron-beam heater (EBH) which allowed cathode temperature alteration at fixed arc current. The increasing in EBH power caused the decrease in the arc voltage U_a at the constant current I. A water-cooled steel disk, 26 mm thick with a centered 15 mm hole, was used as an arc anode. The diameter of the anode was 130 mm. The distance between electrodes was about 30 mm. The cathode temperature was measured by a brightness-temperature pyrometer. Estimated difference between measured and average temperature of the cathode surface was less than 3%. In experiments the cathode temperature was in a range of 1.9-2.0 kK. At these temperatures gadolinium saturated vapor density changes within the limits of 10^{14} -$5*10^{14}$ cm^{-3} [7].

After-anode plasma parameters (electron temperature, plasma potential and average charge of heavy particles) were measured by Langmuir and condensation probes. Ion energy spectrum was registered by a retarding field analyzer (RFA) that consists of a multi-grid probe with two 0.02x0.02 mm mesh size grids and a collector plate. The electron reflecting grid was biased at -60 V, and collector potential was varied from -30 to +30 V. The distance between both grids and collector was 0.5-1 mm. The diameter of the collector was about 30 mm. Note that in the described configuration this RFA could operate at plasma densities in the range of $10^9 - 10^{11}$ cm^{-3}.

RFA was placed at the distance of 20 cm above the anode along the discharge axis. A Langmuir probe was located close to the RFA outlet for plasma potential measurement. I-V curve for the Langmuir probe was processed using a standard method [8]. These data were used for RFA I-V curve processing and correction of ion energy spectrum shift on plasma potential where the analyzer was located.

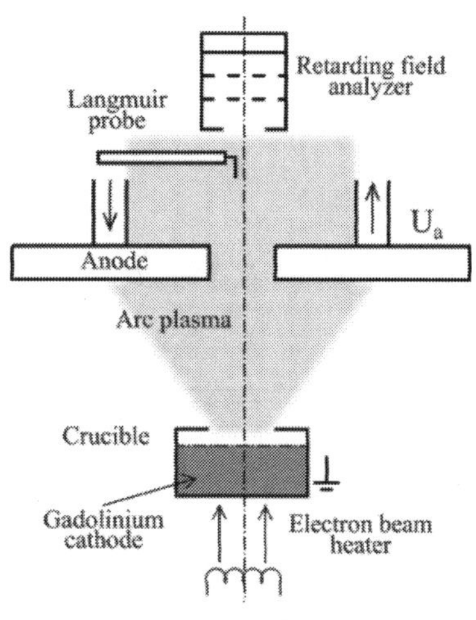

Figure 1. Experimental setup.

3. Experimental results

Typical plasma potential and electron temperature dependencies on arc voltage measured by a Langmuir probe at arc current I = 50 A are shown in figure 2. Similar results were obtained for other arc currents. At $I = 30$ A plasma potential monotonically rises from 4 to 6 V, and electron temperature increases from 0.5 to 1.7 eV. At $I = 75$ A the increase in the arc voltage from 5 to 12 V causes the plasma potential growth from 3 to 4 V and electron temperature increase from 0.5 eV to 1.3 eV.

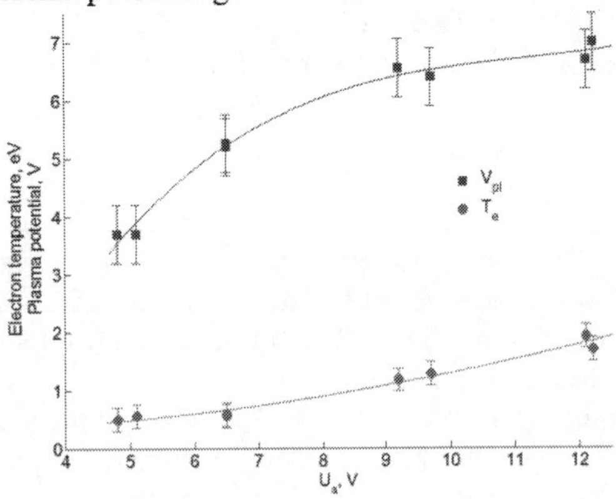

Figure 2. Plasma potential and electron temperature dependencies on arc voltage at a distance of 20 cm above the anode at arc current $I = 50$ A.

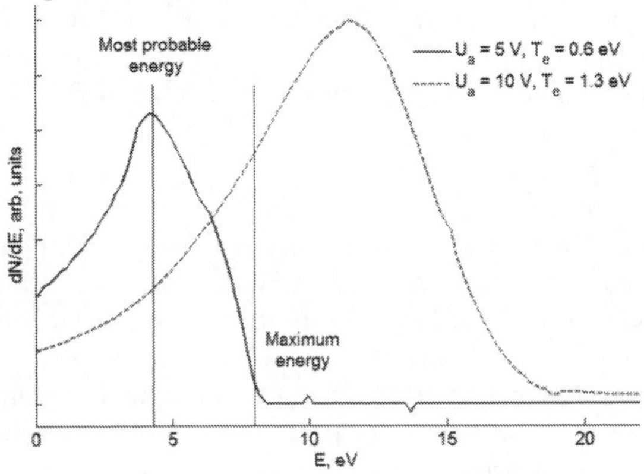

Figure 3. Ion energy distribution functions, $I = 50$ A.

Figure 3 shows the measured ion energy distributions at arc current $I = 50$ A. Figure 4 demonstrates the ion energies measurements (excluding ion charge state) in dependence on arc voltage at currents of 30, 50, 75 A. In different regimes values of most probable ion energies are in range of 3 – 20 eV, which significantly exceeds thermal energy of evaporated gadolinium atoms. Plasma density was determined from measured ion energy and ion saturation current of the Langmuir probe. At the distance of 20 cm above the anode at arc currents from 30 to 75 A the density of plasma was in a range of $3 \cdot 10^{9} - 1.5 \cdot 10^{10}$ cm^{-3}.

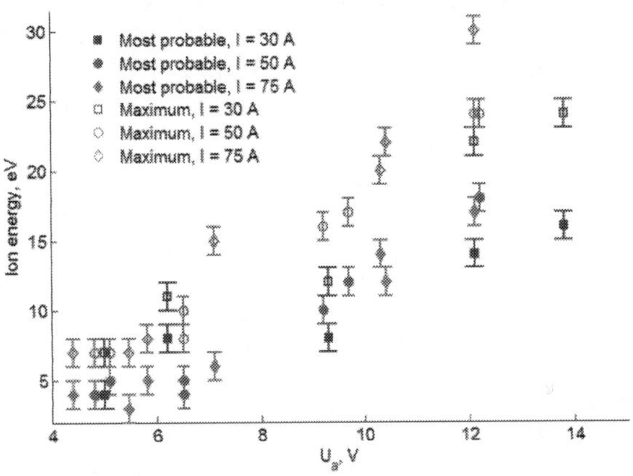

Figure 4. Most probable and maximum ion energies at the distance of 20 cm above the anode (excluding ion charge state).

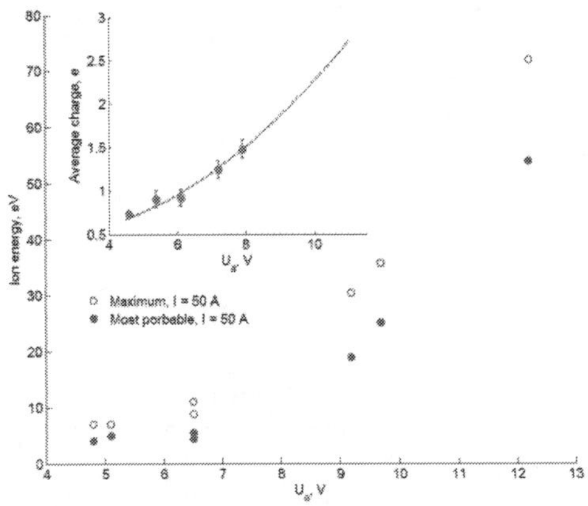

Figure 5. True plasma ion energies at arc current of 50 A. Average ion charge of gadolinium plasma.

Energy values showed in figure 4 were obtained excluding the ion charge state. According to work [3] in the studied discharge the double charged gadolinium ions appear in plasma at arc voltages above 6 V. In figure 5 there is a true ion energy dependence considering its average charge at arc current of 50 A [3]. The approximation of average ion charge versus arc voltage from [3] is shown in the upper right corner of figure 5.

According to [9] ion kinetic energies in a contracted vacuum arc on a gadolinium cathode are higher than our mean values and reach 50 eV at arc voltage of 20 V and electron temperature of 1.7 eV. The same inequality in values of ion energies was observed in [10] where both arc regimes were studied on a carbon cathode – cathodic spot arc and cathodic distributed arc. By [10] most probable ion energy value in a cathodic spot arc was 25 eV (at arc voltage of 20 V) and in cathodic distributed arc it was 20 eV (at arc voltage of 30 V). Arc current in both cases was equal to 60 A and carbon ions remained mostly single charged.

4. Estimations of after anode arc plasma parameters

Calculation of plasma jet parameters is needed for practical application. The whole area of arc plasma flow can be divided into three regions. The first region covers a discharge gap where supersonic plasma jet is formed. The second region is an anode hole and the third region extends from the anode to a collector which receives a plasma jet. One of the important distinctions of plasma processes in these regions is a nature of current changing. In the first region it remains constant along the region length. In the second region current is consumed by the anode and therefore it decreases toward the anode outlet. In the third region current can be practically absent. Exactly such a case was realized in our experiments. The leakage current on chamber walls was less than 2 mA, and apparently its main part was formed between the cathode and anode where plasma density is high enough.

As current changing nature in described significantly different regions, they can be considered separately and obtained solutions can be connected on the borders. In [11] two-temperature flow the model of fully ionized gadolinium plasma with fixed ion charge z_i was described. A quasi-one-dimensional approach for averaged values over the flow cross section was used. The considered discharge geometry was close to the one used in our experiment. According to [11] as moving away from the cathode the pressure in plasma jet decreases faster than magnetic forces which contract it. After jet pinching its cross section $F(x)$ is determined by discharge parameters. The cross section F(x) dependency on discharge parameters could be established as a result of jet pinching. Pinching of plasma jet leads to a qualitative change of plasma flow behavior. If the effect of magnetic forces is neglected the diffusive charge transfer dominates as moving away from the cathode – plasma potential decreases and electron temperature strives for zero. Pinching causes a weak rise of plasma potential and electron temperature.

In this work we used a model [11] for calculation and observed only after anode plasma. Boundary conditions for this region are specified by plasma parameters in the discharge gap and after anode area. After the anode current in plasma is absent, therefore adiabatic expansion of plasma cloud takes place, and ion kinetic energy increases on account of electron enthalpy.

Further it is considered that plasma stream after anode is supersonic (Mach number M_0 is greater than unit) and it expands in a channel with a specified cross section $F(x)$. Ionization and recombination processes in plasma can be neglected, so ion charge z_i remains constant and the ratio of electron T_e and ion temperature T_i along the jet does not change.

Changing of Mach number along the channel length M can be found from the solution of the differential equation [11]:

$$\frac{M^2-1}{M^2(1+M^2/3)}\frac{dM^2}{dx}=\frac{2}{F}\frac{dF}{dx} \tag{1}$$

Solution of (1) is:

$$\frac{F(x)}{F_0}=\frac{M_0}{(1+M_0^2/3)^2}\frac{(1+M^2/3)^2}{M} \tag{2}$$

Relation $F(x)=F_0(1+x^2tg^2\alpha)$ was considered as a channel cross section where F_0 is an area of anode outlet and α is a jet opening angle. From (2) using selected stream geometry a distribution of all main plasma parameters can be obtained.

Electron temperature along the stream is:

$$\frac{T_e(x)}{T_{e0}} = \frac{1+M_0^2/3}{1+M^2/3} \tag{3}$$

Ion kinetic energy W_i [11] is:

$$W_i = \frac{m_i u_i^2}{2} = \frac{5}{6} M^2 (T_i + z_i T_e) \tag{4}$$

Its changing along the channel length is:

$$\frac{W_i(x)}{W_{i0}} = \frac{M^2}{M_0^2} \frac{1+M_0^2/3}{1+M^2/3} \tag{5}$$

Electron density is:

$$\frac{n_e(x)}{n_{e0}} = \frac{F_0}{F(x)} \frac{M_0}{M} \sqrt{\frac{1+M^2/3}{1+M_0^2/3}} \tag{6}$$

Plasma potential distribution along the jet length is:

$$\varphi(x) = \varphi_0 - \frac{5}{2\mathrm{e}}(T_{e0} - T_e(x)) = \varphi_0 - \frac{5}{6e} T_{eo} \frac{M^2 - M_0^2}{1+M^2/3} \tag{7}$$

where φ_0 is plasma potential in anode outlet which is approximately equal to anode potential.

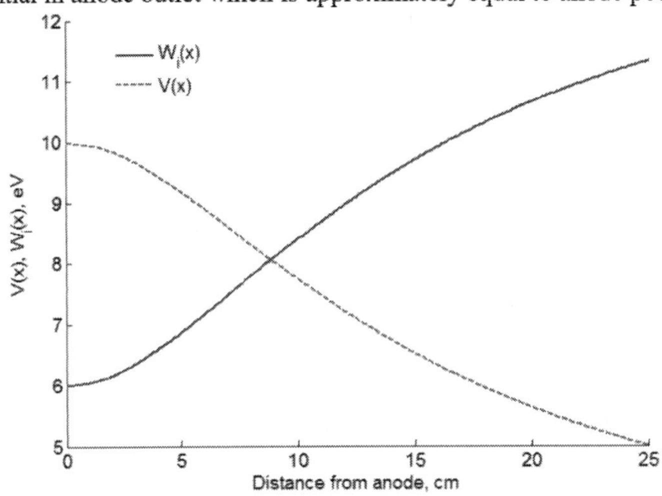

Figure 6. Calculated distributions of plasma potential *V(x)* and ion kinetic energy $W_i(x)$ in after anode area.

After anode plasma parameter distributions obtained with the help of (1)-(7) are shown in figures 6 and 7. Values of free parameters M_0 = 1.5, *tg* α = 0.05, T_{i0} = 0.2 eV were selected to coincide with experimental data at the distance of 20 cm above the anode. Value φ_0 = 10 B is selected for the corresponding arc regime, T_{e0} = 3 eV is selected according to the measured electron temperature within the discharge gap from [3].

5. Conclusion

In conclusion let us list our main results. Ion energy distribution of after anode plasma in a vacuum arc with DCS on a gadolinium cathode was measured. It was concluded that external cathode heating allows to change ion energy in the range of 4-70 eV at arc voltage of 4-14 V, under arc current of 30-75 A. It was noted that in a vacuum arc with DCS ion energies are less than in a contracted arc.

Overall, obtained experimental results on plasma jet characteristics satisfy requirements of plasma separation technology at low arc voltages [1].

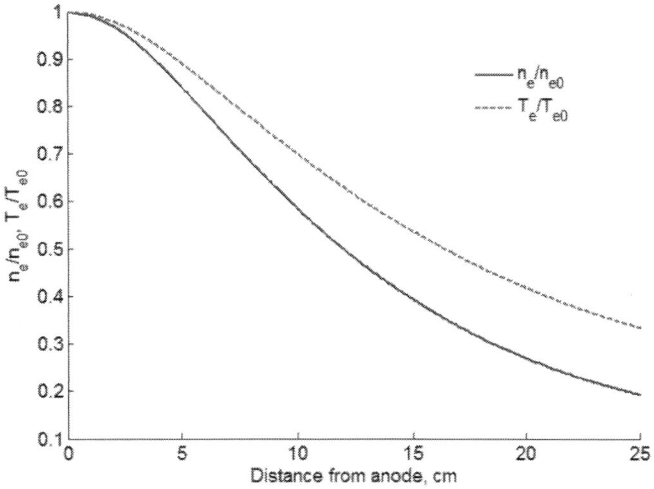

Figure 7. Calculated distributions of electron density $n_e(x)$ and electron temperature $T_e(x)$ in after anode area.

Experimental data were used to analyze adiabatic plasma expansion through the anode outlet. Calculated results allowed us to estimate distributions of plasma potential, plasma density, ion kinetic energy and electron temperature along the discharge axis in the whole area behind the anode. These data could be used in the development of plasma sources on the base of a vacuum arc with DCS.

References

[1] Vorona N A, Gavrikov A V, Samokhin A A, Smirnov V P and Khomyakov Yu S 2015 *Phys. Atom. Nuclei* **78** 1624

[2] Vasin A I, Dorodnov A M and Petrosov V A 1979 *Sov. Tech. Phys. Lett.* **5** 634-6

[3] Amirov R K, Vorona N A, Gavrikov A V, Lizyakin G D, Polishchuk V P, Samoilov I S, Smirnov V P, Usmanov R A and Yartsev I M *2015 Plasma Physics Reports* **41** 808–13

[4] Smirnov V P, Samokhin A A, Vorona N A and Gavrikov A V 2013 *Plasma Physics Reports* **39** 456–66

[5] Samokhin A A, Smirnov V P, Gavrikov A V and Vorona N A 2016 *Tech. Phys.* **61** 283–89

[6] Yushkov G Yu, Anders A, Oks E M and Brown I G 2000 *J. Appl. Phys.* **88** 5618

[7] Grigoriev I and Meylikhov E (eds) 1991 Physical Quantities. Handbook (Moscow: Energoatomizdat)

[8] Raizer Yu P 1991 *Gas Discharge Physics* (Berlin: Springer)

[9] Anders A and Yushkov G Yu 2002 *J. Appl. Phys.* **91** 4824

[10] Veerasamy V S, Amaratunga G A J, Weiler M, Park J S and Milne W I 1994 *Surface and Coatings Technology*, **68/69** 301–08

[11] Bronin S Ya, Polishchuk V P, Sychev P E, Shabashov V I and Yartsev I M 1993 *High Temp.* **31** 26–32

Acknowledgments

We are grateful to V. I. Kiselev for his important help in the preparation and realization of the experiment. The investigation is supported by the Russian Science Foundation (no. 14-29-00231).

Propagation of atmospheric pressure helium plasma jet into ambient air at laminar gas flow

M Pinchuk[1,2], O Stepanova[2], N Kurakina[1] and V Spodobin[1,2]

[1] Laboratory of Gas-Dynamic Systems, Institute for Electrophysics and Electric Power of Russian Academy of Sciences, St. Petersburg, 191186, Russia
[2] Faculty of Physics, Saint Petersburg State University, St. Petersburg, 199034, Russia

E-mail: o.m.stepanova@spbu.ru

Abstract. The formation of an atmospheric pressure plasma jet (APPJ) in a gas flow passing through the discharge gap depends on both gas-dynamic properties and electrophysical parameters of the plasma jet generator. The paper presents the results of experimental and numerical study of the propagation of the APPJ in a laminar flow of helium. A dielectric-barrier discharge (DBD) generated inside a quartz tube equipped with a coaxial electrode system, which provided gas passing through it, served as a plasma source. The transition of the laminar regime of gas flow into turbulent one was controlled by the photography of a formed plasma jet. The corresponding gas outlet velocity and Reynolds numbers were revealed experimentally and were used to simulate gas dynamics with OpenFOAM software. The data of the numerical simulation suggest that the length of plasma jet at the unvarying electrophysical parameters of DBD strongly depends on the mole fraction of ambient air in a helium flow, which is established along the direction of gas flow.

1. Introduction

Study of the “plasma – biological objects” interactions has become one of the main current trends of research on gas discharge plasma technologies [1-5]. There are many important reasons for considering low-temperature plasma as a new breakthrough instrument in biology [1], medicine [1-5], food [6] and agriculture [7] industries. The most rapidly spreading tools for producing low-temperature plasma at atmospheric pressure are plasma jet sources based on a dielectric-barrier discharge, in particular, generators of helium or argon jets propagating in ambient air.

To introduce such plasma sources into existing technological procedures and to create new technologies based on atmospheric pressure plasma jets (APPJ), a number of optimization problems should be solved. At first, we should be able to predict the distance of the propagation of a plasma jet into the surrounding air.

Some authors point out that a limited distance of plasma jet propagation (ionization wave propagation) in the ambient air is determined with the concentration of plasma supporting gas [8, 9], but they do not mention the influence of applied voltage. When the plasma supporting gas is injected from the generator into air, its mixing with the air occurs, and its concentration along the length of plasma jet decreases. The propagation of ionization waves is limited by the certain value of air admixture. The data suggesting this have been obtained for plasma jets generated in inert gases flows directed along the upward vertical at the applying of rectangular unipolar high-voltage pulses [8] and sinusoidal high voltage [9].

The given paper presents the first results of checking conclusions of [8, 9] for a helium plasma jet generated by sinusoidal voltage in a gas flow directed along upward and downward verticals.

2. Experimental set-up

To generate a plasma jet, we used a discharge cell with axisymmetric electrode system "inner wire – outer ring" which is presented in Figure 1.

Figure 1. The sketch of the discharge cell. *1)* quartz tube; *2)* grounded electrode; *3)* high-voltage electrode; *4)* gas flow pipe fitting).

As a dielectric barrier, quartz tubes with the inner diameters of 5.58 and 7.49 mm and the thickness of walls of 1 mm were used. The inner electrode of copper wire 1.5 mm in diameter was placed inside the tube along its center line at the distance of 7.5 mm from the edge of the tube. The outer electrode was a 5-mm-wide copper-foil strip wrapped around the tube at the distance of 5 mm from its edge.

Sinusoidal high voltage pulses with the peak-to-peak magnitude of 4 kV and frequency of 21 kHz was applied to the inner electrode. A schematic diagram of the experimental set-up with the system of electrical measurements is shown in Figure 2. The voltage signals from the electrodes were recorded with a capacitor voltage divider with a division ratio of 1:1350, while current signals were recorded with a non-inductive resistive shunt of 89 Ω. The signals were recorded by Agilent DSO-X 2014A oscilloscope (Agilent Technologies, USA).

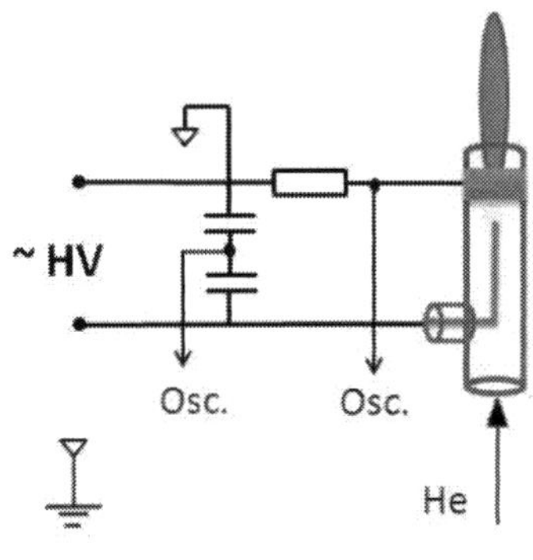

Figure 2. A schematic diagram of the experimental set-up with the system of electrical measurements.

As a plasma-supporting gas, high-purity helium was used. Gas flow rate was controlled by a Bronkhorst EL-Flow F-201AC flowmeter in a range of 0-45 l/min. Some additional details of the gas flow system can be found in [10].

To estimate the experimental length of a generated plasma jet, it was photographed using Nikon D80 DSLR equipped with Sigma 28-200 Macro lens.

3. Brief description of computational model

A helium laminar jet submerged into air was simulated using OpenFoam 2.1.1 program package. To optimize the calculations and take into account an axial symmetry of the problem, a segment of 50×300 mm^2 with the angle of 4 degrees was considered with different density of computational mesh $(0.3\text{-}1)\times0.5$ mm^2. Physical parameters of a helium flow (fraction of air, density, heat conductivity, viscosity) were given according to the industrial certificate of gas production.

The solver *twoLiquidMixingFoam* of OpenFOAM was modified in accordance with the problem. Unsteady-state equations of gas dynamics for an axially-symmetric helium jet submerged into air are as follow:

1) the mass conservation law (continuity equation):

$$\frac{\partial \rho \alpha_i}{\partial t} + \nabla \cdot \left(\rho \vec{\vartheta} \alpha_i\right) = -\nabla \cdot \left(-\rho D \nabla \alpha_i\right), \quad (1)$$

where α_i is the mass fraction of each species ($i = 1$ is for helium species; $i = 0$ is for air species); ρ is the density of the gas mixture; $\vec{\vartheta}$ is the gas mixture velocity ; D is the diffusion coefficient for species in the gas mixture (for the laminar regime of the gas flow it depends on the gradient of component concentration);

2) motion equation:

$$\frac{\partial \rho \vec{\vartheta}}{\partial t} + \nabla \cdot \left(\rho \vec{\vartheta} \vec{\vartheta}\right) = -\nabla p + \nabla \cdot \left(\tilde{\tau}\right) + \rho \vec{g}, \quad (2)$$

where p is pressure; $\tilde{\tau}$ is the stress-tensor with due consideration of the force of viscous tangent friction; $\rho \vec{g}$ is gravitational force.

The calculation was carried out for the laminar regime of a helium flow in air at different helium flow rate.

4. Preliminary experimental results and discussion

The calculation results of helium concentration profiles were compared with the photos of generated APPJ. The matching of simulation results with the experimental findings are depicted in Figures 3 and 4 for the gas flow direction along upward and downward verticals, correspondingly. The parameters of voltage supply were set as follows: peak-to-peak voltage was 4 kV, pulse frequency was 21 kHz. The discharge current was up to 15 mA.

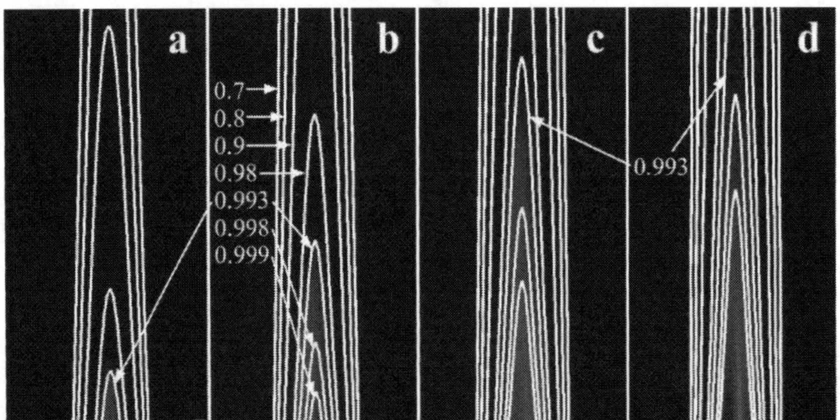

Figure 3. Photos of the plasma jet generated in a laminar helium flow directed along upward verticals through the tube of 7.49 mm in a diameter at the gas flow rates of a) 1.36, b) 2.09, c) 3.03, and c) 4.08 m/s. White lines are helium concentration profiles obtained by computational simulation.

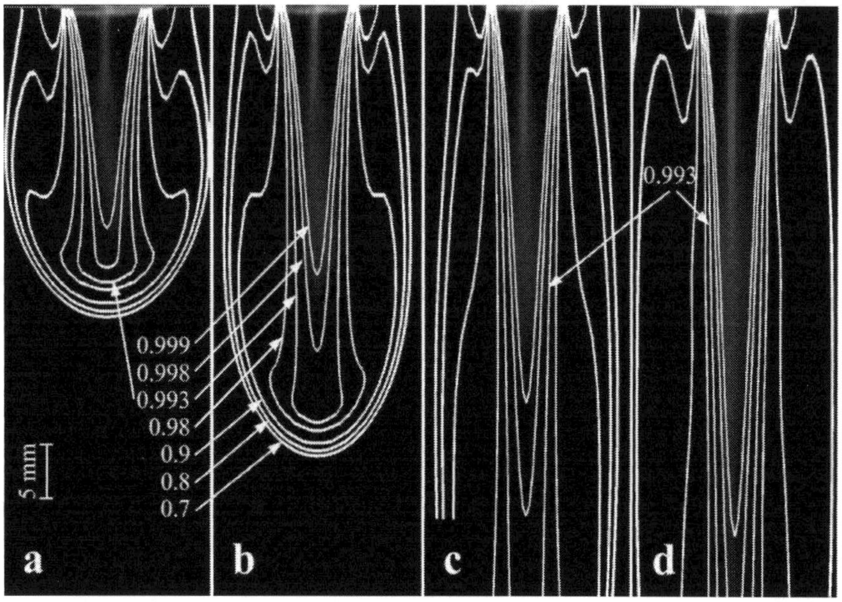

Figure 4. Photos of the plasma jet generated in a laminar helium flow directed along downward vertical through the tube of 7.49 mm in a diameter at the gas flow rates of a) 1.31, b) 1.73, c) 2.5, d) and 3.34 m/s. White lines are helium concentration profiles obtained by computational simulation.

Dependences of the dimensionless length of APPJ for the gas flow directed upward and downward through the tubes of different diameters versus Reynolds number are shown in Figures 5 and 6. Reynolds number is $\mathrm{Re} = \frac{\rho_{He} \cdot \vartheta \cdot d}{\mu_{He}}$, where ρ_{He} and μ_{He} are the density ($\frac{kg}{m^3}$) and dynamic viscosity ($\frac{\mu N \cdot s}{m^2}$) of helium at atmospheric pressure and temperature of 20 °C, respectively; ϑ is gas outlet velocity ($\frac{m}{s}$); d is the inner diameter of the discharge tube (m).

The APPJ length increases as a linear function of a gas flow rate under laminar regime. For Reynolds number exceeding 200-300, a laminar gas flow begins to transfer into a turbulent one, and the deviation of dependence of APPJ length from the linear function appears. Under the developed turbulent regime (Reynolds numbers is more than 500) for both directions of gas flow and both diameters of the tubes, the dimensionless APPJ lengths were approximately of 4. Under this regime, air effectively mixes with helium, and the length of mixture plasma jet does not depend on the gas flow orientation in space.

Comparison of the experimental and calculated data showed that for the laminar regime the expansion boundary of APPJ is limited by the helium concentration of 0.987-0.996 at the discharge generation by a peak-to-peak voltage of 4 kV with frequency of 21 kHz (Figures 5 and 6).

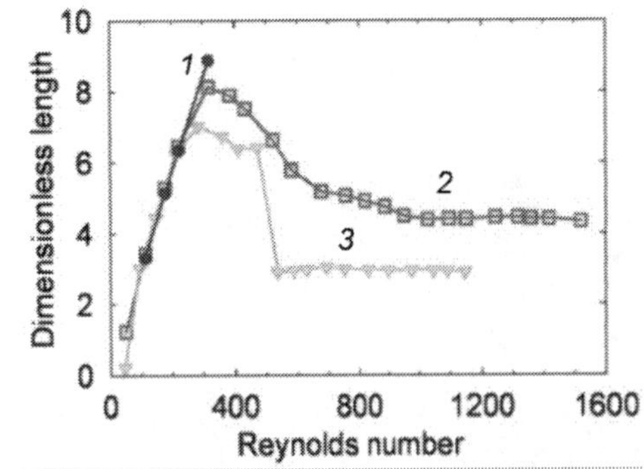

Figure 5. Dimensionless plasma jet length (Ljet/d) versus Reynolds number (Re) for helium flow along upward vertical through the tube of 5.58 (curve *2*) and 7.49 (curve *3*) mm in a diameter in comparison with the simulation results (curve *1*).

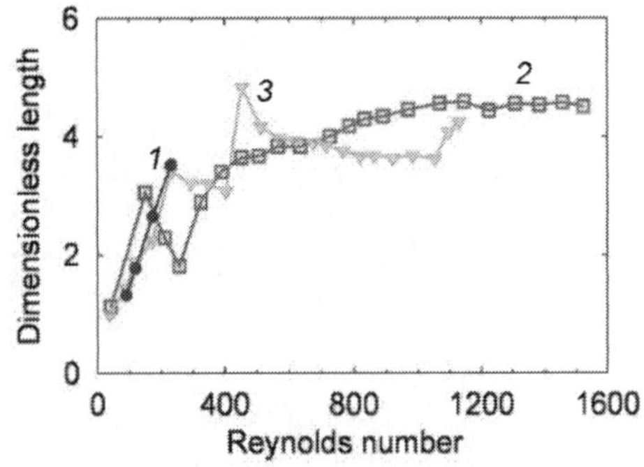

Figure 6. Dimensionless plasma jet length (Ljet/d) versus Reynolds number (Re) for helium flow directed along downward vertical through the tube of 5.58 (curve *2*) and 7.49 (curve *3*) mm in a diameter in comparison with the simulation results (curve *1*).

We should note that the plasma jet propagation is connected with the propagation of ionization waves (guided streamers). Thus, the air impurity influences APPJ length, since ballast air interrupts the resonant transmission of radiation by the ionization wave.

For the other modes of the applied voltage, APPJ length versus gas flow rate is slightly different, but the general trend corresponds to the data given above.

5. Conclusion

According to the series of presented data, the limit boundary of APPJ expansion in helium at laminar regime is determined by the achievement of the certain volume concentration of air impurity, which gradually increases with the distance from the tube outlet.

For the laminar regime, the propagation area of guided streamers corresponds to the boundary of the relative helium volume concentration of 0.987-0.996 for discharge at peak-to-peak voltage of 4 kV and frequency of 21 kHz. At the same time, the length of APPJ linearly increases with the rising of gas flow rate.

Some calculations in comparison to more extensive experimental data are planned to be carried out.

References

[1] Tendero C, Tixier Ch, Tristant P, Desmaison J, Leprince Ph 2006 *Spectrochimica Acta Part B* **61** 2-30.

[2] Shintani H, Sakudo A, Burke P and McDonnell G 2010 *Experimental and Therapeutic Medicine* **1** 731-738.

[3] Fridman A and Friedman G 2013 *Plasma Medicine* (Philadelphia: WILEY Drexel University) p 545.

[4] Graves David B 2014 *Physics of Plasmas* **21** 080901.

[5] Laroussi M 2015 *IEEE Transactions on Plasma Science* **43(3)** 703-712.

[6] Misra N N, Tiwari B K, Raghavarao, Cullen P J 2011 *Food Engineering Reviews* **3** 159–170.

[7] Dobrin D, Magureanu M, Bogdan N, Mandache, Ionita M-D 2015 *Innovative Food Science and Emerging Technologies* **29** 255-260.

[8] Xiong R, Xiong Q, Nikiforov A Yu, Vanraes P, Leys Ch 2012 *Journal of Applied Physics* **112** 033305.

[9] Li Q, Li J-T, Zhu W-Ch, Zhu X-M, and Pu Y-K 2009 *Applied Physics Letters* **95** 141502.

[10] Pinchuk M, Kudryavtsev A, Stepanova O, Subbotin D 2015 *Proc. Int. Conf. on Plasma Physics and Plasma Technology (Minsk)* vol 1 (Minsk: Belarus/Kovcheg) pp 51-54.

Acknowledgments
This work was partially supported by Saint Petersburg State University (grant no. 0.37.218.2016) and the Russian Foundation for Basic Research (grant no. 16-08-00870).

IOP Conf. Series: Journal of Physics: Conf. Series **830** (2017) 012061 doi:10.1088/1742-6596/830/1/012061

Combined impact features for laser plasma generation

E Loktionov, Yu Protasov, V Telekh

Bauman Moscow State Technical University, Moscow, 105005, Russia

E-mail: stcpe@bmstu.ru

Abstract. Laser-induced plasma has been considered for multiple applications by the moment, and its characteristics strongly depend on laser radiation parameters. Reaching demanded values for the latter might be rather costly, but, in certain cases, similar or even better results could be reached in case of additional impact (optical, electric, magnetic, corpuscular, mechanical etc.). Combined impact effects are mainly based on target properties or interaction mechanism change, and found to decrease plasma generation thresholds by orders of magnitude, improve energy efficiency significantly, and also broaden the range of plasma parameters. Application area, efficiency and optimal regimes for laser plasma generation at such combined impact have been considered. Analysis based on published data and own experiments was performed for both target material and induced plasma flows. Criterial parameters have been suggested to characterize both combined impact and response to it. The data on plasma generation thresholds, controlled parameters, working media supply systems and recovery rate of droplets are very important for technology setups, including those for material modification.

1. Introduction

Control of plasma parameters is crucial for surface modification technology. Competitive plasma setups should provide high performance, be energy efficient and consist of reliable and inexpensive components. The analysis of data published on combined generation and control of laser plasma properties reveals several ways for performance improvement: modification of material properties for the most efficient laser radiation absorption and energy conversion; reduction of losses by proper energy transfer form and means; plume collimation and acceleration without energy consumption.

Laser impact efficiency is usually characterized by ablation threshold W_a (minimum radiation fluence enough for macroscopic material removal), ablation rate h (crater depth increase after pulse), specific mass flow rate $\Delta m/E$. Laser plasma generation efficiency is determined by specific impulse $I_{sp}\approx\langle v\rangle/g$ ($\langle v\rangle$ – is mass averaged rate), momentum coupling coefficient $C_m\approx\Delta m\langle v\rangle/E$, laser to kinetic energy efficiency $\eta\approx\Delta m\langle v\rangle^2/2E$ and some other parameters [1, 2], and can be improved by additional impact on the material or plume. For example, the following additional impacts led to increased laser plasma generation efficiency: material heating to fusion or glass transition temperature; material UV irradiation leading to excitation of chromophores and breaking of bonds; material surface covering with ~1 mm gel layer; multiple irradiation of the same spot until deep narrow crater formation (or preliminary manufacturing of those, e.g., wide blind pores); attachment of a cylindrical nozzle with a diameter ~2.4 times bigger than laser focal spot; attachment of a plate capacitor with divergent electrodes narrowing towards the ends, and irradiation with the focus closer to the cathode; application of a constant magnetic field (>0.15 T) normal to the irradiated surface; radiation focusing at two closely localized points under two-pulse impact.

More promising ways for the improvement of combined impact efficiency are the following: local modification of material optical properties at the irradiated area; laser impact on ferrofluids at magnetic field and liquids confined in pored matrices; combined impact of laser and charged particle beams; radial confinement of laser plasma plume. Investigation of these ways of laser-induced plasma generation could provide new data on laser-matter interaction. Additional impact on working medium can also improve the performance of storage, transportation and dosing subsystems of plasma technology setups.

2. Review of previous works

Material heating can be performed electrically, optically or chemically. For some materials optimal temperature rates have been discovered (depending on efficiency parameters), those coming from the change in material properties, intensity of physical and chemical processes (e.g., hardness change or oxide films formation). One of convenient ways to control temperature is the change of interpulse delay between heating and acting irradiation. It has been shown experimentally that energy deposited per amount of removed material decreases. Since only surface layer needs to be heated, inductors and microwave heaters seem to be very promising.

Under double pulse UV impact on PMMA (with resonance absorption at 308 nm) a very pronounced synergistic effect has been discovered caused by excitation of chromophore groups and breaking of intermolecular bonds (followed by formation of oligomers) [3]. The most pronounced effect was observed under simultaneous impact that is considered as the evidence of photo processes. Except oligomerization, UV radiation can induce condensation, crystallization and polymerization (the latter can be also induced under electron beam impact and heating). Transient modification of material properties occurs at ultrashort prepulse irradiation [4], which makes quartz opaque to visible radiation due to excitation of electrons until relaxation or local damage. The structure of some materials can be also sensitive to magnetic or electric field due to orientation of dipoles.

Solid or liquid confining layers are also used (both at front and back side of the target) for the combined impact induced reduction of energy losses, plasma compression and heating at initial stages (also due to efficient absorption of laser radiation), increase of the duration of plasma-target interaction. Laser impact energy can be reduced for such layered targets, and the momentum coupling coefficient is multiplied significantly due to detonation [5]. If two solid surfaces are divided by a liquid layer, the effect becomes even more pronounced at high radiation intensities because of matter density increase at the interface. Confining layer optimal characteristics can not be evaluated theoretically (tried to though [6]), and can be obtained only experimentally for specific target, cover and laser parameters.

Efficiency of different nozzles (cylinder, wedge, cone, hemisphere, paraboloid) in laser plasma injectors has been studied extensively [7]. The results demonstrate, that momentum coupling coefficient reaches its maximum at wedge [8] or cone angles less than 15° [9], slightly converging nozzle can be also effective [10]. Inputs of shockwave and ablation plume are found analytically to be about equal in a plain nozzle sealed from one side [11]. An optimum for nozzle throat to laser focal spot ratio was found to be 2.2 for cone and 2.4 for cylinder [12]. For deep ablation craters, as a kind of micro-nozzles, it was demonstrated that at reduced radiative losses, plasma temperature is proportional to the crater depth, and it decreases slower. The same relates to particle number density [13], and therefore, to the pressure. These effects become less pronounced at crater diameter increase [14] and depth-to-diameter ratio decrease [15]. Filling micro cavities with a liquid leads to a significant increase of the momentum coupling coefficient proportionally to a liquid layer thickness, but specific impulse decreases due to formation of droplets [16]. Nozzles can be used not only for reactive force increase, but as a reflective optics as well. This is possible with some difficulties though, which are mainly related to wall cooling under radiative flux much stronger than a convective one and to manufacturing of micro-nozzles of demanded quality. Annular focusing could be helpful for resolving these problems, since it leads to cylindrical plume formation [17].

In most cases, ablation plume is ionized significantly, so electric field could be obviously used for its acceleration to reach mass averaged velocity of 10s km/s. Combined laser-electric configurations have been investigated with 3 ways of acceleration: electrostatic, electromagnetic, and electro-thermal. Control of plasma characteristics with laser radiation makes it possible (with corresponding shape of electrodes) to implement all of these types in a single device by setting parameters depending on demanded performance.

Plasma generation efficiency at laser ignition in a plate capacitor with ablating dielectric increased significantly at focal spot positioning closer to a cathode [18]: initial plume speed was the highest at the anode. Moreover, the magnetic field of up to 0.25 T appears in such capacitor [19], and current oscillations lead to the formation of at least 3 ionization waves. Combined geometric (narrow cylindrical channel) and electrostatic impact under negative potential application increased electron number density by side-compression of the plume and slowed down the recombination at late stages (>100 ns); positive potential application led to plasma radiation output at initial stages (<100 ns).

Magnetic impact has been considered for flow axis control [20], acceleration [21], compression [22, 23] and collimation [24]. Those are most pronounced for the high-energy fraction of the plume [25] and lead to the stratification and internal electric field generation [26]. Axial symmetry of the plume (this is characteristic for pure laser impact) disappears in transverse magnetic field [23]. This also leads to flow deceleration [22], since plasma tends to move along the field lines [27], and does not improve gas dynamic performance (but these effects are less pronounced for light particles [28]). Axial magnetic field starts affecting the plume structure from values of ~0.04–0.16 T [29]. For axial field of 4.5 T recoil momentum was found to double [30], most likely, due to plume compression, so the field input is about equal to laser ablation one [31]. Preliminary collimation of the flow (due to a concave surface) under multiple irradiation of the same spot [27] can be significantly improved by 1 T magnetic field. These effects can be obtained using permanent magnets, since no external energy is spent for those both absolute and specific performance can be improved.

A significant plume velocity and ionization rate increase takes place at double-pulse irradiation, but temperature increases slightly as compared to a single impact [32]. If the target surface is covered by a liquid layer, double-pulse impact has a negative effect (as compared to the uncovered target) due to the increased density of the upper layer [33]. There are several time segments characteristic for different physical processes, for which local optima can be obtained at different impact conditions.

The previous paragraph considered collinear double pulse impact. In the orthogonal scheme (second pulse radiation is focused in the plume in parallel to the target surface [34]), plasma flow velocity depends on this focal spot distance from the target surface and decreases due to the interaction of flows induced by separated sources, but leads to highly effective collisional excitation and ionization [35]. In the case of the pre-pulse parallel to the target surface (with the energy which is sub-threshold for the ablation), crater volume increased only slightly and depended on the first pulse energy at 0.2–1000 μs delays (with the maximum effect reached at ~10–30 μs) [36]. Parallel scheme has been also investigated [37]. Unlike pre-pulse orthogonal one with no ablation at the first stage, the shockwave energy in the parallel scheme is higher due to spatial confinement. And unlike the collinear scheme, the second pulse goes through the region free of nanoparticles and clusters, formed by the pre-pulse, which increases the fraction of radiation absorbed by the target. Combination of these factors leads to a significant ablation rate increase; craters do not have so obvious rims as in the pre-pulse orthogonal scheme due to more effective melt drainage.

3. Experimental results

We have investigated experimentally several new ways of combined impact for laser plasma generation: irradiation of ferrofluids in axial magnetic field of 0.05–0.5 T; modification of optical properties of the liquids containing photoactivators and absorbing additives; symmetric radial confinement of the plume; laser-electric breakdown of gases. Detailed descriptions of experimental setups were published elsewhere [38, 39]. All the results were evaluated in terms of plasma generation threshold, ablation mass flow rate, recoil momentum, radiation spectral output, and the specific parameters calculated (specific impulse, momentum coupling coefficient, energy efficiency, quantum yield) [40]. Examples of combined impact benefits are presented in Figure 1.

Laser ablation rate of the ferrofluid (averaged for 10 pulses) was 0.21 g/J, in magnetic field at 140 mT it was 0.16 g/J (decrease was proportional to the viscosity increase); in glass cylindrical nozzle at 480 mT it decreased by 3 orders of magnitude down to 0.2 mg/J, and was about equal to that for solid polymer targets [41]. The nozzle had no such effect at zero magnetic field, since the droplets remained on the walls. With the field applied, the droplets moved back towards the magnet, so only evaporated matter took off. Magnetic field moving along nozzle walls could accelerate retrieval of the droplets. When ferromagnetic (steel) nozzle was used, droplets moved towards its end, since magnetic flux at the edges had the highest value. The increase of the momentum coupling coefficient under the impact on droplets in the magnetic field was ca. 30% (up to 4 mN*s/J), and energy efficiency increased by half (up to 3%), staying at low level that is characteristic for liquids. Droplets stopping is important not only for material saving and mass averaged velocity increase, but also in technology, e.g. for thin films deposition.

The efficiency of combined laser-electric impact on gases has been also investigated (more detailed results are published in this volume). Specific regimes of combined impact were discovered, leading either to the decrease or the increase of breakdown thresholds. Combined impact efficiency rate has been suggested as a relation of optical and electric breakdown threshold values at combined and pure impact. Optimal pressures for combined impact were found to be in the range of 10^2–$3*10^5$ Pa, and the efficiency gradual increase was inversely proportional to the radiation wavelength of 213–1064 nm. These results were predictable theoretically. The comparatively long-living (several microseconds) dense radiating cord was a less expected experimental finding. It was, formed at certain ratios of laser and electric arc induced shockwaves energies [42, 43]. This should increase output of short wavelength radiation, used for surface modification.

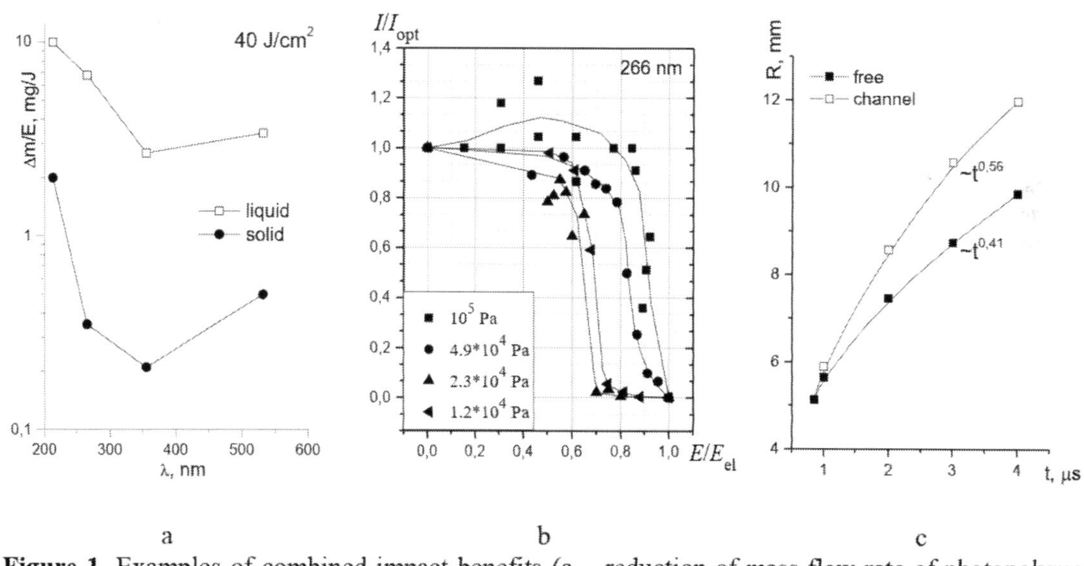

a b c

Figure 1. Examples of combined impact benefits (a – reduction of mass flow rate of photopolymer, b – reduction of breakdown threshold components under laser-electric impact at different pressures of Kr, c – reduction of shock wave deceleration in a square channel).

Spatial confinement effects were investigated using dual-wavelength interferometry and shadowgraphy of ablation plumes for 6 configurations (axial, radial, combined, static and dynamic) at ambient conditions and vacuum, linear, single- and double point focusing [44]. Generation regimes, structure and lifetimes have been evaluated for stagnating, vortex, compressed and radiating areas. Analysis of shockwaves dynamics in relatively narrow (<3 focal spot diameters) transparent channels has revealed the change of shock front dimensionality from 2D ellipsoid to a quasi-1D flat. The shockwave energy remains concentrated at a small surface, unlike fast growing hemispherical front, so the efficiency of material shock treatment can be improved.

4. Conclusion

The analysis of different combined impacts on target materials and laser plasma flows has been performed in terms of efficient energy absorption and reduction of losses, plume acceleration and collimation. Several specific methods for the improvement of laser plasma generation efficiency have been found out from previously published results. Also, several promising ways have been suggested and evaluated. The reduction of mass loss by three orders of magnitude due to droplet phase stop and retrieval has been demonstrated for laser ablation of ferrofluids in axial magnetic field and confined by paramagnetic nozzle. This finding is extremely important for surface modification plasma setups.

References

[1] Loktionov E Y, Ovchinnikov A V, Protasov Y S, Protasov Y Y and Sitnikov D S 2014 *High Temperature* **52** 132-134
[2] Loktionov E Y, Ovchinnikov A V, Protasov Y Y and Sitnikov D S 2010 *Doklady Physics* **55** 446-449
[3] Srinivasan R and Braren B 1990 *J. Appl. Phys.* **68** 1837-1841
[4] Lin C H, Rao Z H, Jiang L, Tsai W J, Wu P H, Chien C W and Tsai H L 2010 *Proc. SPIE* **7585** 75850I
[5] Phipps C R, Seibert Ii D B, Royse R W, King G and Campbell J W 2000 *Proc. SPIE* **4065** 931-938
[6] Zheng Z-Y and et al. 2007 *Chinese Phys. Lett.* **24** 501
[7] Ageev V P, Barchukov A I, Bunkin F V, Gorbunov A A, Hudyakov V M, Konov V I, Korobeinikov V P and Putyatin B V 1981 *Acta Astronautica* **8** 625-641
[8] Han B, Chen J, Shen Z-H, Lu J and Ni X-W 2011 *Opt. Laser Technol.* **43** 604-608
[9] Horisawa H, Tamada K and Kimura I 2006 *Proc. SPIE* **6261** 62612M-10
[10] Ageichik A A, Egorov M S, Rezunkov Y A, Safronov A L and Stepanov V V 2004 *Proc. AIP* **702** 49-60
[11] Li Q, Hong Y, She J, Cao Z and Wang J 2006 *Proc. AIP* **830** 121-129
[12] Li L, Tang Z, Hu X and Peng J 2011 *Proc. AIP* **1402** 258-268
[13] Hou Z, Wang Z, Liu J, Ni W and Li Z 2013 *Opt. Express* **21** 15974-15979
[14] Kumar B, Singh R K and Kumar A 2013 *Phys. Plasmas* **20** 083511
[15] Zeng X, Mao X, Mao S S, Yoo J H, Greif R and Russo R E 2004 *J. Appl. Phys.* **95** 816-822
[16] Zheng Z Y, Zhang J, Zhang Y, Liu F, Chen M, Lu X and Li Y T 2006 *Appl. Phys. A* **85** 441-443
[17] Veloso F, Chuaqui H, Aliaga-Rossel R, Favre M, Mitchell I H and Wyndham E 2006 *Rev. Sci. Instruments* **77** 063506-5
[18] Ayabe T, Horisawa H, Funaki I and Kimura I 2007 *30th Int. Electric Propulsion Conf.* (Florence) IEPC-2008-58
[19] Koizumi H, Noji R, Komurasaki K and Arakawa Y 2007 *Phys. Plasmas* **14** 033506-10
[20] Kajimura Y, Matsuda N, Hayashida K, Maeno A and Nakashima H 2008 *Proc. AIP* **1084** 934-938
[21] Pagano C and Lunney J G 2010 *J. Phys. D: Appl. Phys.* **43** 305202
[22] Harilal S S, Tillack M S, O'Shay B, Bindhu C V and Najmabadi F 2004 *Phys. Rev. E* **69** 026413
[23] Albertazzi B, Béard J, Ciardi A, Vinci T, Albrecht J, Billette J, Burris-Mog T, Chen S N, Da Silva D, Dittrich S, Herrmannsdörfer T, Hirardin B, Kroll F, Nakatsutsumi M, Nitsche S, Riconda C, Romagnagni L, Schlenvoigt H-P, Simond S, Veuillot E, Cowan T E, Portugall O, Pépin H and Fuchs J 2013 *Rev. Sci. Instruments* **84** 043505
[24] Roy A, Murtaza Hassan S, Harilal S S, Endo A, Mocek T and Hassanein A 2014 *Phys. Plasmas* **21** 053106

[25] Chen S N, Gauthier M, Higginson D P, Dorard S, Mangia F, Riquier R, Atzeni S, Marquès J-R and Fuchs J 2014 *Rev. Sci. Instruments* **85** 043504
[26] Ikeda S, Nakajima M, Hasegawa J, Kawamura T and Horioka K 2014 *Nuclear Instruments Meth. Phys. Research A* **733** 103-106
[27] Rubin A, Zaidi S and Miles R. 2012 *50th AIAA Aerospace Sciences Meeting including the New Horizons Forum and Aerospace Exposition* (Nashville) AIAA-2012-197
[28] Li Y, Hu C, Zhang H, Jiang Z and Li Z 2009 *Appl. Opt.* **48** B105-B110
[29] George S 2011*Application of imaging to study the evolution and dynamics of laser produced plasma from solid and thin film Li targets*. (Cochin)
[30] Zaidi S H, Smith T W, Murray R and Miles R B 2005 *29th Int. Electric propulsion Conf.* (Princeton) IEPC-2005-232
[31] Maeno A, Yamamoto N, Nakashima H, Fujioka S, Johzaki T, Mori Y and Sunahara A 2011 *Appl. Phys. Lett.* **99** 071501
[32] De Giacomo A, Dell'Aglio M, Bruno D, Gaudiuso R and De Pascale O 2008 *Spectrochimica Acta Part B: Atom. Spectrosc.* **63** 805-816
[33] Sakka T, Tamura A, Nakajima T, Fukami K and Ogata Y H 2012 *J. Chem. Phys.* **136** 174201
[34] Sánchez Aké C, Sanginés de Castro R, Sobral H and Villagrán-Muniz M 2006 *J. Appl. Phys.* **100** 053305
[35] Sobral H, Sanchez-Ake C and Villagran-Muniz M 2008 *Spectrochimica Acta Part B: Atom. Spectrosc.* **63** 493-497
[36] Sobral H and et al. 2011 *J. Phys. D: Appl. Phys.* **44** 085201
[37] Cristoforetti G, Legnaioli S, Palleschi V, Tognoni E and Benedetti P 2010 *Appl. Phys. A* **98** 219-225
[38] Loktionov E Y 2015 *Applied physics* (5) 12-14
[39] Loktionov E Y 2014 *J. Appl. Spectrosc.* **81** 305-308
[40] Loktionov E Y, Ovchinnikov A V, Protasov Y S, Protasov Y Y and Sitnikov D S 2012 *Optics and Spectroscopy* **112** 631-637
[41] Loktionov E Y, Ovchinnikov A V, Protasov Y Y, Protasov Y S and Sitnikov D S 2012 *J. Appl.d Spectrosc.* **79** 104-112
[42] Loktionov E Y, Pasechnikov N A, Protasov Y S, Protasov Y Y and Telekh V D 2015 *J. Appl. Spectrosc.* **82** 607-613
[43] Loktionov E Y, Pasechnikov N A, Pavlov A V, Protasov Y S and Telekh V D 2015 *J. Phys.: Conf. Ser.* **652** 012043
[44] Loktionov E Y, Protasov Y Y, Telekh V D and Khaziev R R 2013 *Instruments and Experimental Techniques* **56** 46-54

Acknowledgements
Research has been performed using “Beam-M” facility, following the State Mission, and was partially supported by the Russian Foundation for Basic Research (grants Nos. 14-08-01087 and 15-38-20890) and by the President of Russia (grant MK-5407.2016.8).

Experimental Research of Dynamics and Macrostructure of Light Erosion Radiative Plasmodynamic Discharges

A Pavlov, Y Protasov, V Telekh, T Tshepanuk

Bauman Moscow State Technical University, Moscow, 105005, Russia

E-mail: stcpe@bmstu.ru

Abstract. The investigations of dynamics and macrostructure (spatial and temporal distributions of charged particles) of optic-plasmadynamic discharges with light erosion mechanism of plasma formation and generation of accelerated gas–plasma flows of various chemical composition in the wide range of dynamic and energy & power characteristics are presented. The plasma was heated as a result of the thermalization due to shock interaction of high kinetic energy plasma flows with a gas medium, which works as a stopping barrier for such discharges. It was shown that the deceleration of the radially inhomogeneous flow at a deformable gas barrier has an essentially two-dimensional nature: a complex system of conical shock wave forms in a hypersonic plasma flow.

1. Introduction

High current light erosion optic-plasmadynamic discharges (OPDD) are produced due to light erosion mechanism (generation, heating and acceleration of irradiating plasma jets) in which electromagnetic energy is converted to kinetic energy of a quasi stationary irradiating gas–plasma flow. Those are discharges with extremely high specific heat input ($W/S \sim 10^{11}$–10^{13} W/m^2, $Wt_{1/2} > 10^5$–10^7 J, W – power, $t_{1/2}$ – the discharge current half-cycle, S – the area of the midsection to cut electrodes [1], [2]).

Studies of OPDD are stimulated by the prospects of their use as high power and manufacturable sources of UV and VUV spectra for lithographic processes [3], [4], as well as shock wave (SW) and radiating plasma flow generators [1], [5]–[8]. These devices can be used for modeling of thermophysical and optical-plasmadynamic processes in high density radiation plasma flows–matter interactions [9]–[11] (for example, for dry surface cleaning in flat panel displays and printed circuit boards manufacture), in the surface plasma metallurgy for structural surface modification of construction materials exposed to compression plasma flows [12], [13], and also for plasma chemistry applications such as implementation of special conditions for photochemical reactions [9]. One of the fastest growing areas of research in recent years at the interface of physics and medicine is active particles and radicals impact on biological objects (molecules, cells, tissues), for the formation of which gas discharge plasma VUV-radiation is used [9], [11], [14].

Effective radiation plasma dynamic sources requirements are [5]: 1) the light erosion mechanism of radiating gas discharge plasma generation and the shock-wave mechanism of heating; 2) the effective conclusion of photon energy from an emitting area; 3) a technological method of energy accumulation and input.

One possible solution to this problem is an erosion type magneto-plasma compressor (MPC) [15] as a highly efficient generator of the accelerated gas–plasma flow in which the kinetic energy may be converted in different ways into heat energy of shock-compressed plasma, which is accompanied by generation of high-brightness radiation. Structurally, the MPC consists of shaped coaxial electrodes separated by a dielectric sleeve. The plasma-forming substances are products of electrodes erosion and dielectric sleeve ablation. The plasma acceleration is carried out under the action of pondermotive forces arising from the interaction of the discharge current radial component with the azimuthal component of its own magnetic field. The plasma acceleration is accompanied by plasma flow electromagnetic cumulation along the system axis, its own magnetic field of the discharge confines plasma flow in the MHD-compression zone.

The results of the dynamics and macrostructure of optical-plasmadynamic discharges in gases are summarized. In these discharges, the plasma heating is effected by thermalization of the kinetic energy of plasma flows, when the shock wave interaction with a gas medium acts as a barrier.

2. Experimental setup

The radiating plasmadynamic discharges created in the erosive type impulse electromagnetic plasma accelerator with coaxial and edge geometry electrodes (diameter 6 and 20 mm) and an ablative PTFE plasma-forming sleeve were researched (*I*) (fig. 1). The gas discharge chamber (*II*) was pumped before each discharge up to the pressure $p_0 \leq 1$ Pa and filled with gas (Ne, Xe, Ag, air) up to the pressure $p_{g0}=2\cdot10^3$–10^5 Pa. The capacitor storage device (*22*) (7.05 kJ, 28 kV) charged up to the voltage of 15–25 kV was switched with accelerator electrodes of a controlled discharger (*21*), the discharge was periodic with damping (7–9 half-cycles of current); full energy deposition into the discharge was ~ 70% of the stored energy, of which 40% is enclosed in the first half-cycle of the discharge current, current maximum was I_{max} ~ 350–850 кA and duration of half-cycle ~ 3.1–9.6 μs. The polarity of electrodes was selected according to the recommendations [15]: if the central electrode during the first half-cycle had negative polarity, then threshold conditions of plasma focus formation considerably decrease, the main discharge characteristics – working media coefficient of use, degree of flow compression, etc. improve, and the discharge light output increases.

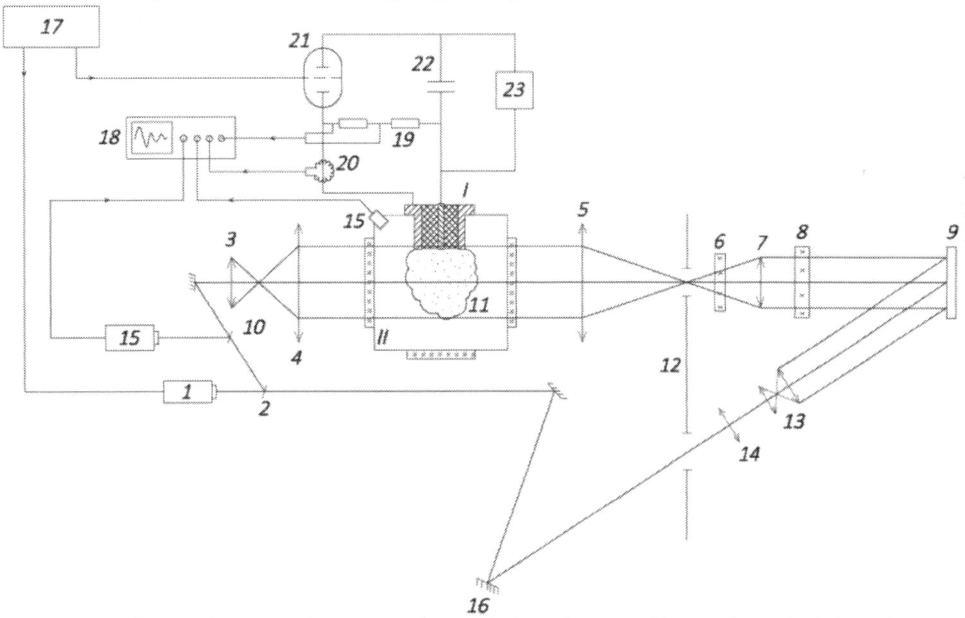

Figure 1. Experimental setup scheme: 1 – laser; 2, 10 – beam splitters; 3, 4, 5, 7, 14 – lenses; 6 – wedge; 8 – optical filter; 9 – screen; 11 – phase object; 12 – pinhole; 13 – beam expander; 15 – photoelements; 16 – mirrors; 17 – pulse generator; 18 – oscilloscope; 19 – voltage divider; 20 – Rogovski coil; 21 – discharger; 22 – capacitor store; 23 – power supply.

Measurements of a discharge current were carried out by Rogovski coil (*20*), working in the current transformer mode. According to current oscillograms (the discharge half-cycle, the relation of the current maximum values in the first and second half-cycles), complete inductivity of a circuit $L \approx 100$ nH and its pure resistance $R \approx 0.03$ Ohm are defined. The maximum value of current in the first half-cycle when the capacitor charging to U=20 kV made $I_{max} \approx 250$ kA. Estimates of the pure resistance of discharge and its inductivity gives: $R_D \approx 0.02$ Ohm, $L_D \approx 15$ nH.

Laser plasma diagnostics was performed using the light field mode of Toepler's schlieren scheme and two-frame holographic interferometry. Single impulses of Nd:YAG laser radiation (*1*) were formed by giving a sequence of synchropulses from a pulse generator (*17*) from the discharge ignition (*21*) and with the required delay (0–15 μs) on the laser electro-optical shutter. The moment of laser operation in relation to the discharge current was controlled by an oscilloscope (*18*) to which channel signals came: from Rogovski coil (*20*), from a voltage divider (*19*), from photoelements (*15*) registering a laser impulse and discharge radiation.

The parallel beam of probe radiation created by the telescope from microlens *3* and lens *4* and had an aperture ~ 150–190 mm. After passing through phase inhomogeneity *11* it was focused by lens *5* (fig. 1) with focal length of 0.75 m on the pinhole with a diameter of 1.8 mm in the screen *12*. Own discharge emission is cut off by the filter (*8*). For obtaining schlieren images mirror *2* was opaque and for interferometry – beam-splitting; in the latter case, the registered screen *9*, besides the object beam, got the reference beam passing through the mirror *16*, which equalized optical path. The telescope *13* and the lens *14* are designed to match the laser radiation spatial structure of both interferometer paths.

The spatial resolution of the laser diagnostic setup was specified by the resolution of the optical system (~50 μm), and temporary one bythe laser pulse duration (~ 50 ns). Minimum detectable by schlieren-images integral concentration gradient for electrons was 10^{18} cm^{-4}, for Ne atoms – $2*10^{19}$ cm^{-4}. For interferometry – the measurement range of quasistationary (in relation to changes during the exposure) electronic concentration distributions N_e have been limited from below by registered shift of periods, and from above by imposing of periods and made for $\int N_e dx$ about $10^{17} - 2*10^{20}$ cm^{-2}. For N_e distributions which are quickly changing in time (for example because of the plasma forming movement or the advanced turbulence), the interference pattern disappeared exceeding $(\partial \int N_e dx)dt$ at the level 10^{25} $cm^{-2}s^{-1}$ (because of bands shift during registration). Visible and near UV light pulses were detected by photoelements.

3. Experimental results

At interaction of the accelerated plasma flow with a gaseous medium, the complex shock wave structure is registered in these experiments. The dynamics of the discharge development can be observed in a series of the schlieren images (fig. 2) and interferograms (fig. 3) for discharge in neon with initial pressure 28.7 and 73.3 kPa. Electron density spatio-temporal distribution (fig.3 d–f) in discharge is received by interferograms treatment (fig.3 a–c).

The zone, characteristic for the plasmadynamic discharges, of the increased density and temperature (MGD-compression zone) is formed before a central accelerator electrode after a gas pushing off; that demonstrates the strong focusing of a flow internal part in an interelectrode interval. Around this zone at different density and chemical composition of gas the conical shock front is registered, and the smaller base of the truncated cone surface is based on the edge of the central electrode (as, apparently, shock wave (SW) formation in this case is due to the interaction of highly focused plasma flow with the edge of the accelerator central electrode). Before the accelerator end faces another conical shock wave converted by the cone big base to the accelerator end is observed (fig. 2 e, f). Narrowing and widening conical shock waves are also observed behind the line of the intersection of these two gas dynamic discontinuities, which, for their part, are reflected by axial flow and the external border of the plasma formation. The originating system of gas dynamic discontinuities reminds the structures which are formed in case of the hyperacoustic expansion in the flooded space of a gas dynamic nozzle [16], but there is no perfect analogy here.

IOP Conf. Series: Journal of Physics: Conf. Series **830** (2017) 012062 doi:10.1088/1742-6596/830/1/012062

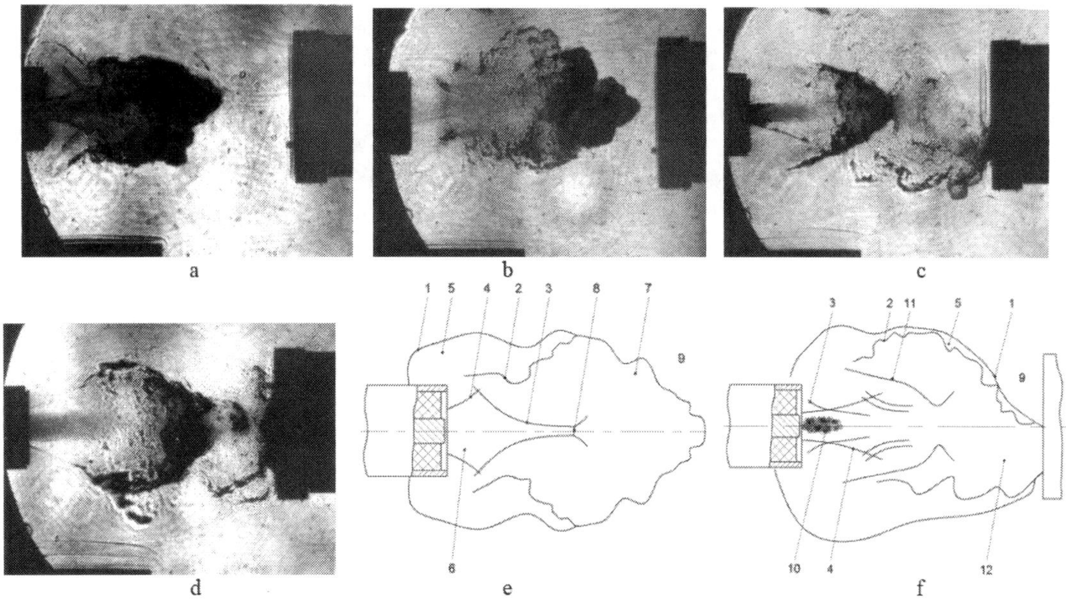

Figure 2. Schlieren photos (a – 3.7, b – 4.2, c – 6.2, d – 7.4 μs) and discharge internal structure (e – 4.2, f – 6.2 μs): 1 – shock wave in gas; 2 – plasma–gas interface; 3 – conical shock wave; 4 – focus shock wave; 5 – impact compressed gas; 6 – erosion plasma flow; 7 – plasma and gas turbulent mixing zone; 8 – Mach's disk; 9 – undisturbed gas; 10 – focus zone; 11 – first and second half-cycles plasma mixing zone and shock front; 12 – first half-cycle collapsing plasma; Ne, P = 28.7 kPa.

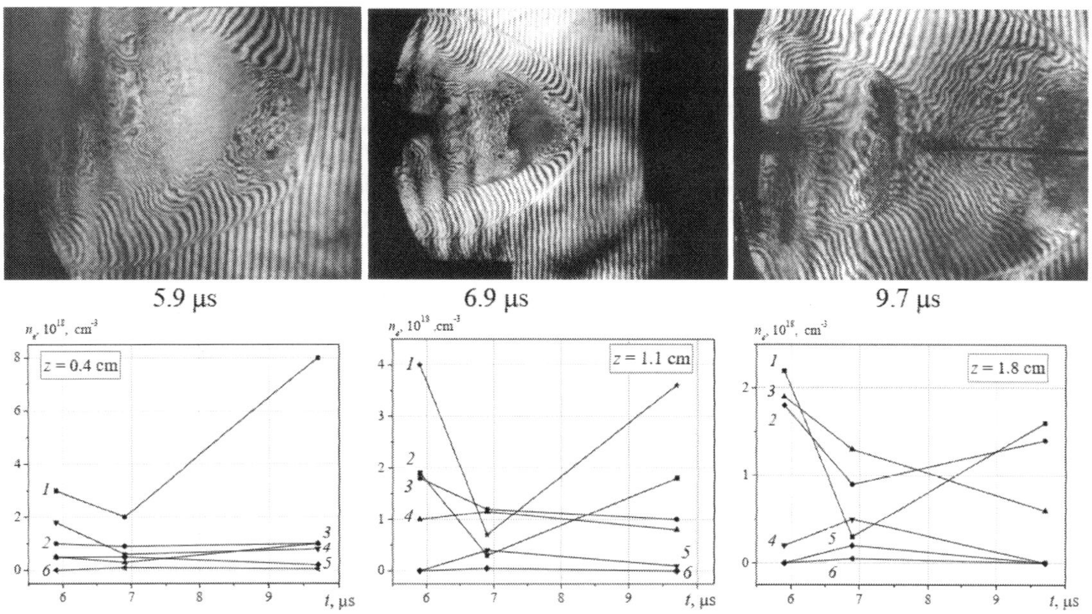

Figure 3. Interferograms (a – 5.9, d – 6.9, c – 9.7 μs) and electron density spatio-temporal distribution in discharge obtained by interferograms treatment (radial sections numbers: 1 – r = 0 cm; 2 – 0.5; 3 – 1.0; 4 – 1.5; 5 – 2.0; 6 – 2.5; z is flow axis); Ne, P = 73.15 kPa.

There is thermalization of a part of plasma elementary volume kinetic energy when passing each shock wave; as a result, in the researched discharge (in comparison with vacuum plasmadynamic discharges [15]) the compression efficiency on the axis significantly increases, characteristic values of flows speed decrease to ~ 30–35 km/s, the radiating capability and plasma temperature grows to ~ 3–6 eV. Gas is affected by light erosion plasma total pressure equal to a gas dynamic flow pressure that will be approved with representation of the plasma piston [17], [18]. The luminescence area consists of some radiating zones located in all plasma structures. In the SW head part the non-uniformly emitting front is watched, which is possible due to non-uniformity of spatio-temporal discharge formation at initial stage and development further of hydrodynamic instabilities. Let us note that the speed of the emitting front, as well as all current macrostructure, weakly depends on a gas sort if initial densities match and is generally specified by initial power and accelerator configuration.

The registered conic gas dynamic discontinuities in plasma at the discharge evolution move up and down the flow and change the slope relative to the axis that cannot be explained only with the movement of a deformable gas barrier. So, at the end of each current half-cycle the first of the specified shock waves departs further from a compression zone, increasing the slope relative to the axis, the second shock wave, on the contrary, moves towards the axis, etc.; such movement of gas dynamic discontinuities is connected with non-stationary character of plasma flows.

Besides, additional complication of interaction pattern is entered by the light erosion plasma forming inertia leading to mismatch between a blow-in of erosive weight and of input energy power deposition in discharge; this strongly influences parameters of flows and shock-compressed plasma [5], [19]. Light radiation from discharge plasma arrives on erosive dielectric and electrodes and during a pause of a discharge current when an electromagnetic acceleration is absent. As a result, during this time in front of the accelerator the light erosion vapors are collected, and the plasma flow of the following half-cycle should punch formed "cork", forcing out vapors to an axis and from the accelerator. The shock wave structure which is formed before the accelerator central electrode at the beginning of the discharge second, third, etc. half-cycles (fig. 2 c, d, 3 f – c) confirms it.

Thus, due to non-stationary character of a power input during the discharge the radiating zones arise and move, therefore the total radiation impulse has the complex time dependence which does not repeat generally the dynamics of a power input. Such flow structure is unstable in relation to turbulization – in the schlieren images (fig. 2 a – d) intensive turbulization in the discharge head part during the first half-cycle is observed. That is accompanied by the perturbation appearance (the size of 0.1–0.5 cm) on the shock wave front propagating in the unperturbed gas. One more type of non-uniformity is observed near contact border plasma–gas. It is a "cellular" structure with a cells size about 0.3–0.5 cm. Observed plasma turbulization significantly affects the spectral–energy characteristics of the OPDD.

Two zones, as it has been established during the studies of spatio-temporal structure, define the discharge light output: a plasma focus zone and plasma–gas shock wave interaction zone. In the acceleration of SW ($\rho_{ro}<10^5$ g/cm^3) during the discharge initial stage (τ < T/8) intensity of plasma focus zone radiation considerably exceeds the brightness of a shock braking zone. Further there is an alignment of radiation intensity from both zones. With transition to the braking mode ($\rho_{ro} > 2\cdot10^5$ g/cm^3) braking shock wave intensity increases and intensity of this zone luminescence is compared to radiation from the plasma focus zone, and at the subsequent moments considerably exceeds it. With gas density growth, the increase in MGD-compression zone radiation intensity is observed, this is related to an intensification of shock wave processes at discharge plasma radial expansion and with photoionization strengtheningleading to reradiation from VUV to more long-wave range.

4. Conclusion

The results of experimental investigation of dynamics and macrostructure (spatial and temporal distributions of charged particles) of a new class of optical-plasmadynamic discharges with light erosion mechanism of plasma formation and accelerated gas–plasma flows generation, with high specific characteristics of a power contribution, in the wide range of dynamic characteristics are presented. The structure of these discharges is investigated using laser diagnostics with high spatial and temporal resolution. The complex system of conical shock waves in hypersonic plasma flow is found. Kinetic energy thermalization has multistage character because of essential not one-dimensionality of accelerated plasma flow with gas interaction. This considerably changes the luminescence body shape and radiating plasma parameters and thus significantly influences discharge light efficiency. The investigated discharges are of interest as sources of powerful short-wave radiation, shock waves and dense multiply ionized plasma.

References

[1] Morozov A I 2000 Radiative Plasmodynamics *Enciclopedia of Low Temperature Plasma. Introduction voluem.* Vol III ed V E Fortov (Moscow: Nauka) P 539
[2] 1991 *Radiative Plasmodynamics* vol 1, ed Yu S Protasov (Moscow: Energoatomizdat)
[3] 2006 *EUV Sources for Lithography* ed V Bakshi (Wash. Bellingham: SPIE Press)
[4] Chkhalo N I and Salashchenko N N 2009 *Nucl. instrum. and methods in phys. research.* A **603(1-2)** 147
[5] Protasov Yu S 2000 Plasma light sources of high spectral brightness *Enciclopedia of Low Temperature Plasma. Introduction voluem.* Vol IV ed V E Fortov (Moscow: Nauka) 232
[6] Gribkov V A 2000 Pulse and not gas discharge methods of plasma generation *Enciclopedia of Low Temperature Plasma. Introduction voluem.* Vol II ed V E Fortov (Moscow: Nauka) P 328-392
[7] Protasov Yu S, Protasov Yu Yu, Suslov V I and Telekh V D 2000 Experimental research of radiative gas dynamic and thermophysical processes in plasma optical converter of laser radiation into AC electrical current. I. *35 Intersociety Energy Conversion Engineering Conference: AIAA paper* **2000-2888**. N.Y.: AIAA.- 7 p.
[8] Protasov Yu Yu, Chuvashev S N and Telekh V D 2002 *Gasdischarge electronics* (Moscow: Bauman Moscow State Technical University Publ.)
[9] Zvereva G N 2012 *Journal of optical technology* **79(8)** 477
[10] Morimoto Y, Sumitomo T, Yoshioka M, Takemura T 2004 *Proc. of IAS (IEEE Industry application society)* p 24
[11] Weltmann K Kindel E Woedtke T Hähnel M Stieber M and Brandenburg R 2010 *Pure Appl. Chem.* **82** 1223
[12] Cherenda N N, Shymanski V I, Uglov V V, Astashinskii V M, Kuz'mitskii A M, Koval' N N, Ivanov Yu F and Teresov A D 2012 *Inorganic Materials: Appl. Research* **3** (5) 365
[13] Cherenda N N and Uglov V V 2010 Modification of Steels Microhardness by Compression Plasma Flows *Handbook of Material Science Research* Ed C Rene and E Turcotte (New York: Nova Science Publ.) p 125-172
[14] Kalghatgi S Kelly C Cerchar E Torabi D Alekseev O Fridman F Friedman G and Azizkhan-Clifford J 2011 *PLos One* **6(1)** 1
[15] Kamrukov A S, Kozlov N P and Protasov Yu S 1984 in *Ion injectors and plasma accelerators* (Moscow: Nauka) p 5–49
[16] Avduevsky V S, Ashtarov E A, Ivanov A V and Pirumov U G 1989 *Gas dynamics of supersonic nonisobaric jets* (Moscow: Mashinostroenie)
[17] Katsnel'son S and Zagorskii A 2001 *J of appl. mechanics and tech. phys.* **42(1)** 13
[18] Kuzenov V V, Polozova T N and Ryzhkov S V 2015 *Problems of Atomic Science and Technology* **4(98)** 49
[19] Kamrukov A S, Kozlov N P, Protasov Yu S and Chuvshev S N 1985 *Technical Physics* **30(3)** 533

Acknowledgements

Research has been performed at "Beam-M" facility, following the government task, and partially supported by the Russian Foundation for Basic Research (grant №№ 14-08-01087).

Energy & mass-charge distribution peculiarities of ion emitted from penning source

N V Mamedov[1], D V Kolodko[2], I A Sorokin[2], I A Kanshin[1] and D N Sinelnikov[2]

[1]Federal State Unitary Enterprise "All-Russian Research Institute of Automatics n.a. N.L. Dukhov", Sushchevskaya st. 22, Moscow, Russia, 127055,
[2]National Research Nuclear University MEPhI (Moscow Engineering Physics Institute), 31 Kashirskoe sh., 115409 Moscow, Russia

E-mail: m_nikitos@mail.ru

Abstract. The optimization of hydrogen Penning sources used, in particular, in plasma chemical processing of materials and DLC deposition, is still very important. Investigations of mass-charge composition of these ion source emitted beams are particular relevant for miniature linear accelerators (neutron flux generators) nowadays. The Penning ion source energy and mass-charge ion distributions are presented. The relation between the discharge current abrupt jumps with increasing plasma density in the discharge center and increasing potential whipping (up to 50% of the anode voltage) is shown. Also the energy spectra in the discharge different modes as the pressure and anode potential functions are presented. It has been revealed that the atomic hydrogen ion concentration is about 5-10%, and it weakly depends on the pressure and the discharge current (in the investigated range from 1 to 10 mTorr and from 50 to 1000 μA) and increases with the anode voltage (up 1 to 3,5 kV).

1. Introduction

As is known, the penning hydrogen ion sources are widely used in plasma chemical processing of materials and deposition of diamond-like thin films, that is why the optimization of operation modes of this kind of sources is still extremely important. The studies of the component composition of the beam extracted from a miniature linear accelerator used to generate neutron fluxes (using D-D or D-T reactions) are also relevant.

Over the last decades, compact neutron generators (NG) are more and more used in various fields of science and technics due to their reliability, radiation safety and easy operation [1]. One of the main elements of NG is the neutron tube (NT) – vacuum neutron tube (VNT) and gas-filled neutron tube (GNT). Despite the obvious advantages of GNT (wide frequency range of neutron pulses generation, the stable neutron flux throughout the lifetime) the increase in the neutron yield and the lifetime are still extremely relevant. As is known, the neutron yield is affected by two main factors – the characteristics of the ion source (IS) (usually used penning type) and properties of target materials. The various options for optimizing the geometric parameters of the bit cell source to increase the discharge current and, as a consequence, to increase the drawn ion current are shown in [2-5]. High neutron output is limited also by the low atomic ions concentration of deuterium (or tritium) when using penning ion source as compared with the molecular one. In [5, 6] the effect of electrode materials on the discharge characteristics and the mass-charge composition has been studied. For example, it is shown that the use of an aluminium with a boron nitride layer as an electrode material increases the proton component ratio up to 16%, but also leads to discharge instabilities such as film sputtering.

As one knows, a low number of atomic ions for the hydrogen discharges is a consequence of the low electron density and short life time of molecular ions H_2^+ [5,7]. The lifetime of ions is directly related to the recombination coefficient, i.e. the materials of the electrode surface [5]. Thus, the increased electron density at least in a local region of the discharge (for example, by changing the discharge combustion mode) can significantly change the mass-charge ratio of the extracted ion beam.

2. Experimental setup description

The high vacuum setup designed in the NRNU MEPhI for the penning ion source main parameters measurements is shown in figure 1. The source (1), fixed by Teflon rings (2), is isolated from the ground potential of the setup. Thus, in addition to the anode potential (Ud < 4kV) the bias potential relative to ground (Uex < 15kV) can be set to the source casing.

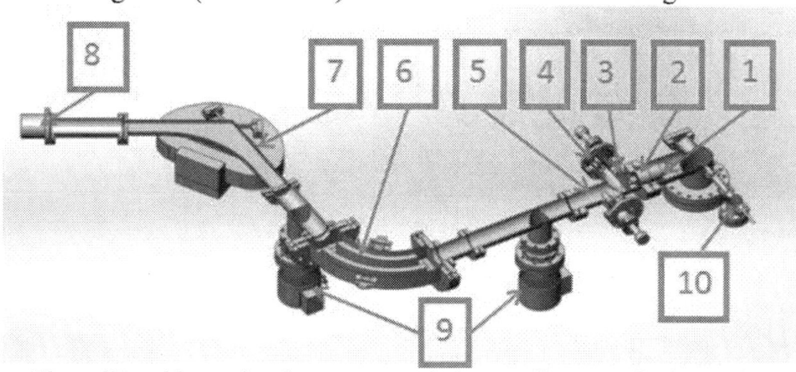

Figure 1 Schematic of the experimental setup (description in the text).

Plate (3) with a circular aperture 1,6 mm diameter is located on the precious linear motion feedthrough (4) in front of the ion source exit orifice at a 5 mm distance. Plate is grounded through an ammeter, which measures the total current of the extracted beam. It is the accelerating electrode at the same time. Measuring probe 2.5x3 mm^2 (5) is mounted at a 50 mm distance from the plate on the same feedthrough. Changing the position of the diaphragm (3) and the probe (5) and measuring the current by ammeter, the spatial distribution of the ion beam and emittance are determined.

Henceforth the extracted ion beam enters the quarter-spherical electrostatic deflector (6) with the energy resolution from 0.8% to 1.5% [8]. The ions passing energy analyzer are registered by the Faraday Cup, (installed on a movable input) or fly into a magnetic mass-monochromator (7). This device permits one to separate the ions with mass to charge ratios of up to $M/Z = 40$; maximum beam energy of up to 10 keV and mass resolution $M/\Delta M \leq 30$ (for 5keV ion beam) [9]. After that the ion beam current is measured by secondary electronic multiplier VEU-1A (8). The recovery of the energy spectra (and similarly mass spectra) takes place automatically in the control program according to the method described in [10].

The vacuum system consists of two turbomolecular pumps Pfeiffer HiPace 80 (9). Working gas is introduced directly into the source volume through a Teflon tube with the precious leak (10). The working pressure in the interaction chamber, (as measured by the pressure sensor Pfeiffer PKR 251,) was of an order of $10^{-3} - 10^{-4}$ Torr, the maximum residual vacuum of the order of $\sim 10^{-6}$ Torr. The gas pressure in the source is calculated according to the Pfeiffer gauge PKR 251 data taking into account the conductivity of the source output aperture and the pumping speed.

The size of the investigated ion source is about 20xØ15mm. Magnetic field is uniform in the center of the cell and has a value of 100 mT. Comsol Simulation of potential distribution on its axis depending on the anode voltage U_d was done.

3. Experimental results and discussion

Before the beginning of the experiments the setup has been evacuated during 3 hours to the ultimate vacuum. Then during 10 to 30 minutes the source has been heated and degassed on exposure to its own discharge (the anode voltage $U_d = 2$ kV, the pressure P = 5 mTorr and the discharge current of the order of $I_d = 600$ μA).

Current-voltage characteristics of the discharge at different pressures inside the source are shown in figure 2. As can be seen from the plot data the discharge characteristic shows sharp jumps at pressures above 5 mTorr and then the discharge current increases slowly and smoothly provided the anode voltage increases. To explain this feature one can use the results of the [11]. Because the ion density is proportional to the gas density, there exists the critical pressure at which the ion density on the inner boundary of the plasma sheath becomes comparable with the electron density. Under the assumption of the electron density finiteness the pressure increase above this level leads to the disappearance of the anodic sheath.

It has been shown [11] that the dependence of the critical pressure on the anode voltage is expressed by a power function with an index between -0.5 and -1. Thus, for certain combinations of magnetic field, pressure and anode voltage at the moment of the transition to another mode of the discharge the potential in the cathode region decreases. In the central region of the anode a dense plasma forms. This explanation qualitatively agrees with experiments [12] that have photos of the discharge at various anode voltages during the current-voltage characteristics measurements. Those photos show that the current jumps correlate well with the change in the shape and area of the discharge.

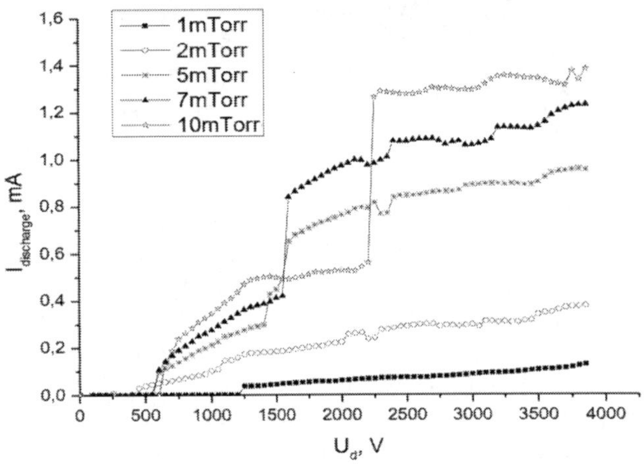

Figure 2. Current-voltage characteristics of the discharge at different pressures inside the source.

A set of extracted ions energy spectra at different voltage on the anode (in this case the source case has been grounded) and at different pressures is presented in Fig. 3-5. Values at the maximum energy of the spectrum are indicated on top of the energy spectra. As can be seen on the "low pressure" graphs (figure 3) the spectra have an asymmetric shape. The sharp growth is observed at the left of the maximum, whereas at the right – the exponential decline (high energy tail). These data are qualitatively similar to Ar energy spectra [13, 14]. Thus, the general form of the spectra (figure 3) can be interpreted as follows. Along the source axis in the discharge region there is no apparent potential difference. All ions generated in this area gain about the same energy reaching the output aperture of the ion source. Thus, due to the anode potential whipping on the axis of the source the energy spectrum of emitted ions shifts to low energy region. "Exponential fall tail" can be explained by the fact that the ion extraction area is not a point and has a finite radial dimension. Therefore, part of ions produced at different equipotential within the boundary layer of the plasma, acquire an energy different from the plasma volume ion one.

As can be seen in figures 4-5 in the transition region at pressures P ≥ 5 mtorr the energy spectra are shifted to the far low-energy region. There is a noticeable broadening of the spectra, the peak intensity maximum of the spectrum decreases. There are additional low-energy peaks with intensity by order of magnitude smaller than the main peak (see figure 4). The obtained results are similar to the data from [15]. An additional experiment has been carried on to prove that the additional peaks are associated with the ions also emitted from the ion source. The source body has been biased relative to the earth (up to 1kV). As a result, the whole energy spectrum as a whole shifts to the right by the applied accelerating voltage magnitude.

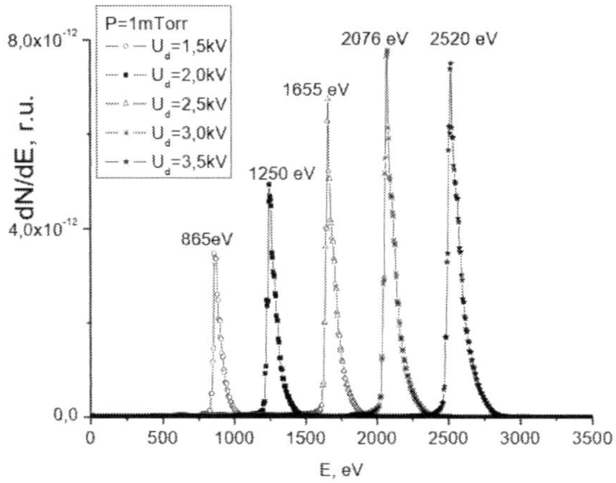

Figure 3. Energy spectra of extracted ions at different anode voltage at pressure P = 1 мTorr.

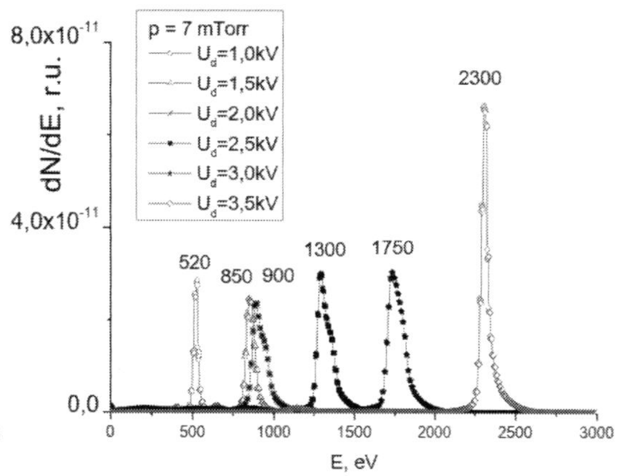

Figure 4. Energy spectra of extracted ions at different anode voltage at pressure P = 7 мTorr.

In the transition region (at pressure P = 7 mTorr and the anode voltage 1.5-2.0 kV) the discharge can "jump" from one combustion mode to another and back again that is demonstrated by change in the shape and position of the energy spectrum. It should be noted that "two hump" shape of energy spectra appearances at the discharge voltage of 1.5-2.0 kV (see figure 4, 5). With further increase in voltage discharge spectra become more monoenergetic once again (see figure 5), with the only difference from the spectra of figure 3 that in figure 5 spectra there are low-energy peaks.

The potential whipping dependence (the difference between the applied voltage to the anode and the maximum in the resulting energy spectrum $\Delta U = U_d - E_{max}/e$) on the anode voltage at different pressures is shown in figure 6. It is seen that at low pressures (P < 5m Torr) the potential whipping grows linearly with increasing applied voltage, whereas at higher pressures a sharp increase in the magnitude occurs and then this value reaches the plateau.

Based on [11], photos [12] and the measured energy spectra, one can conclude that after the ignition discharge continuously goes out from T mode (Townsend region) to LMF mode (low magnetic field), i.e. at small *B* and increasing gas density (pressure). The potential distribution is distorted by a cloud of negative charge existing throughout all volume of the discharge cell. There is a transition mode (TM) at high pressures (P ≥ 5 mtorr), but the transition to the high pressure mode (HP) is not observed. Cathode layer forms; central plasma is still present; the potential whipping increases.

An overview of the extracted ion mass spectrum at a pressure of 7 mtorr and the discharge voltage 3.5 kV without additional accelerating voltage is presented in figure 7. Each peak is signed with corresponding weight. As can be seen, typical component composition ions emitted from penning source are H^+ ~6%, H_2^+ ~90, 5%, $H_{2\text{-}1}^+$ ~1,4%, H_2^+(energy) ~0,2%, H_3^+ ~0,3%, OH^+ ~0,2%, H_2O^+~0,8%, N^+ ~0,08%, O^+~0,09%. $H_{2\text{-}1}^+$ - molecular ions, collapsed during beam transport before entering the magnetic mass separator. H_2^+- (energy) peak of the molecular ions, whose energy spectrum is shifted to the left of the main peak and has an order of magnitude lower intensity (as, for example, in figure 8). The amount of impurity ions (water, oxygen, and nitrogen) is less than 1.5% in all experimental data.

The dependences of the atomic/molecular ions extracted from the ion sources ratio ($\delta = \frac{H^+}{H^+ + H_2^+}$) on the anode voltage at different working gas pressure based on the obtained data are made. As can be seen from figure 8, the atomic ions fraction is in the range of 5-10%. As a whole there are no fundamental differences of the atomic/molecular ratio on the pressure and discharge current in the specified range. The atomic/molecular ratio increases with increasing anode voltage.

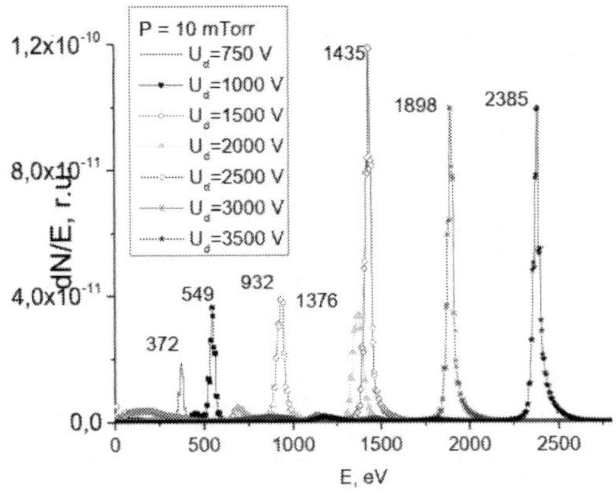

Figure 5. Energy spectra of extracted ions at different anode voltage at pressure P= 10 мTorr.

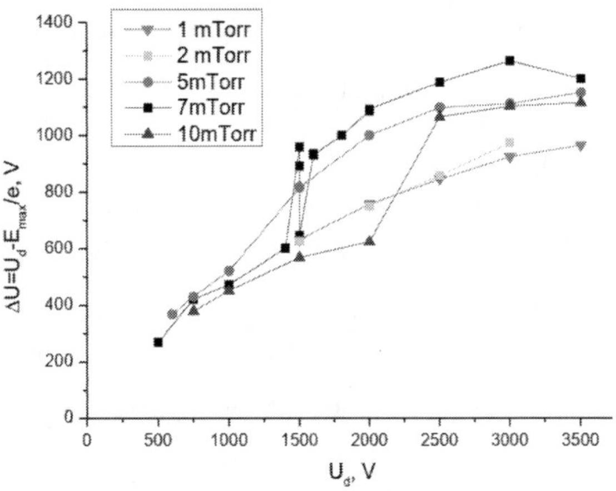

Figure 6. The potential whipping dependence on the anode voltage at different pressures.

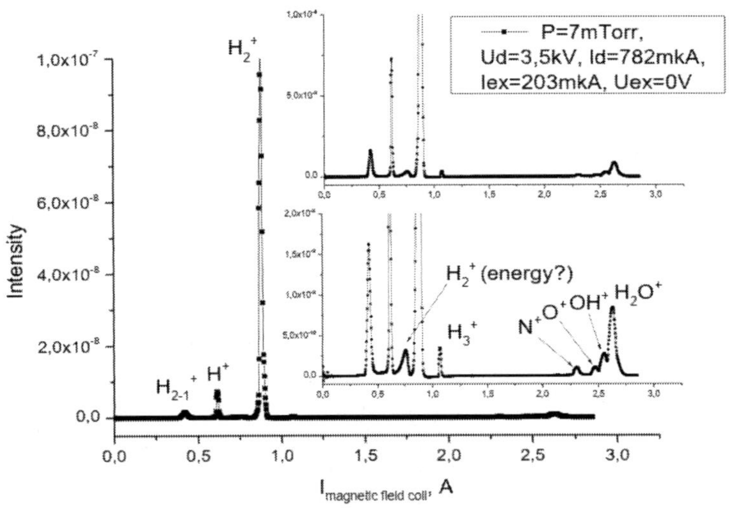

Figure 7. Review of the mass spectrum at a pressure of 7 mtorr and the discharge voltage 3.5 kV.

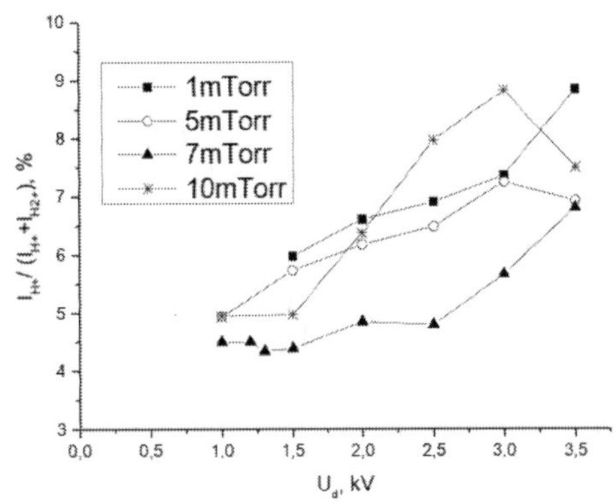

Figure 8. The dependences of atomic/molecular ratios of ions extracted from the penning sources on the anode voltage at different working gas pressure.

4. Conclusion

Investigations of energy distributions and molecular composition of the ions emitted from a penning source are carried out. Current-voltage characteristics for different working gas pressure are measured. It is shown that at pressures above 5 mTorr there are abrupt fluctuations of the discharge current, and then the smooth slow growth with the anode voltage increase. Current 'jumps' are explained by the plasma density and potential whipping increase in the center of the discharge. Shapes of the energy spectra depending on the mode of discharge are investigated. The values of the potential inside the source at different pressures and anode voltages are defined. Component composition of the extracted ion beam is measured. The atomic ions fraction is in the range of 5-10%, depends weakly on pressure and discharge current (in the investigated range from 1 to 10 mTorr and from 50 to 1000 μa) and increases with increasing anode voltage (from 1 to 3.5 kV).

References

[1] Valkovic V 2016 *14 MeV Neutrons. Physics and Applications* (CRC Press Taylor&Francis Group, Boca Raton, London, New York), 500 p
[2] Das B K, et al. 2012 *Nuclear Instruments and Methods in Physics Research A* **669** 19
[3] Liu W, et al. 2014 *Nuclear Instruments and methods in Physics Research A*, **768** 120
[4] Rovey J L, Ruzic B P, Houlahan T J 2007 *Rev. of Sci. Instrum.* **78** 1
[5] Sy A, Ji Q, .Persaud A, Waldmann O, Schenkel T 2012 *Rev. of Sci. Instrum.* **83** 1
[6] Ludewight B, et al. 2011 *Neutron Source R&D – from Compact Generators to Neutron Beams* (Proceedings of FNDA)
[7] Chen F K 1984 *J. Appl. Phys.* **56** 3191
[8] Mamedov N V, Kurnaev V A, Sinelnikov D N, Kolodko D V 2015 *Instruments and Experimental Techniques* **58(1)** 43
[9] Kolodko D V, et al. 2014 *Proceedings of 20th International Workshop on Beam Dynamics and Optimization* (Saint Petersburg), p. 87
[10] Kurnaev V A., Urusov V A 2010 *Technical Physics Letters* **36(5)** 454
[11] Schuurman W 1967 *Physica* **36** 136
[12] Mamedov N V, Schitov N N, Kanshin I A 2015 *Physical-Chemical Kinetics in Gas Dynamics* **16(4)** 590
[13] Rohwer P, et al. 1982 *Nuclear Instruments and methods* **204** 245
[14] Rohwer P, et al. 1983 *Nuclear Instruments and methods* **211** 543
[15] Nagy J L 1965 *Nuclear Instruments and methods* **32** 229

Acknowledgments

These studies have been supported by RFBR grant №12-02-13510-OFI_M_RA and have been done in NRNU MEPhI in collaboration with VNIIA. The authors would like to thank Dr. N.N. Schitov from Dukhov All-Russia Research Institute of Automatics for the help and useful discussions.

Mass spectroscopy of the ion flux produced during inductively coupled plasma nitriding process

D V Kolodko, A V Kaziev, D G Ageychenkov, E A Meshcheryakova, A A Pisarev, and A V Tumarkin

National Research Nuclear University MEPhI (Moscow Engineering Physics Institute), 31 Kashirskoe sh., 115409 Moscow, Russia

E-mail: kaziev@plasma.mephi.ru

Abstract. Ion fluxes on the surface of sample embedded in inductively coupled plasma have been studied in conditions typical for titanium alloy nitriding: total pressure 0.44 Pa, Ar/N_2 = 70%/30%, and RF power 1500 W. The gas composition was independently monitored by the quadrupole analyser. The ion fluxes were sampled using a specially designed electrostatic extractor and then analysed with a magnetic sector mass-separator. The extractor design allowed us to apply a bias voltage to the plasma facing electrode thus imitating interaction of ions with the surface during the plasma processing. The ion fluxes of Ar^+, N_2^+, and N^+ on the surface were measured. The mass spectroscopy diagnostics unit is suitable for extensive ion content studies in the plasma technology facilities.

1.Introduction

The effectivity of technological plasma processing devices strongly depends on the plasma parameters such as plasma density, electron temperature, plasma potential, and ion composition. The latter is of substantial importance since many of the surface modification processes in plasma depend on mass and energy spectra of ions bombarding the substrate, and on their amounts. For example, the fluxes of atomic nitrogen are essential for plasma nitriding of titanium alloys and steels. This is also important for the plasma ion immersion implantation (PIII) technique, where the N^+/N_2^+ ratio affects the energy of incident atoms and hence their penetration depth. It is therefore vital to study the behaviour of ion fractions incident on the material surface during the plasma processing. In our previous paper [1] we have discussed the methods of ion flux analysis for the technological plasma devices. The only reliable way to perform such measurements is to use the charged particles analysis technique, and there are several reports on this topic [1–4].

We have recently described a magnetic sector mass spectrometry unit for analysing fluxes of ions with different masses in the inductively coupled plasma (ICP) reactor used for nitriding [1]. It was shown that the mass resolution of the spectrometer is well-suited to control mass spectra that correspond to ion fluxes incident on the samples embedded in plasma. Such monitoring is a promising diagnostic tool for better characterisation of technological plasmas.

Here, for a given plasma treatment regime, we compare data obtained by optical emission spectroscopy, gas contents diagnostics, and by the ion mass analyses. Absolute ion fluxes produced in the ICP nitriding process were obtained.

2.Experimental

The experiments were carried out in an ICP facility used for surface modification [5] equipped with a mass analysing unit [1]. The ICP discharge was excited by a flat circular antenna Beams&Plasmas RPG-250 cooled by water and protected by a quartz shield. The longitudinal magnetic field was produced by electromagnetic coils. The RF power was supplied by COMET cito 1330 series generator.

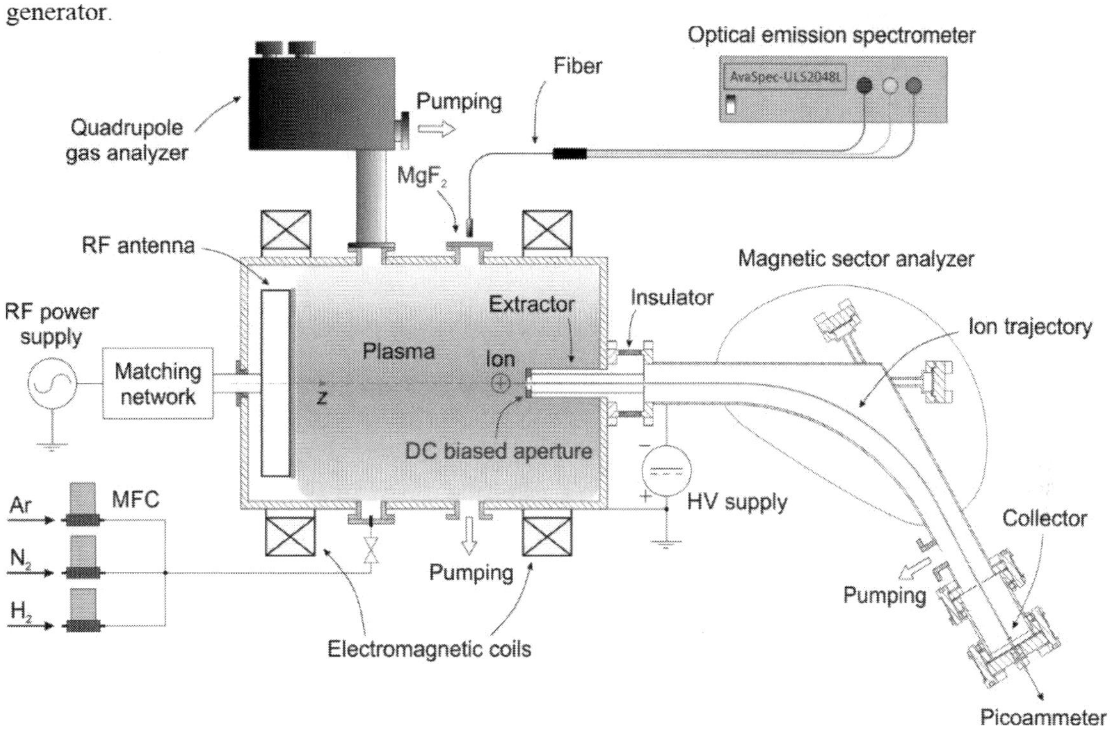

Figure 1. Experimental setup scheme.

The ICP vacuum chamber was evacuated to a base pressure of 1×10^{-4} Pa with a turbomolecular pump backed by a dry scroll pump. Argon and nitrogen mixture (Ar/N_2 = 70%/30%) was used as a working gas. The gas components ratio was regulated by mass-flow controllers (MFC).

The gas mixture composition was continuously monitored with a differentially pumped Extorr XT-100 quadrupole residual gas analyser (GA). Low working pressure in the GA was maintained by a regulating valve. The optical emission spectroscopy unit comprising three Avantes AvaSpec ULS2048L channels was used to qualitatively determine the plasma composition.

The mass analysing unit was attached coaxially to the ICP chamber in order to extract and mass-filter the ions from plasma. It consisted of the extractor followed by the magnetic sector mass analyser (MA). The MA configuration and operating parameters were described in detail previously [1]. The extractor, which was immersed into plasma, was a two-electrode electrostatic lens with circular apertures. The entrance aperture and the extracting aperture were 1 mm and 1.5 mm in diameter, respectively. The entrance aperture imitated the surface of a sample subjected to plasma treatment. It was electrically insulated from the rest of the unit and connected to a dedicated electrical vacuum feedthrough. Thus, the bias could be applied to the entrance electrode, and the total current could be measured at various biases. The relative currents of ion species with masses M_i measured in the MA collector circuit were then used to calculate the current densities of ions of each kind, and eventually to obtain the absolute values of ion fluxes.

In a typical regime for titanium alloy nitriding in the ICP reactor, a fixed ratio of Ar and N_2 partial pressures was used: p_{Ar} = 0.31 Pa, and p_{N2} = 0.13 Pa. The forward RF power was 1.5 kW, and the reflected power did not exceed 2%. The fixed DC bias voltage applied to the entrance aperture of the extractor was $U_b = -300$ V.

3.Results and discussion

The optical emission spectrum for the discussed discharge regime is shown in figure 2. The most intense lines correspond to Ar^+, Ar, N_2, N_2^+, N, N^+, OH, H, and NH. The relative intensities of N and N^+ emission lines are small compared with those of N_2 and N_2^+.

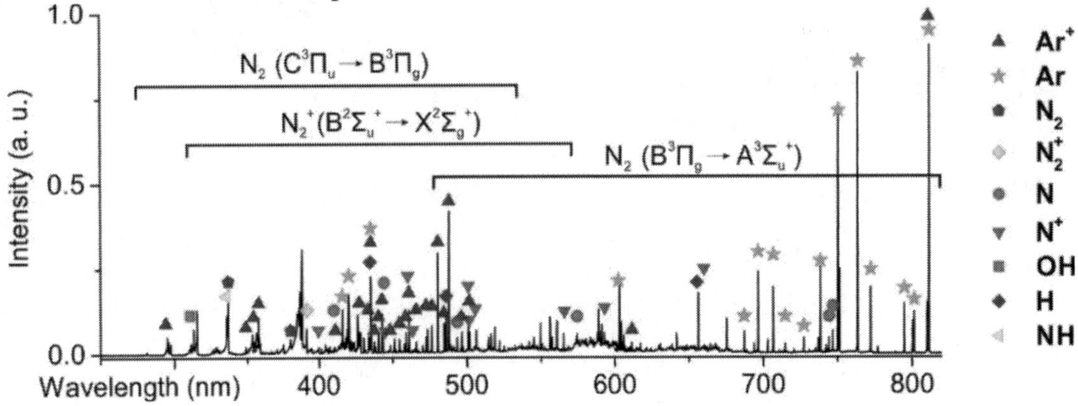

Figure 2. Optical emission spectrum of the ICP discharge operated in the Ar–N_2 mixture (RF power 1 kW, total pressure 0.44 Pa).

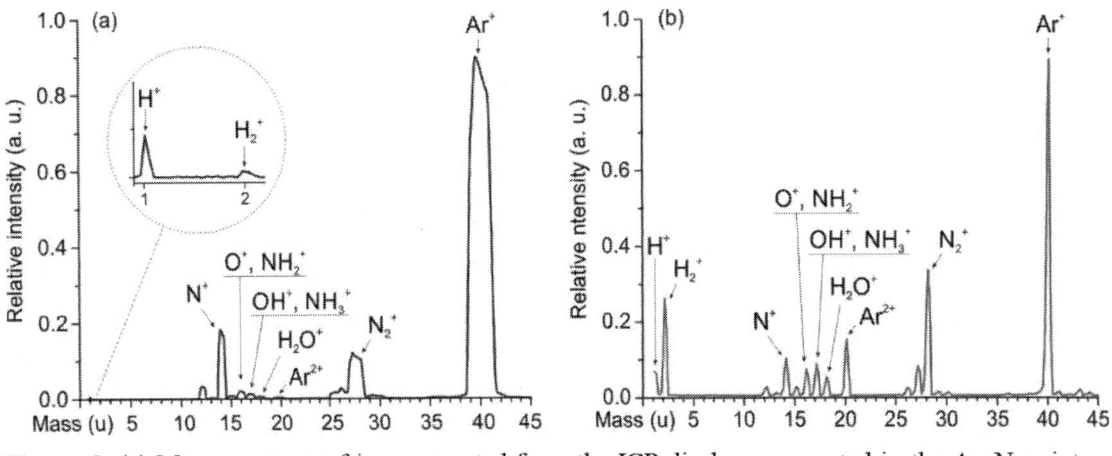

Figure 3. (a) Mass spectrum of ions extracted from the ICP discharge operated in the Ar–N_2 mixture (RF power 1.5 kW, total pressure 0.44 Pa, $U_b = -300$ V). (b) Mass spectrum of gas species registered by the gas analyser under the same conditions.

Figure 3a demonstrates the spectrum of ions, extracted from plasma, obtained using the magnetic mass analyser. Figure 3b demonstrates mass spectrum of gases in the chamber recorded with quadrupole gas analyser (GA), which represents the mass distribution of ions produced inside the GA by its own ionisation source. In the MA spectrum, the signals of Ar^+, N^+, N_2^+ are the most intense, and the signal of atomic ions N^+ exceeds that of molecular ions N_2^+. The GA spectrum demonstrates presence of Ar, N, and N_2, as well as NH_n compounds, the latter with relative intensities considerably higher than those in the MA spectrum. It is generally accepted that the NH_n species are mainly generated through the surface reactions on the vacuum chamber walls [6]. This might account for high relative content of NH_n^+ generated in GA ionizer. The differences in the N^+/N_2^+ ratios as well as in Ar^{2+} relative intensities are supposedly due to different conditions of dissociation and ionisation in MA and GA, since MA deals with the ions produced in the plasma reactor, while GA detects ions produced by its ionisation source. The presence of the hydrogen species in the GA spectrum is due to relatively poor pumping efficiency of light gases by the differential pumping system of GA.

One should notice that both the MA and GA spectra show significant fraction of atomic nitrogen ions while the OES lines corresponding to N and N^+ are relatively weak. The sensitivity of optical diagnostics is therefore insufficient for reliable determination of atomic nitrogen fraction in our ICP process.

Under $U_b = -300$ V, the total current measured at the entrance aperture of the extractor was 25 mA. The collecting area was about 20 cm^2. Taking into account the relative ion contents measured with MA (see figure 3a), the fluxes of ions arriving to the extractor can be evaluated: $\Gamma_{Ar+} = 5.7\times10^{15}$ $cm^{-2}s^{-1}$, $\Gamma_{N+} = 1.3\times10^{15}$ $cm^{-2}s^{-1}$, and $\Gamma_{N2+} = 9.0\times10^{14}$ $cm^{-2}s^{-1}$. The discussed diagnostic technique therefore enables to determine fluxes and fluences for ions incident on the substrate during plasma processing of metals.

4.Conclusion

The magnetic sector mass analysing unit [1] was used to determine the absolute fluxes of ions incident on the biased surface immersed into the inductively coupled Ar–N_2 plasma. Under fixed ICP nitriding conditions, the absolute ion fluxes of Ar^+, N^+, and N_2^+ ions have been evaluated: $\Gamma_{Ar+} = 5.7\times10^{15}$ $cm^{-2}s^{-1}$, $\Gamma_{N+} = 1.3\times10^{15}$ $cm^{-2}s^{-1}$, and $\Gamma_{N2+} = 9.0\times10^{14}$ $cm^{-2}s^{-1}$. It was demonstrated that monitoring the gas species contents with a conventional quadrupole gas analyser is insufficient for comprehensive characterisation of the plasma processing procedure. Particularly, the ratio Γ_{N+}/Γ_{N2+}, which is essential for plasma nitriding investigation and implementation, can be yielded only by dedicated mass analyses of the ion fluxes.

References

[1] Kolodko D V, Kaziev A V, Ageychenkov D G, and Pisarev A A 2016 *J. Phys.: Conf. Ser.* **747** 012018

[2] Kim G H, Rim G H, and Nikiforov S A 2001 *Surf. Coat. Technol.* **136** 255

[3] Zolotukhin D B, Tyunkov A V, Yushkov Yu G, and Oks E M 2015 *Rev. Sci. Instrum.* **86** 123301

[4] Gushenets V I, Bugaev A S, Oks E M, Kulevoy T V, Hershcovitch A, and Brown I G 2008 *Rev. Sci. Instrum.* **79** 02B701

[5] Meshcheryakova E, Zibrov M, Kaziev A, Khodachenko G, and Pisarev A 2015 *Phys. Procedia* **71** 121

[6] Sode M, Jacob W, Schwarz-Selinger T, and Kersten H 2015 *J. Appl. Phys.* **117** 083303

Acknowledgments

The work was supported by the Ministry of Education and Science of the Russian Federation (contract No. 14.Y26.31.0008) and by National Research Nuclear University MEPhI in the framework of the Russian Academic Excellence Project (contract No. 02.a03.21.0005, 27.08.2013).

IOP Conf. Series: Journal of Physics: Conf. Series **830** (2017) 012065 doi:10.1088/1742-6596/830/1/012065

Study of parameters of a facility generating compressive plasma flows

A Ya Leyvi

South Ural State University, pr. Lenina 76, Chelyabinsk, 454080 Russia

E-mail: leiviai@susu.ru

Abstract. The prosperity of plasma technologies stimulates making of a facility generating compressive plasma flows at the South Ural State University. The facility is a compact-geometry magnetoplasma compressor with the following parameters: stored energy up to 15 kJ, voltage of a bank from 3 to 5 kV; nitrogen, air, and other gases can serve as its operating gas. The investigation of parameters of the facility showed the following parameters of compressive plasma flows: impulse duration of up to 120 μs, discharge current of 50-120 kA, speed of plasma flow of 15-30 km/s. By contrast to the available facilities, the parameters of the developed facility can be adjusted in a wide range of voltage from 2 kV to 10 kV, its design permits generating CPF in horizontal and vertical positions.

1. Introduction

Recently much attention has been given to the development of methods of treatment of structural materials by high-intensity beams of charged particles and plasma with the density of power flow of ~ 10^5–10^8 W/cm^2, since the treatment allows enhancing performance properties of treated items. Radiation technologies are featured by a series of advantages as against methods conventionally used in machinery: the possibility to treat high-precision parts of complicated shapes, short treatment time, easy automation of the process, etc. Hence, investigations of processes of radiation effects on a substance aimed at the development of novel materials are a crucial task of the modern materials science.

The accumulated data on explored interaction of high-intensity charged particles and plasma with a solid body prove the prosperity of newly-developed radiation technologies for processing of different structural materials [1].

Lately issues of the treatment of materials by plasma have received primary consideration, since plasma treatment has a number of unanswerable advantages against laser technologies. Firstly, the plasma treatment of structural materials is much cheaper than laser technologies and traditional chemical methods. Moreover, plasma facilities have high capacity, i.e., high rate of modification of a unit area of the material surface per a unit of time. If a fine powder is introduced in plasma, a range of possible coatings is enlarged, if compared with evaporation of the material with other methods for further deposition. The electron-plasma treatment is an advanced subject [2]. In this case the adhesion of a plasma-applied coating is strengthened, when further exposed to an electron beam.

There are a variety of methods used to generate plasma for technological purposes. The most widely used are methods based on plasma of vacuum-arc charges [3,4] or of a discharge glowing in gas [5]; plasma generated when a material is evaporated under powerful energy flows (laser emission, electron or ion beams) and plasma is formed in a facility generating Compressive Plasma Flows (CPF)

[6]. The CPF [7] is featured by high density (up to 10^{18} cm^{-3}), velocity of plasma flow (10^3-10^5 m/s), temperature (from 2 to 5 eV), duration of exposure from 80 to 200 μs, allowing to settle a wide range of scientific and practical tasks of the exposure of various materials to high-intensity energy flows [8,9].

However, despite of a great promise in the CPF application the problem analysis demonstrates that theoretical and experimental investigations of the CPF interaction with a solid body are insufficiently explored. As a rule, the available publications mainly address particular technological tasks [9,10] and do not deeply analyse physical nature of the obtained results. Thus, with a bulk of experimental data on the CPF interaction with a solid body, there appears an urgent need in systematic experimental and theoretical study of the problem.

Although an idea on the CPF generation was formulated in 60-s of the last century [11], the number of facilities for such technological tasks is extremely low.

Due to the need in the systematic study of the CPF interaction with a substance a facility generating Compressive Plasma Flows was developed at South Ural State University. By contrast to the available facilities, parameters of the developed facility can be adjusted in a wide range of voltage from 2 kV to 10 kV, its design permits generating CPF in horizontal and vertical positions.

2. Experimental facility

The CPF generating facility is a magnetoplasma compressor (MPC) [7]. It consists of three key components: a power source, a vacuum chamber and an MPC.

The power source has two main blocks - a charging device and an energy storage. The charging device is designed to charge an impulse sectional bank of capacitors (the energy storage). The charging device is a bridge inverter with a retardation coil. The bank of capacitors has the current of 0.7 A and voltage of 5 kV.

The bank with the total margin up to 15 kJ consists of 12 impulse capacitors K41-I7 (capacity 100 μF, nominal voltage 5 kV). The capacitors are connected into one sectional bank of capacitors with a copper bus 4x40 mm. 12 impulse capacitors are structurally packed in two sections, 6 capacitors each to enlarge voltage up to 10 kV. The battery is discharged into the MPC through an ignitron IRT-6. The ignitron can be controlled.

The vacuum chamber is made of a sheet of organic glass 16 mm thick. The chamber internal size is 250x250x500 mm. Flanges from stainless steel 12X18H10T are placed in end faces of the chamber. The flanges are screwed up with each other by studs M8 from stainless steel 12X18H10T. The front flange has 8 holes, M8 in diameter, symmetrically spaced in a circle 50 mm in diameter, which are designed to house electrodes of an anode, and a central orifice 22 mm in diameter to place a cathode pack.

There are 8 electrodes of the anode (copper rods 8 mm in diameter 14 mm long) screwed in the front flange), an electrode of the cathode (a frustum with diameters 30 mm and 60 mm 50 mm long with an axial orifice (diverter)) made of copper inside the vacuum chamber (a discharge chamber of the facility).

The cathode isolator is made of fluorine plastic 4. The front flange has a hole ½ inch in diameter to supply the plasma generating gas.

The CPF generating facility operates as follows. Air is evacuated from the vacuum chamber with a vacuum pump up to 100 Pa, then the chamber is filled with the plasma generating gas up to the required pressure (from 500 Pa to 2 kPa). The plasma generating gas is filled from a balloon through a gear reducer. The pressure inside the discharge chamber is controlled with vacuum meters. The facility can operate with various plasma generating gases and their mixtures at different pressures. The arc discharge is initiated in the discharge chamber with the controlled ignitron IRT-6. The ignitron is switched on with an impulse lasting 10-20 μs with 2 kV voltage from a discharge control device between the cathode and the discharge control electrode. When the ignitron is switched on, high voltage is supplied from accumulating capacitors to electrodes housed in the discharge chamber.

3. Investigation of the CPF generating facility parameters

The discharge current was taken with a current meter based on the Rogovski belt principle. The discharge current shape and impulse parameters were measured with an oscillograph Rigol DS1052E in control sockets of a high-voltage rectifier and are shown in Fig.1a. The voltage level in accumulating capacitors was taken with a kilovoltmeter, while the voltage shape was recorded with an oscillograph (Fig.2b) in control sockets of the high-voltage rectifier.

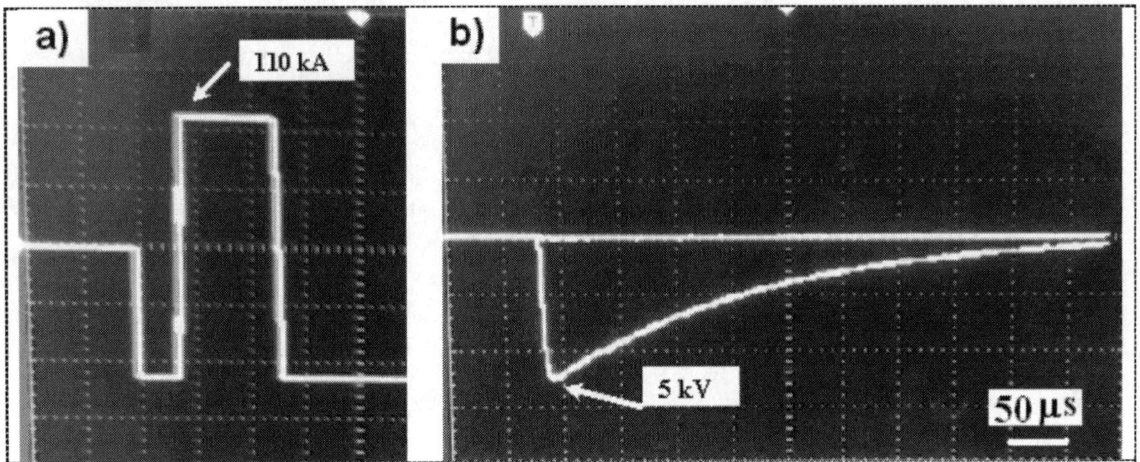

Figure 1. Oscillograms of the discharge current (a) and voltage (b).

The CPF velocity was measured using a passing-over time method. For this two pairs of electrodes were placed at l=30 cm, a pair of electrodes were s=1 cm apart. The plasma flow, passing over the pair of electrodes, closed it. The corresponding oscillograms are illustrated in Fig.2. The first pair closure corresponded to -5V. The second pair was closed at +12V. As it can be seen in the Figure, the total time of closure of the electrodes is approximately 100 µs, and time within which the plasma flow passes over a 30-cm distance amounts to 20 µs. Thus, the impulse duration results in 100 µs (at nominal voltage 4.2 kV), and the plasma flow velocity amounts to 15 km/s.

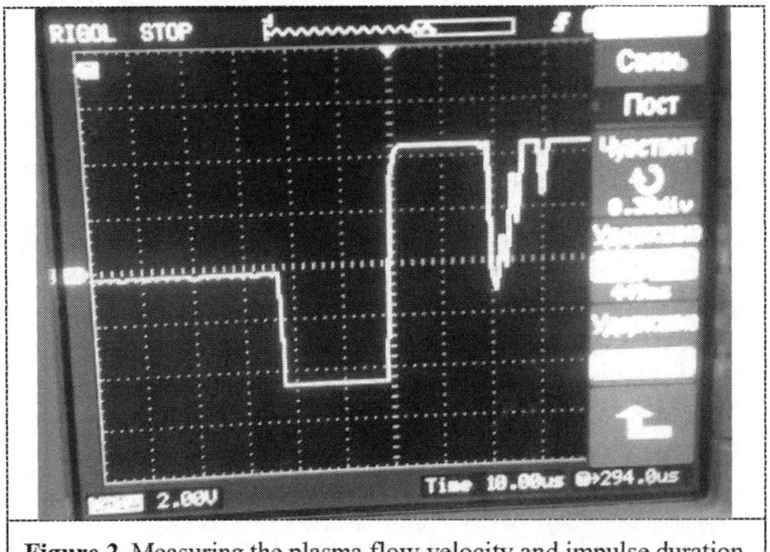

Figure 2. Measuring the plasma flow velocity and impulse duration.

In the course of the experiment specifications of the designed CPF generating facility were defined. The facility parameters are summarized in Table 1.

Table 1. The facility parameters

Parameters	**Value**
Plasma generating gas	Nitrogen, hydrogen, argon, air
Capacity of the accumulating capacitor	1200 μF
Voltage of the accumulating capacitor	2…5 (10) kV
Stored energy, maximum value	15 kJ
Discharge current	50…120 kA
Impulse duration of discharge current	up to 120 μs
Plasma velocity of the compressive flow	15…30 km/s
Length of compressive plasma flow	8…12 cm
Plasma flow diameter	1…2 cm
Pressure in vacuum chamber	from 40 Pa to 101.3 kPa
Pump output	72 l/min
Water rate (to cool the ignitron)	72 l/min

4. Experimentally studied CPF interactions with a substance

The experimental facility was used for experiments with modified surfaces of samples. For the experiment we took samples from steel 3 (50x50 mm in section, 3 mm thick) and glass (100x100 mm in section, 3 mm thick). The geometry sizes of the samples were selected so that they exceeded the plasma flow diameter with due account to the plasma scattering in radial direction along the surface.

After processing we visually observed a non-treated area; an area where the plasma flow falls normally, the area is featured by the formation of craters and the increased surface relief; an area where the plasma flow scatters radially along the surface, the area is featured by a wavy relief elongated in direction of the plasma flow scattering.

Fig.3 presents the surface of steel 3 before and after it was treated with the CPF. The initial roughness was Ra=1 μm, upon treatment with the CPF at density of the absorbed energy 20 J/cm2 it can be noted that the surface roughness increased, in accordance with the profilometry data it reached the value Ra=10 μm.

The foreign inclusions can be also noted in Fig. 3 upon the sample treatment. The foreign inclusions appeared since the cathode material got damaged and precipitated as drops on the sample. The regime of the plasma generator was observed when relative positions of the cathodes and anodes became unsymmetrical.

In Fig.4 similar results are given for the CPF treated glass. Before the CPF treatment the glass surface was smooth, upon the treatment the surface damage can be visually seen as a system of cracks with rounded edges due to melting. The craters, which surfaces are free of cracks, were also observed.

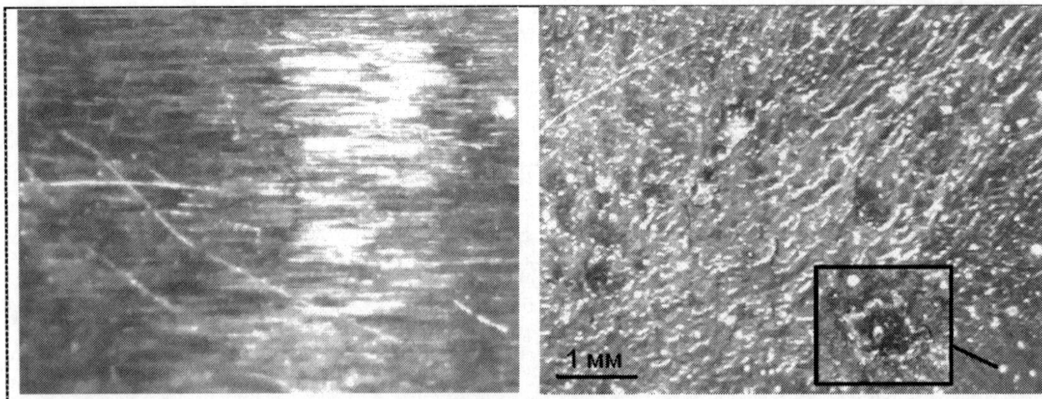

Figure 3. The surface of steel 3 before (to the left) and after (to the right) the CPF treatment.

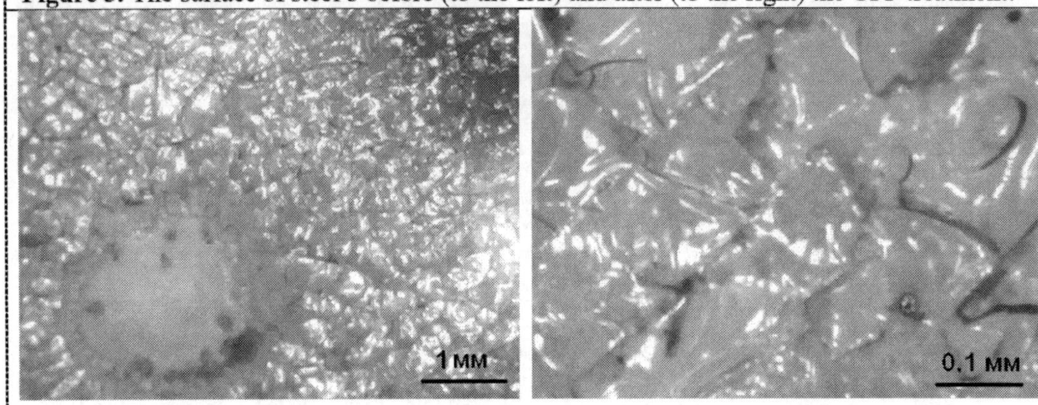

Figure 4. The glass surface upon the CPF treatment.

5. Conclusion

The experimental facility generating CPF was made and its parameters were investigated. The investigation of parameters of the facility showed the following parameters of compressive plasma flows: impulse duration of up to 120 μs, discharge current of 50-120 kA, speed of plasma flow of 15-30 km/s. By contrast to the available facilities, parameters of the developed facility can be adjusted in a wide range of voltage from 2 kV to 10 kV, its design permits generating CPF in horizontal and vertical positions.

The surface modifications caused by the CPF treatment were experimentally explored. It was shown that upon the CPF treatment a system of cracks with rounded edges due to melting appeared on the glass surface.

References

[1] 2006 *Materials Surface Processing by Directed Energy Techniques* (Edited by Y. Pauleau. Elsevier p 722

[2] Leyvi A Ya, Yalovets A P and Kozhevnikov D G 2016 *Materials Science Forum* **843** 73

[3] Ryabchikov A I, Stepanov I B, Eremin S E, and Sivin D O 2008 QUASIRIBBON VACUUM ARC ION SOURCE "RADUGA-6" *Proceedings of 9-th International Conference on Modification of Materials with Particle Beams and Plasma Flows (Tomsk, 21-26 September 2008)* pp 11-14

[4] Shugurov V V 2008 Extended Arc Plasmagenerator PINK-P *Proceedings of 9-th International Conference on Modification of Materials with Particle Beams and Plasma Flows (Tomsk, 21-26 September 2008)* pp 27-30
[5] Koval N N, Schanin P M, Lopatin I V, et al. 2002 Vacuum Ion-Plasma installation "Duet" *Proceedings of the 6th Int. Conf. on Modification of Materials with Particle Beams and Plasma Flows. Tomsk, 2002* pp 112-116
[6] Uglov V V, Anishchik V M, Cherenda N N, Sveshnikov Yu V, Astashynski V M, Kostyukevich E A, Kuzmitski A M, Askerko V V 2008 *Surf. Coat. Technol.* **202** 2439
[7] Uglov V V, Anishchik V M, Astashynski V V, Astashynski V M, Ananin S I, Askerko V V, Kostyukevich E A, Kuz'mitski A M, Kvasov N T, Danilyuk A L 2002 *Surface and Coatings Technology* **158 –159** 273
[8] Cherenda N N, Uglov V V, Kuleshov A K, Astashynski V M, Kuzmitski A M 2016 *Vacuum* **129** 170
[9] Cherenda N N, Kuleshov A K, Shymanski V I, Uglov V V, Bibik N V, Rusalski D P, Astashynski V M, Kuzmitski A M *High Temperature Material Processes* **19(2)** 153
[10] Cherenda N N, Smilgin A A, Uglov V V, Astashynski V M, Kuzmitski A M, Remnev G E 2014 *Surface & Coatings Technology* **255** 79
[11] Morozov A I 2006 Vvedenie v plazmodinamiky (Moscow: FIZMATLIT ISBN 5-9221-0681-3) P 576

Acknowledgements
The research was supported in part by the Ministry of Education and Science of the Russian Federation (contract no. 1030 (2014105-GZ) for support of research at the South Ural State University).

Low temperature ion nitriding of Ti-6Al-4V alloy

R D Agzamov, K N Ramazanov, A F Tagirov, E L Vardanyan

Ufa State Aviation Technical University, Department of Mechanical Engineering, ul. Karla Marksa 12, Ufa, Russia

E-mail: vipt127@yandex.ru

Abstract. The paper presents the results of nitriding the titanium alloy VT6 (Ti-6Al-4V) in a non-self-maintained high-current discharge and in a glow discharge in different regimes. Nitriding was conducted in pure nitrogen and in the mixture of nitrogen and argon. It is shown that microhardness of the samples increased; however, the efficiency of surface modification depends on the modification conditions.

1. Introduction

Currently, titanium alloys play an important role in aviation and space industry due to heat resistance, high specific strength, high corrosion resistance, low density and non-magnetization.

However, along with the positive characteristics such alloys also demonstrate disadvantageous properties: low surface hardness and poor tribological characteristics [1, 2].

Surface modification can improve physical and mechanical properties of titanium alloys. The use of traditional gas nitriding in this case is impossible because this type of treatment requires temperatures in the range of 800-900 °C. Ion nitriding is one of the most efficient methods of surface treatment that increase hardness, wear resistance, and corrosion resistance. The temperature of nitriding of titanium alloys can be reduced in glow discharge and non-self-maintained arc discharge plasmas. Low temperature nitriding can preserve high quality of surface and exclude deformation of final products [3,4,5].

The aim of this work is to study the impact of low temperature ion nitriding on the microhardness and depth of the hardened Ti-6Al-4V layer.

2. Experimental approach

The Ti-6Al-4V alloy with chemical composition presented in Table 1 was used as the initial material for the investigations.

Table 1. Chemical composition of steel [wt%]

Fe	C	Si	V	N	Ti	Al	Zr	O	H
0.3	0.1	0.15	3.5–5.3	0.05	86.485–91.2	5.3–6.8	0.3	0.2	0.015

Literature analysis has shown that gas composition influences the hardness of a surface [5]. Nitriding of the titanium alloy was carried out in the mixture of argon and nitrogen of different concentrations. It is also known [1-5] that the depth of a nitrided layer is substantially impacted by temperature and time of exposure.

IOP Conf. Series: Journal of Physics: Conf. Series **830** (2017) 012066 doi:10.1088/1742-6596/830/1/012066

In the study, several samples of VT6 titanium alloy (Ti-6Al-4V) were subjected to nitriding in a non-self-maintained high-current arc discharge using PINK plasma source [6] with an incandescent cathode installed in NNV-6.6-I1 installation [6] under the following conditions (Table 2):

Table 2. The treatment regimes on installation NNV-6.6-I1

Sample No.	Temperature of treatment [°C]	Treatment time [h]	Composition of the working fluid
1	550	3	60% Ar, 40% N_2
2	500	1	20%Ar, 80%N_2
3	400	1	20%Ar, 80%N_2
4	500	3	100%N_2
5	400	1	100%N_2

The other samples were subjected to nitriding in a glow discharge on modernized ELU-5M installation under the following conditions (Table 3):

Table 3. The treatment regimes on installation ELU-5M

Sample No.	Temperature of treatment [°C]	Treatment time [h]	Composition of the working fluid
1	500	3	85%Ar, 15%N_2
2	500	3	85%Ar, 10%N_2, 5%H_2
3	500	1	85%Ar, 10%N_2, 5%H_2
4	450	1	85%Ar, 15%N_2

The initial microhardness of the VT6 specimen surface subjected to quenching and aging was 330 kgf/mm^2. The uniformity of microhardness distribution over the surface was determined by measuring the microhardness diagonally at regular intervals with PMT-3M microhardness tester. The depth of the modified layers was measured in cross sections at angle $\alpha \approx 6°$. A visual analysis of the samples was used to investigate color uniformity against the initial state in order to identify possible surface defects.

3. Results and discussion

The visual analysis showed that after ion nitriding in a non-self-maintained high-current arc discharge, the samples appeared dull with a barely noticeable golden hue (Figure 1a). The samples treated in a glow discharge acquired a golden hue with slight color heterogeneity (Figure 1b).

Surface microhardness tests showed that the hardness increased in all samples in the range of 348-672 kgf/mm^2. Surface microhardness for the samples treated with non-self-maintained arc discharge and glow discharge is shown in Tables 4 and 5, respectively.

Table 4. Surface microhardness of the samples before and after ion nitriding

Regime	Microhardness before nitriding [kgf/mm^2]	Microhardness after nitriding [kgf/mm^2]	Hardening [%]
Sample 1		672	104
Sample 2		528	60
Sample 3	330	373	13
Sample 4		493	49
Sample 5		348	6

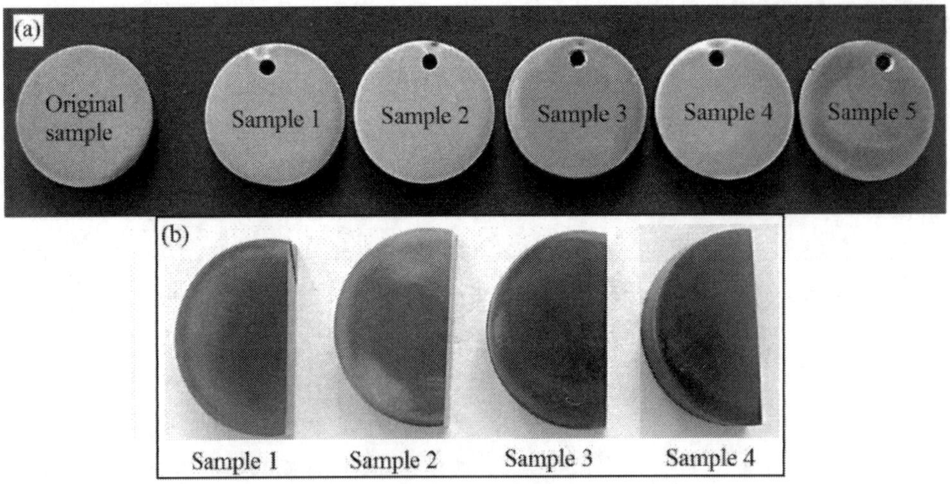

Figure 1. General view of the samples before and after ion nitriding: a) samples treated in non-self-maintained high-current discharge; b) samples treated in glow discharge.

Table 5. Surface microhardness of the samples before and after ion nitriding

Regime	Microhardness before nitriding [kgf/mm2]	Microhardness after nitriding [kgf/mm2]	Hardening [%]
Sample 1	330	353	7
Sample 2		389	17.8
Sample 3		390	18
Sample 4		389	17.9
Sample 5		393	19
Sample 6		395	20

The increase of the microhardness for the samples treated in glow discharge is less significant as compared to those treated in non-self-maintained arc discharge.

Changes in the profile along the microhardness depth of the samples processed in non-self-maintained arc discharge are shown in Figure 2. The maximum surface microhardness of nitrided samples of 672 kgf/mm^2 was found in Sample 1. The minimum surface microhardness of nitrided samples of 348 kgf/mm^2 was found in Sample 5. Microhardness in Samples 2, 3 and 4 was 528, 373 and 493 kgf/mm^2, respectively. Thus, the surface hardness and growth kinetics of the nitrided layers were influenced by nitriding temperature at the greatest [7, 8]. The proportions of gas composition do not have any significant impact on hardness and depth of the nitrided layers.

Hardness-depth changes of the samples after nitriding in a glow discharge are shown in Figure 3. The hardness on the surfaces of all samples that were ion nitrided in glow discharge was about 390 kgf/mm^2. The maximum depth of the hardened layer was found in Sample 2 and reached 17 μm. The dependencies in Figure 3 show that the temperature change of 50 °C had no significant effect on the surface hardness. Addition of hydrogen increases the depth of the modified layer. This is probably due to the fact that hydrogen reacts with oxygen, which results in reduced oxygen concentration in the surface layer and increased nitrogen adsorption sites on the surface. This in turn enhances nitrogen in the sample volume and increases the hardness of the surface layer [6, 9].

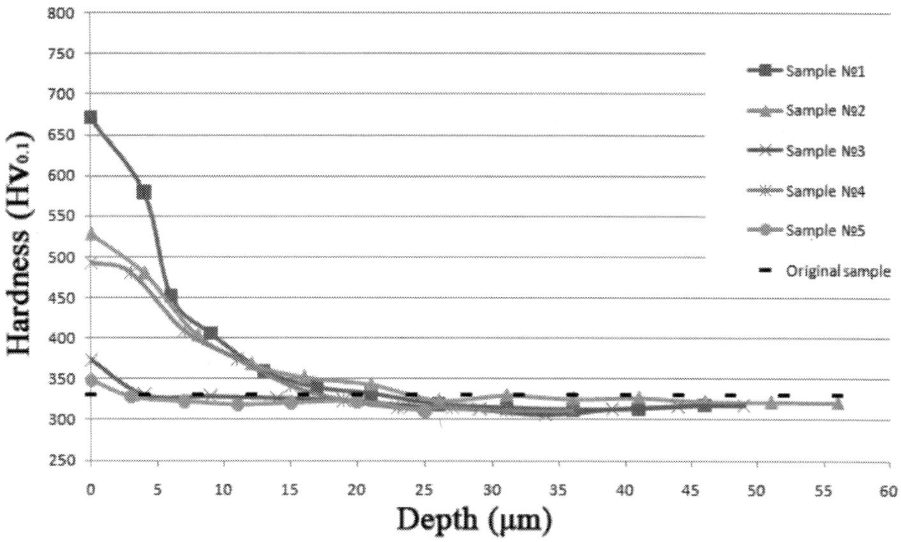

Figure 2. Hardness-depth changes of samples after nitriding in a non-self-maintained arc discharge.

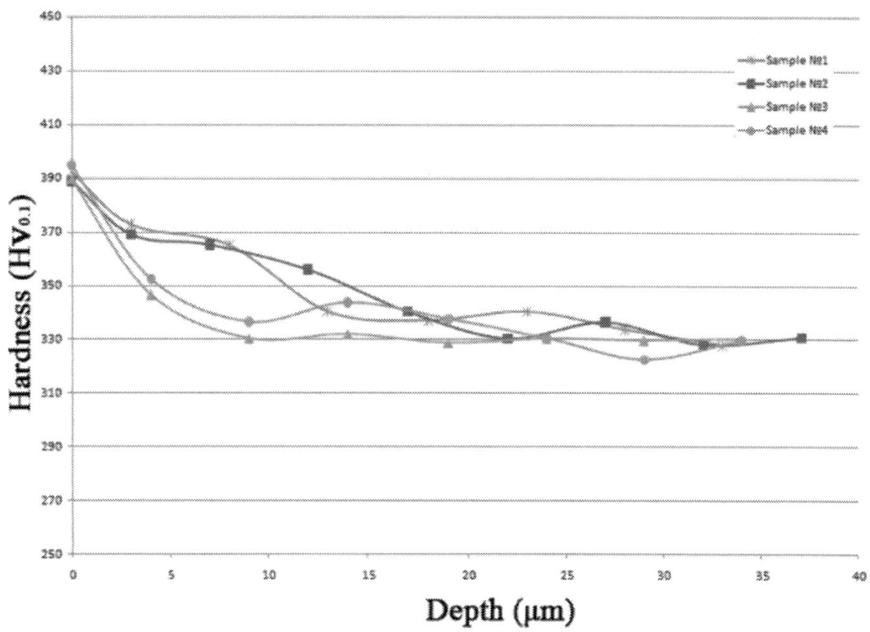

Figure 3. Hardness-depth changes of samples after nitriding in glow discharge.

4. Conclusion

Samples that were subjected to ion nitriding demonstrate sufficiently even microhardness distribution on the surface and in the volume of the material without sharp transitions to the core.

The study showed that ion nitriding can be efficient in improving microhardness. The microhardness and depth of a surface layer are greatly influenced by temperature.

Visual study showed that non-self-maintained high-current discharge nitriding leads to only slight change of sample color. The samples processed in a glow discharge acquired a golden color due to the appearance of titanium nitride on their surfaces.

References

[1] Ramazanov K N, Zolotov I V, Khusainov Yu G and Khusnutdinov R F 2015 Improving the operating properties of parts of titanium alloys by surface hardening in high density plasma of glow discharge *Journal of Physics: Conference Series* **652** 012055

[2] Budilov V, Agzamov R and Ramazanov K 2007 Ion nitriding in a hollow cathode glow discharge *Metal Science and Heat Treatment* **7** 25

[3] Arzamasov B, Bratukhin A, Yeliseev Yu and Panajoti T 1999 *Ion thermochemical treatment of alloys* (Moscow: Publishing House of Bauman Moscow State Technical University) p 400

[4] Checulin B, Ushkov S, Razuvaeva I and Goldfain V 1977 *Titanium alloys in machine industry* (Leningrad: Machine industry) p 247

[5] Vershynin D and Smoljakova M 2011 Effect of structural and phase composition on properties of titanium alloys at low temperature nitriding in a gas mixture *International conference on Interaction of Radiation with Solids* 32

[6] Berlin Ye, Koval N and Seidman L 2012 *Plasma thermochemical treatment of steel products* (Moscow: Tehnosfera) p 464

[7] Malinov L, Zecheva A and Sha V 2004 Dependence of microstructure and properties of commercial titanium alloys on parameters of nitriding from gas phase *Metal Science and Heat Treatment* **7** 21

[8] Pogreluk I Nitriding of titanium alloys 2000 *International conference* **4** 182

[9] Smoljakova M and Vershinin D 2011 Influence of gas mixture composition on titanium structure and properties in low temperature nitriding *Perspektivnye materialy* 478

IOP Conf. Series: Journal of Physics: Conf. Series **830** (2017) 012067 doi:10.1088/1742-6596/830/1/012067

Atomistic simulation of structural damage during ion irradiation of iron single crystals

A V Korchuganov, K P Zolnikov, D S Kryzhevich

Institute of Strength Physics and Materials Science of the Siberian Branch of the Russian Academy of Sciences, 2/4 pr. Akademicheskii, Tomsk, 634055, Russia

E-mail: avkor@ispms.ru

Abstract. The evolution of atomic displacement cascades initiated near free surfaces with different crystallographic orientations in bcc iron specimens was studied on the base of the molecular dynamics approach. The craters surrounded by adatom mounds were formed in the case of the (111) surface irradiation. The dislocation loops consisted of vacancies were generated after the (110) surface irradiation. The dislocation Burgers vector was a/2 <111> or a <100>. It was shown that the type of structural damage is determined by the anisotropy of propagation of shock waves generated by atomic displacement cascades. For energies of the atomic displacement cascade lower than 20 keV the number of adatoms and survived point defects was higher for specimen with the (110) free surface due to the different character of surface damage. The increase in the cascade energy up to 20 keV results in formation of almost the equal number of survived point defects for the (110) and (111) free surfaces as displacement cascades were developed on larger distance from the irradiated surfaces.

1. Introduction

Ion irradiation has a wide area of application from scientific research to industrial processes. One of the most important is the surface modification of material. Generally, the study of structural changes in the surface region of the irradiated material is conducted for ion energies from tens to hundreds of keV. For such energies ions penetrate in the material to a depth of few tens of nanometers. It is known that collision of ion with lattice atoms leads to generation of atomic displacement cascades near the surface of the material. The peculiarities of generation and evolution of atomic displacement cascades in the surface region during ion bombardment are actively investigated [1-6]. It was shown that the free surfaces have a significant impact on the evolution of atomic displacement cascades. In particular, the number of survived point defects in crystallites with the irradiated surface may be two orders of magnitude greater than the corresponding number in bulk samples [5,6]. If a cascade is generated in the volume of the material the fraction of self-interstitial atoms (SIAs) in the formed clusters is higher than the fraction of vacancies. In the case of the near-surface cascade the fraction of clustered vacancies is always higher than the fraction of SIAs. This is due to the fact that most of the SIAs compared to the vacancies escape to the free surface and form adatom islands on it. The free surfaces influence not only the number and position of survived point defects, but also the type of the formed damage near the surface. In particular, the formation of craters and dislocation loops in the surface region after the ion irradiation was found in the experimental [1-4] and theoretical studies [5,6]. Experiments on Xe^{+} ion irradiation of In, Ag, Pb, Au showed that the number of craters per ion increases with an increase in the material density [1]. Dislocation loops in the irradiated metals are

observed in experiments [2-4] and computer simulations [5,6]. For example, a greater number of loops was indicated in the thick iron sample, as compared to the bulk material, irradiated with Fe+ ions of energy 500 keV [2]. Detected loops were identified as interstitial and mostly had the Burgers vector a <100>. Vacancy loops of two types a / 2 <111> and a <100> were formed near the free surface in molybdenum irradiated with ions Sb + [3]. Experimental data on the iron and Fe-Cr alloy irradiated with Fe+ and Xe+ ions of energies 100-150 keV also showed the presence of two dislocation types in samples: a / 2 <111> and a <100> [4]. Similar loops were observed in molecular dynamics simulations when atomic displacement cascades were generated in iron specimens near the (100) and (110) free surfaces [5,6]. Irradiation was carried out with an energy of iron atoms ranging from 25 to 100 keV.
It can be expected that the irradiation of the samples with lower energies will lead to larger structural changes in the surface region, as a cascade will develop closer to the surface. This should also increase the influence of the free surface orientation on the number and character of the distribution of the generated defects. In this connection, the molecular dynamics simulations were conducted to study the features of structural changes in the iron samples during Fe+ irradiation of free surfaces with different crystallographic orientations.

2. Methodology

Simulation of atomic displacement cascades formed under ion irradiation was performed using a software package LAMMPS [7]. Interatomic interaction was described by the many-body potential [8] constructed in the Finnis-Sinclair approximation. The objects of the study were cube-shaped iron crystallites with 20 nm edges. One of the crystallite faces with the indices (110) or (111) was subjected to irradiation. Similar to [9-11] in order to simulate the collision of an ion with the crystallite the momentum directed in the volume of the sample normal to the free surface was given to the one of surface atoms – the primary knocked atom (PKA). The PKA energy in various calculations was ranged from 1 to 20 keV. The temperature of the samples was 300 K. The face opposite to the irradiated surface was fixed in the direction of PKA momentum. Periodic boundary conditions were used in other directions. The determination of the position of point defects (vacancies and SIAs) in the irradiated sample was based on the calculation of the occupancy of the Wigner-Seitz cells. To identify the structure of extended defects and determine the type of generated dislocations Common Neighbor Analisys [12] and Dislocation Extraction Algorithm [13] were used.

3. Results and discussion

The simulation showed that the number of survived vacancies in the bulk is always exceeds the number of SIAs for both free surfaces. This is due to the higher mobility of SIAs, which escape to the free surface and form islands of adatoms on it. The dependence of the number of vacancies and SIAs in the bulk on the PKA energy for the considered free surfaces is presented in Fig.1. The figure shows that in the sample with the (110) surface the number of vacancies and SIAs is larger than in the sample with the (111) surface. This is due to different morphology of the damage caused by cascades on surfaces. The crater surrounded by adatom islands is formed when a PKA with energy of 5 keV or more is generated at the (111) surface (Fig. 3a). In the case of the (110) surface a crater is not formed, although adatom islands are present. They have a larger area and more ordered structure. In the volume of the crystallite near the surface a large cluster is formed, which is a vacancy-type dislocation loop with the Burgers vector a<100> or a/2<111> (Fig. 3b). It should be noted that the local increase of the atomic volume may lead to the formation of a new phase in the material [14, 15].

In the present work the size of the crater on the (111) surface and the vacancy cluster near the (110) surface was estimated and expressed in the number of vacancies. Calculations showed that the sizes of crater and vacancy loop coincide within statistical error. The average size of the crater and dislocation loop as a function of the PKA energy is shown in Fig. 2. Note that the generation of the PKA on the free surface leads to the sputtering of some atoms from the crystallite and escape of atoms to the

IOP Conf. Series: Journal of Physics: Conf. Series **830** (2017) 012067 doi:10.1088/1742-6596/830/1/012067

surface of the specimen. Their amount increases with increasing the PKA energy. The number of adatoms and sputtered atoms is always higher for the (111) surface. The formation of greater number of sputtered atoms and adatoms during irradiation of the (111) surface explains the lower amount of SIAs in samples with a given surface. For the considered surfaces the difference in the amount of adatoms and the sputtered atoms increases with the growth of the PKA energy and disappears at 20 keV. This is because the main part of the cascade develops at a greater distance from the surface. Therefore, the effect of the irradiated surface orientation on the number of formed defects is reduced.

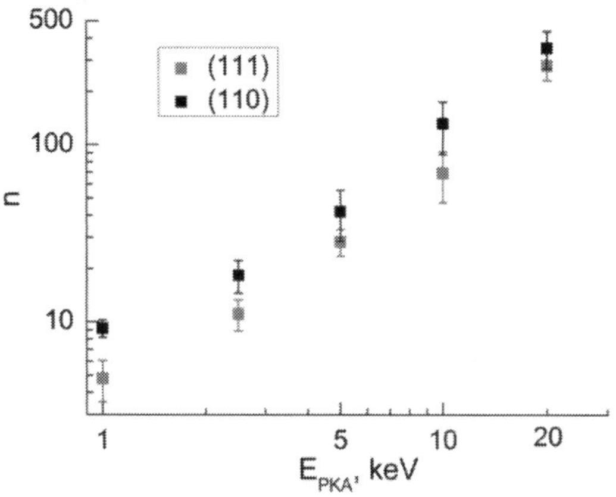

Figure 1. The number of survived point defects in the volume of the crystallite versus the PKA energy for cascades near the free surfaces (110) and (111).

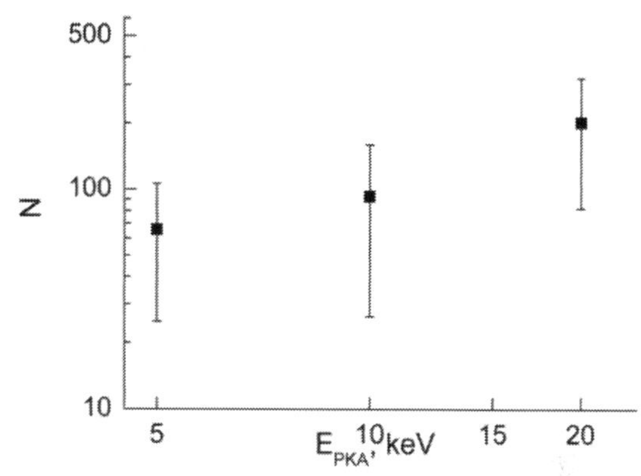

Figure 2. The average size of the crater on the (111) surface and vacancy loop in the sample with the (110) surface (in vacancy count) versus the PKA energy.

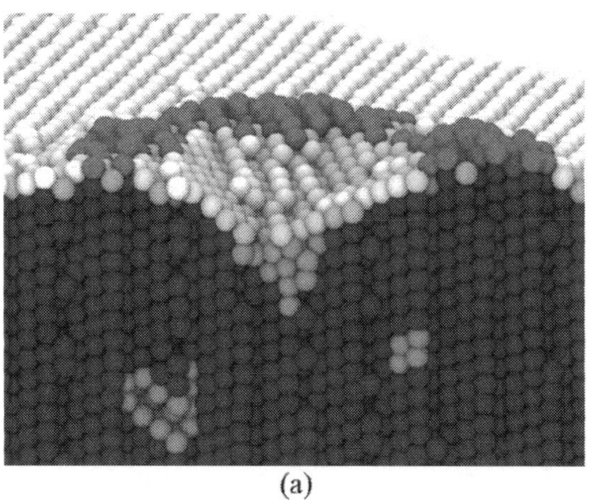
(a)

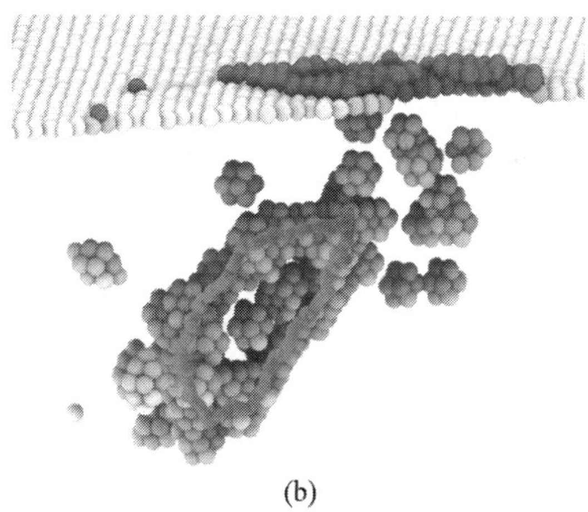
(b)

Figure 3. Slices of crystallites containing the crater on the (111) surface (a) and the dislocation loop a<100> under the (110) surface (b). Red color marks adatoms, blue - atoms having the bcc local environment, grey - atoms with a distorted local environment. In figure b atoms with the bcc local environment are not shown, the dislocation with Burgers vector a<100> is highlighted in pink.

Different character of the damage caused by the generation of the PKA on free surfaces is connected with anisotropy of motion of shock waves formed by atomic displacement cascades in the volume of the crystallite [16]. The waves propagating along close-packed directions <111> have the highest amplitude [17]. Under the irradiation of the (111) surface one of these waves reaches the free surface with the amplitude sufficient to form a crater. When the PKA is generated on the (110) surface the shock wave propagating in the direction normal to the free surface has the amplitude insufficient for the formation of the crater. However, the passage of this wave leads to the escape of large amount of SIAs to the (110) surface. As a result, a large vacancy cluster representing a dislocation loop is formed in the sample.

The formation of a crater on the (111) surface characterized by a denser packing of the atomic planes, compared with the (110) surface, is in good agreement with the experimental data [1]. Authors found that the irradiation of denser materials with ions results in more frequent formation of craters at the free surfaces. It was shown in [6] that the probability of a crater formation increases with decreasing a distance from the atomic displacement cascade to the free surface. This is because the amplitude of the shock waves generated by the atomic displacement cascade is quickly attenuates with the distance traveled.

Calculations have shown that atomic displacement cascades generate shock waves approximately at the same distance from the considered free surfaces. Therefore, the packing of atomic layers parallel to the free surface plays an important role in the formation of the crater or the dislocation loop.

4. Conclusion

The evolution of atomic displacement cascades initiated near the free surface differs significantly from the cascade behavior in the bulk. The type of resulting damage of the irradiated free surface largely depends on its crystallographic orientation. Craters are formed after generation of displacement cascade on the (111) free surface. Dislocation loops of vacancy type are nucleated near the irradiated (110) surface. As a result, the specimen with this surface contains the higher number of survived point defects than the specimen with the (111) surface. Difference in the number of survived point defects vanishes with increasing cascade energy. Formation of a crater or dislocation loop is connected with anisotropic character of motion of shock waves generated by atomic displacement cascades. This anisotropy is caused by different packing of atomic planes parallel to the irradiated free surfaces.

References

[1] Donnelly S E and Birtcher R C 1999 *Philos. Mag.* A **79** 133–45
[2] Prokhodtseva A, Décamps B and Schäublin R 2013 *J. Nucl. Mater.* **442** S786–9
[3] English C A and Jenkins M L 2010 *Philos. Mag.* **90** 821–43
[4] Jenkins M L, Yao Z, Hernández-Mayoral M and Kirk M A 2009 *J. Nucl. Mater.* **389** 197–202
[5] Aliaga M J, Schäublin R, Löffler J F and Caturla M J 2015 *Acta Mater.* **101** 22–30
[6] Osetsky Y N, Calder A F and Stoller R E 2015 *Curr. Opin. Solid State Mater. Sci.* **19**(5) 277–86
[7] Plimpton S 1995 *J. Comp. Phys.* **117** 1–19
[8] Mendelev M I, Han S, Srolovitz D J, Ackland G J, Sun D Y and Asta M 2003 *Philos. Mag.* A **83** 3977–94
[9] Zolnikov K P, Korchuganov A V, Kryzhevich D S, Chernov V M and Psakhie S G 2015 *Nucl. Instr. Meth.* B **352** 43–6
[10] Psakhie S G, Zolnikov K P, Kryzhevich D S, Zheleznyakov A V and Chernov V M 2009 *Phys. Mesomech.* **12**(1–2) 20–8
[11] Psakhie S G, Zolnikov K P, Kryzhevich D S, Zheleznyakov A V and Chernov V M 2009 *Crystallogr. Rep.* **54** 1002–10
[12] Honeycutt J D and Andersen H C 1987 *J. Phys. Chem.* **91** 4950–63
[13] Stukowski A and Albe K 2010 *Modelling Simul. Mater. Sci. Eng.* **18**(8) 085001
[14] Zolnikov K P, Psakhie S G and Panin V E 1986 *J. Phys. F: Met. Phys.* **16**(8) 1145–52
[15] Psakhie S G, Kryzhevich D S and Zolnikov K P 2012 *Tech. Phys. Lett.* **38** 634–7
[16] Zol'nikov K P, Uvarov T Y and Psakh'e S G 2001 *Tech. Phys. Lett.* **27** 263–5
[17] Korchuganov A V, Zolnikov K P, Kryzhevich D S, Chernov V M and Psakhie S G 2015 *Nucl. Instr. Meth.* B **352** 39–42

Acknowledgments
The work was carried out at the support of the Programme of fundamental research of state academies of sciences for 2013-2020.

Features of plastic deformation nucleation in the elastically loaded aluminium crystallites during irradiation

D S Kryzhevich[1, 2], A V Korchuganov[1, 2], K P Zolnikov[1, 2], S Y Korostelev[1]

[1] Institute of Strength Physics and Materials Science of Siberian Branch Russian Academy of Sciences, pr. Akademicheskii 2/4, Tomsk, 634055, Russia
[2] National Research Tomsk State University, 36 Lenin Ave., Tomsk, 634050, Russia

E-mail: kryzhev@ispms.ru

Abstract. The simulation of peculiarities of the surface layer reconstruction in the crystallites of aluminum after the ion bombardment and the copper film irradiated by the electron beam is carried out. The performed calculations are based on the molecular dynamics method. It is shown that the orientation of the irradiated surface and preliminary elastic deformation have a significant impact on features of atomic structure formation in the ion-modified layer in aluminum. Weak structural changes in the surface layer are observed at the irradiation of the {100} surfaces. Sufficiently great number of stacking faults is formed under irradiation of the {111} and {110} surfaces. It is shown that heating by the electron beam of the {110} surface in the copper film leads to the formation of stacking faults. It is shown that the preliminary elastic deformation of the material lowers the energy of the irradiation, at which formation of structural defects and fragmentation of crystallites of aluminum and copper will take place.

1.Introduction

Surface treatment by ion and electron fluxes can significantly increase the strength, durability and corrosion resistance of metallic materials, increase their biological compatibility [1-3]. Change of physical-mechanical properties under irradiation is determined by the development of a number of processes. Doping the surface region of the material with chemical elements from the ion beam is performed during ion irradiation. Ion-beam and electronic processing significantly alter the defect system and can lead to fragmentation of the microstructure of the irradiated surface. In this case, the thickness of the fragmented layer essentially depends on the power density, the impact duration and the degree of preliminary elastic deformation of the material. So, the thickness of the fragmented layer can reach several tens of microns under irradiation of the material by energetic ion beams of high energy (>3 MeV). In the case of using ion beams with the energy less than 1 MeV the thickness of the fragmented layer is reduced to a few microns. The thickness of the modified layer (the layer with the changed defect structure) may exceed hundreds of microns. At small values of the ion beam energy due to the effect of the long-range the thickness of the modified surface layer can significantly exceed not only the depth of the mean free path of incident ions, but the average grain size. For the fragmentation of polycrystalline materials under irradiation, as in the case of plastic deformation, the occurrence of high elastic stresses which are localized at the grain boundaries and their junctions is necessary [4,5]. It is known that strong elastic fields localized in the surface region are formed upon irradiation of metals and alloys by high energy fluxes of charged particles (ions and electrons). The process of a fragmentation may lead to the formation of the new nonequilibrium micro- and nanoscale structures in the surface layers of irradiated metals and alloys. Several competing mechanisms for relaxation of the material which are associated with the phase and structural transformations, fragmentation, plastic deformation in the subsurface of the material take place under ion fluxes action [6-9]. Obtaining reliable experimental information about the structural response of the material under irradiation is connected with significant difficulties, due to the small temporal and spatial scales. An effective approach to overcome these difficulties and to obtain detailed information about dynamics of change of the microstructure during irradiation of a material is a computer simulation [10-12].

In this work, the atomic structure changes of the surface layers of aluminum crystallites after bombardment by low-energy particles were studied. Also the influence of the irradiated surface orientation on the peculiarities of the atomic structure formation in the ion-modified layer was revealed. The characteristics of structural changes in a thin film of copper irradiated by the electron beam were also studied in this work.

Formulated problems were solved on the basis of the molecular dynamics method. The simulation was performed using the software package LAMMPS [13]. Interatomic interaction potential for aluminum was described in the framework of the embedded atom method [14]. Crystallite sizes ranged from 75 000 to 1 000 000 atoms. Initial temperature of the crystallites was 300 K. Irradiated surfaces had crystallographic indices {100}, {110} and {111}. Aluminum atoms were used as the incident particles. The velocity of incident atoms was directed normal to the free surface. Its value was equal to 210 Å/ps, which corresponds approximately to 60 eV. The number of incident atoms was 540 (Fig. 1 a). Simulation of structural changes in the surface layers at irradiation was carried out for mechanically unloaded and elastically uniaxially pre-streched in [110] direction crystallites.

A simulated copper film had a form of parallelepiped with a size of 8x8x20 nm. Free boundary conditions were applied in the direction normal to the smallest cross-section, and periodic boundary conditions were set in the other directions. The power density of the electron beam was ~ 0.03 eV/(atom*nsec).

2.Computational results and discussion

The calculations have shown that as in the case of mechanical loading [15] the orientation of the irradiated surface has a significant effect on the characteristics of structural changes in the surface layer of aluminum crystallite. So, the {100} surfaces were the most resistant to structural rearrangements upon irradiation. A local change of atomic environment symmetry took place only for the individual atoms of the {100} surface layers, which were involved in the atomic displacement cascades (Fig. 1 b). However, such rearrangements of atomic configurations did not lead to the formation of structural defects (dislocations, stacking faults, interstitial dumbbells, etc.). Note that pre-elastic deformation of the samples prior to irradiation had no significant effect on the character of the structural changes in the surface layer. In the case of the {111} surface irradiation the intrinsic and extrinsic stacking faults were formed in the surface region in the (111) planes even in the undeformed crystallites.

Preliminary deformation led to the formation of stacking faults in adjacent directions (Fig. 1). At the irradiation of the {110} surface there were atoms with changed symmetry of their atomic environment, some of which restored the symmetry of local environment, and the other part formed the stacking faults (Fig. 2 a). The size of the defects grows with the increase of the specimen elastic pre-streching. Stacking faults were formed in adjacent planes and interacted with each other forming vacancy chains (Fig. 2 b, c). This process slowed down the accumulation of defects. Fig. 3a, b clearly shows how the structural response of the irradiated crystallite changed with increasing degree of the sample elastic pre-streching from 4 to 5%.

The simulation results showed that irradiation of the crystallites was accompanied by intensive sputtering of atoms and atomic clusters away from the irradiated surface. The total number of sputtered atoms exceeded the number of implanted atoms. Note that under the same irradiation conditions the process of sputtering was the most active for the {111} surface.

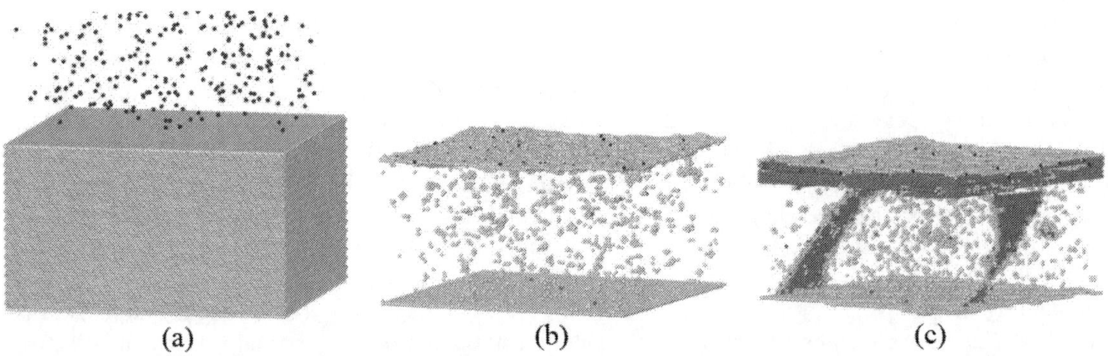

Figure 1. The initial structure of the simulated system (irradiated sample and incident atoms) (a). Structure of the samples deformed to 4 % after irradiation of {100} (b) and {111} (c) surfaces. The blue color indicates incident atoms; atoms with fcc, hcp and uncertain symmetry of local environment are marked in green, red and grey, respectively. Atoms with fcc symmetry of local environment are not shown in Fig. (b) and (c).

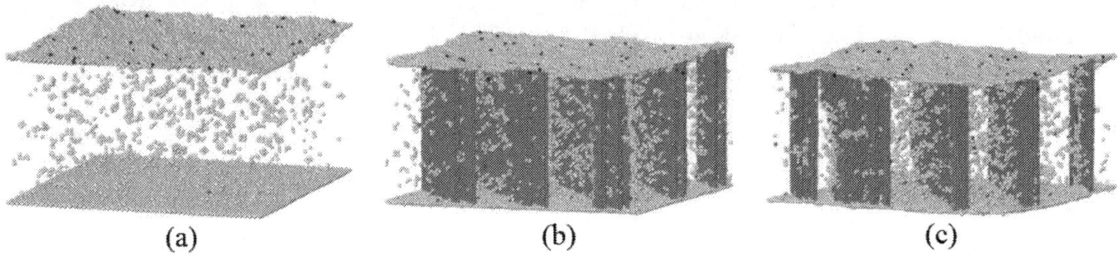

Figure 2. Structure of the simulated samples after irradiation of the {110} surface. Pre-stretching: a) 0 %; b) 4 %; c) 5 %. Red and grey colors indicate atoms with hcp and uncertain symmetry of local environment, respectively. The atoms with fcc symmetry of local environment are not shown.

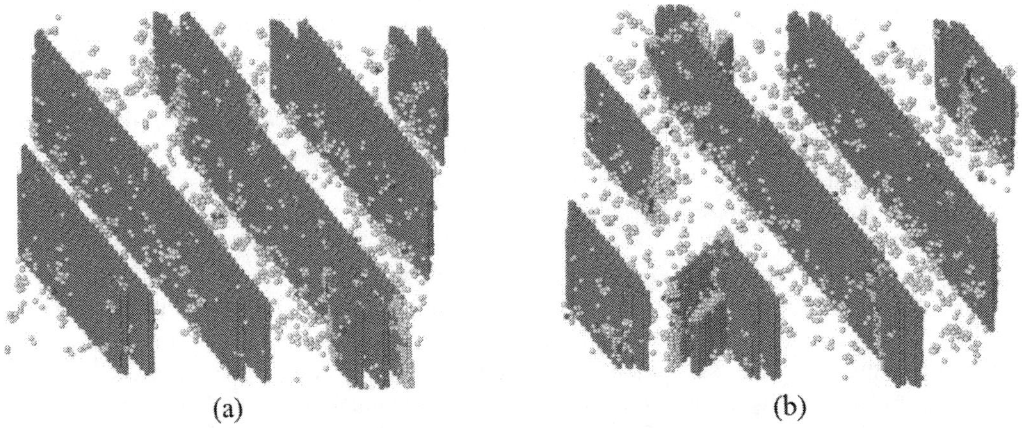

Figure 3. Projection of defect structure of the sample with irradiated {110} surface on (110) plane. Pre-stretching: a) 4%; b) 5%.

Kinetic temperature calculation showed that for the given irradiation intensity the surface layer may be heated to a temperature above the melting point. Kinetic temperature versus time for the {110} surface layer with thickness of 8.5 lattice parameters is shown in Fig. 4. After irradiation the temperature of the surface layer quickly decreased due to the applied thermostat and was characterized by well-pronounced oscillations. The nature of the temperature oscillations in the surface layer as well as the sputtering of atoms and atomic clusters from the irradiated surface is associated with the generation of shock waves caused by rapid heating of the surface layer [12]. The frequency of the oscillations well correlates with the time of the passage of shock waves of the distance from the loaded face to the back surface (from which they are reflected) and backwards.

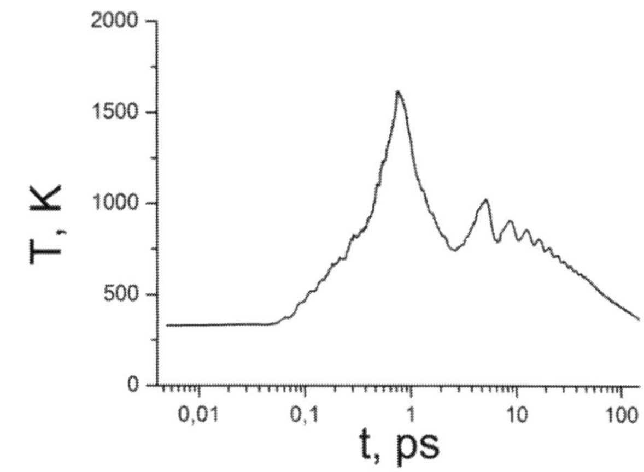

Figure 4. The temperature versus time for the {110} surface layer of the sample pre-stretched to 4%.

The simulation results of the electron beam impact on the copper film show that up to the melting point the heating rate was proportional to the density of the heat source power. The spatial distribution of temperature was homogeneous in the sample. Further heating of the material leads to a change in the spatial distribution of temperature in the sample. This is due to the fact that the liquid phase starts to nucleate in the region of the free surfaces. The analysis of the change of the phase state of the film allows allocating two regions: the region with the crystalline state and the region with the liquid phase separated by interphase boundary, consisting of 4-6 atomic planes. The local environment of individual atoms changes its fcc symmetry on the hcp one at the process of heating of the film with the {110} free surface. As a result stacking faults are formed in the film. The material structure after such local transformation is shown in Fig.5.

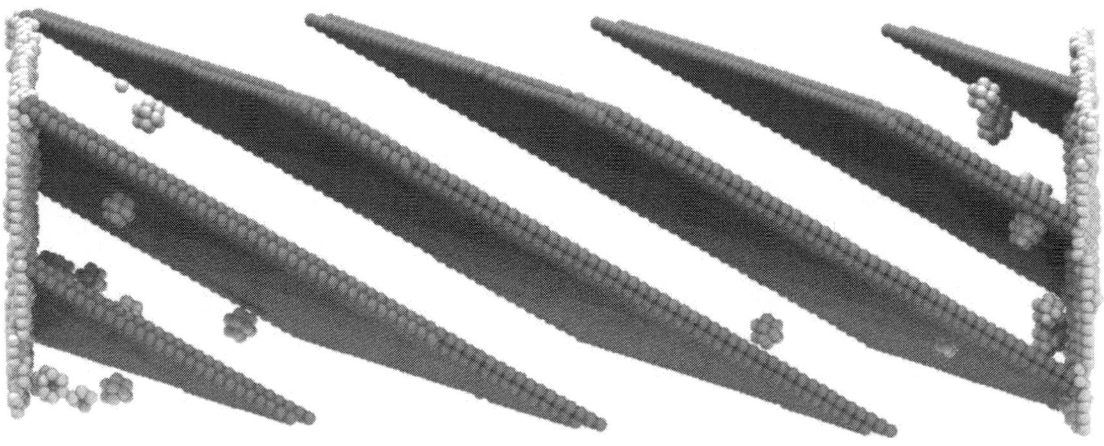

Figure 5. Relaxed structure of copper film after the impact by an electron beam. Red color corresponds to stacking faults, grey – atoms with undefined symmetry of the local environment. Atoms with local fcc symmetry are not shown.

3.Conclusion

The simulation results showed that the formation of defects during ion irradiation of crystalline aluminium and electronic heating of a copper crystalline film largely depends on the crystallographic orientation of the irradiated surface. It was revealed that the {100} surfaces are the most resistant to structural changes. A large number of stacking faults are formed at the irradiation of the {111} and {110} surfaces. The increase in the energy of the incident beams and the preliminary elastic deformation can lead to fragmentation of the surface layers of irradiated materials. The grains in the surface layer having the above mentioned crystallographic orientation to the direction of irradiation and excess elastic stress will be subjected to fragmentation.

References

[1] Meisner S N, Meisner L L, Lotkov A I and Tverdokhlebova A V 2015 *Steel Transl.* **45(4)** 258
[2] Gleiter H Nanocrystalline materials 1969 *Progr. Mater. Sci.* **33** 223
[3] Psakhie S G, Zolnikov K P, Kryzhevich D S, Zheleznyakov A V and Chernov V M 2009 *Crystallogr. Rep.* **54** 1002
[4] Meisner S N 2014 *Russ. Phys. J.* **57**(3) 403–10
[5] Zolnikov K P, Psakhe S G and Panin V E 1986 *J. Phys. F: Met. Phys.* **16(8)** 1145
[6] Psakhie S G, Kryzhevich D S and Zolnikov K P 2012 *Tech. Phys. Lett.* **38** 634
[7] Zolnikov K P, Korchuganov A V, Kryzhevich D S, Chernov V M and Psakhie S G 2015 *Nucl. Instr. Meth.* B **352** 43
[8] Psakhie S G, Zolnikov K P, Kryzhevich D S, Abdrashitov A V and Lerner M I 2010 *Phys. Mesomech.* **13(3-4)** 184
[9] Psakh'e S G, Zol'nikov K P and Saraev D Y 1998 *Tech. Phys. Lett.* **24(2)** 99
[10] Psakhie S G, Zolnikov K P, Kryzhevich D S, Zheleznyakov A V and Chernov V M 2009 *Phys. Mesomech.* **12(1–2)** 20–8
[11] Psakh'e S G and Zol'nikov K P 1998 *Combust. Explos. Shock Waves* **34(3)** 366
[12] Korchuganov A V, Zolnikov K P, Kryzhevich D S, Chernov V M and Psakhie S G 2015 *Nucl. Instr. Meth. B* **352** 39
[13] Plimpton S 1995 *J. Comp. Phys.* **117** 1
[14] Mendelev M I, Kramer M J, Becker C A and Asta M 2008 *Philos. Mag.* **88(12)** 1723
[15] Zol'nikov K P, Uvarov K P and Psakh'e S G 2001 *Tech. Phys. Lett.* **27** 263

Acknowledgments

The work is supported by the Program of Presidium of the Russian academy of sciences No. 13 "Thermophysics of high energy densities".

IOP Conf. Series: Journal of Physics: Conf. Series **830** (2017) 012069 doi:10.1088/1742-6596/830/1/012069

Numerical study of heating and evaporation processes of quartz particles in RF inductively coupled plasma

Yu M Grishin and Long Miao

Bauman Moscow State Technical University, 2-ya Baumanskaya str., bld. 5, 1, Moscow, 105005, Russia

E-mail: 691632748@qq.com

Abstract. Numerical simulations of heat and evaporation processes of quartz particles in Ar radio frequency inductively coupled plasma (ICP) are investigated. The quartz particles are supplied by the carrier gas into the ICP within gas-cooling. It is shown that with the increase of amplitude of discharge current above critical value there is a toroidal vortex in the ICP torch at the first coil. The conditions for the formation of vortex and the parameters of the vortex tube have been evaluated and determined. The influence of vortex, discharge current, coil numbers and feed rate of carrier gas on the evaporation efficiency of quartz particles have been demonstrated. It was found that the optimal discharge current is close to the critical value when the quartz particles with initial sizes up to 130 μm can be fully vaporized in the ICP torch with thermal power of 10kW. The heat and evaporation processes of quartz particles in the ICP torch have significant importance for the study of one-step plasma chemical reaction method directly producing silicon from silicide (SiO_2) in the argon-hydrogen plasma.

1. Introduction

Currently, in connection with the development of various new techniques [1-4], the interest in the inductively coupled plasma (ICP) steadily increases [5, 6]. The method [1] of one-step plasma chemical reaction directly producing silicon from silicide (SiO_2) in the argon-hydrogen ICP plasma torch is proposed and investigated. In our study the solid particles of SiO_2 are introduced by hydrogen into argon plasma, which is created by the ICP torch. There are three different physic-chemical processes occurring consistently in the heterogeneous flows: 1) evaporation and decomposition of SiO_2 solid particles with the formation of silicon and oxygen atoms; 2) non-equilibrium gas-phase chemical reaction with the binding of oxygen and hydrogen atoms, but without formation of silicon monoxide SiO; 3) condensation of silicon vapor with the formation of polycrystalline silicon particles or films.

Analysis of the results [1] showed that the high purity of the polycrystalline silicon can be acquired by using the electrodeless torches of induction frequency plasma with the fulfillment of optimal heat and evaporation conditions for the solid particles of SiO_2. At present, a wide range of theoretical and experimental studies is needed for the design and optimization of structural and operating parameters of the ICP torches. In this paper, the results of numerical simulation of the heat and evaporation processes of quartz particles in the ICP plasma torch with axial gas-cooling are presented.

2. Physical and mathematical model

It is assumed that heating and evaporation of quartz particles happens in sparse heterogeneous plasma flows. The fundamental equations included Maxwell's equations, plasma dynamics equations with the account of influence of electromagnetic force and Joule heat, as well as the equations of movement, heat and mass transfer of particles.

The scheme and geometric parameters of ICP plasma torch with axial gases feeds are shown in Figure 1. In our work, the elementary sizes of ICP plasma torch are set as: $Z_2 = 50$ mm, $Z_{inf} = 63$ mm, $Z_{inl} = 123$ mm, $Z_4 = 400$ mm, $R_1 = 1.7$ mm, $R_2 = 18.8$ mm, $R_3 = 25$ mm, $R_4 = 33$ mm, $d_{coil} = 6$ mm, $\delta_1 = 3$ mm, $\delta_2 = 2.2$ mm, $\delta_3 = 3.5$ mm. The feed rates of central and sheath gases are $G_2 = 1.02 \times 10^{-4}$ kg/s and $G_3 = 1.41 \times 10^{-3}$ kg/s. However, the feed rate of carrier gas ranges from $G_{1min} = 0.87 \times 10^{-6}$ kg/s to $G_{1max} = 0.42 \times 10^{-4}$ kg/s and the length of the injection probe Z_1 ranges from 3 to 50 mm.

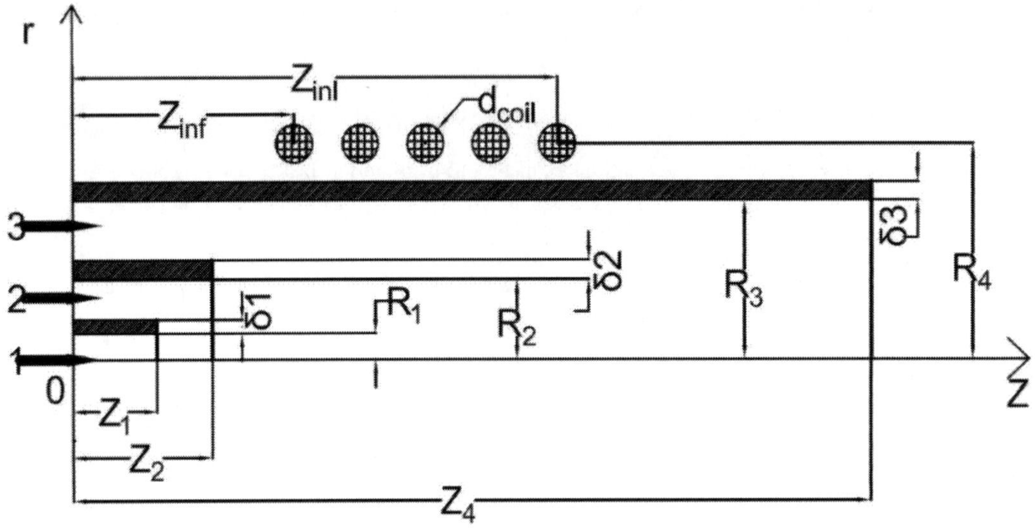

Figure 1. Scheme and geometric parameters of the ICP plasma torch: 1) injection probe of carrier gas and quartz particles, 2) channel of central gas, 3) channel of sheath gas.

The coils ($N = 2$, 3 or 5) are located at a fixed length $l_{coil} = Z_{inl} - Z_{inf} = 60$ mm along the ICP quartz pipe. The amplitude of the discharge current J_{coil}, varying sinusoidally with frequency $\omega = 3$ MHz was chosen in the range 20–170 A.

In this model, it is assumed that the spiral inductor can be regarded as a system of cylindrically symmetric parallel coils. The states of gas flows at the inlets to the ICP channels are azimuthally symmetrical and stationary. With the fulfillment of these assumptions, the electromagnetic and gas dynamics equations can be solved in a two-dimensional cylindrical coordinate system (r, z) (2D-model).

In the present study, the calculation of gas-dynamic parameters of the laminar ($Re < 10^3$) subsonic axisymmetric stationary plasma flows is carried out on the basis of plasma dynamics equations [7], in which the electromagnetic forces and Joule heat (energy loss by radiation is negligible) have been taken into account. Argon is considered as the carrier, central and sheath gas. The thermal parameters of argon plasma are determined as approximation of local thermodynamic equilibrium (LTE) from known data [8].

At the inlets ($Z = 0$) the temperature ($T = 300$K) and the axial component of the velocities of central ($U_{z2} = 0.054$ m/s) and sheath gas ($U_{z3} = 2.1$ m/s) are set constant. The axial component of the carrier gas U_{z1} ranges from 0.054 m/s to 2.1 m/s. At the outlet $Z = Z_4$, constant pressure $P = P_{atm} = 10^5$ Pa is specified.

The calculations of electromagnetic field parameters, intensity of the Joule heat and electromagnetic forces are made on the basis of Maxwell's equations, which is written by the magnetic vector potential [9]. The boundary conditions for the magnetic vector potential are given in [10].

The calculations of the movement, heat and mass transfer of quartz particles in plasma flows with steady distribution of velocity and temperature are carried out by the assumption that: the quartz particles have spherical shape with initial diameter d_0 = 50 - 200 μm; the interactions between quartz particles are neglected and the temperature in the particles is equally distributed. The equation system includes:

–momentum equation of quartz particles in plasma flows with a due consideration of aerodynamic drag force, gravity and virtual mass force;

–energy equation of quartz particles, taking into account the convective heat transfer from plasma and the latent heat resulting from the evaporation of particles.

The evaporation of particles is simulated within the framework of diffusion model, in which the coefficient of heat and mass transfer is calculated by Ranz-Marshall formula [11]. The velocity and entry angle β at the outlet of injection probe (relatively to the axis) are determined by the solution of momentum equation of quartz particles. The thermal parameters of quartz particles are determined from the known data [8].

Numerical solution of the equations system was made in ANSYS using finite volume method. The grid model was built in ANSYS ICEM package using HEXA_8 block of hexagonal structure. Detailed descriptions of the application of physical and mathematical model and numerical solution are presented in [12-16].

3. Results and Discussion

It is found that the maximal temperature in the ICP torch ranges from 8 to 10 kK when the total feed rates of carrier, central and sheath gases $G_1+G_2+G_3 \approx 1.55 \times 10^{-3}$ kg/s and the amplitude of discharge current J_{coil} ranges from 20 to 170 A. In order to ensure the normal thermal state of quartz pipe (wall temperature less than 800-900 K), the discharge current J_{coil} in ICP inductor with three and five coils should not exceed 160-170 A and 120-130 A so that the thermal power in the ICP torch does not exceed 10 kW.

It was established that there can be two different kinds of plasma flow patterns in the ICP torch, which appear alternately with the change of discharge current J_{coil}. When the J_{coil} is above the critical value J_{cr} there is a toroidal vortex (Figure 2) in ICP torch at the first coil. The formation of vortex is connected with the appearance of high pressure zone around the ICP axis because of the influence of axial component of electromagnetic force. With the increase of the discharge current, the intensity of the vortex becomes higher. However, when J_{coil} is below the critical value J_{cr}, the electromagnetic force decreases, and the high pressure zone disappears. As a result, the toroidal vortex is not formed. Under this kind of plasma flow pattern, the central and carrier gases flow almost along the ICP axis into high temperature zone.

When the geometry parameters of the ICP-torch and gases feed rates are defined, the critical value J_{cr} are determined by the specific number of coils $n = N/\ell_{coil}$. For two (n = 0.33 cm^{-1}), three (n = 0.5 cm^{-1}) and five (n = 0.83 cm^{-1}) coils, J_{cr} is equal to 110-120 A, 80-90 A and 40-50 A, respectively.

The intensity of vortex depends both on the discharge current, and on the feed rate of carrier gas. The reduce of feed rate of carrier gas from 0.42×10^{-4} kg/s (Figure 2, upper half-plane) to 0.87×10^{-6} kg/s (Figure 2, lower half-plane) can lead to a substantial increase of vortex area and intensity. However, it was found that the length of injection probe Z_1 have low influence on the vortex locations and intensity.

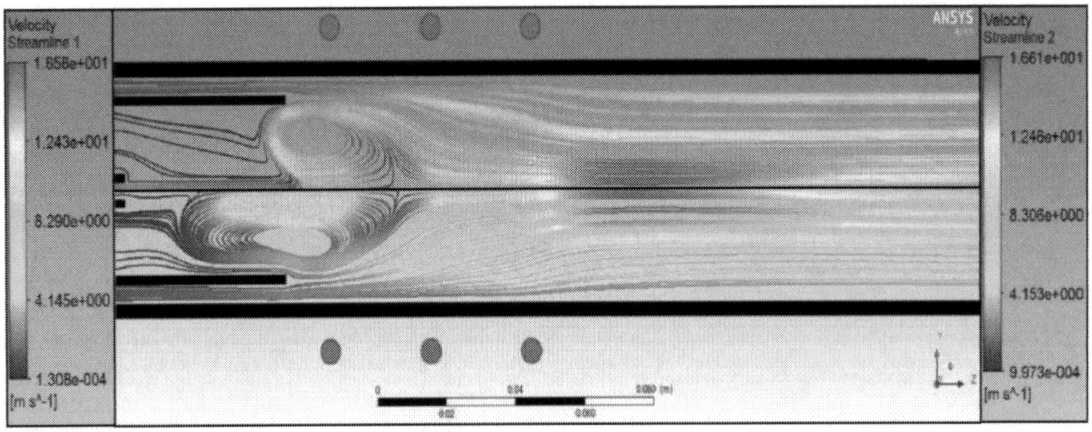

Figure 2. The streamlines in the channels of ICP torch (N = 3, Z_1 = 3 mm) under J_{coil} =170 A and different feed rates of carrier gas: upper half-plane- G_{1max} = 0.42×10^{-4} kg/s ($U_{Z1} = 2.1$ m/s, $U_{Z2} =$ 0.054 m/s); lower half-plane –G_{1min} = 0.87×10^{-6} kg/s (U_{z1} = U_{z2} = 0.054 m/s).

The evaporation efficiencies of particles $\varepsilon = 1-(d_k/\ d_0)$ (d_k is the diameter of particles at the ICP torch outlet $Z= Z_4$, d_0 is the diameter of particles at the injection probe inlet) depends on the discharge current J_{coil}, plasma flow patterns, initial particle diameter d_0 and entry angle of particles β at the outlet of injection probes. It has been studied that when the particles are transported into the injection probe under the cone angle of 45° at the inlet of injection probe, the range of entry angles β at the outlet of injection probe depends on the initial diameter of particles: β = 0–10° for particles with $d_0 \leq 100$ μm, β = 0–25° for particles with $d_0 \leq 200$ μm.

As discussed above, the fine particles ($d_0 \leq 100$ μm) are transported into the ICP torch under entry angle $\beta \leq 10°$. They fly through the cavity of the vortex or by the streamlines into the high temperature zone. The trajectories of such particles have a spindle shape. Investigations show that particles with diameter $d_0 \leq 100$ μm can be fully vaporized before they fly out of the ICP within coils with the number $N \geq 3$ and discharge currents J_{coil} about 40–50 A.

It was found that large particles (150 μm < d_0 ≤ 200 μm) preserve their velocity at the outlet of injection probe when they fly in the ICP torch despite the different kinds of plasma flow patterns because of their large inertia. The evaporation efficiency ε of these particles weakly depends on the discharge current J_{coil} and does not exceed 0.2 for the ICP with $N \leq 3$ and J_{coil} ≤170 A.

The results show that the evaporation efficiency of particles with d_0 from 100 < ≤150 μm and entry angle of particles $\beta \leq 10°$ have maximal values when the discharge current J_{coil} is close to the critical value J_{cr} (Figure 3, solid lines). There can be several reasons for this.

First, when the discharge current is below the critical value J_{cr} there is no vortex in the ICP torch. The particles fly directly through the high-temperature zones. With the increase of discharge current, the evaporation efficiency ε becomes higher due to the rise of plasma temperature and the intensification of heat and mass transfer processes in the ICP torch.

Second, when the discharge current is above the critical value J_{cr}, the increase of discharge current can significantly promote the plasma velocity along the axis (by 2 times, then $J_{coil} < J_{cr}$) because of the formation of vortex in the ICP torch, which respectively leads to a remarkable reduction of particle resident times in the high temperature zone. As a result, the evaporation efficiency ε drops. When $\beta \geq 10°$ the evaporation efficiency of particles with 100 μm< d_0 ≤150 μm monotonically increases with the growth of discharge current (Figure 3, dashed lines).

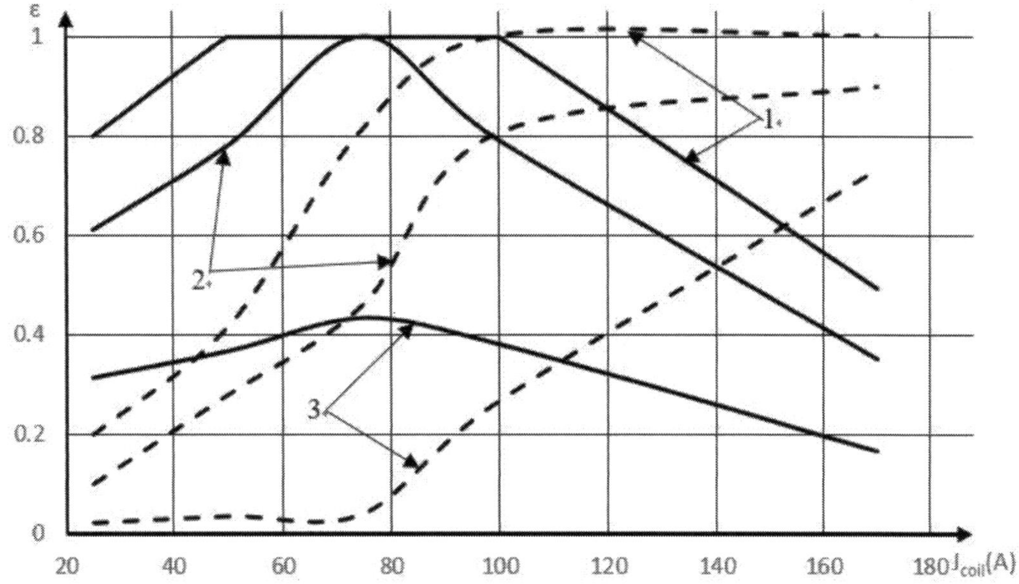

Figure 3. The dependence of evaporation efficiency of particles on discharge current under different initial diameters of particles, which are supplied into the ICP torch (N = 3, $U_{Z1} = 2.1$ m/s, $U_{Z2} = 0.054$ m/s) under axial direction of injection (solid lines) and cone injection angle 45° (at the inlet of injection probe) (dashed lines): 1) d_0= 120 μm; 2) d_0=130 μm; 3) d_0=150 μm.

4. Conclusion

There are two different kinds of plasma flow patterns in the ICP torch depending on the discharge current J_{coil}. When the discharge current J_{coil} is above the critical value J_{cr}, there is a toroidal flow of vortex in the ICP torch at the first coil. However, if the discharge current J_{coil} is below J_{cr}, the carrier and central gases flow almost axially into the high temperature zone (without a vortex).

The evaporation efficiency of particles depends on the discharge current J_{coil}, plasma flow patterns and initial diameter of particles d_0. First, the fine particles ($d_0 \leq 100$ μm) can be completely vaporized in the ICP torch with the number of coils $N \geq 3$ and discharge current J_{coil} about 40-50 A. Second, for the particles with 100 μm< d_0 ≤150 μm, the evaporation efficiency ε is maximal when the discharge current J_{coil} is near the critical value J_{cr} (for $\beta < 10°$). In the case they are injected at $\beta > 10°$, the evaporation efficiency of particles monotonically increases with the growth of discharge current. Third, for the large particles (150 μm< d_0 ≤200 μm), the evaporation efficiency does not exceed 0.2 under the allowable working states (thermal power in ICP ≤ 10 kW).

The ICP torches with gas-cooling and energy power up to 10 kW can be effectively used for heating and evaporating silica particles with maximum size no greater than 150 μm. The processes are significant for the plasma-chemical method for producing silicon from quartz particles in argon-hydrogen plasma [17].

References

[1] Grishin Yu M, Kozlov N P and Skryabin A S 2012 *High Temperature* **50** 459
[2] Boulos M I 1997 *High Temperature Material Processes* **1** 17
[3] Novikov I N and Kruchinin A M 2014 *Technical Physics Letters* **40** 920
[4] Borisov L A, Grishin Yu M, Gulin E N, A. D. Kairyak, Kozlov N P and Kutyrev M V 2007 High Temperature **45** 708
[5] Bapat A, Perrey C R, Campbell S A, Carter C B and Kortshagen U 2003 *Journal of Applied Physics* **94** 1969

[6] Morsli M E and Proulx P 2007 *Journal of Physics D: Applied Physics* **40** 380
[7] Morozov A I 2012 *Introduction to plasma dynamics* (CRC Press)
[8] Belov G V 2002 *Thermodynamic modeling: methods, algorithms and programs* (Moscow: Science World) (in Russian)
[9] Holik E F 2008 *Simulation results of an inductively-coupled RF plasma torch in two and three dimensions for producing a metal matrix composite for nuclear fuel cladding* (Texas: Master Theses)
[10] Grishin Yu M and Miao Long 2016 *Science and Education* **5** (in Russian)
[11] Aissa A, Abdelouahab M, Noureddine A, Ganaoui M E, and Pateyron B 2015 *Thermal Science* **19**(5) 1521
[12] Grishin Yu M and Miao Long 2016 *Science and Education* **8** (in Russian)
[13] Grishin Yu M and Miao Long 2016 *Applied Physics* **4** 33
[14] Kuzenov V V, Ryzhkov S V and Shumaev V V 2015 *Problems of Atomic Science and Technology* **1** (95) 97
[15] Kuzenov V V, Polozova T N and Ryzhkov S V 2015 *Problems of Atomic Science and Technology* **4** (98) 49
[16] Ryzhkov S V 2015 *Sustainable Cities and Society* **14** 313
[17] Grishin Yu M, Kozlov N P and Skryabin A S 2016 *High Temperature* **54** 619

IOP Conf. Series: Journal of Physics: Conf. Series **830** (2017) 012070 doi:10.1088/1742-6596/830/1/012070

Modeling of the electron-beam boriding in the system Fe-B-C-O_2

D E Dasheev and N N Smirnyagina

Institute of Physical Materials Science SB RAS, 670047, Ulan-Ude, Russia

E-mail: fokter@mail.ru

Abstract. This paper reviews the conditions of iron borides formation and simulation of surface layers saturation depending on the stoichiometry of original components. Temperature fields have been investigated as well, which form certain phases in accordance with the pressure in the chamber and the power of the electron beam. A thermodynamic study of phase equilibria in Fe-B-C-O systems has been performed. This was done in order to optimize conditions for forming functional layers on the surface of iron-carbon alloys as a result of electron beam boriding in vacuum. Furthermore, strength characteristics of iron boride layers have been determined. Then these layers obtained by different methods and using various source components have been thoroughly compared with each other during the analysis.

1. Introduction

There are many methods and techniques that exist for improving the qualitative properties of surface layers, such as strength, wear resistance, hardness, ductility, corrosion resistance etc. Modification of metals' and alloys' surface allows increasing the service life and reliability of various parts of machines and tools. Along with traditional and modern machine building methods of obtaining protective coatings, which increase the hardiness and wear resistance of tools, such as carbonization, nitriding, carbonitriding, boriding, chroming etc., the beam technology, for example laser and electron beam processing, increasingly prevails in recent years [1]. Thus, obtaining products with optimized functional properties and structure for specified conditions is some kind of challenge for our present-day industries. Therefore, such method as treatment with concentrated energy flux which generates locally and globally heterogeneous body structure can be referred to one of the most promising ways in this field. Using the electron beam as a heat source can significantly enhance the ability of surface modification of metals and alloys [2].

This article describes thermodynamic study of Fe-B-C-O phase equilibrium in a high vacuum. Thermal processes at electron beam treatment of different power have been analyzed and simulated. The article also deals with the gradual construction of mathematical model of thermal fields arising due to sample exposure by the electron beam.

2. Experimental part

2.1. Modeling phase equilibrium and revealing of fields of crystallization borides and carbides in systems Fe-B-C-O_2

In the present study used interface of a program complex TERRA [3]. Calculations were carried out in the range from 300 to 2073 K at a variation of the general pressure in system in a range from 10^5 to 10^-

[3] Pa. In calculations considered following phases: oxides FeO, Fe_2O_3, Fe_3O_4, B_2O_3; carbides B_4C, Fe_3C; borides FeB, Fe_2B.

2.2. Experimental technique

During the experiment we used the electron beam installation of special design which is equipped with a powerful electron beam projector (gun) on thermal cathodes [4]. During this experiment we used metal samples (St 3, 45) in the form of cylinders of 15 mm diameter and 7 mm in height. The surface of a sample was coated by various stoichiometric compositions such as Fe_2O_3:3B:3C, Fe_2O_3:2B:3C. Then this coated surface was exposed to highly concentrated steams of energy (20 keV) which initiated the SHS process accompanied by the high heat release. Electron beam treatment was carried out in vacuum not higher than 2×10^{-3} Pa with electron beam power of 250-450 W for 1-3 minutes. Finally, we obtain solid combustion products, particularly iron borides.

X-ray phase analysis was carried out on the Phaser 2D Bruker (Cu $K\alpha_1$ – radiation). The microstructure of layers was investigated by METAM PB-22 microscope with the program NEXSYS Image Expert. The Vikers (HV) microhardness measurements are made by pressing a diamond indenter, of a specified shape, into the surface with a known force.

3. Results and their discussion

3.1. The analysis of thermal fields and phase equilibria

The thermodynamics of phase equilibria in Fe-B-C-O systems has been studied in order to optimize conditions for forming functional layers on the surface of titanium and iron-carbon alloys as a result of electron-beam boriding in vacuum.

Thermal properties and the character of Fe_2B and FeB boride dissociation have been simulated depending on the total pressure in the system. Thus, the temperature and pressure influence on its behavior was determined. For example, an interaction of Fe_2O_3 with various boronizing components (B_2O_3, B_4C, B) at pressure of 10^5 Pa begins at temperature of 1300-1600 K, as well as at pressure of 10^{-2}-10^{-3} Pa the temperature falls down to 850-900 K. It is found that in mixtures with B_4C and B the phase transformation should originally occur in condensed state with B_2O_3 liquid phase formation. The layers have a clear boundary between the base and the layer itself. The thickness of layers is 200-350 µm and microhardness comes to 3500 MPa.

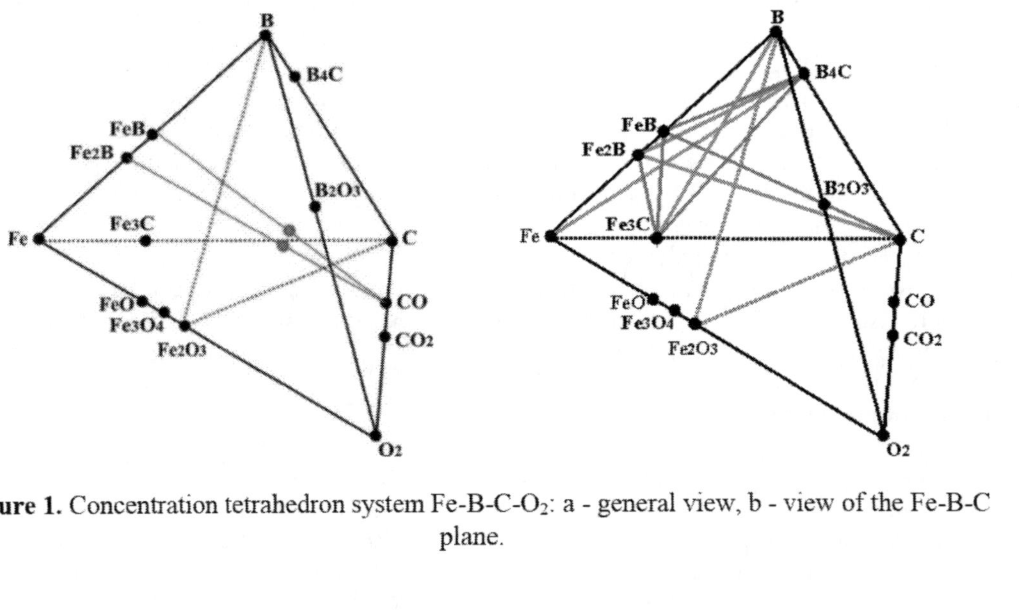

Figure 1. Concentration tetrahedron system Fe-B-C-O_2: a - general view, b - view of the Fe-B-C plane.

Fig. 1a shows the concentrated tetrahedron of Fe-B-C-O_2 system and Fig. 1b displays all possible cuts in Fe-B-C ternary system. This ternary system was the subject of a detailed study because the electron beam boriding uses amorphous boron B and boron carbide B_4C as boronizing components. The phase equilibria in Fe-B-C have been studied. The crystallization fields of all possible phases have been defined, as well as temperature and pressure influence on its behavior has been determined. (fig. 2). It shows that Fe-B_4C cut is not quasi-binary.

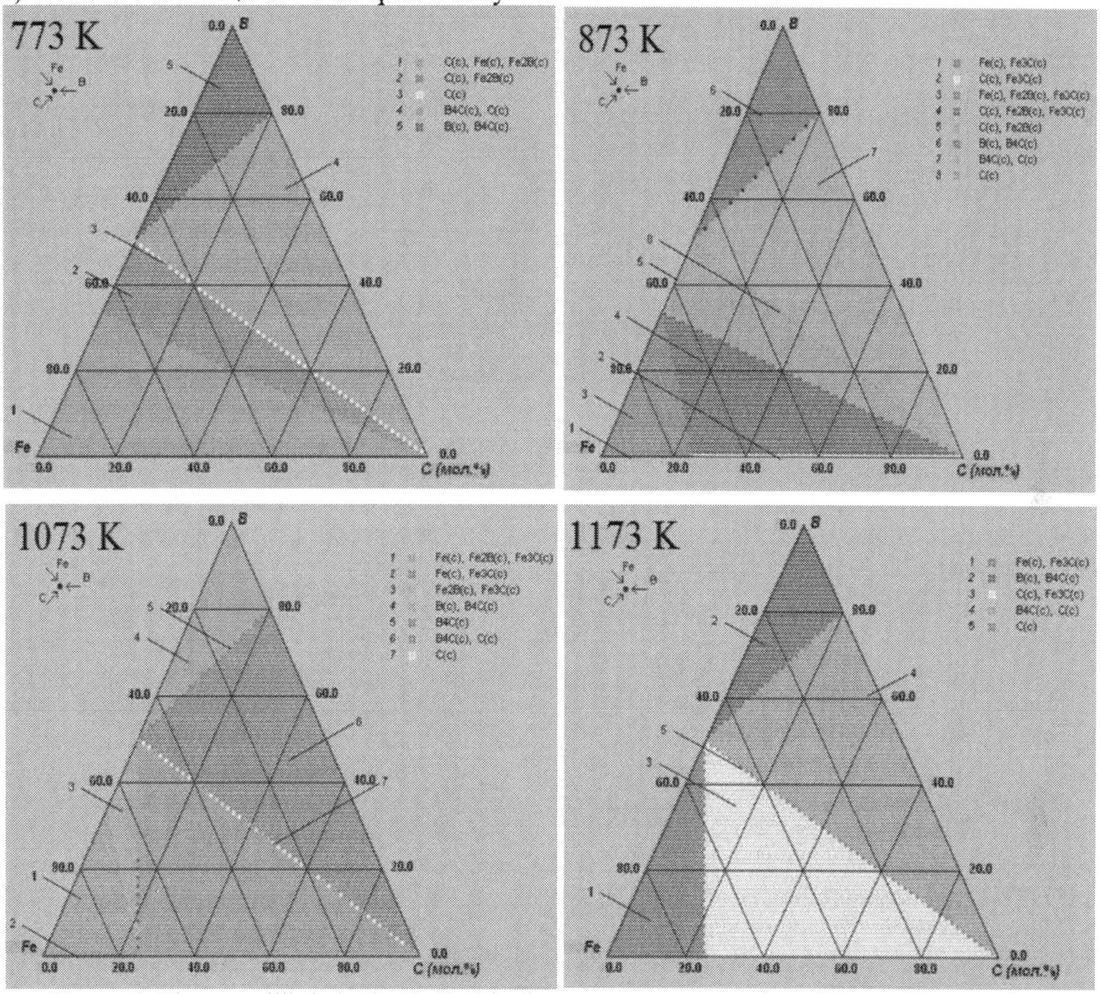

Figure 2. Phase equilibria in the Fe-B-C in the temperature range from 773 to 1173 K under a pressure of 10^{-3} Pa.

Fig. 3 shows the isotherms in Fe-B-C systems at pressure of 10^{-3} Pa which indicate significant release of the energy, that initiates SHS interaction process of Fe with B_4C resulting in the formation of iron carbide as a secondary phase with respect to the metal base. This may occur in the structure of the boride layer obtained by electron beam boriding. Hence, such studies of the boride layer structure support this assumption.

3.2. Boride Layers after Electron Beam Boriding

It is established that at electron beam boriding on a metal surface brightly expressed layers of the thickness about 350÷360 microns (the daub from amorphous boron) and depths up 100÷110 microns (the daub from B_4C) will be formed. In both cases the fixed boundary between the layer and metal basic is founded. In comparison with the metal basic the layers have lower speed of the etching that testifies to them of considerably high corrosion stability.

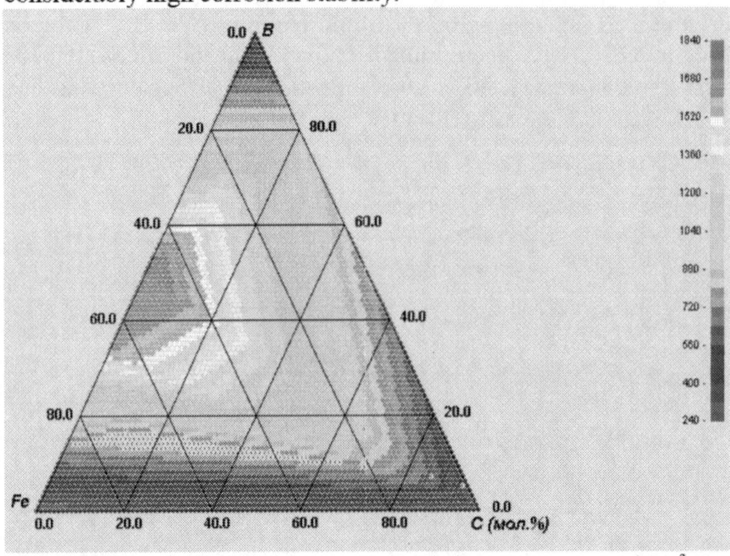

Figure 3. The isotherms in the system Fe-B-C at a pressure of 10^{-3} Pa.

The transition zone after electron beam boriding was not observed and the legible boundary between a layer and base metal was observable (Fig. 4.a, b). The layer consists of rounded crystals are settling down on a surface and a eutectic.

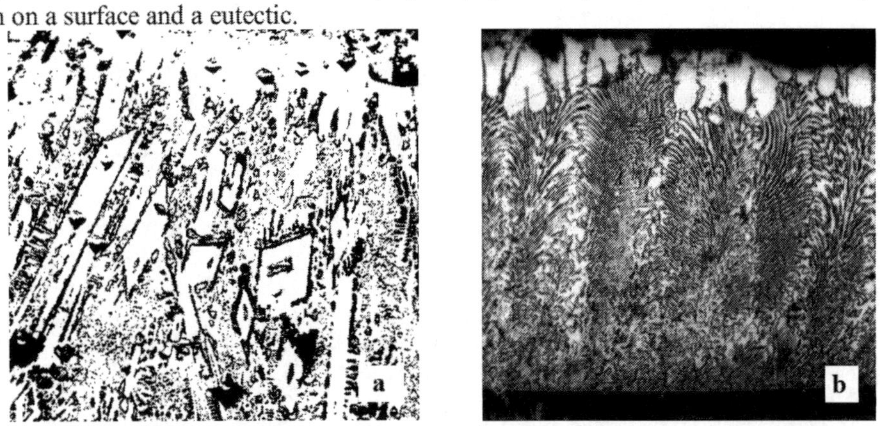

Figure 4. Layers boride microstructure formed on steel 45 surface (× 500):
a – electron beam boriding-daub from amorphous boron; b – from B_4C.

The X-ray diffraction analysis is established that layers contain the iron borides Fe_2B and FeB. The relative maintenance of these borides has depended from daub composition. In case of an amorphous boron forest it FeB and B_4C-Fe_2B. Besides on X-ray diffraction patterns there are the lines of different intensity belonging to the cementite Fe_3C and ferrite α-Fe.

The boride layer formed from daub B_4C (Fig. 4.b) consists from round off engagements, which locating on the layer surfaces and eutectic. The micro hardness values 820÷840 and 510÷530 MPa for the layer surfaces and eutectic were obtained. The rounds of engagements are primary crystals of borides that answers entropic criterion of stability of the crystals limited form at the crystallization in conditions close to equilibrium. According to this criterion, if the value of the entropy of fusion (ΔS) is less than 2 kcal/mol × K to the crystals are rounded. Obtained in work, the values of the entropy of melting for iron boride Fe_2B is ΔS = 1,5 kcal/mol × K. In turn, the rounded forms borides determine the shape of the eutectic crystals.

The boride layer formed from daub with amorphous boron has other structure (Fig. 4.a). It consists of particles of the various forms: rhombic, prismatic, dendritically. On layer surface the continuous light film with needles, directed deep into of a sample is placed. Microhardness of film makes up 1200÷1250 HV. Inside this film the rare (1-2) large inclusions with micro hardness 1750÷1820 MPa is meet. Under the film there are the primary crystals and eutectic with micro hardness 840÷880 MPa and 500÷540 MPa, accordingly.

3.3. Self-propagating high-temperature synthesis (SHS) of iron borides

Among the electron-beam technologies is worth allocating a vacuum electron-beam welding, in which are implemented at the same time the processes of powder metallurgy and classical metallurgy in the electron beam directly on the sample. For this technology, typically use of thermoreacting powder mixtures, in which can be implemented a SHS and process liquid phase sintering with using products of synthesis. Thermoreacting powder mixture contains different components that reduce the temperature of interaction. By lowering the pressure in the chamber is also reduced the temperature of formation of borides. FeB and Fe_2B layers were synthesized from a reaction mixture containing the boron carbide B_4C, oxide Fe_2O_3 , carbon C and the organic binder. Electron beam technologies give credit for a vacuum electron-beam surfacing where we can observe simultaneous processes of powder metallurgy and classical micrometallurgy in the electronic beam directly on the parts. This method is characterized by using heat-sensitive powder mixtures with possible self-propagating high-temperature synthesis (SHS) and liquid-phase sintering by using chemicals. Heat-sensitive powder mixtures contain different components that allow reducing the interaction temperatures. The temperature of boride formation is also reduced due to pressure drop down in the chamber. Fe_2B and FeB layers have been synthesized from a reaction mixture containing boron carbide B_4C, Fe_2O_3 oxide, C carbon and the organic binder.

As mentioned above the decrease of the total pressure in the chamber accompanied by temperature reduction allows forming iron borides. When the pressure is 10^{-3} Pa, the temperature for generating iron borides is~900 K which allows to form the layer without melting the steel surface but during the beam exposure accompanied by a large amount of energy due to SHS, the thin melt layer (5-7 μm) appears on a metal surface. This melted layer is embedded by boride particles which are evenly distributed throughout the whole surface. After the beam exposure the crystallization process results in the formation of a dendroid layer. Dendrites grow along the heat transfer perpendicular to thermal fields. (Figure 5).

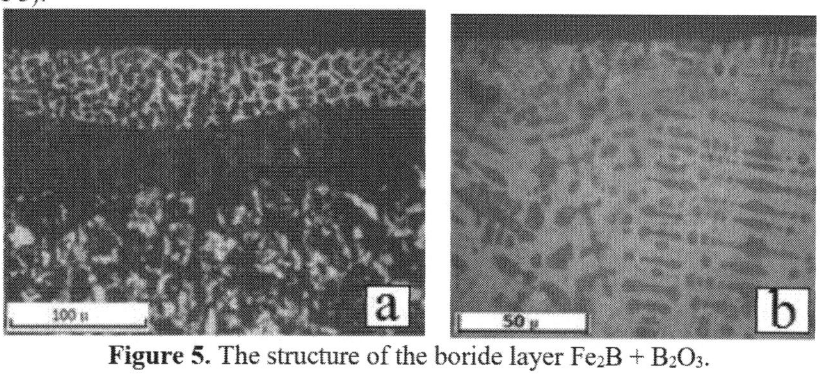

Figure 5. The structure of the boride layer $Fe_2B + B_2O_3$.

The study of layers' microhardness obtained by this technique proves that the hardest borides are FeB of 1200-1500 MPa. Microhardness of Fe_2B borides – 1100-1300 MPa. Separate particles located on the surface of a layer are the hardest with its microhardness of 3000-3500 MPa (Figure 6). The thickness of this microlayer is about 10 μm. Microhardness was measured using PMT-3 microhardness tester according to Vickers method.

The analysis showed that in the layer there are a phase of iron borides FeB and Fe_2B.X-ray diffraction is shown in figure 7. Dendrites are ferritic inclusions with parameters cells, a = 0.2821 nm). In the original steel 45 ferrite has a cubic volume-centered cell with a = 0.2869 nm. The use of high-resolution diffractometer D8 enabled to detect the x-ray lines reflexes plane (110) ferrite owned metal base. Also, found reflection from 20 % intensity, which, in our opinion, is the ferrite, which is formed during crystallization narrow melted near-surface zone.

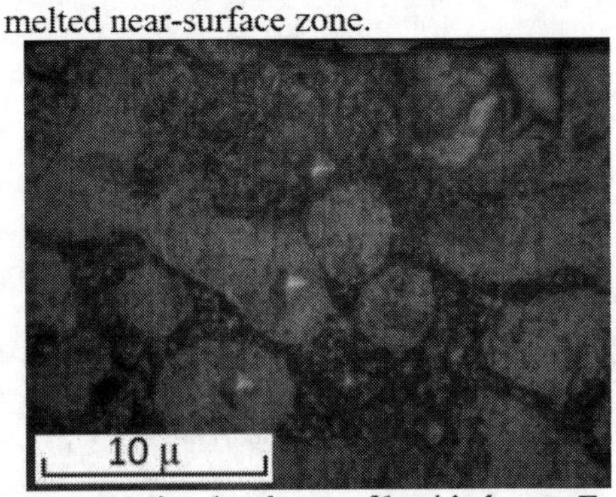

Figure 6. Microhardness of boride layers Fe_2B.

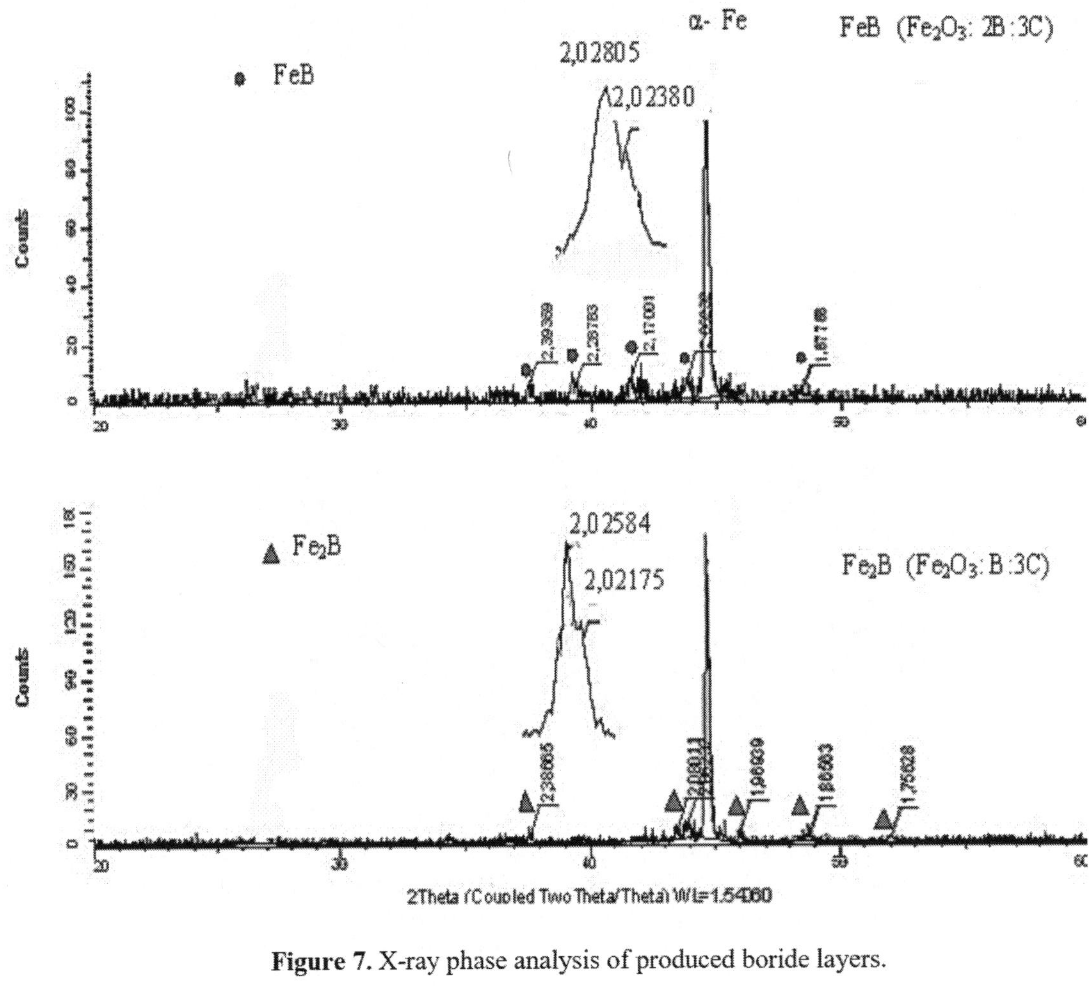

Figure 7. X-ray phase analysis of produced boride layers.

The study of layers structure confirmed this method of coating on the base of iron borides. Microstructure shows the principles of crystallization of the boride layer. Photos made by METAM PB-22 microscope with NEXSYS Image Expert provide a clear image of the visible boundary between the layer and the base as well as the zone of thermal exposure by the electron beam. The thickness of the whole layer runs up to 250 μm.

4. Conclusion

This paper presents the thermodynamic model of iron boride formation with the completed analysis of thermal processes and phase equilibria of given structures. It also provides investigations of strength properties of obtained iron boride layers, that prove the significant increase of microhardness and wear resistance of studied materials.

References

[1] Dasheev D E, Smirnyagina N N 2012 *Izvestia vuz. Fizika* **55(12-2)** 51
[2] Polukonova A E, Smirnyagina N N, Dasheev D E 2015 *Modelling of heat process in processing of low-carbon steel by focused electron beam. Innovative technologies in science and education* (Proceedings of the 4th international scientific-practical conference) p. 118.
[3] Trusov B G 2002 TERRA software package for calculation of plasma chemical processes (*Materials of III Intern. Symp. on Theoretical and Applied Plasma Chemistry*. Ples.) pp. 217-218.
[4] Dasheev D E, et al. 2015 *Journal of Physics: Conference Series* **652** 012002

Acknowledgments

This work was supported by the Russian Foundation for Basic Research, projects No. 15-48-04217-r_sibir_a and No. 16-48-030875-r_a.

Numerical study of mass transfer of a material under intense flows of high-speed electrons and plasma

A Ya Leyvi[1,2], **A P Yalovets**[1]

[1]South Ural State University, pr. Lenina 76, Chelyabinsk, 454080 Russia
[2]LLC NPP EFOM, st. Severnay 56, Chelyabinsk, 454902 Russia

E-mail: leiviai@susu.ru

Abstract. The paper presents the mechanisms of mass transfer in a material under compressive plasma flows. The work contains numerical studies on mass loss from the surface of a sample treated by Compressive Plasma Flow, in the first case, and by High-Current Electron Beam, in the second case. The mass ablated from the material surface was computed with BETAIN software. The effects of plasma flow pressure on mass transfer of the material and processes of formation of the molten layer and erosion of mass from the material surface were investigated. It was revealed that plasma flow pressure formed a shock-compressed layer on the sample, and molten liquid was displaced from the center to peripheral regions. If sample size is less than the diameter of plasma spot, the melt is displaced out of the sample. This approach can explain experimental values of molten thickness and mass loss.

1. Introduction

Nowadays, the treatment of materials with intense energy flows (ion beams, low-energy high-current electron beams or compressive plasma flows) used to modify properties of a material is one of the promising methods in modification of structural materials [1]. This processing increases mechanical strength, microhardness, wear-resistance. The significant factor of the change of forming material properties is mass transfer occurring in the treated material.

Mass transfer in metal samples under low-energy high-current electron beams (LEHCEB) is well studied experimentally and theoretically in [1-4]. As shown in [4], the main mass transfer mechanisms under (LEHCEB) irradiation are thermodiffusion, Richtmyer–Meshkov instability (RMI) [5, 6], and thermocapillary convection [4, 7, 8].

A simple estimation of the role of thermodiffusion shows that over the metal melt lifetime (several tens of microseconds), the diffusion layer depth does not exceed 1 μm, which is much smaller than the mixing depth (10–20 μm) detected experimentally [2, 3]. Thus, the mass transfer by thermodiffusion takes place in material under intense energy flows, but this is not the main mass transfer mechanism.

Refs. [4, 7 ,8] studied mechanisms of mass transfer during LEHCEB irradiation. It was shown that the main mechanism of mass transfer under LEHCEB treatment is thermocapillary convection.

Main mechanisms of mass transfer in material under LEHCEB irradiation were revealed, while in the case of processing by CPF, the results of numerical simulation strongly differ from experimental data. The paper considers the following mechanisms of mass transfer: thermodiffusion, thermocapillary convection and influence of plasma flow pressure on mass transfer.

IOP Conf. Series: Journal of Physics: Conf. Series **830** (2017) 012071 doi:10.1088/1742-6596/830/1/012071

The purpose of this work is to determine the influence of convection on mass transfer under CPF processing and to perform numerical study of the mass loss from the surface of a sample treated by LEHCEB and CPF.

2. Numerical results

Let us consider the role of convection in mass transfer of material exposed to CPF. The surface layers melt under CPF with the following parameters: discharge duration was about 100 μs, plasma flow velocity was (1-3)×10^4 m/s, plasma pressure on surface was up to 15 MPa [9, 10], and molten depth was up to 10-20 μm [11].

When a target is nonuniformly heated in its depth, a convective flow can appear. The following two types of convection are possible in a conducting material: thermocapillary convection and thermogravitational convection (which is caused by the acceleration of an irradiated medium). As shown in ref. [4], under heating metal target by LEHCEB, thermocapillary convection develops, while the role of thermogravitational one is negligible.

Thermocapillary convection can develop, when there is temperature gradient directing into target from free surface. Energy release of CPF is surface source, unlike the case of the energy release of LEHCEB. Therefore, the temperature gradient pointing in opposite direction, to the free surface, has stabilizing effect on volume flow. Temperature perturbations on an irradiated surface are always present, even at an ideally uniform beam current density, because of the surface microrelief. This can result in increased volume flow in the surface layer, namely, thermocapillary convection.

Let us consider the melt with the depth of 30 μm in 2D Cartesian coordinate system and temperature of T = 2200 K (which corresponds to experimental data) to study the role of thermocapillary convection. Non-uniform surface heating is simulated by heating of the round area of 2 μm in the center of the calculated domain with CPF heat flux (Fig. 1). Let us use the domain of 30x30 μm. On the boundary of melt-solid there is non-slip condition. On the one hand, the heat flux goes into inner layers of target due to heat conduction. On the other hand, surface layer gets heat from plasma flow. As surface tension is a temperature function, thermocapillary force arises. It directs from hotter zone to colder one.

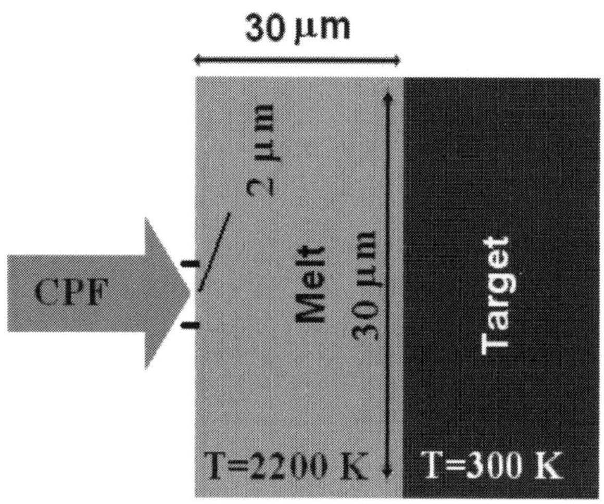

Figure 1. Schematic illustration of numerical experiments on the interaction of plasma flow with the target.

Numerical simulation was conducted using the model presented in [4]. Fig. 2 presents results of the numerical research of temperature field at 50 μs without and with thermocapillary force. As shown in the figure, thermocapillary force induces formation of two vortexes which mix material efficiently and, as a result, temperature field becomes more uniform and temperature gradient directing to free surface decreases. Thus, non-uniform surface heating of the target excites volume flow that causes more uniform temperature volume distribution.

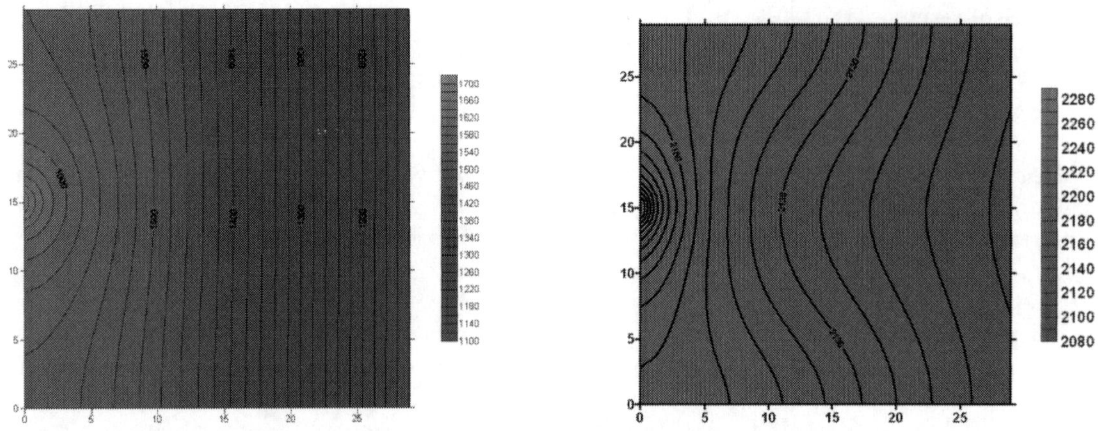

Figure 2. The results of the numerical research of temperature field at 50 μs without thermocapillary force (left) and with it (right).

The numerical calculation of melt depth and mass loss dependencies on absorbed energy density was run out by BETAIN software [12]. CPF processing was taken into account as heat flux on the free surface. Fig. 3 demonstrates the results of the numerical research of dependence of mass loss on absorbed energy density under CPF processing and LEHCEB irradiation.

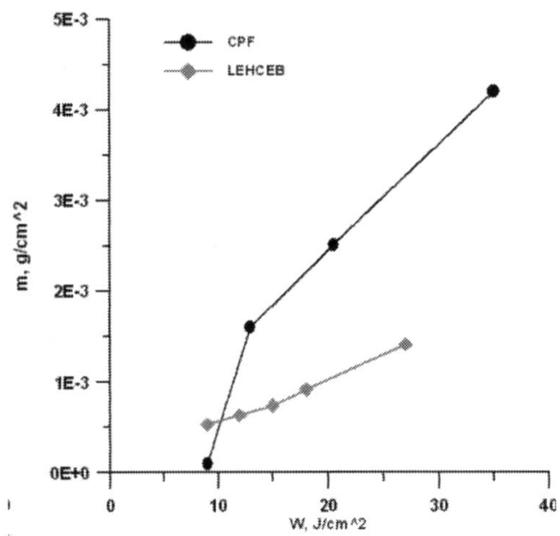

Figure 3. The results of the numerical research of dependence of mass loss on absorbed energy density under CPF processing and LEHCEB irradiation.

The melt depth determined experimentally during LEHCEB with absorbed energy density of $W = 20$ J/cm^2 was up to 20 μm, while the melt thickness specified by conductivity did not exceed 5 μm. Consideration of thermocapillary convection results in the decrease of melt thickness. Thus, due to thermocapillary convection, the surface molten layers have more uniform temperature, despite the increase of temperature gradient on the melt-solid boundary. Such gradient is less than heat flux on the free surface by CPF. And as a result, there is an intense ablation of thicker molten layer.

Such difference of experimental data from numerical results can be explained by the fact that the influence of shock compressed layer on material dynamics is not considered in the simulation of CPF processing. Taking the above mentioned into account, let us suppose that the target layer melts under CPF processing, then molten liquid displaces from the area of high plasma pressure to peripheral regions due to the influence of shock-compressed layer on material. If sample size is less than the diameter of plasma spot, the melt is displaced out of the sample.

Mass displacement tends to outcrop deeper material layer exposed to CPF. Such dynamics of material heating ensures deeper melting of target than that due to heat conductivity. Surface tension levels the melt until material crystallization after CPF treatment end.

The numerical study of the influence of shock-compressed layer on melt dynamic was carried out to confirm proposed mechanism of relief formation depicted in Fig. 4. For this purpose, the model [13] of potential flow of incompressible liquid in 2D Cartesian coordinate system has been applied. Molten volume was simulated by half-infinite media with plane surface. Mass density, surface tension and coefficient of viscosity were taken from [14] at temperature of 1800 K. Linear dimensions correspond to experimental data which results are shown in Fig. 5. Numerical calculation was performed for 200 μs which is about the life time of melt, while shock-compressed layer was taken into account during CPF processing, namely during 80 μs.

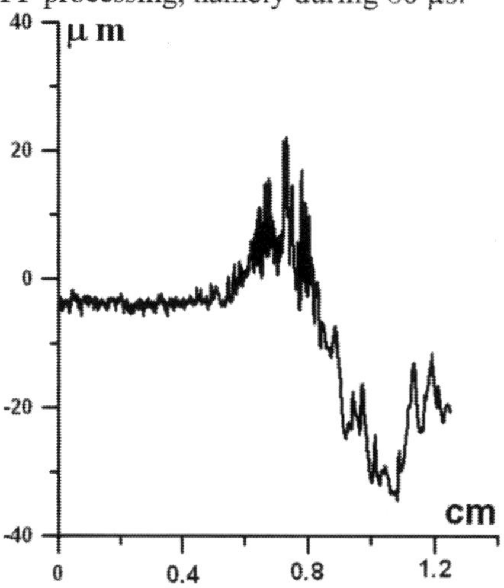

Figure 4. The sample surface exposed to CPF with the deposed film [15].

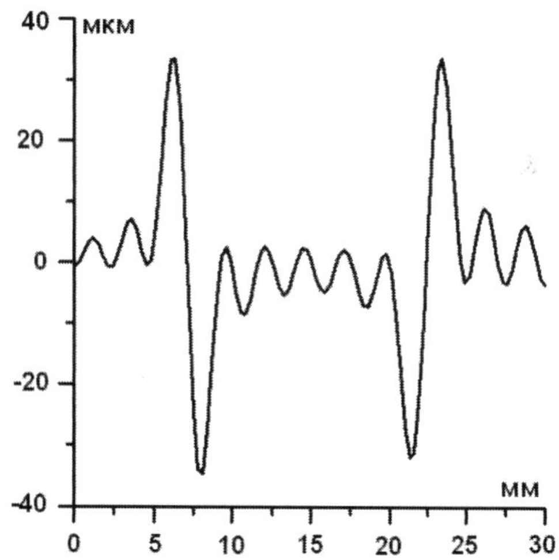

Figure 5. The numerical sample surface profile after CPF processing.

The calculation shows capillary waves developing on the sample surface under CPF processing, while after CPF processing end, there is perturbation amplitude growth due to Richtmyer–Meshkov instability until crystallization.

Fig. 5. represents numerical sample surface profile after CPF processing. As figure shows wave-like relief forms at CPF processing. A breastwork arises due to plasma pressure forming shock-compressed layer that is observed experimentally. The numerical research has revealed that breastwork width depends on pressure and linear dimensions of the region, which is subjected to action of shock-compressed layer. Calculated amplitude of perturbation under crystallization depending on conditions (plasma flow pressure, melt lifetime) is from 10 to 40 μm, which is in accordance with experimental melt thickness.

3. Conclusion

In the paper, the numerical research on mass loss from target surface under treatment by compressive plasma flows and high current electron beam has been conducted.

It was numerically shown that non-uniform heating of target surface results in the development of volume flow, which made temperature more uniform in molten volume.

It was revealed that plasma flow pressure-formed shock-compressed layer on sample and molten liquid is displaced from center to peripheral regions. If sample size is less than the diameter of plasma spot, the melt is displaced out of the sample. This approach can explain experimental values of molten thickness and mass loss.

References

[1] 2006 *Materials Surface Processing by Directed Energy Techniques* (Edited by Y. Pauleau. Elsevier p 722

[2] Mueller G., Bluhm H., Heinzel A., Schumacher G., Strauss D., Weisenburger A., Zimmermann F., Engelko V., and Notchovnaya N. 2002 *in Proceedings of the 6th International on Modification of Materials with Particle Beams and Plasma Flows, Tomsk*, p. 328.

[3] Rotshtein V. P., Markov A. B., Ivanov Yu. F., Karlik K. V., Uglov B, V., Kuleshov A. K., Novitskaya M. V., Dub S. N., Pauleau Y., Thiery F., and Shulepov I. A., 2004 *in Proceedings of the 7th International Conference on Modification of Materials with Particle Beams and Plasma Flows, Tomsk* p. 258.

[4] Volkov N.B., Leivi A.Ya., Talala K.A., Yalovets A.P. 2010 *Technical Physics* 2010 **Vol. 55** No 4 pp 484–490

[5] 1969 Meshkov E. E., Izv. (Akad. Nauk SSSR, Mekh. Zhidk. Gaza, No. 5,) p 151

[6] Richtmyer R. D. (1960) *Commun. Pure Appl. Math.* 13 297

[7] Eidel'man E. D. 1995 *Usp. Fiz. Nauk* **165** 1279

[8] 1972 Gershuni G. Z. and Zhukhovitskii E. M. Convective Stability of an Uncompressed Fluid (Nauka, Moscow) p 285 [in Russian].

[9] Uglov V.V., Anishchik V.M., Cherenda N.N., Sveshnikov Yu.V., Astashynski V.M., Kostyukevich E.A., Kuzmitski A.M., Askerko V.V. 2008 The formation of a tungsten containing surface layer in a carbon steel by compression plasma flow. *Surf. Coat. Technol.* **202** pp 2439–2442

[10] Uglov V.V., Anishchik V.M., Astashynski V.V., Astashynski V.M., Ananin S.I., Askerko V.V., Kostyukevich E.A., Kuz'mitski A.M., Kvasov N.T., Danilyuk A.L. 2002 The effect of dense compression plasma flow on silicon surface *Surface and Coatings Technology* **158 –159** pp 273–276

[11] Cherenda N.N., Uglov V.V., Kuleshov A.K., Astashynski V.M., Kuzmitski A.M. 2016 Surface nitriding and alloying of steels with Ti and Nb atoms by compression plasma flows treatment *Vacuum 129* pp 170-177

[12] Yalovets A. P. and Mayer A. E. 2002 *in Proceedings of the 6th International on Modification of Materials with Particle Beams and Plasma Flows, Tomsk* p 297.

[13] Leivi A.Ya., Yalovets A. P., Krasnikov V. S.. 2013 The Liquid-Phase Mass Transfer of Material in the Film–Substrate System Exposed to Intense Energy Flows. *High Temperature Material Processes* **vol 17(1)** pp 15–23

[14] 1991 Fizicheskie vilichiny. Spravochnik (Grigirev I S Meylihova E Z Energoatomizdat) p 1234 [in Russian].

[15] Leyvi A.Ya., Yalovets A.P. and Kozhevnikov D.G. 2016 Applying compressive plasma flows to improve adhesion in a film-support system *Materials Science Forum* **Vol. 843** pp 73-77 doi:10.4028/www.scientific.net/MSF.843.73

Acknowledgements

The research was partially supported by the Ministry of Education and Science of the Russian Federation (contract no. 1030 (2014105-GZ) for support of research at the South Ural State University).

Multiscale models of metal behaviour and structural change under the action of high-current electron irradiation

A E Mayer, V S Krasnikov, P N Mayer, V V Pogorelko

Department of Physics, Chelyabinsk State University, Bratiev Kashirinykh Street 129, Chelyabinsk 454001, Russia

E-mail: mayer@csu.ru

Abstract. We present our models of the tensile fracture of metals in the solid and molten states, the melting and the plastic deformation of the solid metals. Also we discuss implementation of these models for simulation of the high current electron beam impact on metals. The models are constructed in the following way: the atomistic simulations are used at the first stage for investigation of dynamics and kinetics of structural defects in material (voids, dislocations, melting cites); equations describing evolution of such defects are constructed, verified, and their parameters are identified by means of comparison with the atomistic simulation result; finally, the defects evolution equations are incorporated into the continuum model of the substance behaviour on the macroscopic scale. The obtained continuum models with accounting of defects subsystems are tested in comparison with the experimental results known from literature. The proposed models not only allow one to describe the metal behaviour under the conditions of intensive electron irradiation, but they also allow one to determine the structural changes in the irradiated material.

1. Introduction

Impact of the high-current electron beam [1-4] on a metal provokes a variety of non-equilibrium dynamic processes. The main effect of the beam is in the rapid heating of substance inside the energy absorption area accompanied by melting. The range of the fast electron reaches half a millimetre or more for the electron energy of about 1 MeV [1-3]. Using the value of 5 km/s as a typical sound velocity in solid metals, one gets the mechanical relaxation time [4] of about 100 ns. Therefore, an electron irradiation pulse with the duration of about several tens of picoseconds produces heating in an almost isochoric mode similar to the powerful femtosecond laser pulses [5,6]. The isochoric heating increases pressure up to several tens of gigapascals [1-3] at the typical beam parameters. The subsequent expansion of the heated and stressed surface layer produces a compression pulse consisting of a shock wave followed by a rarefaction wave and propagating deep into the irradiated metal. The compression pulse induces the dynamic plastic deformation with a substantial change of the material microstructure [3], and the pulse reflection from the rear free surface of the irradiated sample can lead to the tensile (spall) fracture [1-3,7]. The melt inside the energy absorption area also reveals a rich physical behaviour. The melt expends initially due to the pressure gradient, and then by inertia. The expansion by inertia transfers the melt into the metastable state at a negative pressure, which leads to the tensile fracture of melt by means of cavitations [8,9]. A complete mathematical model of the high current electron beam interaction with metals should take into account all the mentioned processes. Among other things, this model is required for interpretation of the electron beam experiments [1-3].

2. Mathematical model

We use the multiscale approach for construction of the mathematical model. Atomistic simulations are used for the investigation of the elementary processes in the defect subsystems, such as growth of voids or melting sites, motion of dislocations etc. It allows us to define the form and parameters of the kinetic equation for defect ensembles. The obtained regularities are generalized on the continuum level, which is applicable for modeling of the dynamics of the macroscopic volumes of metal under irradiation. Continuum model supposes an averaged description of the lattice defects in the form of various parameter fields (scalar density of dislocations, concentration and size of voids or melting sites). For description of the non-equilibrium melting, it is supposed that each substance element consists of a mixture of the solid and molten parts.

The continuum model consists of a standard set of the continuum mechanics equations supplemented by models for specific physical processes and properties. The wide-range equations of state [10,11] are used for calculation of the pressure and temperature as functions of density and internal energy. The electron beam action is described by means of the energy release function D, which is calculated from the solution of the fast electron transport problem using the method [12]. The molecular dynamic (MD) simulations are performed with the help of LAMMPS [13] and various interatomic potentials based on the embedded atom model. These simulations are used for construction and verification of the plasticity, fracture and melting models [8,14-18]. The fracture model [8,16-18] describes the nucleation, growth and coalescence of voids including the late stage with the predomination of the voids interaction [9]. The plasticity model [14,19,20] describes the plastic deformation tensor increment as a result of the dislocation motion with velocity determined by the acting stresses; the dislocation kinetics takes into account generation, annihilation and immobilisation of dislocations. The continuum model of melting [21] is based on the description of the current phase state by means of fields of concentration and size of the melting sites; the site size is supposed to be much less than the typical spatial scale of the considered problem; so, the interfaces are not traced explicitly, but an averaged picture is considered instead. Two model problem statements are used in MD simulations of melting for determination of the model equations and coefficients: (i) reaching of the equilibrium in an isolated system with a flat interface between solid and liquid; (ii) melting of an initially homogeneous monocrystal heated by a source with a constant power.

Verification of the model had been performed by means of comparison of the modelling results both with the MD simulations and the experimental data known from literature. For verification of the plasticity and fracture models, we compare the modelling results with the high-velocity impact experiments, which are the standard test of the dynamic properties of various materials.

3. Application for the high-current electron irradiation problem

The continuum model is applied to the calculation of the stainless steel dynamics under the action of the high-current electron beam of the SINUS-7 accelerator (Tomsk, Russia, Institute of High Current Electronics SB RAS) [1-3]. The beam parameters are the following: the maximal energy of fast electrons is 1.35 MeV, the pulse duration is 45 ns (full-width on half-maximum), the maximal power density in the center of the beam is 34 GW/cm^2, and the incident energy density is 1.2 kJ/cm^2; the experimentally determined oscillograms of the beam current and the accelerating voltage are used [3].

Figure 1 shows the spatial distributions of the stress tensor components in the consequent moments of time. The electron beam acts on the left side of the sample and induces an intensive heating up to 8000 K, and the following melting and ablation of matter inside the energy absorption area (the fast expanding part of the sample). The thickness of the energy absorption area is about 1 mm; therefore, the heating is close to an isochoric one. Compressive stress reaches 33 GPa at 50 ns from the irradiation beginning [3]. The expansion of the heated material becomes substantial starting at approximately 50-100 ns from the beginning of the irradiation; the compressive stresses rapidly decrease thereafter. The running shock wave is formed with the initial amplitude of 17 GPa and with a duration of the compression pulse of 0.2 μs. The duration of the compression pulse is restricted by the unloading wave, which propagates from the free surface behind the shock wave. An additional unloading wave originates from the side unloading and converges to the target axis. Interference of two unloading waves creates an expansion wave with tensile stresses following the shock wave; the values of tensile stresses grow up as this unloading wave converges to the sample axis. The expansion wave reaches the target axis at the moment when the shock wave passes the distance of about the beam radius. The shock wave amplitude decreases with propagation deep into the target and the duration of the compression pulse increases due to the interactions with the following expansion wave. An elastic precursor with the amplitude of 1 GPa propagated at the head of the shock wave.

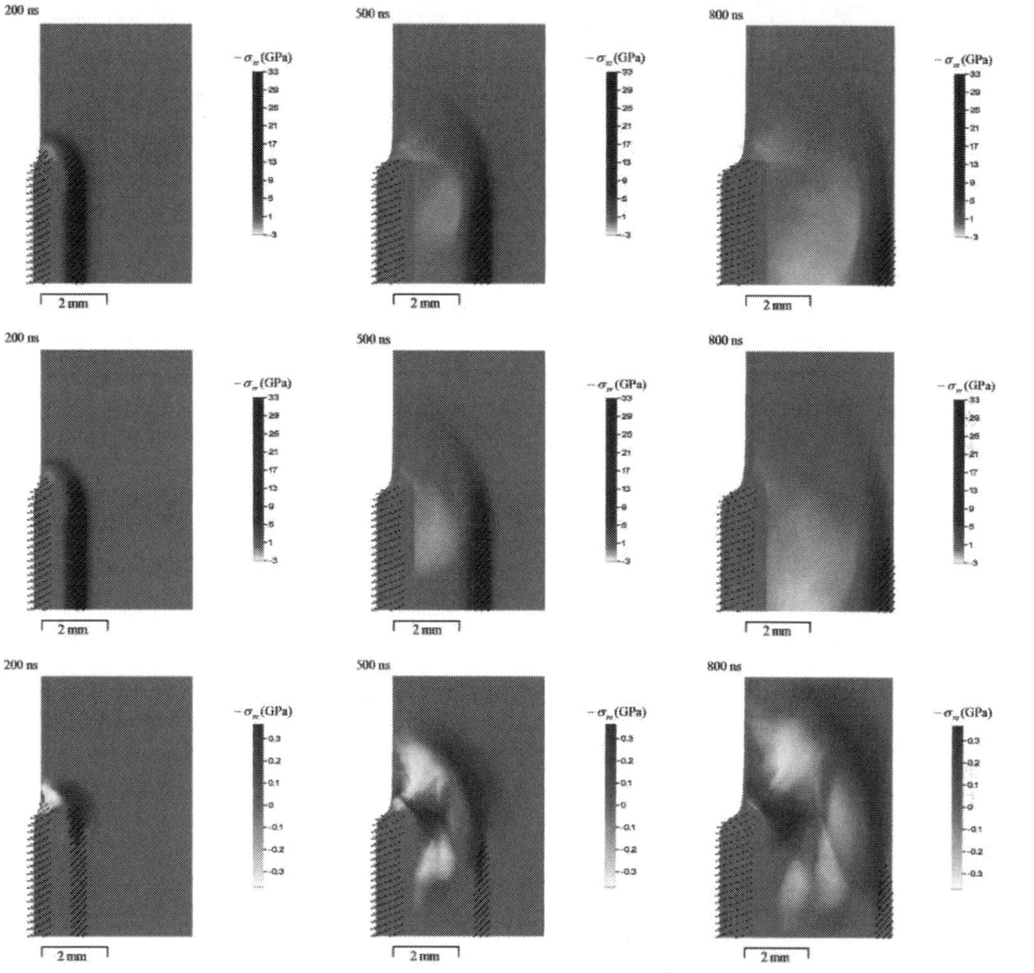

Figure 1. Calculated distributions [3] of the stress tensor components in stainless steel sample with the thickness of 4.5 mm in consequent moments of time (200, 500 and 800 ns) after beginning of the irradiation; stresses are shown with negative sign. Electron beam (SINUS-7 accelerator) irradiates the left side of the target; axial direction is horizontal one and radial direction is vertical one. Little arrows show the substance velocity vectors.

Calculated spatial distributions of the volume fracture of voids for the various sample thicknesses are presented in figure 2. One can see a distinct fracture area near the rear surface that is formed at the shock wave reflection. For the targets thinner than 6.5 mm, the condition of complete fracture is realized within this area, while only an incomplete fracture takes place for the thicker samples. Although, for thicknesses 9 mm, the maximal tensile stress along the sample axis exceeds the spall strength, the decrease of the acting stresses with radius prevents the complete fracture. The expected positions of the main cracks or the incomplete fracture zones are schematically shown in figure 3, which are determined as widest in radial direction cross-sections within the fracture area. The distances between such cross-section and rear surface are the calculated thicknesses of the spalled layer, which are plotted in figure 6 along with the experimental results taken from [3]. Both experimental and calculated dependences are close to linear functions and correlate with each other.

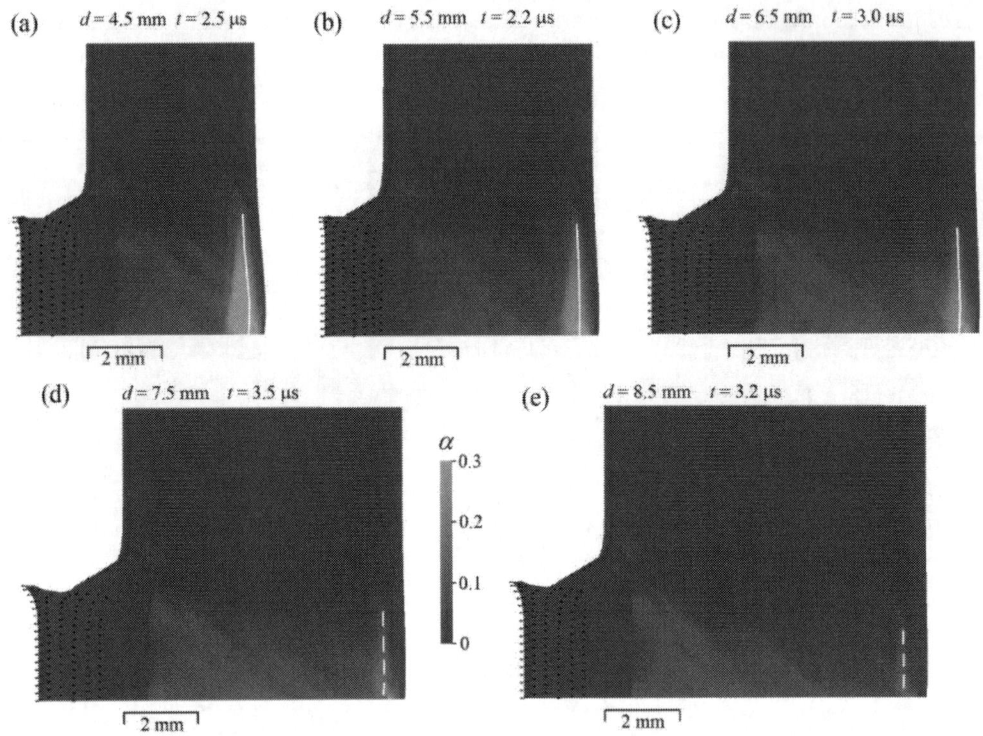

Figure 2. Calculated distributions of the volume fraction of voids α in samples with various thicknesses made of stainless steel and irradiated by the high-current electron beam (SINUS-7 accelerator [1-3]). (a,b,c) solid lines show the expected position of the main crack, (d,e) dashed lines show the expected position of the zone of incomplete fracture. Expanding cloud of ablated material is from the left. Little arrows on the left sides show the substance velocity vectors for the expanding ablated matter.

Influence of the non-equilibrium melting [21] on the temperature distribution in the irradiated aluminum is shown in figure 4, which presents the spatial distributions of the substance temperature T, the melting temperature T_m depending on the local pressure and the volume fraction $(1-\alpha_m)$ of solid phase for time moments 25 and 30 from the beginning of irradiation.

The considered electron beam heats the aluminum layer 2 mm in thickness; spatial distribution of the energy release power is non-uniform which leads to the non-uniform temperature distributions. The energy release power reaches 0.08 PW/kg in maximum which leads to the heating rate of about 70 K/ns. Complete melting occurs in the central part of the energy absorption area to the moment of time of 25 ns. Temperature of the uniform melt begins to exceed T_m in this central area, while it is close to T_m at the edges of the complete melting zone. Two regions of partial melting adjoin the complete melting zone from both sides. $T \approx T_m$ in the internal parts of these partial melting regions which implies reaching the phase equilibrium; increase in the melt volume fraction goes here to the extent of energy supply by the beam. Aluminum is overheated $T > T_m$ within the external parts of the partial melting zones; the melt volume fraction goes here at the expanse of both the energy transfer from by beam and the leaving the metastable overheated state. Areas of overheated solid metal adjoin the partial melting zones. The overheating reaches $T - T_m \approx 100\,K$. The left zone of partial melting disappears at the time moment of 30 ns; aluminum becomes completely melted up to the exposed free surface. The right zone of partial melting moves deep into the target in the course of time; the molten layer thickness increases till the beam action termination.

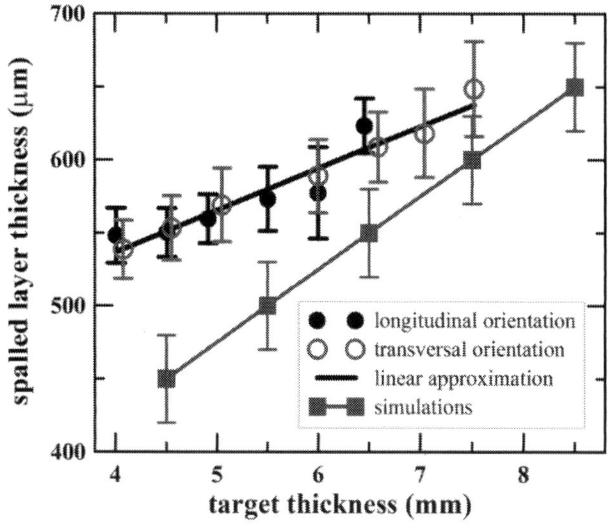

Figure 3. Experimental [3] and calculated dependencies of the spalled layer thickness on the sample thickness for stainless steel. Experimental points correspond to both the longitudinal and the transversal orientations of the δ-ferrite interlayers relative to the shock wave propagation direction [3].

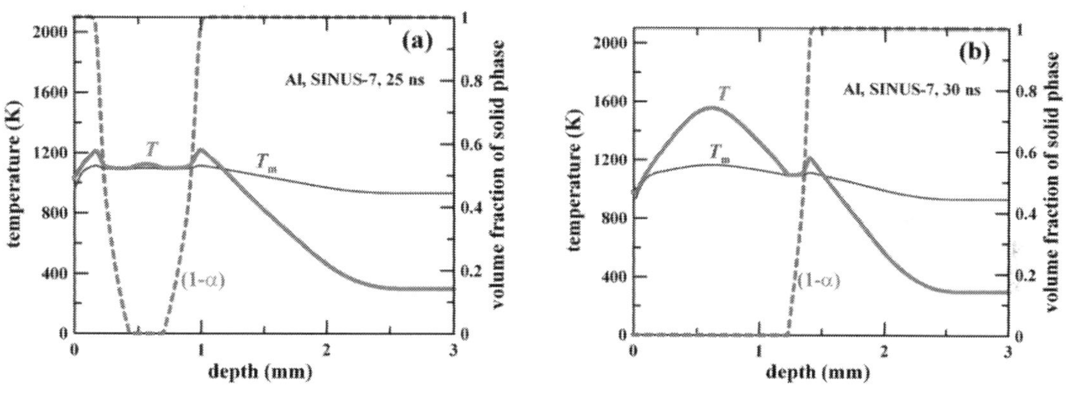

Figure 4. Spatial distributions of the substance temperature T (thick solid lines), the melting temperature T_{m} (thin solid lines) and the volume fraction $(1-\alpha_m)$ of solid phase (dash lines) in the consequent moments of time: irradiation of aluminum by the high-current electron beam with parameters of SINUS-7 accelerator. Initial target thickness is 3 mm; the electrons of the beam fall normally on the left surface of target.

4. Conclusion

Here we present our models of the tensile fracture of metals in the solid and molten states, the melting and the plastic deformation of the solid metals. Also we discuss implementation of these models for simulation of the high current electron beam action on metals. The models are constructed in the following way: the atomistic simulations are used at the first stage for investigation of dynamics and kinetics of structural defects in material (voids, dislocations, melting cites); equations describing the evolution of such defects are constructed, verified, and their parameters are identified by means of comparison with the atomistic simulation result; finally, the defects evolution equations are incorporated into the continuum model of the substance behaviour on the macroscopic scale. The obtained continuum models with accounting of defects subsystems are tested in comparison with the experimental results known from literature. The proposed models not only allow one to describe the metal behaviour under the conditions of intensive electron irradiation, but they also allow one to determine the structural changes in the irradiated material.

References

[1] Dudarev E F, Kashin O A, Markov A B, Mayer A E, Tabachenko A N, Girsova N V, Bakach G P, Kitsanov S A, Zhorovkov M F, Skosyrskii A B and Pochivalova G P 2011 *Rus. Phys. J.* **54** 713–720
[2] Dudarev E F, Markov A B, Mayer A E, Bakach G P, Tabachenko A N, Kashin O A, Pochivalova G P, Skosyrskii A B, Kitsanov S A, Zhorovkov M F and Yakovlev E V 2013 *Rus. Phys. J.* **55** 1451–1457
[3] Gnyusov S F, Rotshtein V P, Mayer A E, Rostov V V, Gunin A V, Khishchenko K V and Levashov P R 2016 *Int. J. Fract.* **199** 59–70
[4] Yalovets A 1997 *J. Appl. Mech. Tech. Phys.* **38** 137–150
[5] Ashitkov S I, Komarov P S, Struleva E V, Agranat M B, and Kanel G I 2015 *JETP Lett.* **101** 276–281
[6] Ashitkov S I, Komarov P S, Struleva E V, Agranat M B, Kanel G I and Khishchenko K V 2015 *J. Phys.: Conf. Ser.* **653** 012001
[7] Mayer A E and Krasnikov V S 2011 *Eng. Fract. Mech.* **78** 1306–1316
[8] Mayer A E and Mayer P N 2015 *J. Appl. Phys.* **118** 035903
[9] Mayer P N and Mayer A E 2016 *J. Appl. Phys.* **120** 075901
[10] Kolgatin S N and Khachatur'yants A V 1982 *Teplofiz. Vys. Temp.* **20** 90–94
[11] Fortov V E, Khishchenko K V, Levashov P R and Lomonosov I V 1998 *Nucl. Instrum. Meth. Phys. Res. A.* **415** 604–608
[12] Evdokimov O B and Yalovets A P 1974 *Nucl. Sci. Eng.* **55** 67–75
[13] Plimpton S 1995 *J. Comp. Phys.* **117** 1–19
[14] Krasnikov V S, Kuksin A Yu, Mayer A E and Yanilkin A V 2010 *Phys. Solid State* **52** 1386–1396
[15] Yanilkin A V, Krasnikov V S, Kuksin A Yu and Mayer A E 2014 *Int. J. Plast.* **55** 94–107
[16] Mayer A E, Mayer P N, Krasnikov V S and Voronin D S 2015 *J. Phys.: Conf. Ser.* **653** 012093
[17] Krasnikov V S and Mayer A E 2015 *Int. J. Plast.* **74** 75–91
[18] Pogorelko V V and Mayer A E 2015 *Mater. Sci.: Eng.: A* **642** 351–359
[19] Krasnikov V S, Mayer A E and Yalovets A P 2011 *Int. J. Plast.* **27** 1294–1308
[20] Mayer A E, Khishchenko K V, Levashov P R and Mayer P N 2013 *J. Appl. Phys.* **113** 193508
[21] Mayer A E and Krasnikov V S 2015 Atomistic and continuum modeling of non-equilibrium melting of aluminum (Preprint arXiv:1512.07891v1 [cond-mat.mtrl-sci])

Acknowledgments

The development of the multiscale fracture model is supported by the Russian Science Foundation (Project No. 14-11-00538). The development of the plasticity model is supported by the Ministry of Education and Science of the Russian Federation (competitive part of State Task of NIR CSU No. 3.1334.2014/K) and by the grant from the President of the Russian Federation (Project No. MD-7481.2016.1). The investigation of the foamed melt dynamics is supported by the grant from the President of the Russian Federation (Project No. MK-9111.2016.8). The development of the non-equilibrium melting model is supported by the Russian Foundation for Basic Research (Project No. 15-32-21039).

Modelling of local ion nitriding in a glow discharge with hollow cathode

V Budilov, K Ramazanov, Yu Khusainov

Ufa State Aviation Technical University, Department of Mechanical Engineering, 12 K. Marx Str., Ufa, Russia

E-mail: ramazanovkn@gmail.com

Abstract. The paper presents the results of computer calculations of glow discharge plasma parameters in a hollow cathode zone and modeling of thermal and diffusion processes at local ion nitriding with a hollow cathode. The proposed model of a glow discharge with a hollow cathode with sufficient accuracy allowed to describe the distribution of plasma parameters in a cathode void. Values of plasma parameters in a cathode void formed by a mesh screen and cathode surface were obtained via the probe method. It was found that the use of hollow cathode effect allows to increase the concentration of ions near the treated surface by 1.5 times. The suggested computer model allows to predict the distribution of the temperature field and depth of a diffusion layer at local ion nitriding with a hollow cathode for various configurations and sizes.

1.Introduction

At present the majority of machine-building enterprises apply the process of ion nitriding in a glow discharge [1,2] to harden the surface of manufactured components and goods. In most cases components used in conditions of local wear are subject to surface hardening, therefore it is often sufficient to treat their working surface, which is exposed to severe wear. It should also be noted that the process of ion nitriding is expensive, power-intensive and still time-consuming [3].

Among the variety of intensification methods it is important to focus on local ion nitriding with a hollow cathode. The use of a hollow cathode during ion nitriding in the areas of severe wear of components allows to reduce the treatment period by 2.5 times [4].

Alongside with intensification the treatment itself becomes complicated. In case of local ion nitriding with a hollow cathode the heating rate of a component under the technological screen is higher than in other areas. Hence, there is uneven distribution of temperature within the component during the whole treatment period. Besides, there is no sufficient data on physical and chemical processes that take place in a cathode void of plasma and their influence on a surface of the treated material. Therefore, better understanding of processes that occur in the surface layer of a material during its treatment will enable the increase of nitriding efficiency by specifying the best technological modes.

This paper presents the results of computer calculations of glow discharge plasma parameters in a hollow cathode zone and modeling of thermal and diffusion processes at local ion nitriding with a hollow cathode.

2.Experimental approach

The calculation model consisting of two disk electrodes placed in a cylindrical chamber was used to determine the glow discharge plasma parameters in the area of hollow cathode. At the distance of 5 mm from the cathode the disk mesh with 1 mm cell size (Fig. 1) was placed. Plasma parameters were defined according to the following equations [5]:

$$\frac{\partial}{\partial t}(n_e)+\nabla\cdot\Gamma_e=R_e-(\mathbf{u}\cdot\nabla)n_e \tag{1}$$

where n_e is the concentration of electrons, R_e is the volume sources and electron sinks, Γ_e is the electron flow vector, μ_e is the electron mobility;

$$\rho\frac{\partial}{\partial t}(w_k)+\rho(\mathbf{u}\cdot\nabla)w_k=\nabla\cdot\mathbf{j}_k+R_k \tag{2}$$

where j_k is thediffusion flux vector, R_k is the specified speed for heavy particles (atoms and ions), u is the average flow speed, ρ is the density of mixture and w_k is the mass content of particles in a mixture.

The theoretical model of local ion nitriding accommodates both thermal and diffusion processes. The process of diffusion is described by the Fick's equation [6]:

$$\frac{\partial c}{\partial \tau}+\nabla\cdot(-D\nabla c)=0 \tag{3}$$

where D is the diffusion coefficient; c is the concentration of a saturant; τ is the time of diffusion saturation; x is a coordinate.

The diffusion coefficient D is found in accordance with Einstein's (Arrhenius's) equation [6]:

$$D=D_0\exp\left(-\frac{E_a}{RT}\right) \tag{4}$$

where D_0 is the preexponential factor; R is an absolute gas constant; E_a is the activation energy; T is the temperature.

To describe the temperature field the model uses the following thermal conductivity equation:

$$\rho\cdot C\frac{\partial T}{\partial \tau}+\nabla\cdot(-k\nabla T)=Q \tag{5}$$

where ρ is the material density; C is the specific heat capacity; τ is time; k is the thermal conductivity ratio; Q is a specific thermal source.

Series of experiments on an upgraded ELU-5M set-up were conducted to check the validity of local ion nitriding model. Figure 2 shows the scheme of the experiment.

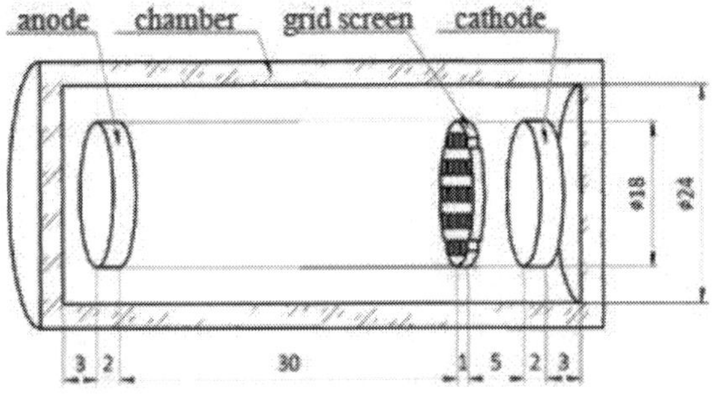

Figure 1. Calculation model to define plasma parameters.

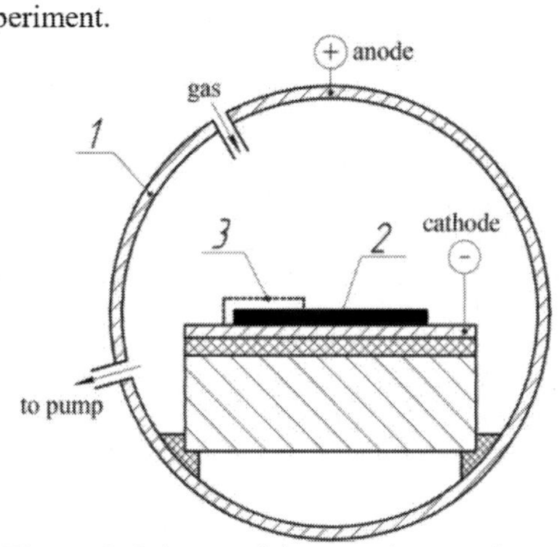

Figure 2. Scheme of the experiment using ELU-5M set-up:
1 – vacuum chamber, 2 – sample,
3 – screen to create a hollow cathode effect.

IOP Conf. Series: Journal of Physics: Conf. Series **830** (2017) 012073 doi:10.1088/1742-6596/830/1/012073

3.Results and discussion

As a result of model calculations the distribution curve of ion concentration in a cathode void was obtained. Figure 3 shows the distributions of ion concentration between a cathode and a mesh screen in axial direction of a discharge obtained through modeling and experiments.

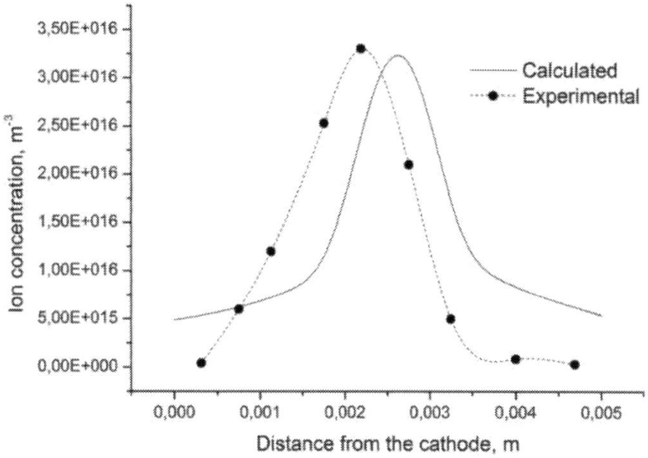

Figure 3. Concentration of ions in cathode void: solid line – calculated data, dotted line – experimental data.

Dependences are identical with slight deviation in values, which may be explained by a number of factors. It should be taken into account that during processing of probe volt-ampere characteristics it was assumed that electron energy distribution in a void is the Maxwell distribution, and heating of mixture and electrodes due to surface bombardment with charged particles was not considered during modeling.

The maximum concentration of ions was observed in the center of a void and made 3.2×10^{16} m^{-3}. Hence, the application of a hollow cathode effect during ion nitriding allows to increase the concentration of ions near the treated surface by 1.5 times.

The distribution of a floating potential in a void (Fig. 4) obtained as a result of modeling also gave acceptable similarity with experimental data.

The width of potential fall area amounted to 1-1.5 mm. At the same time the width of the area on the side of a mesh is bigger, which is explained by the distortion of electric field configuration due to holes present on the screen.

As a result of calculations of a computer model the distribution of a temperature field on the surface of a gear wheel was obtained at local ion nitriding with a hollow cathode (Fig. 5) during 8 hours. At the same time the temperature of teeth surface made ~ 550 °C.

The maximum temperature difference between various sections of a 200 mm gear wheel nitraded with/without a hollow cathode amounts to ~ 35 °C (Fig. 5). Thus, the configuration and geometry of the treated component distribution impact the temperature field at local ion nitriding with a hollow cathode.

Figure 6 shows the diagrams of the treated surface in two points A and B obtained through modeling and experiments, which correspond to the maximum and minimum temperature of a surface (Fig.5).

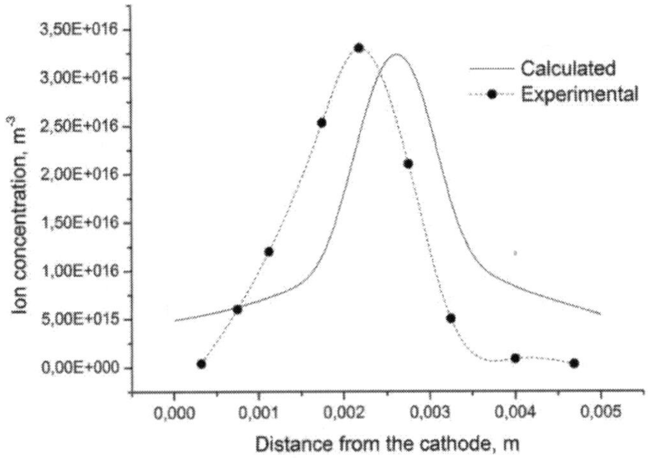

Figure 4. Distribution of a potential in a void: solid line – calculated data, dotted line – experimental data.

The analysis of obtained dependences showed that after heating the component and beginning of an operating mode the temperature remains stable and equals °C T=550 until the treatment is finished. This is caused by the steady-state equilibrium between thermal energy input from gas ions and its output from the surface to the external environment. The heating time necessary to heat the treated surface up to the saturation temperature made about 40 min (while the heating rate made 0.2 °C/sec).

Figure 7 shows the estimated and experimental time dependences of temperature in point A during cooling of a component upon completion of local ion nitriding with a hollow cathode.

The diagram of component cooling is of great practical importance and allows to choose the holding period for loading after completion of local ion nitriding with a hollow cathode for components of various sizes and configurations. The given dependences show that the treated surface of a component is cooled down to the temperature of 250 °C (523K) within approximately 30 min.

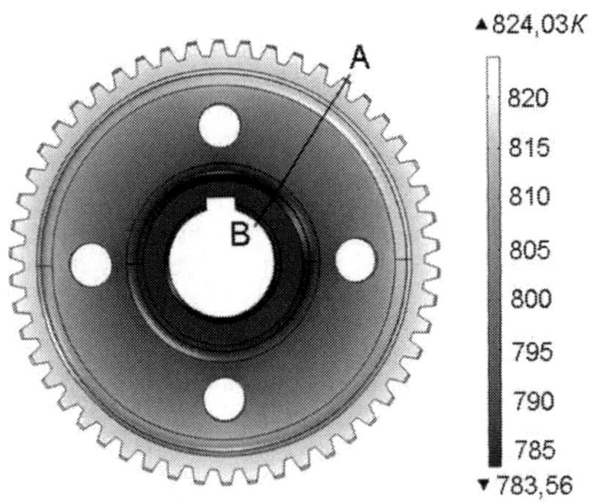

Figure 5. Computational distribution of temperature on gear wheel surface at local ion nitriding with a hollow cathode.

Figure 8 shows the concentration curve of a saturant along the depth of a diffusion layer for locally nitraded area with a hollow cathode that was obtained through a computational model.

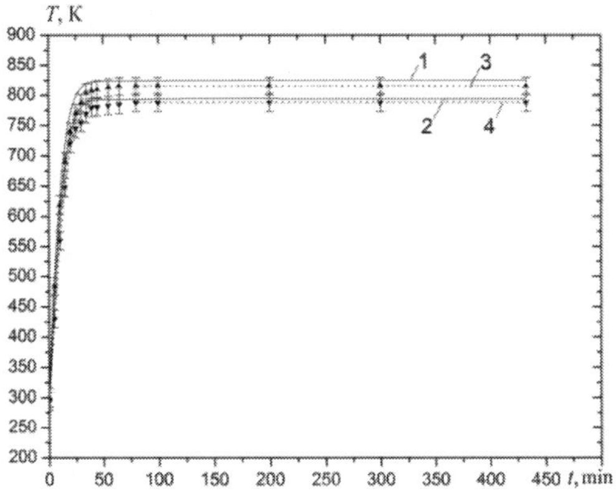

Figure 6. Diagram of surface heating in points A and B during local ion nitriding with a hollow cathode: *1* – master curve, (point A); *2* – master curve, (point B); *3* – measured curve (point A); *4* – measured curve (point B).

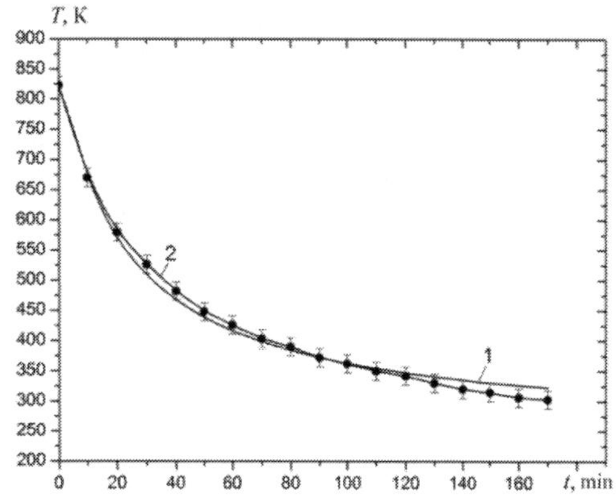

Figure 7. Diagram of the treated component surface cooling (point A): *1* – master curve; *2* – measured curve.

The analysis of the obtained dependence shows that the depth of a hardened layer varies from 180 to 220 microns. Ion nitriding with a hollow cathode was carried out at the following modes: *T =550 °C, P=60 Pa, U=500 V, t=8 h.*

Microhardness was measured by depth of the nitraded layer at the section of this sample (Fig. 9) in order to define the depth of the hardened layer.

Measurements of microhardness along the sample section showed that the depth of the hardened layer is within the range of up to 220 microns. Thus, experimental data correlate well with the calculation results.

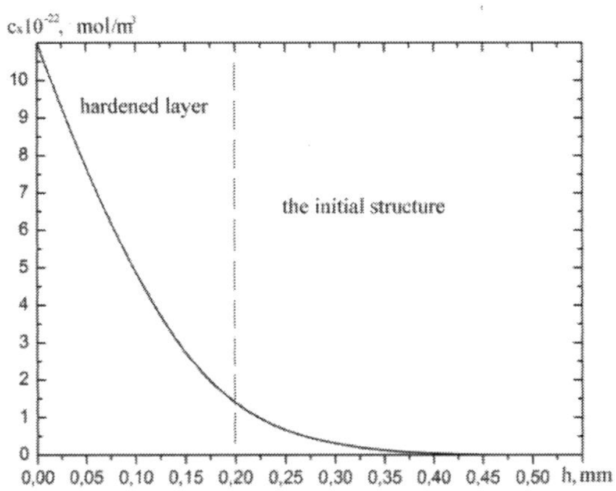

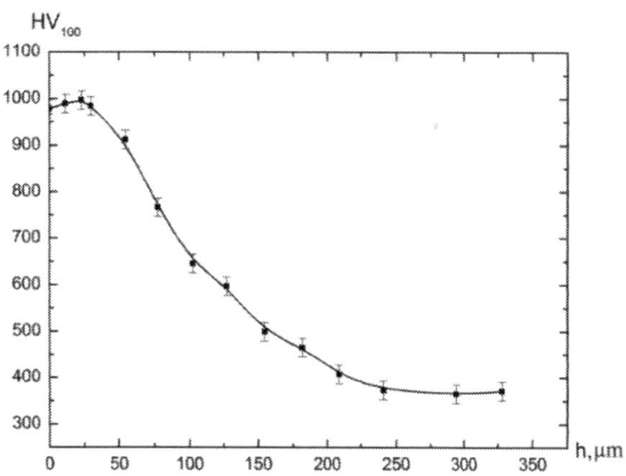

Figure 8. Estimated concentration change of a saturant (nitrogen) along the depth of a diffusion layer of steel 38XMYUA: *T*=550°C, *t*=8 h.

Figure 9. Change of microhardness along the depth of a nitraded layer of steel 38XMYUA:*T*=550°C, *P*=60 Pa, *U*=500 V, *t*=8 h.

4.Conclusion

The following conclusions can be drawn.

- The model of a glow discharge with a hollow cathode with sufficient accuracy describing the distribution of plasma parameters in a cathode void was established.
- The experimental values of plasma parameters in a cathode void were obtained.

- The use of a hollow cathode effect in areas of intense wear of a component allows to achieve the increase in ion concentration near the treated surface by 1.5 times.
- The computer model allowing to predict the distribution of the temperature field and depth of a diffusion layer at local ion nitriding with a hollow cathode for components of various configuration and size was suggested. The computer model also allows to determine the concentration of a saturant by the depth of a diffusion layer.
- It was found that the nitraded surface of 200 mm gear wheel is heated up to the temperature of diffusion saturation (550 °C) within approximately ~ 40 min, following which the surface temperature remains invariable until the treatment is completed. This is caused by the established balance between thermal energy supply from high-energy ions of a working gas and its withdrawal through radiation from a surface into the external environment. At the same time the maximum difference of temperatures between areas nitraded with/without a hollow cathode makes ~ 35 °C.

References

[1] Zinchenko V M 2001 *Engineering of gear wheel surface by chemical heat treatment methods* (Moscow: Publishing House of Bauman MSTU) p 303
[2] Kozlovskiy I S 1970 *Chemical heat treatment of gear wheels* (Moscow: Mechanical engineering) p 232
[3] Budilov V V, Agzamov R D and Ramazanov K N 2007 *MiTOM* **7** 25
[4] Budilov V V, Ramazanov K N, Khusainov Yu G and Zolotov I V 2014 *USATU Bulletin* **18** 32
[5] Granovsky V L 1971 *Electric current in gas. Steady current* (Moscow: Nauka) p 543
[6] Kogan Ya D, Kolachev B A, Levinskiy Yu V 1987 *Force constants of metals with gases (Reference book)* (M.: Metallurgy) p 368

IOP Conf. Series: Journal of Physics: Conf. Series **830** (2017) 012074 doi:10.1088/1742-6596/830/1/012074

Influence of martensitic steel ultrafine-grained structure on diffusion processes at low-temperature ion nitriding

K N Ramazanov, R S Esipov, E L Vardanyan, R D Agzamov

Ufa State Aviation Technical University, Department of Mechanical Engineering, 12 Karla Marksa Str., Ufa, Russia

E-mail: kamram@rambler.ru

Abstract. The paper presents research results on the influence of martensitic steel ultrafine-grained (UFG) structure on diffusion processes at low-temperature ion nitriding. UFG structure was obtained through severe plastic deformation by torsion (SPDT) of coarse-grained (CG) steel samples. Ion nitriding was performed under the constant voltage mode during 4 hours at a temperature of 420 °C. Optical microscope was used to study the microstructure of a sample. Thickness of a modified layer was defined according to measurements of microhardness by depth. The analysis of research results showed that UFG steel structure enhances effective intensification of diffusion processes by 1.7 times, which in its turn contributes to low-temperature modification of a surface. At the same time, surface microhardness of UFG steel after treatment amounted to 1,200 HV and 525 HV of CG steel correspondingly.

1.Introduction

At present various methods of surface structural and phase modification based on diffusion processes, such as ion nitriding, are used to improve surface tribological properties of construction steel. However, such methods can only be applied at high temperatures (550-680 °C) and prolonged exposure time (up to 24 hours) [1,2]. As a result of such treatment, parts with complex geometry are subject to buckling, and the surface is scattered by saturating ions [3]. Besides, at temperatures over 600 °C the diffusive zone depth may decrease due to the reduction of ion current rate under the influence of thermionic emission from the cathode surface [1]. The aforesaid disadvantages may be cleared out by reducing the treatment temperature, which, however may lead to the essential delay of diffusion processes, increase of treatment cycle duration and rise of energy costs.

It is known [4,5] that ultrafine-grained (UFG) steel structure obtained via severe plastic deformation (SPD) is characterized by the grain size of 100-300 nm and non-equilibrium grain boundaries with enhanced concentration of vacancies and dislocations. Such crystalline defects have the stimulating impact on the progression of diffusing element atoms into the material [6-8]. Thus, UFG structure of a material allows to intensify the process of diffusion saturation at low-temperature structural and phase modification.

The purpose of this paper is to study the influence of martensitic steel structure on diffusion saturation at low-temperature ion nitriding.

2.Experimental approach

Table 1 shows chemical composition of martensitic steel 13X11N2V2MF-SH, which was used as a basic material in the research. Steel was subject to thermal treatment at 1,050 °C during 1 hour with subsequent air cooling and tempering at 800 °C within 1 hour. As a result of such treatment the microhardness of a material made 340 HV.

Table 1. Chemical composition of steel 13X11N2V2MF-SH, weight %

C	Cr	Ni	W	Mo	Mn	Si	V	Cu	Fe
0.134	11.4	1.66	1.77	0.46	0.44	0.32	0.21	0.12	residual

Discoidal samples with UFG and coarse-grained (CG) structure of 20 mm in diameter and 1 mm thickness were prepared for further ion nitriding. UFG structure was obtained via severe plastic deformation by torsion (SPDT) (Fig. 1) at the temperature of 300 °C and pressure of P=6 GPa with 10 revolutions per minute. Microhardness of steel surface after SPDT made 750 HV.

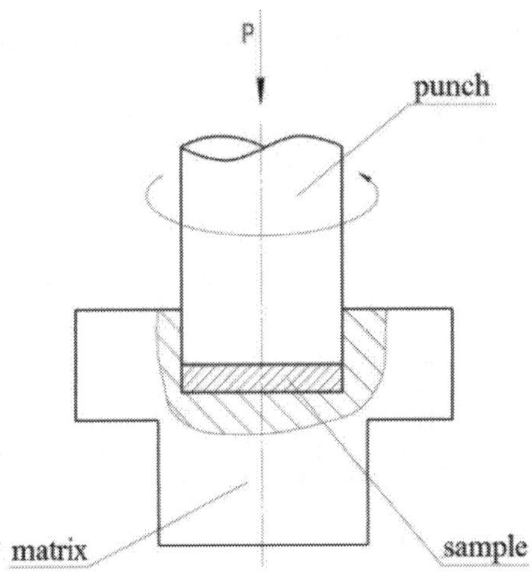

Figure 1. SPDT for martensitic steel sample.

Ion nitriding was performed on the modernized unit ELU-5M [9] according to scheme shown in Figure 2. Samples with CG and UFG structures were subject to treatment in equal conditions during 4 hours at 420 °C in nitrogen-containing gaseous mixture (50%N_2+45%Ar+5%H_2) at the pressure of 150 Pa. Small amount of hydrogen in gaseous mixture is needed to deactivate oxygen, which prevents nitrogen diffusion to get into the material thus forming iron oxides on a surface [3].

In order to assess thickness of a nitrided layer, microhardness by depth was measured using microhardness tester Duramin-1/-2 under the load of 10 g during 10 seconds. Net thickness of a diffusion zone was determined according to hardness distribution curve up to the hardness value of non-nitrided material.

The structure of a nitrided layer was studied against angled laps using an optical microscope Olympus GX51.

3.Results and discussion

Figures 3a and 3b show microhardness distribution curves in cross section of samples after ion nitriding during 4 hours at 420 °C and at the pressure of 150 Pa. The results of surface microhardness measurements are presented in Table 2.

IOP Conf. Series: Journal of Physics: Conf. Series **830** (2017) 012074 doi:10.1088/1742-6596/830/1/012074

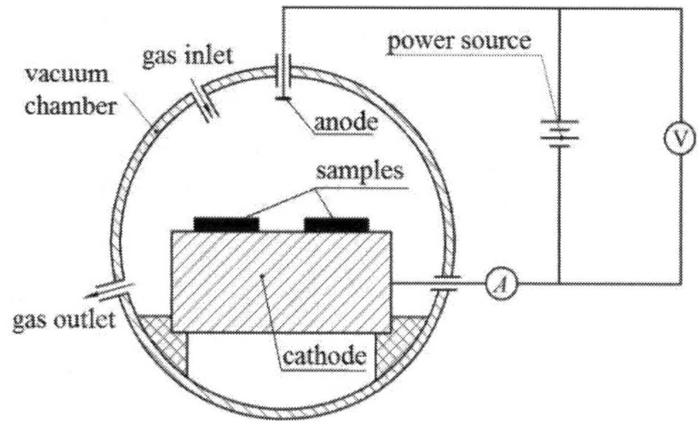

Figure 2. ELU-5M unit for ion nitriding.

The analysis of microhardness distribution curves in cross section after ion nitriding (Fig. 3) shows smooth decrease in hardness from the surface to the basic material for all samples under study. At the same time the net thickness of the modified layer of UFG steel is approximately 1.7 times higher than the same value of the CG steel. Surface microhardness of UFG steel was about 1,250 HV that justifies high concentration of nitrides on a surface [6,7]. Surface microhardness of CG steel made about 525 HV.

Table 2. Surface microhardness

Initial state, $HV_{0.1}$	Surface microhardness of UFG samples, $HV_{0.1}$	Surface microhardness of CG samples after nitriding, $HV_{0.1}$	Surface microhardness of UFG samples after nitriding, $HV_{0.1}$
340±20	750±20	525±20	1250±20

Higher thickness of UFG steel nitrided layer is caused by intense diffusion processes due to multiple grain boundaries and crystalline defects that contribute to the progression of nitrogen atoms into the material at nitriding temperature of 420 °C [6,7]. High increase of UFG steel surface microhardness leads to the formation of γ', ε and α'-phases in the surface layer of a material [10]. At the same time, due to multiple defects of UFG steel the formation of phases is heterogeneous [7].

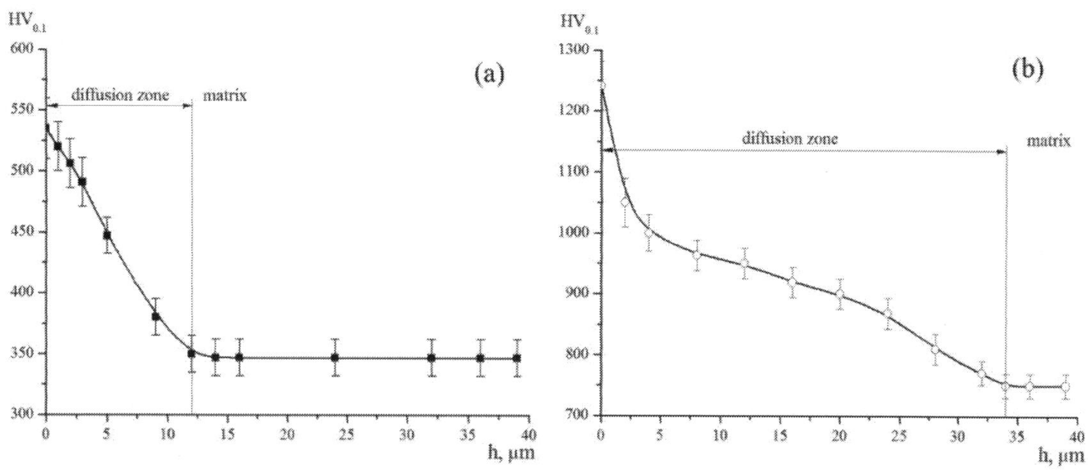

Figure 3. Microhardness distribution curves in cross section of samples after ion nitriding at 420 °C during 4 hours: (a) CG steel, (b) UFG steel.

Figures 4a and 4b show optical images of samples microstructure after ion nitriding at 420 °C during 4 hours.

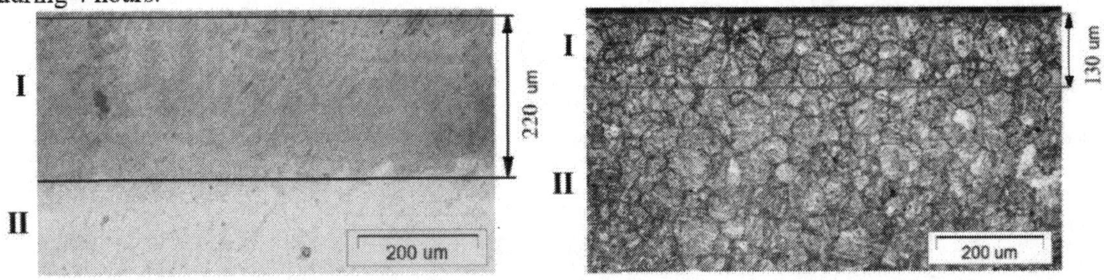

Figure 4. Microstructure of samples at angle laps after ion nitriding at 420 °C during 4 hours: (a) UFG steel, (b) CG steel.

Diffusion zone – I and the zone of a basic material – II were identified on optical images of sample microstructures. The total thickness of UFG steel modified layer was about 220 microns. Thus the transition time between diffusion zone and the zone of a basic material is smooth, which is justified by measurements of microhardness. Due to its fine structure it was impossible to identify the grain boundaries of a UFG sample (Fig. 4a). During metallographic study CG steels were characterized by enhanced etchability along the grain boundaries with about 130 microns thickness, which correlates with microhardness measurement.

The analysis of optical images of sample microstructures showed that UFG structure contributes to the effective increase in speed of diffusive processes at low-temperature ion nitriding during 4 hours at a temperature of 420 °C, thus allowing to reduce the duration of treatment by 1.7 times.

4.Conclusion

The study of the influence of the UFG structure of martensitic steel on diffusion processes at low-temperature ion nitriding proved that preliminary treatment by pressure that forms UFG structure via SPD contributes to the effective increase in the modified layer speed by 1.7 times at 420 °C and within 4 hours.

It was also defined that UFG structure with high volume ratio of non-equilibrium grain boundaries initiates nitrogen grain boundary diffusion mechanisms. At the same time the surface layer is characterized by heterogeneous formations of γ', ε and α'-phases that define high values of surface microhardness after treatment.

References

[1] Lakhtin Yu M and Kogan Ya D 1976 *Nitriding of steel* (Moscow: Mechanical Engineering) p 256
[2] Lakhtin Yu M and Krymsky Yu N 1968 *Physical processes at ion nitriding (Protective coatings of metals* vol 2*)* (Kiev: Naukova Dumka) pp 225-229
[3] Ramazanov K N, Ishmukhametov D Z, Sadkova N S 2011 *Ion nitriding in glow discharge non-homogenios plasma* vol 15 (Ufa: USATU) p 67
[4] Valiev R Z, Islamgaliev R K and Alexandrov I V 2000 *Progress in Materials Science* **45** 103
[5] Zhilyaev A P and Langdon T G 2008 *Progress in Materials Science* **53** 893
[6] Tong W P, Han Z, Wang L M, Lu J and Lu K 2008 *Surf. Coat. Technol.* **202** 4957
[7] Li Y, Wang L, Zhang D and Shen L 2010 *Appl. Surf. Sci.* **257** 979
[8] Ferkel H, Glatzer M, Estrin Y and Valiev R Z 2002 *Scripta Mater.* **46** 623
[9] Budilov V V, Ramazanov K N, Zolotov I V and Khusainov Yu G 2014 *Reinforcement technologies and coatings* **12** 27
[10] Ferkel H, Glatzer M, Estrin Y, Valiev R Z, Blawert C, Mordike B L 2003 *Materials Science and Engineering* **348** 100

IOP Conf. Series: Journal of Physics: Conf. Series **830** (2017) 012075 doi:10.1088/1742-6596/830/1/012075

Research of influence of mobile cathodic stains of the vacuum arc for reception of the adjustable roughness of metal surfaces

V N Anikeev and M Yu Dokukin

Bauman State Technical University, 2-nd Baumanskaya 5, Moscow, 105005, Russia

E-mail: anikeev4032@yandex.ru

Abstract. In the modern technics there is a requirement in micro- and macrorough surfaces of products for improvement of their operational characteristics (improvement of adhesive properties of various coverings, decrease in deterioration of rubbing details because of the best deduction of greasing, increase of the heat exchanging coefficient from a surface, stimulation of adhesive processes on sites of contact to a bone fabric of medical implants in stomatology and orthopedy etc.). In the given work the modes of reception regulated micro- and macrorough surfaces on samples from a titanic alloy and stainless steel by electrothermal influence of moving cathodic stains in the vacuum arc discharge are investigated. Chaotically moving stains, possessing high specific power allocation (~ 10^7 W/cm^2), "scan" the difficult design of a product, including "shadow" sites, doing rough its blanket. The sizes of roughnesses are regulated by a current and time of influence of the discharge, pressure in the vacuum chamber and a number of other parameters. The scheme of experimental device, photo and the characteristic of rough surfaces and technological modes of their reception are resulted.

1. Introduction

In various areas of modern science and technology there is a need in micro- and macrorough surfaces of products of the improvement of their operational characteristics.

Let us list a number of such areas:

-improvement of adhesive properties of various coverings, including paint and varnish.

-Improvement of ergonomic and decorative characteristics in demanded cases, for example, giving of a necessary roughness of a contacted surface of a product, as in the known technological process of mechanical knurling.

-Decrease in deterioration of rubbing details (the minimum deterioration appears not on smooth surfaces, and at some optimum values of roughnesses because of the best deduction of greasing) [1].

-Increase in the heat-removing coefficient from a surface at the increase in its roughness (an intensification of heat exchange at the expense of increase in the general heat-removing surface and the changes of characteristics of a heat-carrier current) [2].

-Improvement of hydro - and aerodynamic characteristics of the flow around bodies by liquids and gases with a stream separation in certain ranges of Reynolds numbers [3].

-Stimulation of adhesive processes on sites of contact to a bone fabric of medical implants in stomatology and orthopedy (the rough well moistened surface of an implant containing roughnesses, fabrics proportional to bio cages is necessary) [4–5].

Implants are made of biocompatible titanic alloys and stainless steels. Their sizes are in a range from several millimeters to several centimeters. There is a considerable quantity of designs of

IOP Conf. Series: Journal of Physics: Conf. Series **830** (2017) 012075 doi:10.1088/1742-6596/830/1/012075

implants. For example, only in tooth implantology cylindrical, cone-shaped, lamellar, disk, screw, hollow, frame and other designs are applied. The rough surface of an implant should not contain sharp edges and easily separated particles [5].

Now such surfaces receive, basically, methods of powder metallurgy, sand-streaming and plasma spraying. These methods have lacks. For example, at the plasma spraying there is a problem of adhesion of a covering (Figure 1) and difficulty of processing of products with the difficult form (hollows, shadow sites).

a)

b)

Figure 1. Macrorough surface obtained by plasma spraying : a – optical zoom x8, b – optical zoom x56.

2. Method of the reception of an adjustable roughness on metal surfaces

In the works devoted to studying electrode processes in vacuum arc electric discharges (look, for example, fundamental works [6–8]) the formation of rough electrode surfaces as a result of action of cathodic (or anode) moving stains is mentioned. In work [9] the reception possibility of micro- and macroroughnesses on metal surfaces by means of a vacuum arc has also been shown. But the characteristics of a received roughness and management possibilities in its parameters were not studied. In a number of works [10–16] the analysis is carried out and the theoretical methods of forecasting of a received roughness are offered.

The novelty of the given work is that in it the experimental research of vacuum arc modes for reception of the necessary (set) roughness of surfaces of products by electrothermal action of moving cathodic stains was carried out. Chaotically moving stains, possessing high power allocation (~ 10^7

W·cm^{-2}), "scan" a product configuration, doing rough its blanket. As cathodic stains, unlike anode, also clear a surface from various pollutions (sites with the smaller exit work of a material – the places of oxides, carbides, nitrides, etc.) in this work, the modes when samples (products) were cathodes in discharges were investigated only.

3. The description of the experimental device and results of research

Experimental research of the reception of roughnesses on metal surfaces was carried out on the device which scheme is presented on Figure 2. The device consists of the vacuum chamber 1 in which the processed sample (cathode) 2, the coaxial anode 3 and the unit of discharge initiation 4 are located. The chamber is pumped out by the system of rotational pumps to pressure of 10^{-2} mm Hg and the necessary working pressure is maintained in it.

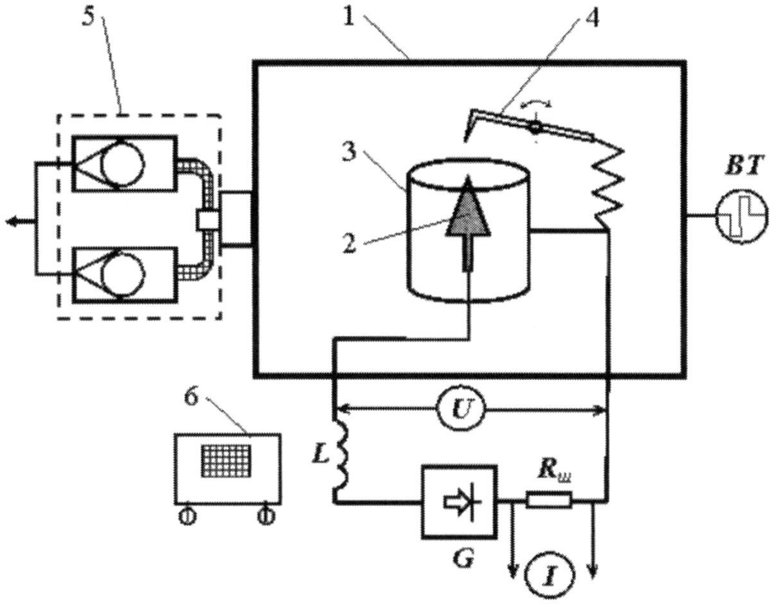

Figure 2. Scheme of the experimental device: 1 – vacuum chamber, 2 – sample (cathode), 3 – anode, 4 – unit of arc discharge initiation, 5 – pumping system, 6 – oscillography, *BT* – thermopair gauge, *G* – rectifier, *L* – inductance, $R_ш$ – shunt.

In a number of investigated modes the chamber was filled with argon to a demanded pressure. The pressure in the chamber was monitored by the vacuummetr with thermopair gauge *BT*. Power supplies of the arc discharge was carried out from the rectifier *G* of thyristor type with the inductance consistently included in a chain $L \approx 10$ mH which provided necessary stabilization of an arc. Arc initiation was carried out by a cathode contact with an additional mobile electrode 4. The current *I* and voltage *U* of the discharge was controlled by analog devices; in addition for the purpose of studying of temporary discharge characteristics filing of oscillograms of current and voltage on the memorable oscillograph was carried out.

Samples for research were made of titanic alloy VT-5 (Grade 6 US) and had working surfaces of two kinds: a cylindrical surface in diameter of 11 mm and length 15 mm and a conic surface with a diameter of the basis 12 mm and height 15 mm. The quality of working surfaces was provided not lower than R_a 2,5.

As the experiments have shown, the received roughness depended on a mode of burning of the discharge on a processed surface which was defined by the pressure in the chamber p, current I and processing time τ.

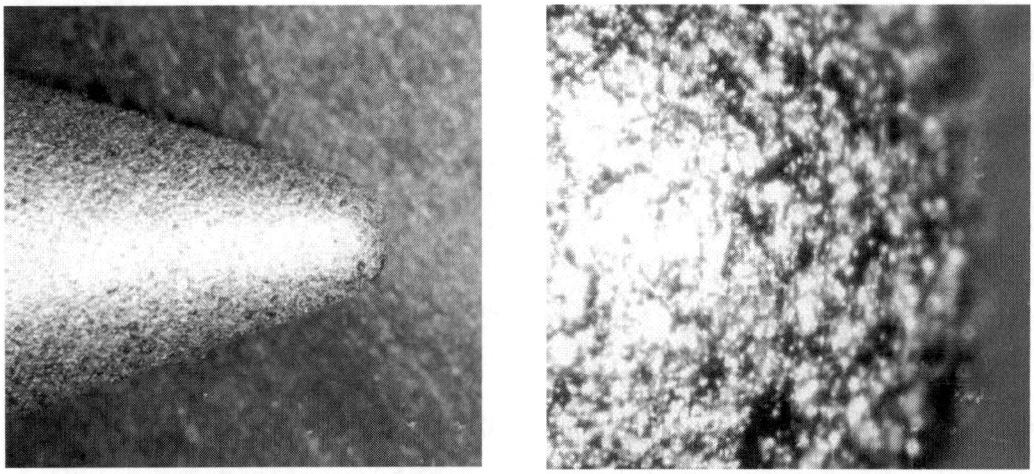

a) b)

Figure 3. Minimum roughness surface obtained by the vacuum arc method: a – optical zoom x8, b – optical zoom x56.

In Figure 3 and Figure 4 the photos of conic titanic samples with characteristic results of processing of their surfaces are presented. The minimum roughness (R_a <10 μm) has been received in a mode: p = (1,5 – 3,0)10^{-2} mm Hg, I = 7 – 9 A, U = 21 – 23 V, $\tau \approx$ 20 s (6 impulses duration on 2 – 4 s). The maximum roughness (R_a > 10 μm) has been reached in a mode: $p = 9 \cdot 10^{-3} - 3 \cdot 10^{-2}$ mm Hg, I = 20 – 25 A, U = 20 – 21 V, τ = 75 s.

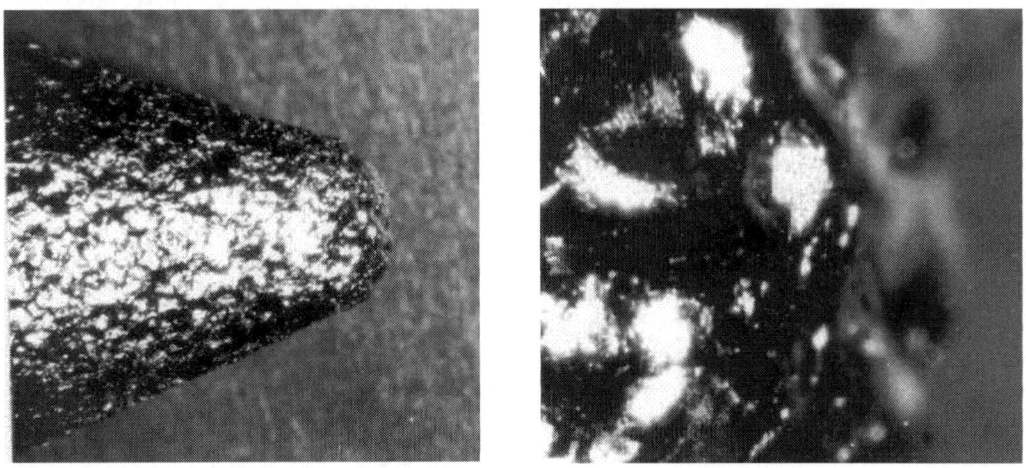

a) b)

Figure 4. Maximum roughness surface obtained by the vacuum arc method: a – optical zoom x8, b – optical zoom x56.

In Figure 5 the results of measurements of a roughness on a reference technique of definition of a R_a index received on the «Profilometr 296» device (a range of measurements 0,02 – 10 μm, the price of division of a scale 0,004 μm, a class of accuracy 2) for conic and cylindrical samples are presented.

Characteristic oscillograms of a current and voltage of the discharge for modes with small (Figure 6) and big (Figure 7) roughness are presented below.

IOP Conf. Series: Journal of Physics: Conf. Series **830** (2017) 012075 doi:10.1088/1742-6596/830/1/012075

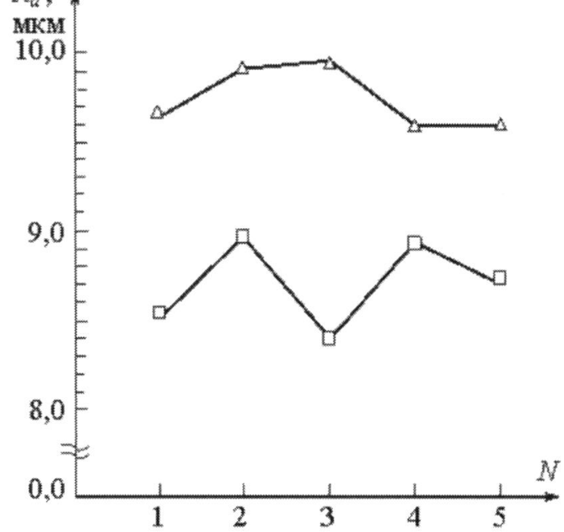

Figure 5. Profilegrams of the processed samples: Δ – cylindrical, □ – conic.

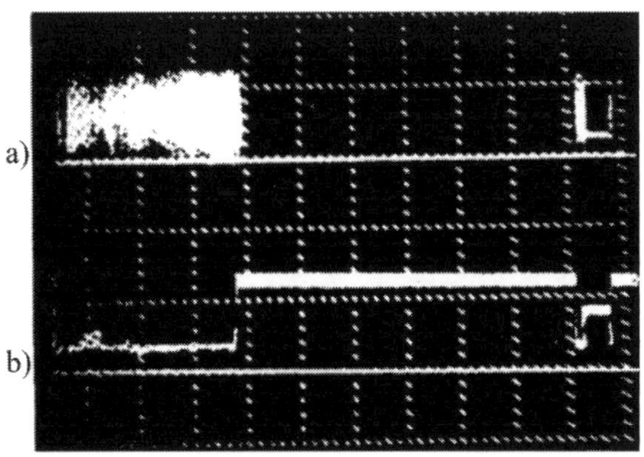

Figure 6. Oscillograms of a current (a) and voltage (b) of the discharge for a small roughness mode: μ_I = 40 A/div, μ_U = 50 V/div, μ_τ = 200 ms/div.

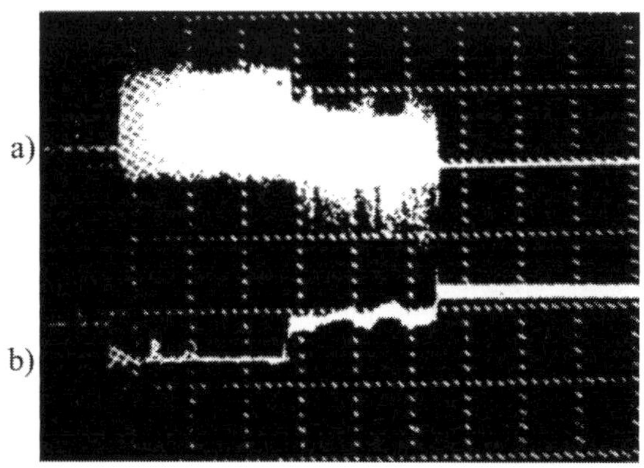

Figure 7. Oscillograms of a current (a) and voltage (b) of the discharge for a big roughness mode: μ_I = 40 A/div, μ_U = 50 V/div, μ_τ = 500 ms/div.

4. Conclusion

The method of reception of an adjustable roughness on metal surfaces by electrothermal action of mobile cathodic stains in the vacuum arc discharge is offered. Results of processing of various surfaces under the form by means of the given method are presented.

The main advantages of the method are:

–maintenance of a blanket roughness of the basic material of a product;

–possibility of the reception of a homogeneous rough surface of a product with a complex configuration (screw, hollow, frame) owing to a specific character of movement of the cathodic stains getting on remote sites of a surface.

Electrothermal action of the cathodic stains in addition clears a surface from various pollutions.

IOP Conf. Series: Journal of Physics: Conf. Series **830** (2017) 012075 doi:10.1088/1742-6596/830/1/012075

References

[1] Kostetsky B I 1979 *Friction, deterioration and greasing in mashines* (Kiev: Technics), in Russian

[2] Fomenko L A and Lovtsova L G 2010 The heat exchange at heating of a surface with an artificial submillimetric roughness *The Surface: x-ray, synchrotron and neutron researches* 109–112, in Russian

[3] Rogovoy Yu A 2005 *Research of influence of an external surface roughness of underwater objects on their hydrodynamic characteristics* (Dis. … Cand. Tech. Sci.: 05.08.01 S-Pb), in Russian

[4] Wil'iams D F and Rouf R 1978 *Implantaty in surgery* (Moscow: Medicine), in Russian

[5] Bukaev M F, Surov A O and Surov O N 2004 *Dentalnaja implantology* (Almaty: Curiosity), in Russian

[6] 1982 *Vacuum arcs. Theory and applications* ed J M Lafferty (Moscow: The World), in Russian

[7] Mesyats G A 2000 *Ektony in the vacuum discarge* (Moscow: The Science), in Russian

[8] Mesyats G A 2011 Explosive electronic emission (Moscow: Fizmatlit), in Russian

[9] Anikeev V N and Dokukin M Yu 2008 *Contr. Papers V Int. Symp. on theoretical and applied plasma chemistry (Ivanovo, on September, 3-8th 2008)* vol 2 (Ivanovo: ISCTU) p 422–425

[10] Benardos P G and Vosniacos G–C 2003 Predicting surface roughness in machining: a review *Int. J. of Machine Tools and Manufacture* **43** 833

[11] Hecker R L and Liang S Y 2003 *Int. J. of Machine Tools and Manufacture* **43** 755

[12] Kuzenov V V, Ryzhkov S V and Shumaev V V 2015 *Problems of Atomic Science and Technology* **4** (98) 53

[13] Kuzenov V V, Polozova T N and Ryzhkov S V 2015 *Problems of Atomic Science and Technology* **4** (98) 49

[14] Kuzenov V V, Ryzhkov S V and Shumaev V V 2015 *Problems of Atomic Science and Technology* **1** (95) 97

[15] Ryzhkov S V 2015 *Sustainable Cities and Society* **14** 313

[16] Aliev I N, Dokukin M Yu and Samedova Z A 2016 *Herald of the Bauman Moscow State Technical University. Natural Sciences Series* **4** 14

Refractory phases synthesis at the surface microalloying using a wide aperture electron beam

S V Fedorov and Min Htet Swe

MSTU "STANKIN", Russia, 127055, Moscow, Vadkovskiy per., 1

E-mail: av288291@akado.ru

Abstract. The experimental results prove the ability to realize technology of chemical heat treatment of some materials by surface microalloying using a wide-aperture low-energy high-current electron beam. Such layers were produced due to initiating exothermic chemical self-propagating high-temperature reactions in the thermal explosion mode between the base and the thin film covered on the base. New phase compounds in reaction products were found.

1. Introduction

Today there are many methods of surface modification of a wide range of materials with surface alloying. The most popular are the methods including chemical and thermal processing (CTP). They include technologies connected to diffuse layer saturation. Usually such processes are carried out as final operation in technological procedure of instrument production, except for the cases when the surface strengthened with CTP is covered with abrasion resistant coatings [1].

Creation of an alloyed surface layer with increased wear-resistance using self-propagating high-temperature synthesis (SHS), based on the use of internal energy of the source reagents, may be an alternative to the traditional methods of chemical and thermal processing. As a rule, reagents in SHS processes are used in the form of fine-grained powders, liquids and gases. Such processes are known in systems with powder and liquid, gas suspension and multilayered films. In this article we describe the peculiarities of producing microalloying subsurface layers in the systems consisting of a massive object with a thin film cover.

The basis of the process is the synthesis of stable compounds in the surface layer of the product by initiating chemical reaction between metals of IV–V groups with non-metals of the 2nd period. At that, the product should contain such non-metals in its structure or to be saturated by them by producing solid solution or some instable compounds. Metal in the form of the layer is applied on the product surface, for instance, with the use of magnetron sputtering. Then the reaction of self-propagating high-temperature synthesis is initiated in the mode of thermal explosion by impulse heating of the product surface. In some cases, interconnection between metal coatings and substrate produces intermetallic phases.

During creating SHS system it is possible to use rather wide spectrum of the reagents which are chemically active at high temperature [2]. Simultaneously other substances can be used as fillers or dilatants including those, which take part in synthesis process as by-products of the reaction. In this case not only chemical nature of the reagents is important but also the thermal effect of the reaction, heat flow conditions and kinetics of phase and structural transformations.

2. Microalloying of the surface samples

The main method of SHS reaction initiating is approaching heat impulse with combustion wave along interphase boundary [3]. The process is more effective in case of heating all surface of the object in the mode of thermal explosion. Evaluation of diffusion coefficient in such reactions is up to 10^{10} times more that it is usually in solid phase. In case of high thermal losses caused by intensive heat removal of the substrate the process can become less stable. However, with increasing temperature of the surface the speed of SHS front will be increasing exponentially.

In our case material examples were subjected to impulses series of a wide aperture low energy high current electron beam (LEHCEB), to initiate exothermic chemical reactions both in liquid and in solid phases, between metal film and nitrogen, carbon or aluminum which are present in the content of base in the free form or in the content of compounds.

The processing was carried out in RITM-SP unit which is a combination of the LEHCEB RITM, and two magnetron sputtering systems were placed in a single vacuum chamber. The unit allows carrying out sputtering of films made of various materials onto the surface of the product and further liquid phase mixing of film materials and the substrate with intensive electron beam impulse [4], which is called microalloying. LEHCEB generation includes electron emission, creating of beam in a plasma diode and its transportation in plasma channel. Use of such generation scheme allows getting a beam with about 5 microsecond length and with current density up to 10^5 A/cm^2 at accelerating voltage from 15 to 30 kW. The area of processing is about 50 cm^2.

During impulse electron processing surface layer covered with an alloying layer is subject to cyclical heating. During quick heating with electron beam with 10^6 K/s speed and its further quick cooling (10^4...10^9 K/s) its external layers cool quicker than the middle that is why there appear tension stresses [5]. The source of excitation of stress waves can be thermo elasticity, caused by the sudden expansion of thin surface layer heated to high temperatures, as well as recoil impulse caused by material vaporizing in the exposure zone and further recession of vapors. The temperature in the subsurface layer reaches the levels which exceeds melting of system compounds. Evaluation of the the impulse surface heating at discharge energy density about 4.5 J/cm^2 results in the value approaching 4000 °C. Heat emission caused by exothermal chemical reaction between a thin film and surface layer allows to produce up to 0.5 J/cm^2 additionally but with certain time delay. After passing of the elastic wave at the moment of relaxation of elastic tensions a burst-like creation of structural defects occurs there, which enhances the process of substance transfer referred to as remote action effect.

3. Microalloying of solid alloy samples with $Nb_{70}Hf_{22}Ti_8$ composition

In this section we describe microalloying processes of the materials surface using the reaction of niobium-hafnium carbide creation in the complex carbide on the surface of carbide phase of hard alloy acting as a donor of carbon atoms, including free atoms.

For experiments H13A inserts, produced by Sandvik Coromant, were used as base material. Thin film (thickness 150 – 250 nm) was applied to the samples by means of magnetron sputtering. The content of the film is NbHfTi alloy. The experiment was carried out to obtain a layer from wear resistant nonstoichiometric carbides with face centered cubic lattice at the next processing stage [6].

A high speed melting and solidification process allows creating unusual structures in hard alloy. And, although according to modern concepts the greatest influence on the strength of a hard alloy having the status of the binder, structure changes of which often can lead to softening, also the changing structure of carbide phases exerts influence here.

Here we are dealing with deformation hardening caused by the elastic wave. However, due to short duration of the process (~ 10^{-5}s) and thermal inertia, heating caused by compression and viscous friction, most likely, is not a physical factor defining substance performance at such conditions. The main part in this case should play mechanic activating of high rate physical and chemical processes within the substance which are present both in liquid or in solid phases.

Covering of the instrument surface by carbide forming elements as – Nb, Hf and Ti before processing with an electron beam allows to produce multiphase carbide structure due to microalloying during SHS. The external layer is saturated with hard-melting carbide phases of MC type, which keep a small form and are homogenously distributed due to extremely high speed of cooling. Moreover, chemical potential of the system changes. The appearance in the melt in the neighborhood with more refractory melting particles leads to the sharp increase in the interphase surface and carbide formation reaction. Due to additional energy release during SHS, there is enough time to produce explosion chemical reactions. Moreover, such compounds as γ-WC, TiC, NbC and HfC have an opportunity to create non stoichiometry structures, with a range of stoichiometry varying from 0.5 to 0.97 which means that effects, connected to surface decarburization, can be eluded.

Here we should notice the role of hafnium carbide, having one of most high formation heat (252 kJ/mol). Impact by the electron beam on an alloy film where hafnium is absent, does not result in the formation of homogenous structure without micro fractures. Internal thermal energy of chemical interaction of source reagents is not enough.

X-ray analysis of crystal structure proves phase content changes, and on the surface of the samples characteristic structure appears, which proves successful SHS in each case (figure. 1b). Powerful thermal impact of LEHTEB on the surface of H13A hard alloy insert covered with film made of NbHfTi alloy about 200 nm thick results in creation of face-centered cubic arrangement of carbide phase, identified as (Nb, Hf, Ti) C_X. γ-WC phase reflection with lattice constant of 0.424 nm are appears. We should also consider one-type FCC structure of niobium-hafnium carbide and of high temperature tungsten carbide. The result of the X-ray analysis depicting the above mentioned specifics features is presented at figure 1b. X-ray diffraction analysis was performed with an X-ray diffraction meter PAN analytical Empyrean Series 2 (CoK_α).

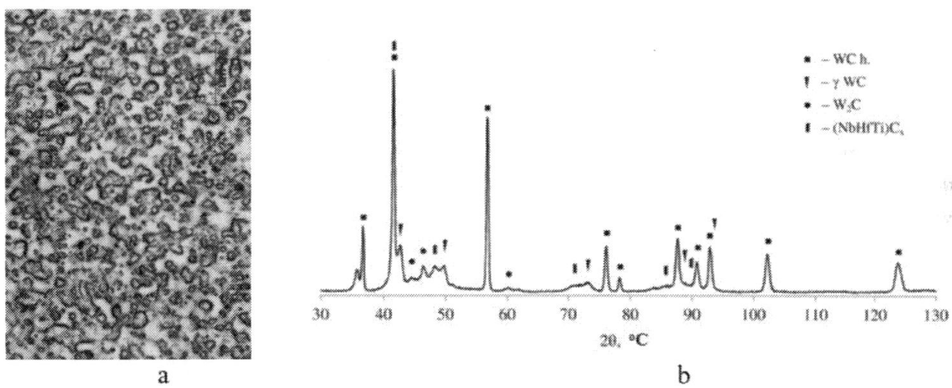

a b

Figure 1. a) Structure of solid alloy surface after microalloying with LEHTEB by $Nb_{70}Hf_{22}Ti_8$, b) X-ray diffraction analysis from H13A sheet surface after microalloying with $Nb_{70}Hf_{22}Ti_8$ alloy.

The specific feature of sub-surface area is formation of crystallographic texture caused by heat transfer during cooling. Comparison of spectrums received during symmetrical and asymmetrical analysis showed the change in intensity of reflection of two main phases of face centered cubic lattice connected to crystalloid texture {100} which is parallel to the object surface.

Influence of geometrical factor should be noticed as well. During the use of impulse electron beams energy losses for backscattering depend on atomic size of the target, energy and incidence angle of electron beams and reach the level up 8 to 50 %. In case if angle of the incidence of an electron beam is about 30°, maximal value of temperature distribution will be transferred to the surface, and in case of further angle decreasing the temperature gradient in subsurface layer increases [7]. This means that homogeneity of heating of subsurface area would be influenced by surface roughness of the irradiated object, especially in case if its value comparable or more than penetration distance of the electrons to the object.

It was noticed that on polished samples of hard alloy when roughness of the surface is 4-5 times less than penetration distance jf the electrons the reaction does not flow right which has negative influence on the properties of a forming material. At that, change of angle of incidence of the electron beam to the surface up to 30° can initiate the reaction. The process flows most effectively at roughness of unprocessed surface varying from 0.5–1.5 µm R_A.

The concentration of Nb, Hf and W in interlayer proves mixing of solid alloy with alloying cover as a result of high temperature synthesis reaction. Maximal thickness which allows to produce modified SHS structure is 3-4 µm. Multiple initiating of the process practically does not change initial microstructure but it can have negative impact on the substrate due to thermal cycling. To make the modification to the fullest extent, as a rule, it is enough to perform a series of five or six pulses.

At the bottom the layer is created due to the contact melting of tungsten carbide grains. The thickness of contact melting zone increases when it is closer to the surface, but the size of initial carbide grains decreases.

On the other hand, the structure of the modified layer transfers to the structure of source solid alloy. In this area we can observe zone liquation of cobalt. The cobalt content in the transfer zone equals 2-3 µm, which is four times more than its content in the source alloy. The occurrence of such liquation can be described by the fact that cobalt phase acting as crystallizing phase transfers with liquid phase to the boundary of thermal influence area. The displacement process is also influenced by the significant difference in the specific density of carbides and liquid cobalt. Existence of such liquation can lead to cracking of the object surface and the most probable risk is maximal cobalt content in subsurface layer.

4. Microalloying of preliminary nitrified rapid cutting steel

We should analyze the process of zirconium nitride formation where instable ferric nitrides of preliminary nitrified rapid cutting steel acted as a nitrogen atom donor.

To carry out the experiments preliminary nitrified high speed steel (HSS) S6-5-2 sheets at the depths of 50 µm with two stage vacuum arcing type discharge [8] as a base material were used.

In figure 2 the influence of LEHTEB impulses series with energy density 4.5 J/cm^2 and about 5 µs lengths to the surface of nitrified steel sample is shown. The heating effect of the electron beam sufficiently below an upper layer of metal not only melted, but also began to evaporate, exposing the carbide component. Irradiation by LEHTEB causes dissociation of ferric nitrides, especially ε-phase; great amount of residual austenite is created on the surface (figure 2b).

After covering the samples by magnetron thin Zr film 150 – 250 nm thick, and further irradiation with an electron beam we have managed to initiate SHS reaction of ZrN formation.

Due to the formation of hard melting nitride film metal vaporizing significantly decreases and the structure becomes finely dispersed. The formation of the nitride phase is proved by the results of X-ray structural analysis (figure 2c). Moreover, we should note that in the latter case the content of residual austenite in subsurface layer significantly decreases.

It can be noted that Zr is discovered in subsurface layer about 2 µm thick.

5. Microalloying of aluminum alloy with titanium and zirconium

Formation of intermetallic phases can be analyzed at the example of aluminum interaction with titan coatings. For electron beam microalloying alloy AlCu4Mg1 was chosen. On flat samples made of these alloy Ti covering were placed in order to get the layer saturated with intermetallic phases of Al-Ti systems after electron beam processing during SHS.

In figure 3 we can see the result of the impact made by an electron beam with the energy density of 4.0 J/cm^2 on Al alloy of the surface covered with titanium film about 0.2 µm thick. Optical microscopy shows formation of dendrite type crystals of intermetallic phase which is proven by X-ray structural analysis. It confirms the intermetallic phase formation Al_3Ti (structural types tI8, a=0.348

nm, c=0.858 nm and tI32/30, a=0.3875 nm, c=3.384 nm, the melting point of 1390 °C). The energy density increased by 4.5 J/ cm^2 causes the vaporizing of the covered film along with the aluminum surface layer. The intermetallic phase content decreases drastically in this case.

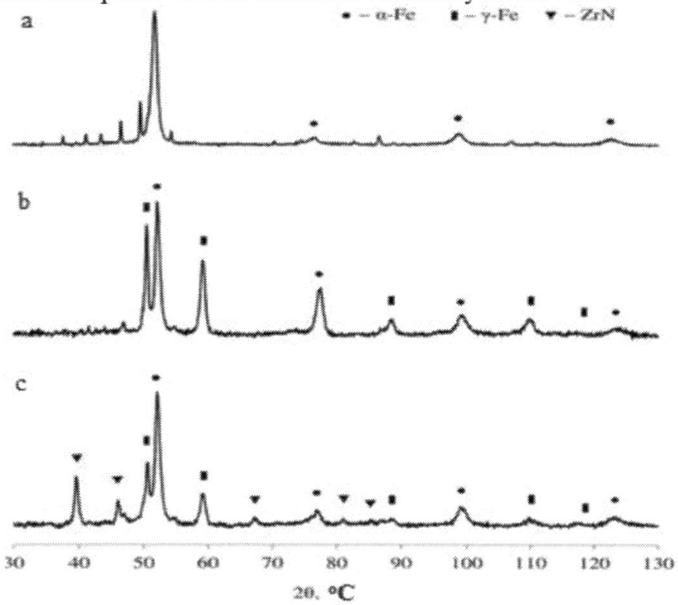

Figure 2. a) Diffract gram (CoK$_\alpha$) of a nitrified HSS P6M5 surface, b) the same after LEHTEB, c) the same after applying of Zr film on the surface before irradiation.

The "polishing" effect is vividly shown on aluminium alloys. The alloy processing, namely the intense melting with the partial vaporizing of the surface layer, causes its refining, its structure crushed to sub-micron crystals and smoothing the micro relief. The initial roughness equal to 2.5 μm has reduced to 0.8 μm.

The essential practical features that the observed effects possess are their stability and reproducibility.

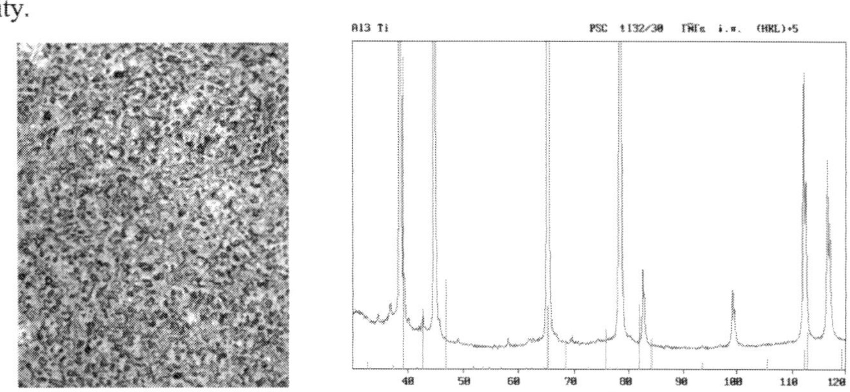

a b

Figure 3. Crystal formation of the intermetallic titanium alluminide Al_3Ti on the aluminium alloy surface due to electron beam processing conducted on the covered titanium film. Surface optical image; b) Diffract gram (CuK$_\alpha$).

6. Conclusion

The obtained experimental results confirm a possibility for the formation layers modified by microalloying on the surfaces of the wide variety of materials. The problem is the increasing the thermal stability of non equilibrium structural and phase states that ensure to achieve unique physical and strength properties. These layers were prepared due to chemical reactions initiation conducted between the alloy and its thin film coated on its surface. The formation of new phase components was detected in the reaction products in this case.

The structure formation in the sub-surface material layer arises in our case from the impulse nature of the conducted processing in the microsecond range. The key factors influenced the microalloying are electron beam energy, which is determined by accelerating voltage, and film thickness. The interdependence that occurs between the modified layer and accelerating voltage has an extremal nature. The electron beam irradiation with the lack of energy is unable to initiate the SHS, but its abundance leads to the extreme film vaporizing.

The processing is to be conducted providing the coating thickness equal to half-value penetration distance into the substrate material. It goes up to about 200 nm in our case. At that, it is preferable to have more or less uniform heating of the sub-surface layer by electrons that the geometrical factor enables. The initial roughness correlates with the electron penetration distance into the substrate.

The microalloying finds its application. In particular, the aforesaid processing recommended before conducting the coating enables to make influence on cutting tool depletion ensuring the fact that its resource will be increased [9].

References

[1] Vereschaka A A 2013 *Advanced Materials Research* **712-715** 347

[2] Rogachev A S and Mukasyan A S 2014 *Combustion for material synthesis* (NY: CRC Press Reference)

[3] Myagkov V G, Bykova L E and Bondarenko G N 1999 *JETF* **115(5)** 1756 (in Russian)

[4] Markov A B, Yakovlev E V and Petrov V I 2013 *IEEE Transactions on Plasma Science* **41** 2177

[5] Proskurovsky I, Rotshtein V P, Ozur G E, Ivanov Yu F and Markov A B 2000 *Surface and Coatings Technology* **125(1-3)** 49

[6] Grigor'ev S N, Fedorov S V, Pavlov M D, Okun'kova A A and Ye Min Soe 2013 *Journal of Friction and Wear* **34** 14

[7] Poate J M, Foti G and Jacobson D C 1986 *Surface Modification and Alloying by Laser, Ion and Electron Beams* (Plenum press, NY and London).

[8] Grigor'ev S N, Metel A S and Fedorov S V 2012 *Metal Science and Heat Treatment* **54** 8

[9] Fedorov S V, Pavlov M D and Okun'kova A A 2013 *Journal of Friction and Wear* **34** 190

Acknowledgements

This project was supported by the grant No. 14-29-00297 of the Russian Science Foundation. The work is carried out on the equipment of the Center of collective use of MSTU "STANKIN".

IOP Conf. Series: Journal of Physics: Conf. Series **830** (2017) 012077 doi:10.1088/1742-6596/830/1/012077

Modification of optical and electrical properties of SnO_2 under the influence of argon ion beam

S Umnov, O Asainov and V Temenkov

Tomsk Polytechnic University, 634050, Lenin 2, bdg. 11, Tomsk, Russia

E-mail: usp@tpu.ru

Abstract. Thin films of tin oxide were deposited on the glass substrates at a room temperature using reactive magnetron sputtering. The ratio between O_2/Ar and the discharge voltage is maintained in such a mode when the deposited films are dielectrics. After the deposition, the films were irradiated with an argon ions beam. The modification of the optical and electrical properties of the films depending on the irradiation time was studied. Optical properties of the films were analyzed in the range of 300-1100 nm using photometry and structural X-ray diffraction. The diffractometric research showed that the films, deposited on a substrate, had a crystal structure, and after argon ions irradiation they became quasi-crystalline (amorphous). It was found that the modification in the transmittance was correlated with modifications in the meaning of surface resistance. The dielectrics films SnO_2 with increasing exposure time became conductive and then the electrical resistance decreased and reached a minimum at 13.2 seconds. Then resistance films began to increase.

1. Introduction

The transparent conductive oxide films are widely used in various industries [1-4]. These optically transparent electrodes are used in displays, solar panels, photoelectric devices, touch panels, etc. One of them is tin oxide, which has found wide practical application. In some practical applications, there is no need to simultaneously achieve the high value of transmittance and low resistance. In such cases it is possible to use an undoped tin oxide. Pure tin oxide has conductivity and transparency. This is due to the existence of intrinsic point defects that act as vacancies [5,6]. Annealing deposited films leads to the modification of the optical and electrical properties. The ratio of oxygen to stannum changed under SnO_2 film annealing at atmosphere and it leads to a decrease in their resistance [7,8]. In this case, the transmission and resistance depend on the duration of the annealing. Annealing the tin oxide films in a vacuum results in a significant reduction of resistance [9]. In work [10] it is shown that when optimizating the oxygen content and the substrate temperature one can receive films with improved transmittance and resistance values. Tin oxide can be in the form of SnO_2 and SnO. Under heating, these phases interact with one another that leads to the modification of films properties [9,11,12]. The effect of films annealing in various atmospheres (air, vacuum or hydrogen atmosphere) is studied in work [9]. The correlation was found between the structural and optical properties of SnO_2 films. In the work [11] tin oxide films were annealed at different time period, however the correlation between the electrical specific resistivity and optical and structural properties was not found. In works [13], the effect of oxygen content in a reactive gas mixture on the properties of the deposited tin oxide films was studied. It is shown that films consist of an amorphous and crystal phase [14]. The formation of textured films in the course of the deposited of tin oxides and its transformation into polycrystalline

SnO_2 at a low temperature (~ 200 ° C) is investigated in [15]. Stoichiometry of films depending on the oxygen content and its influence on electric properties were investigated in [16]. The influence of deposited rate of a film on specific electric resistance has been studied in [17]. The mechanism of conductivity of pure oxide tin films deposited at various temperatures of a substrate has been investigated in work [18]. In work [19] the influence of the application of an external constant magnetic field on resistance and an optical transmission of the deposited films are investigated. Oxygen negative ions of the magnetron discharge [20] also have an impact on the films properties. The influence of ionic stimulation in the course of films deposition on optical properties of a film is considered in [21]. The properties modification of already deposited films happens at the irradiation by an ionic beam of already deposited films [22,23,24]. In this research, the tin oxide films were deposited by a reactive magnetron dispersion, and the films after sedimentation were irradiated by a bunch of argon ions.

2. Experimental details

Thin films of tin oxide were deposited on the glass substrates using reactive magnetron sputtering. The size of Sn target (99.5%) was 700x100x5 mm^3. The sputtering was carried out at a working pressure of 0.3 Pa in the gas mix O_2 and Ar (O_2 – 60%, Ar – 40%). The voltage of the magnetron discharge was ~ 275 V. Under these conditions the deposited films were dielectric. The power of the discharge at sedimentation was stabilized at the level 1.2 kW. The heating of substrates was not carried out. The substrates were placed on a carrier that was able to carry out a reciprocating motion under the magnetron and the ion source in a vacuum chamber. The distance between a sample and a target was 8 cm. Speed movement of the carriage was 3.62 cm/sec. After the deposition, tin oxide films were irradiated with argon ions. The ions beam was located at an angle 90° to a substrate surface. Ions were generated by a source of ions with the closed drift of electrons at a voltage 2800 V and discharge current 0.28 A. The current density on a surface of samples was $2.4 \cdot 10^{-3}$ A/cm^2. However, the pressure in a chamber was 0.166 Pa. The thickness of the deposited films was measured using profilometer MicroMeasure 3D Station. For measuring a spectrum of a transmission in the range of 300-1100 nanometers the spectrophotometer SF-256 was used. The structure of films was studied on the x-ray diffractometer XRD6000 . The average size of crystallites was estimated by half of width of x-ray diffraction lines.

3. Results and discussion

Dielectric films of tin oxide were deposited on glass substrates. For study of the film structure, SnO_2 coatings with thickness of 800 nm were deposited. To measure optical and electrical characteristics, the film thickness was 200 nm.

Figure 1 shows the diffractogram films of tin oxide before and after irradiation by Ar ions. It is found that the films SnO_2 after magnetron deposition have crystal structure with shares of amorphous (quasicrystal). The analysis of XRD data showed that the deposited films consist of pure SnO_2 tetragonal structure without the presence of other phases – SnO and Sn. Similar results were observed in work [21]. Moreover, the presence of other phases of tin oxide in films was not revealed. The size of crystallites makes ~ 10 nm. After irradiation of structure became quasicrystal (amorphous).

In Fig. 2 the transmission measurements of SnO_2 films at various periods of irradiation time are presented. According to the figure, the transmission T (λ) depending on irradiation time has different nature. On the impact at 0 to 13.2 sec., the dependences T (λ) have similar qualitative nature.

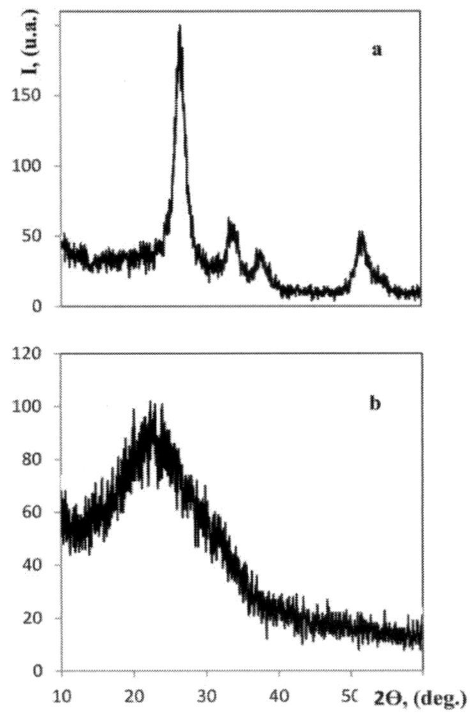

Figure 1. XRD diagram samples of SnO_2: a- before irradiation: b-after irradiation.

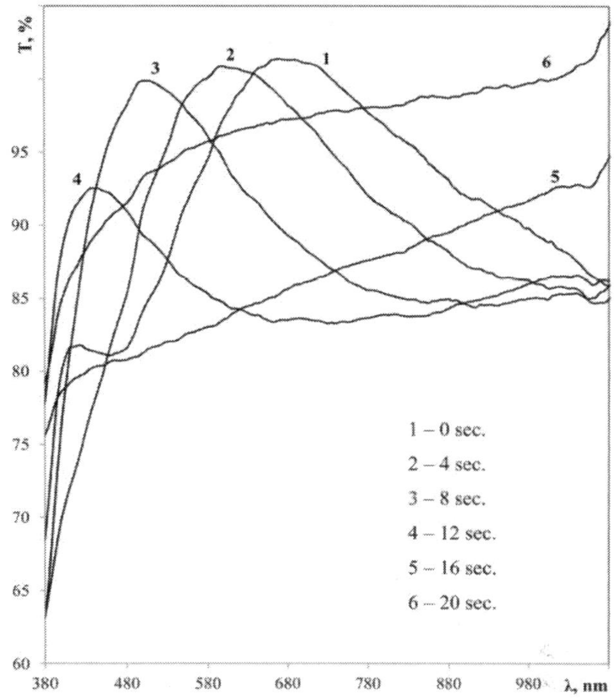

Figure 2. Dependence of the transmission on the irradiation time.

In this time range, the maximum of transmission is existed. The shift of the transmission maximum to the area of short waves during ionic beam exposure time less than 13.2 second is probably connected with the formation of SnO_2 polycrystalline structure [12] of nanodisperse sizes.

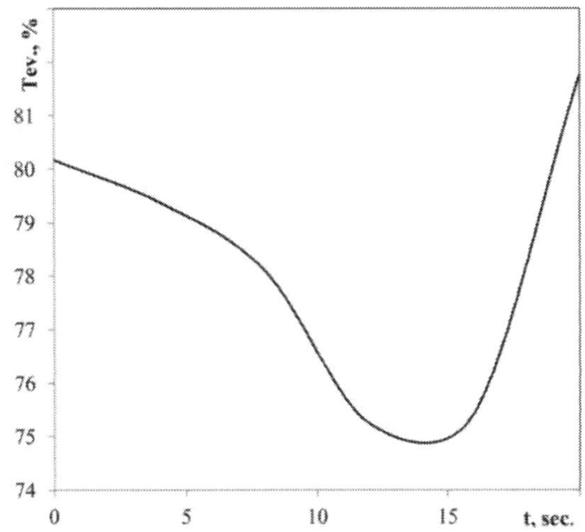

Figure 3. Dependence of the average value of the transmission on the irradiation time.

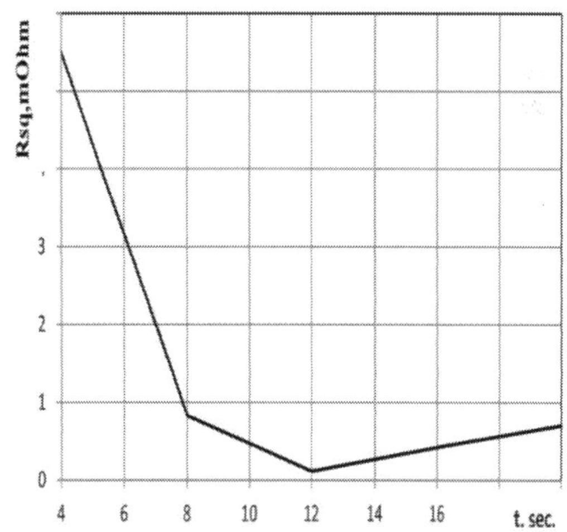

Figure 4. Dependence of surface resistance on the irradiation time.

On the ionic beam impact less than 13.2 sec., the values of a transmission maximum are changed slightly which is perhaps connected with a minor change in a stoichiometric structure of a film. The film transmission decreases under the ion irradiation more than 13.2 sec. This is due to the

IOP Conf. Series: Journal of Physics: Conf. Series **830** (2017) 012077 doi:10.1088/1742-6596/830/1/012077

fact that on the ions influence a SnO_2 dissociation with formation of metal phase takes place [12]. At ion irradiation time more than 13.2 sec., the dependence T (λ) has different character – a rather gradual transmission growth depending on the wavelength. However, the transmittance of SnO_2 films grows with the future increase of ion irradiation time.

Fig. 3 shows the average value change of samples transmission of a deposited SnO_2 film in the range of 380-1100 nm depending on Ar ions irradiation time. As it is seen in Fig. 3, at first the transmission decreases with the increase under the ions irradiation and then increases. The transmission minimum is observed after the irradiation during 13.2 sec. The further growth takes place mostly due to the transmission improvement in a long-wave area relatively to the transmission maximum.

Fig. 4 demonstrates the measurements of a film surface resistance depending on the influence time of an ion beam. It correlates with the change of average transmission. Initially, the film resistance decreases and reaches the minimum value at the argon ions influence time of ~ 13.2 sec. As it has been noted above, at such ion beam influence time the content of small particles of metal tin reaches the highest value in the surface layer of a film. As a result, the film resistance decreases [10] since metal tin particles participate in conductivity. Further impact on an ion beam film leads to the metal component dispersion that reduces its contribution to conductivity, and as a consequence the film resistance increases.

4. Conclusions

Pure SnO_2 dielectric polycrystalline films were deposited at a room temperature by means of reactive magnetron sputtering. After deposition, SnO_2 films were irradiated by argon ions generated by an ion source of with the closed electrons drift. After the irradiation, the film became quasicrystal. It was found that at the increase in ions influence time there is a slightly gradual decrease in average transmission value (from 80 to 75%), which reaches a minimum at ~ 13.2 sec. The further ion beam influence leads to the transmission increase up to 82%. The correlation between optical and electric properties of SnO_2 films is observed. Thus, at first with the increase in Ar ions influence time on the film the film electric resistance decreases and then increases.

Therefore, the optical and electric properties of deposited films can be corrected by means of an argon ions beam impact on the SnO_2 deposited film.

References

[1] Harrison P G, M.J. Willett M J 1988 *Nature* **332** 337
[2] Batzill M, Diebold U 2005 *Surf. Sci.* **79** 47
[3] Zhao W, M. Zhang M, Ai Z et al 2014 *J. Phys. Chem. C* **118** 23117
[4] Qian J, Liu P, Xiao Y et al 2009 *Adv. Mater.* 21 3663
[5] Jarzebski Z M, Marton J P 1976 *J. Electrochem. Soc.: Reviews and news* **123** 199
[6] Kılıç C, Zunger A 2002 *Phys. Rev. Lett.* 2002 **88** 095501
[7] Alterkop B, Parkansky N, Goldsmith S and Boxman R L 2003 *J. Phys. D: Appl. Phys.* **36** 552
[8] Kim I H, Ko J H, Kim D et al 2006 *Thin Solid Films* **515** 2475
[9] Mukashev B M , Tokmoldin S Zh, Beisenkhanov N B et al 2005 2005 *Mater. Sci. Eng: B* **18** 164
[10] Abhijit De and Swati Ray 1991 *J. Phys. D: Appl. Phys.* **24** 719
[11] Mukhamedshina D M, Beisenkhanov N B, Mit K A 2006 *Thin Solid Films* **495** 316
[12] Karapatnitski I A, Mit K A, Mukhamedshina D M, et al 2002 *Surf. Coat. Technol.* **151-152** 76
[13] Goodchild R G, Webb J B and Williams D F 1985 *J. Appl. Phys.* **57** 2308
[14] Gubbins M A, Casey V, Newcomb S B 2002 *Thin Solid Films* **405** 270
[15] Mukashev B N, et al 2007 *Superlattices and Microstruct.* 42 103
[16] Lee W H, Son H C, Moon H S et al 2000 *J. Power Sources* 89 102
[17] Shamala K S, Murthy L C S, Narasimha Rao K 2004 *Bull. Mater. Sci.* **27** 295
[18] Shikha Bansal, Dinesh K Pandya, Subhash C Kashyap 2012 *Thin Solid Films* **524** 30
[19] Tadatsugu Minami, Hidehito Nanto and Shinzo Takata 1988 *Japan. J. Appl. Phys.* **27** L287
[20] Klaus Ellmer and Thomas Welzel 2012 *J. Materials Research* **27** 765

[21] Won-Kook Choi, Hyung-Jin Jung and Seok-Keun Koh 1996 *J. Vac. Sci. Technol.* A **14** 359
[22]Asainov O, Umnov S, Shinin A 2015 *Journal of Physics: Conference Series* **652** 012046
[23]Asainov O, Umnov S, Temenkov V 2016 *Journal of Physics: Conference Series* **669** 012059
[24]Umnov S, Asainov O, Temenkov V 2016 *Journal of Physics: Conference Series* **124** 012148

Ion-plasma nitriding of austenitic steel in a low-pressure low-frequency inductive discharge with ferrite core

M V Isupov[1], V A Pinaev[1], D O Mul[2], N S Belousova[2]

[1]Kutateladze institute of thermophysics SB RAS, Lavrentiev av. 1, Novosibisrsk, 630090, Russia
[2]Novosibisrsk State Technical University, Karl Marx av. 20, Novosibirsk, 630073, Russia

E-mail: isupovmikhail@yandex.ru

Abstract. An experimental investigation of ion-plasma nitriding of austenitic stainless steel AISI 321 in a low-frequency (100 kHz) nitrogen inductive discharge has been performed for the nitrogen pressure of 7 Pa, nitrogen ion densities of 10^{10}–10^{11} cm^{-3}, sample temperatures of 440–590 °C, the densities of current on the sample surface of 1.2–3.3 mA/cm^2, sample biases of -500 and -750 V. The time of ion-plasma treatment was 20 and 60 min. It is shown that even for the short (20 min.) ion-plasma treatment in the low-frequency inductive discharge, formation of nitrided layers with the thickness of up to 40 μm and microhardness of up to 9 GPa is observed.

1. Introduction

New plasma nitriding technologies based on the use of "external" low-pressure, high-density plasma sources for remote generation of nitrogen ions allow to enhance the speed of nitrided layer formation increasing its thickness and give more flexibility to control processing parameters [1, 2]. Thus, increasing the lifetime and efficiency of such plasma sources is an important task. Inductively coupled plasma sources (ICP) having high lifetime and high ion density at low pressures seem to be a good choice for ion-plasma treatment but have a few disadvantages: high current frequency (~1–10 MHz), low power factor of ICP coil (cosφ<1) and thus high power losses in the ICP coil and matching network [3]. Using closed ferrite cores to improve magnetic coupling between the inductor and plasma in ICP allows to create new low-pressure, high density inductively coupled plasma sources with a low current frequency (~10–100 kHz) and a high power factor of ICP coil (cosφ≈1), thus with a good power transfer from power supply into plasma [3]. Such low-frequency inductive discharges could be used in many areas of plasma processing [3], including ion-plasma nitriding technologies. The aim of this work is to investigate the process of ion-plasma nitriding in the low-pressure, low-frequency inductive discharge improved with ferrite cores for the case of austenitic stainless steel, which is quite hard to be nitrided [1, 2].

2. Experimental setup

A scheme of experimental setup is shown in figure 1. Gas discharge chamber 1 is made of stainless steel water cooled sections with the inner diameter of 230 mm. The sections are sealed and dielectrically separated with silicon rubber gaskets. The length of the chamber is 1 meter. A narrow (ID of 40 mm) U-shaped part 2 of the discharge chamber, along with the main part of the discharge

IOP Conf. Series: Journal of Physics: Conf. Series **830** (2017) 012078 doi:10.1088/1742-6596/830/1/012078

chamber 1, forms a toroidal current path for the low frequency inductive discharge. Ferrite cores 3 significantly enhance magnetic coupling between the inductive discharge and inductor 4, so a power supply with a current frequency of 50–100 kHz 5 can be used for discharge generation. The power supply is connected to the inductor through a matching network 6. Rogowsky coil 7 is used to measure the discharge current. Voltmeter 8 is used to measure the discharge voltage and to determine the electric field strength in the discharge chamber. MKS baratron 626a 9 is used to determine the nitrogen pressure. Double electric probe 10 is used to determine the electron density of plasma. Stainless steel samples 11 with the size of 20x20x1.5 mm (12X18H10T austenitic stainless steel, an analog of AISI 321) are placed inside the discharge chamber. The sample temperature is measured with a thermocouple 12. Sample bias is regulated with a DC power supply 13. To pump out the discharge chamber, a fore pump 14 is used.

To measure microhardness of nitrided layers, cross-sections of samples are prepared. Microhardness is measured with a Wolpert micro vickers tester 402MVD with a load of 10 g. To determine the microhardness, a few measurements are done in a middle part of a layer, and then an averaged value of microhardness is calculated. The microstructure analysis is carried out by X-ray diffraction (ARL X'TRA). An optical microscope (Carl Zeiss Axio Observer A1m) is used to observe the morphology and thickness of the nitrided layers.

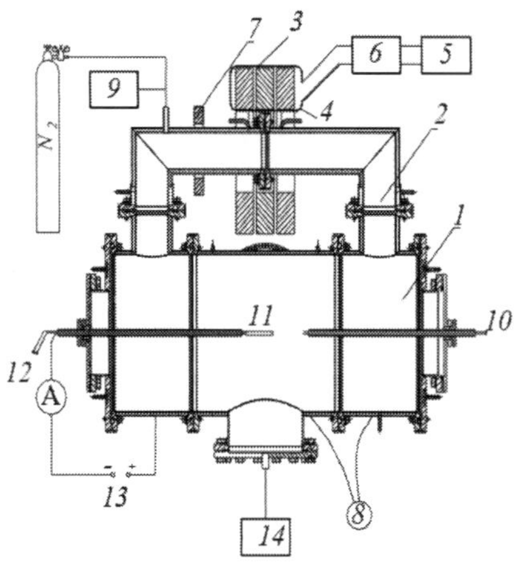

Figure 1. Experimental setup.
1 – Gas discharge chamber (internal diameter of 230 mm), 2 – U-shaped part of the discharge chamber (ID of 40 mm), 3 – ferrite cores, 4 – inductor (ICP coil), 5 – power supply (100 kHz, 500 V), 6 – matching network (variable LC circuit), 7 – current transformer (Rogowsky coil), 8 – voltmeter, 9 – MKS baratron 626a, 10 – double electric probe, 11 – stainless steel sample, 12 – thermocouple, 13 – DC power supply, 14 – forevacuum pump.

3. Experimental results and discussion

Ion-plasma nitriding of the stainless steel samples has been performed for the nitrogen pressure of 7 Pa. The current of the low-frequency inductive discharge was varied in the range of 7–60 A (discharge power of 1300–6400 W), which corresponds to the ion densities in the range of 10^{10}–10^{11} cm^{-3}. These ion densities are comparable with the ion density in a hot-cathode gas arc (nitrogen pressures of 0.1–0.6 Pa, discharge current of 100 A), which was used for plasma nitriding of stainless steels 12X18H10T [1] and AISI 304 [2]. It is necessary to underline that the absence of electrodes in the low-frequency inductive discharge removes limitation on the discharge current strength and allows to increase the current strength and nitrogen ion density up to significantly higher levels.

Figures 2, 3 show the optical micrographs of a cross-section of nitrided samples, treated for 20 min. under DC sample bias of -500 V and nitriding temperatures of 470–590 °C. Even for the short time of treatment, nitrided layers with the thickness of about 20–40 μm and microhardness of up to 8–9 GPa are observed. For comparison, hot-cathode arc treatment [1] for the same time under nitriding temperature of 520 °C caused only insignificant (up to 2.5 GPa) hardening of the surface layer, and the nitrided layer thickness was only 9 μm. This difference may be caused by a lower sample bias used in [1] (-300 V) and a lower current density on the sample surface respectively [4]. It is necessary to underline that short time nitriding of austenitic stainless steel using a "standard" DC glow discharge treatment produces only very thin nitrided layers of a few micrometers [5].

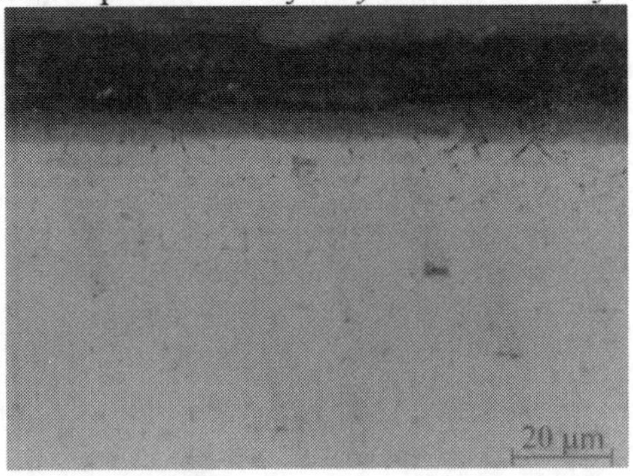

Figure 2. Cross-section of a nitrided sample, treatment time 20 min. Nitriding temperature 470 °C, sample bias -500 V, sample current density 3.2 mA/cm^2, layer microhardness 6.8 GPa.

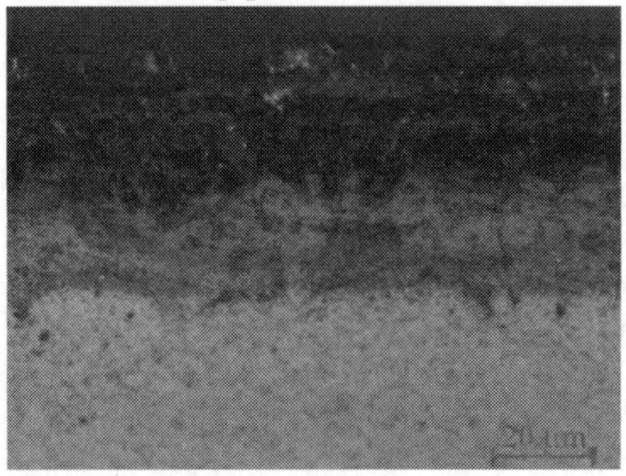

Figure 3. Cross-section of a nitrided sample, treatment time 20 min. Nitriding temperature 590 °C, sample bias -500 V, sample current density 3.3 mA/cm^2, layer microhardness 8.1 GPa.

Figures 4, 5 show the optical micrographs of a cross-section of nitrided samples treated for 60 min. under DC sample bias of -750 V and nitriding temperatures of 440–565 °C. For 1 hour time of treatment, nitrided layers with the thickness of up to 33 µm and microhardness of up to 13 GPa are observed. In this case, the sample current density was a bit lower (1.2–2.4 mA/cm^2), even for a higher sample bias, due to a lower discharge current and nitrogen ion density respectively.

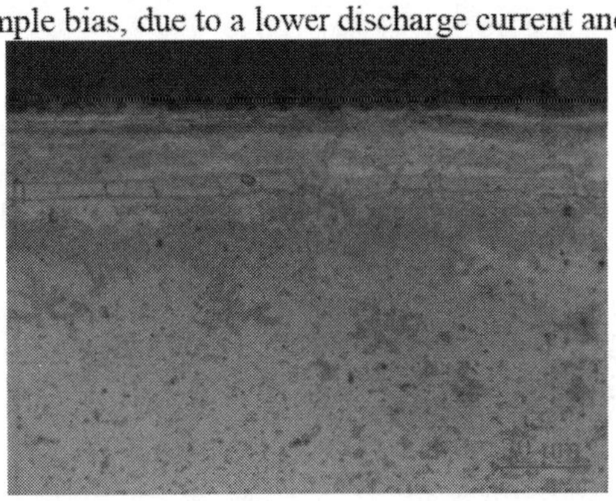

Figure 4. Cross-section of a nitrided sample, treatment time 60 min. Nitriding temperature 440 °C, sample bias -750 V, sample current density 1.2 mA/cm^2, layer microhardness 7.4 GPa.

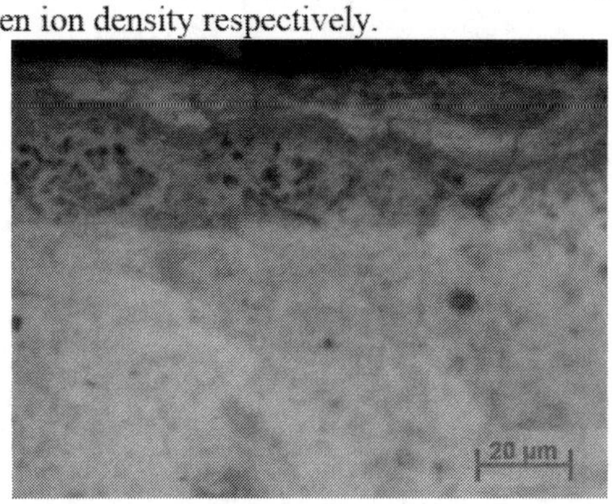

Figure 5. Cross-section of a nitrided sample, treatment time 60 min. Nitriding temperature 565 °C, sample bias -750 V, sample current density 2.4 mA/cm^2, layer microhardness 10.3 GPa.

For comparison, hot-cathode arc treatment [1] for the same time under DC sample bias of -300 V and nitriding temperature of 520 °C formed a nitrided layer with the thickness of 24 µm and microhardness of about 9 GPa (stainless steel 12X18H10T). For stainless steel AISI 304 treated under nitriding temperature of 580 °C, sample bias of -700 V and sample current density of 3.2 mA/cm^2, the nitrided layer with the thickness of 48 µm and microhardness of about 12 GPa was observed [2]. DC glow discharge nitriding during 1 hour under the same temperatures produces layers with the thickness of about 10 µm [5] and microhardness of about 10 GPa. Thus, comparing the low-frequency ICP nitriding with the treatment in the hot-cathode gas arc [1, 2] and the "standard" DC glow discharge nitriding [5], we observe a significant enhancement in the speed of nitriding of austenitic stainless steels for the case of "external" high-density plasma sources. In the case of short-time treatment (20 min.), the best results are achieved for the low-frequency ICP nitriding, probably due to a high current density (up to 3.3 mA/cm^2) on the sample surface. The same sample current density was achieved in [2], but unfortunately the authors in [2] did not present data for 20 min. treatment.

Figure 6 shows X-ray diffraction patterns of the surface of treated samples for various treatment conditions (nitriding temperature, sample bias, current density on the sample surface, and time of treatment). Also, the resulting layer microhardness is shown.

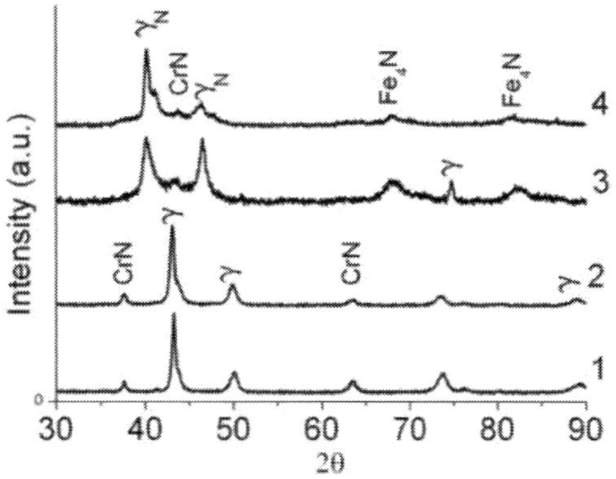

Figure 6. The X-ray diffraction patterns of treated samples.
1 – 470 °C, -500 V, 3.2 mA/cm^2, 20 min., 6.8 GPa;
2 – 590 °C, -500 V, 3.3 mA/cm^2, 20 min., 8.1 GPa;
3 – 440 °C, -750 V, 1.2 mA/cm^2, 60 min., 7.4 GPa;
4 – 530 °C, -750 V, 2.0 mA/cm^2, 60 min., 12.9 GPa.

As it is seen from figure 6, in the case of the short-time treatment, X-ray diffraction patterns have peaks related to the austenitic phase (γ) with a face centred cubic structure, and peaks related to chromium nitride. The presence of the austenitic phase indicates that nitriding process is not completed. With increasing the treatment time up to 60 min. and nitriding temperature up to 530 °C, the peaks of austenitic phase completely disappear. For 60 min. treatment, X-ray diffraction patterns have peaks related to the nitrogen expanded austenitic phase (γ_N), iron nitride Fe_4N and chromium nitride CrN (the same peaks were found in [2]). It is followed by the increase in the layer microhardness (from 7–9 GPa for 20 min. treatment up to 10–13 GPa for 60 min treatment).

4. Conclusion

A new method of ion-plasma nitriding based on the use of the low-pressure low-frequency inductively coupled plasma enhanced with ferrite cores as an "external" source of nitrogen ions has been tested. Austenitic stainless steel 12X18H10T (an analog of AISI 321) was treated in the temperature range of 440–590 °C, sample biases of -500 and -750 V, current densities on the sample surface of 1.2–3.3 mA/cm^2, for 20 and 60 min. It is found that even the short (20 min) ion-plasma treatment results in the formation of the nitrided layers with the thickness of about 20–40 μm and microhardness of about 7–9 GPa (depending on the nitriding temperature). The significant enhancement in the speed of nitriding (in comparison with other methods) can be explained by a high nitrogen ion density (~10^{11} cm^{-3}) in the low-frequency ICP, and high current density on the sample surface respectively. The further increase in the ion density (up to 10^{12}–10^{13} cm^{-3}) is still possible, and it may result in a further increase in the nitriding speed.

References

[1] Korotaev A D, Ovchinnikov S V, Tymencev A N, Pinzhin Yu P, Goncharenko I M, Koval N N and Schanin P I 2004 *Physics and Chemistry of Materials Treatment* **1** 22–7
[2] Zhao Y, Yu B, Dong L, Du H and Xiao J 2012 *Surface & Coatings Technology* **210** 90–6
[3] Godyak V 2013 *J. Phys. D: Appl. Phys.* **46** 283001
[4] Wang Q, Ba D, Ming Y, Xu L and Guo D 2014 *Plasma Science and Technology* **16** 960–3
[5] Li Y, Wang Z and Wang L 2014 *Applied Surface Science* **298** 243–50

Acknowledgments

This work was fulfilled in the framework of the budget project III.18.2.2.

The Corrosion Behavior of Ni-Based Superalloy Irradiated by Intense Pulsed Ion Beam

H W Zhong[1], J Zhang[1], X Yu[1], J Shen[1], G Y Liang[1], X J Cui[1], X F Zhang[1], G L Zhang[1], S Yan[2], X Y Le[1]

[1]School of Physics and Nuclear Energy Engineering, Beihang University, No.37 Xueyuan Rd, Beijing, 100191, P.R. China,

[2]Institute of Heavy Ion Physics, Peking University, No.201 Chengfu Rd, Beijing, 100871, P.R. China

E-mail: xyle@buaa.edu.cn

Abstract: Intense pulsed ion beam (IPIB) has been extensively used in the surface strengthening of metal materials in the past decades. Quite a lot of these materials need to be operated in the corrosive environment. Therefore, it is of significance to research the corrosion behavior of metal materials after IPIB irradiation. In this work, the corrosion behavior of Ni-based superalloy irradiated by IPIB in NaCl solution was studied. Compared to the original samples, after IPIB irradiation the corrosion resistance of samples were improved, and craters were formed on the surface of Ni-based super alloys. It is found that micro-areas with craters would be eroded prior to other areas in caustic solutions. The analysis revealed that craters play an important role in the corrosion process of the metal after irradiation. This research would help understand the influence of craters induce by IPIB on the operational performance of metal materials.

1. Introduction

Intense pulsed ion beam (IPIB) is a kind of flash heat sources, which has been extensively used in material surface modification, especially for engineering materials [1]. Because of the short pulse duration and high energy density of the beam, after IPIB irradiation the temperature of the marital surface will rise rapidly (heating rate 10^8 -10^{11} K/s) accompanied by the shock waves being ignited by the thermal shock in the target [2]. In past works, it has been shown that the corrosion resistance of some materials such as steels, magnesium alloys can be increased by IPIB irradiation. There are many explanations to comprehend the improvement of corrosion resistance for materials after irradiation, including grain refinement, the ablation of alloy components with low melting point, the formation of amorphous structure and so on [3,4]. However, the corrosion behavior of micro region in irradiated material is not clear.

Ni-based superalloy is commonly used in engines vanes, cartridge receiver and other aviation parts due to its excellent property of high strength and oxidation resistance [5]. They always have to be in touch with corrosive liquids during manufacture and operation [6]. Once the corrosion happened on the material surface, the surface properties of materials will deteriorate rapidly. Therefore, the corrosion resistance of Ni-based superalloy is very important for its performance.

In this work, we studied the electrochemistry corrosion behavior of Ni-based superalloy before and after IPIB irradiation and explored the mechanism of electrochemistry corrosion process. The result will be helpful for the application of Ni-based superalloys.

2. Experiments

2.1 Materials

The compositions of Ni-based superalloy used in this study are shown in Table 1. The dimension of samples is 10×10×3 mm. Before irradiation, the samples were polished by 3000# silicon carbide sandpaper using ultrasonic cleaning in alcohol.

Table 1. Composition of Ni-based superalloy samples (in wt.%).

Cr	Fe	Mo	Mn	Al	Ti	S	C	Ni
17.0~20.0	≤4.0	4.0~5.0	≤0.5	1.0~1.5	2.2~2.8	≤0.01	≤0.08	based

2.2 Irradiation parameters

Irradiation experiments were performed on TEMP-4M accelerator at Tomsk polytechnic university. The parameters used in this work are as follows: ion beam composition of 70% C^{n+} and 30% H^+, accelerating voltage of 250kV, energy density of 2 J/cm^2 and pulse duration of 60 ns. The morphology of the samples was studied in Nova NanoSem 430 scanning electron microscope and the analysis of the element was carried out by the affiliated EDS accessory.

2.3 Electrochemistry test

Irradiation experiments were performed on TEMP-4M accelerator at Tomsk polytechnic university. The parameters used in this work are as follows: ion beam composition of 70% Cn+ and 30% H^+, accelerating voltage of 250kV, energy density of 2 J/cm^2 and pulse duration of 60 ns. The morphology of the samples was studied in Nova NanoSem 430 scanning electron microscope and the analysis of the element was carried out by the affiliated energy dispersive spectroscopy (EDS) accessory.

3. Results

3.1 Surface morphology after IPIB irradiation

Fig.1 displayed the change of sample's surface morphology after IPIB irradiation. As shown in Figure 1. (a), there are wispy scratches left on the original sample after polishing. After IPIB irradiation, these scratches began to smooth and disappear owing to the effect of melting and flowing of the surface material. In the meanwhile, craters appeared (Figure 1. b). The result of EDS (Table.2) revealed that

Mg, S, C and other impurity elements gathered on the center of crater. It has been proved that the eruption of alloy components with a low melting point can generate craters on material surface [7]. Cracks were also found in the center of craters (Figure 1. c), which were the result from the thermal stress induced by energy deposition and shrinkage during surface material solidification.

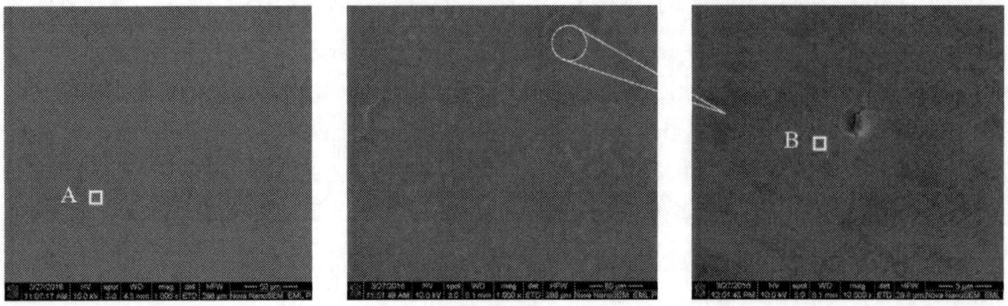

Figure 1. Surface morphology of samples: original sample (a); 2.0J/cm^2, 1 shot (b); enlarged image of crater (c)

Table. 2 Compositions of region marked in Fig. 1 (*Atom %).*

Region	C	O	Ti	Cr	Ni	Al	Mo	Mg	S
A	5.99	2.39	2.50	18.81	67.27	1.00	2.03	——	——
B	22.24	30.67	3.10	4.82	11.65	——	——	20.85	6.66

3.2 Electrochemistry test result

Figure 2 shows the polarization curves of samples in 3.5 wt. % NaCl solutions. The shapes of curves compared, the similar variation tendency revealed that the mainly eroded components between the original and irradiated samples were similar. The test results were given out Table. 3 and the corrosion current density was conducted by Tafel extrapolation. The increase of corrosion potential and decrease of corrosion current density illustrated that the corrosion resistance of Ni-based superalloy was improved after IPIB irradiation. The improvement of corrosion resistance can be caused by the grain refinement induced by IPIB irradiation.

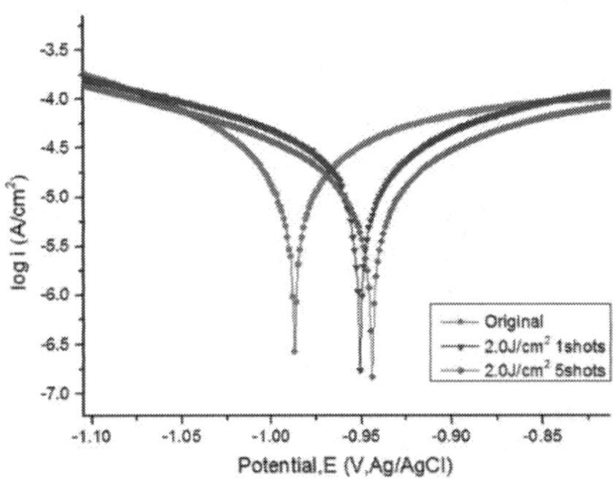

Figure 2. The polarization curves of original, 1shot and 5shots with 2.0J/cm^2 samples.

3.3 Morphology of samples after corrosion

Fig. 3 shows the morphology of the original sample after 5 min corrosion. Large cracks and corrosion damage appeared and the granular corrosion products were found. Compared with the original sample, the corrosion behavior of the irradiated sample was non-uniform (Fig. 4). Clearly, craters would be prior to be eroded than other regions and a deeply corrosion hole appeared in the center of the crater (Fig. 5). This phenomenon can be specified by three major factors. Firstly, the crack in the center of the crater will accelerate the corrosion process. The specific surface area of the micro-region increased due to the cracks. Moreover, narrow space inside cracks were insufficient for the adequate exchange of ions in the solution. Secondly, impurity elements Mg, S, C which were concentrated on the center of craters were more active than other alloy elements. So, they would precede to be oxidized and dissolved into solution. Thirdly, the curvature of the edge of the crater was larger than other regions, it meant that charges in the solution would be relative gathered to promote the corrosion process.

Table. 3 Comparison of electrochemistry corrosion parameters of samples.

Sample	Corrosion Potential（V）	The Logarithmic of Corrosion Current Density (A/cm^2)
original	-0.987	-4.547
2.0J/cm^2, 1shot	-0.951	-4.585
2.0J/cm^2, 5shot	-0.944	-4.652

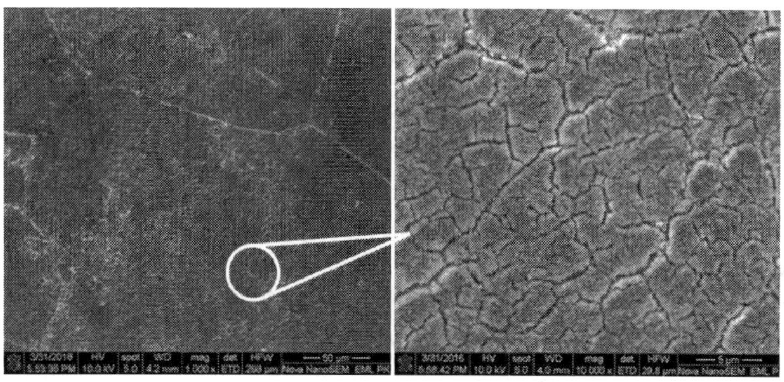

Figure 3. The morphology of the original sample after 5 min electrochemistry corrosion. (The right one is the enlarged image on the left).

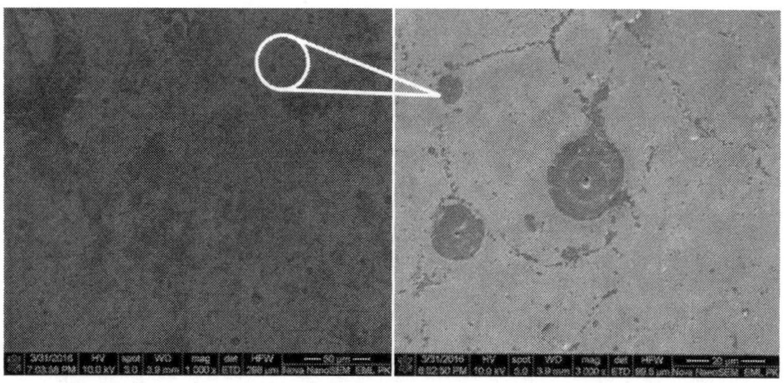

Figure 4. The morphology of the sample irradiated by 2J/cm^2, 1shot and 5 min electrochemistry corrosion. (The right one is the enlarged image of the left)

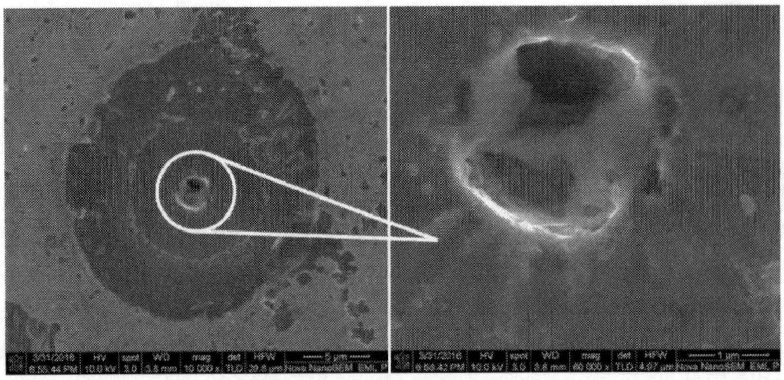

Figure 5. Enlarged images of craters after corrosion. (The right one is the enlarged image on the left)

4. Conclusion

Under the irradiation of IPIB, craters appear on the surface of the Ni-based superalloy. Through the analysis of electrochemistry corrosion process, it is found that the corrosion resistance in 3.5 wt. % NaCl solution of Ni-based superalloy will be improved after IPIB irradiation. In the meanwhile, the corrosion behavior becomes non-uniform compared with the original sample due to that fact the craters will be the first regions to be eroded.

Acknowledgements

This research is supported by National Natural Science Foundation of China No. 11175012 and National Magnetic Confinement Fusion Program (Grant No. 2013GB109004).

We sincerely express the gratitude to the support from Mr. Pavlov, Prof. Remnev and Lab No.1, High Technology Physics Institute, Tomsk Polytechnic University, Russia.

References

[1] Remnev G E *et al* 1999 *Surface & Coatings Technology* **114** 206
[2] Rej D J *et al* 1997. *J Vac Sci Technol* A **15** 1089
[3] Zhang X D *et al* 2011 *Surface & Coatings Technology* **206** 295
[4] Ma X et al 2014 *Applied Surface Science* **311** 567
[5] Bradley E F 1988 *Asm International Metals Park Oh* 2008
[6] Rongvaux J M 1995 *Heat treatment process for a NI-based superalloy* (US, US 5417782 A)
[7] Zhu X P *et al* 2003 *Surface & Coatings Technology* **173** 105
[8] Sun W F *et al* 2006 *High Power Laser & Particle Beams* **18** 2082

Effect of structural steel ion plasma nitriding on material durability in pulsed high magnetic fields

A V Spirin, V I Krutikov, D S Koleukh, A S Mamaev, S N Paranin, N V Gavrilov and A S Kaigorodov

Institute of Electrophysics Ural Branch of the Russian Academy of Sciences, Yekaterinburg, Russia

E-mail: spirin@iep.uran.ru

Abstract. The work was aimed to study the influence of plasma nitriding on electrical and mechanical properties of structural steels and their durability in pulsed high magnetic field. The plates and cylindrical magnetic flux concentrators were made of several steel grades (30KhGS, 40Kh, 50KhGA, 38Kh2MYuA, and U8A), heat-treated, and subjected to the low-temperature (400, 500°C) plasma nitriding. Electrical and mechanical properties of materials, phase composition of steel surface layer, microstructure and microhardness profiles were investigated on the plates before and after plasma treatment. Microstructure and microhardness profiles across the subsurface layer of plasma treated and untreated concentrators applied for high magnetic field generation were also studied. Magnetic field of 50 T under tens of microseconds in duration inside the flux concentrators was generated by long-life outer coil.

1. Introduction

As is known, tool coils (inductors) for magnetic pulse welding (MPW) of hard metals have poor durability under the application of magnetic fields higher than 30 – 40 T. The features of pulsed high magnetic field generation and MPW process are well described in the literature [1-3]. During MPW metallic workpieces are placed inside an inductor and rapidly deformed under the action of electromagnetic Lorentz force caused by the magnetic field and eddy current induced in the workpiece surface. At magnetic field of 40 T the acting pressure is 0.6 GPa.

MPW mostly employs singe-turn coils due to their easy manufacturing and high strength. Such a coil corresponds to a massive conductor having a cylindrical hole and through-the-wall slit as shown in Figure 1a. It allows flowing circular current around the hole and generating an axial magnetic flux. Figure 1b demonstrates another design that represents a pulse transformer modification. It contains a primary massive coil and a secondary coil usually called as a magnetic flux concentrator or a field shaper. This design is more preferable in practice because a damaged secondary coil can be quickly replaced; nevertheless this construction has lower efficiency of magnetic energy delivery.

When a pulsed current flows through an inductor (e.g. at a capacitor bank discharge), it tends to concentrate inside a thin surface layer of the inductor (skin-effect). In real conductors having finite electrical conductivity some diffusion of magnetic flux induced by pulsed current into conductor interior takes place. To restrict magnetic field diffusion the current pulse duration (half-period) must be as short as tens of microseconds. In such a fast rising magnetic field inductor material due to the current concentration inside the skin-layer suffers high thermo-mechanical stresses induced by electromagnetic force and Joule heating. These are the main factors which cause inductor damage.

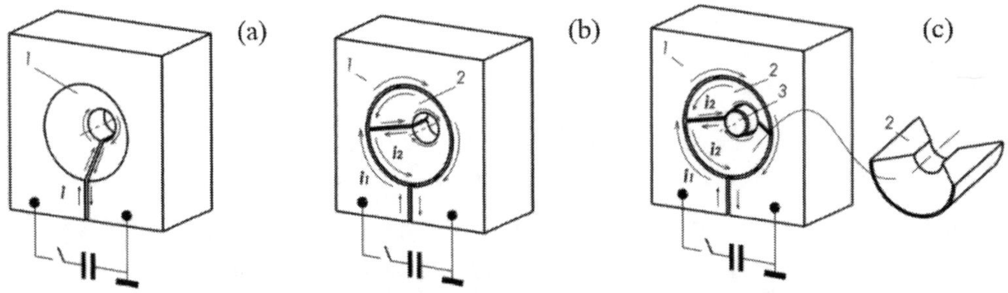

Figure 1. Diagram of a single-turn inductor (a) and inductors with single (b) and duplex (c) flux concentrator: 1 – primary coil, 2 – secondary coil (magnetic flux concentrator), 3 – rod.

In the literature one can find papers devoted to the theoretical analysis of thermo-mechanical stresses in thick coils, e.g. [4,5]. It was shown that the thermo-mechanical stresses in fast rising magnetic fields usually exceed the magnetic field pressure applied. Some solutions of magnetic field diffusion equations indicate at least two ways to reduce the Joule heating of the conductor surface [6,7]. The first way is to change the waveform of magnetic field pulse. The current pulse waveforms [sin, (1-cos)] which are the most common in pulse power technology provide higher surface energy release than linearly increasing magnetic field. However, this waveform is not applicable for MPW. The second way is to use a conductor whose conductivity increases with depth allowing to reduce Joule heating at the surface by the factor of 2 – 4 compared to uniform conductivity case [6,7].

Thus, modifying the inductor current-carrying layer, changing its conductivity and strength, may be the useful way to improve inductor performance. Nowadays structural steels are alternative inductor materials instead of strong copper-based alloys due to lower cost, higher strength, and applicability to modification by, e.g. plasma nitriding. This process for steel modification is well known in the literature [8,9], but specific application of steel for inductor production leaves open the questions on electrical properties and inductor durability.

The work is aimed to study the influence of plasma nitriding on electrical and mechanical properties of structural steels and their durability in pulsed high magnetic field of 50 T in amplitude and about 6 μs in rising time. In the present work we used the coil assemblage with a duplex concentrator (figure 1c) considering its easier nitriding and characterization of concentrator inner surface after each magnetic field pulse.

2. Experimental

2.1. Materials and sample preparation

Several domestic structural steel grades were chosen for study according to their appropriate thermal and mechanical properties. Nomenclature and chemical composition of steels are listed in Table 1. They include 30KhGS, 40Kh, 50KhGA, 38Kh2MYuA with medium carbon content of 0.3 – 0.5% and chromium content of 0.8 – 1.7%. A high carbon instrumental steel U8A was also used.

Three types of samples were prepared. First, thick enough plates used for the nitriding study, with dimensions of 10x20x2 mm. Second, thin plates used for the electrical conductivity measurements, with dimensions of (6 – 8)x(25 – 30)x(0.35 – 0.50) mm. Third, cylindrical flux concentrators with cone ends (field shapers) having 30 and 10 mm outer and inner diameters, 30 mm length, and inner channel length of 10 mm were used for generating the high magnetic field. The samples were studied after quenching with tempering and after nitriding the quenched samples. The plates and concentrators quenching was performed in the temperature range of 850 – 900°C for medium carbon steels and at 780°C for U8A. The tempering was performed in air at 400 and 200°C during 2 hours for untreated reference and to be nitrided samples. Some thin plates were subjected to isothermal annealing in vacuum at 770°C for U8A and 850°C for the tests to characterize the equilibrium conductivity.

Ion plasma treatment was performed using an electron source with a plasma cathode based on low-pressure glow discharge described in detail elsewhere [9]. Electrons emitted by the plasma cathode were accelerated in a double layer of a space charge formed in meshes of the plasma cathode grid. Initial energy of fast electrons was determined by the voltage of 150 V applied between the grid and plasma chamber. Gas ionization by the electron beam provided plasma generation. The working chamber was evacuated by a turbomolecular pump to the residual pressure of 0.002 Pa. The argon-nitrogen mixture with the Ar/N_2 ratio of 0.8 and the total pressure of 3 Pa was used. The gas ions accelerated by bias voltage (200 V) applied to the specimen provided a sample surface bombardment and heating as well. The measured ion beam current density was 2 – 3 mA/cm^2. The temperature of the specimens was controlled by a thermocouple and was fixed at 400 or 500°C by adjusting the beam current in the range of 2.0 – 3.3 A. Figure 2 shows arrays of the plates and the concentrators for nitriding. The duration of nitriding process was 4 hours both for the thick plates and the concentrators. The samples for resistivity measurements were subjected to a double-sided nitriding at 500°C during 5 h.

Table 1. Data sheet chemical composition.

Grades	Content, wt. %								
	C	Cr	Mn	Si	Al	Mo	Cu	Ni	Fe
30KhGS	0.31	1.01	1.01	1.14		0.01	0.23	0.08	bal.
40Kh	0.38	0.82	0.57	0.26		0.02	0.22	0.11	bal.
50KhGA	0.49	1.06	0.93	0.23			0.16	0.13	bal.
38Kh2MYuA	0.38	1.69	0.48	0.22	0.86	0.19	0.12	0.22	bal.
U8A	0.80	0.27	0.21	0.22			0.21	0.11	bal.

Table note: *in steel designation: Kh – chromium, G – manganese, S – silicon, M – molybdenum, Yu – aluminum.*

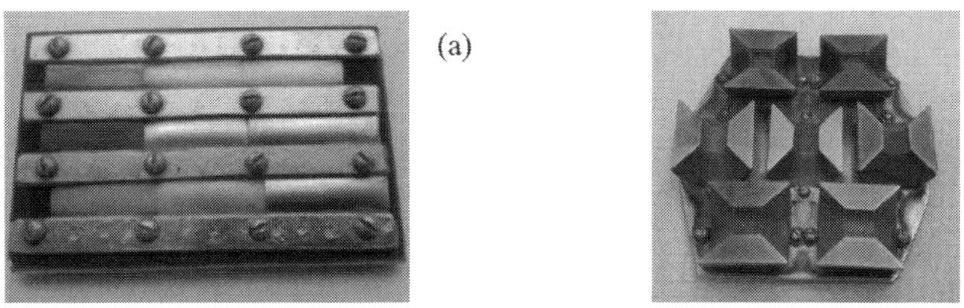

(a) (b)

Figure 2. View of arrays of the plates (a) and flux concentrators (b) for nitriding.

2.2. *Electrical conductivity measurement*

The resistivity of the thin plates was measured by 4-probe method using digital DC milliohm meter GOM-802 (Good Will Instrument) at ambient temperature of 25°C. Special brass contactors with fixed distance between the potential probes were applied to measure the resistance of the narrow thin plates. The long edges of nitrided samples were cut after treatment to remove untreated regions. Curvature of the nitrided samples was taken into account at resistivity evaluation. The accuracy of resistance measuring (9 times per every sample) at 1 A test current was 0.5%. The accuracy of sample geometry and potential probes distance measurements was accordingly 0.2 and 0.3%. The summary error for resistivity was established at 0.8%. The resistivity was measured on three types of thin plates: after annealing, after quenching, and after nitriding.

2.3. *Magnetic flux concentrators (field shapers) testing*

The experiment aimed to investigate the effect of fast-rising high magnetic field on steel properties was carried out using a pulsed current generator based on a capacitor storage (C = 435 μF, U_{max} = 25 kV). The generator short circuit current peaks at 1.5 MA with 5 – 6 μs rising time. In our experiments the generator was loaded to a single-turn inductor with magnetic flux concentrator as shown on figure 3 (on the photo the concentrator and rod are slightly extracted). We used the duplex concentrator considering its easier nitriding and characterization of inner surface after each current pulse. Both parts of duplex concentrators were well insulated from each other and from the primary coil using the Kapton® tape (0.05 mm thick) having high electrical and thermal strength. A tempered beryllium bronze rod of 9 mm in diameter was inserted inside the channel to increase magnetic flux density. The self-inductance of the coil assemblage was about 16 nH.

To establish the magnetic field of 50 T in amplitude a circular gap arranged pick-up coil was used. The magnetic field in the space between the concentrator and the rod in their middle plane was measured at different rod diameters. Figure 4a demonstrates the peak field to peak current ratio (inductor geometrical factor) depending on the rod diameter. It is seen that different current load results in the geometrical factor variation due to Joule heating. Using the dependence at the highest current, the geometrical factor of 45 T/MA was determined for working rod diameter. This factor value was used to calculate the working current to be 1050 kA.

The figure 4b demonstrates the time dependence of discharge current measured by a Rogovsky coil and magnetic field measured by the pick-up coil with the rod of 8.5 mm in diameter at 13 kV generator charge voltage. The magnetic field rising time and half period were 5.5 and 12.5 µs, respectively. After the testing at different pulse numbers: zero pulses (initial state), 5 pulses, and after damage, the parts of concentrators were cut across the middle plane and polished for examination.

Figure 3. Coil assemblage: 1 – primary coil, 2 – magnetic flux concentrator, 3 – rod.

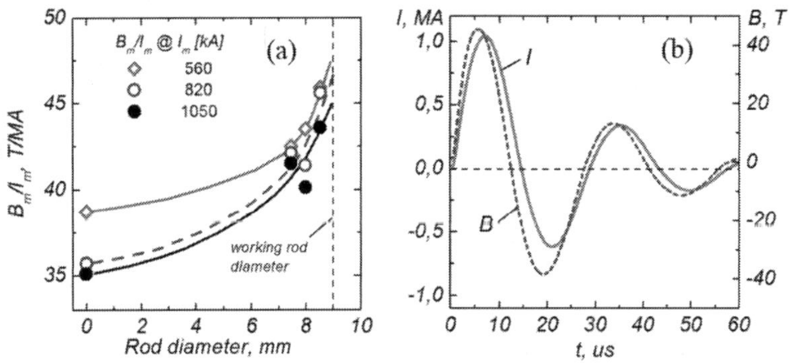

Figure 4. Inductor geometrical factor vs. rod diameter at different current load (a) and time dependence of current and magnetic field at U_o = 13 kV (C = 435 µF).

2.4. Micromechanical testing

Micromechanical testing was performed using Nanotest 600 instrument with a diamond indentor. The profiles of Vickers microhardness Hv were measured both on cross-sectioned polished flat plates, through-the-thickness, and on flux concentrators towards the radial direction. Indentor load of 100 mN was used to provide the spatial resolution of tens of micrometers.

2.5. Structural and microstructural investigation

Phase composition of nitrided thick plates surface was studied using X-ray diffractometer D8 Discover (Bruker AXS) with Cu K$\alpha_{1,2}$ radiation and graphite monochromator. A Rietveld program TOPAS 3 was used to analyze the X-ray patterns. Polished and chemically etched (nitric acid alcohol solution) cross sections of the plates and the concentrators were analyzed by Olympus BX41RF metallograph.

3. Results and Discussion

3.1. Micromechanical and structural properties of the flat samples

The microhardness profiles measured on nitrided thick plates are presented in figure 5. The modified layer of 70 – 100 µm and 150 – 200 µm thick was obtained after the nitriding at 400 and 500°C respectively for all the steels except U8A.The steels with 0.8 – 1%Cr have similar Hv profiles. The value of Hv at the surface was 6.8 – 7.4 GPa and 8.6 – 8.9 GPa after nitriding at 400 and 500°C, respectively. The high surface saturation with nitrogen was inherent to the higher chromium steel 38Kh2MYuA (1.7% Cr) that resulted in significant increase in surface microhardness that was 11.3 and 13.5 GPa after nitriding at 400 and 500°C, respectively. Weak nitriding effect was obtained on high carbon steel U8A with its near-surface Hv value only 30% higher than that in the bulk. Hatched rectangles on both graphs represent the range of Hv for all the steels before nitriding. One can conclude that nitriding process temperature resulted in decreasing the bulk strength of steels due to the tempering process. Thus, the temperature of 500°C was suggested to be the maximal (optimal) to produce flux concentrators with modified layer of 150 μm thick in the average and strength enough bulk material that is important for magnetic pulse tool.

X-ray analysis revealed that the main surface phase at 400°C nitriding temperature was α-Fe (cubic, S.G.: Im-3m) with the highest content of 86 – 96 wt.% for the steels 30KhGS, 40Kh, and 50KhGA and the Fe_2N phase in the rest. In contrast to them the 38Kh2MYuA steel at this nitriding temperature contained the Fe_2N as high as 62 wt.% and 13 wt.% of $FeN_{0.88}$ (tetragonal, S.G.: I4/mmm). X-ray data for 500°C nitriding temperature are listed in table 2.

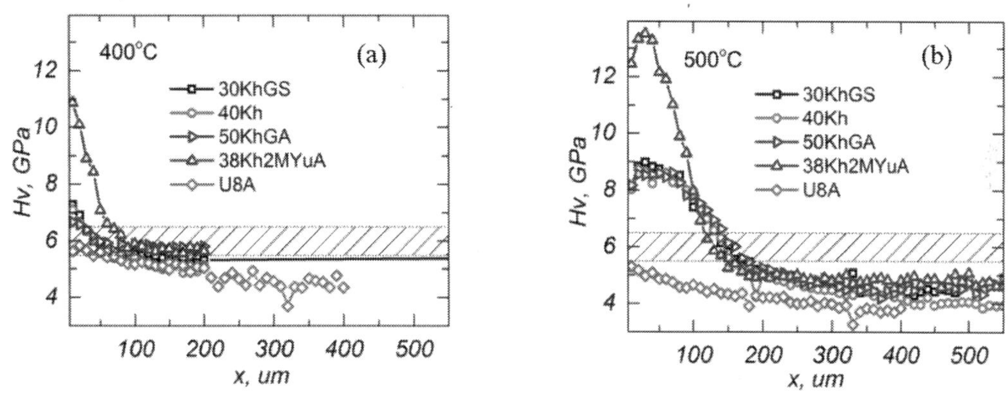

Figure 5. Vickers microhardness profiles on cross-sectioned plates after nitriding at 400°C (a) and 500°C (b). Indentor load was 100 mN.

Table 2. Steel surface phase composition after nitriding at 500°C.

Grades	Content, wt.%		
	Fe_2N (orthorhombic, S.G.: Pbcn)	γ-Fe_4N (cubic, S.G.: Pm-3m	ε-Fe_3N_{1+x} (hexagonal, S.G.: P6322)
30KhGS	96	4	-
40Kh	54	11	35
50KhGA	83	17	-
38Kh2MYuA	44	26	30
U8A	49	51	-

3.2. Electrical properties of steel samples

Table 3 summarizes the resistivity of steels measured on thin plates for three states: after annealing, after quenching with subsequent tempering at the temperature of 200°C/2 h, and after plasma nitriding the quenched and tempered samples. Copper grade Cu-DHP was as a reference material: the measured resistivity, 2.05 ± 0.05 μOhm·cm, was in good agreement with Cu-DHP data sheet.

With the reference to Table 1 it can be concluded that the resistivity of the steels after annealing and quenching is greatly influenced by the chromium content. However, after the quenching the resistivity increases accordingly to carbon content as wel,l which is clear from the ratio ρ_q/ρ_a. It is probably associated with the recrystallization of ferric carbide precipitates of small size on the grain boundaries during the samples tempering. An unexpected result was obtained on nitrided samples for which an enhanced conductivity was observed. It becomes clear from the ratio ρ_n/ρ_a. The conductivity was occurred to be up to 27% higher than that of the equilibrium one. In general, the effect was as high as the chromium content was low except for 38Kh2MYuA. This increase in conductivity is difficult to attribute to any certain phase in the "iron-nitrogen" system since the nitrided plates have gradient structure according to the X-ray data: thin surface regions with iron nitride phases and a bulk region which is a solid solution of nitrogen in iron base. The chromium and carbon binding by the nitrogen is also a possible reason of the conductivity increase. To explore this effect a low carbon (0.06%) and chromium (<0.1%) steel must be examined.

Table 3. Resistivity of the steels before and after plasma treatment.

Grades	Resistivity@25°C, 10^{-8} Ohm·m			ρ_q / ρ_a	ρ_n / ρ_a
	ρ_a (annealed)	ρ_q (quenched)	ρ_n (nitrided)		
30KhGS	35.6±0.6	42.5±0.3	31.4±0.3	1.19	0.88
40Kh	24.5±0.3	31.6±0.4	20.9±0.4	1.29	0.85
50KhGA	24.5±0.2	34.1±0.3	21.5±0.2	1.39	0.88
38Kh2MYuA	36.7±0.6	44.3±0.3	28.8±0.3	1.21	0.78
U8A	20.4±0.2	27.3±0.3	20.1±0.3	1.34	0.98

Figure 6a shows a view of the nitrided plates (from the left: 30KhGS, 40Kh, 50KhGA, 38Kh2MYuA, and U8A). One can see that the samples bending of different degree takes place after the nitriding. At that, correlation between the curvature and the chromium content can also be found. The higher chromium content resulted in stronger bending in contrast to U8A plate which was almost flat. The degree of nitrogen saturation through the plate thickness was indirectly confirmed by the Hv profile. The profiles for the steels 50KhGA, 40Kh, and 38Kh2MYuA are presented in figure 6b.

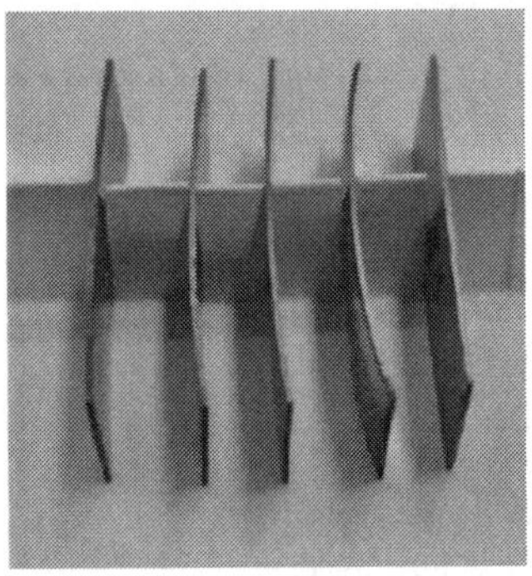
(a)

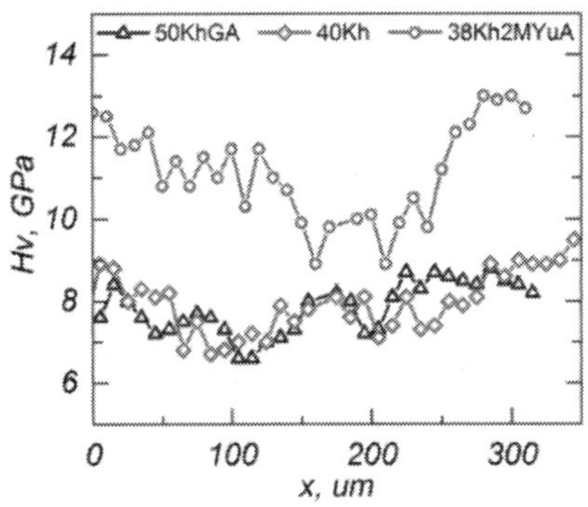

(b)

Figure 6. View of thin plates for resistivity measurements (a) and Hv profiles through-the-thick (b).

3.3. Micromechanical and microstructural properties of flux concentrators of 30KhGS steel before and after magnetic field generation

The preliminary results of 30KhGS steel flux concentrators durability testing and investigation of micromechanical and microstructural properties evolution are presented here. Plasma treated and untreated concentrators without magnetic field treatment were taken as reference samples.

Figure 7a represents a cross midsection photo of the nitrided (1N.1) and untreated (1.2) concentrators parts after damage, i.e. after the formation of visual longitudinal radial cracks. The figure 7b shows the microhardness radial profiles measured near the inner channel surface for three different untreated half concentrators: "initial" – without magnetic field generation, after 5 pulses of high magnetic field and after damage (47 pulses). The figure 7c shows the same for three nitrided half concentrators. It is seen that initially the untreated sample (figure 7b) has practically uniform Hv distribution. After the following pulses the surface strength gradually decreases due to the tempering effect with the bulk strength maintenance. On plasma treated sample an expected Hv radial distribution in the initial state is observed. The behavior of Hv(r) after 5 and 15 pulses is practically the same. Despite of lower microhardness, which was detected for the sample after 5 pulses, the modified layer is characterized by a high heat-resistance.

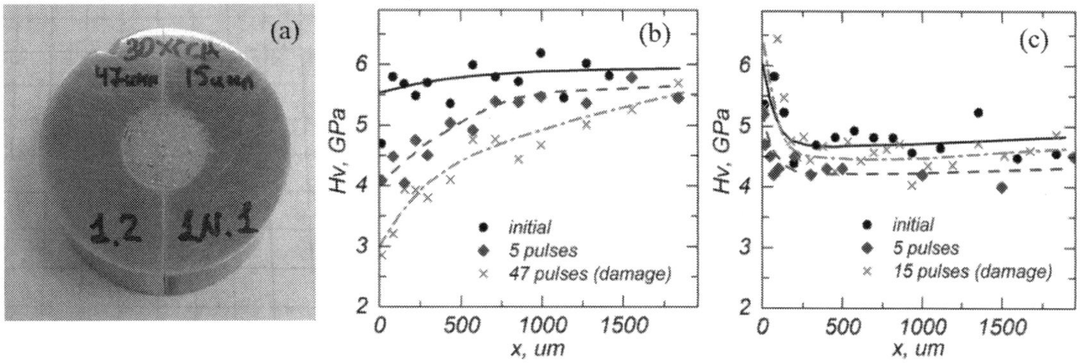

Figure 7. View of the cross midsections of nitrided and untreated concentrators (a) and Hv radial profiles of untreated (b) and nitrided (c) concentrators.

It was observed that the formation of a hard nitrided surface layer, which is brittle enough, results in rapid crack growth. The crack formation mechanism is the same as described in [2]. Due to the high strength of nitrided layer the cracks with sharp edges are formed during the skin-layer cooling under the action of tensile stresses. Further increase in crack depth is enhanced by the so-called "saw effect" and accompanied by plasma sparks ejection. In untreated concentrators longitudinal corrugations at inner channel surface was formed due to the plastic deformation, especially at the channel edges. It was directly confirmed by the surface viewing after several pulses. Finally, the number of ultimate pulses for plasma treated 30KhGS steel concentrator was 15 in contrast to untreated one which was tested 47 times and the crack nuclei formation occurred only at channel edges.

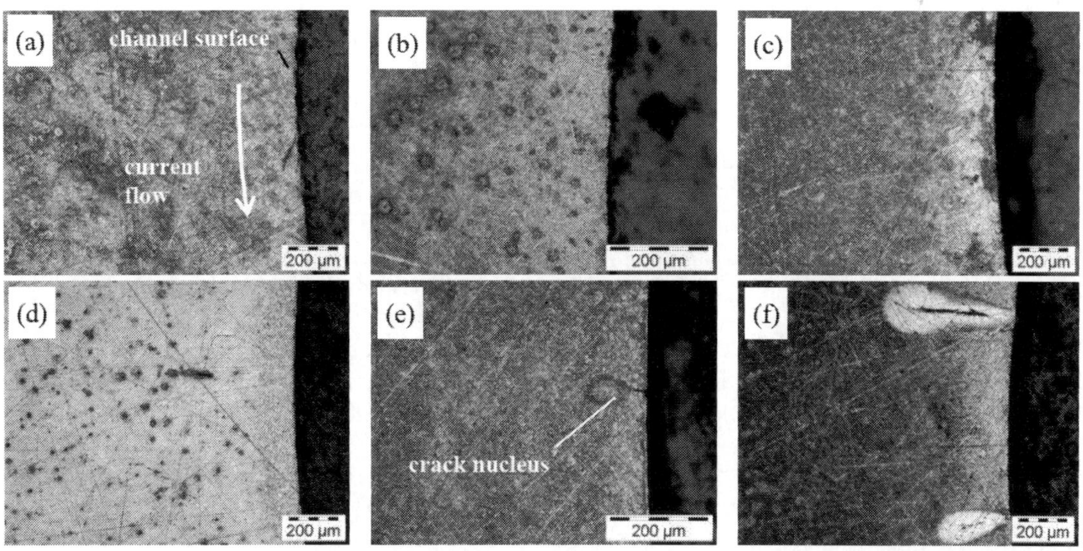

Figure 8. Microimages of inner channel region of untreated (a – c) and nitrated (d – f) concentrators at different numbers of 50 T magnetic field pulses: a, d – zero, b, e – 5, c – 15, f – 47.

Figure 8 represents the microimages of cross midsections of the plasma treated and untreated concentrators close to the channel surface. The steel structure after the quenching is determined to be mainly acicular martensite. Under the magnetic field treatment, pulse by pulse, the martensite phase undergoes a transformation at a surface layer of 150 – 200 μm thick (heat-affected area). In plasma treated samples the modified layer of about 170 μm thick is well defined and characterized by grains refinement due to the precipitation of nitride phases and iron base saturation with nitrogen. No significant microstructure transformation was detected for these samples under the magnetic field treatment. Only the cracks nuclei and cracks enhanced through the melting and spark plasma ejection were observed after 5 and 15 pulses, respectively.

4. Conclusion

The influence of low-temperature (400, 500°C) plasma nitriding on electrical and mechanical properties of structural steels, including 30KhGS, 50KhGA, 40Kh, 38Kh2MYuA, and U8A, has been investigated. The examination of the samples treated in argon-nitrogen plasma revealed the common behavior of microhardness profiles. The modified layers with thicknesses of 70 – 100 μm and 150 – 200 μm have been obtained after nitriding at 400 and 500°C respectively for all the steels except the high carbon steel U8A. The annealed steels resistivity was determined in the range of 20 – 37 μOhm·cm depending on the steel grade. An increase in resistivity was established for all the steels after the quenching; at that, the higher carbon content resulted in higher resistivity increase by 1.19 – 1.34 times. An increase in conductivity was observed after the steels ion plasma nitriding. The conductivity was occurred to be up to 27% higher than that of the equilibrium one. The durability tests of 30KhGS magnetic flux concentrators in pulsed high magnetic field of 50 T in amplitude and 5.5 μs in rising time have been performed. The ultimate number of pulses for nitrided 30KhGS steel concentrator was 15 in contrast to untreated one that withstood 47 pulses without significant cracking. Micromechanical and microstructural properties evolution of plasma treated and untreated concentrators of 30KhGS have been studied as well. It was observed that the decrease of surface strength occurred from pulse to pulse in quenched samples. On the contrary, quenched steel surface modification by the nitriding resulted in stability of microstructure owing to higher heat-resistance. The formation of hard surface layer in nitrided concentrator which is brittle enough resulted in rapid crack growth. The further experiments on durability of 30KhGS steel concentrator having an enhanced surface conductivity and strength under the action of moderate of 30 T magnetic fields (a fatigue failures testing) will allow understanding the applicability of surface modification by plasma nitriding.

References

[1] Knoepfel H 1970 *Pulsed High Magnetic Fields* (Amsterdam, North-Holland)
[2] Shneerson G A, Dolotenko M I and Krivosheev S I 2014 *Strong and Superstrong Pulsed Magnetic Fields Generation* (Walter De Gruyter Incorporated)
[3] Kapil A, Sharma A 2015 *Journal of Cleaner Production* **100** 35
[4] Titkov V V 1991 *Jornal of Technical Physics* **61** (4) 54
[5] Karpova I M, Titkov V V 1994 *Jornal of Technical Physics* **64** (7) 137
[6] Shneerson G A 1992 *Technical Physics Letters* **18** (6) 18
[7] Chaikovsky S A, Oreshkin V I, Datsko I M et al 2014 *Physics of Plasmas* **21** 042706
[8] Rolin'ski E 2015 *Plasma-assisted nitriding and nitrocarburizing of steel and other ferrous alloys Thermochemical Surface Engineering of Steels* ed E J Mittemeijer and A J Marcel (Amsterdam: Woodhead Publishing) chapter 11 pp 413–457
[9] Gavrilov N V, Mamaev A S and Chukin A V 2016 *Technical Physics Letters* **42** (5) 491

Acknowledgments

The authors are thankful to Ph.D. A.I. Medvedev for X-ray analysis. The study was performed within the state assignment No. 0389-2014-0002, RFBR project (No. 16-08-00919-a), and Fundamental research program of UB of RAS (No. 15-17-2-27, No. 15-17-2-28).

IOP Conf. Series: Journal of Physics: Conf. Series **830** (2017) 012081 doi:10.1088/1742-6596/830/1/012081

Some properties of near-surface layer of graded-gap MBE HgCdTe after boron ion implantation

A V Voitsekhovskii[1], S N Nesmelov[1], S M Dzyadukh[1], I I Izhnin[1, 2]

[1]National Research Tomsk State University, Tomsk, Russia
[2]Scientific Research Company "Carat", Lviv, Ukraine

E-mail: bonespirit@sibmail.com

Abstract. The effect of ion implantation of boron ions with an energy of 100 keV and a dose of $(1\text{-}6)\times10^{15}$ cm^{-2} in the MBE HgCdTe films on the characteristics of the MIS structures based on these films was investigated. The changes of the conductivity type in the near-surface layer of HgCdTe after ion implantation of boron and etching by ions of argon were detected. The concentrations of the major charge carriers in the near-surface layer of the epitaxial films after ion implantation and after ion etching were close to 5.88×10^{16} cm^{-3} and 2.47×10^{17} cm^{-3}, respectively.

1. Objects of research

The narrow-gap HgCdTe semiconductor is widely used to create highly-sensitive infrared detectors [1]. Ion implantation method is often used when creating HgCdTe photodiodes. The admittance measurements for MIS structures allow to study the changes in the properties of the near-surface semiconductor layer after ion irradiation.

The structures based on n-$Hg_{0.775}Cd_{0.225}Te$ were grown by MBE on the GaAs (013) substrates in Rzhanov Institute of Semiconductor Physics SB RAS. Graded-gap layers with high content of CdTe were grown on both sides of the working layer. The CdTe composition distribution at thickness of the epitaxial film $Hg_{1-x}Cd_xTe$ measured by an automatic ellipsometer in the process of growing an epitaxial film is shown in figure 1. The hole concentration at 78 K was 8.6×10^{15} cm^{-3}. The hetero-epitaxial film was divided into three parts. The first part was used to make a control sample. The second part was subjected to ion etching during 5 minutes with ion energy of 0.5 keV at current density of 0.1 mA/cm^2. Implantation of boron was carried out in the third part with ion energy of 100 keV and dose of $(1\text{-}6)\times10^{15}$ cm^{-3}. MIS structure based on all films parts were created by deposition of SiO_2/Si_3N_4 insulator and indium electrodes. The inset of figure 1 schematically shows the sequence of the layers in the fabricated MIS structures.

The measurements were performed with an automated setup for spectroscopy of the nanoheterostructure admittance on the basis of non-optical cryostat Janis and immittance meter Agilent E4980A (temperature range of 8-475 K, the range of the measured impedance of 0 - 1 GOhm with the resolution on the capacitance and normalized on frequency conductance to 1 fF , the range of voltage bias of -40-+ 40 V, the frequency range of 20 Hz - 2 MHz). In measurements, the voltage change from negative values to positive ones was taken as the forward direction of sweep, and the voltage change from positive values to negative ones was taken as the reverse direction of sweep.

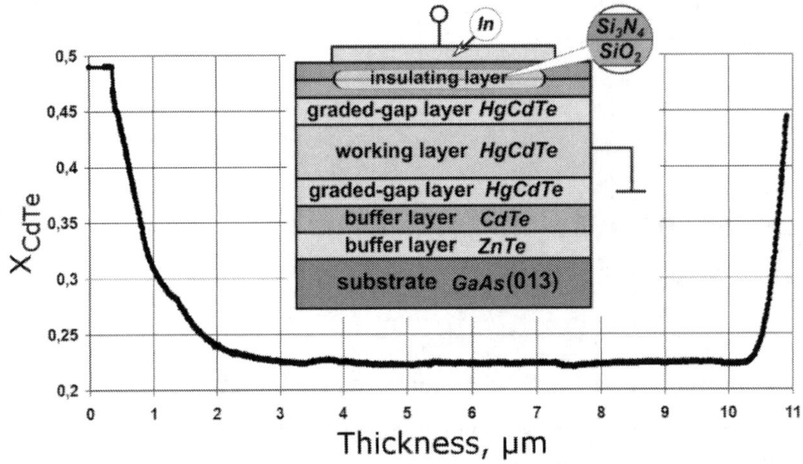

Figure 1. The CdTe composition distribution on thickness of the epitaxial film $Hg_{1-x}Cd_xTe$, measured by an automatic ellipsometer in the process of growing an epitaxial film. The inset in the figure shows the sequence of the layers in the fabricated MIS structures.

2. Results of the study

Figure 2 shows the capacitance-voltage (CV) characteristics of the MIS structure based on the initial film measured at a temperature of 77 K. Figure 2 shows that the CV characteristics correspond to the *p*-type semiconductor. The CV characteristics at a frequency of 1 MHz are almost a high-frequency behavior. For all measurements of the capacitance and conductance, the effect of the series resistance of the epitaxial film bulk was excluded [2]. For MIS structures based on *p*-HgCdTe series resistance of epitaxial film bulk is usually significant due to the low mobility of the majority charge carriers (holes). At a frequency of 10 kHz CV characteristics show the low-frequency behavior, the dip of the low-frequency CV characteristic has a large depth due to the presence of the graded-gap near-surface with a high composition of CdTe [3]. Figure 3 shows the CV characteristics of MIS structure based on the HgCdTe film after ion implantation of boron measured at 77 K at a forward voltage sweep.

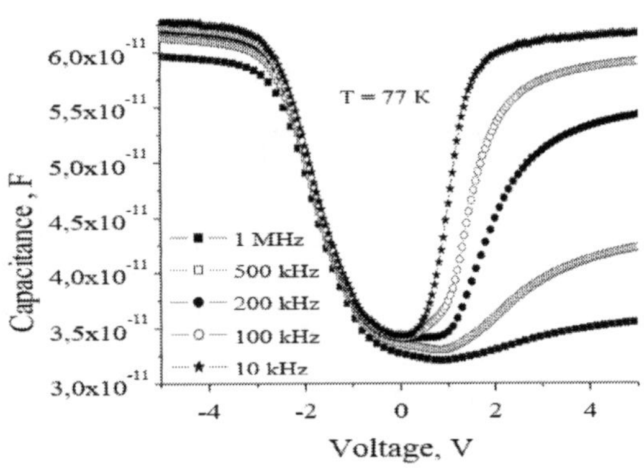

Figure 2. CV characteristics of MIS structure based on the initial HgCdTe film measured at a temperature of 77 K at different frequencies of the test signal.

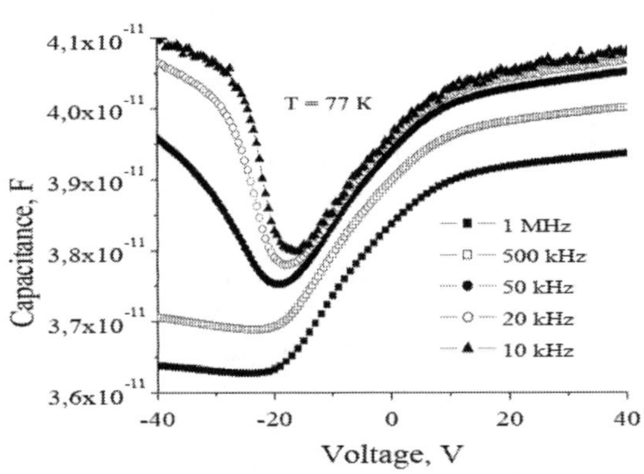

Figure 3. CV characteristics of MIS structure based on the HgCdTe film after ion implantation measured at a temperature of 10 K and 77 K at different frequencies of the test signal.

Figure 3 shows the change in the type of conductivity of the semiconductor film due to ion implantation of boron. It may be noted that the majority charge carrier concentration after ion implantation of boron has increased sharply leading to a low modulation of capacitance in the CV characteristic. Figure 4 shows the CV characteristics of MIS structure based on $Hg_{1-x}Cd_xTe$ films after etching by ions of argon measured at 77 K at different frequencies at the forward sweep of voltage. Figure 4 shows that the change of the conductivity type of epitaxial film also occurred in the case of ion etching.

Modulation of capacitance on the CV characteristics for the MIS structure based on the film after ion etching is more than for the structure based on the film after ion implantation. It indicates that the concentration of the majority charge carriers after ion etching has also increased, but not so significantly how after ion implantation of boron. The value of a slope angle for $1/C^2(V)$ dependency where C is MIS structure capacitance in the depletion mode at a frequency of 1 MHz (figure 5) was used to estimate the concentration of major charge carriers for the test samples at 77 K [4]. Defined by this method the concentration of majority carriers of charge in the near-surface layer was 2.57×10^{15} cm^{-3} for the structure based on the initial film. For the structure based on the film after ion etching the concentration of the majority charge carriers amounted to 5.88×10^{16} cm^{-3}, and for the structure after ion implantation of boron the concentration has reached the value 2.47×10^{17} cm^{-3}.

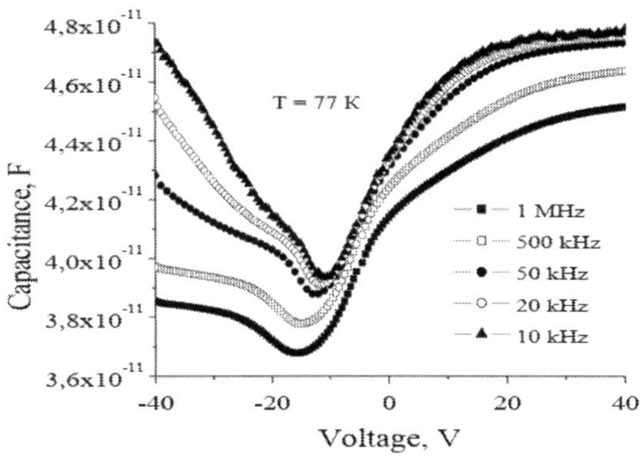

Figure 4. CV characteristics of MIS structure based on HgCdTe film after ion etching measured at a temperature of 77 K at different frequencies of the test signal.

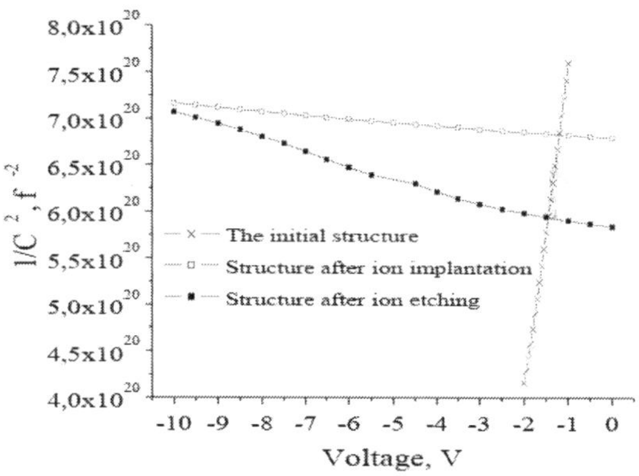

Figure 5. The slope of the $1/C^2$ in the depletion mode at a frequency of 1 MHz at temperature 77 K for all three structures.

Figure 6 shows the dependencies of the series resistance of epitaxial film bulk on reciprocal temperature at a frequency of 50 kHz for structures based on the initial film and also films after ion implantation and ion etching. Figure 6 indicates that the initial structure has the greatest value of series resistance of epitaxial film over the whole temperature range. A large value of epitaxial film resistance is typical for structures based on films with p-type conductivity. This occurs because of the low mobility of holes in the HgCdTe layer. The structure based on the film after ion implantation has the smallest value of series resistance of the epitaxial film. This is due to the fact that as a result of implantation and annealing there appears a high concentration of electrons in a sufficiently thick layer of the semiconductor film. Unlike the implantation - ion etching also is changing the conductivity of the film, but in this case, the electron concentration was not so high. It is also likely that the penetration of boron ions in the epitaxial film during ion etching $Hg_{1-x}Cd_xTe$ occurs to a lesser depth as compared with implantation.

Figure 7 shows the dependencies of the differential resistance of the space charge region (SCR) in strong inversion mode on the reciprocal temperature at a frequency of 50 kHz for structures based on the initial film and also films after ion implantation and ion etching. Figure 7 indicates that up to a temperature of about 25 K the structure based on the film after ion etching has a maximum value of the differential resistance of the SCR. However, the pace of the fall of the differential resistance at higher temperatures for a given structure results in the fact that after 30 K values of its value becomes smaller than for the other structures. The rate of the decline of the differential resistance with increasing temperature for the original structure and structure created on the film after ion implantation are similar up to a temperature of 70 K. In a whole temperature range the values of differential resistance of SCR for original structure do not exceed the values for the structure after ion implantation.

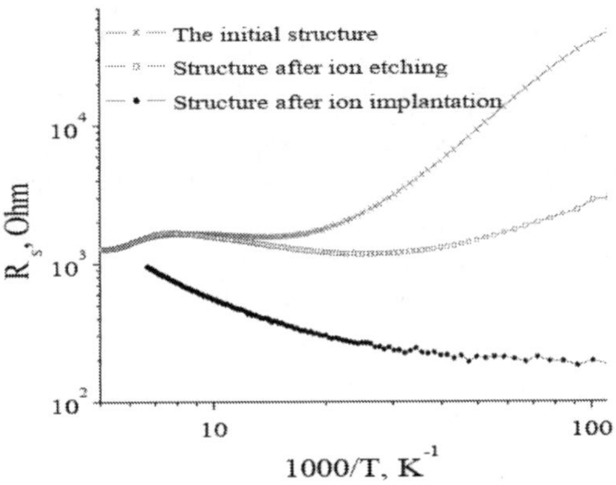

Figure 6. The dependencies of the series resistance of epitaxial film bulk on reciprocal temperature at a frequency of 50 kHz.

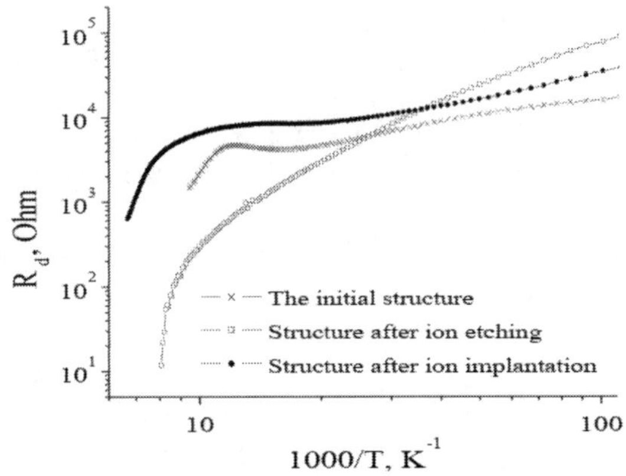

Figure 7. The dependencies of the differential resistance of the SCR in strong inversion mode on reciprocal temperature at a frequency of 50 kHz.

3. Conclusions

The conductivity type in the near-surface layer of HgCdTe changes after both radiation impacts. The electron concentration after implantation was much higher than after etching. The electron concentration determined from the CV characteristics was equal to 5.88×10^{16} cm^{-3} and 2.47×10^{17} cm^{-3} for the ion etching and ion implantation, respectively.

Ion implantation leads to a significant reduction in the resistance of epitaxial film bulk compared to ion etching under the same distance from the rear electrode to the indium field electrode. The question of the dependence of the differential resistance of the SCR on the temperature for different types of impact requires an additional study.

References

[1] Rogalsky A 2011 *Infrared detectors 2nd ed.* New-York CRC Press, Taylor and Francis Group 876
[2] Voitsekhovskii A, Nesmelov S, Dzyadukh S 2005 *Izvestiia VUZov. Fizika* **6** 31
[3] Voitsekhovskii A, Nesmelov S, Dzyadukh S 2012 *Thin Solid Films* **522** 261
[4] Gelder W, Nicollian E 1971 *Journal of Electrochemical Society* **118** 138

Acknowledgement

The study was performed at financial support by RFBR and the Administration of Tomsk region as part of a research project № 16-42-700759.

IOP Conf. Series: Journal of Physics: Conf. Series **830** (2017) 012082 doi:10.1088/1742-6596/830/1/012082

Influence of a high-frequency pulsed nanosecond diffusion discharge in the nitrogen atmosphere on the electrical characteristics of a CdHgTe epitaxial films

D Grigoryev[1,2], A Voitsekhovskii[1], A Korotaev[1], D Lyapunov[1], K Lozovoy[1], V Tarasenko[2], M Shulepov[2], M Erofeev[2], V Ripenko[3], S Dvoretskii[1,4], N Mikhailov[4]

[1] Tomsk State University, Russia, 634050, Tomsk, 36 Lenin Av.
[2] Institute of High Current Electronics, Russia, 634055, Tomsk, 2/3 Akademichesky
[3] Tomsk Polytechnic University, Russia, 634050, Tomsk, 30 Lenin Av.
[4] Semiconductor Physics SB RAS, Russia, 630090, Novosibirsk, 13 Lavrentyeva

denn.grig@mail.tsu.ru

Abstract. The effect of a high-frequency nanosecond volume discharge forming in an inhomogeneous electrical field at atmospheric pressure on the CdHgTe (CMT) epitaxial films is studied. The measurement of the electrophysical parameters of the CMT specimens upon irradiation shows that that the action of pulses of nanosecond volume discharge leads to changes in the electrophysical properties of CMT epitaxial films due to formation of a near-surface high-conductivity layer of the n-type conduction. The preliminary results show that it is possible to use such actions in the development of technologies for the controlled change of the properties of CMT narrow-band solid solutions and production of structures heterogeneous with respect to conduction.

1. Introduction

Gas discharges of various types are widely used for the modification of material properties. The authors have demonstrated the possibility of the formation of a volume discharge at atmospheric pressure. Such a volume discharge is something like a corona discharge. In both cases, the electrodes need special configuration allowing to create a strongly non-uniform electric field in the discharge gap. Therefore, at least one of the electrodes should have a small radius of curvature. The formation of a corona discharge occurs efficiently under reduced pressure, and to obtain the corona discharge under high pressure it is necessary to supply voltage of hundreds of kilovolts. At the same time discharge quickly passes into the spark stage, a breakdown and the evaporation of the electrode material occur.

In recent years, a wide interest is focused on the study of the properties of a volume diffuse discharge induced by a beam of runaway electron avalanches (REP DD) and its application for modification of the properties of various materials [1]. A feature of this discharge is that it is weakly dependent on the ambient pressure (can be quite easily formed at high pressure). It is not necessary for its formation to collect the ions and electrons in the discharge gap, as is the case of the corona discharge. Also there is no need for pre-ionization of the gap by any third-party source. Under the influence of such a discharge it is possible to realize high power density of energy deposition (up to 800 MW/cm^3). At the same time the runaway electron beam with a current amplitude behind the foil of tens to hundreds of amperes is generated from the discharge plasma. The beam current pulse duration at half-height is not higher than 100 ps. In addition, a feature of such discharges is the combined effect of plasma of a dense nanosecond discharge with power density of the energy deposition of hundreds of megawatts per cubic centimeter, ultrashort electron beam with a broad energy spectrum, and the optical radiation of the plasma discharge in different spectral ranges. REP DD is a pulse discharge due to the nature of its formation. In this case most of the time of the existence of discharge, an external voltage, applied to the discharge gap, is small (less than 10 kW) or does not exist. That is, the discharge exists until the end of the recombination of ions. The value of the accelerating voltage in the gap is insufficient for effective ion implantation. So, during the existence of the accelerating voltage only those ions formed by the flown wave of runaway electrons are introduced, which were directly near the surface. Therefore, the impact of REP DD modifies only the surface layers of the material samples, not deeper than 10-200 nm, depending on the material.

Ternary semiconductor compound mercury cadmium telluride ($Cd_xHg_{1-x}Te$, CdHgTe, CMT, x – material composition) is currently one of the basic materials for the creation of highly efficient photodetectors operating in the infrared region [2]. For temperature imaging the most commonly used region of wavelengths is 8-14 μm, which corresponds to the $Cd_xHg_{1-x}Te$ composition with x = 0.22. Despite the great advances made in the technology of photodetectors based on CMT, the search for new and effective methods for the controlled management of the properties of the original material is still an urgent task. The first experimental data on the effect of the REP DD on properties of epitaxial CMT are presented in [3-5]. The frequency of exposure pulses of volume discharge was 1 Hz (low frequency pulsed volume discharge). This paper presents the first experimental results of the effect of the high frequency volume nanosecond discharge in a nitrogen atmosphere on the electrical properties of epitaxial CdHgTe material. Selection of gas discharge media is determined by the necessity to compare the obtained results with the results of the impact of low-frequency pulsed volume discharge in air at atmospheric pressure.

2. Samples and experimental techniques

Two series of samples of CMT epitaxial films of p- and n-type grown by molecular beam epitaxy at the Institute of Semiconductor Physics SB RAS in Novosibirsk were prepared for conducting research. The material composition of the working layer of epitaxial films was x = 0.22.

The experiments were performed using a specially designed setups and discharge chamber. The design of the diffuse plasma generator that allows us to modify the surface of the flat circular samples with a diameter up to 10 mm is shown in figure 1. Point cathode 1 with a small radius of curvature, made of tool steel, was pressed into caprolon insulator 2. To control the operation of the generator quartz windows 4 with a diameter of 20 mm were made in a metal case 3. The voltage across the discharge gap was measured by capacitive divider 5 and the current of discharge was measured by shunt 6 made of a low-inductive chip resistors. The working gas was supplied into the discharge chamber through the opening 7. In the bottom of the housing 3 the outlet openings 8 are located. The laminar gas flow directed along the cathode was formed by caprolon conical nozzle 9. The distance from a point cathode to the plane anode 10 with a modifiable sample has the ability to be customized in the range of 1-20 mm. Figure 1b shows a photograph of a glow of discharge plasma in a stream of nitrogen at a pumping speed of 5 liters / min, the voltage pulse repetition frequency of 40 Hz and electrode gap of 12 mm. The cathode is located on the top.

An NPG-18/3500N high-voltage pulse-periodic generator was used as a voltage source, from which voltage pulses with an amplitude in the incident wave of up to 20 kV, a full width at half-maximum (FWHM) of 18 ns, and a rise time of 4 ns were supplied to the cathode. The discharge voltage was measured using a capacitive divider, and the discharge current was measured using a shunt connected between the anode and the chamber wall.

The $Cd_xHg_{1-x}Te$ specimens were irradiated in the pulsed-periodic mode at the pulse repetition rate 1200 Hz. The action was realized through 5, 10 and 20 minutes in an atmosphere of nitrogen. The nitrogen flow was 1.5 L/min. Measurements of the electro-physical parameters of the samples before and after irradiation were made at the temperature of liquid nitrogen using the Hall Electromotive

Force (EMF) method in the Van-der-Pau configuration. The measurements were performed at a constant current flowing through the sample (I = 1 mA) for the two directions of the current and two directions of a constant magnetic field. The surface of the irradiated samples was studied using atomic force microscope (AFM) "Ntegra Prima" (by NT-MDT) at ambient conditions.

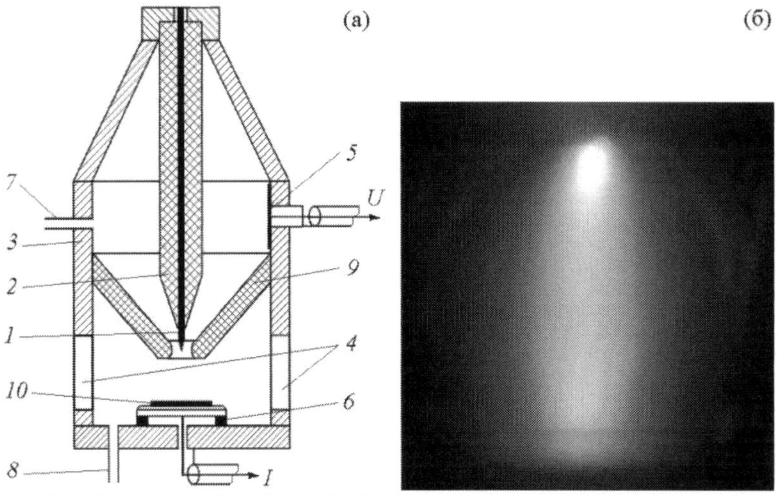

Figure 1. Discharge chamber (a) and photograph of a glow of discharge plasma (b).

3. Results and discussion

The study of the surface structure of initial and irradiated epitaxial films with an atomic force microscope revealed that the surface quality of the samples after exposure to a volume discharge is not changed. Surface roughness slightly increases from 1.7 to 2.1 nm.

Measurements of the electrical parameters of the samples of CMT epitaxial films after exposure to volume discharge pulses have shown that for all samples an increase in the conductivity of the material is observed.

As shown in figure 2 for p-type samples, a sharp change in the character of the behavior of the field dependence of the Hall coefficient is observed. For the initial sample alternating dependence of the Hall coefficient with a dot of inversion 0.2 T is observed. The irradiated material is characterized by the behavior of the field dependence which is characteristic for n-type material. At the same time samples of CMT epitaxial films are characterized by low values of electron mobility $\sim 2 \cdot 10^3$ $cm^2V^{-1}s^{-1}$, which is two orders of magnitude lower than the values corresponding to epitaxial material of n-type conductivity of high quality.

The observed change in the behavior of the field dependence of the Hall coefficient and low values of charge carrier mobility for the irradiated material allows us to conclude that in the process of impact on samples of epitaxial films of the volume discharge, formation of a layer with a high concentration of electrons on the surface or in the near-surface region of the material occurs. Conductivity of this layer is such that it bypasses the bulk of the epitaxial film during the Hall effect measurements. This conclusion is confirmed by the values of the electrical parameters for the irradiated material of n-type conductivity (figure 3). The magnitude of the Hall coefficient decreases with increasing the irradiation time, while the value of the mobility of the charge carriers is not substantially changed. In addition, the experimental data are in good agreement with the results of the impact of low-frequency discharge in air at atmospheric pressure on the electrical parameters of epitaxial CMT [3-5].

However, the features of the impact of high-frequency volume discharge compared to the low-frequency are observed. Various changes in the value of the Hall coefficient over time of the impact of the volume discharge pulses is observed for investigated samples. For example, for the material of n-type the conductivity increase in exposure time from 5 to 20 minutes leads to a monotonic decrease of the

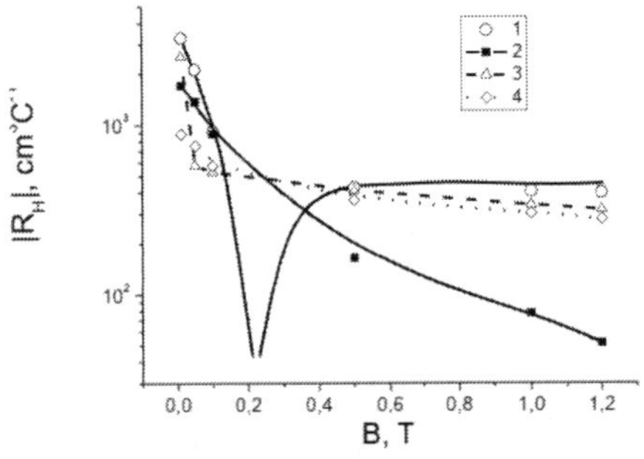

Figure 2. The field dependence of the Hall coefficient for samples of p-CdHgTe epitaxial films before (1) and after exposure to a volume discharge. The time of exposure (minutes): (2) - 5, (3) – 10, (4) – 20.

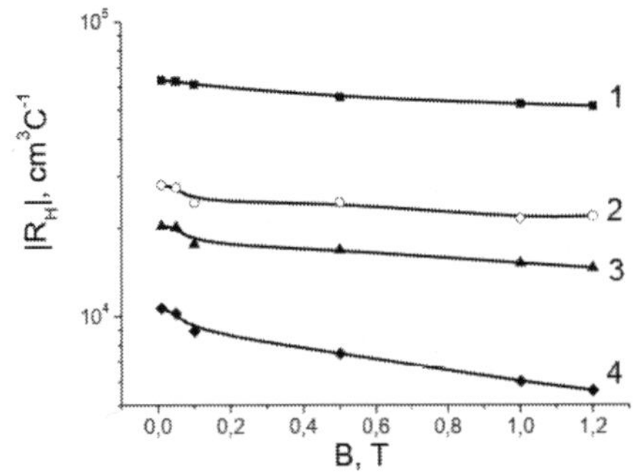

Figure 3. The field dependence of the Hall coefficient for samples of n-CdHgTe epitaxial films before (1) and after exposure to a volume discharge. The time of exposure (minutes): (2) - 5, (3) – 10, (4) – 20.

Hall coefficient value from $3\cdot10^4$ to $6\cdot10^3$ cm^{3}C^{-1}. While for the samples with p-type conductivity the Hall coefficient value increases with increasing the irradiation time. In addition, the stability study of the electrical parameters of the irradiated samples showed that the relaxation of the electrical parameter values over time is observed. Moreover, the character of relaxation for samples of different conductivity type is different. For p-type samples the Hall coefficient value is doubled for a period of 30 days. At the same time for the material of n-type conductivity the Hall coefficient value changes nearly ten times.

4. Conclusion

Thus, the obtained experimental data show that the effect of high-frequency pulses of nanosecond volume discharge in nitrogen atmosphere leads to a change in electrical properties of epitaxial CdHgTe, which is caused by the formation of the highly conductive surface layer of n-type conductivity during the process of exposure. The observed experimental data are in good agreement with the results obtained under the influence of low-frequency volume discharge in air at atmospheric pressure. At the same time revealed features of the high-frequency exposure (presence of relaxation of the electrical parameters of the irradiated samples, the dependence of the dynamics of the Hall coefficient values change over irradiation time on the type of conductivity of the material) allow us to make the assumption that the mechanisms of low- and high-frequency impact of volume nanosecond discharge on the electrical parameters are different.

References

[1] Tarasenko V F 2014 *Runaway electrons preionized diffuse discharges* (Nova Science Pub. Inc.)
[2] Rogalski A 2012 *Opto-electronics Review* **20** 279
[3] Voitsekhovskii A V, Grigor'ev D V, Korotaev A G et al 2012 *Russian Physics Journal* **54** 1152
[4] Grigoryev D V, Voitsekhovskii A V, Lozovoy K A et al 2015 *Journal of Physics: Conference Series* **652** 012025
[5] Grigoryev D V, Voitsekhovskii A V, Lozovoy K A et al 2015 *Proceedings of SPIE* **9810** 98100U

Acknowledgments

This research was supported by grant (№ 8.2.10.2015) from "The Tomsk State University Academic D.I. Mendeleev Fund Program" and grant (№14-29-00052) from Russian Science Foundation.

IOP Conf. Series: Journal of Physics: Conf. Series **830** (2017) 012083 doi:10.1088/1742-6596/830/1/012083

Electrical characteristics of epitaxial MCT after As^+ implantation

A Voitsekhovskii[1], I Izhnin[1,3], A Korotaev[1], D Lyapunov[1], S Dvoretskii[2], P Smirnov[2]

[1]National Research Tomsk State University, Tomsk, 634050, Russia
[2]A.V. Rzhanov Institute of Semiconductor Physics, Siberian Branch of the Russian Academy of Sciences, Novosibirsk, 630090, Russia
[3]R&D Institute for Materials SRC "Carat", Lviv, 79031, Ukraine

E-mail: dima.lyapunov_1992@mail.ru

Abstract. In this work we studied the characteristics of MBE MCT films after the introduction of different energies As+ with different doses of irradiation. Some of the samples were subjected to post-implantation annealing. Electrical characteristics of the samples were determined from Hall measurements. Voltage-current characteristics of the structures were also measured. Activation As and modification of the characteristics of MCT outside the implanted layer after annealing have been detected. Also we found differences in the p-n junction depths and electrically active defects profiles.

1. Introduction

At present molecular beam epitaxial (MBE) layers of mercury-cadmium-telluride HgCdTe (MCT) solid solutions are widely used to create multi-element semiconductor detectors of IR radiation providing signal processing directly in the focal region [1]. Along with the study of the initial properties of epitaxial films of MCT grown by MBE, forming a p-n junction by ion implantation is an urgent problem.

For various reasons (in particular, to reduce the dark currents) there appears more advantageous manufacturing photodiode structure of "p-on-n" type. In [2] it was shown that in such structure dark currents become smaller by two orders of magnitude. This is due to low mobility and lifetime of holes that there are minority carriers for the n-base of a photodiode. Reducing dark current increases the photosensitivity of such a structure. But for such structures smaller yields of detectors (5% less) and a higher level of inhomogeneities per unit of area of photodiode array are typical. This is due, primarily, to the lack of understanding of mechanisms of radiation defect formation and the formation of a p-n junction in such structures. These mechanisms underlie the processes of manufacturing "p-on-n" structures.

The aim of this work was to study the characteristics of MBE MCT films, after the introduction of different energies As+ with different doses of irradiation.

2. Samples and measurement procedure

The epitaxial films were grown at the Semiconductor Physics Institute of the Siberian Branch of the Russian Academy of Sciences. After growth, the films had n-type conductivity, for conversion to p-type conductivity, the films were annealed in a neutral atmosphere of hydrogen or helium. The parameters of the samples are given in Table 1.

Arsenic ion implantation was carried out at the Institute of Semiconductor Physics in Novosibirsk on the plant ion implantation. Irradiation of arsenic ions was performed at room temperature in a dosage range 10^{13}-10^{15} cm^{-2}. The parameters of the ion beam are as follows: energy ions $E_1 = 190$ keV (samples 1-5 series), $E_2 = 350$ keV for sample 5 series). The ion current density is $j = 0.06$ $\mu A\times cm^{-2}$. As shown in ref. [2], for these ion currents we can neglect heating of the sample during the implantation process. After the implantation for the fifth series of samples of the two-stage activation annealing was conducted.

Table 1. Parameters of the samples.

series	d, μm	p, cm^{-3}	μ_p, $cm^2V^{-1}s^{-1}$
1	7.56	$1.5\cdot10^{16}$	334.444
2	7.06	$2.7\cdot10^{16}$	200.116
3	15.51	$1.2\cdot10^{16}$	428.118
4	14.7	$1.57\cdot10^{16}$	462.414
5 (n-type)	7.41	$7.4\cdot10^{14}$	9500

Measurements of the electrophysical parameters of the samples were made at the temperature of liquid nitrogen using the Hall method in the Van-der-Pau configuration. The electron concentration distribution as a function of semiconductor depth was determined by the method of differential Hall measurements using an etching process. The error in the Hall and conductivity measurements did not exceed 2-3%. The error in the volume concentration and mobility of charge carrier measurements is mainly determined by the accuracy in the measurement of the thickness of the removed layer and did not exceed 10-12%.

To determine the depth of the p-n transition 3 samples were selected which produced mesa structure. The height of the mesa was changed with the step of 1 micron by chemical etching. Measurements of current-voltage characteristics of the samples were performed using the automated spectroscopy admittance nanoheterostructures equipment. It permits quantitative measurements of electrical parameters of materials and nanostructures and microelectronics.

3. Experimental results and discussion

Measurement of the electro-physical parameters of the samples after implantation shows that As ion irradiation resulted in a conversion of the conduction type, with the formation of a highly alloyed n^+-layer in the surface of the material due to the generation of radiation defects, which is characteristic of MCT. The dependence of the layer charge carrier concentration N_S on the radiation dose Φ is shown in Figure 1.

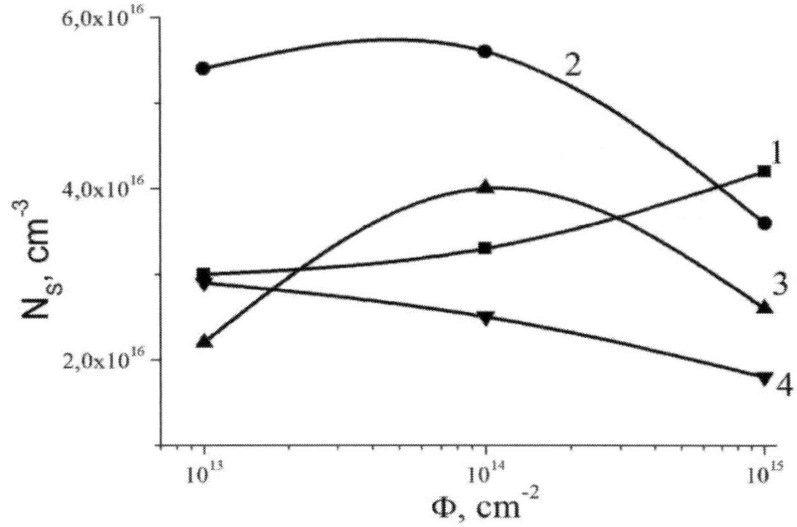

Figure 1. The dependence of the electron concentration N_S on the radiation dose. Curves numbers correspond to the samples of series.

The figure shows that the concentration decrease is observed at higher doses of radiation, which is contrary to the argon implantation data [3]. This decline can be attributed to the greater disorder of the implanted layer by applying a high-energy and heavier (by 1.9 times in comparison with argon ions) arsenic ions.

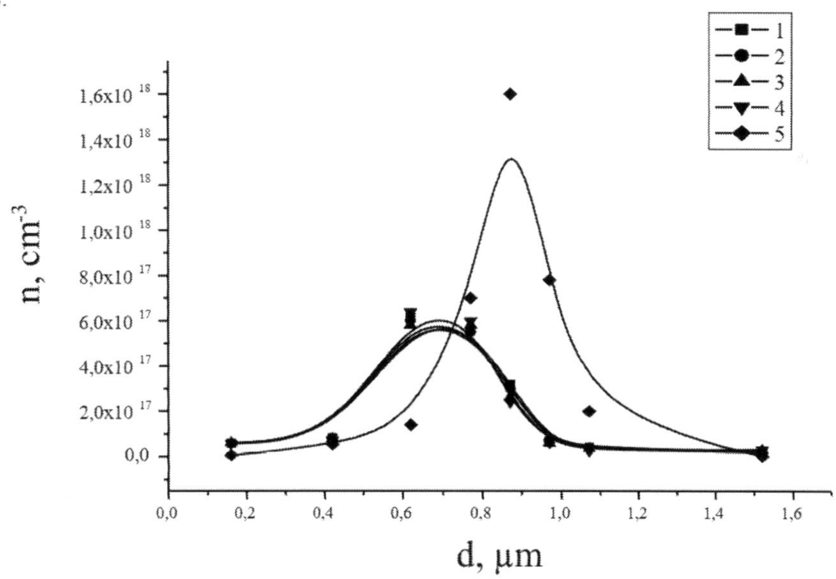

Figure 2. The dependence of the electron concentration n on the depth. Curves numbers correspond to the samples of series. $\Phi = 10^{14}$ cm^{-2}.

The profiles of the carrier concentration are shown in Figure 2. It is evident that for samples a starting p-type layer does not change the profile of the electrically active defects. It is typical for ion implantation in MCT. The profile depths are substantially more than R_p of implanted ions [4]. The concentration at the maximum of the distribution curve is about 10^{18} cm^{-3}. For a sample of the original n-type subjected to irradiation in addition, post-annealing electric profile lies somewhat deeper, which corresponds to the data in ref. [5]. Thus, it can be argued that defects introduced by irradiation are being eliminated, while the implanted impurity is being activated during the annealing process. It should be noted that the parameters of the samples after etching of 1.5 microns of material have virtually returned to the value prior to irradiation.

CVC are shown in Figures 3-5. Up to a depth of 3 μm all CVC have an ohmic character. For samples 5 series diode current-voltage characteristic appears at a depth of 3 μm (Figures 3, 4), which corresponds to a depth profile of arsenic after implantation and post implantation annealing [5]. Therefore, the formation of a p-n junction can be attributed to the activation of arsenic in the annealing process in this case.

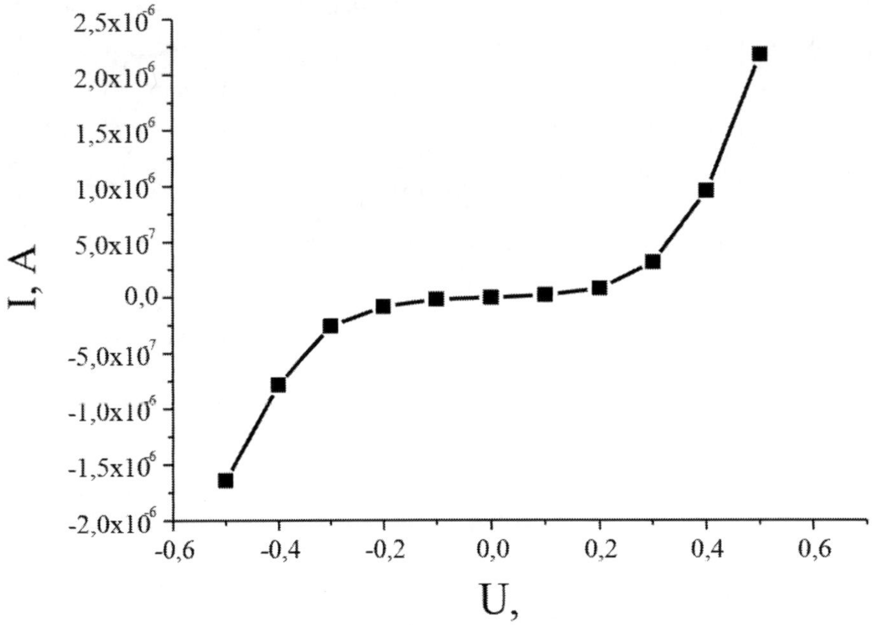

Figure 3. The current-voltage characteristic of the sample number 5 after 350 keV As ions implantation and post-annealing and etching of 3 μm.

It should be noted that the depth of the p-n junction does not depend on the energy of the implanted ions, while the electrical characteristics are different. This difference may be explained by the different concentration of activated arsenic in the region of the p-n junction.

For sample number 4 a diode type current-voltage characteristic is found only at a depth of 10 microns, which is far beyond the electrical profile and implanted arsenic profiles (Figure 2). Probably, the formation of the p-n junction is due to the deep diffusion of radiation-induced defects, the concentration of which cannot be detected by electrical measurements.

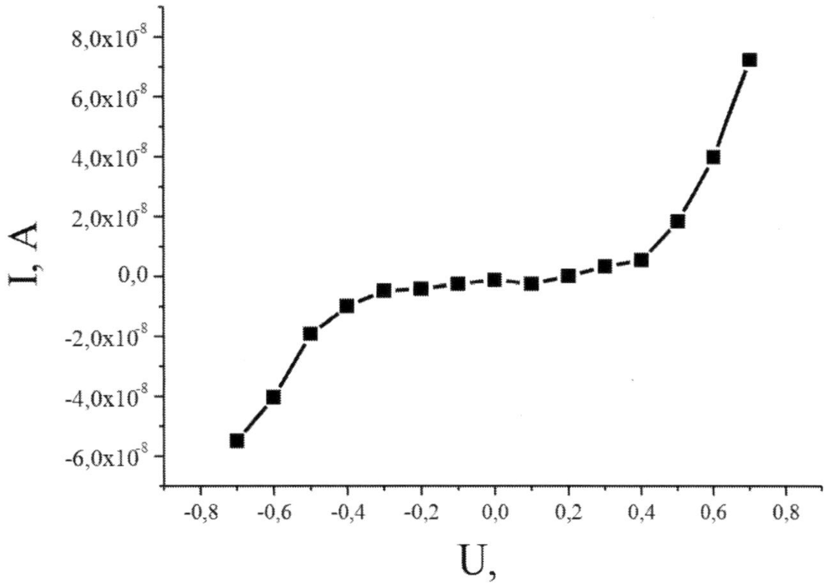

Figure 4. The current-voltage characteristic of the sample number 5 after 190 keV As ions implantation and post-annealing and etching of 3 μm.

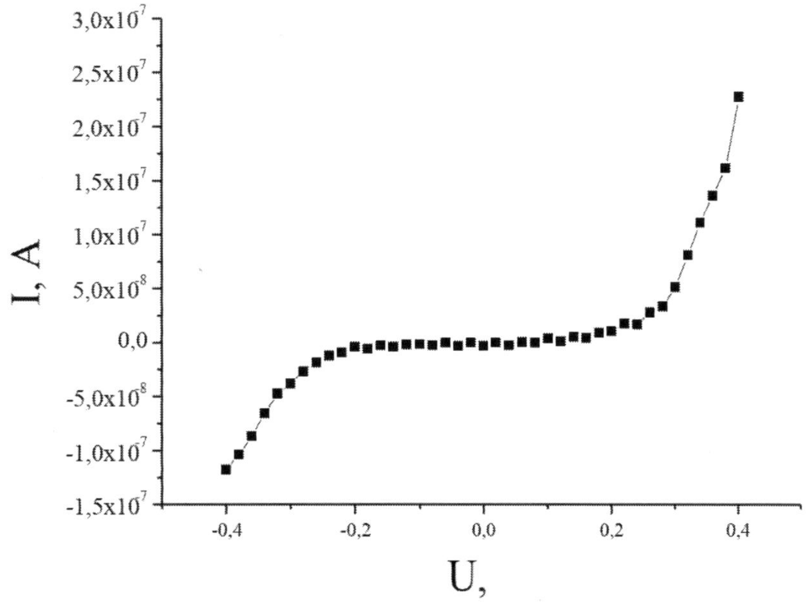

Figure 5. The current-voltage characteristic of the sample number 4 after 190 keV As ions implantation and etching of 10 μm.

4. Conclusion

Measurements of the electrical parameters showed the classic pattern similar to MCT samples after ion implantation. The influence of the presence of a variable-gap layer on the surface of the material and the substrate on the ion implantation results was shown.

It was noted that there is a decrease in the concentration of the layer at high doses, which can be explained by greater disorder of the implanted layer through the use of high-arsenic and heavy ions.

Significant differences in the p-n junctions depths for the obtained structures and in the measured profiles of distribution of electrically active defects are found. It is proposed to explain this distinction by the presence of the "tails" with low concentration of electrical defects and impurities.

References

[1] Rogalski A 2011 *Infrared Detectors* (CRC Press, Boca Raton)
[2] Mollard L., Destefanis G., Baier N., Rothman J., Ballet P., Zanatta J.P., Tchagaspanian M., Papon A.M., Bourgeois G., Barnes J.P., Pautet C. 2009 *J. Electron. Mat.* **38**(8) 1805
[3] Destefanis G.L. 1988 *J. Cryst. Growth.* **86** 700
[4] Wang L., Zhang L.H. 2000 *J. Electron. Mat.* **29**(6) 873
[5] Shin S.H., Arias J.M., Zandian M., Pasko J.G., Bubulac L.O., De Wames R.E. 1995 *J. Electron. Mat.* **24**(5) 615

Technology of treatment of building materials with the plasma torch

R R Galeev, R T Nasibullin, A Sh Sadriev and R Sh Sadriev

Kazan Federal University, Naberezhnye Chelny, Russian Federation

E-mail: asibullin.ramil@mail.ru

Abstract. The paper describes the solutions of the building materials heat conduction problems – heating source power and size definition based on temperature measurements on a material surface and on layers, heating source characteristics and surface temperature identification by temperature measurements on a treated area. The basic problem of the building concrete block heating by mobile Gauss type thermal source are addressed. As a result there are given formulae for an estimation of the plasmatron useful power and maximal temperature on the depth $z \geq 1$ mm with accuracy enough for a practical appliance. Also values of the main technological parameters during the treatment are recommended and the technology of concrete block finish is shown.

1. Introduction

Today electric arc treatment of materials is widely used in various industries: metallurgy, engineering, construction, chemical industry, etc. [1-3]

Most plasma devices make the low-temperature plasma by gas heating in the electric arc [1. 2]. For high-speed plasma (more 90 m/sec) flow and high consumptions linear blowed arc plasmatrons are usually used. In such plasma torches the high powers are implemented at high voltages and relatively low currents (up to 1000 A). In this case, the plasma flame has a small size. The extended sources of heat are required for ensuring the necessary productivity in some practical cases (e.g., melting of construction details; plasma treatment of dielectric materials). These sources have the form of a single pressed arc or a wide plasma flame. They are characterized by high enthalpy and low-speed of gas flow (< 50 m/s).

The thermal processes of heating of material by means of high-concentrated energy flows are investigated mainly for metals [3]. Plasma heating of ceramic and building materials is investigated much less. This determined the aim of comprehensive theoretical and experimental research of heat transfer processes during electric arc treatment of materials.

In this article we investigated heat processes in the surface layer of the concrete slab during plasma treatment.

2. Theoretical statements

As we know the heat treatment of silicate materials with a powerful heat source occurs usually at high speed, and is characterized by great values of Peclet number. Consequently, if we analyze the stationary problem of a temperature field in a moving slab at the coordinates associated with the surface heat source with intensity q(x), we focus on two main factors: the speed of movement of the heat source and the heat conductivity on depth of the heated material. Thus, we have the following simplified stationary problem:

$$v\frac{\partial T}{\partial x}+x\frac{\partial^2 T}{\partial x^2}=0 \tag{1}$$

$$T(\infty)=0, -\lambda\frac{\partial T}{\partial z}=q(x) \tag{2}$$

where

$$q(x)=q_0 e^{-\frac{x^2}{r^2}}$$

The solution of this problem is:

$$T=\frac{q_0}{\lambda}\sqrt{\frac{\pi\chi^x}{v}}\cdot e^{\frac{z^2\vartheta}{4\chi^x}-\frac{x^2}{r^2}} \tag{3}$$

On condition that: $\frac{\chi^x}{v}>0$. The surface temperature is easily determined from equation (3) when z = 0.

$$T_n(x)=\frac{q_0}{\lambda}\sqrt{\frac{\pi\chi^x}{v}}\cdot e^{-\frac{x^2}{r_0^2}} \tag{4}$$

After a detailed analysis of the equation (3) we have two asymptotics for maximal temperature by depth z: linear asymptotic and hyperbolic asymptotic.

Linear asymptotic ($Dp < 1$):

$$T_{MAX}(z)=A-B\cdot z \tag{5}$$

where

$$A=\frac{q_0}{\lambda}\sqrt{\frac{\pi\chi^x r_0}{2ve^{0.5}}},\ B=\frac{q_0}{\lambda e^{0.25}}$$

Hyperbolic asymptotic ($Dp > 1$):

$$T_{MAX}(z)=\frac{C}{z} \tag{6}$$

where

$$C=\frac{q_0\chi r_0}{\lambda v}\sqrt{\frac{2}{e}}$$

Dp-criterion:

$$Dp=\frac{z}{2}\sqrt{\frac{v}{\chi r_0}}$$

Dp-criterion allows to delimit the linear asymptotic (near-surface layer) from the hyperbolic asymptotic (technologically significant layer). If the plasma torch with a characteristic size $r_0 = 0.05$ m moves along the surface of the concrete slab at a speed of $v = 0.3$ m/s, then from the equality the

$Dp = 1$ we obtain $z \approx 0.5$ mm. I.e., at a depth of $z < 0.5$ mm the maximal temperature decreases linearly, and at a depth of $z > 0.5$ mm the maximal temperature decreases hyperbolically.

We can use these temperature asymptotics for solving an inverse problem - definition of power and the characteristic size of the heat source on measurements of temperature at different depths of the heated material.

3. Research results

As input parameters we use a thermal characteristics of concrete [4-6]: $\lambda = 1.45$ W/(m·°C); $\chi = 0.8 \cdot 10^{-6}$ m²/s; $T_{MEL} = 1700$ °C; $v = 0.3$ m/s. The melting temperature of the concrete approximately coincides with a surface temperature $T_S = T_{MEL}$ (at $z = 0$). Then from equation (5) we obtain the equality:

$$\frac{q_0}{\lambda}\sqrt{\frac{\pi\chi r_0}{2ve^{0.5}}} = 1700 \qquad (7)$$

We will find a value (C) in a technologically significant layer ($z > 0.5$ mm) after processing data with least squares method for hyperbolic function (7)

$$C = \frac{q_0 \chi r_0}{\lambda v}\sqrt{\frac{2}{e}} = 0.4582 \qquad (8)$$

Power of the Gaussian surface heat source is determined by the expression:

$$P = q_0 \pi r_0^2$$

Then, we find the power and the characteristic size of the heat source from the integrated solutions of equations (7) and (8): $P \approx 42{,}7$ kW, $r_0 \approx 0.051$ m.

Next we consider another inverse problem: finding the temperature on the surface T_S of the heated slabs from the available data $T_{MAX}(z)$. In this case it is necessary to use more complicated asymptotic instead of a hyperbolic asymptotic (6), namely:

$$T_{MAX}(z) = \frac{C}{z}\left(1 - \frac{D}{z^2} + 6\frac{D^2}{z^4}\right) \qquad (9)$$

where

$$C = \frac{q_0 \chi r_0}{\lambda v}\sqrt{\frac{2}{e}},\ D = \sqrt{2}\frac{\chi r_0}{v}$$

Then we use nonlinear least squares method and find values of the parameters: C = 0.4260 m·°C, D = $1.48 \cdot 10^{-7}$ m².

We obtain the following equation:

$$\begin{cases} \dfrac{q_0 \chi r_0}{\lambda v}\sqrt{\dfrac{2}{e}} = 0.427 \\ \sqrt{2}\dfrac{\chi r_0}{v} = 1.48 \cdot 10^{-7} \end{cases} \qquad (10)$$

From this system we determine the characteristic size of the source $r_0 \approx 0.04$ m and its power $P \approx 40$ kW. We use the linear asymptotics (5) at $z = 0$ to calculate the temperature on the surface. The result is $T_S \approx 1750$ °C.

The obtained surface temperature well combined with the melting temperature of the concrete (1700 °C). The values of r_0 and of P are lower about 10 – 15 % than the real characteristics of the plasma torch. If we consider the extreme sensitivity of inverse problems to inaccuracies of input data, the increase in the number of unknown parameters from two to three under the same experimental initial data, this result can be considered quite satisfactory. We can recommend values of the main technological parameters, with taking into account the results of the research to ensure the quality of the treated surface and the stable mode of operation of the plasma torch in the treatment of concrete slabs: $G = (1500 - 2500)$ kg/s; $I = (1200 - 1500)$ A; $U = (70 - 100)$ V; $H = (0.01 - 0.03)$ m; $v = (0.2 - 0.3)$ m/s.

Temperature research for the heated surface of the concrete slab shows us that the temperature is not dangerous for the material and doe not exceed 142 °C. The Figure 1 shows the heating curve during continuous heating of concrete slab. Each heating passage was shifted on half of the width of the plasma flame.

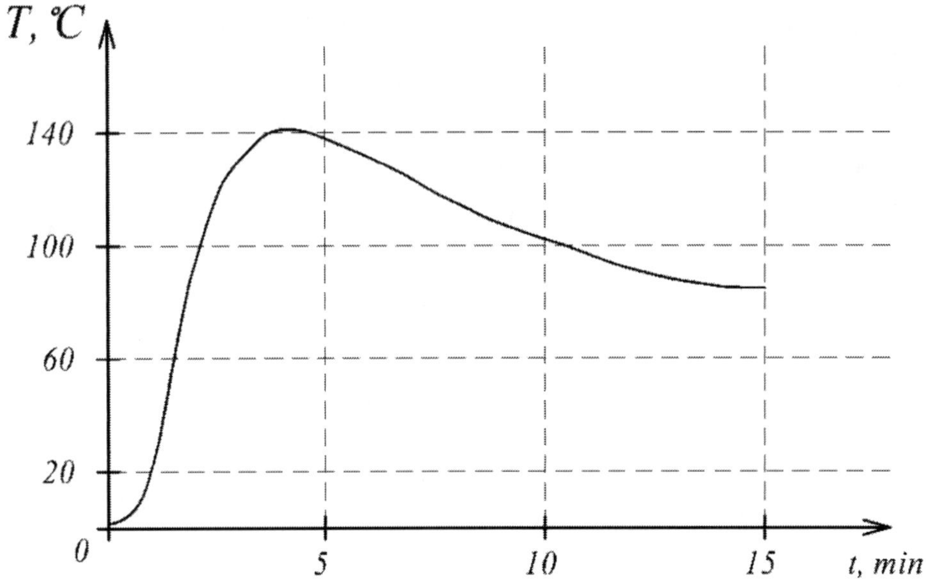

Figure 1. The heating curve during continuous heating of concrete slab.

4. Conclusion

We have received the technology for treatment of concrete slabs by plasma surface reflow method, that is based on the research results which include the selection of the base material, decorative blends, salts, powders, methods of forming products, the determination of plasma treatment conditions, test the properties of the coatings, composition of the technology process of the plasma surface reflow of building materials, etc.

It is necessary to highlight most important aspects of the technology which was obtained during the research and allow to achieve a high quality of products treatment.

It is desirable to use light quartz sand with a particle size of 3-5 mm during plasma treatment as a placeholder subsurface layer. It is also possible to use crushed white glasses with the same particle size in admixture with sand of usual composition (main placeholders). With this composition of subsurface layer the staining properties of a pigment will be better, and striation from the treatment plasma torch will be less noticeable.

The material of building product does not overheat due to high speed of plasma treatment (plasma torch speed 0.2 – 0.3 m/s). It allows us to exclude a special heat insulating layer.

It is preferable to make the treatment of product «face-up» during plasma treatment. In this case the surface is well cleaned and does not have grease residues. If the product is treated «face-down», the surface will have densified structure. It is not so easy to clean such a surface and include residues of grease that make the surface black during the treatment.

IOP Conf. Series: Journal of Physics: Conf. Series **830** (2017) 012084 doi:10.1088/1742-6596/830/1/012084

References

[1] Morzov A I 2008 *Introduction of plasma dynamics* (Moscow: Fizmatlit)
[2] Zhukov M F 1987 Theory of thermal electric arc plasma (Novosibirsk)
[3] Gaysin F M, Nasibullin R T, Sadriev A Sh, Sadriev R Sh and Sarimov L R The practice of features of high-frequency discharges in the technological cycle 2015 *Scientific and Technical Volga region Bulletin* **5** 184
[4] Zotkin A G 2012 *Concrete and concrete constructions* (Moscow: Phoenix)
[5] Nevill A M 1972 *Properties of concrete* (Moscow: Publishing house of literature on construction)
[6] Galeev R R and Abdrakhmanova L A 2014 Development of compounds and physico-chemical principles for the production of efficient building materials use of wastes of different nature *Abstracts of papers of 66 All-Russian scientific conference on problems of architecture and construction* 28

IOP Conf. Series: Journal of Physics: Conf. Series **830** (2017) 012085 doi:10.1088/1742-6596/830/1/012085

Legitimacies of the growth of oxide films on ion-modified samples from the alloy E110 at high-temperature oxidation

B A Kalin, N V Volkov, R A Valikov, A S Yashin and D N Ignatiyev

National Research Nuclear University MEPhI (Moscow Engineering Physics Institute)
31 Kashirskoe shosse, Moscow 115409, Russia

E-mail: nvvolkov@mail.ru

Abstract. The possibility to obtain barrier layers during the surface modification by ion-plasma to decrease the oxidation and hydrogenation rate of fuel claddings from zirconium alloys at various operation conditions is studied in this paper. Fuel claddings from the alloy E110 obtained on the outer surface of fuel elements after ion polishing are determined on the basis of calculations and experiments performed: the average ion energy in a beam is 3-4 keV, the ion current is 0.4-0.7 mA cm^{-2}, and the total irradiation dose is 5×10^{18} ion cm^{-2}. Cleaning and polishing of samples from the alloy E110 were carried out on the installation ILUR-03 by a radial beam of argon ions with a wide energy spectrum. Thin films (up to 100 nm) of alloying elements – Fe, Al, Mo, Y, Mg, and Cr were obtained on the polished surface by magnetron deposition and then implanted into the near-surface layer under the influence of a radial beam of argon ions. Study of the element composition of cleaned and modified near-surface layers was performed by X-ray microanalysis. A part of the samples was preliminarily autoclaved in distilled water at the temperature of 350 °C and the pressure of 17 MPa for 500 h to obtain a protective layer from an oxide film enriched by implanted elements. It is shown that alloying elements are mostly located at the oxide-metal boundary and in the oxide.

1. Introduction

The surface modification of materials by ion beams is one of the most effective and precise methods of increasing the service properties of articles. For example, fuel claddings made from the zirconium alloy E110 with an ion-modified surface can exhibit a much higher corrosion resistance in reactor conditions (water medium, the temperature is 350 °C, and the pressure is 17 MPa) [1-3]. This is due to the formation of dense oxide films at the initial stages of oxidation. Ion polishing of the surface and alloying of the near-surface layer of the material by ion mixing can promote an increase of the protective properties of the oxide. The study of the oxide growth on an ion-modified surface in conditions of intensive high-temperature oxidation corresponding to the emergency regimes of the reactor (steam-water medium, the temperature is up to 1200 °C) is of particular interest. The integrity of the fuel cladding and the working capacity of the fuel element directly affect the safety of the reactor in these regimes.

2. Materials and tools

Fragments of fuel element tubes made from the alloy E110 (Zr-1% Nb) on the basis of sponge and having an optimized composition were used as samples for the research. The outer diameter of the samples was 9.15 mm, their wall thickness was 0.65 mm, and their length was up to 500 mm.

The outer surface of the samples was treated on the installation ILUR-03 by a radial beam of argon ions with an average energy of 3-4 keV. The ion current was equal to 0.4-0.7 mA cm^{-2} and the total irradiation dose was 5×10^{18} ion cm^{-2}. The treatment of the samples included the removal of impurities of the sorbed material in the near-surface layer, polishing of the surface up to a roughness better than R_a=0.3 μm, and alloying by ion mixing of the atoms of films preliminarily deposited on the surface of samples using magnetrons. Fe, Al, Mo, Y, Mg, and Cr of a purity better than 99.99 at.% were used as targets for magnetron sputtering. The film thickness did not exceed 150 nm. The scheme of the samples' treatment on the installation ILUR-03 is shown in Figure 1.

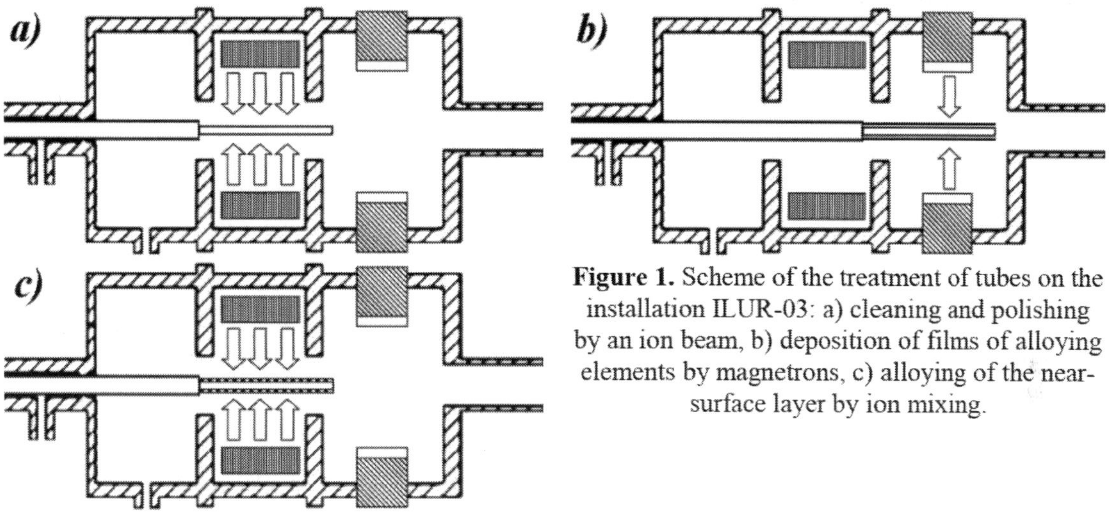

Figure 1. Scheme of the treatment of tubes on the installation ILUR-03: a) cleaning and polishing by an ion beam, b) deposition of films of alloying elements by magnetrons, c) alloying of the near-surface layer by ion mixing.

The prepared samples were exposed to corrosion tests in a water-steam environment at the temperature of 1200 °C for 100, 200, and 500 s. A part of the samples was preliminarily autoclaved in water at the temperature of 350 °C and the pressure of 17 MPa for 30 and 500 h. The oxidation degree of fuel claddings was estimated using a specific overweight Δm or a local depth of oxidation (LDO):

$$\Delta m = 920 \times \exp(\frac{10410}{T}) \times \sqrt{t}, \quad (1)$$

where Δm is the specific overweight of oxygen, mg cm^{-2}; T is the temperature, K; and t is the time of oxidation, s.The LDO-parameter is determined experimentally or by calculation as the ratio of the total thickness of an equivalent zirconium layer, which would react with the steam, to the initial thickness of the fuel cladding:

$$LDO = N \times \delta_e \times \delta_0 \times 100\%, \quad (2)$$

where LDO is the local oxidation depth, %; N is the coefficient that takes into consideration the two-sided oxidation of the fuel cladding, N=2; δ_e is the equivalent layer thickness (the calculated thickness of zirconium which would transfer to ZrO_2); δ_0 is the initial sample thickness. The samples were studied by optical and scanning ion microscopy, the element composition was studied by X-ray microanalysis.

3. Results and discussion

The appearance of the samples before and after the high-temperature oxidation is shown in Figure 2. The oxide film on all the samples has a smooth dark gray color with no visible signs of damage or crumbling.

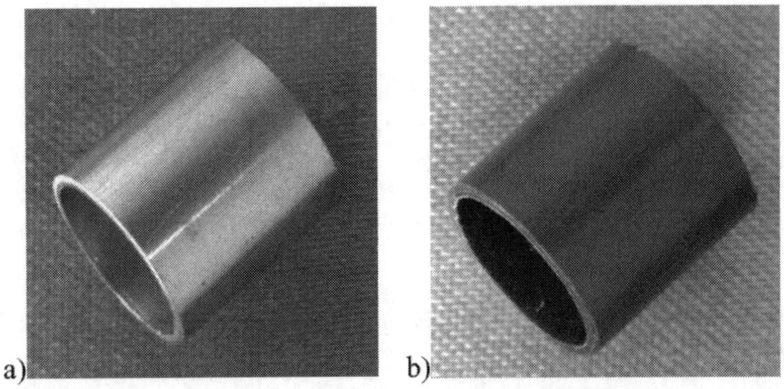

Figure 2. Photos of the samples of fuel claddings with a diameter of 9.15 mm and a length of 10 mm made from the alloy E110: a) before and b) after the corrosion tests.

Figure 3 shows the dependence of the LDO-parameter on the oxidation time of samples with various states of the surface at the temperature of 1200 °C. As the analysis showed, the oxidation kinetics of the samples in the initial state complies to a parabolic law. The surface treatment of tubes by an ion beam results in a change in the character of dependencies. For this time range of the tube after the ion polishing and alloying by Al, lower LDO-values are observed. At that, a slowdown of the oxide growth on the alloyed sample is observed at the initial stage, which is probably due to the inhibition of the film growth owing to the redistribution of Al atoms in the oxide-metal system.

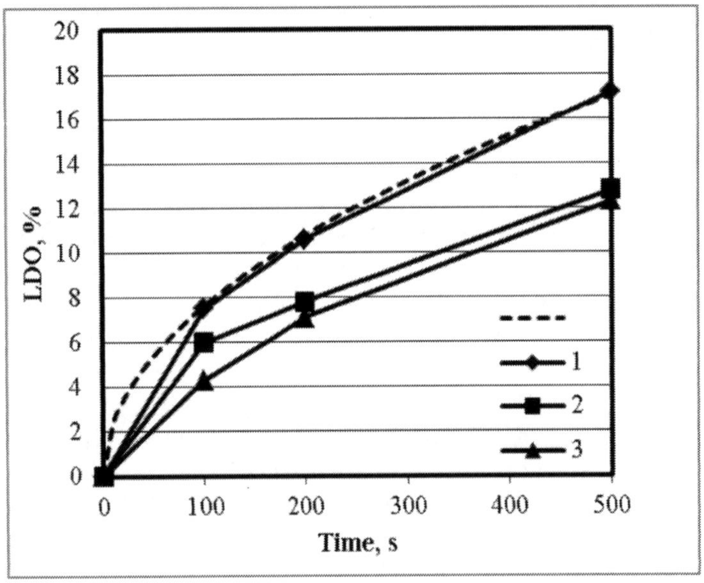

Figure 3. Dependence of the LDO-parameter of samples on the oxidation time at 1200 °C: 1) in the initial state; 2) after the ion polishing; 3) after the surface alloying by Al atoms. The dotted line shows the trend of the parabolic dependence of oxidation for the samples in the initial state.

The effect of the preliminary autoclaving of fuel tubes on the kinetics character of high-temperature oxidation of the material was studied. As can be seen from the data presented in Figure 4, the oxide film grown under autoclave conditions on the surface of samples in the initial state slows down the process of high-temperature corrosion, which is in agreement with the data of other researchers [4]. The samples with an ion-alloyed surface that were preliminarily oxidized show bigger values of the LDO-parameter in comparison with tubes that were not exposed to autoclaving, which can be seen in Figure 5. These results are presumably connected with the evolution of the modified layer of a material in the process of oxidation at 350 °C. At this stage, Al atoms are redistributed in the oxide-metal system and suppress the oxide film growth in the specified temperature range.

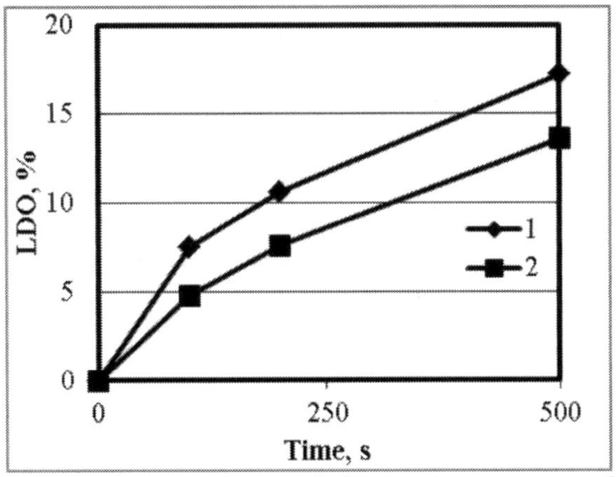

Figure 4. Dependence of the LDO-parameter of samples on the oxidation time at 1200 °C: 1) in the initial state; 2) after preliminary autoclaving at 350 °C for 500 h.

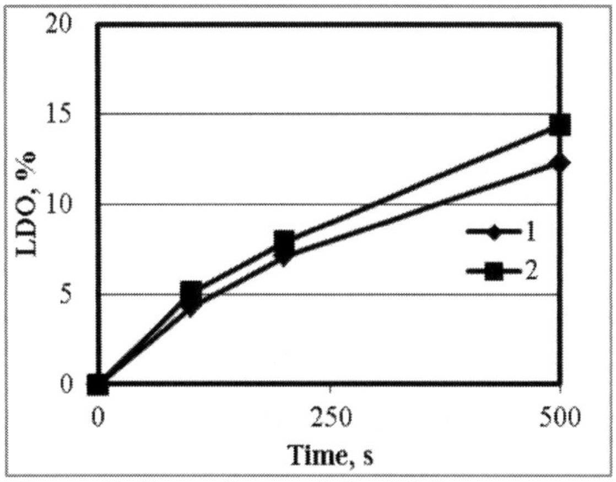

Figure 5. Dependence of the LDO-parameter of samples after ion alloying by Al atoms on the oxidation time at 1200 °C: 1) without preliminary autoclaving; 2) after oxidation at 350 °C for 30 h.

4. Conclusion

The behavior of fuel tubes from the alloy E110 on the basis of sponge has been studied in the process of high-temperature oxidation. It has been shown that the ion modification of the surface using the regimes of polishing and alloying of the material by ion mixing of preliminarily deposited films can suppress the growth rate of the oxide at temperatures up to 1200 °C.

References

[1] Valikov R A, Yashin A S, Yakutkina T V, Kalin B A, Volkov N V, Krivobokov V P, Yanin S N, Asainov O Kh and Yuriev Yu N 2015 *J. of Physics: Conf. Ser.* **652** 012068

[2] Kalin B A, Volkov N V, Valikov R A, Yashin A S, Krivobokov V P, Yanin S N, Asainov O Kh and Yuriev Yu N 2016 *IOP Conf. Series: Materials Science and Engineering* **130** 012051

[3] Kalin B A, Volkov N V, Valikov R A and Yakutkina T V 2016 *IOP Conf. Series: Materials Science and Engineering* **130** 012005

[4] Guilbert S, Lacote P, Montigny G, Duriez C, Desquines J and Grandjean C 2013 *Conference: 17th International Symposium on Zirconium in the Nuclear Industry*

Irradiation of Fe-Mn Supersaturated Solid Solution with Ions of Various Atomic Masses (Ar^+, Xe^+) and Analysis of the Role of Nanosized Dynamic Effects in the Activation Processes of Long-Range Type

V V Ovchinnikov[1,2], **F F Makhin'ko**[1], **V A Semionkin**[1,2] **and** K **V Shalomov**[1,2]

[1]Institute of Electrophysics, Ural Branch of Russian Academy of Sciences, Amundsena street 106, Yekaterinburg, 620016, Russia
[2]Ural Federal Technical University named after the First President of Russia B.E. Yeltsin, Mira street 19, Yekaterinburg, 620002, Russia

E-mail: viae05@rambler.ru

Abstract. A multiple increase in the atom mobility in metastable supersaturated (quenched from 850 °C) Fe-8.16 at % Mn solid solution is detected at temperatures less than 250°C under irradiation with 5-keV Ar^+ and Xe^+ ions of different masses. The irradiation-induced atom redistribution in the entire volume of foils 30 μm thick at a projected Ar^+ and Xe^+ ion ranges as much as 20-30 nm only is found and studied by the transmission Mössbauer spectroscopy. Long-range effects at low irradiation doses and anomalously low temperatures are attributed to "radiation shaking" of metastable media with post-cascade solitary waves in contrast to thermally stimulated radiation-enhanced processes in the narrow nanoscale near-surface layers of the alloy. It has been shown that heavier Xe^+ ions at higher irradiation doses have a stronger impact on the solid solution than Ar^+ ions.

1. Introduction

In work [1] it has been shown that upon irradiation of materials with heavy ions of low and medium energy (from a few tens to several hundreds of keV), the source of defects is only a surface layer less than 1 μm thick rather than entire volume of material in contrast to severe plastic deformation.

In the case of neutron irradiation, defects in material are accumulated slowly due to the high penetrating ability of neutrons (neutron interaction cross section is by 6-7 orders of magnitude smaller than that of ions). This is the reason of why primary recoil atoms and cascades of atomic displacements formed by these atoms under neutron irradiation are separated by distances much greater than those under ion irradiation when accumulated fluences (cm^{-2}) are the same.

Nevertheless, low-dose and long-range effects under neutron and ion irradiation are well-known. These effects manifest themselves by dramatic changes in the structure and properties of materials at a small number (< 0.001) of displacements per atom and by the fact that the distance at which the changes occur is several orders of magnitude greater than the projected range of implanted ions (or primary recoil atoms under low-dose neutron irradiation).

These effects seem to represent two sides of the same coin, and cannot only be explained by relatively slow processes, for example, by radiation-enhanced or radiation-induced diffusion, which well represent radiation swelling and radiation-induced creep in the framework of the classical radiation physics.

It is easy to estimate, as it has been done in [1], that the diffusion length of vacancies is certainly not more than 1 μm for a few seconds of irradiation at room temperature in metals with a melting temperature of T_{melt}>1000 K. There is a large number of studies, which indicate dramatic changes in the structure and properties in the entire volume of materials irradiated with accelerated ions for several seconds. These changes occur without heating and other obvious factors causing such effects at a depth from several tens of **micrometers** to several **millimeters** [2-4]. Low-dose and long-range effects in highly-defected quenched and highly-deformed materials with numerous traps and sinks for the radiation-induced defects also cannot be explained by the role of more mobile interstitial atoms.

It is suggested in [2-4] that the role of nanosized radiation-dynamic effects under cascade-forming corpuscular radiation is significant and crucial in some cases. Nanoregions of cascades of atomic displacements are zones of explosive energy release. The temperature of the cascade regions thermalized for 10^{-12} s is 5000-6000 K and above (the rate of energy release is comparable to that during nuclear explosion). A sharp increase in the temperature and pressure [5] in these regions causes the emission of powerful elastic and shock **postcascade** solitary waves.

It is known that a nonideal state of solid solutions, which often manifests itself near the temperature and concentration phase transitions, can have a significant effect on the properties of materials.

The thermodynamics equations imply that the tendency of solid solutions to the atomic ordering and concentration segregation, at a certain ratio of the energy of interatomic interactions between components can be characterized by an extremely low critical temperature, when the diffusion mobility of the atoms is extremely small. Nevertheless, there are instances where some alloys after long-term service (over hundreds and thousands of hours) at relatively low temperatures exhibit changes in the properties. As a result of special investigations it was found that this is due to intraphase processes, for example, the concentration separation of solid solutions into substitutional impurities [6]. The reason for the changes in the properties of alloys can also be a short or long-range atomic ordering formed in them.

New potential for the control of the atomic structure of materials at low temperatures (due to a sharp increase in the atom mobility throughout the volume of the material [3] far beyond the above calculated diffusion length) is offered owing to the discovery of the mentioned radiation-dynamic effects caused by radiation shaking of condensed matter with post-cascade shock waves under ionizing radiation.

As shown in [5, 7], the temperature and pressure in the regions of nanosized thermal peaks strongly depend on the mass and energy of bombarding ions. The density of the released energy in the dense cascades of atomic displacement increases with ion energy reduction and an increase in the ion mass. So-called metastable media with increased stored energy in a state corresponding to intermediate free energy minimum rather than the absolute one are in a greater extent exposed to external exposure. In particular, this refers to quenched alloys.

In the present study, we investigated the effect of low-energy Ar^+ (39.95 amu) and Xe^+ (131.29 amu) ions (~5 keV), which differ from each other in their atomic mass, on the quenched iron alloy with 8.16 at % Mn (having a tendency to atomic separation) at low temperature (250° C), when the diffusion of substitutional impurities is almost frozen. At the same time, it is shown [2-4] that the atomic mobility in these alloys at lower temperatures can be repeatedly increased by ion irradiation. This refers to the near-surface layers of irradiated materials, the length of which is several orders of magnitude greater than the projected range R_p of the implanted ions (and reaches 10^5 R_p or more [4]). Concentration profiles of the Ar^+ and Xe^+ ions are shown in Figure 1. As can be seen, the ranges of the ions under consideration are extremely small and less than 0.02-0.03 μm, which is about 10^3 times smaller than the thickness of the studied iron–manganese alloy foils under ion irradiation.

X-ray diffraction is ineffective for the analysis of the processes taking place due to similarity of electronic scattering factor of Mn and Fe, which are neighbors in the periodic table, low manganese concentration, nanosized intraphase separation, and the precipitation of the second phase (austenite). Therefore, we used the Mössbauer effect, as in [8], for the analysis of these processes.

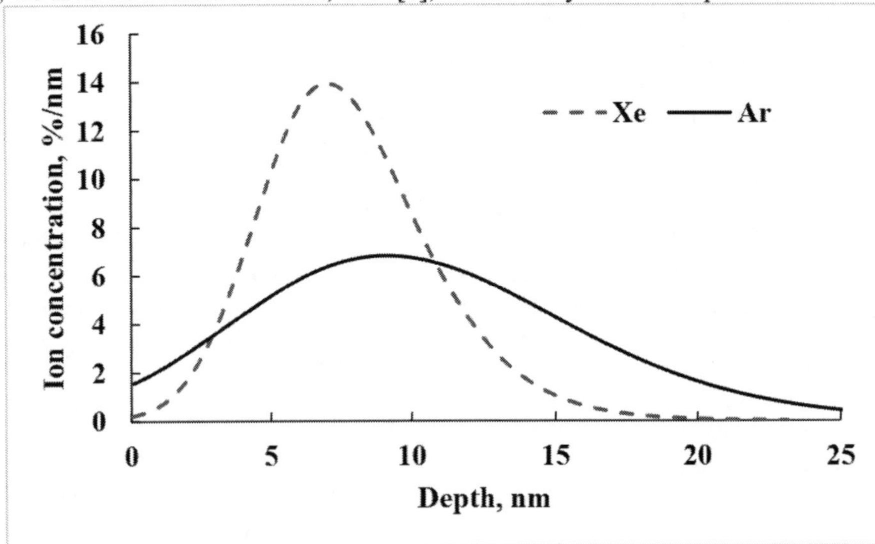

Figure 1. Concentration profiles of the Ar^+ and Xe^+ ions in the Fe-Mn alloy (calculated using kinetic Boltzmann equations [5] method).

2. Experimental

The Fe – 8.16 at % Mn alloy under study according to the chemical analysis had the following content of impurities: 0.070 C, 0.020 N, 0.008 S, and 0.010 wt % P. The alloy was melted in an induction furnace in the argon atmosphere. The ingots were forged to obtain rods 6 × 12 mm in cross section and rolled at 1000°C to form sheets 0.5 mm thick. Pretreatment consisted of normalizing (900°C), annealing (600°C), and sample quenching from 850°C and from γ region (corresponding to the equilibrium fcc γ phase) in a 15% solution of NaCl. After quenching the alloy has a bcc α'-martensite structure with 3-5% of the retained γ phase (austenite). Thin foils for Mössbauer studies were mechanically and electrolytically polished in electrolyte (12 g Cr_2O_3 and 100 mL H_3PO_4). The thickness of the samples was 25 μm.

The samples were irradiated in a continuous mode using an ILM-1 ion beam implanter equipped with a PULSAR-1M ion source based on a low-pressure glow discharge with a hollow cold cathode [9], which also can operate in a pulsed-periodic mode. The samples were irradiated with Ar^+ and Xe^+ ion beams at an energy of 5 keV and current densities from 150 to 170 μA/cm^2 (the current density was continuously varied to achieve the desired temperature). The samples were hanged with thin threads with a low heat conductivity, so that heat removing from a sample was carried out exclusively by radiation. This made it possible to predict the temperature of the sample theoretically [10] (using the emissivity factor chosen in advance from the experiment). In the course of irradiation, the sample temperature was controlled with the help of a thin chromel-alumel thermocouple connected with an Advantech Adam 4000 automated system designed for digital signal registration.

Mössbauer studies were performed using an SM-2201 automated Mössbauer spectrometer under constant acceleration conditions. Isotope ^{57}Co served as the source of γ quanta in Rh. The speed scale of the Mössbauer spectra was calibrated with respect to pure iron.

3. Results and discussions

Mössbauer spectra of the Fe – 8.16 at % Mn alloy after irradiation with low argon and xenon ion fluences (E = 5 keV, j = 150-170 μA/cm^2) significantly change (Figure 2). In particular, the intensity of the components of the external peaks vary, which correspond to the presence of l_1=0 and l_1=1 of atoms

Mn near atom Fe. Component $l_1=0$ decreases, whereas $l_1=1$ increases (indicated by arrows in Figures. 2b, 2c), which clearly indicates the redistribution of atoms in the solid solution [11].

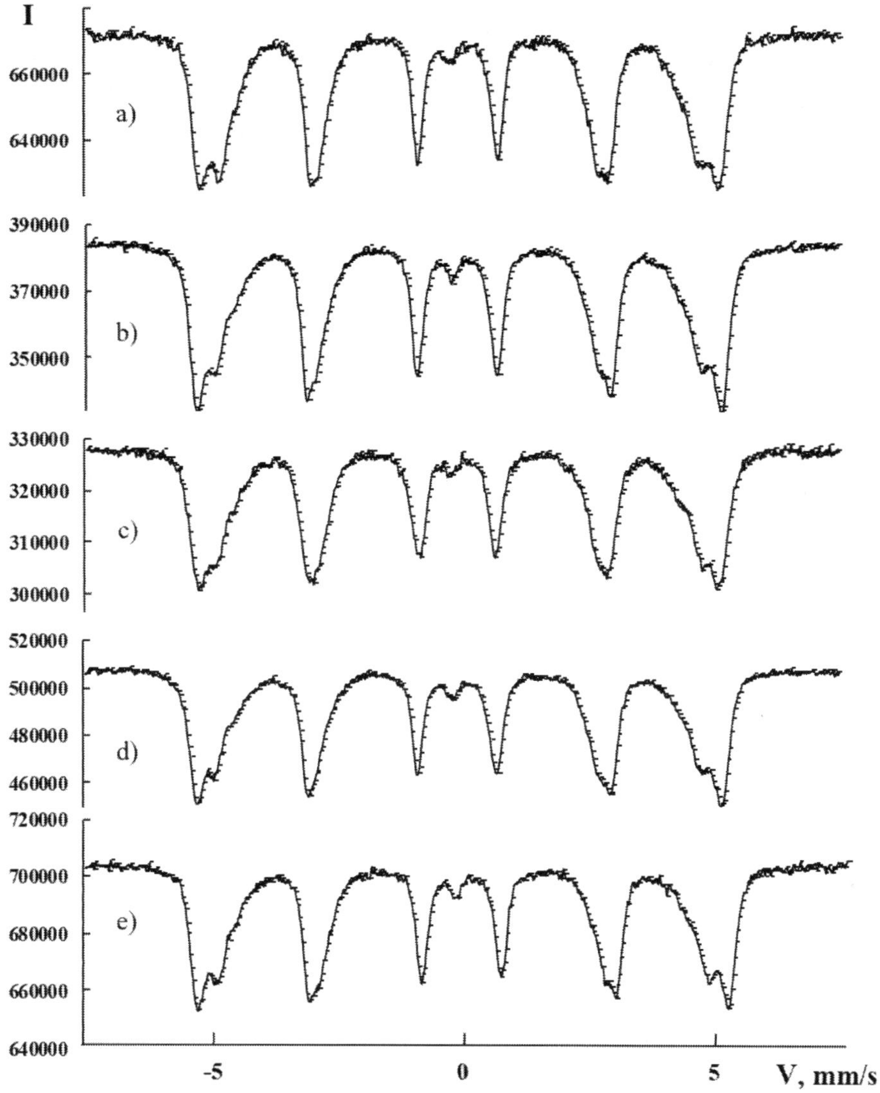

Figure 2. Mössbauer spectra of the samples of the Fe – 8.16 at % Mn alloy after (a) quenching and irradiation at the following various fluences of (b, c) Ar^+ and (d, e) Xe^+ ions: (b, d) heating to 250 °C (without holding); (c, e) heating to 250 °C (for 20 s); dotted lines correspond to experimental data.

The short-range order parameter according to Cowley [12] in the first coordination sphere α ($\alpha = -\frac{\varepsilon_{ab}}{c_a c_b}$, here *a* - Fe, *b* - Mn, ε_{ab} – pair correlation parameter [13]) was determination by analyzing the external shape of the peaks, which corresponded to such nuclear transitions as +1/2 → +3/2 and -1/2 → -3/2, and the Mössbauer spectra (since they have the highest intensity and resolution), taking into account electric and magnetic interactions in the radius of the first coordination sphere using the method described in [11, 13], where a method for calculating a statistical error of the parameter α by referring the structural matrix is described. This approach is justified because, according to [14], the effect of the second, the third, and other remote coordination spheres on the parameters of hyperfine interactions is relatively weak for the above-mentioned concentration of Mn.

The values of parameter α, obtained from the processing of the left and right Mössbauer spectra peaks within the error were zero for all quenched samples.

The intensities of Mössbauer lines after irradiation with Ar^+ ions significantly change already at a fluence of ~ $1 \cdot 10^{16}$ cm^{-2} (exposure time was only ~10 s). During this time the temperature of the samples reaches a stationary temperature of 250°C, the samples were treated without holding at this temperature. This means that the alloy under Ar^+ and Xe^+ion irradiation acquires short-range atomic separation for a very short period of time. The value of short-range atomic order α_1 thus becomes positive (α = 0.17-0.18) both for Ar^+ and for Xe^+ (see Figure 3). A further increase in the irradiation time to 30 s (fluence of ~ $3 \cdot 10^{16}$ cm^{-2}) causes a decrease in the short-range order parameter to α = 0.16 in the case of Ar^+ ion irradiation and to α = 0.12 in the case of Xe^+ ion irradiation.

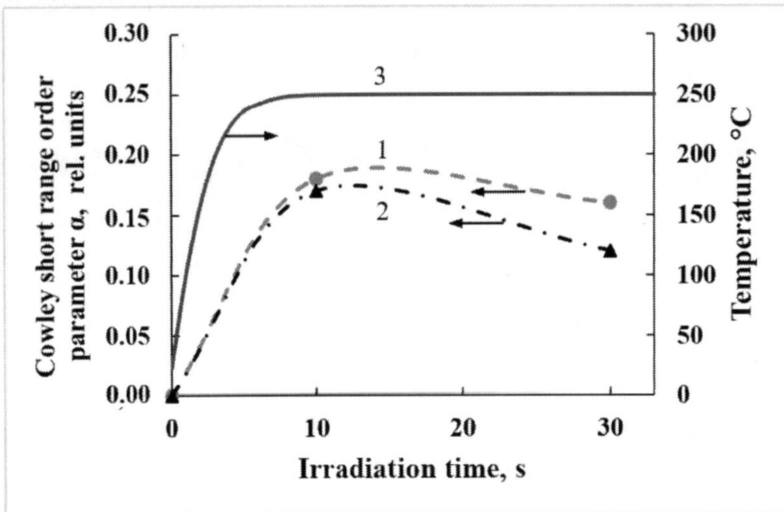

Figure 3. Cowley parameter of short-range atomic order as a function of the irradiation time: (*1*) irradiation with Ar^+ and (*2*) Xe^+ ions. Curve *3* indicates the temperature of the samples as a function of the irradiation time under given irradiation conditions.

This can be explained by the fact that solid solution depleted with manganese at low temperatures may tend to not only short-range separation, but also to short-range atomic ordering found in [8]. In contrast to the separation, this leads to the change in the intensities of spectra components corresponding to $l_1 = 0$ and $l_1 = 1$ impurity atoms in the immediate vicinity of iron atoms.

That is, during the first seconds of irradiation, the solid solution is separated into zones depleted and enriched with manganese atoms. Thereafter, the atomic ordering can occur in the formed zones, leading to an increase in the effective concentration of manganese in the first coordination sphere of iron atoms. The latter indicates the complex ratio of the energy of the pair interaction aa, bb and ab in the iron-manganese system. It may be necessary to refine the model of the atomic redistribution in the iron-manganese alloy, and, consequently, the model approximating the Mössbauer spectra to describe this process in detail. In particular, this concerns not only the effect of the first coordination sphere, as was done in [8] and in this work, but also more distant coordination spheres (at least the 2nd coordination sphere).

Data of [8] somewhat differ from those of the present study, which may be explained by the different composition of alloys used (in particular, the difference in the concentration of interstitial impurities). The alloy used in the present work contains more interstitial impurities than in [8], which is manifested in the presence of a small amount of retained austenite in the quenched alloy (central peak in the center of the Mössbauer spectra, Figure 2).

It should be noted that parameter α in the case of heavier Xe^+ ions more significantly decreases (i.e., an increase in the effective concentration of Mn in the in the nearest neighborhood of Fe), which is in agreement with a large energy release in dense cascades of atomic displacements generated by Xe^+ ions, compared to those generated by Ar^+ ions.

4. Conclusion

Our investigation is an indirect proof of the fact that nanosized dynamic effects (appeared due to the formation of thermal peaks) play the significant role in the effect of accelerated Ar^+ и Xe^+ ions on the structural state of the metastable quenched Fe – 8.16 at % Mn alloy. The temperature of the thermal spikes was measured in [5, 7]. Indeed, the short-range atomic ordering of the short-range separation type is formed at the projected range of the used ions less than 0.02-0.03 nm and temperatures less than 250°C, which are insufficient for the redistribution processes of substitutional atoms in the target volume (25 μm thick) due to thermally and radiation-stimulated diffusion, for the first 10 s of irradiation. The ordering degree somewhat decreases (substantially more for heavier Xe^+ ions) at a further increase in the fluence; which seems to be caused by the complicated structure of atomic ordering, to reliably identify the type of which additional data are required.

In this connection, it seems interesting to study in detail the regularities of the processes using various types of ions and irradiation modes and theoretical models for describing the experiments.

References

[1] Gushchina N V, Ovchinnikov V V and Mücklich A 2016 *Phys. Status Solidi B* **253** 770
[2] Ovchinnikov V V 1994 *Proceedings XVI International Symposium on Discharges and Electrical Insulation in Vacuum* (Moscow-St. Petersburg. SPIE) **2259** p. 605
[3] Ovchinnikov V V, Chernoborodov V I and Ignatenko Yu.G. 1995 *Nucl. Instrum. and Meth. in Phys. Res.* B **103** 313
[4] Ovchinnikov V V 2008 *Phys. Usp.* **51** 955
[5] Ovchinnikov V V Makhin'ko F F and Solomonov V I 2015 *Journal of Physics: Conference Series* **652** 012070
[6] Ovchinnikov V V, Zvigintsev N V, Mikhailov S B and Orel S S 1981 *Fiz. Met. Metalloved.* **51** 815
[7] Ovchinnikov V V, Makhin'ko F F, Solomonov V I, Gushchina N V and Kaigorodova O A 2012 *Technical Physics Letters* **38** 86
[8] Ovchinnikov V V, Gushchina N V and Ovchinnikov S V 2015 *Physics of Metals and Metallography* **116** 1294
[9] Gavrilov N V, Mesyats G A, Nikulin S P, Radkovskii G V, Eklind A, Perry A J and Treglio J R. 2003 *J. Vac. Sci. Technol.* **A14** 1050.
[10] Abdrashitov V G and Ryzhov V V 1993 *Surface Physics, chemistry, mechanics* **2** 106
[11] Ovchinnikov V V, Bogachev I N and Charushnikova G A 1975 *Fiz. Met. Metalloved.* **39** 1270
[12] Cowley J M 1950 *Phys. Rev.* **77** 5 669
[13] Ovchinnikov V V 2006 *Mössbauer Analysis of the Atomic and Magnetic Structure of Alloys* (Cambridge International Science Publishing. Cambridge. UK)
[14] Litvinov V S, Ovchinnikov V V, Dovgopol S P and Karakishev S D 1979 *Fiz. Met. Metalloved.* **47** 96

Acknowledgments

This work was supported by the Russian Scientific Foundation, project no. 15-19-10054.

Investigation of ceramics based on Cu-Sn powder obtained by plasma dynamic method

Yu Shanenkova, A Sivkov, A Ivashutenko and A Tsimmerman

National Research Tomsk Polytechnic University, 30 Lenin av., Tomsk, 634050, Russia

E-mail: julia_kolganova@mail.ru

Abstract. Composites based on copper matrix are of a great interest in various applications. Copper-tin alloys are intensively investigated due to their thermal and chemical stability in combination with good mechanical properties. This work shows the possibility to obtain Cu-Sn ceramics by spark plasma sintering using nanoscale powders consisting of copper and tin, synthesized by plasma dynamic method. This method is implemented by using a coaxial magnetoplasma accelerator with copper electrodes and adding the solid precursor (tin) in the accelerator before carrying out the synthesis process. The synthesized Cu-Sn powders were investigated by X-Ray diffractometry and transmission electron microscopy. It was determined that the final material consists of phase Cu41Sn11. Using this product, the bulk ceramics samples were obtained by spark plasma sintering at different temperatures (150 °C, 250 °C and 500 °C). The changes in microstructure of copper-tin ceramics in dependence on the sintering temperature were also studied. After analyzing all ceramics samples by X-Ray diffractometry and scanning electron microscopy methods, it was found that the optimal temperature for sintering Cu-Sn ceramics, which was made of the powder synthesized by a plasma dynamic method, was equal to 250 °C at pressure 60 MPa. At these conditions, the ceramics sample had the lowest porosity with the smallest grain size.

1. Introduction

Sliding bearings are widely-used in rotating mechanism of different machines. The lifetime of their work strongly affects the stable operation of the responsible mechanisms. The most common their failure cause is not a breakage but the wear and damage of the working surfaces in their parts. Thus, the solution of the problems connected with the increase in the lifetime of machines depends on improving the durability and reliability of friction units and, in particular, sliding bearings [1].

There is a group of soft metals, so-called babbit, which are used as materials for bearings [2]. These species include tin (Sn), lead (Pb), cadmium (Cd), antimony (Sb) and zinc Zn, which are characterized by the presence of solid structural components in the plastic matrix. However, these materials have low fatigue resistance. Therefore, there is a necessity to introduce different additives into material composition, used for producing the bearings.

Earlier work [3] describes a method for improving the conformability of the friction pair by using a coating made of soft anti-friction metals such as indium, copper and silver. The most known method for depositing such coatings is a method of electroerosion alloying. Another way to solve this problem is a pre-activation of copper-tin powders to improve the tribological characteristics of Cu-Sn alloy [4]. The authors of the work [5] study the change of tribological properties in materials based on copper-tin with or without the use of lead at hot pressing of samples. Furthermore, the use of intermetallic pair copper-tin can more efficiently and selectively recover the CO2 to CO in a wide range of potentials [6].

This paper shows a new unique method for obtaining materials based on tin-copper in a high-speed plasma jet generated by a coaxial magnetoplasma accelerator with copper electrodes [7], as well as the creation on their basis of bulk samples using a spark plasma sintering installation. Change in the density and porosity of the obtained bulk samples based on Cu-Sn, as well as their chemical and phase composition were studied with increasing sintering temperature.

2. Experimental

Plasma-dynamic method is based on a system with a coaxial magnetoplasma accelerator, which is supplied from the capacitive energy storage with the charging voltage up to 5 kV and the charging capacity to 28.8 mF. In such a way, the charging energy can have a value up to 360 kJ [8-11]. A sketch-map of the coaxial magnetoplasma accelerator and the principle of its operation are shown in figure 1.

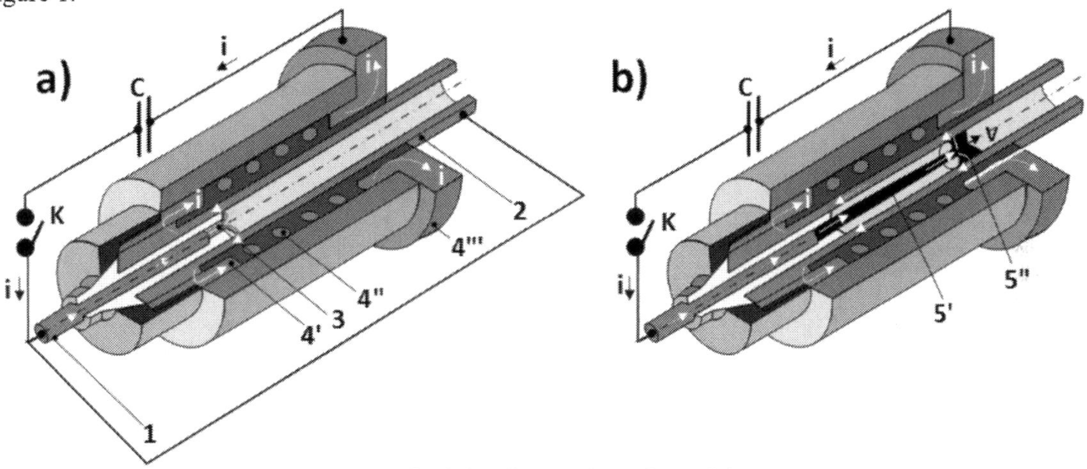

Figure 1. Sketch, elements and principle of operation of coaxial magnetoplasma accelerator: a) initial state; b) state at working: 1 - Central electrode; 2 - Electrode-barrel; 3 - Plasma formation zone; 4 - Inductor (4' – contact cylinder; 4" – solenoid; 4''' – contact flange); 5 - Plasma structure of high-current discharge (5' – Plasma Z-pinch harness; 5" – round plasma jumper).

Previous works have demonstrated the possibility to synthesize the nanodispersed powders with different stoichiometric and phase composition [8] using the accelerator. In order to produce the powdered materials based on copper and tin the coaxial accelerator with copper electrodes was used in order to obtain Cu precursor. Before starting the experiment, tin-containing precursor (chips of tin) was put in the plasma formation zone. A working chamber-reactor, which were used to implement plasma-dynamic process, was preliminary evacuated and filled with inert gas (argon) to prevent the saturation of the resulting material by oxygen (avoid product oxidation).

After closing the power keys (K), the discharge current starts to flow through the current channel. After rising the current up to some level, it creates a plasma arc jumper in the plasma formation zone. Due to the high velocity and temperature of plasma copper material is eroded from the wall of the accelerating channel. This copper interacts with tin, which was put into accelerator and converted to plasma state by increasing current. Thus Cu-Sn-containing plasma exits from accelerating channel into the space of the chamber-reactor, where plasma-chemical reaction is implemented. The synthesized product settles down to the wall of the chamber. After one hour, the chamber is opened and the product is collected.

Figure 2 shows the photogram of the plasma flow. The video record was carried out using a Photron Fastcam SA1.1 high-speed camera through the special protective glass. When the capacitive energy storage has the parameters as mentioned above, at the exit of the accelerator the flow speed is maximum and reaches up to 4.5 km/s.

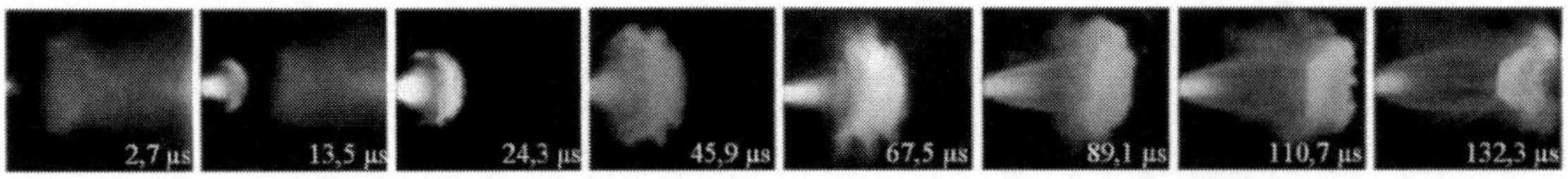

Figure 2. Photogram of plasma flow movement in working chamber.

After collecting the product of a plasma-dynamic synthesis, it was studied by means of X-ray diffractometry (XRD), scanning electron microscopy (SEM) and transmission electron microscopy (TEM) methods. Scanning electron microscopy (SEM) images were recorded using a Hitachi TM-3000 scanning microscope with carbon substrate. X-Ray diffraction (XRD) analysis was carried out using a Shimadzu XRD 7000S diffractometer (CuKα-radiation). The database PDF-2 and the software PowderCell 2.4 were used to make full-profile structural analysis and estimative calculations. Transmission electron microscopy (TEM) images of the product were made using a Philips CM-12 microscope.

After that, the obtained nanodispersed Cu-Sn powder was sintered using the installation for spark plasma sintering (SPS 10-4 Advanced Technology). According to literature data [12], the initial sintering conditions were chosen. The resulting product of a plasma-dynamic synthesis was pre-sifted through a sieve (500 mesh). After that, it was laid in a graphite mold under a pressure of 60 MPa and was sintered at a temperature of 150 °C under high vacuum. After receiving the first ceramics sample, it was decided to raise the temperature up to 250 °C (sample 2) and then up to 500 °C (sample 3), which were at the same pressure and in the same atmosphere.

3. Results and discussion

The results of XRD analysis of the product obtained by plasma-dynamic synthesis, as well as reference cards of the estimated phases are shown in figure 3. It is possible to see that the composition of the synthesized material contains such phases like copper Cu (card No. 4-836), the intermetallic copper-tin $Cu_{41}Sn_{11}$ (card No. 30-510) and lead Pb (card No. 4-686). The appearance of lead in the product can be explained by the fact that tin, which was put in the plasma formation zone, was not a pure product. It has in its composition 10% of lead. The main phase in the product is copper, due to its erosion wear from the surface of the accelerating channel, and the excess mass of submitted material. Also there is the required phase of the intermetallic compound $Cu_{41}Sn_{11}$. Moreover, it is worth noting that the most intense peaks at 43-44 and 48-50 degrees are sufficiently broadened. Upon closer examination, their separation into two peaks is clearly seen. These broadened peaks include the reflections of both phases (copper and intermetallic $Cu_{41}Sn_{11}$).

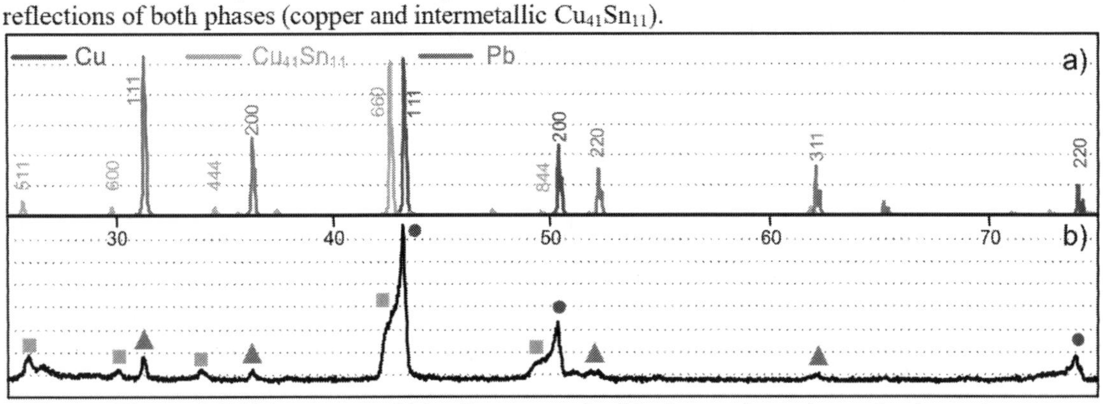

Figure 3. XRD patterns of (a) reference cards, (b) plasma dynamic synthesis product.

In order to investigate the particle morphologies a transmission electron microscope analysis was carried out, the results of which are presented in figure 4. This figure shows that the electronic microdiffraction from the selected area (SAED) has a point-ring character. The phases found at decrypting this diffraction are in a good agreement with XRD data. The bright-field image shows a typical cluster of particles in the product. Dark-field images are obtained in the light of diffracted beams by shifting the diaphragm aperture in the area of corresponding reflexes. The indices of the reflecting crystal planes are indicated on dark-field images and their numbers are indicated on SAED.

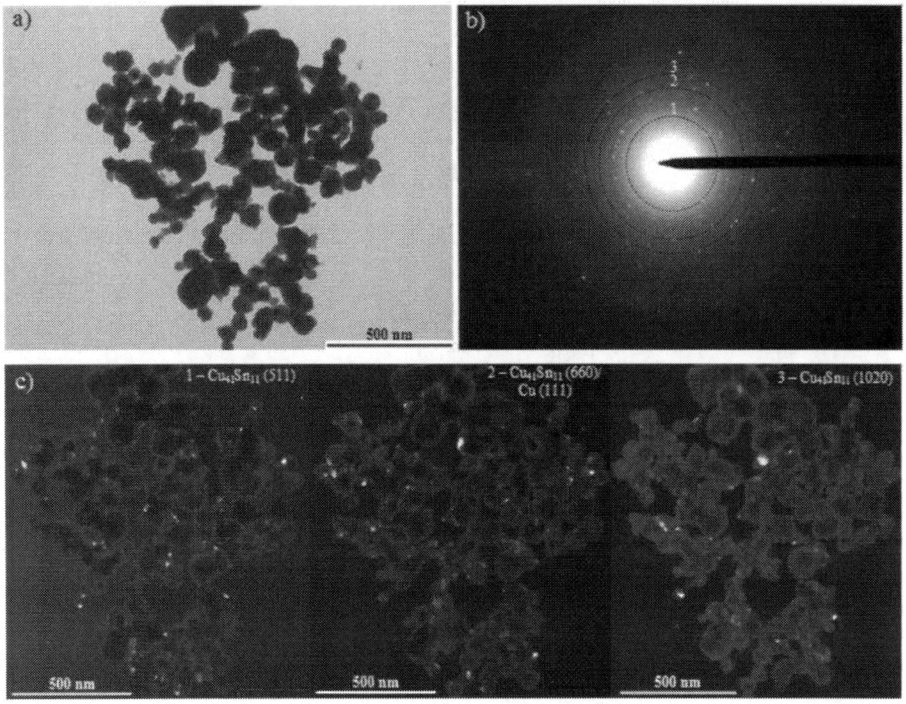

Figure 4. TEM-images of synthesized product: a) bright-field image; b) SAED; c) dark-field images in the light of different reflexes.

The use of transmission microscopy allows accurately finding that the particle sizes in the material range from about 10 nm to ~ 200 nm. The reason for this is the pulsed nature of the plasma dynamic process, during which the current and the electrodischarge plasma parameters change in a very wide range. Such broad particle size distribution can have a positive effect, when creating the bulk materials by spark plasma sintering. During sintering and compacting the fine fraction fills the space between the larger particles that provides obtaining the high-density ceramics.

XRD patterns of samples sintered at different temperatures are presented in figure 5. XRD pattern of the bulk metal material obtained at 150 °C (figure 5b) is almost identical to the diffraction pattern of the initial powder (figure 3b) both at the location of the main peaks and at the intensity of their phases in comparison with each other. Some changes in intensity are seen with increasing in sintering temperature of the bulk material up to 250 °C. In this case, there is an increase in the yield of the phase of the intermetallic compound (figure 5c). However, such increase in sintering temperature on 100 °C does not show significant changes in the product structure. Dramatic changes in XRD pattern are visible for the product sintered at a temperature of 500 °C. The peaks, which are responsible for a phase of the intermetallic compound $Cu_{41}Sn_{11}$, almost disappear, while reflexes of lead increase dramatically. It can be explained by a known phase diagram in Cu-Sn system. At temperatures about 500 °C, tin dissolves in copper, and the sensitivity of XRD analysis becomes not enough to clearly identify intermetallic compound.

The results of scanning electron microscopy are shown in figures 6-8. At sintering the powdered product at a pressure of 60 MPa and a temperature of 150 °C (figure 6), a huge number of pores in the obtained sample is seen even with a slight magnification (1.0k). This is also confirmed at the magnification of 5.0k for a more detailed analysis. Moreover, at this magnification it is clear that the grain size varies from several microns to 40 microns. According to this, it can be concluded that the obtained bulk sample under such conditions has not a sufficient quality due to the significant number of pores. The presence of such amount of pores can be explained by too low sintering temperature of the sample.

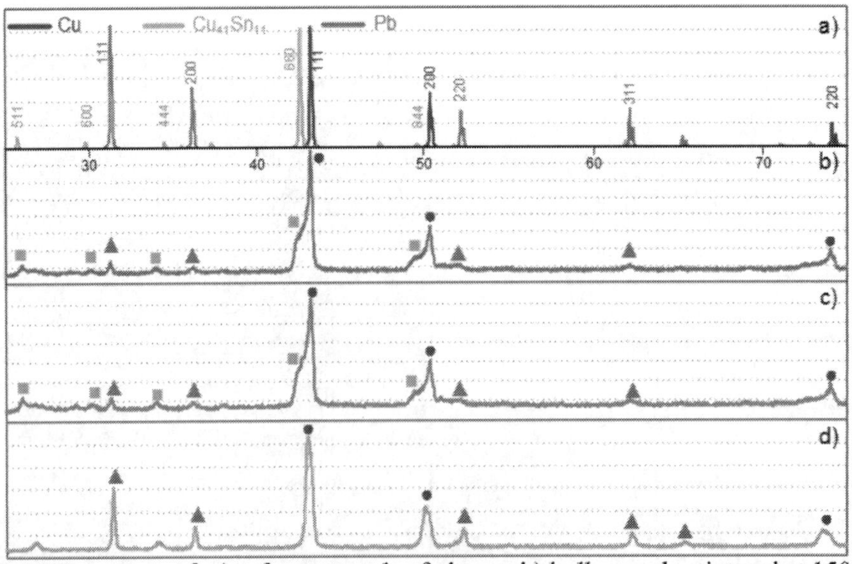

Figure 5. XRD patterns of: a) reference cards of phases; b) bulk sample, sintered at 150 °C; c) bulk sample, sintered at 250 °C; d) bulk sample, sintered at 500 °C.

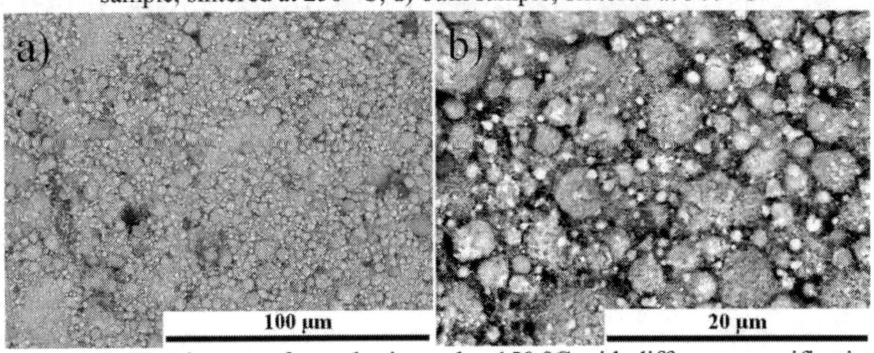

Figure 6. SEM-images of sample sintered at 150 °C with different magnification.

The further increase in the sintering temperature leads to a decreasing in pore amount in the sample and the creation of grain boundaries (figure 7). Nonetheless, the quality of the bulk metal sample was not satisfactory. Thus, the decision was made to increase the value of the sintering temperature up to 500 °C (figure 8).

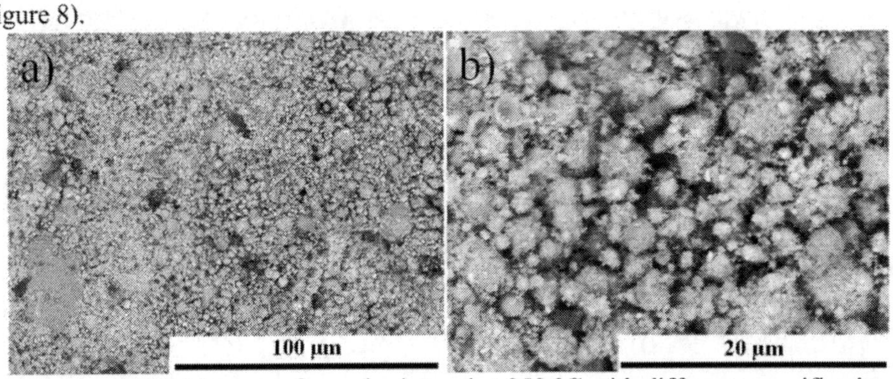

Figure 7. SEM-images of sample sintered at 250 °C with different magnification.

Figure 8a shows that the obtained material has a less porous structure than the previous samples. This product mainly consists of grains of gray color (~ 95 volume %) that are likely to be copper (according to XRD). However, there are impregnations (white color), which fill material in the intergranular space, reducing the number of pores. The white material seems to be the intermetallic lead, because no reflexes of intermetallic $Cu_{41}Sn_{11}$ were found according to XRD data. At higher magnification (x5.0k), it is shown that the grain size varies from 1-2 μm to 20 μm.

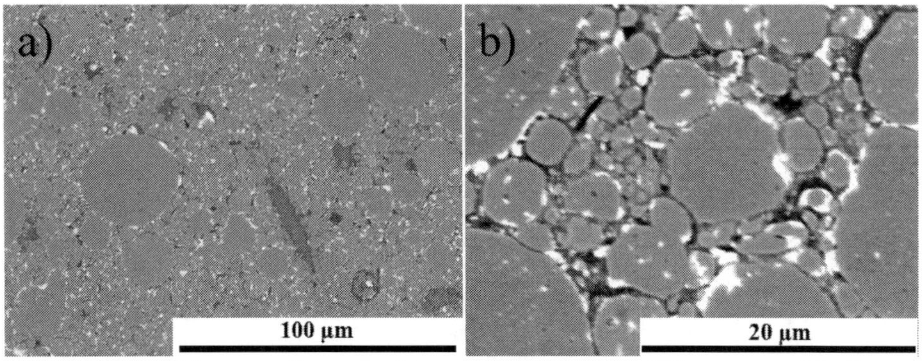

Figure 8. SEM-images of sample sintered at 500 °C with different magnification.

4. Conclusion

As a result of this work, the powder based on copper and tin was obtained using a high-power pulsed system based on the coaxial magnetoplasma accelerator with copper electrodes. Obtaining the intermetallic compound is confirmed by XRD results and analysis by transmission electron microscopy. According to these data, the main phases in the synthesized product are copper and intermetallic compound Cu41Sn11. The particle sizes vary from 10 to 200 nm.

The results on spark plasma sintering show the possibility to obtain the bulk material on the basis of such synthesized copper-tin powder. The influence of the sintering temperature on the resulting products shows that ceramics, obtained by using plasma dynamic synthesis product Cu-Sn, can be sintered even at 150 °C. However, the mean pressure and low temperature did not allow achieving the high quality of bulk material. The increase in the temperature up to 500 °C reduces the number of pores, thereby increases the sample density. However, the increase in sintering temperature at the same time led to the decrease in the content of intermetallic copper-tin phase. Thus, further research is required to find the optimal conditions of sintering for obtaining high-quality bulk samples, which will save the initial phase composition.

Acknowledgements

This work was supported by the Russian Science Foundation (grant № 15-19-00049).

Reference

[1] Lebedev O V et al. 2009 *Nauchnyj zhurnal Bratskogo gosudarstvennogo universiteta:Sistemy, metody, tehnologii* **2** 20-23
[2] Guo X, Zhang G, Li W, Gao Y, Liao H, Coddet C 2009 *Appl. Surf. Sci.* **255(6)** 3822
[3] Tarelnik V B, Martzinovskij V S, Dzuba A V 2013 *Kompressornoe i jenergeticheskoe mashinostroenie* **1** 49
[4] Vityaz P A at al. 2014 *Vestnik Vitebskogo gosudarstvennogo universiteta* **26** 110
[5] Popescu I N, Ghiţă C, Bratu V, Navarro G P 2013 *Appl. Surf. Sci.* **285** 72
[6] Sarfraz S, Garcia-Esparza A T, Jedid A, Cavallo L, Takanabe K 2016 *ACS Catalysis* **6(5)** 2842
[7] Sivkov A, Shanenkova Y, Saigash A, Shanenkov I 2016 *Surf. Coat. Technol.* **292** 63
[8] Sivkov A, Ivashutenko A, Shanenkova Y and Shanenkov I. 2016 *Adv. Powder Technol.* **27(4)** 1506
[9] Sivkov A, Saigash A and Kolganova Y 2013 *Russian Electrical Engineering* **84(8)** 418
[10] Sivkov A, Saygash A, Kolganova J, Shanenkov I 2014 *IOP Conference Series: Materials Science and Engineering* **66**(1) 012048
[11] Kuzenov V V, Ryzhkov S V, Frolko P A, Shumaev V V 2015 *Trudy MAI* **82** 1
[12] Ileskina G S and Psarev V I 1980 *Powder Metallurgy and Metal Ceramics* **19(7)** 463

IOP Conf. Series: Journal of Physics: Conf. Series **830** (2017) 012088 doi:10.1088/1742-6596/830/1/012088

Computer Analysis of the Field Ion Images of Platinum Irradiated with Ar^+ (E = 30 keV) Ions

V A Ivchenko[1,2] and V V Ovchinnikov[1,2]

[1]Institute of Electrophysics, Ural Branch of Russian Academy of Sciences, Amundsena street 106, Yekaterinburg, 620016, Russia
[2]Ural Federal Technical University named after the First President of Russia B.N. Yeltsin, Mira street 19, Yekaterinburg, 620002, Russia

E-mail: viae05@rambler.ru

Abstract. The work presents the field ion microscopy images of the surface atomic layers of irradiated platinum (layer-by-layer from 1st to 5th and more distant). A specially developed algorithm and software were used for image analysis to study the effect of Ar^+ ions (E = 30 keV) on the subsurface atomic structure of pure platinum. The coordinates of atoms, the brightness of their images, and their sizes were determined. The curves of the distribution of the brightness and atomic radii were built. It is shown that the width of the distributions significantly decreases when moving from the damaged surface deep into the platinum bulk, whereas the number of atoms in the field ion microscopy images increases. Field ion microscopy images with only those atoms whose brightness (or radius) is in certain range were synthesized (these atoms can be highlighted in the ionic field microscopy images). It has been established that damaged zones include the images of atoms with both abnormally low and abnormally high radii. The principal possibility of reconstruction of the 3D-picture of the irradiated metal at a temperature of liquid nitrogen ($T \sim 77$ K) is shown.

1. Introduction

A problem of paramount importance in the radiation physics of solids is a detailed study of the effect of different types of ionizing radiation on structural-phase state of irradiated media. The topical problems in developing the methods of modification of structure and properties of materials by the beams of accelerated ions are related to the depth of occurring structural changes and phase transformations as well as to defining the types of the radiation-induced defects and the peculiarities of their distribution in the bulk of material. In case of cascade-forming irradiation, the interest lies in experimental study of the structure of the atomic displacement cascades passage zones, the so-called "depleted zones" [1-3], containing different types of surviving defects after the stages of thermal spikes and intra-cascade diffusion.

Field ion microscopy (FIM) is an in-situ experimental method allowing the nature of radiation-induced lattice damage to be studied at an atomic level. Just as importantly, the method, along with visualization of an atomic-pure surface of irradiated materials, also allows the controlled removal of surface atoms under an electric field at cryogenic temperatures.

Field ion micropictures include images of thousands of atoms located at both the regular lattice sites and in the areas containing different types of defects. The source data for quantitative analysis of a lattice state would be the data on coordinates, brightness, dimensions, as well as features of surface atom image shapes, depending on the depth of location of layers under exposure to different external factors.

As a result, a series of dependences may be derived. A promising method in analyzing defect structure and atoms dynamics is reconstruction of the imaged atomic size and brightness distribution functions for subsurface and more remote layers. The most important fundamental task is digital reconstruction of a 3D structure.

This paper presents the results of analysis of field ion images of pure platinum surface atomic layers after irradiation with Ar^+ (E = 30 keV) ions with application of an algorithm and a computer software developed specifically for this purpose on the basis of the mathematical pattern recognition techniques.

2. Experimental

The object under investigation was 99.99% pure polycrystalline platinum. The specimens for exposure to irradiation and subsequent studying using a field ion microscope were needles (with 30-50 nm tip point radius of curvature) made from wire blanks by the method of electrochemical polishing. Irradiation of a needle-shaped platinum emitter preliminarily certified under a field ion microscope was carried out with singly-charged accelerated (to E = 30 keV) Ar^+ ions to a fluence of $F = 10^{16}$ ion/cm^2, at 150 μA/cm^2 ion current density (T = 70°C at needle basis).

The irradiated specimen tip was again placed under a microscope and, by registration with a photo or video camera of surface field ion micropictures at controlled removal of surface atoms, experimental data were collected for further analysis of the subsurface volume. The surface micropictures were registered so as to allow a minimum number of evaporated atoms in the intervals between shots. In this paper the intervals between the first five micropictures were the times of evaporation under an electric field of one atomic layer (001). Such procedure is necessary for precision study of lattice defects appearing under the action of accelerated Ar^+ ions.

The field ion microscope was outfitted with a microchannel ion-electron converter enhancing 10^4 times the brightness of ion images. Liquid nitrogen served as a coolant (T = 77 K), and spectrally pure neon was used as an imaging gas.

3. Results and discussions

Figure 1*a* shows an ionic micropicture taken of a platinum specimen certified prior to the irradiation. After exposure of the metal to an argon-ion beam, Figure 1*b*, an ionic contrast was registered which is not typical for an ionic micropicture of a defect-free metal. Micropictures of platinum surface irradiated with 30-keV argon ions to fluence $F = 10^{16}$ cm^{-2} (ion/cm^2) display breaks of a (regular before irradiation) circular pattern of ionic contrast. These are the irregularities in the contrast circular pattern that are interpreted as crystal structure defects. The defect type is identified in the course of the controlled removal of surface atoms under an electric field.

In the course of surface atoms removal, a fairly large number of microphotographs of the irradiated metal surface were obtained (each layer from 1st to 88th), sufficient for detailed description of the subsurface volume.

The algorithm of identifying single atoms and fixing their coordinates is based on application of the maximum of brightness function $Y(x, y)$ criteria (where x and y are image coordinates). The first- order $\frac{\partial Y}{\partial y}, \frac{\partial Y}{\partial x}$, also the second-order $\frac{\partial^2 Y}{\partial x^2}, \frac{\partial^2 Y}{\partial y^2}$, derivatives were analyzed in finite difference representations with account for the discrete nature of digital images.

Atomic image coordinates, brightness, dimensions were determined and atomic brightness and radii values distribution functions were desined with the use of computer software. After that the field ion images of all atoms, or only of the atoms with the brightness or radii lying within the preset limits, were synthesized (colour marking of different groups of atoms in the ion field images is feasible).

IOP Conf. Series: Journal of Physics: Conf. Series **830** (2017) 012088 doi:10.1088/1742-6596/830/1/012088

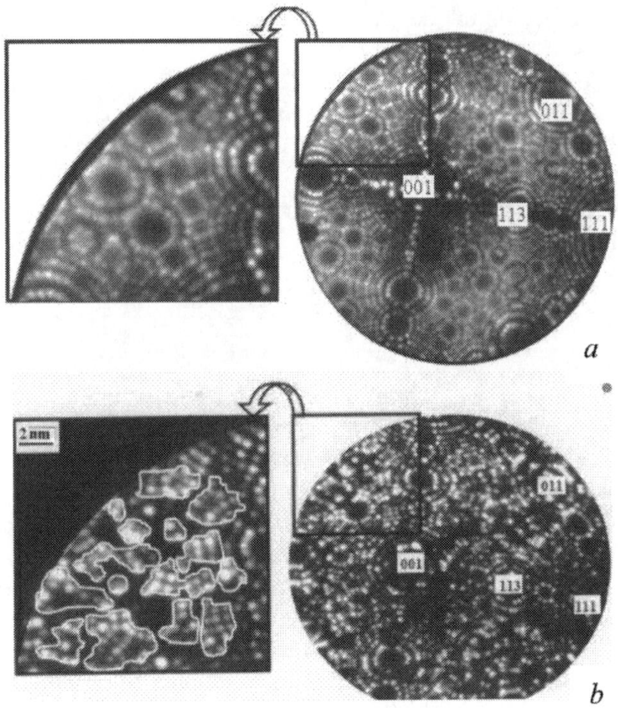

Figure 1. Pt neon images: *a* – ionic contrast of certified crystal; *b* – ionic contrast of surface after irradiation with Ar^+ ions ($F = 10^{16}$ ion/cm^2).

Figure 2 shows the field ion images of atomic layers of pure Pt irradiated with 30-keV argon ions, depending on their location depth.

As a result of these images computer processing, coordinates of all atoms, their brightness and radii were determined. For describing brightness, a scale of brightness from 0 to 255 units was used. Digital field-ion images transformed to black-and-white patterns were processed.

Figure 3 shows an example of application of the developed software in the analysis of a field-ion image of the first subsurface layer of pure platinum irradiated with a beam of accelerated ions under the above-described conditions.

Figure 4 shows the radii distribution functions $f(R)$ (probability density) for the images of atoms of the 1st, 5th and 81st from surface atomic layers of irradiated platinum. As can be seen, as it approaches the irradiated surface, the distributions width increases significantly. A similar conclusion may be made for the computed brightness distribution functions $f(Y)$ (Figure 5).

It should be noted that the number of imaged atoms in the micropictures increases with distance of the atomic layer from the irradiated surface, varying from 1560 in the 1th layer to 2014 in the 81st layer located at maximum depth in the least damaged zone corresponding to an almost ideal lattice. This correlates to a lower defect ratio of deeper layers (can be assumed that the radius of the tip of emitter remaining practically unvaried in the course of the controlled removal of at least the first dozen atomic layers).

Using the imaged atoms brightness and radii distribution functions, highlighting the zones of the strongest change of field images resulting from ion irradiation may be attempted. Assuming that micropictures of the atomic layers exposed to irradiation include the images of both damaged and undamaged sections of a lattice, and subtracting function $kf_{81}(R)$ with selectable "weight" ($k \leq 1$) from function $f_1(R)$ (where the lower indices denote the layer numbers), we obtain the distribution function for the damaged sections. The latter include images of the atoms with both anomalously low and

IOP Conf. Series: Journal of Physics: Conf. Series **830** (2017) 012088 doi:10.1088/1742-6596/830/1/012088

anomalously high radii (Figure 6). The sections with anomalously low radii have been formed in a significantly greater number.

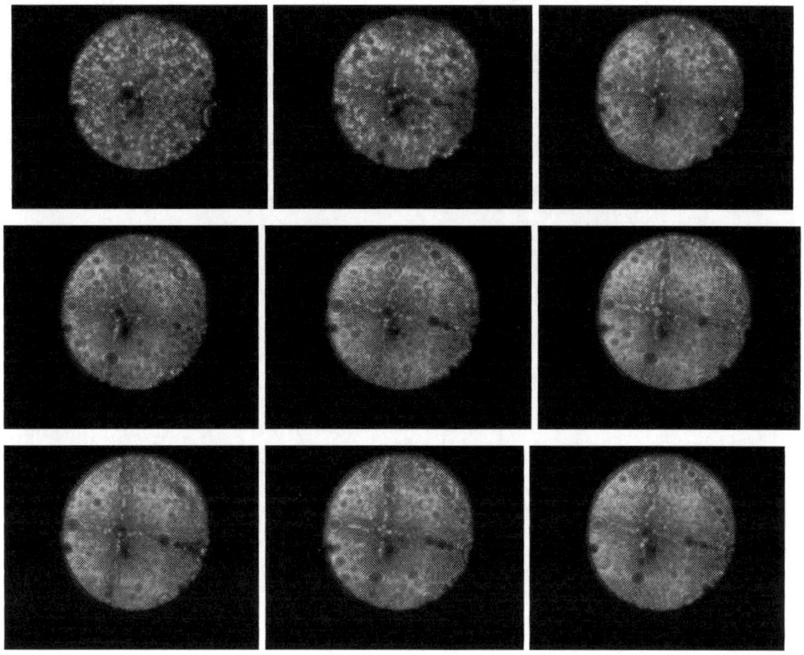

Figure 2. Field ion images of subsurface atomic layers of pure platinum irradiated with Ar^+ ions (E = 20 keV, j = 150 μA/cm^2, F = 3.5 10^{15} cm^{-2}): *1, 2, 3, 4, 5* – 1st, 2nd, 3rd, 4th and 5th surface atomic layers; *6, 7, 8, 9* – 9th, 10th, 11th and 12th layers (count made by evaporation under field of one atomic layer (001)).

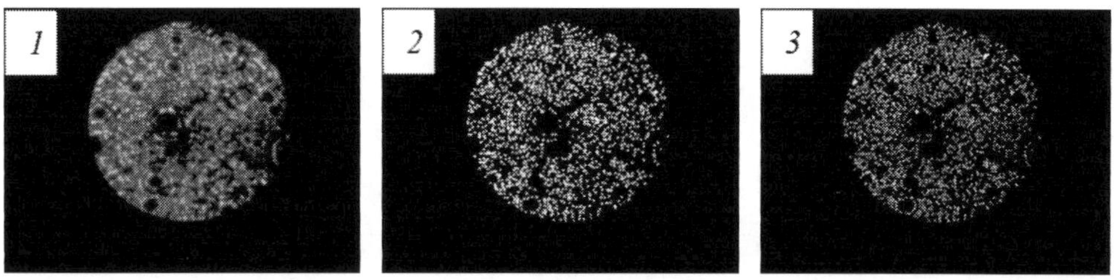

Figure 3. Field ion images of platinum emitter first subsurface layer (R ~ 30 nm) (*a*): *1* – experiment; *2* – reconstructed image with defined atomic coordinates (radii and brightness of all imaged atoms are fixed, R = 2 ps, Y = 255 units); *3* – synthesized image based on atomic coordinates recognition and computation of atomic brightness and effective radii.

Figure 7 shows the reconstructed image of a picture corresponding to the distribution functions for the 1st and 81st layers (*a* and *b*), and to two sections of difference function $f_1(R) - kf_{81}(R)$ (*c* and *d*) with anomalously low and anomalously high radii values.

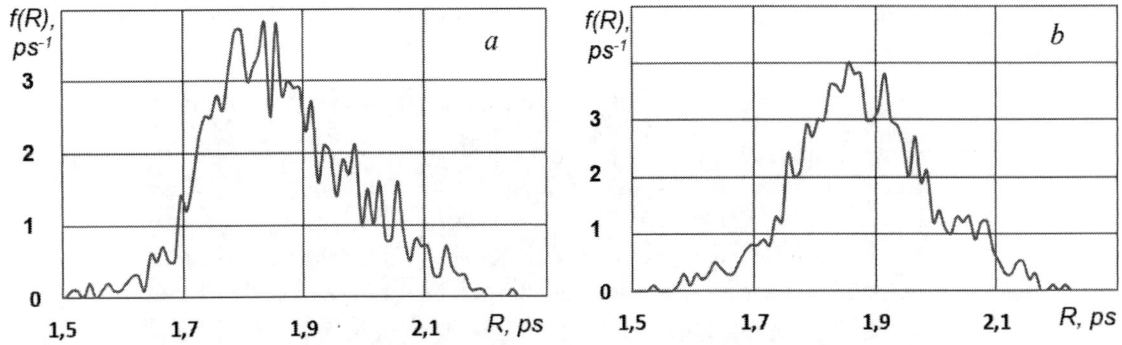

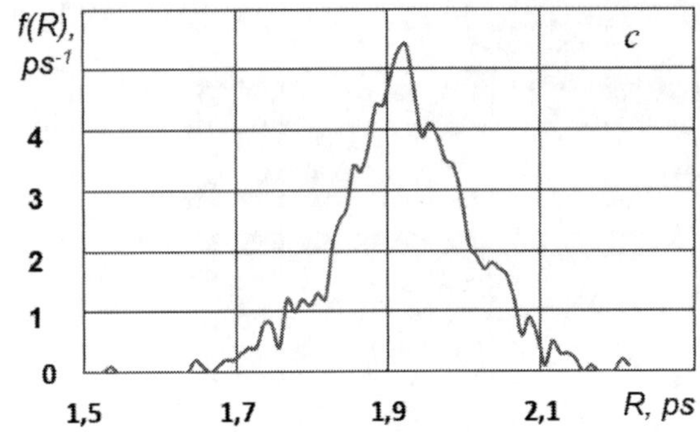

Figure 4. Imaged atoms radii *R* (ps) distribution functions (probability densities *f(R)*) for irradiated platinum: *a* – first subsurface layer, *b* – fifth subsurface layer, *c* – 81st layer.

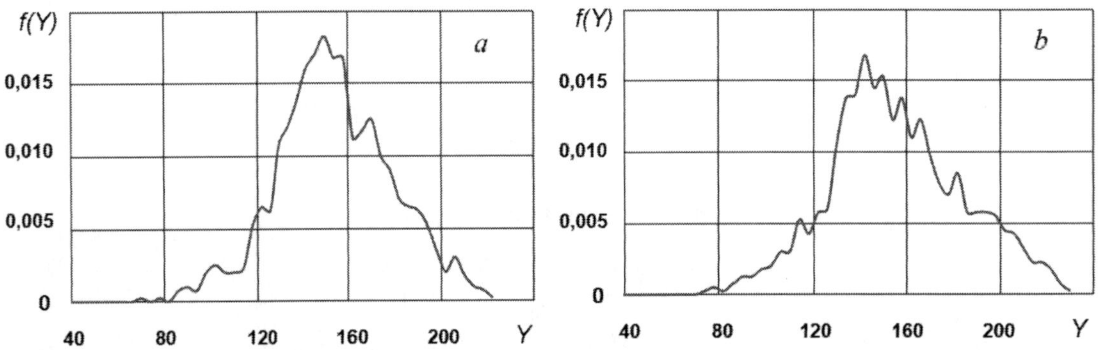

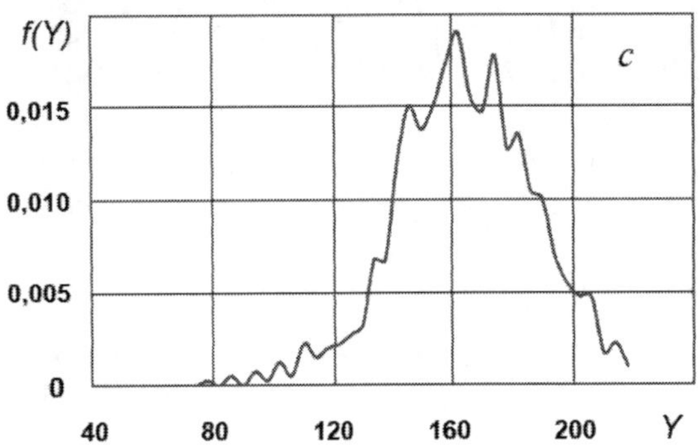

Figure 5. Imaged atoms brightness distribution functions (probability densities *f(Y)*) for irradiated platinum: *a* – first subsurface layer, *b* – fifth subsurface layer, *c* – 81st layer.

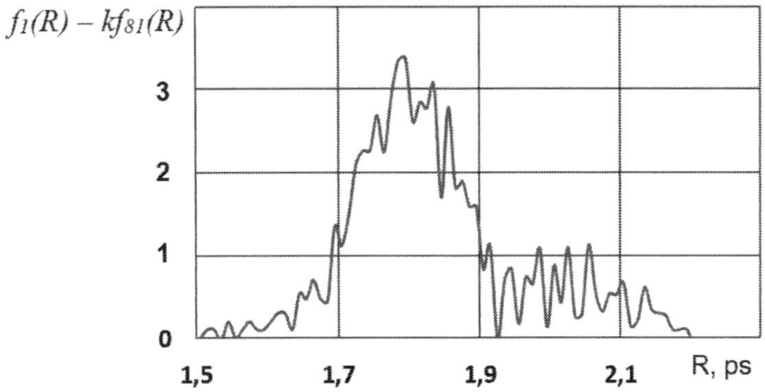

Figure 6. Difference curve $f_1(R)-kf_{81}(R)$ characterizing modified areas of field ion image obtained from the irradiated platinum first subsurface layer ($k = 0.43$).

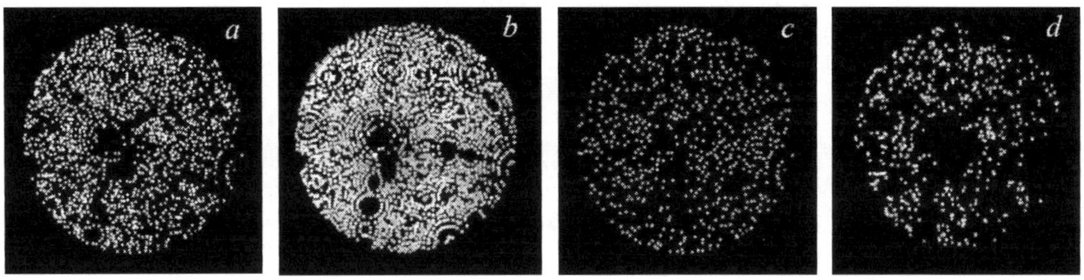

Figure 7. Synthesized field ion images: 1st, 81st atomic layers *(a, b)*; areas with low (*c*) and high (*d*) radii values in first layer ($Y = 255$).

4. Conclusion

Thus, in the present work we have tested the software for analysis of field ion images of an atomic-pure surface aimed at studying the nature of the effect of Ar^+ ($E = 30$ keV) ions on the atomic structure of subsurface atomic layers in pure platinum (99.99 Pt). The distribution functions for brightness and effective radii values of the imaged atoms have been derived.

As further research, the reconstruction of a 3D picture of defect layers of irradiated platinum (as well as other metals, e.g., W) at arbitrary emitter sections is proposed, potentially displaying several tens of atoms, based on software-derived numerical arrays of such atoms coordinates.

References

[1] Boowkett K M and Smith D A 1970 *Field Ion Microscopy Defects in Crystalline Solids* (Amsterdam, London: North Holland Publishing Company)

[2] Muller E W and Tsong T T 1973 *Field ion microscopy, Field ionization and Field evaporation* (Oxford: Pergamon press)

[3] Kirsanov V V Suvorov A L and Trushin Yu V 1985 *Processes of radiation defect formation in metals* (Moscow: Energoatomizdat)

Acknowledgments

This work was performed as part of State Task No. 0389-2014-0002 and it was supported in part by RFBR Grant No. 16-08-00615 and it was supported by the Act 211 of the Government of the Russian Federation (Agreement No. 02.A03.21.0006).

IOP Conf. Series: Journal of Physics: Conf. Series **830** (2017) 012089 doi:10.1088/1742-6596/830/1/012089

Effect of ion beam treatment (Ar^+, E= 30 keV) on the microstructure of titanium alloys

N V Gushchina[1], V V Ovchinnikov[1,2], F F Makhin'ko[1] and S A Linnik[1]

[1]Institute of Electrophysics, Ural Branch of Russian Academy of Sciences, Amundsena street 106, Yekaterinburg, 620016, Russia
[2]Ural Federal Technical University named after the First President of Russia B.N. Yeltsin, Mira street 19, Yekaterinburg, 620002, Russia

E-mail: viae05@rambler.ru

Abstract. The effect of continuous Ar^+ ion beams (E = 30 keV, j = 400 μA/cm^2; F = $5\cdot10^{16}$ cm^{-2}) on the hardness and microstructure of the Gr2-titanium-based and Ti-6Al-4V alloys after cold working and subsequent recrystallization annealing is studied. It has been found that the initial microstructure of the annealed alloys in the equilibrium state, which consisted of recrystallized equiaxed grains, does not change after irradiation; thus the hardness of the irradiated samples is comparable with that of the initial samples. The initial fine fiber structure of the deformed Ti-6Al-4V alloy changes insignificantly after both annealing (T = 790°C, 30 min) and ion irradiation under used conditions. In contrast to that, the effect of rapid radiation annealing of the cold-worked titanium Gr2 alloy with beams of accelerated ions of inert gas at low temperatures (lower by 130°C than the conventional annealing temperature of these alloys at 680°C) and for a shorter time (9 min instead of 35 min, τ_{irr} ~ 20 s) has been revealed. The result is the formation of a uniform recrystallized structure with equiaxed fine grains 5-10 μm in size in the entire volume of the 3-mm-thick samples, despite the fact that the average projected range of 30-keV Ar^+ ions in titanium is only 20 nm.

1. Introduction

Examples of the successful use of accelerated ion beams in industrial processes to modify the electrical, magnetic, mechanical, contact-chemical, and other material properties have emerged in recent years [1-10]. A number of technological ion sources [2, 11, 12] capable of generating ion beams of various chemical elements with energies from a few to several tens of keV are developed and successfully used in manufacture.

However, the main problem of accelerated-ion-beam-based technologies used to modify the physical and mechanical properties of metals and alloys is a little depth of the impact zone. In normal conditions it is limited by a zone of the accelerated ion range, the length of which at ion energies obtained in the technological accelerators ($<10^6$ eV) does not exceed several tens or, in the best case, several hundred nanometers [13].

Authors of many published works dedicated to irradiation of metals and alloys with accelerated ion beams studied structural changes in the ion range zone and tried to find a way to increase the ion penetration depth using heating (i.e., thermally-stimulated processes) and increased ion energy. However, all these attempts to use high-energy ions (> 10 MeV) and powerful ion beams in order to increase the ion penetration depth deeper than several tens of micrometers, which is not sufficient for a great number of commercial applications, have failed.

The use of radiation-dynamic long-range effects [7-10] increases substantially the depth of the modified zone under ion-beam processing. These effects are caused by the generation of nanosized post-cascade shock waves (during evolution of dense cascades of atomic collisions) and propagation of these waves in a medium rather than by the doping of thin surface layers and the formation of radiation defects in them. These waves can initiate the plastic flow of the material on their front and phase transformations that are similar to combustion and detonation phenomena; because of this, changes in the structure and properties can be observed at the submillimeter and millimeter depths [8-10].

In particular, it was established in our works studying the influence that ion beams have on the structure and properties of various metastable media [10, 14] that under certain circumstances ion bombardment can be an alternative to furnace annealing. Radiation annealing in alloys occurs at significantly lower temperatures (by 150-200 K) and energy consumption, and this process takes less time, as compared with isothermal furnace annealing. The giant depth of the influence of accelerated ions (from 3 to 8 millimeters for commercial aluminum alloys [10, 14-16]) in the course of radiation annealing (at one-side irradiation) should be noticed. On the basis of these data, we obtained a patent for a method of treatment of commercial Al-Mg, Al-Cu-Mn-Mg, and Al-Li-Cu-Mg aluminum alloys with gas ion beams (Ar^+) with an energy of 40 keV, instead of prolonged (for 1-2 h) intermediate furnace annealing at increased temperatures (320-400 °C) [17].

It is of interest to study the possible application of radiation-dynamic effects under ion irradiation for the targeted modification of the structure and properties of other metals and structural materials based on them.

The aim of the work is to study the effect of Ar^+ ions with energy of 30 keV on the structure and properties of titanium-based Gr2 and Ti-6Al-4V alloys. In contrast to iron and aluminum, titanium as a basis for structural materials, owing to its small density (4.505 g/cm^3), large specific strength, incredible high corrosion resistance, and considerable strength at increased temperatures shows an advantage in aerospace engineering.

2. Experimental

The following alloys were used as the objects of the study: commercial titanium of Grade 2 (α alloy) and Ti-6Al-4V alloy (α+β martensite alloy). The titanium alloy sheets ~ 3 mm thick in different initial states were prepared at PAO "VSMPO-AVISMA Corporation" (Verkhnyaya Salda, Sverdlovsk region): after cold working and recrystallization annealing (Gr2: 680°C, 35 min; Ti-6Al-4V: 790°C, 30 min). The chemical compositions of the alloys are listed in Table 1. Unlike aluminum and its alloys, titanium has a high melting temperature and other peculiarities. For example, it has the ability to exist in two polymorphic states, namely, α titanium and β titanium, transitioning into one another at ~ 882.5 °C [18]. Owing to the polymorphic α↔β transformation, a sufficiently large variety of structures can be obtained in titanium-based alloys. This, in turn, ensures a change in mechanical properties in a wide range.

The exposure to accelerated ion beams with an energy of 30 keV was carried out using an ILM-1 ion beam implanter equipped with a PULSAR-1M ion source based on a low-pressure glow discharge with a hollow cold cathode [11] in a continuous mode. A line-focus beam 20×100 mm^2 in cross section, under which plate-like samples of the alloys with dimensions of 90 × 210 × 3 mm^3 were moved back and forth at a speed of 1 cm/s, was cut out from Ar^+ ion beam of circular cross section using a collimator. The following beam parameters were specified: ion energy E = 30 keV, ion current density j = 400 μA/cm^2, and fluence $F = 5\cdot10^{16}$ cm^{-2}. The total time of treatment was approximately 9 min but the pure fluence accumulation time was about 20 s only. In the course of irradiation, the sample temperature was controlled with the help of a thin chromel-alumel thermocouple connected with an Advantech Adam 4000 automated system for digital signal registration. The maximum temperature of the samples was no higher than 550°C.

IOP Conf. Series: Journal of Physics: Conf. Series **830** (2017) 012089 doi:10.1088/1742-6596/830/1/012089

Brinell hardness and Rockwell hardness were measured using a TSh-2M and a TK-2M hardness testing machines, respectively, on both irradiated and nonirradiated sample sides. The metallographic analysis of the initial and irradiated samples was performed using a Neophot-30 optical microscope in the longitudinal and transverse directions over the entire cross section of a sample.

Table 1. Chemical composition of the studied titanium alloys.

Content of elements, wt. %							
Ti	Fe	O	H	N	C	Al	V
Gr2							
balance	0.3	0.25	0.015	0.03	0.1	-	-
Ti-6Al-4V							
balance	0.4	0.2	0.015	0.05	0.1	5.5-6.75	3.5-4.5

3. Results and discussion

The hardness of the samples of the titanium Gr2 and Ti-6Al-4V alloys in the various initial states and after irradiation with continuous Ar^+ ion beams are listed in Table 2.

The results listed in Table 2 suggest that the hardness of the *annealed* Gr2 and Ti-6Al-4V alloys remains the same after Ar^+ ion irradiation under used conditions. The resulting effect of ion irradiation on the deformed samples of the same alloys is different. Thus, irradiation decreases the hardness of the deformed Gr2 alloy to the values close to those of the annealed samples of this alloy. The hardness of the deformed Ti-6Al-4V alloy remains the same after irradiation.

Table 2. The hardness of the samples of the titanium Gr2 and Ti-6Al-4V alloys in the initial state and after irradiation with continuous Ar^+ ion beams.

Alloy	Alloy processing	HB		HRC	
		Side 1*	Side 2	Side 1	Side 2
Gr2	Cold working	244 ± 5	248 ± 5	26.5 ± 0.5	28.6 ± 0.5
	Cold working + irradiation (Ar^+, E = 30 keV, j = 400 μA/cm^2; $F = 5\cdot10^{16}$cm^{-2}; $\tau_{irr} \sim 20$ s)	195 ± 5	189 ± 5	16.3 ± 0.5	15.3 ± 0.5
	Annealing at T = 680°C for 35 min air cooling	172 ± 5	167 ± 5	14.0 ± 0.5	12.0 ± 0.5
	Annealing + irradiation (Ar^+, E = 30 keV, j = 400 μA/cm^2, $F = 5\cdot10^{16}$ cm^{-2})	168 ± 5	167 ± 5	12.5 ± 0.5	13.5 ± 0.5
Ti-6Al-4V	Cold working	323 ± 5	318 ± 5	38.4 ± 0.5	38.1 ± 0.5
	Cold working + irradiation (Ar^+, E = 30 keV, j = 400 μA/cm^2; $F = 5\cdot10^{16}$cm^{-2})	330 ± 5	329 ± 5	38.6 ± 0.5	38.3 ± 0.5
	Annealing at T = 790°C for 30 min, air cooling	322 ± 5	328 ± 5	39.2 ± 0.5	39.7 ± 0.5
	Annealing + irradiation (Ar^+, E = 30 keV, j = 400 μA/cm^2, $F = 5\cdot10^{16}$ cm^{-2})	334 ± 5	334 ± 5	39.4 ± 0.5	39.5 ± 0.5
Note: *1 – irradiated side; 2 – nonirradiated side.					

Figure 2 demonstrates the microstructure of the Gr2 and Ti-6Al-4V alloys in various initial states. The cold-worked alloys exhibit a fine fiber structure (Figures 1a, 1c). The microstructure of the Gr2 alloy sheets 3 mm thick after annealing consists of recrystallized equiaxed grains, which are finer in the center over the transverse section than those along both surfaces. The grain size in the center of transverse section is ~ 25-40 μm (Figure 1b). In the Ti-6Al-4V alloy after annealing under used conditions (Table 2), a fine fiber structure retains (Figure 1d), but the fibrous structure becomes less pronounced.

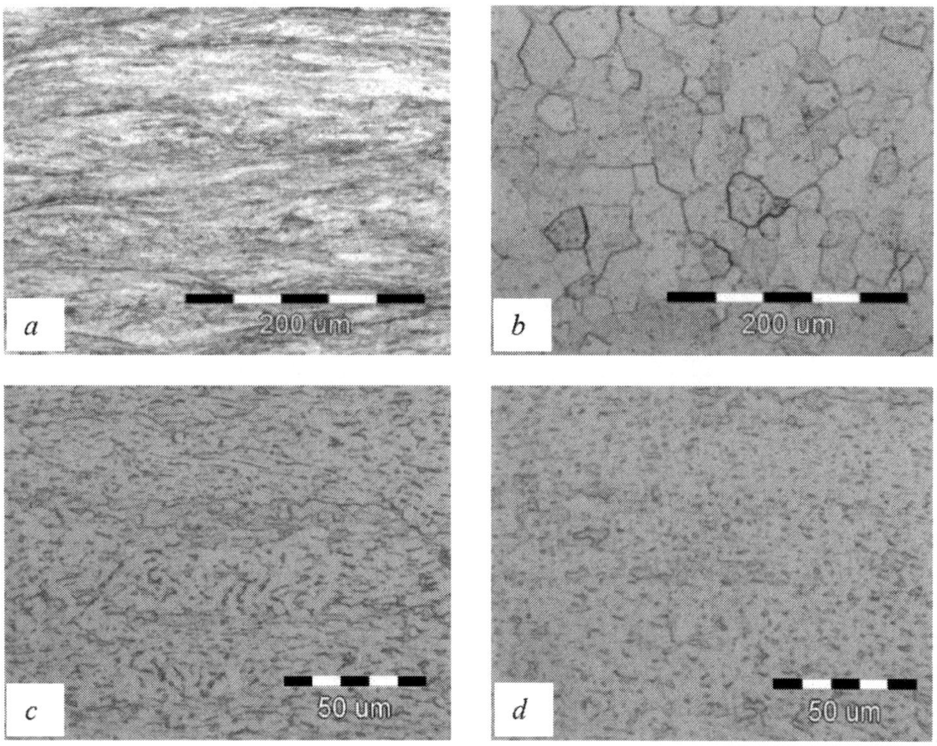

Figure 1. Microstructure of the sheets ~ 3 mm thick made of the (a, b) Gr2 and (c, d) Ti-6Al-4V alloys after (a, c) cold working and (b, d) recrystallization annealing; the central part of the samples over the cross section.

The irradiation of the deformed Gr2 alloy initiates recrystallization processes resulting in the formation of equiaxed fine grains 5-10 μm in size in the entire volume of the samples. Figure 1 demonstrates the microstructure of the Gr2 alloy over the cross section of sample 3 mm thick after ion irradiation. It can be seen that the irradiation-annealing-induced recrystallized structure is the same on the irradiated side, in the central part of the sample, and on the nonirradiated side. The average projected range of Ar^+ ions with an energy of 30 keV in titanium is only ~ 20 nm [13]. In this case, the processes observed over the large depth are caused by the radiation-dynamic effects (related to the propagation of nanosized post-cascade shock waves in medium) [10].

It should be noted that radiation-induced annealing of the deformed Gr2 alloy occurs during heating of the samples (without holding) to a temperature of 550°C. This is lower by 130°C than the conventional annealing temperature of these alloys (680°C). Furthermore, the irradiation annealing takes place for a shorter time (9 min instead of 35 min). Pure time of radiation annealing is ~20 s. The size of the recrystallized grains after ion beam treatment is smaller than that formed during furnace recrystallization annealing, which, as a rule, have a positive effect on the processing plasticity of the alloy.

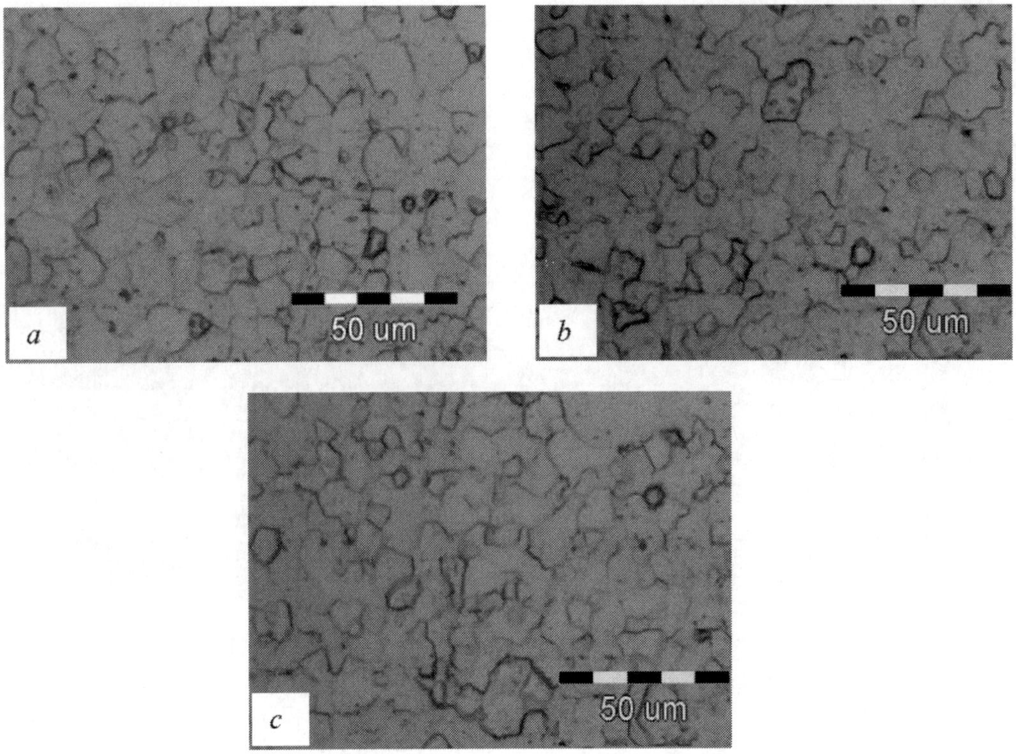

Figure 2. Microstructure of the Gr2 alloy after ion irradiation over the cross section of a sample 3 mm thick: (a) irradiated side, (b) central part of the sample thickness, and (c) nonirradiated side.

After irradiation of the deformed Ti-6Al-4V alloy, the initial fine fiber structure is retained. It is well known that alloyed titanium alloys, in particular, Ti-6Al-4V, at the same deformation degree have a higher recrystallization temperature than Gr2, since such alloying elements as Al and V markedly increase this temperature. It should be noted that the deformed microstructure has remained also after full recrystallization annealing at 790°C for 30 min (see Table 2). Two-phase titanium alloys are characterized by the fact that a marked change in the microstructure can be observed only during isothermal annealing [18], which includes alloy heating to relatively high temperatures of ~ 900-950°C, which are sufficient for recrystallization processes. Moreover, slow cooling to temperatures, which provide a high stability of the β phase and the final cooling in air to room temperature, are required.

Irradiation of the *annealed* Gr2 and Ti-6Al-4V samples under used conditions does not lead to the change in the microstructure. The samples are in an equilibrium state after annealing in contrast to the deformed metal, which possesses an increased internal energy and is thermodynamically unstable. Therefore *radiation-dynamic* effects (observed at low and very *low fluences*) do not show up.

Long-range effects of the other nature [4] are observed under *high-dose* irradiation (10^{17}-10^{18} cm^{-2}) of the annealed materials. They are related with the accumulation of high *static* stresses created by implantation of impurities (capable to exceed the yield strength of the material) and the formation of dislocation loops in the surface layer. These stresses ensure the movement of the generated dislocations deep into material (to few tens of μm). Such effect has not been observed in our case, since the fluence was relatively low.

4. Conclusion

It was established that the initial microstructure, which consisted of recrystallized equiaxed grains, was retained after Ar^+ ion irradiation of the *annealed* titanium Gr2 and Ti-6Al-4V alloys under used conditions (E = 30 keV, j = 400 μA/cm^2; $F = 5 \cdot 10^{16}$ cm^{-2}). The hardness of the irradiated samples was comparable with that of the initial samples. In the case of the irradiation of *cold-worked* Ti-6Al-4V alloy samples, the initial structure and the hardness remained the same. On the contrary, irradiation of the *cold-worked* Gr2 alloy samples 3 mm thick with the initial fine fiber structure reduced its hardness (to the same value on both the irradiated and nonirradiated sample sides). This was associated with the recrystallization in the entire volume of samples, resulting in the formation of fine equiaxed grains of 5–10 μm in size. The average projected range of 30-keV Ar^+ ions in titanium was ~ 20 nm only, which is more than 10^5 times greater than the depth of the impact.

Radiation-induced annealing of the deformed Gr2 alloy occurred during heating of the samples (without holding) to a temperature of 550°C. This is lower by 130°C than the conventional annealing temperature of these alloys (680°C). The treatment time decreased from 35 to 9 min (τ_{irr} ~ 20 s). Thus, this work shows the possibility of rapid radiation annealing of the titanium Gr2 alloy with beams of accelerated ions of inert gas at low temperatures at the expense of radiation-dynamic effects.

References

[1] Didenko A N, Ligachev A E and Kurakin I B 1987 *The impact of beams of charged particles on the surface of metals and alloys* (Moscow: Energoatomizdat)
[2] Gribkov V A, Grigorev F I, Kalin B A and Yakushin V L 2001 *Promising radiation-beam material processing technologies* (Moscow: Kruglyi god)
[3] Anishchik V M and Uglov V V 2003 *Ion- and plasma-beam modification of tool materials* (Minsk: BGU)
[4] Didenko A N, Sharkeev Yu P, Kozlov E V and Ryabchikov A I 2004 *Long-range effects in ion-implanted metallic materials* (Tomsk: NTL)
[5] Gokul Lakshmi S, Tamilselvi S, Rajendran N and Arivuoli D 2004 *J. of applied electrochemistry* **34** 3 271
[6] Bratushka S N and Malikov L V 2011 *Probl. At. Sci. Technol.* **6** 126
[7] Kreindel Yu E and Ovchinnikov V V 1990 *Vacuum* **42** 81
[8] Ovchinnikov V V, Chernoborodov V I and Ignatenko Yu G 1995 *Nucl. Instrum. and Meth. in Phys.Res.* **B 103** 313
[9] Ovchinnikov V V, Goloborodsky B Yu, Gushchina N V et al. 2006 *Appl. Phys.* A. 2006. **83** 83
[10] Ovchinnikov V V 2008 *Phys. Usp.* **51** 955
[11] Gavrilov N V, Mesyats G A, Nikulin S P, Radkovskii G V, Eklind A, Perry A J and Treglio J R. 2003 *J. Vac. Sci. Technol.* **A14** 1050
[12] Gavrilov N V and Kamenetskikh A S 2004 *Review of Scientific Instruments* **75** 5 1875
[13] Biersack J P and Haggmark L G 1980 *Nucl. Instr. & Meth.* **174** 257
[14] Ovchinnikov V V, Gavrilov N V, Gushchina N V et al 2010 *Russian Metallurgy (Metally)* **3** 207
[15] Ovchinnikov V V, Gushchina N V, Makhinko F F et al 2008 *The Physics of Metals and Metallography* **105** 375
[16] Ovchinnikov V V, Goloborodsky B Yu, Gushchina N V et al. 2014 *Izv. Vyssh. Uchebn. Zaved., Fiz.* **57** 10/3 206
[17] Method for producing flat products made of aluminium alloys: Patent RF № 2363755 / Ovchinnikov V V, Gavrilov N V, Gushchina N V, Shkolnikov A R, Mozharovsky S M, Filippov A V Published 10.08.2009. Bulletin № 22. Priority 08.12.2006
[18] Gorynin I V and Chechulin B B 1990 *Titanium in engineering* (Moscow: Mashinostroenie)

Acknowledgments

This work was supported by the Russian Scientific Foundation, project no. 15-19-10054.

Dynamics of titanium surface characteristics after its treatment by runaway electron preionized diffuse discharge

M A Shulepov, M V Erofeev, V S Ripenko and V F Tarasenko

Institute of High Current Electronics SB RAS, Tomsk, Russia

E-mail: MixShlp@yandex.ru

Abstract. Ageing changes of the elementary composition of titanium surface (VT-1) treated by runaway electron preionized diffuse discharge in nitrogen at atmospheric pressure have been studied. Dependence of the elementary composition of titanium surface layers on surface free energy has been shown.

1. Introduction

Runaway electron preionized diffuse discharge (REP DD) has some distinguished properties such as the possibilities to uniformly generate discharges in different gases and their mixtures in a wide range of pressures without contraction [1,2] as well as short duration and low discharge input power which allows to modify near-surface layers (several tens of nanometers [3]) of the materials placed into discharge plasma. The possibility of thin surface layer modification without damages [3,4] is very important for nanotechnology.

The main types of metal surface modification by REP DD are oxidizing and cleaning from the carbon contaminations [5,6]. Oxidizing of the metal near-surface layers forms the protective coating against the influence of the ambient environment. Due to that fact titanium is the corrosion resistive metal. It is also a known method of the titanium surface hardening due to increasing the thickness of the oxidized layer by titanium heating in oxygen atmosphere. The main disadvantage of this method is a difficulty to control growth of the oxidized layer because of the changes of the oxygen diffusion rate with the titanium heating. Moreover, increasing the oxidized layer leads to corrosive cracking under mechanical action, destruction of the surface layer, and propagation of the oxidizer into the metal depth with time, that is why the formation of the thin protective coating on the titanium surface is an important task.

In this paper we present the ageing results of the titanium surface, which was modified by cold plasma of REP DD. Four titanium alloy (VT-1) specimens were used to determine the datum point of the elementary composition changes with time. Before the diffuse discharge treatment all specimens had been polished until mirror surface (roughness Ra = 0,079 μm) and cleaned in ultrasonic bath with alcohol.

2. Experimental setup and methods of measurements

In the experiments, we used a voltage pulse generator GIN-100-1 (Antares Ltd., Russia) with voltage amplitude of 50 kV and pulse repetition frequency (PRF) up to 800 Hz. The REP DD was generated into interelectrode gap 1 (Figure 1a) by applying voltage pulse to a sharp-end cathode 2. A treated specimen was placed on a plane anode 3, the number of discharge shots was 80000. The distance

IOP Conf. Series: Journal of Physics: Conf. Series **830** (2017) 012090 doi:10.1088/1742-6596/830/1/012090

between the cathode and anode was 12 mm. REP DD was initiated in nitrogen flow rate of 2 sl/min. Figure 2b shows the photograph of the REP DD at PRF of 40 Hz.

The changes in concentration of the main chemical elements in the surface layers were measured with Auger-electron spectrometer Shkhuna-2 (NPO "Electron", Russia) on 2, 27 and 63 days after treatment. The specimens were stored in the ambient environment covered from dust.

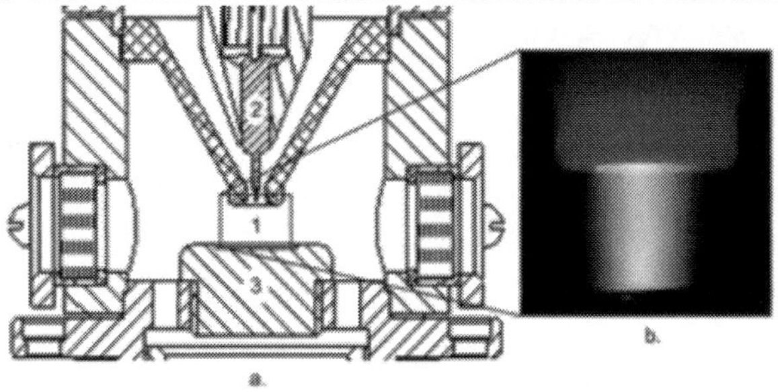

Figure 1. Scheme of the discharge chamber. a: 1-interelectrode gap, 2 – sharp-end cathode, 3- plane anode; b: photograph of the REP DD

3. Results and discussion

Figure 2 shows the results of the elementary composition on the titanium surface specimens.

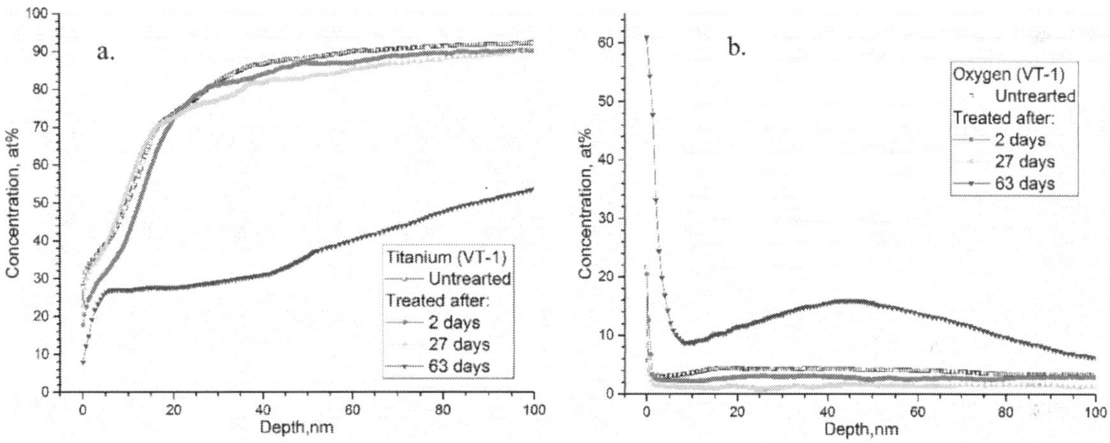

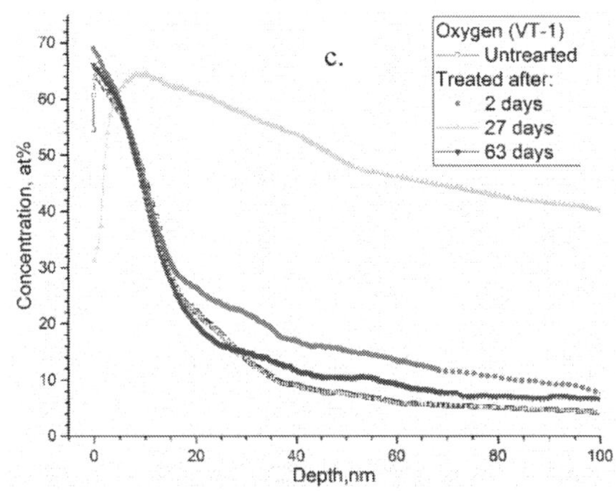

Figure 2. In-depth concentration profiles of titanium (a), carbon (b) and oxygen (c) of untreated (initial) specimen and modified specimens by 80000 REP DD pulses at different period of time after treatment

It is seen from the figure that the surface cleaning (figure 2b) after plasma treatment allows the oxygen from the ambient air intensively to diffuse into surface layers of the specimens even during the treatment process (figure 2c). As can be seen from figure 2c, in the treated specimen after 2 days, the oxygen concentration in the top surface layer at the depth up to 30 nm is 10 at. % higher than that in the untreated specimen. The changes of the carbon concentration (figure 2b) show that top surface layers were cleaned from the carbon-contained contaminants during ultrasonic cleaning, but at a depth larger than 5 nm specimen was cleaned by REP DD.

During the next 27 days, oxygen concentration in the surface layers was practically unchanged due to all chemical bonds occupied by oxygen. Nevertheless, oxygen atoms continue to diffuse into the metal bulk, while carbon concentration remains unchanged.

However, the significant changes occurred after 63 days after treatment. Impurities concentration reached 90 at. % on the surface, and at the depth less than 10 nm, the main share of the impurities are carbon-contained compounds. At a depth larger than 10 nm, more than 60 at. % belongs to the oxygen. Carbon concentration deeper than 120 nm also increased and reached 15 at. % at the depth of 5 nm.
In addition, surface free energy (SFE) of the titanium alloy VT-1 before and after discharge treatment at different moments of time was evaluated through water contact angle measurements according to the Owens-Wendt method [7] from equation (1) and (2):

$$(1+\cos\theta_{L1})\gamma_{L1} = 2(\gamma^{d}_{L1}\cdot\gamma^{d}_{S})1/2 + 2(\gamma^{p}_{L1}\cdot\gamma^{p}_{S})1/2 \quad (1)$$
$$(1+\cos\theta_{L2})\gamma_{L2} = 2(\gamma^{d}_{L2}\cdot\gamma^{d}_{S})1/2 + 2(\gamma^{p}_{L2}\cdot\gamma^{p}_{S})1/2 \quad (2)$$

where θ_{L1} и θ_{L2} – water contact angles of control liquds, γ^{d}_{L1}, γ^{d}_{L2} – dispersive SFE of control liquids, γ^{p}_{L1}, γ^{p}_{L2} – polar SFE of control liquids, γ_{L1}, γ_{L2} – overall SFE of control liquids, γ^{d}_{S} и γ^{p}_{S} – dispersive and polar SFE of titanium specimen. Deionized water and glycerine was used as the control liquids.

Figures 3 and 4 show the dependences of SFE, polar and dispersive components from the number of discharge pulses and ageing time.

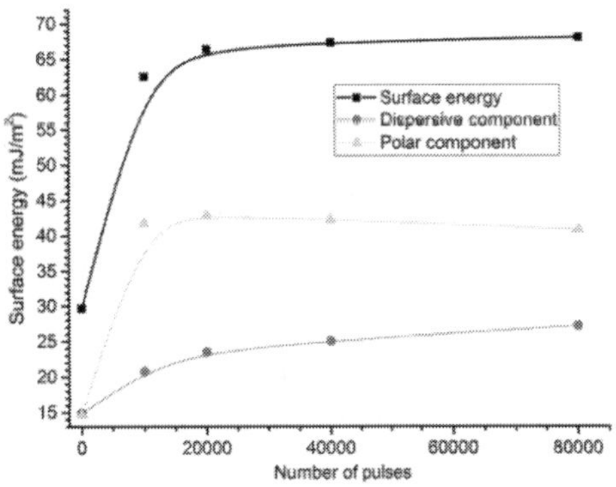

Figure 3. Titanium alloy VT-1 surface free energy and its components versus number of REP DD pulses.

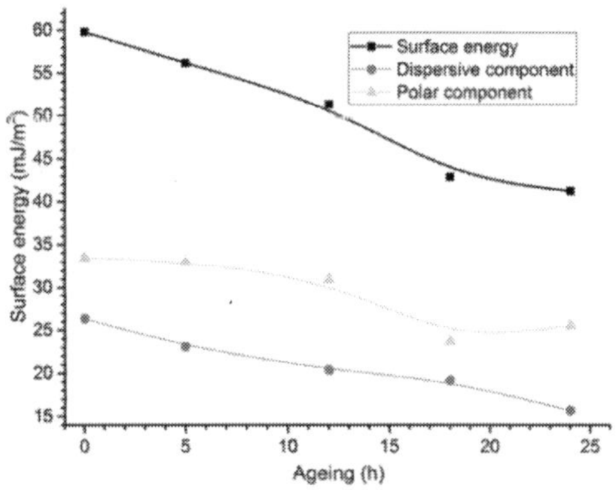

Figure 4. Titanium alloy VT-1 surface free energy and its components versus ageing at the ambient air.

Experiments of SFE measurements have shown activation of titanium surface under the REP DD treatment. As can be seen from figure 3, SFE increased 2 times after the treatment by 20000 discharge pulses. Herewith dispersive component of the SFE increased constantly while polar component after 20000 pulses decreased. As it is known, dispersive component includes van der Waals force and others unspecific interactions (for instance, Coulomb interaction); polar component includes strong interaction and hydrogen bonds. The increase of the dispersion component of SFE can be explained by increasing the oxygen adsorption into titanium bulk along with increasing the surface temperature under discharge action. Decreasing the hydrogen adsorption due to increasing surface temperature [8] and formation of titanium oxide in the top surface layers lead to a decrease in the polar component. However, titanium SFE decreased 1.5 times in 24 hours after plasma treatment (figure 4). It should be noted that the value of dispersive component drops is faster than that of the polar component. Oxide layer formed under the REP DD action does not allow oxygen to penetrate into the depth of titanium, while it cannot protect the titanium surface from the hydrogen penetration.

Formation on the surface of the big amount of carbon-contained compounds can be explained by interaction of CO and CO_2 from the air atmosphere with water film formed in reaction of oxygen with hydrogen on the titanium surface. At room temperature, CO_2 is adsorbed by water with formation of carbonic acid H_2CO_3 [9], which is unstable in the water solutions, and dissociates to hydrocarbonate HCO_3 due to reverse hydrolysis. In the end, on the metal surface at room temperature in the ambient air, the formation of the complex equilibrium system could be possible:

$$CO_2 \uparrow \overset{H_2O}{\longleftrightarrow} CO_2 \cdot H_2O \leftrightarrow H_2CO_3 \overset{-H^+}{\longleftrightarrow} H_2CO_3^- \overset{-H^+}{\longleftrightarrow} CO_3^{2-}$$

Unfortunately, Auger-electron spectroscopy (AES) method does not detect hydrogen at a depth up to hundreds of nanometers, nevertheless, the presence of such reaction on the surface of the titanium alloy referenced in [10], which described well the obtained experimental results.

4. Conclusion

Thus, the formation of the protective oxide film with thickness of 20 nm under the action of REP DD was confirmed by the AES depth profiles. This film protects titanium from the oxygen adsorption into the bulk depth at least during the month after plasma treatment. Experimental results of surface free energy before and after REP DD action show physical-chemical processes of titanium ageing.

References

[1] Tarasenko V F, editor. 2014 *Runaway Electrons Preionized Diffuse Discharges*. (New York: Nova Science)
[2] Tarasenko V F, Orlovskii V M and Shunailov S A 2003 *Russian physics journal* **46(3)** 325
[3] Shulepov M A, Akhmadeev Yu Kh, Tarasenko V F, Kolubaeva Yu A, Krysina O V and Kostyrya I D 2011 *Russian Physics Journal* **53(12)** 1290
[4] Tarasenko VF, Baksht EK and Burachenko AG 2015 *Russian Physics Journal* **58(12)** 1702
[5] Tarasenko V F, Shulepov M A and Erofeev M V 2015 *Phys. Atom. Nuclei.* **78** 1670
[6] Tarasenko V F, Erofeev MV and Shulepov M A 2015 Cleaning and modification of the near-surface layers of metals under the action of runaway electron preionized diffuse discharge *Plasma Sciences (ICOPS)* DOI: 10.1109/PLASMA.2015.7179729
[7] Owens D K and Wendt R C 1969 *Journal of Applied Polymer Science* **13** 17
[8] Yakimenko L M 1977 *Electrode materials in applied electrochemistry* (Moscow: Chemistry) p.264
[9] Nijland O J 1990 *Organic chemistry* (Moscow: High school) p. 751.
[10] Lidin R A, Molochko V A and Andreeva L L 2007 *Reactions of inorganic substances: Handbook* (Moscow: Drofa) p. 637

Acknowledgements

The work was supported by the Russian Science Foundation (project #14-29-00052).

IOP Conf. Series: Journal of Physics: Conf. Series **830** (2017) 012091 doi:10.1088/1742-6596/830/1/012091

Modelling of process formation of the nanocomposite TiN-Cu layers received by vacuum-arc evaporation of Ti and magnetron sputtering of Cu

D B-D Tsyrenov, A P Semenov, N N Smirnyagina.

Institute of Physical Materials Science SB RAS, Sakhyanovoy 6, Ulan-Ude, 670047, Russia

E-mail: dmitriyzak@mail.ru

Abstract. Modeling a TiN-Cu layers deposition process on a fused silica substrate under given conditions is carried out in this work. Calculation formulas which allow to determine the films thickness and their uniformity, with a substrate holder being located at an angle of 45 degrees to the normals of mutually perpendicular planes of the evaporated titanium cathode and the magnetron sputtering copper cathode, are given. The results of this work will be used to analyze the distribution velocity of the substance condensation flow and the character of the formed composite layers depending on the geometry of the cathode-substrate system.

1.Introduction

The studies connected with the development of new technologies for making composite layers with good plasticity and high hardness are of great interest. In this regard, creating composite coatings TiN-Cu by combining two discharge processes (vacuum arc evaporation of titanium in a nitrogen-containing plasma and ion-plasma sputtering of the copper target) is proposed. The design features of the installation realizing this approach, various forms of condensation surfaces, the heterogeneity of the streams density distribution and their characteristics in the working volume – they all put geometric factors in a number of insufficiently studied parameters significantly affecting the properties of the received coatings [1].

There are a number of reports [1-5] on calculating the parameters of a coating process (current density, deposition rate and the coating thickness) by the method of vacuum-arc evaporating or the method of magnetron sputtering at different points of a vacuum chamber. However, they are not suitable for calculations when the substrate holder is positioned at an angle of 45 degrees to the normals of the evaporated titanium cathode and the sputtering copper magnetron cathode planes perpendicular to one another. At the same time the properties of the obtained coatings are largely determined by the deposition rate and the character of their distribution on the condensing surface and the layer thickness.

This report is devoted to studying the condensation distribution speed of substance flows generated by a vacuum arc evaporator and a planar magnetron, the thickness of the obtained coatings and the character of the formed layers depending on the geometry of the cathode – substrate system.

2.Model description

We will make a deposition process model for the case when the target and the substrate are not parallel, but are coaxial. In this case the sputtering angle and the angle of condensation are not equal, i.e. $\varphi \neq \phi$. Calculations of the film thickness were made according to the angle of condensation, and it was 45 degrees. At the same time we will make the following assumptions: I – the sprayed and evaporated atoms have a directional movement and are distributed in space according to the cosine law; II – the atomized and vaporized atoms do not collide with each other and with the atoms of the working gas, III – the sprayed and evaporated atoms are condensed at the point of impingement with the substrate.

In our calculations the expression for magnetron sputtering (2) was used. The expression (10) was used to determine the film's thickness in case of the flat disk evaporator sizes being final with the radius of R'.

The geometric scheme of the model «target -substrate» is shown in figure 1.

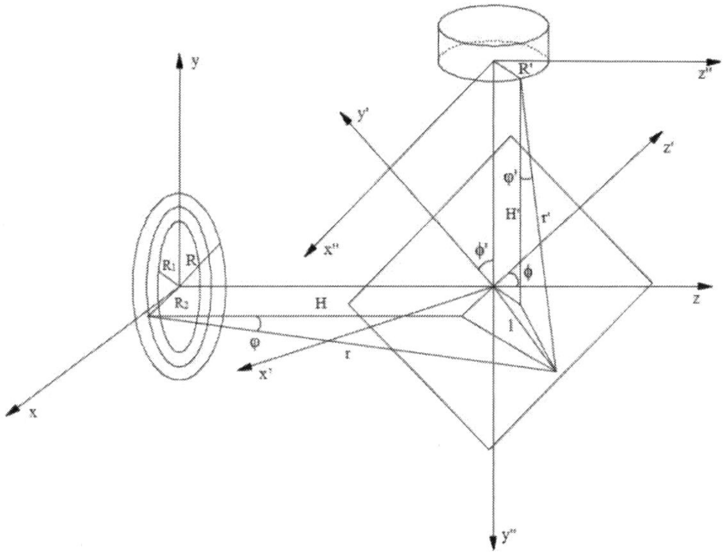

Figure 1. The geometric scheme of the model

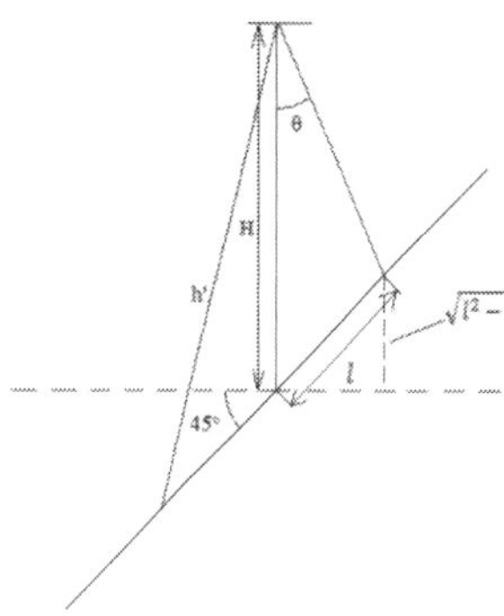

Figure 2. The geometric scheme of the distance changes between the cathode and the substrate

Method of a magnetron sputtering. In general, the thickness of the film at any point of the substrate is described by the expression:

$$h = \frac{V}{\pi \cdot r^2} \cos\varphi \cdot \cos\phi \cdot t \tag{1}$$

where V is the sputtering rate in thickness;
φ is the angle between the normal of the sputtering surface and the sputtering direction;
ϕ is the angle between the normal of the substrate's surface and the direction of deposition;
r is the distance between the sputtering element and the deposition point;
t is sputtering time.

We will use numerical integration of expression (2) in the calculations [6].

$$h = 2 \cdot V \cdot t \int_{R_1}^{R_2} \frac{H^2 \cdot R}{\left[\left(H^2 + R^2 - l^2\right)^2 + \left(2 \cdot l \cdot H\right)^2\right]^{3/2}} dR \tag{2}$$

where H is the distance between the cathode face and the processed surface;

l is a coordinate along the surface of the substrate (the distance between the center of the substrate and a surface processed);
R is a cathode radius.

In figure 2 the change in the distance between the cathode and the substrate is schematically shown if the condensation angle is 45 degrees for magnetron sputtering and vacuum arc evaporating.

The distance H between the cathode end and the substrate element changes in accordance with expression (3), depending on the positive or negative values of l.

$$H' = H' \pm \sqrt{l^2 - (l \cdot \cos 45°)^2} \quad (3)$$

The sputtering rate is a quantitative characteristic of the materials ionic sputtering process which is defined as:

$$V = \frac{h}{t} \quad (4)$$

where h is the thickness of the sputtered material;
t is sputtering time.

It is possible to use the expression (4) for calculating the sputtering rate:

$$V_{sput} = \frac{j_i \cdot S \cdot M_a}{e \cdot N_a \cdot \rho} \quad (5)$$

where j_i is the ionic current density in the cross section perpendicular to the direction of falling ions;
S is a coefficient of material sputtering;
M_a is the atomic mass of the target atoms;
e is an elementary charge;
N_a is Avogadro's number;
ρ is thematerial density.

According to Sigmund's theory for amorphous and polycrystalline materials with the energy of ions less than 1 keV the sputtering coefficient is defined with the help of the following expression [6]:

$$S = \frac{3}{\pi^2}\alpha \frac{M_i \cdot M_a}{(M_i + M_a)^2} \cdot \frac{E_i}{2 \cdot E_{sub}} \quad (6)$$

where M_i and M_a are atomic masses of ions and atoms of the target;
E_i is falling ions energy;
E_{sub} is the sublimation energy of the target atoms [6];
α is a dimensionless parameter depending on the M_i/M_a.
The ionic current density is calculated according to the formula:

$$j_i = \frac{I}{S_r} \quad (7)$$

where j_i is the ionic current density in the cross section perpendicular to the direction of falling ions;
I is the discharge current;
S_r is the area of the sputtering ring. It is defined by the following expression:

$$S_k = \pi \cdot (R^2 - (R - b)^2) = \pi \cdot b(2R - b) \quad (8)$$

where, R is thesputtering radius;
b is the width of the sputtering ring;

Method of a vacuum-arc evaporation. When a sufficiently high negative potential is applied to the treated surface which is in plasma the ionic current of saturation is supplied to the surface. The magnitude of this ionic current in nonequilibrium plasma is given by the formula of Bom [7].

$$j_{H',l} = \frac{\mu_p \cdot I_a \cdot \overline{z_i}}{\pi \cdot m_i \cdot \left(R'^2 + H'^2 \right)} \tag{9}$$

where μ_p is the coefficient of cathode erosion [10];
I_a is the arc current;
m_i is the mass of condensed ion;
$R'$is the cathode radius;
$H'$is the distance between the cathode end and the treatment surface;
$\overline{z_i}$ is the average charging number of ions.
Depending on technological modes of the deposition process it is possible to calculate the ionic current density at an arbitrary point of the vacuum chamber by the formula [3]:

$$j_{H',l} = \frac{\mu_p \cdot I_a \cdot \overline{z_i}}{2\pi \cdot m_i \cdot R'^2 \cdot} \left[1 + \frac{R'^2 - H'^2 - l^2}{\sqrt{\left(R'^2 + H'^2 + l^2 \right)^2 - 4R'^2 \cdot l^2}} \right] \tag{10}$$

where l is a coordinate along the surface of the substrate (the distance from the center of the substrate to the treated surface).
According to figure 2 the distance H between the cathode end and the substrate element changes in accordance with expression (3), depending on the positive or negative values of l.

We will use the following expression to determine the coating layer thickness at an arbitrary point of a vacuum chamber:

$$h = V_c \cdot t \tag{11}$$

where V_c is the coating condensation rate;
t is the deposition time;
The coating condensation rate is defined from a ratio:

$$V_c = \frac{m_i}{\overline{z_i} \cdot e \cdot \rho} \cdot \sum_{i=1}^{n} \left(\alpha_c \cdot j_{H',l} - S_{sp} \cdot j_{H',l} \right) \tag{12}$$

where ρ is the unit weight of the condensed substance;
α_c is the condensation coefficient;
S_{sp} is the sputtering coefficient.

While coatings being deposited, the substrate sputtering coefficient will aspire to zero, so it may be neglected. At the same time the condensation coefficient may be accepted equal to 1. Then the equation (12) will take the following form:

$$V_c = \frac{m_i \cdot \alpha_c \cdot i_{H',l}}{\overline{z_i} \cdot e \cdot \rho} \tag{13}$$

3.Results and discussion

According to the model described above the calculation algorithms are realized in the software package Mathcad:
- layer thickness when the substrate holder is located at an angle of 45 degrees to the normals of the evaporated titanium cathode and sputtering copper magnetron cathode mutually perpendicular planes;
- the dependence of the deposited layer thickness on the offset relative to the cathode axis and the cathode-substrate distance, with the graph plotting (fig.3, 4).

We should say that the calculations were made for the separate work of the magnetron and vacuum-arc evaporator. Further the case of the evaporator and the magnetron simultaneous operation will be studied when the layer condensation will be fulfilled with two overlapping streams of the substance atoms.

The distribution of the deposited film thickness depending on the plate radius and distance to the cathode is the result of the calculations in the software package Mathkad. These distributions are shown on the following graphs:

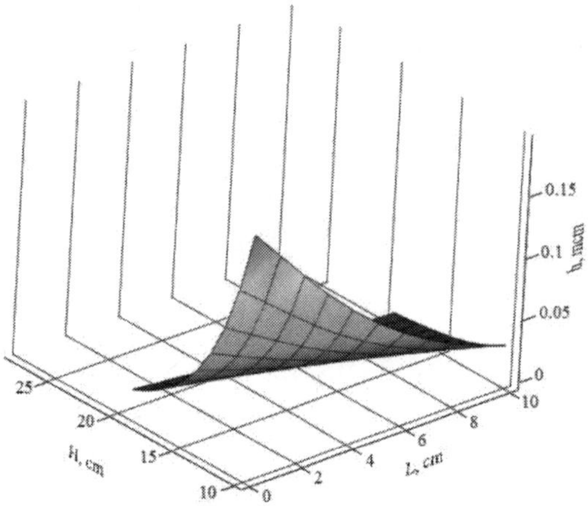

Figure 3. Distribution on thickness of a deposited layer from the offset relative to the cathode axis and distance to magnetron cathode

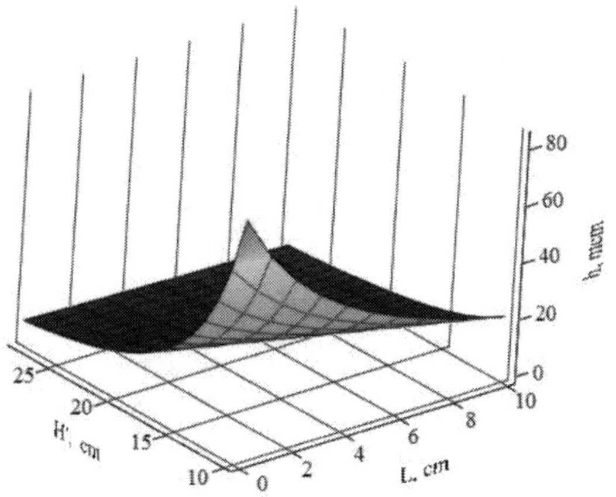

Figure 4. Distribution on thickness of a deposited layer from the offset relative to the cathode axis and distance to evaporator cathode

The calculation was carried out for coating deposition on the VU-1M installation. In Table 1 the process parameters, equipment characteristics and constants used in the calculations are shown.

Table 1. Values of the parameters and constants used in calculations

t, min	S (Cu), atom/ion	M_a (Cu), g/mol	e, C	N_a, atom/mol	ρ, g/cm^3	$M_н$, g/mol	α	E_{sub}, eV	L, cm
20	1,503	63,5	1,6x10^{-19}	6,023x10^{23}	8,96	40	0,33	3,56	5
R, cm	I, A	H, cm	μ_p, kg/C	m_i, kg	z_i	α_k	R_k, cm	I_d, A	H', cm
6	4	10..19	53*10^{-9}	79,5*10^{-27}	1,79	1	3	90	10..28

Experimental studies on deposition of layers by magnetron sputtering and vacuum arc evaporation have been conducted previously [8, 9]. In the case of vacuum-arc evaporation (H=28 cm L=0 cm) the layers thickness was 5-7 μm and in the case of magnetron sputtering (H=19 cm L=0 cm) it was 0.05-0.08 μm. The layer thickness depended on the deposition time t, arc current of evaporator I_∂, ion current density of magnetron I and the target material. The graphs show that the experimental data are in agreement with the calculated values for the vacuum arc evaporation. According to the figures one can see a significant effect of the geometric parameters of the cathode-substrate system on the uniformity of the coating thickness.

4.Conclusion

The method of calculating thicknesses of layers for separate processes of magnetron sputtering and vacuum arc evaporation in a plasma-chemical reactor is offered. Unlike the existing methods of calculation, this method takes into account the location of the substrate holder at an angle of 45

IOP Conf. Series: Journal of Physics: Conf. Series **830** (2017) 012091 doi:10.1088/1742-6596/830/1/012091

degrees to the normals of mutually perpendicular planes of a titanium evaporator cathode and a copper cathode of magnetron. Automated algorithms of calculating parameters and dependences of the layer thickness on the offset from the cathode axis and cathode-substrate distance have been carried out. This will allow to use the proposed calculation method to assess the film distribution uniformity along the radius of the substrate.

References

[1] Kunchenko Y V Kunchenko V V 2005 *ФІП PSE* **3(3-4)** 199
[2] Kireev R M Yagafarov I I 2012 *USATU bulletin* **16(1)** 88
[3] Budylov V V Kireev R M Shechtman S R 2007 *Technology of a vacuum ion-plasma processing*: Tutorial. (Moscow: MAI) pp 155
[4] Voevodin A A Lyubimov V V Erokhin A L 1992 *Issues of atomic science and technology. Series: Physics of radiation damage and radiation materials science* **1(62)** 121
[5] Golosov D A Melnikov S N Dostanko A P 2012 *Electronic treatment of materials* **48(1)** 63
[6] Nikonenko V A *Mathematical modeling of technological processes*: Modeling in MathCAD software environment Practical work MISaA pp 5-11
[7] Budilov V V 1993 *Physical principles of vacuum plasma technology of coatings deposition*: Tutorial USATU p 74
[8] Tsyrenov D B.-D Semenov A P Sushentsov N I 2009 *Proceedings of III Int. seminar name Kreyndel* pp 129-132
[9] Tsyrenov D B-D Semenov A P 2008 *Proceedings 9th International Conference on Modification of Materials with Particle Beam and Plasma Flows* pp 80–83
[10] Kostrizhitsky A I *Reference book of the installations operator for coatings deposition in vacuum* Mechanical engineering (Moscow) p 176

Acknowledgments

The work was performed within the State task of Russian FARO (project No. 0336-2014-0001) and partially supported by the Russian Foundation for Basic Research, project 15-48-04086-r_sibir_a.

Study of diffusion influence on plasma channel while transporting a low-energy high intensity electron beam in the low pressure gas

I L Zvigintsev and V P Grigoriev

National Research Tomsk Polytechnic University, Tomsk, Russian Federation

E-mail: Zvigintsev@yandex.ru

Abstract. This work studies the mathematical model of plasma channel. We consider the beam current in the range of 100-400 A and the external magnetic field in the range of 100-1500 G. It is shown that plasma channel expands under the influence of diffusion. The channel expansion is inversely proportional to the external magnetic field magnitude.

1. Basic equations

Gas ionization in the drift space is caused by the gas atom ionization by fast beam electrons as well as by fields on the front edge of the beam. The change in plasma channel parameters such as an electron density, plasma temperature and plasma conductivity will depend on the gas pressure, geometry of the drift tube and beam parameters. This will lead to the plasma current appearance, the current neutralization of the beam and conditions of the beam transport change. The plasma current changes the magnetic field of the beam, and this current must be taken into account while calculating the vector potential.

We consider a model that describes the beam transport in gas at pressure of $p < 0.1$ Torr under the full charge neutralization of the beam. An additional model of initial conditions was used to obtain the full charge neutralization.

Let us assume that an electron beam with radius r_b is injected into a plane $z = 0$ along the axis z of the drift tube with radius R_c. The basic heterogeneity is associated with beam and plasma density distribution along the radius because the beam transit time through the drift tube is much smaller than the beam rise time. The beam and plasma fields in the cylindrical coordinate system are described by a nonlinear equation:

$$\frac{1}{r}\frac{\partial}{\partial r}\left(r\frac{\partial A_z}{\partial r}\right) = -\frac{4\pi}{c}\left(j_{bz} + j_{pz}\right), \tag{1}$$

where A_z – vector potential, which satisfies the boundary conditions $A_z(r=R_c) = \partial A_z/\partial r|_{r=0} = 0$ and determines an electric field $E_z = -(1/c)(\partial A_z/\partial t)$ and a magnetic field $B_\theta = -\partial A_z/\partial r$; $j_{bz} = ev_b n_b$ – beam current density; j_{pz} – plasma current density; c – speed of light; e – elementary electric charge; v_b – velocity of beam electrons; n_b – beam density. Let us suppose the drift tube has an ideal wall conductivity. The drift tube has ground potentials on its ends. The plasma current is equal to zero at the reference time. The initial condition for the vector potential is derived from the equation (1).

The interaction among the plasma current density, the electric field and the plasma conductivity is determined by the equation:

$$\frac{1}{\nu_{ef}}\frac{\partial j_{pz}}{\partial t} = \sigma E_z - j_{pz}, \tag{2}$$

where $\sigma = e^2 n_e / (m_e \nu_{ef})$ – plasma conductivity; m_e – electron mass; n_e – plasma electron density; $\nu_{ef} = \nu_{ea} + \nu_{ei}^{(1)} + \nu_{ei}^{(2)}$ – effective collision frequency of plasma electrons and heavy particles; ν_{ea} – collision frequency of plasma electrons and gas atoms; $\nu_{ei}^{(l)} = 1.45\cdot10^{-6} n_i^{(l)} T_e^{-3/2} \ln(2.4\cdot10^{20} T_e^3 / n_i^{(l)})$ – collision frequency of plasma electrons and ions; $n_i^{(l)}$ – the atomic (l=1) and molecular (l=2) ion density; T_e – electron plasma temperature in eV [1]; $\nu_{ea} = 3.7\cdot10^{-8} n_g T_e$ for argon, $\nu_{ea} = 4.4\cdot10^{-8} n_g T_e^{0.5}$ for helium [2]; n_g – gas density.

The balance of particles is described by the following equations:

$$\frac{\partial n_i^{(1)}}{\partial t} = \langle\sigma_{ib} v_b\rangle n_b n_g + \langle\sigma_{ie} v_e\rangle n_e n_g - K_k n_i^{(1)} n_g^2 - \alpha_{r1} n_i^{(1)} n_e^2 - \ldots$$
$$\ldots - \alpha_{r2} n_i^{(1)} n_e n_g - \alpha_{r3} n_i^{(1)} n_e + \mathrm{div}\left(D_{A\perp}\,\mathrm{grad}\, n_i^{(1)}\right), \tag{3}$$

$$\frac{\partial n_i^{(2)}}{\partial t} = K_k n_i^{(1)} n_g^2 - \alpha_{rd} n_i^{(2)} n_e, \tag{4}$$

$$\frac{\partial n_g}{\partial t} = -\langle\sigma_{ib} v_b\rangle n_b n_g - \langle\sigma_{ie} v_e\rangle n_e n_g - K_k n_i^{(1)} n_g^2 + \ldots$$
$$\ldots + \alpha_{r1} n_i^{(1)} n_e^2 + \alpha_{r2} n_i^{(1)} n_e n_g + \alpha_{r3} n_i^{(1)} n_e + 2\alpha_{rd} n_i^{(2)} n_e, \tag{5}$$

$$n_e = n_i^{(1)} + n_i^{(2)} - n_b, \tag{6}$$

where σ_{ib}, σ_{ie} – ionization cross-sections by beam and plasma electrons [3]; v_e – velocity of plasma electrons; α_{r1}, α_{r2}, α_{r3}, α_{rd} – recombination coefficients: impact, neutral third particle, radiation and dissociative; K_k – conversion coefficient of atomic ions into molecular ones; $D_{A\perp}$ – ambipolar diffusion coefficient. Boundary conditions for the plasma ion density and the ambipolar diffusion coefficient are similar to the vector potential boundary conditions. The recombination coefficients for argon atoms are determined by formulas: $\alpha_{r1} = 4.7\cdot10^{-26} T_e^{-9/2}$ cm^6/s, $\alpha_{r2} = 1.09\cdot10^{-30} T_e^{-5/2}$ cm^6/s, $\alpha_{r3} = 2.7\cdot10^{-13} T_e^{-3/4}$ cm^3/s, $\alpha_{rd} = 9.1\cdot10^{-7}(0.026/T_e)^{0.61}$ cm^3/s. The recombination coefficients for helium atoms are determined by formulas: $\alpha_{r1} = 3.6\cdot10^{-27} T_e^{-9/2}$ cm^6/s, $\alpha_{r2} = 1.07\cdot10^{-30} T_e^{-5/2}$ cm^6/s, $\alpha_{r3} = 6.45\cdot10^{-14} T_e^{-3/4}$ cm^3/s, $\alpha_{rd} = 7\cdot10^{-7}(0.026/T_e)^{0.61}$ cm^3/s [4, 5]. The conversion coefficients are: $K_k = 2.5\cdot10^{-31}$ cm^6/s for argon, $K_k = 5.77\cdot10^{-32}$ cm^6/s for helium [6]. The ambipolar diffusion coefficient is calculated by:

$$D_{A\perp} = \frac{c^2}{\dfrac{\Omega_e^2 + \nu_{ea}^2}{\nu_{ea}} + \dfrac{m_i}{m_e}\dfrac{4\Omega_i^2 + \nu_{ia}^2}{2\nu_{ia}}}\frac{T_e + T_i}{E_0}, \tag{7}$$

where T_i – plasma ion temperature in eV; Ω_e, Ω_i – cyclotron frequencies of the plasma electrons and ions; m_i – ion mass; $\nu_{ia} = 4.6\cdot10^{-10} n_g T_i^{0.5}$ – collision frequency of plasma ions and gas atoms [1].

The connection of the plasma electron temperature and the electric field that includes both elastic and non-elastic collisions of plasma electrons with heavy particles, is determined by [7]:

$$\frac{\partial T_e}{\partial t} = \frac{2}{3}\frac{E_0 r_0}{m_e \nu_{ef}}\left(E_z^2 - p^2\left(\frac{T_e}{k}\right)^8\right), \tag{8}$$

where $k = 11.65$ for argon and $k = 18$ for helium [2]; r_0 – classical electron radius; E_0 – electron mass energy equivalent in eV.

As the aforementioned model describes the beam transport under the full charge neutralization, it is necessary to determine the initial conditions for the plasma density at various beam injection modes.

IOP Conf. Series: Journal of Physics: Conf. Series **830** (2017) 012092 doi:10.1088/1742-6596/830/1/012092

2. Initial conditions

When the electron beam is injected into a neutral gas, the basic processes of the plasma channel are the beam passing current ionization and the plasma electron ionization in the space-charge field of the beam:

$$\frac{dn_i}{dt} = \sigma_{ib} n_g \frac{j_q}{e} + \sigma_{ie} n_g v_e n_e , \tag{9}$$

$$\frac{dn_e}{dt} = \frac{dn_i}{dt} - \frac{n_e}{\tau_e} , \tag{10}$$

$$n_g = 3.5 \cdot 10^{16} p - n_i , \tag{11}$$

where n_i – plasma ion density; j_q – passing current density; $\tau_e = L / v_e$ – mean time of plasma electron liberation at the drift tube; L – length of the drift tube.

When the beam is injected in the neutral gas, there is no space-charge neutralization of the beam. When an external magnetic field B_z is strong, the beam is kept from the transverse separation. The losses of the transported beam current are associated with the virtual cathode when

$$I_b \geq I_{pr} , \tag{12}$$

where I_b – injected beam current; I_{pr} – passing current. The virtual cathode (VC) arises at a distance of $d = 1.5 \cdot 10^{-3} U^{3/4} / j_b^{0.5}$ from the injection plane, where j_b – injected beam current density; U – VC voltage. Only one part of the injected beam current passes through the drift tube. It does not exceed the limiting current, which equals

$$I_{pr} = I_{pr}^{(0)} / (1 - f) , \tag{13}$$

$$I_{pr}^{(0)} = \frac{2}{9} I_A^{(0)} \beta^3 \left(1 + \frac{4}{3} \ln \left(\frac{R_c}{r_b} \right) \right)^{-3/2} , \tag{14}$$

where $I_{pr}^{(0)}$ – limiting vacuum current [8], $f = n_i / n_b$ – charge neutralization of the beam; $I_A^{(0)}$ – Alfven critical current; β – relative electron velocity. Formula (14) gives satisfactory results when R_c / r_b changes from 1 to 10. VC disappears when the inequality (12) is false, and the beam current starts to pass to a full degree.

The passing beam electrons ionize the gas. As a result, the degree of the beam charge neutralization increases with time. This leads to an increase of the passing current. The full charge neutralization is achieved at the moment, when f is equal to one. The values of the beam and plasma parameters at this moment are taken as initial conditions.

In this work, we describe the initial conditions when there is a strong external magnetic field. The given initial conditions model is a separate part of the total ionization and plasma channel program. This model takes into account the transient mode. A similar model for the high-current beams in the absence of an external magnetic field is presented in work [9].

3. Simulated results

We consider a uniform beam injection into the drift tube filled with the inert gas. The beam parameters correspond to experiment [10]: $r_b = 2$ cm, $R_c = 4.1$ cm, $L = 20$ cm. The beam profile [10] (current −300 A, pulse duration – 145 μs, leading edge pulse time – 20 μs) is shown in figure 1.

The slow change of the beam current is the cause of the plasma current smallness and the electric field strength of the total current (fig. 1, 2). At the edge of the injected current pulse, the plasma current is nearly non-existent. This occurs as the pulse duration is big enough and the fall in the plasma current is quite fast. At the falling edge, the plasma current changes its direction and increases the total current. The result does not depend on the external magnetic field.

The electric and the magnetic fields become lower at smaller values of the beam current maximum (fig. 2). The magnetic field has a common radius distribution throughout the entire pulse – it is equal to zero on the axis and has the maximum value on the beam boundary.

The figure 3 shows leading edge erosion of the beam due to the formation of VC. At a higher pressure, the virtual cathode disappears or is not formed. It is proved that the virtual cathode is not formed in helium when the pressure is $3\cdot10^{-3}$ Torr and it is formed when pressure is 10^{-3} Torr. Other parameters influence the virtual cathode too.

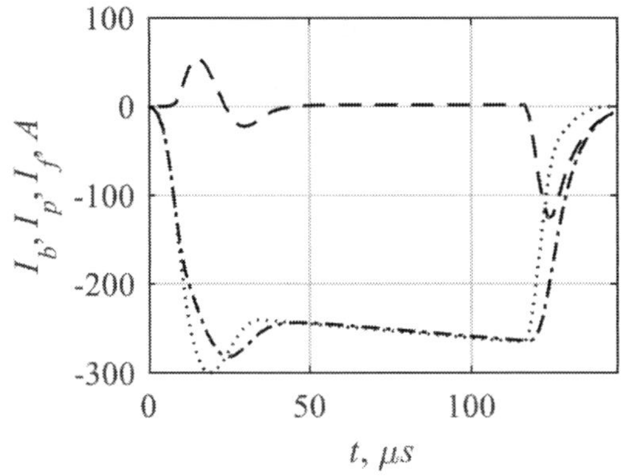

Figure 1. The time dependences of the plasma current (dashed line), the beam current (dotted line) and the total current (dash-and-dot line). $p = 10^{-3}$ Torr, $\varepsilon_b = 30$ keV, $I_{b0} = -300$ A, $B_z = 100, 300, 1500$ G, Ar.

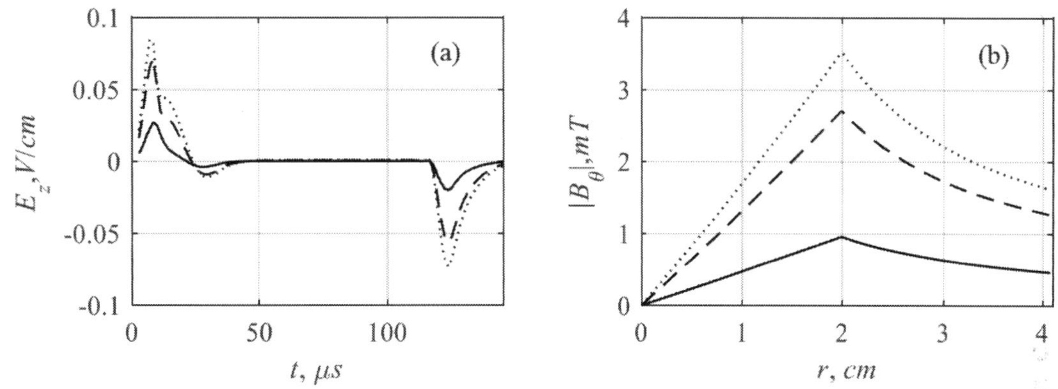

Figure 2. (a) The time dependences of the electric field strength of the total current on the axis; (b) The radial dependences of the magnetic field of the total current at the pulse peak. $p = 10^{-3}$ Torr, $\varepsilon_b = 30$ keV, $B_z = 300$ G, Ar. The solid line – $I_{b0} = -100$ A, the dashed line – $I_{b0} = -300$ A, the dotted line – $I_{b0} = -400$ A.

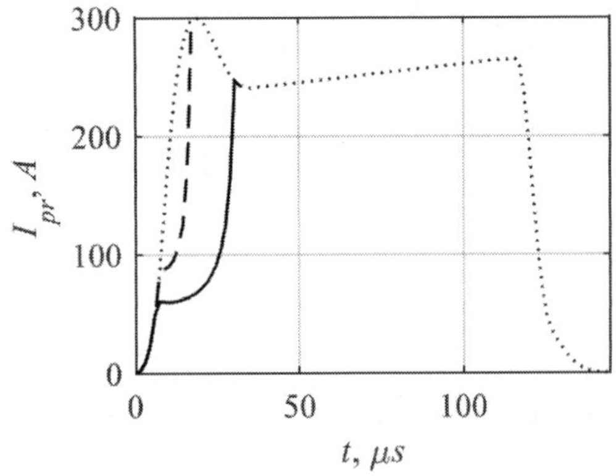

Figure 3. The leading edge erosion of the beam. He, $\varepsilon_b = 30$ keV, $I_{b0} = -300$ A, $B_z = 1500$ G. The solid line – $p = 5\cdot10^{-4}$ Torr, the dashed line – $p = 10^{-3}$ Torr, the dotted line – $p = 3\cdot10^{-3}$ Torr, the injected beam current.

The argon ionization cross-sections are higher than helium. As a result, the argon is ionized faster than helium. In argon, the full gas ionization occurs quickly. The figure 4 shows the dependence of diffusion on the external magnetic field. It is seen that there is no diffusion in argon, and helium, when the external magnetic field equals 1.5 kG. The plasma channel expands rapidly when the external magnetic field equals 100 G.

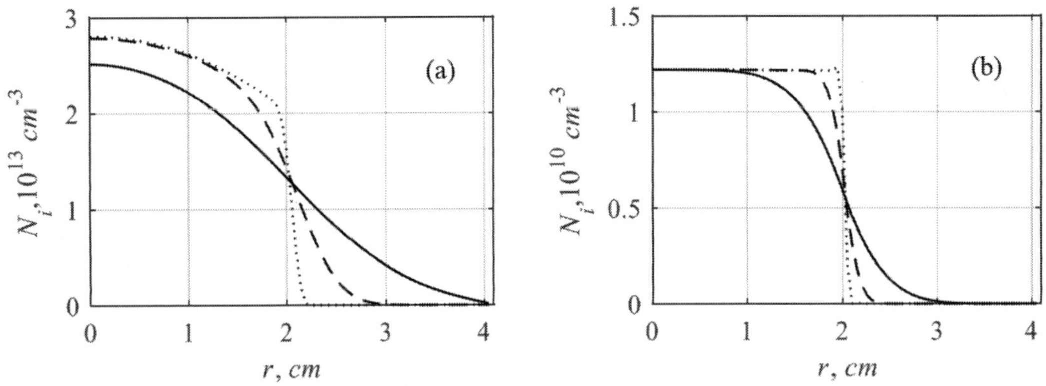

Figure 4. The radial dependences of the plasma ion density at the pulse peak. $I_{b0} = -100$ A, $p = 10^{-3}$ Torr, $\varepsilon_b = 30$ keV. The solid line – $B_z = 100$ G, the dashed line – $B_z = 300$ G, the dotted line – $B_z = 1500$ G. (a) Ar, (b) He.

4. Conclusion

The developed model allows calculating the plasma current, the electric and magnetic fields, the diffusion of the plasma channel, the plasma electron temperature, the gas ionization and recombination, the leading edge erosion of the beam, etc. The model is suitable for a wide range of parameters: the gas pressure is 10^{-4}-10^{2} Torr, the beam energy is 10-30 keV, the beam current is 100 A-15 kA and the microsecond impulses.

References

[1] Golant V E, Zilinski A P and Sakharov S A 1977 *Fundamentals of Plasma Physics* (Moscow: Atomizdat) p 384

[2] Zakharov A V 1987 *Investigation of the gas parameters influence on transportation of a high-intensity electron beam in the absence of external fields* (Tomsk) p 126

[3] Gryzinski M 1965 *Phys. Rev.* **138** 336

[4] McDaniel I and Nigan U 1986 *Gas Lasers* (Moscow: Mir) p 548
[5] Mitchner M and Kruger Ch 1976 *Partially Ionized Gases* (Moscow: Mir) p 496
[6] Vriez C P and Oskam H J 1980 *Phys. Rev.* A **22** 1429
[7] Grigoriev V P and Zvigintsev I L 2011 *Izvestia TPU* **319** 88
[8] Didenko A N, Grigoriev V P and Usov Y P 1977 *Power Electron Beam and Their Applications* (Moscow: Atomizdat) p 280
[9] Zvigintsev I L and Grigoriev V P 2012 *Russian Physics Journal* **55** 460
[10] Grigoryev S V, Astrelin V T, Devjatkov N V, Kandaurov I V, Koval N N, Kozyrev A V, Moskvin P V and Teresov A D 2010 *16th Int. Symp. on High Current Electronics: Proc.* (Tomsk) pp 19-22

Creation of the nanostructured superhard and heatproof layers on die steels under influence of intensive electron beams in vacuum

A Milonov, B Danzheevand N Smirnyagina

Institute of Physical Materials Science SB RAS, 670047, Ulan-Ude, Russia

E-mail: terwer81@mail.ru

Abstract: Studying and comparing microstructure and microhardness of boride layers formed on die D2 steel by using different methods – electron beam borating with continuous and pulsed beams under vacuum. The created layers have a heterogeneous structure combining hard and plastic components resulting in decrease of boride layer embrittlement.

1. Introduction

Today in the conditions of ever-evolving technology there are growing requirements for the material strength of machine, device and tool components, especially with regard to their high temperature strength and heat resistance. Transition metal borides have high fusing temperatures (above 2,000°C) and hardness values and are quite oxidation-proof which makes them of particular interest in terms of coating formation basis. Boride layers have high physical-and-mechanical parameters. Layer microhardness can reach 20,000 MPa provided that these microhardness values can prevail until temperatures reach ~ 600-700°C which allows to apply boriding for improving wear resistance of products made of die D2 steel and operating under high temperatures [1-3].

Table 1. Chemical composition of D2 steel material.

Chemical element	%
Carbon (C)	1.45–1.65
Vanadium (V)	0.15–0.30
Silicon (Si)	0.1–0.4
Copper (Cu), maximum	0.3
Manganese (Mn)	0.15–0.45
Molybdenum (Mo)	0.4–0.6
Nickel (Ni), maximum	0.35
Phosphorus (P), maximum	0.03
Chromium (Cr)	11.0–12.5
Sulphur(S), maximum	0.03

D2 steel is cold deformation die steel with a high content of chrome and inclusions of molybdenum (0.5% on average) and vanadium (0.2% on average). D2 steel has high heat resistance, strength, through-hardening capability, hardening capacity and wear resistance. D2 steel microhardness is 2.560

MPa. This steel is also processable, can easily be machined and mechanically shaped and acceptably grinded.

As is well known, steels with 1.25-1.45% C content refer to ledeburitic class steels, i.e. contain carbide eutectics as-cast and have quench hardness of 8.41-9.07 MPa. These steels contain a high amount of carbide-forming elements and an increased amount (20%) of chrome carbides (Cr_7C_3, $Cr_{23}C_6$).

Since chrome (Cr) represents an alloying element of cold deformation die steel it improves machining (cutting) properties and wear resistance and enhances steel hardness and through-hardening capability which is particularly important for fabricating large male dies and die blocks.

Wolfram (W) is introduced into steel to improve hardness, wear resistance and through-hardening capability which also enhances the cutting capability of the tool.

Vanadium (V) contained in die steels is present in VC carbide and solid solutions. Vanadium significantly reduces the die steel responsivity to overheating, enhances steel heat resistance and improves the proeutectoid constituent particle distribution. With 0.3–0.5 % content of vanadium steel strength and plasticity will be significantly higher compared to those of high vanadium steels.

Molybdenum (Mo) is introduced into high-chromium steel to enhance its ductility and improve through-hardening capability. At the same time molybdenum has a negative effect on scaling resistance and due to this reason its content is limited to 1.4–1.8 %.

Therefore, high-chromium D2 steel is applicable for fabricating dies, male dies and rollers with 4.56...5.80 MPa hardness and under the operating temperature of up to 700 °C.

In the context of this study layers were obtained as a result of electron-beam borating (EBB) [4]. Saturating coating with 0.5-1 mm thickness was applied to the prepared surface of samples. The coating composition included boron carbide B_4C and an organic binder.

2. Experimental details

Vanadium boride synthesis was conducted on the surface of D2 die steel. The samples were prepared by applying the coating to the prepared steel surface. The coating composition included mixtures of oxide V_2O_3, amorphous boron and carbon using 1:1 proportions and an organic binder – 1:10 BPH-6 solvent acetone adhesive.

The samples were heated by means of electron beams using pulse conditions with the following parameters: accelerating voltage - U=24 kV; beam current - I_r=63 A. Processing was carried out with pulse length t=20 μs; number of pulses N=1800; beam current pulse repetition frequency f=6 Hz. Vacuum vessel pressure: 5×10^{-2} Pa. [5,6]

Continuous electron beam heating for 2-5 minutes with specific capacity of 2-2.5 W/sm^2. Residual pressure in the vacuum vessel did not exceed 2×10^{-3} Pa.

The boride layers were analyzed by x-ray diffraction (Advance D8 Bruker, Cu Kα- radiation). The samples microstructure was observed using a metallographic microscopy METAM RB-21 (completed by digital chamber VEC-335 and program complex ImageExpert Pro 3.0) and scan electron microscopy LEO 1430 VP. Microhardness was measured by using PMT-3 microhardness tester at a loading 0.5 H [7].

3. Results and their discussion

Metallographic analysis discovered that the structures of D2 steel surface layers obtained as a result of pulse and continuous electron beam borating (EBB) were different (Fig.1).

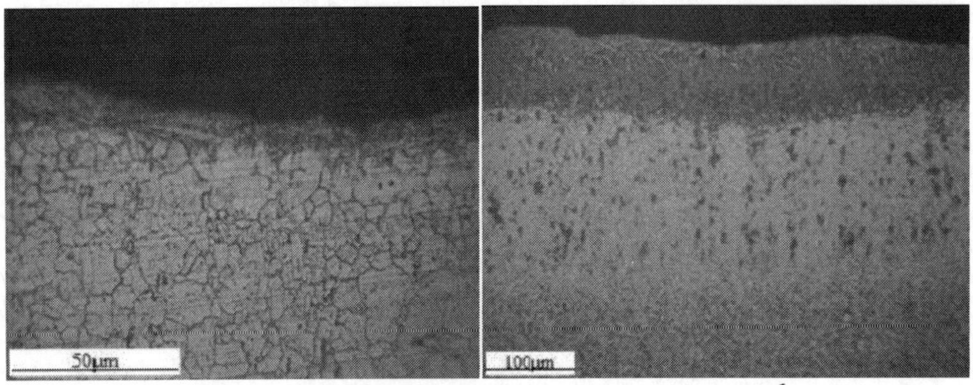

Figure 1. Structure of layer on the basis of VB_2 on D2 steel: a – impulsive electron beam, b – continuous electron beam.

After applying EBB the transition zone is missing and there is a distinct visible border between the layer and main part of the steel. The layer is composed of rounded crystals located on the surface or throughout and eutectics (Fig.1). Layer thickness is 20-30 and up to 250-300 µm after pulse and continuous electron beam bonding, respectively. Small layer thickness obtained after pulse electron beam borating is attributable to the fact that the beam picks off the resulting products of self-propagating high-temperature synthesis before D2 steel surface melting.

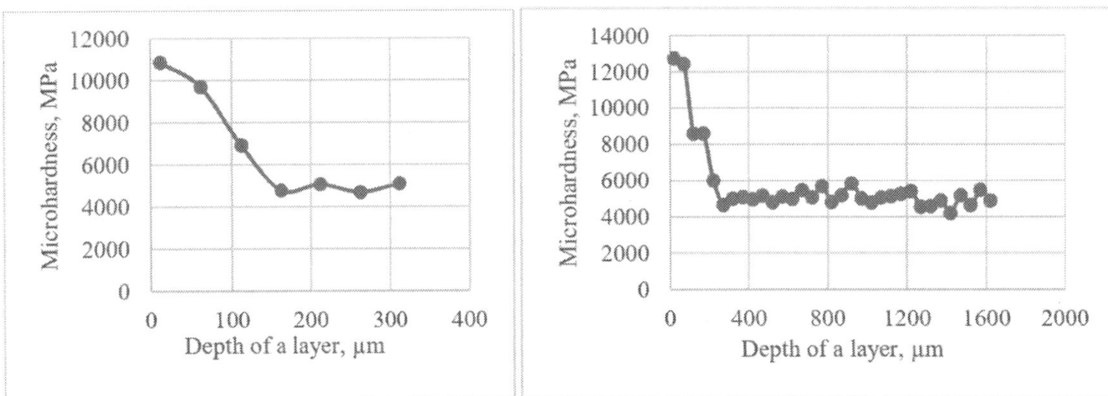

Figure 2. VB_2 layer microhardness on D2 steel: a - pulse electron beam, b - continuous electron beam.

Measurement applied to the hardness of vanadium boride layers with 30-50 µm spacing discovered its uneven distribution across the thickness (Fig. 2). However, every examined sample revealed regular microhardness distribution depending on layer thickness. Particular very rare inclusions have HV≈20000 MPa and are located in near-surface layer zones. The layers are characterized with the most complex disordered structure. Increase in microhardness of the basis to HV≈5000 MPa is attributable to the fact that it has been strengthened by using electron beams.

Boride layers were tested for thermal resistance and high-temperature strength. For this purpose all samples were heated inside a KO-14 resistance furnace until reaching particular temperatures and soaked for 2 hours to reach equilibrium.

Figure 3 shows a graph of weight variance with increase in temperature. It can be seen that when the temperature is increased to 900 °C and above the weight starts to decrease. This is because of layer oxidization and breakup which is supported by research of microstructure and microhardness (Fig. 4, 5).

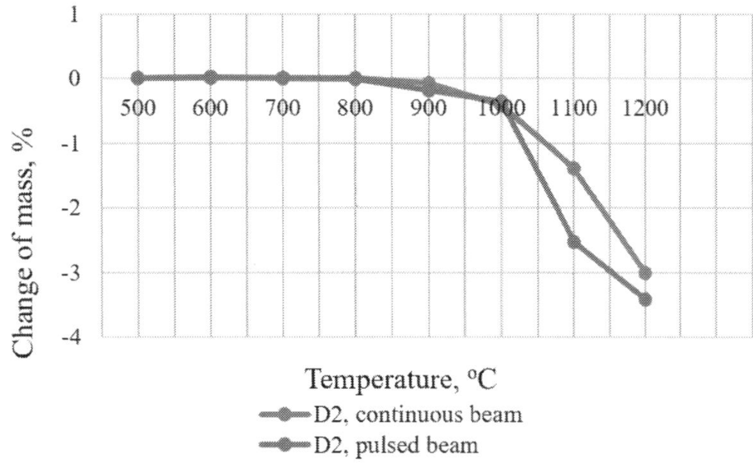

Figure 3. Change of mass with the height of temperature.

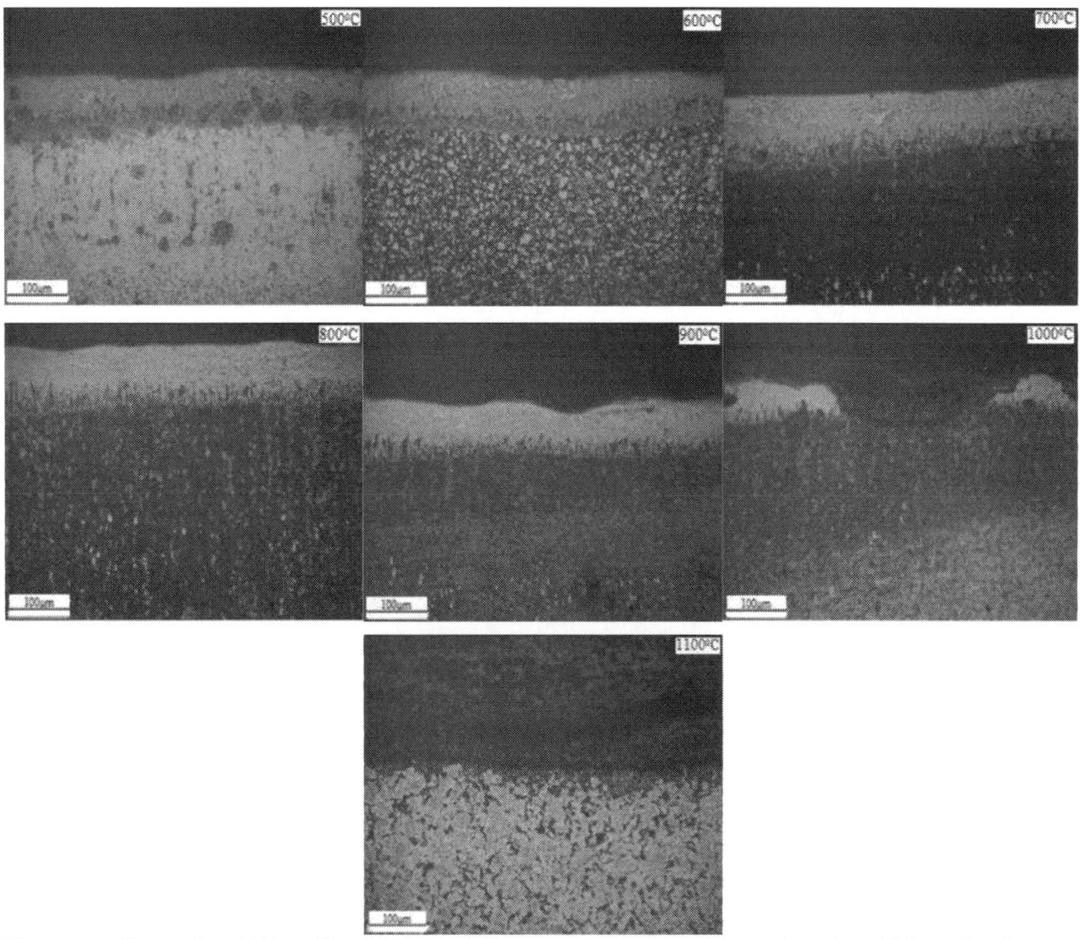

Figure 4. Thermal stability of boride layers after treatment a continuous bunch on D2 steel at heating on air.

The examination of microstructure shows that the thicknesses of boride layers obtained in both cases gradually start to decrease when the temperature is 700°C and above (Fig. 4).

Further rise of the heating temperature to 1100 °C in the open air results in a complete breakup of bonded layers (Fig. 4). This is most intensively observed in the layers formed during pulse beam borating, where the layer loses its weight and completely breaks down when temperatures exceed 900-1100 °C (Fig. 4).

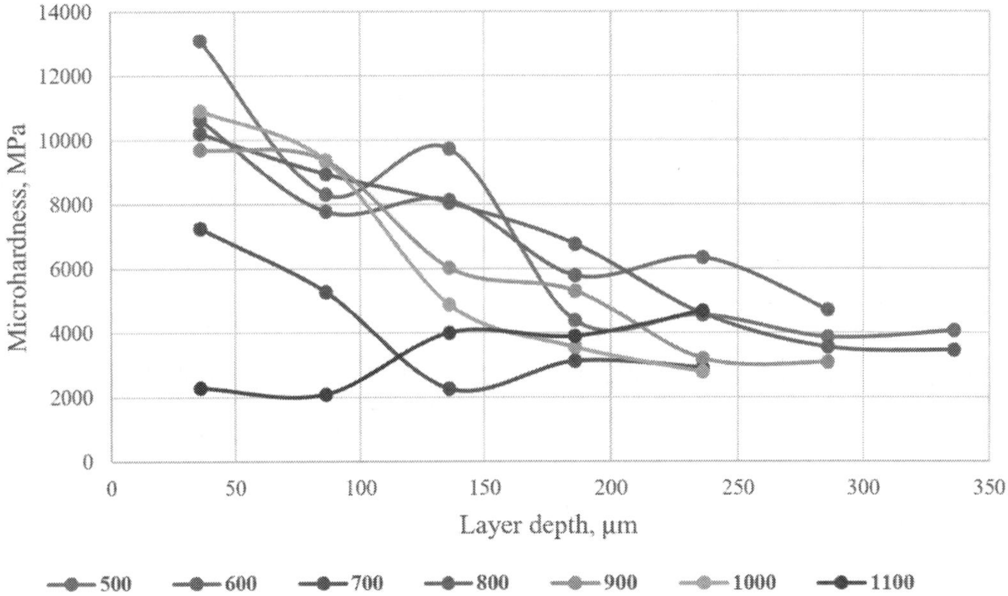

Figure 5. Microhardness the boride of layers when heating on air.

The research allows to make a conclusion regarding EBB application for hardening cutting tools, etc. exposed to heat up to high temperatures during the operation without a significant reduction of performance properties.

It is known that alongside with high hardness and wear resistance the boride layers also have a significant disadvantage, i.e. high embrittlement. The conducted research showed that the use of electronic heating allowed to reduce embrittlement and to improve plasticity. After electron beam borating the layers become more plastic compared to after solid phase borating. Also, after electron beam borating the layers have a heterogeneous structure combining hard (brittle) and more plastic structural components. Such a combination partly explains the absence of thermal cracks during the heating of boride layers to high temperatures.

4. Conclusion

1.Therefore, the electron-beam processing under vacuum resulted in the formation of vanadium boride layers which have an uneven structure across the thickness, contain different phases and as a result have an uneven distribution of physical-and-mechanical properties (for example, microhardness). Hard particles with 5-7 μm size are located in plastic eutectics. The layer surface has maximum values of microhardness.

2.The examination of microhardness of boride layers after pulse and continuous electron-beam borating allows to make a conclusion about the use of both borating methods for hardening cutting tools, etc. exposed to heat up to high temperatures during the operation without a significant reduction of performance properties.

References

[1] Hocking M, Vasantasri W, Sidky P 2000 *Metallic and ceramic coatings: Obtaining, properties and applications* (Moscow: Mir, trans. from English,) p 518

[2] Labunets V F, Voroshnin L G, Kindarchuk M V 1989 *Wear resistance of boride coatings* (Kiev: Publishing house "Technique") p 204

[3] Kolubaev A V, Tarasov S Y, Trusova G B, et all 1994 *Ferrous metals* **7** 49

[4] Semenov A P, Smirnyagina N N, Sizov I I 2001 *Technology of metals* **4** 32

[5] Koval N N, Shchanin P M, Vintizenko L G, Tolkachev V S 2005 *PTE* ***1*** 135

[6] Koval N N, Devyatkov V N, Grigoriev S V, Sochugov N Ch 2006 Plasma electronogen of Solo *Works of the II International Kreyndelevsky seminar "A plasma issue electronics engineering"* (Russia, Ulan-Ude) pp 79-85

[7] Milonov A S, Danzheev B A, Smirnyagina N N, Dasheev D E, Kim T B and Semenov A P 2015 *Journal of Physics: Conf. Ser.* **652** 012010

Acknowledgments

This work was supported in part by the Russian Foundation for Basic Research, project№ 16-48-030875 r_a.

IOP Conf. Series: Journal of Physics: Conf. Series **830** (2017) 012094 doi:10.1088/1742-6596/830/1/012094

Nitriding of VT3-1 titanium alloy in a glow discharge with hollow cathode

V.V. Budilov, K.N. Ramazanov, I.V. Zolotov, Yu. G. Khusainov, E.L. Vardanyan

Ufa State Aviation Technical University, Department of Mechanical Engineering, 12 K. Marx Str., Ufa, Russia, e-mail: ramazanovkn@gmail.com

Abstract. The paper presents research results of VT3-1 titanium alloy samples subjected to nitriding in glow discharge with a hollow cathode. The influence of working gas composition on the thickness of a hardened layer at nitriding in glow discharge with a hollow cathode was studied. It was found that the increase of argon up to 80-90% in gaseous mixture during nitriding in glow discharge with a hollow cathode leads to the increase of the hardened layer thickness to ~ 90 microns and surface microhardness to ~ 850 $HV_{0,1}$, and contributes to a smaller gradient of the surface layer microhardness. The hardened layer on the surface of VT3-1 titanium alloy obtained upon nitriding in glow discharge with a hollow cathode is less fragile and resists plastic deformation more efficiently in comparison with the hardened layer obtained via traditional ion nitriding.

1. Introduction

At present titanium alloys are widely used in aviation industry, which is specified by unique combination of their physical, mechanical and technological properties: rather low density, high strength-to-weight ratio, corrosion resistance, thermal integrity. However their application is limited by a number of disadvantages related to surface properties of titanium and its alloys. Low hardness and, as a result, wear resistance of a surface, tendency to sticking, high friction coefficient alongside with overwhelming amount of materials limit the use of titanium alloys in production of components subject to intense friction and surface wear [1, 2].

The paper suggests using glow discharge with hollow cathode to form high density plasma [3–6] that ensures efficient ion etching of a nitride film formed on a surface of a treated titanium alloy that in its turn leads to the intensification of ion nitriding. The purpose of this paper is to study the influence of nitriding in glow discharge with a hollow cathode on the surface structure and properties of VT3-1 titanium alloy.

2. Experimental approach

Figure 1 shows the layout of an upgraded set-up ELU5-M used for titanium alloy nitriding in glow discharge with a hollow cathode. In order to form a cathode void over the sample a screen from titanium alloy with holes was installed and placed above the same current potential (Fig. 1). As a result high density plasma is formed in a cathode void, which leads to intense nitrogen saturation of a surface.

Two-phase VT3-1 titanium alloy (Ti–6Al–2Mo–1Cr), which is used in production of build-in and stamp-welded components, is able to operate long times at 400–500 °C. The samples were subject to preliminary quenching and aging.

Samples were studied via the scratch test using CSM Scratch Tester to define the influence of nitriding in glow discharge with a hollow cathode on properties of a surface layer of VT3-1 titanium alloy. Diamond pyramid hardness was used as an indenter. The applied load was changed with the linear movement of indenter from 0.03 N to 30 N at the speed of 12 N/min. The path length of the indenter was 5 mm, the speed of its movement was 2 mm/min.

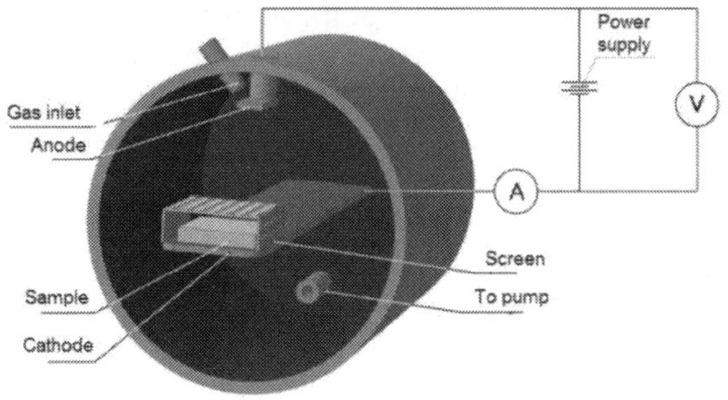

Figure 1. Experimental set-up for sample nitriding in glow discharge with hollow cathode

3. Results and discussion

To define the influence of gaseous mixture composition on the thickness of a hardened layer of VT3-1 titanium alloy, samples were subjected to nitriding in glow discharge with a hollow cathode at a temperature of 800 °C during 4 hours in gaseous mixtures with various contents of nitrogen and argon. Figure 2 shows microstructures of sample sections from VT3-1 titanium alloy obtained using an optical microscope.

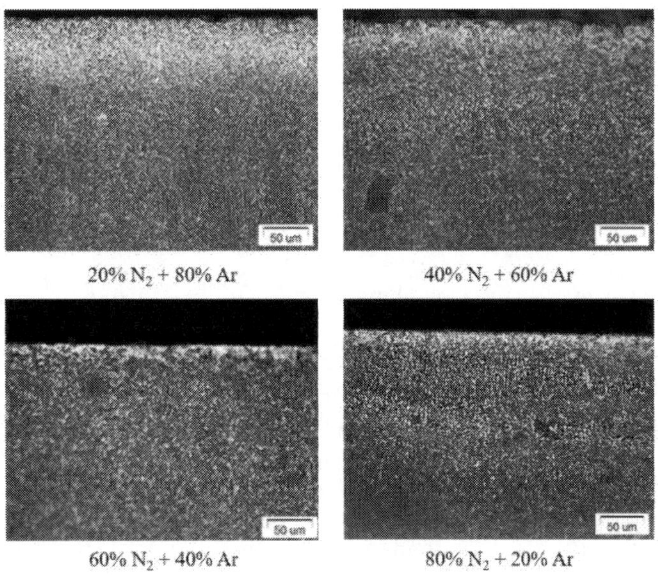

Figure 2. Microstructures of sample sections from VT3-1 titanium alloy subjected to nitriding in glow discharge with a hollow cathode at 800 °C during 4 hours in gaseous mixtures with various contents of nitrogen and argon

The analysis of the obtained microstructure images showed that samples with light poorly etched layer that were nitraded in gaseous mixture with higher content of argon have higher thickness. The thickness of a light layer of samples nitrided in gaseous mixture containing 80% Ar and 20% N_2 makes about 70 microns; then the layer is smoothly transformed into the microstructure of a basis while samples nitraded in gaseous mixtures with higher content of Ar and N_2 are presented as films with the thickness of several microns.

Data obtained as a result of the analysis allow to make a conclusion that nitriding in glow discharge with a hollow cathode in gaseous mixtures with low content of nitrogen in VT3-1 alloy leads to the formation of a distinct light diffusion layer of about 70 microns thick which is smoothly transferred into the microstructure of a basis. The increase of nitrogen in gaseous mixture of up to 60–80% leads to the formation of a thin light non-etched film on the surface of alloys.

Microhardness profiles by layer depth obtained via measurements of microhardness on section surfaces were obtained in order to assess the thickness of the reached hardened layer after nitriding in glow discharge with a hollow cathode in gaseous mixtures with various contents of nitrogen and argon. Dependences of microhardness on distance from the surface of samples of VT3-1 titanium alloy obtained through measurements are shown in Figure 3.

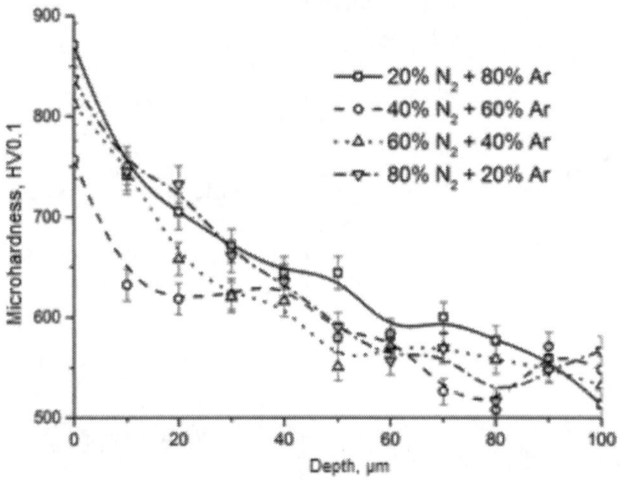

Figure 3. Distribution of microhardness of VT3-1 titanium alloy surface layers nitrided in glow discharge with hollow cathode in gaseous mixtures with various contents of nitrogen and argon within 4 hours

The analysis of obtained distributions showed that nitriding in glow discharge with a hollow cathode of both alloys in gaseous mixture with 20% nitrogen allows to receive the biggest values of both the surface microhardness and thickness of the hardened layer. The thickness of this layer after nitriding during 4 hours amounts to 70-90 microns, microhardness smoothly decreases to values of microhardness of a basis that is one of the main requirements imposed on the nitrided layer.

The obtained experimental data and its further analysis demonstrate that nitriding of VT3-1 alloy in glow discharge with a hollow cathode is the most efficient in gaseous mixtures with low (15-20%) concentration of nitrogen. Despite a high saturant gradient, nitriding in mixtures with high content of nitrogen is less efficient as it leads to the formation of a nitride film thus creating a barrier for nitrogen diffusion deep into the material. High content of argon in gaseous mixture leads to intense etching of a nitride film and, as a result, to the formation of big diffusion layers. Besides, high content of argon in gaseous mixtures ensures more intense heating up to the working temperature and nitriding at lower burning voltage of the discharge.

Figure 4 shows dependences of acoustic emission on the imposed load obtained as a result of a scratch test applied for all samples. The analysis of obtained dependences showed that active breakdown of a surface layer of a sample nitrided in traditional glow discharge begins at the smallest imposed loads while when the sample was nitrided in glow discharge with a hollow cathode, the wave peaks of acoustic emission were registered starting from the imposed load of 1.8 N. This fact allows to make a conclusion that the surface of titanium alloys nitrated in glow discharge with a hollow cathode has bigger hardness and abrasive wear resistance in comparison with the surface nitraded in traditional glow discharge.

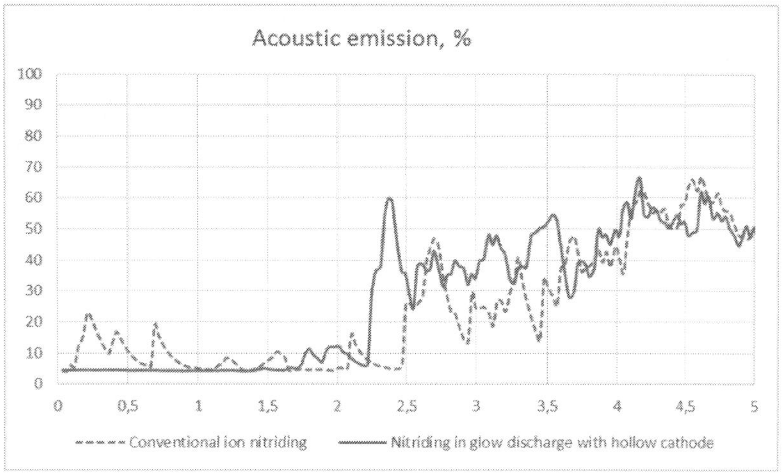

Figure 4. Acoustic emission obtained during a scratch test of VT3-1 titanium alloy samples nitrided in glow discharge

Figure 5 shows images of scratches of the studied samples obtained using an optical microscope.

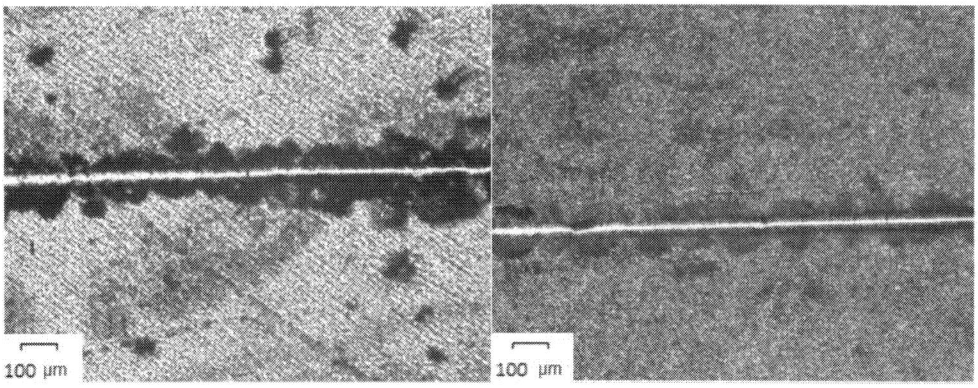

a b

Figure 5. Images of scratches formed on VT3-1 titanium alloys following the scratch test, a – nitriding in traditional glow discharge, b – nitriding in glow discharge with hollow cathode

The analysis of images (Fig. 5) revealed chips at the edge of scratches of a surface nitride layer, typical for brittle failure, and the size of the width of such chips for samples nitraded in glow discharge with a hollow cathode is smaller, which may indicate smaller fragility of the surface layer of these samples.

4. Conclusion

As a result of the research it was found out that the increase in argon content in gaseous mixtures of up to 80–90% during nitriding of VT3-1 titanium alloy in glow discharge with a hollow cathode during 4 hours at 800 °C leads to the increase in thickness of the hardened layer up to ~ 90 microns and surface microhardness up to ~ 850 $HV_{0.1}$, and contributes to the formation of a smaller microhardness gradient of a surface layer.

The hardened layer on the surface of VT3-1 titanium alloy obtained upon nitriding in glow discharge with a hollow cathode is less fragile and resists plastic deformation more efficiently in comparison with the hardened layer obtained via traditional ion nitriding.

References

[1] Zwicker U. 1979 *Titanium and its alloys*, translated from German (Moscow: Metallurgy)
[2] Ilyin A.A., Kolachev B.A. and Polkin I.S. 2009 *Titanium alloys. Composition, structure, properties: Reference book* (Moscow: VILS-MATI)
[3] Budilov V.V., Agzamov R.D. and Ramazanov K.N. 2007 *MiTOM* **7** 25
[4] Ramazanov K.N., Zolotov I.V., Khusainov Yu.G. and Khusnutdinov R.F. 2015 *Journal of physics: Conference series* **652** 012055
[5] Akhmadeev Yu.H., Ivanov Yu.F., Koval N.N., Lopatin I.V. and Schanin P.M. 2008 *Surface. X-ray, synchrotron and neutron study* **2** 108
[6] Lopatin I.V., Akhmadeev Yu.H. 2011 *Scientific News of Belgorod State University. Series: Mathematics. Physics* **22** 5 180–186

Hardening Roll Surface by Plasma Nitriding with Subsequent Hardfacing

A Pesin[1], D Pustovoytov[1], R Vafin[2], I Yagafarov[2], E Vardanyan[2]

[1]Nosov Magnitogorsk State Technical University, Magnitogorsk - 455000, Russia
[2]Ufa State Aviation Technical University, Ufa - 450000, Russia

Email: vafinrk@mail.ru

Abstract. The wear of the surface layer of rolls after ion nitriding in glow discharge, followed by a coating of TiN - TiAlN plasma arc are studied and simulated. stress-strain state of the material rolls under asymmetric rolling with ultra-high shear deformations is simulated. The effect of thermal fields, formed upon contact of the tool and a deformable sheet, the structure of aluminum alloys, are considered.

1. Introduction

The purposeful asymmetry in the process of plastic deformation is gained due to the roll velocity misalignment by proportions from 1:2 to 1:4, when the cold rolling is provided under the high contact friction with severe single deformations (no less than 50%) [1]. The asymmetry results in the significant increase of the friction force which provides the high shear deformations of the material, and in the surface wear of rolls.

Currently, the development of the new ion-plasma techniques for hardening the roll surface which provide the purposeful change of the structure and phase composition of the surface layer and the physical and mechanical characteristics of the tool [2,3] is an urgent issue. The stimulating techniques of ion – nitriding with the crossed electric and magnetic fields [4-8] due to the charged particles densification followed by the TiN-TiAlN coating are considered to be perspective. The material modification is achieved through the increase in the number of the ionizing events, the ion mixing, sputtering, flash heat, atom and molecule deposition, defect formation, chemical interaction, and the radioactive diffusion stimulation.

2. Materials and Research Methods

The numerical simulation of the asymmetric thin sheet rolling was performed as a two-dimensional problem using the DEFORM 2D software. Also, the metal heating was considered, so the thermal and mechanical problems were solved as well. To solve the thermophysical problem, the thermophysical properties of the Al 5083 alloy were predefined as follows: the heat conductivity coefficient was 180.2 N/s/°C, heat capacity coefficient was 2.433 N/mm^2/°C, emissivity was 0.7. Also, the thermophysical properties of the roll material were predefined – AISI D2 from the DEFORM 2D library: the heat conductivity coefficient was 50.71 N/s/°C; the heat capacity was

3,81 N/mm²/°C. The curve of the Al 5083 yield was predefined from the DEFORM 2D library within the temperature band 20 — 500 °C.

The experimental setup to research the glow discharge characteristics in crossed electrical and magnetic fields was developed using the commercial ELU-5 equipment. The setup uses the pulsed power supply with the duty cycle S=80 % and the frequency of 50kHz.

The vacuum chamber was equipped with a standard magnetron with a length of 450 mm, a height of 50 mm, and a width of 100 mm and the magnetic field induction of 0.03 T, which was attached to the regular place of the vacuum chamber. As the working gas, argon as well as the mixture of nitrogen, argon and acetylene in the proportion of N_2 50% - 80%, Ar 25% - 10%, C_2H_2 25% - 10% were used. After the pressure got below 2 Pa, the chamber was blown with argon and filled with the working gas. The working gas pressure was altered within 5 - 200 Pa. The experimental circuit of the experimental ion nitriding is displayed in figure 1.

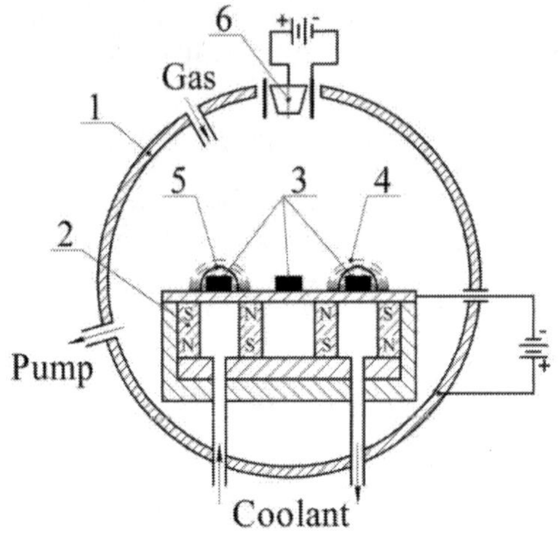

Figure 1. Scheme of the experiment: 1 – vacuum chamber, 2 – magnet system, 3 – samples, 4 - lines of magnetic induction, 5 – high-density nitrogen plasma, 6 – arc plasma source

The research was conducted with the cylindrical-shaped samples of the shear steel P6M5 and X12, each was 12 mm in diameter and 4 mm high. The samples were mechanically polished before nitriding. After nitriding, the samples were coated by TiN and TiAlN compositions using the NNV-6.6-I1 machine. The process sequence was as follows. Firstly, the sample surface was cleaned and activated with argon plasma for 20 min when the bias voltage of -700 V was applied to the specimens. When cleaning, the samples were heated up to 440 °C. Then, they were coated with the TiN and TiAlN compositions. The following coating modes were applied: the nitrogen pressure was 0.11 – 0.13 Pa, the discharge current was 15A supported by altering the heating current of the emission cathode, the arc current of the titan cathode was 90A, of the aluminum one 60A. The stoichiometric compound of the TiN and TiAlN coating was obtained checking the arc current of the titan and aluminum cathodes. Simultaneously, the bias voltage of -200 V was applied to the materials. The temperature of the specimens under coating did not exceed 500 °C.

3. Results and Discussion

The analysis of the numerical simulation in DEFORM 2D revealed that, comparing to the symmetric distortion, the rolling force decreased by 2.4 – 3.2 times under 50 – 75% spellerizing and the misalignment of roll velocities Vi/V_2 = 2 — 4. However, there was a significant torque increase on the rolls: by 1.5 – 3.5 times on the bottom roll, by 1.1 – 2.6 times on the top roll. The deformation heating of Al 5083 alloy in the deformation zone which reached 345 °C when £ is 75 %, Vi is 100 mm/s, Vi/V_2 is

4, and p is 0,4, strongly impacts the force parameters of the asymmetric rolling. When decreasing the rolling velocity down to 1 mm/s, the metal heating does not exceed 50 °C, the rolling force, however, increases from 4 to 7 kN by 1.75 times (per 1 mm of the sheet width), the rolling torque on the bottom roll increases by 1.8 times, from 137kN to 245 kN (per 1 mm of the sheet width), and by 1.96 times on the top roll: from 85 kN to 167 kN (per 1 mm of the sheet width).

The results of the electron microscopical study of samples demonstrate that the nitriding in crossed electrical and magnetic fields of P6M5 and X12 steel results in the formation of nitride layer under which a diffusion sublayer with microtexture is formed.

When nitriding the P6M5 steel in a glow discharge plasma for 4 hours, a hardened layer of 200 μm is formed (figure 2). It is known [2] that the hardened layer height reaches 25 μm after 3 hours of gas nitriding.

The analysis of X12 microtexture revealed the thin nitride zone of 10 – 15 μm (figure 3) with minimal grain boundary nitride precipitate in the diffusion zone. The nitride layer in the images of microtexture looks to be completely texture-free, it includes the iron nitride and the alloying element nitride. Both the nitride and diffusion layers observed microscopically are considered as the common thickness of the nitrogenized layer. The research of the microtexture and the phase composition of the diffusion sublayer (figure 7) revealed the nitrogenous solid solution of the parent metal, its nitrides and the nitrides of alloying elements [7]. During the alpha-phase diffusion saturation, the CrN phases in the shape of fine substances of 1 – 2 μm are separated (figure 3).

There is no clear boundary between the nitride layer and the diffusion sublayer in the microtexture images. The transition from the nitride layer to sublayers is smooth, that is one of the key requirements to the nitrogenized layer [3].

The thickness of the nitride layer in X12 steel was 10 μm after the 4 hour nitriding in crossed electrical and magnetic fields (figure 3) and in P6M5 it was 80 μm (figure 2). The high percentage of Cr in X12 steel provides the thinning of the hardened layer, as may be supposed, due to the chrome nitride layer which prevents the nitrogen diffusion as well as the diffusion layer formation.

There is no parent phase on the sample surface after nitriding in the crossed electrical and magnetic fields. Instead of that, a range of phases is formed as a result of interaction of the glow discharge plasma with the matrix material.

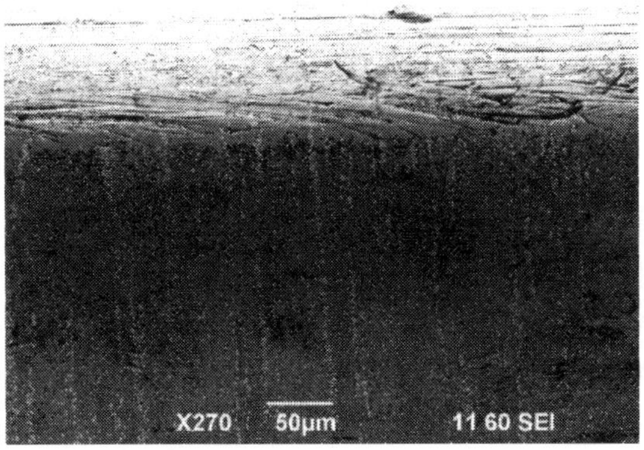

Figure 2. The section microtexture of P6M5 steel after 4 hour nitriding in crossed electrical and magnetic fields, where p is 80Pa

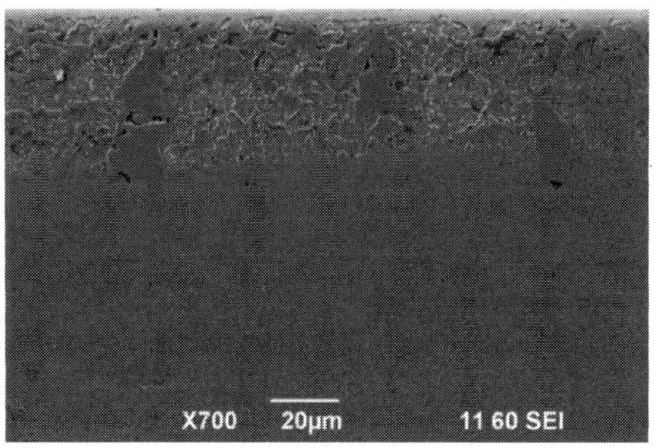

Figure 3. The section microtexture of X12 steel after 4 hour nitriding in crossed electrical and magnetic fields, where p is 80Pa

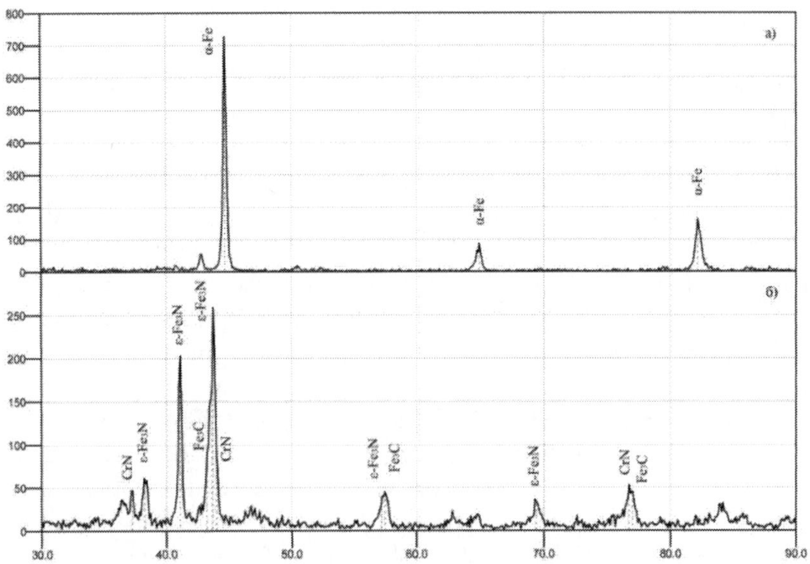

Figure 4. X-Ray diffraction patterns of the P6M5 surface layers before (a) and after the 4 hour nitriding in crossed electrical and magnetic fields, where p is 80Pa (b)

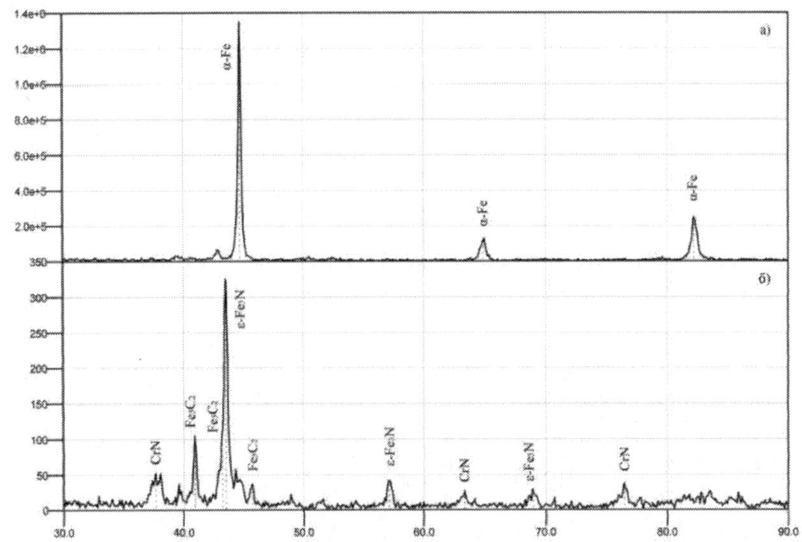

Figure 5. X-ray diffraction patterns of the X12 surface layers before (a) and after the 4 hour nitriding in crossed electrical and magnetic fields, where p is 80Pa (b)

The X-ray pattern (figure 4) of the P6M5 surface after the ion nitriding revealed that around the 40°—50° angles, there is a broadening at the $Fe_3(N,C)$ and CrN peak bases which can be caused by several phases with the similar values of interplanar distance and the retained compression stress.

The analysis of the X12 surface after the nitriding in crossed electrical and magnetic fields demonstrated the reflexes of the Fe_3N ε-phase as well as the phases which was composed of CrN. The significant fall in intensity of the α-iron peaks (figure 5) after nitriding indicates the decreasing α-phase fraction in the surface layer as a result of forming the Fe_3N ε-phase. After the treatment in the nitrogen – argon – acetylene mixture with proportions of N_2 50% - 80%, Ar 25% - 10%, C_2H_2 25% - 10%, the sample

IOP Conf. Series: Journal of Physics: Conf. Series **830** (2017) 012095 doi:10.1088/1742-6596/830/1/012095

surfaces revealed the reaction to iron carbide (Fe_5C_2) as well. Using the nitrogen – argon – acetylene mixture with proportions of N_2 50% - 80%, Ar 25% - 10%, C_2H_2 25% - 10% in nitriding provides the deactivation of the oxygen remained and the diffusion saturation.

After the 180 minutes of coating deposition, the coating of the TiN composition of 9 μm (figure 6) as well as the TiAlN coating with thickness of 6 μm (figure 7). Then, the materials were cooled to 100 °C under vacuum. The micro-hardness dependence on depth is presented in Figure 8.

The high wearing is provided by the TiN and TiAlN coating as well as by hard flexible nitrogenized layers under coating. Moreover, smoothing the sharp transition between the "soft" material and the hard coating contributed to the wearing quality and decreased the hardness gradient between two different textures as well.

Compared to the conventional nitriding, the advantage of the ion nitriding in magnetic field in the microhardness distribution through the obtained layer is obvious. The thickness of the hardened layer was increased from 54 μm to 84 μm, i.e. by 1.5 times. The samples placed in magnetic field demonstrated smoother microhardness depth distribution (figure 8a).

Figure 6. The P6M5 microtexture after the ion nitriding in glow discharge with magnetic field and TiN coating

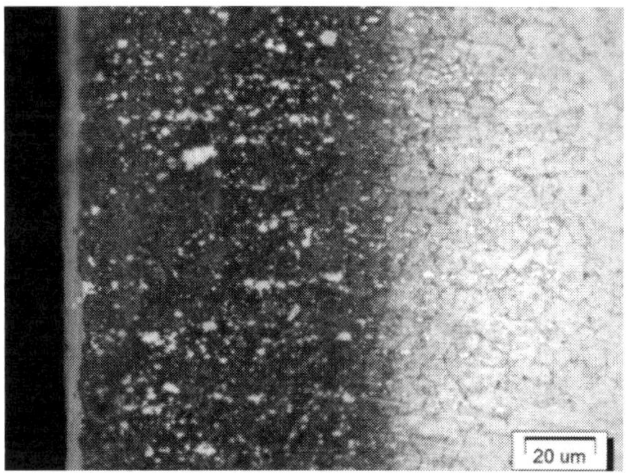

Figure 7. The P6M5 microtexture after the ion nitriding in glow discharge with magnetic field and TiAlN coating

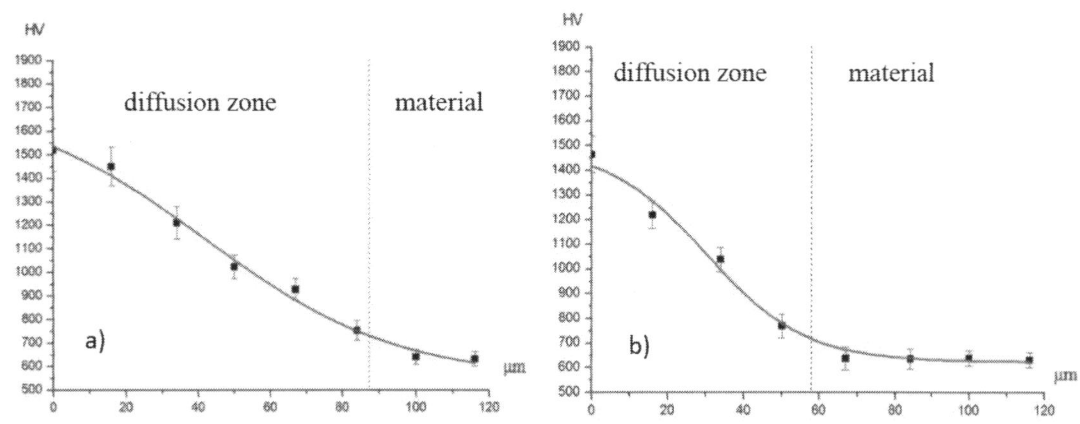

Figure 8. The micro-hardness dependence on depth in the P6M5 samples after the ion nitriding under 500°C: *a* – in magnetic field (*P*=44 Pa); *b* – in the absence of magnetic field (*P*=120 Pa)

4. Conclusion

- The analysis of the numerical simulation in DEFORM 2D revealed that comparing to the symmetric distortion the rolling force decreased by 2.4 – 3.2 times when 50 – 75% spellerizing and the misalignment of roll velocities $V_1/V_2 = 2 — 4$. However, there was a significant torque increase on the rolls: by 1.5 – 3.5 times on the bottom roll, by 1.1 – 2.6 times om the top roll.
- The crossed electrical and magnetic fields provide the energy and density increase of the charged particles current to the treated surface.
- It was experimentally proved that the 4 hour ion nitriding in the crossed electrical and magnetic fields of the P6M5 and X12 steel under pressure of 80 Pa resulted in the surface modified layer which was compounded of Fe_3N, CrN nitride phases of increased hardness. The CrN phase appeared as fine particles of 1-2 nm.
- The crossed electrical and magnetic fields increase the thickness of the hardfaced layer of the P6M5 sample up to 200 μm for 4 hours in comparison with 25 μm for 3 hours under gas nitriding.
- The ion nitriding in glow discharge with magnetic field is an efficient technique to prevent the TiN and TiAlN coating from excessive plastic deformation of the support.

Acknowledgments
The study was supported by the grant from the Russian Science Foundation (#15-19-10030)

References

[1] Ji Y.H., Park J.J. 2009 *Materials Science and Engineering* **499** 14
[2] Goncharenko I.M. 2000 Evolution of the structure and phase composition of hardened 4140 steel in the process of plasma nitriding *5th International Conference on Modification of Materials with Particle Beams and Plasma Flows* (Tomsk: Institute of Physics and Technology TPU) pp 330–333
[3] Edenhofer B. 1976 *Metal and Material Technological* **8(8)** 421
[4] Arnell R.D., Kelly P.J. 1999 *Surface and Coatings Technology* **112** 170
[5] Brading H.J., Morton P.H, Earweaker G. 1992 *Surface engineering* **8(3)** 206
[6] Bradley J.W., Arnell R.D., Armour D.G. 1997 *Surf. and Coat. Technol.* **197** 538
[7] Vafin R.K., Ramazanov K.N. 2010 Surface Modification of Tool Steel during Ion Nitriding in Magnetic Field *10th International Conference on Modification of Materials with Particle Beams and Plasma Flows* (Tomsk: Publishing House of the IOA SB RAS) pp 458–461
[8] Koval N.N. 2000 Elion nitriding of steels *5th International Conference on Modification of Materials with Particle Beams and Plasma Flows* (Tomsk: Institute of Physics and Technology TPU) pp 327–329

The gradient structure of the NiTi surface layers subjected to tantalum ion beam alloying

S L Girsova[1], T M Poletika[1], L L Meisner[1,2] and E Yu Schmidt[2]

[1]Institute of Strength Physics and Material Science SB RAS, 2/4 Akademichesky ave., Tomsk, 634055, Russia
[2]National Research Tomsk State University, 36 Lenin ave., Tomsk, 634050, Russia

poletm@ispms.tsc.ru

Abstract. The NiTi shape memory alloy has been modified by ion implantation with Ta to improve the surface and biological properties. The elemental and phase composition and structure of the surface and near-surface layers of NiTi specimens after the Ta ion implantation with the fluency D = 3 × 10^{17} cm^{-2} and D = 6 × 10^{17} cm^{-2} are examined. The methods of Auger electron spectroscopy (AES), transmission electron microscopy (TEM), and electron dispersion analysis (EDS) are used. It is found that a nonuniform distribution of elements along the depth of the surface layer after the ion implantation of NiTi specimens, regardless of the regime, is accompanied by the formation of a number of sublayer structures.

1. Introduction

Shape memory alloys are known to enjoy wide applications, e.g. NiTi alloy is employed in biomedical industry for the manufacture of implants. To overcome the problem of Ni release and to improve the biological, chemical, and mechanical properties, many surface treatment techniques are used [1]. Thus, the ion beam implantation gained recognition as an effective method for the deposition of thin protective layers on the surface of NiTi alloys [2]. A complex gradient structure would form in surface layers of as-treated NiTi alloys; a significant change would also occur in their phase composition. The mechanisms involved are still unclear. This might be due to the lack of reliable experimental evidence for structural changes occurring in the surface and near-surface layers of treated materials. The aim of the given study is to investigate the surface layer structure of as-treated material at the nanoscale levels. The study is carried out for test specimens made of NiTi alloy, which were modified by the Ta-ion beam implantation; the surface layer of as-treated alloy is found to possess an enhanced biocompatibility and satisfactory X-ray contrast range [3, 4]. The ion beam implantation would cause oxide layers, TiO_2 included, to form on the surface of Ti and Ti-based alloys, which is liable to enhance the biocompatibility, chemical stability, non-toxicity, and absorptive capacity of as-treated material [2, 4].

2. Experimental

The study was carried out on test specimens of NiTi alloy (the alloy composition is $Ni_{50.5}Ti_{49.5}$; the alloy structure is B2). The specimen surface was polished before the treatment. The Ta-ion beam modification was performed with a DIANA-3 ion implanter unit using pulsed Ta-ion beams; the fluency $D = 3 \times 10^{17}$ cm^{-2} and $D = 6 \times 10^{17}$ cm^{-2}; the average accelerating voltage was 80 kV and the pulse repetition frequency was 50 Hz. By ion implantation, the specimens were heated to the temperature ≤473 K. Using the method of Auger electron spectroscopy (AES), the layer-by-layer analysis was made for the element composition of the modified specimen surface using a SHUNA-2 (NR TPU, Russia). The defect structure studies were carried out with the aid of transmission electron microscopy (TEM) on JEM 2100 (JEOL) at the accelerating voltage of 200 kV. The cross-section specimens were prepared by the ion thinning method on EM 09100IS (JEOL). The equipment used in the study and mentioned hereinafter was provided by the Shared Use Center of the ISPMS SB RAS (Tomsk, Russia).

3. Element concentration profiles (AES)

Figure 1 illustrates profiles obtained for near-surface layers of as-treated specimens by AES. Irrespective of the dose, oxide layers enriched with Ti and O atoms form on the specimen surface. O and C atoms are adsorbed from the residual atmosphere of the vacuum chamber of the ion implanter. The oxide layer thickness is 10 nm and 30 nm for the fluency $D = 3\times10^{17}$ cm^{-2} and $D = 6\times10^{17}$ cm^{-2}, respectively. Oxide layers were found to contain such concentrations of O and Ti that are sufficient for the formation of stoichiometric dioxide TiO_2.

The important parameter characterizing the state of NiTi surface layers is the content of Ni atoms. The Ni concentration profiles obtained for oxide sublayers are displaced with depth. The above results indicate that the as-treated specimen surface is depleted of Ni. The AES data presented in Figure 1 suggests that the latter effect might be due to Ni-O bonds, being more readily reduced relative to Ti-O bonds [5]; besides, the preferential sputtering of Ni atoms, which is four times higher than that of Ti, might be responsible for the same effect [6]. Due to surface layers being depleted of titanium, Ti atoms are accumulated in areas with a low content of oxygen and titanium. Hence, in the concentration profiles obtained for the ion-modified layer the maxima of Ti and Ni are shifted relative to one another. The Ta concentration profile obtained for oxide layers is also displaced with depth (Figure 1). The maximum Ta content is found to occur at the depth 20÷30 and 50 ÷ 60 nm in going down to the lower boundary of oxides, respectively, for the regimes $D = 3\times10^{17}$ cm^{-2} and $D = 6\times10^{17}$ cm^{-2}. The ion-modified surface layer for both as-treated specimens has the thickness 80 nm and 100 nm.

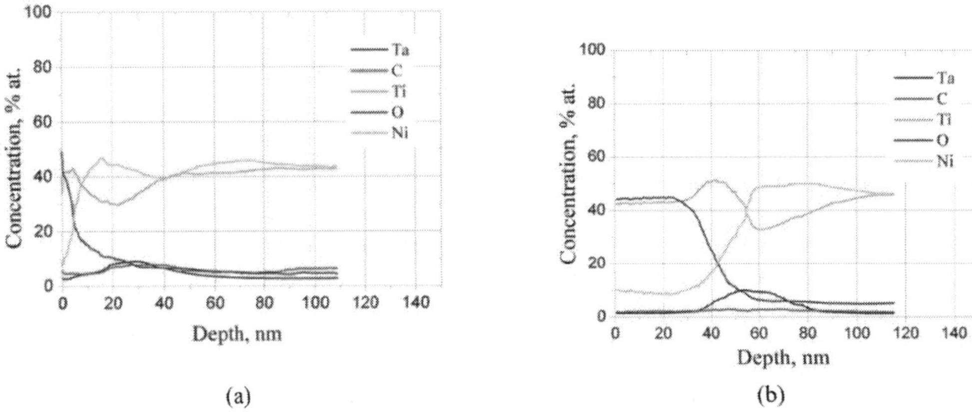

Figure 1. Element concentration profiles within surface layers of NiTi specimens with implanted Ta: fluency $D = 3\times10^{17}$ cm^{-2} (a) and fluency $D = 6\times10^{17}$ cm^{-2} (b).

4. Surface layers structure (TEM)

Using the TEM technique, the surface layer structure is examined. It is found that a nonuniform distribution of elements along the depth of surface layers after the Ta ion implantation, regardless of the regime, is accompanied by the formation of a number of different sublayers. Thus, two main layers are found, i.e. the upper oxide layer and the lower-lying layer of the material in the amorphous state.

The surface layer structure of NiTi specimens subjected to the Ta-ion implantation is illustrated in Figure 2. The upper oxide sublayers are 20÷30 nm thick. The diffraction analysis data suggests that oxide layers preferably include TiO_2 modifications, i.e. brookite and rutile. For specimens with a higher irradiation dose $D = 6\times10^{17}$ cm^{-2}, the oxide layer structure is more complex (Figure 2). It is seen that there are two layers: the light outer layer and the dark one adjacent to the lower boundary of the oxide. The observed contrast suggests that the dark crystalline underlayer contains atoms of heavier element tantalum. We can assume that this is due to the linear dependence of both the depth and the width of the Ta concentration profile of the ion treatment regime. Thus, with increasing irradiation dose, the maximum implantable element expands with overlapping O and Ta distribution profiles, which implies the existence of nanoparticles of Ta oxides. Indeed, an increase in the dose of irradiation is accompanied by a rise of temperature and the number of radiation-induced defects, thus increasing the diffusion mobility of heavier atoms. However, there is no reliable experimental TEM data confirming the presence of tantalum oxide in implanted layers of NiTi.

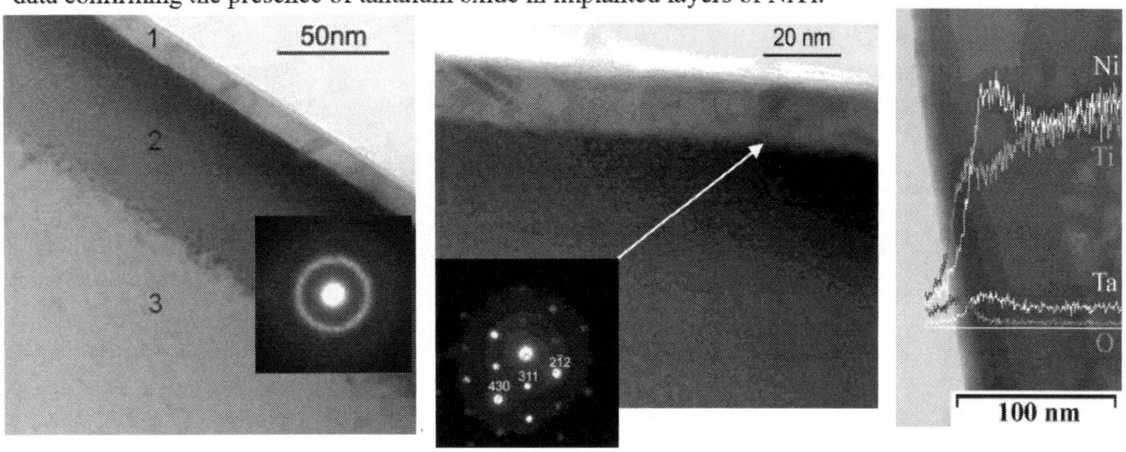

(a) (b) (c)

Figure 2. TEM images of surface layers of NiTi subjected to the Ta-ion implantation with the fluency $D = 3\times10^{17}$ cm^{-2}: (a) a bright-field image of surface layers and a nanodiffraction (NBD) pattern of the amorphous ion-modified layer (the probe on the specimen size 30 nm); 1) the oxide layer (nanocomposite ceramics); 2) the amorphous layer; 3) the near-surface crystalline layer; (b) NBD pattern of the near-surface oxide layer, zone axis [-345] of TiO_2 brookite (indicated by the arrow); (c) EDS data obtained for the surface layer, which reveals a layer containing localized Ta atoms.

The amorphous layer has a thickness of 80 nm and 100 nm for the regimes $D = 3\times10^{17}$ cm^{-2} and $D = 6\times10^{17}$ cm^{-2}, respectively (Figure 2a). The NBD pattern obtained for the latter near-surface amorphous layers of both implanted NiTi specimens has a characteristic broad diffuse halo corresponding to reflection {110} B2, which exhibits a short-range order (Figure 2a). It is believed that distortion-type amorphization takes place, thus causing distortion of the original NiTi crystal at the scale level of a unit cell [7]. It should be noted that, regardless of the dose, the variation in the TEM image contrast suggests that the amorphous material layer has a darker zone 30÷40 nm thick and a lower-lying lighter zone (Figure 2a). According to the sorption *Z*-contrast (here *Z* is the atomic number of elements, $Z_{Ta} = 73$), the element tantalum is localized in the former zone. In support of this view, the energy dispersion analysis is made for the amorphous layer, which suggests that a Ta-enriched layer really exists (Figure 2c). It was found that the thickness of this layer is independent of the ion treatment regime.

A thin defect sublayer is found to occur between the modified amorphous layer and the substrate for both ion-treated specimens; it is characterized by an inhomogeneous TEM contrast (Figures 2 and 3). Generally, such an intermediate layer has a high internal stress level and contains all kinds of defects, e.g. point defects, dislocation loops, etc., as well as nanophases on the base of alloy components [8]. Due to the formation of an ion-modified (amorphous) surface layer, the material volume would change significantly. This would cause significant dilatation and an increase in the compressive stress level in the near-surface layer, thereby initiating plastic yielding of the material [2]. The effect of dilation is observed for the near-surface layer of NiTi alloy subjected to the ion implantation treatment to the depth ≥10 μm [9]. It should be noted that this effect becomes stronger with increasing irradiation dose. The effect is manifested in the formation of dislocations, dislocation loops, and network substructures (Figure 2e). Modified near-surface layers have low dislocation density, which is largely due to recovery processes initiated by high temperature. The overall picture is diverse and inhomogeneous; in particular, the deformation structure varies significantly. Thus, traces of the martensitic R-phase are observed in the form of individual lamellae or groups of lamellae. The formation of the R-phase should be due to the shifting temperature interval of martensitic transformations as well as the suppression of the B19′ phase formation in thin near-surface layers by the implantation treatment [2].

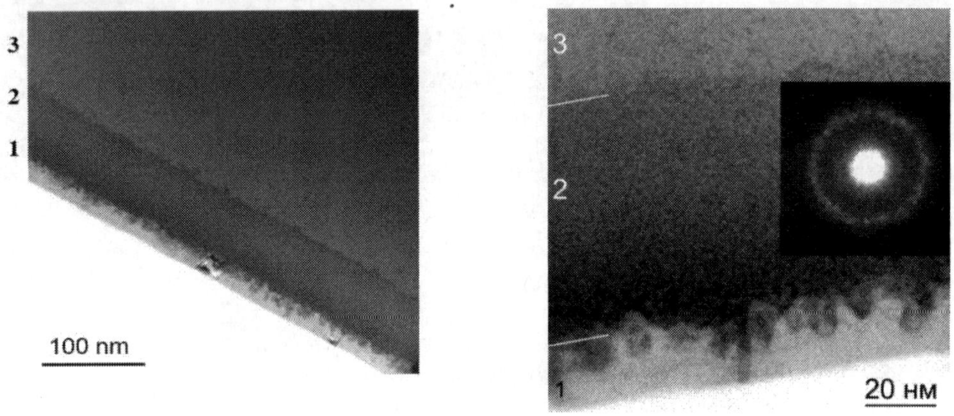

(a) (b)

Figure 3. TEM images of surface layers of NiTi alloy subjected to the Ta-ion implantation with $D = 6 \times 10^{17}$ cm^{-2}: (a) a bright-field image; the oxide layer enriched with Ta is indicated by the arrow; (b) a bright-field image of the amorphous layer; an NBD pattern of the amorphous layer (the probe on the specimen size 25 nm); 1— the oxide layer (nanocomposite ceramics); 2 — the amorphous layer; 3— the near-surface crystalline layer.

4. Conclusion

It is found that a nonuniform distribution of elements along the depth of the surface layer of NiTi after ion implantation, regardless of the regime, is accompanied by the formation of a number of different nanolayer structures. Thus, two main layers are found: the oxide layer and the ion-modified amorphous layer. The oxide layer contains preferably titanium oxide TiO_2. With increasing irradiation dose, we observe an increase in the thickness of the oxide layer and changes in its structure, resulting in the formation of two-layer nanocomposite ceramics. Ta atoms are localized in the lower-lying amorphous layer ≤100 nm thick. With increasing irradiation dose, the maximum implant element is not only shifted into the specimen, but also expands with overlapping O and Ta distribution profiles. This is due to the linear dependence of both the depth and the width of the concentration profile of Ta in the ion treatment regime. The AES and EDS data show a significant loss in the Ni oxide layer, which is an important factor when using NiTi in medicine as implants. The effect of dilation was manifested in the formation of dislocations and dislocation loops and network substructures.

References

[1] Otsuka K and Ren X 2005 *Prog. Mater. Sci.* **50** 511
[2] Pogrebnjak A D and Bratushka S N 2013 *Russian Chem. Rev.* **82** 1135
[3] Cheng Y, Cai W, Li H T and Zheng Y F 2006 *J. Mater. Sci.* **41** 4961
[4] Li Y, Zhao T, Wei S, Xiang Y and Chen H 2010 *Mater. Sci. and Eng. C.* **30** 1227
[5] Kucharski S, Levintant-Zayonts N, Luckner J 2014 *Mater. and Design* **56** 671
[6] Tan L, Crone W C 2002 *Acta Mater.* **50** 4449
[7] Parkhomenko V D and Dubinin S F 2012 *Phys. Metals and Metallogr.* **113** 641
[8] Schmidt B 2007 *Radiation Effects & Defects in Solids* **162** 171
[9] Meisner L L, Poletika T M, Girsova S L, *et al* 2016 *Vacuum* **129** 126

Acknowledgments

The investigations of the structure of the surface and near-surface layers of NiTi specimens after Ta ion implantation were supported by the Russian Academy of Sciences in the framework of the Program of Basic Research for years 2013–2020, project No. 23.2.1. The investigations of the elemental and phase composition and the surface structure of initial NiTi specimens were supported by the Russian Science Foundation (project No. 15-13-00023).

Formation of Ti-Ta-based surface alloy on TiNi SMA substrate from thin films by pulsed electron-beam melting

L L Meisner[1,4], A B Markov[2], G E Ozur[2], V P Rotshtein[1,3], E V Yakovlev[2], S N Meisner[1], T M Poletika[1], S L Girsova[1], V O Semin[1,4], Yu P Mironov[1]

[1]Institute of Strength Physics and Materials Science SB RAS, Tomsk, Russia
[2]Institute of High Current Electronics, SB RAS, Tomsk, Russia
[3]Tomsk State Pedagogical University, Tomsk, Russia
[4]National Research Tomsk State University, Tomsk, Russia

E-mail: llm@ispms.tsc.ru

Abstract. TiNi shape memory alloys (SMAs) are unique metallic biomaterials due to combination of superelastisity and high corrosion resistance. Important factors limiting biomedical applications of TiNi SMAs are a danger of toxic Ni release into the adjacent tissues, as well as insufficient level of X-ray visibility. In this paper, the method for fabrication of protective Ni-free surface alloy of thickness ~1 μm of near $Ti_{70}Ta_{30}$ composition on TiNi SMA substrate has been successfully realized. The method is based on multiple alternation of magnetron co-deposition of $Ti_{70}Ta_{30}$ thin (50 nm) films and their liquid-phase mixing with the TiNi substrate by microsecond low-energy, high current electron beam (≤15 keV, ~2 J/cm^2) using setup RITM-SP (Microsplav, Russia). It was found by AES, XRD, SEM/EDS and HRTEM/EDS examinations, that Ti-Ta surface alloy has an increased X-ray visibility and gradient multiphase amorphous-nanocrystalline structure containing nanopores.

1. Introduction

Successful biomedical applications of TiNi alloys exhibiting shape memory effect and superelasticity (SME-SE) are associated with their ability to reversible return a substantial predetermined strain (8–12%) when heated to body temperature [1, 2]. Important factors limiting these applications, particularly, in cardiovascular surgery, are a danger of release of toxic Ni into the adjacent tissues [3,4], as well as their insufficient level of X-ray visibility [5].

The optimum solution of these problems seems to produce a protective surface layer up to several micrometers in thickness which would possess not only excellent biochemical resistance, but also high biomechanical properties under cyclic loading in aggressive biological media. This approach can be realized by producing a surface alloy which would have mechanical properties (elastic modulus and SME-SE characteristics) similar to these of TiNi substrate, but free of nickel. Surface alloys, as known, are produced by pulsed melting of an alloying film with a substrate and/or by ion implantation [6]. That is why, surface alloys, compared to conventional PVD and CVD-based coatings, do not present problems associated with poor adhesion and delamination. In this respect, the Ti-Ta system, whose alloys possess high-temperature SME (HTSME), looks very promising. The martensitic transformation (MT) temperature range of these alloys is 50–100 °C above ambient temperature [7], so, they are in martensitic states with elastic characteristics comparable to these of muscle tissues at human body temperature. In addition, alloying of titanium alloys, including TiNi SMAs, with tantalum improves their corrosion resistance [8], biocompatibility [9], and X-ray visibility [10].

This work aims to fabricate and characterize the novel thin (1–2 μm) protective surface layers based on Ni-free alloys with superelastic properties on TiNi substrates. The paper presents the results of studies of elemental and phase compositions as well as structure of ~1-μm-thick Ti-Ta thin-film coating and Ti-Ta-based surface alloy fabricated on TiNi substrate by magnetron co-deposition of Ti-Ta films followed by pulsed melting by microsecond low-energy, high-current electron beam (LEHCEB) [11].

2. Experimental

Fabrication of Ti-Ta-based surface alloy was performed in united vacuum cycle using modified automatic setup RITM-SP equipped with LEHCEB source, DC-magnetron sputtering unit and common vacuum chamber evacuated by turbo-molecular pump [12]. Magnetron unit consisted of three magnetrons arranged symmetrically on a circle, two of which contained Ti (VT1-0, Russia) and Ta (GIRMET, Russia) targets of 75 mm diameter. The angle between magnetron axis and the normal to substrate holder was 28°. The co-deposition of $Ti_{70}Ta_{30}$ (at.%) thin films was carried out at room temperature at Ar working gas of 0.1 Pa in a vacuum chamber with a base pressure of $\sim 10^{-3}$ Pa. 560 and 160 W power were applied to Ti and Ta targets, respectively, which corresponded to the film co-deposition rate of 1.8 ± 0.3 nm/s. The substrate was rotated with the rate of 20 rpm during the co-deposition to obtain a homogeneous film composition over the substrate.

Substrates with the dimensions of 15×10×1 mm were made of commercial (VIM) TiNi alloy (MATEK-SMA Ltd., Russia). Its chemical composition was as follows (wt%): 55.08 Ti, 0.051 Ni, 0.03 C, 0.002 O; and the transformation temperature $A_s = 303$ K. The methods for preparing TiNi substrates as well as their microstructural characterization are given in [13]. To prevent the local delamination of a Ti-Ta film due to cratering during the pulsed melting [13], the substrate was pre-irradiated with LEHCEB in the following mode: pulse duration of 2–2.5 μs, maximum electron energy of 25 keV, energy density $E_s = 3.8\pm0.7$ J/cm^2; beam diameter of ~60 mm, number of pulses $n = 32$, and repetition rate of 1 pulse/5 s. After this pre-treatment, TiNi substrate was installed by manipulator alternately along the magnetron unit axis for co-deposition of 50-nm-thick $Ti_{70}Ta_{30}$ film, and along the LEHCEB axis for pulsed melting of film/substrate system. Pulsed melting was performed in the following mode: maximum electron energy of 15 keV, $E_s = 2\pm0.2$ J/cm^2; $n = 5$.

The thickness of $Ti_{70}Ta_{30}$ film, energy density and number of pulses in each cycle of "co-deposition + pulsed melting" were chosen based on the two main requirements: (i) a thickness of the melted layer of a film/substrate system should be a little more than ~1 μm; (ii) a liquid-phase diffusion depth of film components into substrate must be, at least, several times greater than original film thickness to provide a diffused transition between Ti-Ta surface alloy and TiNi substrate. The number of cycles of "co-deposition + pulsed melting" was $N = 20$; thus, the predicted thickness of $Ti_{70}Ta_{30}$ surface alloy was ~1 μm. The energy density of LEHCEB was preliminarily controlled by calorimeter with an accuracy no worse than ±15%.

Elemental depth profile analysis of test samples was performed by Auger electron spectroscopy (AES) (Shkhuna-2, Russia) at sputtering rate of analyzed near-surface layer of 2÷3 nm/min and 3 keV Ar+ ion beam diameter of 1 mm; maximal analyzed depth is 400 nm. The average elemental compositions of Ti-Ta coating and surface alloy were determined with the INCA Energy Dispersive Spectrometer (EDS) (Oxford Instruments, UK) installed on scanning [EVO 50 (Zeiss, Germany)] and transmission (JEM 2100; the electron-probe diameter was ~12 nm) electron microscopes (SEM, TEM). X-Ray diffraction (XRD) analysis was carried out on DRON-7 diffractometer (Burevestnik, Russia) with Co-Kα radiation in symmetric and asymmetric Bragg diffraction geometry. XRD patterns in asymmetric geometry were obtained at glancing incidence angles α = 3, 5, 8 and 15°. TEM experiments were performed on JEM 2100 electron microscope (JEOL, Japan) at accelerating voltage of 200 kV by the method of thin foils. Thin foils were made by ion thinning using EM 09100IS device

(JEOL, Japan) from thin (0.3 mm) plates, spark-cut perpendicularly to the plane of the samples. This enabled studying the microstructure at different depths below the surface.

3. Results and discussion

3.1. Elemental composition

3.1.1. Ti-Ta thin-film coating. According to AES data, as-co-deposited ~1-μm-thick Ti-Ta coating was oxygen-enriched up to a depth of ~50 nm. This is because the Ti/Ta films absorb residual oxygen from vacuum chamber. The coating was uniform in composition at the depths $x \geq 50$ nm, and their composition was $Ti_{\sim 80}Ta_{\sim 20}$ (at.%) at x = 200 nm. Similar measurements were performed by SEM/EDS analysis to clarify the coating composition. Figures 1a, 1b and 1c shown the results of SEM/EDS surface analysis of the coating at accelerating voltage of 10 kV, wherein, taking into account the density of Ti-Ta coating/alloy, the thickness of the analyzed surface layer is ~0.5 μm [14]. According to data obtained by averaging over 6 EDS measurements, the coating composition is $Ti_{\sim 70}Ta_{\sim 30}$ (at.%) in accordance with the prediction. The content of O and C impurities in the coating is less than the limit of detection (LOD), i.e. ≤4-5 at.%.

3.1.2. Ti-Ta-based surface alloy. According to AES data, the ~50 nm-thick near-surface layer of Ti-Ta surface alloy, as that of Ti-Ta coating, is enriched by oxygen, which delivered into the melt from residual atmosphere of vacuum chamber during processing. The surface alloy is homogeneous in composition at depths $x \approx 50 \div 350$ nm, but, in contrast to the coating, it contains nickel, and its composition is $Ti_{40}Ta_{39}Ni_{11}O_5C_5$ (at.%) at intermediate depth of 200 nm. In turn, SEM/EDS analysis has shown (Figure 1e,f,g) that surface alloy, as co-deposited coating, has an average composition of $Ti_{\sim 70}Ta_{\sim 30}$ (at.%) over thickness of ~0.5 μm. The presence of nickel (according to AES data) may be caused by its lower melting point, as compared to Ti and Ta, so it may segregate towards the surface from the molten substrate at the solidification front during multiple cycles of pulsed melting/solidification.

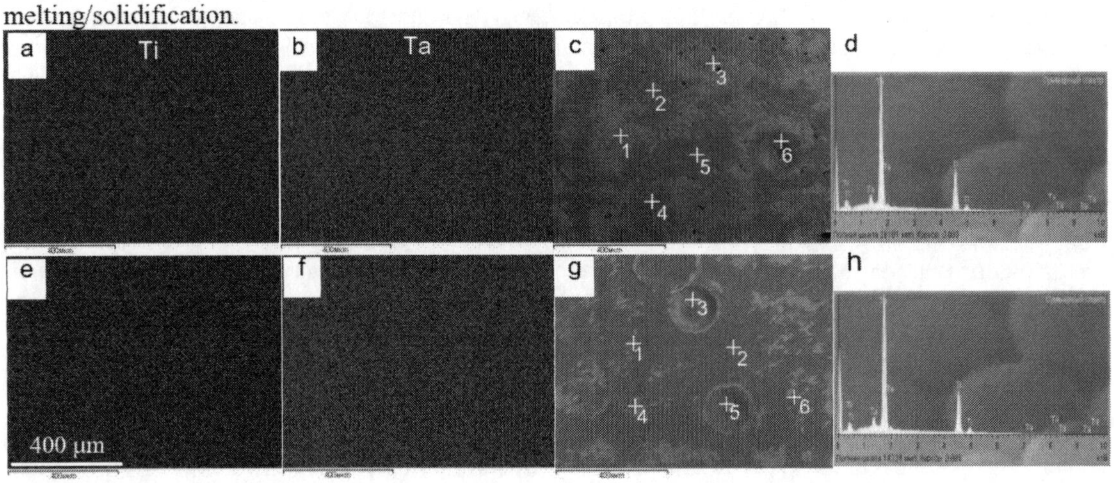

Figure 1. Results of SEM/EDS surface analysis for Ti-Ta coating (a-d) and Ti-Ta based surface alloy (e-h): EDS map distributions of Ti (a, e) and Ta (b, f); SEM images (c, g); EDS spectrums (10 keV) (d, h) obtained by averaging over 6 points shown in (c) and (g), correspondently.

3.2. Phase composition

Figure 2 shows XRD patterns of TiNi substrate in the initial state, the substrate with co-deposited Ti-Ta coating and the substrate with Ti-Ta surface alloy. Analysis of XRD patterns given in Figure 2b, showed that Ti-Ta coating has multiphase structure composed of a bcc-phase based on β-Ti [15], β-

(Ti,Ta) [16], and α-Ta [17]. It follows from an analysis of the location of additional reflexes, that the coating can contain also one-component phases based on deposited Ti and Ta in their low-symmetric modifications: α-Ti with hexagonal structure [18] and β-Ta with tetragonal structure [19]. Significant broadening of main XRD reflexes indicates that the coating has nanocrystalline (domain) structure. The Ti-Ta-based surface alloy, as it follows from Figure 2c, has also multiphase structure, in which the main phase, most likely, is based on disordered Ti-Ta solid solution with amorphous-nanocrystalline structure.

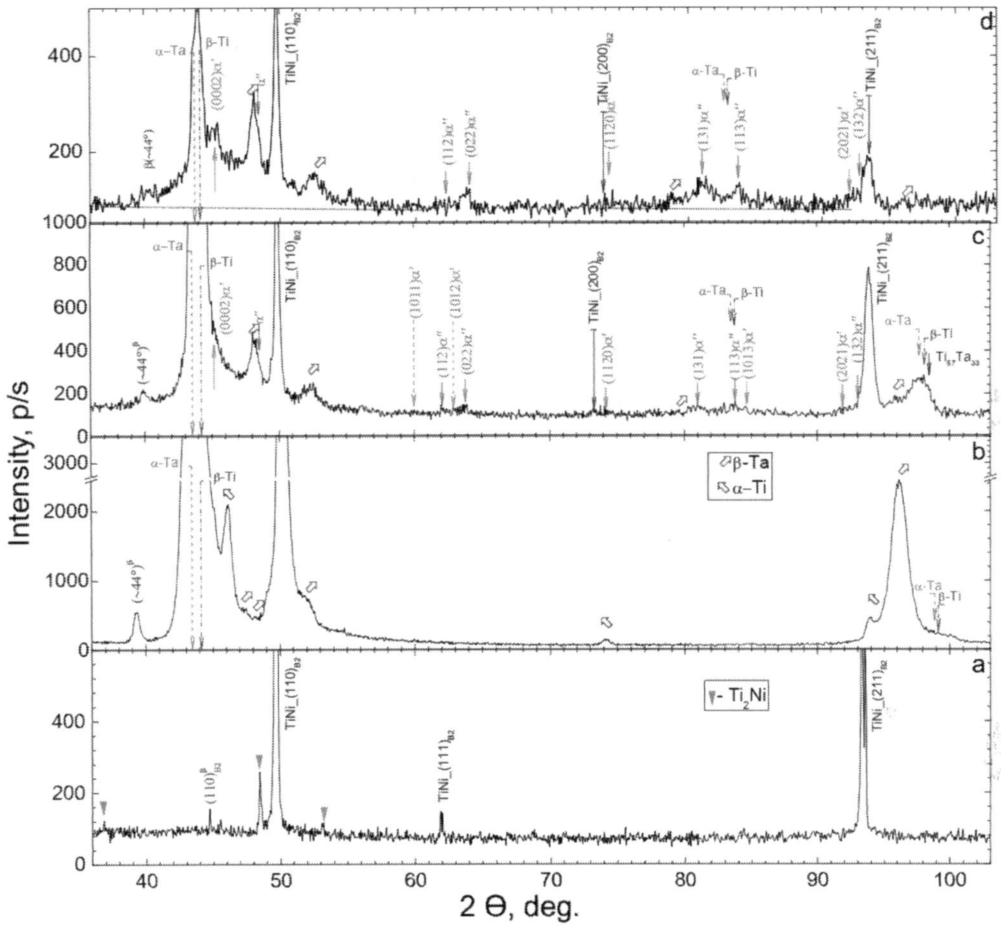

Figure 2. XRD patterns of TiNi substrate (a), Ti-Ta coating (b) and Ti-Ta surface alloy (c, d), obtained in symmetric (a, b, c) and asymmetric (d) Bragg diffraction geometries; (d) glancing incidence angle α=15°. DSRs are located above the tangential (dotted) lines (d) in angular ranges of 40÷60° and 75÷90°.

The large number of clearly identified, although low-intensity reflexes is evidenced by the presence of martensitic α′- and α″ - phases with composition close to $Ti_{70}Ta_{30}$ (at.%) [16], as well as β-Ta-phase [19] in the surface layers. In turn, the broadening of corresponding reflexes point to the nanocrystalline (domain) structure of these phases.

XRD patterns, corresponding to the coating and to the surface alloy (Figures 2b and 2c), contain diffuse scattering regions (DSRs) in angular ranges of 40÷60° (Figure 2b,c,d) and 75÷90° (Figures 2c and 2d), indicating the presence of relatively large volume fraction (~5 vol.%) of quasi-amorphous phases. Angular positions of gravity centers of DSRs correspond to those of main XRD reflexes of bcc phases based on pure Ti and Ta as well as on Ti-Ta system. This means that the short-range order in the atom arrangement in the quasi-amorphous and bcc phases is identical. From this, it follows that crystallization from the amorphous state will probably occur by the growth of bcc phases.

The XRD reflections from the B2 phase of TiNi substrate either absent (Figure 2b) or have very low intensity (Figure 2c). Taking into account symmetric Bragg diffraction geometry, this means that both Ti-Ta coating, and Ti-Ta surface alloy of thickness ~1 μm absorb X-rays efficiently, and this is a reliable indication of their increased X-Ray visibility as compared with TiNi substrate.

The cross-section TEM examination of Ta-Ti-based surface alloy (Figure 3a) showed that its thickness is ~1 μm, which agrees with the expected value. Near the interface of surface alloy/substrate, an equiaxed polycrystalline structure with a average grain size of ~50 nm is formed. Surface alloy has slightly pronounced columnar structure, indicating on directional solidification in the process of formation. There are numerous nanosized inclusions having essentially spherical shape and different contrast (from white to dark gray). Inclusions are distributed non-uniformly in depth: increased inclusion density is observed near the interface of surface alloy/substrate.

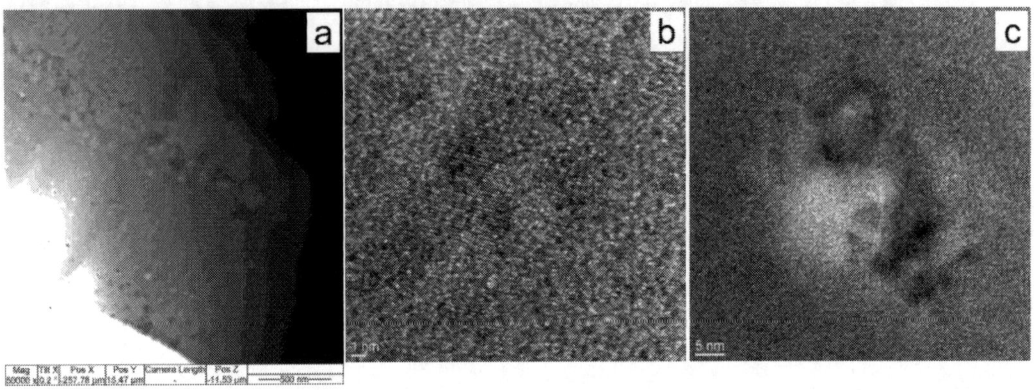

Figure 3. Bright field TEM/HRTEM images of Ti-Ta-based surface alloy: (a) cross section view; (b) nanocrystalline particles embedded into the amorphous matrix; (c) nanopore inside the amorphous matrix.

The high-resolution TEM (HRTEM) analysis revealed a significant volume fraction of mixed amorphous-nanocrystalline structure in the surface alloy, signs of which were identified by XRD analysis. It is seen from Figure 3b, that individual nanocrystals with the size ≤10 nm are surrounded by regions with amorphous structure. The structure in a vicinity of typical white inclusion (presumably, nanopore) is shown in Figure 3c. One can see, that a conglomerate of nanocrystals with dark gray contrast is distributed along the boundary of white spherical precipitate, and, partly, inside it. This picture can be interpreted as a preferential nucleation and growth of nanocrystals on free surface (nanopore wall) in the matrix phase of the surface alloy. This result, which requires a separate study, can clarify the mechanisms of amorphization and crystallization of materials rapidly quenched from the melt

According to TEM/EDS data, white inclusions contain oxygen (≥30 at.%) and argon (working gas). Chemical composition and shape of such inclusions (Figure 3a) indicate that they represent nanopores. Gray inclusions represent probably nanosized precipitates based on Ti, Ta, Ni (e.g., α'-$Ti_{\sim 70}Ta_{\sim 30}$, Ti_2Ni, Ta_2Ni). Different contrast of these particles is associated with their different chemical composition.

4. Conclusion

The method for fabrication of Ni-free surface alloy of thickness ~1 μm of near $Ti_{70}Ta_{30}$ composition on TiNi SMA substrate has been successfully realized. The method is based on multiple alternation of magnetron co-deposition of $Ti_{70}Ta_{30}$ thin films and their liquid-phase mixing with TiNi substrate by

IOP Conf. Series: Journal of Physics: Conf. Series **830** (2017) 012097 doi:10.1088/1742-6596/830/1/012097

microsecond LEHCEB. The Ti-Ta-based surface alloy has an increased X-Ray visibility and gradient multiphase amorphous-nanocrystalline structure containing nanopores. The immediate objectives are: (i) detailed studies of concentration depth profiles of the surface alloy; (ii) more detailed studies of the evolution of phase composition and atomic structure in depth by methods of grazing incidence X-Ray diffraction and HRTEM.

References

[1] *Shape Memory Alloys for Biomedical Applications* ed Yoneyama T and Miyazaki S (CRC Press, Boca Raton, FL) 2009
[2] *Shape Memory and Superelastic Alloys: Applications and Technologies* ed Yamauchi K, Ohkata I, Tsuchiya K, Miyazaki S (Cambridge, UK: Woodhead Publishing) 2011
[3] Ryhänen J, Niemi E, Serlo W, Niemelä E, Sandvik P, Pernu H, Salo T 1997 *J. Biomed. Mater. Res.* **35 (4)** 451
[4] Shabalovskaya S, Humbeeck J Van 2009 *Shape Memory Alloys for BiomedicalApplications* ed Yoneyama T and Miyazaki S (FL, USA: CRC Press, Boca Raton) pp 194–227
[5] Duerig T W, Tolomeo D E, Wholey M 2000 *Min. Invas. Ter. & Allied Technol* **9(3/4)** 235
[6] *Surface Modification and Alloying: by Laser, Ion, and Electron Beams* ed Poate J M, Foti G, Jacobson D C (New York: Plenum Press) 1983
[7] Buenconsejo P J S, Kim H Y, Hosoda H, Miyazaki S 2009 *Acta Materialia* **57(4)** 1068
[8] Martel Fuentes O, Mirza Rosca J, Rodriguez Galindo E, Baltes L S, Tierean M H 2013 *Metalurgia International* **18** 18
[9] Capellato P, Riedel N A, Williams J D, Machado J P, Popat K C, Claro A P 2013 *Engineering* **5** 707
[10] O'Brien B, Stinson J, Carroll W 2008 *Acta Biomaterialia* **4(5)** 1553
[11] Ozur G E, Proskurovsky D I, Rotshtein V P, Markov A B 2003 *Laser Part. Beams* **21(2)** 157
[12] Markov A B, Mikov A V, Ozur G E, Padei A G 2011 *Instrum. Exp. Tech.* **54(6)** 862
[13] Meisner L L, Markov A B, Proskurovsky D I, Rotshtein V P, Ozur G E, Meisner S N, Yakovlev E V, Poletika T M, Girsova S L, Semin V O 2016 *Surf. & Coat. Techn.* **302** 495
[14] Goldstein J I, Newbury D E, Echlin P, Joy D C, Lyman C E, Lifshin E, Sawyer L C, Michael J R 2012 *Scanning Electron Microscopy and X-Ray Microanalysis* (New York and London: Plenum Press, ISBN 0306472929) p 84, 240, 294, 295
[15] http://www.icdd.com/products/pdf2.htm. PDF # 44-1288.
[16] Dobromyslov A V, Dolgikh G V, Dutkevich Ya, Trenogina T L 2009 *The Physics of Metals and Metallography* **107(5)** 502
[17] http://www.icdd.com/products/pdf2.htm. PDF # 4-788.
[18] http://www.icdd.com/products/pdf2.htm. PDF #5-682.
[19] http://www.icdd.com/products/pdf2.htm. PDF #25-1280.

Acknowledgement

This work was funded by Russian Science Foundation (project No. 15-13-00023 of May 18, 2015). The SEM/EDS, TEM and XRD examinations were carried out using the equipment of the Shared Use Centre "Nanotech" of the ISPMS SB RAS.

Coating synthesis on dielectric substrates assisted by pulsed beams of high-energy gas atoms

S N Grigoriev, Yu A Melnik and A S Metel

Moscow State University of Technology "STANKIN", 1 Vadkovsky per., Moscow GSP-4, 127055, Russia

E-mail: a.metel@stankin.ru

Abstract. Titanium nitride and aluminum nitride coatings have been deposited on glass and aluminum oxide substrates in a flow of metal atoms accompanied by high-energy gas atoms. The metal atoms are produced due to sputtering of a flat rectangular magnetron target. The gas atoms with energy up to 25 keV are produced due to charge exchange collisions of ions extracted from the magnetron discharge plasma and accelerated by high-voltage pulses applied to a flat grid parallel to the target. The metal atoms pass through the grid and deposit on the substrate. Conjunction of their trajectories with those of gas atoms bombarding the growing coating enables the coating synthesis on complex-shape dielectric products planetary rotating inside the vacuum chamber. Mixing high-energy gas atoms of the coating and substrate atoms substantially improves the coating adhesion.

1. Introduction

Plasma- and beam-assisted deposition methods are widely used for production of various functional coatings on the surface of machine parts and tools [1]. Properties of a growing coating depend on the energy transferred to the atoms condensing on its surface. This energy may be supplied due to the product heating. When the energy is supplied by accelerated ions bombarding the product, the equilibrium heating of the coating and the product is replaced by non-equilibrium atomic-scale heating [2]. In this case, the energy is independent of the substrate temperature. It enables coating deposition on heat-sensitive materials and wide-range regulation of the coating properties.

Under energy of ions rising up to hundreds and thousands of volts, their sputtering efficiency monotonically rises, and the ions are able to sputter all deposited atoms. For this reason, the coating modification by high-energy ions is usually carried out in pulsed regimes. Due to ion mixing of the product and coating atoms, the width of the interface may be comparable with the coating thickness, thus ensuring perfect adhesion [3-5].

When a conductive coating is deposited on a conductive product, it can be bombarded with ions extracted from the discharge plasma and accelerated by a negative DC or pulsed bias voltage applied to the product. Nevertheless, the modification of dielectric coatings or any coatings on dielectric products can be carried out only with the use of broad beam sources of accelerated ions [6] or fast atoms [7-9], because it is impossible to apply high negative voltage to dielectrics.

When a complex shaped product is being coated, some parts of the product surface shadow the others during its rotation inside the chamber, and at some time intervals, the accelerated particles do not bombard the growing coating at all. This is because accelerated particles from a beam source [10] and metal atoms, for instance those from the source of metal atoms [11], converge on the product from different directions. There is no shadowing only when the source [12] has common emissive grid for both metal atoms and accelerated gas ions moving to the product along the same trajectories. This source significantly improves adhesion of thin films deposited on dielectrics when the growing film is bombarded by pulsed beams of fast gas atoms with energy of several keV. To ensure adhesion of 0.1-mm-thick superhard coatings on dielectric ceramic products, it is necessary to create the same physical conditions of their synthesis, as in [3-5]. Therefore, the energy of fast neutral gas atoms must be heightened up to ~25 keV and more.

It is impossible to produce high-energy ions using the source [12], because of electrical breakdowns at 20-kV amplitude of pulses applied to the anode of hollow cathode glow discharge used for generation of plasma emitter, and problems with the discharge ignition at high voltage [13].

It was shown in [7] that the beams of fast neutral atoms and molecules can be produced due to charge exchange collisions of accelerated ions in space charge sheaths near negatively biased grid immersed in low-pressure gas discharge plasma. In this case, the beam parameters depend mainly on the pressure and do not depend on the gas discharge type. It is possible, for instance, to use the magnetron discharge plasma as ion emitter and to produce pulsed beams of high-energy gas atoms bombarding the growing coating due to application of high-voltage pulses to the grid parallel to the planar magnetron target. Similarly to [12], the target material atoms taking part in the coating synthesis will enter the chamber through the same grid together with high-energy gas atoms. This work is dedicated to the development of the magnetron sputtering system, generating pulsed beams of high-energy gas atoms bombarding the growing coating.

2. Experimental setup

Figure 1 depicts the scheme of the experimental setup used for the deposition of coatings bombarded by pulsed beams of high-energy gas atoms.

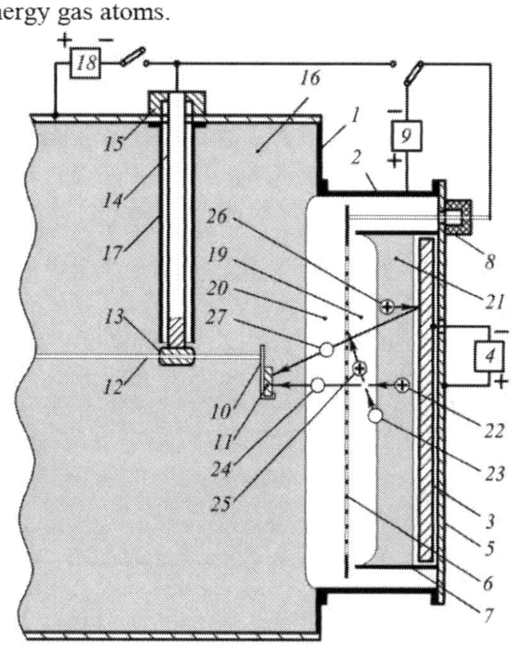

Figure 1. Scheme of the experimental setup.

On vacuum chamber *1*, rectangular 130-mm-deep hollow case *2* is installed with an inner cross-section of 600×200 mm. At the case bottom, a planar magnetron is mounted with water-cooled 150-mm-wide and 370-mm-long titanium target *3*. Magnetron power supply *4* is connected between target *3* and grounded flange *5*, which serves as the magnetron discharge anode. The power supply ensures stabilized current in the target circuit ranging from 1 to 8 A at voltage of up to 650 V between the anode and the target. Inside case *2* grid *6* is placed. The width of the gaps between grid *6* and the inner surfaces of case *2* amounts to 10 mm. The system design allows variation of the distance between target *3* and grid *6* from 60 to 100 mm. Using feed-through *8* the grid is connected to high-voltage pulse generator *9*. The generator allows regulation of the pulse amplitude from 3 to 30 kV, the pulse width from 5 to 50 μs, and the repetition rate from 5 to 50 Hz. Holder *10* of substrates *11* is mounted on horizontal rod *12* passing through guide bush *13*, which rotates on cylindrical electrode *14* of feed-through *15*. The holder design allows varying the distance between substrate *11* and target *3* from zero to 300 mm.

After the power supply *4* is switched on, the discharge is ignited and the chamber is filled with glow of the discharge plasma (Figure 2), which is most intense near the target surface.

Figure 2. Photograph of the discharge plasma glow inside the chamber at pressure 0.3 Pa of argon and nitrogen (15%) mixture and current in the magnetron target circuit of 8 A.

When a high-voltage pulse is applied to grid *6*, near its surface sheaths *19* and *20* of positive space charge are formed separating the discharge plasma into two parts. During the pulse, plasma *16* filling chamber *1* is not involved in the discharge. It is cut off from plasma *21* of the ongoing magnetron discharge located between target *3* and grid *6*. The ions *22* from plasma *21* are accelerated by the grid; due to collisions in sheaths *19* and *20* with gas molecules *23*, they turn into fast neutral molecules *24*. All ions *25* resulting from the charge exchange collisions are collected by grid *6*.

3. Experimental results

Substrate *11* was placed on holder *10*, and on its surface a 0.5-mm-thick titanium mask was fixed. After the coating deposition, the mask was removed, and height of the step between the open surface of the substrate and its surface covered with the mask was measured using the DektakXT profilometer (Bruker Nano Inc., USA). The titanium deposition rate was determined by dividing the height of the step by the deposition time. In the absence of grid *6* and at the current in the target circuit of 8 A, it amounted to 2.5 μm/h at the distance from the target of 200 mm; to 9.5 μm/h, at the distance of 100 mm and to 17 μm/h, at the distance of 50 mm.

After grid *6* (Figure 1) with the transparency of 70 % was installed at the distance of 60 mm from target *3*, the titanium coating deposition rate—at argon pressure of 0.4 Pa and current in the target circuit of 8 A—on a substrate distanced from the target at 90 mm diminished from 10 to 7 μm/h, i.e. by 30%. When nitrogen (15%) was added to argon, the deposition rate diminished more than 2 times, and within 2 hours a 6.5-μm-thick golden titanium nitride coating with the microhardness of 2200 HV was synthesized on the substrate.

When high-voltage pulses were applied to grid *6*, near both surfaces of the grid sheaths of positive space charge were formed, where ions were accelerated and high-energy atoms were produced. The sheaths width depends on the pulse amplitude and the ion current density. To determine the current density of ions *22* from plasma *21* (Figure 1) located between grid *6* and target *3*, dependencies of the grid current I_g on the grid voltage U_g were obtained (Figure 3).

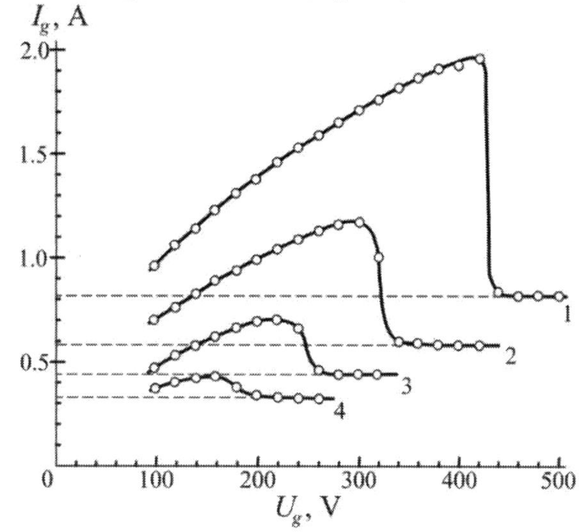

Figure 3. Dependencies of current I_g on voltage U_g between the chamber and the grid in the grid circuit at the pressure of argon and nitrogen (15%) mixture of 0.4 Pa, current in the target circuit of 4 A and distance between the grid and the target of 70 mm (*1*), 80 mm (*2*) 90 mm (*3*) and 100 mm (*4*).

The current in the grid circuit rises when the grid voltage increases. But at a voltage of hundreds of volts it dramatically reduces approximately by 2 times and with a further increase in the voltage it keeps the constant value I_{gi}. After the current drop, glow intensity of plasma *16* in the chamber sharply diminishes, and the ion current in the circuit of the probe installed in the chamber reduces by ten times. It can be explained by the fact that the grid prevents electrons of discharge plasma *21* from penetrating into chamber *1*. In this case, the current in the grid circuit I_{gi} is equal to the total current of ions extracted from plasma *21* and secondary electrons emitted by the grid. Since at voltages lower than 600 V, the coefficient of ion-electron emission does not exceed 0.1 [14], the secondary electron current can be neglected, and I_{gi} may be considered as average current of ions arriving from plasma *21* to grid *6*. Figure 4 shows the dependencies of current I_{gi} in the grid circuit on the distance h between the grid and the target at currents in the target circuit ranging from 4 to 8 A.

Before the deposition with application to the grid of high-voltage pulses, no activation and heating of the substrates has been carried out, hence the formation of the transitional layer between the substrate and the coating started at room temperature. At the pressure of the argon and nitrogen (15%) mixture of 0.7 Pa, the current of 8 A in the target circuit and the amplitude of 20 kV of the voltage pulses applied to the grid, a brown 5.8-μm-thick nitride coating with the microhardness of 2700 HV was synthesized within 2 hours on the initially cold substrate made of hard alloy. Evaluation of its adhesion using a scratch tester, showed that critical loads for the coatings synthesized using 20-kV pulses are several times higher than that of conventional nitride coatings, synthesized at a constant bias voltage of 100 V of the substrate.

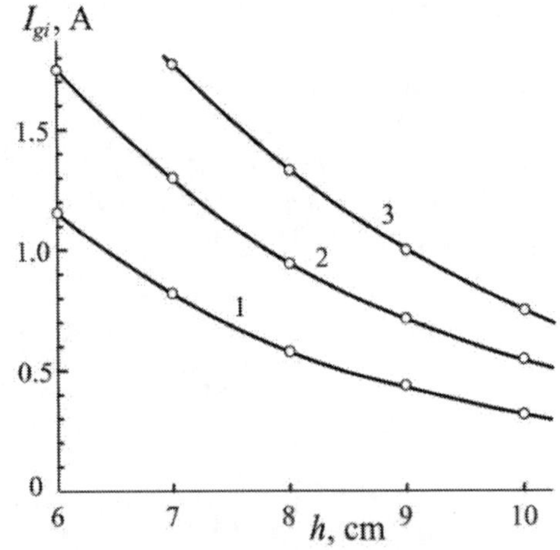

Figure 4. Dependence of current I_{gi} on the distance h between the grid and the target in the circuit of the grid preventing the magnetron discharge plasma from penetrating into the chamber, at the current in the target circuit 4 (*1*), 6 (*2*) and 8 A (*3*).

When the amplitude of pulses applied to the grid is reduced to 15 kV, the coating parameters remain virtually unchanged. However, at voltages lower than 10 kV, the microhardness decreases down to 2200 HV, the adhesion is considerably reduced, and at the amplitude of 5 kV, delamination of the coating is already observed.

4. Discussion

The most striking difference of coatings deposited using high-voltage pulses applied to the grid is their perfect adhesion despite the interface formation at room temperature without any preheating and activation. The adhesion is improved due to bombardment of the substrate by high-energy gas atoms. In the surface layer with a thickness ranging from 20 to 100 nm, each of these atoms triggers dislodgement of hundreds of atoms from the crystal lattices of the substrate and the growing coating. As a result, atoms of the substrate and the coating are mixing together. This is the reason of the interface width increase and the adhesion improvement.

The energy of atoms produced due to charge exchange collisions of ions accelerated in sheaths *19* and *20* (Figure 1) is determined by the ratio of the ion charge exchange length, depending on their energy and gas pressure, to the width of the sheath, which depends on the voltage applied to grid *6* and the current density of ions *22* accelerated from plasma *21*. When current in the target circuit is equal to 8 A, the current of ions *22* in the grid circuit is equal to I_{gi} = 2.3 A, and the average ion current density j on the grid with the surface area of 0.072 m^2 is equal to 32 A/m^2.

From the Child-Langmuir law [15], at j = 32 A/m^2 and the pulse amplitude U = 25000 V, the sheath width is equal to 0.03 m. The fast molecule energy corresponds to the potential of the point, where the charge exchange collision took place [7]. Therefore, it is distributed continuously from zero to the value eU corresponding to the pulse amplitude. At room temperature and pressure of p = 0.1 Pa, the density of the gas molecules is equal to $n = 2.5\times10^{19}$m^{-3}. The charge exchange cross section of 25-keV argon ions is equal to $\sigma = 10^{-19}$ m^2 [15, 16]. Therefore, the charge exchange length of these ions $\lambda = 1/(n\sigma) \sim 0.4$ m exceeds by ten times the width of the sheath. When they pass through the sheath, only a few of accelerated ions are converted into neutral molecules with an energy $\sim eU$. The number of such molecules rises with pressure increase to $p \sim 0.5$ Pa, when the charge exchange length $\lambda \sim 0.08$ m exceeds the sheath width approximately by 2 times [20]. At $p \sim 1$ Pa λ is less than the width of the layer ~ 0.3 m and the maximum energy of molecules is reduced, since most of the ions are converted into fast molecules, even before they reached the grid. The experimentally observed deterioration in adhesion of the coatings at the gas pressure lower than 0.7 Pa can be explained by a decrease in the energy and the number of high-energy gas molecules.

5. Conclusion

1. Bombardment of growing coating by neutral molecules with energy up to 20 keV creates the same physical conditions for its synthesis on the products made of metals and dielectric ceramics, as bombardment by ions with equal energy. This enables the synthesis of coatings with enhanced adhesion and microhardness on products rotating inside the chamber and made of both conductive and dielectric materials.
2. The reason of the coating adhesion improvement can be the interface width increase, caused by mixing of the product with atoms of the growing coating by gas atoms with energy up to 25 keV in the very beginning of the coating deposition.

References

[1] Grigoriev S and Metel A 2004 Plasma- and beam-assisted deposition methods *Nanostructured thin films and nanodispersion strengthened coatings* ed A A Voevodin, A A Shtansky and E A Levashov (Boston, Dordrecht, London: Kluwer Academic Publishers) pp 147–154
[2] Musil J 1998 *Vacuum* **50** 363
[3] Ruset C and Grigore E 2002 *Surf. Coat. Technol.* **156** 159
[4] Ruset C, Grigore E, Collins G A, Short K T, Rossi F, Gibson N, Dong H and Bell T 2003 *Surf. Coat. Technol.* **174/175** 698
[5] Grigore E, Ruset C, Short K T, Hoeft D, Dong H, Li X Y and Bell T 2005 *Surf. Coat. Technol.* **200** 744
[6] Kaufman H R 1990 *Rev. Sci. Instrum.* **61** 230
[7] Metel A S 2012 *Plasma Phys. Rep.* **38** 254
[8] Metel A S, Grigoriev S N, Melnik Yu A and Bolbukov V P 2012 *Instrum. Exp. Tech.* **55** 122
[9] Metel A S, Melnik Yu A and Panin V V 2011 *Plasma Phys. Rep.* **37** 357
[10] Cakir A F, Metel A, Urgen M and Grigoriev S 2000 *Galvanotechnik* **91** 768
[11] Grigoriev S N, Melnik Yu A, Metel A S, Panin V V and Prudnikov V V 2009 *Instrum. Exp.Tech.* **52** 731
[12] Metel A, Bolbukov V, Volosova M, Grigoriev S and Melnik Yu 2013 *Surf. Coat. Technol.* **225** 34
[13] Metel A S 1986 *Sov. Phys.–Tech. Phys.* **31** 1395
[14] Kaminsky M 1965 *Atomic and ionic impact phenomena on metal surfaces* (Berlin, Heidelberg, New York: Springer–Verlag)
[15] Langmuir I 1929 *Phys. Rev.* **33** 954
[16] Phelps A V 1991 *J. Phys. Chem. Ref. Data* **20** 557
[17] Massey H S W, Burhop E H S 1952 *Electronic and ionic impact phenomena* (Oxford: Clarendon Press)

Acknowledgements

This project was supported by Grant of the Russian Science Foundation No. 14-29-00297.

IOP Conf. Series: Journal of Physics: Conf. Series **830** (2017) 012099 doi:10.1088/1742-6596/830/1/012099

Improvement of the magnetron sputtered coating adhesion through pulsed bombardment by high-energy ions

Yu A Melnik and A S Metel

Moscow State University of Technology "STANKIN" 1 Vadkovsky per., Moscow GSP-4, 127055, Russia

E-mail: a.metel@stankin.ru

Abstract. Comparative study of titanium nitride deposition has been carried out, the growing coating being uninterruptedly bombarded by 100-eV ions or ions accelerated by high-voltage pulses applied to the substrate. The study revealed that microhardness of coatings synthesized using 25-kV pulses rises from 21 GPa to 29 GPa when percentage of nitrogen in the mixture with argon increases from 15% to 20%. With a further increase of nitrogen percentage to 30%, the microhardness slightly diminishes to 27 GPa. In contrast to golden coatings synthesized at low-voltage substrate biasing, the color of titanium nitride coating produced using high-voltage pulses is brown. The most striking difference of coating deposited using high-voltage pulses applied to the substrate is its perfect adhesion despite the interface formation at the room temperature without any preheating and activation. The adhesion characterization using a scratch-tester has revealed that critical loads of coatings synthesized using 25-kV pulses are several times higher than those of conventional nitride coatings synthesized at uninterrupted substrate biasing of 100 V. When the pulse amplitude diminishes to 5 kV, the critical loads and microhardness of the coating decrease to conventional values.

1. Introduction

For production of various functional coatings on the surface of machine parts and tools, plasma- and beam-assisted deposition methods are widely used [1]. The growing coating properties depend on the energy transferred to the atoms condensing on the product surface. When this energy is supplied by accelerated ions bombarding of the product, the equilibrium heating of the coating and product is replaced by a non-equilibrium atomic scale heating [2], the coating properties depend on the ion energy and are independent of the substrate temperature. This enables the coating deposition on heat sensitive materials and a wide-range regulation of the coating properties.

For instance, the density of Mg films deposited using a Mg^+ ion beam grows with the ion energy from 83% of the bulk material density at the energy of 24 eV to its maximal value (about 100% of the bulk material density) at the ion energy of 48 eV and then falls down to 50% and 15% with the energy increasing to, respectively, 120 eV and 300 eV [3]. At the same time, the Mg-film adhesion rises with the ion energy monotonically due to deeper penetration of accelerated particles into the substrate.

When slow atoms are condensing on the substrate and the growing film is bombarded by accelerated ions, the film properties depend not only on the energy of ions, but also on the ratio of their flow density on the substrate surface to that of condensing atoms. At the ion energy rising up to hundreds and thousands of volts, their sputtering efficiency monotonically rises, and the ions are able to sputter all deposited atoms. For this reason the film modification with high-energy particles is

Published under licence by IOP Publishing Ltd

IOP Conf. Series: Journal of Physics: Conf. Series **830** (2017) 012099 doi:10.1088/1742-6596/830/1/012099

usually carried out in pulsed regimes [4, 5]. This work is dedicated to the investigation of the coating adhesion dependence on pulsed ion bombardment.

2. Experimental setup

Figure 1 presents schematic of the experimental setup used for the deposition of coatings with pulsed bombardment by high-energy ions. On vacuum chamber *1* is mounted a planar magnetron with water-cooled 150-mm-wide and 370-mm-long titanium target *2*. Magnetron power supply *3* is connected between target *2* and grounded flange *4*, which serves as the magnetron discharge anode. The power supply ensures stabilized current in the target circuit ranging from 1 to 8 A at the discharge voltage up to 650 V between the anode and the target.

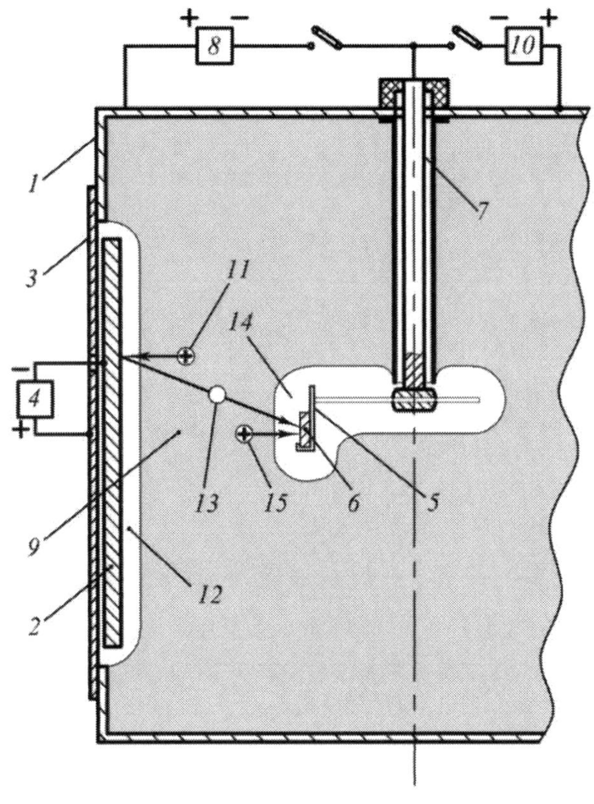

Figure 1. Schematic of experimental setup. *1* – vacuum chamber, *2* – target, *3* – anode, *4* – magnetron power supply, *5*– holder, *6* – substrate, *7* – feed-through, *8* – high-voltage pulse generator, *9* – plasma, *10* – DC power supply, *11*, *15* – ions, *12*, *14* – positive space charge sheaths, *13* – sputtered atom.

Holder *5* of substrates *6* is mounted on a horizontal rod passing through a guide bush, which rotates on high-voltage feed-through *7*. It is possible to connect high-voltage pulse generator *8* to feed-through *7* and to apply the pulses directly to substrate *6*. The generator allows regulation of the pulse amplitude from 3 to 30 kV, the pulse width from 5 to 50 μs, and the repetition rate from 5 to 50 Hz. To reduce the current of ions from plasma *9* filling, the chamber in the circuit of the feed-through its surface is covered with a grounded cylindrical screen. The holder design allows variation of the distance between substrate *6* and target *2* from zero to 300 mm. A negative voltage ranging from 100 to 1200 V can be applied to substrate *6* from DC power supply *10*.

After power supply *4* is switched on and the magnetron discharge is ignited, the chamber is filled with glow of discharge plasma *9*, which is most intensive near the target surface. Ions *11* accelerated in sheath *12* of positive space charge bombard target *2*, and sputtered atoms *13* of the target material deposit on substrate *6*. When a high-voltage pulse is supplied to substrate *6*, sheath *14* of positive space charge is formed around holder *5* and accelerated in the sheath high-energy ions *15* bombard the coating growing on the substrate.

3. Experimental results

To appreciate the plasma density distribution a 120-mm-high and 160-mm-wide flat electrode was installed in parallel to the target surface. The electrode was attached to feed-through *7* instead of substrates holder *5* and negatively biased to 50 V. Currents in its circuit at argon pressure of 0.3 Pa, distance to the target h = 40 mm and currents in the target circuit of 3 A, 5 A and 8 A amounted to 0.6 A, 1.15 A and 1.65 A, respectively (solid curves in Figure 2). When the distance h rises to 100 mm, these currents nearly linearly decrease to 0.06 A, 0.14 A and 0.24 A. When the distance runs up to 150 mm, the currents amount to 0.02 A, 0.04 A and 0.08 A, respectively, and are only slightly reduced with the further increase of the distance.

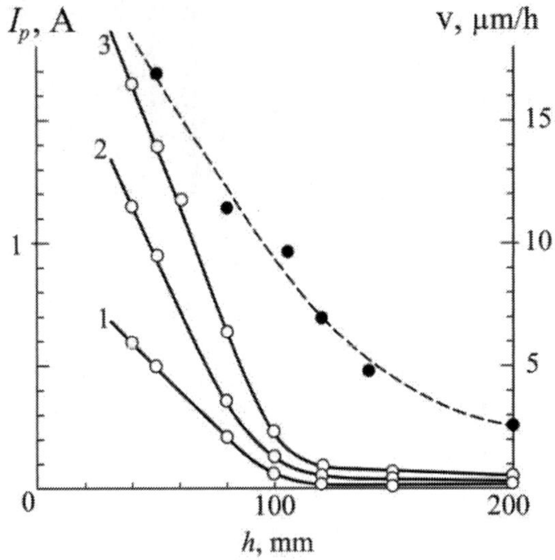

Figure 2. Dependence on the distance h from the target of titanium coating deposition rate v at current of 8 A in the target circuit (dashed line) and current in the probe circuit I_p (solid lines) at the current in the target circuit of 3 (1) 5 (2) and 8 A (3).

The dashed curve in Figure 2 presents dependence on the distance to the target of the titanium coating deposition rate on flat substrates made of high speed steel. Substrate *6* was placed on holder *5*, and on its surface facing target *2*, a mask was fixed. After 5-minute-long etching of the substrate and the mask with argon ions, extracted from the magnetron plasma and accelerated by negative bias voltage of 1200 V applied from power supply *10*, titanium coatings were deposited on their surfaces.

During the deposition process, the bias voltage of the substrate amounted to 100 V. After the 30-minute-long coating deposition the mask was removed, and height of the step between the open surface of the substrate and its surface covered with the mask, was measured using the DektakXT profilometer (manufactured by Bruker Nano, Inc., USA). The titanium deposition rate was determined by dividing the height of the step by the deposition time. At the current in the target circuit of 8 A, it amounts to 2.5 μm/h at the distance from the target of 200 mm, rises to 9.5 μm/h at the distance of 100 mm and to 17 μm/h at the distance of 50 mm. Measurements have shown that at the distance of 120 mm from the target, the titanium deposition rate amounts to 7 μm/h and is independent of the argon pressure in the range from 0.2 to 2 Pa. When reducing the current in the target circuit, the deposition rate is decreasing proportionally to the current.

After addition of nitrogen to argon (15%), the blue glow of the discharge plasma changes its color to pink (Figure 3), and the coating deposition rate is reduced by 2 times. Increasing the nitrogen content in the gas mixture up to 20% reduces the deposition rate by 2.5 times, and addition of 30% nitrogen reduces the rate by 3 times. In all cases, a golden hard coating of titanium nitride is synthesized on the substrates. With the reduction of the nitrogen content in the gas mixture to 10%, synthesis of titanium nitride stops, and on the substrate is deposited titanium coating. In this case, the color of the discharge plasma is again light blue, and the deposition rate increases to the values measured during the discharge in argon without nitrogen addition (dashed curve in Figure 2). When the current in the target circuit amounts to 8 A and the pressure of argon mixed with nitrogen (15%) is equal to 0.4 Pa on the substrate made of high speed steel and distanced from the target at 90 mm is deposited within 1 hour a 5.6-μm-thick titanium nitride coating with microhardness of 2200 HV25.

Figure 3. Photograph of the substrate holder and the magnetron discharge plasma glow.

In contrast to the above coating deposition at the negative voltage of 100 V after the substrate pretreatment by argon ions with energy of 1200 eV, no pretreatment has been carried out before the deposition with pulsed bombardment of the substrate by high-energy ions. At the pressure of argon and nitrogen mixture ranging from 0.2 to 0.8 Pa, first high-voltage pulses with specified amplitude, width and repetition rate are applied to the substrate, and only after that the magnetron discharge with specified current of 2 to 8 A is activated. The substrate with initial temperature equal to the room temperature immediately becomes immersed in the gas discharge plasma. Titanium atoms arrive to its surface and synthesis of the nitride coating starts, the growing coating being bombarded by argon and nitrogen ions accelerated by high-voltage pulses. Only a few minutes later the substrate temperature measured by IP 140 IMPAC pyrometers (manufactured by LumaSense Technologies, Inc., USA) grows up to 200-400 °C due to the substrate bombardment by high-energy ions and heating by the discharge radiation.

When high-voltage pulses are applied to the substrate, the coating deposition rate is a little lower as compared to the deposition at the bias voltage of 100 V. At the argon pressure of 0.4 Pa, current in the target circuit of 8 A, pulses amplitude of 25 kV, their width of 40 μs and their repetition rate of 50 Hz, the titanium coating deposition rate at a distance h = 90 mm from the target amounted to 10 μm/h. At the same pressure and current, addition to argon of nitrogen reduced the coating deposition rate down to 5 μm/h.

In contrast to the golden coatings synthesized on the substrate at a constant voltage of 100 V, titanium nitride coatings produced using high-voltage pulses are brown. Measurements with the use of POLYVAR-MET durometer (manufactured by Reichert-Jung, Austria) showed that microhardness of 5-μm-thick coatings on hard alloy substrates, synthesized using 25-kV pulses grows from 2100 HV25 to 2900 HV25 with increase in the nitrogen content from 15% to 20%. With a further increase in the nitrogen content to 30%, it slightly diminishes to 2700 HV25. However, the increase in microhardness of coatings with increasing the nitrogen content occurs not regularly. Analysis of the gas medium and plasma composition in the vacuum chamber by means of the EQP energy-mass analyzer (manufactured by HidenAnalitical Ltd., England) revealed that increase of the microhardness requires a sufficiently low content of water vapor and, accordingly, of hydrogen and oxygen ions.

Evaluation of adhesion using a scratch tester, showed that critical loads for the coatings synthesized using 25-kV pulses are several times higher than for conventional nitride coatings synthesized at a constant bias voltage. When, however, the pulse amplitude diminishes from 25 to 5 kV, the critical loads and microhardness decrease to conventional values.

4. Discussions

The experimental results showed that usually titanium coatings delaminate if deposited without pre-heating and activation of the substrate by argon ions with energy up to 1200 eV. To ensure an admissible adhesion of titanium nitride coating, it is also necessary to deposit an adhesive titanium sublayer before the coating synthesis. The most striking difference of coatings deposited using high-voltage pulses applied to the substrate is their perfect adhesion despite the formation of the transitional layer (interface) at room temperature without any preheating and activation. The adhesion was improved due to bombardment of the metal substrate by ions with an energy ranging from 5 to 25 keV, which penetrate the substrate surface in the form of neutral atoms. In the surface layer with a thickness, respectively, from 20 to 100 nm, each of these ions triggers dislodgement of 100 to 500 atoms, respectively, from the crystal lattices of the substrate and the growing coating. As a result, atoms of the substrate and atoms of the coating mix together. This is the reason of the interface width increase and the adhesion improvement.

5. Conclusion

Pulsed bombardment of growing coating by high-energy ions results in the coating adhesion improvement due to increase in the interface width, caused by mixing by the ions of the substrate and coating atoms in the very beginning of the coating deposition.

References

[1] Grigoriev S and Metel A 2004 Plasma- and beam-assisted deposition methods *Nanostructured thin films and nanodispersion strengthened coatings* ed A A Voevodin, A A Shtansky and E A Levashov (Boston, Dordrecht, London: Kluwer Academic Publishers) pp 147–154
[2] Musil J 1998 *Vacuum* **50** 363
[3] Amano J and Lawson R 1977 *J. Vac. Sci. Technol.* **14** 695
[4] Ruset C and Grigore E 2002 *Surf. Coat. Technol.* **156** 159
[5] Metel A S, Bolbukov V P, Volosova M A, Grigoriev S N and Melnik Yu A 2014 *Instrum. Exp. Tech.* **57** 345

Acknowledgements

This project was supported by Grant No. 14-29-00297 of the Russian Science Foundation.

Product surface hardening in non-self-sustained glow discharge plasma before synthesis of superhard coatings

P S Krasnov, A S Metel and H A Nay

Moscow State University of Technology "STANKIN", 1 Vadkovsky per., Moscow GSP-4, 127055, Russia

E-mail: a.metel@stankin.ru

Abstract. Before the synthesis of superhard coating, the product surface is hardened by means of plasma nitriding, which prevents the surface deformations and the coating brittle rupture. The product heating by ions accelerated from plasma by applied to the product bias voltage leads to overheating and blunting of the product sharp edges. To prevent the blunting, it is proposed to heat the products with a broad beam of fast nitrogen molecules. The beam injection into a working vacuum chamber results in filling of the chamber with quite homogeneous plasma suitable for nitriding. Immersion in the plasma of the electrode and heightening of its potential up to 50–100 V initiate a non-self-sustained glow discharge between the electrode and the chamber. It enhances the plasma density by an order of magnitude and reduces its spatial nonuniformity down to 5–10%. When a cutting tool is isolated from the chamber, it is bombarded by plasma ions with an energy corresponding to its floating potential, which is lower than the sputtering threshold. Hence, the sharp edges are sputtered only by fast nitrogen molecules with the same rate as other parts of the tool surface. This leads to sharpening of the cutting tools instead of blunting.

1. Introduction

Ion-plasma processing is widely used in industry for strengthening the surface layer of various products [1]. The useful life of any product can be substantially increased using hard wear-resistant coating. This is due to the coating hardness of about 25 GPa [2], which exceeds the bulk material hardness by many times. Necessary for synthesis of titanium nitride and other hard coatings metal vapor is produced by means of vacuum arc evaporation [3] and magnetron sputtering [4]. To prevent the formation of droplets in the coatings, the vacuum arc plasma can be filtered [5]. The density of magnetron discharge plasma near the product surface and the sputtering rate are substantially enhanced when a pulsed DC magnetron [6] is used. Metal vapor can be also produced due to sputtering a target at the bottom of a hollow cathode [7].

It was shown in [8] that under certain physical conditions, superhard coatings with hardness of 40–80 GPa and ultrahard coatings with hardness of 80–105 GPa can be synthesized. The survey of those conditions is given in [9]. Usually, the coating thickness amounts to a few microns. When its hardness is higher than the hardness of bulk material by an order of magnitude, the thin coating can be broken by external loads as an egg shell or a thin ice on the water surface. Performance and useful life of the coated product can be substantially improved when the thin coating is deposited on a strengthened surface layer of the product, the layer being by an order of magnitude thicker than the coating.

IOP Conf. Series: Journal of Physics: Conf. Series **830** (2017) 012100 doi:10.1088/1742-6596/830/1/012100

Strengthening of the surface can be carried out, for instance, by means of nitriding. Nitrided surface layer exhibits a high loadbearing capacity and fatigue strength, thus preventing elastic and plastic deformations of the product as well as brittle rupture of synthesized coating [10].

Combined processing of substrates made of high-speed steel in vacuum arc discharge plasma, which included nitriding and subsequent deposition of titanium nitride coatings, has been successfully carried out in [11]. Drawbacks of the arc method are soft metal droplets in the hard coatings, non-practical consumption of the target material and nonuniformity of the discharge plasma. To eliminate the droplets and the plasma nonuniformity, it was proposed in [12, 13] to produce the nitrogen plasma in hollow cathode glow discharge [14, 15] using the working vacuum chamber as the hollow cathode.

When a product immersed in any plasma is heated by ions accelerated from the plasma by negative voltage applied to the product, the ion current density on its surface depends on the product shape. It can be 10 to 100 times higher on its sharp edges than in the cavities. During 1-hour-long pre-nitriding of cutting tools, the thickness of the surface layer sputtered from the tool cutting edge by the ions can exceed the mean thickness of the sputtered tool material by 10 to 100 times and amount to 20–30 μm. When initial radius of the cutting edge is equal to 15 μm, nitriding can result in an increase of the edge radius up to 45 μm, thus making the tool dull.

Preventing the tool blunting requires homogeneity of the tool surface sputtering. This paper is dedicated to the production of a uniform nitrogen plasma and heating of the tools immersed in the plasma by a broad beam of fast nitrogen molecules [16, 17].

2. Experimental setup

Figure 1 presents the scheme of 500-mm-diameter vacuum chamber *1* equipped with a beam source of fast neutral molecules, which comprises 400-mm-diameter and 120-mm-deep titanium hollow cathode *2*, 40-mm-diameter disc anode *3* and 300-mm-diameter emissive grid *4* made of 2-mm-thick titanium sheet.

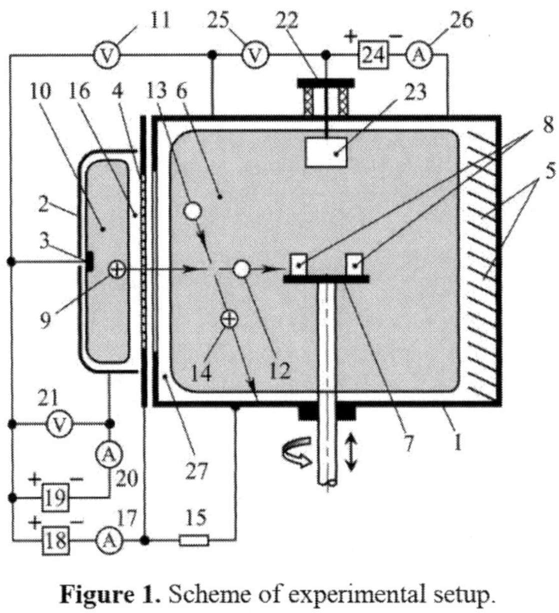

Figure 1. Scheme of experimental setup.

Distance between the centers of 4.6-mm-diameter orifices of the grid amounts to 5 mm and the grid transparency amounts to $\eta = 0.8$. Distance between grid *4* and set of plates *5*, which cover the opposite wall of chamber *1* preventing it from sputtering by fast molecules, amounts to 700 mm. A flat probe (not shown in Figure 1) is introduced into chamber *1* to measure parameters of nitrogen plasma *6*. Through the chamber bottom rotating table *7* is introduced for placing products *8* to be processed in plasma 6 by fast neutral molecule beam.

IOP Conf. Series: Journal of Physics: Conf. Series **830** (2017) 012100 doi:10.1088/1742-6596/830/1/012100

Ions *9* are accelerated between plasma emitter *10*, filling cathode *2* and plasma *6*, potential of the latter being by 2–3 V higher than the ground potential of chamber *1*. For this reason, the energy of accelerated ions corresponds to measured voltage between anode *3* and chamber *1* with kilovoltmeter *11*. At the gas pressure $p = 0.2 - 1$ Pa and the distance of 0.1 m from grid *4* most of accelerated ions *9* turn into fast neutral molecules *12* due to charge exchange collisions [18] with gas molecules *13*. The current of slow ions *14* in the circuit of chamber *1* through feedback resistor *15* with 3-kΩ resistance induces negative bias voltage of 100–200 V of the grid, thus preventing penetration of electrons from plasma *6* through grid *4* into positive space charge sheath *16* between grid *4* and plasma emitter *10*.

Neglecting the secondary electron emission from grid *4* we may assume that beam current is equal to the measured with ammeter *17* sum of currents in chamber *1* and grid *4* circuits multiplied by the grid transparency $\eta = 0.8$. The energy of ions *9* is regulated using high-voltage power supply *18*, the beam current is regulated using power supply *19* between anode *3* and cathode *2*. Ion current in the circuit of cathode *2* is measured with ammeter *20*, and cathode potential drop is controlled with voltmeter *21*. Using feedthrough *22*, a 2-mm-thick flat electrode *23* is immersed in plasma *6*, its surface area amounting to 0.03 m^2. The electrode is connected to positive pole of power supply *24*; its bias voltage can be regulated from zero up to 500 V and measured with voltmeter *25*, the current in its circuit being measured with ammeter *26*.

3. Experimental results

Figure 2 presents the dependence of ion current density j_i on the probe surface at the beam axis at the probe distance x from grid *4* at nitrogen pressure $p = 0.2$ Pa, equivalent beam current $I_b = 0.8$ A and energy of molecules E_b ranging from 1 to 4 keV. The dashed curve presents $j_i(x)$ at $I_b = 0.8$ A, $E_b = 4$ keV and gas pressures $p = 0.5$ Pa. As the plasma density n is proportional to the ion current density, the presented in Figure 2 distributions show that n reaches its maximum at distance $x = 0.05$–0.1 m from the grid surface and monotonically diminishes with a further increase of the distance.

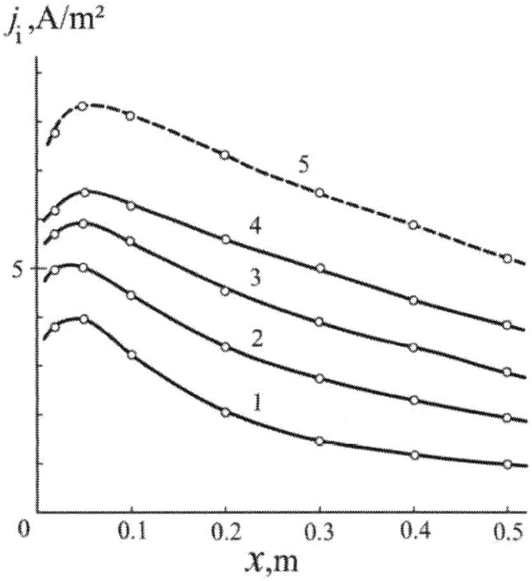

Figure 2. Dependence of ion current density j_i at the beam axis at distance x between the probe and the grid at nitrogen pressure $p = 0.2$ Pa, beam equivalent current $I_b = 0.8$ A and energy of molecules $E_b = 1$ (solid curve 1), 2 (2), 3 (3) and 4 (4) keV, as well as at $I_b = 0.8$ A, $E_b = 4$ keV and $p = 0.5$ (dashed curve 5).

At the molecule energy of 1 keV, the axial distribution of plasma density n is highly nonuniform and exhibits a fourfold decrease in n when x is increasing from zero to 0.5 m. After the molecule energy rises to 4 keV, the plasma nonuniformity does not exceed ±30%. At the same energy of fast molecules, the plasma density increases, when the nitrogen pressure p rises from 0.2 Pa to 0.5 Pa.

The plasma density can be substantially increased and its homogeneity improved using a non-self-sustained glow discharge with electrostatic confinement of electrons [19] between chamber *1* and electrode *23* immersed in plasma *6* filling the chamber. Figure 3 presents dependences of current I in the circuit of electrode *23* on its bias voltage U at $p = 0.5$ Pa, beam current $I_b = 0.1$ A (solid curves), $I_b = 0.2$ A (dashed curves) and energy of molecules $E_b = 1$ (1), 2 (2) and 4 keV (3). When the voltage U exceeds 20 V, current I of non-self-sustained discharge and velocity dI/dU both grow due to the gas ionization by electrons emitted by the chamber *1* and accelerated in the sheath *27* up to energy eU exceeding the ionization threshold $E_i = 15.6$ eV of nitrogen [21]. At one and the same U value velocity dI/dU is rising when the beam equivalent current I_b and/or its energy E_b are growing (Figure 3). At $U > 150$ V electrons produce in the chamber more ions than fast neutral molecules of the beam.

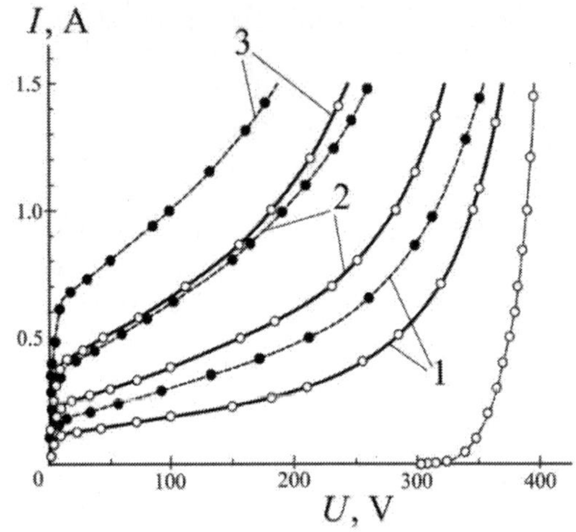

Figure 3. Current-voltage characteristics (CVC) of non-self-sustained discharge at $p = 0.5$ Pa, beam current $I_b = 0.1$ A (solid curves), $I_b = 0.2$ A (dashed curves) and energy of molecules E_b = 1 (1), 2 (2), 4 keV (3) and CVC of the self-sustained discharge (dash-dot curve) at $p = 0.5$ Pa.

Dash-dot curve in Figure 3 presents current-voltage characteristic of self-sustained glow discharge between its anode *23* and chamber *1* playing the role of hollow cathode, when the beam source is switched off. It shows that only at the voltage U exceeding 300 V the discharge supported by the beam does not expire after the beam source is switched off.

Radial distributions of ion current density j_i in the discharge plasma demonstrate an increase of plasma density by an order of magnitude with increasing voltage U of the non-self-sustained discharge and a decrease of plasma nonuniformity down to ±5%, which is quite suitable for the products nitriding.

Ten cutting plates *8* (Figure 1) made of high-speed steel were positioned in the center of chamber *1* on rotating table *7*. They were heated by the beam of fast neutral nitrogen molecules with 4-keV energy and 0.4-A equivalent current up to 500 °C and soaked in plasma *6* of non-self-sustained discharge at voltage $U = 100$ V and nitrogen pressure $p = 0.5$ Pa. After the one-hour-long processing, the initial microhardness 860 HV of the plates increased up to 1350 HV. Usual for plasma immersion ion nitriding increase of the mean radius of cutting edges was not observed. On the contrary, the radius even decreased from 20±3 μm down to 17±3 μm.

4. Discussion

The data presented above prove that injection of a broad beam of fast neutral nitrogen molecules into a vacuum chamber results in production of quite nonuniform plasma. The plasma density grows with the beam equivalent current, energy and nitrogen pressure. This fact reveals the leading role of fast neutral molecules in gas ionization. At an adequate pressure, the current of ions produced by fast molecules [20] may exceed two times and more equivalent current of the beam.

Nonuniformity of the beam-produced plasma can be readily improved using non-self-sustained glow discharge with electrostatic confinement of electrons [19]. At the discharge voltage *U* exceeding 150 V fast electrons many times return back to the cathode sheath *27* and produce therein new electrons capable of ionizing the gas. Multiplication of fast electrons in the cathode sheath [14] is the basic process of electrostatic trap effect in glow discharge. It results in ionization intensity growth, plasma density increase by an order of magnitude and its nonuniformity decrease to ±5%.

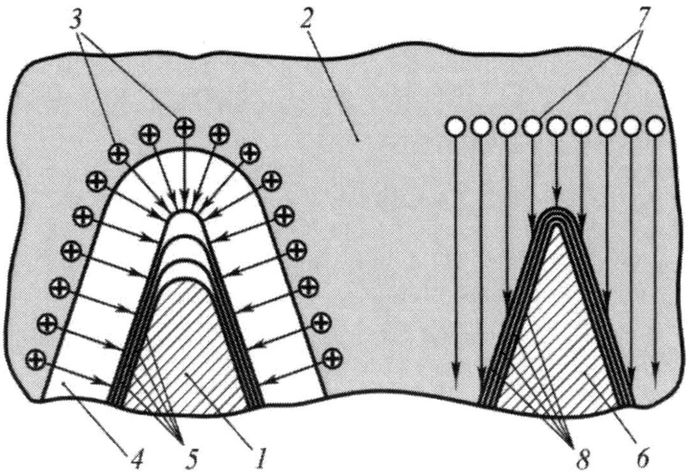

Figure 4. Scheme of the knife sputtering in homogeneous plasma.

1 and *6* – knifes; *2* – plasma; *3* – ions; *4* – positive space charge sheath; *5* – sputtered by ions knife surface in three equal time intervals; *7* – fast molecules; *8* – the knife surface sputtered by fast molecule beam in three equal time intervals.

Figure 4 can explain why the radius of cutting edge does not increase after nitriding in beam-produced plasma. When steel knife *1* is immersed in homogeneous plasma *2* and negatively biased, it is heated by ions *3* extracted from plasma *2* and accelerated in sheath *4* of positive space charge. At a bias voltage of 100–1000 V, the sheath width exceeds 1 mm. For this reason, the radius of cutting edge ~10 μm is hundred times lower than the radius of plasma surface surrounding the knife-edge and current density of ions sputtering the edge is hundred times higher than on the rest surface of the knife. Intensive sputtering results in a substantial increase of the edge radius. On the contrary homogeneous sputtering of the knife by fast molecules results in a decrease of the edge radius.

5. Conclusion

Nitrogen plasma can be produced inside a working vacuum chamber by the injection of a broad fast neutral molecule beam. In contrast to conventional plasma immersion ion nitriding, the above new method of tool strengthening in plasma produced by a broad beam of fast neutral molecules results in sharpening of the tools instead of blunting.

References

[1] Grigoriev S and Metel A 2004 Plasma- and beam-assisted deposition methods *Nanostructured thin films and nanodispersion strengthened coatings* ed A A Voevodin, A A Shtansky and E A Levashov (Boston, Dordrecht, London: Kluwer Academic Publishers) pp 147–154

[2] Zhitomirsky V N, Grimberg I, Rapoport L, Boxman R L, Travitzky N A, Goldsmith S and Weiss B Z 2000 *Surf. Coat. Technol.* **133-134** 114

[3] Boxman R L and Zhitomirsky V N 2006 *Rev. Sci. Instrum.* **77** (2006) 021101/1

[4] Musil J, Rajsky A, Bell A J, Matous J, Cepera M and Zeman J 1996 *J. Vac. Sci. Technol.* A **14** 2187

[5] Boxman R L, Zhitomirsky V N, Alterkop B, Gidalevich E, Beilis I, Keidar M and Goldsmith S 1996 *Surf. Coat. Technol.* **86-87** 243

[6] Musil J, Lestina J, Vlcek J and Tolg T 2001 *J. Vac. Sci. Technol.* A **19** 420

[7] Grigoriev S N, Melnik Yu A, Metel A S, Panin V V and V.V. Prudnikov V V 2009 *Instrum. Exp. Tech.* **52** 731

[8] Veprek S, Mannling H D, Karvankova P and Prochazka J 2006 *Surf. Coat. Technol.* **200** 3876

[9] Veprek S 2013 *J. Vac. Sci. Technol.* **A 31** 050822/1

[10] Avelar-Batista J C, Spain E, Housden J, Matthews A and Fuentes G G 2005 *Surf. Coat. Technol.* **200** 1954

[11] Grigoriev S N, Metel A S and Fedorov S V 2012 *Met. Sci. Heat Treat.* **54** 8

[12] Metel A 2002 *Surf. Coat. Technol.* **156** 38

[13] Metel A S, Grigoriev S N, Melnik Yu A and Prudnikov V V 2011 *Plasma Phys. Rep.* **37** 628

[14] Metel A S 1985 *Sov. Phys.–Tech. Phys.* **30** 1133

[15] Metel A S 1986 *Sov. Phys.–Tech. Phys.* **31** 1395

[16] Metel A S, Grigoriev S N, Melnik Yu A and Bolbukov V P 2012 *Instrum. Exp. Tech.* **55** 122

[17] Metel A S 2012 *Plasma Phys. Rep.* **38** 254

[18] E.W. McDaniel E W 1964 *Collision phenomena in ionized gases* (New York, London, Sydney: John Wiley & Sons, Inc)

[19] Metel A S, Melnik Yu A and Panin V V 2011 *Plasma Phys. Rep.* **37** 357

[20] A.V. Phelps A V 1991 *J. Phys. Chem. Ref. Data* **20** 557

[21] Massey H S W, Burhop E H S 1952 *Electronic and ionic impact phenomena* (Oxford: Clarendon Press)

[22] Kaminsky M 1965 *Atomic and ionic impact phenomena on metal surfaces* (Berlin, Heidelberg, New York: Springer–Verlag)

Acknowledgements

This project was supported by Grant No. 14-29-00297 of the Russian Science Foundation.

IOP Conf. Series: Journal of Physics: Conf. Series **830** (2017) 012101 doi:10.1088/1742-6596/830/1/012101

Calcium phosphate coatings modified with zinc– or copper–incorporation on Ti–40Nb alloy

E G Komarova[1], M B Sedelnikova[1], Yu P Sharkeev[1], A A Kazakbaeva[2], I A Glukhov[1] and M A Khimich[1,2]

[1]Institute of Strength Physics and Materials Science of SB RAS, 2/4 Akademichesky ave., Tomsk, 634055, Russia
[2]National Research Tomsk State University, 36 Lenina ave., Tomsk, 634050, Russia

E-mail: katerina@ispms.tsc.ru

Abstract. The influence of the microarc oxidation parameters and electrolyte composition on the structure, properties and composition of CaP coatings modified with Zn– or Cu–incorporation on the Ti–40mas.%Nb (Ti–40Nb) alloy was investigated. The linear growth of thickness, roughness, and size of structural elements with process voltage increasing has been revealed. It was shown that the CaP coatings have the low contact angles with liquids and, consequently, high free surface energy. This indicates a high hydrophilicity of the coatings. X–ray diffraction analysis showed that the coatings have X–ray amorphous structure. The increase of the process voltage leads to the formation of such crystalline phases as $CaHPO_4$ and β–$Ca_2P_2O_7$ in the coatings. The maximum Ca/P atomic ratio was equal to 0.4, and Zn or Cu contents was equal to 0.3 or 0.2 at.%, respectively.

1. Introduction

Materials used for implants should have not only high strength properties (yield strength, ultimate strength, fatigue strength, microhardness, wear resistance, etc.), but also low elastic properties, such as elastic modulus. Elastic modulus is an important biofunctional characteristic of implant materials. An elastic modulus of titanium and its alloys ranges from 100 to 120 GPa, which is significantly higher than that of bone tissue which ranges from 15 to 55 GPa [1]. Accordingly, the most promising research area in medical materials science is the application of β–Ti alloys with a low elastic modulus, for example, Ti–Nb or Ti–Nb–Zr systems. Thus, niobium alloyed titanium (~ 40 mas.% Nb) could decrease the elastic modulus to 55–60 GPa [2]. To provide reliable osseointegration of bone tissue–implant interface integration, the implant surface should possess specific morphology and porosity. To solve this task, as well as to impart bioactive properties, metal surfaces are covered with bioactive calcium phosphate (CaP) coatings. There are various deposition methods for production of biocoatings on the metal implant. During the last few decades, the microarc oxidation (MAO) method is being widely used as electrical and chemical surface treatment for the formation of oxide and CaP coatings on metals [3]. Such coatings have a wide spectrum of physical and chemical properties, including high corrosion resistance, wear resistance, hardness, chemical stability in aggressive media, etc.

In addition, a serious problem in biomedicine is bacterial infection of medical implants. Bacterial infection is a result of bacteria adhesion to the implant surface and formation of a biofilm that sometimes cannot be destroyed by antibacterial drugs coming from outside [4]. Therefore, at present the investigations aimed at synthesis of biomaterials with antibacterial properties and at the study of bactericidal properties of implanted materials and patterns of interaction of antibacterial agents with surrounding tissues of the organism are of great interest. Bactericidal and anti–inflammatory actions of implanted materials are mainly caused by the presence in their structure of certain chemical microelements possessing natural antiseptic properties. It is well known that zinc (Zn^{2+}) and copper (Cu^{2+}) in the form of free ions possess strong slowing down activity and strong antimicrobial activity against various bacteria. Zn and Cu are necessary cofactors for enzymes participating in synthesis of various components of the bone matrix [5]. In addition, Zn is an important microelement of the human body and plays an important role in various biological functions, including DNA synthesis, activity of enzymes, metabolism of nucleic acids, biomineralization, and hormonal activity. It is well known that the large Cu deficiency causes heavy skeleton disorders [6]. Addition of Zn and Cu microdosages within biocoatings will allow the directed antimicrobial activity to be obtained for a long time and the danger of growth of pathogenic microorganisms to be minimized. Incorporation of the above–indicated modifying additives will promote balancing between microelement concentrations of biocoatings and bone structures as well as improving conditions for implant adaptation to organism environments. The aim of the present work was to produce the Zn and Cu–incorporated CaP coatings on Ti–40Nb alloy by the MAO method under different process voltages and investigate their structure, composition, and physical and chemical properties.

2. Materials and Methods

The Ti–40Nb alloy was produced by the method of electron arc melting with nonconsumable electrode [7]. The specimens were cut from Ti–40Nb alloy billets and their sizes were 10×10×1 mm^3. Specimens were prepared with silicon–carbide papers of 120, 480, 600, 1200 grit, respectively. Then specimens were ultrasonically cleaned (Elmasonic, Germany) in distilled water and ethanol for 10 min and further dried in air. Previously it was [8] reported that to carry out MAO method for deposition of CaP coatings on specimens, the Micro–arc–3.0 technique was used. The Zn– or Cu–incorporated CaP coatings were deposited from an aqueous solution prepared from 30wt.% phosphoric acid (H_3PO_4), 6wt.% Zn– or Cu–substituted hydroxyapatite (HA), and 9 wt.% carbonate calcium ($CaCO_3$) in the anode regime [9]. Zn– or Cu–substituted HA ($Ca_{9.9}Zn_{0.1}(PO_4)_6(OH)_2$ or $Ca_{9.9}Cu_{0.1}(PO_4)_6(OH)_2$) was produced by mechanochemical synthesis in the AGO–3 planetary mill at the Institute of Solid State Chemistry and Mechanochemistry SB RAS, Novosibirsk [10]. In the previous works [3, 9] we found the optimal MAO parameters for CaP coating deposition on the Ti such as pulse frequency of 50 Hz, pulse duration of 100 μs, process duration of 10 min, and voltage in the range of 200–300 V. In this work, we used the same MAO parameters for the coating synthesis on Ti–40Nb alloy.

The surface morphology and cross–sectional structure of the CaP coatings were observed by scanning electron microscopy (SEM, LEO EVO 50, Zeiss, and Philips SEM 515) and transmission electron microscopy (TEM, JEOL JEM–2010, Japan) in the "Nanotech" center at ISPMS SB RAS, and Scientific Educational Center at NR TSU. For TEM analysis replica of CaP coating were prepared by removing the CaP particles from the substrate. In addition, the elemental composition of the coatings was analyzed by energy–dispersive X–ray spectroscopy (EDX, INCA, Oxford Instruments) in conjunction with the SEM system. The size of structural elements of the coatings is measured by the secant method according to ASTM E1382–9 and DD ENV 1071–5. The chemical composition was determined by X–ray diffraction (XRD, DRON 7, Burevestnik, in "Nanotech" center at ISPMS SB RAS) in the angular range 2θ = 10–90° with a scanning step 0.03 with Co K_α radiation. The average roughness (*Ra*) was estimated with a Hommel–Etamic T1000 profilometer. To measure the adhesion strength of the coatings two cylinders were glued to both sides of the samples with coating and fixed in grips in order to be tested under tension (Instron–1185, Great Britain). The adhesion strength is the maximum stress required to tear the cylinder off the coating. It was measured as $\delta_A = F / S$, where F is the breakout force of coating from the substrate and S is the area of separation [8].

3. Results and Discussion

Figure 1 shows the pulse current density of MAO process against the process duration for the deposition of Zn– or Cu–incorporated CaP (Zn–CaP or Cu–CaP) coatings on Ti–40Nb under different applied voltages. The presence of steplike pulses is demonstrated by identical dependences of current densities of deposition processes as Zn–CaP as Cu–CaP coatings. This indicates the pulse nature of microplasma discharges. The substance transport in the arc and electrolyte components precipitation on the substrate occur under these discharges. Figure 1 shows that the current density increases with the voltage increment. This is caused by the increase in the microplasma discharges intensity. In addition the current density decreases with the microarc treatment time increasing. This is due to the formation of the coating with a low electrical conductivity on the surface of the specimen (dielectric CaP layer).

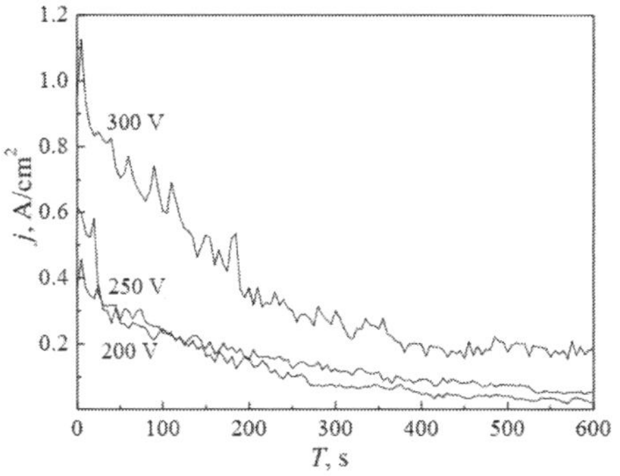

Figure 1. Current density against the process duration for deposition of Zn– or Cu–CaP coatings under different applied voltages.

Figure 2 shows SEM–images of cross–sectional structure of the Zn– and Cu–CaP coatings deposited under different oxidation voltages. It is seen that the coatings have gradient structure with many pores. With increasing the oxidation voltage, the thickness and sizes of pores increase. Further SEM investigations showed that the surface morphology of the Zn– and Cu–CaP coatings is identical and represented by the structural elements of spheroidal form with pores (Figure 3, a, d). The same surface morphology with spheres and pores were obtained for CaP coatings on the Ti in the previous works [3, 9]. The process of spheres and pores formation on the coating surface is similar to the process of formation and collapse of electrolyte–suspension "bubble" in the area of microplasma discharges.

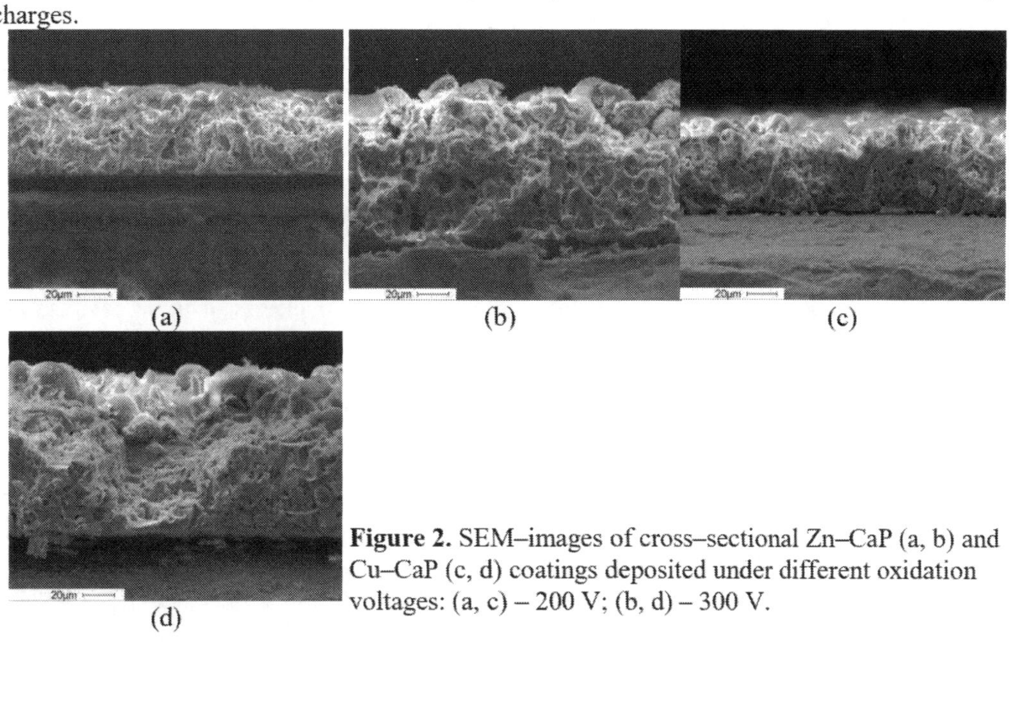

Figure 2. SEM–images of cross–sectional Zn–CaP (a, b) and Cu–CaP (c, d) coatings deposited under different oxidation voltages: (a, c) – 200 V; (b, d) – 300 V.

The measurements of the structural elements (spheres and pores) sizes showed that the histograms of the element size distribution have unimodal character for all coatings (Figure 3, b, c, e, f). The size of structural elements depends substantially on the MAO parameters, in particular, on the applied voltage. Zn– and Cu–CaP coatings start to form at the initial process voltage of 200 V. In this case, the average sizes of spheres and pores for both coatings are 17.7 and 4.4 μm, consequently (Figure 3, a–c). The increase in the MAO voltage to 300 V leads to the growth of spheres and pores to 25.9 and 7.2 μm, respectively, and to the destruction of some structural elements. In this case, the plate–like crystals of the new substance are formed on the surface of fragments and hemispheres (Figure 3, d).

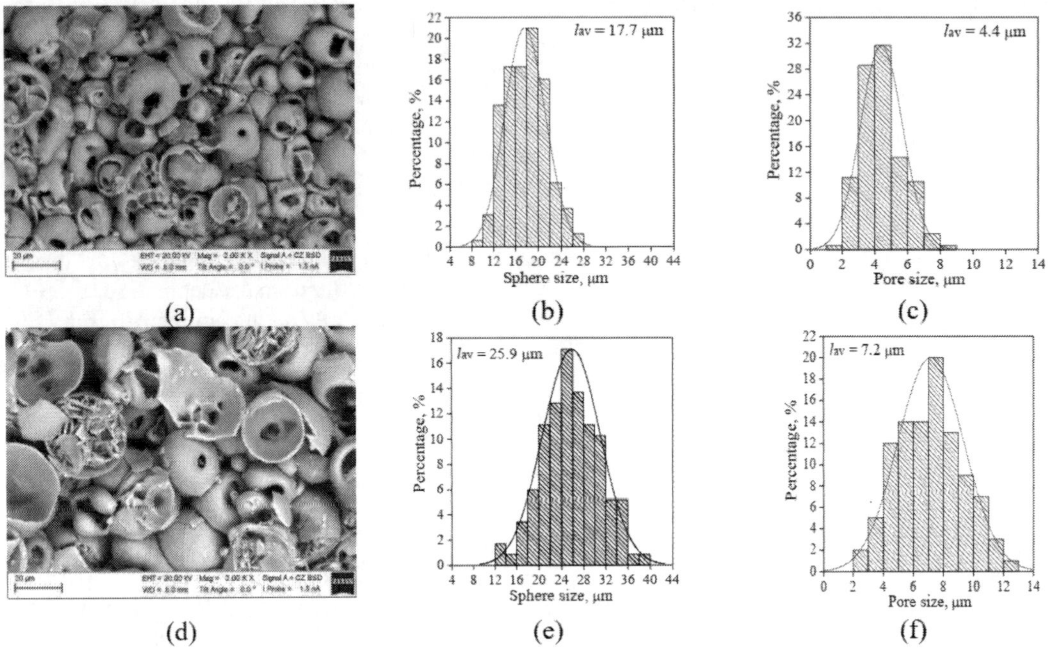

Figure 3. SEM–images (a, d) and histograms of structural element distribution by sizes (b, c, e, f) for Zn– or Cu–CaP coatings deposited under oxidation voltages: (a–c) – 200 V; (b–f) – 300 V.

It was found that with increasing of the process voltage the thickness and roughness of the Zn–CaP and Cu–CaP coatings increase linearly from 50 to 110 μm and from 3 to 6 μm, respectively (Figure 4, a). Khlusov et al [11] reported about novel concepts of "niche–relief" and "niche–voltage" for stem cells and supposed that the average roughness in the range of 2.5–5.0 μm is optimal for successful stem cells adhesion to the coating surface and their further differentiation into the bone tissue. Also, Sammons et al [12] reported that the microporous rough surfaces improved osseointegration of the implants. The change in the surface morphology, topography and thickness of these coatings under increase of the process voltage is connected with increasing of microplasma discharge intensity. However, such important characteristic of the coatings as the adhesion strength to substrate has a low values and decreases from 10 to 4 MPa with increasing of the process voltage (Figure 4, b). This is due to the increase in the coating thickness and size of pores. Unfortunately, it is faulty adhesion for medical applications because according to the ISO 13779–4, for medical devices the adhesion strength of coatings to substrate must be not less than 15 MPa. In this regard, to increase the coating adhesion, the surface was preliminary treated before the coating deposition using sandblasting and subsequent chemical etching [8]. Such pretreatment allows to increase the adhesion strength twice.

Results of XRD analysis showed that the Zn–CaP and Cu–CaP coatings deposited under the process voltages of 200–250 V are in the X–ray amorphous state (Figure 5). Low intensive reflexes corresponding to the β–calcium pyrophosphate (β–$Ca_2P_2O_7$) and monetite ($CaHPO_4$) phases are presented in XRD patterns. With increasing the oxidation voltage to 300 V the intensive crystalline monetite phase forms in the coatings. Monetite exhibits osseoinductive properties like other acid calcium phosphates, which "etches" bone apatite and induces desorption of specific osteoinductive BMP–type proteins [13]. In addition, monetite dissolves easily in the body fluids and transforms into hydroxiapatite at subsequent mineralization. The XRD results are in agreement with the SEM results. Figures 3 and 4 illustrate that the plate–like crystals are typical for monetite crystals.

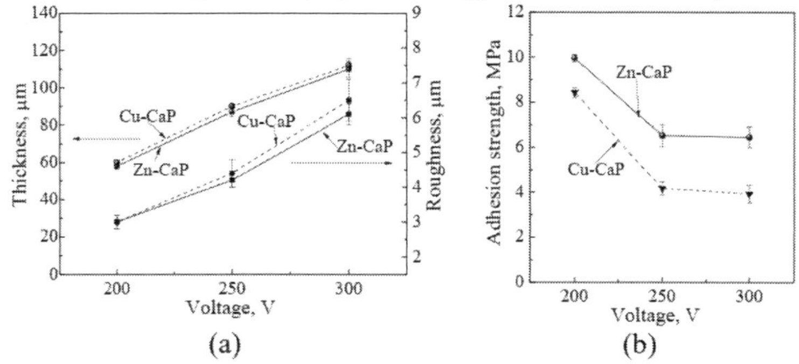

Figure 4. Graphs of thickness and roughness (a), and adhesion strength (b) of Zn–CaP and Cu–CaP coatings against the applied voltages.

Figure 6 shows bright–field (BF) and dark–field (DF) TEM–images with selected–area diffraction (SAD) patterns for particles of Zn– or Cu–CaP coatings. The TEM–study reveals the amorphous structure of coatings. It was observed in the SAD–patterns with two diffusion halos (Figures 6, c, f). It confirms the XRD results (Figure 5) indicating the X–ray amorphous structure.

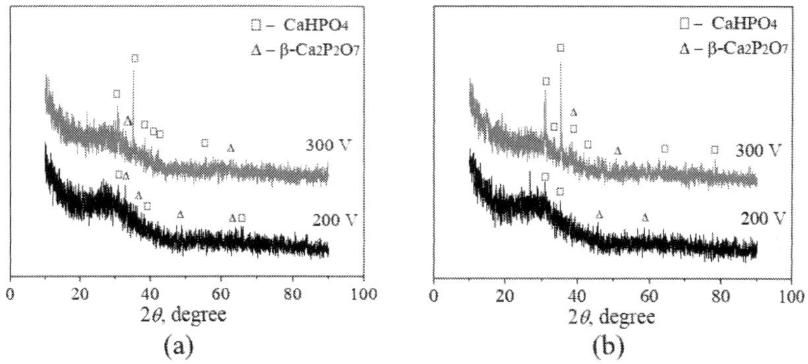

Figure 5. XRD–spectra of Zn–CaP (a) and Cu–CaP (b) coatings deposited under different process voltages.

The EDX microanalysis performed from surface and cross–sectional profile of Zn–CaP and Cu–CaP coatings revealed the elemental composition with the following elements: Ca (5.7–9.8 at.%), P (15.8–25.2 at.%), O (51.2–63.3 at.%), Ti (5.8–9.8 at.%), Nb (2.2–4.2 at.%), Zn or Cu (0.1–0.4 at.%). The larger amount of Ti than Nb in the coatings can be connected with predominant Ti content in amorphous state. The EDX results showed that the maximum Zn and Cu incorporation of 0.3 and 0.2 at.% corresponds to the coatings deposited under oxidation voltage of 250 V. With increasing the process voltage and consequently the temperature in the microplasma discharge region, the process of Ca^{2+} ion deposition from the electrolyte is intensified and the Ca/P ratio increases [3]. The maximum Ca/P ratio of 0.5 is obtained for both coatings deposited under the process voltage of 300 V.

4. Conclusion

Thus, the investigations of the morphology, structure, and properties of Zn– and Cu–CaP coatings on Ti–40Nb alloy formed by MAO method in electrolyte based on substituted HA under different applied voltage in the range of 200–300 V were performed. It was found that the thickness and roughness of the coatings, and the average size of the structural elements (spheres and pores) grow linearly with increasing the MAO voltage. In this case, the coating adhesion strength to substrate decreases. It was established by the XRD that the coatings deposited under voltages of 200–250 V have X–ray amorphous structure. Increase of the voltage to 300 V leads to the formation in the coatings of crystalline phases such as $CaHPO_4$ and β–$Ca_2P_2O_7$. The EDX analysis showed that maximum contents of 0.3 at.% Zn or 0.2 at.% Cu were found in the coatings deposited under the voltage of 200–250 V. Maximum Ca/P ratio of 0.5 was obtained for both coatings deposited under 300 V.

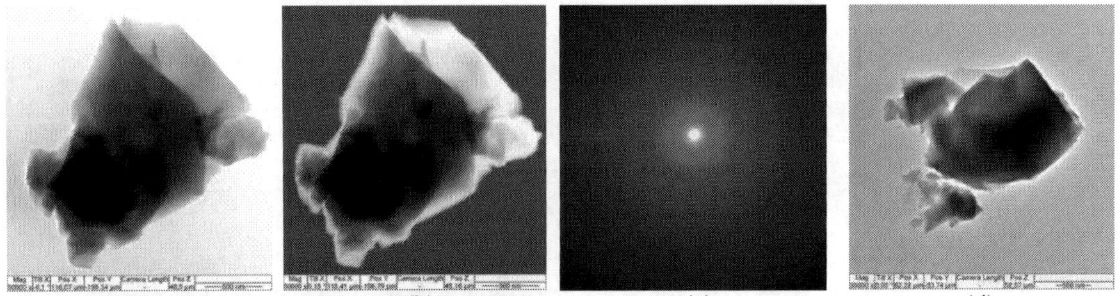

(a) (b) (c) (d)

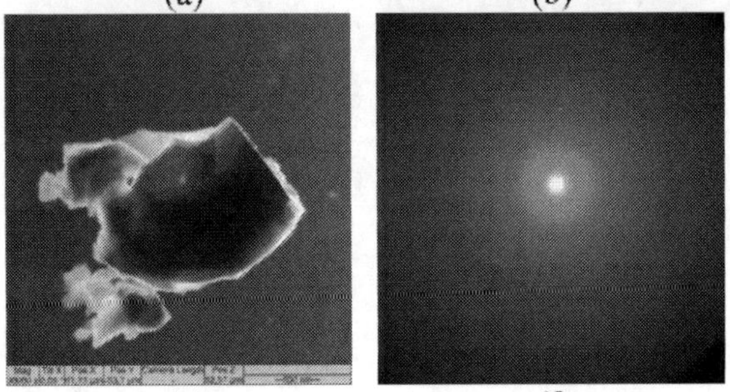

(e) (f)

Figure 6. BF–TEM (a, d) and DF–TEM (b, e) images and SAD patterns (c, f) for particles of Zn– or Cu–CaP coatings deposited under the voltages of 200 V (a–c) and 300 V (d–f).

References

[1] Gepreel M A H and Niinomi M 2013 *J. Mech. Behav. Biomed. Mat.* **20** 407
[2] Lee C M, Ju C P and Chern Lin J H 2002 *J. Oral Rehab.* **29** 314
[3] Sedelnikova M B, Komarova E G and Sharkeev Yu P 2016 *Key Eng. Mater.* **695** 144
[4] Gallo J, Holinka M and Moucha C S 2014 *Int. J. Mol Sci.* **15** 13849–80
[5] Hu H, Zhang W, Jiang X, Liu X and Ding C 2012 *Acta Biomat.* **8** 904
[6] Lowe N M, Fraser W D and Jackson M J 2002 *Proc. Nut. Soc.* **61** 181
[7] Sharkeev Yu P, et all 2016 *Key Eng. Mater.* **683** 174
[8] Sharkeev Yu P *et al* 2014 *Biocomposites based on calcium–phosphate coatings, nanostructured and ultrafine–grained bioinert metals, their biocompatibility and biodegradation* (Tomsk: Publishing House of Tomsk State University)
[9] Komarova E G, et all V V 2015 *Izv. Vuz. Fiz.* **58** 117–21
[10] Chaikina M V, et all 2014 *Euro. J. Inorg. Chem.* **2014** 4803
[11] Khlusov I A, et all 2013 *IFMBE Proc.* **38** 99
[12] Sammons R L, Lumbikanonda N, Gross M and Cantzler P 2005 *Clin. Oral Impl. Res.* **16** 657
[13] Safronova T V and Putlyaev V I 2013 *Nanosist. Fiz. Khim. Matem.* **4** 24

Acknowledgments

The work has been partially financially supported by the Fundamental Research Program SB RAS, Project No.III.23.2.5; the RFBR, Grant No.15–03–07659; and "UMNIK" Program, Grant No. 5256GU1/2014. Authors wishing to acknowledge assistance and encouragement from Prof. M.V. Chaikina (ISSCM SB RAS, Novosibirsk, Russia), Prof. Q. Zhu (GRINM, Beijing, China), A.I. Tolmachev, P.V. Uvarkin, and T.V. Tolkacheva (ISPMS SB RAS, Tomsk, Russia).

The dynamic sublayers for improving the adhesion of CVD diamond films on copper

A V Gaydaychuk, S A Linnik and V V Okhotnikov

National Research Tomsk Polytechnic University, 30 Lenin Avenue, 634050, Tomsk, Russia

E-mail: gaydaychuk@tpu.ru

Abstract. Polycrystalline diamond films were fabricated on copper substrates by a multi-step process comprised of physical vapor deposition of Al-based composite interlayer on Cu substrate and depositing continuous diamond film on composite interlayer by plasma assisted chemical vapor deposition (PACVD). The interface characteristics and adhesion strength were investigated by Auger electron spectroscopy, Raman analysis, calotest and indentation test. The results show that the continuous film without cracks is successfully obtained. The improved adhesion between diamond film and substrate results from the low residual stress in the film due to their compensation by Al interlayer in the sample cooling process.

1. Introduction

Diamond is one of the best candidates for thermal spreading materials due to its excellent thermal conductivity, chemical stability, low dielectric constant and high dielectric strength [1, 2]. However, the high thermal conductivity of diamond by itself is not enough to dissipate heat, as the heat capacity of diamond is limited [3]. Copper is a cheap material for thermal spreading with average thermal conductance and large heat capacity. Therefore, chemical vapor deposition (CVD) of diamond film on copper substrate has attracted great attention.

The main problem of diamond–copper heat spreader is adhesion failure. The poor adhesion of diamond film coatings has been attributed to the high residual stress in the coatings and the weak interfaces between the coatings and Cu substrates [3, 4]. In addition to reducing the residual stresses in diamond coatings, many efforts, including preseeding, scratching, and adding an interlayer, have been made to enhance the interfacial adhesion of the diamond/copper systems [5–9]. Appropriate interlayer between the diamond film and Cu substrate can effectively improve the interfacial adhesion. The interlayer is expected to perform the following three main functions: as adhesive layer, to enhance diamond/copper bonding strength; as diffusion barrier, to protect the detrimental element, which catalyzes diamond to graphite and consequently leads to the coating delamination; and as nucleation layer, to enhance diamond nucleation.

In this paper, Al-based composite interlayer for improving diamond–copper adhesion strength was investigated.

2. Experimental details

There are two methods to improve diamond/copper adhesion strength: method related to preparation of the substrate (preseeding, scratching etc.) and method directly associated with diamond film (residual stress reduction and interlayer adding). The first type of method initially ineffective because at a certain thickness of the diamond film (due to multiple internal stresses increase) regardless of diamond/substrate adhesion magnitude the film will peel off. For this reason, the second method (Al-based composite interlayer adding) was chosen.

Fusible Al was selected as composite interlayer base material. It must perform diamond film internal stress relaxation during the cooling process from growth to room temperature. Second material of the Al-based composite interlayer is cooper (as C diffusion barrier). The third one is Cr (as nucleation layer).

Cooper samples 20 mm in diameter and 5 mm thick were used as substrates. These were polished with a series of SiC sandpapers from 400 down to 1200 grit and ultrasonically cleaned in acetone for 10 min. For subsequent ion etching and deposition of Al-based composite interlayer, a PVD setup was used which was designed and manufactured in Tomsk Polytechnic University (Figure 1). Ion etching and interlayer deposition was performed in one continuous vacuum cycle. The main operating parameters are listed in Table 1. Two types of samples were prepared: 1) Al (520 nm)/Cu (80 nm)/Cr (210 nm); 2) Al (520 nm)/Cu (250 nm)/Cr (210 nm).

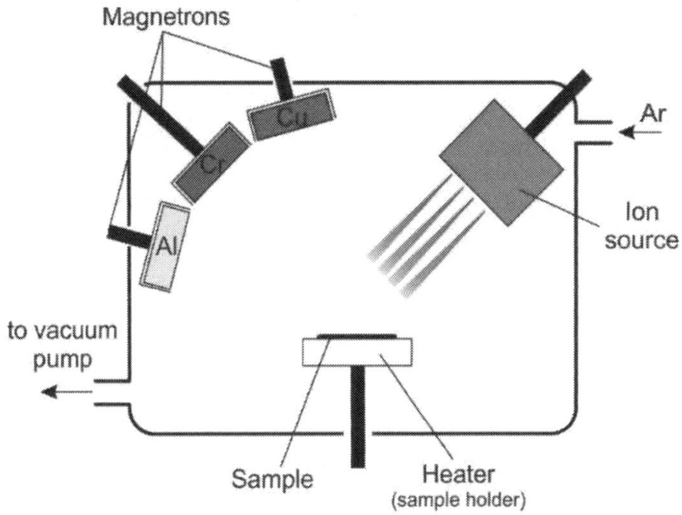

Figure 1. Schematic illustration of the PVD setup.

Polycrystalline diamond films deposited on the Al-based composite interlayer applied on Cu substrate was carried out by a glow discharge plasma-assisted chemical vapor deposition system. The details were described elsewhere [10, 11]. The deposition parameters employed included the discharge current of 10 A, the gas pressure of 13.3 kPa, the CH_4 concentration in H_2 of 6%, the substrate temperature of 800–850 °C and the deposition time of 30 min.

Table 1. The main operating parameters.

	Al	Cu	Cr
Magnetron current [A]	4	2	2.3
Deposition rate [nm/min]	86	41.5	105
Working pressure [kPa]	10^{-3}	10^{-3}	10^{-3}
Ion etching (Ar^+ ions) [keV]	3.5	3.5	3.5

IOP Conf. Series: Journal of Physics: Conf. Series **830** (2017) 012102 doi:10.1088/1742-6596/830/1/012102

3. Results and discussion

Calotest was used to control the thickness of the Al-based composite interlayer (Figure 2). The micrograph shows three clearly visible interlayer: 1) Al, 2) Cu, 3) Cr. Calotest performed without diamond film.

The phase composition and the purity degree of the synthesized diamond films were determined using the most accurate method, i.e. Raman spectroscopy (NanoScan Technology Centaur I HR spectrometer). Figure 3 shows the Raman spectrum obtained in the area of the solid film. It is close to the spectrum characteristic to monocrystalline diamond. The line width, which characterizes sp^3-carbon (1333 cm^{-1}), is less than 50 cm^{-1}, which indicates the absence of amorphous sp^3-carbon. The lines of the parasitic inclusions typical for polycrystalline diamond [12] (lines of trans-PA (1450 cm^{-1}) and G band of graphite (1560 cm^{-1})) are also not observed in the spectrum. Thus, the deposited films consist of high-quality diamond material.

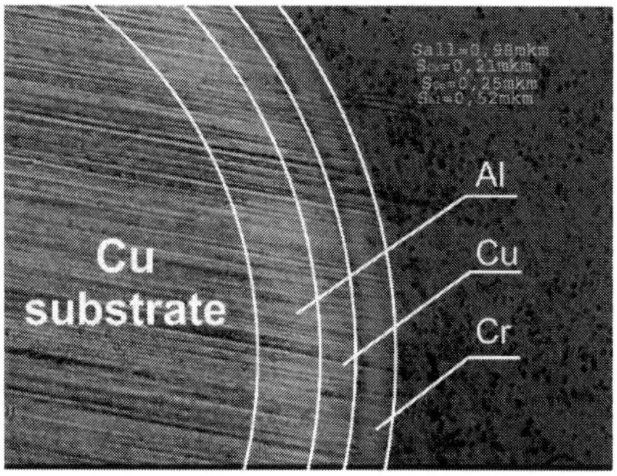

Figure 2. Optical micrograph of film thickness from calotest sample.

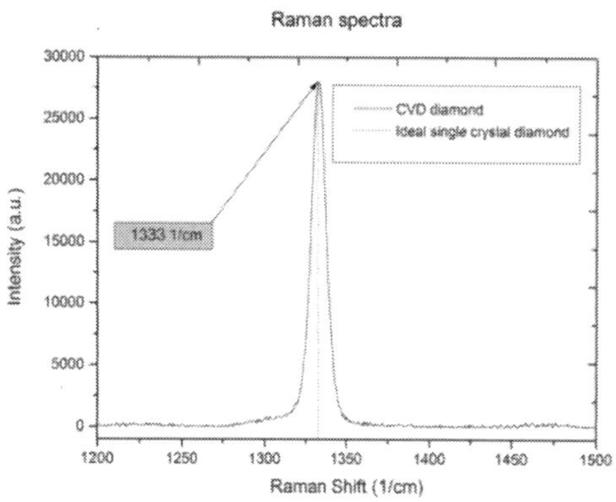

Figure 3. The Raman spectrum of deposited diamond film (Al (520 nm)/ Cu (250 nm)/ Cr (210 nm)).

Nevertheless, the shift in the peak from normal peak position of natural diamond (1332 cm^{-1}) indicates that the film is under compressive stress. According to Wan-qi QIU et al. [13] the compressive stress can be calculated by the following equation:

$$\sigma = -0.567(v_m - v_o) \quad (1)$$

where v_m and v_o are wavenumbers obtained from as-prepared diamond film and natural diamond, respectively. Using this equation, the compressive stress in the diamond film grown on Al-based composite interlayer is calculated to be 0.57 GPa, which is much lower than those in Refs. [14, 15]. The second sample compressive stress were also calculated (equation (1)) and amounted to 1.6 GPa.

It is widely known that diamond film deposited on cooper substrate usually crack or peel off during post cooling process due to the poor bonding strength and the large mismatch of thermal expansion coefficients between diamond and copper [16]. For a comparative analysis of diamond/cooper adhesion strength, a microhardnes tester with spherical indenter with d = 180 μm and 15 kg load was used. The microstructures of the film after test are shown in Figure 4. The diamond film delamination regions was not observed in the case of 250 nm Cu interlayer thickness (Figure 4a). When the thickness of Cu interlayer is 80 nm, delamination region about 1 mm in diameter is present (Figure 4b). This suggests that the diamond /cooper adhesion strength decreases.

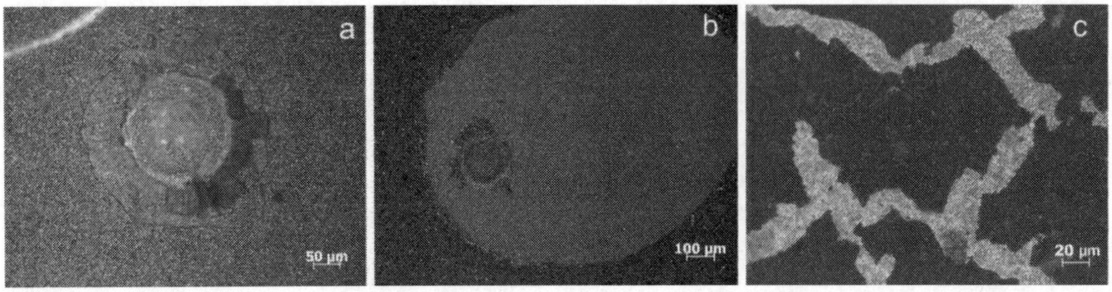

Figure 4. Optical image of indentation imprint: a) sample with 250 nm Cu interlayer in Al based composite interlayer; b) sample with 80 nm Cu interlayer in Al based composite interlayer; and c) sample without Al based composite interlayer.

For a detailed study of Cu interlayer thickness influence on the diamond/cooper adhesive strength, element composition of Al based composite interlayer was investigated (Figure 5).

AES pattern shows the presence of carbon in the interlayer of aluminum in the case of 80 nm Cu interlayer (Figure 5b). At the same time, the 250 nm Cu interlayer completely blocks carbon diffusion in Al (Figure 5a). This explains the enhancement of the diamond film adhesion on a sample with a 250 nm copper interlayer. If thickness of the Cu is sufficient, Al retains its properties and compensates the compressive stresses occurring in the diamond film in the postcooling process. Otherwise, the resulting aluminum carbide has a higher melting point than aluminum, which reduces the efficiency of stress compensation. With high probability at the Cu/Al-based composite interlayer interface, Al/Cu intermetallic compound is formed. It also increases the adhesion strength of the system Cu/Al based composite interlayer/polycrystalline diamond film.

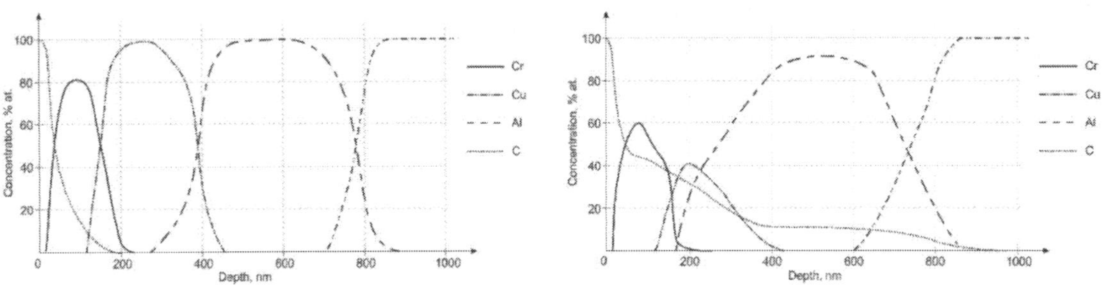

Figure 5. Auger electron spectroscopy pattern of obtained samples: a) Al (520 nm)/ Cu (250 nm)/ Cr (210 nm); b) Al (520 nm)/ Cu (80 nm)/ Cr (210 nm).

4. Conclusion

During this work, the continuous polycrystalline diamond films of high purity were deposited on Al-based composite interlayer coated on copper. The improved adhesion between diamond film and Cu substrate results from the internal stress of the diamond film compensated by fusible Al interlayer and Al/Cu intermetallic compound formation at the Cu/Al-based composite interlayer interface.

References

[1] Liao M Y and Koide Y 2011 *Crit. Rev. Solid State Mater. Sci.* **36** 66
[2] May P W 2008 *Science* **319** 1490
[3] Liu X, Wei Q, Yu Z and Yin D 2012 *Phys. Status Solidi A* **209** No. 11 2217
[4] Yang T F 2007 *Surf. Coat. Technol.* **202** 1208
[5] Fan Q F, Pereira E and Gracio J 1999 *Journal of Materials Research* **13** 2787
[6] Liu X Z, Wei Q P, Zhai H and Yu Z M 2013 *Transactions of Nonferrous Metals Society of China* **23** 667

[7] Fan Q H, Fernandes A, Pereira E and Gracio J 1999 *Vacuum* **52** 163
[8] Ali N, Ahmed W, Rego C A and Fan Q H 2000 *Diamond and Related Materials* **9** 1464
[9] Battabyal M, Beffort O, Kleiner S, Vaucher S and Rohr L 2008 *Diamond and Related Materials* **17** 1438
[10] Linnik S A, Gaydaychuk A V 2013 *Diamond and Related materials* **32** 43
[11] Linnik S A, Gaydaychuk A V 2015 *Materials Letters* **139** 389
[12] Linnik S A, Gaydaychuk A V and Baryshnikov E Y 2015 *Materials Today: Proceedings* **3** 138
[13] Qiu W Q, Hu Z G, Liu Z W, Zeng D C and Zhou K S 2014 *Trans. Nonferrous Met. Soc. China* **24** 758
[14] Hartsell M L, Plano L S 1994 *Journal of Materials Research* **9** 921
[15] Narayan J, Chen X 1992 *Journal of Applied Physics* **71** 3795
[16] Fan Q H, Gracio J and Pereira E 1997 *Diamond and Related Materials* **6** 422

Acknowledgments
This work was financially supported by the Russian Foundation for Basic Research (research project No. 16-32-00008 Mol_a).

The influence of repetitively pulsed plasma immersion low energy ion implantation on TiN coating formation and properties

D O Sivin, P S Ananin, S V Dektyarev, A I Ryabchikov and A E Shevelev

National Research Tomsk Polytechnic University, 30 Lenin Avenue, 634050, Tomsk, Russia

E-mail: sivin@tpu.ru

Abstract. Application of high frequency short pulse plasma immersion low energy ion implantation for titanium nitride coating deposition using vacuum arc metal plasma and hot-cathode gas-discharge plasma on R6M5 alloy was investigated. Implementation of negative repetitively pulsed bias with bias amplitude 2 kV, pulse duration 5 μs and pulse frequency 10^5 Hz leads to 6.2-fold decrease of vacuum arc macroparticle surface density for macroparticles with diameter less than 0.5 μm. Ion sputtering due coating deposition reduces the production rate approximately by 30%. It was found that with bias amplitude range from 1.1 to 1.4 kV and pulse duration 5 μs yields to formation of coatings with local hardness up to 40 GPa. This paper presents the results of experimental studies of adhesion strength, tribological properties and surface morphology of deposited TiN coatings.

1. Introduction

Hard thin-film nitrides coatings formed by Physical Vapor Deposition (PVD) are widely used to improve wear-resistant, anticorrosive properties of cutting tools, machine parts and etc. Among these materials the most widely studied is titanium nitride (TiN). Typically, the standard method of TiN forming involves the use of plasma of vacuum arc discharge. Vacuum arc has undeniable advantages such as high degree of ionization, high coating growth rate, the ability to generate plasma of almost any conductive material [1–3]. The drawback of the vacuum arc is associated with the presence of debris metal particles, with sizes from nanometers up to hundreds of micrometers [3], often refer as macroparticles (MP). Deposition of the MPs on the treated surface significantly impair the functional characteristics of obtained coatings.

Several different approaches were proposed and used to reduce the MP density in the deposited coatings [4–7]. One of the methods based on the application of a negative DC bias to a substrate, immersed in vacuum arc plasma [4, 5]. The authors of these papers mentioned that the number of MPs on titanium nitride coating can be reduced by approximately 3–4 fold with the increasing of negative bias potential up to -1 kV. In other Refs [8, 9], the authors used high-frequency short-pulse negative biasing during the deposition of unfiltered vacuum-arc plasma. The application of pulsed negative bias (with pulse duration of microseconds) allowed increasing the amplitude of the negative bias up to 2–3 kV. The quantitative reduction of the surface density of aluminum or titanium MPs by several orders of magnitude was demonstrated.

IOP Conf. Series: Journal of Physics: Conf. Series **830** (2017) 012103 doi:10.1088/1742-6596/830/1/012103

This work studies the regularities of synthesis of TiN coatings with lower MP-content and improved physical-mechanical properties using vacuum-arc plasma deposition and negative high-frequency short-pulse biasing of the sample.

2. Experimental setup

The coating synthesis were carried out on an experimental setup presented in figure 1. The working pressure in vacuum arc chamber was 10^{-3} Pa. The unit is equipped with an industrial vacuum-arc source of metal plasma, and gas plasma source "PINK". Titanium VT-1 was used as the cathode material. Arc current was equal to 100 A and the density of ionic current saturation of metal plasma near the surface of the substrate was approximately 4.4 mA/cm^2.

Experimental samples were made of R6M5 high speed steel and were mounted on a massive holder located in the center of the vacuum chamber. The distance from the sample to the cathode of vacuum-arc evaporator was 40 cm. The surfaces of the substrates were polished to roughness $R_a \sim 0.25$ μm.

The negative bias was formed by the high-frequency short-pulse negative bias generator with following parameters: pulse duration τ = 1–9 μs, pulse repetition rate f = 10^5 Hz, bias potential amplitude φ = 0.25–3 kV.

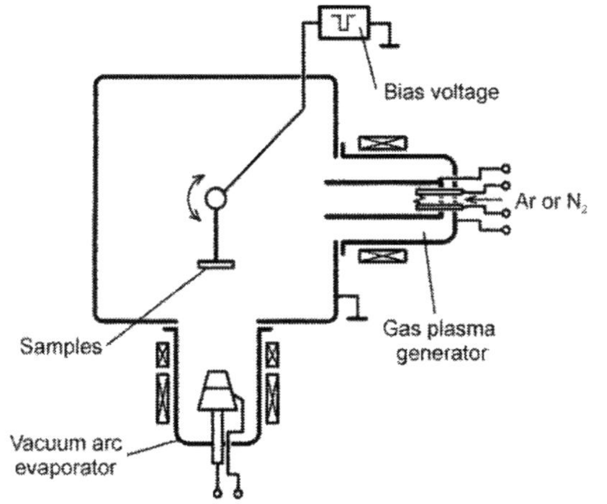

Figure 1. The experimental setup scheme.

Before the synthesis of nitride coatings in a single vacuum cycle ion cleaning, surface activation of the samples and their heating were carried out, using gas-discharge plasma of argon at the pressure of the working gas of 1 Pa and high-frequency short-pulse biasing with amplitude up to 1.2 kV. As a result of ion bombardment, the surface layer of the samples was activated to ensure the best adhesion between the coating and the substrate, and heated to a temperature of 300–400 °C. In addition, the formation of the nitride coating was preceded by 5 min deposition of the transition titanium layer. The deposition of the nitride coating was performed for 40 min. The pressure of the working gas (nitrogen) was chosen in the range 0.1–2.4 Pa. The parameters of the pulsed bias potential were varied in the following ranges: pulse amplitude, 0.9–2.0 kV; pulse duration, 3–5 μs.

The surface morphology and the macroparticle surface density were studied using electron scanning microscope "Hitachi TM-3400S" and three-dimensional profilometer "STEL MicroMeasure 3D Station". To obtain statistically reliable results, for each experimental point, the total area for macroparticles quantification was 270×210 μm.

The element composition of the formed coating analyzed by Auger electron spectroscopy (AES). The hardness was measured using "CSM Nano-Hardness Tester". The Vickers indenter load was 100 mN. The adhesion properties examined using "CSM Micro-Scratch Tester". In the scratch test, a Rockwell diamond indenter with the diameter of 20 μm was used. The tribological properties were investigated by "CSM HighTemperature Tribometer". Wear studies were carried out by the "ball-on-disc" method using a ball with the diameter of 3 mm at the normal load of 2 N.

3. Results and discussion

Prior to the TiN coating deposition with high-frequency short-pulse negative bias, a series of experiments with a DC negative bias of 0.3 kV was performed to determine the optimal pressure of the working gas (nitrogen) for the formation of the coating with the best properties. Figure 2 presents the results on the hardness and critical load studies of the obtained coatings versus nitrogen pressure in the vacuum chamber. The coatings with the best properties were synthesized at the pressure of 0.6 Pa. Therefore, all subsequent experiments on the coating formation were carried out at the same pressure.

The experiments on the formation of TiN coatings with different parameters of negative high-frequency short-pulse bias were done. Table 1 presents the modes of the TiN coating synthesis and main physical and mechanical characteristics.

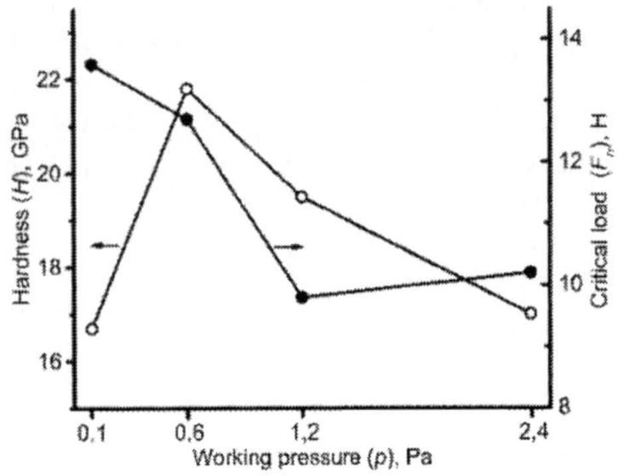

Figure 2. The influence of nitrogen pressure on hardness and adhesion of TiN coatings formed with DC negative bias of 0.3 kV.

Table 1. The modes of synthesis and physical properties of the TiN coatings.

N	Coating synthesis parameters				Characteristics of the deposited coatings				
	$-\varphi$ [V]	τ [μs]	f [Hz]	$P(N_2)$ [P]a	Thickness (s) [μm]	Hardness (H) [GPa]	Critical load (F_n) [N]	Mean friction coeff. (μ)	Wear [relative unit]
1	0.3	D.C.		0.6	5.2	21.8	12.5	–	–
2	0.9	3	10^5	0.6	5.3	18.7	11.8	–	–
3	0.9	4			5.0	20.5	9.5	0.22	0.69
4	0.9	5			5.0	23.2	10.2	0.22	0.69
5	1.1	5			4.9	28.1	16.0	0.27	0.66
6	1.4	5			4.4	27.5	13.0	0.22	0.57
7	2.0	5			3.6	19.5	8.4	0.25	0.62
8	Initial sample HSS (R6M5)				–	16.0	–	0.42	1.0

Figure 3 shows SEM images of the coating surfaces depending on the deposition conditions. The microphotographs demonstrate a slight decrease in the number of small MPs on the surface of the substrates with increase in the bias amplitude. Figure 4 demonstrates the size distribution of MP for some typical modes of deposition. The application of high frequency negative bias with a frequency of 10^5 Hz, amplitude of 2 kV and duration of 5 μs enables the 6.2-fold reduction of the number of MPs with diameter less than 0.5 μm. The total number of MPs is decreased 2.8-fold. In addition, the SEM images in Figures 3c and 3d show a significant erosion of the coating surface, which may be associated with intensive ion sputtering at the high bias amplitudes.

Figure 5 presents the thickness of coatings formed under various conditions. According to the data, the increase in the amplitude of the bias up to 2 kV leads to a decrease in the coating growth rate by approximately 30%.

IOP Conf. Series: Journal of Physics: Conf. Series **830** (2017) 012103 doi:10.1088/1742-6596/830/1/012103

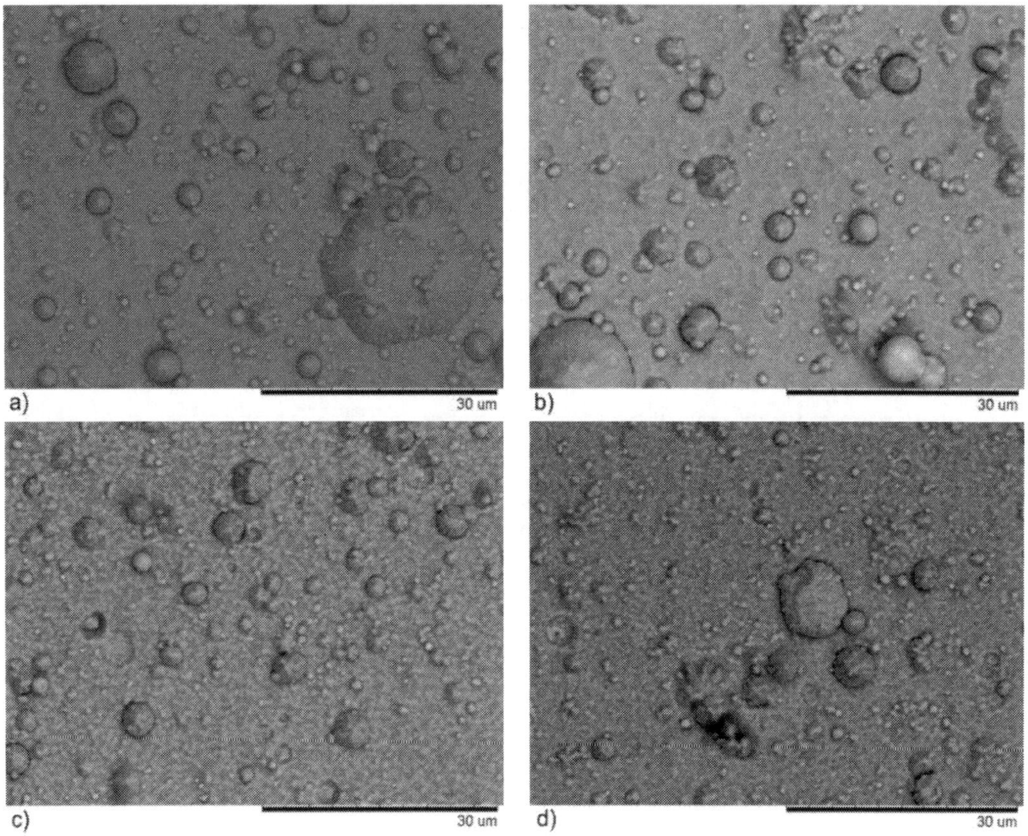

Figure 3. Microphotographs of the TiN coating surface with the deposited MPs in different modes of deposition: a) at anode potential of 0 V; b) at negative DC bias of 0.3 kV and negative pulsed bias (f = 105 Hz, τ = 5 μs) with an amplitude c) φ = −1.1 kV; d) φ = −2.0 kV.

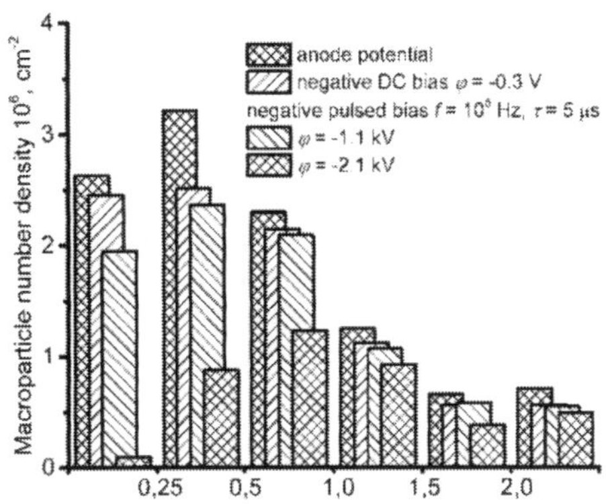

Figure 4. The number of counted MPs as a function of diameter for the case of anode potential, negative DC bias of 0.3 kV and negative pulsed bias (f = 105 Hz, τ = 5 μs) with amplitudes of 1.1 kV and 2.0 kV.

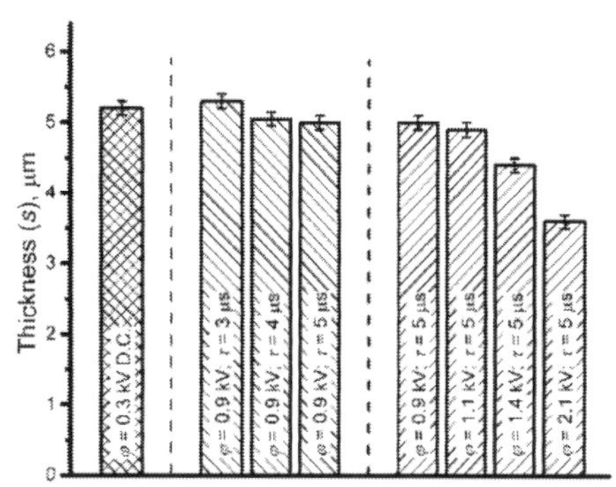

Figure 5. The TiN coating thickness obtained under various modes of deposition.

The results of the elemental composition studies of the coatings deposited in the experiments with a pulsed negative bias of the various amplitudes are presented in Table 2. The average concentration of nitrogen is 41 at.%, and concentration of titanium is 52 at.%. With an increasing amplitude, a slight decrease in the concentration of nitrogen is observed, probably due to the decomposition of titanium nitride at high temperatures (>700 °C).

Table 2. Element composition of TiN coatings.

N	Coating synthesis parameters				Element concentration [at.%]		
	$-\varphi$ [V]	τ [μs]	f [Hz]	$P(N_2)$ [Pa]	Ti	N	other
4	0.9	5	10^5	0.6	51.4	42.1	6.5
5	1.1	5			49.7	44.2	6.1
6	1.4	5			53.6	40.7	5.7
7	2.0	5			53.8	38.2	8.0

The obtained nitride coating has high hardness, exceeding the hardness of the default HSS sample. The average values for hardness of the formed films are presented in Table 1. The highest hardness of about 28 GPa is observed in TiN coatings synthesized with a negative pulsed bias with a pulse duration of 5 μs and amplitudes of 1.1 kV and 1.4 kV, the lowest hardness of 18.7 GPa in TiN coating with the following bias parameters: φ = 0.9 kV, τ = 3 μs. A large number of performed indentations (49 measurements for each sample) revealed the presence of superhard phases (≥40 GPa). Figure 6 shows the results of surface hardness measurements of different experimental samples in a form of histogram showing the relative frequency of the superhard coatings observation.

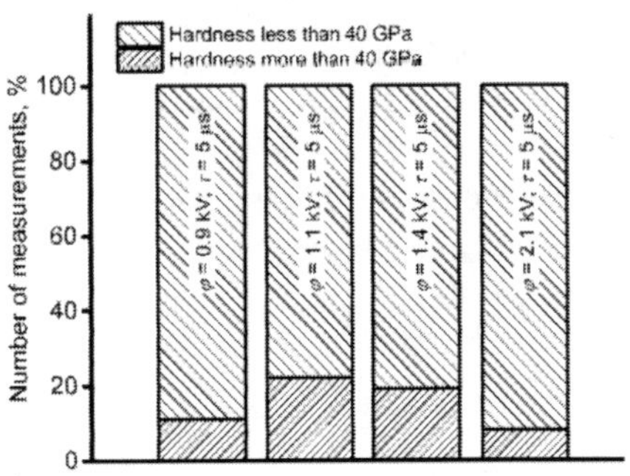

Figure 6. Histogram of the number of hardness measurements with values less or more than 40 GPa on coatings obtained with various negative bias parameters.

Analysis of the data showed that with the application of pulsed negative bias with a pulse duration of 5 μs and amplitudes of 1.1 kV and 1.4 kV, approximately 20% of the total indentations expose the maximum value of hardness (≥40 GPa) distinctive for the superhard TiN phases. Further increase in the bias potential leads to a decrease of this parameter down to 8%. One of the possible explanations is the growth of the crystalline titanium nitride grain with the prolonged exposure of high temperatures. The Young's modulus of the obtained nitride coatings is in the range of 350-680 GPa.

Figure 7 shows the dependence of the dynamic friction coefficient versus the number of cycles of the tribological test for the R6M5 steel samples and for the samples with a TiN coating synthesized at different parameters of repetitively pulsed negative bias. During the test, the friction coefficient remains at the level of the average values of 0.22–0.27. Table 1 presents the average values of a friction coefficient and relative wear of the experimental samples. The mean coefficient of friction in the tribological tests for TiN coating is approximately 1.7-times less than that in the test with the default sample material. Figure 8 presents the data on the wear of the original sample material and some of the TiN coatings. The maximum wear resistance was observed on the samples modified with ion-plasma treatment at bias amplitude of 1.4 kV and a pulse length of 5 μs, which is 1.8-fold higher than that of the initial material. The minimum wear resistance of the coating deposited with the different bias parameters is also 1.5-fold higher than wear resistance of the default HSS sample.

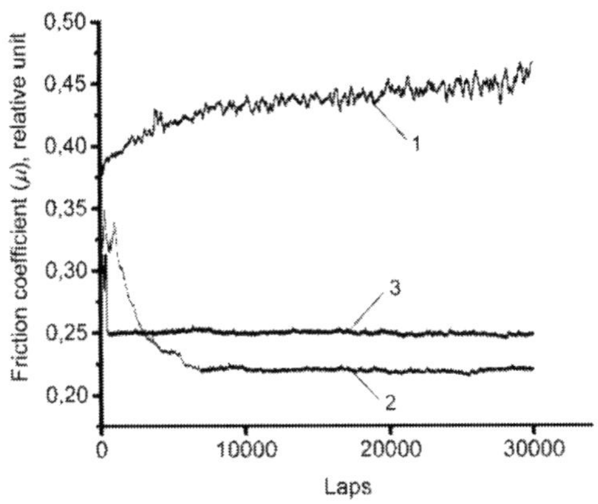

Figure 7. The dynamic friction coefficient: *1)* R6M5 alloy, TiN coating obtained with negative pulsed biasing (f = 105 Hz, τ = 5 μs) with an amplitudes of *2)* φ = –1.1 kV, *3)* φ = –2.0 kV.

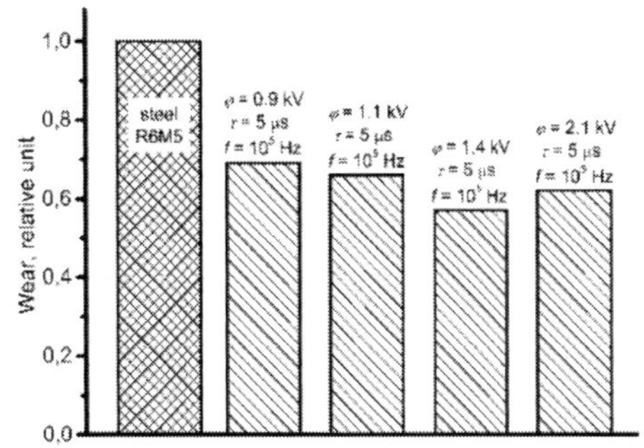

Figure 8. Wear of TiN coatings obtained at different bias parameters, with respect to the wear of the original sample of R6M5 steel.

The results of the scratch test of TiN coatings are presented in Table 1. Analysis of the dynamics on changes of friction forces and sensor data from the acoustic emission allowed determining the corresponding critical load of the coating separation from the substrate. The coating with high adhesion and a critical load of 16 N was obtained with a negative pulsed bias potential with the pulse duration of 5 μs and amplitude of 1.1 kV. The critical load of this sample is 1.9-fold greater than the critical load (8.4 N) of the sample with the smallest adhesion (sample N7 in Table 1).

4. Conclusion

In the process of synthesis of titanium nitride by vacuum-arc discharge, the application of negative high-frequency short-pulse negative bias with the frequency of 10^5 Hz with amplitude of 2 kV and duration of 5 μs provides the 6.2-fold reduction in the number of vacuum arc macroparticles (with the diameter less than 0.5 μm) on the surface of the coating. The use of pulsed bias with a pulse duration of 5 μs and amplitudes of 1.1 kV and 1.4 kV ensure the formation of TiN coatings with the hardness of 28 GPa with the best wear resistance and adhesion.

References

[1] Martin P J, *et al.* 2005 *Surf. Coat. Technol.* **200** 2228
[2] Flink A, *et al.* 2005 *Surf. Coat. Technol.* **200** 1535
[3] Boxman R L, Sanders D M and Martin P J 2009 *Handbook of Vacuum Arc Science and Technology: Fundamentals and Applications* ed Martin P J (N.J: Noyes Publications) p 936
[4] Tau C N, Koh E S and Akari K 1990 *Surf. Coating Techno.l* **43–44** 324
[5] Keidar M, Aharonov R and Beilis I I 1999 *J. Sci. Technol.* **A17 (5)** 3067
[6] Keidar M and Beilis I I 1999 *IEEE Trans. Plasma Sci.* **27 (3)** 810
[7] Ryabchikov A I, *et al.* 2006 *Rev. Sci. Instrum.* **77** 03B516
[8] Ryabchikov A I, Sivin D O and Bumagina A I 2014 *Applied Surface Science* **305** 487
[9] Ryabchikov A I, Sivin D O and Bumagina A I 2014 *Applied Surface Science* **310** 130

Acknowledgements
This work was performed within the State Mission «Science» of the Ministry of Education and Science of the Russian Federation.

TiC-based coatings deposition using electric discharged plasma

I Rahmatullin, A Sivkov, D Gerasimov, A Ivashutenko and J Shanenkova

National research Tomsk polytechnic university, 30 Lenin Avenue, 634050, Tomsk, Russia

E-mail: riam@tpu.ru

Abstract. The aim of this work was to research the possibility of using coaxial magneto plasma accelerator for TiC-coatings deposition on steel substrates. As a result, coatings with 0.01 m^2 area was deposited. They were researched using XRD, SEM; also, the nanohardness on cross section of coating was measured. The influence of energy and carbon load on phase content, average hardness and microstructure is shown. It is established that the finest microstructure and average nanohardness is 15.3 GPa are achieved at energy W = 46.7 kJ and carbon load of 2.0 grams.

1. Introduction

TiC-based coatings are characterized as wear resistant, hard, strength, corrosion and oxidation resistant material [1]. Moreover, TiC is refractory and thermally stable material, which has low friction coefficient and high thermal and electric conductivity [2]. TiC-based coatings can be deposited using different techniques such as laser cladding [1], laser assisted CVD [2] or plasma assisted CVD [3, 4], ion plating [5–7] and PVD [8].

There are certain problems with TiC deposition methods in spite of wide variety of them. They are low adhesion, residual stress, deposition time and substrate preheating to high temperature 500-1050 °C [3].

One of the possible ways to improve this is the implementation of coaxial magneto plasma accelerator (CMPA) generating supersonic jet of high current plasma. Previous investigations demonstrate possibility of using that kind accelerator for TiC/Ti-coatings on copper substrate with a high degree of adhesion [9]. In this work, TiC-coatings were deposited on steel substrates. The average hardness at cross section of coatings reached 15.3 GPa and were achieved at maximum energy of process W = 46.7 kJ. The influence of carbon load and energy output on coating hardness are shown.

2. Experimental

TiC-based coatings were deposited on the steel substrates using high voltage high current coaxial magneto plasma accelerator with power supply from capacity energy storage. CMPA unit is shown in Figure 1. CMPA is designed as a two coaxial titanium electrodes. One of them is central electrode placed in barrel electrode with insulated gap between them. Insulated gap plays a role of plasma structure forming channel. Electrodes are placed into the inductor which stabilizes and accelerates arc discharge after breakdown of electrodes gap. Carbon load was filled into the gap and provided initial

IOP Conf. Series: Journal of Physics: Conf. Series **830** (2017) 012104 doi:10.1088/1742-6596/830/1/012104

electric conductivity for breakdown at a switching time. Thus, titanium electrodes were eroded under high current plasma flow, and the eroded material was reacting with filled carbon.

Carbon load was varied in the range of 0.4–2.0 g. Distance between the end of the barrel electrode and the substrate was 150 mm and was constant in all experiments. Area of the substrates was 0.01 m^2. Charging voltage and capacity in all experiments was 2.8 kV and 14.4 mF, respectively. Summary of initial experimental data is presented in Table 1. Energy parameters at process of synthesis were recorded using Tektronix TDS 2012B digital oscilloscope. Obtained waveforms were calculated respectively to transformation coefficients of Rogowski transformer and resistive voltage divider.

Deposited coatings were researched by methods of X-Ray diffractometry (XRD) using Shimadzu XRD6000S diffractometer. Samples were prepared as polished cross sections, and nanoharndess was measured on them using NANO Hardness Tester NHT-S under 300 mN load. After indentation, the samples were etched using a mixture of ammonium nitrate (65%), hydrofluoric acid HF (40%) and glycerin (87%). Etching mixture was applied to the polished side during 20 seconds and was washed by water flow. The microstructure of the etched samples was investigated using scanning electron microscope Hitachi TM3000 and Quanta 200 3D.

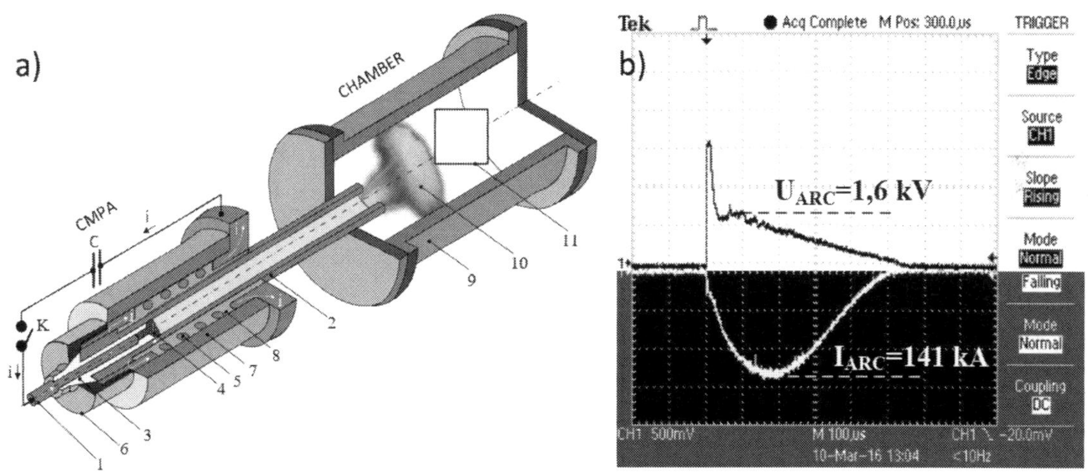

Figure 1. CMPA design (a) and waveforms of the process (b). *1* – central electrode; *2* – barrel electrode; *3* – central electrode insulator; *4* – carbon-filled gap; *5* – inductor; *6* – steel case; *7* – CMPA body; *8* – insulation; *9* – chamber; *10* – plasma jet; *11* – substrate.

3. Results and discussion

As a result, coated samples for different deposition conditions were obtained. Figure 2 presents typical coated samples obtained at different conditions respectively to the Table 1. Coatings change exterior from metallic to matted according to increasing of carbon load. Moreover, the relief of the coatings becomes more uniform.

Table 1. Initial experimental data and energy parameters of the deposition process.

#	m_C [g]	U_{ch} [kV]	C_{ch} [mF]	U_{arc} [kV]	I_{arc} [kA]	P_{arc} [MW]	W [kJ]
1	0.4	2.5	14.4	0.8	140	122	24
2	0.5	2.8	14.4	1.3	133	132	35.7
3	1.5	2.8	14.4	1.55	117	145	42.0
4	2.0	2.8	14.4	1.6	141	180	46.7

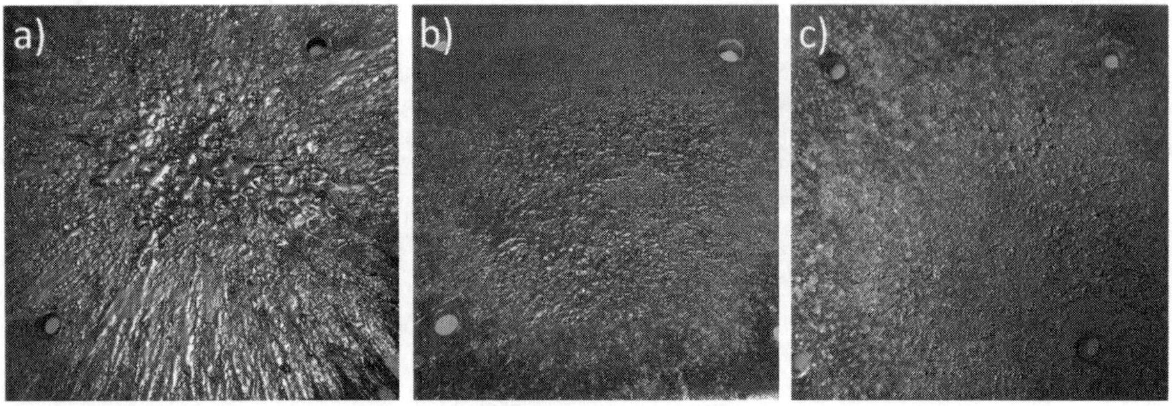

Figure 2. Substrate surface after deposition under different carbon load: a) 0.5 g; b) 1.5 g; c) 2.0 g.

XRD-investigations demonstrate TiC fraction increase with carbon load increase. The level of released energy influence on TiC proportion also. Besides TiC in coating Ti, Fe and Fe_3C also exist. XRD-pattern in Figure 3 shows the absence of carbon in the coating. The presence of Ti is caused by the erosion of electrodes and shortage of carbon load. Iron appears due to the contact of high temperature plasma jet with substrate and consequent mixing, and the reaction between iron and carbon occurs with iron carbide synthesis.

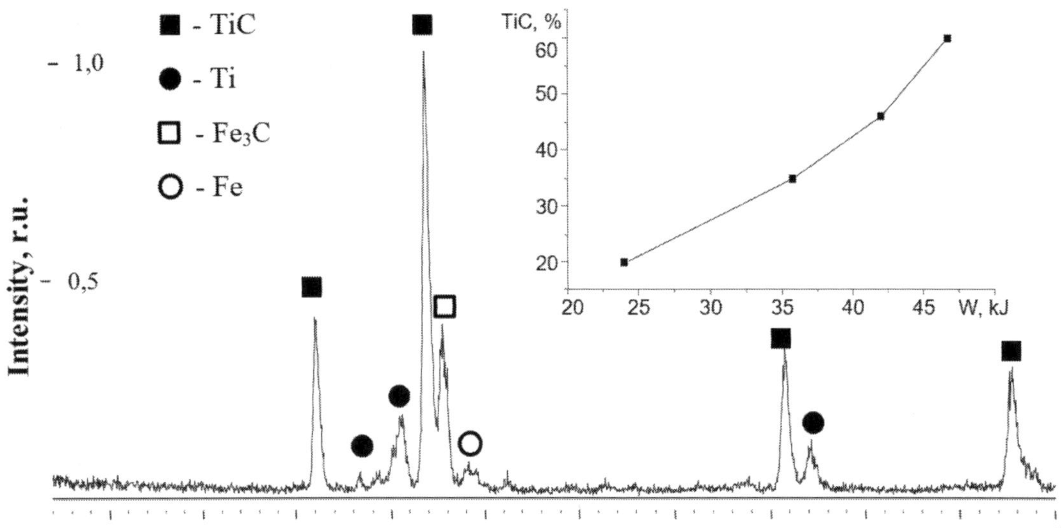

Figure 3. XRD-pattern of coating obtained in experiment 4 (Table 1) and influence of energy on TiC fraction.

Nanohardness measuring on polished cross section of coating shows upward trend of average hardness with an increase of released energy level, carbon load and proportion of TiC in coating as demonstrated in Figures 4a, 4b and 4c.

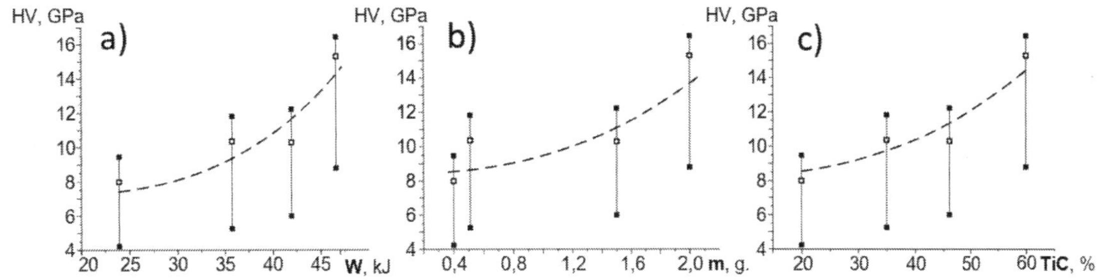

Figure 4. Influence of experimental conditions on the average nanohardness of the coatings.

Etched cross-sections of deposited coatings were investigated using scanning electron microscopy, which allowed identifying their microstructure presented in Figure 5. The thickness of coatings varies in the range of 30–300 μm and grows with increasing of carbon load and released energy. SEM results demonstrate polycrystalline structure of the coatings. Microstructure and grain boundaries become more uniform and ordered, while carbon load and energy of the process are increasing. The finest microstructure was reached in experiment with m = 2.0 grams and W = 46.7 kJ as it can be seen Figure 5c. On the basis of the image, it can be argued about uniform distribution of TiC grains in Ti-matrix, which allow reaching the highest level of nanohardness for described conditions.

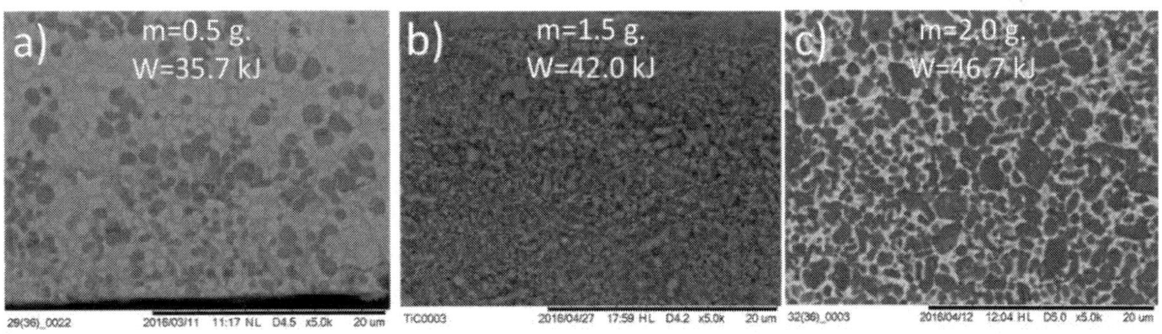

Figure 5. Microstructure SEM-pictures of coatings deposited at different conditions.

4. Conclusion

Investigation results demonstrate the possibility of using coaxial magneto plasma accelerator of supersonic electrodischarged plasma jet for TiC- based coatings deposition with the thickness up to 300 μm in a short time cycle about 500 μs. Influence of released energy and carbon load mass on coating thickness, microstructure and hardness is established.

References

[1] Savalani M, Ng C, Li Q and Man H 2012 *Applied Surface Science* **258**(7) 3173
[2] Yansheng G, Rong T and Takashi G 2013 *Materials Research Bulletin* **48**(8) 2766
[3] Shanaghia A, Rouhaghdama A, Ahangaranic S, Chub P and Farahania T 2012 *Applied Surface Science* **258**(7) 3051
[4] El Mel A, Angleraud B, Gautron E, Granier A and Tessier P 2010 *Surface and Coatings Technology* **204**(12–13) 1880
[5] Guojun Z, Bin L, Bailing J, Fuxue Y and Dichun C 2009 *Applied Surface Science* **255**(21) 8788
[6] El Mel A, Angleraud B, Gautron E, Granier A and Tessier P 2011 *Thin Solid Films* **519**(12) 3982

[7] Longoa G, Girasolea M, Pompeoa G, Cricentia A, Misianob C, Acclavioc A, Tizzonid A, Mazzolae L, Santinid P et al 2012 *Surface and Coatings Technology* **204**(16–17) 2605
[8] Saoulaa N, Hendaa K, Kesrib R, Shrivastavac S, Erasmusc R and Cominsc J 2012 *Acta Physica Polonica A* **121** 172
[9] Sivkov A, Shanenkov I, Pak A, Gerasimov D and Shanenkova Y 2016 *Surface and Coatings Technology* **291** 1

Acknowledgments
This work was supported by the Russian Scientific Foundation (grant No. 15-19-00049).

Effect of different methods of preliminary surface treatment and magnetron sputtering on the adhesion of Si coatings

D P Borisov, V A Slabodchikov and V M Kuznetsov

Tomsk State University, Tomsk 634050, Russia

E-mail: borengin@mail.ru

Abstract. The paper presents research results on the adhesion of Si coatings deposited by magnetron sputtering on NiTi substrates after preliminary surface treatment (cleaning and activation) with low-energy ion beams and gas discharge plasma. The adhesion properties of the coatings obtained by two methods are analyzed and compared using data of scratch and spherical abrasion tests.

1. Introduction

One of the promising implant materials for medicine, e.g. for treatment of vascular diseases, is nickel-titanium (NiTi) alloy [1], featuring high strength and elastoplastic properties (shape memory effect and superelasticity) and applicable for manufacturing self-expanding intravascular stents, including peripheral, occluders, cava filters, cardiac valves, clips, and clamps.

The main requirement on such implants is their biocompatibility to provide the growth of endothelial cells on an implant and its successful implantation. The anticorrosion properties of NiTi implants, their biocompatibility, and the cell adhesion to their surface can be increased through surface modification with Si atoms [2] either by Si doping to a depth of tens of nanometers or by deposition of a Si coating of tens of nanometers [3,4].

When depositing a Si coating on a NiTi implant, it is required to provide its high adhesion to the implant surface, thus precluding its separation during the operation of the implant in a body, for which the implant surface should be chemically clean and free of any impurities (adsorbed residual gases, residues of organic and inorganic solvents) that can form intermediate layers between the coating and the implant surface and affect the coating adhesion up to its loss. There are different methods of preliminary surface treatment or cleaning: for example, with ion beams and with gas discharge plasma which provide, in addition to efficient desorption of contaminants, a large number of point defects in surface layers, enhanced diffusion, and hence, better coating adhesion.

Here we report on a comparative study to assess how the coating adhesion is influenced by two widely used types of preliminary surface cleaning: with ion beams and with gas discharge plasma at a negative bias.

2. Experimental equipment and research technique

The test material was NiTi with a Ni content of 50.9 at% applicable for manufacturing intravascular implants. The NiTi specimens were square plates of dimensions 10×10 mm and thickness 1 mm polished to an average surface roughness $R_a = 0.05$ μm and deposited with Si coatings ≈600 nm thick by magnetron sputtering of Si targets. Before coating deposition, the specimens were treated with ion beams and gas discharge plasma to compare the effect of two methods on the coating adhesion.

The experiments with ion beam pretreatment were performed on a Kvant ion-magnetron setup with a vacuum chamber of dimensions 600×600×600 mm at the Laboratory of Nanotechnology and Materials Science of Coatings of the Institute of Strength Physics and Materials Science SB RAS (Russia). On the upper flange at the center of the chamber, there was a resistance heater (molybdenum). The setup comprised a vacuum arc ion source and a circular planar magnetron sputtering system located on the side flanges of the vacuum chamber at an angle of 120° to each other.

Before placing in the vacuum chamber, the specimens were cleaned in an ultrasonic bath successively with benzine and ethanol for 20 mins and were dried. Then, the specimens were placed in the vacuum chamber and the chamber was pumped by a cryogenic pump to an ultimate residual pressure of 0.003 Pa.

The ion source operating in continuous mode produced a Ni ion flow (Ni^+) at an accelerating voltage of 900 V with an ion current density to a substrate of 6 mA/cm^2. The substrates were located perpendicular to the ion source and were spaced from it by 220 mm. The time of ion beam pretreatment was 2 mins during which the radiation dose at the NiTi substrates reached $4.5 \cdot 10^{21}$ ion/cm^2. Simultaneously, the Mo heater heated the substrates to 300 °C.

After ion beam pretreatment, the substrates were deposited with a Si coating on the magnetron sputtering system with a Si target of diameter 100 mm located parallel to the substrates at a distance of 60 mm. The time interval between ion beam irradiation and coating deposition was 1–2 min. The average power of the magnetron sputtering system was 1 kW at a pulse frequency of 50 kHz and pulse duration of 16 μs. The working gas was argon at a pressure of 0.3 Pa; the negative bias applied to the substrate was 100 V, and the substrate temperature was 300 °C.

The experiments with gas discharge plasma pretreatment were performed on a SPRUT technological vacuum plasma setup developed at Tomsk State University (Russia) [5].

The main plasma device of the setup is its plasma generator with thermionic (hot) cathode units located on the diametrically opposite flanges of a cylindrical vacuum chamber measuring 0.7 m^3 in volume. Applying a voltage between the hot cathodes and grounded vacuum chamber with gas supply to the cathode cavities at the same flow rate causes the ignition of a non-self-sustained arc discharge the operation of which is ensured by thermionic emission from the hot cathodes. The working volume of the chamber is thus uniformly filled with "shadow-free" bulk gas plasma (e.g., Ar plasma) which provides efficient treatment (cleaning, etching, heating, etc.) of all sides of articles immersed in the plasma. The density of the gas discharge plasma produced by the generator can be varied from 10^8 to 10^{11} cm^{-3} over wide operating pressure range (0.13–0.67 Pa) by varying the discharge current from 10 to 250 A. The setup includes four magnetron sputtering systems equally spaced over the lateral surface of the vacuum chamber pumped with a cryogenic pump rated at 5000 l/s.

Once the NiTi specimens were placed in the vacuum chamber of the SPRUT setup, the chamber was pumped to an ultimate residual vacuum of $6 \cdot 10^{-4}$ Pa and was checked for possible air leak-in. The leak-in permissible for vacuum plasma treatment was no more than $3.4 \cdot 10^{-5}$ Pa·m^3/s (0.02 sccm); this value is determined by specifications of the setup and excludes the presence of excess gas impurities (oxygen, carbon, etc.) in the vacuum chamber. The specimens were subjected to finish surface cleaning in the argon plasma in which they were heated to the desired temperature by varying both their negative bias and the plasma density. The gas discharge plasma produced in the setup provides highly efficient ion cleaning such that the only preliminary treatment of the specimens outside the vacuum was their rinsing in benzine and ethanol. In the vacuum chamber of the SPRUT setup, the specimens were arranged on special holders, allowing accurate measurements of their temperature with a chromel-alumel thermocouple.

The treatment of the NiTi specimens on the SPRUT setup included ion plasma cleaning of their surface in the gas discharge plasma of high-purity argon (99.998%) at a discharge current of 20 A and operating pressure of 0.3 Pa. For extraction of Ar ions to the specimen surface, the specimens were biased with respect to the anode (vacuum chamber) by applying a negative pulsed bias with a pulse duration of 17 μs, pulse repetition frequency of 30 kHz, and amplitude of up to 400 V. Thus, the ion current density at the specimen surface during the pulses was 0.3 mA/cm^2, ensuring efficient surface cleaning and heating of the specimens by Ar ions to T = 90 °C in 30 min.

The same bias parameters but with an amplitude of 200 V were used for deposition of Si for which the plasma generator was turned off and four magnetron sputtering systems with pure silicon targets were simultaneously turned on at the same Ar pressure in the vacuum chamber. The operation of four unbalanced magnetron sputtering systems with a diffused power of 0.2 kW at each target ensured the generation of plasma containing Ar and Si ions in the specimen region at a distance of 440 mm from the systems. At the bias used, the ion current density from the plasma to the specimen surface was 0.4 mA/cm^2. During deposition for 90 min, the specimen temperature increased to T=150 °C.

The properties of the coatings deposited after two types of preliminary surface treatment were studied on analytical equipment of the Material Properties Measurements Centre of the Institute of Physics and Technology at Tomsk Polytechnic University (Russia). The adhesion properties of the coatings were studied in scratch tests on a CSEM micro scratch tester with increasing load on an indenter with a rounding-off radius of 100 µm throughout its motion (track). The critical load in newtons (adhesion strength) at which the coating was separated from the substrate was determined from no less than six scratch tracks for each of the specimen by visual observation in an optical microscope. The coating thickness was measured using a rotating sphere on a Calotest device (CSM Instruments) with a Nikon microscope. The same method was used to analyze the character of fracture and the quality of coating adhesion.

3. Results and discussion

Figures 1 and 2 show images of the tracks produced by the scratch tester in studying the adhesion strength of the coatings deposited after pretreatment with ion beams and gas discharge plasma, respectively. It is seen that the critical load for coating separation (adhesion strength) in two cases is almost the same. The onset of fracture and separation of the coatings in the form of islands (Figure 1a) and bands (Figure 2a) is observed at an indenter load of 16 N.

The indenter load at which the coatings are fully separated (critical load) is equal, on average, to 19 N for ion beam treatment (Figure 1b) and to 20 N for discharge plasma treatment (Figure 2b). The results of scratch testing demonstrate that the coatings deposited after gas discharge plasma treatment on the SPRUT setup have higher adhesion than the coatings deposited after ion beam treatment, though the difference is very slight. By and large, the coatings of both types feature high adhesion properties compared, e.g., to antifriction coatings [6] whose adhesion estimated with the same device is no greater than 10 N.

It is seen from Figures 1 and 2 that the coatings obtained by two methods differ in their fracture (separation). The coatings deposited after gas discharge plasma treatment are separated from the substrates as such (Figure 2). In the specimens coated after ion beam treatment, there is a film between the substrate and the coating (Figure 1). The film has likely the same thickness throughout the interface and is separated in portions baring the substrate surface.

The foregoing allows the conclusion that the substrate surface is strongly activated by high-dose Ni ion implantation, and this, together with the high substrate temperature after the process (300 °C), time interval between ion activation and deposition, and uncontrollable composition of the gas atmosphere, can lead to the formation of an oxide film. The adhesion of the oxide film to the substrate is higher than the adhesion of the Si coating to the film. As can be understood from Figure 1, the Si coating was stripped at lower indenter loads before the film was detected, and the critical load 16 and 19 N determines the adhesion strength of the intermediate oxide film to the substrate in view of its successive separation in portions of equal thickness. The above conclusion is confirmed by spherical abrasion tests demonstrating different patterns of abrasion for two coating types at the same coating thickness (0.6 µm). Figure 3a shows an image of the abraded Si coating which was deposited after Ni ion implantation. It is seen that the abrasion reveals a dark layer with a "ragged" boundary between the Si coating and the substrate.

IOP Conf. Series: Journal of Physics: Conf. Series **830** (2017) 012105 doi:10.1088/1742-6596/830/1/012105

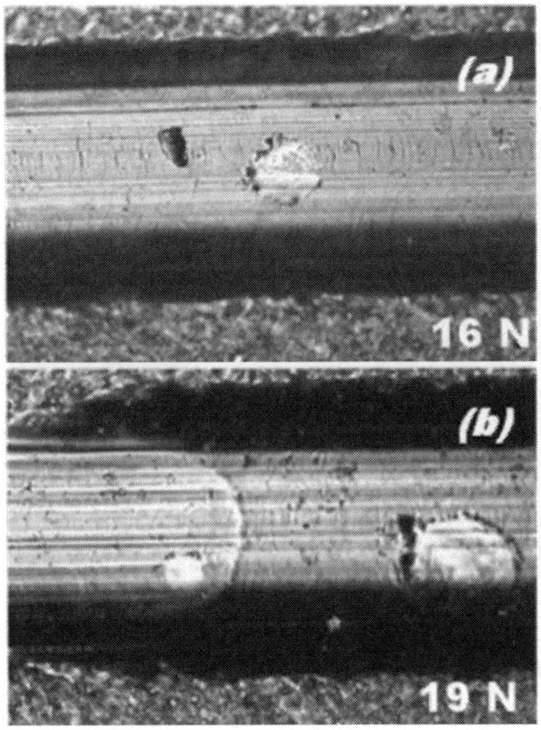

Figure 1. Indenter tracks for the Si coatings deposited after pretreatment of the substrates with ion beams.

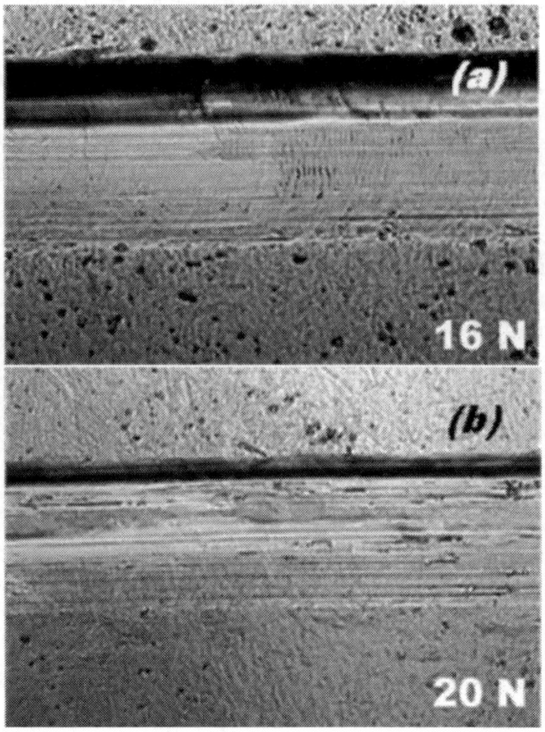

Figure 2. Indenter tracks for the Si coatings deposited after pretreatment of the substrates with gas discharge plasma.

At the same time, the coating deposited after preliminary plasma treatment on the SPRUT setup (Figure 3b) reveals a perfectly smooth abraded crater of uniform color and regular circular shape with tight attachment of the coating to the substrate.

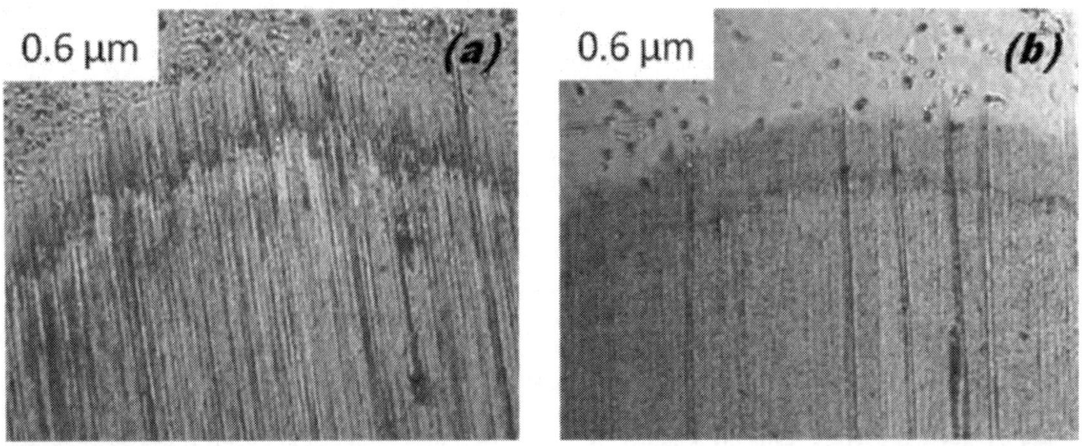

Figure 3. Images of the spherically abraded Si coatings (Calotest) on the NiTi substrates pretreated with ion beams *(a)* and gas discharge plasma *(b)*.

The results of spherical abrasion tests also demonstrate that of the two methods of coating deposition, the SPRUT setup provides higher adhesion at the same surface roughness of the specimens before pretreatment and coating deposition. The most probable cause of the high adhesion of the ion plasma coatings deposited on the SPRUT setup consist in stringent requirements on vacuum plasma processes, low residual pressures in the vacuum chamber, and low leak-in for operation with high-purity argon, and these factors provide highly efficient surface cleaning without any contamination by oxygen, carbon, and their compounds.

4. Conclusion

By and large, both methods of deposition of Si coatings on NiTi substrates demonstrate high coating adhesion. The higher quality and the higher adhesion of the coatings deposited with preliminary plasma cleaning and surface activation of the substrates suggests efficiency of this method of coating deposition. In particular, the plasma method offers advantages over ion beam irradiation for treatment of articles of complex configuration, and this opens up great opportunities for its use in developing efficient technologies of treatment, including treatment of medical materials.

References

[1] Stoeckel D, Pelton A, and Duerig T 2004 *Eur. Radiol.* **2**(14) 292
[2] Meisner L L, Lotkov A I, Matveeva V A et al. 2012 *Adv. Mater. Sci. Eng.* doi: 10.1155/2012/706094
[3] Psakhie S G, Lotkov A I, Meisner L L et al. 2013 *Rus. Phys. J.* **55**(9) 1063
[4] Kashin O A, Borisov D P, Lotkov A I et al. 2015 *AIP Conf. Proc.* **1683** 020077
[5] Borisov D P, Koval N N, Kuznetsov V M et al. 2013 *IEEE Trans. Plasma Sci.* **41** 2183
[6] Potekaev A I, Savostikov V M, Tabachenko A N et al. 2015 *J. of Phys.: Conf. ser.* **652** 012058

Acknowledgments

This study (state task No.11.1655.2014/K) was supported by the Ministry of Education and Science of the Russian Federation.

IOP Conf. Series: Journal of Physics: Conf. Series **830** (2017) 012106 doi:10.1088/1742-6596/830/1/012106

Solid oxide fuel cell anode surface modification by magnetron sputtering of NiO/YSZ thin film

A A Solovyev[1,2], A V Shipilova[1,2], I V Ionov[1,2], E A Smolyanskiy[1], A L Lauk[1], A N Kovalchuk[1,2], G E Remnev[1] and A M Lebedynskiy[1]

[1] Tomsk Polytechnic University, 30 Lenin Ave., Tomsk, 634050, Russia
[2] Institute of High Current Electronics SB RAS, 2/3 Akademichesky Ave., Tomsk, 634055, Russia

E-mail: andrewsol@mail.ru

Abstract. NiO/ZrO_2-Y_2O_3 (NiO/YSZ) anode functional layers (AFL) with 16-60 vol.% of NiO were deposited onto NiO/YSZ anode substrates by magnetron sputtering, followed by annealing in air at 1200 °C. The optimal deposition conditions for NiO/YSZ were determined. NiO content in the films was varied by changing the oxygen flow rate during the sputtering process. The microstructure and phase composition of NiO/YSZ anode functional layer were studied by SEM and XRD methods. Anode functional layers were fully crystallized and comprised of grains up to 500 nm in diameter after reduction in hydrogen. Anode-supported solid oxide fuel cells (SOFC) with the diameter of 20 mm including the magnetron sputtered AFL, 4-microns thick YSZ electrolyte and $La_{0.6}Sr_{0.4}Co_{0.2}Fe_{0.8}O_3$/$Ce_{0.9}Gd_{0.1}O_2$ (LSCF/CGO) cathode were fabricated and tested. Electrochemical properties of the single fuel cells were investigated as a function of NiO volume content in AFL and AFL thickness.

1. Introduction

A ceramic-metal (cermet) composite of Ni and yttria-stabilized zirconia (YSZ) is the most common anode material of solid oxide fuel cells (SOFC) operating at a temperature of 800-1000 °C. However, the desire to create a commercial SOFC requires to mitigate the problem, associated with a high operating temperature. In turn, the lower operating temperature would increase polarization of electrodes and ohmic resistance of electrolyte. The ohmic resistance can be decreased either by using ultra-thin electrolytes, or new materials with higher ionic conductivity at lower temperatures [1, 2]. Polarization losses of the anode can be reduced by creating new alternative materials or by nanostructuring commonly used NiO-YSZ cermet [3, 4]. The second approach seems to be the most promising for improving the SOFC characteristics. First, conventional NiO/YSZ cermet is cheaper than other materials, chemically stable in a reducing atmosphere at high temperatures and has thermal expansion coefficient close to that of YSZ electrolyte [5]. Second, it is known that the electrochemical reaction at the anode occurs at a three-phase boundary (TPB), the area of contact between the metal, the electrolyte and the working gas. Therefore, the electrode reaction rate and, hence, its efficiency is determined by the TPB length. The formation of nanoporous anode structure will significantly increase the TPB length. At that, the nanostructuring of the whole anode substrate is not required, since the coarse anode facilitates the rapid diffusion of fuel gas to the active reaction area and removal of reaction products out from the anode [6]. Formation of the nanostructured anode functional layer

(AFL) will lead to an increase of the TPB length of an anode and to a decrease of activation polarization of the anode [7].

Among many methods used to prepare the anode functional layer, the greatest success in terms of the nanostructures formation was achieved by methods of physical vapor deposition (PVD), such as pulsed laser deposition [8], magnetron sputtering [9, 10], electron beam evaporation [11] and others. Among all, magnetron sputtering is characterized by several advantages. By changing deposition parameters such as substrate temperature, working pressure, angle of incidence of the deposited particles, etc. both dense and porous coating structures can be formed. It is also possible to adjust the number, size and shape of pores in the deposited films.

In our recent work [12], we have used magnetron deposition technique to fabricate nanocomposite NiO/YSZ anode layer onto YSZ electrolyte support. As a result of nanocomposite NiO/YSZ anode deposition, the power density of the fuel cell was increased by two times, as compared to a cell with the anode fabricated by conventional screen printing technique. But such cells had high ohmic losses due to the large electrolyte thickness. Therefore, in the present work, we suggest forming the nanoporous AFL for the anode-supported SOFC with a thin electrolyte. AFL should also serve as an interface layer between the highly porous anode substrate and thin gas-tight electrolyte.

2. Methodology and materials

NiO/YSZ thin films were deposited onto NiO/YSZ substrates by reactive magnetron co-sputtering of Ni (99.995% purity) and Zr/Y (85/15 at.%, 99% purity) targets. The substrates with a diameter of 20 mm and 0.4-mm-thick were laser cut from commercial porous NiO/YSZ anodes for SOFCs (Ningbo SOFCMAN Energy Technology Co., China). The Ni and Zr-Y targets were inclined at an angle of 45° to the substrate holder. The vacuum chamber was evacuated to the base pressure of 0.01 Pa using a turbo-molecular pump. The substrates were heated up to 370 °C and the temperature was maintained during the sputtering process. Argon and oxygen were introduced into the vacuum chamber with the flow rate of 30 sccm and 9–16.5 sccm, correspondingly, at a working pressure of 0.2 Pa. To improve films adhesion before the deposition, the substrate surface was cleaned for 30 seconds by ion beam at 1000 V and 15 mA generated by an ion source with a closed electron drift. Prior to deposition, the targets were pre-sputtered for two minutes with a shutter covering the substrates in order to remove any contaminant on the surface of the targets and to ensure stable sputtering conditions. Ni target was supplied with a pulsed DC power supply (Applied Electronics, Ltd., Russia). The average power was 350-500 W at a frequency of 80 kHz. The bipolar pulse powering (Applied Electronics, Ltd., Russia) was used for Zr/Y target sputtering with pulse repetition frequency of 75 kHz, positive pulse duration of 4 μs and average power of 1500 W. After deposition, the samples were kept in the vacuum chamber until substrate temperature dropped to 150 °C.

The obtained samples were annealed in air at 1200°C for 1 hour to prevent Ni agglomeration in the film during subsequent use under hydrogen atmosphere. The heating/cooling rate was 200 °C/h. Undesirable massive metal phase agglomeration was observed after the reduction or thermal treatments of the cermet composites in the form of the thin film regardless of the deposition methods [13]. Ni agglomeration leads to the loss of connectivity and poor high temperature stability of electrodes. The method of suppression of Ni agglomeration in Ni/YSZ films by post-annealing was first used by Noh et. al. [14].

To study the electrochemical activity of NiO/YSZ, thin films SOFCs including YSZ electrolyte and $La_{0.6}Sr_{0.4}Co_{0.2}Fe_{0.8}O_3/Ce_{0.9}Gd_{0.1}O_2$ (LSCF/CGO) cathode were manufactured. 4 μm-thick YSZ electrolyte was deposited by reactive magnetron sputtering of Zr/Y target. Sputtering was also carried out using a bipolar pulsed power supply. We have previously shown that the bipolar pulsed magnetron sputtering can significantly increase the deposition rate of oxide coatings, as compared with unipolar pulsed DC or DC mode [15].

The LSCF/CGO cathode layer was prepared on the electrolyte surface by a screen printing technique using $La_{0.6}Sr_{0.4}Co_{0.2}Fe_{0.8}O_3/Ce_{0.9}Gd_{0.1}O$ paste (CERA-FC Co., Korea). The cathode area was 1 cm^2. Cathode was sintered at 800°C during the SOFC electrochemical test.

Electrochemical tests were performed using ProboStat^{TM} device (NorECs, Norway). Details about the single cell test bench and testing procedures were given in our previous work [16]. The scanning electron microscope Quanta 3D 200 was used to study the cross section of the deposited films and to determine their thickness. The NiO content in the film was determined by X-ray diffractometry (XRD-6000 diffractometer with Cu K_α radiation). To minimize the contribution related to the substrate, the technique of X-ray diffraction at grazing incidence was used.

3. Results and discussion

In general, conventional SOFC anodes are made by mixing NiO and YSZ at a certain ratio. It is desirable to have NiO content as high as possible to improve the performance of the anodes. But high NiO content makes it impossible to match its thermal expansion to that of the YSZ. Change of the NiO content in the NiO/YSZ film deposited by co-sputtering of two targets is possible either by varying the power supplied to the Ni target (at constant gas flow rates and constant power to Zr/Y target), or by changing the oxygen flow rate (at constant power supplied to both targets). The second method was chosen in this work. The main problem of the simultaneous reactive sputtering of different targets consist in the fact that the increase in the discharge power at one of them results in a decrease in the partial pressure of oxygen in the chamber, which, in its turn, causes the change in the sputtering conditions of the other target. It is known that the deposition rate during reactive magnetron sputtering strongly depends on the target surface conditions. The target can become oxidized (poisoned) by the oxygen which reduces the amount of sputtered material. Therefore, the oxygen flow rate has to be precisely controlled for better results.

The deposition rate (μm/h) and NiO content in the film are highly dependent on the oxygen flow rate and deposition time at the growth conditions as shown in Figure 1. At oxygen flow rate of 9 sccm, the deposition rate of NiO/YSZ film is relatively high and is equal to 30 μm/h. This is transition region between metal and oxide sputtering modes. Above 9 sccm, deposition rate starts to fall abruptly but after 12 sccm the fall becomes slow. This means the process enters the oxide sputtering mode. In this mode, targets were oxidized, and the deposition rate was noticeably lower (3.6-8.6 μm/h).

NiO content in the film tends to decrease with increasing oxygen flow rate. A similar relationship was observed in [17] at sputtering of a single Ni/Zr/Y (68.8/26.6/4.6 wt.%) target. However, as can be seen from Figure 1, at the same oxygen flow rate and constant power, the NiO/YSZ film may contain different amount of NiO phase. This is because with increasing coating thickness, and, hence, deposition time there is a gradual decrease in the oxygen partial pressure in the chamber. This leads to the fact that the oxidation state of the sputtering targets is changed. In order to exclude this phenomenon, it is necessary to control the oxygen partial pressure, for example, by measuring the plasma composition by optical emission spectroscopy and controlling the oxygen flow rate by feedback control system [18].

Figure 2 presents XRD patterns of NiO/YSZ films with NiO content of 36, 45 and 62 vol.%. These samples were annealed at 1200 °C before measurements. The films exhibit a crystalline structure with clear indicated peaks. The intensity of NiO peaks increases with the decreasing oxygen flow rate, indicating an increase in the NiO content in the deposited films. According to data (curve 3) presented in Figure 2, the (400) YSZ reflection in X-ray pattern of the film with 62 vol.% NiO disappeared whereas the (311) NiO reflection intensity considerably increased. Interestingly, the increasing NiO content in the films leads to (111) YSZ reflection intensity increase, while intensity of YSZ (200) reflection decreases. Thus, textural evolution of YSZ phase is observed which is the transition from [200] out-of-plane preferential orientation to [111]. Lamas et. al. [19] showed that in magnetron sputtered YSZ films change of the orientation from [200] to a complete [111] out-of-plane orientation is due to the competition between the fastest growth direction and the lowest surface energy.

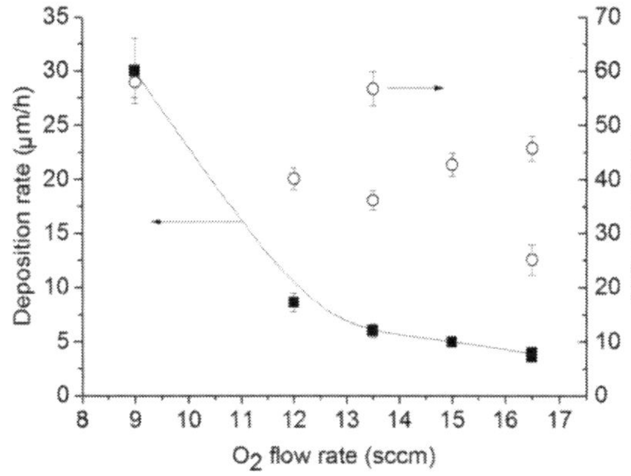

Figure 1. Effect of oxygen flow rate on deposition rate and NiO content of the NiO/YSZ films.

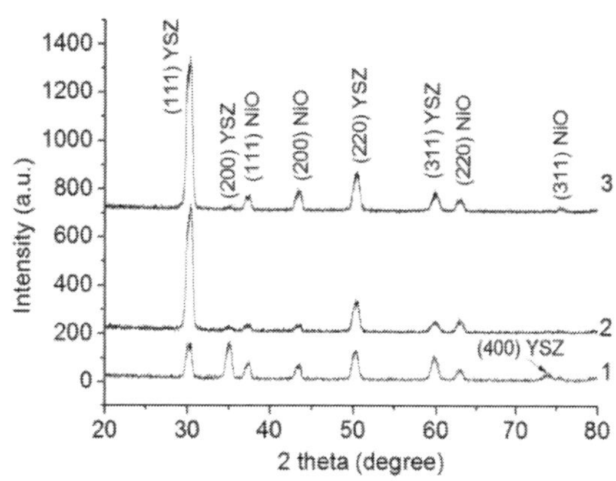

Figure 2. X-ray diffraction patterns of the magnetron sputtered NiO/YSZ films with different NiO content: 1) 36 vol.%, 2) 45 vol.%, 3) 62 vol.%.

The dependence of the NiO/YSZ film microstructure on the NiO content is shown in Figure 3.

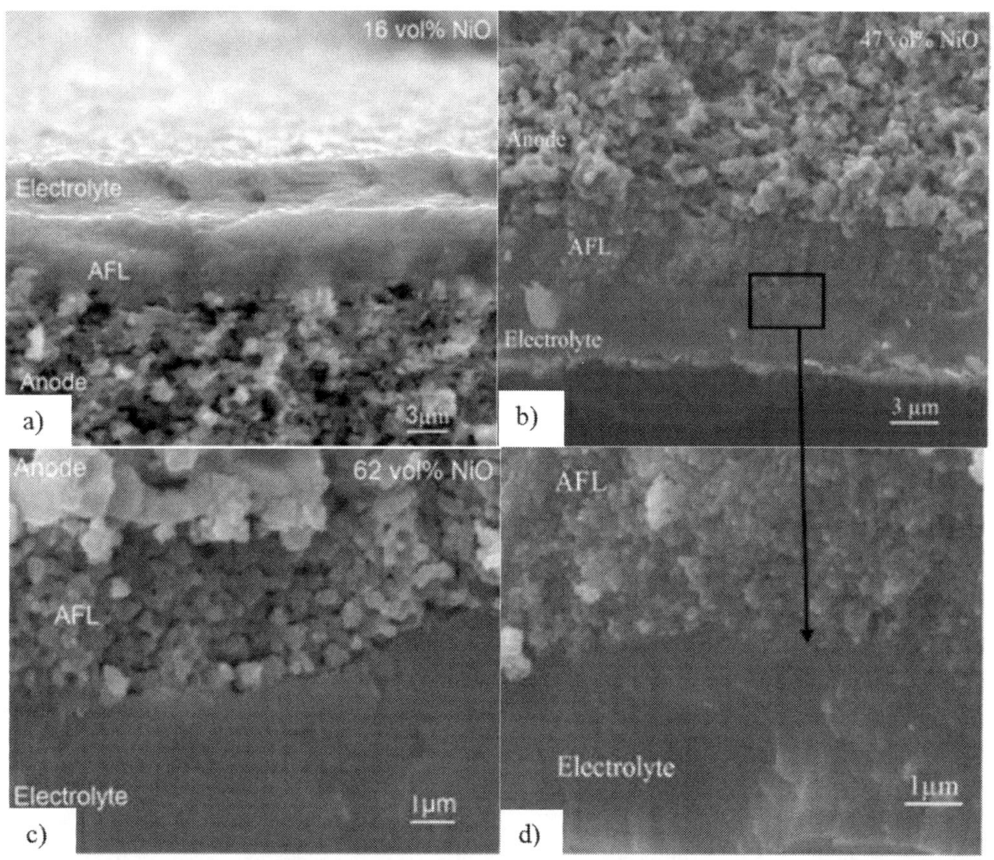

Figure 3. Cross sectional SEM images of the SOFCs with nanostructured anode functional layer with NiO content: a) 16 vol.%; b) 47 vol.%; c) 62 vol.% and d) is the magnified image of the electrolyte and anode interface marked in Figure 3b. SEM images were obtained after cells testing and NiO reduction in hydrogen.

IOP Conf. Series: Journal of Physics: Conf. Series **830** (2017) 012106 doi:10.1088/1742-6596/830/1/012106

It can be seen that AFL with the NiO content of 16 vol.% has the microstructure, which is slightly different from the microstructure of the YSZ electrolyte. A clear boundary between the NiO/YSZ film and the electrolyte is not distinguishable because AFL has quite dense structure. The porosity of NiO/YSZ layer, as expected, increases with increasing NiO content in the layer up to 62 vol.%. For the sample with 47 vol.% of NiO in AFL, the pore size is in the range of 50-100 nm, and the grain size is less than 200 nm. Pore and grain size for the sample with 62 vol.% of NiO is about 50-400 nm, and 150-600 nm, respectively. For all samples, there is a good contact of the AFL both with an anode substrate and an electrolyte layer. The change in anode substrate surface porosity and roughness by the formation of the NiO/YSZ AFL is clearly observable where the porous and rough anode surface (Figure 3) displays relatively large voids formed from large NiO and YSZ particles. AFL has a relatively smooth and uniform surface for subsequent deposition of a high quality YSZ electrolyte layer. The analysis of the microstructure of the YSZ electrolyte films showed that the nanoporous anode functional layer provides the conditions for the formation of thin homogeneous and dense YSZ electrolyte films without columnar grains (Figure 3), which is typical for magnetron sputtered films.

In order to estimate the impact of the anode functional layer on the fuel cell performance, the voltage-current (V-I) and power-current (P-I) characteristics of the anode-supported SOFC with different AFLs were measured at 700 °C (Figure 4). AFLs with different NiO content and thickness were used. The results presented in Figure 4 shown that power density increases almost twice with increasing of NiO content in the AFL from 16 up to 60 vol.%. Also, the thickness of AFL significantly impacts the fuel cell performance. The optimum thickness of deposited nanoporous layer from the point of view of power density is equal to about 4 μm. At the larger thickness of the AFL the diffusion losses begin to increase due to the small size of the gas transport channels. At the layer thickness of 1-2 μm the length of the three-phase boundary on the anode increases slightly and therefore, power density of the cells is significantly lower.

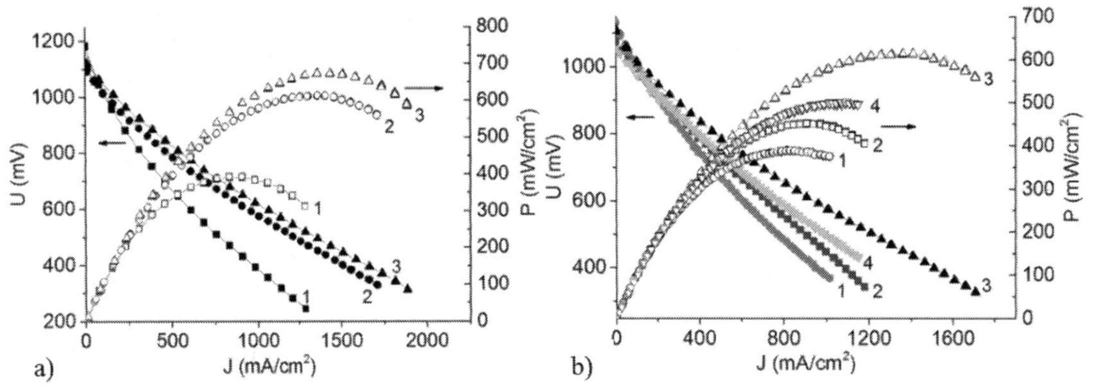

Figure 4. V-I and P-I characteristics of the anode-supported SOFC with magnetron sputtered NiO/YSZ AFL measured at 700 °C: a) AFL with different NiO content, 1) 16 vol.%, 2) 47 vol.%, 3) 60 vol.%, and thickness of 3.5 μm; b) AFL with different thickness, 1) 1.3 μm, 2) 2.2 μm, 3) 4.3 μm, 4) 9.5 μm, and NiO content of 50 vol.%.

4. Conclusion

NiO/YSZ thin films with different NiO content and thickness were deposited by reactive magnetron co-sputtering of Ni and Zr-Y targets. The films were annealed at 1200 °C in air to prevent Ni agglomeration in hydrogen atmosphere. NiO/YSZ films with NiO content from 16 to 60 vol.% and thickness from 1.3 to 9.5 μm were deposited and tested as anode functional layers of anode-supported SOFCs. It was shown that the increase of NiO content in the film from 16 up to 60 vol.% leads to a nearly twofold increase of SOFC power density. It was also found that SOFC maximum power density was achieved at the NiO/YSZ film thickness of 4 μm.

References

[1] Souza S, Visco S, De Jonghe L 1997 *Solid State Ionics* **98** 57-61
[2] Xia C, Liu M 2001 *Solid State Ionics* **144** 249-55
[3] Kennouche D, Hong J, Noh H, Son J, Barnett S 2014 *Phys. Chem. Chem. Phys.* **16(29)** 15249
[4] Noh H, Yoon K, Kim B, Je H, Lee H, Lee J, Son J 2014 *J. Power Sources* **247** 105-11
[5] Zhu W, Deevi S 2003 *Materials Science and Engineering A* **362** 228-39
[6] Tsipis E, Kharton V 2008 *J. Solid State Electrochem.* **12**, 1367-91
[7] Chen K, Chen X, Lü Z, Ai N, Huang X, Su W 2008 *Electrochim. Acta.* **53** 7825-30
[8] Noh H, Son J, Lee H, Song H, Lee H, Lee J 2009 *J. Electrochem. Soc.* **156(12)** 1484
[9] Garcia-Garcia F, Yubero F, González-Elipe A, Balomenou S, Tsiplakides D, Petrakopoulou I, Lambert R 2015 *Int. J. Hydrogen Energy* **40(23)** 7382-87
[10] Garcia-Garcia F, Yubero F, Espinós J, González-Elipe A, Lambert R 2016 *J. Power Sources* **324** 679-86
[11] Meng B, Sun Y, He X, Li M 2008 *Mater. Sci. Technol.* **24** 997-1001
[12] Solov'ev A, Sochugov N, Ionov I, Shipilova A, Koval'chuk A 2014 *Russ. J. Electrochem.* **50** 647-55
[13] Jou S, Wu T 2008 *J. Phys. Chem. Solids* **69** 2804-12
[14] Noh H, Son J, Lee H, Ji H, Lee J, Lee H 2010 *J. Eur. Ceram. Soc.* **30** 3415-23
[15] Solovyev A, Sochugov N, Rabotkin S, Shipilova A, Ionov I, Kovalchuk A, Borduleva A 2014 *Appl. Surf. Sci.* **310** 272-7
[16] Solov'ev A, Sochugov N, Shipilova A, Efimova K, Tumashevskaya A 2011 *Russ. J. Electrochem.* **47** 494-502
[17] Rezugina E, Thomann A, Hidalgo H, Brault P, Dolique V, Tessier Y 2010 *Surf. Coatings Technol.* **204** 2376-80
[18] Hirohata K, Nishi Y, Tsukamoto N, Oka N, Sato Y, Yamamoto I, Shigesato Y 2010 *Thin Solid Films* **518** 2980-3
[19] Lamas J, Leroy W, Depla D 2012 *Thin Solid Films* **520(14)** 4782-5

Acknowledgments

The research was carried out with the financial support of the Russian Science Foundation (grant No. 16-19-00089).

Study of the generation of a weakly ionized medium for spraying of biocompatible coatings by two-stage pulsed plasma set

P A Tsygankov, A S Skriabin, E Yu Loktionov, K Yu Kuvaev, and R I Chelmodeev

Department of Power Engineering, Bauman Moscow State Technical University, Baumanskaya 2-ya 5/1, Moscow, 105005, Russia

E-mail: terra107@yandex.ru

Abstract. The paper is devoted to the experimental study of the generation of the weakly ionized high velocity heterophasic flows by gas detonation in acetylene-oxygen mixtures. Such flow can contain fine hydroxyapapite (HOA) particles with an average diameter of $d_p \leq 100$ μm for spraying of biocompatible coatings. The characteristic velocity (≈2.3…2.5 km/s) and pressure (≈3.1…3.8 MPa) of the detonation waves were defined and the attenuation dependences of those parameters were stated. The structure of the detonation flows was visualized and the chemical composition was estimated. The samples of the HOA coatings (with a thickness of 80…100 μm) were prepared on the carbon nanocomposites and their properties were characterized by SEM, EDX and XRD. The possibility of the using of such flow as a working fluid for electromagnetic acceleration of HOA was offered by the two-stage pulsed plasma set.

1. Introduction

Nowadays the using of the two-stage pulsed plasma sets [1] for coating spraying systems is under consideration. A weakly ionized (with a degree of ionization of $\approx 10^{-4} \ldots 10^{-3}$) detonation wave (DW) is formed by gaseous detonation [2] within a cylindrical barrel at the first stage. Additional acceleration of the detonation products takes place due to discharging of capacities into this wave during the second stage within a coaxial electrode unit. Owing to Joule heating generation of a conductive plasma channel with a temperature of 20…30 kK occurs. The channel expends [3] with a supersonic velocity and generate of a shock wave. The using of the preliminary detonation allows decreasing energy consumption for this acceleration within the coaxial electrode unit. The particles which were injected before the wave front are accelerated up to velocities of 1.4…2.0 km/s and transferred to a substrate where they form a coating.

Note that the complications of the flows generation processes with non organic high-molecular particles can limit the efficiency. As an example of this method spraying biocompatible coatings [4] from hydroxyapatite (HOA) $Ca_{10}(PO_4)_6(OH)_2$ was studied in this paper. As substrates the carbon nanocomposites [5] were used. The detonation with the regime of weak wave generation can decrease the dynamic loads on to the substrate and avoid its crash.

2. Experimental set

Based on commercial CCDS 2000 equipment [2] the experimental set working with C_2H_2 and O_2 mixtures as the fuel was designed (see figure 1). The facility "Beam-M" [6] was used for the flow characterization. During the experiments the fuel composition was stoichiometric. The generation of the detonation flows was occurred within a cylindrical barrel (1) with the diameter of d=1.6 cm by a spark (2). The set was equipped by a gas feed system (3). The moving of non-dusty flows and heterophasic (with an average diameter of HOA particles of d_p≈70 nm and 100 μm) flows was studied. An injection of HOA was realized into the direction normal to the barrel axe by a pneumatic feeder (4). A volume flow rate of HOA was about of 0.3…1.0 mm^3 per a shot. Piezometric sensors were used for control of the flow dynamics. The one (5) was built in the barrel at the injection zone and the other (6) was embedded into a substrate (7). The distance δ between the muzzle and the substrate was varied from 2.5 to 23 cm. The distance l between the injection zone and the muzzle was l=40 cm. The data recording from the sensors was realized by a storage oscilloscope (8) with the using of a syncronizing signal from the spark. The flow structure was visualized by a laser Schlieren photography. A Nd:YAG laser (10) with the wavelength of λ=532 nm was applied as a source of the radiation that was fixed by a high-speed camera (11). The polarizing filter (12) was used for the regulation of an intensity of the laser radiation. The system consisting of an interference filter (13) with λ_{max}=532 nm and Δλ=10 nm and a diaphragm (14) was used for cutoff of the flow (9) radiation and visualization of flow disturbances. Producing of a collimated laser beam was done by lenses (15) and (16). Delay of a picture capturing by the camera was set relativity the sensor (5) by BNC 575 synchronizing unit (17). A part of the flow radiation was transmitted by prism (18) to Solar SC125 spectrometer (19) for its characterization.

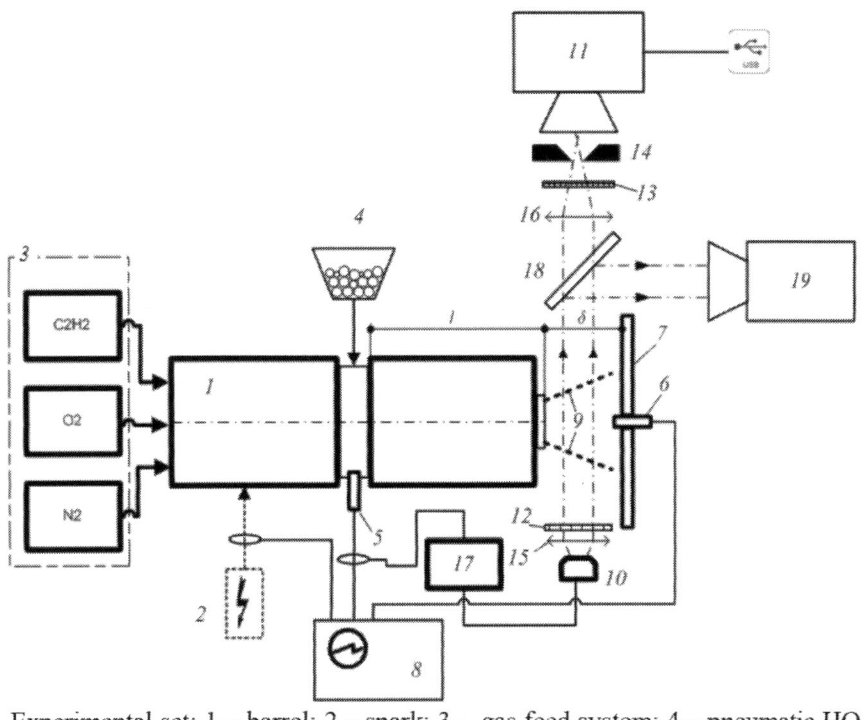

Figure 1. Experimental set: 1 – barrel; 2 – spark; 3 - gas feed system; 4 – pneumatic HOA feeder; 5 and 6 – piezometric sensors; 7 – substrate; 8 – oscilloscope; 9 – detonation flow; 10 - Nd:YAG laser; 11 – high-speed camera; 12 - polarizing filter; 13 - interference filter; 14 – diaphragm; 15 and 16 – lenses; 17 - BNC 575 synchronizing unit; 18 – prism; 19 - Solar SC125 spectrometer.

IOP Conf. Series: Journal of Physics: Conf. Series **830** (2017) 012107 doi:10.1088/1742-6596/830/1/012107

3. Results and discussion

Spectroscopy (in the wavelength range of 150…1250 nm) of the non-dusty flows at δ/l=3.13 showed (see figure 2) the presence of the intensive lines of atoms and first ions of C, O and atomic H. In a case of HOA injection with $d_p \approx 70$ nm the specter contained also further the lines of atoms and ions of P and Ca. Furthermore, increasing of the intensity of atomic O was detected. Such circumstances evidence about partial pyrolysis of HOA.

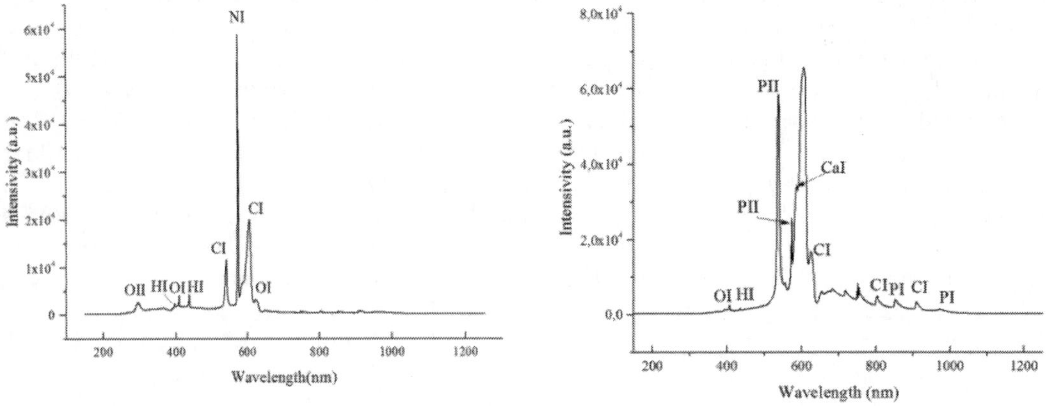

a b

Figure 2. Spectra of non-dusty flow (a) and heterophasic flow (b) with $d_p \approx 70$ nm.

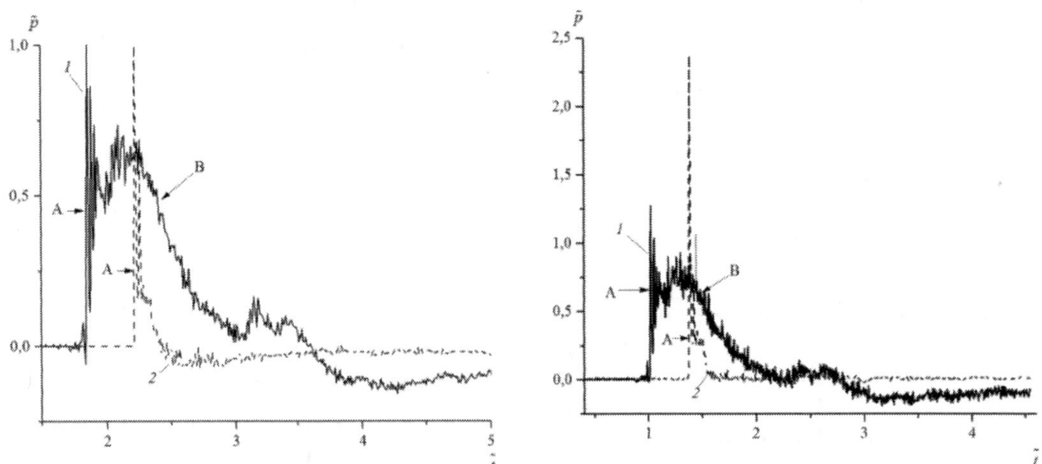

a b

Figure 3. Dynamics of pressure for non-dusty flow (a) and heterophasic flow with $d_p \approx 70$ nm (b): 1 - $\tilde{p}(\tilde{t})$ at the injection zone; 2 - $\tilde{p}(\tilde{t})$ at the distance of δ/l from the muzzle.

Data from the piezometric sensor (5) demonstrated (see figure 3) registration of the strong DW (A) which propagates and compresses the non-disturbed fuel-oxygen mixture. Thereby the initiation of intensive combustion occurs which was indicated as pressure decreasing (B). In figure 4 the dependences of dimensionless pressure $\tilde{p}(\tilde{t})$ are presented. Here $\tilde{p} = p / p_0$ and $\tilde{t} = t / t_0$ is dimensionless time, where $p_0 = 3.1$ MPa and $t_0 = 440$ μs. It was shown that the amplitude $\tilde{p}(\tilde{t})$ of the heterophasic DW within the injection zone is ≈1.1…1.25 times higher than of non-dusty flows because of the turbulization.

The sensor (6) demonstrated that DW transforms into a strong shock wave (SW) at the muzzle. The SW attenuates with δ/l increasing and becomes weak at $\delta/l>7$. The characteristic values of heterophasic flow velocities ($D_0\approx2.3…2.5$ km/s) at the muzzle were estimated. Note that for the flows with HOA the pressure pulse is 1.5…3.5 times higher owning to the action of HOA particles on to the substrate.

Schlieren photography demonstrated the features of structure and dynamics of the non-dusty flows and the heterophasic flows with HOA particles ($d_p\approx70$ nm). It was shown the absence of influence of such particles on the DW structure at the provided volume flow rates. In figure 4 the photos of the detonation in flowing (from left to right) are presented at the varied delay time t_d. The diffraction of the wave front (1) at the muzzle (2), the shock-compressed mixture (3) of the fuel and oxygen and high brightness region (4) of combustion were visualized.

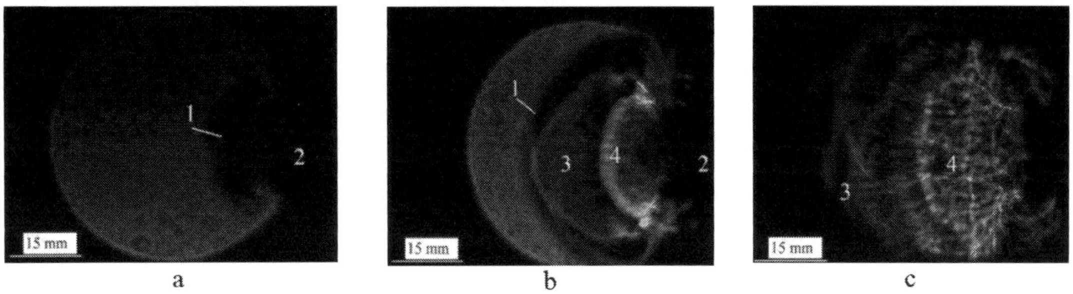

Figure 4. Dynamics of the flow structure at varied t_d: 1 – wave front; 2 – muzzle; 3 – shock-compressed layer and 4 – combustion region; a - t_d=170 μs; b – 210 μs; and c – 220 μs

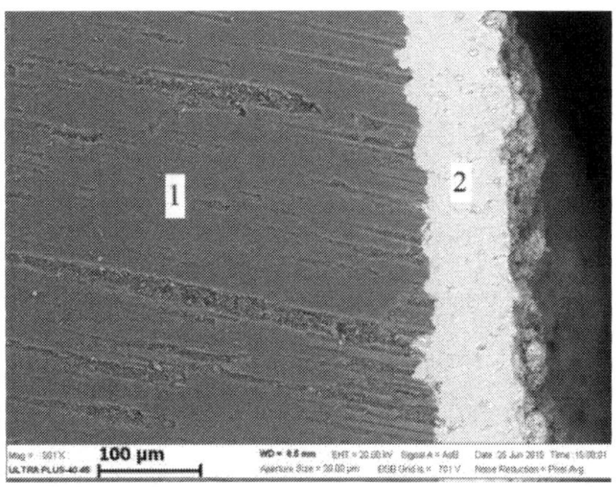

Figure 5. SEM image of the sample: 1 – carbon substrate; 2 – HOA film

With the using of the set the samples of the coatings (with thickness of 80…100 μm) on the carbon substrates were prepared. The coating morphology (see figure 5) was studied by SEM with Zeiss Ultra plus 55. This has a close packed structure without visible defects. EDX and XRD (DRON 3M diffractometer) showed that the coatings are polycrystalline with Ca and P concentration ratio of ≈1.67 that corresponds to raw HOA.

4. Conclusion

The interaction of the detonation flows with fine HOA particles was studied. It was shown that the produced flow is high velocity weak ionized medium. This medium is available for electromagnetic acceleration within the coaxial electrode unit of the two-stage pulsed plasma set for spraying of biocompatible coatings.

References

[1] Pogrebnjak A D and Tyurin Yu N 2005 *Usp. Fiz. Nauk* **48**(5) 515 – 544
[2] Nikolaev A Yu *et al* 2003 *Combust. Explo. Shock+.* **39**(4) 382-410
[3] Alekseev Yu A and Kazeev M N 1981 *Fiz. Plasmy* **7**(5) 1084-1095
[4] Nosenko V *et al* 2015 *Nanoscale Res. Lett.* **10** 464-469
[5] Gordeev S K *et al* 2002 Porouse composite for bone replacment and its preparation method *Russia patent* 2181600
[6] Loktionov E Yu *et al* 2013 *Instrum. Exp. Tech+.* **56**(1) 46-54.

Acknowledgments

This work was performed at the facility "Beam-M" (RFMEFI59014X0001), supported by the State Task of the Russian Ministry of Education and Science. Authors thank Sc. Doctors V.Yu. Ulianitskiy, N.S. Gavryushenko and N.A. Medik for the discussion and «NTM+» Ltd for the equipment.

Argon trapping in the depositing metal coating

L B Begrambekov, A V Grunin, Ya A Sadovskiy, N A Puntakov, N S Utkov, A M Zaharov

National Research Nuclear University MEPhI, 115409 Russia, Moscow, Kashirskoe sh.,31

E-mail: grunin@plasma.mephi.ru

Abstract. The paper has investigated conditions and parameters of argon trapping in molybdenum, tantalum and tungsten layers during their deposition on tungsten substrate by the atoms sputtered from the respective targets in argon plasma. The substrate temperature during deposition was 1273 K. The rate of deposition was 1 µm/h. It was shown that electron irradiation of the deposited layer with the beam intensity of 4 mA/cm^2 initiated argon trapping in tungsten and tantalum coating with approx. 2 x 10^{27} at/m^3 and 8 x 10^{26} at/cm^3, respectively, but did not stimulate argon trapping in the molybdenum layers. Features of argon trapping in the tungsten coating and its release are investigated in detail.

1. Introduction

Argon trapping in solids during ion and plasma irradiation, as well as in surface films deposited in plasma, has been mentioned in a number of articles [1-4]. The authors of paper [1] have reported argon trapping in tungsten during its irradiation by argon ion beam. Argon trapping and retention in tungsten irradiated by argon plasma ions was observed in [3]. Trapping of argon during coating deposition in plasma, which included tungsten and molybdenum layers, has been found in paper [2], while argon has been detected in carbon layers, irradiated by argon plasma after its formation.

It should be also noted that argon seeding into the divertor plasma is proposed as an option to reduce flux density of divertor irradiation for future ITER and DEMO thermonuclear facilities. Trapping of argon in the plasma-facing materials and deposited layers with its uncontrolled release is still a problem that has not been solved yet.

The development of surface properties modification by argon ion and argon plasma irradiation, as well as the solution of argon trapping in the plasma-facing nuclear fusion facility elements, is hindered by a lack of knowledge in the field of trapping and retention of argon and other noble gases in materials. The research of these processes is the problem presented in this work.

2. The choice of research subject and experiment conditions

In all papers listed above, argon trapping in metals [1–3] and carbon [4] occurred when the surface irradiation by argon atoms at thermal velocities was accompanied by the irradiation with accelerated argon ions (or ion beam as in [3]) or accelerated argon plasma ions and, consecutively, fast recharge atoms [1, 2, 4]. It can be assumed then that argon trapping is stimulated both by processes originating from kinetic energy of fast irradiating particles and by excitation of electron sub-system of surface layers by ion bombardment of the surface. At this stage, it was decided to limit the research to thermal velocity argon trapping in the metallic layer during deposition under fast electron irradiation. In this case, electron sub-system excitation occurs, and argon atom implantation into the surface due to kinetic energy is excluded.

Tungsten, tantalum and molybdenum layers of equal thickness were deposited on tungsten substrate in argon plasma. Argon trapping in these materials during deposition was studied. This choice of materials for deposition allowed comparing the trapping for different metals with high melting points, the same type of crystal lattice (BCC) and the same lattice parameter a = 0.3 nm, but with a number of distinct characteristics. Atomic numbers of tantalum and tungsten are similar, but tantalum has a remarkable oxide layer on the surface. Oxidation properties of tungsten and molybdenum are close, however, the atomic number of molybdenum is half of that of tungsten. The deposition process of aforementioned surface films was conducted with and without electron irradiation of the coating during deposition. The research on argon trapping by the deposited film for different electron flux intensity, deposition rate, substrate temperature and oxygen presence in plasma was then conducted for tungsten films.

3. Experimental

The coatings were deposited on the "COating Deposition and MAterial Testing" (CODMAT) rig. The CODMAT rig allows irradiating materials with high-density ion or electron fluxes and depositing coatings (Airapetov et al. [5]). The experiments were conducted with 15×7×1 mm tungsten substrates rinsed in an ultrasonic ethanol bath and then irradiated by argon ions before the deposition to provide proper coating adhesion.

The deposited layer was formed by atoms sputtered by ions of argon plasma from corresponding targets (tungsten, molybdenum and tantalum). The construction of the experimental setup allows irradiating the deposited film with electrons emitting from ion irradiated targets and accelerating towards the substrate. The use of anti-dynatron electrode allows varying or fully suppressing the electron flux from the target due to ion-electron emission.

The pressure of residual gas in the stand did not exceed 10^{-5} Pa, pressure of plasma-generated gas (argon) was equal to $4x10^{-2}$ Pa. The energy of target sputtering ions was 7 keV for most experiments. The deposition rate was 1 μm/h, and the thickness of the deposited film was 2.5 μm. The temperature of substrate should be taken as 1200 K unless specified.

The amount of argon trapped in the films was measured by the method of Thermal Desorption Spectrotry (TDS) on the Multifunctional Investigation Complex for Mass Analysis (MICMA) (Begrambekov et al. [6]). The TDS analysis was conducted at the temperatures of up to 1550 K with the heating rate of 5 K/s. The sample was heated by the radiation from tungsten spiral directly below it.

The amount of gas trapped was also determined by Energy-Dispersive X-ray Spectrometry (EDS) via Oxford X-act console installed on the Tescan Vega 3 Scanning Electron Microscope. The thickness of the deposited films was 2.5 μm, which significantly exceeded the maximum depth of the EDS analysis. FEI Scios dual-beam device was used for making a cross-section of the film and studying its structure.

4. Results and discussion

Deposition of tungsten, tantalum and molybdenum coatings on tungsten substrate with the absence of electron radiation was conducted during the first set of experiments. In this case, the surface of the sample was irradiated by the flux of argon atoms with thermal velocities as well as fast recharge atoms. The rate of deposition of the layers was equal to 1 μm/h, and the deposition process lasted for 2.5 hours. The EDS analysis has shown the absence of trapped argon in all three samples.

In the next set of experiments, the rate of layer deposition and experiment duration were unchanged, but the coating deposition was accompanied by electron radiation. The flux density of electrons during the deposition was 4 mA/cm^2, the electron energy was 7 keV. The EDS analysis of the deposited layer has shown approx. 2×10^{27} at/m^3 and 8×10^{26} at/m^3 of argon in the near-surface area for tungsten and tantalum, respectively. There was no argon in molybdenum. It can be asserted then that the electron radiation stimulated the trapping of argon in the tungsten and tantalum layers during deposition, but not in molybdenum.

The explanation of varying ability of molybdenum, tantalum and tungsten to trap argon requires additional analysis. We shall only note that the obtained result correlates with the ratio of substrate temperature during the deposition to melting points of metals, equal to 0.41 for molybdenum, 0.36 for tantalum and 0.31 for tungsten. The argon atom mobility and, consequentially, re-emission flux from the coatings could increase with the ratio getting higher.

The dependence of argon trapping in the coating during deposition from the experimental conditions was ascertained on tungsten coatings.

To reveal argon distribution throughout the substrate and determine the presence of argon in the substrate, a cross-section was cut in the sample down to the near-surface layer of the substrate. The cutting was conducted on the FEI Scios dual-beam setup with a gallium ion beam. To improve the quality of the cross-section, an additional layer of platinum was deposited on the cavity about 0.5×0.5×0.8 μm in size (Figure 1). The electron beam was scanning the depth of the deposited coating and the near-surface layer of the substrate. The analysis has shown a uniform distribution of argon throughout the deposited coating and its absence in the substrate.

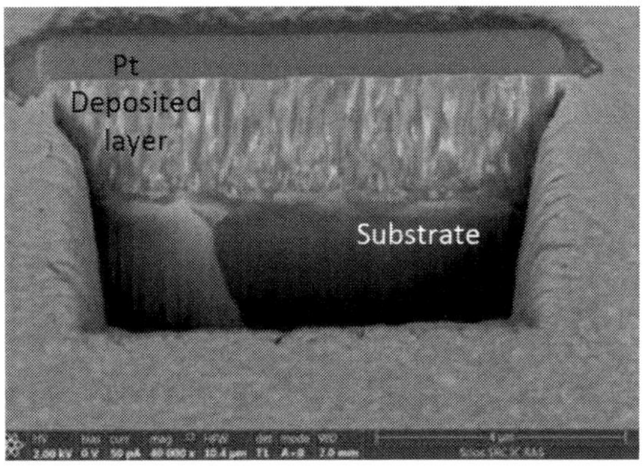

Figure 1. Cross-section of the deposited film.

The EDS analysis of the tungsten layer surface deposited in the same conditions, but irradiated with electron flux of halved intensity (approx. 2 mA/cm^2), has shown the decrease of argon content in the analysis zone to nearly halved amount versus the first experiment (approx. 1×10^{27} at/m^3). This experiment has confirmed the suggestion on the dependence of argon trapping on electron radiation accompanying the coating deposition.

The use of ion-electron emission from the target for electron irradiation of the deposited film enabled us to measure argon trapping with differing deposition rates, but constant ratio of flux of tungsten atoms for deposition to electrons. This was accomplished by varying the flux of ions sputtering the target, while keeping their energy constant. The deposition rate was changed in the range of 0.5 to 2.5 μm/h, which was in accordance to the change of electron flux of 2-10 mA/hr.

The results of the EDS analysis have shown that the amount of argon trapped in the deposited layer is kept constant throughout the entire range of deposition rates. To determine the influence of oxygen in plasma on argon trapping, a series of experiments was conducted with tungsten deposition on a substrate and oxygen content in plasma varying from 4% to 10%. The EDS analysis has shown that the increase of oxygen concentration in plasma led to the increase of oxygen in the deposited layer, reaching 7 at.% with the concentration of oxygen in plasma equaling to 7 at.%. At the same time, the amount or trapped argon was kept constant at the level of the coating deposited with no oxygen present in plasma.

For thermal desorption of argon from the coating, the substrate was heated by the irradiation from the heated flat tungsten spiral installed under the bottom of the sample. In the first experiment, the

IOP Conf. Series: Journal of Physics: Conf. Series **830** (2017) 012108 doi:10.1088/1742-6596/830/1/012108

temperature of the sample was raised at the rate of 5 K/s. The thermal desorption spectrum is shown in Figure 2.

Obviously, the peak of desorption is observed at approx. 1370 K. The amount of desorbed argon from the sample was equal to 1.6×1020 at/m^2. After the TDS measurements, exfoliation of the film from the substrate was found. The comparison of the results measured by EDS of argon in exfoliated fragments with argon in the sample before the heating has shown that about 10% of argon was liberated from the sample during the heating. This result can be explained by the temperature of the film reducing quickly after exfoliation to the point not sufficient for argon desorption.

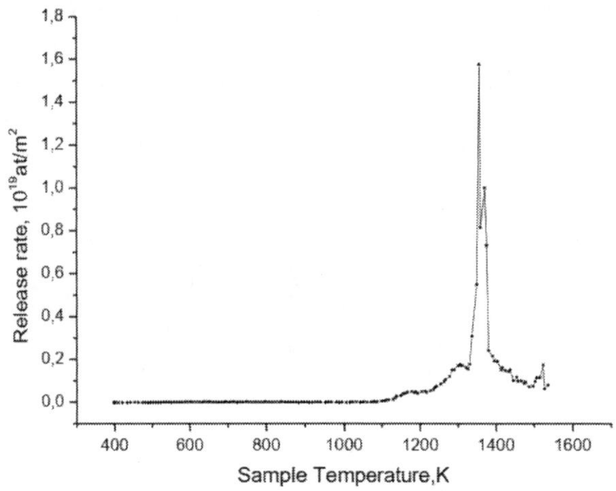

Figure 2. TDS spectra for tungsten film.

Slow heating (1 K/s) and 1-hour exposure of sample to the temperature of 1473 K was performed in the second experiment. After such heating, the EDS analysis has shown the absence of argon in the deposited coating. The cross-section of the heated sample (Figure 3) has shown exfoliation of the film from the substrate. Such defects can happen as a result of stress appearing in the coating during argon desorption or due to argon accumulation along the boundary between the deposited film and the substrate.

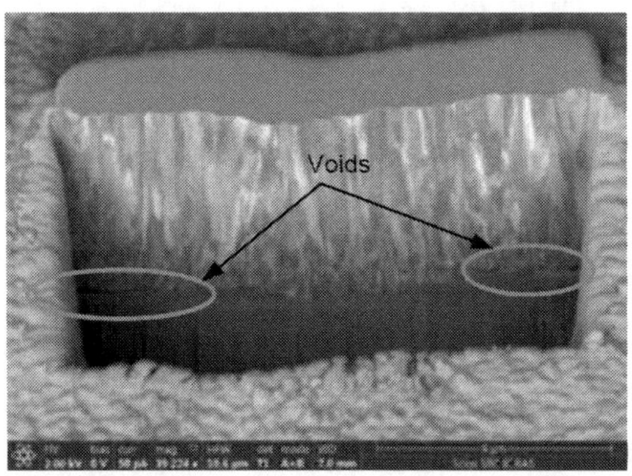

Figure 3. Cross-section of the heated sample.

A significant amount of argon in the deposited layer allows for making an assumption that most of it is held in microscopic pores formed during the crystallization of the layer. The size of such pores, supposedly, does not exceed 20 nm, which prevents them from being observed using the methods of analysis used in this research. The pressure of argon in such pores should balance out the pressure of surface tension. The value of surface energy $\varepsilon = 4$ J/cm^2[8] was used to calculate the pressure of surface tension in a pore, equaling to 600 MPa[7] for a pore of 20 nm in diameter and argon being in a supercritical fluid state. With the diameter of the pore $d \leq 1.3$ nm, the pressure of surface tension would exceed 6 GPa. In such pores, according to paper [7], solid argon is expected to be formed.

5. Conclusion

The paper has investigated conditions and parameters of argon trapping in molybdenum, tantalum and tungsten layers during their deposition on tungsten by the atoms sputtered from the respective targets in argon plasma. The substrate temperature during deposition was 1200 K. The rate of deposition was equal to 1 μm/h.

It was shown that electron irradiation of the deposited layer with the intensity of 4 mA/cm^2 has stimulated argon trapping in the tungsten and tantalum coating layers with approx. 2x10^{27} at/m^3 and 8x10^{26} at/m^3, respectively, but did not lead to argon trapping in molybdenum. A suggestion was made that the obtained result correlates with the ratios of temperatures of substrates during the deposition process to the melting points of molybdenum, tantalum and tungsten, which are 0.41, 0.36 and 0.31, respectively.

The research on the specifications of argon trapping in the coating layers during deposition has allowed making the following conclusions. Argon is uniformly distributed throughout the entire depth of coating and does not penetrate tungsten substrate. The amount of trapped argon is approximately proportional to the intensity of electron radiation. The concentration of argon in the deposited film was kept constant with constant ratio of deposited tungsten atom flux to electron flux, respectively, in the range of 0.5 to 2.5 μm/h and 2-10 mA/hr. The amount of trapped argon did not change with the addition of oxygen into plasma of up to 10 at.%. The argon desorbed from the deposited film during heating for 1 hour at 1473 K. It is assumed that most of argon is held in microscopic pores formed during crystallization of the deposited coating. The size of the pores does not exceed 20 nm, with argon inside being in a state of supercritical fluid.

References

[1] Barvinok V, Ryasnyi A 2009 *Izv. Sam. Ak. Nauk* **11№3(2)** 378-381 [In Russian]
[2] Wang P, et al. 2014 *Physica Scripta* **T159** 014046.
[3] Dhal S, el al. *Radiation Effects and Defects in Solids* **170(2)** 144–151.
[4] Xie, J., & Komvopoulos, K. (2016) *Journal of Applied Physics* **119(9)** 095304.
[5] Ayrapetov A, el al. 2016 *Journal of Physics: Conference Series* **700** 012041.
[6] Begrambekov L, et al. *Nuclear Instruments and Methods in Physics Research Section B: Beam Interactions with Materials and Atoms* **315** 110–116.
[7] Kataoka, Y., & Yamada, Y. 2013 *Journal of Computer Chemistry* **12(2)** 101–108.
[8] Lassner E, Schubert W-D 1999. Tungsten: Properties, Chemistry, Technology of the Element, Alloys, and Chemical Compounds.

Acknowledgments

This work was supported by Competitiveness Growth Program of the Federal Autonomous Educational Institution of Higher Professional Education National Research Nuclear University MEPhI and by the Ministry of Science and Education of the Russian Federation (agreement No. 14.575.21.0049 as of 27.06.14).

The authors also wish to thank doctor Artemov V V for his assistance in the work on the FEI Scios dual-beam device.

IOP Conf. Series: Journal of Physics: Conf. Series **830** (2017) 012109 doi:10.1088/1742-6596/830/1/012109

Sand-blasting treatment as a way to improve the adhesion strength of hydroxyapatite coating on titanium implant

I Grubova[1], T Priamushko[1], M Surmeneva[1], O Korneva[2], M Epple[3], O Prymak[3], R Surmenev[1]

[1]Department of Theoretical and Experimental Physics, National Research Tomsk Polytechnic University, 634050 Tomsk, Russia
[2]Material Properties Measurements Centre, National Research Tomsk Polytechnic University, 634050 Tomsk, Russia
[3]Inorganic Chemistry and Center for Nanointegration Duisburg-Essen (CeNIDE), University of Duisburg-Essen, 45117 Essen, Germany

E-mail: rsurmenev@gmail.com

Abstract. In the current study, the effect of corundum particle sizes (50 and 250–320 μm) used for sand–blasting on the structure, roughness, wettability, mechanical properties, and adhesion of radio frequency magnetron hydroxyapatite coating deposited on treated titanium substrate is studied. Morphology analysis revealed that pretreatment uniformly deforms the surface and induces the formation of pits, which size depends linearly on the grit size. The deposited coatings (Ca/P was in a range of 1.75-1.79) are homogeneous and repeat the relief of the substrate (mean roughness Ra is 1.9±0.1 (250–320 μm) and 0.8±0.1 μm (50 μm)). Texture coefficient calculations revealed the predominant (002) growth texture of hydroxyapatite coatings. The resistance of the coating to plastic deformation and the surface hardening were significantly higher for coatings formed on sand blasted with particle size of 50 μm. Scratch test have shown the significant improvement of wear resistance and lower friction coefficient of coatings for smoother samples. Dynamic contact angle measurements revealed the hydrophilic properties of the hydroxyapatite coating. Thus, sand–blasting of titanium with corundum powder having the size of 50 μm prior to the deposition of RF magnetron coating is recommended for the medical applications intended to improve the bonding between the substrate and coating.

1. Introduction

Since the early 1980s, the excellent mechanical properties of titanium (Ti) substrates have been commonly utilized in the biomedical field for the compensation of poor mechanical properties of porous bulk calcium phosphate (CaP) materials [1, 2], which are widely accepted as a biocompatible materials chemically similar to the mineral component of bones and hard tissues.

The CaP-coated Ti implants have been widely employed in the orthopedic and dental fields due to their excellent biocompatibility [3, 4], which in turn leads to the development of good interfacial strength between the implant and bone [3, 4].

However, in order to maintain long-term stability, CaP-coated Ti implants have to demonstrate not only the good bone regeneration process, but also the high bonding strength at the interface between the coating and metallic substrate. Bonding strength should be high enough to sustain possible fatigue stress during the surgical operation or after implantation.

Altering the metallic surface texture, namely, the implant roughness, via different pretreatment techniques or/and their combination is the most commonly used and relatively inexpensive way that can help in tackling above mentioned challenges as the substrate properties play an important role in obtaining stable coating, which in turn leads to the effective implant–tissue interaction and osseointegration.

It is known, that the mechanical properties of biomaterials are strongly governed by the film fabrication method and substrate characteristics. According to E. Mohseni et al, the sputtering technique has the highest adhesion of coating to the substrate as compared to other methods, which can be attributed to the sputter cleaning and ion bombardment processes [5].

The main sputtering technique employed today in the application of CaP coatings on medical implants is radio frequency (RF) magnetron sputtering. This technique allows producing thick, uniform, dense and pore-free films with high bonding strength to the metal implants [6, 7].

Nowadays, prior to RF-magnetron spray procedure, the Ti substrates are normally pretreated by grinding (GR), sand–blasting (SB) and acid–etching (AE) [7, 8] to remove the surface impurities and roughened the surface in order to improve the surface characteristics, such as hardness, Young's modulus and adhesion of the CaP coatings to substrate, thus increasing life expectancy of implants. Our previous study have assessed the effects of SB at different pressures on grain size, mechanical properties and surface wettability of RF magnetron silver-containing hydroxyapatite (HA) coating and revealed that the coating microstructure could be designed by controlling the pre-treated Ti surface topography [7]. However, the type, size and geometry of the sand particles are also important factors determining the result of the treatment. Therefore, in the current study, the effect of SB particle size (50 and 250–320 μm) on the surface structure, roughness, wettability, mechanical properties, and adhesion failure at the interface of RF magnetron deposited HA coating is studied.

2. Materials and methods

The commercially available technically pure Ti plates (10 x 10 x 2 mm) with the following chemical composition (in wt%): 0.30 Fe, 0.10 Si, 0.10 C, 0.05 N, 0.25 O, 0.0015 H and balance Ti were chosen as a material for the modification. Two groups of the Ti substrate were prepared using SB treatment with 250-320 μm and 50 μm of Al_2O_3 particles designated below as group 1 and group 2, respectively. Then all the samples were AE using a 1:2:2.5 mixture of HF (40 %), HNO_3 (66 %) and distilled water for 10 s in a laminar flow hood. Subsequently, the samples were washed using ultrasonic bath (Eurosonic Micro) sequentially during 10 min in acetone, ethanol and deionised water and dried in an autoclave at 70–80 °C.

The thin coatings (<800 nm) were deposited using a commercially available setup with an RF generator COMDEL (USA, 13.56 MHz). The samples were rotated in the deposition chamber. The precursor-powder of HA ($Ca_{10}(PO_4)_6(OH)_2$, $Ca/P = 1.67 \pm 0.02$) was synthesized using a mechanochemical activation method. The preparation and investigation of the target are reported in detail elsewhere [6].

Optical ellipsometry (Ellipse 1891-S AG) was used to determine the thickness of the deposited coatings. Surface morphology and chemical composition were investigated via scanning electron microscopy (SEM) using a Quanta 400 instrument (FEI, USA) equipped with an energy-dispersive X-ray analysis (EDX analysis system Genesis 4000, SUTW-Si (Li) detector) operated under high vacuum. The surface roughness was studied by Dektak 6M Stylus Surface Profilometer (Veeco Instrument Inc., USA). Surface phase composition was investigated by X-ray diffractometer (XRD) (D8 Advance, Bruker, Germany). The wettability of the samples was studied via the static and dynamic drop method using deionized water (Easy Drop Instrument (Kruess, Germany)). Mechanical properties of the samples were studied by Nano Hardness Tester «NanoScan-3D» using a Berkovich indenter at the load of 2 mN. For each sample, 30 indentations were performed. Based on the elastic deformation theory, Young's modulus (E) and nanohardness (H) values were determined in accordance with Oliver–Pharr method. Wear resistance was evaluated using the plastic resistance parameter (H/E ratio) as well as resistance of the material to plastic deformation (H^3/E^2 ratio).

Adhesion properties of the coatings were determined by Micro-Scratch Tester (Micro-Scratch Tester MST-S-AX-0000).

3. Experimental results and discussion

The appearance of the Ti surface was significantly altered by multi-surface pretreatment for all the groups, as shown in Figures 1a and 1b. SEM analysis of pretreated Ti substrates revealed that the modification via the combination of SB and AE uniformly deforms the surface and induces the formation of homogeneously distributed micro-pits, which size depends linearly on the grit size of corundum powder. Thus, the combination of these pretreatment methods leads to the formation of textured micro-relief on Ti surface, which application as a substrate for the deposition of nanostructured coatings can decrease micro-tension and increase the effective surface area, which in turns is very promising for osseointegration.

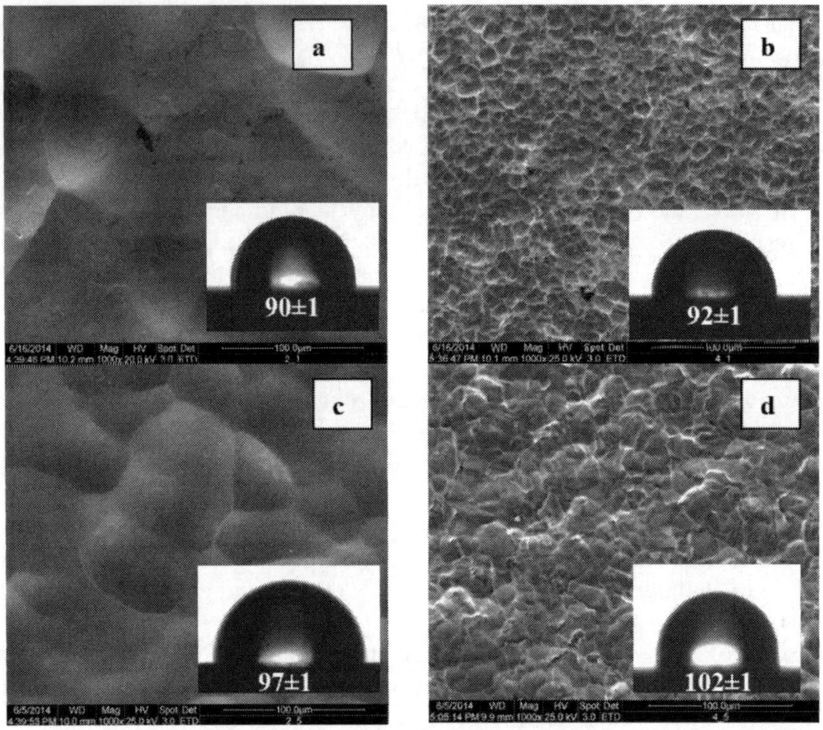

Figure 1. Typical SEM images showing the Ti surface morphology after SB and AE (a–b) (the insets are optical images of a water droplet), and after the deposition of the HA coatings on the nano-patterned Ti substrates (c–d) (the insets are optical images of a water droplet). For the surfaces (a, c) and (b, d), the grit size of Al_2O_3 were 250-320 μm and 50 μm, respectively.

Different powder size at SB also resulted in obtaining various surface roughness characteristics of substrates. A significantly higher surface roughness of group 1 compared with group 2 was observed after SB (Table 1). Chemical etching of the titanium surface was selective and led to the smoothing of initial blasted topography, which in turn led to the decrease in roughness (Table 1).

The resulting deposited coatings are homogeneous with unclear grain boundaries and repeat the relief of the initial Ti substrate (Figures 1c and 1d). HA coatings have no significant influence on the parameters of surface roughness. The mean roughness R_a of HA-coated Ti substrates was 1.94±0.14 and 0.76±0.02 μm for group 1 and group 2, respectively (Table 1).

Table 1. Effect of the surface treatment procedure on the surface roughness parameters and mechanical properties of the samples.

Parameters	SB		SB + AE		SB + AE + HA coating	
	Group 1	Group 2	Group 1	Group 2	Group 1	Group 2
R_a, (μm)	6.6 ± 0.5	1.3 ± 0.4	1.8 ± 0.1	0.96 ± 0.04	1.94 ± 0.14	0.76 ± 0.02
R_q (μm)	8.4 ± 0.6	1.8 ± 0.1	2.4 ± 0.2	1.21 ± 0.04	2.57 ± 0.23	0.97 ± 0.03
R_z (μm)	36.7 ± 2.4	12.2 ± 1.8	12.4 ± 1.2	7.43 ± 0.20	14.61 ± 1.96	5.57 ± 0.25
H (GPa)	5.8 ± 0.3	3.9 ± 0.3	3.1 ± 0.2	3.09 ± 0.19	12.23 ± 0.61	15.20 ± 0.66
E (GPa)	108.7 ± 11.7	86.8 ± 13.5	74.6 ± 0.2	80.87 ± 11.47	118.41 ± 5.85	147.11 ± 15.93
H/E	0.053	0.044	0.041	0.038	0.101	0.101
H^3/E^2 (GPa)	0.016	0.008	0.005	0.005	0.131	0.164
h_c (nm)	101 ± 6	127 ± 10	145 ± 6	130 ± 5	74 ± 2	71 ± 3

EDX analysis revealed that the Ca/P ratio of the coatings for both regimes of SB was in a range of 1.75-1.79, which is close to that reported for CaP coatings deposited via RF magnetron sputtering [7].

XRD patterns for all of the coatings are mainly composed of crystalline HA with (002), (211) and (300) Bragg peaks and several peaks corresponding to the substrate (Figure 2). Blasting with 50 and 250-320 μm corundum particles resulted in lesser average crystallite sizes of HA estimated from the XRD patterns and were 66±1 and 96±2 nm, respectively (Table 2). The formation of HA grains with c-axis is preferentially aligned in the direction perpendicular to the substrate. This is evidenced by the calculations of the texture coefficient (*TChkl)* [8], which revealed the predominant (002) growth texture for all groups of HA coatings (Table 2). The determined unit cell parameters were in good agreement with that of HA standard (ICDD, #09-432, $a = b = 0.9418$ nm, $c = 0.6884$ nm (space group *P63/m*) (Table 2).

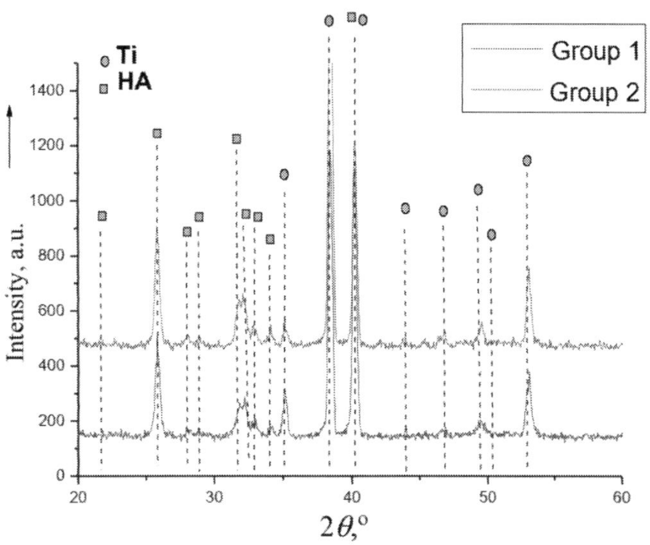

Figure 2. The typical XRD patterns of HA coatings deposited onto the surface of sand blasted and acid etched Ti substrates. The size of Al_2O_3 at SB was: a) 250-300 μm and b) 50 μm.

Nanoindentation results indicated intensive surface hardening after nanocrystalline HA coating deposition (Table 1). Nanohardness of the HA films in the case of substrate treated with the smaller particle size was 15.20±0.66 GPa, while with the higher particle size it was 12.23±0.61 GPa. Based on the data obtained before for pure HA coatings deposited on the surface of untreated Ti (nanohardness ~10 GPa [9]), we can assume that the substrate surface microstructure strongly affected the mechanical properties of HA films. It can be seen from the obtained data that H/E ratio correlated with the mechanical performance under load (i.e., deformation for tribological applications) [10] was significantly higher as compared with the non-coated samples, due to an increased hardness of the HA coating. The resistance of the material to plastic deformation for the coating (H^3/E^2 parameter proposed by Johnson [11]) in the case of smaller abrasive particle size (0.164 GPa) was significantly higher than for the larger one (0.131 GPa) (Table 1). Thus, it can be concluded that the mechanical properties of the coating can be enhanced via variation of the particle size during blasting.

Table 2. Effect of Ti surface treatment on the phase composition, lattice parameters, crystallite size, element composition, texture coefficient of deposited coatings and wettability, where θ_{adv} is advancing contact angle.

Group	Phases	Lattice parameters (Å)	Crystallite size (nm)	Ca/P	TC_{002}	TC_{112}	TC_{300}	θ_{adv} (°)	$\Delta\theta$ (°)
1	Ti	a = 2.931, c = 4.646	214	1.75	1.7 ± 0.1	0.7 ± 0.1	0.5 ± 0.1	85 ± 2	35
	HA	a = 9.468, c = 6.787	66						
2	Ti	a = 2.949, c = 4.691	335	1.79	2.1 ± 0.1	0.6 ± 0.1	0.4 ± 0.1	87 ± 1	36
	HA	a = 9.411, c = 6.876	96						

In the present paper, the main task in the pretreatments of Ti surface prior to coating deposition was to enhance the bonding strength of the coating to substrate. Therefore, the adhesion strength at the interface was the most important characteristic, which was measured via scratch test method. Scratch test results revealed that the deposited HA coatings exhibited improved wear resistance and lower friction coefficient, especially in the case of 50 μm corundum particles.

Figures 3a and 3b show the typical scratch test diagram of the friction coefficient change as a function of the scratch length for the HA-coated sample. The clear jumps of the friction coefficient from ~0.29 to ~0.65 and from ~0.40 to ~0.70 in arbitrary units are visible for group 1 and group 2, respectively (Figures 3a and 3b). The evolution of the coating failure can be divided into three stages. At low loads (under ~2.6 N), cracking on the coating trackside was evident (Figures 3e and 3f, insets). As the load gradually increased, delamination at the trackside occurred. Eventually, the coating was delaminated from the substrate along the scratch path when the load increased up to ~2.60 and to ~3.14 N for group 1 and group 2, respectively.

The surface of Ti after SB followed by AE revealed almost the same static contact angle values (figure 2 c–d insets). The water contact angles for groups 1 and group 2 treated surfaces were 90 ± 1° and 92 ± 1°, respectively. The HA coating deposition process resulted in an increase in the water contact angle, that is, 97± 1° and 102 ± 1° for groups 1 and 2, respectively (Figures 2e and 2f, insets). Although the XRD patterns for all groups of the coatings are mainly composed of HA phase, some differences in the morphology of the HA films were observed. Therefore, surface chemistry gives the most contribution to the wettability of the HA coating than the surface texture. Thus, the deposited film is hydrophobic in static because of the formation of the multi-scale surface. However, dynamic contact angle measurements revealed the hydrophilic properties of the deposited HA coating due to a

IOP Conf. Series: Journal of Physics: Conf. Series **830** (2017) 012109 doi:10.1088/1742-6596/830/1/012109

higher hysteresis as compared with uncoated Ti surface. In this respect, we assume that HA coating can decrease stress-shielding effect and improve the fixation of an implant in bone.

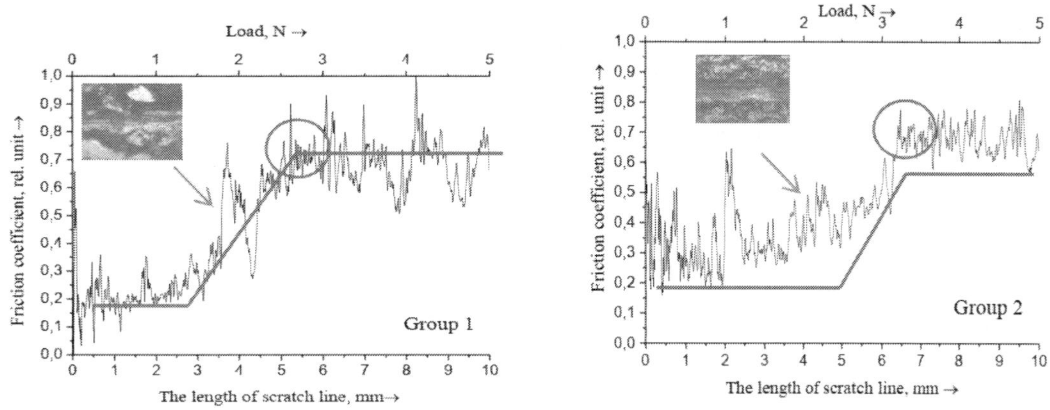

Figure 3. Results of scratch testing the HA coating deposited on the treated Ti substrates. The size of Al_2O_3 at SB was: a) 250-300 μm and b) 50 μm.

4. Conclusion

The results of the present study demonstrate that the predetermined surface texture of Ti substrates plays a crucial role in achieving the desired mechanical properties of RF magnetron deposited HA coatings (800 nm). Combination of SB and AE pretreatment methods led to the formation of the textured micro-relief on Ti surface with the mean roughnesses R_a of 1.94±0.14 and 0.76±0.02 μm, for group 1 and group 2, respectively. Nanoindentation results indicated an intensive surface hardening after nanocrystalline HA coating deposition with the (002) texture. Nanohardness and Young's modulus of the coatings were 12.23±0.61 GPa, 118.41±5.85 GPa and 15.20±0.66 GPa, 147.11±15.93 GPa for groups 1 and group 2, respectively. The resistance of the material to plastic deformation (H^3/E^2 ratio) for the HA coating corresponding to group 2 (0.101 and 0.164 GPa, respectively) were significantly higher than that of the coated samples of group 1 (0.101 and 0.131 GPa). Scratch test results revealed that the deposited HA coatings exhibited improved wear resistance and lower friction coefficient.

Eventually, the coatings were delaminated from the substrates along the scratch path when the load increased up to ~2.60 and to ~3.14 N for group 1 and group 2, respectively. Obtained data revealed that HA surface chemistry gives more contribution to the wettability of the HA coating, than the surface texture. Dynamic contact angle measurements have shown the hydrophilic properties of the deposited HA coating due to a higher hysteresis as compared with uncoated Ti surface. Thus, SB of Ti with corundum powder having the particle size of 50 μm prior to the deposition of RF magnetron coating is recommended for medical applications intended to improve the bonding between the substrate and coating, which in turn can decrease stress-shielding effect and improve the fixation of an implant in bone.

References

[1] Ducheyne P, Hench L L, Kagan I I A, Martens M, Bursens A and Mulier J C 1980 Effect of hydroxyapatite impregnation on skeletal bonding of porous coated implants *J. Biomed. Mater. Res.* **14** 225-237

[2] Jarcho M 1981 Calcium phosphate ceramics as hard tissue prosthetics *Clin. Orthop. Relat. Res.* **157** 259-278

[3] Chen W, Liu Y, Courtney H S, Bettenga M, Agrawal C M, Bumgardner J D and Ong J L 2006 In vitro anti-bacterial and biological properties of magnetron co-sputtered silver-containing hydroxyapatite coating *Biomaterials* **27** 5512-5517

[4] Qu J, Lu X, Li D, Ding Y, Leng Y, Weng J, Qu S, Feng B and Watari F 2011 Silver/hydroxyapatite composite coatings on porous titanium surfaces by sol–gel method *J. Biomed. Mater. Res. B* **97** 40-48

[5] Mohseni E, Zalnezhad E and Bushroa A R 2014 Comparative investigation on the adhesion of hydroxyapatite coating on Ti–6Al–4V implant: a review paper *Int. J. Adhes. Adhes.* **48** 238-257

[6] Surmeneva M A, Chaikina M V, Zaikovskiy V I, Pichugin V F, Prymak O, Epple M and Surmenev R A 2013 The structure of an RF-magnetron sputter-deposited silicate-containing hydroxyapatite-based coating investigated by high-resolution techniques *Surf. Coat. Technol.* **218** 39-46

[7] Grubova I Y, Surmeneva M A, Ivanova A A, Kravchuk K, Prymak O, Epple M, Buck V and Surmenev R A 2015 The effect of patterned titanium substrates on the properties of silverdoped hydroxyapatite coatings Surf. Coat. Technol. 276 595-601

[8] Perry A J 1986 The structure and colour of some nitride coatings *Thin Solid Films* **135** 73-85

[9] Pichugin V F, Surmenev R A, Shesterikov E V, Ryabtseva M A, Eshenko E V, Tverdokhlebov S I, Prymak O and Epple M 2008 The preparation of calcium phosphate coatings on titanium and nickel-titanium by rf-magnetron sputtered deposition: composition, structure and micromechanical properties *Surf. Coat. Technol.* **202** 3913-3920

[10] Leyland A and Matthews A 2000 On the significance of the H/E ratio in wear control: a nanocomposite coating approach to optimised tribological behavior *Wear* **246** 1-11

[11] Johnson K L 1985 *Contact Mechanics* (London: Cambridge University Press) p 464

Acknowledgments
The authors would like to express their gratitude to Mr. T.M. Mukhametkaliyev for the help with the fabrication of thin HA films and to Prof. V. Buck for the assistance with the experiments. This research was supported by the "P.L.U.S." scholarship funded by National Research Tomsk Polytechnic University, President's fellowships MK-7907.2016.8, Russian Fund for Basic Research (15-08-08652), State Mission NAUKA (Science) (#1359).

IOP Conf. Series: Journal of Physics: Conf. Series **830** (2017) 012110 doi:10.1088/1742-6596/830/1/012110

Phase equilibrium in system Ti-Si-C-B and synthesis of MAX phase layers in vacuum under the influence of electron beam

N N Smirnyagina[1,2], V M Khaltanova[1,2], D E Dasheev[1] and A E Lapina[1]

[1]Institute of Physical Materials Science SB RAS, 670047, Ulan-Ude, Russia
[2]Buryat State Universities, 670000, Ulan-Ude Russia

E-mail: smirnyagina09@mail.ru

Abstract. Composite layers on the basis of carbides and borides the titan and silicon on titanic alloy VT-1 are generated at diffused saturation by electron beam treatment in vacuum. Formation in a composite of MAX phase Ti_3SiC_2 is shown. Thermodynamic research of phase equilibrium in systems Ti-Si-C and Ti-B-C in the conditions of high vacuum is executed. The thermodynamics, formation mechanisms of superfirm layers borides and carbides of the titan and silicon are investigated.

1. Introduction

Last year's interests to the titan-silicon-boron-carbon alloys increases, as they, possess ease, high durability, excellent electro conducting and stability at the raised temperatures. The development of new compositions according to phase diagrams of the Ti-Si-B-C system is one of the main ways to obtain and investigate new titanium alloys with improved properties. By present time of the MAX phases (double carbides of silicon and titan Ti_3SiC_2, $Ti_5Si_3C_x$) are synthesized by various methods in a nanostructural status in the form of layers, films or ceramic.

The crystal structure of thin layers and physical and chemical, mechanical properties etc. are investigated [1-4]. Ti_3SiC_2 is a rare example of so called "plastic ceramics". It combines properties typical of ceramic compounds, such as high melting temperature, thermal and chemical stability, with mechanical properties of metallic systems. The main difficulty while synthesizing Ti_3SiC_2 is the loss of silicon during reaction. The reaction features depend on the Ti_3SiC_2 crystal structure that consists of titanium carbide blocks divided by silicon layers. The inter-atomic dimension analysis shows that they are larger for silicon in Ti_3SiC_2 than for other compounds Ti-Si-C. At the same time, Ti-C dimensions are close to those of titanium carbide. The weak bond of silicon in the Ti_3SiC_2 lattice can be the cause of its loss in thermal reactions.

In [5] the conclusion is drawn on synthesis possibility complex carbide phases, analogues of MAX phases in system Ti-Si-C as a result of high-intensity influences, for example electron beams.

To improve properties of the material, information about the phase diagram and thermodynamic properties in the Ti–Si–C system is essential [6]. The authors strongly recommend the simultaneous use of thermodynamic calculations and experiments in order to establish technologically important phase diagrams highly efficiently.

Certainly, modelling phase equilibrium and revealing a field of borides and carbides in ternary systems Ti-Si-S and Ti-B-C represents to crystallization certain scientific interest. It is

IOP Conf. Series: Journal of Physics: Conf. Series **830** (2017) 012110 doi:10.1088/1742-6596/830/1/012110

physicochemical basis for working out of technology of functional coating formation at electron beam processing in vacuum.

In the present study revealed optimum conditions of formation the MAX phases at electron beam processing of titanic alloys.

2. Experimental procedures

Modeling phase equilibrium in ternary systems Ti-Si-S and Ti-B-C were carried out. In the present study used interface of a program complex TERRA [7]. Calculations are spent in a temperature interval 300-4500 K at a variation of the general pressure in system in a range from 10^5 to 10^{-3} Pa.

In calculations considered following phases: silicide's Ti_3Si, Ti_5Si_3, Ti_5Si_4, TiSi и $TiSi_2$; carbides B_4C, SiC, TiC; borides TiB, TiB_2 and Ti_3SiC_2, $Ti_5Si_3C_x$.

Layers of composites formed on preliminary prepared surface of the samples made of titanic alloy VT-1 in the form of square plates in the size 15x15 mm and height of 7 mm. It was used sating daubs. Sating daubs contained reactionary a component (Si:2 C) and the organic binding. As the organic binding applied a solution 1:10 glue BF-6 in acetone. Electron beam processing was carried out in a vacuum at most $2x10^{-3}$ Pa at power electron beam W = 250-450 W for 1-3 min.

X-ray phase analysis was carried out on the Bruker Phaser 2D diffractometer with a Cu Ka radiation source of a wavelength of 1.504 Å over a 2θ range from 20° to 100°. The microstructure of layers was investigated by METAM PB-22 microscope with the program NEXSYS Image Expert. The microhardness of composite layers was measured by indentation technique using a Vicker's indenter.

3. Results and discussion

Phase equilibrium and revealing of crystallization fields of borides and carbides of the titan and silicon in ternary systems Ti-Si-C and Ti-B-C and their analysis representations about mechanisms and laws of education borides and carbides in the conditions of vacuum allow systematizing and expanding modeling. On the figure 1 the concentration tetrahedron of system Ti-Si-B-C is presented.

Phase equilibrium and revealing of crystallization fields of borides and carbides of the titan and silicon in ternary systems Ti-Si-C and Ti-B-C and their analysis representations about mechanisms and laws of education borides and carbides in the conditions of vacuum allow systematizing and expanding modeling. On the figure 1 the concentration tetrahedron of system Ti-Si-B-C is presented.

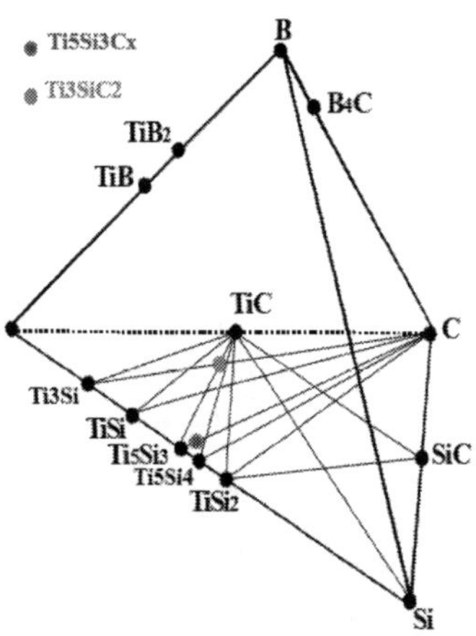

Figure 1. Concentration tetrahedron of system Ti-Si-B-C.

Phase equilibrium in system Ti-Si-C was calculated. On the figure 2 are presented phase balance in system Ti-Si-C at pressure $P=10^{-3}$ Pa. Areas of co-existing phases are allocated. It is shown, that the titan silicide's titans of various compositions form quasi-binary cuts with titan carbide, instead of with silicon carbide.

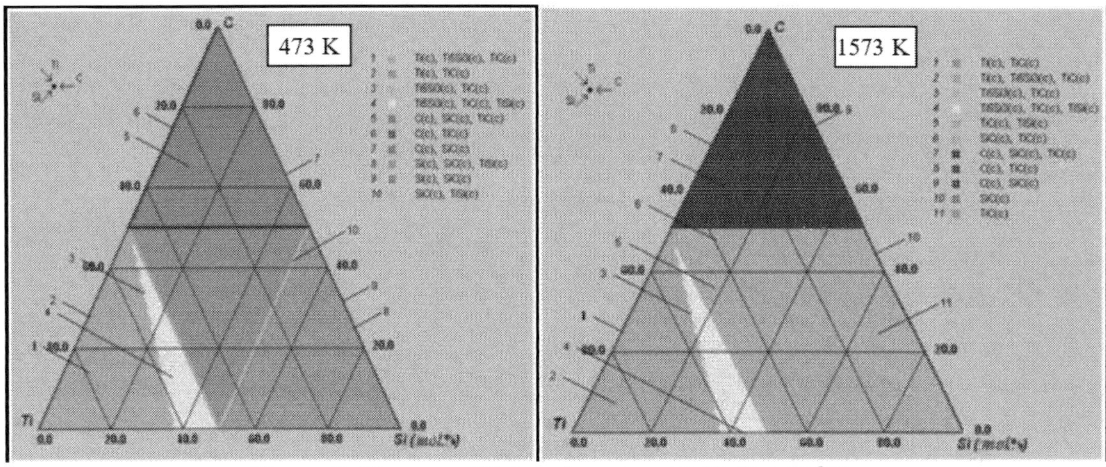

Figure 2. Phase equilibrium in system Ti-Si-C at 10^{-3} Pa.

Phase balances in system Ti-B-C were studied. In this system cut Ti-B_4C is not binary. At him there can be various phases from ternary system Ti- B-C. Phase balance (isothermal sections at 1000 K) in system Ti-B-C are presented on the figure 3 at pressure 10^{-3} Pa, besides on the figure 4 are presented isotherms of this system at discussed pressure. Formation of ternary compounds is not revealed.

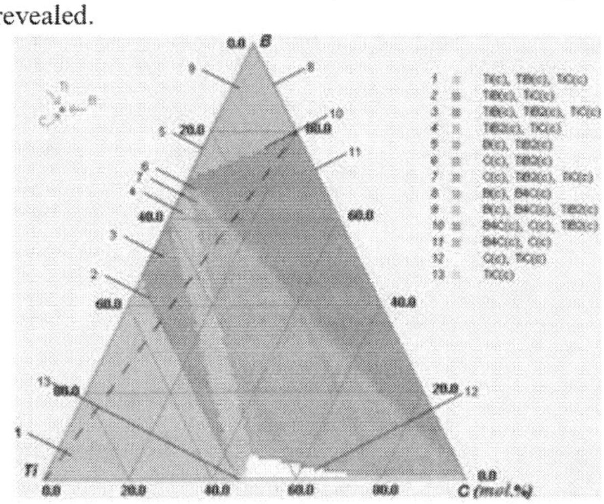

Figure 3. Phase equilibrium in system Ti-B-C at T=1000 K (P=10^{-3} Pa).

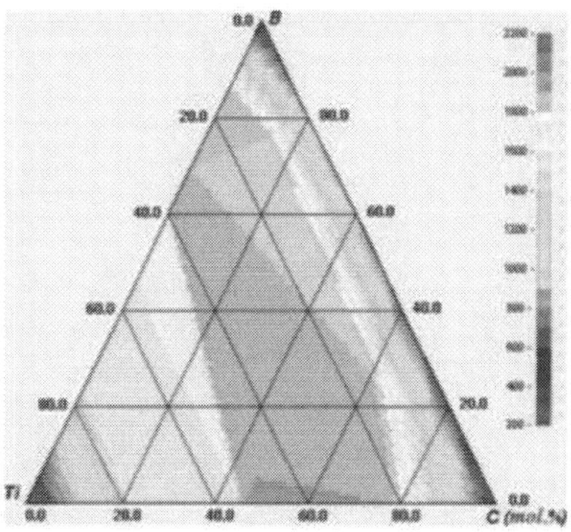

Figure 4. Isotherms in system Ti-B-C at P=10^{-3} Pa.

According to thermodynamic calculations the coexistence of various areas of crystallization is possible: B_4C; C; C+TiB_2; B_4C+C+TiB_2; C+TiB_2+TiC; TiC, TiB_2+TiC, TiB+TiB_2+TiC; TiB+TiC and Ti +TiB (at pressure below 1 Pa). At atmospheric pressure at heats (above 3000K) areas single-phase boride TiB_2, boron, their joint presence, and three-phase area with B, C and TiB_2 are found out.

Formation processes of the carbides and borides titan proceed with allocation of a significant amount of energy, thereby raising temperature in system to 2000÷2150 K (P=10^5 Pa) and 1600÷1725 K (P=10^{-3} Pa).

Titanic alloy VT-1 was carried out by power electron beam. The diffraction results were compared with the results published by the International Centre for Diffraction Data (ICDD), and the measured values were close to the d values published by ICDD in the Powder Diffraction File (PDF). Initial alloy VT-1 was represented a mix of two forms of the titan (α - and β-Ti). Titanic alloy VT-1 after

IOP Conf. Series: Journal of Physics: Conf. Series **830** (2017) 012110 doi:10.1088/1742-6596/830/1/012110

electron beam treatment was also many phases. Lattice parameters of cells α-Ti: a=0,29504 *nm* and c=0,46833 *nm* (P63/mmc, PDF 03-065-33 62), and parameters of a cubic cell β-Ti: a=0,3306 (Im-3m, PDF 00-044-1288).

Composite layers on the basis of titan and silicon carbides were formed. Layers of 80-100 microns are formed. Figure 5 showed the microstructures of the overview cross-section of the composite layers in sample. It was noted that there was a metallurgical combination between the coating and VT-1 substrate in sample 1. A layers structure is eutectic.

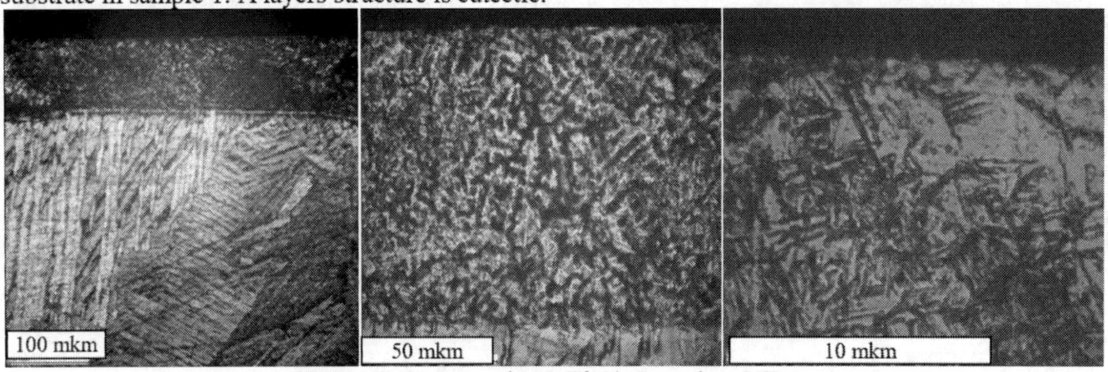

Figure 5. Structure layer Ti-Si-C on titan VT-1.

According to XRD studies the composite layer are containing 76,5 mas. % crystalline and 23,5 mas. % amorphous phases. Thus the metal basis consists from 84,1 mas. % β- titan with and 4,1 mas. % α-titan. The layer contains double carbide Ti_3SiC_2 and silicon carbide β-SiC (PDF01-073-1665, F43m), with a cubic cell a=0,4358 nm.

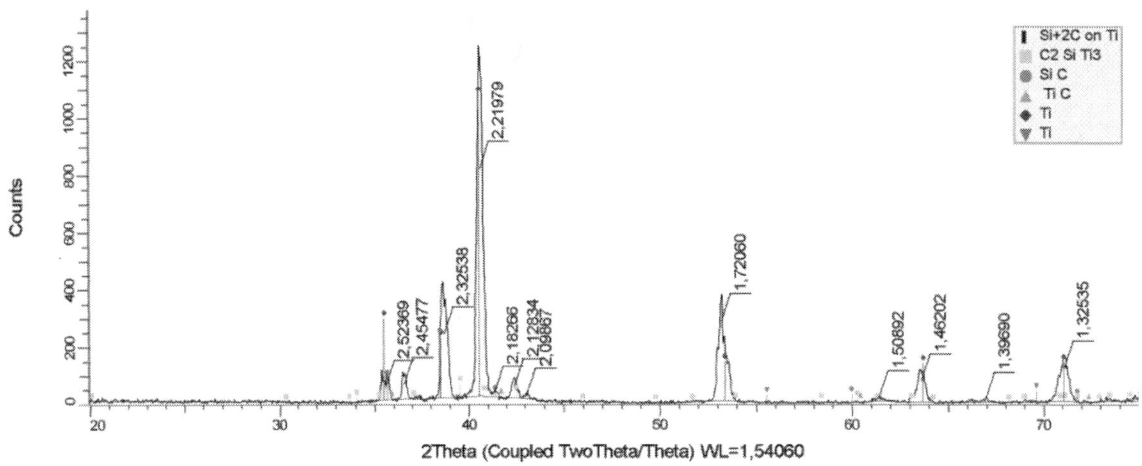

Figure 6. XRD pattern of composite layers Ti-Si-C on titan alloy VT-1.

Solid state boriding from sating daub of amorphous boron were carried out. On XRD patterns the disordered phases of titan α-Ti and boride titan TiB_2 (PDF 01-073-2148) are presented There is no reflexes β-Ti that testifies to heats on a surface of a titan alloy at electron beam boriding. The considerable thickness of the modified layer (more than 700-900 microns) is observed.

Attempt of synthesis Ti_6Si_2B from sating daub of composition Si-B-2C is made. Layers of thickness 30-80 microns are formed.

Measurement of microhardness of composite layers has shown achievement of values to 10 GPa.

4. Conclusion

In summary, preliminary results of electron beam boriding of titan alloy VT-1 are presented. The thermodynamic investigation of equilibrium in systems Ti-B-C-Si is carried out with the purpose of formation conditions optimization of functional layers on a surface of titan alloys as a result of electron beam boriding in vacuum. The thickness of MAX phase Ti_3SiC_2 coating is received 100 μm. Composite layers will consist from Ti_3SiC_2. Research by means of a metallographic microscope has shown that carbide borides layers formation is difficult physical and chemical process. Their formation is difficult physical and chemical process. The structure, phase structure and strength characteristics are investigated.

References

[1] Korosteleva E N, Pribytkov G A and Gurskikh A V 2009 *Powder Metallurgy and Metal Ceramics* **48** 8
[2] Vahlas C, Chevalier P Y, and Blanquet E. 1989 *CALPHAD* **13(3)** 273
[3] Debashis Bandyopadhyay 2004 *Journal of Phase and Diffusion* **25(5)** 415
[4] Yong Du, Julius C, Schuster, Hans J. Seifert and Fritz 2000 *Journal of the American Ceramic Society* **83(1)** 197
[4] Ivanov Iu F, Teresov A D, Soskova N A 2012 *Izvestiia vuzov. Fizika* **55(5/2)** 130
[5] Smetkin A A, Maiorova Iu K 2015 *Bulletin PNRPU Mechanical engineering, materials science* **17(4)** 120
[6] Wakelkamp W J J, van Loo F J J, and Metselaar R 1991 *J Eur Ceram Soc* **8(1)** 35
[7] Trusov B G 2002 *TERRA software package for calculation of plasma chemical processes* (Materials of III Intern. Symp. on Theoretical and Applied Plasma Chemistry. Ples) pp. 217-218

Acknowledgements

This work was supported by the RFBR, project 15-48-04217-r-sibir-a.

Coating of a substrate with surface preliminarily treated with intensive flows of high-speed electrons and plasma using a magnetron

A Ya Leyvi[1,2], **A P Yalovets**[1]

[1]South Ural State University, pr. Lenina 76, Chelyabinsk, 454080 Russia
[2]LLC NPP EFOM, st. Severnay 56, Chelyabinsk, 454902 Russia

E-mail: leiviai@susu.ru

Abstract. This paper presents results of the research in influence of preliminary surface treatment on adhesion of film to substrate (St8 or brass- target, CU, Ni-film). The preliminary treatment has been conducted by two methods: first, by compressive plasma flow with charge duration ~100μs, plasma formation rate (15-20)×10^3 m/s; second, by low-energy high-current electron beams with pulse duration 2-3 μs and electron energy up to 30 keV. The investigation shows the strong influence of preliminary sample treatment and processing parameters on adhesion of film to substrate and final roughness. Experimental investigations have revealed the best adhesion of film to substrate, and the smoothest film is corresponding to the mode with preliminary LEHCEB irradiation with electron energy of 25, 30 keV. It was shown that the alternation of deposition with LEHCEB irradiation has resulted in large-scale surface smoothing: the surface has become glassy whereas craters of 10-20 μm have formed. It was shown that the magnetron-covered samples can withstand the saline mist for twice as long as samples with galvanic coatings.

1. Introduction

Presently, the treatment of materials by intense energy flows [1-5] (ion beams, low-energy high-current electron beams or compressive plasma flows) has become an effective tool for modification of their properties and for improving their operational characteristics such as microhardness, wear-resistance, change of the surface relief, adhesion of film to a substrate.

It is well known [6] that many factors such as surface roughness [7], surface impurities [8] (contamination, oxide films, adsorbed gases), film deposition methods and others have influence on adhesion of film to substrate. According to [6, 7], surface roughness affects film adhesive interaction both during film deposition and its separation. The adhesive strength is directly proportional to real contact area. The real contact area can be increased by extending surface roughness. It should be noted that the adhesion strength is determined by the size of protrusions and depressions on a surface and their shape. As declared in [9, 10], the surface relief roughness of a sample enlarges after treatment by compressive plasma flow (CPF).

Ref [11] has shown that the treatment by CPF and low-energy high-current electron beams (LEHCEB) increases roughness and, as a result, the adhesion of the film to the substrate. In this case, real contact area treated by CPF increases by 5-10% as compared with that treated by LEHCEB. As a result, the adhesive property of film-substrate system after CPF processing is better than that after LEHCEB treatment.

However, it is well known that film roughness depends on substrate roughness, whereas most of technologies show increased requirements to the film roughness. Ref [11-16] have revealed that LEHCEB treatment can tend to smooth the relief. Therefore, LEHCEB is used as a tool to smooth relief in the research.

The paper is devoted to study the influence of preliminary treatment of substrate surface on adhesion of the film-support system.

2. Experimental methods

Samples of brass and carbon steel 8 were taken. Copper and nickel films were applied by magnetron deposition, which was conducted in current stabilization mode; the current was 3 A and overall deposition time ranged from 30 to 60 min.

LEHCEB processing was made by the electron beam generation facility with the pulse duration range of 1-2 μs and electron energy varying from 5 to 30 keV. Electron processing has conducted in smoothing regime (roughness decrease) with the energy up to 20 keV, and in the regime of roughness increase (crater formation) at 25 and 30 keV.

There were two kinds of samples: first, brass and steel 8 samples with deposed nickel films; second, steel 8 samples with copper film. The samples of both types were exposed to preliminary CPF treatment.

The samples of both kinds were also exposed to LEHCEB with the pulse duration of 1-2 μs, electron energy of 25 and 30 keV and 50 impulses. According to [11, 12], such preliminary treatment results in surface layer melting, intense ablation and, as a consequence, surface cleaning from oxides, and increased roughness of the surface.

The film was applied by magnetron deposition in two modes. First, single-film coating with consequent LEHCEB irradiation and alternation of the film magnetron deposition with LEHCEB processing. In the single-coating mode, electron irradiation was carried out with pulse duration of 1-2 μs, electron energy 10, 15 and 20 keV and 20 impulses.

In the other mode, the deposition carried out for 10 minutes. Then, the covered film-substrate system was exposed to LEHCEB with pulse duration of 1-2 μs, electron energy 10-15 keV and 50 impulses. Consequently, the film was deposited with following LEHCEB irradiation. There were 6 equal cycles. Electron energy and deposition time were determined in the case of LEHCEB irradiation. The surface layer of film-substrate melted without intense ablation as compared with smoothing mode.

3. Results and discussion

One of the steel 8 samples after CPF and LEHCEB processing is depicted in Fig.1. The processing was run in roughening mode. The sample treated by CPF has visual distinguished region with the radius of 1.5-2 cm with complex relief. The LEHCEB irradiation of material has resulted in crater formation on the sample surface. Figure 2 illustrates the sample surface with copper film. Film deposition was alternated with LEHCEB irradiation. The figure demonstrates that the substrate is rougher than the film. Thus, the alternative deposition with LEHCEB processing has caused large-scale smoothing and the surface has become glassy, whereas there are 10-20-μm craters.

Fig.3. shows the first type of samples before and after nickel film deposition. The main purpose of nickel film covering of brass samples is the improvement of adhesive properties in comparison with galvanic covering. The samples with magnetron and galvanic coatings were subjected to saline mist. The results of the study have revealed that magnetron coatings can withstand saline mist for twice as long time as compared to galvanic ones.

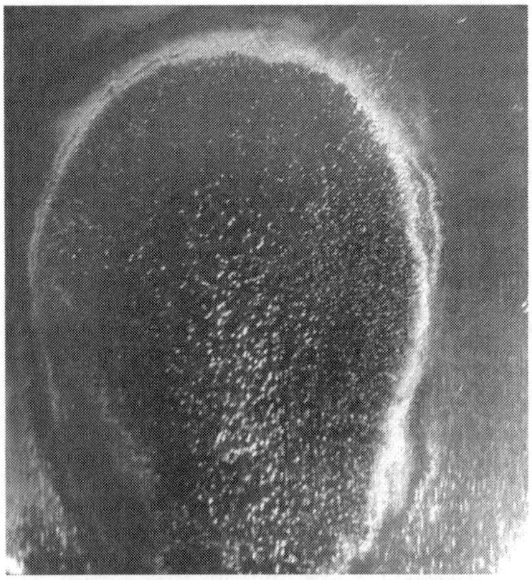

Figure 1. The sample surface after CPF and LEHCEB processing.

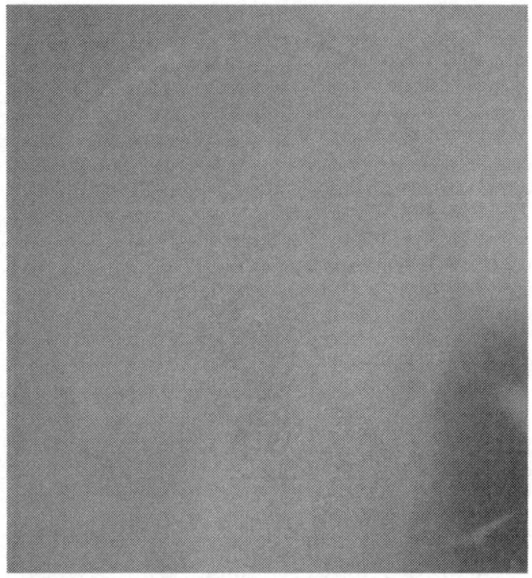

Figure 2. The sample surface with copper film after CPF and LEHCEB processing.

The experiments have demonstrated the best adhesion of the film to the substrate. The smoothest film corresponds to the regime with preliminary LEHCEB irradiation with electron energy of 25 and 30 keV and alternating 10-minute deposition by LEHCEB irradiation by 10-15-keV electrons.

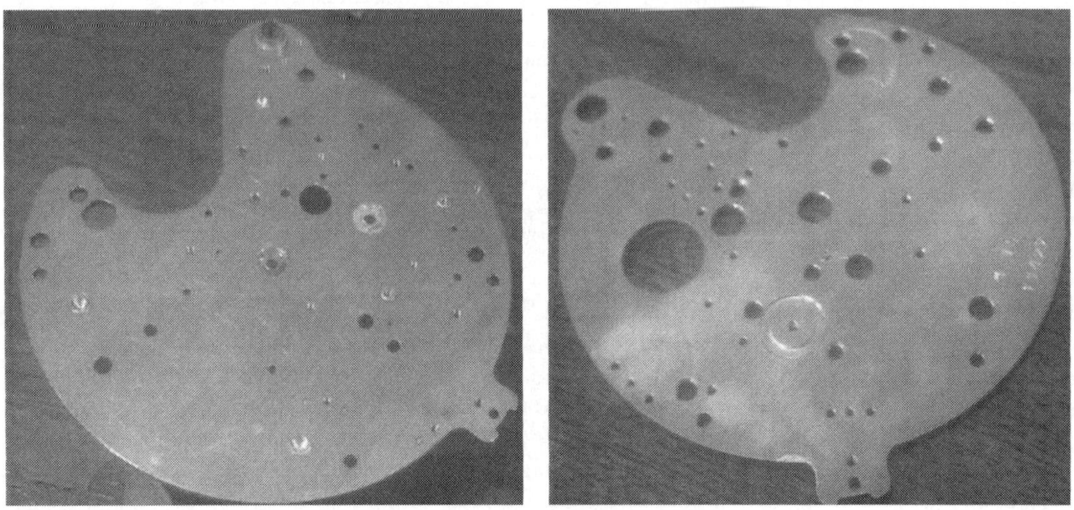

Figure 3. The sample surface before (left) and after (right) nickel film deposition.

4. Conclusion

The research of the influence of preliminary treatment of substrate surface on adhesive property of film-substrate system and final film roughness was conducted.

It was shown that the alternation of deposition with LEHCEB irradiation have resulted in large-scale surface smoothing; the surface has become glassy, whereas 10-20-μm craters have formed.

It was shown that the magnetron coated samples can stay in saline mist for the twice as long as compared to galvanic coatings.

Experimental investigations have revealed the best adhesion of the film to the substrate, and the smoothest film corresponds to the mode with preliminary LEHCEB irradiation with electron energy of 25 and 30 keV.

References

[1] Pauleau Y 2006 *Materials Surface Processing by Directed Energy Techniques* (Elsevier Publishing) p 744

[2] Shulov V A , Novikov A S , Paikin A G , Ryabchikov A I 2007 *Surface and Coatings Technology* **201 19–20** 8105

[3] Rej D J, Davis H A, Nastasi M, Olson J C, Peterson E J, Reiswig R D, Walter K C, Stinnett R W, Remnev G E, Struts V K 1997 *Nuclear Instruments and Methods in Physics Research Section B: Beam Interactions with Materials and Atoms* **127–128** 987

[4] Anishchik V M, Uglov V V, Astashynski V V, Astashynski V M, Ananin S I, Kostyukevich E A, Kuzmitski A M, Kvasov N T, Danilyuk A L, Rumianceva I N 2003 *Vacuum* **70 2–3** 269

[5] Uglov V V, Anishchik V M, Cherenda N N, Sveshnikov Yu V, Astashynski V M, Kostyukevich E A, Kuzmitski A M, Askerko V V 2008 *Surf. Coat. Technol.* **202** 2439

[6] Kinloch A J 1987 Adhesion and Adhesives: Science and Technology (Springer Science & Business Media) p 441

[7] Soo Hong Kim, Sun Woong Na, N.-E. Lee, Yun Woo Nam, Young-Ho Kim. 2005 *Surface & Coatings Technology* **200** 2072

[8] Ju Dy Lim, Pui Mun Lee, Daniel Min Woo Rhee, Kam Chew Leong, Zhong Chen. 2015 *Applied Surface Science* **355** 509

[9] Juárez-Moreno J A, Ávila-Ortega A, Oliva A I, Avilés F, Cauich-Rodríguez J V 2015 *Applied Surface Science* **349** 763

[10] Ghalmi Zahira, Farzaneh Masoud 2015 *Cold Regions Science and Technology* **115** 42

[11] Leyvi A Ya, Yalovets A P and Kozhevnikov D G 2016 *Materials Science Forum* **Vol. 843** 73

[12] Krasnikov V S, Leyvi A Ya, Mayer A E, Yalovets A P 2007 *Technical Physics* **52** (4) 431

[13] Raharjo P, Uemura K, Okada A, Uno Y 2004 *Proc. of the 7th Int. Conf. on Modification of Materials with Particle Beams and Plasma Flows* (Tomsk: IAO Publishing) p 263

[14] Raharjo P, Uemura K, Okada A, Uno Y 2004 *Proc. of the 7th Int. Conf. on Modification of Materials with Particle Beams and Plasma Flows* (Tomsk: IAO Publishing) *p* 267

[15] Shulov V A, Novikov A S, Paikin A G, Belov A B, Lvov A F, Remnev G E 2007 *Surface and Coatings Technology* **201 19–20** 8654

Acknowledgements

The research was partially supported by the Ministry of Education and Science of the Russian Federation (contract no. 1030 (2014105-GZ) for support of research at the South Ural State University).

IOP Conf. Series: Journal of Physics: Conf. Series **830** (2017) 012112 doi:10.1088/1742-6596/830/1/012112

Influence of microstructure and surface topography on the electrical conductivity of Cu and Ag thin films obtained by magnetron sputtering

D A Polonyankin, A I Blesman and D V Postnikov

Omsk State Technical University (OmSTU), 11 Mira Avenue, Omsk, 644050, Russia

E-mail: nano@omgtu.ru

Abstract. Conductive thin films formation by copper and silver magnetron sputtering is one of high technological areas for industrial production of solar energy converters, energy–saving coatings, flat panel displays and touch control panels because of their high electrical and optical properties. Surface roughness and porosity, average grain size, internal stresses, orientation and crystal lattice type, the crystallinity degree are the main physical properties of metal films affecting their electrical resistivity and conductivity. Depending on the film thickness, the dominant conduction mechanism can affect bulk conductivity due to the flow of electron gas, and grain boundary conductivity. The present investigation assesses the effect of microstructure and surface topography on the electrical conductivity of magnetron sputtered Cu and Ag thin films using X–ray diffraction analysis, scanning electron and laser interference microscopy. The highest specific conductivity (78.3 MS m^{-1} and 84.2 MS m^{-1}, respectively, for copper and silver films at the thickness of 350 nm) were obtained with the minimum values of roughness and grain size as well as a high degree of lattice structuredness.

1. Introduction

Designing insulating, conducting and semiconducting films from metals and their oxides onto substrates is one of the high technological areas in microelectronics and optical industries due to their distinctive electrical and optical properties [1].

In the field of electronic and optical devices, silver is an exceptional material because of its reflectivity which reaches 97.7% at a wavelength of 500 nm and the smallest resistivity among all the metals [2]. High reflecting ability in the visible range of wavelengths as well as large electrical conductivity cause the widespread pure silver application for optical reflectors and for producing electrical conductors in microprocessor technology. However, the impact of environmental factors (temperature, humidity) and the aggressive exposure of chlorine and sulfur reduce the performance of silver coatings [3].

Copper metal coatings are widely used in microelectronics because of its low electrical resistivity (1.67 μΩ cm at room temperature), small degree of atom surface migration and high melting point [4]. Nanostructured thin copper films formed on silicon substrates by magnetron sputtering are used for protective electromagnetic screens production [5]. Application of copper magnetron sputtering for integrated microcircuits manufacture is caused by its high thermal and electrical conductivity (only silver has a better electrical conductivity at a higher material cost) [6]. Thin copper films sputtered on porous silicon substrate with high optical and electrical properties are widely used in production of LEDs, optical fibers and filter, photovoltaic diodes and different type of sensors [7].

The review of published data on the problem of thin copper and silver films formation by magnetron sputtering allows us to formulate the following statements. Conductive copper and silver coatings can be divided into two groups: the first one contains ultrathin films with the thickness less than 40 nm, which exhibit size effect property [8], and the second one contains films related to submicron scale with the thickness ranging from 50 to 800 nm [9] having grain-boundary mechanism of conductivity, which also obeys the classic laws of conductivity theory (Ohm's law, resistivity dependence on temperature and conductor geometry, etc.). According to film with the thickness related to submicron region, the dominant conduction mechanism can impact bulk conductivity due to the electron gas flow and grain boundary conductivity [10]. The most significant physical, chemical and structural properties of films affecting their electrical conductivity include surface roughness [11], its porosity [12], adhesion force and energy [13], average grain size [8, 14], internal stresses [15], crystal lattice type and coating crystallinity degree [16, 17]. The physical and electrical properties of films considerably depend on such process parameters of their production as temperature [2], substrate material [11], chamber pressure [18], sputtering type (high–frequency [9, 19] or DC [17] sputtering) and power [20], deposition rate [11], electromagnetic field configuration and target area [21].

Thus, magnetron sputtering is extensively used for surface modification of products demanded in various industries, while thin–film coating design is the basic part or a constituent of the production technology for protective, wear–resistant, low–friction, anti–corrosion and decorative copper and silver coating with special optical and electrical properties [22]. Thus, obtaining Cu and Ag conductive coatings by magnetron sputtering is a topical problem of modern materials science and nanotechnology due to significant application value of these method and materials in technologies of thin–film coatings production with high physical, chemical and performance characteristics.

The object of this study is to conduct a comparative research of microstructure and surface topography influence on electrical conductivity of Cu and Ag thin films produced under identical conditions by RF magnetron sputtering having the thickness ranging from 100 nm to 500 nm with 50-nm step.

2. Experimental details

For Cu and Ag films preparation, ADVAVAC VSM–200 magnetron setup was used. Thin Cu and Ag films with the thickness ranging from 100 nm to 500 nm and 50 nm step were sputtered on soda–lime silicate glass substrate at high vacuum RF mode with the source power of 100 W under $(3.2\div3.4)\cdot10^{-3}$ mbar argon pressure and pre-ionic chamber cleaning for 10 minutes. The selection of the thickness range is conditioned by the reason that such thickness is far from the size effect region [9] and bulk properties are expected. Thin films were sputtered with pure argon (Ar) by using Cu and Ag targets (99.95% purity, with 50-mm diameter).

Cu and Ag qualitative and quantitative analysis was performed by scanning electron microscopy (JEOL JCM–5700) and energy dispersive spectroscopy (EDS). XRD investigation of silver and copper films with the thickness in the range from 100 nm to 500 nm and 50 nm step was carried out by a sliding X–ray beam method (fixed tube angle θ=5°) in Bragg–Brentano configuration with 2θ from 40° to 80° for Cu films and from 30° to 80° for Ag films and Cu–Kα radiation ($\lambda = 0.154$ nm). Shimadzu XRD–7000 diffractometer allowed obtaining values of 2θ angle and corresponding full–width at half–maximum (FWHM) for all identified peaks.

Surface roughness analysis was provided by laser interference microscopy (LIM) performed in dependence on the film thickness. Laser interference microscope MIM–340 (Shvabe) is designed for non–contact and non–destructive submicron measurements in nanometer region and is equipped with 405 nm wavelength violet laser.

3. Results and discussion

SEM images and EDS results of Ag and Cu films with 350 nm thicknesses are shown in Figures 1 and 2, respectively. Results of quantitative EDS analysis demonstrate that Ag and Cu films do not contain impurity atoms, which is also confirmed by XRD results.

As seen from mixed XRD pattern presented in Figure 3 and 4, it contains the peaks reflected by Ag and Cu crystal lattice and peaks related to planes with [(111), (200), (220) and (311)] Miller indices was detected. The intensity of (111) peaks surpasses manifold the intensities of other peaks. The shape of (111) peak indicates a high crystallinity degree of sputtered Ag and Cu films. Thus, we can conclude that Ag and Cu films with the thickness ranging from 100 nm to 500 nm possess polycrystalline face–centered cubic structure of bulk materials and their crystallographic plane orientation preferentially has (111) texture. Applying sliding X–ray beam allowed us to detect scattering radiation only from Ag and Cu, because the film thickness is greater than X–ray penetration depth, and the signal coming from the substrate is not observed.

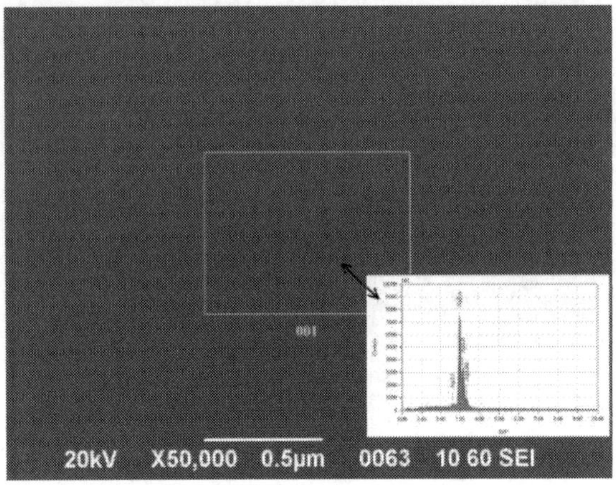

Figure 1. SEM image and EDS results of Ag film with 350 nm thickness.

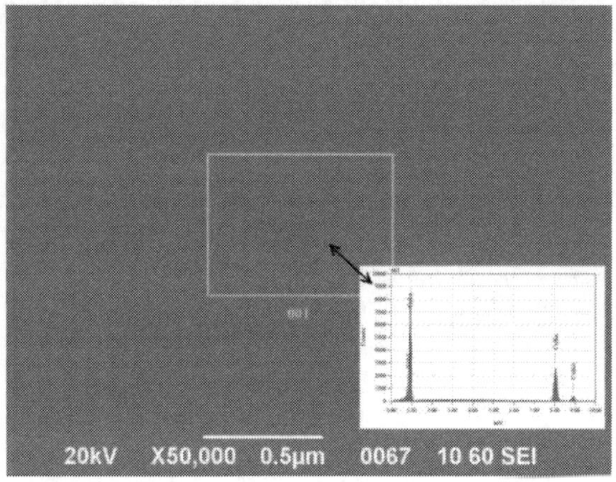

Figure 2. SEM image and EDS results of Cu film with 350 nm thickness.

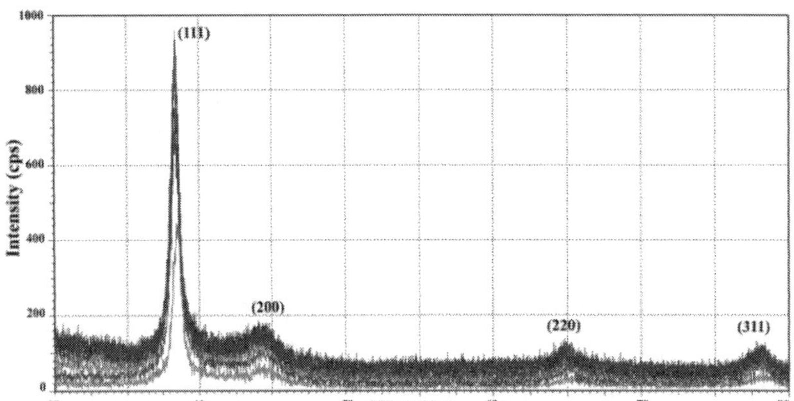

Figure 3. Mixed XRD pattern of Ag films with the thickness in the range from 100 nm to 500 nm and 50 nm step.

3D LIM images and surface roughness profiles of Ag and Cu films with 350 nm thicknesses are shown in Figures 5 and 6, correspondingly. RMS values of surface roughness were calculated by MIM–Soft program for the base length of 80 nm after bonding together four surface profiles by MIM–Series software for each film thickness value. The average roughness of Ag films ranges from 3.59 nm (for 350-nm film) to 4.86 nm (for 100-nm film), while the average roughness of Cu films ranges from 5.44 nm (for 350-nm film) to 7.69 nm (for 500-nm film) depending on the film thickness (Figure 7).

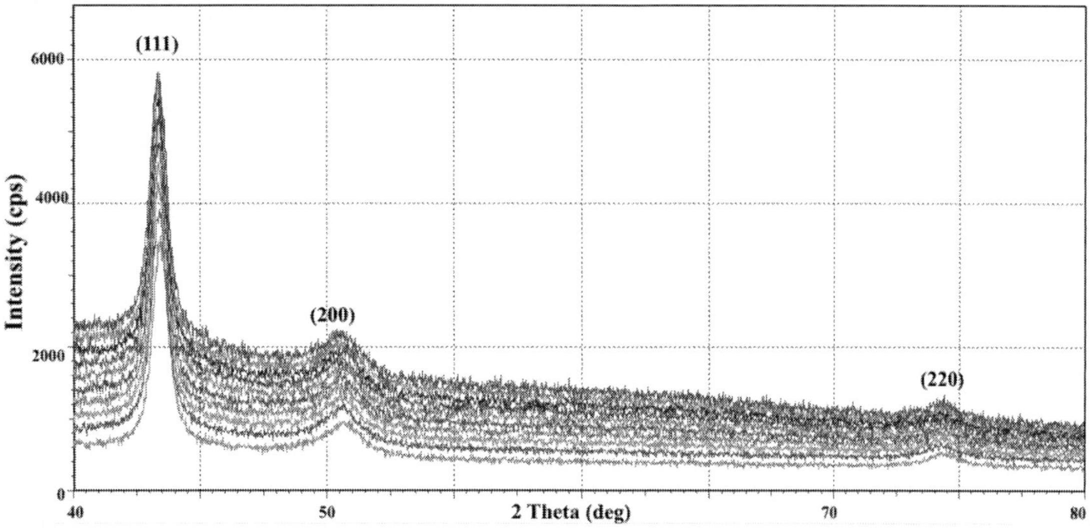

Figure 4. Mixed XRD pattern of Cu films with thickness in the range from 100 nm to 500 nm and 50 nm step.

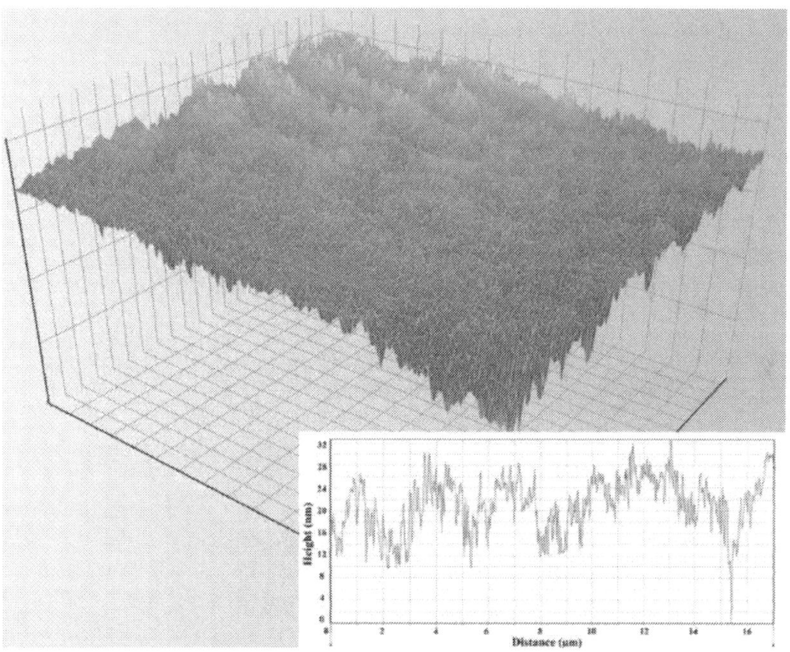

Figure 5. 3D LIM image and surface roughness profile of 350-nm Ag film.

On the basis of XRD data, according to Debye–Scherrer formula, the average grain size was calculated in dependence on the film thickness (see Figure 7). The average grain size of Ag films ranges from the minimum 10.02 nm (for 350-nm film) to maximum of 11.12 nm (for 500-nm film). The average grain size of Cu films ranges from the minimum of 11.43 nm (for 350-nm film) to maximum of 13.24 nm (for 500-nm film). The roughness and the average grain size dependences on the film thickness can be explained by island mechanism, when the force and energy of adhesion to the substrate is less than the mutual binding energy of the deposited atoms, and film properties with the thickness growth become close to that of bulk material.

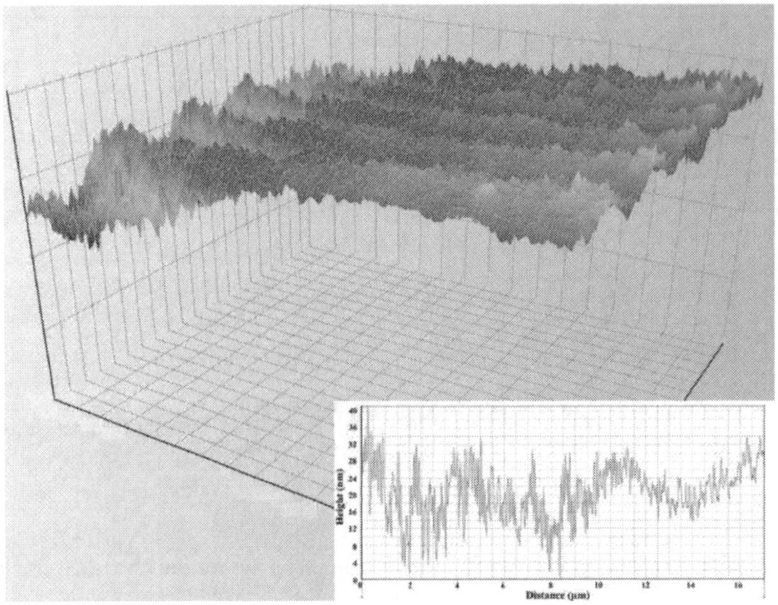

Figure 6. 3D LIM image and surface roughness profile of 350-nm Cu film.

Copper and silver resistivity was measured by ohmmeter for all values of films thickness from 100 to 500 nm with 50 nm step, formed on the substrate of the cover glass (with the length and the width of 7.5 and 2.5 mm, respectively), as well as their specific conductivity was calculated. The dependencies of specific conductivity on film thickness are shown in Figure 8.

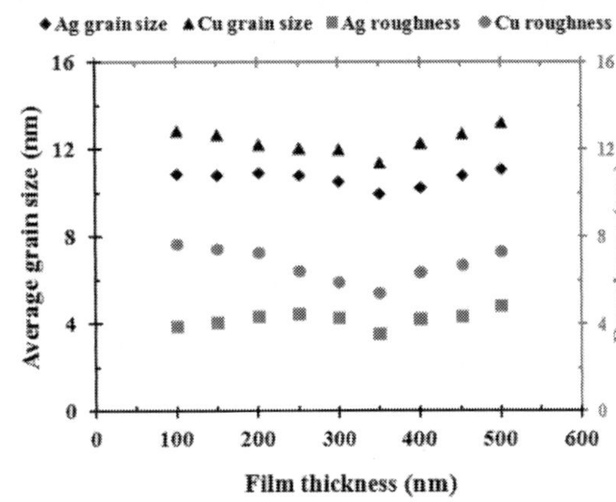

Figure 7. The average grain size and roughness vs. Cu and Ag film thickness.

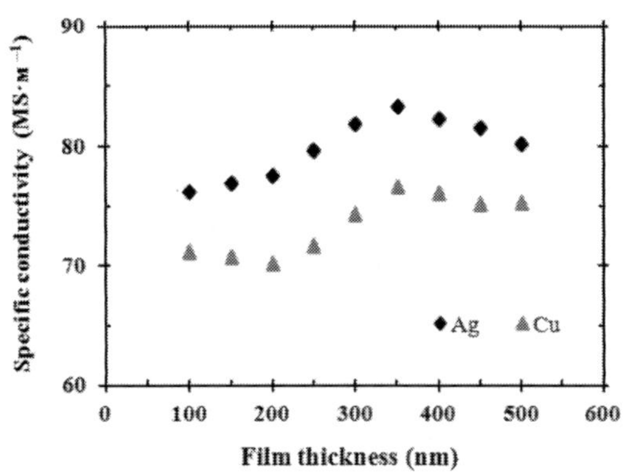

Figure 8. Specific conductivity vs. Cu and Ag films thickness.

Specific conductivity of Ag ranges from the minimum of 76.2 MS m^{-1} (for 100-nm film) to the maximum of 84.2 MS m^{-1} (for 350-nm film), while the specific conductivity of Cu films ranges from 70.2 MS m^{-1} (for 200-nm film) to 78.3 MS m^{-1} (for 350-nm film).

4. Conclusion

Cu and Ag films produced by RF magnetron sputtering having the thickness ranging from 100 nm to 500 nm and with 50 nm step according to XRD results possess polycrystalline face–centered cubic structure of bulk materials and their crystallographic plane orientation preferentially have (111) texture.

The average roughness of Ag films ranges from 3.59 nm (for 350-nm film) to 4.86 nm (for 100-nm film), while the average roughness of Cu films ranges from 5.44 nm (for 350-nm film) to 7.69 nm (for 500-nm film) in dependence on the film thickness. The average grain size of Ag films ranges from the minimum of 10.02 nm (for 350-nm film) to the maximum of 11.12 nm (for 500-nm film). The average grain size of Cu films ranges from the minimum of 11.43 nm (for 350-nm film) to the maximum of 13.24 nm (for 500-nm film). The roughness and the average grain size dependencies from the film thickness can be explained by island mechanism, when the force and energy of adhesion to the substrate is less than the mutual binding energy of the deposited atoms, and film properties with the thickness growth become close to the bulk material.

The highest specific conductivity (78.3 MS m^{-1} and 84.2 MS m^{-1}, respectively for copper and silver films at the thickness of 350 nm) were obtained with the minimum values of roughness and grain size as well as a high degree of lattice structuredness.

References

[1] Gordon R 1997 *J. Non–Cryst. Solids* **218** 81
[2] Suzuki T, Abe Y, Kawamura M, Sasaki K, Shouzu T and Kawamata K 2002 *Vacuum* **66** 501
[3] Nakanishi Y, Kato K, Omoto H and Yonekura M 2013 *Thin Solid Films* **532** 141
[4] Cao B, Yang T R, Li G P, Cho S J and Kim H 2012 *Advanced Materials Research* **430** 419
[5] Desideri D, Natali M, Cavallin T and Maschio A 2014 *IET Science, Measurement & Technology* **8** 1
[6] Mech K, Kowalik R and Żabiński P 2011 *Archives of Metallurgy and Materials* **56** 903
[7] Ansari Z A and Hong C L 2002 *Mater. Sci. Eng., B* **90** 103
[8] Ke Y, Zahid F, Timoshevskii V, Xia K, Gall D and Guo H 2009 *Phys. Rev. B* **79** 155406
[9] Marechal N, Quesnel E and Pauleau Y 1994 *Thin Solid Films* **241** 34
[10] Rossnagel S M and Kuan T S 2004 *J. Vac. Sci. Technol. B* **22** 240
[11] Yang J, Huang Y and Xu K 2007 *Surf. Coat. Technol.* **201** 5574
[12] Carton O, Ghaymouni J, Lejeune M and Zeinert A 2013 *Journal of Spectroscopy* **2013** 1
[13] Tkachenko E A, Postnikov D V, Blesman A I and Polonyankin D A 2016 *IOP Conf. Ser.: Mater. Sci. Eng.* **110** 012009
[14] Sursaeva V G, Gottstein G and Shvindlerman L S 2016 *Scripta Mater.* **116** 91
[15] Sun X, Hong R, Hou H, Fan Z and Shao J 2007 *Thin Solid Films* **515** 6962
[16] Jiang H 1998 *J. Vac. Sci. Technol. A* **16** 3376
[17] Chan K–Y, Luo P–Q, Zhou Z–B, Tou T–Y and Teo B–S 2009 *Appl. Surf. Sci.* **255** 5186
[18] Del Re M, Gouttebaron R, Dauchot J P, Leclère P, Lazzaroni R, Wautelet M and Hecq M 2002 *Surf. Coat. Technol.* **151** 86
[19] Jiang H, Klemmer T, Barnard J, Doyle W and Payzant E 1998 *Thin Solid Films* **315** 13
[20] Chan K–Y and Teo B–S 2007 *Microelectron. J.* **38** 60
[21] Promros N, Sittimart P, Patanoo N, Kongnithichalerm S, Horprathum M, Bhathumnavin W and Paosawatyanyong B 2016 *Key Eng. Mater.* **675** 193
[22] Kelly P J and Arnell R D 2000 *Vacuum* **141** 159

IOP Conf. Series: Journal of Physics: Conf. Series **830** (2017) 012113 doi:10.1088/1742-6596/830/1/012113

Influence of substrate structure on adhesion behavior of TiN and $TiAl_3$/TiAlN coatings deposited by vacuum arc plasma

E Vardanyan, K Ramazanov, I Yagafarov, V Budilov, R Agzamov

Ufa State Aviation Technical University, Ufa - 450000, Russia

E-mail: vardanyaned@gmail.com

Abstract. This paper presents the study of the influence of a substrate structure on the adhesion behavior of monolayer coatings TiN and multilayer coatings $TiAl_3$ / TiAlN deposited by vacuum arc plasma. Martensitic steel with a coarse-grained (CG) and ultrafine-grained (UFG) structure was used for samples. The samples were subjected to ion nitriding in a glow discharge before coating deposition. Adhesion of the coatings was examined with CSM ScratchTEST. For samples with different structures, critical load was defined at which microcracks are formed in the coatings.

1. Introduction

Durability and reliability of gas turbine engines (GTE) are largely determined by load-carrying capacity of compressor blades. The blades are vital and highly loaded parts that experience considerable alternating and cyclic loads of high frequency during operation. Gas turbine blades typically brake due to erosion, corrosion and fatigue damage [1]. Research of the past two decades showed that mechanical and fatigue properties of metals and alloys can be effectively improved by the formation of UFG nanostructures with submicron and nanocrystalline grains by intense plastic deformation (IPD) [2]. Thus, materials for GTE compressor blades should possess a UFG structure [2]. A promising approach to protect GTE compressor blades is to apply multilayer coatings of periodically arranged nanometer layers of different materials that have high mechanical properties and performance characteristics [3-5]. A range of coatings with different chemical composition (TiN, Ti-Al-N, Ti-C-N, Ti-Zr-N, ZrO_2) and structure (mono- and multilayer) is used to protect GTE compressor blades. Among them multilayer Ti/Al coatings are of great practical interest due to high corrosion resistance in aggressive environments, strength, and low specific weight. The Ti-Al-N system also forms [6,7] a large number of different intermetallic compounds with unique physical and mechanical properties.

One of the most important characteristics of coatings is their adhesion to substrates. Most studies of film deposition analyze not only their structure, phase composition, mechanical characteristics, etc., but also their adhesive properties. Adhesion is commonly assessed by separation of a coating from the substrate and scratching. Separation is mainly applied to thick films, while thin films are typically subjected to scathing [8]. The latter is used in ~70% of applications. While separation defines adhesion in MPa, in scratch tests adhesion is characterized by the critical load at which a film delaminates. ASTM assesses adhesion of coatings against their critical load at which the coating spalls off [9].

The aim of this study is to investigate the influence of martensitic steel structure on the adhesive properties of intermetallic Ti-Al coatings.

2. Experimental approach

Intermetallic Ti-Al coatings were deposited in the installation NNV-6.6-I1. Samples of martensitic steel EI-961Sh were placed in a vacuum chamber. Technological regimes were changed in the following ranges: chamber pressure $P = 10^{-1}–10^{-2}$ Pa; arc current $I = 60-120$ A; processing time $t = 60$ min.

The coatings were deposited with simultaneous spraying of one-component cathodes from Al and Ti in the conditions of rotation of the table around its axis. During the deposition, the samples moved through four areas: 1) where the samples were exposed to an Al ion flux; 2) where the samples were exposed to an Al and Ti ion flux (transition area); 3) where the samples were exposed to a Ti ion flux; and 4) where the samples were not exposed to deposition (shadow zone) [9]. The resulting coatings contained different grades of intermetallic phases.

To study the influence of the substrate structure (coarse or ultrafine grain structure) on the adhesion characteristics of the coatings, the samples were exposed to ion nitriding and subsequent deposition of intermetallic Ti-Al coatings onto martensitic steel (see Table 1).

Table 1. Treatment of samples

Sample	Substrate structure and treatment
1	CG + nitriding+ TiN
2	UFG + nitriding + TiN
3	CG + nitriding + $TiAl_3$/TiAlN
4	UFG + nitriding + $TiAl_3$/TiAlN

A CSM Instruments scratch tester [9] (Figures 1 and 2) was used to determine adhesion power and scratch resistance and identify the coating failure mechanism. The surface of the coatings was scratched with a diamond spherical indenter of the Rockwell C type with a curvature radius of 200 mm at a continuously increasing load. A computer was used to register the applied load (F_N, H), the penetration depth of the indenter (Pd, μm) and acoustic emission (AE, dB), which reflected the destruction, delamination or spalling of the coating. After that, the load was released and the indenter moved in the opposite direction to measure the restored scratch depth (Rd, μm). The tests were performed in the following conditions: the indenter was increasingly loaded at 0.3-30 N; the indentation rate was 2 mm/min; the scratch's length was 5 mm; the load application rate was 11.88 N/min; the frequency of the discrete signal was 60 Hz, and the power of the acoustic emission was 9 dB. The minimum (critical) loads LC, at which first cracks appeared, were specified in the tests.

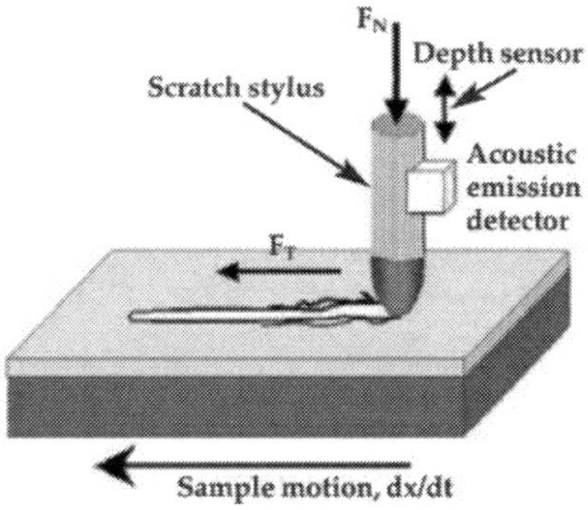

Figure 1. Scratch adhesion test schematic (from ASTM C1624) [9].

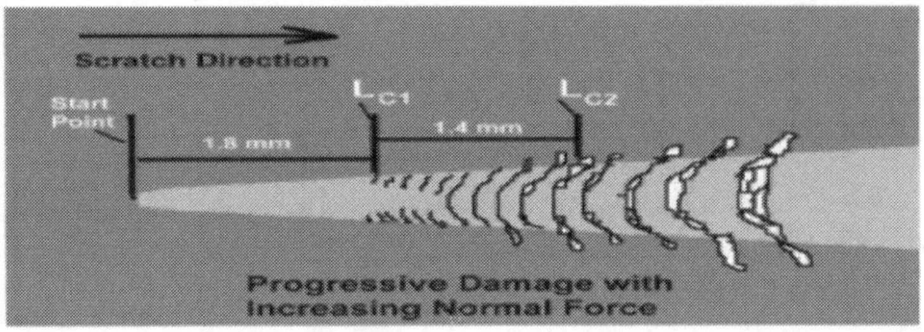

Figure 2. Formation of microcracks at an increasing load [9].

3. Results and discussion

Coating thickness was 10-10.5 μm, while the thickness of individual layers in multilayer coatings $TiAl_3$ / TiAlN was ~ 0.8...1 μm (Figure 3).

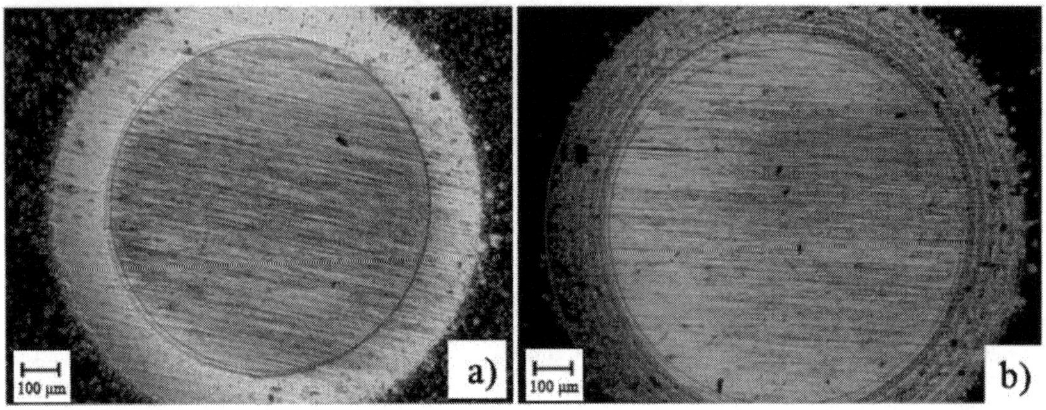

Figure 3. Measurement of coating thickness for a) TiN at 10x magnification; b) $TiAl_3$/TiAlN at 10x magnification.

CG and UFG samples of martensitic steel with TiN and $TiAl_3$/TiAlN coatings processed with different technologies (Table 1) were investigated with sclerometry (Table 1). Optical images of the scratch tracks are shown in Figure 4.

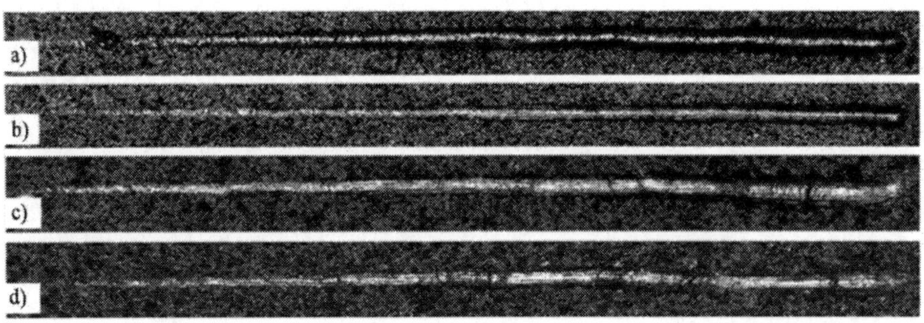

Figure 4. Optical images of the scratch tracks on treated samples: a) CG + ion nitriding + TiN; b) UFG + ion nitriding + TiN; c) CG + ion nitriding + $TiAl_3$/TiAlN; d) UFG + ion nitriding + $TiAl_3$/TiAlN.

The indenter penetration depth is given in Figure 5.

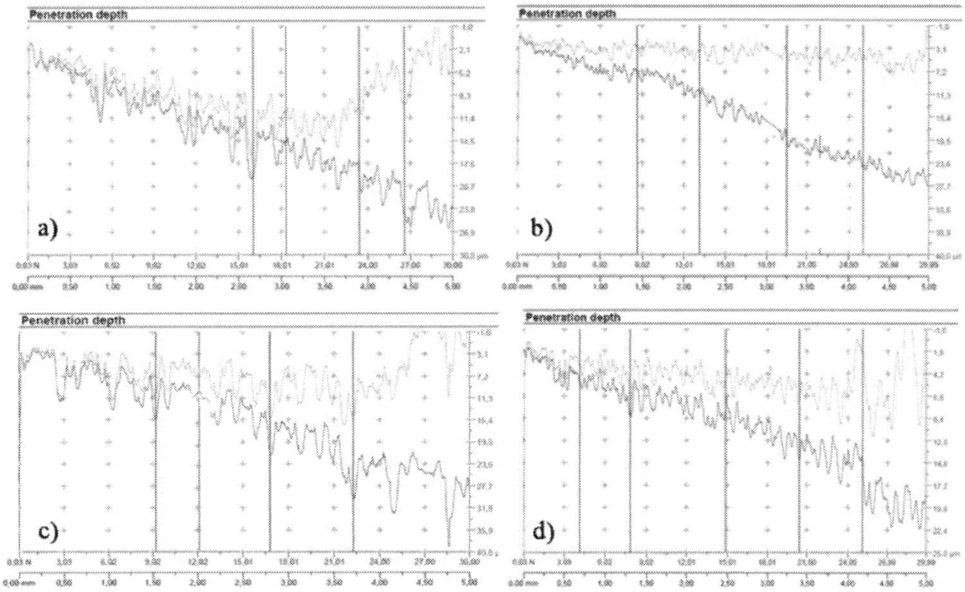

Figure 5. Indenter penetration depth: a) CG + ion nitriding + TiN; b) UFG + ion nitriding + TiN; c) CG + ion nitriding + $TiAl_3$/TiAlN; d) UFG + ion nitriding + $TiAl_3$/TiAlN.

The results show that the indenter penetration depth under the load of 30 N and the scratch depth after the load was released differ during the transition from CG structure (Figures 5a and 5c) to UFG structure (Figures 5b and 5d). Scratching of samples with monolayer TiN coatings showed that the depth of loaded indenter penetration for CG and UFG structure is similar and amounts to ~25-27 μm. However, the track depth after release of the load for samples with the CG structure is about 15 μm, which exceeds the thickness of the coating. Thus, the TiN coatings on the samples with the CG structure destroyed completely. At the same time, the depth of the track after release of the load was 7.5 μm and the coatings remained for the samples with the UFG structure. The diagram for the indenter penetration depth demonstrates that elastoplastic recovery regions in samples with the UFG structure are significantly greater than those in the samples with the CG structure.

The study of the samples with multilayer $TiAl_3$/TiAlN coatings showed that the loaded penetration depth of the indenter is 32 μm and 22 μm for the specimens with the CG and UFG structure, respectively. The measurements after the load release showed that for the samples with CG structure the track depth was about 15 μm, which also exceeds the thickness of the coating. On the contrary, for the samples with UFG structure the depth of the track was about 7 μm.

Scratch tests of the coated samples (Figures 5 and 6) show that the area of the elastic recovery in the samples with the CG structure is significantly narrower than that of the samples with the UFG structure for both TiN and $TiAl_3$/TiAlN coatings. This finding indicates a larger mechanical performance of the samples with the UFG structure. Table 2 gives optical determination of the critical load at which microcracks occur in the coatings.

Table 2. Critical load of microcracks formation

Treatment	1	2	3	4
Critical load [N]	16	19	16.8	22

The study of the samples with the UFG structure of the substrate material coated with TiN and $TiAl_3$/TiAlN showed that microcrack formation starts at the critical load of 19 N and 22 N, respectively.

4. Conclusions

The comparative analysis of the scratch tests carried out for the deposited martensitic steel proved that:

- all samples demonstrate high adhesion to substrates. There is no delamination of the coatings at the maximum load F = 30 N;
- the critical load, at which microcracks are formed is 16-22 N for the samples with the UFG structure. This value is significantly greater than that of the samples with the CG structure (15-19 N) for both types of coatings (TiN and $TiAl_3$/TiAlN).

Thus, the formation of UFG structure in a substrate improves adhesion strength of the coatings.

References

[1] Nihamkin M Sh, Ravens LV, Konev I P 2006 *Bulletin engine* **3** 93
[2] Feuerstein A, Kleyman A 2009 *Surface and Coatings Technology* **204** 1092
[3] HetmaEczyk M, Swadlba L, Mendala B 1996 *Surface Treatment Conference materials, III Nationwide Science Conference* **9** 240
[4] Urbahs A, Savkovs K, Urbaha M, Kurjanovics I 2012 *Nanodevices and Nanomaterials for Ecological Security. NATO Science for Peace and Security Series B: Physics and Biophysics* 307
[5] Zhirkov I, Oks E, Rosen J 2015 *J. Appl. Phys.* **117** 213
[6] Musil J, Stupka P 2015 *Int. J. Nanomanufacturing* **11** 78
[7] Budilov V et al 2004 *Mat. Sci. and Eng A* **375-377** 656
[8] Budilov V, Vardanyan E 2015 *Journal of Physics: Conference Series* **652** 012053
[9] Lunev M, Nemashkalo O 2010 *FIP the PSE* **8**(1) 64
[10] Stephen T, Gonczy A 2005 *Int. J. Appl. Ceram. Technol.* **2** 422

Acknowledgments

The authors are grateful to the Russian Science Foundation for support of this work through the project No. 15-19-00144.

IOP Conf. Series: Journal of Physics: Conf. Series **830** (2017) 012114 doi:10.1088/1742-6596/830/1/012114

Investigation of influence of structure and $TiAl_3$/TiAlN intermetallic coatings on the corrosion behavior of martensitic steels

E Vardanyan, K Ramazanov, I Yagafarov, A Khamzina, R Agzamov

Ufa State Aviation Technical University, Ufa - 450000, Russia

Email: vardanyaned@gmail.com

Abstract. The paper considers several approaches to protect martensitic steels with ultrafine-grain (UFG) structure in aggressive environments. Scanning electron microscopy was used to study the microstructure and composition of steel substrates and coatings. The samples were also subjected to corrosion tests. Regularities of corrosion behavior were specified for the UFG steels. The samples were subjected to ion nitriding in a glow discharge and deposition of protective $TiAl_3$/TiAlN coatings in vacuum arc discharge plasma. Corrosion rates were identified for different treatments.

1. Introduction

Recently, bulk nanostructured materials of the grain size of 100-300 nm with unique structures and properties were developed. The materials change such fundamental characteristics as the Debye and Curie temperature, saturation magnetization, etc. Currently, particular attention is given to the method of producing such materials and namely severe plastic deformation (SPD) because it allows creating high volume semi-finished nanostructured products in the form of bars, thin foils and wires for various applications. At the present time we also observe transition from laboratory SPD experiments to designing pilot industrial technology based on continuous processes [1,2].

Until now, great progress has been made in the development of physical principles that improve properties in the study of structural models and evolution upon annealing, the study of some fundamental deformation characteristics, namely, magnetic properties, elasticity, internal friction, grain boundary diffusion, etc. However, the issue of corrosion behavior of such materials remains understudied, as existing data are insignificant and contradictory.

Properties of nanocrystalline materials cannot be predicted from the properties of corresponding coarse-grain analogues. The imperfections of the crystal lattice in metals and alloys due to deformation and high defect ratio in grain-boundary areas contribute to a local increase in the velocity of dissolution of these zones, which ultimately leads to accelerated dissolution. Therefore, there is a need in approaches to improve resistance to corrosion of such materials, which include ion nitriding, ion-plasma coating, combined methods, passivation, oxidation, chemical and electrochemical polishing.

The study of corrosion properties for coatings with intermetallic phases (TiAl, $TiAl_3$) are given in [3-7]. Literature review showed that Ti-Al intermetallic coatings that contain $TiAl_3$+TiAl phases demonstrate good corrosion resistance and can be considered for the protection of UFG martensitic steel.

The aim of this work is to identify regularities in corrosion behavior of CG and UFG martensitic steel in relation to the type of its processing.

2. Experimental approach

Specimens of martensitic steel EI-961SH with coarse-grain and ultrafine-grain structure were used as experimental materials.

Samples of 20 mm in diameter and 1 mm thick were subjected to severe plastic deformation by torsion (SPDT) under the pressure of 6 GPa with 10 twists to produce the UFG structure.

Intermetallic Ti-Al coatings were applied in the installation NNV-6.6-I1. Samples of martensitic steel EI-961SH were placed in a vacuum chamber. Technological spraying regimes changed in the following ranges: chamber pressure $P = 10^{-1}–10^{-2}$ Pa; arc current $I = 60$-120 A; processing time $t = 60$ min [9].

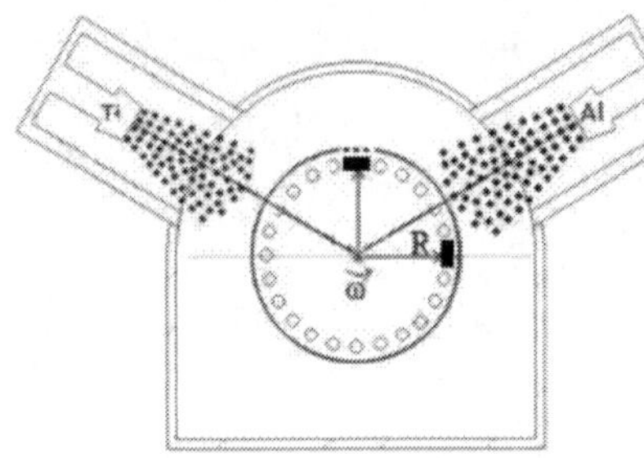

Figure 1. Scheme of the experiment.

Ion nitriding was carried out in a modernized installation ELU-5M. Argon and a mixture of nitrogen, argon, acetylene (N_2 75% + Ar 20% + C_2H_2 5%) were used as a working gas. The pressure of the working gas varied in the range of 5...200 Pa. [10].

Corrosion tests were carried out according to GOST 9.905-2007. The tests were performed in artificially created conditions that simulated climatic effects of the atmosphere; the corrosive medium was composed of 3% NaCl.

The conductivity of the solutions was measured with Dulcometer LFWS 1C2 with the cell constant of 1.0 cm^{-1} and the measuring range of 10-200 *S/cm. The temperature was controlled automatically with the integrated potentiometer Pt 100. Conductivity was measured with a conductivity probe that was inserted into test solutions.

The Dulcometer was set up in the following order: a series of standard KCl solutions with known electrical conductivity was prepared; the probe was dipped into the electrolyte and conductivity was recorded using a special button after 3-5 min.; then the Dulcometer measured electrical conductivity of the test solution in the automatic mode.

Corrosion resistance was measured by gravimetry.

3. Results and discussion

The microstructure and elemental composition of the surface in UFG samples as well as the coating layers were investigated by scanning electron microscopy (Figure 2, Figure 3).

Table 1 shows the results of measuring the atomic ratio of the chemical elements in different points of the coating (Figure 3).

Analysis of the chemical composition showed the maximum content of nitrogen in dark areas (points 1, 3, 5 in Figure 3) and reduced content of nitrogen in light areas (points 2, 4 in Figure 3), which correspond to intermetallic layers. With the transition from the upper to the lower layers of the coating, an iron fracture increases, which could move from the investigated steel. The increased nitrogen content in point 6 as compared to points 2 and 4 is due to the nitrided layer that enters the scanning region.

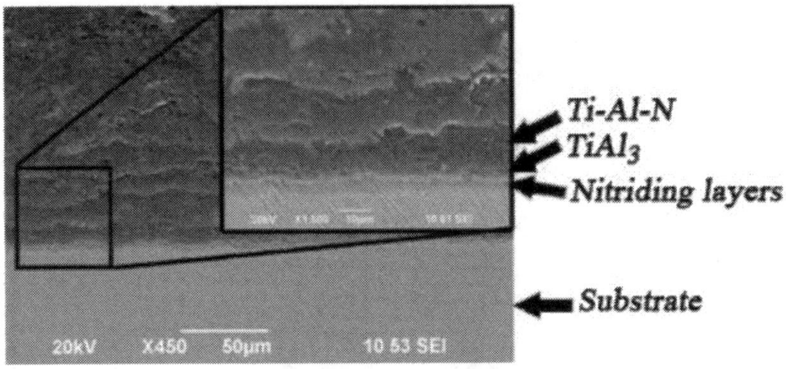

Figure 2. UFG steel microstructure after ion nitriding and subsequent coating with $TiAl_3$/TiAlN.

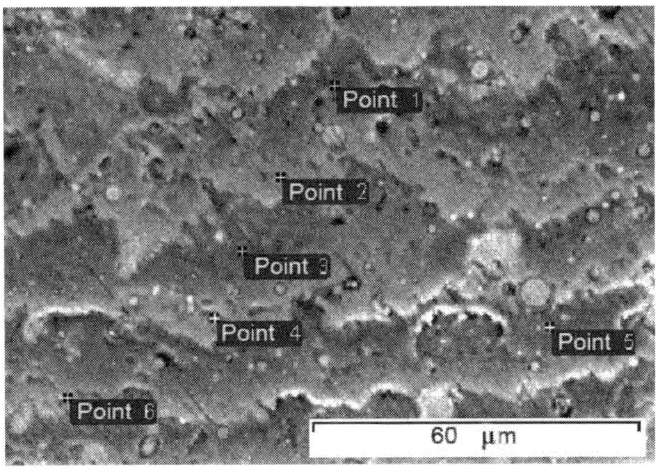

Figure 3. Microstructure of the multilayer coating and points of measuring the chemical composition: points 1, 3, 5 – TiAlN; 2, 4, 6 – $TiAl_3$.

There are no literature data on the effect of ion modification followed by deposition of ion-plasma coatings on the corrosion properties of UFG martensitic steels. The present study investigated the corrosion behavior of UFG martensitic steels subjected to ion-plasma treatment: ion nitriding followed by deposition of the $TiAl_3$/TiAlN coating.

Table 1. Atomic ratio of chemical elements in different points of $TiAl_3$/TiAlN coating

Points	Atomic ratio, %			
	N	Al	Ti	Fe
1	35.12	22.8	40.7	0.4
2	27.27	26.7	45.4	0.4
3	39.77	22.2	36.6	0.5
4	27.4	22.1	48.5	0.8
5	34.34	21.5	42.9	0.7
6	33.15	21.5	43.1	0.9

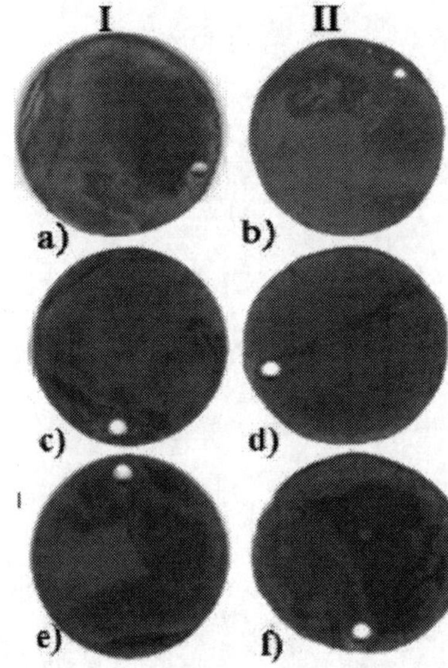

Figure 4. Surface of martensitic steel samples after corrosion tests: I CG structure; II UFG structure; a, b) initial state; c, d) + $TiAl_3/TiAlN$ coatings; e, f) + ion nitriding + $TiAl_3/TiAlN$ coating.

A comparison of corrosion rates for steel EI961 without subsequent coating in a 3% NaCl electrolyte showed that the corrosion rate for the CG structure is 2.5 times lower than that for the UFG structure. High activity of UFG steel is due to a significant decrease in grain size, elongated boundaries and high density of dislocations at the grain boundaries, which together lead to accelerated dissolution in interaction with the activating external environment. Steel corrosion mainly associates with oxygen depolarization. The resulting primary products Fe^{+2} and OH^- react to form white ferric hydroxide (II), which is oxidized by atmospheric oxygen to form brown ferric hydroxide (III):

$$4Fe(OH)_2 + O_2 + 2H_2O \rightarrow 4Fe(OH)_3.$$

Further transformation produces rust, i.e. complex hydrated ferric oxides:

$$nFe_2O_3 + mFeO = qH_2O.$$

Corrosion rate is controlled by the rate of oxygen diffusion to its surface in asodium chloride solution.

Comparison of the corrosion rates for CG and UFG steel coated with $TiAl_3/TiAlN$ showed a significant reduction in the corrosion rate (2 times for UFG steel and 4.7 times for CG steel). Significant corrosion effect with the deceleration of the corrosion rate by 2-5 times has been achieved. Increased corrosion resistance of UFG samples coated with $TiAl_3/TiAlN$ is probably related to the presence of the first layer of nanocrystalline iron in the multilayer coating followed by a layer of titanium aluminum nitride with ultrafine grains and high lattice curvature. This unusual microstructure of the coating allows high-corrosion properties as compared to CG substrates.

The studies revealed that the samples subjected to ion nitriding + $TiAl_3/TiAlN$ coating are more resistant to corrosion than the samples subjected to $TiAl_3/TiAlN$ coating only. Comparison of corrosion rates for pre-nitrided UFG samples is 3 times higher than the corresponding value for the samples that were not subjected to nitriding prior to coating and 14 times higher compared to uncoated samples. Obviously, corrosion resistance of the samples subjected to ion nitriding and subsequent deposition of coatings is higher than that of the samples that were not nitrided, in particular due to the gradient (as opposed to abrupt and step-like) transition of the substrate core (product) parameters to the characteristics of the coating. Increased corrosion resistance was confirmed by continuity of the coating (Fig. 4) after the corrosion tests as compared to the other samples.

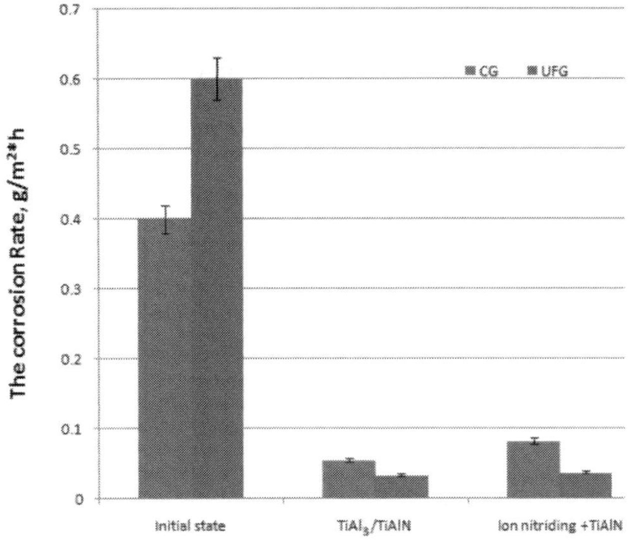

Figure 5. Corrosion rate for CG and UFG steels.

Figure 4(c) shows that the sample surface is characterized with shallow pitting of 0.01 mm depth. Pitting or corrosion holes in martensitic steels often have microscopic dimensions, are surrounded by reddish-brown or multi-colored corrosion products, often have circular deposits of corrosion products around the holes. In martensitic steel pitting corrosion is caused by chloride ions that penetrate through the passive steel layer in some areas.

Analysis of the surface microstructure of the coated specimens (Fig. 4 (d)) showed no corrosion; instead, the coating changed its color to dark gray. Thus, the surface of UFG samples experiences less significant corrosion attacks.

4. Conclusion

Corrosion tests of martensitic steels with coarse-grain and ultrafine-grain structure revealed that:

- Corrosion resistance of CG martensitic steels is 1.5 times higher as compared to the UFG samples. Therefore, the impact of the structure on the corrosion resistance should be taken into account when using UFG martensitic steels.
- Deposition of protective coatings $TiAl_3/TiAlN$ onto martensitic steels may increase the corrosion resistance of the CG materials by 8 times and of the UFG materials by 20 times.
- Ion nitriding prior to deposition of protective coatings can improve corrosion resistance only for UFG materials (22 times).

References

[1] Vaiev R Z, Islamgaliev R K 2000 *Progress in Materials Science* **45** 103
[2] Zhilyaev A P, Langdon T G 2008 *Progress in Materials Science* **53** 893
[3] Musil J, Stupka P 2015 *Int. J. Nanomanufacturing* **11** 78
[4] Smialek J L 1993 *Corrosion Science* **35** 1199
[5] Goral M et al 2011 *Intermetallics* **19** 744.
[6] Du H L et al 2003 *Oxidation of Metals* **60** 29
[7] Miyake M, Tajikara S, Hirato T 2011 *Surface & Coatings Technology* **205** 5141.
[8] Budilov V V et al 2004 *Mat. Sci. and Eng A* **375-377** 656.
[9] Budilov V V, Vardanyan E L 2015 *Journal of Physics: Conference Series* **652** 012053

[10] Ramazanov K N, Khusainov Yu G, Zolotov I V 2015 *Journal of Physics: Conference Series* **652** 012055

Acknowledgments
The authors are grateful to the Russian Science Foundation for support of this work through the project No. 15-19-00144.

Computer simulation of metal wire explosion under high rate heating

K P Zolnikov[1,2], D S Kryzhevich[1,2], A V Korchuganov[1,2]

[1]Institute of Strength Physics and Materials Science of the Siberian Branch of the Russian Academy of Sciences, 2/4 pr. Akademicheskii, Tomsk, 634055, Russia
[2]National Research Tomsk State University, 36 Lenin Ave., Tomsk, 634050, Russia

E-mail: kost@ispms.ru

Abstract. Synchronous electric explosion of metal wires and synthesis of bicomponent nanoparticles were investigated on the base of molecular dynamics method. Copper and nickel nanosized crystallites of cylindrical shape were chosen as conductors for explosion. The embedded atom approximation was used for calculation of the interatomic interactions. The agglomeration process after explosion metal wires was the main mechanism for particle synthesis. The distribution of chemical elements was non-uniform over the cross section of the bicomponent particles. The copper concentration in the surface region was higher than in the bulk of the synthesized particle. By varying the loading parameters (heating temperature, the distance between the wires) one can control the size and internal structure of the synthesized bicomponent nanoparticles. The obtained results showed that the method of molecular dynamics can be effectively used to determine the optimal technological mode of nanoparticle synthesis on the base of electric explosion of metal wires.

1. Introduction

The electrical explosion of metal wires is one of the most promising technological approaches for the synthesis of nanoparticles of a given composition and internal structure. This approach allows synthesizing particles with complex internal structure consisting of several metallic and non-metallic phases which significantly influences their physical and mechanical properties [1–4]. The metal wire explosion occurs as follows: when a high-density electric pulse (10^6–10^9 A/cm^2) is sent through a metallic wire, it is rapidly heated, melted, and then it explodes. The explosion products disperse into gaseous atmosphere with the formation of particles. This approach allows synthesizing a wide range of metals, oxides, nitrides and other powders with complex internal structure. It was shown that the internal structure of nanopowders influences their physical, chemical and mechanical properties [5–8]. Experimental study of the processes in electric explosion of wires has significant difficulties connected with ultrahigh temperature and high-velocity processes of explosion. It should be noted that the use of the particle method in its various representations is promising for describing structural and phase transformations, generation of charged clusters, formation of gas phase and dispersion particles under the electric explosion of wires [9, 10].

The purpose of this paper was the investigation of the synthesis of bicomponent particles as a result of simultaneous electric explosion of copper and nickel wires. The influence of the distance between metal wires and viscosity of surrounding medium on characteristics of the synthesized particles and the distribution of chemical elements inside of them were studied.

2. Results of simulation and their discussion

Investigations in the paper were carried out using the molecular dynamics method [11–13]. The potentials calculated in the framework of the embedded atom method were used to describe interatomic interaction [14]. These potentials allow calculating with good accuracy the surface properties, energy of structural defects, elastic characteristics and other properties that are necessary for a correct simulation of electric explosion.

Copper and nickel specimens of cylindrical shape were chosen as conductors for explosion. Each simulated wire consisted of 110 000 atoms, the height of a cylindrical crystallite was about 50÷60 and the diameter was about 25÷30 lattice parameters. Each specimen consisted of two grains. In view of the small size of the simulated wires, they had a shape of rectangular prisms. Periodic boundary conditions were used along the cylinder axis, and the free surface was simulated in other directions. Loading was applied in the following steps: the system was kept at the temperature of 1000 K, and then copper and nickel wires were rapidly heated up to 7000 K and 9000 K, respectively. The thermostat was applied to the simulated system after 100 ps after explosion. The distance between the wires in different calculations varied within the range from 40 to 260 Å.

The high-rate heating resulted in explosive disintegration of the wires accompanied by the formation of nanosized particles (atomic clusters) and gaseous phase. The cluster size was determined by assuming that atoms belong to one cluster, if the distance between them is shorter than the radius of the second coordination sphere in a perfect lattice close to the melting point. The cluster size was defined by the number of atoms in it. The cluster of minimum size was assumed to contain no less than 13 atoms because the first coordination sphere in the fcc lattice consists of 12 atoms.

The analysis of simulation results shows that after the wire has been heated, the process of dispersion occurs by stages. At the first stage of dispersion process, the average interatomic distance rapidly increases; however, the thermal expansion of the specimens causes no loss of continuity. At the next stage, fast fracture processes occur in the specimens, which involve formation of clusters of different sizes and intensive surface evaporation of atoms. The fracture leads to abrupt decrease in the temperature of the simulated system. This is due to the fact that a significant part of the kinetic energy of the specimens is expended to break atomic bonds.

The change of the number of clusters and the number of atoms in the gas phase in the simulated system as a function of time is shown in Fig. 1. The figure shows that the number of the synthesized clusters in 70 ps starts to saturate. The fraction of the gas phase in the simulated system grows continuously until the beginning of cooling (100 ps), and then decreases due to deposition of atoms on the surface of forming clusters.

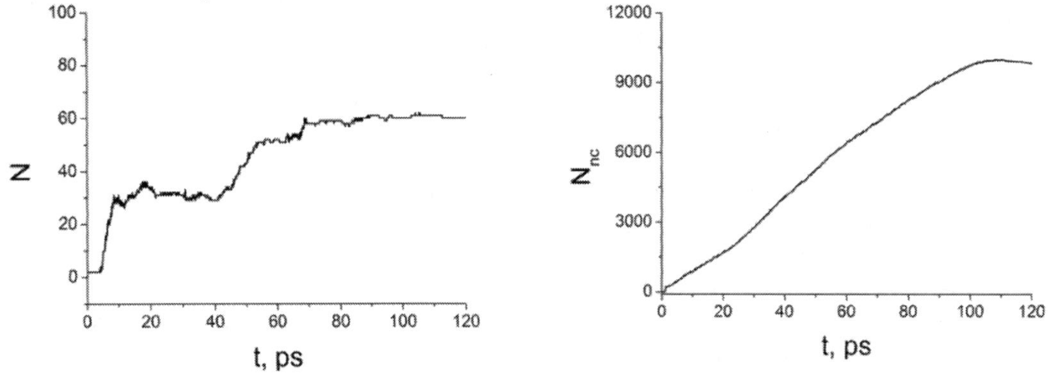

(a) (b)

Figure 1. The dependencies of the cluster number (a) and number of atoms in the gas phase (b) on time. The distance between the wires before detonation was 80 lattice parameters.

Figure 2. Structure projection on plane normal to cylinder axis of exploded system at different time points after explosion: a) 30 ps; b) 100 ps. The distance between the wires before the explosion was 80 lattice parameters.

Note that setting a high temperature heating allowed for "reasonable" computational time (in the framework of molecular dynamics method) to describe the dispersion of the simulated wires and the particle synthesis. The structure projection of simulated copper-nickel system at different points in time after the explosion is shown in Fig. 2. The figure shows that the bicomponent particles were formed in the process of dispersing (copper atoms are shown in red, nickel – in blue).

The results show that the distance between the dispersing wires has a significant influence on the number of generated clusters, their composition and structure, as well as the fraction of gas phase, which is formed in the process of explosion. The total number of formed clusters and number of bicomponent clusters as a function of the distance between the wires after relaxation process are shown in Fig. 3. The figure clearly shows that for the simulated system there is optimal distance interval, at which maximum number of bicomponent particles is synthesized. This interval corresponds to 40-160 lattice parameters between the wires before loading. It can be assumed that the optimum distance between the wires for the synthesis of bicomponent particles will depend largely on the wire thickness, their form and, to a lesser extent on the mode of heating and environmental properties, at which dispersion takes place.

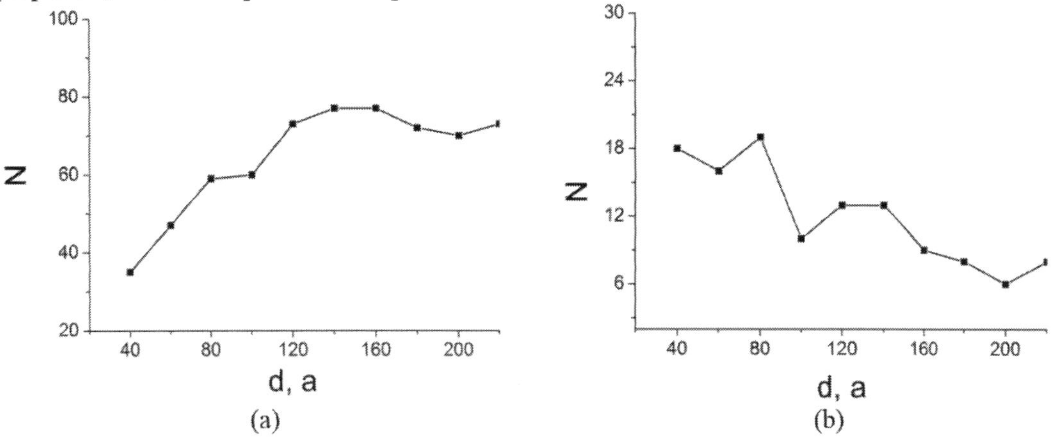

Figure 3. The total number of (a) formed clusters and (b) bicomponent clusters versus the distance between the wires (a is lattice parameter).

The cluster distribution by their size and component composition depending on the distance between the wires at the end of the calculations was analyzed. It was found that a large number of clusters with a high concentration of the second component was formed for small distances between the dispersing wires. It is obvious that the evolution of the simulated system closer to equilibrium state and particle formation will continue with the slowing rates for a longer time intervals. In view of the limited computing resources, the evolution of the system towards equilibrium cannot be described within the framework of molecular dynamics method without using some approximations. A quite efficient approach is the use of viscoelastic boundary conditions that simulate the properties of the environment in which metal wires are dispersed, and the increase of the integration step at lowering temperature of the simulated system.

Typical particle distribution after the application of viscoelastic boundary conditions and cooling the dispersed system down to 2000 K is presented in Fig.4. The analysis of simulation results showed that the basic mechanism of particle formation is the agglomeration of smaller clusters, but not the deposition of atoms from the gas phase on the particle surface. It is seen in Fig. 4 that the chemical composition of the formed particles varies in a wide interval. Moreover, the chemical composition along the cross section of the bicomponent particles varies strongly. The concentration of copper atoms near the surface of bicomponent particles is much higher than in the bulk of the particles (Fig. 5).

It should be noted that temperature dynamics of the simulated system has a feature. The temperature abruptly decreases after high-rate heating. This behavior of the simulated system is connected with fracture processes of the wires and particle formation. The process of the wire fracture is accompanied by an increase in the free surface area of the simulated system and leads to the transition of a significant part of the kinetic energy into potential energy. The decrease in the intensity of the thermal pulse loading leads to the formation of particles of larger size. The calculations showed that the wire heating at high-rate electric pulse can lead to a significant increase in their volume without discontinuity (the jump of the atomic volume was about 9%). Such behavior of the crystalline wires might be related to a lower rate of accommodation processes in the internal structure as compared to the heating rate.

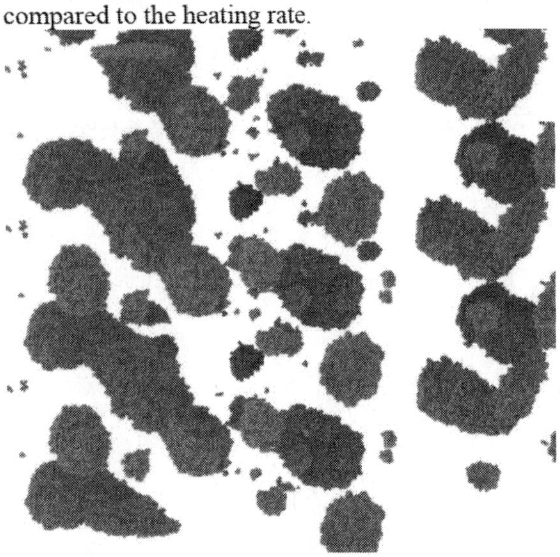

Figure 4. Structure of dispersed copper–nickel system in 150 ps. The distance between wires before the explosion was 80 lattice parameters.

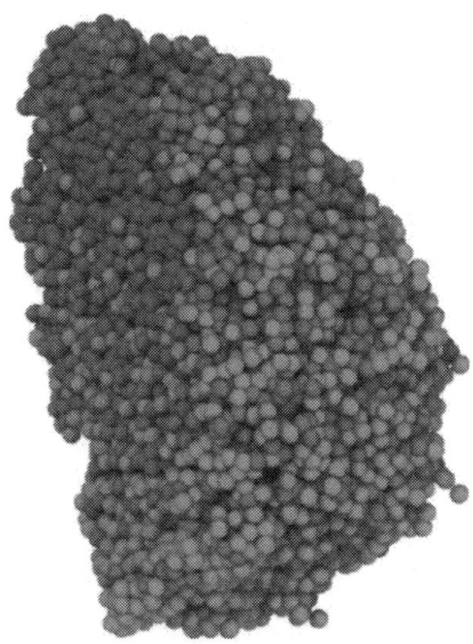

Figure 5. The cross-section of the bicomponent cluster (the copper atoms are marked in red, the nickel atoms in blue). The cluster contains around 10 000 atoms, the concentration of copper in it being 30%.

3. Conclusion

The results of simulation show that there are two mechanisms of particle formation at metal wires explosion: agglomeration and deposition from the gas on the particle surface. The first one is agglomeration of small clusters and it prevails over the process of atom deposition from the gas on the free surface of the particles. The chemical composition of the formed particles is heterogeneous along their cross section. So, the concentration of copper atoms near the particle surface is higher than in the volume. It was found that the distance between dispersing wires greatly affects the process of particle formation. On the basis of the performed calculations one can conclude that the molecular dynamics metod can be used effectively to determine the optimal loading conditions of wire explosion to synthesize particles with desired composition and size.

References

[1] Jonson R, Siegel B 1970 *Rev. Sci. Instr.* **42(6)** 854
[2] Bennett F 1968 High-temperature Exploding Wires *Progress in High Temperature Physics and Chemestry* vol 2 ed C Rouse (N.Y.: Pergamon Press) p 463
[3] Psakhie S G, Zolnikov K P, Kryzhevich D S, Abdrashitov A V and Lerner M I 2010 *Phys. Mesomech.* **13(3-4)** 184
[4] Abdrashitov A V, Kryzhevich D S, Zolnikov K P and Psakhie S G 2010 *Procedia Eng.* **2(1)** 1589
[5] Zol'nikov K P, Uvarov T Y and Psakh'e S G 2001 *Tech. Phys. Lett.* **27** 263
[6] Psakhe S G, Zolnikov K P and Korostelev S Y 1995 *Pisma v Zhurnal Tekhnicheskoi Fiziki* **21(13)** 1
[7] Psakhie S G, Kryzhevich D S and Zolnikov K P 2012 *Tech. Phys. Lett.* **38** 634
[8] Zolnikov K P, Psakhie S G and Panin V E 1986 *J. Phys. F: Met. Phys.* **16(8)** 1145
[9] Psakhie S G, Zolnikov K P, Skorentsev L F, Kryzhevich D S and Abdrashitov A V 2008 *Phys. Plasmas* **15(5)** 053701
[10] Psakhie S G, Zolnikov K P, Kryzhevich D S, Zheleznyakov A V and Chernov V M 2009 *Crystallogr. Rep.* **54** 1002
[11] Zolnikov K P, Korchuganov A V, Kryzhevich D S, Chernov V M and Psakhie S G 2015 *Nucl. Instr. Meth.* B **352** 43
[12] Psakhie S G, Zolnikov K P, Kryzhevich D S, Zheleznyakov A V and Chernov V M 2009 *Phys. Mesomech.* **12(1–2)** 20
[13] Zolnikov K P, Korchuganov A V, Kryzhevich D S, Chernov V M and Psakhie S G 2015 *Nucl. Instr. Meth.* B **352** 39
[14] Bonny G, Pasianot R C, Castin N and Malerba L 2009 *Philos. Mag.* **89** 3531

Acknowledgments

The work was carried out at the support of the Tomsk State University Competitiveness Improvement Program.

Plasma dynamic synthesis of iron oxides in a discharge plasma jet with possibility to control final phase composition

I Shanenkov, A Sivkov, A Ivashutenko and M Gukov

Institute of Power Engineering, Tomsk Polytechnic University, Lenin av. 30, Tomsk, Russian Federation

E-mail: Swordi@list.ru

Abstract. Magnetite Fe_3O_4 and epsilon iron oxide ε-Fe_2O_3, having excellent frequency characteristics and high electrical resistivity, are considered as the most promising phases among all iron oxides for high-frequency equipment in order to increase the working frequency of the data transmission. Despite the large number of existing methods for synthesizing these materials, many of them do not provide both of these phases. In opposite to these methods, the plasma dynamic synthesis can provide the synthesis of necessary phases in a one-step process. The process is implemented in an electrodischarge iron-containing plasma jet, which interacts with gaseous precursor (oxygen). The use of plasma jet allows obtaining nanoscale powdered products. This work shows the results of the experiment series, where the influence of initial energy parameters on the final phase composition was studied. It is found that the plasma dynamic synthesis allows obtaining both magnetite Fe_3O_4 and epsilon phase ε-Fe_2O_3 during one short-term process (less than 1 ms). It is also established that the final phase composition strongly depends on the initial parameters of the system. The increased energy parameters lead to the formation of the product with predominant content of epsilon phase, while lower parameters allow synthesizing magnetite phase. Thus, by changing energy parameters, it is possible to control the final composition in the considered system.

1. Introduction

Magnetic iron oxides are of a great interest in research and industry. They have numerous applications in different areas, ranging from engineering (e.g., magnetic recording media or magnetic seals) to biomedical applications (e.g., magnetic resonance imaging, drug delivery or hyperthermia) and their appealing novel properties [1–5]. Among various magnetic iron oxides, those that exhibit soft magnetic behavior with excellent frequency characteristics and high electrical resistivity are strongly required as core materials used in magnetic components for the ceaseless increase of the working frequency of the data transmission and multiple accesses in computer, mobile and blue-tooth devices [6,7]. The most known iron oxide phases, having outstanding characteristics of electromagnetic radiation (EMR) absorption, are magnetite Fe_3O_4 and relatively new epsilon phase ε-Fe_2O_3.

In the case of magnetite, the best EMR absorption properties are shown by hollow spheres of Fe_3O_4 [8,9]. They can provide a strong attenuation (up to –40 dB) of electromagnetic (EM) signal in a wide range of frequencies (from 2 to 18 GHz). Moreover, the hollow structure of these spheres allows obtaining low-weight coatings and samples, which protect from EMR. Absorbing materials based on epsilon phase can effectively attenuate the EM signal in a range of frequencies from 110 to 183 GHz, which makes them very prospective for using in military applications. The main problem is the difficulties with its synthesis. It is known that this phase is thermally unstable (converts to hematite at temperatures about 750 °C) and can exist only in nanodispersed state (also converts to hematite when interacts with other particles due to low surface energy) [10].

In the past decades, preparations of magnetic iron oxide nanoparticles have been extensively studied, especially in synthesis chemistry. One has witnessed various methods (e.g., solid state reactions [11]; high energy ball milling [12]; sol-gel [13]; chemical co-precipitation [14]; conventional two-step synthesis [15], and so on) that have yielded various kinds of magnetic iron oxide nanoparticles. Nonetheless, not all of these methods allow synthesizing both ε-Fe_2O_3 and Fe_3O_4 phases and those which do suffer from such disadvantages as the process duration (usually, it is multi-step process, which can continue up to 24 hours or more), the necessity to use a large amount of expensive precursors and the low yield of the final product [16]. Differing from the preparation methods mentioned above, an alternative approach—plasma dynamic synthesis [17-19]—can provide synthesis of necessary phases in one-step process.

The aim of this work is to synthesize the soft magnetic phases of iron oxide by implementing the process in a discharge plasma jet using a coaxial magnetoplasma accelerator (CMPA). It is shown that the plasma dynamic method allows obtaining ultradispersed powders of iron oxide including both ε-Fe_2O_3 and Fe_3O_4 phases. It is found that the system energy parameters directly influence the ratio between these phases in the final product. It is suggested that the rate of current rise plays the key role in obtaining the powder with high content of epsilon phase by plasma dynamic method.

2. Experimental

Both the design and the function of CMPA-based system can be explained using Figure 1. In order to synthesize iron oxide powders, the system based on iron electrodes (central electrode 1 and barrel-electrode 6) was used. Electrodes were separated by a fiberglass insulator 2. All this assembly was held by metal cases 3 and placed into solenoid 5. Before starting the process, the capacitive energy storage *C* was charged. In the first moment of time (t_0), when the power keys 7 were closed, the energy from capacitive storage was released and the process was started. Current flew through electrodes and explosive Fe wires. When its value became large enough, Fe wires exploded forming the arc discharged plasma structure. This plasma structure was accelerated by forces of conductive and inductive electrodynamics. Iron, which was necessary for synthesis, was accumulated from the central electrode and inner walls of electrode-barrel due to their electro erosion. The gas precursor (O_2) was added into the working chamber, which was preliminary evacuated, through the gas collector.

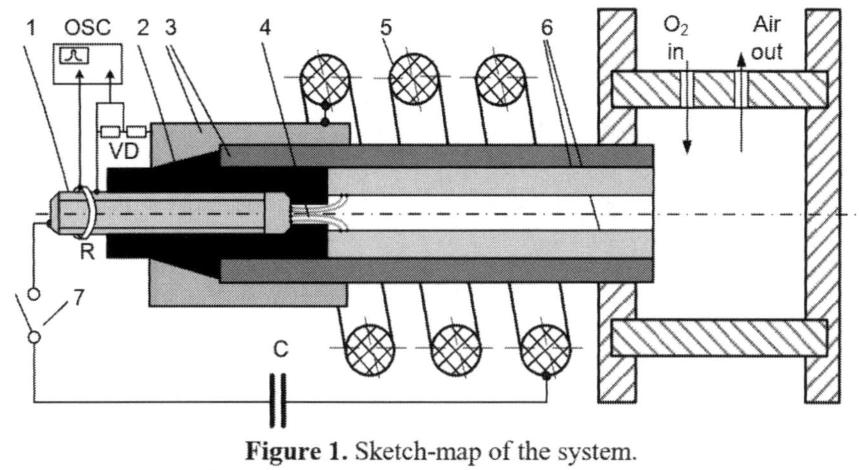

Figure 1. Sketch-map of the system.

During the process, such energy parameters as the discharge current and voltage between electrodes were recorded using Rogowski coil *R* and ohmic voltage divider *VD*, respectively. Both signals were recorded by Tektronix 2012b oscilloscope. Based on the current and voltage data, the discharge power and input energy were calculated.

A series of experiments was carried out in order to investigate the possibility to control the phase composition in the resulting powders by changing energy parameters of the system. The main data of experiments are shown in Table 1. It is worth noting that experiment 3 was carried out using without using Fe explosive wires. In order to initiate the discharge, carbon spray was used to create a thin conductive layer on the surface of the insulator's inner surface.

Synthesized powders were investigated using such methods as X-Ray diffractometry (XRD) and scanning electron microscopy (SEM). The XRD analysis was carried out using Shimadzu XRD7000S diffractometer with Shimadzu CM-3121 counter monochromator. In order to provide qualitative and quantitative analyses of XRD patterns, PowderCell 2.4 software and PDF2+ database were used. SEM analysis was carried out using Hitachi TM3000 microscope.

Table 1. Initial energy parameters of experiments and mass of products.

	Experiment 1	**Experiment 2**	**Experiment 3**
Charging voltage U_c [kV]	3.22	2.50	2.50
Charging capacity C_c [mF]	14.4	14.4	28.8
Charging energy W_c [kJ]	75.0	45.0	90.0
Input energy W [kJ]	57.8	24.2	67.2
W/W_c [%]	77.1	56.3	74.6
Eroded mass m_e [g]	9.0	2.5	7.1
Product mass m_{prod} [g]	7.3	2.0	6.6

3. Results and Discussions

As a result of each experiment, oscillograms of the voltage *v(t)*, current *i(t)*, power *p(t)* and energy *w(t)* were recorded (Figure 2). As it can be seen from oscillograms of current, the process had an aperiodic nature in all experiments. Such process feature can be explained by increasing the system`s active resistance since a certain time moment. This leads both to an increase in the ratio between active resistance and inductive resistance, which in this system is practically unchanged, and to the transition to the aperiodic discharge. The main difference between experiments with Fe explosive wires and carbon spray was in the character of plasma structure formation. In the case with Fe wires, which had the resistance of few mOhm, in the first moment of time, after closing power keys, there was an immediate rise of current. It led to the explosion of Fe wires with the formation of plasma structure. In the case of carbon spray, there was a step of about 25 μs, when the capacitive energy storage discharged on the high active resistance (~5 kOhm) of the thin layer. An aperiodic nature was also seen for this discharge. When the carbon layer was heated enough under the influence of current (~10 kA), the arc discharge started to burn, and the process was similar to experiments with Fe wires.

As can be seen from Table 1, the initially selected energy parameters significantly influenced the energy characteristics of the process. For example, at the minimum charging energy (experiment 2) the magnitude of the main parameters was lower than in other experiments. The maximum current value was equal to 113 kA, and the magnitude of the input energy was 24.2 kJ, while the efficiency of its conversion was 56%. The higher initial energy parameters (experiments 1 and 3), the higher the maximum current (up to ~225 kA) and the energy conversion efficiency (~75%). The increase in current led to an increase in the mass of eroded material and in the mass of the final product.

The formation process of ultra- and nanodispersed particles in the CMPA-based system was considered in detail earlier [20]. The higher current leads to the increase in the temperature and plasma flow speed. In turn, this influences the increase in the temperature gradient and the speed of the material sputtering from the front of the bow shock wave in the chamber-reactor. Both of these factors have a significant effect on the cooling rate and the final dispersivity of the material. It is known that the epsilon phase ε-Fe_2O_3 can exist only when its particle sizes are less than 200 nm. Thus, the increase in the energy parameters of the experiment and, in particular, the increase in the current level have to directly impact the phase composition of the final synthesized samples.

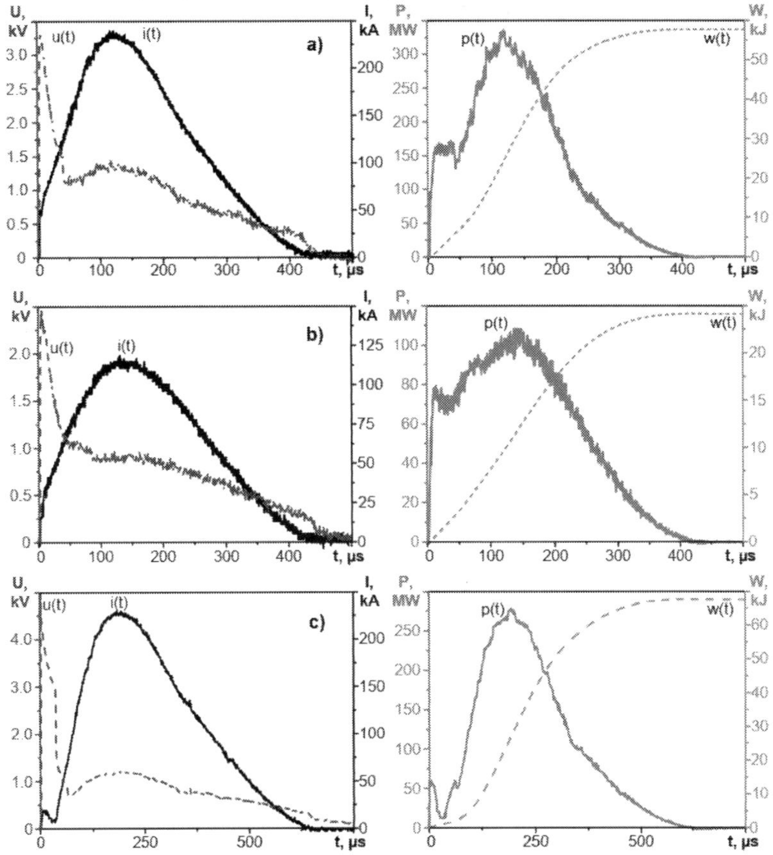

Figure 2. Oscillograms of current i(t), voltage u(t), discharge power p(t) and input energy w(t), recorded during implementing the experiments: a) exp. 1; b) exp. 2; c) exp. 3.

All synthesized samples were studied using X-Ray diffractometry (XRD) method. XRD patterns of products are shown in Figure 3. Table 2 contains the results of the quantitative analysis of phase composition carried out using a PowderCell 2.4 software and PDF2+ database, together with the additional data obtained after a more detailed study of oscillograms (pulse time, max current, max voltage and current rising edge). The comparison of XRD patterns shows that in the experiment with low energy parameters (experiment 2), the peaks corresponding to hematite and epsilon phases are low-intensity and present in the form of traces. In turn, magnetite peaks have the greatest intensity and are more clearly seen in comparison with XRD patterns for products of other experiments. As for the other two experiments with high energy parameters, they are characterized by the presence of well-seen maxima of epsilon and hematite phases. The presence of hematite in the final product is quite likely due to the fact that the synthesized nanoparticles ε-Fe_2O_3 have a low surface energy, and in the process of sputtering from the front of the bow shock wave they can interact with each other. This, ultimately, leads to their growth and transition to the α-Fe_2O_3 phase.

A more detailed analysis of oscillograms allowed establishing the relationship between the level of the current, the front of its rising and the final phase composition (Table 2). It was found that the higher level of discharge current influence the yield of the epsilon phase in the final product. Such feature seems to be natural due to the fact that the larger value of the discharge current causes a greater temperature gradient, which, in turn, leads to an increase in the cooling rate and the formation of smaller particles. Because ε-Fe_2O_3 can exist only, when its particles less than 200 nm, the obtaining of higher energy parameters is necessary condition for the considered system to obtain the product with higher yield of ε-Fe_2O_3 phase. In fact, when the maximum current value is equal to 113.0 kA (exp. 2), the ε-Fe_2O_3 content is 13.0%. With increase in current level up to ~225 kA, the content of epsilon phase is also increased up to 37.0% (exp. 1) and 50.0% (exp. 3).

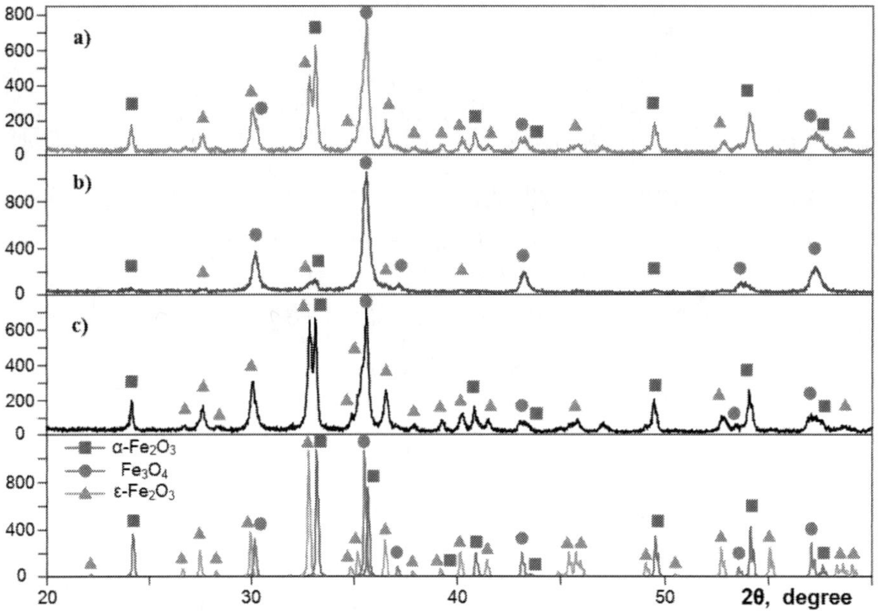

Figure 3. XRD patterns of the products synthesized in different experiments: a) 1; b) 2; c – 3); as well as reference models of iron oxide phases from PDF2+ database.

Table 2. Energy parameters of experiments and phase composition of products

		Experiment 1	Experiment 2	Experiment 3
Energy parameters	**Charging voltage** U_c [kV]	3.22	2.45	2.50
	Charging capacity C_c [mF]	14.4	14.4	28.8
	Pulse time t_p [μs]	405.0	415.0	610.0
	Max current I_m [kA]	231.0	113.0	225.5
	Max voltage U_m [kV]	1.39	0.94	1.21
	Input energy W [kJ]	57.8	24.2	67.2
	Rate of current rise di/dt [a.u.]	1.80	0.94	2.12
Phase composition	**ε-Fe_2O_3** [wt%]	36.7	13.0	50.0
	Fe_3O_4 [wt%]	36.0	76.5	24.0
	α-Fe_2O_3 [wt%]	27.3	10.5	26.0

In spite of being almost equal maximum discharge currents, experiments 1 and 3 differ from each other with the content of epsilon phase. This fact can be explained by the difference in the rate of the current rise. According to Table 2, this parameter is equal to 2.12 for experiment 3 and 1.80 for experiment 1. The rate of current rise shows the speed of energy input, which influences the speed of the plasma flow. In turn, the speed of plasma flow directly influences the speed of material sputtering, which is necessary for obtaining the nanoscale particles. Thus, when the rate of current rise is higher, the synthesized product has a less particle size that leads to formation of the powder with the predominant content of epsilon phase. In opposite to experiments with high energy parameters, the experiment, where the maximum current is equal to 113.0 kA, showed the high yield of magnetite Fe_3O_4. This seems to be natural due to the same reasons discussed above. When the current and its rise rate are decreased, there is a reduction of main plasma parameters that negatively influences the conditions for epsilon phase formation. In this case, magnetite becomes a dominant phase in the product. Thus, the CMPA-based system allows not only synthesizing soft magnetic phases of iron oxide but also controlling their ratio in the final product. This control is implemented by changing the initial energy parameters of the capacitive energy storage, which is used for power supplying the CMPA, and by changing the way of the plasma structure formation. Despite the found dependence on the maximum level of discharge current, there is a limit value for energy parameters, which cannot be exceeded, in order to provide the reliable work of the system.

4. Conclusion

Soft magnetic phases of iron oxide (magnetite Fe_3O_4 and epsilon ε-Fe_2O_3) can be synthesized in one-step short-term (less than 1 ms) process using the plasma dynamic method in the system based on the coaxial magneto plasma accelerator. The synthesis is implemented in an electric-discharge iron-containing plasma jet flowing into oxygen atmosphere. It is found that the energy parameters of the process sufficiently influence the phase composition of the final product. The content of epsilon phase strongly depends on the level of discharge current and rate of its rising. The higher these parameters can be reached, the more yield of the epsilon phase can be obtained. In opposite, at lower energy parameters Fe_3O_4 phase dominates in the final product. Thus, the CMPA-based system allows controlling the final phase composition by changing the energy parameters.

References

[1] Majetich S and Sachan M 2006 *J. Phys. D. Appl. Phys.* **39** R407
[2] Leslie-Pelecky D L and Rieke R D 1996 *Chem. Mater.* **8** 1770
[3] Frey N A, Peng S, Cheng K and Sun S 2009 *Chem. Soc. Rev.* **38** 2532
[4] Lu A H, Salabas E L and Schuth F 2007 *Angew. Chemie - Int. Ed.* **46** 1222
[5] Estrader M, Roca a G and Nogués J 2015 *Phys. Rep.* **553** 1
[6] Pardavi-Horvath M 2000 *J. Magn. Magn. Mater.* **215** 171
[7] Toneguzzo P, Viau G, Acher O, Fiévet-Vincent F and Fiévet F 1998 *Adv. Mater.* **10** 1032
[8] Wei J, Liu J and Li S 2007 *J. Magn. Magn. Mater.* **312** 414
[9] Liu Y, Cui T, Li Y, Zhao Y, Ye Y, Wu W and Tong G 2016 *Mater. Chem. Phys.* **173** 152
[10] Tucek J, Zboril R, Namai A and Ohkoshi S I 2010 *Chem. Mater.* **22** 6483
[11] Lu J, Yang S, Ng K M, Su C-H, Yeh C-S, Wu Y-N and Shieh D-B 2007 *Nanotechnology* **18** 289001
[12] Munoz J E, Cervantes J, Esparza R and Rosas G 2007 *J. Nanoparticle Res.* **9** 945
[13] da Costa G M, De Grave E, de Bakker P M a. and Vandenberghe R E 1994 *J. Solid State Chem.* **113** 405
[14] Khalil M I 2015 *Arab. J. Chem.* **8** 279
[15] Lind K, Kresse M, Debus N P and Müller R H 2002 *J. Drug Target.* **10** 221
[16] Syue M R, Wei F J, Chou C S and Fu C M 2011 *J. Appl. Phys.* **109** 101
[17] Sivkov A A, Ivashutenko A S, Nazarenko O B, Saigash A S, Pak A Y and Kolganova Y L 2016 *Inorg. Mater. Appl. Res.* **7** 354
[18] Shanenkov I I, Sivkov A A, Pak A Y and Kolganova Y L 2014 *Adv. Mater. Res.* **1040** 813
[19] Kuzenov V V, Polozova T N and Ryzhkov S V 2015 *Problems of Atomic Science and Technology* **4 (98)** 49.
[20] Sivkov A, Naiden E, Ivashutenko A and Shanenkov I 2016 *J. Magn. Magn. Mater.* **405** 158

Acknowledgements
This work was supported by the Russian Science Foundation (grant No. 15-19-00049).

Plasma dynamic synthesis of ultradispersed zinc oxide and sintering ceramics on its basis by SPS method

Yu Shanenkova, A Sivkov, A Ivashutenko, I Shanenkov and K Firsov
National Research Tomsk Polytechnic University, 30 Lenin av., Tomsk, 634050, Russia

E-mail: julia_kolganova@mail.ru

Abstract. Zinc oxide is a well-known semiconductor material having good electrical, optical and catalytic properties. It can be used in different areas from cosmetics to drug delivery and biosensors. The synthesis of nanosized zinc oxide is an urgent task for obtaining ZnO-based ceramics with enhanced physical properties. This work shows the possibility to implement the plasma dynamic synthesis of zinc oxide in one short-term process (less than 1 ms) using an electrodischarge zinc-containing plasma jet, flowing into oxygen atmosphere. It allows synthesizing a mono-crystalline powder with particle size distribution from tens to hundred nanometers. The synthesized powdered product is investigated using by X-Ray diffractometry (XRD), scanning electron microscopy and high-resolution transmission electron microscopy. According to XRD, the obtained product consists of hexagonal zinc oxide with lattice parameters a = b = 3.24982 Å, c = 5.20661 Å that is clearly confirmed by microscopy data. This powder was used to produce a bulk ceramics sample on its basis by spark plasma sintering. The influence of sintering parameters on the structure of the resulting sample was studied. The optimal parameters were found which allows obtaining the more dense ceramics with a better microstructure. It was also found that the absence of exposure time after reaching the working temperature and pressure allows decreasing the porosity of ceramics.

1. Introduction

It is well known that zinc oxide is a promising material because of its properties such as high electron mobility, good transparency, wide bandgap, strong luminescence [1]. In addition, the last studies of this material have allowed opening good catalytic properties of ZnO [2-7]. All of these unique characteristics make zinc oxide useful for using in different production areas such as nanobiology, nanomedicine, piezoelectric generators, optoelectronic and photovoltaic devices [8-13].

The properties of zinc oxide strongly depend on its structure, morphology, aspect ratio, size and orientation of the particles [14-16]. ZnO crystallites with the different particle morphology, for example, spherical, prism, stars, rods and others, have been previously studied [17-23]. Nonetheless, the hexagonal structure of zinc oxide is of great scientific interest [24, 25].

Except the material structure, the particle size is also an important factor for ZnO. Nanodispersed powders are widely used in science and technology because of their unique properties (low density and high specific surface features) [26]. In spite of this, in some scientific areas it has a special interest in obtaining a particular particle forms with a wide distribution from several tens to hundreds of nanometers [27]. For example, in order to create a high density ceramic, it is necessary to introduce particles of smaller fraction (nanoparticles) in a micron or submicron powder [28].

This paper shows a unique method for obtaining nanodispersed zinc oxide using a system based on a coaxial magneto plasma accelerator. The main advantages of this method are the high speed of the process, as well as the high purity of the synthesized product, which has the desired structure and particle size distribution. This work reports the plasma dynamic synthesis that allows obtaining ultradispersed zinc oxide in the one short cycle (up to 10^{-3}) of CMPA work. Using XRD and HRTEM it is found that the synthesized product consists of single-crystalline hexagonal ZnO particles with the necessary stoichiometry. The particle size distribution histogram allows affirming that the product has the broad distribution up to 350 nm. It is also shown that the synthesized product can be used for sintering bulk samples using Spark Plasma Sintering (SPS) technology. The density of the obtained ceramics sample is up to 99%.

2. Experimental

The previous article [29] showed a scheme and the operation principle of coaxial magnetoplasma accelerator with zinc electrodes [30,31]. Using this accelerator, the nanodispersed ZnO powder was synthesized. After collecting the product of a plasma-dynamic synthesis, it was studied by means of X-ray diffractometry (XRD), scanning electron microscopy (SEM) and transmission electron microscopy (TEM) methods. After that, this product was sintered, using the setup for spark plasma sintering (Advanced Technology SPS 10-4).

Scanning electron microscopy (SEM) images were recorded using a Hitachi TM-3000 scanning microscope with carbon substrate. X-Ray diffraction (XRD) analysis was carried out using a Shimadzu XRD 7000S diffractometer (CuKα-radiation) with the counter monochromator Shimadzu CM 3121. The database PDF-2 and the software PowderCell 2.4 were used to make full-profile structural analysis and estimative calculations. Transmission electron microscopy (TEM) images of the product were made using a Philips CM-12 microscope. The particle size distribution histogram was built by measuring the size of several hundred particles in accordance with data of a number of bright-field TEM images. High resolution transmission electron microscopy (HRTEM) was carried out using a Jeol JSM-7500FA microscope.

3. Results and discussion

It was experimentally found that almost pure crystalline zinc oxide ZnO in the considered system is formed when the volume oxygen (O_2) concentration in a gas mixture is over than 80%. This is confirmed by the XRD pattern of the initial product (fig. 1). Its analysis showed the closeness of the synthesized powder to the structural model of ZnO of the hexagonal crystal system (space group $P6_3mc$ (no.186)) with the lattice parameters a=b=3.24982 Å, c=5.20661 Å (card number 36-1451) [32]. A full-profile structural analysis allowed estimating the average value of the coherent-scattering regions (CSR) for the crystalline phase ZnO, which was equal to ~124.0 nm, as well as the average level of internal microscopic strains in the crystal structure ($\Delta d/d=0.043\cdot10^{-3}$). The last parameter indicates the weakly strained state of the single-crystalline particle structure in the synthesized powder. It is also a reason for a good agreement between the calculated values of the lattice parameters a=b=3.24970 Å, c=5.20270 Å with the previously mentioned parameters, which are closest to the structural model of the zinc oxide. Such a structure for even nanoscale ZnO particles became evident after analyzing the data of transmission electron microscopy (fig. 2a). The high density of ZnO particle ($\rho = 5.61$ g/cm^3) allowed obtaining high-quality brightfield and darkfield TEM images.

An analysis of SEM and TEM images showed that the particles had a broad enough size distribution from 10 nm to 500 nm. However, in order to get the more detailed information on the size of the particles, the distribution of the particle size was calculated, using a number of bright-field TEM-images, and the corresponding histogram was built (figure 2b). It is possible to see that particles have a wide size distribution up to 350 nm. The reason for this is the pulsed nature of the plasma-dynamic processes, during which the formed plasma flew out from the accelerator channel in free space. The regulation of plasma parameters at this stage seems to be not possible. At the end of the process, when the energy parameters of plasma and its temperature decrease, the cooling rate drops sharply that leads to an increase in particle size. Nonetheless, about 85% of ZnO particles have sizes not more than 150 nm that is in a good agreement with the average value of CSR (XRD data).

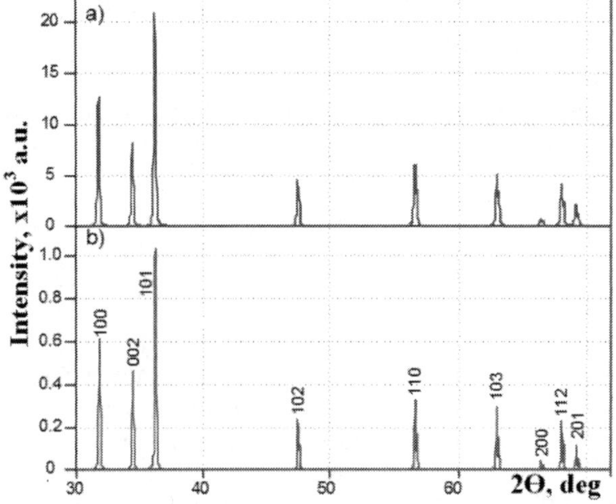

Figure 1. Typical XRD patterns a) for product of plasma dynamic synthesis in Zn-O system; b) structural model of hexagonal zinc oxide.

Judging by TEM images, the single-crystalline particles have an almost perfect crystallographic design in the form of hexagonal rods, as it is shown in the inset of figure 2. Regardless of the overall crystallite sizes, in this and similar clusters there are the particles, the length of which is either greater than the width of the lateral faces or less. The comparison of bright-field and dark-field images in the light of diffracted beams (corresponding reflexes are pointed by the Miller indices) shows a full compliance of reflecting plane systems to the theoretical model.

The marked features of the particle size distribution and the size ratio of their facets are connected with the pulse nature of the plasma-dynamic synthesis process. A very wide range of current change (from 0 up to 100 kA) provides the same wide range of changes of pressure and temperature (from room to approximately 10^4 K) in the discharge plasma, as well as the plasma jet velocity and velocity of spraying the liquid phase of the synthesized material in the chamber space. This influences the material crystallization rate, the size and shape of the crystallites. A detailed analysis of these processes requires additional special studies and is not the subject of this work.

It is known that for obtaining the high-quality ceramics with the micron and submicron structure by spark plasma sintering (SPS) it is necessary to introduce the finer fraction in quantity of 45 % in the main micron or submicron powder. In plasma-dynamic synthesis powders, this condition is automatically satisfied because of a uniform distribution the nanodispersed fraction in the total mass. While the powder is being compacted, the fine fraction fills the space between the larger particles, providing obtaining the high-density ceramics.

Due to the absence of the experience, the first experiments on the SPS-compaction of ultrafine ZnO powders were carried out in the mode, the main parameters of which were determined according to the published data [26]. The powder was compacted without any preliminary preparation in a vacuum. A graphite mold and punches with a diameter of 20 mm were used. A heating unipolar pulse current was switched on after compressing the dispersed sample between the punches with a pressure of 60 MPa. The temperature was raised stepwise with a rate of 50 °C/min to 600 °C, then at a rate of 10 °C/min to 800 °C. After reaching the preset maximum temperature, the temperature lowering was carried out at a rate of 50 °C/min to the room temperature. Only after the moment, when the temperature became equal to the room temperature, the pressure was removed from the sample.

The material density, measured by the hydrostatic method, was equal to $\rho = 4.5$ g/cm^3 (80%), that indicates its high porosity. The SEM-images of the chip (figure 3) show the high heterogeneity of the bulk material structure (fig. 3a). Its main mass has a nanoporous structure, formed by the ZnO particle-crystallites, which are united by crosspieces, which can be formed during spark sintering (fig. 2). In the sintered material, there are larger rounded particles, which are present in the original powder too.

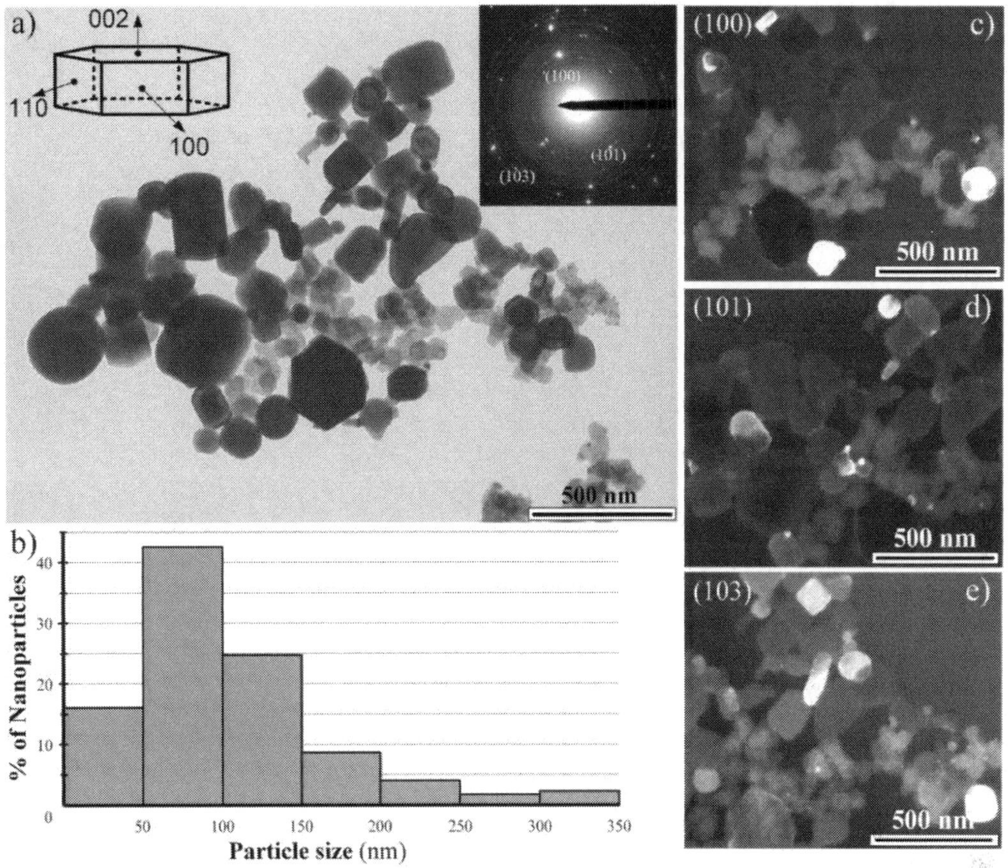

Figure 2. a) TEM data for initial product of plasma dynamic synthesis, consisting of single-crystalline particles of zinc oxide; b) histogram of powder particle size c)-e) dark-field images in the light of diffracted beams.

Figure 3. SEM-images of surface chip of ceramics sample: a) x1.0k, b) x20.0k, c) x10.0k.

In the sintered material, an obvious inhomogeneity is created by numerous focal defects of the rounded shape, with sizes up to several tens of microns and a typical coarse-grained structure (fig. 3b). This structure is formed by densely packed monolithic grains with sizes up to ~5.0 μm. These features indicate a liquid-phase mechanism of the formation of point defects. Such particles, having such sizes

IOP Conf. Series: Journal of Physics: Conf. Series **830** (2017) 012117 doi:10.1088/1742-6596/830/1/012117

and shapes, do not exist in the initial ZnO powder (figure 3). However, the dense agglomerates of nanoscale particles have even larger sizes. These dense agglomerates may be the reason for forming point defects during SPS-compaction, because there can be a local increase in both the density of the heating current and the realized energy. This leads to the melting of nanoparticles and the formation of large rounded focal points of the liquid ZnO phase. The melt crystallization with the formation of close-packed micron-sized grains occurs, when the heating current and cooling of the pressing are reduced.

Analysis of the results of the first experiments on the SPS-compaction showed the necessity to disaggregate powders and to optimize the heating mode. In this regard, before the next experiment, the product of the plasma-dynamic synthesis was subjected to the heat treatment in air at 200 °C for 1 hour with the followed processing in a ball mill (corundum crucible and a ball) at 30 minutes. The SPS process was also carried out in a graphite mold at a constant pressure of 50 MPa.

In the presented experiment, the heating mode was substantially changed. The speed of temperature growth was ~800 °C/min with the maximum level of 1100 °C. The process was carried out without isothermal exposure and with the natural cooling of the sample under pressure. As a result, the sample of ZnO-ceramics had the high density ρ = 5.6 g/cm^3 (99.9%) that influenced its high thermal conductivity (λ = 32.0 W/m·K). Studies of ceramics sample by scanning electron microscopy (SEM) showed the absence of focal macro defects (figure 4) that should be considered as a positive result of disaggregating the initial ZnO powder.

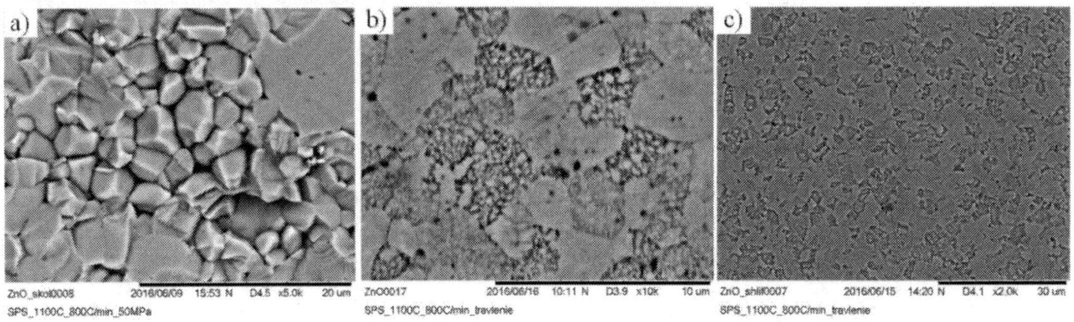

Figure 4. SEM-images of microstructure of obtained SPS-ceramics:
a) transverse chip of sample b) etching surface for 1 second c) etching surface for 30 seconds.

The transverse chip of the sample (figure 4a) passes through the volume of grains that indicates the strength of grain boundaries. The sizes of the largest grains do not exceed ~5.0 μm. Numerically insignificant large (open) pores with size up to ~1.0 μm were generally formed because of a non-completion of the triple grain boundaries. In addition, in the bodies of some large grains, there are small (less than 0.5 microns) closed pores. In the SEM-image, it is seen that surfaces of the chip of the smaller part of the grains have a layered nature, which seems to be caused by internal substructure of the grains.

Figure 4b shows an image of the sample flat surface, which is normal to the pressing direction, after a short etching for 1.0 seconds in a 5 percent solution of glacial acetic acid at different magnifications. The image is made in the mode of detecting back scattered electrons (COMPO), which gives the contrast in density of the material. It is seen that a large part of the surface cross-section (75%) consists of cuts of dense grains with a relatively light contrast. A smaller part of the surface area consists of grains, the cut surfaces of which were subjected to severe etching, which provided the development of the internal substructure. It is formed by particles, which consist of the densest material and give the brightest contrast.

The longer etching of the cross-section for 30 seconds does not lead to an increase in the total area of the material etching. However, it shows (fig. 4c) that strongly etched grains of ZnO ceramics microstructure consist of parallel micron-sized crystals with needle shape, which is a characteristic for zinc oxide of the hexagonal crystal structure. In addition, SEM-image seeing the substructure of the dense grain, which is almost impervious to strong etching. This and similar SEM-images show the cuts (in the plane of cross-section) of elongated ZnO nanocrystals, Z-axis of which in the crystallographic direction (002) (figure 2a) nearly coincide with the pressing direction.

4. Conclusion

Thus, using a system based on a coaxial magnetoplasma accelerator and an installation for the spark plasma sintering, a bulk sample of SPS-ceramics based on hexagonal ZnO with the monocrystalline structure was obtained. The studies of its microstructure allow concluding that the obtained ceramics is quite dense with no visible focal defects. Such a structure of the sample is provided by the fact that the synthesized product of plasma-dynamic synthesis contains both nanoscale and micron size particles which is a positive moment for creating the ceramics.

Reference

[1] Duong T T, Do Q N, Pham A T and Nguyen D C 2016 *J. Alloys Compd.* **686** 854-858
[2] Yan H Q, He R R, Yang P D 2003 *J. Pham, Adv. Mater.* **15(5)** 402-405
[3] Pourmortazavi S M, et al. 2015 *J. Molec. Struct.* **1099** 232-238
[4] Lanje A, et al. 2013 *Adv. Powder Technol.* **24** 331-335
[5] Lu P, Huang S, Chen Y, Chiueh L, Shih D 2015 *J. Food Drug Anal.* **23** 587-594
[6] Zarrinkhameh M, Zendehnam A, Hosseini S M 2015 *J. Ind. Eng. Chem.* **30** 295–301
[7] Sedlák J, et al. T 2015 *Adv. Powder Technol.* **26** 1064-1071
[8] Moezzi A, McDonagh A M, Cortie M B 2012 *Chem. Eng. J.* **185** 1-22
[9] Roy D, Fendler J. 2004 *Adv. Mater.* **16(6)** 479-508
[10] Law M, Greene L, Johnson J, Saykally R and Yang P. 2005 *Nature materials* **4(6)** 455-459
[11] Vanmaekelbergh D, Van Vugt L K 2011 *Nanoscale* **3(7)** 2783-2800
[12] Liu Z, Song H, Yu L and Yang L 2005 *Appl. Phys. Lett.* **86(11)** 3109
[13] Zhao M H, Wang Z L and Mao S X 2004 *Nano Lett.* **4(4)** 587-590
[14] Hong R, Li J, Chen L, Liu D, Li H, Zheng Y and Ding J 2009 *Powder Technol.* **189(3)** 426-432
[15] Wang H et al. 2013 *Adv. Powder Technol.* **24(3)** 599-604
[16] Jagadish C and Pearton S J 2011 *Zinc oxide bulk, thin films and nanostructures: processing, properties, and applications* (Elsevier)
[17] Dao V D, Choi Y, Yong K, Larina L L and Choi H S 2015 *Carbon* **84** 383-389
[18] Liu X, et al. 2009 *J. Am. Chem. Soc.* **131(42)** 15106-15107
[19] Zhang J, Zhao B, Pan Z, Gu M, Punnoose A 2015 *Cryst. Growth Des.* **15(7)** 3144-3149
[20] Xu S, Li Z H, Wang Q, Cao L J, He T M and Zou G T 2008 *J. Alloys Compd.* **465(1)** 56-60
[21] Jitianu M and Goia D V 2007 *J. Colloid Interface Sci.* **309(1)** 78-85
[22] Anand V and Srivastava V C 2015 *J. Alloys Comp.* **636** 288–292
[23] Xu S, Li Z H, Wang Q, Cao L J, He T M and Zou G T 2008 *J. Alloys Comp.* **465(1)** 56-60
[24] Wang V, Ma D, Jia W and Ji W 2012 *Solid State Commun.* **152(22)** 2045-2048
[25] Topsakal M, Cahangirov S, Bekaroglu E and Ciraci S 2009 *Phys. Rev. B* **80(23)** 235119
[26] Rahimi-Nasrabadi M, et al. M 2015 *J. Mol. Struct.* **1083** 229-235
[27] Simonelli G, Arancibia E L 2015 *J. Molec. Liquids* **211** 742-746
[28] Xu Z et al. 2016 *Ceram. Int.* **42(13)** 14350-14354
[29] Sivkov A, et al. 2016 *Adv. Powder Technol.* **27(4)** 1506–1513
[30] Sivkov A, Saigash, A and Kolganova Y 2013 *Russian Electrical Engineering* **84(8)** 418-421
[31] Kuzenov V V, Polozova T N, Ryzhkov S V 2015 *Problems of Atomic Science and Technology* **4** 98
[32] Sivkov A, Shanenkova Y, Saigash A and Shanenkov I 2016 *Surf. Coat. Technol.* **292** 63-71
[33] Damonte L C, et al. 2004 *Powder Tech.* **148** 15–19

Acknowledgements
This work was supported by the Russian Science Foundation (grant № 15-19-00049).

The peculiarities of CVD diamond coatings synthesis in abnormal glow discharge plasma using repetitively-pulsed mode

S A Linnik, A V Gaydaychuk and V V Okhotnikov

National Research Tomsk Polytechnic University, 30, Lenin Avenue, 634050, Tomsk, Russia

E-mail: stepan_lin@mail.ru

Abstract. We report about the features of polycrystalline diamond coatings CVD synthesis in repetitively-pulsed plasma of abnormal glow discharge. The discharge burning time was varied from 0.5 to 10 ms with proportional pauses. The dependences of deposited diamond films growth rate on the durations of the discharge burning and pauses are presented. The mutual influence of two plasma filaments on each other and onto the substrate has unequivocally established. Raman spectroscopy, X-ray diffractometry and SEM were used for identification of phase composition and microstructure of deposited films. Implementation simplicity and reliability of the proposed discharge system may find application in diamond film deposition industries.

1. Introduction

The increase in CVD diamond films growth rate is the subject of intensive research around the world because of their unique characteristics, such as hardness, wear resistance, chemical inertness and thermal conductivity [1-3]. Despite the fact that all diamond CVD reactors are based on the same principles, each particular reactor type has its own deposition rate, applied power, working pressure and other parameters. Among the various diamond films deposition methods, there are those such as microwave plasma (MPCVD) [4], hot filament (HFCVD) [5], arc-jet [6] and glow discharge [7,8] CVD. Among the other methods, glow discharge CVD is considered to be one of the most effective diamond deposition method because of its simplicity, high growth rate (up to 10 μm/h [9,10]) and scalability.

In last papers [9,11], we described the construction of a new AC abnormal glow discharge CVD system based on one plasma line, which enabled the deposition of diamond films on the substrate sized 35x200 mm. But the problem of construction of the glow discharge CVD reactor with several plasma lines is still topical, due to the lack of data on the mutual influence of several plasma lines.

In this paper, we report about the peculiarities of CVD diamond coatings synthesis in two AC abnormal glow discharge plasma lines discharge system using repetitively-pulsed mode. The discharge characteristics and the system construction are described. The structure, phase composition and growth rates of produced diamond films are examined. The mutual influence of two plasma lines on each other and onto the substrate has unequivocally established.

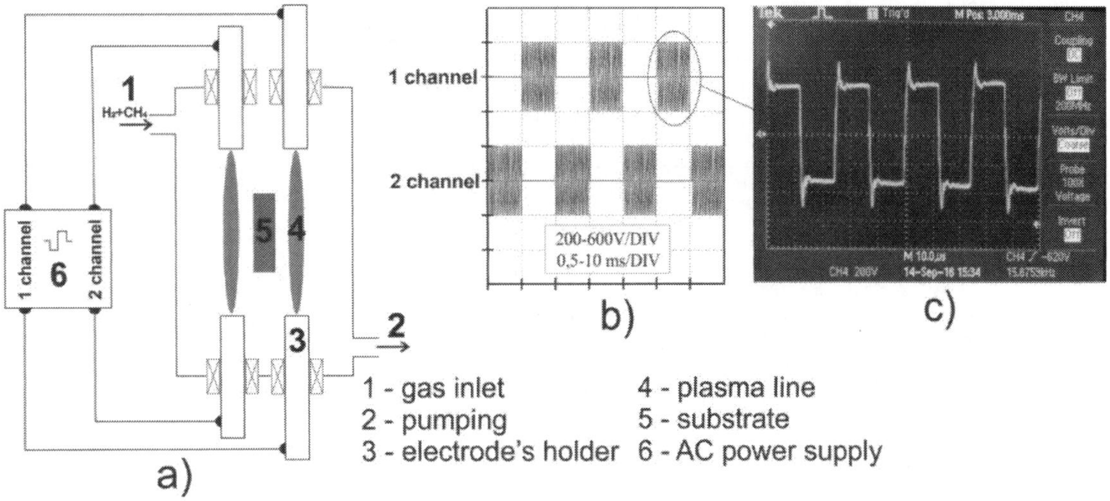

Figure 1. Scheme of the experimental arrangement (a), combined time-voltage diagram of two plasma lines burning (b) and voltage oscillogram of one channel (c).

2. Experimental

Figure 1 shows the scheme of the experimental setup for diamond film deposition in AC glow discharge using two plasma lines. The plasma lines burning mode could vary from parallel (both plasma lines were working simultaneously without interruption) to repetitively-pulsed type (plasma lines were working alternately with a frequency of channel switching from 50 to 1000 Hz). The waveform of each channel burning voltage was represented (as in work [9]) with unipolar rectangular pulses and a short igniter voltage impulse at the initiation of each pulse.

In our experiments the power supply with the operating current up to 20 A and voltage up to 1000 V was used. The amplitude of ignition pulses reached 1.2 kV. The discharge system consisted of four cooled electrode holders, in which the electrodes made of tungsten were fixed. The spacing between two plasma lines also could be varied in the range from 20 to 100 mm, and the interelectrode distance could reach as much as 20 cm. The gas mixture with methane content from 3 to 6 % (be the volume flow) was used as a working atmosphere. The total flow rate of the precursor gasses could change from 50 to 300 ml/min. The substrates (molybdenum rods and plates) were placed on the cooled substrate holder between the plasma lines. The pressure in the deposition chamber could range from 20 to 100 Torr and it was controlled by a diaphragm gauge (Televac 1EP/N). The substrate temperature during deposition was maintained at 850±25 °C using an IR pyrometer. During the deposition substrates were at ground potential.

The surface morphologies and phase composition of the films were analyzed by scanning electron microscopy (SEM) using an FEI Quanta 3D SEM without the sputtering of any conductive film. The samples were characterized by X-ray diffraction (XRD) using Cu - Ka radiation on a Shimadzu XRD-6000 diffractometer. Raman spectra were recorded on a NanoScan Technology Centaur I HR spectrometer.

3. Results and discussion

Figure 2 illustrates that discharge behavior strongly depends on the frequency of channels switching, but the discharge is stable at frequency of 50 Hz as well as at 1 kHz. At low pressures in CVD reactor (less than 25 Torr) the volume form of discharge between each pair of electrodes occurs, and electrodes do not have a significant influence on each other (see figure 2a). At higher pressures, the discharge is constricted in the cord (plasma line) (see figure 2b-g) and its behavior changes significantly. Plasma lines begin to exert the influence on each other with decreasing of downtime (i.e.

effectively by increasing the frequency of channel switching). It is expressed in bending of the plasma lines towards each other (see figure 2d). The reason of this can be high temperature gas area (and therefore with lower dielectric strength) which generated with discharge plasma. It is clear that lifetime of this area continues for some time after the extinction of the discharge.

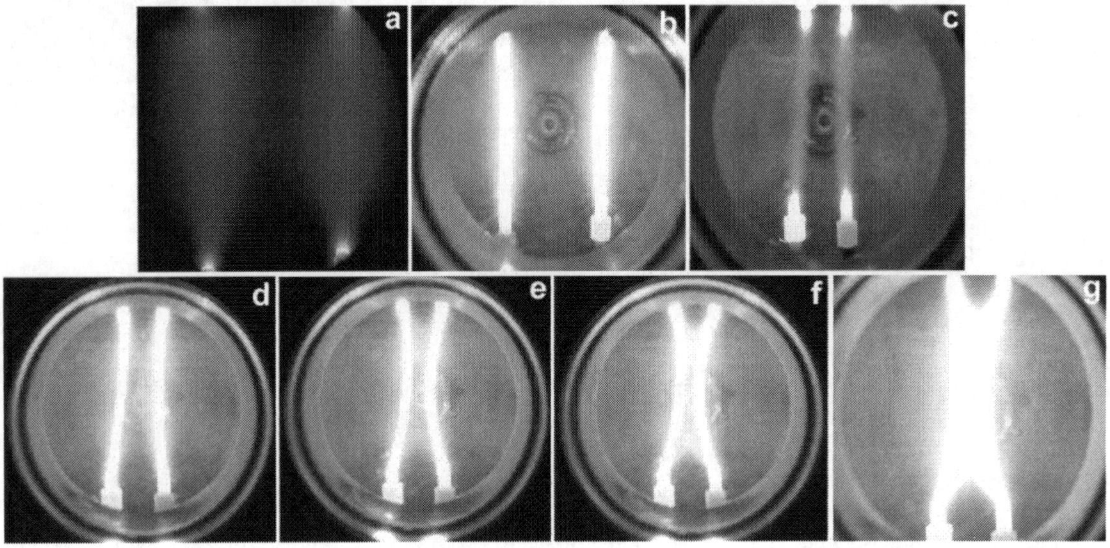

Figure 2. Colour photographs of two plasma lines burning taken through the viewport: (a) volume discharge at 20 Torr in pure hydrogen at a switching frequency of 1 kHz, (b) constricted discharge at 60 Torr in H_2/CH_4 atmosphere at a switching frequency of 100 Hz, (c) constricted discharge at 40 Torr in pure hydrogen at a switching frequency of 100 Hz, (d) increasing interconnection between plasma lines at a switching frequency of 300 Hz (60 Torr; H_2/CH_4 atmosphere), (e) more significant interconnection between plasma lines at a switching frequency of 500 Hz (60 Torr; H_2/CH_4 atmosphere), (f) at 700 Hz, (g) at 1 kHz.

For a while after first plasma line attenuation the second plasma line can be more easily ignited not in a straight line between electrodes (the shortest distance), but in a curved arc towards the plasma line. As a result, at high frequencies of channel switching (i.e. in a state approaching a parallel mode of combustion) the common combustion region is formed (see figure 2f,g). It represents a superposition of two plasma lines. This phenomenon may be very useful for the development of CVD reactors for large area deposition on the flat area of substrates (because the uniformity of the coatings will increase with combining of plasma lines), but in this case the deposition on milling tools (3D substrate) can be problematic.

Since the main purpose of this discharge system is the deposition of diamond coatings we conducted a series of experiments on the deposition of these coatings on the samples of various shapes and sizes with different frequency of channel switching. As a result it was found that in a pulsed mode the optimal gas atmosphere for sustained diamond growth is similar to a parallel mode of plasma combustion (methane content must be no more than 10 %). Any correlation of this parameter on the switching frequency was not observed. At the same time, it was found that the deposition rate in a pulsed mode, essentially the same as in the parallel mode of combustion, despite the half as much power consumption (calculated to one plasma line). Table 1 shows the main deposition parameters of CVD diamond films, obtained in different discharge modes (50 Hz, 500 Hz, 1 kHz). It is seen that the difference in deposition rates comparable to the error in the measurement of the substrate temperature (which directly influences the deposition rate). This fact of the absence of the effect of reducing the power input to the discharge on the coating deposition rate is quite atypical. For example for microwave CVD reactors the direct correlation between the deposition rate and increasing the injected power was observed [12]. This phenomenon can be explained by different mechanisms and intensities of active radical generation in an anomalous glow and microwave discharges. In the

Table 1. The deposition parameters of CVD diamond films obtained in different discharge modes.

Discharge mode	Discharge power per one plasma line, W	Reactor pressure, Torr	H_2/CH_4 ratio	Substrate temperature, °C	Diamond film growth rate, μm/h
Constant (40 kHz without modulation)	5.5	60±1	50:3	850±25	7.5±0.2
50 Hz modulation	2.75	60±1	50:3	850±25	7.0±0.2
500 Hz modulation	2.75	60±1	50:3	850±25	7.1±0.2
1000 Hz modulation	2.75	60±1	50:3	850±25	7.4±0.2

plasma line (quasiarc discharge) the area with very high temperature (over 5000 °C) is generated, while the temperature in microwave discharge plasma under the same pressure does not exceed 2500 ° C [13]. At higher temperatures more active particles (radicals, unstable molecules) required for diamond gas phase growth are generated [14] and lifetime of such particles after discharge attenuation also directly depends on the temperature.

Figure 3a shows the SEM images of the 25 μm textured diamond film with well-faceted microcrystallites without amorphous non-diamond impurities. Therefore, to determine a change in the phase composition of the film surface at different stages of deposition, Raman spectra were obtained (see figure 3b). The typical Raman lines of diamond and amorphous (sp^2) carbon (1332 cm^{-1} and 1450 cm^{-1}) are observed. The XRD diffractogram obtained from the front surface of the diamond films is shown in figure 3c. It contains crystalline diamond diffraction peaks at 44.0, 75.5 and 91.8°, corresponding to the {111}, {220} and {311} planes, respectively. There are no lines of other phases which confirms the purity of the diamond film.

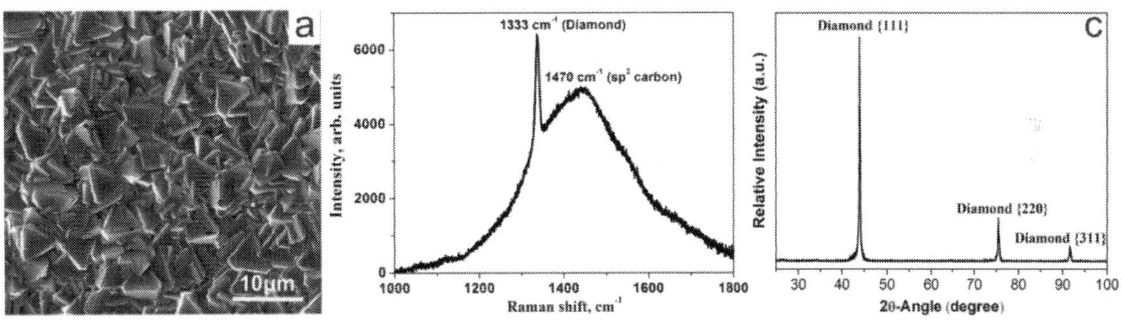

Figure 3. The data about structure and phase composition of obtained diamond coatings: SEM image of 25 μm film (a), Raman spectra (b) and x-ray diffraction pattern (c).

4. Conclusion

We consider the features of polycrystalline diamond coatings CVD synthesis in repetitively-pulsed plasma of abnormal glow discharge. The discharge burning time was varied from 0.5 to 10 ms with proportional pauses. The dependences of deposited diamond films growth rate on the durations of the discharge burning and pauses are presented. The mutual influence of two plasma filaments on each other and onto the substrate has unequivocally established. Raman spectroscopy, X-ray diffractometry and SEM were used for identification of phase composition and microstructure of deposited films. Implementation simplicity and reliability of the proposed discharge system may find application in diamond film deposition industries.

References

[1] Williams O A 2001 *Diam. Relat. Mater.* **20** 621
[2] May P W 2000 *Philos. Trans. R. Soc. Lond. A* **358** 473
[3] Seelmann-Eggebert M, Meisen P, Schaudel F, Koidl P, Vescan A, Leier H 2001 *Diam. Relat. Mater.* **10** 744
[4] Liang Q, Yan C, Lai J, Meng Y, Krasnicki S, Shu H, Mao H, Hemley R J 2014 *Cryst. Growth Des.* **14** 3234
[5] Matsumoto S, Sato Y, Kano M, Setaka N 1982 *Jpn. J. Appl. Phys.* **21** 182
[6] Han Q Y, Or T W, Lu Z P et all 1991 *Proceedings of the Second International Symposium on Diamond Materials* **91** 115
[7] Chae K –W, Baik Y –J, Pack J –W, Lee W –S 2010 *Diam. Relat. Mater.* **19** 1168
[8] Suzuki K, Sawabe A, Yasuda H et all 1987 *Appl. Phys. Lett.* **50** 728
[9] Linnik S A, Gaydaychuk A V 2013 *Diam. Relat. Mater.* **32** 43
[10] Linnik S A, Gaydaychuk A V 2014 *Vacuum* **103** 28
[11] Linnik S A, Gaidaichuk A V 2012 *Tech. Phys. Lett.* **38** 258
[12] http://sekidiamond.com/microwave-plasma/products/AX6600.html
[13] Hassouni K, Grotjohn T A, Gicquel A 1999 *J. Appl. Phys.* **86** 134
[14] Corat E J, Goodwin D G 1993 *J. Appl. Phys.* **74** 2021

Acknowledgments

This work was supported financially by Russian President's grant № MK-5516.2016.8 and by the Russian Foundation for Basic Research (research project № 16-32-60018 Mol_a_dk).

IOP Conf. Series: Journal of Physics: Conf. Series **830** (2017) 012119 doi:10.1088/1742-6596/830/1/012119

Effect of precursor mass on product phase composition in plasma dynamic synthesis of tungsten carbide

K N Shatrova, A A Sivkov, I I Shanenkov and A S Saigash

Institute of Power Engineering, National Research Tomsk Polytechnic University, Lenin av., 30, Tomsk 634050, Russian Federation

E-mail: Swordi@list.ru

Abstract. An interest in WC_{1-x} cubic tungsten carbide results from its catalytic properties similar to those of platinum group metals and the synergistic effect between WC_{1-x} and Pt in reactions of hydrogen evolution and hydrogen oxidation. However, according to the phase diagram of the W–C system, the cubic phase WC_{1-x} only exists in a narrow range of temperature stability (about 2798–3058 K), which makes it difficult for being obtained. To date, there are different methods for synthesizing tungsten carbide powder with a low content of cubic phase that complicates the study of WC_{1-x} properties. A direct plasma dynamic synthesis is known as one of the promising methods to produce WC_{1-x}. The aim of this work is to find the optimal amount of tungsten precursor to obtain cubic tungsten carbide with a high purity by plasma dynamic method. The synthesized products were examined by X-ray diffraction (XRD) and transmission electron microscopy (TEM). The XRD patterns showed that the main phase was cubic tungsten carbide with negligible content of hexagonal tungsten carbide W_2C and pure tungsten W. According to a quantitative analysis of synthesized products, which were obtained using masses of initial tungsten equal to 1.0, 0.7, 0.6 and 0.5 gram, the yield of WC_{1-x} phase was 84, 89, 95 and 92 wt%, respectively. The results of TEM displayed that the synthesized powders consist of crystallites, having the size less than 100 nm (WC_{1-x}), and a carbon matrix. This carbon was not detected in XRD due to its presence as an amorphous phase.

1. Introduction

Tungsten carbides play an important role in industry because of their high melting point, high Vickers hardness, high Young's modulus, high fracture toughness, low thermal expansion coefficient and good wear resistance [1]. Due to these properties, tungsten carbides are widely used for creating cutting tools, wear resistant materials, drilling tools, etc. [2-5]. Moreover, electrocatalytic properties of tungsten carbides are of utmost interest since this chemical compound may be used as a platinum catalyst substrate. It allows reducing the fuel cell cost and enhancing fuel cell efficiency [1, 2].

It is known that hexagonal tungsten carbide WC is the dominant phase in the W-C system. Nevertheless, there are two more phases in the system: W_2C and WC_{1-x}. According to a recent theoretical analysis [4], cubic WC_{1-x} has the density of states close to the Fermi level twofold greater than that of W_2C and 6 times greater than that of WC. It is the reason to suggest that WC_{1-x} can be the most active catalyst among the three phases of tungsten carbides. Therefore, cubic tungsten carbide WC_{1-x} is now the object of scientific research in the whole world. The problem is yet, that the cubic phase of WC_{1-x} tungsten carbide only exists in a narrow range of temperature stability, which makes it more difficult to obtain as compared to the hexagonal WC and W_2C [3].

To date, in a number of articles [5-9] authors present different approaches to cubic tungsten carbide production while plasma dynamic synthesis shows the best results from the point of cubic phase output [10]. In this paper, investigation was carried out to determine the optimal precursor amount to obtain a greater WC_{1-x} percentage in a product by direct plasma dynamic synthesis.

2. Experimental

The proposed method is based on the use of a high-current pulsed coaxial magnetoplasma accelerator [11] (CMPA) with graphite electrodes [12]. The CMPA-based system includes three main elements such as capacitive energy storage, CMPA and working chamber. The capacitive energy storage consists of 24 sections with 1.2 mF capacitor banks. This allows varying the energy, which is supplied to CMPA during the process. In a set of experiments, capacity of the energy storage was 6 mF and charging voltage of capacitor banks was 3 kV. At chosen energy parameters, the value of energy, which was accumulated in the energy storage, was equal to 27 kJ. This allowed reaching the value of discharge current up to ~130 kA, discharge voltage 1.1 kV and discharged power ~120 MW.

In the set of four experiments, the tungsten powder of 1.0, 0.7, 0.6 and 0.5 gram (average particle size was 1 micron) was used as a precursor, which was placed in the plasma formation zone of the accelerator. Under the influence of a rising discharge current, this precursor was converted to the plasma state and involved into the plasma flow movement. The second precursor was accumulated from the walls of the accelerating channel, made of graphite. Thus, in the moment, when the plasma flow exited the accelerating channel, it included both tungsten and carbon in ionized state. The final powder (tungsten carbide) was synthesized in a super-high-speed tungsten-carbon plasma jet flowing into the working chamber filled with argon atmosphere at normal conditions. The methodology of the experiments was described in detail in our previous works [10, 11]. In order to investigate, how the mass of tungsten precursor influences the yield of cubic tungsten carbide, all synthesized samples were carefully collected and studied without any additional treatment.

The synthesized products were analyzed by X-ray diffraction (XRD) method using a Shimadzu XRD7000 diffractometer (CuKα-radiation, $\lambda = 1.54$ Å) with a graphite monochromator Shimadzu CM-3121 at the pace of 0.02 degrees and exposure time of 1 second. Qualitative analysis was conducted using the standards of the structural data base PDF2+. The size of coherent-scattering regions (CSR) was estimated according to Debye-Scherrer equation. The morphology of the powder was studied by a transmission electron microscope (TEM, Philips CM12) operating at 200 kV and high-resolution TEM (HRTEM, JEOL JEM 2200F).

3. Results and discussion

Figure 1 shows X-ray diffraction patterns of the synthesized products at the initial tungsten fill of 1.0 g (1), 0.7 g (2), 0.6 g (3) and 0.5 g (4). According to the XRD patterns, at the tungsten masses equal to 1.0 g and 0.7 g, the synthesized products consist not only of the main phase (cubic WC_{1-x}), but also include the hexagonal phases of tungsten carbide W_2C and pure tungsten W. Moreover, W_2C and W peaks are quite intense. The appearance of these phases can be explained by the lack of carbon ions in comparison with tungsten ones, which present in the plasma in a large amount due to a larger precursor mass. An electro erosion mechanism, which is used to obtain carbon during the process, has a limited value. It means that at the chosen energy conditions there is a necessity to reduce the mass of tungsten precursor, in order to increase the yield of cubic tungsten carbide. This suggestion is confirmed, when the tungsten mass is reduced to 0.6 g and 0.5 g. W peaks disappear and the intensity of W_2C peaks reduces significantly. The percentage phase content and the size of coherent scattering regions (CSR) for the all synthesized powders were calculated by the XRD patterns and summarized in Table 1.

The changes in the phase composition of the synthesized products can be explained using the phase diagram of the W–C system (Figure 2). Evidently, when the initial tungsten weight is 1.0 g and 0.7 g, it is present in excess. Therefore, in the process of cooling the tungsten-carbon plasma down to about 3,000 K the cubic phase WC_{1-x} forms, after which the atomic carbon amount in the plasma is reduced.

In the conditions of insufficient amount of carbon, W_2C is formed and some amount of pure tungsten remains during further cooling (region 1 in Figure 2). When the mass of tungsten is 0.6 g or 0.5 g, initially WC_{1-x} is formed and then the carbon shortage is not so critical as in the first and the second experiments. Therefore, in these cases, W_2C is formed and the unreacted W disappears (region 2 in Figure 2). Absence of pure W traces indicates that such mass fills are most appropriate from the standpoint of the atomic ratio of W/C, which allows getting rid of the presence of "extra" phases. However, exactly tungsten powder of 0.6 g may be optimal for production of cubic tungsten carbide of high purity in the plasma dynamic system, because in this case WC_{1-x} phase reaches 95 wt%.

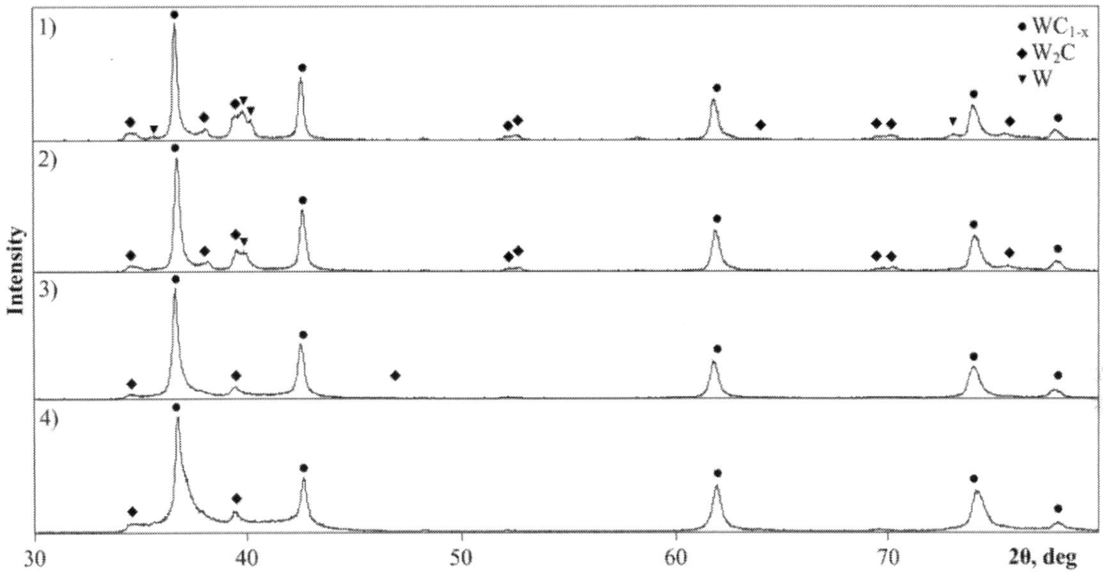

Figure 1. X-ray diffraction patterns of synthesized products.

Table 1. Calculated data of full-profile X-ray diffraction analysis

Number of experiment	WC_{1-x} [%]	W_2C [%]	W [%]	C [%]	CSR [nm]
1	84.0	10.6	4.8	0.6	39.9
2	89.3	7.5	2.5	0.7	32.6
3	95.0	3.7	0.6	0.7	29.6
4	92.2	6.5	0.6	0.7	23.9

Unfortunately, there is a lack of quantitative analysis of materials containing the cubic tungsten carbide phase. Therefore, we estimated the percentage of the phases (wt) by the ratio of the major peak intensities in the XRD patterns to compare our results with the literature data.

In polyporous C@WC_{1-x} composite, intensities of the carbon major peak and the cubic tungsten carbide major peak are practically equal [6]. It means that the cubic tungsten carbide has reached 10–20%. A nanosized tungsten carbide-cobalt composite powder synthesized by thermal plasma contains about 60% WC_{1-x} [7]. After high-intensity pulsed ion beam irradiation of WC-Ni cemented carbides, Zhang et al. have obtained the sample consisting of about 70% WC_{1-x}, 30% WC and Ni traces [8]. In electro-discharge process, nano-structured tungsten carbide powder has been obtained [9]. In this material WC_{1-x} phase has reached about 90%. The powder produced from tungsten powder of 0.6 g, APS 1 micron by the plasma dynamic method proposed in this paper consists of 95% WC_{1-x}. Therefore, this method can be very efficient in obtaining the cubic tungsten carbide phase.

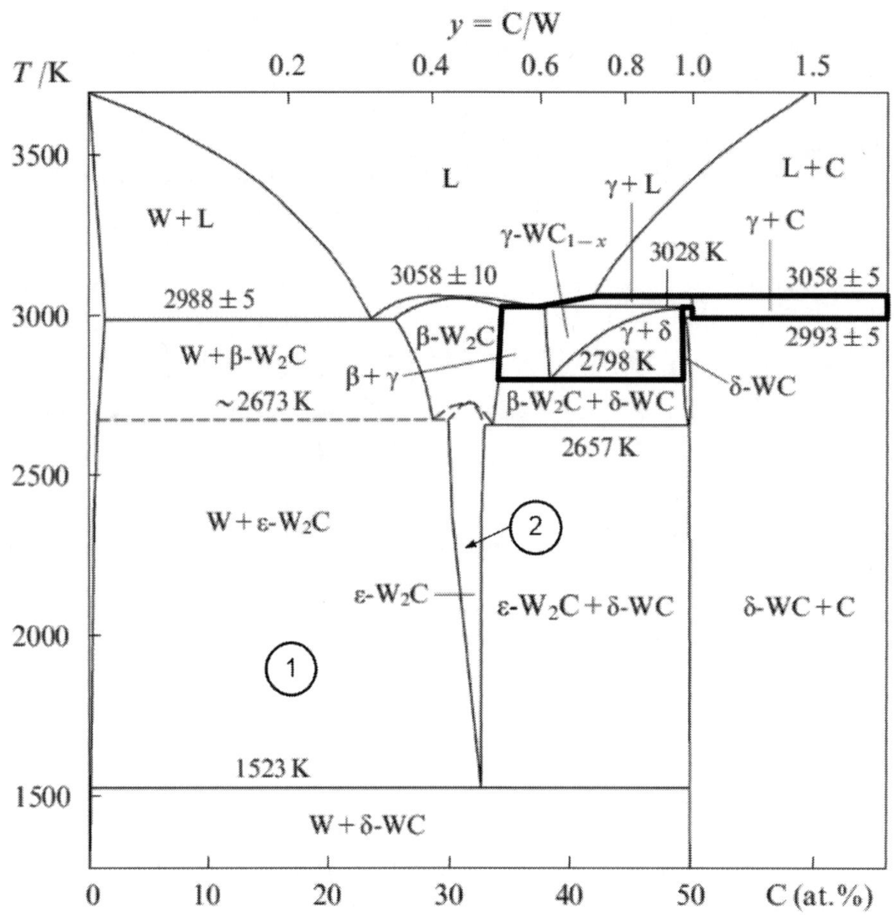

Figure 2. Phase diagram of the W–C system [3].

The results of TEM analysis show that the synthesized powders consist of objects of two types: rounded dark particles with a size of less than 100 nm, corresponding to the cubic phase of tungsten carbide, and graphite matrix (Figure 3a). This carbon is not detected in XRD due to its amorphous state. The particle size obtained from the TEM micrograph is in a good agreement with CSR, calculated from the XRD results. Figure 3b shows the selected-area electron diffraction (SAED) pattern, in which two fuzzy rings of the ultrafine graphite phase and individual point reflexes, belonging WC_{1-x} phase, can be identified. The values of interplanar distances, calculated by HRTEM image (Figure 3c), correspond to one for the phase (ICDD no. 00-020-1316). Crystalline particles of WC_{1-x} phase have a shell with a thickness of about 10.0 nm. It can be argued that the shell consists of crystalline carbon as it shines in the dark-field TEM image in the light of diffracted beam of the carbon structure (Figure 3d).

4. Conclusion

In this paper, the optimal amount of the initial tungsten powder, in order to obtain the cubic phase of high purity by plasma dynamic synthesis, was found. According to the results of X-ray diffraction, cubic tungsten carbide powder with 95 wt% of WC_{1-x} was produced using the tungsten mass of 0.6 g and the charging energy of 27 kJ. The plasma dynamic synthesis, based on using the coaxial magnetoplasma accelerator, may rate as one of the most effective methods for producing the cubic tungsten carbide phase that should outrank the WC and W_2C phases in catalytic activity. This property of the cubic tungsten carbide will be investigated at the next stage of our scientific work.

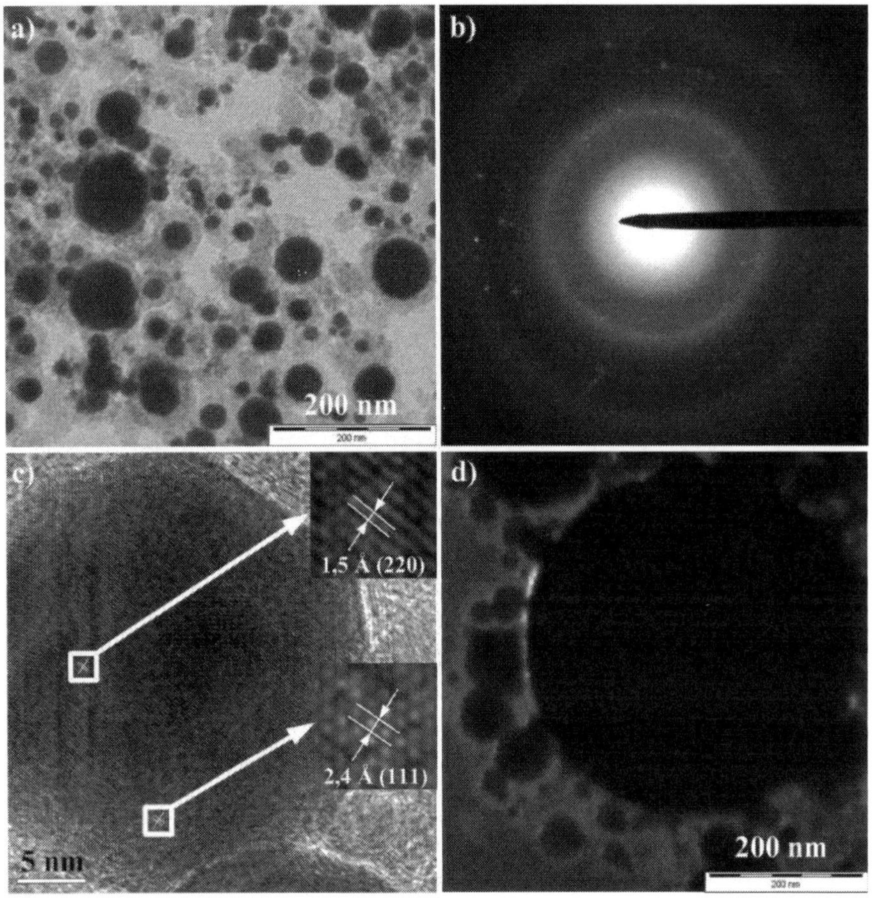

Figure 3. TEM analysis of the plasma dynamic synthesis product: a) bright-field image; b) SAED; c) HRTEM; d) dark-field image.

References

[1] He C, Meng H, Yao X and Shen P K 2012 *Int. J. of Hydrogen Energy* **37** 8154–60
[2] Kurlov A S and Gusev A I 2006 *Inorganic Materials* **42** 121–7
[3] Suetin D V, Shein I R and Ivanovskii A L 2009 *J. of Physics and Chemistry of Solids* **70** 64–71
[4] Gao Y, Song X, Liu X, Wei C, Wang H and Guo G 2013 *Scripta Materialia* **68** 108–10
[5] Zheng H J, Yu A M and Ma C A 2011 *Chinese Chemical Letters* **22** 497–500
[6] Ryu T, Sohn H Y, Hwang K S and Fang Zh Z 2009 *J. of Alloys and Compounds* **481** 274–7
[7] Zhang F G, Zhu X P and Lei M K 2012 *Surface and Coatings Technology* **206** 4146–55
[8] Shabgard M R and Najafabadi A F 2014 *Advanced Powder Technology* **25** 937–45
[9] Pak A, Sivkov A, Shanenkov I, Rahmatullin I and Shatrova K 2015 *Int. J. of Refractory Metals & Hard Materials* **48** 51–5
[10] Sivkov A A and Pak A Ya. RF Patent No. 2431947, H05H 11/00, F41B 6/00, Byull. Izobret., No. 29 (2011) [in Russian].
[11] Kuzenov V V, Polozova T N and Ryzhkov S V 2015 *Problems of Atomic Science and Technology* **4 (98)** 49.
[12] Sivkov A A, Pak A Y, Shanenkov I I, Kolganova Y L and Shatrova K N 2014 *IOP Conference Series: Materials Science and Engineering* **66** 012001.

Acknowledgements
This work was supported by the Russian Science Foundation (grant No. 15-19-00049).

Synthesis of silicon carbide nanopowders in free flowing plasma jet with different energy levels

D Nikitin, A Sivkov, I Rahmatullin and A Ivashutenko

National Research Tomsk Polytechnic University, 634050, Tomsk, Lenin Avenue, 30

E-mail: NikitinDmSr@yandex.ru

Abstract. Silicon carbide (SiC) nanopowders were produced by the synthesis in an electrodischarge plasma jet generated by a high-current pulsed coaxial magnetoplasma accelerator. The present work focuses on the experiments where the obtained hypersonic plasma jet flew into space of the reactor chamber without impact on a target. The energy level of experiments was changed from ~10.0 to ~30.0 kJ. Four experiments were carried out at different energy levels. The powder products synthesized by the plasmadynamic method were studied by such well-known methods: X-ray diffraction (XRD), transmission electron microscopy (TEM). All the powders mainly contain cubic silicon carbide (β-SiC) particles with clear crystal structures and triangular shapes. SiC content reaches its maximum value 95% at the energy level 21.0 kJ, then SiC content is decreased to 70% the energy level 27.8 kJ. The powder crystallites in different experiments have approximately the same average crystallite size because quasistationary time, which allows growing powder crystallites, is absent.

1. Introduction

Silicon carbide (SiC) is one of the most important non-oxide materials and it is a very attractive material for a wide range of industrial applications. This compound has many excellent physical and chemical properties due to strong covalent bonding: superhardness, low density, high mechanical strength, Young's modulus, thermal and corrosion resistance, radiation hardness [1–4].

Silicon carbide is a promising ceramics for various industrial applications even at temperatures above 1000 °C including externally stable ceramics, high temperature semiconductor devices, grinding powder, brake pads, modern nuclear reactors and etc. [5, 6].

Nanomaterials are known to have unique and higher properties than conventional materials. Various silicon carbide nanostructures are widely used as ceramics reinforcement, for creation of nanostructured ceramics, micro- and nanoelectromechanical systems (MEMS and NEMS) [7–9]. It is well known that carbothermal reaction, physical vapor transport (PVT), self-propagation high-temperature synthesis (SHS), chemical vapor deposition (CVD), sol-gel methods, gas-phase reaction method are the most commonly method used to synthesize SiC materials [10–13]. The cost, time and energy expenses are comparatively high in all above synthesized methods. There is a big problem to produce nanosized silicon carbide particles in large volumes.

Nano-SiC synthesis can be realized in the hypersonic plasma of 11–28 kJ. In this work, the synthesis of nanosized silicon carbide powders was implemented by plasmadynamic method. The possibility of SiC nanopowders production by plasmadynamic method was shown in [14, 15]. The aim of this work is to implement nano-SiC synthesis in a free flowing hypersonic plasma jet and to evaluate the effect of plasma energy on nanosized product characteristics.

2. Experimental Conditions

The plasmadynamic method has been already successfully applied for obtaining nanopowders of the Si-C system. This method is based on the use of a coaxial magnetoplasma accelerator invented in Tomsk Polytechnic University. This original installation was used to generate a plasma jet. The schematic diagram of the experimental system is shown in Figure 1. Only the type of a coaxial magnetoplasma accelerator with graphite electrodes [3] could be applied for obtaining silicon carbide nanopowders without different extraneous phases (not Si-C system) associated with electric erosion of electrodes. The fundamentals of the applied synthesis technique are discussed in detail for synthesis of Si-C phases in [14], [15]. The present work focuses on the experiments where the obtained hypersonic plasma jet flew into space of the reactor chamber. Such the experiments are significantly different from the previous works [15] where another form of the flowing plasma jet was used. These works include experiments with the impact of the plasma jet on a copper target installed at a distance of 23 mm from acceleration channel.

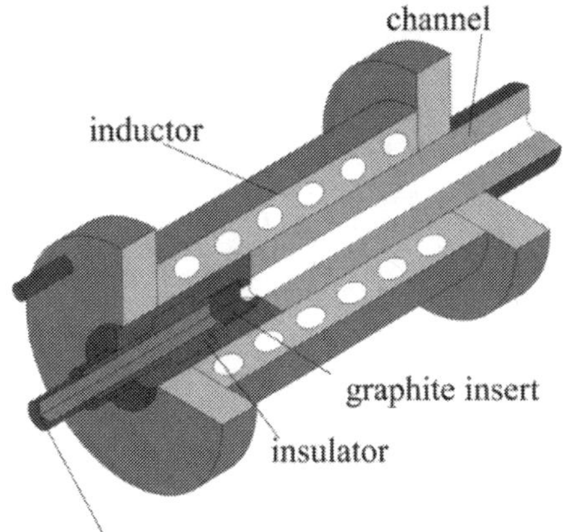

Figure 1. Schematic diagram of the experimental system based on a coaxial magnetoplasma accelerator.

Pulsed power supply of a magnetoplasma accelerator was carried out by a capacitive energy storage with the maximum stored energy W_{max} = 360 kJ. The present study includes the experiments which were carried out at the different levels of accumulated energy W_c and charging voltage U_{ch} from 2.0 kV to 3.5 kV. It was assumed that the variation of plasma jet energy should lead to changing main characteristics of the synthesized product such as the phase composition (percentage content of silicon carbide and attendant phase) and the product dispersion. The capacity of a capacitive energy storage remained constant C = 6 mF.

A mixture of solid precursors in the form of silicon (averaged sizes of particles ~50 μm) of and carbon (amorphous powder) in an approximately stoichiometric ratio was placed to a plasma formation zone. The discharge between the central electrode and the coaxial electrode (accelerated channel) lead to transition of a precursor mixture to a plasma jet. The generated plasma jet was accelerated by the magnetic field of an accelerating channel and an inductor (a solenoid). Table 1 shows the basic electrical and energy parameters of the experiments: operating current I_m, the voltage on the electrodes of magnetoplasma accelerator U_m, the discharge power P_m, the released energy W. The above values were determined by typical oscillograms of working current and voltage on electrodes obtained by a Tektronix TDS1012 oscilloscope. The space of the reactor chamber was filled with inert gas (argon).

The powder product synthesized by the plasmadynamic method was collected from the wall of the reactor chamber after precipitation of suspended particles. The collected powders were studied without any additional preparation by such well-known methods as X-ray diffraction (XRD) (Shimadzu XRD

6000 diffractometer, CuKα-radiation, λ = 0.15406 nm), transmission electron microscopy (TEM) (Philips CM 12 microscope). Qualitative X-ray analysis was performed using PowderCell 2.4 software package based on Rietveld method using PDF4+ database.

Table 1. The average size of nanoparticles and aggregates in suspensions.

Experiments	Electrical and energy parameters				
	U_c [kV]	I_m [kA]	U_m [kV]	P_m [MWt]	W [kJ]
1	2.0	2.5	3.0	3.5	27.8
2	65	76	98	111	21.0
3	12.0	18.8	27.0	36.8	18.8
4	10.5	13.0	19.0	29.7	12.0

3. Results and discussion

The results of the product study by X-ray diffractometry are shown in Figure 2 in the forms of the obtained X-ray diffraction patterns (for two experiment with the released energy $W_1 = 21$ kJ and $W_2 = 12$ kJ). According to the nature of peaks, the product consists of several crystalline phases. The full-profile phase and structural analysis of the X-ray diffraction spectrums of the products are presented in Table 2. All the products include phases corresponding to the following structural models: cubic silicon carbide β-SiC, space group (SPGR) – F-43m {216}; cubic silicon cSi, SPGR – F-43/d-32/m {227}; graphite gC, SPGR – P6-3mc {186}.

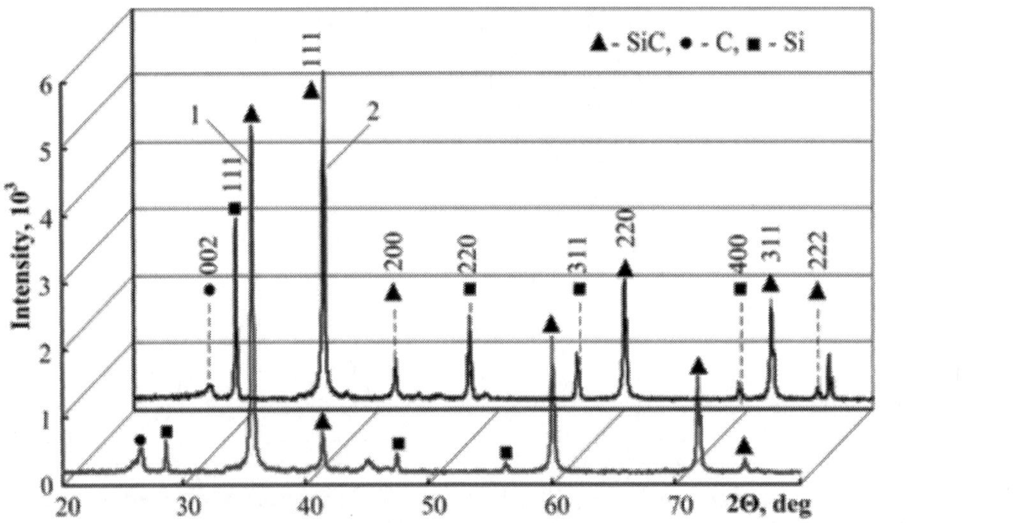

Figure 2. Diffraction patterns of the synthesized powders.

In general, phase composition of the products which were synthesized in the experiments with free flowing plasma jet is similar to the powders obtained by impact of the plasma jet on a copper target [4]. Comparison of the quantitative analysis results allows saying that there is an approximate equality of the silicon carbide content in the products synthesized at the same energy levels. However another tendency appears in relation to the effect of plasma jet energy on phase composition. SiC content reaches its maximum value (ω = 95%) at the energy level W = 21 kJ, then SiC content is decreased down to ω = 70% the energy level W = 27.8 kJ. This tendency can be explained by increase of pT-parameters due to increase of the energy level and increasing wear of the surface of a carbon accelerating channel at the greatest energy. Contamination of the final product by eroded carbon is confirmed by the calculated data presented in Table 2: the main impurity in the product of the experiment at the energy level W= 27.8 kJ is carbon phase, graphite. Increase of the energy level leads to changing the ratio of the silicon and carbon impurities to increase the amount of carbon.

Table 2. The full-profile phase and structural analysis of the X-ray diffraction spectrums of the products.

Phase / *W* [kJ]		SiC	Si	C
1. 12.0	Content [%]	65,0	19.0	16.6
	CSA [nm]	60	40	20
2. 18.8	Content [%]	80.0	6.6	15.7
	CSA [nm]	60	35	10
3. 21.0	Content [%]	95.0	0.6	6.3
	CSA [nm]	50	40	10
4. 27.8	Content [%]	70.0	4.0	24.5
	CSA [nm]	70	80	10

A slight change of the main phase of coherent scattering areas (CSA) within 20 nm should be noted. This phenomenon is related to the form of a flow of hypersonic plasma jet from an accelerating channel. When a plasma jet impacts a copper target, pT-parameters increase in the shock wave induced in the vicinity of the copper target, so the size of silicon carbide crystallites increase. When a plasma jet flows into a reactor chamber freely without any impact on a target, a quasistationary time which allows growing powder crystallites is absent. In the experiments of the last case, powder crystallites have approximately the same average crystallite size, characterized by the value of CSA.

The results of the study by X-ray diffraction are confirmed and complemented by the results of the analysis by transmission electron microscopy (TEM) presented in Figure 3. Figure 3a shows the general bright-field TEM-image typical for of the experiments carried out by free flowing plasma jet. The present image includes the accumulation of crystalline particles. Significant quantity of typical triangular shaped particles in are presented in Figure 3a. The intensive Debye rings corresponding to cubic silicon carbide proves the predominant content of this phase in the product. The selected area electron diffraction (SAED) is presented in Figure 3b. The most striking and the main diffraction reflexes relate to reflecting planes of cubic silicon carbide. Dark-field TEM-image (Figure 3c) was obtained at the diffracted light beam on the packets plane SiC (111). The glow of reflecting planes of particles can be seen. It proves the particles of this fraction relate to silicon carbide phase.

4. Conclusions

Silicon carbide (SiC) nanopowders were produced by the synthesis implemented in an electrodischarge plasma jet generated by a high-current pulsed coaxial magnetoplasma accelerator. The present work focuses on the experiments where the obtained hypersonic plasma jet flew into space of the reactor chamber without impact on a target. The energy level of experiments was changed from ~10.0 to ~30.0 kJ. Four experiments were carried out at different energy levels. The powder products synthesized by the plasmadynamic method were studied by such well-known methods as X-ray diffraction (XRD), transmission electron microscopy (TEM). All the powders mainly contain cubic silicon carbide (β-SiC) particles with clear crystal structures and triangular shapes. SiC content reaches its maximum value 95% at the energy level of 21.0 kJ, then SiC content is decreased to 70% at the energy level of 27.8 kJ. The powder crystallites in different experiments have approximately the same average crystallite size because quasi-stationary time—which allows growing powder crystallites—is absent.

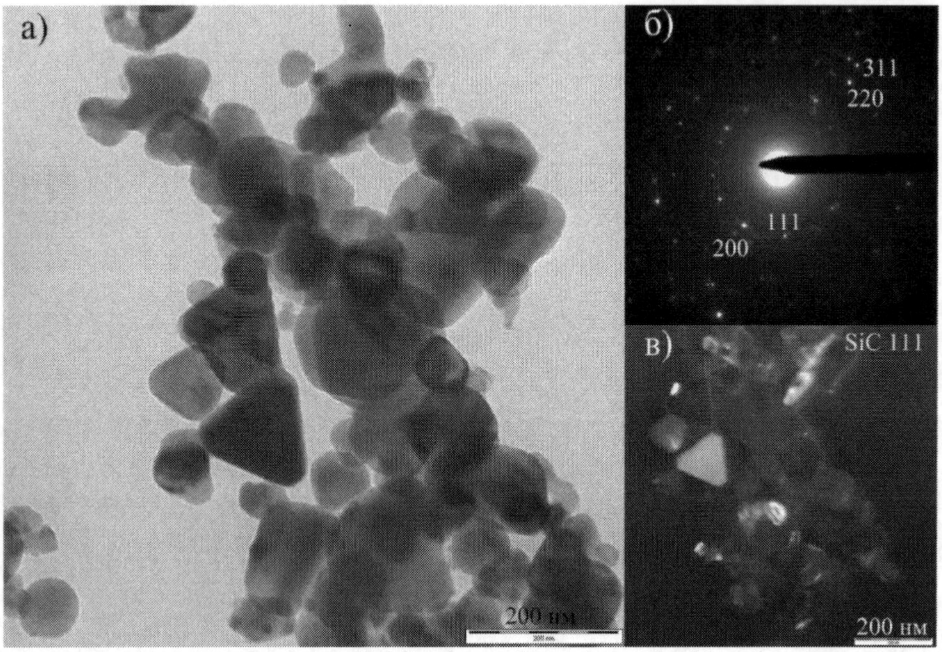

Figure 3. Bright-field TEM-images of the plasmadynamic synthesis product (W = 29.7 kJ).

References

[1] Wesch W 1996 *Nucl. Instrum. Methods Phys. Res., Sect. B.* **116** 305
[2] Harris G L 1995 *Properties of Silicon Carbide* (London: NSPEC)
[3] Gerhardt R 2011 *Properties and Applications of Silicon Carbide* (Rijeka: InTech)
[4] Sarin V K 2014 *Comprehensive Hard Materials. Volume 2, Ceramics* (Oxford: Elsevier Ltd.)
[5] Knippenberg W F 1963 *Phillips Res. Rep.* **18** 1
[6] Van der Berg N G, Malherbe J B, Botha A J, Friedland E 2012 *Appl. Surf. Sci.* **258** 5561
[7] Celebi Efe G, Ipek M, Zeytin S, Bindal C 2012 *Composites Part B* **43** **4** 1813
[8] Moazami-Goudarzi M., Akhlaghi F. 2013 *Powder Technol.* **245** 126
[9] Maboudian R, Carraro C, Senesky D G, Roper C S 2013 *J. Vac. Sci. Technol., A* **31** **5** 1
[10] Rajarao R, Ferreira R, Sadi S H F, Khanna R and Sahajwalla V 2014 *Mater. Lett.* **120** 65
[11] Huczko A, Kurcz M, Dabrowska A, Baranowski P, Bhattarai A and Gierlotka S 2014 *J. Cryst. Growth* **401** 469
[12] Van Laar J H, Slabber J F M, Meyer J P, van der Walt I J, Puts G J and Crouse P L 2015 *Ceram. Int.* **41** **3** **B** 4326
[13] Omidi Z, Ghasemi A and Bakhshi S R 2015 *Ceram. Int.* **41, 4** 5779
[14] Sivkov A A, Nikitin D S, Pak A Y, Rakhmatullin I A 2013 *J. Superhard Mater.* **35** **3** 137
[15] Sivkov A, Nikitin D, Pak A, Rakhmatullin I 2015 *Nanotechnologies Russ.* **10** 34

Acknowledgments

This work was supported by the Russian Science Foundation (project no. 15-19-00049).

Plasmadynamic synthesis of Ti-B nanopowders

D Nikitin, A Sivkov, D Gerasimov and A Evdokimov

National Research Tomsk Polytechnic University,634050, Tomsk, Lenin Avenue, 30

E-mail: NikitinDmSr@yandex.ru

Abstract. Ti-B nanopowders were produced by plasmadynamic synthesis. This method was implemented by the synthesis in an electrodischarge plasma jet generated by a high-current pulsed coaxial magnetoplasma accelerator. Two experiments were carried out using copper and titanium conductors for initiating the plasma jet. The synthesized products were analyzed by several modern techniques including X-ray diffractometry and transmission electron microscopy. The variety of crystalline phases in the products of both Ti-B and Ti-O systems were identified. The most successful results were obtained using titanium conductors. In this case, the product mainly contains titanium boride and titanium diboride. Synthesized powder consists of hexagonal and cubic particles, which were identified as titanium boride and titanium diboride particles, respectively. The method using titanium conductors for initiating the plasma jet is more efficient and simple way for obtaining TiB/TiB_2 nanopowders.

1. Introduction

Nowadays, there is a high demand for materials with high physical and mechanical characteristics. These materials are urgently needed in the industry, particularly in the aerospace field, in the field of heavy engineering, medicine. The use of these materials helps to reduce maintenance and repair costs of technical equipment, increases their service life, allows obtaining easier and cheaper construction, as well as reducing the risk of accidents.

Titanium diboride has high hardness and has high abrasive properties and resistivity to metal melts [1]. It is a refractory material, so it is increasingly used in industry at high temperatures and in aggressive environments [2]. Titanium diboride TiB_2 does not react with molten steel, cast iron, Zn, Cd, Sn, Pb, Bi, weakly reacts with copper at temperatures above 1400 °C [3]. This makes it possible to produce thermocouples and their protective covers, as well as a variety of lining materials and details for metallurgical furnaces. Titanium diboride is used for production of evaporators for molten metals and electrolytic cells lining, as a component of high-temperature alloys, tool materials, wear-resistant surfacing coatings [4]. Borides of titanium are required in the manufacture of jet engine parts, gas turbine blades [5]. The high hardness allows using borides as abrasives providing a higher surface when processing ductile metals and alloys than synthetic diamond grinding [6].

Titanium diboride and boride particles with micro- and nanometer dimensions are known to improve the strength and creep resistance of metal-matrix composites [7]. Nanoceramics based on titanium borides has greater fracture toughness, ductility and ease of densification compared with conventional titanium borides ceramics [7]. Therefore, one of the ways to improve ceramics properties is creation of nanoceramics. The most common methods of TiB_2/TiB nanopowder synthesis are mechanochemical synthesis, self-propagating high-temperature synthesis, sol-gel reduction [8].

The present paper reports a novel method of direct and quick production of nanodispersed Ti-B phases by a plasmadynamic technique. This method has been already successfully used for synthesis of different nanopowders and nanocoatings (including nano-SiC, TiN, WC and etc.) [9], [10].

2. Experimental Conditions

The paper presents the results of experimental and analytical studies on the synthesis of nanocrystalline powders containing in its composition phase diboride and titanium boride. Required conditions for obtaining a compound of titanium and boron are provided in high energy impact of a high-speed (about 10^3 m/s), high-density (>1.0 g/cm^3) and high-temperature (~1000 K) discharge plasma jet. Such a plasma jet was generated by a coaxial magnetoplasma accelerator of a high current (~10^5 A) discharge of Z-pinch type. The process of generating a plasma jet, implementation of the chemical reactions and formation of the final product takes a short time (~1.0 ms). The generated plasma jet was accelerated by the magnetic field of an accelerating channel and an inductor (a solenoid).

The main metal component for the synthesis of metal compounds was introduced into the system by means of electroerosion of titanium from surface of a titanium accelerating channel of magnetoplasma accelerator. The second main component (amorphous boron) in form of powder was placed into the plasma formation zone. Due to the use of a dielectric precursor in the plasma formation zone there was a problem of initiating the plasma bridge between the central electrode and the accelerating channel. Figure 1 shows the unit of a central electrode with applied method initiating a plasma jet. Amorphous boron powder was placed axially into plasma formation zone (4) as a precursor of a chemical reaction. Thin metal conductors (3) were stretched between central electrode (1) and coaxial electrode (2) along the surface of the plasma formation zone in order to initiate arc discharge in dielectric space. Two experiments were carried out using copper (1) and titanium (2) conductors for initiating the plasma jet. Plasma shots were produced in a sealed volume of the cylindrical reactor chamber filled with argon at normal pressure and room temperature.

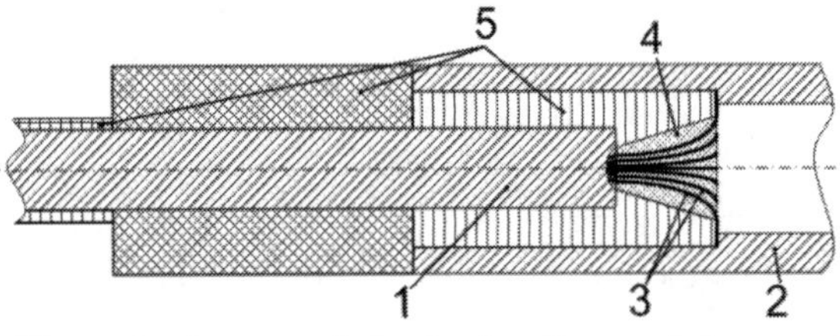

Figure 1. The unit of a central electrode of a coaxial magnetoplasma accelerator: central electrode (1), coaxial electrode (2), copper/titanium conductors (3), plasma formation zone (4), insulators (5).

A magnetoplasma accelerator was supplied by a capacitive energy storage with the maximum stored energy W_{max} = 360 kJ. The electrical and energy parameters of the experiments are presented in Table 1. Figure 2 shows the oscillograms of operating current I, the voltage on the electrodes of magnetoplasma accelerator U, the discharge power P, the released energy W obtained by a Tektronix TDS1012 oscilloscope. According to these waveforms and tables, explosive material of conductors does not practically affect the energy release in the process. Opening a chamber reactor and collecting the powdered product was performed with almost complete deposition of suspended particles on the walls of a chamber. This time is the time of passivation powder in argon atmosphere.

Synthesized in the experiments powder products were analyzed without any preparation by such modern methods as X-ray diffraction (XRD) using Shimadzu XRD 6000 diffractometer (Cu Kα-radiation, λ = 0.15406 nm), transmission electron microscopy (TEM) using Philips CM 12 microscope. Qualitative X-ray analysis was made by PowderCell 2.4 software package using PDF4+ database.

Table 1. Electrical and energy parameters of the experiments.

Exp.	Method of initiating	m_{pow} [g]	I_{max} [A]	U_w [kV]	P_{max} [MW]	W_c [kJ]	W [kJ]
1	Cu-cond.	3.5	120	1.40	155	67.5	33.5
2	Ti-cond.	3.2	140	1.40	201	67.5	38.3

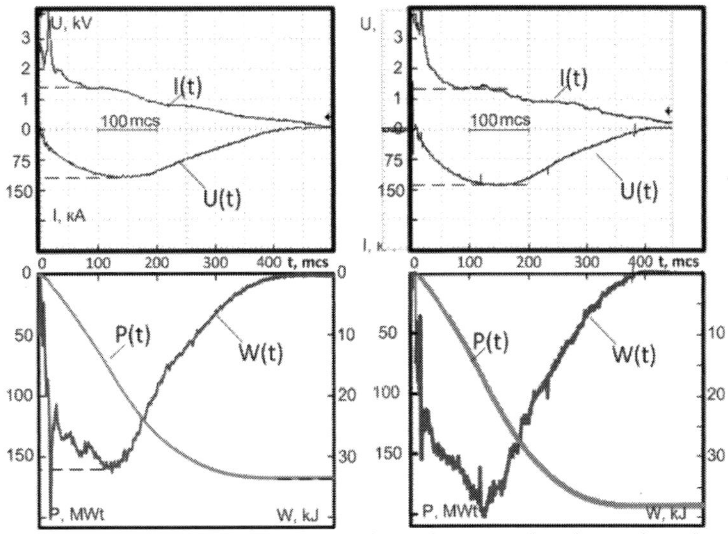

Figure 2. Oscillograms of operating current I, the voltage on the electrodes of magnetoplasma accelerator U, the discharge power P, the released energy W.

3. Results and discussion

The XRD patterns of the products obtained with the use of copper and titanium conductors are shown in Figure 3. The identified peaks are marked in the diffraction patterns. The analysis showed that the synthesis product of experiment 1 (the use of Cu-conductors as a method of initiating) contains crystalline phases: TiB_2 (ICDD Card No. 00-07-0275; hexagonal, space group P6/mmm, No. 191), TiB (ICDD Card No. 00-06-0641; cubic, space group Fm3m, No. 225), Ti (ICDD Card No. 00-01-1197; hexagonal, space group P63/mmc, No. 194), TiO_2, (ICDD Card No. 00-49-1433; orthorhombic, space group Pbnm, No. 62), TiO_2, (ICDD Card No. 00-76-0326; tetragonal, space group P42/mnm, No. 136), TiO_2, (ICDD Card No. 00-82-0514; tetragonal, space group P42/mnm, No. 136), TiB (ICDD Card No. 00-05-0700; orthorhombic, space group Pnma, No. 62). Quantitative XRD analysis was performed only for the synthesis product of the experiment 2 (the use of Ti-conductors as a method of initiating) due to the multicomponent product of the powder synthesized in experiment 1. Quantitative XRD analysis is presented in Table 1.

The variety of crystalline phases in the products of both Ti-B and Ti-O systems were identified. The presence of oxygen in the products in the forms of several titanium oxides TiO_2 (including different crystal systems) were caused by usage of commercially pure boron, which included a small quantity of boron oxide. As a result of the transition of oxygen into the plasma state, titanium—which is more reactive than boron—entered into chemical reactions of compound with oxygen, and the formation of several oxides occurred. The products also contain boride of titanium and diboride of titanium. The preferential content of borides in the products compared with diboride is explained by the certain ratio of boron and titanium for respective phases in accordance with the Ti-B state diagram [11], 20% of boron in the system. The products contain pure titanium, and this fact demonstrates the excess of titanium in the system. Also, the diffraction patterns contain the amorphous halo at low angles characteristic for a product including an amorphous phase. In these cases, such an amorphous phase is pure boron, which was placed into the plasma formation zone. The boron had not been able to react with titanium because of the sharp process of electrical explosion of conductors, and it was rapidly taken out of the chemical reaction zone to the reactor chamber in the unreacted state by a hypersonic plasma jet. It should be noted that the diffraction pattern of experiment 1 has a halo at low angles with larger area indicating a greater weight of amorphous boron in the product. It corresponds to a sharper process of electrical explosion reflected by the oscillograms (Figure 2).

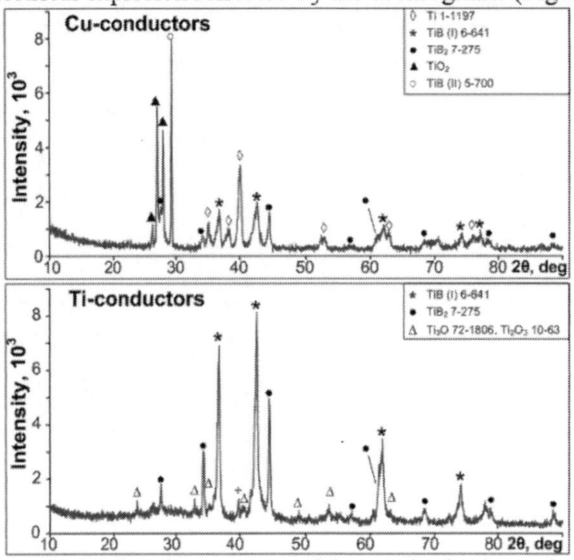

Figure3. Diffraction patterns of the synthesized powders.

Table 2. Quantitative XRD analysis.

	Contents of phases (ω); CSR			R_{wp}/R_{exp}
	TiB_2	TiB/TiN	Ti-hex	
ω [%]	24.50	69.50	6.00	25.8/9.8
CSR [nm]	90.3	47.4	32.5	

A smaller amount of crystalline phase in the product was identified in the experiment 2, as it can be seen from the corresponding diffraction pattern (Fig. 3) and the results of the quantitative XRD analysis (Table 1). The main synthesized phases are hexagonal titanium diboride TiB_2 and cubic titanium boride TiB. The values of goodness S (R_{wp}/R_{exp}), which take values >1, indicate the correctness of calculations. The calculation revealed a high content of titanium boride corresponding to a state diagram at low content of titanium and low temperatures of formation. According to the values of coherent scattering areas (CSA) in Table 2, all the phases are nanodispersed.

The product of experiment 2 was analyzed by transmission electron microscopy (Figure 4). According to the obtained images, the synthesized powder includes crystalline agglomerates of particles with sizes up to several hundred nanometers. Figure 4a shows the bright field image. There are several types of particles in the powdered product. The first type (1) is particles of a round or hexagon shape. The first type can be referred to a space group P6/mmm of titanium diboride. The second type (2) is particles of square shape. The second type can be referred to a space group Fm3m of titanium boride. The selected area electron diffractions (SAED) are presented in Figures 4b and 4d. The most striking and the main diffraction reflexes relate to reflecting planes of titanium diboride and titanium boride. The dark-field images (Figures 4c and 4e) prove accordance of the mentioned types of particle with titanium diboride boride and titanium boride.

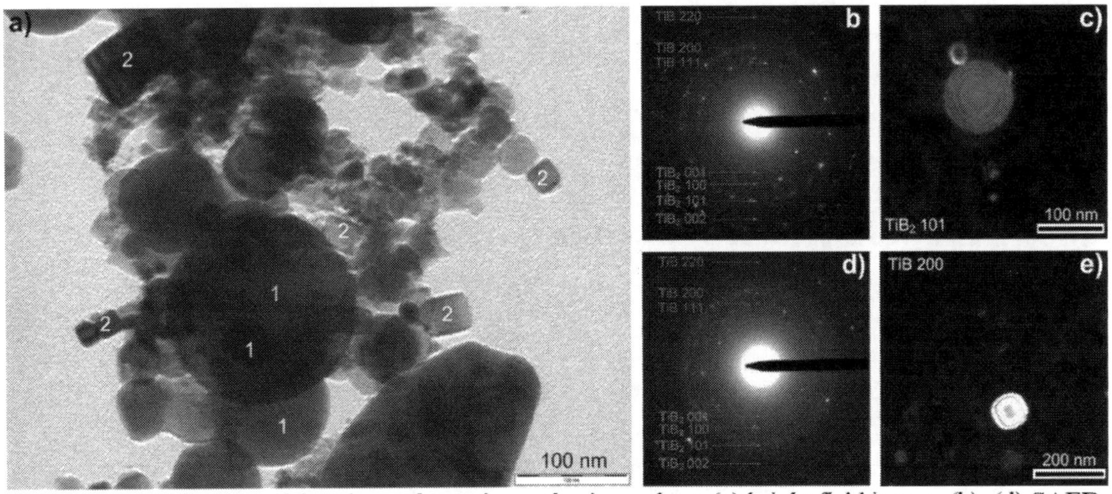

Figure 4. TEM-images of the plasmadynamic synthesis product: (a) bright field image, (b), (d) SAED, (c), (e) dark-field images.

4. Conclusions

Ti-B nanopowders were produced by plasmadynamic synthesis. This method was realized by the synthesis in electrodischarge plasma jet generated by a high-current pulsed coaxial magnetoplasma accelerator. Two experiments were carried out using copper and titanium conductors for initiating the plasma jet. The synthesized products were analyzed by several modern techniques including X-ray diffractometry and transmission electron microscopy. The variety of crystalline phases in the products of both the system Ti-B and Ti-O were identified. The most successful result was obtained using titanium conductors. In this case, the product mainly contains titanium boride and titanium diboride. The synthesized powder consists of hexagonal and cubic particles, which were identified as titanium boride and the titanium diboride particles, respectively. The method using titanium conductors for initiating the plasma jet is a more efficient and simple way for obtaining TiB/TiB_2 nanopowders.

References

[1] Basu B, Raju G B, Suri A K 2006 *Int. Mater. Rev.* **51** 352
[2] Carenco S, Portehault D, Boissiere C, Mezailles N, Sanchez C 2013*Chem. Rev.* **113** 7981
[3] Andrievski R A 2015 *Russ. Chem. Rev.* **84** 540
[4] Ravi Chandran K S, Panda K B, Sahay S S 2004 *JOM* **56** 42
[5] Saito T 2004 *JOM* **56** 33
[6] Munro R G. 2000 *J. Res. Natl. Inst. Stand. Technol.* **105** 709
[7] Bates S E, Buhro W E, Frey C A, et al. 1995 *J. Mater. Res.* **10** 2599
[8] Nozari A, Heshmati-Manesh S, Ataie A 2012 *Int. J. Refract. Met. Hard Mater.* **33** 107
[9] Sivkov A, Nikitin D, Pak A, Rakhmatullin I 2015 *Nanotechnologies Russ.* **10** 34
[10] Pak A, Sivkov A, Shanenkov I, Rahmatullin I, Shatrova K 2015 *Int. J. Refract. Met. Hard Mater.* **48** 51
[11] Murray J L, Liao P K, Spear K E 1986 *Bull. Alloy Phase Diagrams.* **7** 550

Acknowledgments
This work was supported by the Russian Science Foundation (project no. 15-19-00049).

Silica nanoparticles produced by DC arc plasma from a solid raw materials

P V Kosmachev[1], V A Vlasov[1,2], N K Skripnikova[1]

[1]Department of Applied Mechanics and Materials, Tomsk State University of Architecture and Building, Tomsk, 634003, Russia
[2]Depatment of Technical Physics, National Research Tomsk Polytechnic University, Tomsk, 634050, Russia

E-mail: pvkosm@gmail.com

Abstract. Plasma synthesis of SiO_2 nanoparticles in experimental atmospheric pressure plasma reactor on the basis of DC arc plasma generator was presented in this paper. Solid high-silica raw materials such as diatomite from Kamyshlovskoye deposit in Russia, quartzite from Chupinskoye deposit in Russia and milled window glass were used. The obtained nanoparticles were characterized based on their morphology, chemical composition and size distribution. Scanning electron microscopy, laser diffractometry, nitrogen absorption (Brunauer–Emmett–Teller method), X-ray photoelectron spectroscopy and energy-dispersive X-ray spectroscopy were used to characterize the synthesized products. The obtained silica nanoparticles are agglomerated, have spherical shape and primary diameters between 10-300 nm. All samples of synthesized nanopowders were compared with commercial nanopowders.

1. Introduction

At present time due to the increasing need for nanosized oxides with special properties to modify different materials the problem of effective production technology developing is particularly relevant. There are a broad range of nanoparticles synthesis methods such as sol-gel, laser ablation, mechanochemical, plasma and flame processes [1-7]. Each of them has some prospects and constraints, but top-priority direction of investigations is the creation of low-waste and economical technology, such as plasma chemical synthesis. Silica nanopowder is one of the most widely applied additives in industry. As a rule, it helps to increase mechanical and chemical properties of materials [8-12]. But nowadays there are not so many works concerned with using low-cost solid raw materials for production nanosilica.

The paper mainly focuses on the investigation of SiO_2 nanopowder obtained from high-silica raw materials by using energy of cold plasma.

2. Materials and methods

Three types of high-silica were used as raw materials for plasma synthesis of silica nanopowder: quartzite (Chupinskoye deposit, Karelia, Russia), diatomite (Kamyshlovskoye deposit, Sverdlovsk region, Russia) and milled window glass (table 1).

IOP Conf. Series: Journal of Physics: Conf. Series **830** (2017) 012122 doi:10.1088/1742-6596/830/1/012122

All these materials are wide spread and very cheap in comparison with popular for silica nanopowder synthesis chemical precursors. Particle size range of used materials was not more than 2 mm.

Table 1. Content of SiO_2 in raw materials

#	Raw material	SiO_2 (wt. %)
1	Quartzite	99
2	Diatomite	80
3	Glass	75

Plasma synthesis of SiO_2 nanoparticles is carried out in experimental atmospheric pressure plasma reactor on the basis of DC arc plasma generator. The process of synthesis is based on processes of melting, evaporation and sublimation of raw material and next silica nanoparticles condensation from gaseous phase.

To evaluate quality of produced silica nanopowders all samples were compared with commercial ones: "Sky Spring" - nanosilica obtained from Sky Spring Nanomaterials, Inc., Houston- Texas, USA and "Tarcosil" - nanosilica obtained by electron beam way in Novosibirsk, Russia [5].

Different kinds of analyses (PSA, SEM, BET, EDX, XPS) were carried out to get information about sizes, morphology, chemical bonds and composition which influence the properties of material.

3. Results and discussion

Results of laser diffractometry (table 2) show strong agglomeration of silica nanoparticles in each test probe. An ultrasonic bath was used to prevent this factor. Diameters after ultrasonic treatment are ten times smaller but it was impossible to destroy agglomerates of particles during particle size analyses.

Table 2. PSA results

#	Raw material	Volume (%)	Diameter (μm)	
			As is	*Ultrasonic bath*
1.	Quartzite	10	9.0	0.4
		50	75.8	2.1
		90	338.2	42.4
2.	Diatomite	10	10.7	1.7
		50	100.3	10.9
		90	507.9	42.6
3.	Glass	10	41.3	1.5
		50	390.9	23.5
		90	783.9	70.1
4.	Tarcosil	10	-	9.0
		50	-	26.5
		90	-	74.0
5.	Sky Spring	10	-	4.8
		50	-	12.2
		90	-	32.2

Thus sizes of nanoparticles agglomerates from quartzite and diatomite are 42.4 and 42.6 μm which is close to sample "Sky Spring" with size 32.2 μm. Sizes of nanoparticles aggregates from glass (70.1 μm) were close to sample "Tarcosil" (74 μm).

Figure 1 illustrates the morphology and sizes of plasma synthesized nanopowders in comparison with commercial ones obtained after Scanning Electron Microscopy.

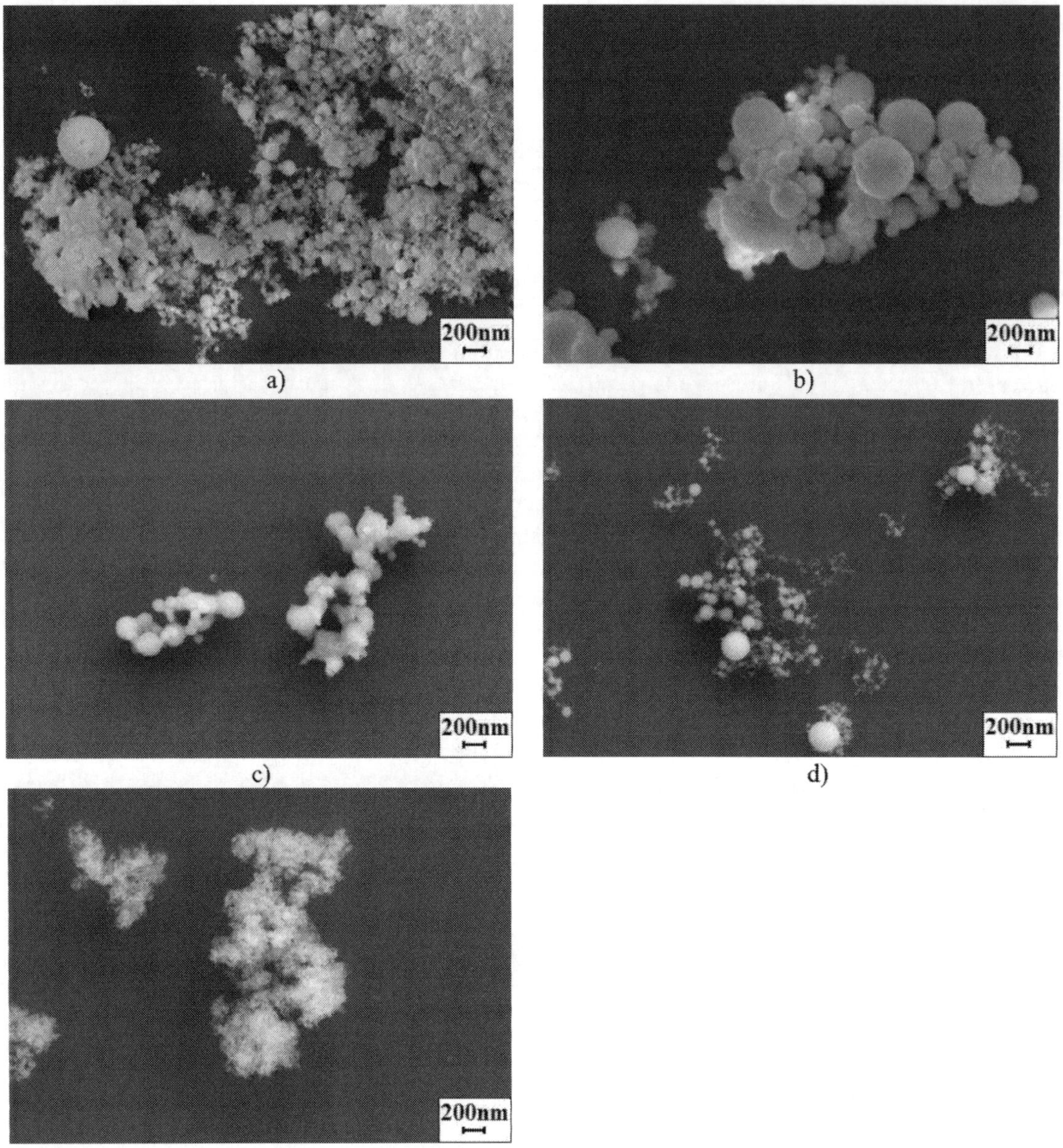

Figure 1. SEM images of nanopowders samples: a – Nanopowder from diatomite, b – nanopowder from quartzite, c – nanopowder from glass, d – Tarcosil, e – Sky Spring.

Investigations have shown that the nanoparticles have a spherical shape, polydisperse and agglomerated. The particle sizes in a range of 10-500 nm. The best result among synthesized nanopowders has the sample from diatomite: it has sizes similar to commercial nanopowders ones. The sample from quartzite has the biggest size among all nanopowders. Sample "Sky Spring" has the smallest dispersion in a range of 10-30 nm.

BET particle diameters (table 3) were calculated based on the measured surface area assuming monodisperse particles [13]. In order to calculate the equivalent diameter, the equation (1) was used:

$$D = \frac{6 \cdot 10^6}{\rho \cdot SSA} \quad (1)$$

Where ρ is bulk density of silica (2200 kg/m3), SSA is a specific surface area in m2/g. and D is the diameter of nanoparticles in nanometers. Plasma synthesized nanopowders were compared with only one commercial nanopowder (Tarcosil) in this analysis because it has comparable particle sizes.

Table 3. Specific surface areas and calculated BET diameters of silica nanoparticles

#	Raw material	SSA ($m^2 g^{-1}$)	Bulk density (g cm^{-3})	BET diameter (nm) (calculated)
1.	Quartzite	11.2	2,20	244
2.	Diatomite	32.2		85
3.	Glass	6.2		440
4.	Tarcosil	31.1		88

Thus SSA of diatomite nanoparticles (32.2 $m^2 g^{-1}$) is close to SSA of Tarcosil nanoparticles (31.1 $m^2 g^{-1}$). Calculated BET diameters were 85 nm and 88 nm correspondingly. Samples from quartzite and glass have smaller SSA (11.2 $m^2 g^{-1}$ and 6.2 $m^2 g^{-1}$) and are consequently bigger BET diameters of particles (244 nm and 440 nm).

Figure 2 demonstrates the elemental composition of nanopowders samples.

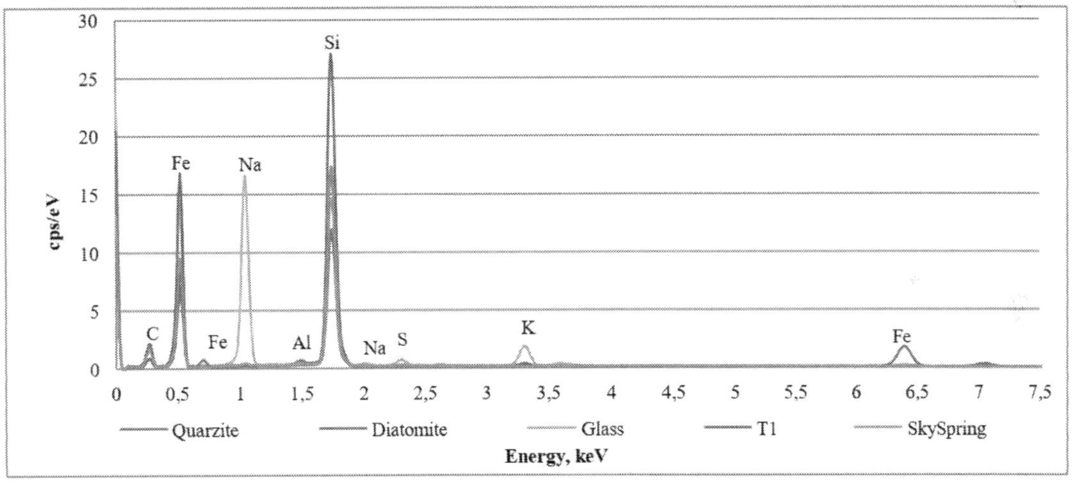

Figure 2. Elemental distribution pattern (EDX).

The pattern of distribution shows that the product contains predominantly the sublimation silicon and oxygen, which indicates representation nanopowder as silicon oxide. There are some impurities such as C, Fe, Na, K in nanosilica samples because a little amount of this element was in raw materials. The presence of carbon was due to the use of graphite as one of the electrodes in plasma system.

Results of XPS (Table 4) show quantitative evaluation of elements content in samples of nanopowders.

Table 4. Elemental composition of silica nanopowders (XPS)

№	Raw material	Element's content, at. %					
		O	Si	C	Fe	Na	K
1.	Quartzite	55.67	31.29	8.85	2.92	1.26	-
2.	Diatomite	59.78	31.15	6.63	0.94	0.81	0.70
3.	Glass	48.18	16.02	16.94	-	17.12	1.74
4.	T1	58.90	32.71	8.38	-	-	-
5.	Sky Spring	61.63	33.67	4.23	-	0.48	-

The highest level of impurities was found in the sample of nanosilica synthesized from milled glass due to its composition.

Conclusion

The possibility to obtain silica nanopowders from solid raw materials by DC arc plasma was shown. All produced samples of nanoparticles have a spherical shape, broad distribution of particle sizes and are strongly agglomerated. There are some impurities (C, Fe, Na, K) in synthesized nanopowders due to its presence in raw materials. Nanopowder from diatomite has nearly the same properties like commercial nanopowders.

Results of the investigation of plasma synthesized silica nanopowders show the possibility of their application in a building sector like additive to materials.

References

[1] Shigeta M., Murphy A.B. 2011 J. Phys. D: Appl. Phys. **44** 174025
[2] Shabanova N A, Sarkisov P D 2004 *Fundamentals of sol-gel technology nanosized silica* (Moscow: Akademkniga)
[3] Zyryanov V V 2008 *Rus. Chem. Rev.* **77** (2) 105–135
[4] Radhip N.R., Pradeep N., Abhishek Appaji M, Varadharajaperumal S. 2015 *J. P. App. Ind. Phys.* **5** (6) 165-172.
[5] Bardakhanov S., Kravets S., Lysenko V., Naumenkov V., Nomoev A., Obanin V., Trufanov D., Shibaev A. 2009 Russ. J. Non-Fer. Met. **50** (4) 383-385
[6] Sazonov R V, Kholodnaya G E, Ponomarev D V, Kaykanov M I 2012 *Izv. Vuz. Fiz.* **55** 72–76
[7] Trufanov D Yu, Nomoev A V, Bardakhanov S P, Sangaa D, Buyantuev M D, Bazarova D Zh 2008 *Vest.NSU: Phys.* **3** 40-46
[8] Messing G.L., Zhang S.C., Jayanthi G.V 1993 *J. Am. Ceram. Soc.* **76** (11) 2707-2726
[9] Sanchez F, Sobolev K 2010 *Constr. Build. Mat.* **24** 2060-2071
[10] Quercia G, Hüsken G, Brouwers H J H 2012 *Cement Concrete Res.***42** 344-357
[11] Sobolev K, Ferrara M 2005 *Am. Ceram. Soc. Bull.* **84** 15-17
[12] Kuli I. et al. 2016 *Am. J. Eng. and App. Sc.* **9** (1) 146-154
[13] A. Abdali 2010 *J. Opt. Adv. M.* 12-3

Acknowledgements

The authors gratefully acknowledge the use of the services and facilities of Innovent Technology Development in Jena.

Plasma-chemical synthesis of carbon nanotubes and fullerenes to create frost-resistant composite building materials

A P Semenov, N N Smirnyagina, B O Tsyrenov, D E Dasheev, Z M Khaltarov

Institute of Physical Materials Science SB RAS, 670047, Ulan-Ude, Russia

E-mail: bulatzsk@gmail.com

Abstract. This paper considers a method of synthesis fullerenes and carbon nanotubes at atmospheric pressure. Carbon evaporates into the plasma arc. The paper discusses the method of synthesis of helium at a pressure of 10^5 Pa. We show the dependence yield of fullerenes and carbon nanotubes from the buffer gas pressure. It has been found that the fullerene yield increased with increasing pressure.The obtained fullerenes and nanotubes find their application in the modification of construction materials. The use of carbon nanomodifiers in the modification of the construction is promising since their introduction significantly improves the physico-mechanical properties using a small quantity of additives. With the introduction of the carbon nanomodifier decrease the porosity of cement stone, which leads to high strength and frost-resistant indicators of the modified cement.

1. Introduction

Currently cement modifiers (concrete)arewidely used, an additive which increases the strength characteristics of the cement stone (or concrete based on it) - compressive strength and flexural, reduces curing and at the same time slows the primary adhesive behavior, reduces water retention, increases frost and m. p. Fullerene is a closed spherical or spheroidal molecule in which the carbon atoms are at the vertices of regular pentagons or hexagons. Because fullerenes have specific physical and mechanical properties, they are used as modifiers for mineral binders. [1]

The effectiveness of carbon nanomaterial technology in cement composite materials and road construction materials depends on the type of nanomaterial. Form, size and contents of fullerenes and carbon nanotubes vary depending on the conditions and method of obtaining. There are various methods of synthesis of fullerenes and nanotubes: laser evaporation of graphite, thermal evaporation of graphite, combustion of hydrocarbons, chemical synthesis, etc. All these methods allow to producefullerenes of different composition, quantity and purity. One of the most productive methods of producing pure fullerenes is the synthesis in arc discharge. In Kirensky Institute of Physics SB RASa method of synthesis of fullerenes in a high-frequency arc discharge at atmospheric pressure was proposed. This technique is comparable to those that donot require to create low pressure, and has the ability to change parameters of the synthesis to obtain the fullerenes of different composition.

2. Experimental part

The set-up which we used was developed at Kirensky Institute of Physics SB RAS [3.4]. Installation diagram is shown in Fig.1. The installation uses the erosion of the electrodes in the arc discharge plasma. The discharge is initiated in the inert gas flow of helium when the current passes through the electrodes 44 kHz. Erosion rods occur in a closed sealed volume, filled with helium.

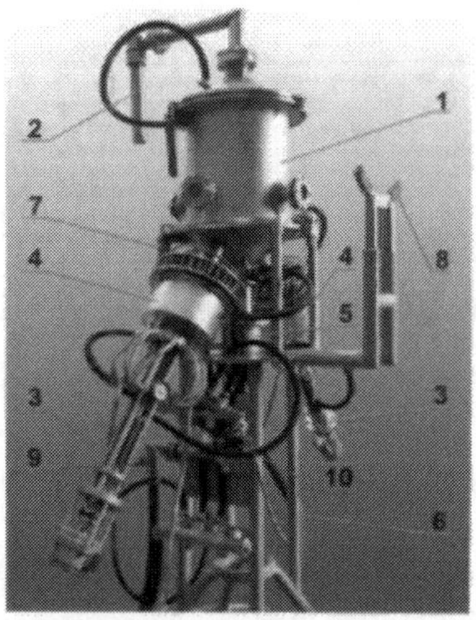

Figure 1. General view of the installation (plasma chemical reactor): 1 - camera for synthesis, 2 - nitrogen trap, 3 - rods, 4 - block matching with the load, 5 drive of the carbon condensation, 6 - stand, 7 - clamps 8 - blank, 9 - valve, 10 - valve control cold water supply.

The synthesized carbon condensate was collected from chamber wall 5. Carbon soot was analyzed on Bruker D2 Phaser diffractometer.

3. Result and discussion

We investigated the dependence of pressure on the yield and composition of the fullerene mixture and the influence of additives on the yield of fullerenes.According to X-ray analysis performed on diffractometer D2 Phaser, we see an increase in the crystalline phase and an increase in yield of higher fullerenes as the pressure increases, reducing the phase corresponding nanotubes and crystalline graphite. Changing the gas pressure fullerene mixture of different compositioncan is obtained.

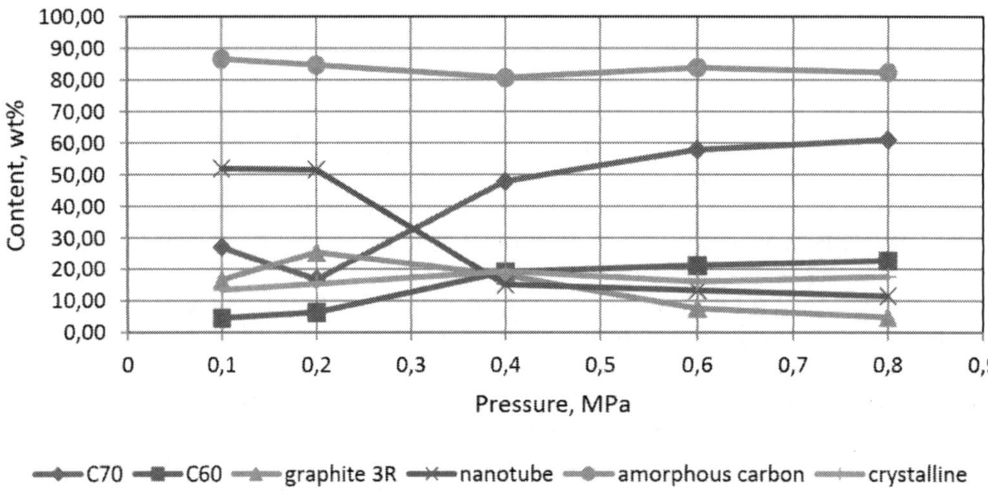

Figure 2. The dependence of the yield and composition of fullerenes on the pressure.

In the synthesis process it is possible to use additives which influence the yield of fullerenes. To investigate this we selected additive type B, Si, Ni(CO), Ni_3B. The dependence of the yield and composition of fullerenes on the pressure is shown in Figure 3. Admixture added to the graphite rod. In the graphite roda hole of 2-3 mm in diameter was drilled. In this hole with the powder added into this hole the synthesis was carried out at a pressure of 0.1 MPa. Depending on the type of doping substances the yield of fullerenes increased, the pressure remained constant.

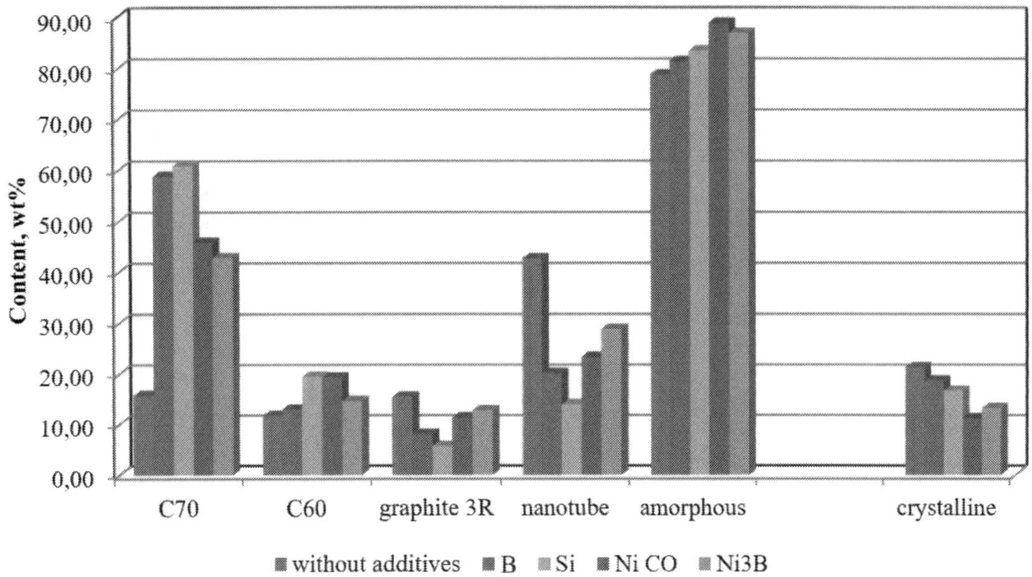

Figure 3. The dependence on the type of fullerene yield additive.

Fullerene mixture is used in the modification of building materials. The addition of fullerene improves the physical and mechanical characteristics when low doses of additives are introduced. This work is particularly relevant for the development of regions with a sharply continental climate, because in the process of modification the strength and frost-resistant properties of building materials with minimal amounts of additives (of the order of 0.001% - 0.01% by weight of cement paste) are improved.

Table 1. The strength properties of cement stone

Composition of the sample	**Tensile strength, MPa**	
	7 days of curing	**28 days of curing**
Check sample	40	61
Sample with CNM 0.01%	38	67
Sample with CNM 0.001%	44	82
C-3® 0.7%	47	50
C-3 0.7% + CNM 0.01%	42	67
C-3 0.7% + CNM 0.001%	58	74
Sika ViscoCrete 5 neu 0.3%	43	63
Sika ViscoCrete 5 neu 0.3% + CNM 0.01%	42	69
Sika ViscoCrete 5 neu 0.3% + CNM 0.001%	40	68

The addition of carbon nanomodifier in an amount of 0.01 % by weight of cement increases the strength by 10% and in the amount of 0.001 % increases the strength by 35 %. With the introduction of the carbon nanomodifier in the cement matrix there is a change in the phase composition, structure and physical and mechanical properties of the cement stone. Carbon nanomodifier changes the structure of the mixing water, creating around their directionally oriented particles of hydrated shell, which leads to the change of rheological characteristics of cement paste. Furthermore, the carbon particles serve as nanomodifier nucleation hydration products of cement, which accelerates the cement hydration and curing, especially in the initial period of hardening. [5]

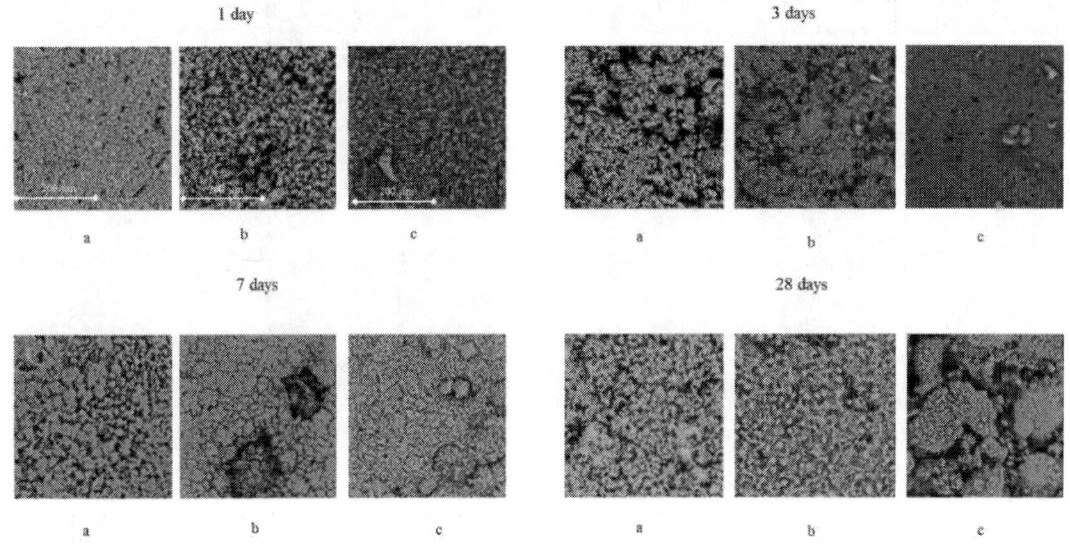

Figure 4. Electron-microscopic analysis cement stone a) check sample b) Portland cement + 0, 01wt. % CNM c) Portland cement + 0, 01wt. % CNM + C-3.

4. Conclusion

This article describes the method of plasma-chemical synthesis of carbon nanomodifiers. We investigated the dependence of the yield of fullerenes and nanotubes on the gas pressure. The increased pressure helium gas increases the yield of higher fullerenes. We also showed the dependence of the carbon content of the condensate on the extension mixture in the synthesis process. Changing the parameters of the process of obtaining nanomodifiers it is possible to obtain carbon soot of different compositions. The obtained carbon nanomaterial finds application in the modification of construction materials, enhancing their strength and frost-resistant characteristics.

References

[1] Eletskii A V, Smirnov B M 1995 *Uspekhi Fizicheskikh Nauk* **165(9)** 977
[2] Kratschmer W, Lamb L D, Fostiropoulos K, Huffman D R 1990 *Nature* **347** 27
[3] Churilov G N 2000 *Instruments and experimental techniques* **43(1)** 1
[4] Churilov G N, Korets A Y, Titarenko Y N 1996 *Journal of Technical Physics* **41** 102
[5] Semenov A P, Semenova I A, Smirnyagina N N, Urkhanva L A, Minyaeva A A, Lkhasaranov S A, Dasheev D E, Khaltarov Z M, Tsyrenov B O, Shestakov N I 2015 *Plasma emission electronics* 241

Acknowledgments

This work was supported by the Program of Russian Academy of Sciences №II.2P «Search basic research in the development of the Russian Arctic» (project №0336-2015-001).

IOP Conf. Series: Journal of Physics: Conf. Series **830** (2017) 012124 doi:10.1088/1742-6596/830/1/012124

Numerical simulation of the effect of laser radiation on matter in an external magnetic field

V V Kuzenov [1,2] and **S V Ryzhkov**[1]

[1] Department of Thermal Physics, Bauman Moscow State Technical University, 2 Baumanskaya Street 5, 1, Moscow 105005, Russia
[2] N.L. Dukhov All-Russian Research Institute of Automatics (State Corporation «ROSATOM» company), Suschevskaya Street 22, Moscow 127055, Russia

E-mail: vik.kuzenov@gmail.com, svryzhkov@bmstu.ru

Abstract. Numerical study of the influence of magnetic field on the plasma compressed by powerful source is carried out. Numerical experiments for the magnetized plasma behavior during and after laser pulse are presented. Convective and thermal parts of a model were tested on several tasks. Spatial distribution of the plasma pressure, temperature, velocity, density, magnetic pressure and laser radiation are obtained in the presence of the external magnetic field.

1. Introduction

The experimental study of thermonuclear plasma should be accompanied by the construction of multilevel radiation and gas dynamic computational modeling that adequately describe the processes in the active zone of fusion reactor. Thus, calculation-theoretical methods are an important element in the development process of the concept of magneto-inertial fusion. A magneto-inertial fusion system is a pulsed thermonuclear device, where a cylindrically or spherically symmetric target is placed in the seed magnetic field and compressed by powerful laser beams or shells of all kinds, including gas, liquid and metal shells, plasma liners formed by the merger of high-speed plasma jets, etc. It is obvious that the developed mathematical models and methods of computing target plasma dynamics of magneto-inertial fusion require in this case obligatory verification on the basis of comparing to the reliable computation and experimental data of the physical experiment.

In this theoretical study, we have calculated the behavior of plasma, due to the interaction of laser pulse with magnetized plasma [1-7]. Nonlinear stability of an axial electric field was taken into account [8], modeling and simulation of laser beams and plasma jet by numerical methods for different applications [9, 10] have been done previously. However, the influence of magnetic fields on power source-matter interaction still requires consideration.

We developed a high accuracy method (eighth order accuracy) for numerical simulation of the laser-matter interaction in the external magnetic field. The effects of magnetic field in plasma compressed by lasers, namely the influence of laser power on magnetized target, are studied.

A numerical method for high resolution developed in this work has improved dissipative and dispersive properties (especially for modeling of turbulent flows) when the multidimensional convective flux discretization is implemented. Quasi-monotone numerical methods do not cause the appearance of false extrema in the spatial distributions (unphysical oscillations developing over time).

The advantage of higher resolution is that high spatial accuracy is obtained with a relatively small number of points of the spatial grid. This approach describes well the sharp and monotone spatial profiles of all gas dynamic parameters. Also note that the developed method is applicable for numerical solving the problem of compression of magnetized plasma by high speed plasma jets.

2. Mathematical model of laser-target interaction with external magnetic field

In general, the subjects of magneto-inertial fusion research are spatial flows from the plasma involving heat and mass transfer, electromagnetic fields, and nuclear reactions. Solving these problems requires the development of new plasmadynamic mathematical models that describe the processes proceeding in thermonuclear plasma, as well as the creation of high-precision numerical methods for solving them, that will allow carrying out with adequate numerical simulation. At the same time, it is expedient to base initial studying of the main physical relationships of magneto-inertial fusion on system of simplified one-dimensional mathematical models.

We will perform the description of the one-dimensional mathematical model of compressing the thermonuclear magneto-inertial fusion targets process for the centrally-symmetric coordinate system. The mathematical model of magneto-inertial fusion targets presented in this work is based on one-dimensional equations of radiation plasma dynamics: Euler system of equations (1), the transfer equation of own broadband radiation (2), the equation of magnetic induction (3), the transfer equation of laser radiation, computational methods for the equations of state and absorption coefficients for laser radiation. This model determines the conditions of creating and achieving self-sustaining fusion reaction:

$$\frac{\partial\rho}{\partial t}+Div(\rho u)=F_\rho,\ F_\rho=-\rho u\frac{(\nu-1)}{r}, \tag{1}$$

$$\frac{\partial(\rho u)}{\partial t}+Div\left(\rho u^2+P\right)=F_{\rho u}+f_r,\ F_{\rho u}=-\rho u^2\frac{(\nu-1)}{r},\ f_r=\frac{1}{c}\left[\vec{j}\times\vec{H}\right]_r,)$$

$$\frac{\partial(\rho E)}{\partial t}+Div\left(\rho Eu+Pu+q_\Sigma\right)=F_E+q_r+Q^e_{Fus},\ q_r=j_rE_r,\ q_\Sigma=q_e+q_i+q_{laz},$$

$$F_E=-(\rho Eu+Pu)\frac{(\nu-1)}{r},\ P=P_e+P_i,\ Div(\ )=\frac{1}{J}\frac{\partial(J\ )}{\partial\xi},\ J=r^{(\nu-1)},$$

where ρ is the density, u is the speed along the radial coordinate r, t is the time, $P=P(\rho,\varepsilon)$ is the static pressure, ε is the specific internal energy, $E=\left(\varepsilon+u^2/2\right)$ is the total energy of gas stream, $\vec{F}=\left(F_\rho,F_{\rho u},F_E\right)$ is the vector of sources, F_ρ is the mass flow density, $F_{\rho u}$ is the momentum flux density, F_E is the laser fluence, q, q_ν are total and spectral radiant fluxes, respectively, T_e and T_i are the electron and ion temperatures in plasma volume ($T=T_e=T_i$), respectively, χ_ν is the spectral absorption constant, f_r is the electromagnetic force, q_r is the energy flow in the electromagnetic field, $q_e=-\lambda_e gradT_e$, $q_i=-\lambda_i gradT_i$, λ_e and λ_i are the thermal conductivity coefficients of electrons and ions, j_r is the specific current, $\vec{H}(r)$ is the magnetic induction vector, P_e is the electron pressure, P_i is the ion pressure, index ν = (1,2) obeys plane and axial symmetry.

The contribution of local (thermonuclear) energy release Q^e_{Fus} to the electronic component of plasma due to the energy transfer to electrons from thermonuclear α–particles can be determined using an approximation formula [11]:

$$Q^e_{Fus}=8{,}483\cdot10^{29}\rho^2\frac{\left(1+0{,}232\cdot T_i^{3/4}\right)\exp\left(-20/T_i^{1/3}\right)}{T_i^{2/3}\sqrt{1+9{,}41\cdot10^{-5}\cdot T_i^{13/4}}},\ \left(\frac{erg}{cm^3\cdot s}\right).$$

The broadband radiation transfer can be considered using multigroup diffusive approximation [12]:

$$\frac{1}{r^n}\frac{d\left(r^n q_v\right)}{dr} + \chi_v c U_v = \chi_v 4\sigma T^4 , \; \frac{c}{3}\frac{dU_v}{dr} + \chi_v q_v = 0 , \tag{2}$$

where q_v and U_v are the spectral flux and apparent density of broadband radiation, respectively; c is the speed of light, v is the frequency group number, χ_v is the spectral absorption constant, n=0 refers to the flat layer, n=1 is applied to the infinite one-dimensional cylinder. Here q_v refers to the radiation flux in the axis r direction.

The equation of magnetic induction taking into account the continuity equation and the conservation law $div\left(\overline{H}\right)=0$) can be written as [13]:

$$\frac{\partial\left(B_z/\rho\right)}{\partial t} + \frac{1}{\mu J}\frac{\partial J\left(\mathrm{v} B_z/\rho\right)}{\partial r} = \frac{c^2}{4\pi\mu\rho J r}\frac{\partial}{\partial r}\left(\frac{Jr}{\sigma}\frac{\partial B_z}{\partial r}\right), \tag{3}$$

The electrical conductivity is determined by the Spitzer formula [14] taking into account possible plasma magnetization:

$$\sigma(z) = 12{,}06\times10^{13}\cdot\frac{T^{3/2}}{\ln\Lambda}\cdot\frac{n_e}{\sum_i n_i z_i^2}\frac{1}{1+\left(\Omega_e\tau_e\right)^2},$$

where z_i and n_i are the average charge and ion density, respectively, $\Omega_e = e\left|\vec{B}\right|/m_e c$ is the electron gyrofrequency, m_e and M_i are the electron and ion masses, respectively.

The thermal conductivity coefficients of electrons and ions $\lambda_{e,I}$, in case plasma is magnetized, can be calculated using the formulas [14].

Parameters of laser radiation along an axis r are found out on the basis of the solution of the transfer equation of laser radiation:

$$\frac{dq_{Las}}{dz} - \chi_\omega q_{Las} = 0, \tag{4}$$

where τ is the half-width at half-maximum pulse duration. An absorption constant of laser radiation χ_ω is determined using a continuum absorption mechanism that is opposite to the braking electron radiation machine in the conditions of the local thermodynamic equilibrium (LTE):

$$\chi_\omega = \begin{cases} \dfrac{4{,}97 g Z_i^2 n_i^{\Sigma} n_e^{\Sigma}}{n_c^2\lambda^2\left(kT_e\right)^{3/2}}\dfrac{1}{\sqrt{1-n_e/n_c}} & , n_e < n_c \\ \infty & , n_e \ge n_c \end{cases},$$

where λ is the wavelength of laser radiation (μm), n_e and n_i are electron and ion densities (cm^{-3}), kT_e is the electron temperature (keV), g is the Gaunt factor [14, 15].

The calculation of the optical $\chi_i(T, \rho)$ and thermodynamic $e(T, \rho)$, $P(T, \rho)$ parameters of the operating environment was carried out in the framework of LTE approximation using the computer system ASTEROID developed by RAS correspondent member S.T. Surzhikov [16]. This system works on Tomas-Fermi's model with quantum and exchange corrections [17, 18] and on average charge model [19-21].

A computational domain and target of magneto-inertial fusion consist of the central part and one coaxial layer. They have a cylindrical shape with the following range of values of initial parameters of the target and medium. The central part of the target (inner radius R = 0.05 cm) is filled with D-T mixture with density $\rho=5\cdot10^{-2}$ g/cm^3 and temperature of T=297 K. It was surrounded by the coaxial layer (outer radius R_c = 0.1 cm) consisting of metal (Al) with density ρ=2.7 g/cm^3 and temperature of T=297 K. The computational domain has an outer radius r=0.2 cm. Thermodynamic parameters of external rarefied medium (Ar) are defined: T=297 K, $\rho=1.29\cdot10^{-3}$ g/cm^3.

3. Results

For a reasonable quantitative assay of physical processes, which can proceed in plasma of a magneto-inertial fusion target, it is necessary to carry out verification of the recommended mathematical model and numerical methods and test the results on the model problems.

Convective parts of a computer model of magneto-inertial fusion targets are tested on the one-dimensional Riemann problem (for some particular initial data, called the Soda problem) about decay of an unstable fracture configuration. The comparison of exact and approximate solutions shows that the difference is no more than one percent.

The "thermal" part of the model was tested on some tasks allowing accurate analytical solutions: heating of the continuous medium [13], filling a flat semi-bounded space $r>0$, by heat flow through the left-hand fixed boundary $r=0$.

Numerical calculations presented in this paper are carried out for the pulsed Nd laser with radiation flux density being within the range from 10^{12} to 10^{14} W/cm^2. Here is a brief description of the obtained results, which geometrical representation is shown in figures 1–4. These results correspond to the following parameters of the mathematical model: $Q^e_{Fus}=0$, laser radiation power density $q_{Las}=2\cdot 10^{14}$ W/cm^2, and exposure time $t_{Las}\in[0,10]$ ns.

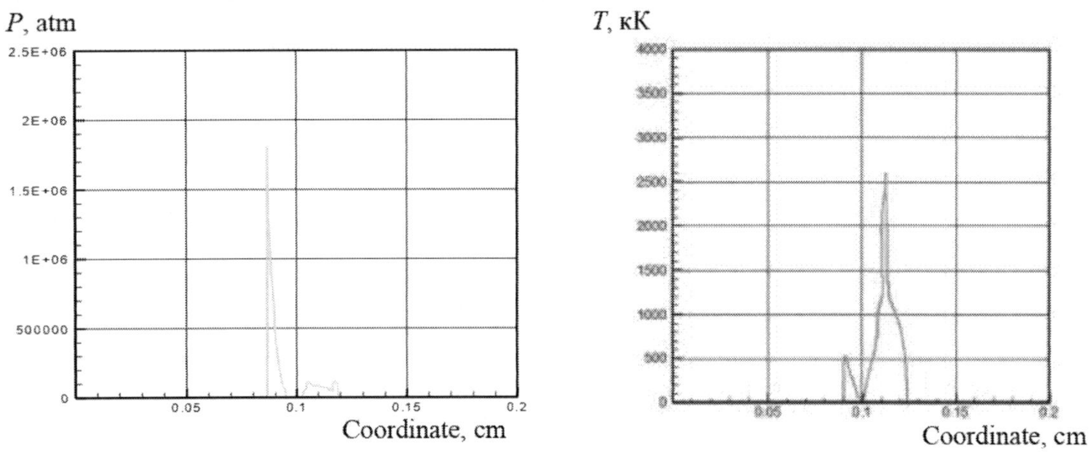

Figure 1. Spatial distribution of the pressure (*left*) and temperature (*right*) at t=2.14 ns (t_{Las}=5 ns).

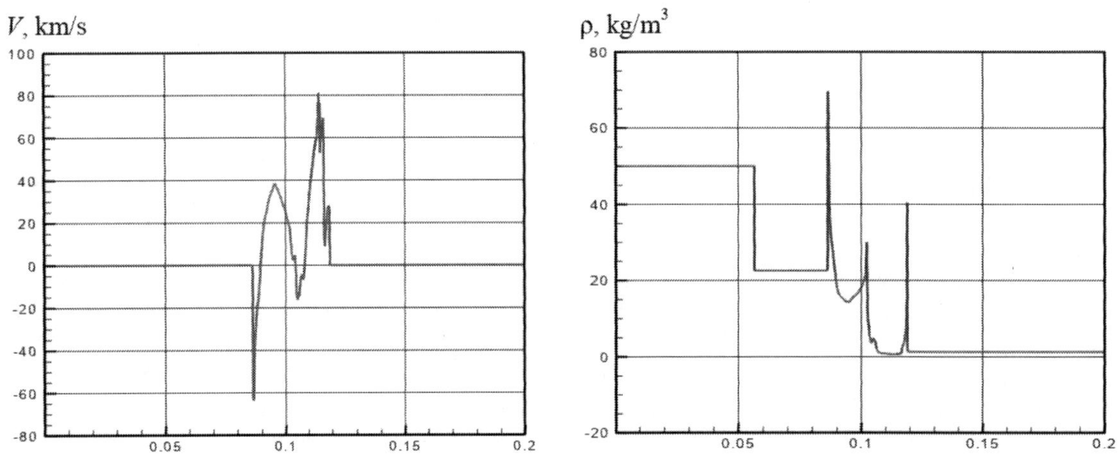

Figure 2. Distribution of plasma velocity (*left*) and plasma density (*right*) at t=2.14 ns along the radial coordinate.

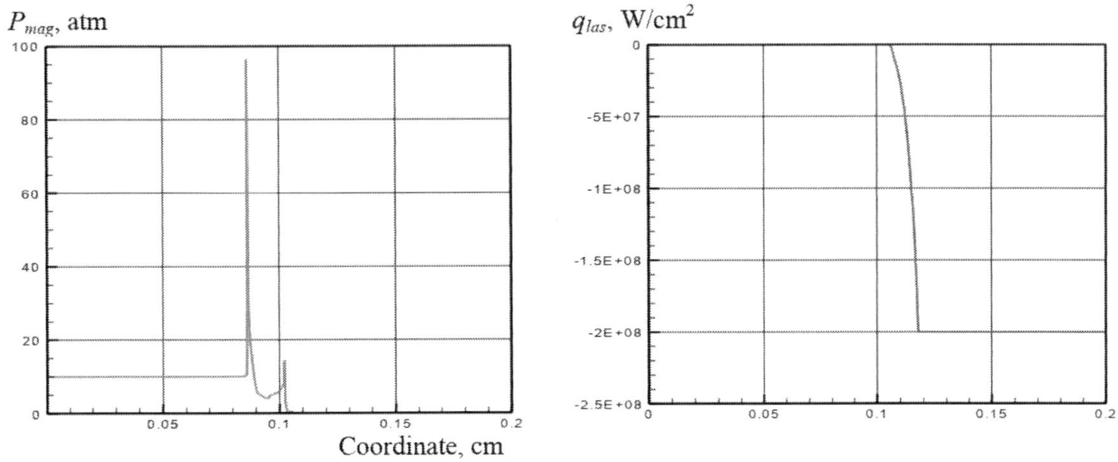

Figure 3. Distribution of the magnetic pressure (*left*) and laser radiation (*right*) at t=2.14 ns (t_{Las}=5 ns).

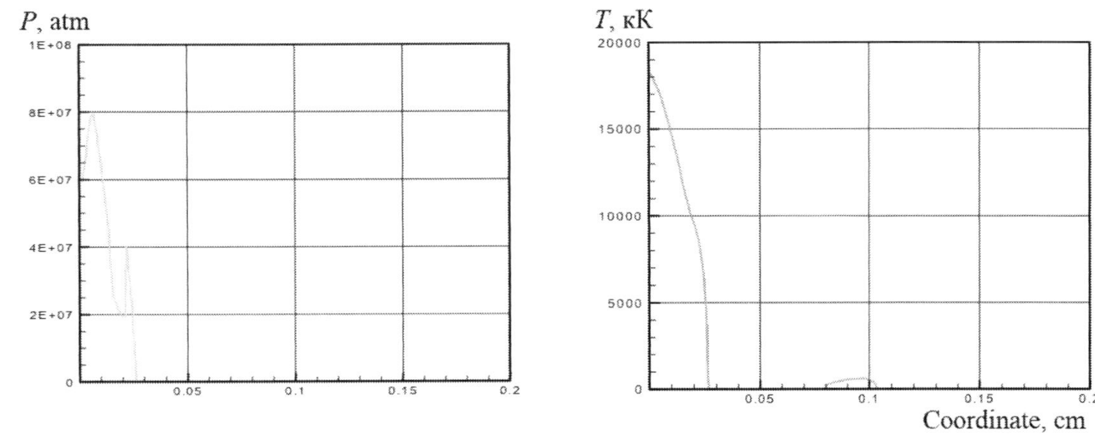

Figure 4. Spatial distribution of the plasma pressure (*left*) and temperature (*right*) at t=5.63 ns (t_{Las}=5 ns).

It follows from the carried-out calculations that the laser compression process of a magneto-inertial fusion target, which is in an external magnetic field concerning time t, can be conditionally presented in the form of three stages: the "initial" compression stage of a magneto-inertial fusion target; the "collapse" stage of a magneto-inertial fusion target; the "expansion" stage of plasma formation.

The distributions of static P and magnetic P_{mag} pressure, plasma temperature T, velocity V, density ρ and laser flux q_{Las}, that correspond to the first "initial" compression stage (t=2.14 ns) of a magneto-inertial fusion target placed in an external magnetic field are given in figures 1-3. The laser pulse duration is 5 ns, the laser radiation flux density is $2 \cdot 10^{14}$ W/cm^2. A thin metallic cylindrical shell material is Al. The target (outer) radius is 0.1 cm, and the inner radius is 0.05 cm.

The plasmadynamic parameters of the collapse stage are defined by the shock wave reflected from the axis of symmetry, which moves towards the outer boundary of the computational domain. At the same time the maximum values of plasma pressure and temperature are observed along the axis of symmetry of the system.

The distributions of the magnetic pressure P_{mag} and combined flux q of self-radiation plasma formation are shown in figures 4-6. These distributions correspond to the "expansion" stage of plasma formation (t=5.63 ns).

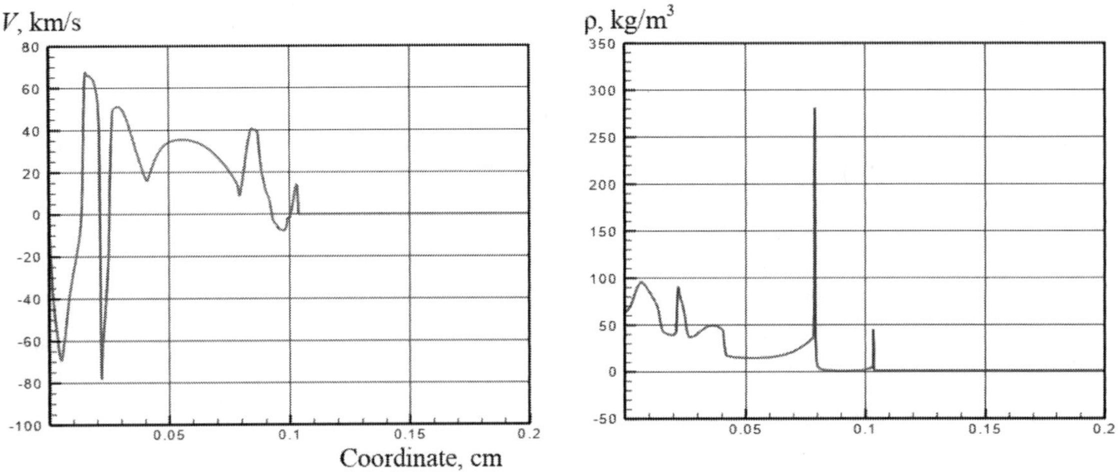

Figure 5. Plasma velocity (*left*) and density (*right*) at t=5.63 ns as a function of radial coordinate.

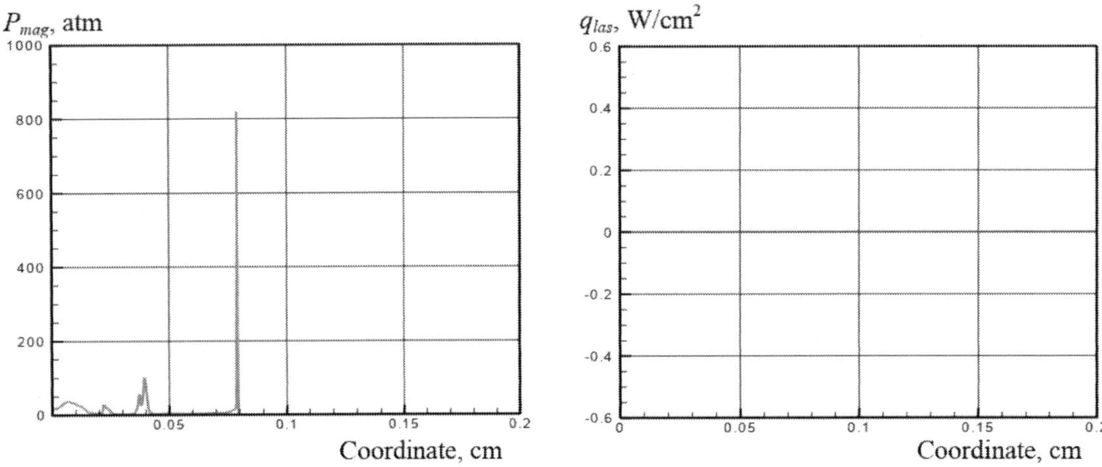

Figure 6. Distribution of the magnetic pressure (*left*) and laser radiation (*right*) at t=5.63 ns (t_{Las}=5 ns).

4. Conclusion

We developed a high accuracy method (eighth order accuracy) for numerical simulation of the laser-matter interaction in the external magnetic field. Quasi monotone compact higher order finite difference scheme provides the improved dissipative and dispersive properties. Numerical study of the influence of magnetic field on plasma compressed by powerful source is carried out. The effects of magnetic field in plasma compressed by lasers, namely the influence of laser power on magnetized target, are studied. Numerical experiments for the magnetized plasma behavior during and after laser pulse are presented.

Acknowledgments

This work was supported by the Ministry of Education and Science of the Russian Federation (Project No 13.5240.2017/БЧ) and Russian Science Foundation (Project No 16-11-10275).

References

[1] Drake R P, Keiter P A, Kuranz C C, Malamud G, Manuel M, Di Stefano C A, Gamboa E J, Krauland C M, MacDonald M J, Wan W C, Young R P, Montgomery D S, Stoeckl C and Froula D H 2016 *J. Phys.: Conf. Ser.* **688** 012016

IOP Conf. Series: Journal of Physics: Conf. Series **830** (2017) 012124 doi:10.1088/1742-6596/830/1/012124

[2] Ryzhkov S V 2010 *Problems of Atomic Science and Technology* **4** 105
[3] Ryzhkov S V 2015 *Sustainable Cities and Society* **14** 313
[4] Kostyukov I Yu and Ryzhkov S V 2011 *Plasma Physics Reports* **37** 1092
[5] Honrubia J J, Murakami M, Mima K, Johzaki T, Sunahara A, Nagatomo H, Fujioka S, Shiraga H and Azechi H 2016 *J. Phys.: Conf. Ser.* **688** 012033
[6] Ryzhkov S V 2014 *Bulletin of the Russian Academy of Sciences. Physics* **78** 456
[7] Ryzhkov S V 2008 *Proc. 35th EPS Conf. on Plasma Physics and Contr. Fusion* ECA **32D** P1.114
[8] Elcoot A E K 2010 *Appl. Math. Model.* **34** 1965
[9] Zhang H, Hu S, Wang G and Zhu J 2007 *Appl. Math. Model.* **31** 1124
[10] Fasano A, Hömberg D and Naumov D 2010 *Appl. Math. Model.* **34** 3831
[11] Danilov I M, Iorish V S and Son E E 2011 *High Temperature* **49** 217
[12] Chetverushkin B N, Ascenzo N D' and Saveliev V I 2014 *Dokl. Math.* **90** 495
[13] Mazhukin V I, Malaphei D A, Matus P P and Samarskii A A 2001 *Comp. Math. Math. Phys.* **41** 379
[14] Braginskii S I 1963 *Questions of plasma theory*, ed M A Leontovich (Atomizdat, Moscow) p 183
[15] Zeldovich Ya B and Raizer Yu P 2002 *Physics of Shock Waves and High-Temperature Hydrodynamic Phenomena* (Dover)
[16] Surzhikov S T 2013 *Computational physics of electric discharges in gas flows* (Walter de Gruyter GmbH)
[17] Kuzenov V V, Ryzhkov S V and Shumaev V V 2015 *Problems of Atomic Science and Technology* **1** 97
[18] Shumaev V V 2016 *Phys. Atom. Nucl.* **79** 1
[19] Kuzenov V V and Ryzhkov S V 2013 *Problems of Atomic Science and Technology* **1** 12
[20] Kuzenov V V and Ryzhkov S V 2013 *Problems of Atomic Science and Technology* **4** 103
[21] Kuzenov V V and Ryzhkov S V 2016 *Bulletin of the Russian Academy of Sciences. Physics* **80** 598

Compositional dependence of thermal, optical and mechanical properties of oxyfluoride glass

I Z Hager[1], H A Othman[1] and D T Valiev[2]

[1]Physics Department, Faculty of Science, Menoufia University, Shebin El-Koom, Egypt
[2]National Research Tomsk Polytechnic University, Tomsk, Russia

E-mail: hosamssl4@yahoo.com

Abstract. Tungsten oxyfluoride glasses are characterized by low phonon energy. This is due to the presence of fluoride ions that have low phonon energy and formation of low phonon energy WO6 units. Oxyfluoride glasses based on WO3–BaF2–RF, where RF= LiF, NaF or mixed (LiF–NaF) have been prepared by melt quenching technique. The density and molar volume of the prepared glasses show a decrease with the increase of RF instead of WO3 content. The glass transition temperature Tg is found to decrease with increasing RF content. The refractive index increases with the addition of heavy polarizable fluorides. The decrease of the elastic moduli and microhardness of these glasses may be due to the decrease in density and the depolymrization effect. The Poisson's ratio increases with increasing RF content due to the structural changes and formation of (NBOs) and (NBF) units. The aim of this work is to prepare a glass host with low phonon energy to be an efficient host with good luminescence properties, when doped with rare earth ions, and to study its structural, thermal, optical and mechanical properties.

1. Introduction

Glasses containing tungsten ions are considered to be very important technological glasses [1]. This is due to their specific thermal, chemical, mechanical and optical properties [1,2]. The addition of WO_3 to the glass enhances its network connectivity [3,4] and this increases its thermal and chemical stability. With their different structural units (WO_4 tetrahedral and WO_6 octahedral units), tungsten ions have a great influence on optical and electrochemical properties of the glasses in which they are incorporated [5-7]. Having different oxidation states (W^{4+}, W^{5+} and W^{6+}), tungsten ions impart specific photo-, thermo- and electrochromic properties to the glass in which they are incorporated so that it could be used in different optoelectronic devices and as a sensor [8]. Addition of WO_3 to glass decreases its phonon energy due to the formation of WO_6. This provides the glass with interesting optical properties such as photochromic and linear absorption effect, and imparts efficient luminescent properties to the glass when it is doped with rare earth ions [9]. At high concentrations of WO_3, WO_6, structural units are linked together through W-O-W bonds that result in the formation of highly polarizable clusters responsible for optical properties [10,11]. Addition of fluoride content affects glass formation and structure and makes promising optical materials for laser hosts. It also helps to decrease the phonon energy, non-radiative losses, reduces the OH content and increases the transparency in the visible region [12]. Rare earth doped tungsten glasses and crystals have excellent properties to be used in different applications such as waveguides, amplifiers, and lasers [13,14].

2. Preparation and measurements

Anhydrous fluorides are commercial products: BaF2 and LiF from Prolabo, NaF from Ridel de Haen and WO_3 from Strem. Calculated amounts of starting materials were mixed and heated up to the melting temperature which is about 650–700 °C. Melting was carried out in a platinum crucible for 30–45 min. Then the melt was poured into a brass mold preheated up to about 200–250 °C. The samples were annealed at near the glass transition temperature for one hour.

The densities of the prepared samples were measured by the Archimedean method using CCl_4 as an immersion liquid. The molar volume was calculated by the equation $V_m = (M/\rho)$, where ρ is the density (g/cm^3) and M is the molecular weight of the glass sample (gm). The error of the density measurements was not more than ±0.001 gm/cm^3.

The characteristic temperatures of the glass (e.g., glass transition temperature T_g, onset crystallization temperature T_x and crystallization temperature T_c) were measured by a differential scanning calorimeter SEIKO DSC 220 at a heating rate of 10 °C.min^{-1} under controlled atmosphere N_2. The average values of the thermal expansion coefficient α were measured by using TMASII Seiko 5200 instrument. Bar samples with a thickness of about 3 mm were used in the measurements. The measurements were performed in the temperature range from 50 °C to near the glass transition temperature of the chosen samples. Then the average values of thermal expansion coefficient were determined in the temperature range of 100–150 °C. The uncertainty of the obtained values was ±3×10−7 °C^{-1}. Heat capacity C_p was measured by a differential scanning calorimeter, DSC-7 PerkinElmer. The samples for heat capacity measurements with a mass of 30–50 mg were enclosed in crimped Al DSC pans. The C_p values of the test glasses were measured from 70 °C to near the glass transition temperature region at 4 K min^{-1} heating rate.

Ultrasonic velocities were measured using KARL DEUTSCH Echograph 1080 ultrasonic flow detector. The ultrasonic longitudinal velocity (VL) was measured using normal longitudinal probe Krautkramer, 5 MHz, and the ultrasonic shear velocity (VS) was measured using normal shear probe Olympus, 2.25 MHz. The velocity was obtained by dividing the round trip distance by the elapsed time. The relative error in the determined values of these velocities was roughly about ±30 m/s.

3. Results and discussion

The obtained data show that the density and molar volume decrease with increasing LiF, NaF and the mixture (LiF-NaF), see Table. 1. The decrease of the density is due to the smaller molecular weight of the added alkali fluorides (LiF=25.94, NaF=41.99) than that of WO_3 (231.85) and also the packing density, V_t, (V_t was calculated according to our previous works [15, 16] and their values are given in Table 1). The decrease in the molar volume is due to the change of the W–O bond length. In addition, the change in the molar volume is the result of the creation of non-bridging oxygen (NBO) and fluorine (NBF). The variation in the density and molar volume with the addition of RF content were attributed to the change of the coordination state of Li^+ and Na^+ ions. It was proposed that at higher concentration Li^+ ion can be coordinated by six F^- ions and behave as a glass-modifier region, and Na^+ ion is coordinated by six F^- ions and acts as a glass-modifier [17].

The transition temperature of the present glasses decreases with increasing LiF, NaF and mixed (LiF-NaF) instead of WO_3 (see Table 1). This is due to weakening of the glass structure. Alkali fluorides have lower T_g which varies inversely with the ion size. Therefore, LiF has unexpectedly most effect in view of its higher bond strength. The decrease of T_g with increasing RF content can be attributed to the breaking of linkages in the glass structure. It may also be due to the replacement of O^{2-} ions with F^- ions that leads to depolymerization of the oxide and as a result, T_g decreases [18]. As previously discussed [19], the structure of the present glass consists of tetrahedral WO_4 and WO_6 octahedral group chains cross-linked by Ba^{2+} or Li^+ and Na^+ ions. The replacement of the transition

IOP Conf. Series: Journal of Physics: Conf. Series **830** (2017) 012125 doi:10.1088/1742-6596/830/1/012125

ion (W) by alkali ion (Li or Na) decreases the degree of cross-linking and thus renders the glass more fluid, therefore the glass transition temperature decreases.

Table 1. The composition, density, molar volume V_m, packing density V_t, cross link density $\bar{n}_c$, F/O ratio, glass transition temperature T_g, crystallization temperature T_c, thermal stability (T_c-T_g), activation energy of crystallization E_a and average linear thermal expansion α of WO_3-BaF_2-RF glass.

Sample Name	WO_3	BaF_2	LiF	NaF	□ $g.cm^{-3}$	V_m cm^3	$V_t x10^{-6}$ m^3	T_g °C	T_c °C	T_c-T_g °C	□$x10^{-7}$ °C^{-1}
		Mol %									
WBL-1	42	30	28		5.83	26.97	0.6151	314	375	61	142
WBL-2	40	30	30		5.77	26.54	0.6141	309	369	60	
WBL-3	38	30	32		5.73	26.00	0.6157	299	346	47	150
WBL-4	36	30	34		5.63	25.73	0.6109	271	382	111	
WBL-5	34	30	36		5.56	25.32	0.7105	263	421	58	163
WBN-1	46	30		24	5.83	29.05	0.6044	352	433	81	152
WBN-2	44	30		26	5.78	28.63	0.6043	343	432	89	
WBN-3	42	30		28	5.73	28.23	0.6037	336	403	67	160
WBN-4	40	30		30	5.67	27.86	0.6024	326	386	60	170
WBNL-1	42	30	10	18	5.78	27.70	0.6094	322	398	76	142
WBNL-2	40	30	10	20	5.70	27.43	0.6059	315	404	89	
WBNL-3	38	30	10	22	5.65	26.99	0.6063	309	375	66	179
WBNL-4	36	30	10	24	5.52	26.94	0.5978	294	357	63	
WBNL-5	34	30	10	26	5.48	26.45	0.5991	293	348	55	193

The decrease in T_g with increasing RF content could be due to the disruption of the network and weakening of the W–O bonds as suggested by the density explanation. Also, the decrease in T_g with the increase in RF content instead of WO_3 content may be due to the decrease in packing density.

The activation energy of crystallization was determined by Chen's equation [20],

$$\ln\left(\frac{T_C^2}{\alpha}\right) = \frac{E_a}{RT_C} \tag{2}$$

where Tc is the crystallization temperature, α is the heating rate (10 Ks^{-1}), E_a is the activation energy of crystallization and R is the universal gas constant. The values of E_a were calculated and given in Table 1. Thermal stability was determined according to:

$$S = T_c - T_g \tag{3}$$

The values of S are given in Table 1. The structure of the present glasses is deformed from WO_4 tetrahedra and WO_6 octahedra group disorderly packs by means of corner-sharing or edge-sharing to form the skeleton of the glass. Ba^{2+}, Li^+ and Na^+ are incorporated into the vacancies of the skeleton and center on the WO_4 and WO_6 anions. The addition of the alkali fluorides RF results in the system of F^- anions which form non-bridging fluorides at one or two corners of the metal oxide polyhedra. This increases the spatial freedom linking between the polyhedral and hence the stability of the glasses is improved. As seen from Table 1, the average thermal expansion increases with the increase of alkali fluoride and mixed alkali fluoride. As RF increases instead of WO_3, the units of each glass become more disconnected. The non-bridge fluorine will be formed at one or two corners of the polyhedral in metal oxide. So that when the sample is heated, the unit volume of each glass expands with increasing RF, which is also confirmed by the decrease in the glass transition temperature T_g as discussed before. In addition, this could be due to the disruption of the network and weakening of W–O bonds that was suggested by density and glass transition temperature. Also, the excess of RF content result in the formation of W–F–W bonds, which provide weak points in the vitreous network, that results in decreased T_g and increased thermal expansion coefficient. The data on specific heat C_P of the present glasses versus temperature during heating in the temperature range from 70 °C up to 170 °C at a rate of 4 °C/min is shown in Figure 1. C_P increases with increased temperature. The C_P values of these glasses are higher than those of oxide glass due to the low vibration frequencies in the oxyfluoride glass, which means that their Debye temperature is also low. It was found that at different temperatures C_P decreases with increasing LiF and the mixed (LiF-NaF), while it increases in case of NaF contents. This is due to the change of the glass structure as mentioned before. The refractive index is one of the interesting optical parameters to design glass for optical devices. The refractive index is changed as follows: n(Na-Li)< n(Na)< n(Li). This may due to the difference in the refractive potential for each alkali fluoride component in the glasses. The larger its value, the larger the contribution to the refractive index of the glass, for example, these values are 0.221, 0.244 for NaF and LiF, respectively [21]. When RF increases instead of WO_3, the denser nature of the glass relatively decreases, which leads to decrease the refractive index [22].

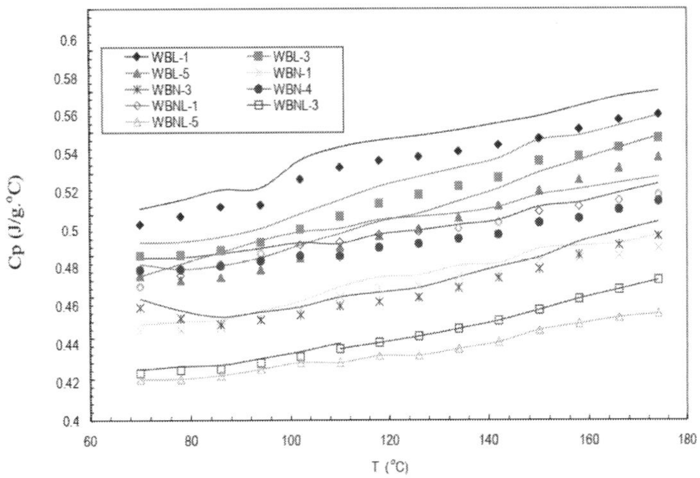

Figure1. Dependence of specific heat Cp on temperature of the prepared glasses.

The addition of heavy polarizable fluorides increases the refractive index and molar refractivity. [23]. This suggests that when the lighter alkali metal is incorporated into the present oxyfluoride glass, the refractive index and molar refractivity will reduce as shown in Figure 2. However, different cations in the glass have different structural effects; those with higher electrical changes and small coordination and bond ionicity due to strong polarization and their refractivities and volumes in fluoride glasses are very different from those in crystals. For alkali substituting, the effect of substitution of an ion with a slightly smaller ionic refractivity is outweighed by a relative decrease in the molar volume [24], ionic refraction of Li^+, Na^+, W^{6+}, O^{2-} and F^- is 0.1, 0.8, 2.6, 3.8 and 2.0, respectively [25].

Finally, the refractive index is decreased due to substitution of alkali contents instead of the present oxide. As a result, lower ionic refractivity in alkali and decreased molar volume of the glass can be observed.

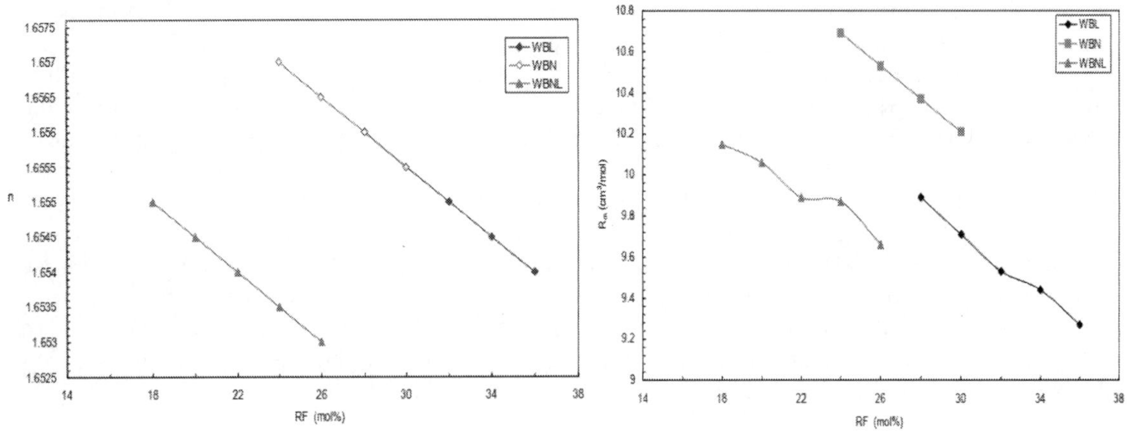

Figure 2. Variation of n & $R_m$with composition for the prepared glasses.

The elastic moduli such as longitudinal L, shear S, Young's modulus E, bulk modulus B and Poisson's ratio σ can be determined as follow:

$$L=\rho V_L^2,\quad S=\rho V_s^2,\quad E=\rho V_s^2\frac{(3V_L^2-4V_s^2)}{(V_L^2-V_s^2)},\quad B=\rho(V_L^2-\frac{4}{3}V_s^2),\quad \sigma=\frac{(V_L^2-V_s^2)}{2(V_L^2+V_s^2)} \qquad (11)$$

where V_L, V_S are longitudinal and transverse velocities of the prepared glasses and ρ is the glass density.

Then the mean ultrasonic velocity (V_{mean}) and the microhardness (H) can be calculated according to:

$$V_{mean}=\left[\frac{(\frac{1}{V_L^3}+\frac{2}{V_S^3})}{3}\right]^{-\frac{1}{3}},\quad H=\frac{(1-2\sigma)E}{6(1+\sigma)} \qquad (12)$$

The longitudinal L, shear S, Young's modulus E, bulk modulus B and themicrohardnessof the prepared glass samples show a decrease with increasing RF content, while the poisson's ratio increases with increasing RF content, see table2.

This may be due to the decrease in density and depolymerisation effect occurred in the glass network with increasing RF content. The increase in the Possion's ratio can be attributed to the structural changes and the formation of nonbridging oxygen units (NBOs) and (NBF) with increasing RF content.

Table 2. Longitudinal modulus, L (GPa), Young's modulus, E (GPa), bulk, B and shear moduli, S (GPa), Poisson's ratio and micro hardness of the present glasses:

Sample Name	L GPa	S GPa	E GPa	B GPa	σ	H GPa
WBL-1	91.085	25.17313	65.905	57.521	0.309	3.205
WBL-2	88.303	24.65226	64.408	55.433	0.306	3.183
WBL-3	84.097	23.06419	60.477	53.344	0.311	2.905
WBL-5	85.219	24.2634	63.132	52.868	0.301	3.219
WBN-1	83.52	22.793	59.825	53.129	0.312	2.852
WBN-2	82.543	21.686	57.331	53.62809	0.322	2.576
WBN-3	78.105	22.981	59.363	47.463	0.292	3.194
WBN-4	77.033	24.238	61.587	44.715	0.270	3.709
WBNL-1	76.422	24.2634	55.668	48.022	0.307	2.743
WBNL-3	81.142	22.769	59.42727	50.78184	0.305	2.961
WBNL-4	79.750	24.609	62.84413	46.93807	0.277	3.661
WBNL-5	76.365	22.895	58.88197	45.8386	0.286	3.268

4. Conclusion

The decrease in density and the molar volume with increasing alkali and mixed alkali fluoride instead of WO_3 is due to the lighter molecular weight of the added LiF and NaF as compared to that of WO_3 and the increase of W–O bond length. The decrease of glass transition temperature, specific heat at constant pressure and the increase of average linear thermal expansion with the increase of alkali and mixed alkali fluoride content are due to the weakening of the glass structure. The refractive index also decreases due to the lower ionic refractivity of the alkali content and the decrease of the molar volume of the prepared glasses. Formation of (NBOs) and (NBF) increases the Possion's ratio for the prepared glasses and the observed decrease in the elastic module and microhardness may be also due to the decrease in the density and the depolymerisation effect that occurred in the glass network with increasing RF content. These glasses have acceptable values of the glass transition, average linear thermal expansion, specific heat and refractive index. As a result, it may be used in many optical and electrical applications.

References

[1] Bih H, Bih L, Manoun B, Azrour M, Elouadi B, Lazor P, *M.* 2010 *J. Cond. Mat.* **12** 2
[2] Araujo, Carla C De, Wenzel Strojek, Long Zhang, Hellmut Eckert, Ga l Poirier, Sidney J. L. Ribeiro, and Younes Messaddeq 2006 *J. Mater. Chem.* **16** 32
[3] Poirier G, Messaddeq Y, Ribeiro S J L, Poulain M 2005 *J. Sol. State Chem.* **178**
[4] Poirier G, Michalowicz A, Messaddeq Y, Ribeiro S J L 2005 *J. Non-Cryst. Solids* **351**
[5] Nalin M, Poirier G, Ribeiro S J L, Messaddeq Y, Cescato L 2007 *J. Non-Cryst. Solids* **353**
[6] Subbalakshmi P, Raghavaiah B V, Balaji Rao R, Veeraiah N 2004 *Eur. Phys. J. Appl.Phys.* **26**
[7] Sambasiva Rao K, Srinivasa Reddy M, Ravi Kumar V, Veeraiah N 2008 *J. Mater.Chem. Phy.* **111**
[8] Lampert C M 1994 *Sol. Energy Mater. Sol. Cells* 32 307
[9] Poirier G, Castro Cassanjes F, Messaddeq Y, Ribeiro S J L 2009 *J. NonCryst. Solids* **355**
[10] Vogel E M, Weber M J, Krol D M 1991 *Phys. Chem. Glasses* **32**
[11] Falcão-Filho E L, de Araújo C B, Manzani D, Messaddeq Y 2009 *J. Appl. Phys.* **106**
[12] Deva Prasad Raju B, Madhukar Reddy C 2012 *Optical Materials* **34**
[13] Conti G N, Berneschi S, Bettinelli M, Brenci M, Chen B, Pelli S, et al. Conti GN, Berneschi S, Bettinelli M, Brenci M, Chen B, Pelli S *et al* 2004 *J. NonCryst. Solids.* pp 345-346
[14] Kuan, Pei-Wen, Kefeng Li, Guang Zhang, Xin Wang, Lei Zhang, Gongxun Bai, Yuenhong

IOP Conf. Series: Journal of Physics: Conf. Series **830** (2017) 012125 doi:10.1088/1742-6596/830/1/012125

Tsang, and Lili Hu. 2013 *Opt. Mater. Express* **3** 6
[15] El-Hofy M and Hager I Z 2000 *Physica status solidi (a)* **182** 2
[16] Hager I Z 2008 *Mat. Chem. & Phys.* **109**
[17] Zhao X and Sakka S 1988 *J. Mater. Sci.* **23**
[18] Coon J and Shelby J E 1990 *Ceram J Am Soc.* **73** 2
[19] Hager I Z, El-Mallawany R and Poulain M 1999 *J. Mater. Sci.* **34**
[20] Chen H S 1978 *Non-Cryst J Solids* **27**
[21] Zhang L and Gan F 1992 *Glass Tech.* **33** 1
[22] Venkateswarlu M Mahamuda S k, Swapna K, Prasad M.V, Rai A S, Babu A M, Shakya S, Prakash G V 2015 *Opt. Mater.* **39**
[23] MacFarlane D R, Newman P J, Zhou Z andJavornicsky J 1993 *J. Non-Cryst. Solids* **161**
[24] Jewell J M and Aggrawal I D 1992 *J. Non-Cryst. Solids* **142**
[25] Korotkov A S and *Atuchin* VV 2008 *Optics Comm.* **281**

OpenMP parallel computing of 2D TiC combustion synthesis process using an explicit finite-volume scheme

A Aoufi

Centre SMS, Laboratoire Georges Friedel, UMR CNRS 5307, EMSE, 158 cours Fauriel, 42023 Saint-Etienne. France

E-mail: aoufi@emse.fr

Abstract. This paper analyzes from a numerical point of view the ignition and propagation of the combustion front during the exothermic TiC combustion synthesis of a material made of pressed titanium and carbide particles when thermophysical properties are either assumed constant or temperature and conversion rate dependent. A two-dimensional cylindrical geometry is considered. The heat supply is prescribed on one, two or three sides of the physical domain. A one-step kinetics is used to describe the reaction Ti+C→TiC in a solid phase and leads to the computation of the conversion rate. A coupling with a non-linear heat equation which takes into account the heat generated by the exothermic kinetics and the two allotropic phase-changes is considered. An explicit finite-volume discretization of the coupled system is constructed and analyzed. Time-step's stability condition is given for a general expression of the thermo-physical characteristics. A discrete maximum principle is reported. Open MP API was used to parallelize the numerical software written in C. An average speedup of three was obtained on an intel quad-core processor i7-2600. The ignition time and the fraction of unreacted material are systematically computed and compared for several heat supply scenario.

1. Introduction

The present paper is devoted to the numerical computation of TiC combustion synthesis [1,2] in a 2D cylindrical coordinate system using an explicit scheme implemented on a multicore architecture using OpenMP API. It is organized as follows. Section two reports the main features of the mathematical modelling of this combustion synthesis process (coupling a partial differential equation with a differential equation) and the expression used for thermophysical parameters (thermal conductivity and heat capacity). Section three outlines the explicit finite-volume discretization of both equations and reports a discrete maximum principle for both equations. Section four explains the main ingredients of the parallelization strategy based upon inserting OpenMP #pragma directives and analyzes the efficiency of the software on a multicore intel i7-2600 cpu. Section five presents some numerical simulation results used to systematically investigate the evolution of ignition curves when thermophysical properties are either constant or temperature and conversion rate dependent. The stability of the combustion wave is investigated thanks to the Zeldovitch number. The contribution of allotropic phase changes is also taken into account numerically. The effect of heat supply on one, two and three faces of the cylinder is also reported. A conclusion summarizes the results presented in this study.

IOP Conf. Series: Journal of Physics: Conf. Series **830** (2017) 012126 doi:10.1088/1742-6596/830/1/012126

2. Mathematical modeling

Temperature T in K, and conversion rate $\xi \in [0,1]$ are the key variables used to describe the evolution of the exothermic equation Ti+C →TiC. The mathematical modeling couples the enthalpy balance with a differential equation.

2.1. *Nonlinear reaction diffusion and boundary condition*

An enthalpy balance written for a cylinder Ω with radius *R* and height *H,* leads to the following non-linear reaction-diffusion equation

$$\rho Cp(T,\xi)\frac{\partial T}{\partial t} - \nabla.(\lambda(T,\xi)\nabla T) = \rho\Delta H k(T)(1-\xi). \quad (1)$$

Along the symmetry axis of the cylinder, the adiabatic boundary condition is prescribed

$$\forall z \in [0,H], \quad -\lambda(T,\xi)\frac{\partial T}{\partial n}(0,z) = 0. \quad (2)$$

Radiative boundary conditions are defined for the three remaining subparts Γ of $\partial\Omega$. The value of $T_{\infty,\Gamma}$ ranges from 300 K (room temperature) to 2400 K. ϵ is the emissivity of the material and σ is the Stefan-Boltzmann constant.

$$-\lambda(T,\xi)\frac{\partial T}{\partial n} = \epsilon\sigma\left(T^4 - T^4_{\infty,\Gamma}\right) \quad (3)$$

The initial condition states that the sample is at room temperature T0 such that

$$\forall x \in \Omega, \quad T(x,0) = T_0 \quad (4)$$

2.2. *Heat capacity*

Assuming that the mass of the system remains constant, a linear mixing law between the heat capacity of reactants *Cp*Ti(*T*), *CpC*(*T*) [3] and the heat capacity of the product *Cp*TiC (*T*) [3] weighted by the conversion rate of the reaction gives (Figure 1)

$$\mathrm{Cp}(T,\xi) = \xi.C_{P_{TiC}}(T,\xi) + (1-\xi).\frac{M_{Ti}.C_{P_{Ti}}(T)+M_C.C_{P_C}(T)}{M_{TiC}} \quad (5)$$

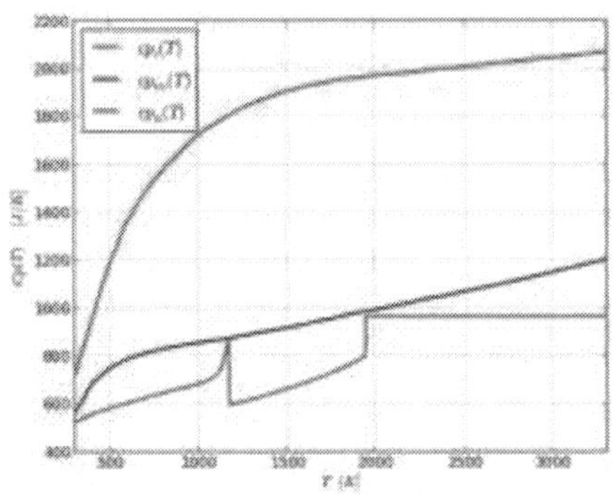

Figure1. Evolution of cptic(T), cpti(T) and cpc(T).

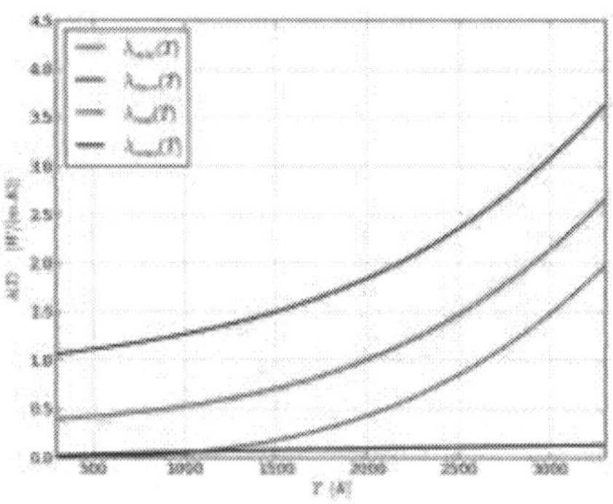

Figure2. Evolution of λ min(T), λ max(T), λ rad(T) and λ conv(T).

2.3. *Thermal conductivity*

The effective thermal conductivity of Ti+C power mixture is the sum of terms taken from [4],[5],[6]. $f_{sl} \in [0,1]$ is the liquid fraction at allotropic phase change occurring at T_f = 1950K. Figure (2) shows the range in which $\lambda(T,\xi,f_{sl})$ varies while omitting $\lambda_{fus}(f_{sl})$ defined only $T = T_f$.

$$\lambda(T,\xi,fsl) = \lambda cond(T,\xi,fsl) + \lambda fus(fsl) + \lambda rad(T) + \lambda conv(T) \quad (6)$$

2.4. *Kinetics*

A first order kinetics with an Arrhenius equation for the rate constant describes Ti+C →TiC. It is given by

$$\frac{d\xi(x,t)}{dt} = k(T(x,t))(1 - \xi(x,t)), \quad k(T) = k_0 e^{-\frac{E^*}{RT(x,t)}}, \quad \xi(x,0) = 0. \tag{7}$$

2.5. *Analysis of the numerical simulations*

The following three quantities are systematically computed in order to investigate the results of the numerical simulations presented in next sections:

- the synthesis temperature *Tsyn*(*x*) ($x \in \Omega$) is the weighted mean of the temperature by the kinetics defined (from [7]) by

$$\forall x \in \Omega, \qquad T_{syn}(x) = \int_0^{+\infty} T(x,t)\frac{d\xi}{dt}(x,t)dt \tag{8}$$

- the normalized fraction of unreacted material *nfum(t)* is defined (from [8])

$$\forall t > 0, \quad nfum(t) = \frac{\|1-\xi(.,t)\|_{L^2(\Omega)}}{\|1\|_{L^2(\Omega)}} \in [0,1\,]. \tag{9}$$

- the maximum of the temperature *Tmax*(*t*) and the mean temperature $\bar{T}(t)$ of $T(.,t)$ over Ω are computed at each time $t > 0$.

3. Finite-volume discretization

The computational domain Ω is decomposed into a set of $NI\times NJ$ rectangular control volumes $\Omega i,j$ with "mass" mi,j . A conservative finite-volume cell-centered approximation is used to discretize the governing equations of the modeling. The handling of each phase change transition during which a liquid fraction appears is done thanks to a specific procedure.

3.1. *Explicit finite-volume discretization of the reaction-diffusion equation*

The integration of equation (1) over a space-time finite volume $\Omega i,j \times [tn, tn+1]$ leads to the discrete equation, where $T_{i,j}^n$ is the mean of T over Ωi,j evaluated at tn

$$m_{i,j}\rho C_p(T_{i,j}^n,\xi_{i,j}^n)\frac{T_{i,j}^{n+1}-T_{i,j}^n}{\Delta t} = E_{i,j}^n(T_{i+1,j}^n - T_{i,j}^n) - W_{i,j}^n(T_{i,j}^n - T_{i-1,j}^n)$$
$$+N_{i,j}^n(T_{i,j+1}^n - T_{i,j}^n) - S_{i,j}^n(T_{i,j}^n - T_{i,j-1}^n) + m_{i,j}\rho\Delta Hk(T_{i,j}^n)(1-\xi_{i,j}^n). \tag{10}$$

Similar discrete equations are written for control volumes intersecting the boundary $\partial\Omega$. Let us define Δt_{rd}^{stab} by

$$\Delta t_{rd}^{stab} = \min_{i,j}\left(\frac{m_{i,j}\rho C_p(T_{i,j}^n,\xi_{i,j}^n)}{E_{i,j}^n+W_{i,j}^n+N_{i,j}^n+S_{i,j}^n}\right). \tag{11}$$

Under the assumption that $0 < \Delta t \leq \Delta tstab$, the following discrete maximum principle properties for the time-explicit discretization ([7],[9] for the time-implicit discretization) are established by induction

(i) $\forall(i,j) \in [1,NX] \times [1,NY]: 0 < T_0 \leq T_{i,j}^n \Rightarrow 0 < T_0 \leq T_{i,j}^{n+1}$.

(ii) $\forall(i,j) \in\,]1,NX[\, \times\,]1,NY[: T_{i,j}^{n+1} \leq T_{i,j}^n + \Delta t\frac{k_0\Delta H}{C_p^{min}}$, where $C_p^{min} = \min_{T>0,\xi\in[0,1]} C_p(T,\xi)$.

It is worth mentioning that when the thermo-physical properties are constant, the time-step for the reaction-diffusion is also constant, therefore the stability condition Δt_{rd}^{stab} is computed once and passed as a parameter to the function advancing the heat transfer the case. When these parameters are temperature and conversion rate dependent, Δt_{rd}^{stab} is recomputed at each step once the positive quantities $E_{i,j}^n, W_{i,j}^n, N_{i,j}^n, S_{i,j}^n$ are evaluated as a function of T and ξ. The computational complexity is therefore a multiple of $NI \times NJ$.

3.2. *Explicit finite-volume discretization of the differential equation*

The integration of equation (7) over a space-time finite volume leads to the discrete equation

$$\xi_{i,j}^{n+1} - \xi_{i,j}^n = \Delta tk(T_{i,j}^n)(1-\xi_{i,j}^n). \tag{12}$$

Let us define $\Delta t_{kin}^{stab} = \min_{i,j}\frac{1}{k(T_{i,j}^n)}$. Under the assumption that $0 < \Delta t \leq \Delta t_{kin}^{stab}$, the following discrete maximum principle properties ([7],[9] for the time-implicit discretization) are established by

induction for $(i,j) \in [1,NX] \times [1,NY]$
(i) $0 \le \xi_{i,j}^{n} \le 1 \Rightarrow 0 \le \xi_{i,j}^{n+1} \le 1$.
(ii) $0 \le \xi_{i,j}^{n} \le \xi_{i,j}^{n+1}$.

3.3. Choice of the time-step

The time-step Δtn is computed thanks to the expression

$$\Delta t_n = \left(\Delta t_{rd}^{stab}, \Delta t_{kin}^{stab}\right). \quad (13)$$

It is worth mentioning that in the numerical simulations presented in next sections, we always observed that $\Delta t_{rd}^{stab} < \Delta t_{kin}^{stab}$. This is explained by the small value of the activation energy E^*.

4. Parallel computing using OpenMP

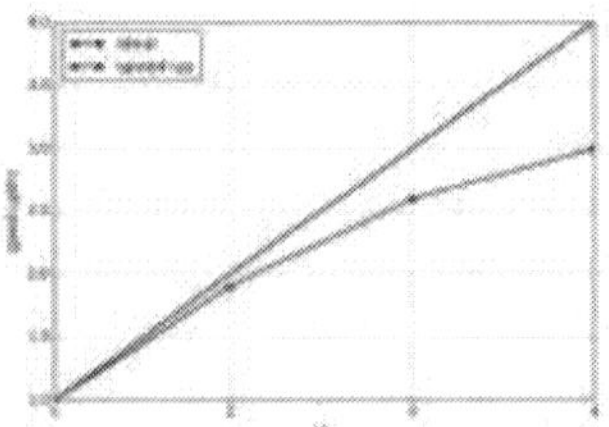

Figure 3. Evolution of speed-up $s_{th} = t(1)/t(th)$, for th=1,...,4.

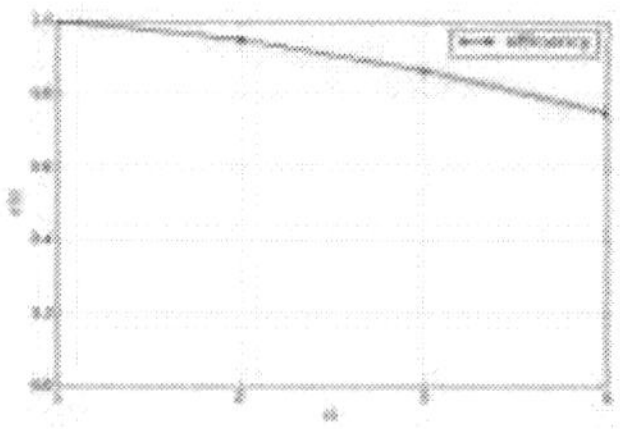

Figure 4. Evolution of efficiency $e_{th} = s_{th}/th$, for th=1,...,4.

The numerical scheme is implemented in C using our library Hepha¨ıstos, compiled wit gnu gcc compiler (version 4.8.2) and run under Ubuntu 14.04 (64 bits) OS. Two dimensional arrays are converted to single dimensional arrays. All loops are parallelized thanks to OpenMP directives manually inserted into the code. Highest level of optimisation activated by an adequate compiler flag is chosen for gcc compiler. This leads to the vectorization of loops using SIMD units of the intel i7-2600 multicore processor.

The value of the time-step for the reaction-diffusion equation is either computed once at the beginning of the program (the values of λ and Cp are constant) or dynamically determined by a reduction operation (the values of λ and Cp are temperature and conversion rate dependent). The efficiency of the program is increased by avoiding redundant computations. All loop invariant quantities are computed once and stored into arrays. Trivial nested loops are parallelized with a specific openmp #pragma directive. The numerical results are checked and are independent from the number of cores. Figure 3 shows the scalability of the numerical code. A speedup of 3 was obtained when the number of threads is 4 for a given mesh. It varies slightly as a function of the mesh. Figure 4 shows that the efficiency of the numerical code is above 0.75 when four cores are used. This means that time spent on transferring datas between cores is much lower than time spent on computations. The implementation is therefore efficient.

5. Numerical simulation

5.1. Stability analysis of the combustion front

Linear stability analysis (from [10]) is based upon the computation of Zeldovich number $Ze = \frac{E^*(T_{ad}-T_0)}{RT_{ad}^2}$. Here $T_{ad} \cong 2900K$ and $E^* \cong 30kcal = 125.4$ kJ, then $Ze = 4.52 < 2(2+\sqrt{5}) \cong 8.4723$ for which a stable propagation with a constant velocity is obtained.

5.2. Key simulation parameters

A simulation is characterized by (*Tz*=0, *Tr*=*R*, *Tz*=*H*) and stopped when *nfum*(*t*) < 10−2, hence the cooling of the material down to room temperature is not computed. R = 0.5cm and H =1 cm. The spatial distribution of *Tsyn*(*x*) is systematically determined for T =1600 K when a) constant properties

IOP Conf. Series: Journal of Physics: Conf. Series **830** (2017) 012126 doi:10.1088/1742-6596/830/1/012126

(c-pc-0), b) variables properties without phase change (v-pc-0) or c) variables properties with phase change (v-pc-1) are considered.

5.3. *Propagation of a longitudinal front*

In this case, (*Tz*=0, *Tr*=*R*, *Tz*=*H*) = (*Tf* , 300, 300), where *Tf* =800, 1600, 2400. A uniform tensorial grid made of 32 × 512 cells is used for the computations. Results are presented in Figure 5, Figure 6, Figure 7 and show that nonlinear expression for $Cp(T,\xi)$ and $\lambda(T, \xi)$ have a leading contribution to the celerity of the synthesis.

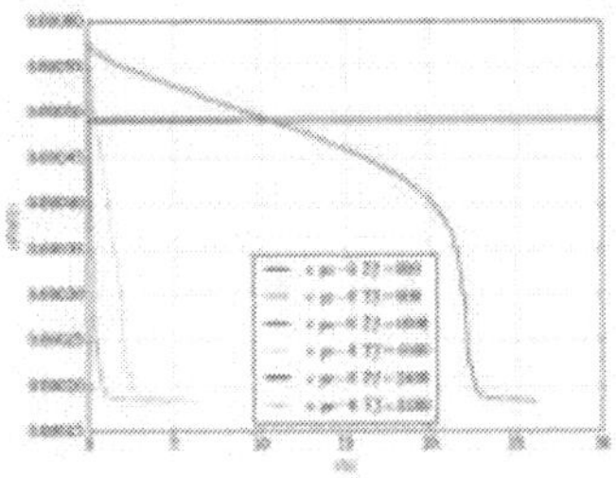

Figure 5. Time evolution of $\Delta t_{rd}^{stab}(t)$

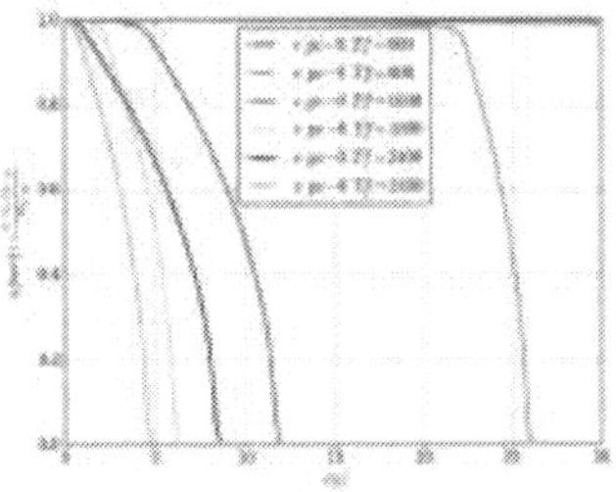

Figure 6. Time evolution of Nfum (t)

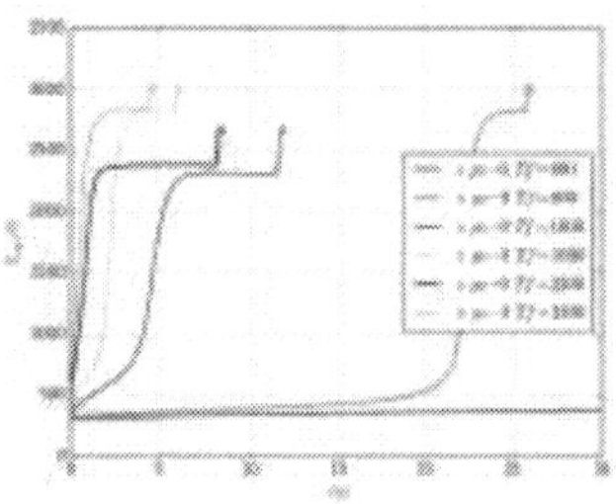

Figure 7. Time evolution of Tmax (t)

5.4. *Propagation of a converging radial front*

In this case, (*Tz*=0, *Tr*=*R*, *Tz*=*H*) = (300, *Tf* , 300), where *Tf* =800 K, 1600 K, 2400 K. A uniform tensorial grid made of 512 × 32 cells is used for the computations. Results are presented in Figure 8, Figure 9, Figure 10, Figure 11, Figure 12. A conclusion similar to the longitudinal case can be established. Phase changes contribution is noticeable graphically.

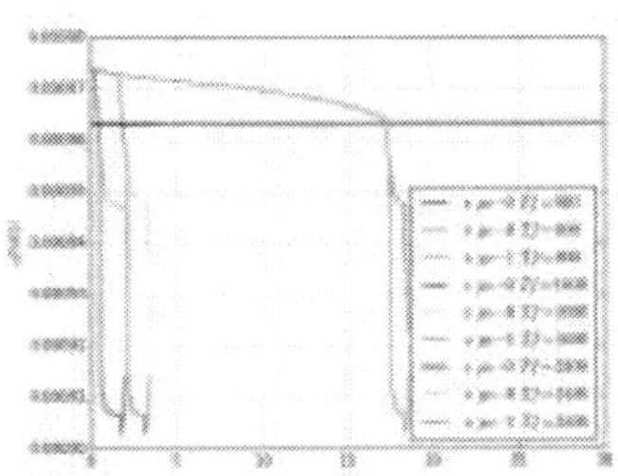

Figure 8. Time evolution of$\Delta t_{rd}^{stab}(t)$

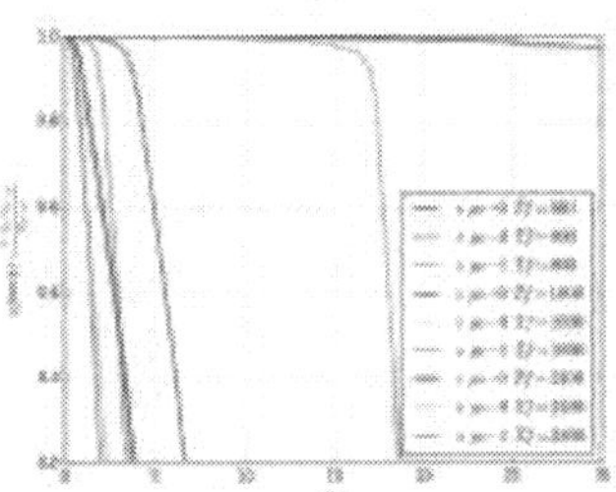

Figure 9. Time evolution of *nfum* (*t*)

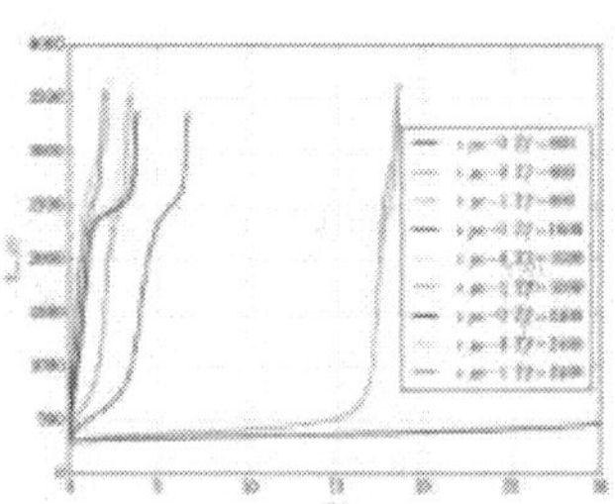

Figure 10. Time evolution of *Tmax* (*t*)

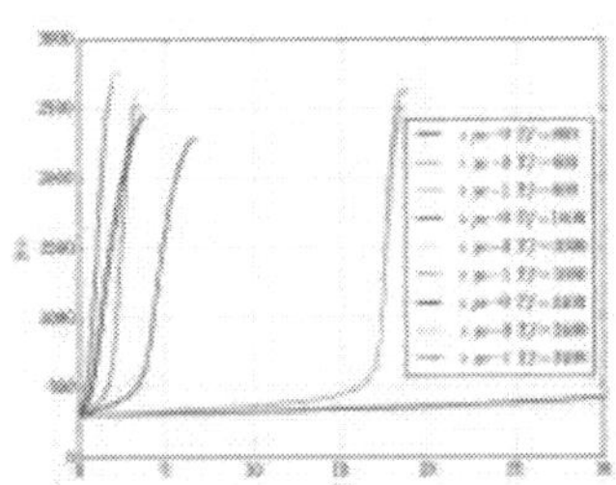

Figure 11. Time evolution of *T* (*t*)

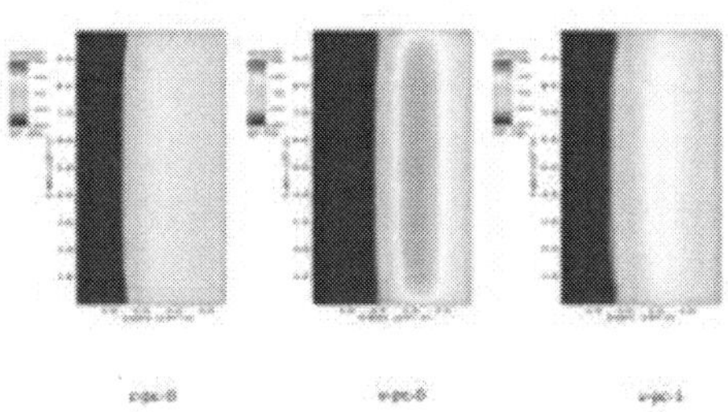

Figure12. Spatial distribution of *Tsyn*(*x*)

IOP Conf. Series: Journal of Physics: Conf. Series **830** (2017) 012126 doi:10.1088/1742-6596/830/1/012126

5.5. Propagation of two longitudinal fronts

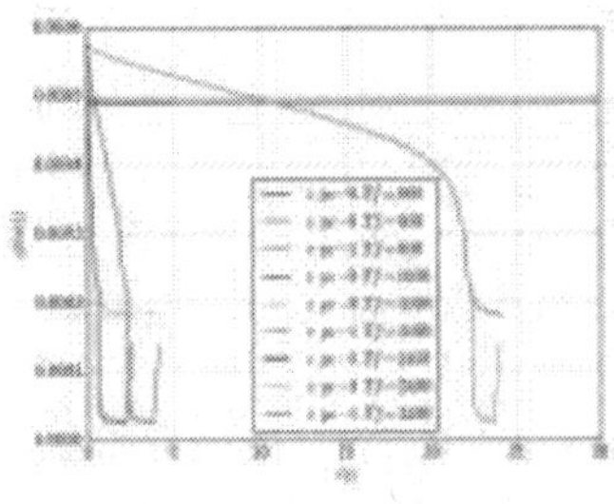

Figure 13. Time evolution of $\Delta t_{rd}^{stab}(t)$

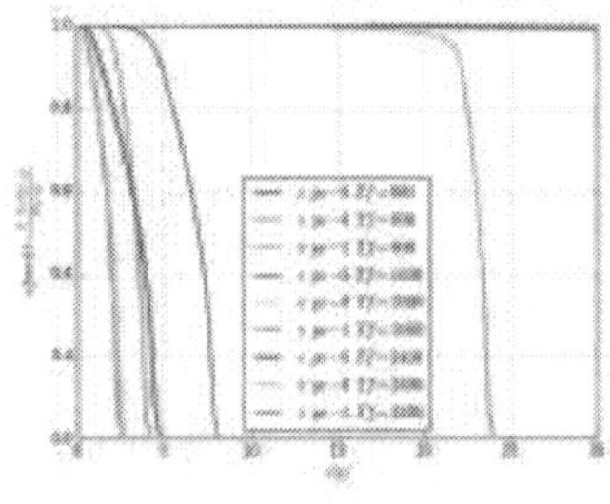

Figure 14. Time evolution of *nfum* (*t*)

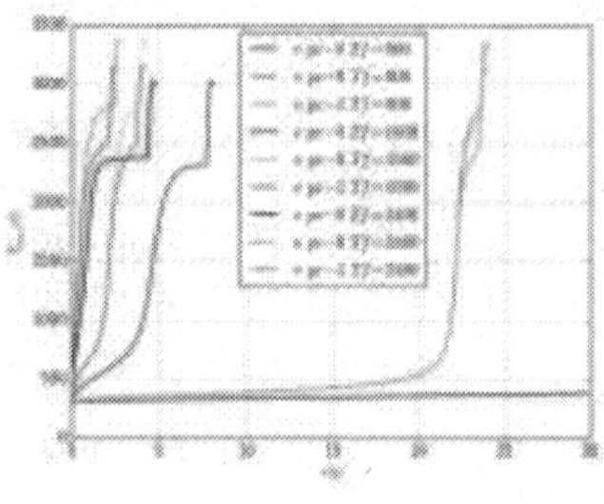

Figure 15. Time evolution of *Tmax* (*t*)

In this case, (*Tz*=0, *Tr*=*R*, *Tz*=*H*) = (*Tf* , 300, *Tf*), where *Tf* =800 K, 1600 K, 2400 *K*. A uniform tensorial grid made of 32 × 512 cells is used for the computations. Results are presented in Figure 13, Figure 14, Figure 15, Figure 16, Figure 17. A similar conclusion can be established. Moreover, along *z* = *H*/2 a melting might occur because *Tmax*(*t*) is locally greater than 3000 K. Latent heat of phase change is therefore needed to accurately capture the effects of this melting.

5.6. Propagation of three converging fronts

In this case, (*Tz*=0, *Tr*=*R*, *Tz*=*H*) = (*Tf*, *Tf*, *Tf*), where *Tf*=800, 1600, 2400. This corresponds to the practical case where the cylindrical sample is placed inside a furnace with uniform temperature *Tf* . A uniform tensorial grid made of 512 × 512 cells is used for the computations.

Results are presented in Figure 18, Figure 19, Figure 20, Figure 21, Figure 22 (notice the symmetries induced by the symmetries of the boundary conditions). The conclusion is similar to those previously obtained.

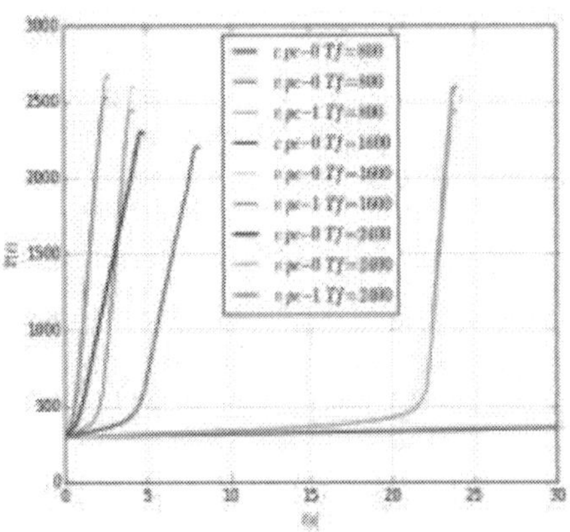

Figure 16. Evolution of $\overline{T}(t)$

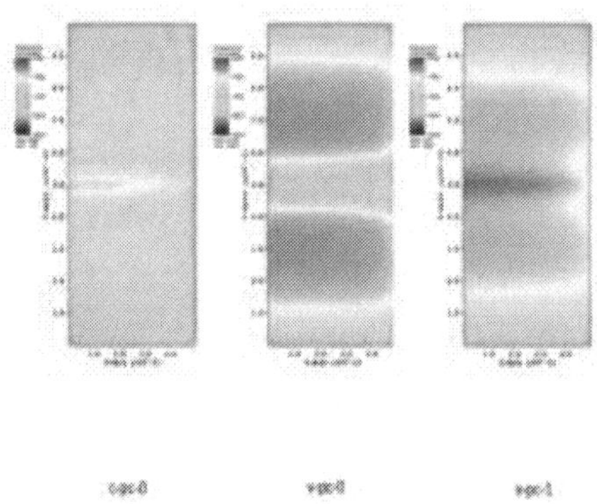

Figure 17. Spatial distribution of *Tsyn*(*x*)

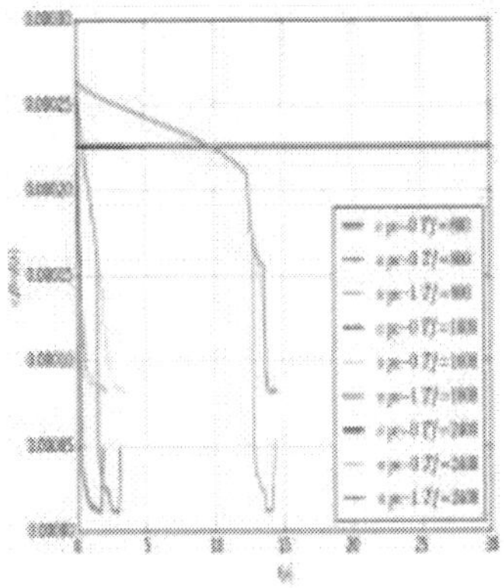

Figure 18. Time evolution of$\Delta t_{rd}^{stab}(t)$

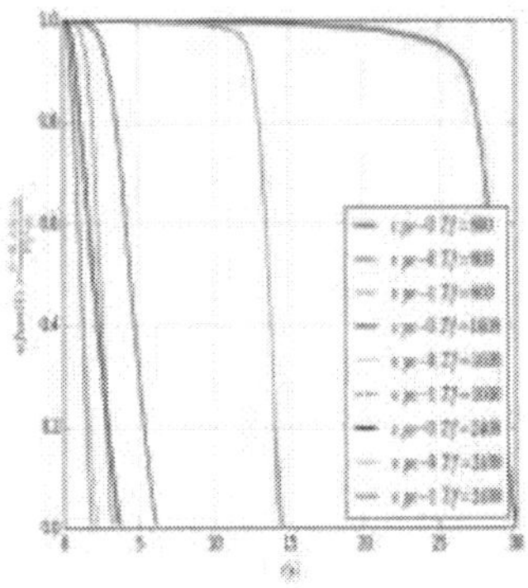

Figure 19. Time evolution of *nfum* (*t*)

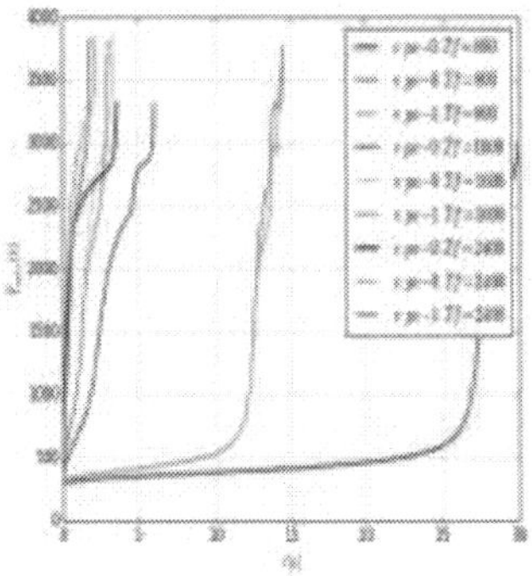

Figure 20. Time evolution of *Tmax* (*t*)

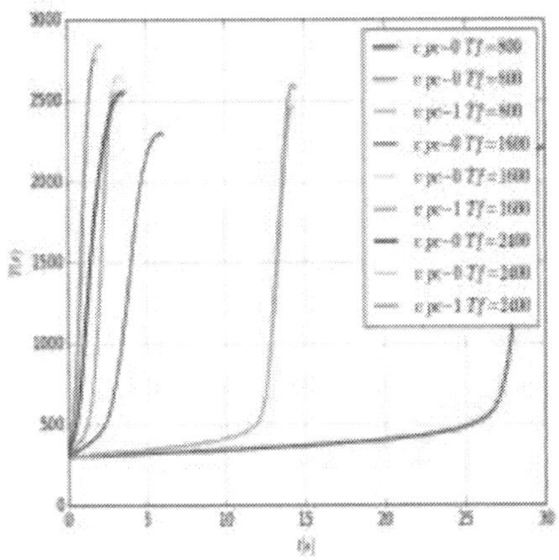

Figure 21. Evolution of $\overline{T}(t)$

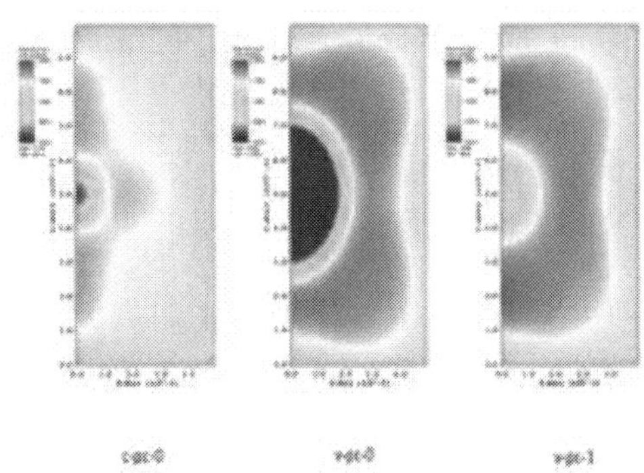

Figure 22. Spatial distribution of *Tsyn*(*x*)

5.7. Ignition curves

Figure 23, Figure 24 and Figure 25 show the ignition (instant τ such that *nfum*(τ)$\cong$ 0.95) curves obtained respectively for two propagating fronts, three propagating fronts and one radially converging front as a function of furnace's temperature *Tf* . Each set of curves, presents a high dependence on *Cp*(*T*, ξ) and λ(*T*, ξ). The contribution of phase changes is negligible (<1%)

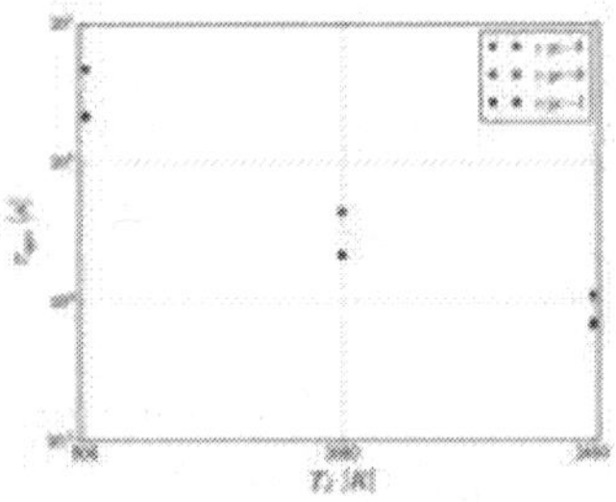

Figure 23. Ignition time for two travelling fronts

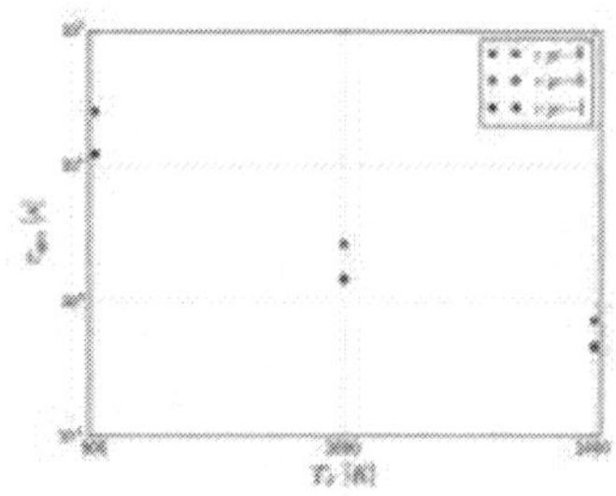

Figure 24. Ignition time for three travelling fronts

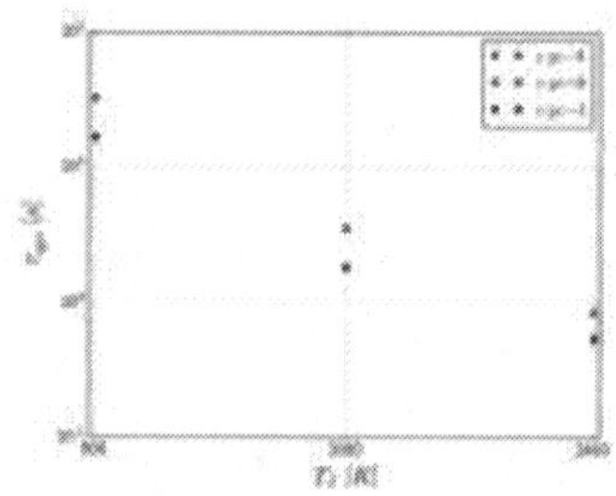

Figure 25. Ignition time for one radially travelling front

6. Conclusions and perspectives

In the present paper we have shown through numerical simulations (thanks to a tuned parallel software using OpenMP and implementing an explicit finite-volume scheme for which a discrete maximum principle was established) that independently of the way external heat supply is applied on part of the boundary $\partial\Omega$, the expressions used for $Cp(T, \xi)$ and $\lambda\ (T\ \xi)$ have a significant influence on TiC combustion synthesis process, quantified by $nfum(t)$ and $Tsyn(x)$. The contribution of the two allotropic phase changes ranges from negligible to minimal.

A similar approach is currently investigated in the case of a cubic sample where the computations are genuinely three dimensional.

References

[1] Merzhanov A G 1969 *Combustion and Flame* **13** pp 143-156
[2] Merzhanov A G 1994 Combustion Science and Technology **1994** pp 307-336
[3] Touloukian S and Dewitt D P 1972 *The TPRC Data Series* IFI Plenum **8**
[4] Vrel D, Lihrmann J M and Tobaly P 1994 *Journal of Materials Synthesis and Processing* **2** pp 179-187
[5] Wakao N and Kato K 1969 *Journal of Chemical Engineering of Japan* **2-1** 24-32
[6] Viljoen H J and Hlavacek V 1991 *AIChE J.* **37** pp 1595-1597
[7] Aoufi A and Dammame G 2006 *Applied Mathematics and Computation* **176** pp 99-116
[8] Aoufi A 2014 *Izvestia vizov physica* **57** pp 16-19
[9] Aoufi A 2004 *Integral Methods in Science and Engineering* Birkhauser, Basel pp 1-6
[10] Bayliss A and Matkowsky B J 1987 *JCP* **71-1** pp 147-168

IOP Conf. Series: Journal of Physics: Conf. Series **830** (2017) 012127 doi:10.1088/1742-6596/830/1/012127

Host and defect-related photoluminescence of structurally disordered $K_3WO_3F_3$ oxyfluoride crystals

A V Kozlov and V A Pustovarov

Ural Federal University, Mira str. 19, 620002 Yekaterinburg, Russia Russia

E-mail: KozlovArtemVladimirovich@gmail.com

Abstract. Spectra of photoluminescence (PL) in the region of 1.5–5.5 eV, PL excitation spectra (3–22 eV), PL decay kinetics and PL temperature dependence were measured for single crystals and ceramics $K_3WO_3F_3$ as well as for ceramics $K_3WO_3F_3$ irradiated by fast electrons. Synchrotron radiation was used for low temperature PL experiments with time resolution. Single crystals are transparent in microwave, visible and near UV range, inter-band transition energy is E_g = 4.3 eV. In $K_3WO_3F_3$, the wide band luminescence in the region of 2.5 eV with the Stokes shift of 1.5 eV with the microsecond decay kinetics is connected with luminescence of triplet self-trapped excitons (STE). This luminescence is formed by electronic transitions in $[WO_3F_3]$ octahedron. Different distortion of KWOF crystal lattice is manifested in the change of the Stokes shift of the STE luminescence band. The 3.2 eV emission band in low-temperature PL spectrum with decay times of 1.8 ns and 11 ns corresponds to singlet STE luminescence. A new 2.9 eV emission band is discovered in low-temperature PL spectrum in the samples irradiated by fast electrons (E = 10 MeV, D = 160 kGy). This emission band is excited not through the intracenter mechanism but through the creation of excitons bound on the defects. It is suggested that it is F-like centers of anionic sublattice induced by the mechanism of elastic collision.

1. Introduction

The group of crystals based on tungstates/molybdates have been of particular interest for over a hundred years. They are widely used in a great range of tasks: in laser technology, as scintillation materials in radiation safety systems, for thermal neutrons detection, double β-decay, etc. In particular, high purity crystals $PbWO_4$ are used as scintillation detectors in the Large Hadron Collider.

The group of oxyfluorides with general formula $A_3MO_3F_3$, where A can be rubidium, cesium, or potassium and M can be tungsten, titanium, niobium or molybdenum, are attractive compounds for developing new noncentrosymmetric crystals having ferroelectric and ferroelastic properties. This is achieved due to the strong distortion of metal-(O,F) polyhedra in the crystal lattice because of different ionicity of metal-O and metal- F bonds. The luminescence spectroscopy can be a sensitive method to study the character of lattice distortion of oxyfluorides.

However, nowadays there are almost no studies dedicated to the study of photoluminescence properties for this crystal group. In [1,2] such research of crystal $K_3WO_3F_3$ was initiated using standard laboratory equipment and time-resolved spectroscopy in a microsecond time range. Moreover, in crystals of this class, there has lately been discovered ultrafast intraband luminescence with a decay time of tens of picoseconds that may have a potential use, for example, in positron emission tomography [3].

The aim of this research is the study of luminescence properties of crystals K_3WO_3F using the method of low-temperature time-resolved spectroscopy, identification of PL excitation mechanisms and defining the role of radiation induced defects of the crystal structure in the processes of energy transformation.

2. Experimental results and discussion

2.1. Samples and experimental details

Three different types of $K_3WO_3F_3$ samples were studied: single crystal, ceramics and ceramics irradiated by fast electrons (the energy of 10 MeV, dose D = 160 kGy). These samples were grown and characterised in the Institute of Geology and Mineralogy SB RAS (Novosibirsk, Russia). Methods of growing and attestation are thoroughly described in [4]. In this context, we will discuss only principal moments. Crystals were grown by the method of solid phase synthesis. Both Raman scattering and XRD methods were used in order to certificate crystals (figure 1 and figure 2). XRD method (figure 2) shows the presence of one pure monoclinic phase with structural parameters **a** = 8.7459(2) Å, **b** = 8.6930(4) Å, **c** = 6.1650(3) Å, β = 135.178(2). In Raman scattering spectrum (figure 1), two regions can be distinguished: the first one in the range from 750 to 1000 cm^{-1} corresponds to stretching vibrations of W–O ion bonds, the second region located under 650 cm^{-1} corresponds to stretching vibrations W–F. Except for the above mentioned certification methods, others were also used for particular chemical analysis. All of them confirmed the adequacy of samples to the monoclinic crystal structure of $K_3WO_3F_3$.

Photoluminescence (PL) and photoluminescence excitation (PLE) spectra were measured using a 400 W deuterium discharge lamp (DDS-400), two double prism DMR-4 type monochromators and a R6358-10 (Hamamatsu) type photomultiplier tube. The PLE spectra were normalized to an equal number of photons incident on the sample, using a yellow lumogene-luminophore with a unit quantum yield in the studied energy range. Under UV-VUV excitation using synchrotron radiation (SR), the time–resolved PL spectra, PL decay kinetics as well as PL excitation spectra were measured on a SUPERLUMI station (channel I, HASYLAB, DESY) [5]. The measurements of PL spectra were carried out using an ARC Spectra Pro–308i monochromator and R6358P Hamamatsu photomultiplier tube. The PL spectra were recorded in two time windows − fast component: a delay relative to the start of the exciting SR-pulse δt_1 = 0.6 ns, span of time windows Δt_1 = 6.0 ns; – slow component: δt_2 = 32 ns, Δt_2 = 48 ns. The time resolution of the entire detection system was 1 ns (FWHM). The temporary interval between excitation SR pulses was equal to 96 ns. The PL excitation spectra were corrected to an equal number of photons incident on the sample recorded using sodium salicylate.

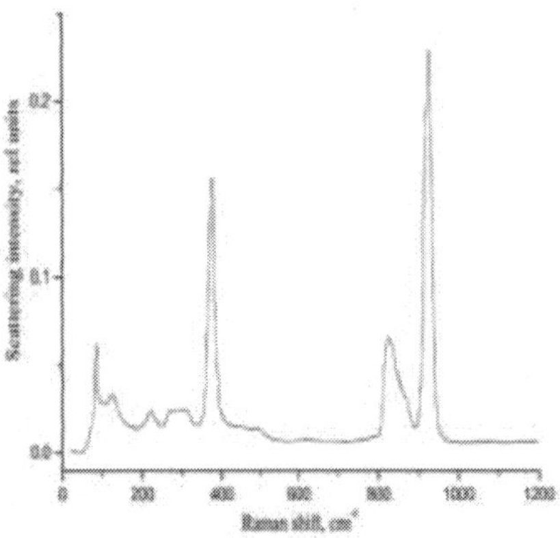

Figure 1. The Raman scattering spectrum of $K_3WO_3F_3$ crystal according to [4, 5].

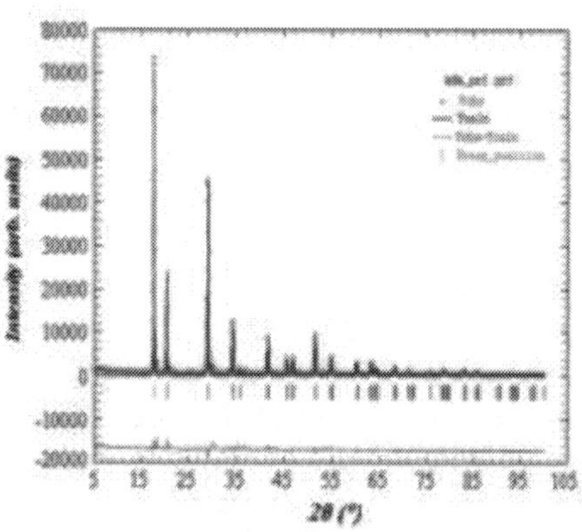

Figure 2. The XRD analysis of $K_3WO_3F_3$ crystal according to [4, 5].

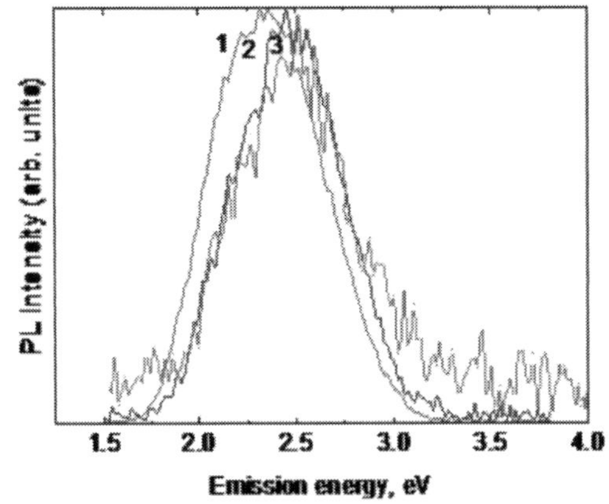

Figure 3. The normalized PL spectra of $K_3WO_3F_3$ at T = 295 K. Excitation energy = 4.11 (1), 4.70 (2) and 5.18 eV (3).

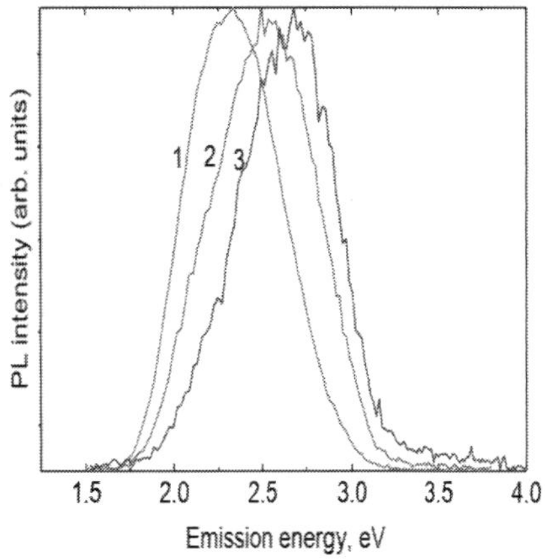

Figure 4. The normalized PL spectra of $K_3WO_3F_3$ at T = 90 K. Excitation energy = 4.11 (1), 4.70 (2) and 5.18 eV (3).

2.2. Luminescence spectroscopy data and discussion

It should be noted that during sample studying there were discovered no significant differences between ceramics and a single crystal $K_3WO_3F_3$. In this connection, in the description of the obtained results, there will be data for ceramics, and in case there are some differences with the single crystal, it will be specified separately.

In figure 3 and figure 4, the normalized PL spectra of $K_3WO_3F_3$, measured using a D_2 discharge lamp at room temperature and T = 90 K under different excitation energy E_{exc} are shown. A broad band in the region of 2.2–2.6 eV characterizes these PL spectra, and its maximum depends on the energy of the excited photons E_{exc}. This broad band emission can be effectively excited only in the long-wave edge of fundamental absorption or in the region of interband transitions. (Recall that the minimum energy of the interband transitions is E_g = 4.3 eV at the room temperature in $K_3WO_3F_3$, according to the data in [5]). This emission is not excited in the region of crystal transparency (figure 5 and figure 6). Moreover, the band of 4.2 eV in the PL excitation spectrum is shifted in the direction of higher energies together with the fundamental absorption edge when crystal temperature drops. Such behavior of the photoluminescence is typical of intrinsic luminescence in tungstate or molybdate crystals. The intrinsic luminescence of tungstates is usually ascribed to the radiative relaxation of exciton-like excitations localized on WO_6 octahedra or WO_4 tetrahedra [7]. However, in KWOF, the PL maximum position depends on the energy of the excited photons E_{exc}. This fact points that at different exciting energies luminescence appears due to electronic transitions in WO_3F_3 octahedron with various local distortions. Based on all above mentioned properties and by analogy with the tungstates crystals, we can say that the photoluminescence in $K_3WO_3F_3$ is associated with the luminescence of self-trapped excitons, i.e. it is not connected with the point defects of crystal structure or an impurity center.

Normalized excitation spectrum recorded in a wide range of energies from 3 to 22 eV at temperature 8.8 K for two emission energies 2.42 eV and 3.20 eV using synchrotron radiation is shown in figure 6. It is seen that the main excitation band is on the edge of the fundamental absorption, next in the region of higher energies (vacuum ultraviolet region), efficiency of excitation is low. It means inefficiency of conditioning transport of charge carries and presence of competing channels of capture with the following nonradiative recombination on the defects or crystal surface.

Photoluminescence spectra measured with time-resolution at T = 8.8 K are presented in figure 7. Using decay kinetics, there were distinguished a fast component – red curve, a slow component – green curve and a time integrated component – black curve.

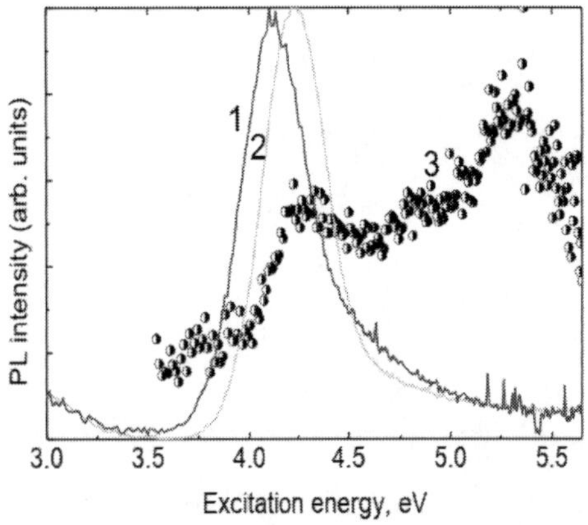

Figure 5. The normalized PL excitation spectra of $K_3WO_3F_3$ measured using a D_2 discharge lamp, T = 295 (1) and 90 K (2, 3). Emission energy E_{emis} = 2.42 (1, 2) and 3.20 eV (3).

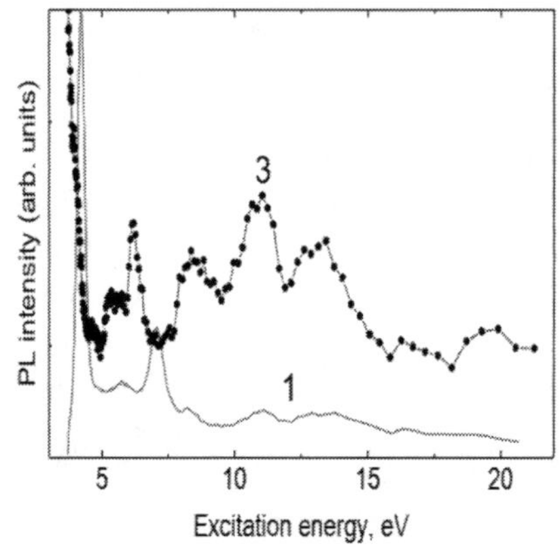

Figure 6. The normalized PL excitation spectra of $K_3WO_3F_3$ measured using synchrotron radiation, T = 8.8 K. Emission energy E_{emis} = 2.42 (1) and 3.20 eV (2).

The parameters of time windows are mentioned in the previous section, furthermore, they are shown in figure 8. The fast component shows well the presence of band 3.2 eV, which previously did not manifest itself in the PL spectra due to low intensity in contrast to the wide band 2.4 eV. As seen in Fig. 3, their band 3.2 eV is well excited by energy 4.2 eV, as well as by energy 5.2 eV. The PL decay kinetics allows us to assume singlet nature of that band. In contrast to the main band 2.4 eV where the decay time is a few tens of microseconds, the 3.2 eV band is characterised by the decay time in the nanosecond region.

Let us discuss more thoroughly excitation mechanisms of luminescence. When excited, there is a high probability of electronic radiative transition from the triplet level to the ground state. However, such transition is prohibited by the selection rules, and as a result, the decay time is a few tens of microseconds. In case when an electron in the excited exciton is on the singlet level, the selection rules allows making a radiative transition to the ground state, the decay kinetics is thousands of times faster in this case. To assess the life time of an exciton in the singlet state, approximation of two exponential was carried out, the results are τ_1 = 1.8 ns и τ_2 = 11 ns. As we suppose, the presence of two different τ values may caused by the distortion of the crystal structure and, as a consequence, by different local surroundings near emission centers.

Further $K_3WO_3F_3$ irradiated by fast electrons was studied. Irradiation by fast electrons involves the creation of point defects due to elastic displacement that will supposedly lead to the creation of capture centers of charge carriers. Such defects affect the intrinsic luminescence and new defect-related luminescence can appear. To evaluate changes in the irradiated crystal, PL emission and PL excitation spectra were measured at the conditions similar to those for the previously unirradiated crystal (figure 4 and figure 5, respectively). At room temperature, no changes were revealed regarding the unirradiated crystal. However, at low temperature, a new emission band in PL spectrum appeared (figure 9, curve 1).

The excitation spectrum of the new PL emission band is presented in figure 10. A new PL band is effectively excited in the region of 4.25 and 5.2 eV. As in case with the unirradiated crystal, a low energy band corresponds to the long-wave edge of the fundamental absorption, i.e. to creation of excitons. It should be noted that the new emission band is not excited in the crystal transparency below 4.0 eV. This allows us to conclude that the new band is excited not through the intracenter mechanism but by the creation of excitons bound on defect. We assume that it is F-like centers of anionic sublattice induced by the mechanism of elastic collision. However, this assumption needs further studies.

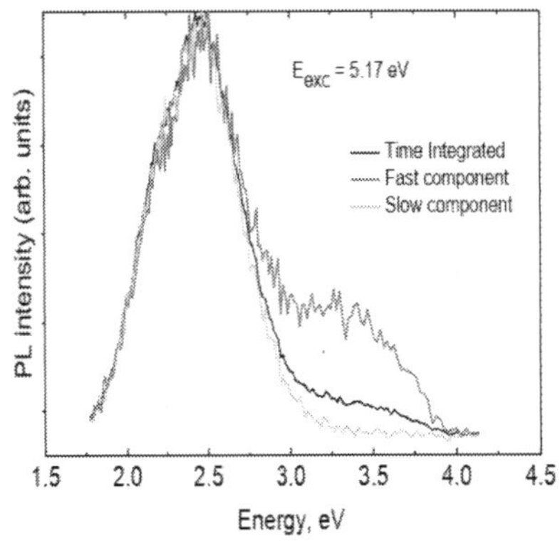

Figure 7. Time-resolved PL spectra of $K_3WO_3F_3$ measured at T=8.8 K.

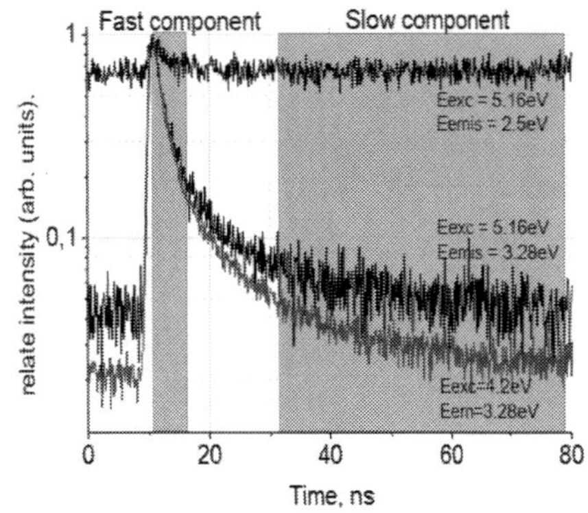

Figure 8. Decay kinetics of $K_3WO_3F_3$ measured at T=8.8 K. Positions of time windows (fast component and slow component) are shown.

The temperature dependence of PL intensity for crystals reference and irradiated by fast electrons is shown in figure. 11. All temperature dependencies are recorded from the temperature 90 K to 400 K. Before irradiation, the temperature dependence in the main band is well described by Mott's law. The quenching starts from T = 150 K, the activation energy is 0.20 eV. However, after irradiation (figure. 11 (2)), it significantly changes and no longer obeys Mott's law. The quenching starts immediately from T = 90 K that indicates the influence created by the irradiation defects to the main band. The 2.9 eV emission band in the PL spectra of the irradiated crystal in its turn has absolutely different temperature dependence, where the process of luminescence quenching starts right from T = 90 K. The activation energy is approximately 0.10 eV. However, the measurements at lower temperatures are needed to examine this temperature dependence in detail.

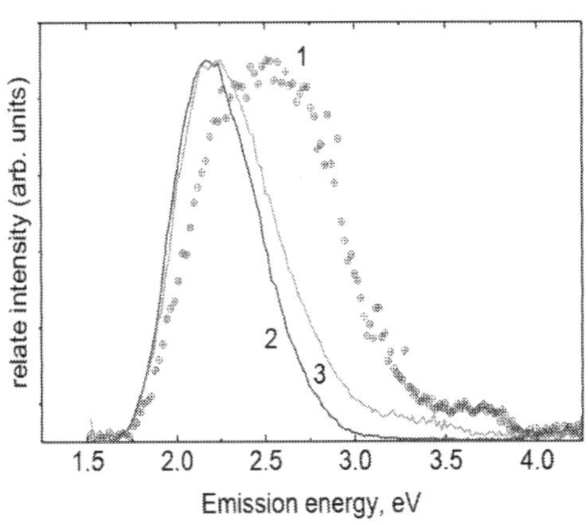

Figure 9. The normalized PL spectra of $K_3WO_3F_3$ irradiated by fast electrons. Excitation energy = 5.2 (1), 4.2 (2) and 4.7 eV (3).

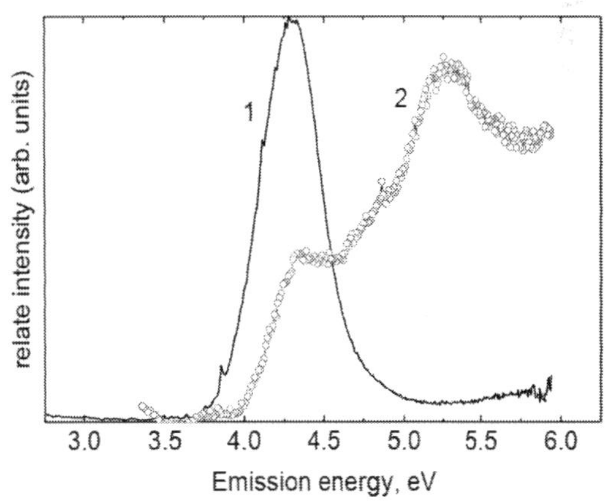

Figure 10. The normalized PL excitation spectra of $K_3WO_3F_3$ irradiated by fast electrons. Emission energy = 2.5 (1) and 2.9 eV (2), T = 90 K.

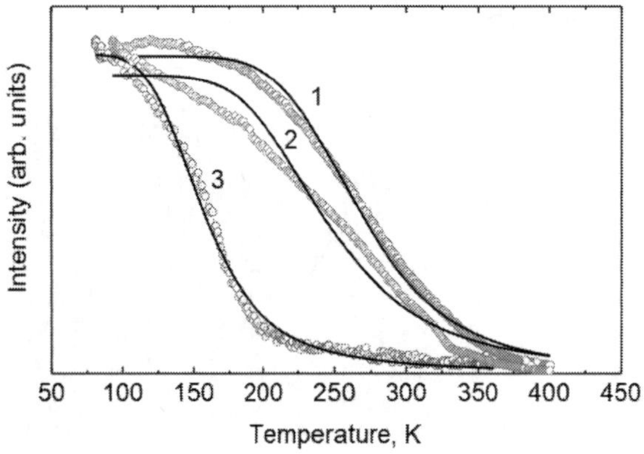

Figure 11. The temperature dependence of PL yield of reference (1) and irradiated by fast electrons (2, 3) $K_3WO_3F_3$ ceramics. E_{exc} = 4.2 eV, E_{mis} = 2.5 eV (1, 2); E_{exc} = 5.08 eV, E_{mis} = 2.9 eV (3). A thin line represents an approximation using the Mott's law, the activation energy shown in figures.

Thus, the PL data show that after irradiation by fast electrons new point defects are formed in the crystal structure. They affect the intrinsic luminescence and create new PL centers at low temperature.

3. Conclusions

The broad band luminescence in the region of 2.5 eV with the Stokes shift of ~ 1.5 eV with the microsecond decay kinetics in $K_3WO_3F_3$ crystals is connected with its intrinsic luminescence – luminescence of STE from the excited triplet states. This luminescence is formed by electronic transitions in $[WO_3F_3]$ octahedron. Different distortion of $K_3WO_3F_3$ crystal lattice is manifested in the change of the Stokes shift of luminescence band of self-trapped excitons.

The 3.2 eV emission band in low-temperature PL spectrum with decay time of 1.8 ns corresponds to electronic transitions from the singlet excited state of STE. Under interband excitation, the PL yield is low that points to ineffective transport of charge carriers due to the competition of non-radiative recombination on point defects.

A new 2.9 eV emission band is discovered in low-temperature PL spectrum in the samples irradiated by fast electrons. This emission band is excited not through the intracenter mechanism but through the creation of excitons bound on defects. It is suggested that it is F-like centers of anionic sublattice induced by the mechanism of elastic collision.

Acknowledgments
Authors are grateful to Dr. L.I. Isaenko for the provided samples. This work has been supported by the Ministry of Education and Science of the Russian Federation (the basic part of the government mandate) as well as by Act 211 Government of the Russian Federation (contract #02.A03.21.0006), Center of Excellence "Radiation and Nuclear Technologies".

References

[1] Omelkov S I, Spassky D A, Pustovarov V A, Kozlov A V and Isaenko L I 2016 *Optical Materials* **58** pp 285–289
[2] Kozlov A V and Pustovarov V A 2016 AIP Conference Proceedings **1767** 020037
[3] Omelkov S I, Nagirnyi V, Vasil'ev A N and Kirm M 2016 *J. of Luminescence* **176** pp 309-317
[4] Molokeev M S, Misyul S V, Fokina V D, Kocharova A G and Aleksandrov K S 2011 *Physics of the Solid State* **53** pp 834–839
[5] Atuchin V V, Isaenko L I, Kesler V G, Lin Z S, Molokeev M S, Yelisseyev A P and Zhurkov S A 2012 *J. of Solid State Chemistry* **187** pp 159–164
[6] Zimmerer G 2007 *Radiation Measurements* **42** pp 859–864
[7] Zhang Y, Holzwarth N A W and Williams R T 1998 *Physical Review* B **57** 12738-50

IOP Conf. Series: Journal of Physics: Conf. Series **830** (2017) 012128 doi:10.1088/1742-6596/830/1/012128

The critical parameters of the thermal explosion micro hot-spot model dependence on the pulse duration

A V Kalenskii [1], A A Zvekov [2] and E V Galkina [1]

[1] Kemerovo State University, Kemerovo 650000, Krasnaya 6, Russia
[2] Federal Research Center of Coal and Coal Chemistry SB RAS, Kemerovo 650000, Sovetsky av. 18, Russia

E-mail: kriger@kemsu.ru

Abstract. The dependencies of critical laser initiation energy density of pentaerythritol tetranitrate (PETN) – aluminum nanoparticles, PETN – cobalt nanoparticles and lead azide – lead nanoparticles on pulse duration were calculated in terms of the refined micro hot-spot model. It was shown that the absorption efficiency of the laser irradiation taken into account makes the initiation criterion change. According to the calculation results, the criterion in the limit of short pulses is energy density matching the experimental data. If the neodymium pulses duration is less than 50 ns, the radius of the nanoparticles with highest temperature varies insignificantly. The expression for the critical hot-spot temperature dependence on the pulse duration was derived. The conclusion was made that the model refining with nanoparticles absorption efficiency dependence on their radius is sufficient for the "small particles' paradox" solution.

1. Introduction

It is essential to change the contemporary used initiation systems to optical ones in order to achieve substantial increasing of the blasting operations safety in the mining industry. In [1,2], the optic detonator stuffed with the primary explosive silver azide was developed but it did not get wide application because of its significant sensitivity to shock and electrical influences. A selectively sensitive to laser initiation composite based on secondary explosive PETN and metal nanoparticles was suggested in [3]. The minimal critical energy density of this explosive compound laser pulse initiation (H) is about 0.7 J/cm^2 [4], which we consider as an encouraging mark on the way of the selectively sensitive optic detonator development. The main processes leading to the explosion initiation are absorption of the laser irradiation by metal nanoparticles, their heating and ignition of the surrounding matrix close to the contact surface nanoparticle – explosive [4,5]. The micro hot-spot model of the explosion initiation with laser irradiation was suggested [6] for the math description of the processes. One of the model assumptions is independence of the nanoparticle absorption efficiency on its size. The absorption cross section was thought to be equal to the geometrical cross section [6]. The critical energy density calculations with pulse duration (t_i) variation were done in terms of the model. In the case of nanoparticles' ensemble the radius of the mostly heated nanoparticle depends on the pulse duration as $t_i^{1/2}$. At the same time, when the pulse duration diminishes, the calculated critical energy density decreases and tends to zero [6]. The experimental dependence $H(t_i)$ measured for lead azide pellets in the limit of the short pulses (t_i<100 ns) does not depend significantly on the pulse duration [6], on the contrary. The theoretical trend $H \to 0$, when $t_i \to 0$ meaning that the optimal nanoparticles are of the small size, was referred to as "the small particles' paradox".

The "small particles' paradox" solution is essential for the physical substantiation of the micro hot-spot model. In [4,7], a modern variant of the model was proposed taking into account the dependence of the absorption efficiency (Q_{abs}) on the nanoparticles' radius (R) and the wavelength, as well as the scattering properties of the nanoparticles. The aim of the present work is to calculate the critical parameters of the lead azide – lead, PETN – Al and PETN – Co composites laser initiation on the pulse duration on the pulse duration in terms of the refined micro hot-spot model keeping in mind the "small particles' paradox" solution for the practically important explosives.

2. Micro hot-spot model

Let us consider the thermal explosion initiation model in the system comprising nanoparticles in the explosive material matrix heated with laser irradiation. The equations describing the conductive heat transport in the composite coupled with the chemical decomposition of the explosive material are as follows [4-7]:

$$\begin{aligned} &\frac{\partial T}{\partial t} = \alpha \cdot \left(\frac{\partial^2 T}{\partial x^2} + \frac{2}{x} \cdot \frac{\partial T}{\partial x} \right) + k_0 \frac{nQ}{c} \cdot \exp\left(-\frac{E}{k_B T} \right), \qquad x > R, \\ &\frac{\partial n}{\partial t} = -k_0 n \cdot \exp\left(-\frac{E}{k_B T} \right), \qquad x > R, \\ &\frac{\partial T}{\partial t} = \alpha_M \cdot \left(\frac{\partial^2 T}{\partial x^2} + \frac{2}{x} \cdot \frac{\partial T}{\partial x} \right), \quad \frac{\partial n}{\partial t} = 0, \quad n_0 = 0, \quad x < R, \end{aligned} \tag{1}$$

the boundary condition at the surface $x = R$ are

$$J - c_M \alpha_M \cdot \left. \frac{\partial T}{\partial x} \right|_{x \to R-0} + c\alpha \cdot \left. \frac{\partial T}{\partial x} \right|_{x \to R+0} = 0 \tag{2}$$

where T is temperature, E is decomposition activation energy, and k_0 is preexponential factor, Q is heat released with the decomposition of a cubic centimetre of the explosive, α and α_M are thermal diffusivities and nanoparticle materials, R is nanoparticle's radius, c and c_M are the volumetric heat capacity of the matrix and inclusion material, n is the share of undecomposed explosive, and $J(t)$ is the absorbed density of the laser pulse radiation power. For the simulation, the following parameters were used: $c = 2.22$ J/(cm^{3}K), $\alpha = 1.1 \cdot 10\text{-}3$ cm^{2}s^{-1}, $E = 165$ kJ/mole, $k_0 = 1.2 \cdot 10^{16}$ s^{-1}, $Q = 9.64$ kJ/cm^3 for PETN; E = 152 kJ/mole, Q =7.42 kJ/cm^3, k_0 = $1 \cdot 10^{13}$ s^{-1}, c = 2.48 J/(cm^{3}K) for lead azide; $c_M = 3.74$ J/(cm$^3 \cdot$K), $\alpha_M = 0.267$ cm^{2}s^{-1} for cobalt; $c_M = 1.46$ J/(cm$^3 \cdot$K)), $\alpha_M = 0.241$ cm^{2}s^{-1} for lead and $c_M = 2.43$ J/(cm$^3 \cdot$K), $\alpha_M = 0.973$ cm^{2}s^{-1} [4-8]. The dependence of the laser pulse radiation power on time is close to the normal distribution function [9]. Taking the maximum intensity of the pulse as a reference time position, we obtain for value $J(t)$ the expression [10]:

$$J(t) = \sqrt{\pi} \cdot Q_{abs} R^2 k_i H_0 \cdot \exp\left(-k_i^2 t^2\right) \tag{3}$$

where k_i is the effective constant corresponding to the pulse full width at half maximum t_i as $k_i = 2(\ln 2)^{1/2} t_i^{-1}$, Q_{abs} is nanoparticle absorption efficiency equal to the ration of the absorption cross section and the geometrical cross section (πR^2). The additional multiplicative constants in equation (3) are sued to norm the integral of $J(t)$ over entire time axis to the value $Q_{abs}\,H_0$.

In the initial variant of the laser initiation micro hot-spot model, it was assumed that Q_{abs} =1 and it does not depend on the nanoparticles' size. In the refined model, we used Mie theory for the spherical nanoparticle absorption efficiency with radius R calculation [7,11] that gave us the $Q_{abs}(R)$ dependence. The critical parameters of the explosion initiation were determined solving the model equations (1)-(3) on the variable step mesh according to the approach thoroughly described elsewhere [5]. The ordinary differential equation set obtained after the dividing of the space into cells was solved by the Runge — Kutta method of 1–5 orders with a variable time step. A relative error at the integration step did not exceed 10^{-12}, whereas the integral relative error estimated by the precision of the energy conservation law fulfilment did not exceed $2.5 \cdot 10^{-6}$.

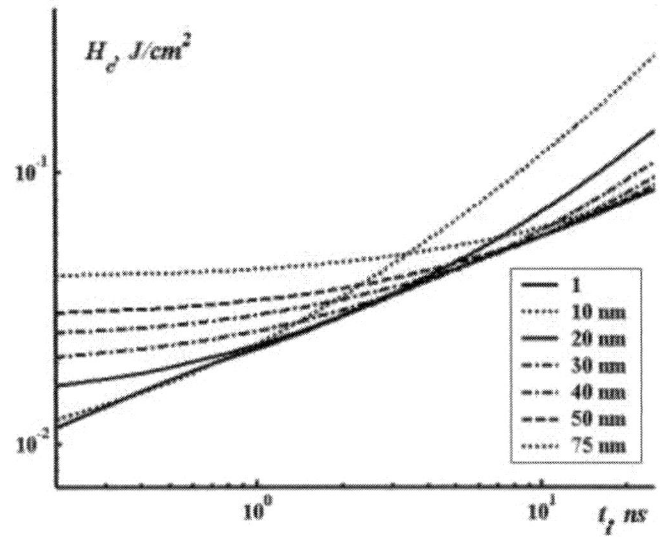

Figure 1. The calculated dependencies of the PETN-Co laser initiation critical energy density on pulse duration assuming that Q_{abs} =1. The Co nanoparticles' radii for each curve are listed on the legend, 1 is the envelop curve.

3. Results and discussion

Figure 1 shows the calculated dependencies of the critical (minimal) energy density of PETN – cobalt nanoparticles composite laser initiation (H) assuming Q_{abs}=1. The pulse duration was varied in the range 0.1 – 50 ns while the curves were calculated for the radii values 10, 20, 30, 40, 50 and 75 nm. At each pulse duration there is a particle with the optimal radius (R_{min}). Its heating to the maximal temperature leads to the self-accelerating reaction hot-spot formation at the minimal laser pulse energy density (H). In the case of nanoparticles ensemble with continuous size distribution, at every pulse duration there is a minimal critical energy density for the entire ensemble $H(t_i)$ presented in figure 1, curve 1. This curve is an envelope curve in the geometrical sense for the dependencies calculated at different radius values. Matching the results of the paper [6] one sees that as the pulse duration decreases the critical energy density of the explosion initiation for the nanoparticles of different size ensemble (envelope) also decreases with the trend $H \to 0$ in the limit $t_i \to 0$. Ignoring the objective laws of the light absorption gives 17-fold critical energy density decrease (from 154 to 8.9 mJ/cm^2) with the pulse duration decreasing from 100 to 0.1 ns. The R_{min} decreases from 140 to 5 nm (almost 30-fold decrease) which is significantly more pronounced than the corresponding diminishing of the critical energy density in this range of pulse durations. The possible pulse duration diminishing gives, obviously, the following decrease of the critical energy density of the explosion initiation.

The dependence of the light absorption efficiency of cobalt nanoparticles in PETN matrix (refractive index of PETN is 1.54) at a wavelength of 1064 nm was calculated using Mie theory in [8]. The applicability of Mie theory for the metal nanoparticles absorption efficiency calculation was proven in [11]. The maximum in $Q_{abs}(R)$ curve with height 1.26 was observed at the radius value

R_m=98 nm. At the radii range $R>R_m$, the absorption efficiency slightly decreases with oscillations. The $Q_{abs}(R)$ dependence is similar to that calculated for the nanoparticles of aluminum, silver, copper, nickel and some other metals in PETN [7, 11-15].

Let us consider the influence of the absorption efficiency dependence on the nanoparticles radius in the limit of short laser pulses. The corresponding calculated dependencies of the laser initiation explosion critical energy density for the cobalt nanoparticles radii 10, 20, 30, 40, 50 and 75 nm on the pulse duration is shown in figure 2. The relative positions of the curves calculated for different radii are affected by both the absorption efficiency and pulse duration. The absorption efficiency value makes the curves calculated for different radii to shifts of the curves upward if the $Q_{abs}(R)<1$ and downward in the opposite case. As a consequence, the envelope curve $H(t_i)$, which has the same meaning as in figure 1, tends to the positive constant in the limit of short pulses. The minimal critical energy density decreases only 3-fold with pulse duration decreasing from 100 to 0.1 ns, giving contrast with the case $Q_{abs}(R)$=1. The radius of the most heated nanoparticle keeps almost constant equal to 85 nm governed by the dependence $Q_{abs}(R)$ in the limit of short pulses. Thus, the absorption efficiency dependence on the nanoparticles radius in the limit of short pulses introduced into the model is sufficient to solve the "small particles' paradox". In the opposite limit of pulses with high durations, the nanoparticles with relatively higher radii are heated most. In this limit, the absorption efficiency dependence on the radius is weak, so the critical energy density of the laser pulse dependence on the pulse duration is close to $t_i^{0.4}$.

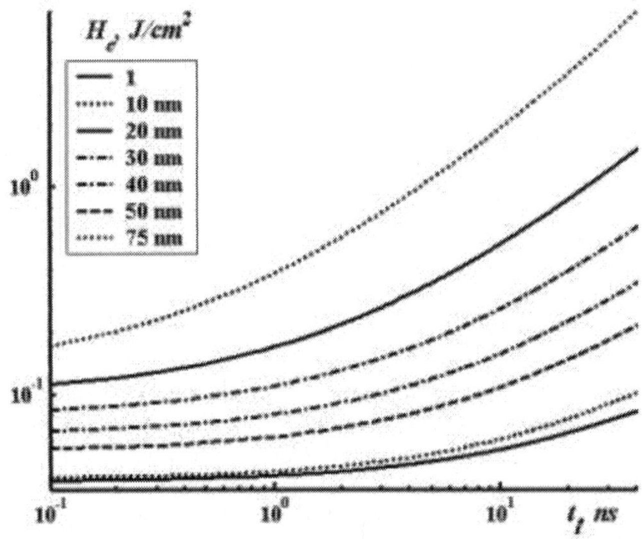

Figure 2. The calculated dependencies of the PETN-Co laser initiation critical energy density on pulse duration with $Q_{abs}(R)$ calculated in terms of Mie theory. The Co nanoparticles' radii for each curve are listed on the legend, 1 is the envelop curve.

The critical hot-spot temperature is another crucial parameter along with critical energy density, which dependence on the pulse duration may give an insight in the critical conditions of the self-accelerating reaction initiation. For that reason we performed the calculations of the hot-spot temperature on pulse duration at the critical energy density values for the aluminium nanoparticles with R=100 nm in PETN matrix. Anticipating the analytical expression for that dependence, the model (1)-(3) reduction to the Nikolay Semenov's model of thermal explosion [16] was done. Let us consider a metal nanoparticle and a surrounding heated layer with thickness h assuming that they have the same temperature as a 'reactor'. The matrix contacting this heated layer could be considered as thermal reservoir for the system cooling.

In terms of these assumptions, the effective heat released after the total explosive consumption inside the 'reactor' is

$$\tilde{Q} = \frac{V}{CV + C_m V_m} \cdot Q, \tag{4}$$

where V is volume of the explosive's heated layer with thickness h, V_m is the nanoparticle volume. The denominator in (4) corresponds to the 'reactors' effective heat capacity. The 'reactors' temperature augmentation concerned on the explosive decomposition is read as

$$\frac{dT}{dt}_+ = \frac{V}{CV + C_m V_m} \cdot Qk_0 \exp\left(-\frac{E}{k_B T}\right) \approx \frac{4\pi R h (R+h)}{CV + C_m V_m} \cdot Qk_0 \exp\left(-\frac{E}{kT}\right). \tag{5}$$

The heat transferred from the 'reactor' to the matrix in a time unit is equal to the heat integrated flux on its boundary. The temperature gradient could be estimated as the difference of the 'reactor' temperature and that of the medium divided by the heated layer thickness. Thus, the temperature diminishing term concerned on the heat conduction is as follows

$$\frac{dT}{dt}_- = \alpha C \cdot \frac{T - T_0}{h} \cdot \frac{4\pi (R+h)^2}{CV + C_m V_m}. \tag{6}$$

Now the task (1)-(3) could be approximately reduced to the Semenov's thermal explosion model using equations (5) and (6) as

$$\frac{dT}{dt} = D \cdot \left[\exp\left(-\frac{E}{k_B T}\right) - \frac{\alpha C}{k_0 Q} \cdot \frac{R+h}{Rh^2} \cdot (T - T_0)\right], \tag{7}$$

where $D = \frac{4\pi k_0 Q h R (R+h)}{CV + C_m V_m}$. The critical conditions of the reaction initiation for that task is

$$\begin{cases} \frac{dT^*}{dt} = D \cdot \left[\exp\left(-\frac{E}{k_B T^*}\right) - \frac{\alpha C}{k_0 Q} \cdot \frac{R+h}{Rh^2} \cdot (T^* - T_0)\right] = 0, \\ d^2 T^* / dt^2 > 0. \end{cases} \tag{8}$$

Where the first line in the equation set (8) we used to derive the dependence of the hot-spot critical temperature on the pulse duration introducing though h value. Assume that one knows the critical temperatures at one of the pulse duration values. Introducing the difference in critical hot-spot temperatures $\delta T = T^* - T_1$, applying the Frank-Kamenetskii approximation [16] and neglecting the small δT value in the second term, one arrives at the expression:

$$T^* = 2T_1 + \frac{k_B T_1^2}{E} \cdot F(h), \tag{9}$$

where $F(h) = \ln\left[\frac{\alpha C}{k_0 Q} \cdot \frac{R+h}{Rh^2} \cdot (T_1 - T_0)\right]$. The thickness of the reacting (heated) layer was estimated as $h = (2\alpha / k_i)^{0.5}$. The dependence of the critical hot-spot temperature calculated using the model (1)-(3) on the $F(h)$ function is shown in figure 3. The calculations with equation (9) were done using the preliminary calculated critical temperature at the pulse duration full width at half maximum 7 ns. The corresponding results are presented in figure 3 as well. The obtained slope coefficient and the intercept on the critical temperature axis match the linear approximation of the results calculated with the full model (1)-(3) with precision 2 and 4%, respectively.

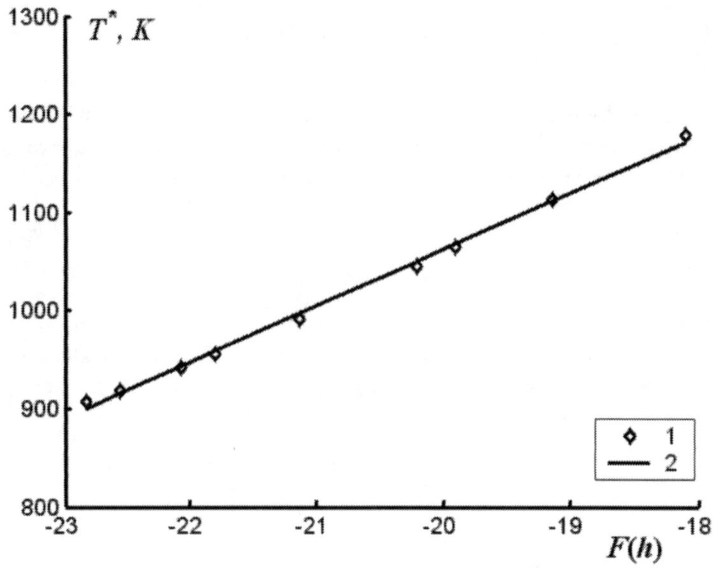

Figure 3. The dependence of the critical temperature on the pulse duration in the coordinates of equation (9).

4. Comparison with experimental data

The critical energy density values were experimentally measured for lead azide with pulse duration variation. Figure 4 shows the dependencies of the critical laser pulse energy densities on the pulse duration calculated in the case of lead azide containing lead nanoparticles. The calculations were done for the radii values 30 nm (1), 50 nm (2), 100 nm (3) and 300 nm (4) using the dependence $Q_{abs}(R)$ obtained previously using Mie theory. As in the case of the PETN-Co system, the absorption efficiency values affects the $H(t_i)$ curves positions shifting them upward when $Q_{abs}<1$. The values for the lead nanoparticles radii 10, 30 and 50 nm inside the lead azide matrix are 0.089, 0.333 and 0.755 respectively [7]. It is evident that critical energy density increases because of Q_{abs} influence for the nanoparticles' with radius 50 nm by 33% and by 200% (3-fold) for the nanoparticles with R=30 nm. As a consequence, the envelope curve $H(t_i)$ tends to the constant limit in the short pulses case. This theoretical result is the direct consequence of the light interaction with metal nanoparticles features. The experimental dependence corrected with the illuminance increasing factor 12 (as in [17]) is presented also in figure 4. One sees that when the pulse duration diminishes, the experimental critical energy density values decrease in the range t_i >200 ns, but tend to a constant limit in the range t_i<100 ns. The theoretical envelope curve for the nanoparticles ensemble shows the critical energy density decreasing with t_i diminishing the range of long pulses with trend to a limit in the case of short pulses matching the experimental data well.

In the lead nanoparticles in the lead azide matrix case, the radius of the nanoparticle that heated most in the limit $t_i \to 0$ is 64 nm at the wavelength 1064 nm. This value is slightly lower than the radius of the nanoparticle that absorbs laser irradiation most effectively. Thus, the hot-spot formation at the threshold conditions occurs on the nanoparticles with radius R_{min} that is not zero. At a wavelength of 1064 nm, the dependencies of the optimal nanoparticle radius calculated assuming Q_{abs} =1 and using the dependence $Q_{abs}(R)$ obtained with Mie theory cross at the pulse duration of 12 ns and R_m=74 nm.

At shorter pulses the optimal radius is determined by the optic properties of the nanoparticles, at longer – by the heat conduction effects mostly.

Decreasing of the absorption efficiency in the limit $R \rightarrow 0$ is typical for every metal. It means that the introduction of the absorption efficiency factor into the hot-spot model of laser initiation is sufficient for the “small particle’s paradox” solution.

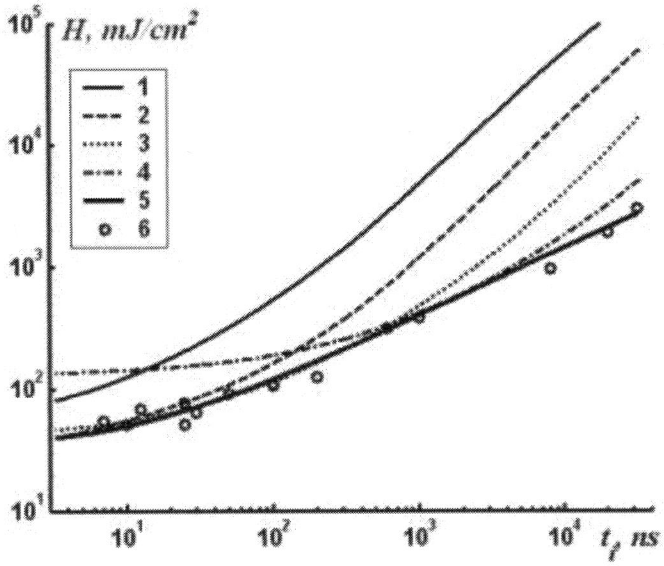

Figure 4. The dependence of the lead azide containing lead nanoparticles critical energy density of the laser pulse on the pulse duration calculated for nanoparticles’ radii 30 (1), 50 (2), 100 (3) and 300 nm (4) using $Q_{abs}(R)$ values obtained with Mie theory; 5 is envelop curve for the nanoparticles ensemble; 6 is experimental data [6] corrected with illuminance increasing factor 12.

5. Conclusion

In terms of the refined micro hot-spot model, the dependencies of critical laser initiation energy density of pentaerythritol tetranitrate (PETN) – aluminum nanoparticles, PETN – cobalt nanoparticles and lead azide – lead nanoparticles on pulse duration were calculated. It was shown that the absorption efficiency of the laser irradiation taken into account makes the initiation criterion change. According to the calculation results, the criterion in the limit of short pulses is energy density matching the experimental data. In the case of cobalt nanoparticles, the optimal nanoparticle radius in the short pulse limit tends to 85 nm. It was concluded that the model refining with nanoparticles absorption efficiency dependence on their radius is sufficient for the “small particles’ paradox” solution. This work was supported by Ministry of Education and Science of the Russian Federation (governmental project No. 2014/64).

References

[1] Kriger V G, Kalenskii A V, Zvekov A A, Anan’eva M V and Borovikova A P 2009 *Russ. J. Phys. Chem.* B **3** pp 636-40

[2] Kriger V G, Kalenskii A V, Zvekov A A and Tsipilev V P 2009 *Combust., Expl., Shock Waves* **45** pp 729-31

[3] Aduev B P, Belokurov G M, Nurmukhametov D R and Nelyubina N V 2012 *Combust., Expl., Shock Waves* **48** pp 361-66

[4] Kalenskii A V, Zvekov A A, Ananeva M V, Zykov I Yu, Kriger V G and Aduev B P 2014 *Combust., Expl., Shock Waves* **50** pp 333-8
[5] Kriger V G, Kalenskii A V, Zvekov A A, Zykov I Yu and Nikitin A P 2013 *Thermophysics Aeromechanics* **20** pp 367-74
[6] Alexandrov E I, Tsipilev V P 1984 *Combust., Expl., Shock Waves* **20** pp 690–4
[7] Kriger V G, Kalenskii A V, Zvekov A A, Zykov I Yu and Aduev B P 2012 *Combust., Expl., Shock Waves* **48** pp 705-8
[8] Ananyeva M V, Kalenskii A V, Zvekov A A, Nikitin A P and Zykov I Yu *2015 Nanosystems: Phys., Chem., Math.* **6** pp 628-36
[9] Kriger V G, Kalenskii A V and Zvekov A A 2010 *Combust., Expl., Shock Waves* **46** pp 60-63
[10] Odincova O V, Ivashenko G J 2015 Int. Sci. Period. 'Modern Fund. Appl. Research' **2** pp 43-8 [in Russian]
[11] Kalenskii A V, Zvekov A A, Nikitin A P, Anan'eva M V and Aduev B P 2015 *Opt. Spectr*. **118**, pp 978-87
[12] Kalenskii A V, Anan'eva M V, Zvekov A A and Zykov I Yu 2015 *Tech. Phys.* **60** pp 437-41
[13] Kalenskii A V, Zvekov A A, Nikitin A P and Anan'eva M V 2015 *Russ. Phys.* J. **58** pp 1098-104
[14] Aduev B P, Nurmukhametov D R, Liskov I Yu, Zvekov A A and Kalenskii A V 2015 Russ. *J. Phys. Chem.* B **9** pp 644-47
[15] Kalenskii A V, Zvekov A A, Nikitin A P and Gazenaur N V 2016 *Russ. Phys. J.* **59** pp 263-72
[16] The Mathematical Theory of Combustion and Explosions 1985 ed Ya Zeldovich (Dordrecht, Netherlands: Kluwer Academic Publishers Group)
[17] Burkina R S, Morozova E Yu and Tsipilev V P *Combust., Expl., Shock Waves* **47** pp 581-900

IOP Conf. Series: Journal of Physics: Conf. Series **830** (2017) 012129 doi:10.1088/1742-6596/830/1/012129

The study of inorganic scintillating materials

G N Dudkin, S I Kuznetsov, V N Padalko and M S Syrtanov

Tomsk Polytechnic University, Tomsk 634050, Russia

E-mail: smit@tpu.ru

Abstract. The procedure for measuring the temporal characteristics and light output of inorganic scintillating materials excited by β -, γ -, and α -particles from radioactive sources is described. Results of measurements of characteristics are presented for ~30 scintillating compounds including cerium-doped yttrium silicate and scandium borate; europium-doped strontium phosphate; cerium-doped strontium silicate, calcium silicate and magnesium calcium silicate, etc. Upon β - and γ -excitation, cerium-doped scandium borate gives the highest light output with a fluorescent lifetime of 40 ± 4 ns. The highest light output for α -excitation was from cerium-doped yttrium aluminum perovskite, with a fluorescent lifetime of 29 ± 3 ns.

1. Introduction

Inorganic scintillating materials are widely used in detectors for diverse applications ranging from high energy physics to nuclear medicine [1-2]. Better understanding of various scintillation mechanisms has led to innovative new materials for both gamma-ray and neutron detection, and the concept of scintillator design and engineering has emerged, whereby materials are optimized according to the scintillation properties needed by specific applications [3]. Although certain incremental improvements may still be achieved in each property, the main challenge then becomes to find a scintillator material that has a correct combination of properties to meet the needs of a given application rather than to find a scintillator with a single outstanding characteristic.

Intensive efforts to find and study new substances with higher light outputs, shorter fluorescent lifetimes, higher transparencies, smaller radiation lengths, etc., still continue [4-11]. The fluorescence decay lifetime of a scintillating substance is usually determined by the statistical method of single-photon counting [12]. The results obtained by this technique are independent of the shape and dimensions of samples under investigation. Measurements of the light output from substances are carried out under conditions that allow comparisons with results from other samples. For this purpose, crystals are grown to have similar dimensions (~1 cm^3). Identical conditions of light collection on a photomultiplier tube are provided, and the relative light yield is determined by the peak height of the total absorption of γ-quanta of known energy [13,14]. However, it is much easier and less expensive to synthesize a new substance in a powdered form. Based on the results of preliminary tests, one may decide whether or not to grow crystals. Thus, in [14], relative scintillation efficiency and fluorescence decay time were studied using a pulsed (τ<1 ns) monochromatic (E_γ = 22 keV) X-ray beam from the Brookhaven National Laboratory synchrotron source. A powdered scintillating compound was placed in a quartz cuvette and its light output was measured by comparing its response to radiation to the response of a bismuth germanate (BGO) crystal allowing for light absorption in the powder and the

IOP Conf. Series: Journal of Physics: Conf. Series **830** (2017) 012129 doi:10.1088/1742-6596/830/1/012129

efficiency of interaction of X-rays with the material. As a result, approximately a hundred new inorganic compounds were investigated within a short period.

Using radioactive sources that are more readily available makes it possible to perform measurements with various types of ionizing radiation. In this work, we measured the characteristics of powdered inorganic scintillators using β-, γ-, and α-sources.

2. Measurement procedure

The samples used in the study were prepared by the deposition of a fractionated luminophore powder from aqueous silicate suspension on a 22 mm diameter glass substrate. The 8–10 μm thick luminophore layer made of a powder with a grain size of 2–3 μm provides, on the one hand, identical conditions of light transmission for different samples and elimination of light loss in the scintillator and, on the other hand, necessary efficiency of detection of γ-photons with energy of ~20 keV.

In the device operating in the statistical mode of single-photon counting (figure 1), sample 1 and radioactive source *3* (^{90}Sr, ^{238}Pu) are placed between two photomultiplier tubes.

One of them, PMT_1 is in optical contact with the sample and detects light pulses whose leading edges give the start mark for intervals measured. The stop signal is provided by the other photomultiplier, PMT_2.PMT-130 photomultipliers with a sensitivity range of 200–650 nm, and good single-electron counting characteristics were used for the measurements. Attenuation of light burst intensity required for single-photon counting on PMT_2 is accomplished with diaphragm 5 and by varying distance *L* from the sample. Since radioactive source 3 prevented the direct passage of light from the sample to PMT_2, mirror 2 and reflector 4 were applied to provide light collection.

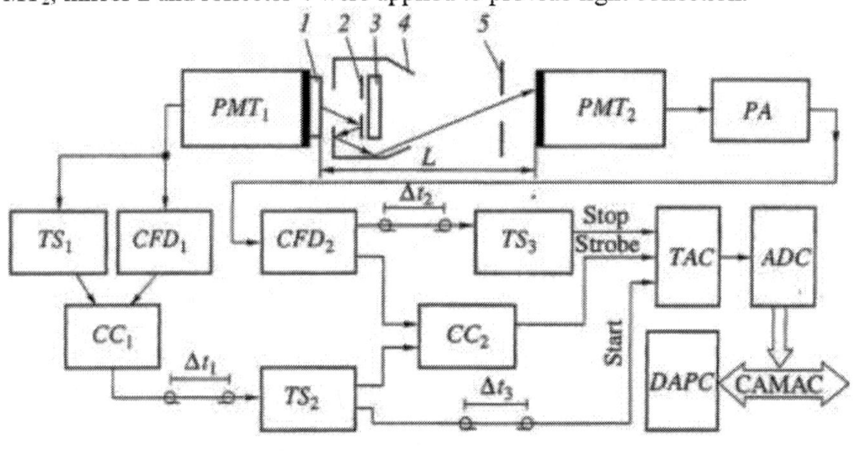

Figure 1. Diagram of the setup for measurements of fluorescence lifetime of scintillating materials. PA is the preamplifier; CFD_1 and CFD_2 are the riding-threshold shapers, TS_1 and TS_2 are the threshold shapers; CC_1 and CC_2 are the coincidence circuits; TAC is the time-to-amplitude converter; ADC is the analog-to-digital converter; DAPC is the data acquisition and processing complex; Δt_1, Δt_2, and Δt_3 are the matching delay lines; (1) sample, (2) mirror, (3) radiation source; (4) reflector, (5) diaphragm.

To form the start signal, a two-threshold detection system is used that includes a constant-fraction discriminator, threshold shaper (TS_1), and coincidence gate. The constant- fraction discriminator with a low (~2 photoelectrons) threshold accomplishes accurate timing of the instant of excitation of the sample. In the case of β-activation, the timing starts from the pulse front of the Cerenkov radiation produced in the glass substrate and determining the front of the overall light pulse. The threshold of shaper TS_1 is established at a level of ~0.5 of the signal peak height.

Pulses from PMT_2 (stop channel) after the amplification with a preamplifier pass to discriminator CFD_2 having the detection threshold in the "valley" of single electron pulse distribution. Relative efficiency of detection of scintillation flashes by this channel is established at a level of 1/20 by the ratio of counts from the coincidence circuits CC_1 and CC_2, thus providing the probability of <2.5% for detecting many-electron events.

The start and stop signals are transmitted to the time-to-amplitude converter (*TAC*) triggered by a strobe pulse from the coincidence circuit CC_2. Timing of signals is accomplished with cable delay lines. Threshold shapers TS_2 and TS_3 are used to restore the front and the amplitude of the start and stop signals, which are distorted by delay lines. Measurements on the luminophore-free glass substrate used as a Cerenkov light emitter gave the time resolution of the device as $\sigma = 0.68$ ns.

Fluorescent lifetime's τ_2 and $\tau_3$are determined as parameters of approximation of time spectrum by two exponentials (figure 2)

The total uncertainty of determination of τ_2 does not exceed 10%. In order to verify the procedure, we measured the fluorescence lifetime of the KS-425-1 commercial luminophore, which is a constituent of a type V-2 electrooptical transducer, upon excitation with monoenergetic electrons. To synchronize the instant of luminophore excitation with the start pulse, a silicon carbide light-emitting diode was used (flash duration <2 ns) which illuminates the photocathode of the electrooptical transducer.

The value measured in this manner agrees, within the limits of accuracy, with the τ_2 and τ_3 values obtained upon excitation with a ^{90}Sr β-source.

To measure the afterglow intensity *F* of a substance, an additional delay ($\Delta\tau_3$) of 5 μs was introduced to the start channel. Time spectra with and without the delay are built-up during the same period, and the *F* value is determined as the ratio of the number of events in these spectra within the interval τ_2 for the given substance. The accuracy of determination of *F* is ~20%.

The light yield from samples is measured by two methods. The first one is based on the use of a ^{90}Sr β-source. For β-particles, the ionization energy loss d*E*/d*x* in a thin layer of a scintillator depends mainly on its density. Relative light yield accurate to within ~20% is determined from the ratio of counts in the start and stop channels (introducing a correction for density of the material).

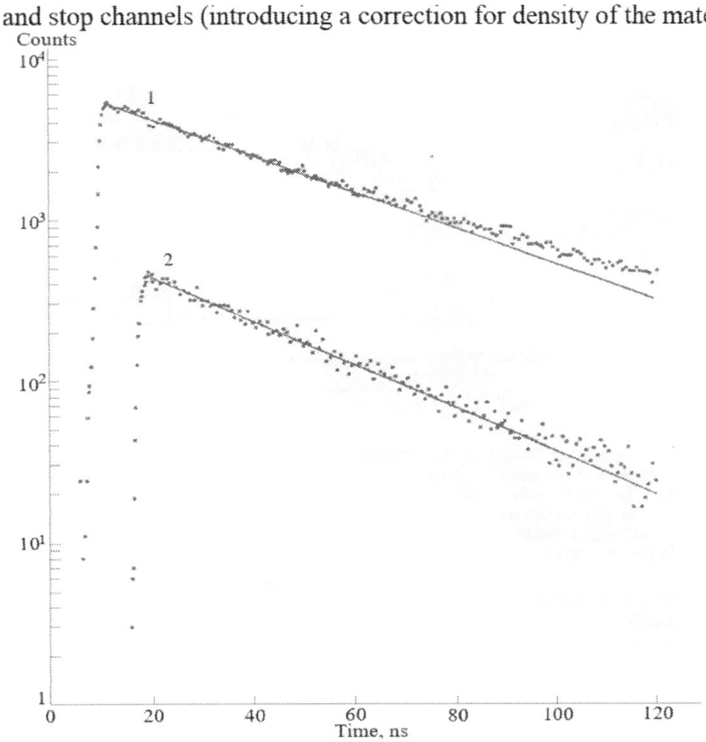

Figure 2. Temporal spectra of fluorescence decay of scintillating materials upon ^{90}Sr β-excitation: (1) KS-425-1 (two-exponential decay law), (2) and $YAlO_3$: Ce (one-exponential decay law).

IOP Conf. Series: Journal of Physics: Conf. Series **830** (2017) 012129 doi:10.1088/1742-6596/830/1/012129

More accurate determination of the light yield is performed with the use of γ-sources of low energies: ^{65}Zn (E_γ= 8.1 keV), ^{113}Sn (E_γ= 24,7keV), and ^{241}Am ($E_{\gamma 1}$ =13.2 keV, $E_{\gamma 2}$ = 17.8 keV, $E_{\gamma 3}$= 59.5 keV) [15]. The relative light yield for α-particles was determined by using a ^{238}Pu source. Because the range of α-particles from this source exceeds the thickness of scintillation screens under study, the mode of total energy absorption is achieved by varying the distance from the α-source to the luminophore so that to obtain the required energy of α-particles at the expense of ionization loss in air.

3. Results

The characteristics of ~30 types of scintillating substances were measured on the apparatus. Tables 1 and 2 list these characteristics for β- or γ-excitation and for α-excitation, respectively. Rows 2–22 in Table 1 and rows 2–8 in Table 2 present the data on luminophores prepared in the Mendeleev University of Chemical Technology (the wavelengths at the maxima of emission spectra were also measured there). The $YAlO_3$: Ce crystal and Scandium borate, produced at the Research Institute of Nuclear Problems of Belarussian State University (Minsk), was kindly presented by the Joint Institute for Nuclear Research (Dubna).

Table 1. Scintillating materials and their characteristics on β-excitation

№	Scintillatortype	λ_m, nm (color)	τ_2, ns τ_3, ns	L_{rel},%	F,%
1	KS-425-1 [Y_2SiO_5 : Ce (1.5 wt%)]	410	38±4 51±4	100	~0.03
2	Y_2SiO_5: Ce + BaF_2	410	39±4	86	0.03
3	Y_2SiO_5: Ce (2 wt %)	380	35±3	29	0.03
4	$2YPO_4$: Ce	365	33±3	5	0.6
5	$Y_{0.8}\cdot La_{0.1}\cdot Cd_{0.1}\cdot PO_4$: Ce	365	39±4	6	0.03
6	$SrO\cdot SiO_2$: Ce (0.04 mol %)	410	48±5	27	0.31
7	$SrO\cdot SiO_2$: Ce (0.04 mol %) with additional calcination at, t = 1250°C for 1 h	410	45±4	5	1
8	$Sr_3(PO_4)_2$: EU_2O_3 (0.6 wt%) t=100°C	425	375±35	89	—
9	$Sr_3(PO_4)_2$: EU_2O_3 (0.6 wt%) t=1200°C	425	395±40	89	—
10	$Sr_3(PO_4)_2$: EU_2O_3 (0.6 wt%) t=1300°C	425	405±40	83	—
11	$Sr_3(PO_4)_2$: EU_2O_3 (1.2 wt%) t =1300°C	425	389±40	44	—
12	3CaO $2SiO_2$: Ce (0.1 mol %)	410	39±4	22	0.3
13	3CaO $2SiO_2$: Ce (0.1 mol%) with additional calcination at, t= 1250°C for 1 h.	410	47±5	4	0.6
14	2CaO MgO· $2SiO_2$: Ce, calcination for 1.5 h, t=1250°C	380	48±5	4	2
15	2CaO MgO· $2SiO_2$: Ce, calcination for 3 h, t= 1250°C	380	47±5	4	0.7
16	$2CaAl_2O_3\cdot SiO_2$: Ce (0.015 mol %)	400	56±5	4	—
17	$CaZrO_3$: Ti	White	276±30	37	—
18	$BaZrO_3$: Ti	White	40±4 959±90	31	—
19	$ScBO_3$: Ce (0.8 mol %)	390	33±3	62	0.03
20	$ScBO_3$: Ce (2.5 mol %)	390	32±3	60	0.02
21	$ScBO_3$: Ce (1.5 mol %) + BaF_2 (0.3 mol %)	390	38±3	67*	0.03
22	$ScBO_3$: Ce (1.5 mol %) + BaF_2 (1 mol %)	390	39±4	75*	0.04
23	$ScBO_3$: Ce (Minsk)	393	40±4	156*	0.03
24	$YAlO_3$: Ce (3 x 3 x 0.1 mm^3 crystal) (Minsk)	347	32±3	98*	0.03
25	CsI(Tl) (3 x 3 x 0.6 mm^3 crystal) (Minsk)	430	710±50	100*	—

IOP Conf. Series: Journal of Physics: Conf. Series **830** (2017) 012129 doi:10.1088/1742-6596/830/1/012129

Table 2. Scintillating materials and their characteristics on α-excitation

№	Scintillator type	λ_m, nm (color)	τ_2, ns τ_3, ns	L_{rel},%	F,%
1	KS-425-1	410	29 ±3	100	0.03
2	Y_2SiO_5: Ce, t = 1350°C	410	28 ±4	69	0.03
3	Y_2SiO_5: Ce, t = 1350°C	410	30±3	116	0.03
4	Y_2SiO_5: Ce + NF_4, t = 1350°C	410	20 ± 2	69	0.02
5	Y_2SiO_5: Ce + BaF_2, t = 1350°C	410	26 ± 2	89	0.02
6	Y_2SiO_5: Ce + BaF_2, t = 1300°C	450	24 ±2	57	0.03
7	Y_2SiO_5: Ce + BaF_2,t = 1350°C,	White	21 ±2	121	0.02
8	$ZrGeO_4$: Ti·P	430	110 ± 10 500 ± 50	75	2
9	KO-425 (Y_2SiO_5: Ce, Gd)	410	45 ±4	72	0.04
10	$ScBO_3$: Ce (Minsk)	393	37 ±4	131	0.03
11	$YAlO_3$: Ce (Minsk)	347 [8]	29 ±3	216	2

4. Conclusion

An analysis of the results shows that the highest light yield upon β-excitation is provided by cerium-doped scandium borate (a sample from Minsk). Its light yield is higher than that of CsI(Tl). For α-excitation, the cerium-doped yttrium aluminum perovskite (YAP:Ce) crystal has unique light yield characteristics. For scintillators intended for the identification of charged particles, the α/β ratio is of practical importance. We determined this ratio for Y_2SiO_5: Ce, $YAlO_3$: Ce, and $ScBO_3$: Ce as 0.19, 0.46, and 0.16, respectively. Tables 1 and 2 also show the dependence of luminescent lifetimes of inorganic scintillators on the type of particles exciting the scintillations. This time is shorter for α- than for β-excitation; the difference is especially pronounced in the case of the KS-425-1 luminophor.

Acknowledgements

The research has been supported by the project VIU_NU_174_2014 in the context of the Competitiveness Improvement Program of the Tomsk Polytechnic University.

References

[1] Gostojić A, Tatischeff V, Kiener J et al 2016 NIM A **832** pp 24–42
[2] Paul Lecoq 2016 NIM A **809** pp 130–139
[3] Lisitsyna L A, Tupitsyna I A and Lisitsyn V M 2015 NIM B **365** pp 278–282
[4] Melcher C L 2005 NIM A **53** pp 6–14
[5] Pania R, Pellegrinia R, Cintia M N et al 2006 NIM A **569** pp 296–300
[6] Carel W E van Eijk *2008 Radiation Protection Dosimetry* **129** pp 13–21
[7] Jung Yeol Yeom, Seiichi Yamamoto, Stephen E Derenzo et al *2013 IEEE transactions on nuclear science* **60**
[8] Joanna Iwanowska, Lukasz Swiderski, Tomasz Szczesniak et al 2013 NIM A **712** pp 34–40
[9] Lucchini M T, Babin V, Bohacek P et al 2016 NIM **816** pp 176–183
[10] Korepanov V I, Lisitsyn V M and Oleshko V I. 2000 *Russian Physics Journal* **43**
[11] Lisitsyn V M, Stepanov S A, Valiev D T *et al* 2016 IOP Conference Series: Materials Science and Engineering **110** 012051
[12] Bollinger L M and Thomas GE 1961 Rev. *Sci. Instrum.* **32** 1044
[13] Moses W W and Derenzo S E 1989 *IEEE Trans. Nucl. Sci.* **36** 173
[14] Derenzo S E, Moses W W, Cahoon J L et al 1990 *IEEE Trans. Nucl. Sci.* **37** 203
[15] Firestone R B and Shirley V S Table of Isotopes, 8th ed. – Wiley New York 1998

IOP Conf. Series: Journal of Physics: Conf. Series **830** (2017) 012130 doi:10.1088/1742-6596/830/1/012130

Using Rouse-Fowler model to describe radiation-induced electrical conductivity of nanocomposite materials

N S Dyuryagina and A P Yalovets

Department of Computer Simulation and Nanotechnology, South-Ural State University (National Research University), 76 Lenin Prospect, Chelyabinsk, 434080, Russian Federation

E-mail: nsdiuriagina@susu.ru

Abstract. Using the Rouse-Fowler (RF) model this work studies the radiation-induced electrical conductivity of a polymer nanocomposite material with spherical nanoparticles against the intensity and exposure time of gamma-ray, concentration and size of nanoparticles. The research has found the energy distribution of localized statesinduced by nanoparticles. The studies were conducted on polymethylmethacrylate (PMMA) with CdS nanoparticles.

1. Introduction

Nanocomposite materials are materials formed by inclusion of nanoparticles into some matrix material. As a result, we can create a new functional material with unique electro physical properties. PMMA+CdS or CdSe nanocomposite allows creation of new types of photo galvanic and optoelectronic devices [1-2]. It is important to consider the possibility of using these devices under increased radiation (space, nuclear engineering, etc.). Therefore, the study of the nanocomposite radiation resistance is an important and relevant task [3]. Particular attention is given to spherical semiconductor nanoparticles such as CdS or CdSe due to the fact that their fluorescence band covers whole visible, near-IR and near-UV bands depending on the particle size. It is known that such size dependent properties are related to quantum confinement effects that are more pronounced with the smaller nanoparticle size. Thus nanocomposites with the nanoparticle size less than 10 nm possess the most interesting electrophysical properties. The maximal realization of electrophysical properties of CdS and CdSe nanoparticles requires their isolation from each other, that is why the nanoparticle concentration normally does not exceed 10 vol.%.

Using the Rouse-Fowler (RF) model, this work studies the radiation-induced electrical conductivity of the polymer nanocomposite exposed to gamma-rays. According to works [4-6], phenomena related to the radiation-induced electrical conductivity of polymers are best described by the Rouse-Fowler model. There are analytic and numerical solutions [7-11] for the model variations where either the spectrum of localization centers (traps) in the bandgap has only one or two states or the distribution of trap energy states follows the exponential law. Nanocomposite materials are characterized by existence of additional centers of localization in the bandgap. The energy distribution of these centers depends on the shape, size and concentration of nanoparticles. Thus the need to describe electrophysical properties of nanocomposites requires the RF model generalization for a certain energy distribution of localization centers in the bandgap.

2. Localization centers

It is known that dielectrics and semiconductors have intrinsic localized states (intrinsic traps) due to various structure defects [7]. Trap properties of many polymers are well understood. For example, such polymers as polymethylmethacrylate (PMMA) have the p-type conductivity and the trap energy distribution follows the exponential low:

$$N(E)=(N_0/kT_1)\exp[-E/kT_1],$$

where N_0 is the full concentration of traps, T_1 is the characteristic temperature of distribution, E isthe trap energy calculated from the valence band top.

Free charge carrier trapping on the energy level (E)is characterized by the capture coefficient $k_E=\upsilon\sigma_a(E)$, where υ is the free charge carrier velocity, $\sigma_a(E)$ is the capture cross-section on the relevant level [7].The thermal release time of trapped charge carriers follows the formula $\tau_E=\nu_0^{-1}\exp(E/kT)$, where ν_0 is the frequency factor. k_E,ν_0 values for a variety of polymers are estimated in works [7-11].

Due to differences in the Fermi levels of materials, the introduction of nanoparticles into matrix materials forms the potential well with additional localized centers just as impurities in semiconductors lead to formation of acceptor and donor states. Following this line of reasoning, localized centers induced by nanoparticles shall be designated as impurity localized centers (impurity traps).

Let us denote the Fermi energies and work functions of the matrix and the particle by $F_M, W_M, F_{pat}, W_{pat}$, respectively. When $F_M < F_{pat}$ ($W_M > W_{pat}$), electrons flow from the nanoparticle to the matrix tending to the energy minimum. As a result, the nanoparticleacquires positive charge $q_{pat}=C_{pat}U_0$, where C_{pat} is the nanoparticle capacitance, $U_0=W_M-W_{pat}$. Electrons held by the field of the positively charged nanoparticle become localized in the potential wellthe depth of which is $-U_0$ from the conductivity band bottom of the matrix. These inclusions are of donor-type. In the opposite case, when $F_M > F_{pat}$ ($W_M < W_{pat}$), the hole potential well forms with the depth U_0 measured from the valance band top. These inclusions are of acceptor-type.

To find the energy distribution of traps it is necessary to solve the steady Schrodinger equation for subatomic particles with the effective mass in the inclusion field. For definiteness let us consider the donor-type inclusion. The potential energy can be written as:

$$U(r)=-U_0 \text{ at } r\le R_{pat} \text{ and } U(r)=-(U_0 R_{pat}/r_D)\cdot\exp\left[-(r-R_{pat})/r_d\right] \text{ at } r>R_{pat}.$$

Here r_D isthe Debye shielding radius at the particle, which is determined so that the number of localized centers is not less than $q_{pat}/2e$ including spin. The Schrodinger equation is solved in spherical coordinates for two areas: $r\le R_{pat}$ and $r>R_{pat}$. Since the problem is one-dimensional, the WKB-method is used to solve the wave equation. For the acceptor inclusion the trap spectrum is sought the same way.

The frequency factor for impurity traps can be estimated in the form: $\nu_0=A_\nu\upsilon_E/2r_E$, where A_ν is the probability of absorbing a phonon by an electron, υ_E is the average velocity of a trapped charge carrier, r_E is the radius of the potential well with the given energy found from the terms: $U(r_E)=E$.

Let us consider the nanocomposite based on PMMA ($W_{PMMA}=4{,}68$ eV) with CdS nanoparticles ($W_{CdS}=6{,}05$ eV). In this nanocomposite, the potential well depth for holes is $U_0=1{,}37$ eV.

Increase of the nanoparticles size leads to increase in the number of energy levels in the potential well and the proportion of deep traps with the large lifetime ($\tau_E\ge10^4$ s). Charge carrier release from deep traps by the thermal motion at $T=300$ K is not possible, but under external exposure, such as ionizing radiation or nanocomposite heating, the deep trap influence on the electrical conductivity increases.

It is important to note that compared to intrinsic localized centers the energy spectrum of impurity centers belongs to a certain particle. Thus, holes trapped on impurity centers and tending to the energy minimum always tend to go to a deeper energy level if it has a vacancy.

3. Ionization of nanocomposite material

To describe the radiation-induced electrical conductivity of materials it is necessary to set the volumetric generation rate of charge carrier pairs (S).

The volumetric generation rates in the matrix and inclusions can be written in the form:

$$S = (1 - f_{pat})\rho D/w_i, \qquad S_{pat} = f_{pat} \cdot (\rho D/w_i)_{pat},$$

where ρ is the mass density of matter, D is the radiation energy absorbed by unit mass of substance per unit time, f_{pat} is the volume ration, w_i is the average energy of electron-hole pair generation, the values with the *pat* index in the brackets are defined for nanoparticles. Generally, for a solid dielectric the value w_i makes several hundreds of eV [7] to exclude pair generation with the small recombination time (less than 10^{-12} s) from consideration.

The charge carrier generation rate in the matrix (S)determines the generation speed of electrons in the conduction band and holes in the valance band. As a rule, ionization of a nanometer-sized inclusion is accompanied by secondary electron removal from it. Such a conclusion can be based on the experimental and theoretic studies of the attenuation length (the distance at which an electron flux is weakened by **e** times) of low energy electrons in different materials [12]. For metals, where the free electron concentration is large and, therefore, the share of electron-electron interaction is large too, the minimal attenuation length makes 3-4 nm and is observed in the range of the secondary electrons energy 10-30 eV. The one in dielectrics is substantially greaterdue to a low probability of inelastic collision between a low energy electron and bound electrons.

As an inclusion is a localization center, electron release from this center leads to occupancy change of the localized state.

In case of the donor inclusion, electron removal leads to increase of its positive charge that results in electron capture by this inclusion from its impurity trap. Thus, ionization of the donor inclusion forms S_{pat} of free electrons and the same number of vacancies on impurity electron states in unit volume per unit time.

In case of the acceptor inclusion, electron removal from it leads to electron transfer with the minimal energy from the matrix to the inclusion, that is hole generation with the energy close to the valence bandtop. Thus, ionization of the acceptor inclusion forms S_{pat} of free electrons and the same number of holes localized on high states in unit volume per unit time.

The dose rates(D) absorbed by materials of the matrix and the inclusion werecalculated for exposure by deceleration radiationquanta generated by electrons with the initial energy of 1 MeV in the wolfram absorber with the thickness of 260 μm. Transfer of electrons is calculated by the method [13]. Using the attenuation coefficient tables [14],transfer of deceleration quanta in the absorber is described by the kinetic equation excluding scattering. At the current density of the electron beam 0.1 kA/cm^2 in the PMMA layer located behind the wolfram target, the absorbed dose rate near the contact boundary is $D_{PMMA} = 1.9 \cdot 10^8$ W/kg. For the CdS layer, it makes $D_{CdS} = 5.58 \cdot 10^8$ W/kg. The consideredirradiation mode at the average energy per ion pair formed $w_i = 100$ eV and $f_{pat} = 0.05$ the volumetric generation rates are $S_{PMMA} = 1.29 \cdot 10^{28}$ m^{-3}s^{-1} and $S_{CdS} = 8.4 \cdot 10^{27}$ m^{-3}s^{-1}. Thus, CdS nanoparticle ionization contributes to PMMA+CdS nanocomposite ionization even more than matrix ionization.

4. The Rouse-Fowler equation system for nanocomposite material

As at this stage of the work only the local conductivity is interesting to us. We assume that ionization is homogeneous in volume, loading effects [6] are not significant and drift of charge carriers can be neglected.It should be noted, that the considered here model and method of its realization are easily generalized to heterogeneous irradiation, loading and drift of charge carriers, and a multidimensional case.

Let us denote the concentrations of free electrons and holes by n, p; n_m, p_l – the concentrations of electrons and holes localized by traps, where $1 \le m \le m_0$, $1 \le l \le l_0$, m_0, l_0 – the numbers of electron and hole states. Let states $1 \le m \le \tilde{m}_0$, $1 \le l \le \tilde{l}_0$ be related to localized states of the matrix (intrinsic localized traps) and $\tilde{m}_0 + 1 \le m \le m_0$, $\tilde{l}_0 + 1 \le l \le l_0$ be related to impurity traps. The other denotations: α – the recombination coefficient, $K_m = k_m(M_m - n_m)$, $K_l = k_l(P_l - p_l)$, k_m, k_l – the capture coefficients on electron and hole traps, M_m, L_l – the concentrations of electron and hole traps, $\tau_{m,l} = \exp(E_{m,l}/kT)/\nu_0$ –the life times of charge carriers on electron and hole traps with the energy $E_{m,l}$,respectively [7].

The equation system describing the dynamics of free and trapped charge carriers has the forms:

$$\dot{n} = S + S_{pat} - \alpha n\left[p + \sum_{l=1}^{l_0} p_l\right] - n\sum_{m=1}^{m_0} K_m + \sum_{m=1}^{m_0} \frac{n_m}{\tau_m}; \quad (1)$$

$$\dot{p} = S + \left(S_{pat} - S'_{pat}\right) - \alpha p\left[n + \sum_{m=1}^{m_0} n_m\right] - p\sum_{l=1}^{l_0} K_l + \sum_{l=1}^{l_0} \frac{p_l}{\tau_l}; \quad (2)$$

$$\dot{n}_m = K_m n - \frac{n_m}{\tau_m} - \alpha p n_m - S'_{pat}\,\delta_{m,m_0}, \quad 1 \le m \le m_0; \quad (3)$$

$$\dot{p}_l = K_l p - \frac{p_l}{\tau_l} - \alpha n p_l + \left(S_{pat} - S'_{pat}\right)\delta_{l,\tilde{l}_0+1}, \quad 1 \le l \le l_0, \quad (4)$$

where $S'_{pat} = S_{pat}$ for donor inclusions, $S'_{pat} = 0$ for acceptor ones.

At the initial time moment, the concentrations of charge carriers have the equilibrium (without irradiation) values $n^0, p^0, n_m^0\ p_l^0$. They can be found by the solving system (1)-(4) with the zero initial conditions for the concentrations of free and trapped charge carriers and without vacancies on impurity traps at $S = S_T$, $S_{pat} = 0$, where S_T is the thermal ionization rate.

The general idea of the system solving (1)-(4) is finding the approximate analytical solution for a certain time interval $\Delta t = t_{k+1} - t_k$ that is chosen to minimize bias of the found analytical solution.

5. Test calculations

The system solving (1)-(4) suggested in this work was tested by comparing our results with the results of the analytic solution of the problem of the dielectric radiation-induced conductivity for the two-trapped model (the Van-Lint model) [7], and with the numerical results for the exponential distribution of traps [9].

Our results fully coincide with the Van-Lint model results, and this fact indicates the adequacy of our model for two traps.

For the exponential distribution of traps, we compared our results with those reported in [9] (figure 1) at the following parameters: $T = 300$ K; $N_0 = 10^{25}$ m^{-3}; $k = 10^{-16}$ m^3/s; $\alpha = 10^{-14}$ m^3/s; $\nu_0 = 10^4$ s^{-1}; $S = 6{,}25 \cdot 10^{19}$ m^{-3}s^{-1}at continuous irradiation. Here, as well as in [9], we consider the dielectric material, which has the *p*-type conductivity with the hole mobility $\mu_p = 10^{-6}$ m^2/V·s. It is clear that the obtained results for the intrinsic radiation-induced conductivity of the polymer coincide well with the results [9].

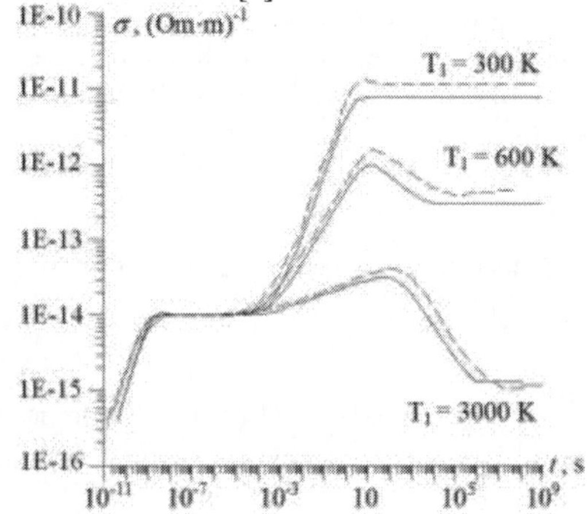

Figure 1. Radiation-induced conductivity $\sigma\,(\mathrm{Om}^{-1}\mathrm{m}^{-1})$ of the polymer matrix with the exponential distribution of traps and various distribution parameters T_1 (K). " " indicates the results from the work [9]; " " plots our results.

6. Radiation-induced conductivity of nanocomposite material

The calculation of the radiation-induced conductivity for pure PMMA and PMMA with CdS nanoparticles against their size, concentration and absorption dose rate is shown in figure 2.

Intrinsic localized centers of PMMA have the exponential distribution with the distribution parameter T_1 = 2150 K at room temperature [11]. The full concentration of intrinsic localized centers is 10^{25} m^{-3}.

The dark conductivity of pure PMMA does not exceed 10^{-20} (Om·m)$^{-1}$.The nanoparticle inclusion leads to formation of additional centers and increase of the nanocomposite dark conductivity. The resulting equilibrium condition is used as initial for the calculation of the radiation-induced conductivity.

During irradiation, when a process of generation dominates over processes of capture and recombination, the radiation-induced conductivity increases to a certain value. This value is kept constant until the end of irradiation by the equilibrium between processes of generation, recombination and capture on traps. After irradiation, the equilibrium is disturbed and active capture of charge carriers by localized centers takes place. As a result, the conductivity drops drastically by four orders. Further behavior of the conductivity is caused by release of localized charge carriers from traps and recombination of these charge carriers.

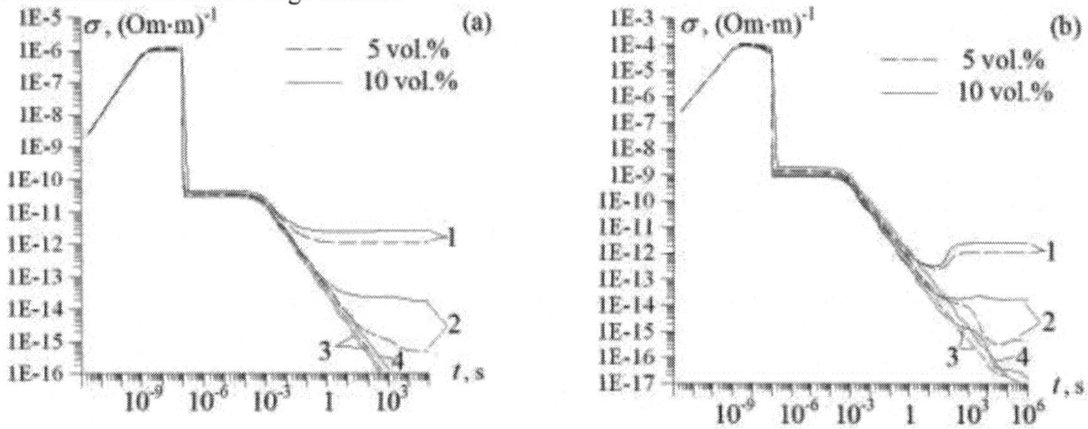

Figure 2. Radiation-induced conductivity of PMMA+CdS nanocomposite with the CdS radius: 2 nm (1); 5 nm (2); 10 nm (3); pure PMMA (4). The absorbed dose rate of the matrix is $D = 10^8$ (a) and 10^{10} W/kg (b). The duration pulse is $\tau_{rad} = 100$ ns.

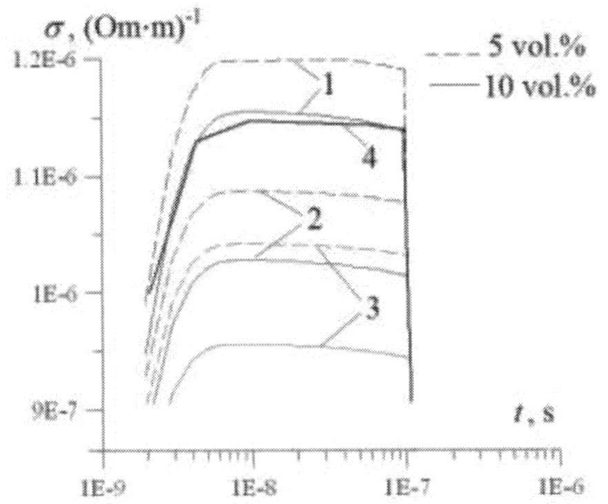

Figure 3. Radiation-induced conductivity of PMMA+CdS nanocomposite during irradiation $t < \tau_{rad} = 10^{-7}$ s against the size and concentration of CdS. The CdS radius is 2 nm (1);5 nm (2);10 nm (3); pure PMMA (4). The absorbed dose rate is $D = 10^8$ W/kg.

During irradiation the conductivity of nanocomposite with CdS nanoparticles with 2 nm in radius exceeds the value of pure PMMA (figure 3). While for nanoparticles with 5 or 10 nm in radius, the electrical conductivity is less than the one of pure PMMA. It is caused by growth of the capture cross-section on impurity centers with increase of the nanoparticles size. Because inclusion of nanoparticles leads to formation of impurity centers, the nanocomposite conductivity decreases with increase of the nanoparticle concentration.

After irradiation,the conductivity returns to its original state of the equilibrium and the relaxation rate depends on the nanoparticle size. Decrease of the relaxation rate with increase of thenanoparticle radius is caused by increase in the number of deep traps. For CdS nanoparticles with 2 nm in radius at $\dot{D}_y \le 10^8$ W/kg, the relaxation time does not exceed 1 s, but for the nanoparticle radius of 5 and 10 nm, it reaches 10^4 s.

7. Conclusion

The test calculations have demonstrated that the suggested method is efficient for solving the Rouse-Fowler equation system for a certain spectrum of traps.

The developed model of nanocomposite ionization considers both matrix ionization and inclusion ionization. Our research has shown that ionization of inclusions changes the concentration of free charge carriers and the occupation of energy states.

Inclusion of CdS nanoparticles into PMMA leads to generation of additional charge carriers and increase of the nanocomposite dark conductivity. The number of impurity centers in PMMA+CdS nanocomposite grows with increase of the nanoparticle size, at that the number of deep traps, charge carrier release from which is possible only under external exposure (ionization, heating), grows too.

During irradiation, the radiation-induced conductivity of nanocomposite depends on the size and concentration of nanoparticles. For CdS nanoparticles with R_{pat} = 2 nm, the electrical conductivity exceeds the value of pure PMMA. While for CdS nanoparticles with $R_{pat} > 5$ nm, the electrical conductivity is less than the one of pure PMMA caused by growth of the capture cross-section on impurity centers with increase of the nanoparticle size.

After irradiation, the conductivity returns to its original state of the equilibrium and the relaxation rate depends on the size and concentration of inclusions. The equilibrium time for the nanocomposite conductivity grows with increase of the nanoparticle radius due to increase in the number of deep traps.

Acknowledgments
This study was supported in part by the Ministry of Education and Science of the Russian Federation (research work of South-Ural State University (Research University) under state assignment no. 1030 (no. 2014105-GZ).

References

[1] Popov V K, Bagratashvili V N, Krotova L I, Rybaltovskii A O, Smith D C, Timashev P S, Yang J, Zavorotnyi Yu S and Howdle S M 2011 A route to diffusion embedding of CdSe/CdS quantum dots in fluoropolymer microparticles *Green Chemistry* **13** pp 2696-2700
[2] Tomczak N, Janczewski D, Han M and Vancso G J 2009 Designer polymer–quantum dot architectures *Progress in Polymer Science* **34** pp 393-430
[3] Davidyuk G Ye, Bozhko V V, Mironchuk G L, Bulatetskaya L V and Kevshin A G 2008 Special features of optical and photoelectric properties of nominally undoped and Cu-doped CdS single crystals *Russian Semiconductors* **42** pp 399-403
[4] Byub R 1962 *Photoconductivity of solids* (Moscow: Foreign Languages Publishing House) p 558
[5] Rose A 1966 *Fundamentals of the photoconduction theory* (Moscow, Mir) p 92
[6] Sessler G 1983 *Electrets* (Moscow: Mir) p 487
[7] Vannikov A V, Matveev V K, Sichkar V K and Tyutnev A P 1982 *Radiation effects in polymers. Electrical properties* (Moscow, Nauka) p 273
[8] Tyutnev A P, Vannikov A V and Minigaleev G S *1989 Radiation electrophysics of organic dielectrics* (Moscow, Energoatomizdat) p 192
[9] Tyutnev A P, Sadovnichii D N and Boev S G 1995 Numerical analyses of the Rose-Fowler model *High Energy Chemistry* **29** pp 115-119
[10] Tyutnev A P, Saenko V S, Pozhidaev E D and Ikhsanov R 2015 Experimental and theoretical studies of radiation-Induced conductivity in spacecraft polymers *IEEE Transaction on Plasma Science* **43** pp 2915-2924
[11] Vaisberg S E, Sichkar V P and Karpov V L 1971 Investigations of electrical conductivity of polymers *Russian Polymer Science* A **13** pp 2502-7
[12] Kovalev V P 1987 *Secondary electrons* (Moscow: Energotomizdat) p 177
[13] Kononov B A, Stepanov Yu M and Yalovets A P 1977 Transfer of fast electrons in layered materials *Atomic Energy* **42** pp 326–328
[14] Nemets O F and Gofman Yu V 1975 Nuclear physics reference book (Kiev, Naukova Dumka) p 416

The Effect of Electric Field on the Explosive Sensitivity of Silver Azide

A P Rodzevich[1a], E G Gazenaur[2b], L V Kuzmina[2c], V I Krasheninin[2c], N V Gazenaur[2d]

[1]Yurga Institute of Technology, TPU Affiliate, 26, Leningradskaya str., Yurga, 652050, Russia
[2]Kemerovo State University, 6, Krasnaya str., Kemerovo, 650000, Russia

[a]nimez@tpu.ru, [b]gazenaur@kemsu.ru, [c]specproc@kemsu.ru, [d]nikhasenauer@gmail.com

Abstract. The effect of a constant contactless electric field on the rate of a chemical reaction in silver azide is explored in this paper. The technology of growing and processing silver azide whiskers in the constant contactless electric field (field intensity was varied in the range from 10^{-3} V/m to 10^{0} V/m) allows supervising their explosive sensitivity, therefore, the results of experiments can be relevant for purposeful controlling the resistance of explosive materials. This paper is one of the first attempts to develop efficient methods to affect the explosive sensitivity of energy-related materials in a weak electric field (up to 10^{-3} V/m).

1. Introduction

Improvement of material resistance to the unsupervised outer factors is one of the most important problems in the present day materials science. This issue is particularly urgent for advancing the resistance of energy-related materials.

Silver azide is a traditional object of research into solid phase chemical reactions. Physical and chemical properties of silver azide are studied quite well [1]. Both slow and explosive decomposition are possible in this material after energy deposition. Silver azide crystals have a great practical value because they can be used as explosive materials with high initiating capacity [1, 2]. Silver azide is used in small-scale primers, in production of explosive materials, commercial cumulative charges, detonating cords, perforating systems, electric detonators, which are in use in oil and gas industries for perforating and pumping-in well sites, as well as in development geophysics as sensors of electromagnetic fields.

The outer energy deposition causes the decomposition process in silver azide crystals, including generation of non-balanced electrons and holes, their transfer to the crystal surface into reaction sites (RS) formed by edge dislocations and Cottrell atmosphere (containing positively charged particles), and a chain reaction in the reaction site as well [2-5].

We have found out in our research that the reactive capacity of silver azide crystals can be controlled efficiently via a contactless electric field [5-7].

The effect of electric field in the solid chemistry is traditionally considered in two aspects: injection and drift of charge carriers, disregarding the polarization. However, polarization depends on the spatial

IOP Conf. Series: Journal of Physics: Conf. Series **830** (2017) 012131 doi:10.1088/1742-6596/830/1/012131

shift of the electron intensity in chemical agents, and it must have the most significant influence on the rate of a chemical reaction.

This paper presents the outcomes of research how the constant contactless electric field (field of polarization) affects explosive decomposition, initiated in silver azide whiskers by the laser pulse or the contact electric field in conditions of mono-polar injection of holes. The possibility to use for these purposes a weak electric field (up to 10^{-3} V/m), simulating real conditions of storage and transport of explosive materials, makes it possible to affect the long-term resistance of energy-rated materials under the action of unsupervised outer factors.

2. Technique of experiment

Experiments required the use of silver azide whiskers grown as described in [7]. Crystallization of silver azide in a weak constant contactless electric field (the field intensity was varied 10^{-3} V/m to 10^{0} V/m), as demonstrated by the outcomes of our experiments [6, 7], allows synthesizing chemically pure mono-disperse crystals with the low concentration of defects, which sizes can be modified via varying the intensity of crystallization field.

The samples were glued at both ends of a high electric resistant mica plate degreased preliminary by ethyl alcohol.

The direct experiment scheme "impact – response" was used for exploring the regularities of explosive decomposition. As a factor initiating explosive decomposition a constant contact electric field was used with intensity of 300 KV/m. The electric field was generated by sources of direct current. Gallium electric contacts were deposited under the microscope onto the most developed crystal face in the central part of the sample (010); the distance between electrodes was 1 μm. The mode of mono-polar injection of holes is activated in these conditions; the holes in the reaction sites force a chemical reaction. The specified electric field intensity makes it possible to observe a transition from the slow decomposition into the explosive one in silver azide whiskers within some minutes (360±20 s).

The explosive decomposition was also initiated by 200 μs to 20 ms pulse with 1070 nm wavelength (energy in the pulse up to 20 J), generated by the ytterbium quasi-continuous fiber laser. The exposure of initiation on the sample surface was identified by the pyroelectric head PE50BF-DIF-C and monitored according to the calibrated photodiode signal. We used only the central laser beam for initiation to support quite homogeneous distribution of actuation intensity over the sample surface [8]. The dispersion of initiating pulse energies did not exceed 3%. We selected one sample of a set to explore the kinetics of explosive decomposition; afterwards it was subjected to initiation by a low energy laser pulse. Unless it had exploded, the pulse energy was increased by 5% and the sample was initiated again. This procedure makes it possible to determine precisely an "own" initiation threshold of a sample in the set. A probabilistic curve of explosion was plotted down on the base of data obtained for approximately 30 crystals.

The explosive sensitivity was determined as explosion delay time. The explosion delay time is measured quite specifically due to the probabilistic character of explosive decomposition, in the area of small fields in particular. The explosion delay to be measured includes a probabilistic component and as the experiments revealed it depends on the "life history" of a sample. The explosion delay time at the intensity of the contact electric field 300 KV/m was accepted as a measurement standard (this field intensity corresponds to the average explosion delay 360 ±20 s under the condition of 80-95 % explosion delay). At least 10 samples were examined for each point of experimental curves. The data of experiments were processed in Microsoft Excel.

The experimental facility shown in figure 1 was used for exploring the effect of a constant contactless electric field on the explosive sensitivity of silver azide whiskers. Polarization fields with the diverse configuration were used for the purpose of experiments (figure 1). The constant contactless electric field intensity was varied from 10^{-3} V/m to 10^{3} V/m.

IOP Conf. Series: Journal of Physics: Conf. Series **830** (2017) 012131 doi:10.1088/1742-6596/830/1/012131

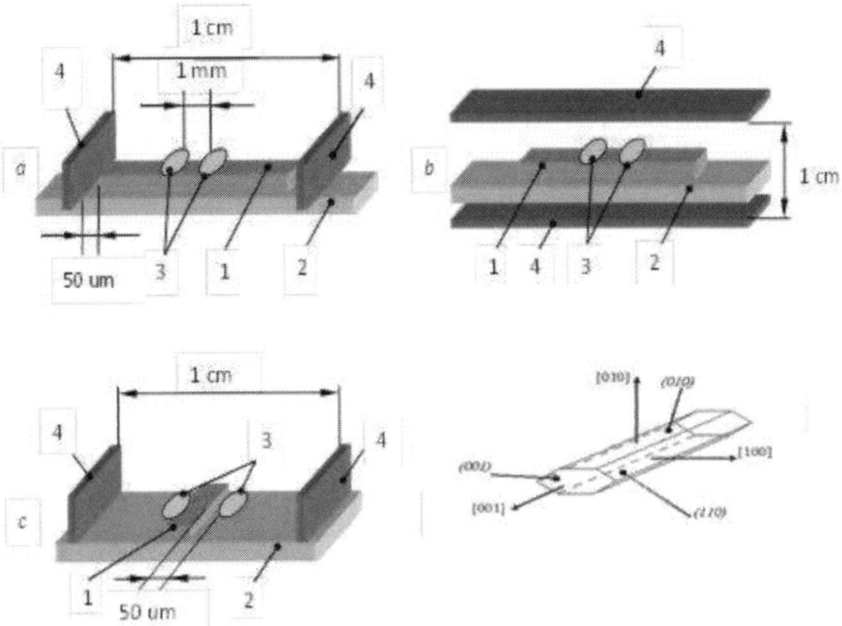

Figure 1. Experimental facility for exploring the explosive decomposition of silver azide samples in: a) a longitudinal contactless electric field, its electric-field vector overlaps with the crystallographic direction [001]; b) and c) a cross contactless electric field, its electric-field vector overlaps with the crystallographic directions [010] and [100].

3. Results and Discussion

The samples of silver azide synthesized via crystallization in the electric field are needle-like habit crystals. Simple forms of a pinacoid {010} and a rhombic prism {110} are recorded on the explored samples. A combined type of forms {010}+m{110} is typical of whiskers synthesized this way. Observing the growth of crystals in the process of crystallization, it was revealed that the silver azide whiskers grow from the top in the direction [001] with the developed face (010) and a side face (110) (see figure 2a). No volume damages seen by the optical microscope were found on the whiskers grown in the electric field, they are optically transparent with perfect faceting (figure 2b).

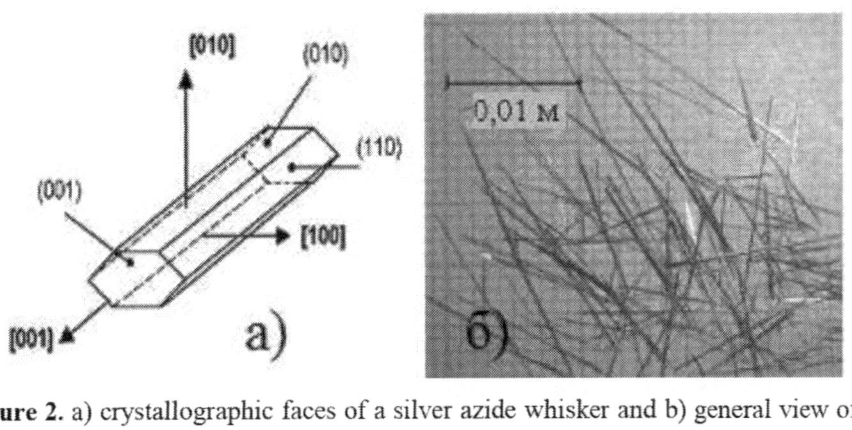

Figure 2. a) crystallographic faces of a silver azide whisker and b) general view of a crystal

The concentration of basic impurities (Fe, Si, Ca, Mg, Al, Na) in the synthesized crystals did not exceed 10^{16} cm^{-3}. The defect structure of the crystals under consideration was analyzed by the X-ray fluorescence method (JEOL JSM-6390 LA, integrated with REM-RMA (РЭМ-РМА) was used for the purpose of the research) and by atomic emission spectrometry (atomic emission spectrometer with inductively coupled plasma iCAP 6500). The density of dislocations in the crystals grown in a weak contactless electric field was determined by the method of etching pits [9] and averaged to $2\cdot10^2$ cm^{-2}.

Having analyzed the explosive decomposition of samples in conditions of laser initiation, we revealed that the crystals grown in the contactless electric field are the most suitable ones for interpretation of experimental results. The crystals grown in the electric field have a narrow probabilistic curve – the width of interval with the most probable explosion divided by the critical energy density does not exceed 0.05; the rate of reaction propagation in the crystal is constant; the obtained kinetic curves are smooth without any pips, therefore, the number of samples per each experimental point can be reduced.

The experimental results into the effect of a contactless electric field (polarization field) of diverse configuration on the explosive sensitivity of silver azide whiskers are provided in figure 3. The straight line in figure 3 represents the explosive sensitivity of crystals without a field of polarization.

The explosive sensitivity of the samples goes down sharply, as one can see in figure 3, simultaneously with the decreasing intensity of the contactless electric field. For instance, the intensity of the contactless electric field below 10^{-2} V/m is in line with the explosion delay time over $6\cdot10^3$ s. We should note that without any regards to mutual configuration of electric fields, in particular in zones with the intensity below 10^{-3} V/m, the curves nearly overlap when passing through the extremum.

As the chemical reaction in the anion sub-lattice of silver azide progresses till a quite stable semi-product N_6 [5, 10] is formed in the sample, the study of its formation in the contactless electric field helps to realize how a field of polarization affects the solid phase decomposition of silver azide whiskers. Furthermore, the dependence of relative dielectric permittivity vs. intensity of the constant contactless electric field was assessed by the method proposed in [11].

The dependences of relative dielectric permittivity ($\varepsilon_{кр}$), explosive sensitivity (figure 4, curve 2), and specific volume N_6 (figure 4, curve 3) on intensity of a constant contactless electric field (longitudinal contactless electric field in figure 1a) are presented in figure 4.

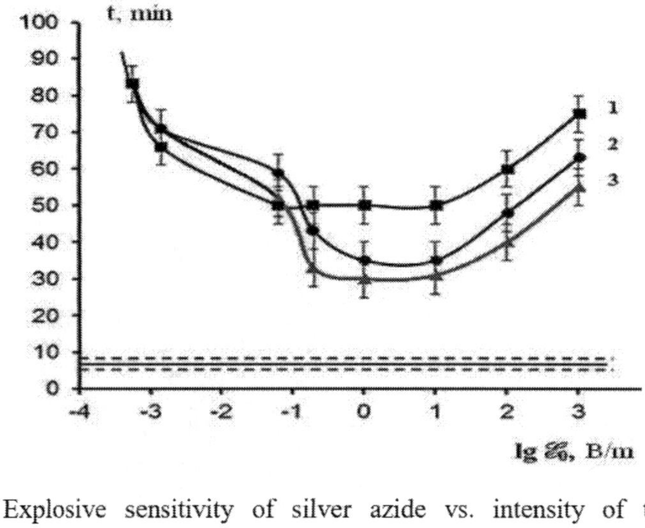

Figure 3. Explosive sensitivity of silver azide vs. intensity of the effecting contactless: 1, 3 – cross electric field, its vector of intensity overlaps with the crystallographic directions [010] and [100], respectively; 2 – longitudinal electric field, its vector of intensity overlaps with the crystallographic direction [001].

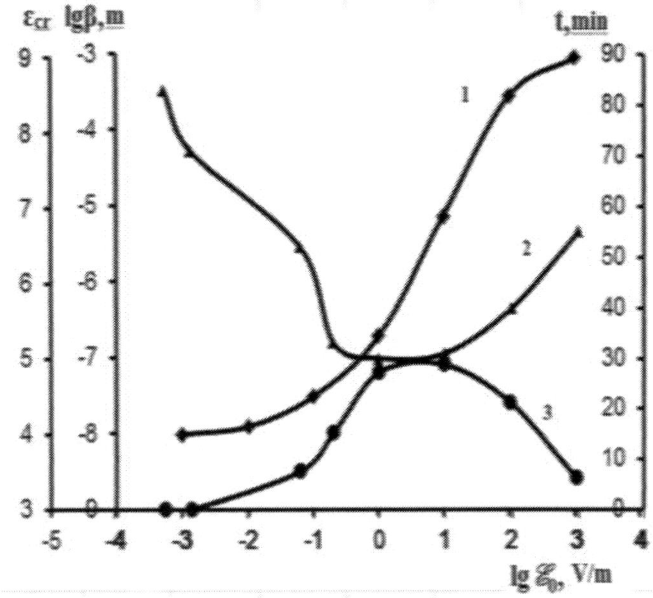

Figure 4. Relative dielectric permittivity $\varepsilon_{кр}$ (1), explosive sensitivity of silver azide crystals (2) and relative volume of semi-product lgβ (3) vs. logarithm of intensity of the constant contactless electric field

The experiments revealed that the amount of the formed semi-product in the contactless electric field drops significantly (the logarithm of the semi-product specific volume averages to 6 m provided that there is no electric field). It is worth noting that the field influences on the amount N_6 in the whole range of its intensity. Therefore, it implies the importance of the effecting field for formation of the semi-product in the process of solid phase decomposition of silver azide crystals.

Considering the intensity of the contactless electric field ranging from 10^0 V/m to 10^3 V/m and the relative dielectric permittivity, an assumption can be made that the mechanism of effecting contactless electric field is a combined one and depends on types of polarization: dipolar orientational, ionic, and electronic ones. However, the effecting electric field seems to be relevant only for the change in electron density in atoms provided that the intensity of polarization field does not exceed 10^{-2} V/m, as a consequence, the semi-product of solid phase decomposition of silver azide whiskers is formed slower. Thus, a weak (below 10^{-2} V/m) contactless electric field influences the rate of chemical reaction due to the electron density of reacting particles drifting along the vector of the electric field intensity, as a result, overlapping of electron clouds is less probable in the course of chemical bound formation. To sum up, polarization in the constant contactless field can be viewed as a method to control the explosive sensitivity of silver azide crystals.

4. Conclusion

The technology of growing and processing silver azide whiskers in the constant contactless field (in the specified range of field intensity) allows controlling their explosive sensitivity, as a result, making it possible to use the obtained experimental data for purposeful supervising the resistance of explosive materials. We note that the time till explosion of samples depends directly on the intensity of the constant contactless electric and magnetic fields in the course of crystallization and on the number of reaction sites in the sample.

The work is performed under support of Russian Foundation for Basic Research, grant №16-03-00313 A "Explosive sensitivity of energy-related materials in outer fields".

References

[1] Walker B F and Fair H D 1977 *Energetic Materials. Physics and chemistry of inorganic azides* (New York: Plenum Press) **1** p 503
[2] Aduev B P, Aluker É D, Belokurov G M, Zakharov Yu A et al 1999 *Journal of Experimental and Theoretical Physics* **89** pp 906-915
[3] Suhushin Yu N, Zakharov Yu A and Ivanov F I 1973 *High Energy Chemistry* **7** pp 261-268
[4] Suhushin Yu N and Zakharov Yu A 1981 *The common regularities of decomposition of solid matters in an electric field* (*In Kinetics and the mechanism of chemical reactions in a solid body*) (Russia: Chernogolovka) pp 152-161
[5] Zakharov V Yu, Krasheninin V I, Gazenaur E G, Gasanov A I and Yakunina V I 2002 *Russian Physics Journal* **45** pp 553-557
[6] Rodzevich A P, Gazenaur E G and Krasheninin V I 2013 *Applied Mechanics and Materials* **379** pp 154-160
[7] Rodzevich A P, Gazenaur E G and Belokurov G M 2014 *Applied Mechanics and Materials* **682** pp 206-209
[8] Kalenskii A V, Zvekov A A, Ananeva M V, Kriger V G, Tsipilev V P and Razin A V 2015 *Combustion, Explosion, and Shock Waves* **351** pp 353-357
[9] Sangwal K 1987 *Etching of crystals: Theory, experiment, and application* (*In S. Amelinckx and J. Nihoul, Eds., Defects in Solids*) (North-Holland: Amsterdam) p 497
[10] Aluker É D, Aduev B P, Zakharov Yu A et al 2006 *Focus on Combustion Research* (New York: Novapublishers) p 55
[11] Rodzevich A P, Gazenaur E G, Kuzmina L V and Krasheninin V I 2014 *Advanced Materials Research* **1040** pp 744-747

Influence of preliminary irradiation by gamma-quanta on development of catastrophic failures during operation of IR-LEDs

A V Gradoboev and A V Simonova

National Research Tomsk Polytechnic University, Tomsk, Russia

E-mail: gradoboev1@mail.ru

Abstract. We have considered the influence of preliminary irradiation by ^{60}Co gamma quanta on development of catastrophic failure during long-term operation of IR-LEDs based upon dual AlGaAs heterostructures. Irradiation was realized in passive power mode of the LEDs prior to aging. Two specific doses of irradiation were used. The first irradiation dose corresponded to the first stage of emission power drop of the LEDs irradiated by gamma-quanta, which is attributed to the radiation-stimulated reconstruction of the initial defect structure of the LED crystal. The second dose corresponded to the second stage of emission power drop, which is attributed solely to the introduction of radiation defects. Aging under long operating time conditions was simulated by a step-stress approach. We have established that the preliminary irradiation by gamma-quanta can be used in manufacturing technology of LEDs to improve their operating parameters on condition that irradiation doses are chosen in strictly provide area of the radiation-stimulated reconstruction of the initial defect structure.

1. Introduction

Nowadays infrared wavelength range light-emitting diodes (IR-LEDs) operate under the conditions of various types of ionizing radiation. It leads to combined influence of ionizing radiation and long-term operation factors. Whereas this influence limits operational reliability of LEDs. In this case, we interpret that the combined influence is the impact of two or more factors spread out over a period of time. It should be noted that today there is insignificant number of works about combined irradiation of semiconductor devices [1-4], while the information about combined influence of long-term operation and ionizing radiation is practically lacking.

In the preliminary investigation we have established that the reliability of the LEDs and, consequently, their lifetime are limited to catastrophic failure (CF). Previously [5] the research of preliminary irradiation by neutrons influence on parameters of IR-LEDs change and CF probability during their further operation have been considered. Moreover only investigation of preliminary irradiation by gamma-quanta influence on emissive power of LEDs decrease during operation has been discussed in [6].

Thus, the purpose of this investigation is to research the influence of preliminary irradiation by ^{60}Co gamma-quanta on CF development in the IR-LEDs during long-term operation. So, in this paper we have researched combined influence of gamma-quanta and long-term operation factors on probability of CF development.

IOP Conf. Series: Journal of Physics: Conf. Series **830** (2017) 012132 doi:10.1088/1742-6596/830/1/012132

2. Materials and methods

The objects of this investigation were industrial LEDs manufactured on the basis of dual AlGaAsheterostructures with 1 μm active layer grown on the monocrystalline n^+-GaAs wafer by means of liquid epitaxy. The crystal size was 450×450 μm^2. Figure 1 represents the structure of the dual AlGaAsheterostructure of the LED.

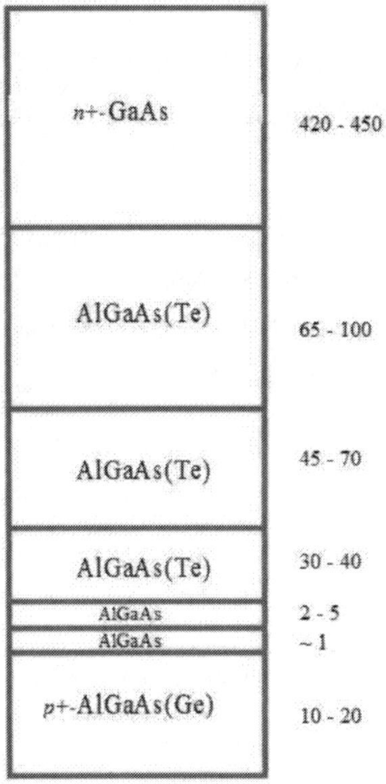

Figure 1. The thickness of the layers of dual AlGaAsheterostructure, microns.

LEDs were manufactured using a standard sandwich technology that involves metallic layer deposition and shaping processes for ohmic contact creation, photolithographic and chemical etching processes for die formation and dicing for wafer division into individual chips.

LEDs had packages and lenses made of an optical compound that was used to form the required angular pattern for output lumen. We emphasize that the preliminary investigation shows us that the optical compound irradiated by gamma-quanta does not lead to changing its optical properties in a given wavelengths range. Consequently, we suggest that the changing of the observable light characteristics of the LEDs is due to the change of characteristics in their active layer only.

In a continuous power mode, the LED forward operating current was I_{op}=50 mA and supply voltage was not over U_{op}=2.0 V. The maximum radiant wavelength was within the range of 0.82…0.90 μm.

Three batches of LEDs were composed of 20 items in each group. The first batch (LED-1) was intended for step-by-step tests without preliminary irradiation. The second batch (LED-2) was destined for preliminary irradiation with a dose equal to $D_{\gamma 1}=5\cdot 10^4$Gy. The third batch (LED-3) was meant for preliminary irradiation with a dose equal to $D_{\gamma 2}=2\cdot 10^6$Gy. Preliminary irradiation was performed with ^{60}Co gamma-quanta while the LEDs were in a passive power mode, i.e. without adding external electric field.

The doses of preliminary irradiation by gamma-quanta were chosen on the basis of the results previously published in [7,8]. We established that decrease of emissive power of the LEDs irradiation by gamma-quanta includes three stages. At the first stage, the fall of emissive power of the LEDs is attributed to the radiation-stimulated reconstruction of the initial defect structure of the LED crystal as evidenced by saturation of this stage as the level of exposure increases. At the second stage, the fall of emissive power of the LEDs during irradiation is attributed solely to the introduction of radiation defects. The second stage proceeded to the third stage, distinguished by the LED transition into the field of low level electron injection with further origination of CF.

These results [7,8] allow us to select the doses for preliminary irradiation by gamma-quanta. The first dose of irradiation was chosen because the emissive power reduction under irradiation corresponds to the first stage of the fall of emissive power of the LEDs. The second dose corresponded to the second stage of emission power drop of the LEDs irradiated by gamma-quanta.

On the other hand, we have established in [9] that the decrease of emissive power of the LEDs during operation can also be described by three stages. At the first stage, the fall of emissive power of the LEDs is attributed to the reconstruction of the initial defect structure of the LED under the influence of step-by-step test factors. At the second stage, the fall of emissive power of the LEDs during long-term operation is attributed solely to the introduction of new defects. The second stage proceeded to the third stage, distinguished by the LED transition into the field of low level electron injection with further origination of CF. To sum up, the observable laws of emissive power decrease of the LEDs under the influence of gamma-quanta and long-term operation factors are identical to a large extent.

Thus, we can obtain information about summing influence of gamma-quanta and long-term operation factors on the probability of CF development.

Consequently, in this work, combined (spread out over a period of time) influence of gamma-quanta and long-term operation factors on probability of CF development are investigated.

For every LED, volt-ampere (*V-I*) characteristics under normal conditions were taken using a measurement complex with spherical photometric integrator which allows measuring the forward voltage within the range of 0…5 V with change of the forward current within the range of 1…500 mA and pitch of at least 1 mA. The error of forward current setting from the given level was ±2% and the error of the forward voltage measurement was ±2%.

The mentioned characteristics of LEDs were obtained at the very beginning and after every stage of the investigation. The results were processed by means of methods of mathematical statistics. Every batch of LEDs under investigation was characterized by average values of measured parameters.

Standard certified equipment was used for the experiments, and the baseplate temperature was 65°C, the increment step of current was I_{step}=+50 mA. The operating current of the first stage was 50 mA. The duration of each stage was t=24 h. Furthermore, each stage of the experiments is distinguished by the temperature of the LED active layer, which depends on ambient temperature (i.e. baseplate temperature), LEDs thermal resistance, and consumed electric power.

In addition, the limit step of accelerated tests did not go beyond catastrophic failure (CF) development. Step-by-step tests stopped when CFs exceeded at least 80% of the LEDs in each batch.

3. Results and discussion

3.1. V-I characteristic of initial LEDs

Firstly, consider the initial characteristics of the LEDs, which were measured before irradiation by gamma-quanta and step-by-step tests. *V-I* characteristic was chosen for control parameter, because changes of *V-I* characteristic preceded CF. Figure 2 represents typical strongly pronounced s-shaped *V-I* characteristic of initial LEDs.

The observed change in *V-I* characteristic (dependence 1, figure 2) of LEDs with supply power increase can be described in the following way. In the area of small voltage (area of small forward current and relatively high resistance of the active layer of LED) typical *V-I* characteristic of the p-n-junction has a relatively small slope (line 2, figure 2). At supply voltage of approximately 1.2 V, the parallel connection of dislocations to the p-n-junction of the LED is observed. It leads to total slope increase of *V-I* characteristic (line 3, figure 2).

Next, an active layer resistance decrease with supply voltage increase is observed. It results in increase of noticeable fall of supply voltage on the series connection ohmic contacts (line 4, figure 2). The reduced slope in the *V-I* characteristic shows the increase of voltage drop in the ohmic contacts.

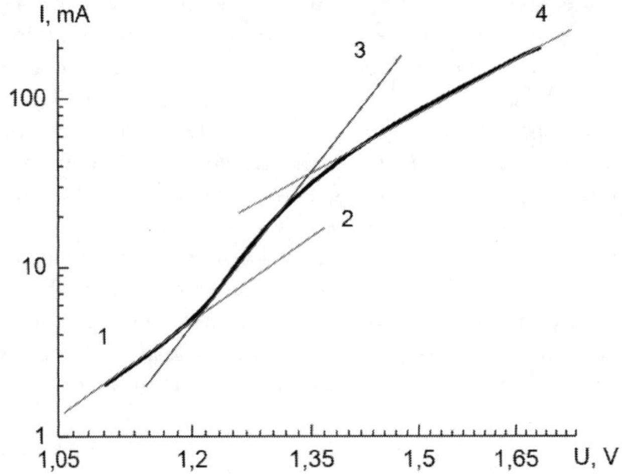

Figure 2. Description of *V-I* characteristic of initial LEDs: 1 is the measured *V-I* characteristic, 2 is model description of *V-I* characteristic before connection of dislocations, 3 is total *V-I* characteristic of active layer of LED and dislocations, 4 is demonstration of ohmic contacts resistance in the high current area.

Previously we have considered the LEDs with s-shaped *V-I* characteristic in more detail in [10].

3.2. V-I characteristic of LED-1 batch

Preliminary investigation shows us that the decrease of emissive power of LEDs does not lead to noticeable change of *V-I* characteristic of LEDs both under irradiation by gamma-quanta and during long-term operation at the influence levels corresponding to the first and second stages of the degradation process.

In addition, previously [9-12] we have established that CF sets measures to reliability of LEDs in these both causes.

Consider the observable changes of *V-I* characteristic and the CF development for the LED-1 batch. We note that the preliminary irradiation was absent for this batch.

Figure 3 illustrates the change of *V-I* characteristic of the LEDs during step-by-step tests.

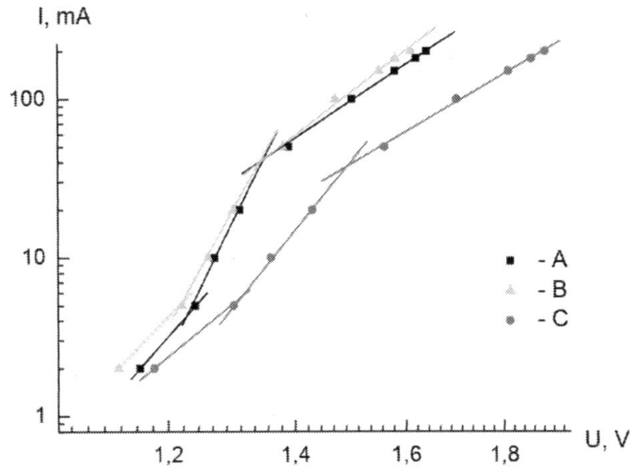

Figure 3. *V-I* characteristics for the LED-1 batch during step-by-step tests: initial (A); during step-by-step tests (B); before the CF development (C).

The above presented results show that slight changes of *V-I* characteristic (dependences A and B, figure 3) have been observed during step-by-step tests. We note that the registered changes of *V-I* characteristic are statistically significant and cannot be attributed to measurement error. At present time, we are unable to explain these changes of *V-I* characteristic.

Major changes of *V-I* characteristic have been observed on the test stage immediately prior to CF (dependence C, figure 3) only. This shift of *V-I* characteristic to the field of higher power supply voltage is accounted to degradation of ohmic contacts resistance, which leads to increase of their ohmic resistance, additional decrease in power supply voltage on the ohmic contacts, their additional heating and CF development at the end.

Consequently, CF probability of LEDs during step-by-step tests is determined by accelerated degradation of ohmic contacts and temperature of active layer of LEDs.

Previously [10-12] we have observed such behavior of *V-I* characteristic of LEDs,when we researched radiation resistance of LEDs.

3.3. V-I characteristic of LED-2 batch

Let us consider the LEDs batches preliminary irradiated by gamma-quanta. Figure 4 describes the changes of *V-I* characteristic for the LED-2 batch. The LEDs were preliminary irradiated by ^{60}Co gamma-quanta with a dose equal to $D_{\gamma 1}=5\cdot 10^4$Gy, which corresponds to the first stage of the emissive power fall under irradiation of LEDs. Figure 4 shows that the strong changes of *V-I* characteristic are absent. We note that the observable minor changes of *V-I* characteristics are probably significant. However, at the moment we are unable to interpret them.

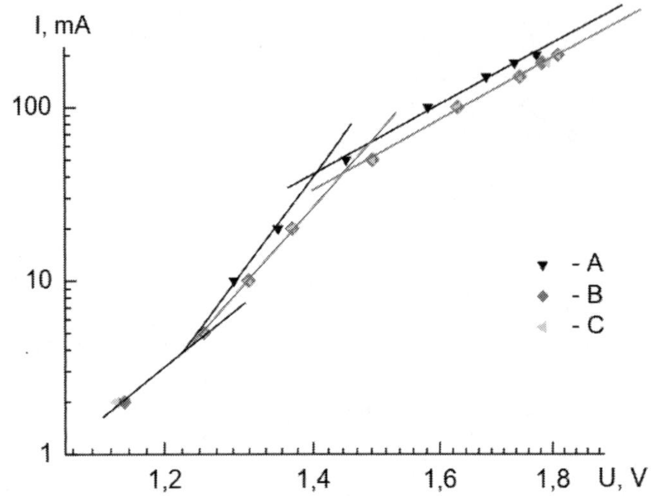

Figure 4. *V-I* characteristics for the LED-2 batch during step-by-step tests: initial (A); during step-by-step tests (B); before the CF development (C).

Furthermore, comparison of the results shown in this figure with the results presented in figure 3 allows us to conclude that probability of CF development is not aligned with accelerated degradation of ohmic contacts in this case. The CF development is determined by temperature of the active layer of LED only.

So, drop-out (local or full) of the stage of the emissive power fall during step-by-step tests due to reconstruction of the initial defect structure under preliminary irradiation by gamma-quanta provides an opportunity toimprove reliability of LEDs by means of ohmic contacts reliability increase. Therefore, such preliminary irradiation by gamma-quanta with the selected dose makes it possible to minimize additional contribution of ohmic contacts to CF probability during step-by-step tests.

3.4. V-I characteristic of LED-3 batch

Consider changes of *V-I* characteristic of the LED-3 batch during step-by-step tests. Figure 5 depicts a noticeable shift of *V-I* characteristic to the field of higher power supply voltage by reason of preliminary irradiation by gamma-quanta.

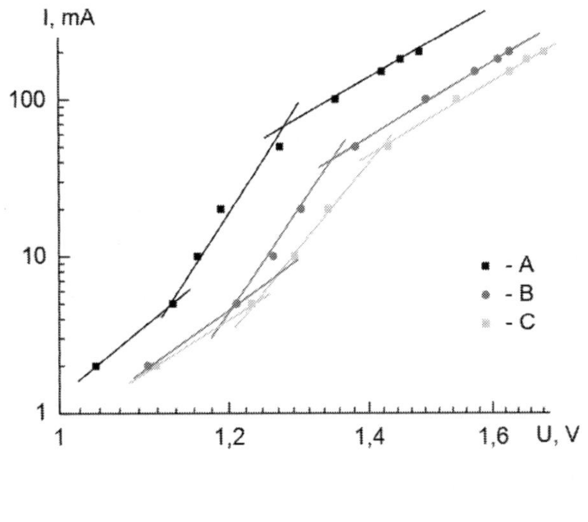

Figure 5. *V-I* characteristics for the LED-3 batch during step-by-step tests: initial (A); during step-by-step tests (B); before the CF development (C).

Thereby, the selected dose of the preliminary irradiation by gamma-quanta leads to accelerated degradation of ohmic contacts. And the decrease of ohmic contacts resistance has been observed at the stage before CF development only. It can be expected that the given process is due to local annealing of radiation-induced defects under appropriate temperatures which are responsible for degradation of ohmic contacts.

In addition, the results represented in figure 5 allow us to suggest the CF development at the earlier stage of the tests.

3.5. Influence of preliminary irradiation on CF probability

Examine the influence of the preliminary irradiation by gamma-quanta on probability of the CF development. Figure 6 demonstrates the dependence of accumulate quantity of CF on the step number of the tests for the LED batches.

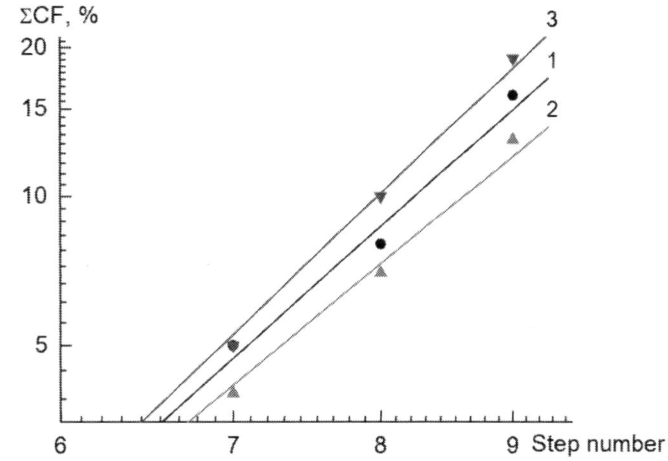

Figure 6. The influence of different doses of preliminary irradiation by gamma-quanta on CF development depending on step number: LED-1 (1); LED-2 (2); LED-3 (3).

This results absolutely confirm the above presented analysis of *V-I* characteristic changes of the investigated LEDs.

Probability of CF development during step-by-step tests is defined by the increase of ohmic contacts resistance and the temperature of the active layer of LED.

Preliminary irradiation by gamma-quanta with doses corresponding to the radiation-stimulated reconstruction of the initial defect structure of the LED crystal improves ohmic contacts resistance to influence of long-term operation factors. Thus, it leads to decrease of the CF probabilityat the early stage of step-by-step tests and increases the reliability of LEDs.

Preliminary irradiation by gamma-quanta with doses corresponding solely to the introduction of radiation defects leads to ohmic contacts degradation and increase of CF probability at the early stage of step-by-step tests. Consequently, the reliability of LEDs drops.

These results allow us to recommend the preliminary irradiation by gamma-quanta in the manufacturing technology of LEDs for improvementof their operational characteristics through irradiation with doses in the strictly regulated field of radiation-stimulated reconstruction of the initial defect structure of the LED crystal.

4. Conclusion

Sum up the main investigation results of preliminary irradiation by ^{60}Co gamma-quanta influence on CF development of IR-LEDs based on AlGaAsheterostructure during long-term operation.

1. The investigated LEDs had s-shaped *V-I* characteristics because of the derived connection of dislocations to the active layer of LED crystal.

2. We have established that the limit step of tests is determined by catastrophic failures due to mechanical destruction of the LED package.

3. The significant changes of *V-I* characteristic are distinctive under influence of both ionizing radiation and long-term operation factors. The observable changes of *V-I* characteristic are declarative of accelerated ohmic contacts degradation of LEDs.

4. CF probability of LEDs during step-by-step tests is determined by accelerated degradation of ohmic contacts and temperature of the active layer of LEDs.

5. Preliminary irradiation by gamma-quanta with doses corresponding to the radiation-stimulated reconstruction of the initial defect structure of the LED crystal improves ohmic contacts resistance to influence of long-term operation factors. Thus, it leads to decrease of the CF probability at the early stage of step-by-step tests and increase of the reliability of LEDs.

6. Preliminary irradiation by gamma-quanta with doses corresponding solely to the introduction of radiation defects leads to ohmic contacts degradation and increase of CF probability at the early stage of step-by-step tests. Consequently, the reliability of LEDs drops.

7. Preliminary irradiation by gamma-quanta can be used in the manufacturing technology of LEDs for improvement of their operational characteristics through irradiation with doses in the strictly regulated field of radiation-stimulated reconstruction of the initial defect structure of the LED crystal.

References

[1] Gradoboev A V and Surzhikov A P The Radiation Resistance Microwave Devices Based on Gallium Arsenide 2005 *Tomsk Polytechnic University* p 277

[2] Martin D I, Ighigeanu D I, Mateescu E N *et al* 2004 *IEEE Transactions on Industry Applications* **40** 41

[3] G. Busatto, V. De Luca, F. Iannuzzo *et al* 2013 *IEEE Transactions on Nuclear Science* **60** 3793

[4] Chaoming Liu, Xingji Li, Jianqun Yang *et al* 2015 *IEEE Transactions on Nuclear Science* **62** 3381

[5] Gradoboev A V, Orlova K N, Asanov I A and Salchak Y A 2014 Influence of the fast-neutron irradiation on the IR-LEDs reliability ISROS *France International Symposium on Reliability of Optoelectronics for Systems* Toulouse

[6] Gradoboev A V, Simonova A V and Orlova K N Influence of irradiation by ^{60}Co gamma-quanta on reliability of IR-LEDs based on AlGaAs heterostructure *Physica Status Solidi C: Current Topics in Solid State Physics*. In Press. DOI: 10.1002/pssc.201600035

[7] Gradoboev A V and Sednev V V 2014 *Izvestia vyzov. Fizika* **57** 20

[8] Gradoboev A V and K N Orlova 2015 *IOP Conf. Series: Materials Science and Engineering* **81**

[9] Gradoboev A V, Simonova A V and Orlova K N 2016 In-service change in radiant power of infrared LEDs *International Siberian Conference on Control and Communications* DOI: 10.1109/SIBCON.2016.7491662

[10] Gradoboev A V, Orlova K N, Asanov I A and Simonova A V 2016 The fast neutron irradiation influence on the AlGaAs IR-LEDs reliability *Microelectronics Reliability* **65** http://dx.doi.org/10.1016/j.microrel.2016.07.143

[11] Gradoboev A V, Asanov I A and Salchak Ya A 2013 *Izvestia vyzov. Fizika* **56** p 111

[12] Gradoboev A V and Orlova K N 2014 *Advanced Material Research* **880** pp 237–241

Local mechanical stress relaxation of Gunn diodes irradiated by protons

A V Gradoboev and E P Tesleva

National Research Tomsk Polytechnic University, 30 Lenin Ave., 634050, Tomsk, Russia

Abstract. The aim of the work is studying the impact of Gunn diodes thermocompression bonding conditions upon their resistance to being radiated with protons of various energies. It was established that the tough conditions of Gunn diodes thermocompression bonding results in local mechanic stresses introduced into the active layer of the device, reduction of electron mobility because of the faults introduction and, subsequently, to reduction of operating current, power of UHF generation, percentage of qualitative units production and general reduction of production efficiency of the devices with required characteristics. Irradiation of Gunn diodes produced under the tough conditions of thermocompression bonding with protons which energy is (40–60) MeV with an absorbed dose of $(1–6)\cdot 10^2$ Gy does not practically reduce the radiation resistance of Gunn diodes produced with application of the given technique. This technique can be recommended for all semiconductor devices on the base of GaAs, which parameters depend significantly upon the mobility of the electrons, to increase the efficiency of production.

1. Introduction

Semiconductor devices are actively applied as the main component base for various electronic devices operating in the upper layers of the Earth atmosphere, near and far space, at various objects with increased radiation background including nuclear power objects [1-3]. In this case, they are exposed to complex (simultaneous) and combined (distributed in time) impact of various kinds of ionizing radiation as well as to various factors of long-time operation. Thus, the given conditions of operation set especially firm requirements for reliability and radiation resistance of semiconductor devices.

In the process of semiconductor and semiconductor based devices production, various technologic defects may be introduced into the active layers of the devices which results in deterioration of the output parameters of the devices, reduction of the percentage of good device output during production as well as in deterioration of their operational characteristics such as radiation resistance and reliability [4].

Currently the works aimed at studying the impact of technological parameters upon the radiation resistance of semiconductor devices are very scarce.

On the other hand, it is known that radiation technologies can be efficiently applied right in the technology of semiconductor devices production [4-6].

Thus, the analysis of available reference data brings us to the conclusion that the research on the impact of technological factors upon the resistance of semiconductor devices to ionizing radiation is relevant.

The aim of the given paper is to study the impact of thermocompression bonding of Gunn diodes upon their resistance to exposure to protons of various energies.

2. Research objects and methods

2.1. Objects of research

We chose industrial Gunn diodes 3A728 [7] as the objects of research. Choosing Gunn diodes as the objects of research is based on their rather high sensitivity to technology factors exposure [8-10]. Gunn diodes under study were produced on the base of one multilevel epitaxial structure of gallium arsenide of n^+-n-n_b^+-n^{++}-type according to standard sandwich technology [11] where n^+ is the below contact heavily doped epitaxial layer; n is working layer; n_b^+ is buffer heavily doped epitaxial layer and n^{++} is heavily doped bulk substrate from volume monocrystalline GaAs. The multilayer epitaxial structure was grown according to the method of gaseous-phase epitaxy within a single technologic process.

When manufacturing Gunn diodes, the following basic technologic methods were applied:

- lost wafer process with the methods of abrasive and chemical polishing;
- ohmic contacts deposition on the base of Au-Ge-Ni with their further formation by thermal treatment;
- formation of active elements with the methods of photolithography and chemical etching;
- dividing the plates into separate crystals with laser scribing.

As a result of completing of the mentioned above operations, we obtained a separate chip (crystal) of Gunn diode containing an active element in the form of a mesostructure with round cross-section on the crystalline underplate with square cross-section.

Then structurally complete Gunn diode was manufactured applying reattachment of the crystal onto the copper chip-carrier (heat removal) with golden coating with implementation of thermocompression bonding. With such construction of Gunn diode the contact was on the side of the heat removal.

Two sets of Gunn diodes were produced: the first set (DG-1) – the optimal conditions of thermocompression bonding; the second set (DG-2) – the tough conditions of thermocompression bonding (increased temperature and pressure). The diodes manufactured this way were put to in-process testing (current training under the increased temperature, thermocycling, etc.) and presorting. The optimal conditions of the thermocompression bonding were selected according to the criteria of minimum process scrap revealed during the in-process testing. It should be noted that for the tough conditions of bonding the increased amount of the process scrap was observed.

Thus, under the tough conditions of thermocompression bonding we observe introduction of local mechanic stresses which result in development of additional faults and deterioration of the electrophysical characteristics of the active layer of the Gunn diode [4,8-10].

From the manufactured sets of Gunn diodes, random series were formed (each including 20 diodes) to be used for the research.

2.2. The methods of Gunn diodes parameters monitoring

Each Gunn diode was characterized in terms of operating current I_{op}, transit-time frequency which corresponds to the maximum value of SHF generation P_{HF} within the frequency range 33.0 – 37.5 GHz and, accordingly, in terms of maximum power of SHF generation P_{HF}. All measurements of Gunn diodes parameters were conducted under the input voltage of 4.0 V.

The listed above SHF characteristics were measured in a corresponding waveguide cell with the power supply used for supplying the required input voltage, rated detector head used for measuring energy and the frequency meter. The waveguide cell was adjusted so that to ensure the maximum possible value of SHF generation power.

The studied series of diodes from the described above sets were characterized by the measured parameters distribution histogram, mean values of parameters and their range. The characteristics of the formed research series for each of the diode sets did not practically differ which definitely allows relating the differences in diode parameter changes during testing to the differences in the technology of their production.

2.3. Methods of research

We measured the described above parameters for each series of the devices before and after the irradiation with protons of various energies. To study the regularities of Gunn diode parameters changing in their relation to exposure increase we applied both single irradiation and successive accumulation of the required exposure level. In this case, the identity of the data obtained for different series allows concluding that measurements of the SHF characteristics of diodes do not result in annealing of radiation defects introduced by irradiation. Then all the observed changes can be undoubtedly related to the differences in the technology of Gunn diodes production. The growth of exposure level was limited to complete diode failure, i.e. extinction of SHF generation.

Further we shall provide only mean values of parameters for all the dependences characterizing changes of the parameters of Gunn diodes from various series related to exposure growth. Besides, in all cases, we are going to analyze the changes of diodes parameters distribution histograms. We applied Isochronous Cyclotron U-240 as the source of proton irradiation as it allows obtaining proton beams with the energy of 8–80 MeV [12]. The level of exposure to protons of different energies was characterized by energy E_p [MeV] and fluence of particles F_p [cm^{-2}]. Energy of protons E_p was calculated with consideration to its losses at the passive elements of the package and the crystal of the Gunn diode. In all cases, the path of protons significantly exceeded the thickness of the active layer of the studied Gunn diode.

As previously [4] it was established that irradiation with protons with the applied energies under the particle fluence over $3 \cdot 10^{10} cm^{-2}s^{-1}$ results in annealing of the part of introduced radiation defects and increase of the temperature of irradiated samples, in the given work we applied particle fluence below $1 \cdot 10^{10} cm^{-2}s^{-1}$.

Thus, irradiation of all the studied series of diodes was conducted under normal climatic conditions and constant control of diode temperature which allows elimination of annealing radiation defects in the process of irradiation.

3. Results and discussion

3.1 The original characteristics of Gunn diodes

First of all, let us consider the original characteristics of Gunn diodes for two studied sets of devices. In figure 1, we present typical distribution histograms of Gunn diodes from the two sets considered above in the series formed for the research according to the SHF generation power. From the presented results we can see significant difference between the studied sets of Gunn diodes in terms of generation power. For the series of Gunn diodes from set DG-1, a typical mean value of power is (144±4) mW and its change range is (120 – 170) mW, while for the typical series of Gunn diodes from set DG-2, the mean value of generation power is (132±5) mW and its change range is (100–160) mW. Similar results are observed for operating currents.

The analysis of the results presented in figure 1 brings us to the conclusion that tough conditions of thermocompression bonding lead to noticeable reduction of the operating current and power of SHF generation of the produced diodes. We also observe a clearly marked left shift of diode distribution histogram in terms of SHF generation – to the region of lower power. In this case, the group of diodes with higher generation power practically disappears and there appears the group of diodes with lower power. The same is obviously demonstrated by the observed differences of the measured power ranges which were presented above. The same difference is also observed for the operating current. It should be also noted that for the diodes from DG-2 set higher frequency of generation is observed corresponding to the maximum of SHF power which value characterizes the transit-time frequency, i.e. the velocity of high-field domain propagation in the active layer of the Gunn diode. Growth of generation transit-time frequency can be explained by the growing resistance of ohmic contacts for Gunn diodes from DG-2 set [4].

It should be noted that during the in-process testing of Gunn diodes from DG-2 set partial restoration of operating current corresponding growth of SHF generation power growth were observed. Similar results were observed when storing diodes from DG-2 set without in-process testing under normal climatic conditions as it is shown in figure 2.

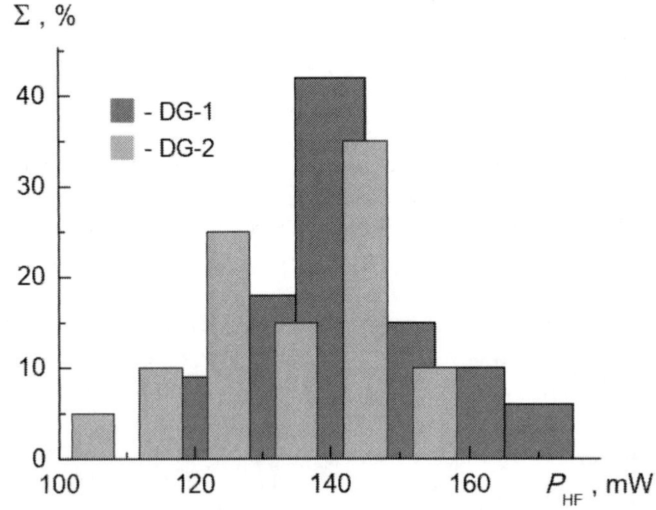

Figure 1. Typical histograms of Gunn diode distribution by power of SHF oscillation in the two batches of Gunn diode.

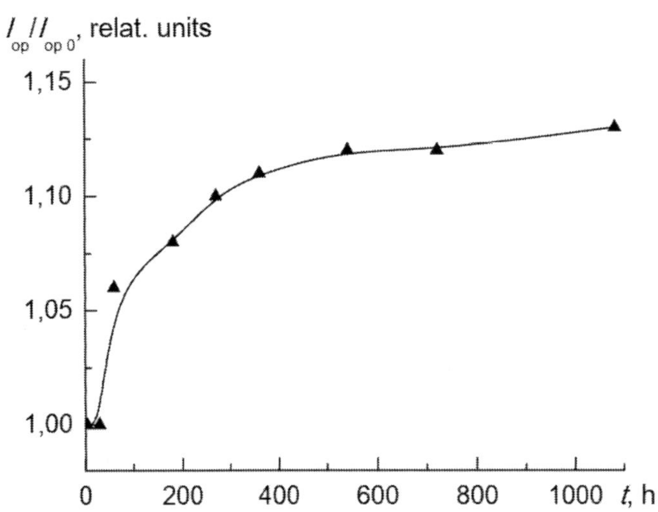

Figure 2. Recovery of operation current of GD-2 without technological tests during storage.

In-process testing and storing Gunn diodes DG-1 without in-process testing under normal climatic conditions did not lead to traceable change of the operating current and, accordingly, to the change of SHF parameters.

On the other hand, storing Gunn diodes of the studied sets undergone the whole complex of in-process testing also did not lead to the traceable change of their parameters and characteristics.

The presented above results of the Gunn diodes analysis allow concluding that the tough conditions of thermocompression bonding lead to deterioration of electrophysical characteristics of their active layer, increase of process scrap and reduction of SHF parameters due to introduction of local mechanic stresses and development of additional defects. The given conclusion agrees with the previously presented results [4, 8-10].

The question about the deterioration of electrophysical characteristics of the active layer of Gunn diodes under the tough conditions of thermocompression bonding still remains open. To answer this

IOP Conf. Series: Journal of Physics: Conf. Series **830** (2017) 012133 doi:10.1088/1742-6596/830/1/012133

question diodes with Schottky barrier were produced according to the same sandwich technology on the base of the epitaxial structure of gallium arsenide of n-n_b^+-n^{++}-type with close characteristics of the active layer. The given structure was grown during single technologic process according to the same technology that was used for growing the epitaxial structure applied in Gunn diodes production. Two sets of diodes with Schottky barrier were produced under identical conditions of thermocompression bonding of diodes into the metal-ceramic package.

Analysis of capacitance-voltage curves of the diodes with Schottky barrier showed that the thermocompression bonding conditions do not influence the concentration and the electron distribution profile in their active layer.

The given results allow concluding that the observed changes of the electrophysical characteristics of the active layer of Gunn diodes depending upon the conditions of the thermocompression bonding are determined by the changes of electron mobility.

3.2. Changes of Gunn diodes parameters under irradiation with protons of various energies

Let us consider the results of the study of proton irradiation influence upon the characteristics of the studied sets of Gunn diodes. Figure 3 shows the changes of the mean value of the operating current under proton irradiation with energy 14.3 MeV and 63 MeV.

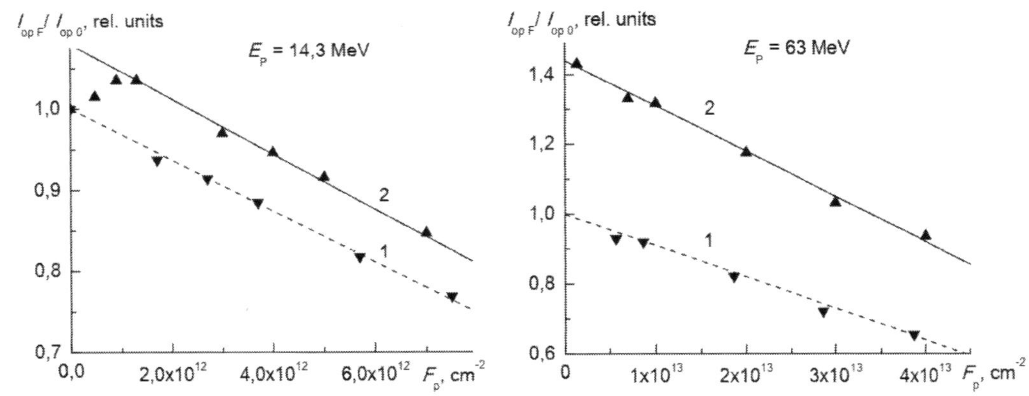

Figure 3. Change of operation current of Gunn diode DG-1 (1) and DG-2 (2) under irradiation by protons with energy of 14.3 MeV и 63 MeV.

For other proton energies, similar results were obtained. From the presented results we can see that for Gunn diodes DG-2 in the region of low levels of exposure restoration of the operating current and, accordingly, growth of SHF generation power are observed as it is shown in figure 4.

At the same time, for Gunn diodes DG-1 irradiated with protons in the region of low exposure levels noticeable changes of the operating current are not observed and for SHF generation power its insignificant growth is observed as it is shown in figure 5.

From figure 4 and figure 5 we see that in both cases irradiation with protons in the region of low exposure levels results in disappearing of diodes with low values of SHF generation power. The observed effect of SHF generation power increase is most significant for the diodes from DG-2 set.

From the results presented in figure 5 we can see that the degree of operating current restoration under irradiation in the region of low exposure levels significantly depends on proton energy which is additionally proved by the results presented in figure 6.

It should be noted that long-time storing of Gunn diodes from DG-2 set leads to noticeable reduction of the degree of operating current restoration under proton irradiation as it is shown in figure 7. In the given case we understand the time of storing under standard climatic conditions as the time between the bonding of the devices and proton irradiation.

On the base of the study results for the original Gunn diodes we can state that the observed operating current restoration of Gunn diodes is determined by restoration of mobility in the region of low levels of proton irradiation.

For different diode sets, different proton energies, the optimal value of proton fluence under which maximum restoration of electron mobility was achieved corresponded to the absorbed dose within (1-6)·10^2 Gy.

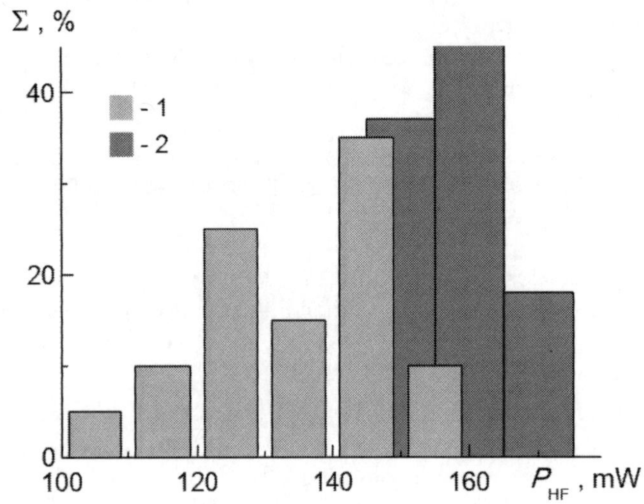

Figure 4. Histograms of Gunn diode DG-2 distribution by power of SHF oscillation before (1) and after (2) irradiation by protons with energy of 63 MeV at low levels of exposure.

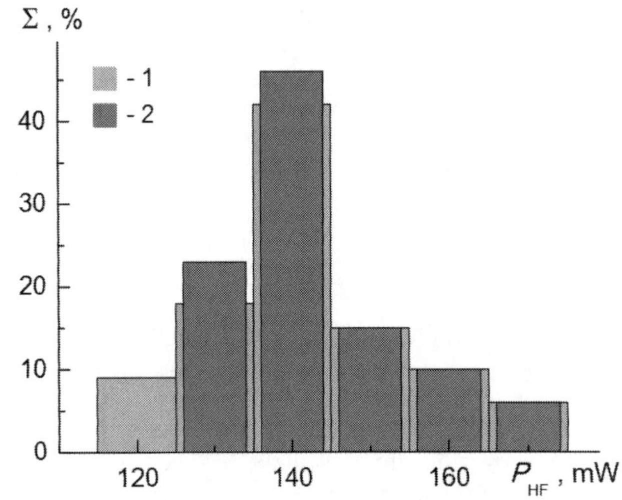

Figure 5. Histograms of Gunn diode DG-1 distribution by power of SHF oscillation before (1) and after (2) irradiation by protons with energy of 63 MeV at low levels of exposure.

Let us sum up the considered above results on proton irradiation of the studied sets of Gunn diodes in the region of low exposure levels.

For Gunn diodes DG-2 produced under tough conditions of thermocompression bonding being irradiated with protons with absorbed dose within (1–6)·10^2 Gy, restoration of electron mobility is observed, which results in growth of the operating current and power of SHF generation.

Thus, irradiation of Gunn diodes with protons with the energy within (40–60) MeV range allows relaxing the local mechanic stresses introduced into the active layers of Gunn diodes under the tough conditions of thermocompression bonding and restore electron mobility in the active layer. Besides, the given exposure can be recommended for treating Gunn diodes with low level of SHF generation power which were discarded after the in-process testing as it allows increasing their SHF generation power and, thus, reduce the amount of process scrap and increase the efficiency of Gunn diodes production.

We should note that relaxation of local mechanic stresses in the semi-conductor devices have been also observed before under other kinds of energy deposition [4,5,13].

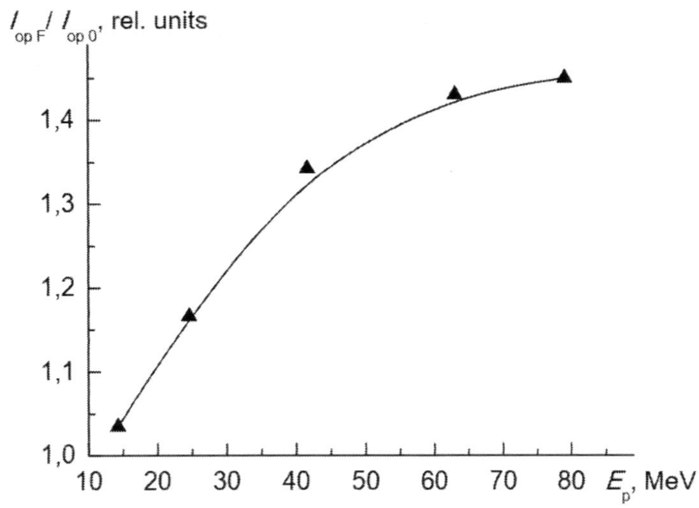

Figure 6. The dependence of average value of reduction degree of operating current of Gunn diode DG-2 on the proton energy.

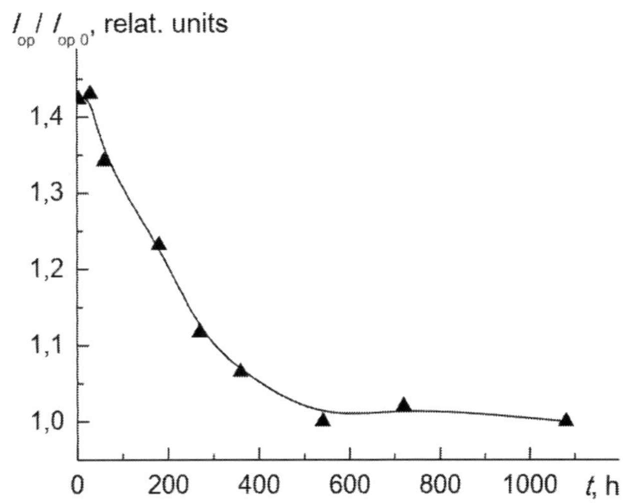

Figure 7. The reduction degree of operating current of Gunn diode DG-2 under irradiation by protons with energy of 63 MeV depending on storage time.

3.3. Resistance of Gunn diodes to irradiation with protons of various energies

Further we shall consider the resistance of the studied sets of Gunn diodes to irradiation with protons of different energies in the range of exposure which leads to their degradation up to complete failure. Here standardization was completed according to the value of the operating current measured after

Gunn diodes bonding.
As irradiation with protons in the region of low exposure can be considered as a technologic operation, then to study the resistance of Gunn diodes to proton irradiation standardization should be done for the values of parameters after the complete production cycle. Figure 8 shows the change of the operating current under irradiation with protons with the energy of 63 MeV for the studies sets of Gunn diodes. Standardization for DG-2 diodes was done for the value of the operating current after preliminary irradiation, and the fluence of protons was taken excluding the fluence of preliminary irradiation.
Similar dependences were obtained for protons of other energies and also for changes of SHF generation power. Results presented in figure 8 show that the Gunn diodes from the sets under study have practically identical resistance to proton irradiation.

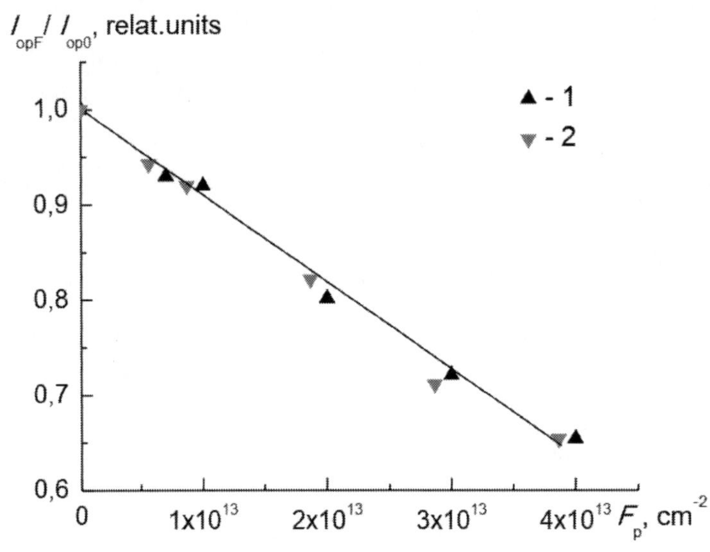

Figure 8. Operating current of Gunn diode DG-1 (1) and DG-2 (2) under irradiation by protons with energy of 63 MeV: symbols are experimental data; lines are results of the calculation of established correlations [4].

In [4], the authors present regularities of electron concentration and their mobility for epitaxial GaAs under irradiation with protons of different energies which can be applied for calculation of Gunn diodes operating current changes. The results of the calculations are presented in figure 8 in the form of a line. It is obvious that there is good correlation between the calculated and the experimental values.

Thus, the obtained experimental data allow concluding that preliminary irradiation with protons in the region of $(1–6)\cdot 10^2$ Gy does not practically reduce the radiation resistance of Gunn diodes produced under the tough conditions of thermocompression bonding. The observed insignificant resistance reduction is explained by insignificant reduction of electron concentration and mobility under preliminary irradiation.

4. Conclusion

In the conclusion, let us sum up the main results obtained in the given work.

1. The technologic factors may lead to significant deterioration of the output parameters and reduction of production efficiency of semi-conductor devices with the required characteristics.
2. Tough conditions of thermocompression bonding of Gunn diodes lead to introduction of local mechanic stresses into the active layer of the device, reduction of electron mobility in the active layer due to introduction of defects and, as a result, to reduction of the operating current level, SSHF

generation power, percentage of quality devices production and general reduction of production efficiency of devices with required characteristics.

3. Irradiation of Gunn diodes produced under the tough conditions of thermocompression bonding with protons with energy of (40–60) MeV and absorbed dose of $(1–6)\cdot 10^2$ Gy allows relaxing the local mechanic stresses without formation of additional defects, restoring electron mobility, level of the operating current and SHF generation power.

4. Preliminary irradiation with protons with the energy of (40–60) MeV and absorbed dose of $(1–6)\cdot 10^2$ Gy does not practically reduce the radiation resistance of Gunn diodes produced with application of the given technique.

5. Preliminary irradiation with protons with the energy of (40–60) MeV and absorbed dose of $(1–6)\cdot 10^2$ Gy may be recommended for treating Gunn diodes which were discarded for the purpose of improving their output parameters and, accordingly, for increasing the percentage of quality devices production.

6. The given technique may be recommended for all semiconductor devices based on GaAs if their parameters depend upon the electron mobility to increase the efficiency of their production.

References

[1] Vavilov V C and Uhin N A 1969 *Radiation effects in semiconductors and semiconductor devices* (Moscow, Atomizdat) p 312
[2] Chaffin R J 1973 *Microwave semiconductor devices: Fundamentals and Radiation Effects* (New York: Willey) p 387
[3] Kulakov V M and Ladygin E A 1980 *Effects of ionizing radiation on the devices of electronic equipment* (Moscow: Sov. Radio) p 244
[4] Gradoboev A V and Surzhikov A P 2005 Radiation resistance of SHF devices based on gallium arsenide (Publishing TPU, Tomsk) p 277
[5] Mamontov A P and Chernov I P 2009 Effect malych doz ioniziryuchego izluchenia (Tomsk, Delteplan) p 288
[6] Korshunov F P *et al* Radiation effects in semiconductor materials and devices technology (Contemporary issues of solid physics: collection of works for the 40th anniversary of IPSS NAS of Belarus and 90-th anniversary of its founder acad. N.N. Sirota, Minsk: Belaruskaya navuka) pp 332 – 364
[7] Semiconductor devices. Super high frequency diodes. Reference book 1992 ed B A Nalivaiko (Tomsk RASKO) p 223
[8] Hasegawa F and Ito H 1972 *Appl. Phys. Lett.* **21** 107
[9] Kotani M 1976 *IEEE Trans. Electron. Dev.* **ED – 23** 567
[10] Mitsui S and Ishihara O 1972 *Japan. J. Appl. Phys.* **11** 603
[11] Haues M and Morgan D Semiconductor devices in the schemes of SHF 1979 ed (Moscow, Mir) p 444
[12] Isochronous Cyclotron U-240 http://www.kinr.kiev.ua/index_en.html
[13] Boltovets N S, Kamalov A B, Kolyadina Ye Yu *et al* 2002 *Letters to TPJ* **28**

Effect of temperature on resistance of LEDs based on AlGaAs heterostructures to ^{60}Co gamma radiation

A V Gradoboev, P V Rubanov and V V Sednev

National Research Tomsk Polytechnic University

Abstract. The paper presents the results obtained in the study of the change in the parameters of IR LEDs based on AlGaAs double heterostructures under ^{60}Co gamma irradiation with regard to irradiation temperature. The study indicated several consecutive stages of LED emissive power lowering under ionizing radiation. Increased temperature during gamma irradiation enhances radiation resistance at the first stage due to radiation-stimulated defect annealing, which reduces relative contribution of the first stage to the overall emissive power lowering. It was found that in exposure at temperature more than 380 K, the first stage of LED emissive power lowering is completely eliminated. At the second stage, increase in resistance is caused by the decreased relative contribution of the less stable first stage to the overall emissive power lowering. The maximum resistance of LEDs to gamma radiation depends on radiation resistance of metal–semiconductor contacts.

1. Introduction

Light-emitting diodes (hereinafter LEDs) are the basis for manufacturing various electronic devices and systems that operate under various conditions, including elevated temperature during exposure to ionizing radiation. A simultaneous (complex) and staggered (combined) effect of ionizing radiation and long operating time may significantly affect both the process of emissive power lowering and reliability of LEDs. Based on general information on radiation defect accumulation and annealing, it can be unambiguously assumed that increased temperature during exposure to ionizing radiation is supposed to decrease the rate of LED parameter degradation due to thermal annealing [1,2]. At present, no information is available on the studies of the effect of irradiation temperature on LED resistance to ionizing radiation.

On the other hand, ionizing radiation can stimulate radiation-enhanced diffusion and accelerate LED degradation [3]. In the first place, the occurrence of radiation-enhanced diffusion effects can be one of the main causes of the accelerated degradation of metal-semiconductor contacts under simultaneous and/or combined action of long-term operation and ionizing radiation. Virtually no information on the effect of ionizing radiation on the performance reliability of semiconductor devices and devices based on them is currently available.

The paper aims to investigate the change in the parameters of IR LEDs based on AlGaAs double heterostructures exposed to ^{60}Co gamma radiation depending on irradiation temperature.

2. Research objects and methods

The objects to study were commercial IR LEDs based on double AlGaAs heterostructures. The heterostructures were produced through liquid-phase epitaxy, and an n-GaAs single crystal was used as a substrate. The thickness of the active layer was about 5 μm.

The emissive power of each of the LEDs was measured in the ball at an operating current of 50 mA before and after irradiation under normal climatic conditions. The obtained measurement results were processed by mathematical statistics. An average value of the measured parameters was calculated for each of the batches of the test LEDs. Each batch contained 20 LEDs. Since the error of the emissive power measurement did not exceed ±2.5%, the spread of the LED emissive power in the batch did not exceed ±10% and it insignificantly increased under irradiation, we do not provide the relations with confidence intervals.

The temperature in the active region of LEDs during gamma irradiation was raised by means of the forward operating current (increased forward current during irradiation results in temperature growth in the active region). The thermal resistance of the LED being known, the active region of the temperature was calculated by the formula

$$T = T_P + R_T \cdot I_{op} \cdot U_{op} \tag{1}$$

where T is the temperature in the LED active region during irradiation; T_p is the temperature of the LED housing measured during irradiation; R_T is thermal resistance of LED; I_{op}, U_{op} are operating current and operating voltage, respectively.

Irradiation was performed in different power supply modes under normal conditions, and the gamma irradiation level was characterized by the absorbed dose D_γ. The temperature of the housing during irradiation was controlled. When estimating the temperature in the active region of LED by relation (1), it was assumed that the thermal resistance of LED did not depend on the temperature and gamma radiation absorbed dose. Thus, the temperature in the LED active region varied in the range of (320 –350) K.

The required irradiation level was provided by a consequent set of the absorbed dose for each of the chosen irradiation temperatures (i.e., for chosen operating current values). Multiple measurements of the tested parameters under intermediate levels of gamma irradiation showed high repeatability of measurement results, which indicated the absence of radiation defect annealing in intermediate measurements.

After irradiation and cooling, LEDs were removed from the irradiation box and the required parameters were measured.

3. Results and discussion

Consider the experimental results. Figure 1 shows the data on emissive power lowering for several LED batches. Each LED batch has its own temperature value in the active region.

The results indicate several consecutive stages of emissive power lowering under ionizing radiation, each of which has its own pattern of LED emissive power lowering. It should be noted that as the irradiation temperature grows, the differences between the identified stages of emissive power lowering become more distinct (in figure 1, the solid arrow shows a shift of the boundary between the first and the second stages of emissive power lowering at increased irradiation temperature).

The studies conducted previously [4] suggest that emissive power lowering at the first stage is caused by radiation reorganization of the defect structure (growth of defects, technological defects, etc.). In other words, this stage of emissive power lowering is finite and limited by the initial defect content. At the second stage, emissive power lowering is caused by introduction of radiation-induced defects. At the third stage, LED turns to the mode of low electron injection into the active LED layer and catastrophic failure occurs (ohmic contact degradation, breakdown and/or burnout of the active layer). This is a physical limit of LED resistance to ionizing radiation. The third stage is not shown in figure 1. The above-described radiation model is fully consistent with the study results published earlier [4, 5]. Consider the first stage of emissive power lowering in more detail. At this stage, LED emissive power lowering can be described by the relation

$$\frac{P_\gamma}{P_0} = \left(k_1 \cdot D_\gamma\right)^\alpha \tag{2}$$

where P_0, P_γ are LED emissive power at a forward current of 50 mA before and after gamma irradiation, respectively; k_1 is the damage coefficient at the first stage, the amount of which does not depend on irradiation temperature and characterizes the resistance of the double AlGaAs heterostructure to gamma radiation at the first stage of emissive power lowering; α is the exponent and its value determines the irradiation temperature contribution to LED emissive power lowering at the first stage.

The results presented in figure 1 show that when temperature grows, the proportionality factor α decreases. This indicates that during LED emissive power lowering caused by radiation-enhanced reorganization of the initial defect structure (the first stage of emissive power lowering), the emissive power is observed to restore as a result of annealing of the defects involved in reorganization of the initial defect structure, i.e. relative contribution of the first stage to the overall process of LED emissive power lowering decreases due to gamma irradiation.

In previous studies conducted to test the performance reliability of these LEDs through step-by-step tests, no facts of emissive power restoration as a result of LED operation at increased temperatures were found [6]. Therefore, it can be argued that the observed annealing of the defects involved in the reorganization of the initial defect structure at the first stage is enhanced by gamma irradiation.

The obtained results were used to determine the dependencies for emissive power lowering at the first stage under gamma irradiation at different temperatures which are well described by relation (2). Some of them are shown in figure 1.

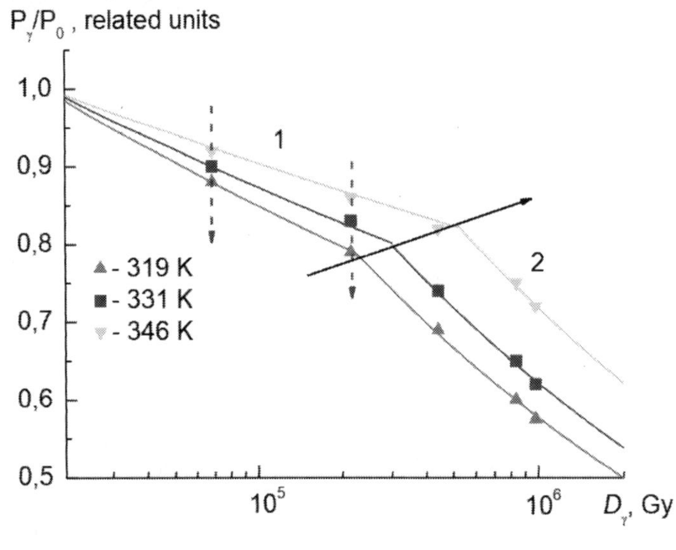

Figure 1. Change in the LED emissive power under gamma irradiation at different temperatures of the active region: 1, 2 are the stages of emissive power lowering; the solid arrow indicates the boundary between stages 1 and 2; vertical arrows show the change in relative contribution of the 1st stage to the overall LED emissive power lowering at different irradiation temperatures.

The experimental data can be used to plot the dependence of α coefficient on irradiation temperature shown in figure 2.

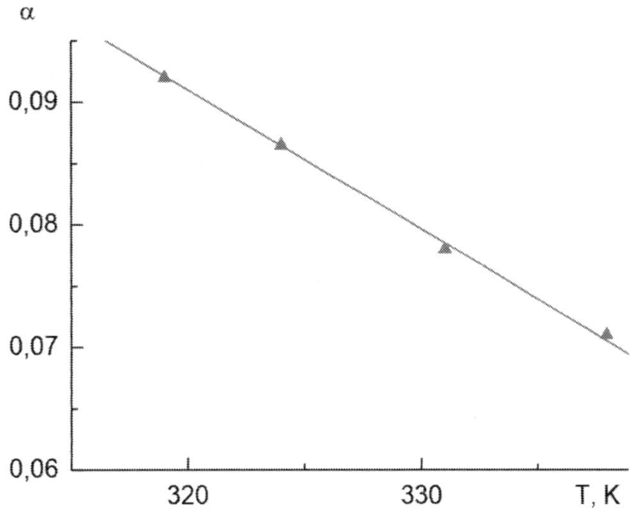

Figure 2. The α coefficient versus the irradiation temperature under gamma irradiation

The determined dependence allows estimation of the temperature which indicates zero contribution of the first stage to emissive power lowering, at $T_{\min 1} \geq 400$ K.

Consider other possible ways to determine the temperature characteristic of the first stage $T_{\min 1}$. If we make sections in the region where only the first stage of emissive power lowering occurs, as shown in figure 1 in dotted downward directed vertical arrows, the dependence of the $(1-P_{\gamma i}/P_0)$ value on the irradiation temperature for each of the sections will indicate the change in the contribution of the first stage of A_{1i} emissive power lowering for the given level of exposure to $D_{\gamma i}$ against irradiation temperature.

Thus, we obtain the results shown in figure 3, which give a similar value of the limit temperature $T_{\min 1} \geq 400$ K.

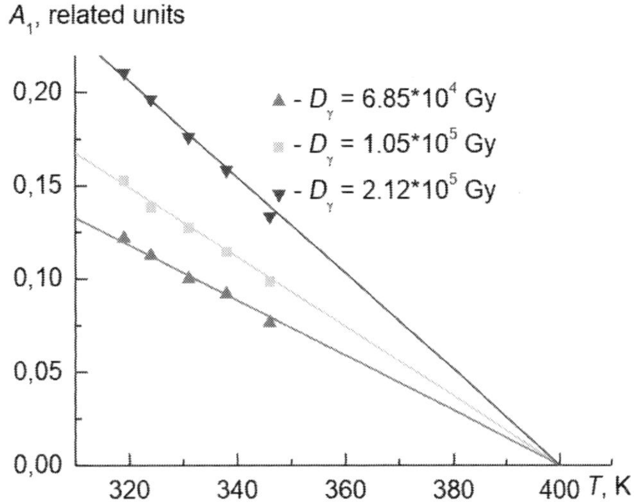

Figure 3. The contribution of the first stage of LED emissive power lowering under gamma irradiation to the overall power lowering versus radiation temperature

IOP Conf. Series: Journal of Physics: Conf. Series **830** (2017) 012134 doi:10.1088/1742-6596/830/1/012134

4. Conclusion

In conclusion, summarize the main results of the studies of emissive power lowering of LEDs based on double AlGaAs heterostructures exposed to ^{60}Co gamma radiation at different temperatures.

1. The observed emissive power lowering of the studied LEDs irradiated with increased gamma doses occurs through three stages:

- at the first stage, LED emissive power lowering is caused by radiation-stimulated reorganization of the initial defect structure;
- at the second stage, LED emissive power lowering is caused by introduction of purely radiation defects;
- for the limit case, the second stage is followed by the third stage, when LED turns to the mode of low electron injection into its active layer.

2. The third stage determines terminal radiation resistance of LEDs. At this stage, catastrophic failures of LEDs caused by mechanical rupture of the metal-semiconductor boundary can be observed.
3. At the first stage of LED emissive power lowering under gamma irradiation at increased temperatures, two different simultaneous radiation-stimulated processes can be observed: radiation-stimulated emissive power lowering due to reorganization of the initial defect structure and radiation-stimulated annealing of defects that cause decrease in LED emissive power lowering.
4. As the irradiation temperature increases, the differences between the first and the second stages of emissive power lowering become increasingly distinct. This demonstrates the validity of the proposed radiation LED model.
5. As the irradiation temperature grows, relative contribution of the first stage to the overall LED emissive power lowering decreases, and at irradiation temperature ≥380 K, the first stage of LED emissive power lowering disappears due to full radiation-stimulated annealing of the defects responsible for LED emissive power lowering at this stage.
6. At the second stage of LED emissive power lowering, the rate of radiation defect introduction does not depend on the irradiation temperature within the studied temperature range of (300–350) K. Increased LED resistance to gamma radiation observed in practice is caused by decreased contribution of the first stage to the overall LED emissive power lowering.
7. For certification testing of LEDs based on AlGaAs heterostructures, a passive mode of power supply is recommended since this mode provides minimal resistance of LEDs to ionizing radiation.

References

[1] Hirunenko L I, Shakhovtsov V I and Shumov V V 1998 *Physics and technology of semiconductors* **32** 2 pp 132-134
[2] Pagava T A 2006 *Physics and technology of semiconductors* **40** 8 pp 919-921
[3] Djafarov T D 1991 Radiatcionno-stimulirovannaia diphphuzya v poluprovodnikah [Radiation-induced diffusion in semiconductors] Moscow Energoatomizdat Publ. 228
[4] Gradoboev A V and Sednev V V 2015 *Applied mechanics and materials* **770** pp 518-521
[5] Gradoboev A V and Orlova K N 2014 *Advanced material research* **880** pp 237-241
[6] Gradoboev A V, Simonova A V and Orlova K N 2015 VI Shkola-konferentsiya molodykh atomshchikov Sibiri: sbornik tezisov dokladov Tomsk
[7] Lampert M and Mark P 1973 Inzhektsionnye toki v tverdykh telakh [Injection currents in solids] Moscow Mir Publ. 416

Luminescence of ultrafine magnesium oxide ceramics with deep traps

S V Nikiforov, V S Kortov, A N Kiryakov and K A Petrovykh

Ural Federal University, Ekaterinburg, Russia

E-mail: s.v.nikiforov@urfu.ru

Abstract. Thermoluminescence (TL) of ultrafine anion-defective magnesium oxide ceramics with a grain size of about 150 nm, which were thermally treated in highly reducing conditions at T=1100–1400 °C, has been studied after high-dose irradiation with a pulse electron beam (130 keV). Thermal treatment was found to cause TL intensity grow, which is associated with an increasing concentration of F-type centers. It was shown that varying occupancy of deep trapping centers changes a yield of the TL peak at 380 K. A hypothesis was made that the traps which are responsible for the TL peak at 380 K mainly have a hole nature in the samples under study.

1. Introduction

Luminescence properties of phosphors are predominantly determined by the processes of charge transfer between different localized states including the ones associated with deep trapping centers [1]. The competing processes in charge trapping by deep traps cause a number of effects essential for applications of many technically important phosphors in particular metal oxides (α-Al_2O_3, SiO_2, BeO). These effects include changes in the luminescence yield (sensitization and desensitization) after radiation treatment [2-5], and non-linear dependences of thermoluminescence (TL) intensity on the absorbed dose of ionizing radiation [3,6]. Moreover, a temperature dependence of competing effect of deep centers can cause luminescence quenching and a resulting drop in the TL yield with an increasing heating rate of the material [7], which should not be neglected in TL dosimetry.

Deep trapping centers are present in such phosphor as magnesium oxide (MgO). An interest in the study of its luminescent properties is due to a possibility of its use as a TL detector of gamma-, UV- and neutron radiations [8-10]. Deep traps in the material are detected through TL measurements in the range of 450–800 K [10,11], and also indirectly via measurements of phototransferred TL [12]. From the data obtained with the use of EPR method, it was found that a number of high-temperature TL peaks are associated with destruction of hole V-type centers connected with cation vacancies and more complex H-centers [10,11]. However, many aspects of the charge transfer processes with deep traps involved, which cause changes in TL sensitivity to ionizing radiation, are little studied and require further research. Improving TL sensitivity is especially important for ultrafine modification of magnesium oxide, as phosphors with small sizes of grains are usually less TL sensitive to radiation in comparison with their bulky analogs [13,14]. Furthermore, the study of TL properties of ultrafine MgO is of special interest because nanophosphors are promising materials for high-dose (over 1 kGy) dosimetry due to their high radiation resistance [13,14].

IOP Conf. Series: Journal of Physics: Conf. Series **830** (2017) 012135 doi:10.1088/1742-6596/830/1/012135

The aim of this work is to study luminescent properties of ultrafine magnesium oxide, which are associated with deep traps.

2. Samples and Experimental Methods

The studied MgO samples were in the form of compacts 5 mm in diameter and 1 mm thick made of nanopowder using the method of uniaxial cold pressing at a unit pressure of 1000 kgf/cm^2. The precursor powder with the grain size of 45–75 nm was produced by "Plasmoterm" (Moscow, Russia) by plasma synthesis method. The proportion of MgO was 99.8 weight%. The powder contained Fe_2O_3 (not more than 0.1%) and SiO_2 (not exceeding 0.1 %) impurities.

To increase the luminescence yield, the precursor compacts were exposed to high temperatures (T=1100–1400 °C for three hours) in vacuum (10^{-3}–10^{-4} Torr) in reducing conditions. The conditions are created due to the presence of carbon as a graphite rod. The reported thermal treatment of the materials based on wide-gap oxides causes increased concentrations of F-type centers in these materials, where these centers are associated with oxygen vacancies which determine luminescent properties of this class of materials [15,16]. It is known that when nanosized materials are treated with high temperatures, nanoparticles stick together to form larger particles [17]. An X-ray diffraction method was used to measure the sizes of nanoparticles in thermally treated MgO compacts. X-ray diffraction studies of nanopowder compacts were carried out with Shimadzu XRD Maxima-7000 diffractometer. Imaging was performed in Cu-K$\alpha_{1,2}$-radiation in scan-step mode with $\Delta(2\theta) = 0.03°$ in the 2θ angle range from 15 to 115°, exposition time in each point was 3 s.

To excite TL, the samples were exposed to a pulsed electron beam from an accelerator (pulse time 2 ns, mean electron energy (130±1) keV, current density 60 A/cm^2) at room temperature. An absorbed dose from one pulse was found experimentally with a film dosimeter PD(F)R-5/50 and was 1.5 kGy/pulse [18]. A source of beta-radiation BIS-10 on the base of $^{90}Sr/^{90}Y$ isotopes with the dose rate of 8 mGy/min was also used in the experiments. TL was measured at linear heating with the rate of 2 K/s by using FEU-130 photomultiplier tube with the maximum spectral sensitivity of 400–420 nm. An experimental facility for measuring TL spectra is described in [19].

3. Results and Discussion

Figure 1 presents X-ray diffraction pattern of magnesium oxide compacts which were annealed at the maximum temperature T=1400 °C. The obtained results show that the samples under study contained a cubic phase (space group $Fm\overline{3}m$). Miller indexes related to different diffraction peaks are also given in figure 1.

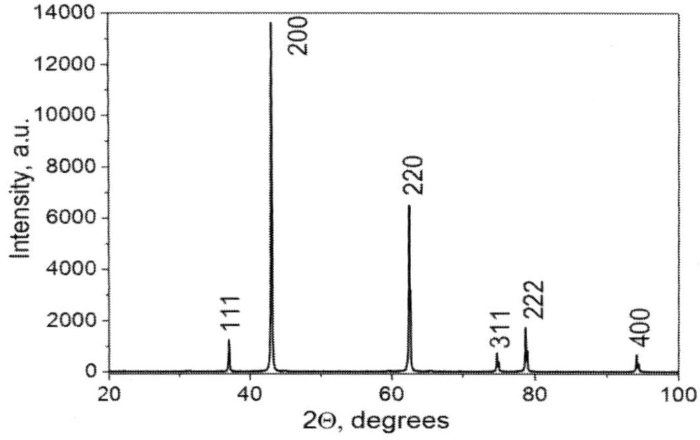

Figure 1. X-ray diffraction pattern of the magnesium oxide compacts thermally treated at 1400 °C

The particle sizes in the samples under study were determined following broadening of the diffraction peaks by using Williamson-Hall method [20]. The observed peak broadening in nanomaterials can be caused by inhomogeneity, breaking of translation symmetry of the crystalline structure and of some elements of point symmetry due to small sizes of the particles. To approximate the profile of the diffraction peaks, the pseudo-Voigt function, which is a weighted superposition of Gaussian and Lorentz functions, was chosen. The peak broadening was determined taking into account the resolution of the device.

Figure 2 shows dependence of the given broadening of peaks on the magnitude of scattering vector for magnesium oxide compacts, which were thermally treated at the maximum temperature (1400°C). It can be seen that broadening grows when the scattering vector becomes longer. This means that broadening is affected not only by particle sizes, but also by micro-deformations of the crystalline lattice. Then, the value of the given broadening β_0 in the crossing point of the straight line with the axis of ordinates was found by extrapolating linear dependence which approximates the experimental points. The particle size was calculated as a reciprocal value of β_0 and was about 150 nm. The results obtained show that thermally treated MgO compacts are ultrafine ceramics.

Figure 3 presents TL curves on a logarithmic scale of the studied MgO compacts (both initial and synthesized at different temperatures) after exposure to a pulsed electron beam (30 kGy). The TL curve of the initial samples features three peaks. Peak A has the maximum at 380 K; it is considered as the main one further on. Peak *B* at 510 K and peak *C* at 630 K, which are associated with deep traps are also seen on the TL curve. The data in figure 3 show that high-temperature annealing in the reducing environment leads to greatly increased TL yield in peak *A* (by more than two orders of magnitude). One can hypothesize that TL intensity grows due to increased concentration of anion defects which create luminescence F-type centers in different charge states. Formation of F- and F^+-centers in the thermally treated compacts under study in our work was verified before by using luminescence spectroscopy [21].

A growing TL yield is also observed in the temperature range of deep traps luminescence (450–670 K) after high-temperature annealing. Emergence of the regions where TL intensity only slightly depends on temperature on the TL curves is an interesting characteristic of TL of the thermally treated samples. To clarify the TL mechanism responsible for such curves, we preliminary measured isothermal decay of the TL samples which were thermally treated at different temperatures [21]. We found that decay kinetics did not depend on the temperature of isothermal exposure, and the TL decay curves were described with hyperbolic function of $I(t)=K/t$-type.

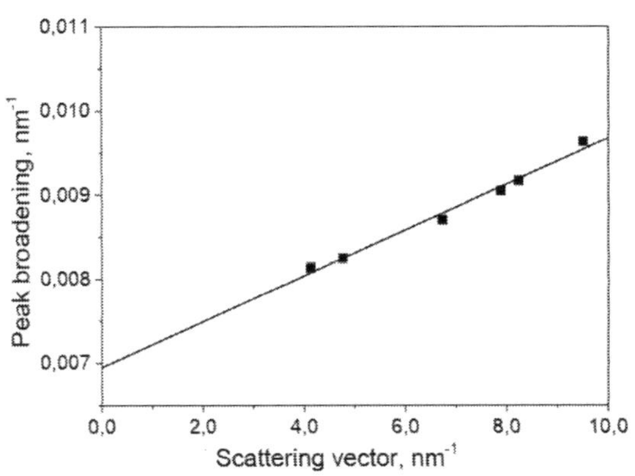

Figure 2. Dependence of the given broadening on the magnitude of the scattering vector for magnesium oxide compacts thermally treated at 1400 °C

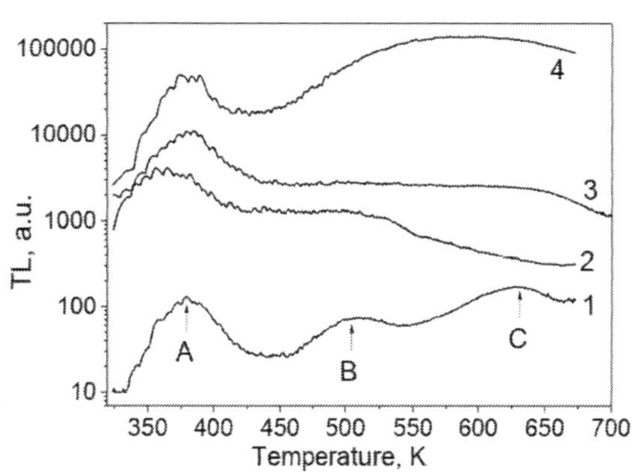

Figure 3. TL of magnesium oxide after exposure to a dose of 30 kGy with a pulsed electron beam: precursor compacts (1); thermally treated compacts at 1200 °C (2); at 1300 °C (3); at 1400 °C (4)

The results obtained confirm the fact of existence of the tunneling mechanism of recombination of charge carriers which were released from the deep traps. The reported mechanism was earlier studied in YPO_4 doped with different impurities [22]. For more details of this effect in magnesium oxide further studies are required.

In addition, we studied an effect of deep traps occupancy on the intensity of the main TL peak at 380 K. MgO ceramics samples thermally treated at 1200°C were used. Deep traps were preliminary filled with 50 pulses of an electron beam (75 kGy) at room temperature. After exposure to an electron beam, the samples were annealed up to 470 K till the traps responsible for the main peak at 380 K were completely emptied. The measurement cycle was as reported below. In the samples with filled deep traps, TL of the main peak was excited with a low dose of beta-radiation (40 mGy) and was recorded at 325–475 K. After that, the samples were annealed up to T_{ann} to partly empty the deep traps. T_{ann} temperature increased with a constant step (20 K) and changed within the range of 470–730 K in every cycle.

Figure 4 shows a dependence of the intensity of the TL peak at 380 K on the temperature of the stepped annealing (curve 1). 1.0 a.u. level is related to the initial TL sensitivity before filling of the deep traps. Curve 2 is TL of the deep traps, which was obtained after thermal cleaning up to T=470 K at a heating rate of 2 K/s to exclude the effect of the TL peak at 380 K. It can be seen that when the deep traps are filled, the intensity of the TL peak at 380 K increases by approximately one order of magnitude. When the deep centers responsible for TL peaks at 510 and 600 K are emptied further, desensitization of TL at 380 K is observed. At T_{ann}=730 K, TL sensitivity returns to its initial value before high-dose irradiation.

This effect can be explained through the model of competing traps capturing carriers of the same sign [23]. Filling of the deep traps decreases the probability of competing trapping of the charge carriers by these traps and leads to a growing number of events of emission recombination on the luminescence centers, which causes sensitization of TL of the main traps responsible for the TL peak at 380 K. Decreasing occupation of the deep centers under stepped annealing, on the contrary, conditions decreasing possibility of radiative transitions and, therefore, desensitization of TL. Such mechanism of changing TL sensitivity, which is associated with a competing effect of deep centers at the stages of irradiation and thermal stimulation, has been observed before in other wide-gap oxide phosphors (Al_2O_3, SiO_2) [2,3,5]. In ultrafine magnesium oxide the effects of TL sensitization/desensitization have been found by us for the first time.

Learning more about the details of the charge transfer processes, the sign of the charges and the nature of the defects associated with the traps of the main TL peak and deep traps in the compacts under discussion in this paper requires further study. What we can state so far is the following.

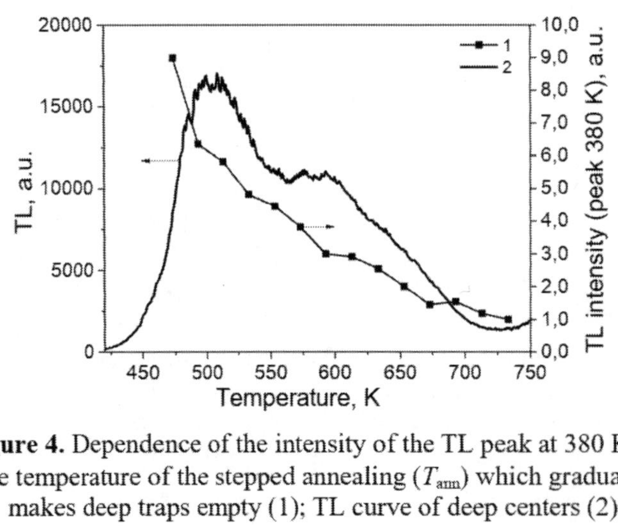

Figure 4. Dependence of the intensity of the TL peak at 380 K on the temperature of the stepped annealing (T_{ann}) which gradually makes deep traps empty (1); TL curve of deep centers (2)

There are data from the literature, which are based on the measurements of EPR spectra, that the TL peak at 380 K is caused by releasing holes from V-type centers [10,11]. These centers are single or associated with impurities cation vacancies which trapped one or several holes [24]. In magnesium oxide such centers can be formed when the oxide is exposed to high-energy particles, which was not observed in our experiments as the energy of the electron beam (130 keV) was much lower than the threshold value necessary to create cation and anion vacancies in MgO according to the knock-out mechanism (330 keV) [25]. There are also literature data on high concentration of similar centers in non-irradiated MgO samples with high proportion of impurities, which is conditioned by a charge-compensation principle [26]. As the samples we examined contained large amount of three-valent iron impurities, one can hypothesize that a chance of formation in them of cation vacancies which form V-type centers is high. When the samples are irradiated, the charge states of these defects change. As a result of further thermal stimulation and destruction of V-type centers, freeing of holes, whose recombination on the luminescence centers causes TL at 380 K, is observed.

Our results of the measurements of the TL spectrum recorded at T=365 K (Figure 5) support the hypothesis that the TL peak at 380 K has the hole nature. It is seen that a band at 3.1 eV (400 nm) caused by relaxation of F^+-centers dominates in the spectrum [27]. Luminescence of F^+-centers in the TL peak at 380 K in the samples under study can be caused by trapping of the holes released from the traps by F-centers which are transformed into excited F^+-centers with their further radiative relaxation: $F+h^+=F^{+}*=F^{+}+h\nu$(3.1 eV). This mechanism has been earlier discussed for TL of anion-defective alumina [2]. In the case of MgO, the presence of the dominating band of F^+-centers may speak for the hole nature of the traps responsible for the TL peak at 380 K, which is in agreement with the data from the literature.

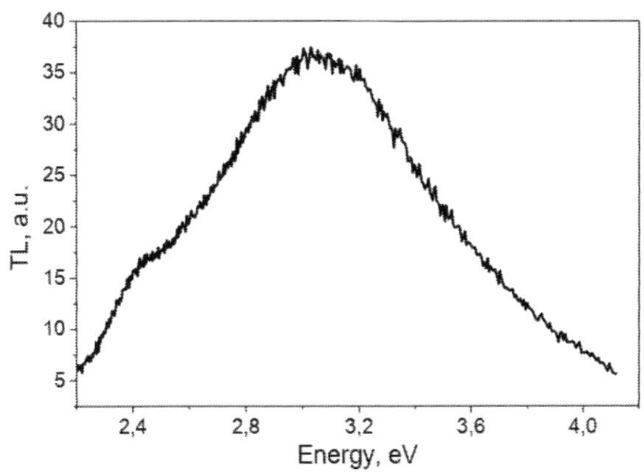

Figure 5. Emission spectrum of the TL peak at 380 K, which was measured at T=365 K

The desensitization effect of TL of this peak under stepped annealing allows hypothesizing on the hole nature of deep traps associated with the TL peaks at 510 and 600 K. Works [10,11] report on the hole nature of the TL peak at 495 K, which is near the peak at 510 K in its temperature position. In these works, the peak at 495 K is associated with V^--centers. There is also an opinion that the TL peak at 615 K is caused by recharging of chromium impurity ions [10].

4. Conclusions

The characteristics of TL associated with deep traps in ultrafine magnesium oxide compacts were studied. It was shown that their thermal treatment in the reducing environment at T=1100–1400 °C make the TL yield grow due to formation of F-type centers and cause the grain size increase up to 150 nm. The effects of changes of the TL yield in the peak at 380 K as a result of varying occupancy of deep traps were found. It was shown that TL sensitization of the peak is caused by a competing interaction of the traps of the main TL peak and deep traps capturing the carriers of same sign. A hypothesis on the predominantly hole nature of the traps responsible for the observed TL peaks of the samples under study was made on the base of the literature data and measurements of the spectral composition of TL.

Acknowledgements
This work has been done as a part of the governmental task of the Ministry of Education and Science of the Russian Federation (Grant N 3.1016.2014/k).

The authors are grateful to professor Aleksandr Lushchik (Tartu University, Estonia) for helpful discussions of the nature of the trapping centers in magnesium oxide.

References
[1] Nikiforov S V and Kortov V S 2014 *JOP:Conf. Ser.* **552** 012042
[2] Yukihara E G *et al* 2003 *Rad. Meas.* **37** pp 627-638
[3] Chen R, Fogel G and Kristianpoller N 1994 *Rad. Meas.* **23** pp 277-279
[4] Murthy K B S *et al* 1978 *J. Phys. D: Appl. Phys.* **11** pp 561-565
[5] Nikiforov S V *et al* 2014 *Rad. Meas.* **71** pp 74-77
[6] Nikiforov S V, Kortov V S and Kazantseva M G 2014 *Phys. of the Sol. St.* **56** pp 554-560
[7] Nikiforov S V, Milman I I and Kortov V S 2001 *Rad. Meas.* **33** pp 547-551
[8] Soliman C 2009 *Rad. Effects and Defects in Solids* **164** pp 257-265
[9] Kortov V S *et al* 1993 *Rad. Prot. Dosim.* **47** pp 273-276
[10] Dolgov S *et al* 2002 *Rad. Prot. Dosim.* **100** pp 127-130
[11] Lushchik A *et al* 2007 *Rad. Meas.* **42** pp 792-797
[12] Sono D A and McKeever S W 2002 *S. Rad. Prot. Dosim.* **100** pp 309-312
[13] Kortov V S 2010 *Rad. Meas.* **45** pp 512-515
[14] Salah N 2011 *Radiat. Phys. and Chem.* **80** pp 1-10
[15] Akselrod M S, Kortov V S and Gorelova E A 1993 *Rad. Prot. Dosim.* **47** pp 159-164
[16] Monge M A *et al* 2000 *Phys. Rev.* B. **62** pp 9299-9304
[17] Rao C N and Cheetham A K 2001 *J. Mater. Chem.* **11** pp 2887-2894
[18] Afanas'ev V N *et al* 2005 *Inst. and Exper. Tech.* **48** pp 641-645
[19] Vokhmintsev A S *et al* 2015 *Meas.* **66** pp 90-94
[20] Hall W H and Williamson G K. 1951 *Proc. Phys. Soc. London. Sect.* B **64** pp 937-946
[21] Nikiforov S V, Kortov V S and Petrov M O 2016 *Rad. Meas.* **90** pp 252-256
[22] Dobrowolska A, Bos A J and Dorenbos P J 2014 *Phys. D: Appl. Phys.* **47** 335301
[23] Chen R, Fogel G and Lee C K 1996 *Rad. Prot. Dosim.* **65** pp 63-68
[24] Dolgov S A *et al* 2011 *Phys. Sol. St.* **53** pp 1244-1252
[25] Lushchik A *et al* 2007 *Phys. Stat. Sol.* **4** pp 1084-1087
[26] Las W C and Stoebe T G 1982 *J. of mat. Science* **17** pp 2585-2593
[27] Rosenblatt G H *et al* 1989 *Phys. Rev.* B **39** pp 10309-10318

Dynamic ignition regime of condensed system by radiate heat flux

V A Arkhipov[1], **N N Zolotorev**[1], **A G Korotkikh**[1,2] **and V T Kuznetsov**[1]

[1] Research Institute of Applied Mathematics and Mechanics, Tomsk State University, 36 Lenin Ave., Tomsk, 634050, Russia
[2] Institute of Power Engineering, Tomsk Polytechnic University, 30 Lenin Ave., Tomsk, 634050, Russia

E-mail: korotkikh@tpu.ru

Abstract. The main ignition characteristics of high-energy materials are the ignition time and critical heat flux allowing evaluation of the critical conditions for ignition, fire and explosive safety for the test solid propellants. The ignition process is typically studied in stationary conditions of heat input at constant temperature of the heating surface, environment or the radiate heat flux on the sample surface. In real conditions, ignition is usually effected at variable time-dependent values of the heat flux. In this case, the heated layer is formed on the sample surface in dynamic conditions and significantly depends on the heat flux change, i.e. increasing or decreasing falling heat flux in the reaction period of the propellant sample. This paper presents a method for measuring the ignition characteristics of a high-energy material sample in initiation of the dynamic radiant heat flux, which includes the measurement of the ignition time when exposed to a sample time varying radiant heat flux given intensity. In case of pyroxyline containing 1 wt. % of soot, it is shown that the ignition times are reduced by 20–50 % depending on the initial value of the radiant flux density in initiation by increasing or decreasing radiant heat flux compared with the stationary conditions of heat supply in the same ambient conditions.

1. Introduction

The features of ignition of condensed substances by means of heating with the external heat flux (conductive, convective, radiant or mixed) are required to develop initiating explosives systems and high-energy materials, to assess fire and explosive features of substances and to solve a number of other practical problems. In this paper, the ignition is a stationary self-sustaining combustion mode after the termination of an external heat flux [1].

At present time, the characteristics of ignition during constant heating by heat flux in the so-called static heating mode are being studied [1, 2]. In real conditions, in particular at the ignition of solid propellant rocket engine [3], combustible materials under fire conditions [4, 5] and etc., the ignition is carried out through heating by heat flux in a dynamic mode, which depends on time.

A theoretical analysis of the dynamic mode of ignition is conducted mainly through a thermal solid-phase model. In this case, the formation of the heated layer in the condensed phase takes place in complicated dynamic conditions, whereby the ignition delay time substantially depends not only on the heat flux density to the sample, but also on the mode of changes in the heat flux induction period [1, 2, 5–9].

IOP Conf. Series: Journal of Physics: Conf. Series **830** (2017) 012136 doi:10.1088/1742-6596/830/1/012136

An experimental research in characteristics of ignition condensed substance by exposure to radiant heat flux is carried out in most of the works with constant heat flux in the induction period [10-17]. Dynamic ignition modes with convective and radiant heat flux in the available literature have been considered in single publications [4, 18-21], apparently, due to the technical complexity of the correctly set experiments.

In [18, 19], the results of ignition of polyvinyl nitrate and pyroxyline are presented. These samples were lit when exposed to the increasing and decreasing convective heat flux created by gaseous products of combustion in a combustion chamber with a nozzle. Due to the large spread of the experimental data on ignition delay time (up to 60%), only quality ignition characteristics were obtained in these studies.

The results of measurements of ignition characteristics for pyroxyline [20, 21] and for timber [4], when exposed to dynamic mode of the heat flux, showed that the same values of the heat flux density for the ignition delay t_{ign} differ for static and dynamic modes of heating.

In these studies in the analysis of experimental data, an average heat flux for dynamic modes of heating was used:

$$\overline{q} = \frac{1}{t_{ign}} \cdot \int_{0}^{t_{ign}} q(t)dt \,, \qquad (1)$$

where $\overline{q}$ is the average value of the variable heat flux density $q(t)$ of the induction period t_{ign}.

As noted in [10], it does not take into account fully the dynamics of change in the temperature profile in the condensed phase. The generalization of the experimental data on the dynamic modes of ignition [4, 18-21] is difficult due to component composition of the samples of condensed systems, methods of research and conditions of the experiment.

The aim of this work is the experimental study of the characteristics of pyroxyline ignition under dynamic modes as a model substance with the known thermal and kinetic characteristics when heated by thermal radiation. To correctly analyze the ignition characteristics, the experiments were carried out in strictly identical conditions using a single methodology for continuous, increasing and decreasing heat flux.

2. Experiment

The experiments were carried out using an optical furnace "Uran-1" [10]. The emitter of the setup consists of an elliptical mirror and a DKsR-10000 xenon lamp (Figure 1) with a nominal heat output of 5.5 kW with radiation A = 0.25–1.85 μm (Figure 2). The influence of the variable heat flux on the delay time of ignition of pyroxyline in the air at atmospheric pressure was studied. The distribution of the radiation in the spectral ranges was as follows: 0.5 kW (9%) in the ultraviolet spectral range, 2 kW (36%) in the visible range, and 3 kW (55%) in the infrared range. The integrated radiant flux (in a broad range of wavelengths) from the DKsR-10000 xenon lamp of Uran-1 was focused by an elliptical mirror into a spot 12 mm in diameter. This capacity corresponds to the integral blackbody radiation at a temperature of 3110 K.

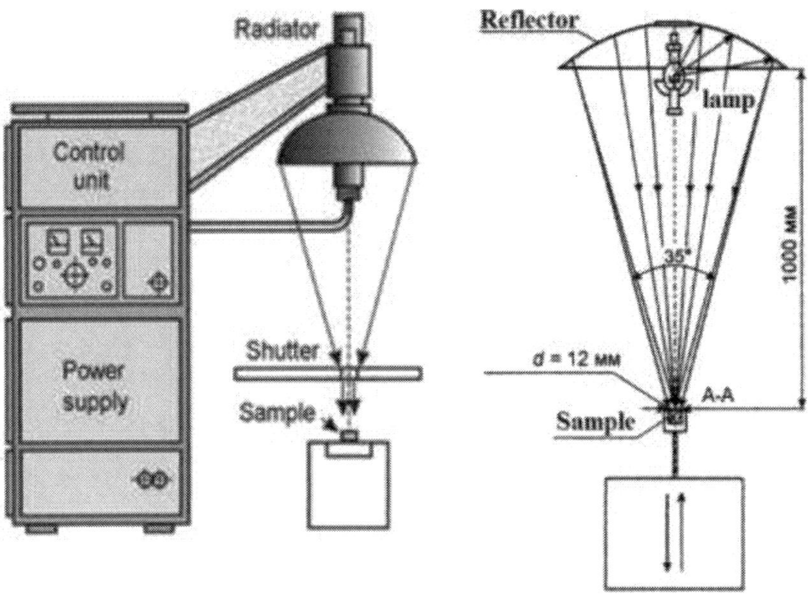

Figure 1. Optical scheme of installation of the integral radiant heat "Uran-1"

In the research, a metal mesh placed perpendicular to the optical axis of the elliptical reflector at a distance of 30–40 cm from the focal plane was used to weaken the power of the radiation. The strength of the current and voltage supplied to the xenon lamp were reduced. The flux density was measured by calorimetry with a relative error not exceeding 10%. We used a copper calorimeter as a disk 10 mm in diameter and 3 mm thick. The surface of the disk that absorbs radiation was covered with lamp black.

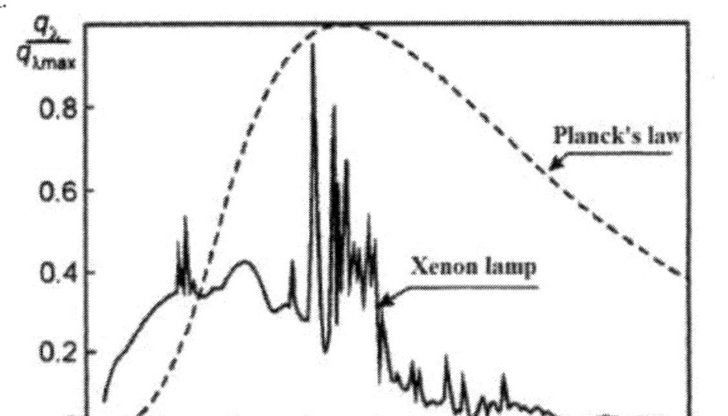

Figure 2. The distribution of the spectral irradiance of the DKsR-10000 xenon lamp and a black body at 3110 K

The pyroxylin samples containing 1 wt. % of carbon black K-354 were used as a model. Carbon black is added to reduce the influence of the optical parameters of the heat source. The diameter and height of the samples was 10 mm, and the density was 1.45 g/cm^2.

The signal of the ignition delay time was recorded by the photodiode FD-9 at the first appearance of flame. When the ignition electrical signals from the photodiodes were recorded by oscilloscope OWON, the data were saved. Indications of the difference between the two signals on the oscilloscope match the ignition delay time.

A dynamic heat flux mode is ensured by moving the test sample while it is heated by radiation heat flux along the optical axis of the elliptical reflector, experimental facility "Uran-1". For this purpose, we used an electromechanical device allowing the sample to move at a speed of 4.75 cm/s The dynamic regime was carried out with a speed of movement of the sample.

In order to make clear the heat flux distribution function the density of heat flux at three or four fixed points on the optical axis of the setup was determined before each measurement of the ignition delay time of pyroxyline. The heat flux density was measured using a micro calorimeter installed in the place of the test sample [7]. The accuracy of the heat flux density measurement does not exceed 10%. Based on the known speed of movement of the sample along the optical axis, it is possible to determine the dynamics of the heat flux density in this experiment.

According to the obtained values of ignition delay time of VEM at different radiation flux, the move distances of the samples at the moment of ignition were calculated. Calibration of measuring the heat flux density of the system by moving the test pyroxylin sample relative to the focal plane of the radiator was carried out in static conditions. A heat flux sensor was placed in the focal plane ($f \approx 1$ m), the exact coordinate of which is determined by the maximum sensor readings ($q = q_f$) for a given supply current value of the xenon lamp and was taken as $x = 0$.

Then the heat flux sensor was moved along the optical axis of the apparatus in the direction from the emitter, and the heat flux density at several fixed points $x_i > 0$ along the coordinate x was measured. Similar measurements were performed to move the sensor of the heat flux toward the emitter. The resulting averaged tabular dependence of $q_i(x_i)$ was approximated by exponential function

$$\overline{q}(x) = \frac{q(x)}{q_f} = \exp(-bx), \quad (2)$$

where $q(x)$ is the heat flux density at distance x from the focal plane; q_f is the heat flux density in the focal plane ($x = 0$).

Figure 3 shows a generalized dependence obtained in the range of $x = (0 \div 17.5)$ cm, $= (10.2 \div 121.2)$ W/cm^2 (coefficient of determination $R^2 = 0.9865$):

$$\overline{q}(x) = \exp(-0.239x) \quad (3)$$

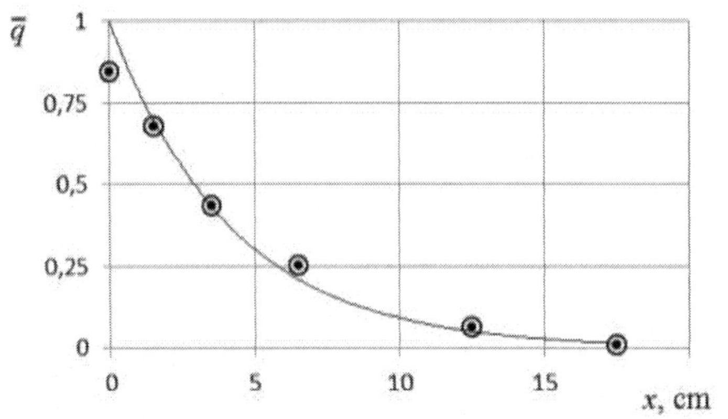

Figure 3. Dependence of $\overline{q}$ on the focal plane distance of the sample

To determine the dynamics of the measurement, the time dependence of the heat flux density (2) can be converted to $\overline{q}$ mean (t), substituting $x = ut$, where u is the displacement speed of the sample relative to the focal plane of the radiator.

When you move the sample from the focal plane in the direction from the emitter (decreasing heat flux)

$$\overline{q}(t) = \exp(-but) \quad (4)$$

and (increasing heat flux) during the reverse movement of

$$\bar{q}(t) = \exp\left[-b(x_0 - ut)\right] \tag{5}$$

Here x_0 is the coordinate of the sample at the initial time $t = 0$.

The dependence of $q(t)$, calculated according to the equations (3), (4) for the conditions of the experiments (u = const = 4.75 cm, x_0 = 17.5 cm, b = 0.239), is shown in Figure 4.

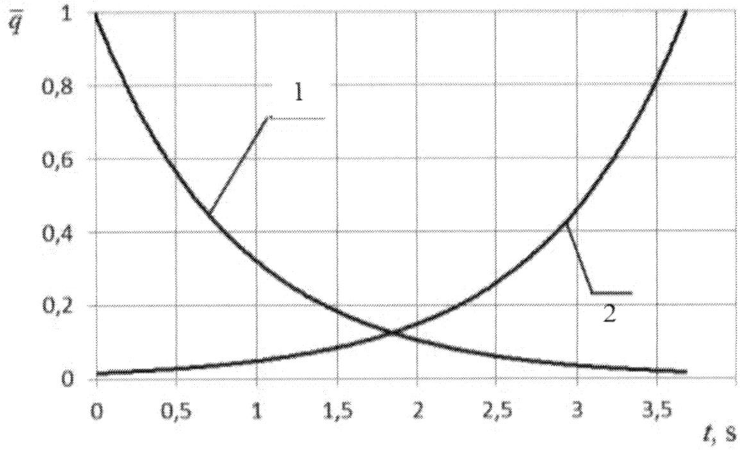

Figure 4. Dependencies $\bar{q}(t)$ for decreasing (1) and increasing (2) heat flux

The movement of the sample from $x = 0$ to x_0 = 17.5 cm and reverse movement from x_0 = 17.5 cm to $x = 0$ was

$$t_0 = x_0/u = 3.684 \text{ s.}$$

3. Experimental results

The experiments were conducted in conditions of increasing heat flux. The experiments differed in the rate of change in the heat flux density during ignition of pyroxylin. The results of the study are shown in Table 1, where a, b are the coefficients of the approximation function, Q is calculated by equation (2). The average values of the heat flux and change in flux $(dq/dt)_{med}$ during the induction period were determined by formulas

$$\tilde{q} = \frac{Q}{t_{ign}}, \quad (dq/dt)_{\text{med}} = \frac{1}{t_{ign}}\left[q(t_{ign}) - q(0)\right]$$

Table 1. Ignition characteristics of pyroxylin with increasing heat flux

Interval q, W/cm^2	a, W/cm^2	b, s^{-1}	t_{ign}, s	Q, J/cm^2	q,W/cm^2	$(dq/dt)_{med}$, W/(cm^2·s)
0.21÷10.50	0.21	1.27	3.08±0.04	8.10	2.63	3.34
0.51÷16.47	0.51	1.25	2.78±0.08	12.77	4.59	5.74
0.99÷19.00	0.99	1.66	1.78±0.06	10.85	6.10	10.12
1.78÷21.24	1.78	1.71	1.45±0.01	13.42	7.85	13.42

Experimental data [1] on pyroxylin ignition characteristics at a constant heat flux q_0 are given in table 2. The value of Q was calculated by the formula

$$Q = q_0 \cdot t_{ign}.$$

IOP Conf. Series: Journal of Physics: Conf. Series **830** (2017) 012136 doi:10.1088/1742-6596/830/1/012136

Table 2. Ignition characteristics of pyroxylin with constant heat flux

q_0, W/cm^2	t_{ign}, s	Q, J/cm^2
3.61	9.1	32.85
5.43	4.06	22.05
13.55	0.70	9.48
17.83	0.42	7.49

Comparison of the characteristics of ignition pyroxylin obtained with increasing heat flux with the experimental data at constant heat flux [11] is shown in Figure 5.

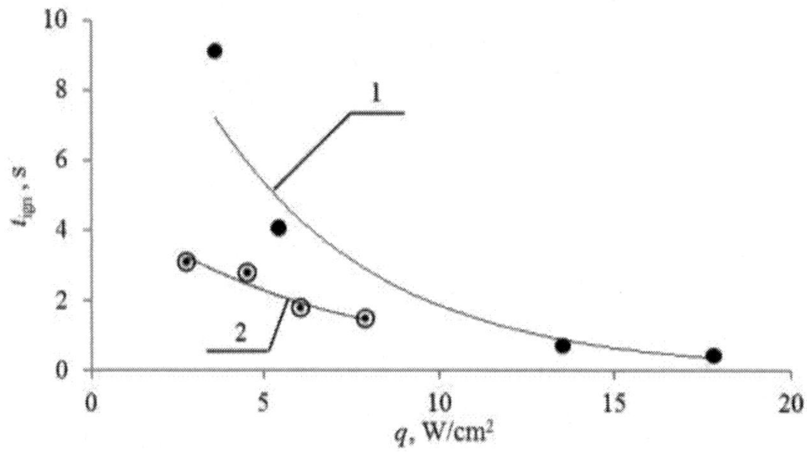

Figure 5. The ignition delay time vs. heat flux density at constant (1) and increasing (2) heat flux

From graphs in figure 5, it is seen that in the studied range q of ignition delay time of pyroxylin at increasing heat flux with an average value is smaller than that at a constant heat flux. With increasing heat flux density, the ignition delay time difference decreases and asymptotically tends to zero when $q > 8$ W/cm^2.

The dependence of the amount of thermal energy Q obtained by means of pyroxylin during the period on the heat flux density in heating with constant (1) and increasing (2) heat flux is shown in figure 6.

In heating the sample with constant heat flux, Q drastically decreases with increasing q_0, and in heating with increasing heat flux, growing dependence $Q(\overline{q})$ is observed. The same kind of Q is dependent on the average rate of the change in the heat fux $(dq/dt)_{med}$ shown in Figure 7.

This effect is apparently due to the particularities of formation of the heated layer of condensed phase in static and dynamic conditions of ignition of pyroxylin..

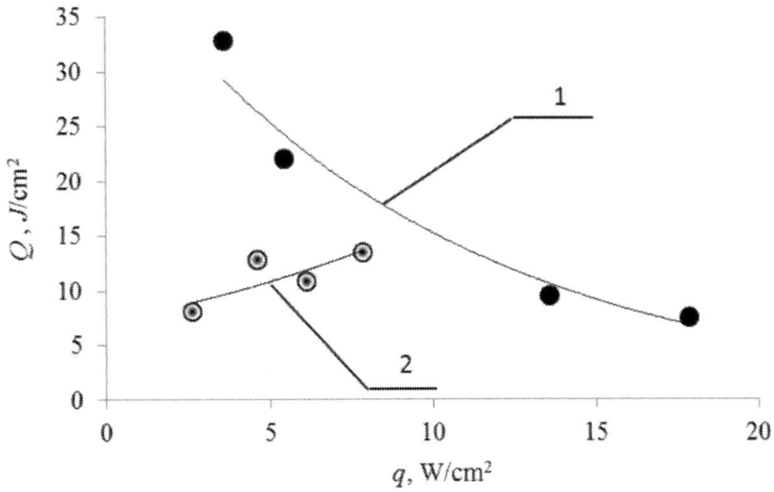

Figure 6. The dependence of energy on the heat flux density at constant (1) and increasing (2) heat flux

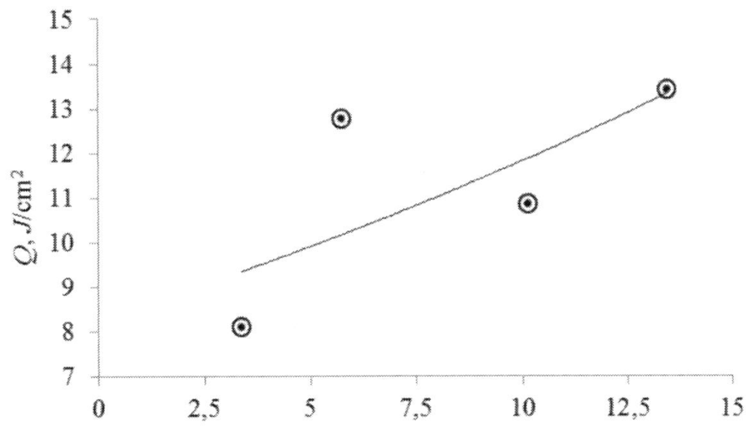

Figure 7. Dependence of energy on the average rate of the heat flux density change

4. Conclusion

A method for measuring the ignition characteristics of high-energy material samples when exposed to dynamic regime radiant heat flux, which includes the measurement of the ignition delay time during exposure of the sample surface to time varying radiant heat flux with the given intensity, was presented. The results of the experimental study of ignition of pyroxyline in heating with increasing radiation flux in the range of (0.2 ÷ 22) W/cm^2 have been presented. The findings suggest that pyroxylin containing 1 wt. % of soot in dynamic heating regime is higher than that in static conditions in the studied range of the heat flux density. It was found that the ignition times are reduced by 20–50 % depending on the initial value of the radiant flux density in initiation by increasing or decreasing heat flux compared with the stationary conditions of heat supply under the same ambient conditions.

Acknowledgement
This work was supported by the Ministry of Education and Science of the Russian Federation under Agreement No. 14.577.21.0157 of 11.28.2014 (unique identifier RFMEFI57714X0157).

References

[1] Vilyunov V N 1984 Ignition theory of condensed substances (Novosibirsk: Nauka)

[2] Merzhanov A G and Averson A E 1971 The present state of the thermal ignition theory: an invited review *J. Combust. and Flame* **16** 1 pp 89-127

[3] Kalinin V V, Kovalev Yu N and Lipanov A M 1986 Nonstationary processes and methods of SRM design nodes (Moscow: Mashinostroenie)

[4] Yang L Z, Guo Z F, Chen X J and Fan W C 2006 Predicting the temperature distribution of wood exposed to a variable heat flux *J. Combust. Science and Tech.* **178** 12 pp 2165-2176

[5] Enaleev R Sh, Krasina I V, Gasilov V S, Chistov Yu S and Tuchkova O A 2013 Wood ignition *J. Vestnik Kazanskogo tekhnolog. un-ta* **11** 10 pp 99-106

[6] Averson A E, Barzikin V V and Merzhanov A G 1968 Dynamical regimes of ignition *J. Combust., Expl., Shock Waves* **4** 1 pp 20-32

[7] Rozenband V I, Averson A E, Barzikin V V and Merzhanov A G 1968 About some laws of dynamical regimes of ignition *J. Combust., Expl.,Shock Waves* **4** 4 pp 493-500

[8] Dik I G and Zurer A B 1987 To problems of ignition of condensed substances with changing heat flow *J. Engineering Physics collection* (Tomsk: Izd-vo Tom. un-ta) pp 21-26

[9] Gusachenko L K, Zarko V E and Richkov A D 2012 Ignition and quenching homogeneous materials energy with light pulse *J. Combust., Expl., Shock Waves* **48** 1 pp 80-88

[10] Mikheev V F, Kovalskiy A A and Khlevnoi S S 1968 Ignition ballistite gunpowder using light radiation *J. Combust., Expl., Shock Waves* **4** 1 pp 3-9

[11] Mikheev V F and Khlevnoi S S 1970 On ignition pyroxyline *J. Combust., Expl., Shock Waves* **6** 2 pp 176-181

[12] DeLuca L, Caveny L H, Ohlemiller T J and Summerfield M 1976 Summerfield M Radiative ignition of double base propellants: Preignition events and source effects *AIAA J.* **14** pp 1111-1152

[13] Vilyunov V N, Kuznetsov V T and Skorik A I 1977 Inflammation pyroxyle using luminous flux of high (Moscow: Nauka)

[14] Arkhipov V A, Korotkikh A G, Kuznetsov V.T and Sinogina E S 2007 Influence of dispersion of metal powders on the characteristics of conductive and radiant ignition of mix composition *J. Chem. phys.* **26** 6 pp 58-67

[15] Atwood A I, Ford K P, Bui D T, Curran P O and Lyle T 2009 Radiant ignition studies of ammonium perchlorate based propellants *Progress in Propulsion Physics* ed L De Luca, C Bonnal, O Haidn and S Frolov (Moscow: TORUS PRESS) pp 121-140

[16] Arkhipov V A, Korotkikh A G, Kuznetsov V T, Razdobreev A A and Evseenko I A 2011 Effect of dispersity alumina powder on compositions ignition characteristics using laser *J. Chem. phys.* **30** 7 pp 68-76

[17] Arkhipov V A, Bondarchuk S S, Korotkikh A G, Kuznetsov V T, Gromov A A, Volkov S A and Revyagin L N 2012 Influence of aluminum dispersity on characteristics of ignition and nonstationary combustion of heterogeneous of condensed substances *J. Combust., Expl., Shock Waves* **48** 5 pp 148-159

[18] Grigoryev Yu M, Lissitzky V I and Merzhanov A G 1967 On ignition of condensed substances in the heated gas *J. Combust., Expl., Shock Waves* **3** 4 pp 512-526

[19] Rozenband V I, Barzikin V V and Merzhanov A G 1968 Ignition of condensed substances convective flow of medium intensity in dynamic conditions *J. Combust., Expl., Shock Waves* **4** 2 pp 171-175

[20] Enaleev R Sh, Mateosov V A, Sinaev K N, Dinavetsky V D and Gainutdinov R Sh 1973 Experimental study of condensed matter ignition dynamic radiant energy supply conditions *Physics of Combustion and Methods of Research* pp 80-86

[21] Arkhipov V A, Korotkikh A G, Kuznetsov V T and Zukov A S 2014 Effect of heat input dynamic mode on the ignition characteristics HEM *J. Izv. Vuzov. Phyzika* **57** 8/2 pp 18-23

[22] Lopatina G, Sasorov V, Spitcyn B and Fedoseev D 1969 *Optical furnace* (Moscow: Metallurgiya Publ.) p 216
[23] Arkhipov V A, Zarko V E, Korotkikh A G, Kuznetsov V T and Razdobreev A A 2015 *Patent RF № 2569641* Bul. 33
[24] Kovalskiy A A, Khlevnoi S S and Mikheev V F 1967 *J. Combust., Expl., Shock Waves* **3** 4 pp 527-541

Ignition by laser radiation and combustion of composite solid propellants with bimetal powders

A G Korotkikh[1,2], V A Arkhipov[2], O G Glotov[3] and N N Zolotorev[2]

[1] Institute of Power Engineering, Tomsk Polytechnic University, 634050, Tomsk, 30 Lenin Ave., Russia
[2] Research Institute of Applied Mathematics and Mechanics, Tomsk State University, 634050, Tomsk, 36 Lenin Ave., Russia
[3] Institute of Chemical Kinetics and Combustion of SB RAS, 630090, Novosibirsk, 3 Institutskaya Str., Russia

E-mail: korotkikh@tpu.ru

Abstract. The use of metal powder (usually aluminum) as a fuel in composite solid propellants (CSPs) for propulsion is the most energy efficient method that allows improvement of combustion characteristics of propellants in the combustion chamber and specific impulse. This paper presents the experimental data of the ignition and combustion processes of CSPs containing Alex aluminum nanopowder and mixtures of Alex/Fe and Alex/B nanopowders. It was found that the introduction of Alex/Fe in CSPs leads to 1.3–1.9 times decrease in the ignition time at q = 55–220 W/cm^2 and to 1.3–1.4 times increase in the burning rate at p = 2.2–7.5 MPa with respect to that for basic CSP with Alex. When introducing Alex/B in CSP, the ignition times are 1.2–1.4 fold decreased, and the burning rate is practically unchanged. However, the agglomeration is significantly enhanced, which is manifested through the increase in the agglomerate particles content in condensed combustion products by a factor of 1.8–2.2 , at 1.6–1.7 fold increase of the agglomerates mean diameter for CSP with Alex/B.

1. Introduction

Modern composite solid propellants (CSPs) contain the oxidizer crystals of ammonium perchlorate (AP), ammonium nitrate (AN) and nitramines (RDX, HMX, CL-20) [1–2], combustible-binder – inert or energetic polymer rubber, and metal fuel – aluminum powder with typical content up to 22 wt.% [3–5]. These systems are used as the energy source of solid rocket motors and gas generators for various purposes. Adjustment of the CSP burning rate is mainly achieved by introduction of the catalysts in propellant formulation, through partial or complete replacement of AP and AN by nitramines, by changes of the excess oxidizer factor, as well as the changed particle size of oxidizer and metal powders.

Metal oxides: MnO_2, Ni_2O_3, Cr_2O_3, MgO, Fe_2O_3, Co_2O_3, CuO, SiO_2, copper, iron, zinc, cadmium and magnesium salts of chromic and metachromic acids, complex cyanides of copper, iron and nickel are used as burning catalysts of CSPs [6–9]. The most common and versatile in this group of catalysts are the systems containing copper, chromium, boron and iron [10–12].

This paper presents the experimental data on ignition, combustion and agglomeration processes of CSPs based on Alex aluminum nanopowder (NP) containing additives of iron and amorphous boron.

2. Experimental

3.1. The CSP samples

We studied the CSP samples on the basis of bidisperse AP (fraction less than 50 μm and 160–315 μm in the ratio 40/60), inert combustible-binder (19.7 wt.%) – butadiene rubber plasticized by transformer oil, and Alex aluminum NP (15.7 wt.%) obtained in argon using electric explosion of conductors. In the second and third compositions of CSPs, 2 wt.% of the Alex NP was partially replaced by 2 wt.% additives of iron and amorphous boron NPs. According to the measurements on the BET analyzer Nova 2200e, the specific surface area of Alex amounts to 7.04 m^2/g in nitrogen, 1.08 m^2/g in iron, and 8.63 m^2/g in amorphous boron. The studied cylindrical samples of CSPs with a diameter of 10 mm and height of 30 mm were produced in the laboratory by extrusion pressing with the subsequent curing. The density of solid samples was in an interval of 1.53–1.59 g/cm^3 depending on the component composition.

3.2. Ignition by laser radiation of CSPs

The ignition study of test CSPs was carried with the use of a set up for radiant heating on the basis of CO_2-laser with the wavelength of 10.6 μm and power of 200 W (Figure 1). Prior to testing, the samples were cut into tablets of 5 mm height. The CSP samples were placed in the hollow cylinder of 10 mm depth made of ebonite for inhibiting the lateral surface and making one-dimensional flow of gasification products.

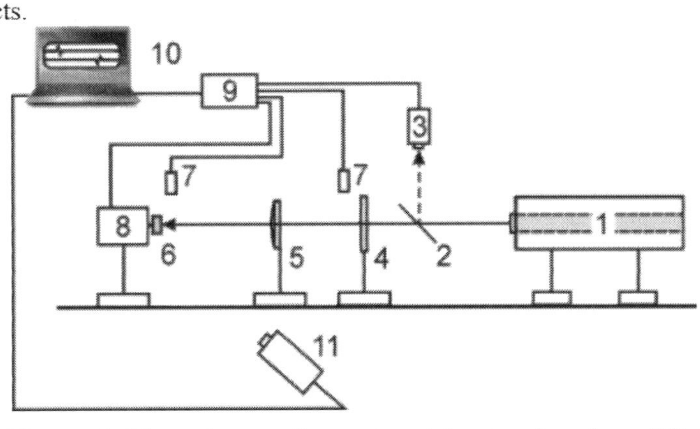

Figure 1. The scheme of experimental setup based on CO_2-laser: CO_2-laser (1); beam-splitting mirror (2); thermoelectric sensor of radiation power (3); shutter (4); lens (5); CSP sample (6); photodiodes (7); transducer of recoil force (8); ADCs (9); PC (11); video camera (11)

Test CSP sample (6) was attached to the substrate of recoil force transducer (8) to record the gasification products outflow from the burning surface. Shutter (4) being opened, radiation was focused onto CSP sample (6) by sodium chloride lens (5). Signals from recoil force transducer (8) and photodiodes (7) were transmitted to L-card-E 14-440 ADC (9) and recorded in personal computer (10), and then processed using LGraph2 software. The time delay of start gasification t_{gas} of the CSP sample was determined as the time interval between the moments of signals change of photodiode near shutter (7) (or a thermocouple installed in the laser beam behind the shutter), and recoil force transducer (8). Photodiode (7) recorded the moment of opening the shutter, transducer (8) recorded the appearance of recoil force signal of gasification products flowing from the front (irradiated) sample surface. The ignition time t_{ign} of CSPs was determined by difference between the moments of signals from two photodiodes (7), one of which recorded the appearance of flame near the end surface of the CSP sample. The relative error of delay times measuring of t_{gas} and t_{ign} was equal to 5–12 % at the value of confidence probability 0.9.

The values of recoil force of gasification products outflow from the end surface of the CSP sample during heating of the reaction layer, ignition and combustion of CSPs were determined using a recoil force transducer [13].

The radiation power and heat flux density of CO_2-laser beam was measured by the thermoelectric sensor of radiation power (3). The maximum radiation power was defined in the center of the laser beam.

3.3. Combustion of CSPs

The study of the effect of iron and amorphous boron additives on the combustion process of CSPs was conducted in blow-through sampling bomb in nitrogen at different pressures [14]. We used the CSP samples with a diameter of 10 mm and height of 25–30 mm in the experiments with sampling of condensed combustion products (CCP). The lateral surface of the samples was inhibited by heat resistant rubber "Solpren" (a styrene/butadiene copolymer). During combustion of Solpren, the carbonaceous particles were formed, the content of which in CCPs did not exceed 1 wt.%.

The experiments were carried out in nitrogen at three pressure levels: 2.2, 4.5, and 7.5 MPa. The CSP burning rate was determined on the basis of the known initial length and the burning time of the sample, which was measured via processing the signal from a pressure sensor installed in a bomb. During combustion of the CSP sample, gaseous and condensed combustion products passed through the metal sieve mesh packet and analytical aerosol filter which capture solid particles of different sizes. The sampled particles of combustion products were divided into two fractions: less than 80 μm and 80–315 μm. Then, the CCPs particles were subjected to morphological, particle size, chemical, and phase analyses.

3. Results and discussion

3.1. Ignition parameters

The values of ignition and gasification times, recoil force of gasification products outflow for the test solid propellants samples under study in atmospheric conditions were determined using a laser setup. Processing of signal recording from the recoil force transducer, the time delays of ignition and gasification, the recoil force outflow of gasification products from the burning surface of CSP samples was performed. The ignition time of CSPs dependences vs. the maximum value of the heat flux density (in the center of the laser beam) were determined (Figure 2).

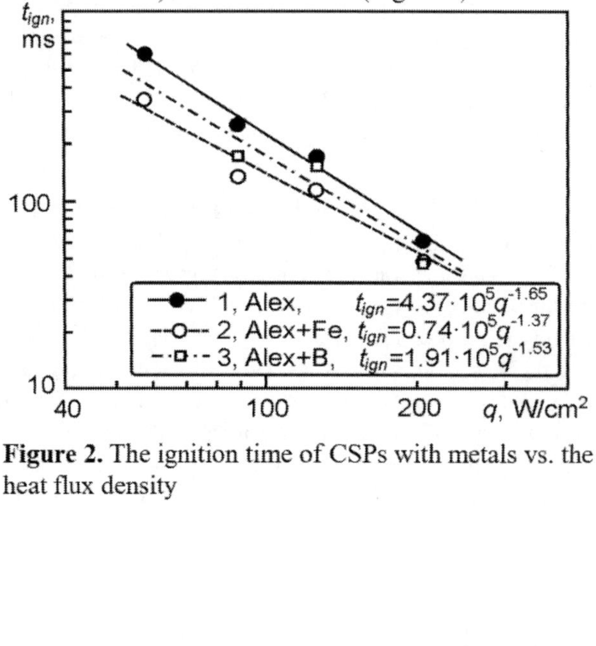

Figure 2. The ignition time of CSPs with metals vs. the heat flux density

Partial replacement of Alex by iron in CSP leads to 1.3–1.9 fold decrease in the ignition time in the range of heat flux density 55–220 W/cm^2. Partial replacement of Alex by boron in CSP leads to 1.2–1.4 fold decrease in the ignition time in the specified range of q. For the CSP sample with Alex, the gasification starts immediately after shutter opening and the gasification time is 40 times less than the ignition time at the heat radiant flux density of q = 65 W/cm^2. For samples with Alex/Fe and Alex/ B, the time moments of ignition start and gasification start coincide. Partial replacement of Alex by iron in CSP in initiation by CO_2-laser leads to 25 fold increase of the gasification time and 1.3 fold increase of the recoil force of gasification products outflow from the burning surface in the period of stationary combustion of CSPs. For partial replacement of Alex by boron in the CSP composition, the gasification time increases 36 times and the recoil force of gasification products outflow increases 1.1 times.

Thus, the additives of iron and boron in the CSPs with Alex allow considerable increase in the gasification time and reduction of the ignition time and also increase in the recoil force of gasification products outflow from the burning surface of CSP samples.

3.2. Combustion and agglomeration parameters

Figure 3 shows the dependencies of the burning rate on pressure for test CSPs in the form

$$u = B \cdot p^n,$$

where [u] in mm/s, [p] in MPa, and also the values of determination coefficient R^2 characterizing the quality of approximation.

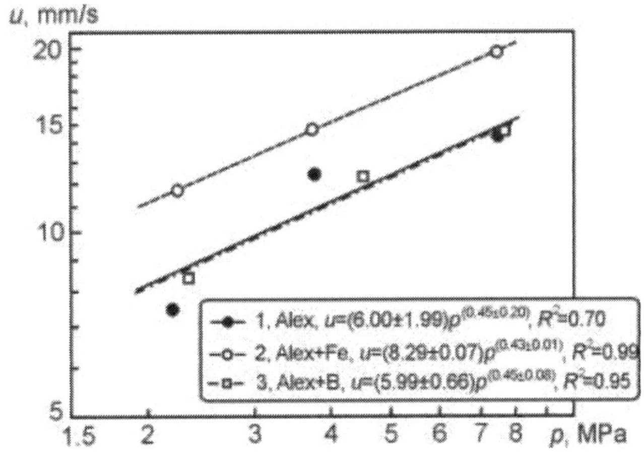

Figure 3. The burning rate of CSPs with metals vs. pressure

As one can see, the dependencies of the burning rate on pressure for CSP with Alex and Alex/B are matched up within the range of experimental error. The CSP with Alex/Fe does not change the exponent of the burning rate law, but leads to significant increase in the burning rate level (from 7.5 to 11.7 mm/s at p = 2.2 MPa and from 14.3 to 19.7 mm/s at p = 7.5 MPa).

A noticeable increase in the burning rate of iron doped CSP can be possibly explained by the iron oxide catalytic effect on the oxidizer decomposition and corresponding increase in the heat release on the burning surface of sample. The thermal analysis data obtained with a thermal analyzer Netzsch STA 449 F3 Jupiter at the heating rate of 10 K/min in argon have shown that the CSP temperature of intensive decomposition beginning is reduced by ~ 20 K upon iron addition. Note also that there is a possibility for thermite reaction between fine aluminum particles and iron oxide (total content in CSP formulation amounts to ~0.16 wt.%) within the subsurface reaction layer resulting in additional release of heat.

Boron additive in CSP formulation does not make catalytic effect on the propellant decomposition. The thermal analysis data shows that the start temperature of intensive decomposition of CSP with Alex/B remains practically unchanged and amounts to 582 K. Consequently, the burning rate of boron doped CSP exhibits the same behavior as original Alex-based CSP.

Figure 4 shows the values of the mean diameter and dimensional parameters of sampled agglomerate particles of CCPs (with $D > 55$ μm). Within the confidence intervals, the scatter of the experimental data on the CCPs parameters (typical error bars) is the same for all tested CSPs.

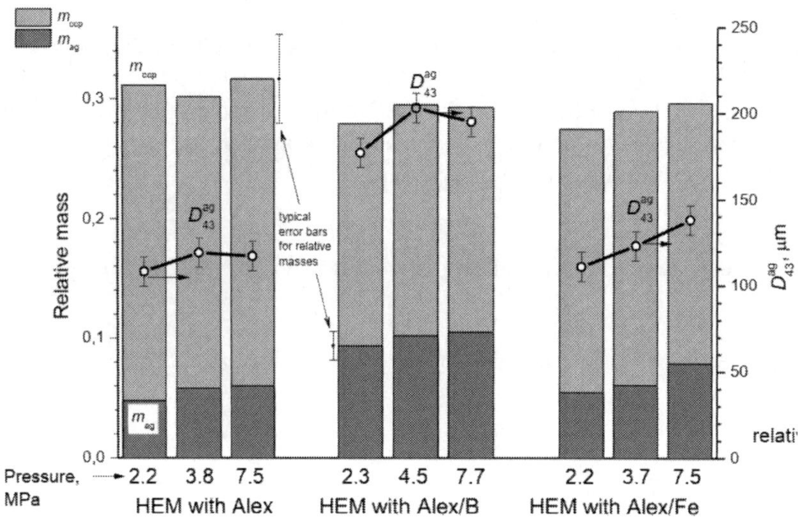

Figure 4. Agglomerate mean diameter and values of agglomerate relative mass m_{ag} and CCPs relative mass m_{ccp} for CSPs with metals

The relative mass of agglomerates m_{ag} (blacked bars in Figure 4) is increased with the growth of ambient pressure for all CSPs. In the case of CSPs with Alex and Alex/Fe, the values of the relative mass parameters of agglomerates m_{ag}, the fraction of metal involved in agglomerate formation and the mean diameter, characterizing the agglomerates properties, are similar. Since the burning rate of CSP with Alex/Fe is higher as compared with CSP with Alex but the agglomerating parameters of CCPs are similar, one can state that agglomeration in the case of CSP with Alex/Fe is conditionally enhanced and this is compensated due to the increase in the burning rate.

For CSP with Alex/B, the relative mass of agglomerates m_{ag}, the fraction of agglomerates in CCPs m_{ag}/m_{ccp} and the fraction of metal involved in the formation of agglomerates are increased. In addition, the agglomerates of CSP with Alex/B are much larger in size (the mean diameter is increased). Taking into account the similar levels of the burning rate, as well as the identity of the geometric structure of CSPs (which is mainly determined by the coarse particles of AP), these facts should be definitely interpreted as an agglomeration enhancement upon introduction of boron.

In all cases, the aluminum content in the fraction of < 80 μm does not exceed 1.3 wt.%. It can be expected that the particles with $D < 55$ μm contain a smaller amount of unburned aluminum. The analysis of the mass size distributions of oxide particles showed that in all cases (except for CSP with Alex at $p = 2.2$ MPa) the basic modes of size distributions are in the same ranges of the particle size analyzer Malvern. Thus, the first mode is observed in the range of 1.9–2.4 μm, most "massive" mode is in the range of 3–11 μm. In the case of CSP with Alex/B and Alex/Fe, the amplitude of the first mode is much larger than in the case of CSP with Alex. The particles formation of micron and submicron size is normally associated with the vapor type combustion of aluminum. Actually, the mass size distributions of fine particles for the studied CSPs have different ratios of mode amplitudes and they are slightly changed with the pressure. The variation of the ratio of modes results in different magnitudes of the mean diameters: for CSPs with Alex, Alex/B and Alex/Fe they are about 6.4 μm, 4.8 μm and 5.9 μm, respectively.

4. Conclusions

1. The effect of iron and amorphous boron nanoadditives in CSP on the basis of AP, butadiene rubber and 15.7 wt.% of aluminum Alex on ignition and combustion characteristics were studied. It was found that the introduction of Alex/Fe in CSP decreases the ignition time by a factor of 1.3–1.9 for the range of heat flux density of 55–220 W/cm^2, increases the recoil force of gasification products outflow from burning surface by a factor of 1.3 in the period of stationary combustion of propellants due to possible catalytic effect, which reduces the beginning temperature of AP high-temperature decomposition in the reaction layer of CSP by 20 °C and interaction of thermite mixture of aluminum and iron particles in the reaction layer of propellants. At the same time, the burning rate of the CSP sample is increased by a factor of 1.3–1.4 in the pressure range of 2.2–7.5 MPa and the agglomeration of metal fuel is slightly increased: the mean diameter of agglomerate particles is increased up to 1.2 fold and the content of agglomerates in the CCPs composition is increased up to 1.4 fold. The content and mean diameter of oxide particles in CCPs are reduced by 16% and 13%, respectively.

2. The introduction of Alex/B in CSP decreases the ignition time by a factor of 1.2–1.4 at q = 55–220 W/cm^2 and increases of the recoil force of gasification products outflow from the burning surface by a factor of 1.1 . In this, the beginning temperature of CSP high-temperature decomposition does not change and equals to ~309 °C. The burning rate is practically unchanged with respect to the CSP sample with Alex. However, the agglomeration is significantly enhanced, which is manifested through 1.8–2.2 fold increase in the agglomerate particles content in CCPs composition, 1.6–1.7 fold increase in the agglomerates mean diameter and 1.6–1.9 fold increase in the metal fraction, which is involved in agglomerates formation. The content and the mean diameter of the oxide particles in CCPs are reduced more significantly than those upon introduction of iron, by a factor of 1.2–1.3 and 1.3–1.4, respectively.

Acknowledgement
This work was supported by the Ministry of Education and Science of the Russian Federation under Agreement No. 14.577.21.0157 of 11.28.2014 (unique identifier RFMEFI57714X0157).

References

[1] Beckstead M W, Puduppakkama K, Thakreb P and Yang V 2007 *Progress in Energy and Combustion Science* **33** pp 497–551
[2] Takahashi K, Oide Sh and Kuwahara T 2013 *Propellants, Explosives, Pyrotechnics* **38** pp 555–562
[3] Sossi A, Duranti E, Manzoni M, Paravan C, DeLuca L T, Vorozhtsov A B, Lerner M I, Rodkevich N G, Gromov A A and Savin E N 2013 *Combustion Science and Technology* **185** pp 17–36
[4] Arkhipov V A and Korotkikh A G 2012 *Combustion and Flame* **159** pp 409–415
[5] Arkhipov V A, Bondarchuk S S, Korotkikh A G, Kuznetsov V T, Gromov A A, Volkov S A and Revyagin L N 2012 *Combustion, Explosion, and Shock Waves* **48** pp 625–635
[6] Arkhipov V A, Bondarchuk S S and Korotkikh A G 2010 *Combustion, Explosion, and Shock Waves* **46** pp 570–577
[7] Shioya S, Kohga M and Naya T 2014 *Combustion and Flame* **161** pp 620–630
[8] Ishitha K and Ramakrishna P A 2014 *Combustion and Flame* **161** pp 2717–2728
[9] Farley C W, Pantoya M L and Levitas V I 2014 *Combustion and Flame* **161** pp 131–1134
[10] Berner M K, Talawar M B, Zarko V E 2013 *Combustion, Explosion, and Shock Waves* **49** pp 625–647
[11] McDonald B A, Rice J R and Kirkham M W 2014 *Combustion and Flame* **161** pp 363–369

[12] Zarko V E and Glotov O G 2013 *Science and Technology of Energetic Materials* **74** pp 139–143
[13] Arkhipov V A, Kiskin A B, Zarko V E, Korotkikh A G 2014 *Combustion, Explosion, and Shock Waves* **50** pp 622–624
[14] Glotov O G, Zyryanov V Ya 1995 *Combustion, Explosion and Shock Waves* **31** pp 72–78

IOP Conf. Series: Journal of Physics: Conf. Series **830** (2017) 012138 doi:10.1088/1742-6596/830/1/012138

The thermostimulated luminescence of radiation defects in KCl, KBr and KI crystals at elastic and plastic deformation

K Shunkeyev, L Myasnikova, A Barmina, N Zhanturina, Sh Sagimbaeva, Z Aimaganbetova and D Sergeyev

Zhubanov Aktobe Regional State University, Aktobe, Kazakhstan

E-mail: shunkeev@rambler.ru

Abstract. The efficiency of radiation defects formation in alkali halide crystals (AHC) was studied by the method of absorption spectroscopy. However, it is not possible to study the deformation-stimulated processes in detail by the absorption spectrum of radiation defects due to the limited sensitivity compared with luminescent spectroscopy. In this regard, thermally stimulated luminescence (TSL) of radiation defects at elastic and plastic deformation was applied in AHC. In the absence of deformation, the dominant peaks in TSL are $\left(X_3^-\right)_{aca}^0$ -centers. After elastic deformation, low temperature peaks of TSL corresponding to F'-, V_K- and V_F-centers became dominant. After plastic deformation, the peaks of TSL corresponding to $\left(X_3^-\right)_{aca}^0$ -centers became dominant. The elastic deformation contributes to the increase in concentration of low-temperature F'-, V_K- and V_F-centers, and the plastic one contributes to that of high temperature $\left(X_3^-\right)_{aca}^0$ -centers (peaks of TSL in KCl at 360K, in KBr at 365K, in KI at 340K), composed by divacancies created by plastic deformation. At elastic deformation, unrelaxed interstitial halogen atoms are converted into V_K- and V_F-centers, and due to this fact the long-range interaction is absent, the result of which are the X_3^- -centers.

1. Introduction

At AHC lattice symmetry lowering by low temperature elastic, high-temperature plastic deformation and by light cations, the increase in the luminescence intensity of self-trapped excitons (STE) occurs [1-3]. This effect is very important for the search of scintillation materials based on AHC. However, in many cases the decay of electronic excitations into radiation defects is not taken into account, what is important in the integration of all kinds of luminescence to increase the quantum yield of the scintillator luminescence. It is known that low temperature uniaxial deformation can reduce the effectiveness of radiation defect formation in AHC [4-7]. However, these experimental results have been obtained on the basis of absorption spectroscopy for recording the optical density of radiation defects, which is not able to take into account the small amount of radiation defects, which are clearly manifested in the TSL.

IOP Conf. Series: Journal of Physics: Conf. Series **830** (2017) 012138 doi:10.1088/1742-6596/830/1/012138

2. Experimental results and discussion

2.1. KCl

The TSL measurement of radiation defects at AHC lattice symmetry lowering is carried out taking into account various disturbing factors. From the change in intensity of TSL, which characterizes the concentration of thermally destroying radiation defects, we can assume the effectiveness of their formation after exposure to external factors (elastic and plastic deformation, cationic impurities of different size, etc.).

Figure 1 shows the TSL of KCl crystals in the absence of deformation, at low temperature elastic deformation and after plastic deformation. From figure 1 (curve 1), it follows that at the absence of deformation the low temperature peaks of TSL of Na^0-, H_A(Na)-, V_K- и V_F were recorded, the maximum destroying temperature of which corresponds to the destruction of the TSL peaks at 125K, 140 K, 205 K and 235 K, respectively, and high temperature peaks of TSL were recorded at temperatures 360 K and 420 K, which are caused by thermal destruction of $V_{4A} = (Cl_3^-)_{aca} K_i^+ Cl_i^-$, $V_4 = (Cl_3^-)^0_{aca}$ and $V_2 = [(Cl_2^-)_2]_{acac}$ centers. At the same time, the dominant peaks are high-temperature ones characteristic of pure crystals, in which the impurity concentration is maximally reduced; the impurity serves as shallow traps for stabilizing mobile interstitial atoms of halogen and unrelaxed holes.

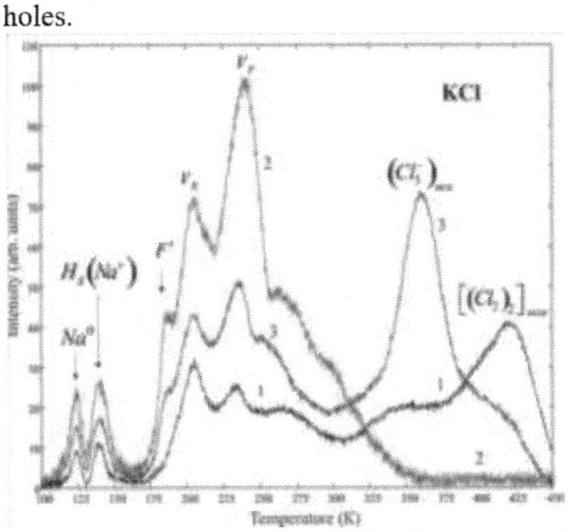

Figure 1. Thermostimulated luminescence of KCl crystals radiated by X-rays in isodose mode (3 mA, 120 kV in equal exposition (90 min) at 90 K) before deformation (1), at low temperature (90 K) elastic deformation (ε=1.2%) (2), after plastic deformation at 300 K (ε=4%) (3).

The low temperature elastic deformation of KCl crystal leads to the increase of intensity peaks of TSL of Na^0-, H_A-, F'-, V_K- and V_F-centers more than 3–4 times (curve 2) in comparison with that of the undeformed crystal (curve 1). At the same time, TSL peaks near 360 K and 420 K disappear, which correspond to high-temperature aggregate formations of halogen V_4 and V_2 centers. In the absence of deformation, the TSL high-temperature peaks are dominant, and after low temperature elastic deformation, low temperature peaks are dominant. The low temperature elastic deformation contributes to the self-trapping of mobile interstitial halogen atoms and unrelaxed holes in regular lattice sites.

Thus, after low temperature elastic deformation by TSL, in KBr crystals we have recorded an increased concentration of F'(185 K)-, V_K(205 K)- and V_F(235 K) centers compared to the original crystal, apparently due to self-trapping and transformation of unrelaxed hole and movable halogen atoms.

Figure 1 (curve 3) represents the TSL of KCl crystal after impact of plastic deformation. Firstly, the intensities of low temperature peaks of TSL (curve 3) reduced compared with the elastically deformed crystal (curve 2), but remain dominant relative to undeformed crystal (curve 1); secondly, the most intense peaks became $(Cl_3^-)^0_{aca}$ and $[(Cl_2^-)_2]_{acac}$-centers, which are destroyed at 360 K and 420 K (curve 3). It is known that $(Cl_3^-)^0_{aca}$-centers are formed by reacting of two interstitial halogen atoms (H-centers) in regular lattice sites, and their efficiency increases significantly if the association of H-centers occurs in divacancies field ($\upsilon_a^+ \upsilon_c^-$), created by plastic deformation.

2.2. *KBr*

From figure 2 (curve 1) we can assume that at the absence of deformation the low temperature peaks of TSL of *F'*-, V_K- and V_F-centers were recorded, the maximum destruction temperature of which corresponds to the TSL peaks at 120 K, 180 K and 235 K, respectively. From high temperature TSL peaks two peaks at temperatures 365 K and 440 K were recorded, which are caused by thermal destruction of $V_4 = (Br_3^-)_{aca} K_i^+ Br_i^-$ and $V_2 = (Br_3^-)^0_{aca}$ -centers.

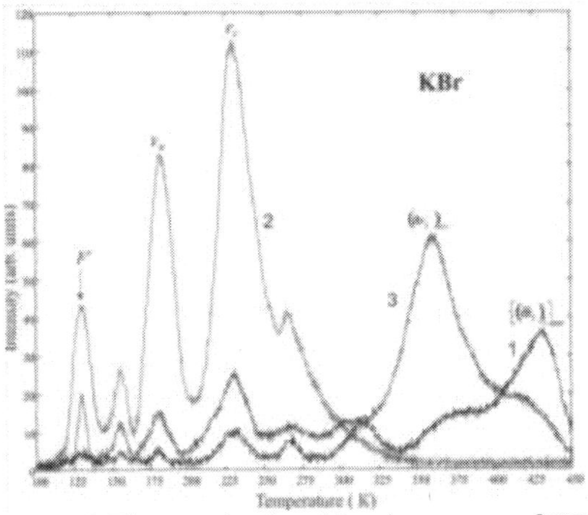

Figure 2. Thermostimulated luminescence of KBr crystals radiated by X-rays in isodose mode (3 mA, 120 kV in equal exposition (90 min) at 90 K) before deformation (1), at low temperature (90 K) elastic deformation (ε=1.2%) (2), after plastic deformation at 300 K (ε=4%) (3).

The low temperature elastic deformation of KBr crystal leads to increase in intensity of TSL peaks of V_K(180 K)- and V_F(240 K)-centers. Thus, the peaks in the region of 365 K and 440 K disappear; they correspond to high-temperature aggregate formations of halogen V_4 and V_2 -centers. Also amplify TSL peaks intensities, located in the area of *F'*-centers destruction temperatures (about 120÷130 K for KBr).

Thus, after low temperature elastic deformation by TSL in KBr crystals we have registered high concentrations of V_K-, V_F- and *F'* -centers compared to those of the original crystal.

The basic pattern of plastic deformation impact is as follows: firstly, the intensities of low temperature V_K-, V_F- and *F'*-peaks decreased in comparison with those of the elastically deformed crystal but remain dominant relative to the undeformed crystal; secondly, the most intense peaks became the peaks of $(Br_3^-)^0_{aca}$ and $[(Br_2^-)_2]_{acac}$ -centers.

2.3. *KI*

In undeformed KI crystal (figure 3, curve 1), the TSL peak in high temperature region at 340 K was recorded; the peak corresponds to thermal destruction of $(I_3^-)^0_{aca}$ -centers, which is absent at low temperature deformation. This means that at uniaxial deformation field as expected, there is no association between *H*-centers (which produces $(I_3^-)^0_{aca}$-centers). Thus the low temperature intense TSL peaks of V_K, V_F, V_{KA} and *F'*- centers were recorded by analogy with KBr and KCl crystals (curve 2).

The effects of plastic deformation on TSL of KI crystal are shown in figure 3 (curve 3). Thus the basic pattern is similar for KBr and KCl crystals, and it is as follows: firstly, the intensities of low-temperature V_K-, V_F- and *F'*-peaks (curve 3) decreased in comparison with those of the elastically deformed crystal (curve 2), but remain dominant relative to those of undeformed crystal (curve 1); secondly, the most intense peaks became $(I_3^-)^0_{aca}$ and $[(I_2^-)_2]_{acac}$-centers, which are destroyed at 365K and 420K (curve 3).

Assuming that at uniaxial compression in KI crystal the concentration of *H*-centers generated by the decay of STE will decrease, apart from inefficiency of their association we can expect the change of their stability channels or conversion into other hole centers, which in interaction with the *F*-centers allow recombination luminescence.

IOP Conf. Series: Journal of Physics: Conf. Series **830** (2017) 012138 doi:10.1088/1742-6596/830/1/012138

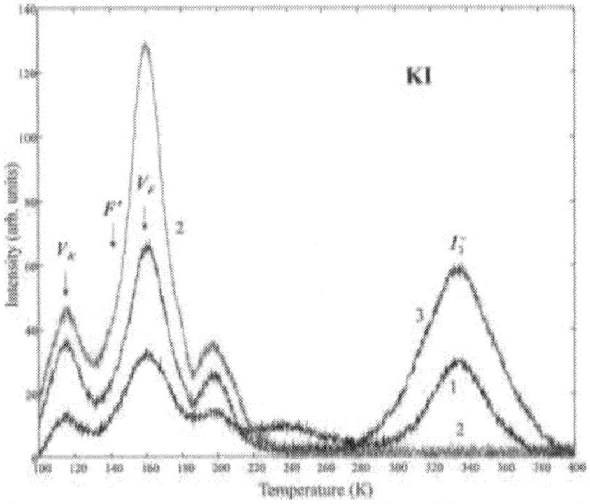

Figure 3.TSL of KI crystals radiated by X-rays in isodose mode (3 mA, 120 kV in equal exposition (90 min) at 90 K) before deformation (1), at low temperature (90 K) elastic deformation (ε=1.2%) (2), after plastic deformation at 300 K (ε=4%) (3).

3. Conclusion

The experimental results on registration of thermally stimulated luminescence of radiation defects in KCl, KBr and KI crystals under the influence of low temperature elastic and plastic deformation allow us to establish the following patterns:

- at low temperature elastic deformation of AHC, the high temperature TSL peaks disappear (in KCl at 360 K, in KBr at 365 K, in KI at 340 K), corresponding to the halogen formations X_3^- and $(X_2)_2$, which are created by interacting of mobile halogen atoms (*H*-centers);

- at low temperature elastic deformation of AHC, the intensity of low temperature TSL peaks of *F'*(120 K)-, V_K(180 K)- and V_F(235 K)-centers increase significantly (10 times more) due to their high concentration;

- at low temperature elastic deformation, redistribution of high temperature TSL peaks intensities (X_3^- and $(X_2)_2$ -centers) in favor of low temperature occurs due to the high concentration of stabilized *F'*-, V_K- and V_F-centers;

- at low temperature elastic deformation, the redistribution of high intensity TSL peaks (X_3^- and $(X_2)_2$ centers) takes place in favor of low temperature ones due to high concentration of stabilized *F'*-, V_K- and V_F-centers;

- after plastic deformation, the effectiveness of X_3^- and $(X_2)_2$-center formation increases (TSL peaks in KCl at 360 K, in KBr at 365 K, in KI at 340 K), which are created by the interaction of two *H*-centers in the field of vacancy formations;

- vacancy formations (divacancies $-\upsilon_a^+\upsilon_c^-$ and quartets of vacancies $-2(\upsilon_a^+\upsilon_c^-)$) play the role of traps for the stabilization of unrelaxed holes and interstitial halogen atoms with the formation of *F'*-, V_K- and V_F-centers;

- increase in TSL peak intensity of self-trapped holes (V_K -centers) is due to the decrease in the mean free path of unrelaxed holes in regular lattice sites.

This work was supported by grants of the Ministry of Education and Science of the Republic of Kazakhstan 4903/GF4 and 4904/GF4.

References

[1] Nishimura H, Tsujimoto T, Nakayama M, Horiguchi T and Kobayashi M 1994 *J. Phys. Soc. Japan* **63** 2818 JPSJ.63.2818

[2] Shunkeyev K, Sarmukhanov E, Barmina A, Myasnikova L, Sagimbaeva Sh and Shunkeyev S 2008 *Phys. Solid State* **50** 1799 S1063783408100016

[3] Lisitsyna L, Kassymkanova R, Esil'baev D and Dauletbekova A 2013 *Phys. Solid State* **55** 1034 S1063783413050211

[4] Shunkeyev K, Zhanturina N, Aimaganbetova Z, Barmina A, Myasnikova L, Sagymbaeva Sh and Sergeyev D 2016 *Low Temp. Phys.* **42** 580 1.4960008

[5] Lushchik A, Lushchik Ch, Vasil'chenko E, Kirm M and Martinson I 2002 *Surf. Rev. Lett.* **9** 299

[6] Shunkeyev K, Sergeyev D, Myasnikova L, Barmina A, Zhanturina N, Shunkeyev S and Aimaganbetova Z 2014 *Russ. Phys. J.* **57** 451 s11182-014-0261-3

[7] Popov A and Balanzat E 2000 *Nucl. Instrum. Methods* B **166** 545

The deformation stimulated luminescence in KCl, KBr and KI crystals

K Shunkeyev[1], D Sergeyev[1], W Drozdowski[2], K Brylev[2], L Myasnikova[1], A Barmina[1], N Zhanturina[1], Sh Sagimbaeva[1] and Z Aimaganbetova[1]

[1]Zhubanov Aktobe Regional State University, Aktobe, Kazakhstan
[2]Nicolaus Copernicus University in Torun, Torun, Poland

E-mail: shunkeev@rambler.ru

Abstract. Currently, strengthening of the intensity of luminescence in alkali halide crystals (AHC) at lattice symmetry lowering is discussed as a promising direction for the development of scintillation detectors [1-3]. In this regard, for the study of anion excitons and radiation defects in the AHC anion sublattice at deformation, the crystals with the same sizes of cations and different sizes of anions were chosen. In the X-ray spectra of KCl at 10 K, the luminescence at 3.88 eV; 3.05 eV and 2.3 eV is clearly visible. The luminescence at 3.05 eV corresponds to the tunneling recharge [*F**, *H*]. Luminescence at 3.88 eV is quenched in the region of thermal destruction of *F*′-centers and characterizes tunneling recharge of *F*′, V_K-centers. In KCl at 90 K, the luminescence of self-trapped excitons (STE) is completely absent. In KBr at deformation not only STE luminescence, but also deformation stimulated luminescence at 3.58 eV were recorded, the last one corresponds to tunneling recharge of *F*′, V_K-centers. In KI crystal at 10 K and 90 K at deformation, only STE luminescence is enhanced. There are no deformation luminescence bands in KI compares with KBr and KCl crystals.

1. Introduction

The luminescence of self-trapped excitons in alkali halide crystals is considered as the intrinsic matrix luminescence, and it currently has been studied in detail [4-7]. The present study is aimed at investigation of the luminescence of electronic excitations in the anion sublattice of AHC matrix.

2. Experimental results and discussion

2.1. KCl

The instrinsic luminescence of KCl crystal due to self-trapped exciton luminescence at 10 K consists of one band with maximum at 2.3 eV with π-polarization (curve 1, figure 1*a*) [8]. The curve 2 of figure 1*a* shows the X-ray spectra of KCl crystal at deformation recorded at 10 K. From these spectra the luminescence bands with maxima at 5.15 eV, 3.88 eV, 3.05 eV and 2.3 eV are clearly visible. It should be noted that in KCl crystal the luminescence with a peak at 5.15 eV has been recorded for the first time and its nature remains unclear, apparently linked to the free exciton radiative relaxation.

Simultaneous exposure of low temperature (10 K) and the uniaxial deformation (ε=1.3%) seem to have created favorable conditions for recording luminescence of free excitons in KCl crystal. The last luminescence band at 2.3 eV refers to the self-trapped exciton radiative relaxation, which due to the temperature quenching has not been recorded at 90 K (curve 2, figure 1*b*).

The luminescence bands at 3.88 eV and 3.05 eV characterize the tunnel recharge between the radiation defects *F'* and V_K, *F** and *H*, respectively. This is evidenced by their temperature dependence. The temperature dependence of the luminescence band at 3.88 eV in KCl crystal shows that it undergoes the maximum quenching at temperature region (120 K) when *F'*-centers are thermally annealed (inset, figure 1*b*). The slow decline of the luminescence intensity at 3.88 eV in the temperature range of 10 to 80K, apparently, is associated with a thermal delocalization of hole radiation defects, which interact with the *F'*-centers. Ranging from 80K to 140K the thermal annealing of *F'*-centers takes place.

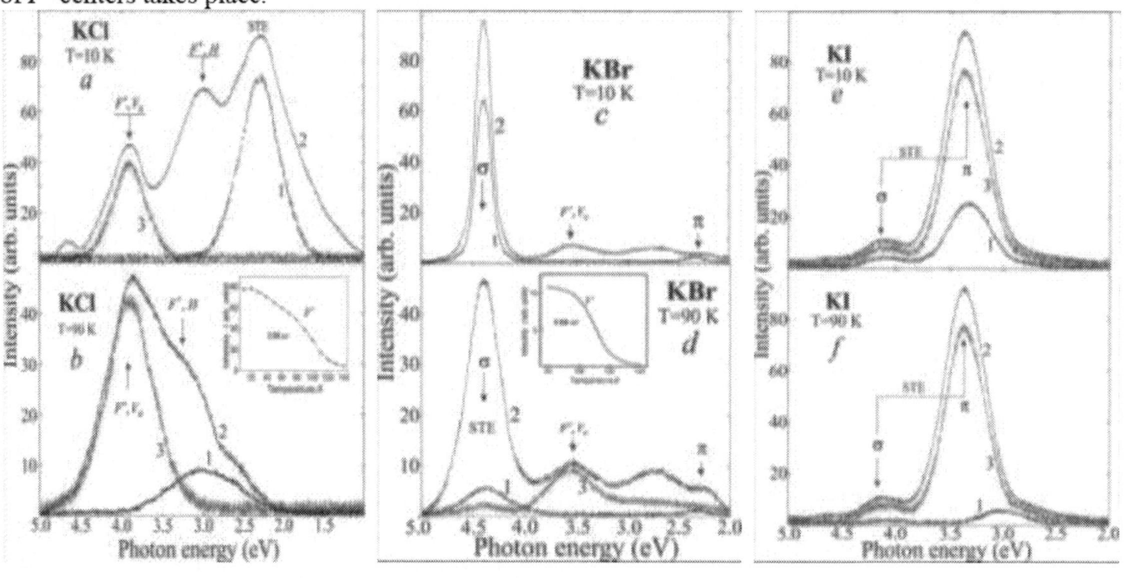

Figure 1. X-ray and tunneling luminescence of KCl, KBr and KI crystals before deformation (1), at deformation (ε=1.3%) (2), tunneling luminescence (3)

From these experimental results we assume that the low temperature elastic deformation of KCl crystal facilitates efficient formation of *F', V_K* and *F, H*-pairs.

Therefore, it is assumed that the effect of luminescence amplification is most likely associated with the tunneling recharge of different charged radiation defects created under the influence of X-ray radiation. Figure 1*a* (curve 3) shows the spectrum of the tunneling luminescence of KCl crystal at low temperature deformation in <100> directions after X-ray irradiation at 10 K.

Algorithm of experiment was carried in the following sequence. After the vacuum obtaining in cryostat, the crystal was cooled to 10 and 90 K, and at this temperature it was deformed uniaxially to the degree ε=2%, and then the crystal was irradiated for one hour by X-ray in mode 3 mA, 120 kV. After the termination of X-ray radiation, the tunneling luminescence of KCl crystal was recorded, i.e., in the absence of the exciting photons.

From figure 1*b* we see the luminescence with the maximum at 3.88 eV is also dominant in the spectrum of tunneling luminescence of KCl crystal deformed at 90 K, indicating the process of recharging between the ground states of stable radiation defects.

From comparison of X-ray spectra (curve 1, figure 1*b*; curve 2, figure 1*a*) and the tunneling luminescence (curve 2, figure 1*b*) of KCl crystal deformed (ε=2%) at low temperature, it is clear that the appearing luminescence band is observed in the tunneling luminescence spectra of these crystals. The contribution increase of the tunneling luminescence in the X-ray spectra of deformed KCl crystal is related to a large number of *F'*-centers and hole V_K-centers and with the increase in the tunneling probability of radiation defects due to oriented action of uniaxial compression.

Therefore, luminescence at 3.88 eV in KCl crystal at low temperature deformation was interpreted as tunneling recharge of radiation defects, most likely, *F'* and V_K -centers.

Such tunnel processes usually occur at temperatures much lower than the liquid nitrogen temperature (4.2 K), as in the case of our experimental conditions. Apparently, the low temperature deformation creates favorable conditions for the formation of interacting pairs of radiation defects, tunneling charge of which is carried out easily.

2.2. KBr

Similar low temperature deformation was carried out for KBr crystals. In KBr crystal compared with KCl crystal, the ionic radii of halogens increase 1.1 times. The figures 1*c* and 1*d* show the luminescent characteristics of KBr crystal.

The intrinsic luminescence of KBr crystal caused by the luminescence of self-trapped exciton at 10 K consists of two bands with maxima at 4.42 eV (σ) and 2.28 eV (π) (curve 1, figure 1*c*) [8]. The curve 2 in figure 1*d* shows that the significant effect is a considerable increase of σ- and π-luminescence of self-trapped exciton and new bands with maxima at 3.58 eV and 2.75 eV (curve 2, figure 1*d*) [9].

From a comparison of X-ray spectra (curve 1, figure 1*d*) and tunneling luminescence (curve 3, figure 1*d*) of KBr crystal deformed (ε=2%) at low temperature it is clear that a new luminescence band at 3.58 eV is observed also in the spectrum of tunneling luminescence. The temperature dependence of the X-ray spectrum of KBr crystal, predeformed at 90 K shows that the observed luminescence band with a maximum at 3.58 eV is most thermally annealed at 110 K, i.e., in the temperature region, when *F'*-centers are thermally destroyed (inset, figure 1*d*). This suggests that one of the partners responsible for tunneling luminescence at 3.58 eV is indeed *F'*-center having the structure ($\upsilon_a^+ e^- e^-$).

2.3. KI

Similar experiments were continued for KI crystal (figure 1*e*, 1*f*). In this case, in KI crystal the ionic radii of halogens compared with KCl crystal increase by 1.2 times. Thus, among KCl, KBr and KI crystals the halogen radius gradually increases according to the following sequence R(Cl)=1.81 Å, R(Br)=1.96 Å and R(I)=2.2 Å.

The intrinsic luminescence of KI crystal caused by the exciton self-trapping luminescence at 10 K consists of two bands with maxima at 4.15 eV (σ) and 3.3 eV (π) (curve 1, figure 1*e*) [8, 10]. In optical creation of electron-hole pairs and at X-ray excitation, the intensity of σ-luminescence at heating begins to be quenched only after 60 K, and at 80 K it decreases only 3 times [2, 11]. While π-luminescence is quenched only at temperatures above 90 K [2]. This means if self-trapped excitons are created, they can be trustworthy recorded at 90 K.

The results of measurements of X-ray spectra at low temperature uniaxial deformation on example of KI crystal are shown in figure 1*f*. We obtain the similar results as in the case with KCl and KBr crystals. It should be emphasized that in comparison with KCl crystal (curve 1, figure 1*a*) in KI crystal the intensity of intrinsic luminescence increases about 100 times at low temperature deformation (curve 2, figure 1*f*).

Luminescence at 3.02 eV increases at compression also, although less efficiently. This band is observed in the phosphorescence spectrum of the deformed KI crystal, and we can assume that it can occur as a result of tunneling recombination between the radiation defects (curve 3, figure 1*f*). By analogy with KCl and KBr, one would assume that it can be associated with the tunneling luminescence between the ground states of *F'*- и V_K-centers. In the same energy region in KI, the exciton self-trapped near impurity of Na^+ (3.0 eV) radiates as well.

3. Conclusion

The experimental results allow us to establish the following patterns of luminescence of KCl, KBr and KI crystals at low temperature uniaxial deformation:

1. The buildup of intensity of self-trapped excitons luminescence and luminescence of tunneling radiation defects.

2. In crystals KCl at low temperature deformation, new luminescence bands with maxima at 3.88 eV and 3.05 eV were recorded, the nature of which is associated with the tunneling recharge of radiation defects – F'- and V_K-centers.

3. In KBr crystals at low temperature deformation in <100> directions, new luminescence bands with maxima at 3.58 and 2.75 eV were recorded. It is assumed that the luminescence at 3.58 eV is associated with the tunnel recharging between the ground states of F'- and V_K-centers.

4. In KI crystals at low temperature deformation, the luminescence of self-trapped excitons with maxima at 4.42 eV and 2.3 eV was recorded.

This work was supported by grants of the Ministry of Education and Science of the Republic of Kazakhstan 4903/ GF4 and 4904/GF4.

References

[1] Lisitsyna L, Kassymkanova R, Esil'baev D and Dauletbekova A 2013 *Phys. Solid State* **55** 1034 S1063783413050211
[2] Shunkeyev K, Sarmukhanov E, Barmina A, Myasnikova L, Sagimbaeva Sh and Shunkeyev S 2008 *Phys. Solid State* **50** 1799 S1063783408100016
[3] Nishimura H, Tsujimoto T, Nakayama M, Horiguchi T and Kobayashi M 1994 J. *Phys. Soc. Japan* **63** 2818 JPSJ.63.2818
[4] Shunkeyev K, Zhanturina N, Aimaganbetova Z, Barmina A, Myasnikova L, Sagymbaeva Sh and Sergeyev D 2016 *Low Temp. Phys.* **42** 580 1.4960008
[5] Lushchik A, Lushchik Ch, Vasil'chenko E, Kirm M and Martinson I 2002 *Surf. Rev. Lett.* **9** 299
[6] Shunkeyev K, Sergeyev D, Myasnikova L, Barmina A, Zhanturina N, Shunkeyev S and Aimaganbetova Z 2014 *Russ. Phys. J.* **57** 451 s11182-014-0261-3
[7] Popov A and Balanzat E 2000 *Nucl. Instrum. Methods* B **166** 545
[8] Song K and Williams R 1994 *Self-trapped excitons* (Berlin: Springer Verlag) 10.1007-978- 3-642 -85236-7
[9] Shunkeyev K, Sarmukhanov E, Barmina A, Myasnikova L and Shunkeyev S 2007 *J. Applied Spectroscopy* **74** 74 s10812-007-0011-1
[10] Babin V, Elango A, Kalder K, Maaroos A, Shunkeyev K, Vasilchenko E and Zazubovich S 1999 *J. Lumin.* **81** 71 S0022-2313(98)00051-9
[11] Shunkeyev K, Myasnikova L, Barmina A and Sagimbaeva Sh 2012 *J. of Phys.: Conf. Series* **400** 042002 1742-6596/400/4/042002

Oxide films state analysis by IR spectroscopy based on the simple oscillator approximation

N V Volkov, T V Yakutkina and V V Karpova

National Research Nuclear University MEPhI (Moscow Engineering Physics Institute) 115409, Russia, Moscow, Kashirskoe highway, 31

E-mail: nvvolkov@mail.ru

Abstract. Stabilization of structure-phase state in a wide temperature range is one of the most important problems of improving properties of oxide compounds. As such, the search of new effective methods for obtaining metal oxides with desired physic-chemical, electro-physical and thermal properties and their control is important and relevant. The aim of this work is identification features state of the oxide films of some metals Be, Al, Fe, Cu, Zr on the metal surface of the polycrystalline samples by infrared spectroscopy. To identify the resonance emission bands the algorithm of IR-spectra processing was developed and realized on the basis of table processor EXCEL-2010, which allow revealing characteristic resonance bands successfully and identification of inorganic chemical compounds. In the frame of simple oscillator model, resonance frequencies of normal vibrations of water and some inorganic compounds: metal oxides – Be, Al, Fe, Cu, Zr were calculated and characteristic frequencies for different states (aggregate, deformation, phase) were specified. By means of IR-spectroscopy fundamental possibility of revealing oxides films on metal substrate features state is shown, that allow development and optimization of the technology for production of the oxide films with desired properties.

1. Introduction

The presence of non-destructive control methods of the state of structural materials, for example, which are used as cladding materials for the fuel elements of VVER-reactors, is important to improve technological processes. One of the most important properties of structural materials is the corrosion resistance that decreases at the violation of the continuity and structure of protective oxide films. To investigate the properties of oxide films after corrosion tests, infrared (IR) spectroscopy, which is a non-destructive control method, can be applied. The aim of the work is to study the features of the state of zirconium alloys oxide films on a metal substrate by IR spectroscopy.

2. Experimental technique

Fragments of Zr-1% Nb tubes with a diameter of 9.1 mm were used as samples for the investigation. The oxidation was carried out in a water-steam environment for four groups of samples under the following conditions: No. 1 (T = 500 °C, t = 30 s); No. 2 (T = 500 °C, t = 300 s); No. 3 (T = 400 °C, t = 600 s); No. 4 (T = 350 °C, t = 1112 h).

The structure of oxide films on a metal substrate was studied by scanning ion microscopy (SIM) using the scanning ion microscope Helios 650 Nanolab. A beam of Ga^{+} ions was used as a probe beam under the following conditions: the ion energy was 30 keV, the ion beam current was 1÷10 nA, the beam cross-section was 10÷30 nm, and the residual gas pressure in the working chamber was below 10^{-8} Pa.

The emission spectra of oxide films were measured using the IR spectrometer Nicolet 6700. Figure 1 shows the scheme of measuring the reflected flux of the "oxide-metal" system, where I_o is the primary incident flux, $I(z=0)$ is the reflected flux from the oxide surface at the angle α, $I(z_t)$ is a part of the reflected flux at the angle α_1 that passed through the oxide layer with a thickness z_t and a refractive index n_1.

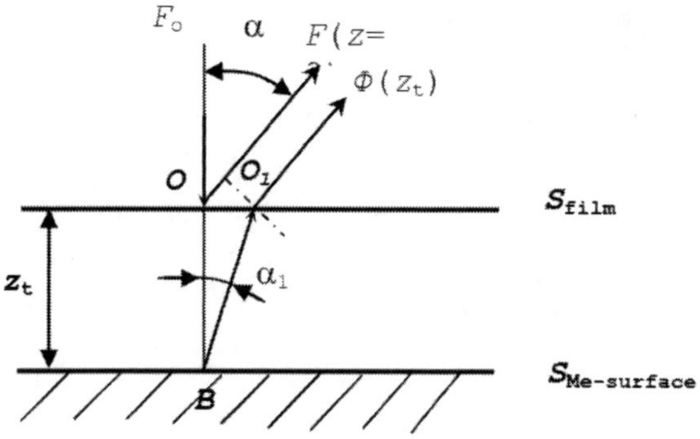

Figure 1. Scheme of forming the reflected flux from a uniform film on a metal surface.

The relationship between the energy of incident, reflected and transmitted electromagnetic waves is described by the following equation

$$R + T + \chi = 1, \quad (1)$$

where R is the reflection coefficient that is characterized by the ratio of the reflected flux to the incident flux $F(R)/F_o$, χ is the radiation absorption coefficient – $F(\chi)/F_o$, and T is the transmission coefficient – $F(T)/F_o$.

On the assumption that the oxide is transparent, the absorption process can be neglected ($\chi \approx 0$) and then the refractive index can be defined as $n = (1+R^{0.5})/(1-R^{0.5})$.

In the field of frequencies of infrared radiation, optical properties are mainly determined by conduction electrons and the bond type of molecular compounds. The resonance frequency of non-deformed molecules can be estimated within the framework of the harmonic oscillator model from the expression:

$$\nu = (1/2\pi) f^{0.5} (M_1^{-1} + M_2^{-1})^{0.5}, \quad (2)$$

where f is the rigidity coefficient of the interatomic bond of the molecule, for example, for diatomic molecules with atomic masses M_1 and M_2, $f = c\nu^2 (M_1^{-1} + M_2^{-1})^{-1}$ at $c = 5.863\times10^{-2}$.

To conveniently process and present IR spectra, the algorithm is implemented on the basis of the table processor of Microsoft Excel 2010. When working with this program, the main peaks of absorption and reflection were revealed. This program is developed for wavelengths in the range from 1.0 to 25.0 μm with a 0.1 μm step. The developed algorithm allows identifying and recognizing the affiliation of the emission peak to an inorganic compound or a group of compounds. The resonance reflection peaks of inorganic compounds in the wavelength range from 1 to 25 μm are extracted and systematized from the literature data that contain more than 2100 types of compounds and cover more than 10000 characteristic resonance lines.

Figure 2 shows the typical emission spectra of the samples after oxidation. The most intensive lines are observed in the spectrum of water (H_2O) for the wavelength ranges of 2.5–3.5, 5.3–7.2 and 13.4–

17.8 μm, and in the spectrum of a hydroxyl group (–OH) for the wavelength ranges of 2.0–3.3 and 3.8–4.5 μm.

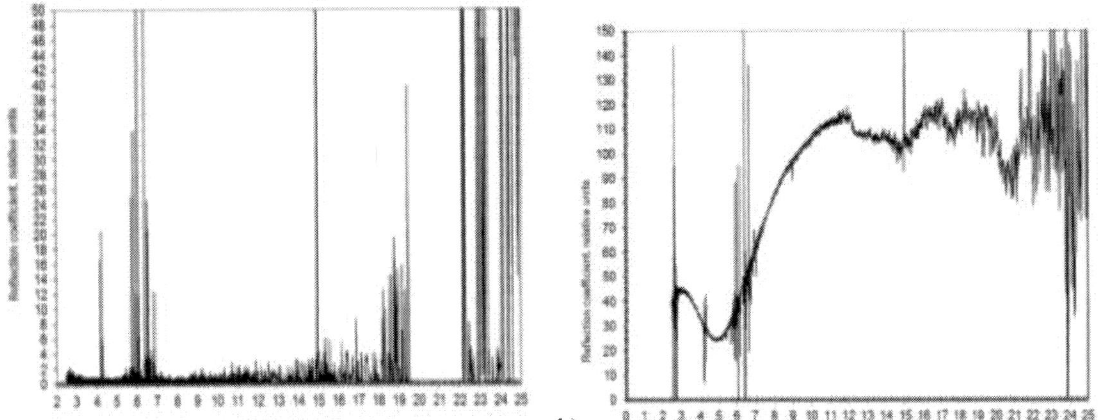

a) b)

Figure 2. Initial IR spectrum of a sample of group No. 1 after oxidation at 500 °C for: a) 30 s and b) 300 s.

3. Results and discussion

The IR spectra obtained were processed using the software package OMNIC-2000 and the developed algorithm based on the table processor of Microsoft Excel-2010, which made it possible to identify the position of the most intensive peaks, a part of which is presented in table 1.

Table 1. Emission peaks in the IR range of the samples after oxidation.

Wave-length, μm	Location of emission peaks for various groups No. 1 T=500 °C, t=30 s		No. 2 T=500 °C, t=300 s		No. 3 T=400 °C, t=600 s		No. 4 T=350 °C, t=1112 h	
	Expe-riment	Calcula-tion	Expe-riment	Calcula-tion	Experi ment	Calcula-tion	Expe-riment	Calcula-tion
	2.71	2.7±0.03	2.65	2.7±0.03	2.74	2.70±0.03	2.71	2.70±0.03
2	2.78	2.7±0.03	-		-		2.79	2.70±0.03
	2.81	2.88±0.03	-		-		-	-
	-	-	2.98	2.94±0.04	2.97	2.94±0.04	2.98	2.94±0.04
	-		-		-		3.21	3.25±0.03
3	-	-	3.95	-	-	-	3.39	-
	-		-		3.96		3.96	-
4	4.75	-	-	-	-	-	4.51	-
	4.98		-	-	-	-	-	-
5	5.04**	-	-	-	5.10**	-	5.08**	-
	5.15**				5.19**	-		
	-	-	-	-	-	-	6.13*	6.13±0.06
6	6.23	6.26±0.06	6.73**	6.5±0.6	-	-	-	-
	-	-	-	-	-	-	-	-
7	7.19	-	-	-	-	-	-	-
	7.23	-	7.23	-	-	-	-	-
10	-	-	-	-	9.85	10.5±0.9	-	-
14	14.22	14.9±1.4	-	-	-	-	-	-

* vibrations of a deformed H2O molecule
** vibrations of a deformed ZrO2 molecule

It follows from the analysis of the position of extracted emission peaks (by an experiment) and comparison with the calculated values (using the harmonic oscillator model) that the appearance of peaks is mainly due to the stretching vibrations of water molecules and zirconium oxide except for vibrations in the wavelength area of 5 and 6 μm (5.04, 5.15, 5.19 and 6.73 μm) that are usually due to incipient internal mechanical stresses.

Table 1 also shows a shift of the peaks of normal vibrations of water that is caused by the distortion of bonds and the formation of new hydrogen bonds influencing the force constant of the valence bonds and angles of the molecule.

The shift of the peaks of normal vibrations of zirconium oxides ZrO, ZrO2-x, and ZrO2 to the high-frequency area usually takes place because of the presence of internal mechanical stresses in the film resulting from the presence of phase transformations with various crystal lattices. If the emission peaks shift to the low-frequency area, it is due to a decrease in the force constant f owing to the loosening of the oxide film structure (the presence of defects in the form of pores and cracks).

4. Conclusion

As a result of the study on the emission spectra of oxide films on a metal surface of zirconium samples after autoclave tests of the samples with the following oxidation regimes: No. 1 (500 °C, 30 s), No. 2 (500 °C, 300 s), No. 3 (400 °C, 600 s), No. 4 (350 °C, 1112 h), it has been found the following:

- The number of emission peaks typical of ZrO and ZrO_2 compounds decreases with increasing the temperature of tests;
- The number of peaks of the oxide film with an oxygen deficiency (ZrO_x) increases with the oxidation time, which, apparently, is due to an increase in the thickness of the protective oxide film and an increase in the oxygen deficiency along its thickness;
- The number of emission peaks of –OH and H_2O compounds decreases with an increase in the time and the temperature of tests.

An algorithm has been developed for processing the spectra to identify the resonance emission peaks. The algorithm has been implemented on the basis of the table processor of Microsoft Excel-2010.

References

[1] Kalin B A, Platonov P A, Tuzov Yu V, IChernov I I and Shtrombakh Ya I 2012 Fizicheskoe materialovedenie: Uchebnik dlya vuzov. Konstruktsionnye materialy yadernoy tekhniki (Moscow: MEPhI) p 672
[2] Douglas D L 1971 The metallurgy of zirconium (Vienna: Atomic Energy Review) p 360
[3] Parfyonov B G, Gerasimov V V and Venediktova G I. 1967 Korroziya tsirkoniya i ego splavov (Moscow: Atomizdat) p 246

Study of deformation dependence on time in polyethylene terephthalate for different static loads and irradiation doses

A I Kupchishin[1,2], V M Lisitsyn[3], M N Niyazov[1] and N A Voronova[1]

[1] Abai Kazakh National Pedagogical University, 13 Dostyk Ave., Almaty, 050010, the Republic of Kazakhstan
[2] al-Farabi Kazakh National University, 71al-Farabi Ave., Almaty, 050012, the Republic of Kazakhstan
[3] Tomsk Polytechnic University, Tomsk 634050, Russia

E-mail: ankupchishin@mail.ru

Abstract. Complex experiments were conducted on deformation dependence on time for different static loads and irradiation doses in polyethylene terephthalate. Curves of time (t) dependence of deformation (ε) show a significant change in the behavior of the material after irradiation. The obtained dependence curves of ε on t for both non-irradiated and irradiated materials are satisfactorily described by cascade-probability model.

1. Introduction

Widespread use of polymeric materials leads to steadily increased volumes of their production. World production of plastics increases two times every fifteen years, and by 2010 it had reached 250 mln. tns. Moreover, the fastest growing market today is polyethylene terephthalate (PET, PETP, polyester). Polyethylene terephthalate is a polymer that has revolutionized the world of packaging, radically changing the situation in the world market in the field of production [1–3].

Materials consisting of PET were developed in the early 1940s and has since proven its versatility of application in various areas of life: in light industry, food industry, and machine-instrument making, mechanical engineering, medicine and pharmaceuticals. In addition, PET is an excellent material for the manufacture of various types of films and containers. High consumer properties of PET containers have provided a rapid growth in the production of packaging for food and beverage products. PET has good heat resistance in the temperature range from –40 °C to +200 °C. A small water absorption provides high stability of properties and equipment sizes. Products made of PET have good mechanical properties, resistant to impact and cracking. PET is stable to diluted acids, oils, alcohols, mineral salts and organic compounds but not to strong alkalis and some solvents. It minimally adsorbs odors and exhibits good gas barrier properties [4–7]. At the same time, the problem of regulations of properties is topical. One of the methods of improving the characteristics of the material is exposure to radiation. Many irradiation types affect the properties of polymeric materials, resulting in a change in their structure depending on various factors [8–10]. Investigation of the effect of charged particles on the properties of polymers is very relevant and allows directed modification of their properties. This work is devoted to the study of deformation dependence on time in polyethylene terephthalate for different static loads and irradiation doses.

2. Methodology

Industrial polyethylene terephthalate 90-μm thick was selected as the test material. The film samples were cut using a special device. The length of the material to be tested was 7 cm and the working part was 5 cm (2 cm of polymer sample were fixed in the clamps), the width was 0.5 cm.

In order to carry out studies, the maximum load at which rupture of the tested polymer occurs had been previously determined using special cargo. For this lavsan material it was 13.5 MPa. Then the strain constituting a certain part of the limit strain was calculated. After that, the sample was mounted in the clamps. The dependence of ε on t was recorded. The timing of the test sample on uniaxial tensile static load was about 50 sec. The data were recorded on a video camera. After that, all the material was processed. The error was 0.1%.

The studies of the dependence of deformation on strain of both non-irradiated and irradiated materials were performed using a modified tensile device based on RU-50 with crosshead speed of 12 cm/min.

Irradiation of samples was carried out using ELU-6 linear electron accelerator with energy of 2 MeV in air. The irradiation dose was equal to 50 and 100 kGy. The samples of the films were placed at a distance of 40 cm from the output window of the accelerator. The material temperature was 23 °C and the relative humidity was 55%.

3. Results and discussion

Figures 1 and 2 show the experimental results on the dependences of deformation ε on time t at different values of static strain: $\sigma_1 = 7$; $\sigma_2 = 10$; $\sigma_3 = 11$; $\sigma_4 = 12$ and $\sigma_5 = 14$ MPa, and the irradiation doses of 0, 50 and 100 kGy.

It was found that the relative elongation of the material depends significantly on the static load, reaching more than 500% of the initial value and becomes a constant over time (Figure 1, curves 1, 3, 5, 7, 9; Figure 2, curves 1, 3, 5).

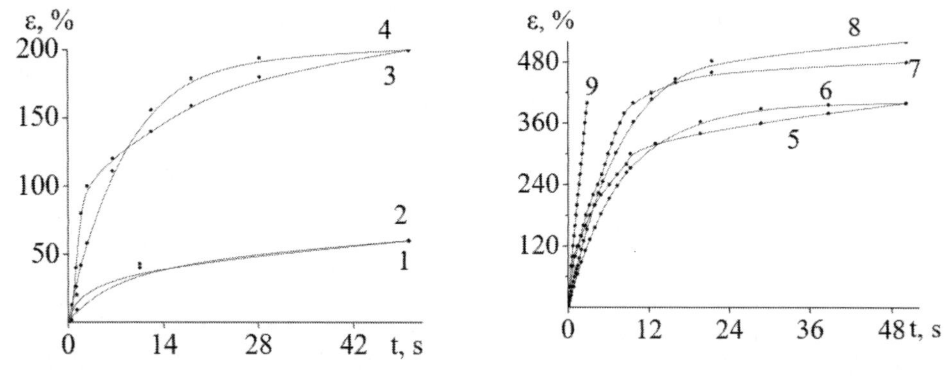

Figure 1. Dependence of deformation on time for non-irradiated polyethylene terephthalate at different static loads experiment at σ = 7, 10, 11, 12, 14 MPa (1, 3, 5, 7, 9); calculation at σ =7, 10, 11, 12 MPa (2, 4, 6, 8)

Calculation of the dependence of ε ***on*** *t* was carried out using the cascade-probability model according to the formula: $\varepsilon = \varepsilon_0[1-\exp(t/t_0)]$. For this experiment, ε_0 is the maximum elongation, t_0 is the time at which $|\varepsilon/\varepsilon_0 - 1|$ decreases by a factor *e*. Table 1 shows the values of t_0 and ε_0 for different values of σ.

Table 1. Dependence of ε_0 and t_0 values on σ

σ, (MPa)	7	10	11	12
ε_0, (%)	60	200	360	520
t_0, (s)	6	8	10	12

As seen from the table, increase in σ value leads to significant increase in ε_0 and t_0. Figure 2 shows that with increasing radiation dose (D), deformation increases sharply at first, and then gradually reaches saturation. Dependence of ε_0 and t_0 parameters on D is shown in Table 2. With increasing σ, Dε increases. According to the results obtained, electron irradiation alters ε by more than 100%.

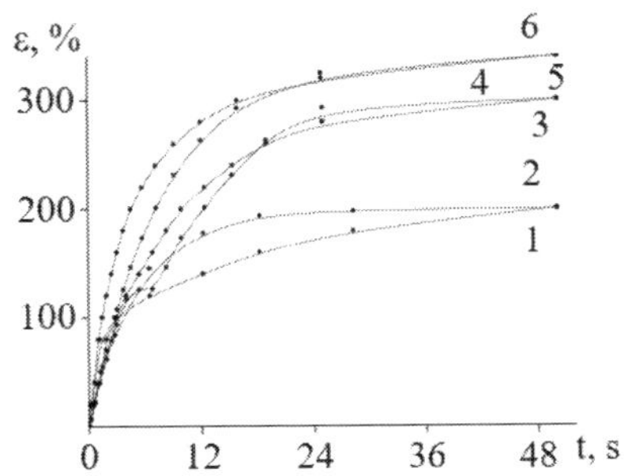

1 – non-irradiated sample; 3 – 50;
5 – 100 kGy, – experiment;
2 – non-irradiated sample, 4 – 50;
6 – 100 kGy, – calculation

Figure 2. Dependence of deformation on time for PET for different doses of electron irradiation at static load of σ = 11 MPa

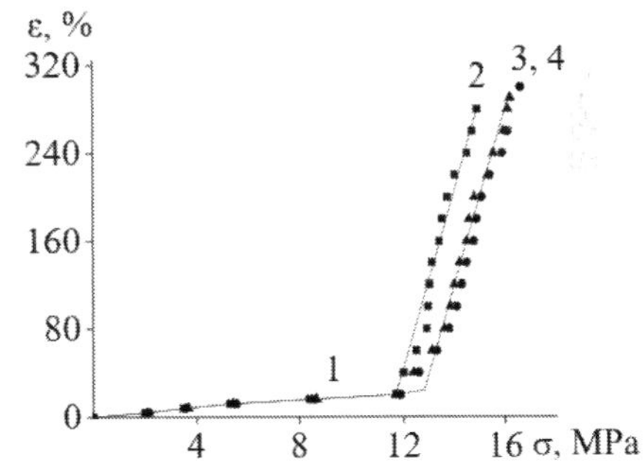

square – non-irradiated sample; circle – 50;
triangle – 100 kGy, – experiment;
1 – non-irradiated sample, 2– 50;
3 – 100 kGy, – calculation

Figure 3. Dependence of deformation on mechanical strain of irradiated and non-irradiated polyethylene terephthalate

Table 2. Dependence of ε_0 and t_0 values on the irradiation dose

D, (kGy)	0	50	100
ε_0, (%)	200	300	340
t_0, (s)	6	8	10

After irradiation by the doses of 50 and 100 kGy, the samples of material gain significant flexibility and begin to tear at deformations and strains greater than those before irradiation. Thus, an increase in tensile strength by 10 and 15 % is observed compared with non-irradiated material (Figure 3, curves 3 and 4). In the strain range of 12–16 MPa, electron irradiation results in a substantial reduction of the deformation (80 %). Lavsan films are characterized by good elasticity, and it is partially reversible. The material contains regularly spaced polar groups, whose function is to enhance the intermolecular interactions and the formation of the material stiffness. According to Figure 3, the experimental curves are satisfactorily described by the linear model [8]. In this case, in the range σ = 0–12 MPa, the experimental data for the non-irradiated and irradiated samples (curve 1) coincide. The change in the angle of inclination of the curve at $\sigma = 12$ MPa is due to the transition from the elastic region of deformation to catastrophic tear of the material. The reduced material strength and increased elongation indicate a significant influence of radiation defects on the structure and mechanical properties of lavsan.

4. Conclusions

1. Complex experiments were conducted on the deformation dependence on time for different static loads and irradiation doses: $\sigma_1 = 7$; $\sigma_2 = 10$; $\sigma_3 = 11$; $\sigma_4 = 12$ and $\sigma_5 = 14$ MPa. It was found that the deformation of the material depends strongly on time and static load.
2. Irradiation of PET samples results in significant improvement in the flexibility, and a significant increase in strain (more than 100% compared with non-irradiated material), which is associated with degradation of the polymer chains. However, strength is virtually unchanged.
3. Experimental dependences of ε on t and ε on σ for both non-irradiated and irradiated materials are satisfactorily described by cascade-probability and linear models.

Acknowledgments
The work was performed as part of the grant of the Ministry of Education and Science of the Republic of Kazakhstan (No. 274).

References

[1] Godovskiy Y K 1970 Successes of chemistry and polymer physics (M.: Khimiya) p 205
[2] Deans J, Vinyard J 1966 Radiation effects in solids (M.: Mir) p 129
[3] Milinchuk V K, Klinshpont E R and Pshezhetskiy S Y 1980 Macroradicals (M.: Khimiya) p 278
[4] Tyutnev A P, Sayenko B C, Pozhidayev Y D, Kostyukov NS 2005 Dielectrics and radiation: Bk. 5: Dielectric properties of the polymers in the fields of ionizing radiation (M.: Nauka) p 453
[5] Krinichnyy V I, Shuvalov V F, Grinberg O Y and Lebedev Y S 1983 Chemical Physics Kinetics and mechanism of formation and modification of polymers **5** pp 621 – 627
[6] Mitrofanov R Y, Chistyakova Y S and Sevodin V P 2006 Processing of PET waste **6** pp 24–29
[7] Wildemann V E 2012 Experimental studies of the properties of materials under complex thermomechanical effects (M.: Fizmatlit) p 203
[8] Kupchishin A I, Taipova B G, Kupchishin A A, Voronova N A, Kirdyashkin V I and Fursa T V 2016 *Catastrophic models of materials destruction IOP Conf. Series: Material Science and Engineering* **110** 012037 pp 1–5
[9] Boos E G, Kupchishin A I, Kupchishin A A, Shmygalev Y V, Shmygaleva T A *et al* 2015 Cascade-probability method. Solution of radiation-physical problems, Boltzmann equations. Relation with Markov chains (Almaty, Abay KazNPU, SRINCTM of al-Farabi KazNU, Kama LLP) p 388
[10] Kupchishin A I, Taipova B G, Kupchishin A A and Kozhamkulov B A 2015 Study on the physical and mechanical properties of composites based on polyimide and polycarbonate *Mechanics of composite materials* **51 1** pp 115–118

Explosive radiation sources for analytical spectroscopy

V I Oleshko[1], S V Plotnikov[2] and A Turlybekuly[2]

[1] Department of Lasers and Lighting Engineering, National Research Tomsk Polytechnic University, Tomsk, 634050, Russia
[2] Department of Engineering and Technical Physics, D. Serikbayev East Kazakhstan State Technical University, Ust-Kamenogorsk, 070004, Kazakhstan

E-mail: aturlybekuly@gmail.com

Abstract. The possibility of using the explosion energy of condensed explosives in solids analytical spectroscopy was investigated. For the atomization and excitation emission spectra of solids ($BaCl_2$ and $SrCl_2$) pressed powders of furazanotetrazinedioxide (FTDO, $C_2N_6O_3$) were used. Initiation of the explosion was induced by the influence of high-current electron beam. The impurity atoms that are the part of FTDO, as well as atoms and molecules formed by the evaporation of $BaCl_2$ and $SrCl_2$ were identified in the spectra of explosion plasma.

1. Introduction

At present, the atomic spectrum analysis is used in various fields of natural science and the production; it allows you to quickly and accurately identify the substance composition. The transfer of the solid substance into the gaseous state can be achieved by exposure to directed concentrated high-power energy flux (DCHPEF) which includes laser radiation and high-current electron beams [1-3]. At such influence, instantaneous vaporization of any substance and simultaneous delivery of all elements, included in the sample, into analytical plasma torch, regardless of their volatility is achieved. In this aspect, explosive radiation sources (ERS) based on the detonation of condensed explosives (explosives) are of interest. High specific energy content of explosives (~5 kJ/g) and a small discharge time allows obtaining high levels of specific heat power and ensure the heating of any material up to tens of thousands degrees. In this case, all substances transfer into the plasma state and emit powerful optical radiation. The possibility of obtaining long-lived plasma formations with full output optical radiation at the level of a few percent of chemical explosives stored energy was showed in [4-5].

With regard to the possibility of ERS application in substance analytical spectroscopy, such studies, to our knowledge, have not been carried out. The purpose of this work is to explore the possibility of using the explosive radiation sources based on the detonation of condensed explosive in atomic and molecular spectroscopy.

2. Experiment

For study objects salts of $BaCl_2$ and $SrCl_2$ were selected, which were placed on the explosives surface. A pulsed nanosecond electron accelerator based on a pulsed high voltage generator GIN-600 connected to a vacuum diode was taken as a source of initiation. Average electron energy in a beam was ~250 keV, maximal current was 3 kA and pulse duration amounted to ~ 12 ns. The scheme of the experimental setup is shown in figure 1 and figure 2.

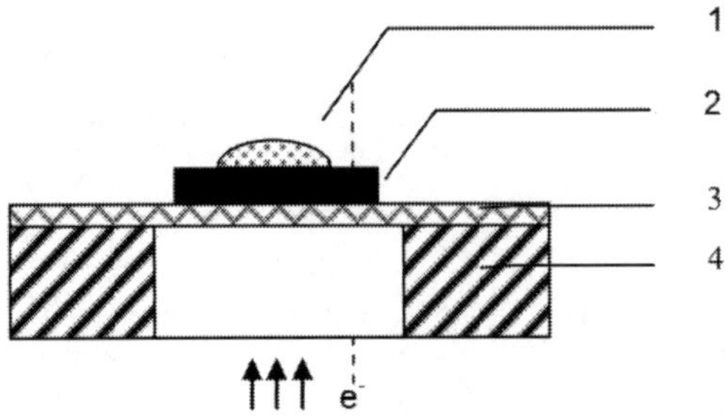

Figure 1. The scheme of explosive initiation by electron beam that were used for measuring spectra of explosion decomposition products: 1–sample, 2–explosives, 3–aluminum foil 10 μm thick, 4 – anod.

Sample 1 was placed on the surface of explosive 2, which was placed on aluminum foil 3 with thickness of 10 μm, and than an initiation by an electron beam through the hole in the anode 4 was made. Two methods were used for recording the explosion spectrum. In the first one, the image of explosion plasma was projected with a lens onto the entrance slit of monochromator MDR-23 and photographed with digital camera SONY α500 mounted on monochromator instead of the exit slit. Spectral resolution of the method amounted to ~0.2 nm. The spectral range of measurements was 400–700 nm.

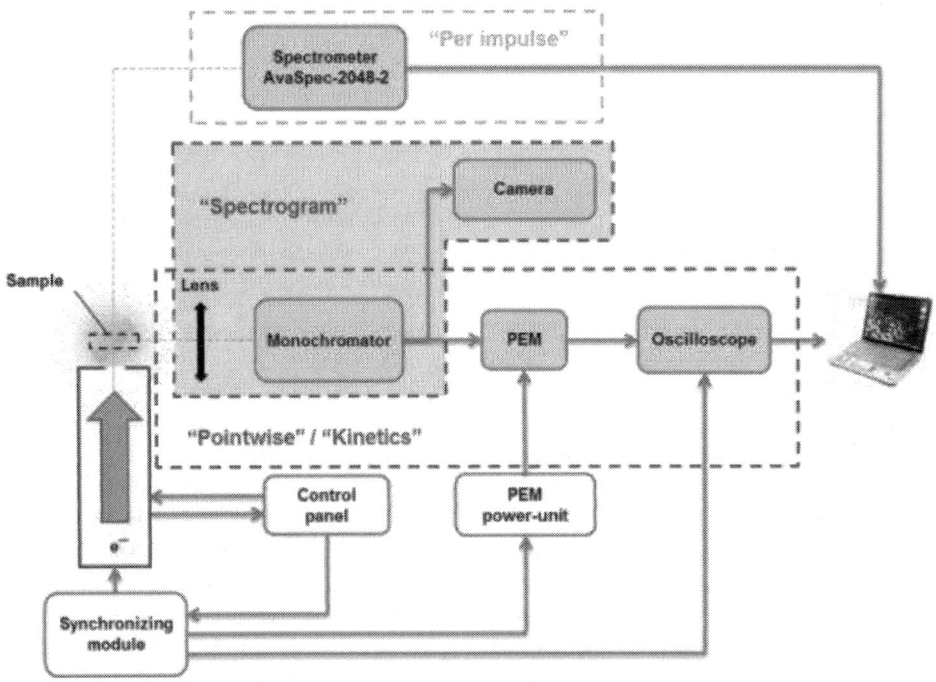

Figure 2. The scheme of experimental setup

Time-integrated fragments of explosion spectrum (Δλ~25 nm) were recorded in spectral response range of camera SONY α500 per one excitation pulse. Atomic spectral lines recorded in explosion plasma were identified using well-known atomic spectra of spectral lamps LSP-1. Those reference spectra were photographed along with explosion spectrum onto the same spectrogram.

In the second method, the integrated spectra of explosion were recorded by means of fiber-optical spectrometer AvaSpec 2048-2. The spectral range of measurements was 220–1100 nm, spectral resolution amounted to ~1.5 nm.

3. Results and discussion

Preliminary, the explosion plasma spectra of "clean" FTDO samples were studied. Figure 3 shows the explosion atomic emission spectra. It can be seen against the background of low-intensity continuous spectrum four lines are distinguished belonging to: sodium (588.9 nm and 589.5 nm), lithium (670.7 nm), and potassium (769.8 nm and 766.4 nm). Sodium doublet line is recorded as one line due to lack of the spectrometer's resolution ability. According to studies [10], alkali metal lines (Na, Li, K) found in the FTDO explosion plasma indicate occurrence of impurities in the samples during manufacture.

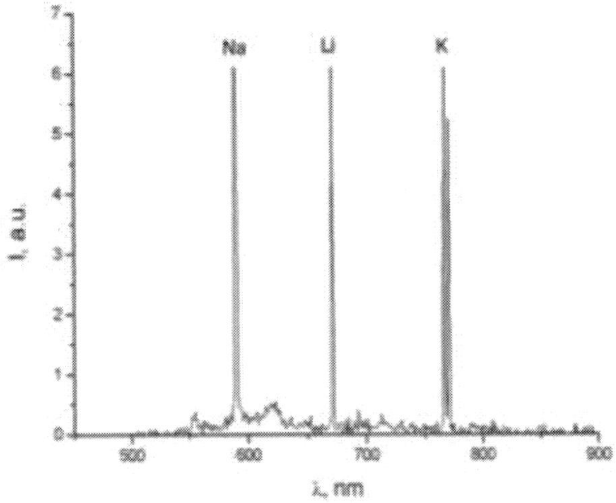

Figure 3. Explosion emission spectrum of pure FTDO disposed on an aluminum substrate at initiating by electron beam.

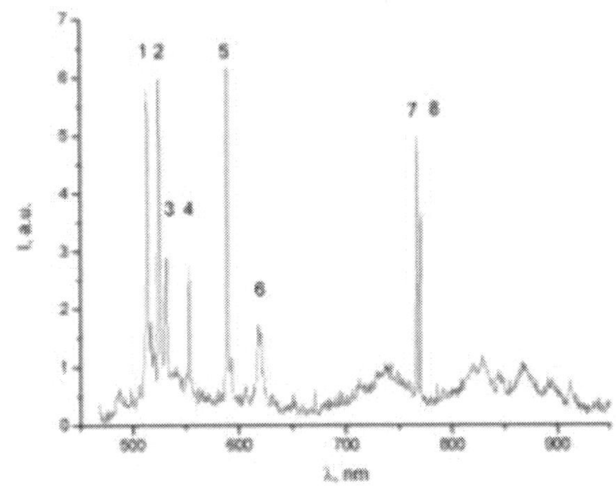

Figure 4. FTDO explosion luminescence spectra with the $BaCl_2$ addition.

FTDO explosion glow spectrum with the addition of barium chloride is shown in figure 4. The figure shows that the explosive emission spectrum contains additional molecular and atomic lines: Ba (553.55 nm), BaCl (513.8 nm and 524.5 nm), CaCl (620.9 nm). Detailed identification of the obtained spectrum is shown in table 1. The spectra of the explosive emission of barium chloride with a high spectral resolution (0.15 nm) were recorded using a digital camera through a monochromator MDR-23 (figure 5). To identify the molecular spectra literature data were used [11-13].

IOP Conf. Series: Journal of Physics: Conf. Series **830** (2017) 012142 doi:10.1088/1742-6596/830/1/012142

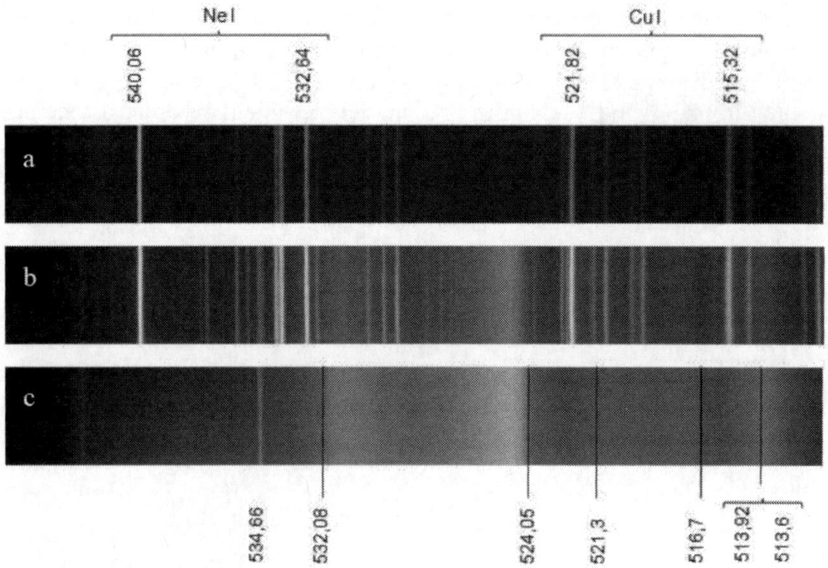

a) the spectra of the LSP-1
b) the FTDO explosion spectra with adding $BaCl_2$ salt against the background LSP-1 spectra
c) the FTDO explosion spectra with adding $BaCl_2$ salt at free explosion products expansion

Figure 6. Explosion glow of $BaCl_2$ and lamps LSP-1 with 0.1 nm resolution.

Figure 6 shows the spectrum of FTDO explosive emission with the addition of strontium chloride. Identification of the obtained spectrum is shown in Table 2. It can be seen that the impurity atoms of the FTDO are present in the explosion spectrum, as spectrum of SrCl molecules formed during explosive evaporation.

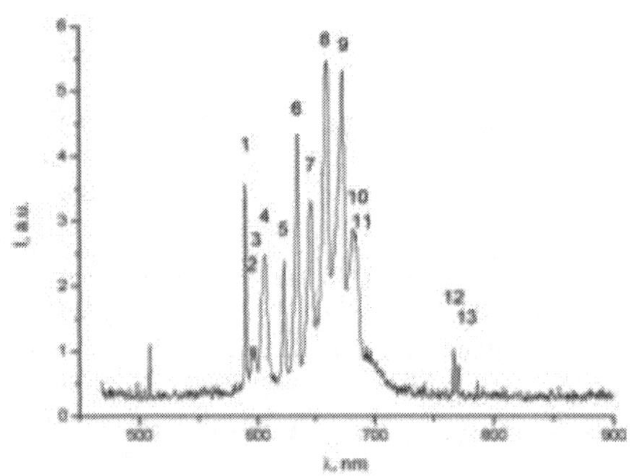

Figure 7. FTDO explosion luminescence spectra with the $SrCl_2$ addition.

The data obtained in the experiments demonstrate the possibility of using explosives as a source of spectra excitation in atomic emission spectrometry and molecular spectral analysis. The main advantage of the explosives use in the spectral analysis is the ability to identify elemental composition of explosives, and other condensed materials. Alkali metal impurities recording in the FTDO emission spectrum shows the high sensitivity of the developed method.

IOP Conf. Series: Journal of Physics: Conf. Series **830** (2017) 012142 doi:10.1088/1742-6596/830/1/012142

Table 1. Explanation of the main FTDO explosion emission spectral lines adding $BaCl_2$ and $SrCl_2$.

	$BaCl_2$	**$SrCl_2$**
№	**Identified line**	**Identified line**
1	BaCl 513,8 [11]	Na 589,593[11]
2	BaCl 524,05 [11]	CaCl 593,4[11]
3	BaCl 532,08 [13]	-
4	Ba553,55 [11]	SrO606,84 [12]
5	Na 589,593[11]	SrCl 623,9[13]
6	CaCl620,1[11]	SrCl 635,8[13]
7	K 766,491[11]	SrOH 646[13]
8	K 769,989[11]	SrOH 659[13]
9		-
10		SrOH 682[13]
11		SrJ 684,77[11]
12		K 766,491[11]
13		K 769,989 [11]

4. Conclusion

The data obtained in the experiments demonstrate the possibility of using explosives as a source of spectra excitation in atomic emission spectrometry and molecular spectral analysis. The main advantage of the explosives use in the spectral analysis is the ability to identify elemental composition of explosives, and other condensed materials. Alkali metal impurities recording in the FTDO emission spectrum shows the high sensitivity of the developed method.

References

[1] Oleshko V I 2012 *The spectral element analysis using powerful electron beams* (Tomsk: Tomsk Polytechnic University) p 96
[2] Oleshko V I, Zarko V E, Lysyk V V, Tsipilev V P and Kalmykov P I 2015 *Technical Physics Letters* **41** pp 10–15
[3] Oleshko V I and Lysyk V V 2012 *Izv.Vuzov.Physics* **55** pp 149-152
[4] Bulanov C C, Esiev R U, Kamrukov A S, Kozlov N P, Morozov M I and Roslyakov I A 2010 *Technical Physics* **80** pp 87–94
[5] Usupaliev U, Usupaliev P U and Shuteev S A 2007 *Technical Physics* **77** pp 50–62

Thermally stimulated processes in Li and Cu doped alkali fluorides irradiated with electron beams of ultra-high dose

Zh K Mamytbekov[1], A N Tcherepanov[1], A I Slesarev[1], M M Kidibaev[2], Q Shi[1,3], K V Ivanovskikh[1], V Yu Ivanov[1], A A Egamberdieva[2] and B V Shulgin[1]

[1]Ural Federal University, 620002, Ekaterinburg, Russia
[2]National Academy of Sciences KR, 720071, Bishkek, Kyrgyz Republic
[3]Taiyuan University of Technology, 030024, Taiyuan, China

E-mail: zzhayloo@gmail.com

Abstract. The thermally stimulated luminescence (TSL) and exoemission (TSE) in Li and Cu doped NaF and LiF single crystals irradiated with electron high energy electron beams of (10 MeV, doses 0.75 and 2 MGy) have been investigated. The results obtained reveal important properties that suggest that the crystals have a sufficient radiation stability and sensitivity for high energy electron beams and are promising for application as high-dose detectors of electron radiation.

1. Introduction

Considerable progress has been made over the last decades in the development and characterization of many kinds of thermoluminescence (TL) phosphors among which alkali fluorides have always been of notable importance [1-4]. Lithium and sodium fluorides are characterized by low effective atomic number (Z_{eff}) that makes them useful for the development of tissue-like thermoluminescence dosimeters (TLDs). The remaining potassium, rubidium, and cesium fluorides are highly hygroscopic and do not attract significant attention.

A drawback of undoped LiF consisted in a complicated TL response that was dependent on its radiation and thermal history. In attempting to solve the problem, a quite significant attention was drawn to the development and study of LiF doped with B, Si, Mg, Cu, Ti and other dopants including lanthanides (see [4-12] and reference therein). Nowadays, commercial TLDs based on LiF:Mg,Ti (TLD-100/600/700) and LiF:Mg,Cu,P (TLD-100H/600H/700H) and their modifications are well accepted in the industry. The second practically important representative of the alkali fluorides family is NaF which thermally stimulated properties have also been extensively studied over the last decades. The development of NaF based TLDs is challenging due to a complex manner of its TL properties influenced by factors like type and energy of ionizing radiation, thermal and radiation dosage history, etc. [1-3, 13-17]. Both LiF and NaF like many others have drawback consisted in a low sensitivity to low doses and an early saturation at higher doses.

Nowadays, we observe rising demand for high-level TL dosimetry materials due to the development of new radiation technologies applicable in materials testing, sterilization and processing, nuclear medicine, etc. Some of them are based on using electron and ion beam facilities as irradiation sources. It was recently shown that LiF based TLD may be prospective for application in extremely wide dose range of up to twelve orders of magnitude (see [18] and references therein). This inspired us to test thermally stimulated processes of NaF and LiF doped with Li and Cu irradiated with ultra-high dose of electron beam. To this end we performed an experimental study of TL and thermally stimulated exoemission (TSEE) glow curves of LiF and NaF doped with Li and Cu irradiated with ultra-high doses of 10 MeV electron beams. The results obtained suggest that the crystals have a sufficient radiation stability and sensitivity for high energy electron beams and can be promising for application as ultra-high-dose detectors of electron radiation.

2. Experimental

Single crystals of NaF:1mol%Li (NaF:Li), NaF:1mol%Li, 0.1mol%Cu (NaF:Li,Cu), NaF:0.1mol%Cu (NaF:Cu), and LiF:0.3mol%Cu (LiF:Cu) were grown in platinum crucibles by the Kiropoulos method [19] at the Institute of Physics of Kyrgyz National Academy of Sciences. The crystals were 5 mm in diameter and about 3 cm long. To perform the measurements the crystals were cut into 1 mm thick slices using a diamond saw and polished. The samples were then examined for their basic luminescence properties in order to confirm the entry of impurity ions into the lattice.

The samples were irradiated with 10 MeV electron beam from a MT-20 microtron (Russia) at Ural Federal University. The samples were mounted on a massive cupper sample holder in order to minimize possible heating. Typically, dose acquired by a sample in one minute was 15-16 kGy that was monitored with SO PD(F)R-5/50 standard fenazin dosimeters (Russia) with uncertainty of up to 10%. The samples were exposed to electron beams for 50 and 133 minutes in order to acquire dose of 0.75 and 2 MGy, respectively.

TL and TSEE glow curves were recorded with an automated exoemission spectrometer [20] equipped with a VEU-6 electron multiplier (Russia) to detect electron emission and a FEU-142 solar-blind photomultiplier (Russia) sensitive in the spectral range of 112–365 nm to detect the luminescence. The setup allows linear heating of samples in 300–800 K range with 0.1–1.0 K/s rate. The measurements were performed in vacuum of ~10^{-5} Torr.

3. Results and discussion

Figure 1 shows TSEE glow curves recorded for NaF:Li and NaF:Li,Cu crystals irradiated by electron beams with 750 kGy and 2 MGy dose. Obviously, the glow curves have a complex structure resulted from superposition of multiple TSEE peaks. The curves were fitted assuming the first order kinetics of thermal processes [21]. The TSEE (and TL) activation energies were calculated using halfwidth of the elementary emission bands obtained from deconvolution of the initial glow curves [22]. The quality of fittings done in the frame of this research was to be satisfactory done if figure-of-merit (FOM) was less than 1%. Kinetics parameters obtained for TSEE glow curves of NaF:Li and NaF:Li,Cu crystals that include peak maxima (T_m), activation energies (E_a), and frequency factor (S) are gathered in table 1 along with the data obtained for LiF:Cu and NaF:Cu considered below.

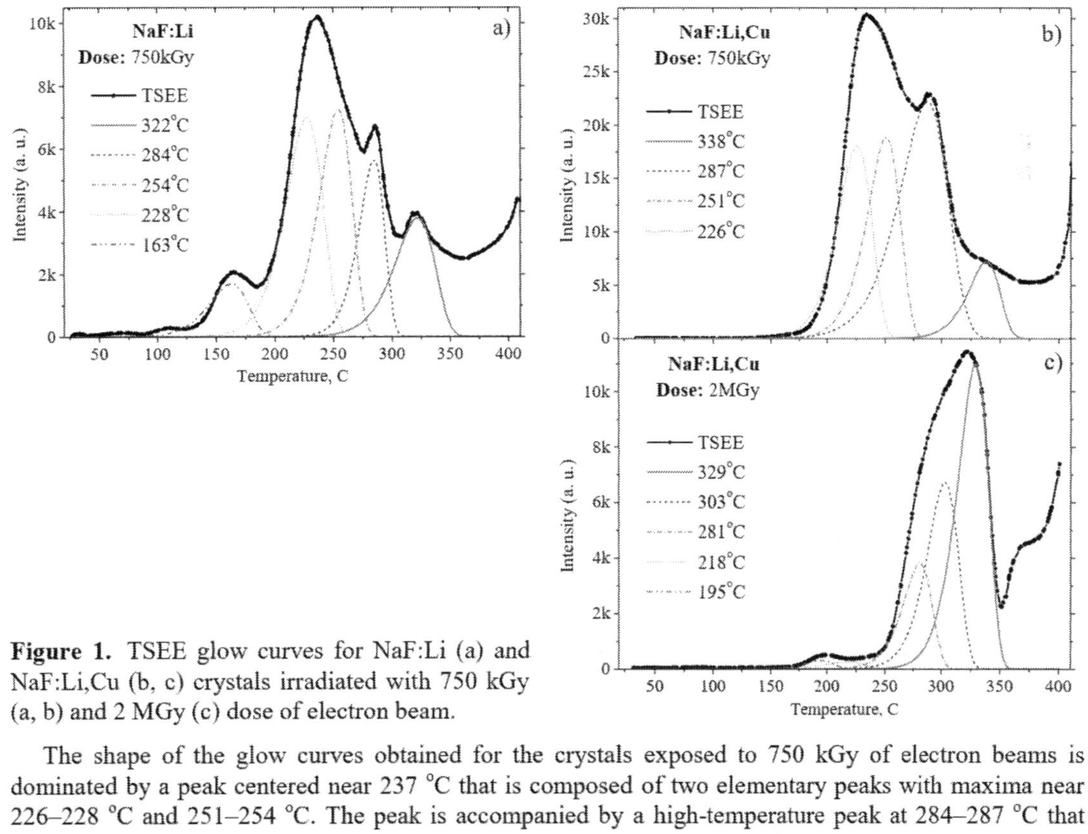

Figure 1. TSEE glow curves for NaF:Li (a) and NaF:Li,Cu (b, c) crystals irradiated with 750 kGy (a, b) and 2 MGy (c) dose of electron beam.

The shape of the glow curves obtained for the crystals exposed to 750 kGy of electron beams is dominated by a peak centered near 237 °C that is composed of two elementary peaks with maxima near 226–228 °C and 251–254 °C. The peak is accompanied by a high-temperature peak at 284–287 °C that becomes more intense in Cu-co-doped crystal. It is worth noting that co-doping with Cu leads to noticeable rise of the TSEE yield. Increasing the dose of irradiation up to 2 MGy leads to dramatic modification of the glow curve that reveals redistribution of populated traps towards deeper ones. Shallow traps responsible for TSEE features detected below 250 °C are not effectively populated at such a high irradiation dose, making less than 1% of the total observed TSEE yield.

Figure 2 shows TL and TSEE glow curves recorded for LiF:Cu and NaF:Cu crystals exposed to 2 MGy dose of electron beams. Both crystals demonstrate well pronounced anticorrelation of the corresponding TL and TSEE glow curves up to about 320 °C. Thermally stimulated processes switch from the domination of electron exoemission to luminescence near 180 °C and 210 °C for LiF:Cu and NaF:Cu crystals, respectively. The TSEE glow curve of LiF:Cu crystal is dominated by the peak at 174 °C. The features located at lower temperatures (peaks at 258, 212, and 160 °C) make less than 15% of the total exoemission yield. NaF:Cu crystal demonstrates a well pronounced TSEE peak at 198 °C. Contribution of a broad TSEE feature centered near 293 °C does not exceed 1%. The TL glow curves of LiF:Cu and NaF:Cu crystals reveal broad non-elementary features spread from about 180 to 300 °C and from 190 to 320 °C, respectively. Both crystals demonstrate rise of TL and TSEE intensity above about 330 °C. We note that the appearance of high-temperature TL features is typical of alkali fluorides irradiated with high and ultra-high dose of ionizing radiation (see [18] and references therein). Kinetics parameters obtained for TSEE and TL glow curves of NaF:Li and NaF:Li,Cu crystals can be found in table 1.

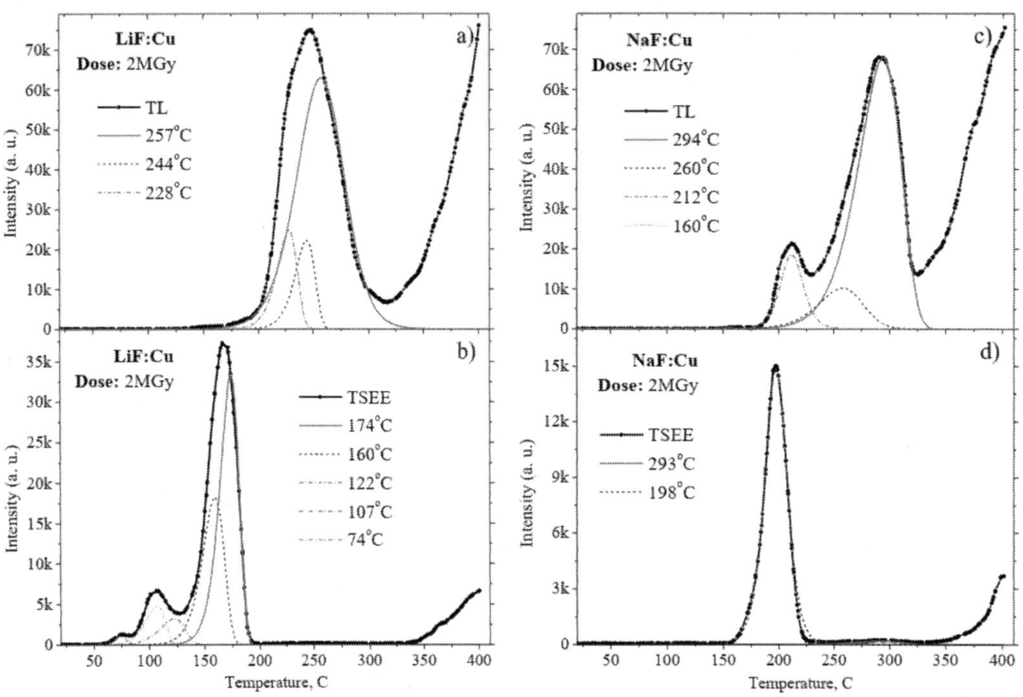

Figure 2. TL (a, c) and TSEE (b, c) glow curves for LiF:Cu (a, b) and NaF:Cu (c, d) crystals irradiated with 2 MGy dose of electron beam.

The above described experimental results can be explained in terms of the model of electronic and ionic excitations in the crystals [23]. The model assumes the presence of a highly mobile electron-hole, ion-ion, and ion-vacancy pairs which are responsible for accumulating of energy released by primary electronic excitation and subsequent transfer of the stored energy to the surface. The latter is known to be enhanced upon high energy electron bombardment [24]. The results documented by some of us earlier in [15-17, 19, 23, 25] along with the results presented in [26, 27] suggest that the energy range of active exoemission in the crystals studied here correlates well with the energy range related to the thermal destruction of F-centers along with aggregated F_2, F_2+ and F_2- centers. We note, that destruction of F-centers may also be connected with their recombination with the hole (H-) centers, which have an increased thermal stability in doped crystals [3, 16, 24]. The TL features related to this kind of thermal processes are likely to contribute to the TL peaks observed around 320–350°C. Obviously, the thermal processes staying behind the TL and TSEE features observed in this work are quite complicated. From the same point of you, we can assume that both the fitting and interpretation of the TL and TSEE glow curves can be ambiguous. One of the reasons for that comes from the fact that the commonly applied fitting method, when a glow curve is considered to be a superposition of several single glow peaks each related to a first-order process, does not account such processes as cascade detrapping or retrapping of carriers and saturation of traps [21, 22]. This peculiarity may results in getting somewhat physically irrelevant kinetics parameters and particularly to extremely high values of the frequency factor [28], that is partly the case in our fitting results. Applying the second-order kinetics or more advanced models is a challenging task that requires additional a priori data on energy structure and concentration of defects. Although defects in alkali halides have been extensively studied so far, understanding of defects due to high and ultra-high dose electron beams is lacking today.

From the practical point of view, it is worth noting that dopants play an important role in formation of trapping and recombination centers responsible for both the TL and TSEE phenomena. Extension of this experimental work would benefit from comprehensive study of TL and TSEE response of the crystals depending on dopant concentration, dose and energy of electron beams.

Table 1. Kinetics parameters for TL and TSEE processes in NaF and LiF doped with Li and Cu irradiated with ultra-high dose electron beams

Sample	Dose[a]	Process	Kinetic parameter[b]	Peaks 1	2	3	4	5
NaF:Li	750	TSEE	T_m	322	284	254	228	163
			E	1.76	2.46	1.71	1.36	0.97
			S	1.4E+13	5.0E+20	4.6E+14	9.0E+11	2.9E+09
NaF:Li,Cu	750	TSEE	T_m	338	287	251	226	
			E	2.39	1.23	1.64	1.46	
			S	1.1E+18	1.7E+09	1.3E+14	1.1E+13	
	2000	TSEE	T_m	329	303	281	218	195
			E	2.39	2.11	2.03	1.37	1.48
			S	2.1E+18	5.6E+16	5.9E+16	2.3E+12	1.8E+14
LiF:Cu	2000	TSEE	T_m	174	160	122	107	74
			E	2.06	1.67	1.22	1.48	1.92
			S	5.1E+21	8.6E+17	8.7E+13	1.3E+18	4.0E+26
		TL	T_m	257	244	228		
			E	1.40	2.61	2.39		
			S	3.0e+11	7.8e+23	3.7e+22		
NaF:Cu	2000	TSEE	T_m	293	198			
			E	1.27	2.56			
			S	2.7e+9	1.1e+26			
		TL	T_m	294	258	212	160	
			E	1.43	1.34	2.87	2.37	
			S	7.3e+10	7.6e+10	2.5e+28	1.8e+26	

[a] Dose is measured in kGy.

[b] Units for T_m, E, and S parameters are °C, eV, and s^{-1}, respectively.

4. Conclusion

Single crystals of NaF:Li, NaF:Li,Cu, NaF:Cu, and LiF:Cu were synthesized using the Kiropoulos method and irradiated with ultra-high dose electron beams of 10 MeV. All the samples revealed good radiation resistance and sensitivity to the irradiation doses applied. The results of TSEE and TL measurements suggest complicated structure of defects in the crystals that can hardly be modelled with the commonly applied model of first-order process. In connection with the observed TL and TSEE glow curves, it is possible to tentatively propose application of the crystals for application in ultra-high dose TL and TSEE dosimetry. This proposition, however, should be further supported by in-depth investigation of TL and TSEE response of the crystals depending on dopant concentration, dose and energy of electron beams.

Acknowledgements

This work was supported by the UrFU Competitiveness Enhancement Program.

References

[1] Sharma J 1956 *Physical Review* **101** pp 1295-1297
[2] Riggin M, Radhakrishna S and Whippey P W 1976 *Phys. Status Solidi A* **37** pp 51-56
[3] Mariani D F and Alvarez Rivas J L 1978 *J Phys. C: Solid State Phys.* **11** pp 3499-3509
[4] Berger T and Hajek M 2008 *Rad. Measur.* **43** pp 146-156
[5] Zimmerman D W, Rhyner C R and Cameron J R 1966 *Health Phys.* **12** pp 525-531
[6] Zimmerman J 1971 *J. Physics C: Solid State Phys.* **4** pp 3277-3291
[7] Taylor G C and Lilley E 1978 *J Phys. D: Appl. Phys.* **11** pp 567-581
[8] Stoebe T G and Watanabe S 1975 *Phys. Status Solidi A* **29** pp 11-29
[9] Nakajima T, Murayama Y, Matsuzawa T and Koyano A 1978 *Nucl. Instr. & Meth.* **157** pp 155-162
[10] Deshmukh B T and Moharil S V 1985 *Bull. Mater. Sci.* **7** pp 427-457
[11] Salah N, Sahare P D and Rupasov A A 2007 *J. Luminescence* **124** pp 357-364
[12] Tang K, Cui H, Zhu H, Liu Z and Fan H 2012 *Radiat. Meas.* **47** pp 185-189
[13] Bhan S 1982 *Phys. Status Solidi A* **69** pp 367-376
[14] Tomita A, Takeyasu T and Fukuda Y 1996 *Rad. Protect. Dosimetry* **65** pp 405-408
[15] Kazakbaeva Z M, Ogorodnikov I N, Kidibaev M M, Alybakov A A and Shul'gin B V 1992 *J. Appl. Spectr.* **56** pp 34-38
[16] Alybakov A A, Gubanova V A, Denisov G S and Umurzakov B S 1984 *Phys. Status Solidi B* **124** K75-K78
[17] Slesarev A I, Zhamangulov A A, Kidibaev M M, Kortov V S and Shul'gin B V 2000 *Tech. Phys. Lett.* **26** pp 386-388
[18] Obryk B, Khoury H J, de Barros V S, Guzzo P L and Bilski P 2014 *Rad. Measur.* **71** pp 25-30
[19] Alybakov A A, Dobrzhanskiy G F and Gubanov V A 1964 *Kristallografiya* **9** pp 940-942
[20] Kortov V S, Isakov G V and Slesarev A I 1996 *Defectoskopiya* **1** pp 50-59
[21] Bos A J 2006 *Rad. Measur.* 41 Suppl. **1** S45-S56
[22] Slesarev A I 2006 *Bulletin of Ural State Technical University* **5** chapter **76** 174-179
[23] Tcherepanov A N, Ivanov V Yu, Koroleva T S and Shulgin B V 2006 *LiF and NaF Bulk, fiber and nanocrystals as detector materials* (Ekaterinburg: USTU Press) ISBN 5-321-00833-7 p 304
[24] Brunsmann U and Scharmann A 1977 *Phys. Status Solidi A* **43** pp 519-528
[25] Shulgin B V, Ivanov V Yu, Tcherepanov A N, Petrov V L, Anipko A V, Neshov F G, Kidibaev M M and Koroleva T S 2007 *Phys. Status Solidi C* **4** chapter 3 pp 1126-1129
[26] Spayser V 1981 *Electron and Ion Spectr. Sol. St.* pp 61-97
[27] Bohun A, Dolejsi J and Czech J 1959 *Phys.* **9** pp 578-589
[28] Chen R, Pagonis V and Lawless J L 2016 *Rad. Measur.* **91** pp 21-27

IOP Conf. Series: Journal of Physics: Conf. Series **830** (2017) 012144 doi:10.1088/1742-6596/830/1/012144

Methods of radiation and environmental studies of build-up areas and their analysis

N K Ryzhakova, K O Shilova, A A Udalov and N V Demyanenko

National Research Tomsk Polytechnic University, Tomsk, Lenin Avenue, 30, Russia

Abstract. The most objective assessment of radon hazard of build-up area can be the amount of radon released from the underlying rocks. In the Russian Federation, the criterion of radon hazard is the radon flux density, measured by the method of accumulation chambers. The paper presents the results of measurements obtained by the method of accumulation chamber using two measuring complexes and the method of "two depths". The measurements were performed on a testing site of Tomsk (Russia); the type of soil is loam. Studies have shown that the considered assessment methods of radon risk have significant drawbacks, because they do not allow reliable determination of the amount of radon coming from the soil surface.

1. Introduction

It was found that one of the most significant sources of radiation is radon and its decay daughter products that create more than half of the dose from all natural sources. In this regard, in the production of engineering surveys of build-up sites, the radiation and environmental studies are conducted, part of which is the assessment of radon hazard of area. However, to date there is no single approach to the assessment of radon hazard of areas. Abroad, the value of "geogenic radon potential" is widely used as a criterion for radon hazard. To determine it, they use different sets of input variables, such as: radium concentration, radon pore activity, gas permeability of soils, features of geological structure of underlying rocks [1,2]. In Russia, the value of radon flux density (RFD) is used to assess radon hazard of areas. The territory is considered safe if *RFD* does not exceed 80 $mBkm^{-2}s^{-1}$. Despite the use of RFD as a criterion of radon hazard, the question of possible influence of measurement methods and procedures on RFD values and value of its variability has not been studied yet. The way of measuring RFD, certificated in the Russian Federation, is method of accumulation chamber. In our study, the RFD measurements were carried out by the method of accumulation chamber using measuring complexes "Alfarad+" and "Camera-01", as well as using the method of "two depths" (TD). TD method was developed in Tomsk Polytechnic University of the Russian Federation [3]. This method is based on the simultaneous measurement of volumetric activity of radon in soil pore space (hereinafter pore activity) at two depths differing twice, and diffusion model of radon transport through soil [4].

2. Measurement methods

In this study, when implementing the accumulation chamber method, measuring systems Alfarad+ and Camera-01 were used. When using complex Alfarad+ (hereinafter AC method) a chamber is installed on the surface of the soil. From this chamber the radon, accumulated during the time of exposure, is transferred to the measuring device. When measuring using complex Camera-01, radon accumulation occurs on activated carbon, followed by measuring its activity according to γ- or β-radiation of short-lived daughter products of radon decay – the so-called method of carbon adsorbers (CA). It should be noted that the method of carbon adsorber can be considered as a modification of the accumulation chamber method, as activated carbon in the adsorber acts as an equivalent volume, in which radon accumulation also occurs.

Practice shows that the accumulation chamber method, used in the Russian Federation, has significant drawbacks. Firstly, the value of RFD measured using accumulation chambers depends on weather conditions, resulting in significant time variations of the measurement results, which can reach 100% or more. Secondly, the accumulation of radon in the chamber or on the adsorber must occur in accordance with the linear law, therefore the exposure time does not usually exceed a few hours. This condition does not allow significant smoothing out of random variations of RFD. AC method has other drawbacks - when processing the results, the exchange processes of diffusion between the air in accumulation chamber and the soil are not taken into account and radon leakage out of the chamber due to insufficient sealing of chamber joint with the soil surface. In the method of CA the impact of atmospheric humidity on the measurement results is not taken into account. This impact could be significant, as carbon actively absorbs water, and radon in its turn dissolves readily in water.

In the study we also used the method of "two depths", wherein the determining of RFD is based on the measurements of pore activity at two small depths, differing twice. In this case, according to the diffusion model of radon transport through uniform soils, the radon flux density can be calculated by formula [3]:

$$RFD = \frac{\lambda \cdot h_1 \cdot A_1 \cdot \eta}{\left(2 - \frac{A_2}{A_1}\right) \cdot \left| \ln\left(\frac{A_2}{A_1} - 1\right) \right|}, \tag{1}$$

where A_1, A_2 – pore activity of radon, measured at depths h_1 and $h_2 = 2h_1$, η - soil porosity; λ – radon decay constant. The condition of soil homogeneity, used in the obtaining of expression (1), is easy to achieve in the case of removing the upper layer of soil with thickness of 0.3...0.5 m at the points of measurement. Method of "two depths" allows obtaining a more reliable estimate of radon flux density than in the case of using accumulation chamber method when measuring RFD, as: a) methods for measuring radon pore activity provide more reliable results than methods for measuring RFD; b) radon pore activity, measured at depths of 0.4...1.2 m, is influenced by the atmosphere in a lesser degree, than radon flux density on the soil surface; c) measurements of VA is carried out in the soil with undisturbed structure, unlike measurements of RFD that are performed on the surface exposed to the strong influence of weather conditions (the surface layer is characterized by a large number of cracks occurring during drying and weathering of soils)

3. Planning of experiment

The measurements of flux density by different methods were carried out in summer 2016 on the territory located in the area of the Lagerny garden of Tomsk. In the study area, at a distance of 6 m from each other, we chose two sites with dimensions of 2m*2m (Figure 2). To provide the uniformity of soil in depth, at each site we removed the upper layer of soil with thickness of ~ 0.5 m. For measurement by TD method, two bore holes with depths of 0.4 and 0.8 m were drilled using a hand drill at each point at a distance of approximately 0.5 m. When measuring radon flux density using measuring complexes Alfarad and Camera–01, the accumulation chambers were installed in the gaps between bore holes (figure 1). In total, we carried out 95 measurements of radon flux density by the method of accumulation chamber, 60 measurements of RFD using activated carbon, 50 measurements of radon flux density using the method of "two depths" and 100 measurements of radon pore activity. Along with the preparation of bore holes, we selected the soil samples using the method of "cutting ring" to determine its porosity.

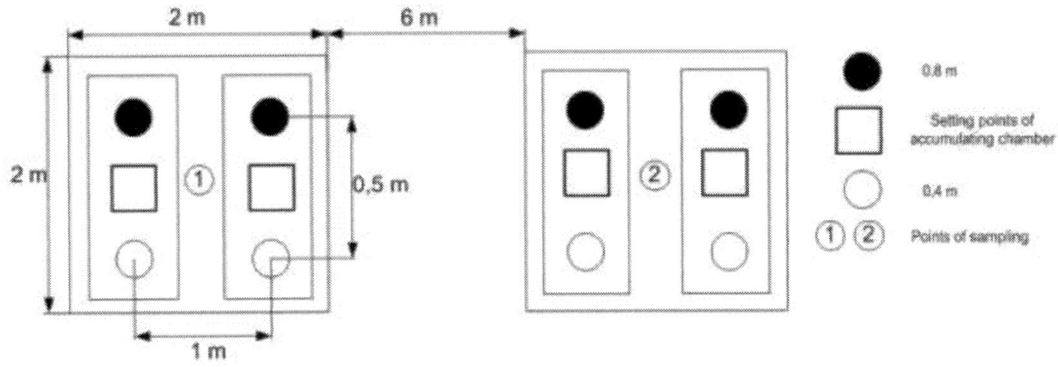

Figure 1. Scheme of measurement points arrangement

4. Measurement techniques

"Cutting ring" method is used to determine the porosity of the soil. The soil samples were selected using a hand drill at depths of 0.1…1.5 m with a step of 10 cm. A cutting ring with an internal volume of 53.2 cm^3 was pressed into selected soil samples, avoiding tool bends. After filling the ring, the soil was undercut flush with the edges and pushed onto the substrate. Then we dried the soli, determined the density of completely dried soil ρ_d and calculated the porosity by the formula: $\eta = 1 - \frac{\rho_d}{\rho_s}$, where ρ_s=2.5g·cm^{-3} – density of mineral soil particles.

To measure the pore activity of radon, we inserted a metal pipe with a diameter of 48 mm into each bore hole, which allowed us to exclude the contribution of radon released from the upper layers of soil, and also served to strengthen the walls of the bore holes. The samplers were dipped into the bore holes on the hanger, the rubber plugs were previously removed from the fittings of samplers (figure 2).

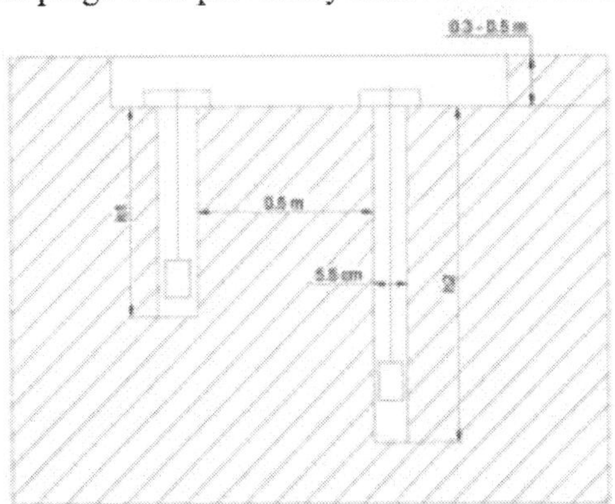

Figure 2. Measurement scheme of radon pore activity

Thereafter, the pipes were sealed with a resilient porous material with thickness of 5 cm, coated with polyethylene film on both sides. To prevent the ingress of rainwater into bore holes, each site was covered with a polyethylene film. The samplers were in bore hole during one day that provided the alignment of radon concentrations in bore holes and samplers. Then the samplers were removed from bore holes and radon pore activity was measured using complex Alfarad+.

In accordance with the procedure of RFD measurements using the measuring complex Alfarad+, the surface on a prepared site was leveled. After installation of accumulation chamber, we carried out the air sampling into the sampler for 20 minutes using a sampling device. After that, the air sampler was connected to a chamber Alfarad+ and volumetric activity of radon was measured in the accumulation chamber (figure 3).

IOP Conf. Series: Journal of Physics: Conf. Series **830** (2017) 012144 doi:10.1088/1742-6596/830/1/012144

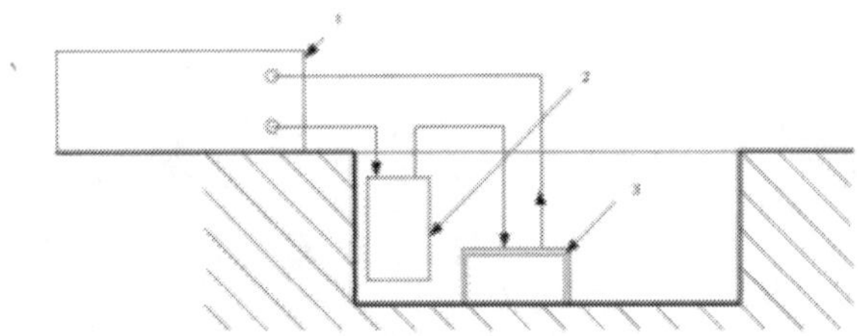

Figure 3. Scheme of air samples measurement to determine RFD. measuring device (1); air sampler (2); accumulation chamber (3)

When measuring using complex Camera 01, we used the method of radon adsorption on activated carbon with an open volume of the chamber (figure 4) and followed by measuring the activity of adsorbed radon in vitro.

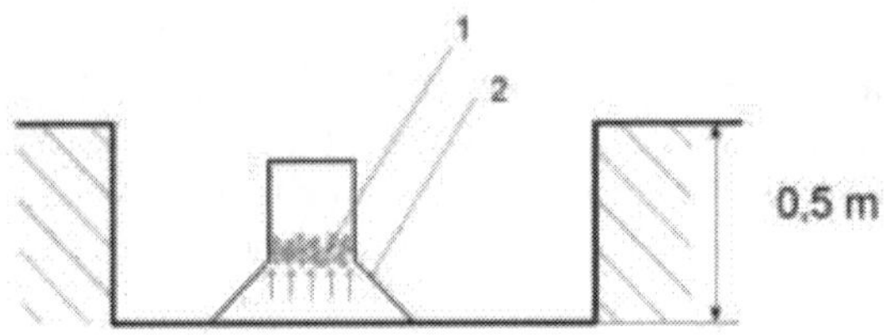

Figure 4. Measurement scheme of radon flux density using carbon adsorbers. carbon (2); accumulation chamber (2).

As the sampling device the accumulation chambers AC-32 were used. We spread a smooth layer of activated carbon onto the grid. Carbon was pre-warmed in the regenerator for one hour at a temperature of 140–160 degrees, and then accumulation chamber was installed on the soil surface (figure 4). In accordance with the service manual of Camera-01, the exposure time can be 1h...10h. When carrying out the measurements, it was noticed that when exposing accumulation chamber more than 2 hours, and at relatively high ambient humidity, the carbon stored in accumulation chamber significantly increased in volume. Therefore it was decided to use the minimum exposure time of ~1h. On expiration of exposure time, the sorbed carbon in vitro was spread into detection block of beta radiation DBR-13 and radon flux density was measured.

5. The results of measurements and their discussion

Table 1 shows the results of measurements of soil porosity depending on the depth. It can be seen that soil porosity varies slightly with the depth, i.e. soil is sufficiently uniform in depth. When calculating RFD according to formula (1), we used the mean value of porosity over all measurements, which was 0.44 of relative units.

Table 1. Porosity of soil at different depths

Depth, m		0.	0.2	0.3	0.4	0.5	0.6	0.7	0.8	0.9	1.0	1.1	1.2	1.3	1.4	1.5
η, (rel. unit.)	1	0.46	0.42	0.43	0.43	0.43	0.42	0.45	0.45	0.43	0.44	0.45	0.44	0.47	0.48	0.45
	2	0.43	0.45	0.50	0.46	0.43	0.40	0.41	0.40	0.43	0.45	0.46	0.41	0.45	0.45	0.42

In total, we carried out 95 measurements of radon flux density by the method of accumulation chamber, 60 measurements of RFD using activated carbon, 50 measurements of radon flux density using the method of "two depths" and 100 measurements of radon pore activity. We have verified the hypothesis for compliance of the measurement results of RFD to normal or lognormal distribution using the criteria of Kolmogorov-Smirnov and Shapiro-Wilk. The verification showed that radon flux density measured by the methods of AC and TD obeys the lognormal distribution, and flux density measured by the method of carbon adsorber obeys the normal distribution. Figure 5 shows the histogram for the measurements results of RFD by different methods.

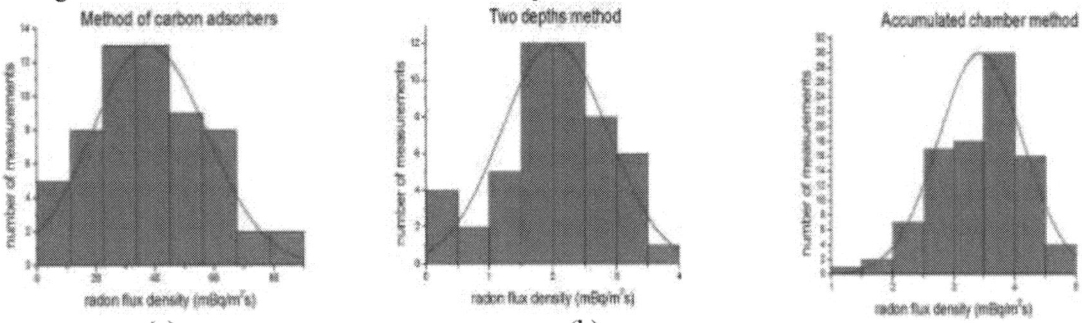

(a) (b) ©

Figure 5. Histogram of measurement results of RFD registered by different methods: (a) carbon adsorber method, (b) "two depths" method, (c) accumulated chamber method.

According to the results of statistical processing for the lognormal distributions, there are calculations of the means, standard deviations and coefficients of variation according to the formulas:

$$\overline{RFD} = \exp\left(\overline{\ln(RFD)}\right)\cdot\exp\left(\frac{\sigma^2_{\ln(RFD)}}{2}\right); \tag{2}$$

$$\sigma^2_{\ln(RFD)} = \frac{\Sigma\left(\ln(RFD)_i - \overline{\ln(RFD)}\right)^2}{(n-1)}; \tag{3}$$

$$\sigma = \sqrt{\left[\exp\left(\overline{\ln(RFD)}\right)\right]^2 \cdot \exp\left(\sigma^2_{\ln(RFD)}\right)\cdot\left[\exp\left(\sigma^2_{\ln(RFD)}\right)-1\right]}. \tag{4}$$

Table 2. The basic results of measurements of pore activity and RFD, determined using the method of "two depths" (TD), accumulation chamber (AC) and carbon adsorber (CA).

Measured value	Range of values $x_{min}\ldots x_{max}$	Mean value $\bar{x}$	Standard deviationσ	Coefficient of variation V, %
$A_{0.4}$ (kBk·m^{-3})	0.3…48	21	10	48
$A_{0.8}$ (kBk·m^{-3})	0.7…69	29	14	49
RFD "TD" (mBk·m^{-2} s^{-1})	2…44	10	10.8	100
RFD "AC" (mBk·m^{-2} s^{-1})	3…96	39	31	80
RFD "CA" (mBk·m^{-2} s^{-1})	2…82	38	19	51

IOP Conf. Series: Journal of Physics: Conf. Series **830** (2017) 012144 doi:10.1088/1742-6596/830/1/012144

Table 2 shows the basic results of measurement: range of values, mean values, standard deviation and coefficients of variation $V = \frac{\sigma}{\bar{x}} \cdot 100\%$, where σ – standard deviation, $\bar{x}$ – mean value.

From the data given in the table, we can see that the mean values of the results of RFD measurements by the methods of AC and CA are practically the same, which indicates the reliability of the means obtained. The coefficient of variation for AC is significantly greater than for CA that is most likely explained by a longer exposure when measuring by CA. Quite unexpected results have been obtained for the method of two depths – the mean value of RFD is four times less, and the scatter in measurement results is greater. During the measurement, it was observed that on the surface of samplers, extracted from bore holes, a large amount of moisture was accumulated. Therefore, in the final period of the study we carried out six measurements of air humidity in each bore hole – relative humidity was 93% on average (much greater than atmospheric humidity – 44%). It is known that radon is readily soluble in water. It can be assumed that radon released from the soil is actively absorbed by moisture, which leads to a decrease in its concentration in the air of soil pores. To obtain the final conclusions about effect of humidity on the measurement results it is necessary to carry out further research.

6. Conclusions

To obtain reliable estimates of the amount of radon coming to the soil surface, it is necessary to use sufficiently large exposure times.

The results of measurements of radon flux density by the methods of CA and "TD" are greatly influenced by air humidity, at which radon exposure is carried out.

Methods and techniques for measurement of flux density and radon pore activity require further study and improvement.

References

[1] Aderhold J, Davydov V Yu, Fedler F, Klausing H, Mistele D, Rotter T, Semchinova O, Stemmer J and Graul J 2001 *J. Cryst. Growth* **222** 701

[1] Neznal M, Neznal M, Matolín M, Barnet I and Mikšová J 2004 *Czech Geological Survey Prague Special Papers* **16** pp 7-47

[2] Kemski J, Siehl A, Stegemann R, and Valdivia-Manchego M 2001 *Science of The Total Environment* **272**(1-3) pp 217–230

[3] Ryzhakova N K 2014 *Appl. Radiation and Isotopes* **91** pp 161–164

[4] Ryzhakova N K 2012 *J. of Appl. Geophysics* **80** pp 151–157

IOP Conf. Series: Journal of Physics: Conf. Series **830** (2017) 012145 doi:10.1088/1742-6596/830/1/012145

Luminescent centers in nanolayers of LiF crystals with embedded silver ions

O I Shipilova[1], V P Dresvyansky[2], E F Martynovich[1,2], A L Rakevich[2], R Yu Shendrik[1,3], V L Paperny[1] and A A Chernich[1]

[1] Irkutsk State University, Irkutsk, 664003, Russia
[2] Irkutsk Branch of the Institute of Laser Physics SB RAS, 664033,Irkutsk, Russia
[3] Institute of Geochemistry SB RAS, Irkutsk, 664033, Russia

E-mail: paperny@math.isu.runnet.ru

Abstract. The paper presents the results of studies of luminescent centers induced by irradiation of LiF crystals with flow of silver ions of a fluence varying in a range of 2×10^{13}–5×10^{17} ions/cm^2 and energy about of 150 keV. Two bands with maxima at 250 nm and 420 nm in the absorption spectra of the exposed crystals were observed. The former was due to the absorption of F colour centers and the latter consisted of F_2 (F_3^+) colour centers band and a plasmon resonance band arose due to the embedded silver ions. The peak of the latter remained stable after annealing the sample at 400 ^{0}C unlike the colour centers bands. The luminescence spectra of the crystals under excitation with a laser irradiation of 375 nm wavelength showed a band with a peak at 450 nm along with the bands with peaks at 530 and 680 nm corresponding to F_3^+ and F_2 colour centers, respectively. The peak of the former reached saturation, while the peaks of the colour centers bands increased with ion fluence increasing. These results show that sub-nanometre sized metal clusters Ag(n+) (n=3-6) are responsible for the observed luminescence, while the Ag particles of a nanometer scale are not luminescent.

1. Introduction

The luminescence of noble metal nanoparticles (NPs) has attracted a lot of attention in the last decades. Their unique optical properties can be applied in different fields such as luminescent labels [1], sensors [2], data storage [3] and fiber lasers [4]. The works in this field largely focus on the studies of optical properties of NPs in liquid media [1, 2] or in glasses, i.e. amorphous media, [3,4]. Along with this, it is of interest to study these properties of NPs produced in crystal media, in particular, in LiF crystals or thin polycrystalline LiF films. This material is remarkable due to an opportunity to produce the so-called colour centers therein. Interest in the study of colour centers in lithium fluoride crystals is supported by numerous practical applications of this material. In particular, the colour centers are used in thermoluminescent detectors of the X-radiation [5]. Also, it is known that the introduction of various impurities (Mg, Cu, etc.) into the LiF matrix increases the sensitivity of the luminescent detectors [6]. Recent studies have shown that such an effect also resulted from doping polycrystalline LiF film with Ag atoms [7]. The objective of this work is to establish changes in the optical properties of the LiF crystal irradiated by high energy silver ions embedded into the crystal matrix and produced both NCs and colour centers.

IOP Conf. Series: Journal of Physics: Conf. Series **830** (2017) 012145 doi:10.1088/1742-6596/830/1/012145

2. Experimental details

2.1. Experimental setup

The LiF samples were irradiated by a flow of silver ions produced with a compact pulsed implanter MEVVA.RU. An ion beam of 200 μs duration and a single pulse fluence of 2×10^{13} ions/cm^2 was extracted by a three-grid accelerating system from a vacuum arc plasma flame and accelerated by a voltage of 50 kV. Since the mean charge state of silver ions in these pulsed discharges is between +2 and +3, we guess that energy of the accelerated ions lies in a range of 100–150 keV. The research in spectral-kinetic characteristics of photoluminescence of lithium fluoride samples irradiated with silver ions was performed with a sensitive scanning confocal fluorescence microscope MicroTime 200 (by PicoQuant Gmbh company) with picosecond time resolution with a spatially-selective time-correlated count single photons. Photoluminescence spectra measured under excitation by picosecond laser with a wavelength of 375 nm were recorded by the spectrometer Ocean Optics 6500. To suppress scattered excitation radiation from the sample, the luminescence was recorded through a filter with a cutoff wavelength of 400 nm. The absorption spectra of the samples were recorded on a spectrophotometer SF-56.

2.2. Results and discussion

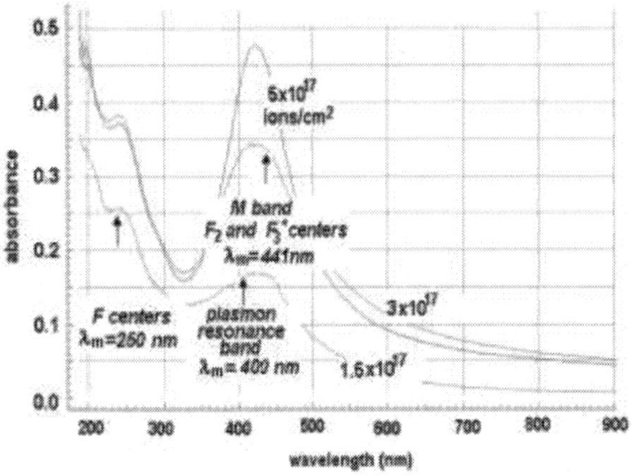

Figure 1. Absorption spectra of LiF crystals irradiated with Ag ions of various fluencies

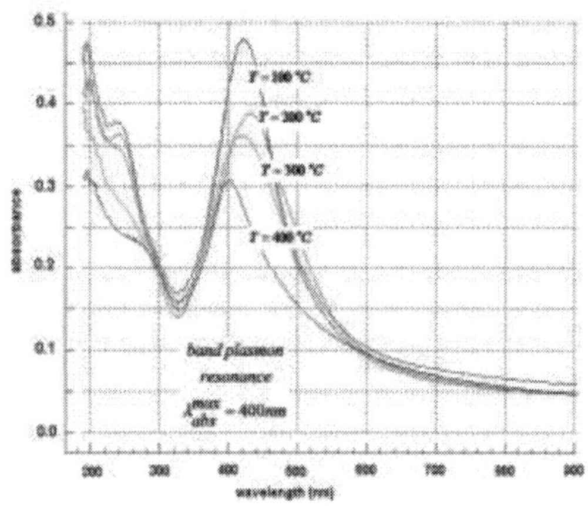

Figure 2. Changes in the absorption spectra of LiF crystals irradiated by Ag ions with fluencies of 3×10^{17} ion/cm^2 during thermal annealing

Figure 1 presents the absorption spectra of LiF crystals irradiated by Ag ions with a few high fluencies. Two bands with maxima at 250 nm and 430 nm wavelengths in the absorption spectra of the exposed crystals were observed. The former was due to the absorption of F colour centers and the latter consisted of F_2 (F_3^+) colour centers band and a plasmon resonance band that arose owing to embedding high energy silver ions into the matrix of the sample and following formation of silver NPs. One can see that increase in the fluencies results in corresponding increase in the absorption bands. Subsequent annealing of the exposed sample at temperature up to 400 ^{0}C results in decrease of the peaks and shift of the latter peak to 400 nm wavelength (see Figure 2). This result demonstrates that this band is thermally stable and, in fact, caused by the silver NPs, while bands due to color centers have disappeared because of the destruction of centers. Note that a marked absorption band can be observed only at sufficiently large fluencies. This means that the embedded ions merge into metallic particles of a nanometer scale only when the ion density in the host matrix is sufficiently large.

Now let us consider the results on studies of luminescent characteristics of the exposed samples. Figure 3 shows the luminescence spectra of LiF crystals irradiated by ion flow of various fluencies. It can be seen that the spectra of the exposed LiF crystals excited by laser radiation of 375 nm wavelength show that in addition to the luminescence bands with maxima at 550 and 680 nm, a luminescence band with a maximum at 460 nm was detected. The two former bands attributed to F_2 and F_3^+ colour centers, respectively, are formed in a thin surface layer by ion irradiation. The luminescence band with a maximum at 460 nm is due to the embedded silver ions. Figure 4 shows also that at low fluencies the peaks of the band due to silver ions are close to peak of the band due to F_2 and F_3^+ colour centers. Note that at high fluencies the latter exceeds significantly the former one.

The curve of the luminescence kinetic depicted in Figure 4a contains components with decay times 13.4 and 5.2 ns, that are close to the characteristic luminescence decay times of F_2 and F_3 centers in LiF crystal, which are of 15 and 8 ns, respectively [8]. In addition, two fast components with the decay times of 0.26 and 1.3 ns are observed. After annealing the sample at 400 ^{0}C for 30 min, the components due to the F_2 and F_3 centers, practically, diminished, while the fast components retained significant value. Taking into account the previous results, one can suppose that these fast luminescent components are due to silver ions.

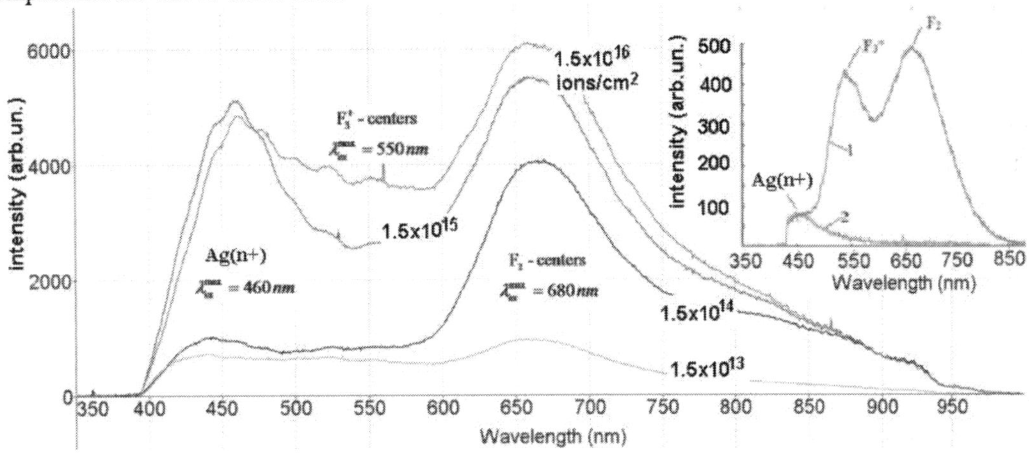

Figure 3. The luminescence spectra LiF crystals irradiated with Ag ions of various fluencies.

The results presented above permitted us to submit the following scenario of behaviour of silver ions imbedded into the host LiF matrix. At low ion fluencies, when the concentration of ions in the matrix is insignificant, these ions combine sub-nanoscale molecular charged clusters (MCs) consisting of n ions (atoms) Ag(n+), here n = 3–6. Under excitation, these clusters produce a luminescent band with peak at 460 nm wavelength and short decay times [9]. The intensity of this band is comparable with intensity of luminescence bands emitted by F_2 and F_3 colour centers resulting from the silver ion exposure of the matrix. When the ion fluencies increase, the concentration of ions embedded in the matrix increases as well. This results in merging these ions into metal particles of a nanometer size (NPs, 10–100 nm) which do not luminesce. These particles are resistant to annealing unlike the F_2 and F_3 colour centers.

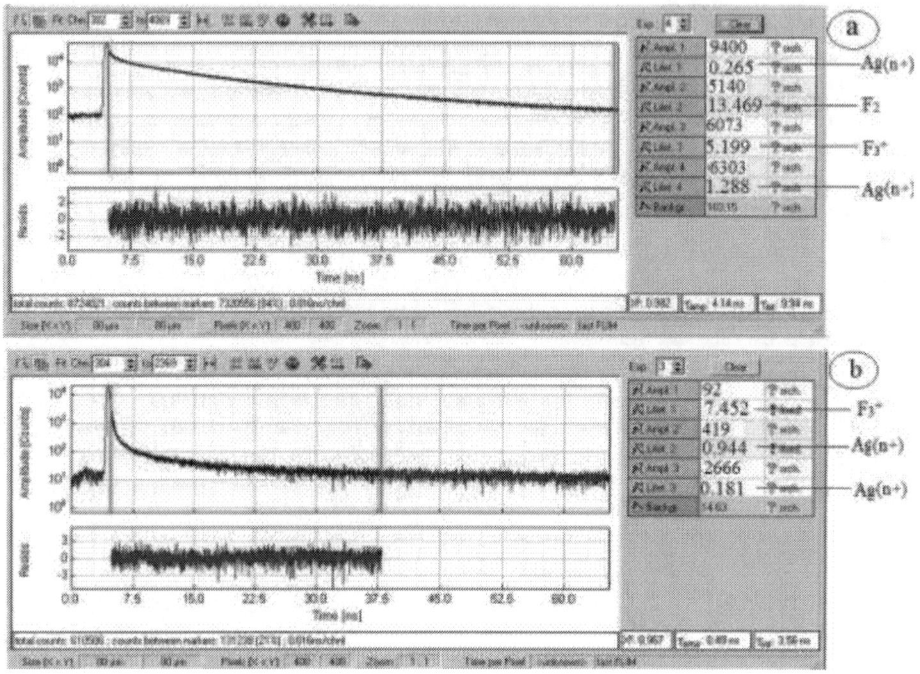

Figure 4. Luminescence decay curve of LiF fluorescence spectrum of the crystal irradiated of fluence of $5\cdot10^{17}$ion/cm^2 before (a) and after (b) annealing at 400 C^0 for 30 min

Conclusion

Thus, the presented results show that implantation of high energy silver ion is an effective method for production of a new type of luminescent medium. By varying the ion fluencies, we can produce silver particle from a sub-nanometer size scale up to several tens of nanometers. These particle types have significantly different luminescent characteristics, namely, while the former ones luminesce with the peak of the band at 460 nm, the latter ones do not luminesce.

References

[1] Zheng J and Dickson R M J 2002 *Am. Chem. Soc.* **124** 13982
[2] Adhikari B and Banerjee A 2010 *Chem. Mater.* **22** 4364
[3] Canioni L *et al* 2008 *Opt. Lett.* **33** 360
[4] Lumeau J and Glebov L 2014 *Appl. Opt.* **53** 7362
[5] Nepomnyashchikh A I, Radzhabov E A and Egranov A V 1984 Color Centers and Luminescence in LiF Crystals (Novosibirsk, Russia: Nauka) [in Russian]
[6] Patil R R and Moharil S V 1995 *J. Phys. Condenced Matters* **7** 9925
[7] Alharbi N A *et al* 2013 *J. Phys. D. Appl.Phys.* **46** 035305
[8] Ivanov N A *et al* 2011 *Technical Physics Letters* **37(** 939
[9] Klyukin D A *et al* 2016 *Opt. Eng.* **55** 067101

The effect of aluminum particles dispersity on characteristics of ammonium perchlorate—aluminum composition laser ignition

V V Medvedev, E V Forat, V P Tsipilev and A N Yakovlev

National Research Tomsk Polytechnic University, 30 Lenin Ave., 634050, Tomsk, Russia

E-mail: forategor@gmail.com

Abstract. Ammonium perchlorate–aluminum compositions taken in stoichiometric ratio were ignited in air and under the cover with 1.06-μm 0.8-ms-long laser pulses. The ignition energy thresholds were measured for samples at various dispersity of Al. The causes of the difference in examined compositions sensitivity to the influence of laser radiation is considered from the perspective of the thermal theory.

1. Introduction

It is known that pyrotechnic mixtures based on ammonium perchlorate (AP) and aluminum can be used to produce pyrocartridges and other ignition means. Laser system can be successfully applied for that kind of devices in the function of igniter [1, 2]. For that reason, it is necessary to understand what characteristics of the mixture affects the ignition process.

It was previously ascertained that ignition characteristics of such mixtures, on top of everything else, can depends on the size of Al particles [3]. The sensitivity to laser pulse (pulse duration 4 ms) comparison between AP mixtures with ultrafine (UF) Al (size of one particle $d = 0.25$ μm) and coarse Al ($d = 80$ μm) was conducted. The energy density threshold (ET) was used as the criterion of ignition. The ET of mixture AP/coarse-Al appears to be significantly higher than ET of AP/UF-Al.

It has been assumed that mixture sensitivity decreases with increasing the size of Al particles. To verify that assumption it is necessary to carry out the experimental research with greater numbers of mixtures of different Al particle size.

2. Materials and methods

In this work we studied the ignition of a stoichiometric AP/Al compositions by laser radiation pulse (wave length λ=1.06 μm) with pulse duration τ_p = 0.8 ms. Analytical-grade ammonium perchlorate was milled in an agate mortar and sieved through a Capron sieve with a mesh size of ~70 x 70 μm. Al powders with average particle size of 0.14; 3; 4; 9 and 80 μm was mixed with AP.

Ultrafine aluminum powder with a mean diameter of 0.14 μm was prepared by electric explosion. It contained less than 93 wt. % of active aluminum. The particles had a spherical shape and a log-normal size distribution [4].

The mixtures samples of ~ 9 mg weight had been placed into a polymethyl methacrylate (PMMA) capsule having 3 mm inner diameter and 3 mm depth. The samples were pressed in the PMMA capsule up to the pressure of 800 kgf/cm^2. The end surfaces of the both open and covered samples were exposed to laser radiation. The covered samples were tested in the absence of gas-dynamic unloading as shown in figure 1.

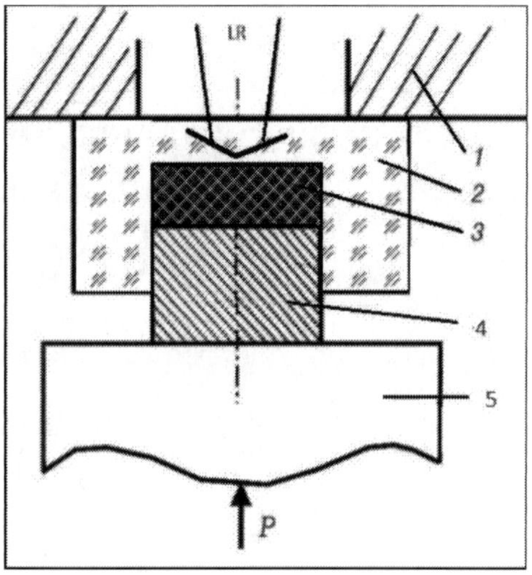

Figure 1. Cross-sectional view of covered sample 1 – Upper press stop; 2 – PMMA capsule; 3 – mixture sample; 4 – lower press stop; 5 – press forcer.

The neodymium laser (wavelength λ=1.06 μm) was used for defining the ET of each sample (table 1). It generates the quasicontinuous laser pulse with the modulation depth not higher than 10 % and duration of 0.8 ms. Laser beam diameter is 2 mm. Non-uniformity of the irradiation was not over than 10 %.

The experimental installation for the laser ignition is schematically presented on figure 2. Further details about installation described in works [5, 6].

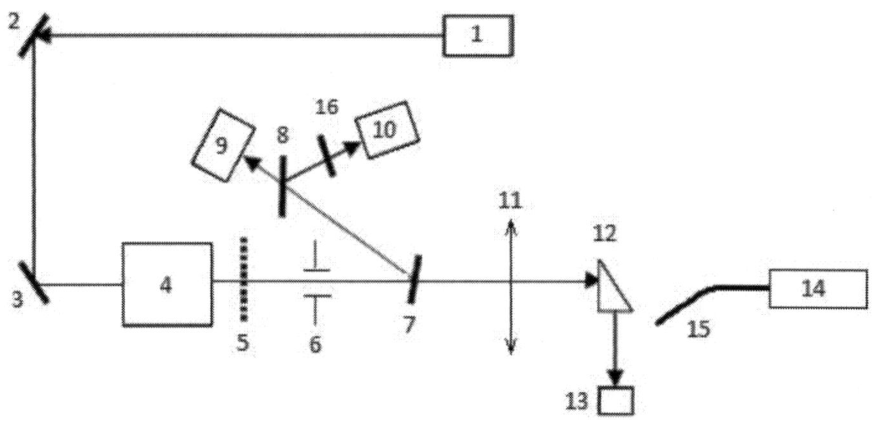

Figure 2. The scheme of the laser ignition experiment of pyrotechnic mixtures: 1 – alignment laser; 2, 3 – rotary mirrors; 4 – pulse laser block; 5, 16 – neutral filters; 6 – diaphragm; 7,8 – beam splitters; 9 – calorimeter; 10, 14 – photodetectors; 11 – lens; 12 – rotary prism; 13 – sample; 15 – fiber optic cable.

The ignition thresholds were determined from the initiation function curve (figure 3). Firstly, the dependence *P(E)* within the range of 0–1 was obtained, where *P* is the ratio between the number of successful ignitions and all attempt; *E* is the energy density of laser radiation. Each point of this function contains 25 experiments with statistic range distribution less than 10 % by the calorimeter results. The ignition threshold H_{50} (or H_{TR}) is a value for which ignition took place with possibility P = 0.5.

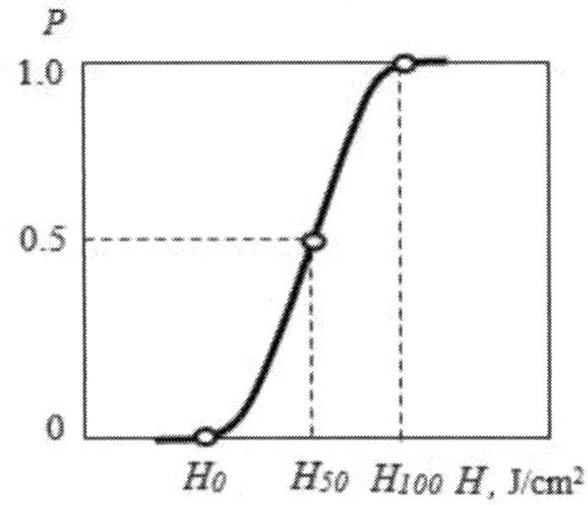

Figure 3. The function of the ignition threshold.

3. Experimental results

AP/UF-Al samples with opened surface burned out with bright and more than 10 cm tall flame jet. The ET is 1.3 J/ cm^2 and the ignition delay amount to 15 ms. That results corresponding the obtained one from the work [3].

From all examined mixture types only AP/UF-Al was able to ignite in case of open surface.

The other mixtures with coarse Al showed no substantial activity even with increase of the energy exposure up to 40 J/cm^2. However, the plasma jet appears on $H \geq 3$ J/ cm^2.

In case of closed surface all mixtures burned out along with blast sound and disruption of PMMA capsule. The ignition delay did not exceed laser pulse duration and decreased with energy level rising. ET are listed in the table 1.

4. Discussion

So the AP/UF Al with open surface is more active than other mixtures that based on coarse Al. In case of covered surface there is an opposite situation. Furthermore, covered samples sensitivity increases with decreasing Al particle size.

To interpret such results, it is necessary to analyze the heating of Al particle and the matrix during the duration of laser pulse. For that purpose, it is enough to be confine by the rough estimate of heating temperature. The two different approaches should be considered.

4.1. Isolated particle approach

It is possible to assume some microzones with isolated from each other Al particles. The particle heating process taking place without heating influence of surrounding Al particles.

In that context the hot spot appears in AP matrix near the Al particle, even in case of high concentration of Al (mixtures AP/UF-Al and AP/Al-10 at table 2).

A coarse estimate of the hot spot temperature ΔT for each particle size during laser pulse τ_u can be found from equation:

$$\Delta T = \frac{F(\rho_0)H_{TR}\cdot 3}{4\cdot c\rho\cdot R_0}\cdot k(R_0,\lambda_0)\cdot\frac{R_0^3}{(R_0+\sqrt{\alpha\tau})^3} \quad (1)$$

$F(\rho_0)$ is the coupling coefficient between light intensity on and under the surface; H_{TR} is the energy density threshold; ρ_0 is the diffusion reflectivity coefficient; ρ is the AP density; c is specific heat; α

is thermal diffusivity of AP matrix; $k(R_0,\lambda_0)=\dfrac{\sigma(R_0,\lambda_0)}{\pi R_0^{\,2}}$ - is relative cross-section of absorption at wavelength λ_0 ; $\sigma(R_0,\lambda_0)$ is the absorption cross-section. According to [3] coefficient ρ_0 equals to 0.35, while value of ρ_0 the coupling coefficient is $F(\rho_0)\approx 3$. For Al particles of considered size range $2R_0$ relative cross-section of absorption is $k(R_0,\lambda_0)\approx 0{,}1$. According to [7] α take a value of 10^{-3} cm^2/s. The results of temperature calculations ΔT_{oz} for different mixtures are listed in the table 1.

Maximum temperature of the hot spot (3500 °C) corresponds to maximum particle size $2R_0$ = 80 μm (AP/Al-1 mixture). Whereby hot spot size is $2R=2R_0+2\sqrt{\alpha\tau}$, take a value about 100 μm. The heating temperature and the enthalpy W ($W=F\cdot H_{TR}\cdot k\cdot\pi\cdot R_0^{\,2}$) decreases along with the particle size $2R_0$ decreasing.

So if $2R_0$ = 80 μm then enthalpy is about 3 mJ and for $2R_0$ = 0,14 μm (2R ~ 10 μm) is 3 × 10^{-10} J, in other words the difference is about 6 orders of magnitude.

The main paradox of that approach is about hot spot with low enthalpy actually are more active than hot spot with much higher enthalpy. That defies the conceptual issues of thermal theory and the common sense.

Hence, the experimental results can't be interpreted from the position of microzone thermal ignition theory. It follows that ignition from the single hot spot, that appears around isolated singe Al particle in such mixtures, is not possible.

Table 1. Energy thresholds and calculation results.

Types of mixture	Typical size Al particle $2R_0$, μm	Energy density threshold for ignition with covered surface H_{TR}, J/cm^2	Energy density threshold for ignition with uncovered surface H_{TR}, J/cm^2	Heating temperature of hot spot around the particle ΔT_{oz}, °C	Hot spot enthalpy W, J
AP/Al -1	80	153	—	3500	~ 3 × 10^{-3}
AP/Al -4	8-10	16.9	—	228	~ 4 × 10^{-6}
AP/Al -8	3-4	11.8	—	86	~ 5 × 10^{-7}
AP/Al -10	2-3	8.3	—	35	~ 8 × 10^{-8}
AP/UF-Al	0.14	3.3	1.3	0.25	~ 3 × 10^{-10}

4.2. Near surface layer approach

The other limiting case has no dependence from sample structure (AP crystal size or Al particle size). Laser radiation is heating the subsurface layer with the depth $Z_l=\dfrac{1}{\mu}+\sqrt{\alpha\tau}$, μ is the absorbance index of mixture. Absorbance index can be found from equation:

$$\mu=\pi R_0^{\,2}\cdot k(R_0,\lambda_0)\cdot C \tag{2}$$

C is Al particle concentration ($C=\dfrac{\eta\cdot\rho_c}{M_{pat}(1-\eta)}$; η is the weight percentage of Al particles; ρ_c is the mixture density; M_{pat} is the mass of one Al particle). The value of μ-index for different mixtures are listed in the table 2. The temperature of heated layer can be found from:

IOP Conf. Series: Journal of Physics: Conf. Series **830** (2017) 012146 doi:10.1088/1742-6596/830/1/012146

$$\Delta T_l = \frac{F \cdot H_{TR}}{Z_l c \rho} \tag{3}$$

$F = 1 - \rho_F$ is absorption coefficient due to Fresnel reflection. The values of heated layer temperature are listed at table 2.

Table 2. The results of layer heating calculations.

Types of mixture	Typical size Al particle $2R_0$, μm	Al particle concentration, particle /cm^3	Layer absorbency index μ, cm^{-1}	Thickness of the heated layer ΔZ, cm	Heated layer temperature ΔT_l, °C
AP/Al - 1	80	$\sim 2 \times 10^6$	9	~ 0.1	765
AP/Al - 4	8-10	$\sim 1 \times 10^9$	74	$\sim 1 \times 10^{-2}$	845
AP/Al - 8	3-4	$\sim 5 \times 10^9$	185	$\sim 6 \times 10^{-3}$	983
AP/Al - 10	2-3	$\sim 1.5 \times 10^{10}$	370	$\sim 4 \times 10^{-3}$	1037
AP/UF-Al	0.14	$\sim 3 \times 10^{13}$	3700	$\sim 1 \times 10^{-3}$	1650

It can be seen that the surface layer temperature increases with decreasing Al particle size alone with decreasing the thickness of heated layer ΔZ.

Nevertheless, the reactivity of the flat form hot spot with heating temperature 1650 °C and layer thickness about 10^{-3} cm (AP/UF-Al mixture) is significantly higher than the reactivity of the another hot spot with heating temperature 765 °C and layer thickness about 10^{-1} cm (AP/Al-1 mixture). This situation corresponds to classic ideas of thermal hot spot ignition and in particular, it corresponds the critical Frank-Kamenetskii parameter [8].

Thus, the increasing mixtures sensitivity along with increasing Al particle size points at thermal macrozone nature of AP/Al mixtures ignition. According to that theory the ignition begins from the hot spot limited by laser beam size, light attenuation depth and the thermal front range into the sample.

The ignition tests of the opened surface samples with coarse Al were not successful. That situation is similar to explosives with gas reaction phase [10]. But experiment result of AP/UF Al mixture ignition invalidate such idea because of lower threshold with open surface rather then covered. That fact points at possibility of condensed phase reaction.

From that position it is easy to explain the sensitivity difference of opened and closed surface of AP/UF-Al samples. In case of opened surface, the additional heating taking place due to Al particles oxidizing by atmospheric oxygen and nitrogen. The oxidation process appears due to high Al particles temperature (about 1650 °C). Perhaps, for mixtures with AP and coarse Al the heated layer temperature is not enough for effective Al particle heating.

In that case the surface closed by transparent dielectric material improves the reaction condition through increasing the reaction velocity along with the pressure into the reaction zone. It can be confirmed by the ignition delay reduction in case of closed surface ignition in comparison with opened surface (figure 4).

IOP Conf. Series: Journal of Physics: Conf. Series **830** (2017) 012146 doi:10.1088/1742-6596/830/1/012146

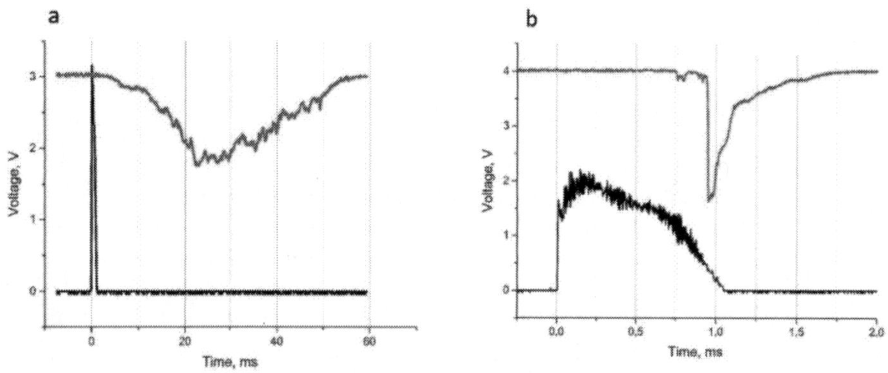

Figure 4. Typical AP/Al mixture ignition oscillogram for a) opened surface; b) closed surface. Upper curve corresponds to the glow of combustion products and lower curve is the laser pulse signal.

5. Conclusions

The energy thresholds for pulse laser ignition of stoichiometric AP/Al mixtures were measured. The mixtures distinguished in the Al particles size. The samples surface was irradiated with and without transparent dielectric.

For laser pulse duration τ_p = 0.8 ms the mixture sensitivity increases with decreasing the specific size of Al particles in the range from 80 μm to 0.1 μm.

The AP/Al mixture ignition begin from the hot spot limited by laser beam size, light attenuation depth and the thermal front range into the sample in a time of laser pulse duration.

The absorbance index, temperature and enthalpy increases along with Al particle size which leads to increasing mixture sensitivity. That parameters were determined to estimate the hot spot reactivity.

It has been suggested that the presence of oxygen and nitrogen facilitate the ignition process only for AP/UF-Al mixture in case of opened surface. The influence of atmospheric oxygen and nitrogen on AP/Coarse Al mixture ignition is not substantial.

In this work the mixtures wherein the AP Crystal size excels the size of Al particles were examined. The mixtures with equal components dispersity should be considered at the next work.

Acknowledgments

This work was financially supported by the Russian Foundation for Basic Research Grant 15-03-05385.

References

[1] Platt W G 1972 U.S. Patent No. 3685392

[2] Williams N P 1977 U.S. Patent No. 4047483

[3] Medvedev V V, Tsipilev V P et al 2005 Conditions of millisecond laser ignition and thermostability for ammonium perchlorate/aluminum mixture *Khimicheskaya fizika* **24** 11 pp 94–96 (in Russian)

[4] Yavorovskiy N A 1996 Ultrafine powders production by electric explosion *Russian Physics Journal* 4 pp 114–136 (in Russian)

[5] Aleksandrov E I and Tsipilev V P 1981 Dimensional effect in the initiation of compressed lead azide by single-pulse laser radiation *Combustion, Explosion, and Shock Waves* **17** 5 pp 550–552

[6] Medvedev V V 2000 A Laser with Adjustable Pulse Duration Based on a Commercial GOS-301 Laser *Instruments and Experimental Techniques* 6 p 807

[7] Rosser W A, Inami S H Thermal Diffusivity of Ammonium Perchlorate *AIAA Journal* **4** 4 p 663-666
[8] Baum F A, Orlenko *et al* 1975 *Explosion Physics Nauka* p 156
[9] Aleksandrov E I, Voznyuk A G, Tsipilev V P 1989 Effect of absorbing impurities on explosive initiation by laser light *Combustion, Explosion, and Shock Waves* 1 pp 1–7
[10] Tsipilev V P, Morozova E U, Skripin A S 2010 Laser initiation of PETN powder in a volumetric compression *Bulletin of the Tomsk Polytechnic University* **317** 4 (in Russian)

PiC code KARAT simulations of Coherent THz Smith-Purcell Radiation from diffraction gratings of various profiles

K P Artyomov[1,2], V V Ryzhov[1,2], A P Potylitsyn[2], L G Sukhikh[2]

[1] Institute of High Current Electronics, Siberian Branch, Russian Academy of Sciences, 2/3 Akademichesky Avenue, Tomsk, Russia
[2] Tomsk Polytechnic University, 30 Lenin Avenue, Tomsk, Russia

E-mail: artyomov@to.hcei.tsc.ru

Abstract. Generation of coherent THz Smith-Purcell radiation by single electron bunch or multi-bunched electron beam was simulated for lamellar, sinusoidal and echelette gratings. The dependences of the CSPR intensity of the corrugation gratings depth were investigated. The angular and spectral characteristics of the CSPR for different profiles of diffraction gratings were obtained. It is shown that in the case of femtosecond multi-bunched electron beam with 10 MeV energy sinusoidal grating with period 292 μm and groove depth 60 μm has the uniform angular distribution with high radiation intensity.

1. Introduction

One of the actual problems of modern accelerator physics is the development of compact sources of THz radiation. Currently there are new applications of coherent THz radiation. The basis of modern THz radiation sources are powerful and bulky gas lasers or compact semiconductor laser systems.

In this paper, we investigate the possibility of developing a compact tunable monochromatic THz CSPR source based on electron accelerator with femtosecond bunched beam. A 10 MeV multi-bunched electron beam with a THz repetition frequency is formed in RF gun with photocathode by a femtosecond laser.

CSPR is generated when a charged particle passes near the diffraction grating. This radiation is described by the dispersion relation, also called Smith-Purcell relation:

$$\lambda = \frac{d}{|n|}\left(\frac{1}{\beta} - \cos\theta\right),$$

where λ is wavelength, d is grating period, n is diffraction order, β is charged particle velocity in units of speed of light, and θ is radiation angle.

There are a number of theoretical and experimental studies of the CSPR generation by multi-bunched electron beams. Using multi-bunched beam allows generation of narrow-band CSPR in super-radiance regime [1,2]. The radiation frequency can be tuned changing the repetition rate of the femtosecond laser pulse in RF gun [3]. In addition, it is possible to change the angle of radiation generation by changing the repetition frequency of micro-bunches. For example, the generation of radiation perpendicular to the grating plane simplifies the radiation output from the installation and recording of its characteristics. Since the radiation is formed on the diffraction grating, its size and the radiation angle will define the spatial size of the wave packet.

The generation of CSPR by THz multi-bunched electron beam from diffracting gratings with an arbitrary profile is a very complex task for which modern numerical simulation methods should be used.

2. Simulation of CSPR characteristics

2.1. Particle-in cell code KARAT

In the present paper, the numerical simulation of the CSPR generation was carried out by PiC-code KARAT [4]. The code allows taking into account the effect of the grating profile on spatial and spectral characteristics of the CSPR and choosing the optimal diffraction grating. KARAT is a fully electromagnetic code based on the particle-in-cell (PiC) method. It has been designed to simulate the dynamics of charged particle beams in their own and external electromagnetic fields in complex geometry. In particular, code KARAT performs well in modeling of the generation of different types of polarized radiation, such as diffraction, transition, Cerenkov radiation and Smith-Purcell radiation [5]. There are 1D, 2D and 3D versions of the code. In all cases, it takes into account the three components of the electromagnetic fields and moments. A finite difference scheme with overstepping on the rectangular shearing grid is used to solve Maxwell's equations.

2.2. Simulation parameters

Numerical simulation of the CSPR generation was carried out for the four types of diffraction gratings (figure 1):

- echelette grating with 45 degrees blaze angle (prof. 1);
- echelette grating with 30 degrees blaze angle (prof. 2);
- sinusoidal grating (prof. 3);
- lamellar grating (prof. 4).

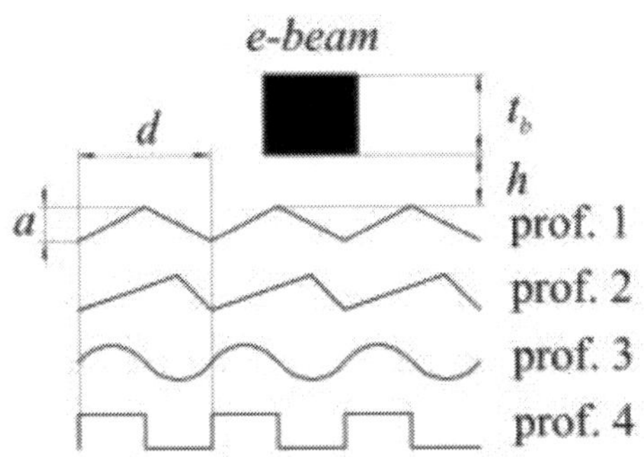

Figure 1. The profiles of diffraction gratings.

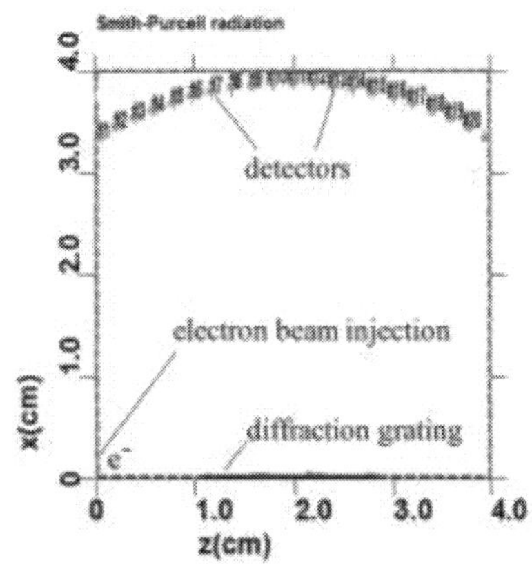

Figure 2. Simulation geometry.

Gratings period $d = 292$ μm provided the generation of THz radiation at an angle $\theta \sim 90$ deg to the direction of motion of the electron beam. The number of grating periods $N_p = 60$. A groove depth a was varied from 0 to 300 μm. The bunch repetition rate $\gamma_b = 1$ THz, number of bunch N_b from 1 to 16, electron energy $E_e = 10$ MeV and beam current $I_b = 75$ A/cm. The number of electrons per micro-bunch $N_e \approx 3 \cdot 10^7$ The longitudinal distribution of electron bunch corresponds to a Gaussian distribution with $\sigma = 55$ fs. In the transverse direction, the electron bunch has a uniform distribution and a thickness of $t_b = 300$ μm. The impact parameter $h = 150$ μm.

The basic geometry is shown in Figure 2. The dimensions of the simulated region were 40×40 mm with grid spacing of $\Delta x = 10$ μm transversely and $\Delta z = 10$ μm longitudinally. Diffraction grating is located at the bottom of the simulated region. The electron beam is injected in a direction parallel to the diffraction grating through the left boundary. The spectral characteristics of the radiation and the spatial and angular dependences are recorded at a maximum distance from the grating.

3. Simulations results

3.1. The dependence of the CSPR intensity of the multi-bunched beam on the diffraction gratings grooves depth

It is known that for a rectangular diffraction grating the dependence of the CSPR intensity of the grating corrugation depth has a periodic character [6]. For sinusoidal, echelette and lamellar diffraction gratings in the groove depth range from 0 to 300 μm code KARAT simulation was carried out.

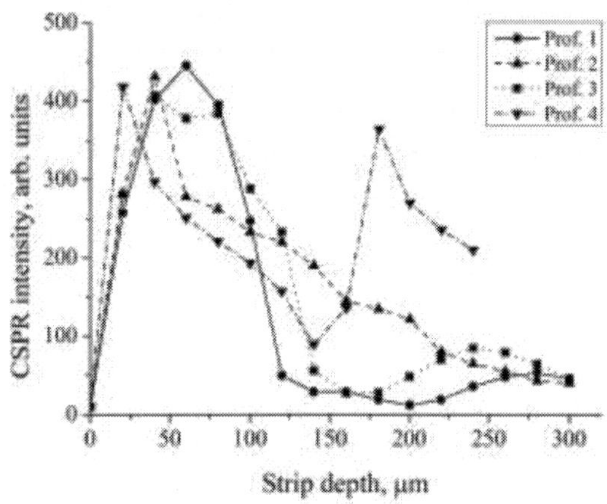

Figure 3. Dependence of the CSPR intensity of the multi-bunched beam on the diffraction gratings grooves depth.

Figure 3 shows the dependences of the radiation intensity of the corrugation depth for all studied diffraction gratings. It is seen that the periodic dependence is observed only for a lamellar grating. The electromagnetic wave enters grating groove gaps and then is reflected from the bottom of the grating and emitted back into space. Because interference of the electromagnetic waves depends on the groove depth, there is a periodic character of the radiation intensity. In sinusoidal and echelette gratings, the groove gaps are narrowed to the base which leads to deterioration of propagation of electromagnetic waves in them. Radiation with different frequencies penetrates into gaps at different depths. The interference conditions are disturbed leading to suppression of radiation with increasing groove depth.

Based on these dependencies, we can conclude that for the generation of THz radiation by diffraction gratings with a period $d = 292$ μm groove depth a should be chosen equal to 60 μm.

3.2. The angular distribution of CSPR by single electron bunch

PiC code KARAT simulation of the CSPR by single electron bunch for four types of gratings was carried out. The depth of the corrugation of the diffraction gratings is chosen on the basis of the results obtained above and was $a = 60$ μm. The angular distribution was recorded in the range of angles from 60 to 120 degrees (figure 2) where the 90 deg corresponds to the perpendicular drawn through the grating center and 120 deg towards the rear with respect to the motion of the beam electrons.

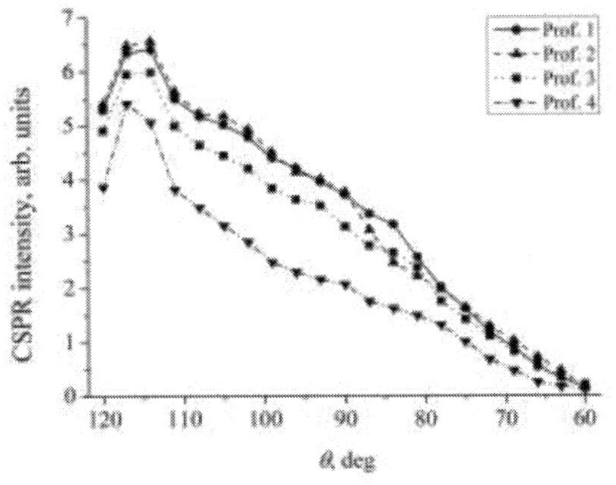

Figure 4. Angular distribution of CSPR by single electron bunch.

Figure 4 shows the angular distribution of CSPR by single electron beam for the four types of diffraction gratings. Figure 4 shows the angular distribution of the radiation intensity for the frequency corresponding to the observation angle. Dependencies increase monotonically with increasing angle of observation. In this case the radiation intensity for a lamellar grating is the lowest. The maximum intensity has the echelette grating. The results agree well with the radiation intensity depending on the grating corrugation depth for multi-bunched electron beam.

3.3. The angular distribution of CSPR by multi-bunched electron beam

The angular distribution of the CSPR by multi-bunched beam is significantly different from the case when single bunch passes near the diffraction grating. For multi-bunched beam as a result of constructive interference a set of plane waves is produced, in our case, propagating perpendicular to the grating with a frequency of 1 THz (figure 5).

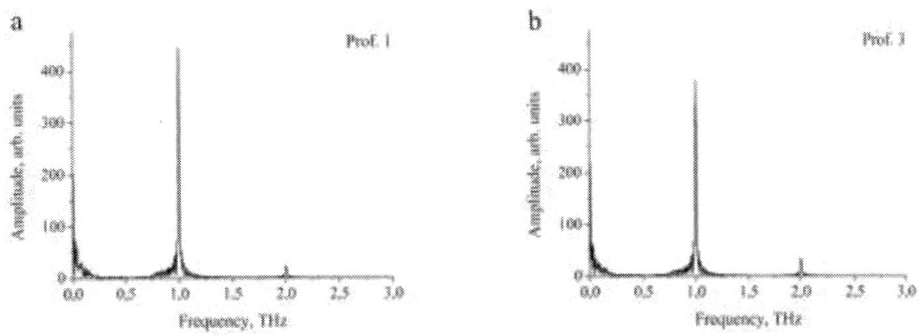

Figure 5. CSPR power spectrum for echelette (a) and sinusoidal (b) diffraction gratings for $a = 60$ μm.

PiC code KARAT simulation of CSPR generation by multi-bunched electron beam consisting of 16 micro-bunches for the diffraction gratings of different profile was carried out. The depth of the corrugation gratings $a = 60$ μm. Figure 6 shows the angular distribution of the CSPR intensity. In this case, CSPR consists of planar wavefronts formed due to constructive interference between successive bunches. The transverse dimensions of the planar wavefronts are determined by the geometrical dimensions of the grating. It should be noted that the efficiency of CSPR generation from the echelette grating is almost two times higher than that for the lamellar grating.

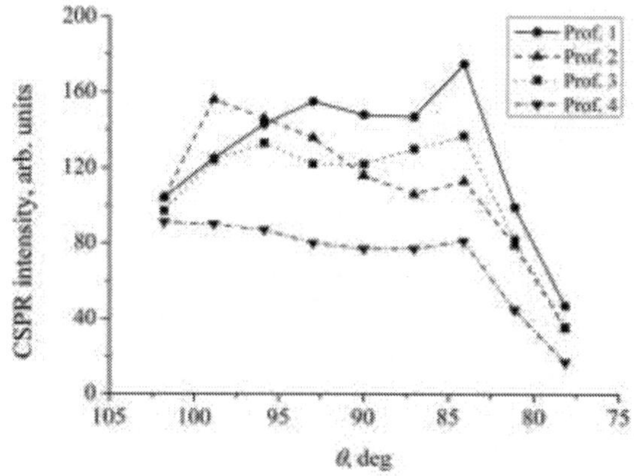

Figure 6. Angular distribution of CSPR by multi-bunched electron beam.

In addition, the shape of the angular distribution depends on the gratings profile. Since the angular distribution for a sinusoidal grating is most uniform at a sufficiently high average intensity. At the same time, the angular distribution for echelette gratings is not symmetric with a significant amplitude difference depending on the observation angle.

4. Conclusion

Numerical simulation by PiC-code KARAT allowed a direct comparison of the efficiency of the CSPR generation from four types of diffraction gratings for single electron bunch and for multi-bunched electron beam. The analysis of the spatial and angular characteristics of radiation leads to the conclusion that the optimal grating for further investigations is the sinusoidal grating with a period d = 292 mm and a depth of corrugation a = 60 mm.

Uniform angular distribution and high intensity of radiation are undoubted advantage over diffraction gratings with echelette profile and, especially, gratings with a lamellar profile.

The work was partly supported by Russian Ministry of Education and Science within the program "Nauka".

References

[1] Sukhikh L G, Potylitsyn A P, Artyomov K P, Aryshev A S, Urakawa J and Karataev P V 2012 Feasibility of THz source based on coherent smith-purcell radiation generated by femtosecond electron bunches in super-radiant regime *IPAC-2012* p 574

[2] Prokop C, Piot P, Lin M C and Stoltz P 2010 Mumerical modeling of a table-top tunable Smith-Purcell terahertz free-electron laser operating in the super-radiant regime *Appl. Phys. Lett.* **96** 151502

[3] Andruszkow J et al. 2000 First observation of self-amplified spontaneous emission in a free-electron laser at 109 nm wavelength *Phys. Rev. Lett.* **85** 3825

[4] Tarakanov V P 1992 User's Manual for Code KARAT (Springfield: BRA)

[5] Artyomov K P, Ryzhov V V, Naumenko G A and Shevelev M V 2012 PIC code KARAT simulation of different types of polarization radiation generated by relativistic electron beam *Journal of Physics: Conference Series* **357** 012022

[6] Shibata Y et al. 1998 Coherent Smith-Purcell radiation in the millimeter-wave region from a short-bunch beam of relativistic electrons *Phys. Rev.* E **57**

IOP Conf. Series: Journal of Physics: Conf. Series **830** (2017) 012148 doi:10.1088/1742-6596/830/1/012148

Discharge-mechanical method of rock breakage

V F Vazhov[1], S Y Datskevich[1], M Y Zhurkov[1], V M Muratov[1] and B Jeffryes[2]

[1] National Research Tomsk Polytechnic University, Tomsk, 634050, Russia
[2] Schlumberger Gould Research, High Cross, Madingley Road, Cambridge CB3 0EL, UK

Abstract. The electric discharge and mechanical technology of hard rock breakage was developed on the ground of mechanical and electrical pulse methods and it was tested for purposes of deep drilling. It was demonstrated that, due to breakage of the rock surface by electric discharges, the rock excavation volume (breakage performance) is significantly improved as compared to conventional mechanical methods.

1. Introduction

Rock breakage is the most labor-intensive and time and money-consuming operation when extracting and processing minerals. Inevitable shortcomings of main conventional mechanical methods of rock breakage are high specific energy consumption, strong dependency of the excavation volume on the rock strength, and rapid wear of a rock destruction tool. An electric discharge (ED) technology of breaking solid dielectric materials, that is invented and being developed in the Tomsk Polytechnic University (Russia), is seen to be promising [1, 2]. Such an ED technology provides the basis for its various applications including rock drilling, rock cutting, rock and ore breaking [3-5], surface processing and finishing [6] and other. The ED method is based on a phenomenon that the electric strength of liquid dielectrics and water is higher than that of solid dielectrics and rocks at a breakdown delay of 1 μs and less [1,7]. An electric breakdown results in a spark channel which penetrates into a solid body and converts the electric energy into energy of mechanical stresses and deformations (crack formation). The ED rock breakage principle allows preventing a lot of disadvantages of conventional mechanical methods due to following reasons: it is a breaking load that effects on the material and there is no need in any-kind mechanical rock destruction tools. When drilling a borehole, a rapid wear of a drill bit leads to significant increase in borehole drilling time and cost, especially it is true for deep and super-deep boreholes, and this retards the geothermal energy development [4].

Each rock breaking method is known to have its own shortcomings and disadvantages that limit fields of its application. Combining various rock breaking methods makes it possible to get new ones coming closer to a universal rock breaking technology [8].

Earlier we suggested an ED method of rock breakage by a continuously moving two-electrode system [9]. Such a method has a lot of advantages against to a fixed multi-electrode system [6] because it reduces the energy loss and the voltage pulse deformation at a pre- breakdown stage and significantly reduces the energy stored by a pulse generator. An ED method of rock breakage by a moving electrode system, when applied to drilling, allows one to implement a cooperative effect on rocks. A joint action of electric discharges and mechanical cutting would result in a higher rock excavation volume due to pre-breaking of rock by electric discharges and that makes its further mechanical disintegration easier.

The present paper shows, for the first time, results of studying the rock breakage by a joint electric discharge and mechanical method in order to determine the cutting forces when processing the surface of a granite unit mechanically and by an electric discharge method.

2. Experimental

In order to study the cutting forces we used a device equipped with a three-component elastic dynamometer with foil resistive strain gages. When processing a rock surface mechanically, cutters with an 8mm-wide main cutting edge and made of tungsten-cobalt hard alloy VK8 were used. A 4mm-deep and 8mm-wide trench was cut in a granite unit (the compression strength $\sigma = 160$ MPa), with a cutting depth of 0.07 mm per turn (Figure 1, a). A 1mm-thick layer of rock was removed off and a main constituent P_z of a cutting force and a normal constituent P_y (pressing-out force) were recorded at a final stage of surface layer removal (Figure 2).

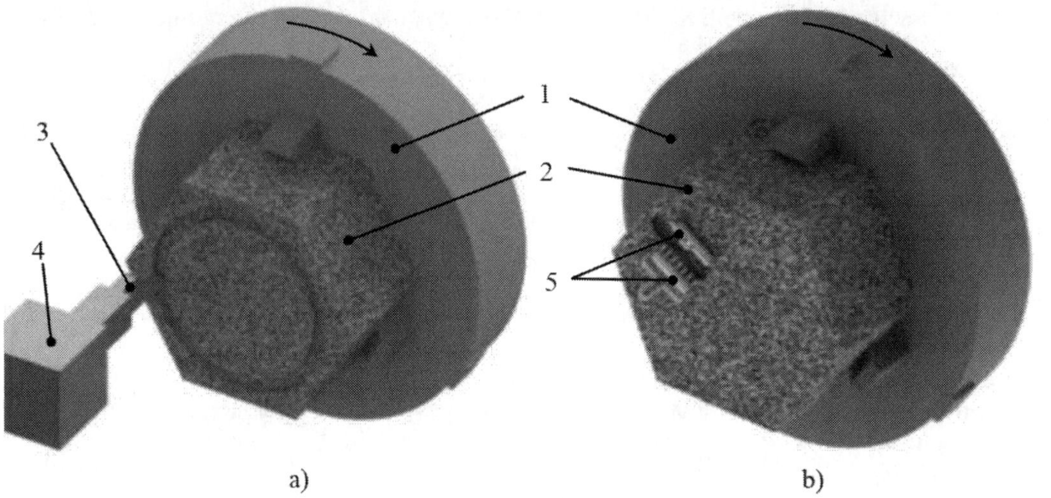

Figure 1. Processing of granite unit. a) mechanically, b) by electric discharges.
1 – rock unit holder; 2 – granite unit; 3 – cutter; 4 – elastic dynamometer, 5 – electrodes

When testing a pre-processed granite unit, a 20mm-gap two-electrode system was moving along the granite surface at a given speed (Figure 1, b). After having treated the granite unit with electric discharges, the rock was exposed to mechanical cutting according to the method described above.

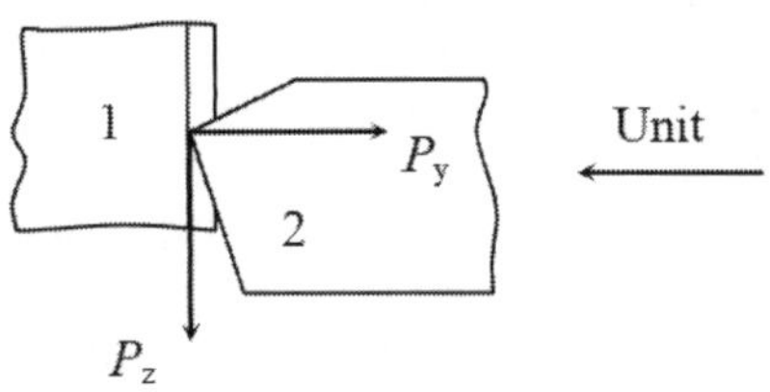

Figure 2. Diagram of forces applied to the cutter
1 – granite unit, 2 – cutter;
P_z – main constituent of cutting force,
P_y – normal constituent of cutting force

3. Results and discussions

Figure 3 shows granite units after being processed mechanically (a) and by a joint electric discharge and mechanical method (b).

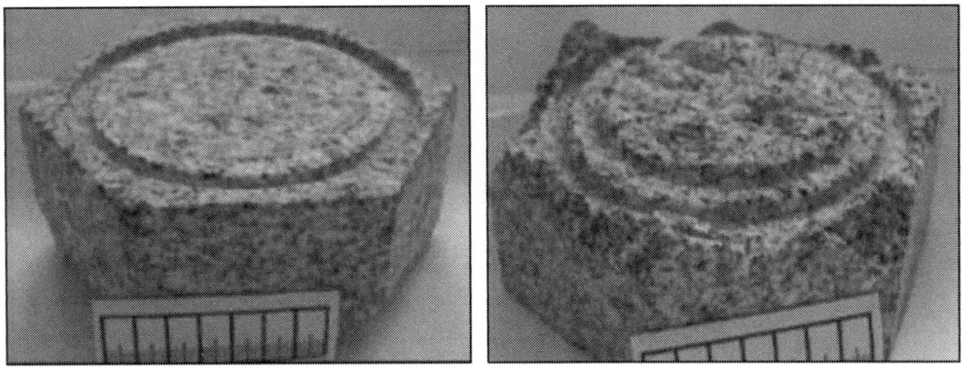

a) b)

Figure 3. Photos of granite units after processing. a) mechanically, b) by a joint electric discharge and mechanical method

Measurements of mechanical cutting forces are given in Table 1. When taking measurements of cutting forces on the rock surface before being exposed to electric discharges, it has been identified that, when 1mm-thick rock surface layers are successively, one after one, removed, the cutting forces do not practically depend on the depth of mechanical impact.

Table 1. Function of the cutting force and depth (without pre-processing by electric discharges)

Mechanical processing depth (mm)	P_z (N)	P_y (N)
1	2950	1550
2	2950	1550
3	2950	1490
4	2900	1530

Figure 4 illustrates results of measuring the cutting forces versus the cutting depth for a granite unit that has been pre-treated by electric discharges. The effect of electric discharge on the cutting forces is observed at a depth of 8 mm and less from the rock surface. Moreover, when removing the first layer off the pre-treated granite unit, the cutting forces appear to be approximately an order of magnitude less than those of when cutting the rock in its original state. The deeper we go, the higher the cutting forces are, because the material is getting less fractured. The depth of penetration of the discharge channel into a solid body is one of important parameters determining the degree and depth of crack formation after pre-processing the rock by electric discharges and the efficiency of rock breakage by the ED technology. Reference [2] states that, by theory, the depth of discharge channel penetration into a solid dielectric is $H = 0.32S$ (S – interelectrode gap). The breakdown channel path has a stochastic nature and the penetration depth, especially in rocks, varies in a rather wide range. Reference [9] covers experimental studies of a depth of discharge channel penetration into a granite unit at a wide range of the interelectrode gap. Whence it follows that an average penetration depth is $H = 6.1$ mm for $S = 20$ mm. Since the energy stored by the pulse generator is released inside the discharge channel as an electric burst, there occurs intensive crack formation inside the rock and rock pieces are split off. The crack length depends on a lot of factors and can reach tens of millimeters in case if cracks cannot come out on the free surface. In our case an average length of cracks is limited by a channel penetration depth since cracks can come out on a free surface. Plasma is released out of the discharge channel, the channel pressure reduces and the process of crack formation stops. Hence, cracks are able to develop inside a granite unit at a mean depth of 6.1 mm, thus, the total depth of rock fracture and crack formation can be 12.2 mm from the rock surface.

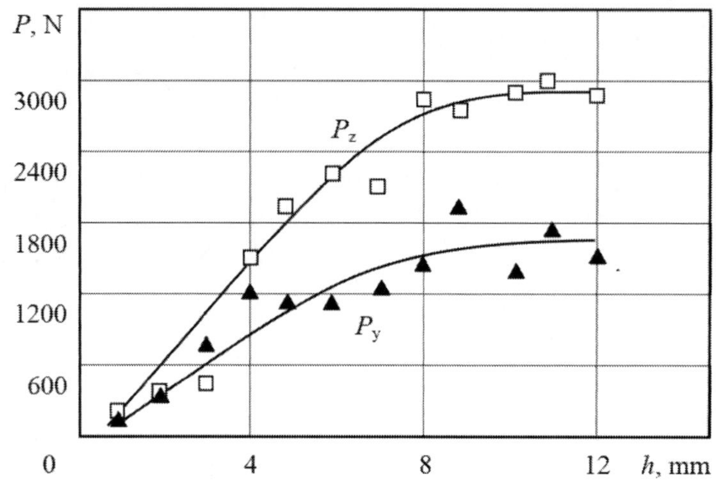

Figure 4. Function of cutting forces vs cutting depth for ED pre-processed surfaces

When cutting the rock at a depth of 8 mm and deeper, we observed the cutting forces corresponding to those of the rock initial condition given in Table 1. Thus, pre-processing of the rock surface by the electric discharge method significantly reduces forces of mechanical cutting that, in its turn, improves the mechanical disintegration performance when, for example, drilling.

More other advantages of the joint method of electrical discharge and mechanical drilling compared to the ED drilling only should also be mentioned. First, mechanical cutters both level an uneven surface of the bottomhole and, at the same time, increase the depth of penetration per turn by getting deeper into a fractured layer of the rock. Second, leveling the bottomhole surface increases the probability of discharge channel penetration into the rock and the depth of its penetration, and it also improves the excavation volume of the rock fractured by the ED method. A greater energy input to the discharge channel and the wider interelectrode gap are assumed to contribute to the greater efficiency of drilling by the joint electric discharge and mechanical technology [10].

4. Conclusion

This work covers research results that have been obtained for the first time and that allow us to make the following conclusion. An electrical burst inside the rock matrix significantly reduces its mechanical strength for shear that, in its turn, improves the mechanical rock breakage performance (excavation volume) by 2 times and more. A joint action of an electrical discharge technology and a mechanical method considerably improves the drilling depth and efficiency as compared to the mechanical drilling only. When drilling by an electrical discharge technology and by a joint, electrical discharge and mechanical, method, the rock strength has considerably less impact on the rock excavation volume than on that by the mechanical drilling only.

The work has been conducted with the financial support from Schlumberger.

References

[1] Vorob'yov A, Vorob'yov G and Chepikov A 1961 Regularities of solid dielectric breakdown at the interface with a liquid dielectric under the voltage pulse impact (Certificate of Invention **107**, 27.07.1999)

[2] Semkin B, Usov Y and Kurets V 1995 Principles of Electric Discharge Breakage of Rocks (Saint petersburg: Nauka) p 276

[3] Voigt M., Großmann S, Speck J and Anders E, 2012 Entwicklung und Erprobung eines Bohrkopfs für Tiefengeothermie basierend auf dem Elektro-Impuls-Verfahren 12. Symposium Energieinnovation (EnInnov 2012, Graz)
[4] Schiegg H, Rødland A, Zhu G and Yuen D 2015 *Journal of Earth Science* **26** pp 037–046
[5] Kurets V, Usov A and Tsukerman V 2002 Electrical Discharge Disintegration of Materials (Apatity: Krasnoyarsk Scientific Center of RAS) p 324
[6] Goldfarb V, Budny R, Dunton A, Shneerson G, Krivosheev S and Adamian Y 1997 11th IEEE International Pulsed Power Conference (Baltimore, USA) pp 1078–1085
[7] Vorob'yov G, Chepikov A and Vazhov V 1998 *Izvestiya vuzov. Fizika* **12** pp 110–113
[8] Protasov Y 2001 Rock fracture (Moscow: Moscow State Technical University) p 453
[9] Vazhov V, Zhurkov M, Lopatin V and Muratov V 2008 *Physical and Technical Issues of Rock Mining* **2** pp 70–77
[10] Vazhov V, Gafarov R, Datskevitch S, Zhurkov M and Muratov V 2010 *J. Appl. Phys.* **80** 8 pp 79–84

IOP Conf. Series: Journal of Physics: Conf. Series **830** (2017) 012149 doi:10.1088/1742-6596/830/1/012149

High-temperature thermoluminescence of anion-deficient corundum and its connection with intrinsic defects and impurities

R N Abashev[1,2], A I Surdo[1,2] and I I Milman[1]

[1] Ural Federal University, 19 Mira Str., 620002 Ekaterinburg, Russia
[2] Institute of Industrial Ecology UB RAS, 20 S. Kovalevskaya Str., 620219 Ekaterinburg, Russia

E-mail: surdo@ecko.uran.ru

Abstract. Anion-deficient corundum and TLD-500 detectors on their basis were studied for a possible correlation between the thermoluminescence (TL) yield in the high-temperature TL peak at 830 K and the transformation of intrinsic defects and contents of impurities. Only 40–60% of TLD-500 detectors were found to have 830 K TL peak. The efficiency of thermally stimulated direct conversion of F and F^+ centers in the 800–930 K region depends on the presence of TL peak at 830 K. Obtained data point to the quenching role of the Ti impurity in the formation of the 830 K TL peak. Also, these data provide indirect evidence of association of the 830 K TL peak with Cr^{3+} ions.

1. Introduction

It was found in [1, 2, 3] that thermoluminescence (TL) response in TLD-500 detectors based on anion-deficient corundum (α-$Al_2O_{3-\delta}$) is dose-proportional in the high-temperature peak at 830 K. In cases where continuous irradiation is used, such response is proportional to a dose up to 2 kGy. At pulsed nanosecond irradiation, it is proportional to a dose up to 6 MGy. In [3], the TL spectrum in the peak at 830 K was also studied to reveal, along with weak luminescence of F^+ centers (anion vacancy with one electron) with $h\nu_{em}$=3.4 eV, intense luminescence with $h\nu_{em}$=4.1 eV whose nature is not known precisely. Similar luminescence was only observed earlier at T≤300 K and was caused by excitons localized at Ti-ions [4]. It is also known that the luminescence of F centers (anion vacancy with two electrons) at T>700 K including the high-temperature peak was thermally quenched. One of the difficulties and still unresolved problems for anion-deficient corundum is also the fact that so far no correlations have been discovered between the impurity composition, concentration of active F^+ and F centers, and the TL yield in the main dosimetric peak at 450 K. For above room temperatures, the research has precisely established only the relationship of the TL yield in the 570–600 K peak with chromium impurity [5]. It is caused mainly by the luminescence of Cr^{3+} ions and is therefore called "chromium" peak. The objective of this study is detection in α-$Al_2O_{3-\delta}$ of possible correlations of TL output in the high-temperature peak at 830 K with the transformation of intrinsic defects and contents of impurities.

2. Samples and experimental techniques

The objects of this study were samples of TLD-500 detectors on the basis of α-$Al_2O_{3-\delta}$ single crystals, some of which were discovered to lack the 830 K TL peak. The impurity composition in the detectors studied was checked by the X-ray fluorescent method using an XRF-1800 spectrometer. The samples were irradiated on UELV-10-10S accelerator with 10 MeV electrons and with a bremsstrahlung by doses of 9 kGy and 300 Gy, respectively [3]. The TL curves were measured at a heating rate of β=2 K/s in a temperature range of T=300–1100 K. The TL signal was recorded by a FEU-142 photomultiplier at a reduced sensitivity to the thermal radiation of the heater, the temperature of which could reach 1200 K. The samples were annealed directly in the spectrophotometers over a temperature range of T_a=300–1400 K with 20 K steps at β=2 K/s. Upon achieving the required T_a, the samples were cooled to 300 K, and their optical absorption (OA) spectra and PL were measured in two parallel experiments by means of a Cary 60 and a Cary Eclipse spectrophotomer, respectively. The concentrations of F and F^+ centers (C_F and C_{F+}, respectively) were determined from OA in the bands with $h\nu_{m1}$=6.1 eV and $h\nu_{m2}$=4.8 eV, respectively, using Smakula's formula. According to [6], oscillator strengths were taken to be equal to 1.3 for the F center and to 0.65 for the F^+ center.

The time-integrated spectra of pulsed cathodoluminescence (PCL) were measured using a LineSpec™ CCD Array Spectrometer over a range of 1.4–5.6 eV with a time gate length of 20 ms for 10 s under excitation with an electron beam from an Arina-02 accelerator (E_e=130 KeV, τ_p=10 ns, W=7 mJ/(cm^2·pulse)).

3. Results and discussion

A study of TL in fairly large batches of TLD-500 detectors revealed the 830 K TL peak in only ~40–60% of the detectors. In the others, the TL peak at 830 K was not observed even after their continuous irradiation with 2–9 kGy doses.

Figure 1a shows changes in C_F (curve 1) and C_{F+} (curve 2) in the samples with the 830 K TL peak depending on T_a. In the annealing temperature range under study, antibatic changes in the $C_F(T_a)$ and $C_{F+}(T_a)$ are observed in them over the entire region of the step annealing procedure. The concentration change of Al_i^+ centers (C_{Al}, curve 3) (interstitial aluminium ion in an octahedral position absorbing and excited at $h\nu_{ex}$=4.1 eV and luminescing at $h\nu_{em1}$=3.8 eV and $h\nu_{em2}$=2.4 eV [7]) demonstrates a similar behavior under annealing to that of the F centers, with a prominent minimum near the main peak. Figure 1a also shows a TL curve ($I_{TL}(T)$, curve 4) in which one can easily discern the main, chromium and high-temperature peaks at 450, 570 and 830 K, respectively. Especially noteworthy is a substantial change in the $C_F(T_a)$, $C_{F+}(T_a)$ and $C_{Al}(T_a)$ in the range of 830–920 K, that is in the high-temperature part of the 830 K TL peak and immediately after it. What we observe here is effective thermally stimulated direct ($F \rightarrow F^+$) conversion of F and F^+ centers. However, comparison of the changes in the C_F and C_{F+} shows that annealing in the range of 830–920 K causes only ~30% of transformed F centers convert into F^+ centers. A similar situation is observed for the reverse conversion ($F^+ \rightarrow F$) of F^+ into F centers at temperatures above 950 K.

Similar evidence of changes in the $C_F(T_a)$ (curve 1) and $C_{F+}(T_a)$ (curve 2) in a sample without a detectable 830 K TL peak is presented in figure 1b. If we compare it with the data for a sample with the 830 K TL peak, one can see that changes in the $C_F(T_a)$ and $C_{F+}(T_a)$ are also observed in the region of 830–920 K, but with a very low efficiency. Nevertheless, as well as in the case of the sample with the 830 K peak, in this case only ~30% of the transformed F centers will be converted into F^+ centers. Note also a substantial increase in the F center concentrations at annealing temperatures above 950 K along with an insignificant decrease in the concentration of F^+ centers.

By way of summarizing the data on changes in the concentrations of F^+ and F centers in the irradiated detectors of both types considered, we can assume that it is F^+ and F centers that actively participate in the formation of the high-temperature peak at 830 K, at least of its high-temperature part.

We therefore then compared the TL curves ($I_{TL}(T)$) in the samples with about similar and essentially different concentrations of anion vacancies C_a defined as $C_a=C_F+C_{F+}$. As may be seen from

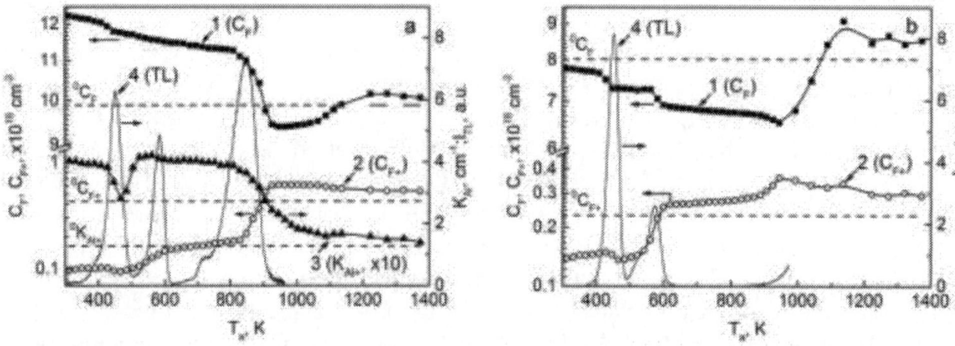

Figure 1. Changes in the concentrations of F (1), F^+ (2) and Al_i^+ (3) centers under step annealing and the TL curve (4) in samples irradiated with a dose of 9 kGy with the 830 K TL peak (part a) and without such a peak (part b).

figure 2 (curves 1 and 2), the two samples with about similar C_a (C_{a1}=9.0·10^{16} cm^{-3} and C_{a2}=9.3·10^{16} cm^{-3}, respectively) have TL yields that are essentially different not only in the main peak but also in the high-temperature one. This pattern for the main peak has been known for a long time and, in our opinion, it is due to its nature. It is one of the varieties of Al_i^+ center that is responsible for the main peak [8]. Thermal stimulation causes a restructuring of this center with the formation of F^+ and F center-bound excitons, which, while relaxing, excite F^+ and F centers.

To elicit the dependence of the TL yield at 830 K on C_a, half of the one of the samples with the 830 K peak was annealed in air at T_a=1800 K for 70 hours. After such thermal treatment, C_a decreased ~7 times (1.3·10^{16} cm^{-3}), the TL yield of the 830 K peak decreased ~2 times, and that of the 450 K peak decreased 15–20 times (figure 2, curve 3).

Thus, an analysis of these data reveals a certain relationship in α-$Al_2O_{3-\delta}$ between the concentration of anionic vacancies and TL yield in the high-temperature peak, but it is not as significant as for the main peak. We may, therefore, assume that certain impurities or other intrinsic centers may be responsible for the TL response in the 830 K peak.

One of the sensitive methods for estimating the contribution of impurities to the relaxation process is cathodoluminescence. Figure 3 shows time-integrated PCL spectra for samples with the 830 K peak (curves 1 and 2, respectively), and samples without such a peak (curves 3 and 4, respectively). The spectra reveal the characteristic luminescence of F^+, F and Al_i^+ centers with $h\nu_{em}$ equal to 3.8, 3.0 and 2.4 eV, respectively, along with sufficiently intensive PCL of Cr^{3+}, Ti^{3+} and Ti^{4+} centers with $h\nu_{em}$ equal to 1.79, 1.7, and 2.2, respectively. The interpretation of PCL for Ti^{3+} and Ti^{4+} centers is based on the data from [9].

The joint analysis of PCL and TL demonstrates that samples with the 830 K peak in which we did not detect PCL of Ti^{3+} centers with $h\nu_{em}$=1.7 eV display an intensive high-temperature peak at 830 K. In samples without the 830 K peak, such PCL is well detectable. Moreover, all spectra display a band with a maximum at 4.26 eV, which is most intensive in sample with the maximum luminescence of Ti^{3+} centers (curve 4).

In order to better understand the nature of the relaxation processes in α-$Al_2O_{3-\delta}$, and in the region of the 830 K TL peak in particular, we studied in the next series of experiments changes in the PL response separately for Al_i^+, Ti^{3+} and Cr^{3+} centers (I_{Al}, I_{Cr} and I_{Ti}) depending on T_a in the 300–1400 K range. The photoluminescence of Al_i^+, Ti^{3+} and Cr^{3+} centers was excited at $h\nu_{ex}$, equal to 4.1, 2.5 and 2.2 eV, respectively, and the PL responses were read out at $h\nu_{em}$, equal to 2.4, 1.7 and 1.79 eV, respectively. This approach ensures a higher sensitivity than OA data in indentifying small changes in the concentrations of Al_i^+, Ti^{3+} and Cr^{3+} centers in corundum under step-annealing, including in the region of the 830 K TL peak.

IOP Conf. Series: Journal of Physics: Conf. Series **830** (2017) 012149 doi:10.1088/1742-6596/830/1/012149

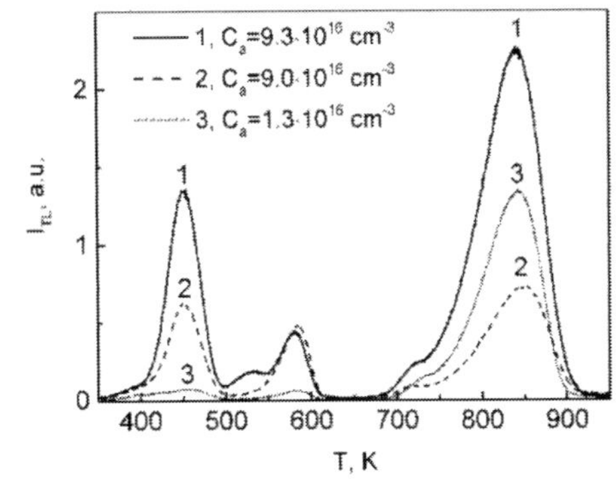

Figure 2. The TL curves of TLD-500 detectors irradiated with a dose of 300 Gy, with C_a: $9.3\cdot10^{16}$ cm^{-3} (1), $9\cdot10^{16}$ cm^{-3} (2), $1.3\cdot10^{16}$ cm^{-3} (3).

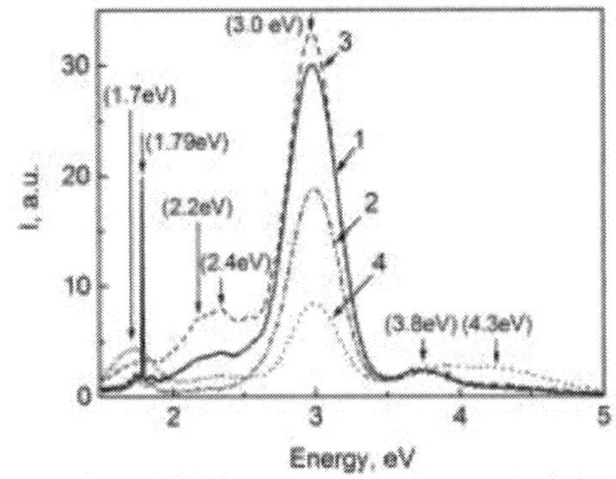

Figure 3. PCL spectra of samples with the 830 K peak (curves 1 and 2) and without such a peak (curves 3 and 4).

Figure 4 shows, for one of samples with the 830 K TL peak (part a) and for sample without such a peak (part b), the dependencies $I_{Al}(T_a)$, $I_{Cr}(T_a)$, $I_{Ti}(T_a)$ and $I_{TL}(T)$ (curves 1–4, respectively). In sample with the 830 K TL peak, the dependence $I_{Al}(T_a)$ is seen to vary in an extreme manner in the regions of the main and high-temperature peaks at 450 and 830 K and immediately after them, while $I_{Cr}(T_a)$ varies symbatically with $I_{TL}(T)$ in the chromium peak at 570 K and antibatically with $I_{TL}(T)$ in the 830 K peak. The extremely low PL response of the Ti^{3+} ions enables one just to identify its possible small increase in the region of the high-temperature peak.

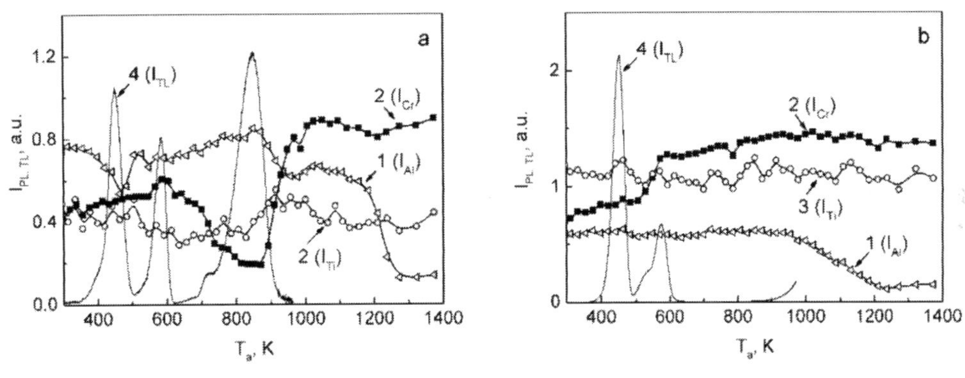

Figure 4. Changes in the intensity of Al_i^+ (1), Ti^{3+} (2) and Cr^{3+} (3) centers PL under step-annealing in samples irradiated with a dose of 9 kGy with the 830 K TL peak (part a) and without such a peak (part b). The figure also has TL curves (4) for samples with the 830 K TL peak (part a) and without such a peak (part b).

In the sample without the peak, the above changes in $I_{Al}(T_a)$, $I_{Cr}(T_a)$ и $I_{Ti}(T_a)$ were not detected in the region of Ta=700–900 K (Figure 4b). Nevertheless, step-annealing can help identify minor changes in $I_{Ti}(T_a)$ (curve 3) and $I_{Cr}(T_a)$ (curve 2) in the main and chromium peaks, respectively.

We may therefore state that in anion-deficient corundum samples with a high-temperature peak impurity and intrinsic Cr^{3+}, Ti^{3+} and Al_i^+ type centers may play a certain part in its formation.

X-ray fluorescent elemental analysis was therefore performed next for specially selected 8 samples of TLD-500 detectors having considerably differing TL yields at 830 K and containing, according to the PCL data, trace amounts of chromium and titanium impurities. The analytical data are summarized in Table 1, which also includes the values of C_a and TL yields in the region of the 830 K peak. Moreover, the data in Table 1 are arranged allowing for the TL yields of the samples in the region of the 830 K peak. For the purpose of TL yield determination, the samples were irradiated with a bremsstrahlung dose of 300 Gy corresponding to the linear region of the dose dependence [3].

An analysis of the data presented in Table 1 shows that the concentration of the chromium impurity in the samples studied is rather small, varying insignificantly, no more than two times. The other impurities found, such as Mg, Si, Ca, Ti and Fe, differ in concentration ~3–4 times as a minimum. The spread of the C_a values is ~8 times. It is important that neither the level of anion deficiency nor the changes in the concentration of each of the impurities found, except for the titanium, correlate with the TL yield in the 830 K peak. We may regard as the most significant result of the analysis the discovery of the quenching effect of the titanium impurity on the TL yield in the high-temprature peak at 830 K. As can be seen, the TL yield in the 830 K peak increases as the concentration of the titanium impurity decreases (C_{Ti}).

Table 1. The result of the X-ray fluorescent elemental analysis in specially selected TLD-500 detector samples

Sample number	Impurity concentration (10^{-4} wt.%)						$C_a \times 10^{-16}$ (cm^{-3})	TL output at 830 K
	Mg	Si	Ca	Ti	Cr	Fe		
1	77	<1	100	<7	78	17	7.8	167036
2	352	429	57	7	98	210	7.1	37532
3	<29	499	64	8	58	370	1.8	29041
4	29	<1	127	12	68	283	9.0	22744
5	190	54	45	13	61	92	8.5	11866
6	30	<1	78	17	73	89	9.3	7493
7	123	1	45	21	81	81	6.3	0
8	88	267	47	29	53	127	14.8	0

4. Conclusion

Thus, our data point to the quenching role of the titanium impurity in the formation of the 830 K high-temperature thermoluminescence peak. Also, the reported data provide indirect evidence of its association with Cr^{3+} ions. However, no reliable data have been obtained as yet on the nature of the UV glow with a maximum at 4.1 eV observed in the 830 K high-temperature peak. According to the data available to us, none of the types of active center in anion-deficient corundum, such as F^+, F, Al_i^+ centers, are additionally excited at temperatures near 830 K. Nevertheless, what is possible for F^+ centers is radiative transition from higher excited states. The interaction between Cr^{3+} and Ti^{3+} ions that we observed can lead to the creation of a multiple component defect, including that with the participation of anion vacancies. Such a complex defect may get restructured in the relaxation process near 830 K, as is the case with Al_i^+ centers in the main dosimetric peak, and with the glow in the region of 4.1 eV.

Acknowledgements

The work was partly supported by RFBR (pr. 16-38-00744) and Act 211 of the Government of the Russian Federation (contract no. 02.A03.21.0006), Center of Excellence "Radiation and Nuclear Technologies".

References

[1] Surdo A I, Milman I I, Abashev R M and Vlasov M I 2014 *Tech. Phys. Lett.* **40** pp 1048-1051
[2] Patent RU № 2570107, 05.11.2015. Abashev R M, Vlasov M I, Milman I I, Moiseykin E V, Sarychev M N, Soloviev S V, Surdo A I and Khokhlov G K (in Russian)
[3] Surdo A I, Abashev R M, Milman I I and Moiseykin E V 2016 *Radiat. Meas.* **90** 192
[4] Mikhailik V B, Kraus H, Balcerzyk M, Czarnacki W, Moszynski M, Mykhaylykc M S and Wahl D 2005 *Nucl. Instr. and Methods Phys. Res. A.* **546** pp 523-534

[5] Bessonova T S, Stanislavskii M P, Sobko A I and Khaimov-Mal'kov V Ya 1977. *J. Appl. Spectrosc.* **27** pp 991-995
[6] Evans B D and Stapelbroek M 1978 *Phys. Rev. B* **18** pp 7089-7098
[7] Springis M J and Valbis J A, 1984 *Phys. Stat. Sol.* **123** pp 335-343
[8] Surdo A I 2007 *Radiat. Meas.* **42** pp 763-766
[9] Yamaga M, Yosida T, Hara S, Kodama N and Henderson B 1994 *J. Appt. Phys.* **75** pp 1111-1117

Luminescence of potassium sulphate crystals activated by Sn^{2+} ions

T Koketai, A Tussupbekova, A Baltabekov and E Mussenova

Karaganda State University named after Academician Y.A. Buketov, Karaganda, Kazakhstan

E-mail: tussupbekova.ak@gmail.com

Abstract. The aim of this paper is to study some spectral-luminescent properties of sulphates activated by tin ions. The paper presents the photoluminescence excitation spectra of K_2SO_4-Sn^{2+} crystal. In this spectrum, peaks are observed at 240 nm and 260 nm. In the long-wavelength wing, the emission band has a "shoulder" indicating the presence of the second emission band. Thus, the presence of two bands of photoluminescence excitation suggests that K_2SO_4 formed two types of the impurity luminescence center differing in the immediate surroundings.

1. Introduction

Potassium sulphate, as well as potassium dihydrophosphate is a crystal with the composite anionic complex. Due to practical application of these crystals in the last decades, specific features of characteristic and impurity electronic excitations [1] have been widely investigated. Potassium sulphate is of interest for the study of luminescence, recombination and radiation processes and associated color centers [2]. It is stable to radiation and the energy of activation is effectively transferred from the sulphate matrix to impurities.

Moreover, a low symmetry of the lattice points of potassium sulphate gives confidence that we can avoid ambiguity in the interpretation of the nature of the electronic states of the color centers. This indicates the relevance of this study.

2. Experimental

The main investigation objects were K_2SO_4 activated by Sn^{2+} ions. K_2SO_4-Sn^{2+} crystals were grown from aqueous solutions. Due to the fact that the salt has reverse $SnSO_4$ solubility, in the stock solution $SnCl_2$ salts were added. The impurity concentration was the same and equal to 0.1 mol%. The thickness of the majority of the objects was approximately 1.5 mm.

Chloride ions are included in the crystal lattice of K_2SO_4 during crystal growth from aqueous solution. Meanwhile K_2SO_4 Cl^- ions do not absorb in the field of transparence of a crystal and do not form luminescence centers excited by X-rays. Sn^{2+} ions belong to the group of mercury-like ions. They, as well as Tl^+, have two s-electrons on the external electronic shell.

IOP Conf. Series: Journal of Physics: Conf. Series **830** (2017) 012150 doi:10.1088/1742-6596/830/1/012150

3. Results and Discussion

Figure 1 shows the radiation spectra of K_2SO_4-Sn^{2+} crystal measured at 80 K. In all spectra of photoluminescence excitation and emission spectra, the necessary correction for the spectral sensitivity of the photomultiplier and the spectral distribution of the source of UV emission is made.

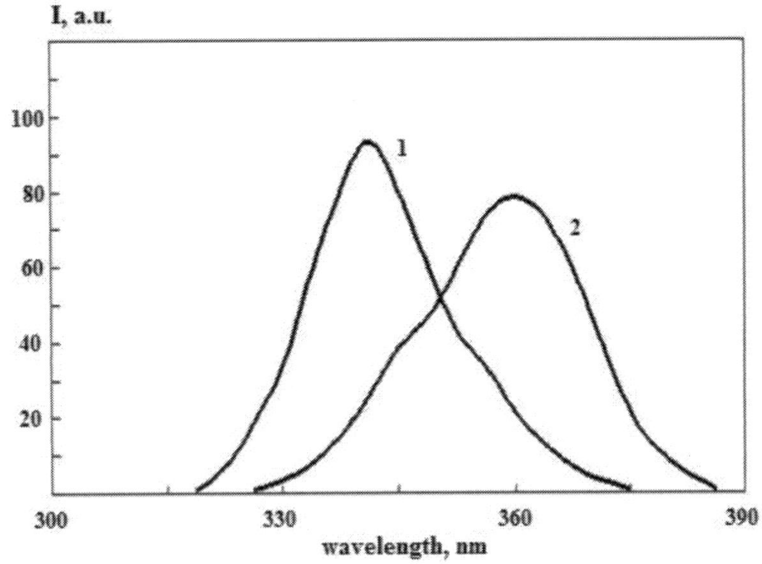

Figure 1. Radiation spectra of K_2SO_4-Sn^{2+} crystal measured at 80K. 1 – in the area of 230 nm excitation, 2 – in the area of 270 nm excitation.

As it is seen in Figure 1, the type of the radiation spectrum is strongly dependent on the position of the excitation band. Curve 1 was obtained in the excitation of the crystal in the region of 230 nm. The maximum of the emission band is observed at 335 nm. In the long-wavelength wing of the emission band there is a "shoulder" indicating the presence of the second emission band. Curve 2 was measured in excitation of the crystal in the region of 270 nm. Upon excitation of the crystal in the field emission maximum is observed at 360 nm. The emission band with a maximum at 335 nm is shown in the short-wave wing spectrum in the form of a "shoulder". Since these emission bands are not observed in the pure crystal of potassium sulfate or potassium sulphate in the crystal activated by chloride ions, therefore, they can be referred to ions Sn^{2+}.

The spectra of excitation of photoluminescence of K_2SO_4-Sn^{2+} crystal measured at 80 K are given in Figure 2. Curve 1 in Figure 1 represents the excitation spectrum measured when observing the emission in the field of 325 nm. Curve 2 is the excitation spectrum measured in observation of the emission in the field of 365 nm. The choice of the observation field is explained by aspiration to reduce overlapping of emission bands. The excitation spectra measured at the temperature of fluid nitrogen have maxima at 240 nm and 260 nm. It is apparent from Figure 1 that the emission band with maximum at 335 nm is mainly excited in the band with maximum at 240 nm, and the emission band with maximum at 360 nm is mainly excited in the band with maximum at 260 nm.

By analogy with the alkaline haloid crystals and the ammonium-haloid crystals activated by mercury-like ions, we connect the excitation bands observed in K_2SO_4-Sn^{2+} with A, the band with the ion Sn^{2+}, replacing the cation in nodal situation [1].

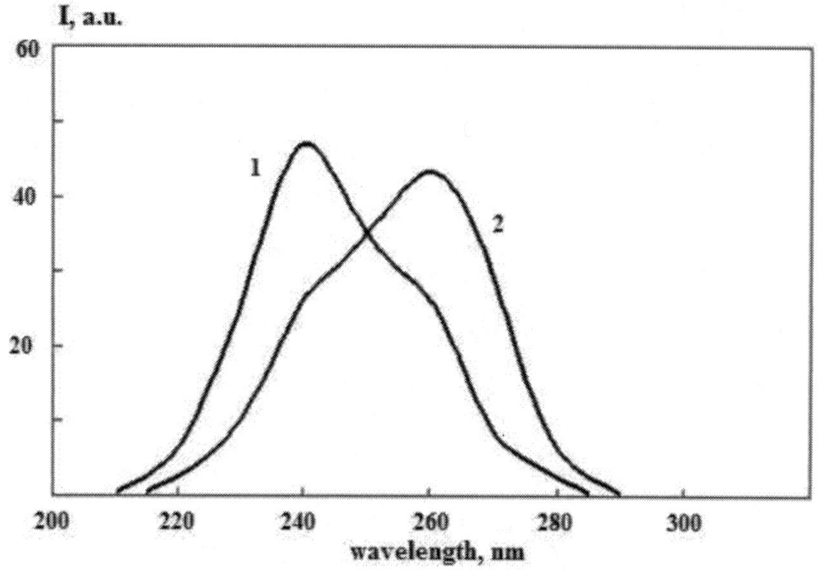

Figure 2. Spectra of excitation of K_2SO_4-Sn^{2+} crystal at temperature 80 K. 1 – emission band in the field of 325 nm, 2 – emission band in the field of 365 nm

Curves of temperature suppression of photoluminescence of K_2SO_4-Sn^{2+} crystal are given in Figure 3. In the figure it is seen that for different emission bands the nature of suppression is different. At increased temperature, K_2SO_4-Sn^{2+} emission spectrum tends to widening.

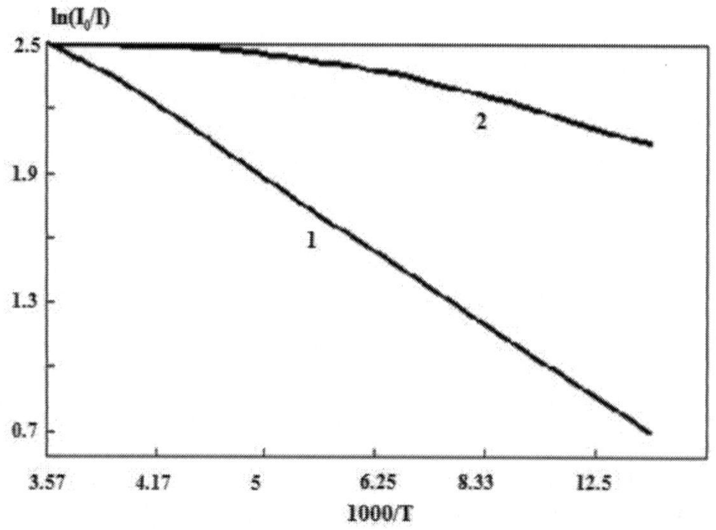

Figure 3. Curves of temperature suppression of photoluminescence of tin in potassium sulphate. 1 – for the emission band with λ_{max}=335 nm, 2 – for the emission band with λ_{max}=360 nm.

From K_2SO_4-Sn^{2+} emission spectra it is obtained that half-width of the emission band with λ_{max}=335 nm at a temperature of 80 K makes 335 nm 0.11 eV, and at 120 K, it is 0.13 eV. From temperature dependences of half-widths of emissions for the band with λ_{max}=335 nm, it is obtained that the efficient oscillation frequency is equal to 2.35×10^{-12} s^{-1}, and for the band with λ_{max}=360 nm, it is 4.6×10^{-12} s^{-1}. Different values of frequencies demonstrate that these luminescence centers either differently interact with a crystal lattice, or have various efficient mass of oscillators. As it is the impurity ion of one nature, different values of frequencies are related to various surrounding [2].

Curves of thermally stimulated luminescence (TSL) of crystals K_2SO_4-Sn^{2+} and K_2SO_4-Pb^{2+} are given in Figure 4. From the figure, it is seen that in addition to own peaks of TSL of the matrix there are luminescence peaks which can be connected with the impurity centers. The curves indicating TSL show that presence of heterovalence impurity ions of tin reduces the light sum of TSL peaks of the matrix in the area of 300 K. This phenomenon is followed by emergence of padding cationic vacancies.

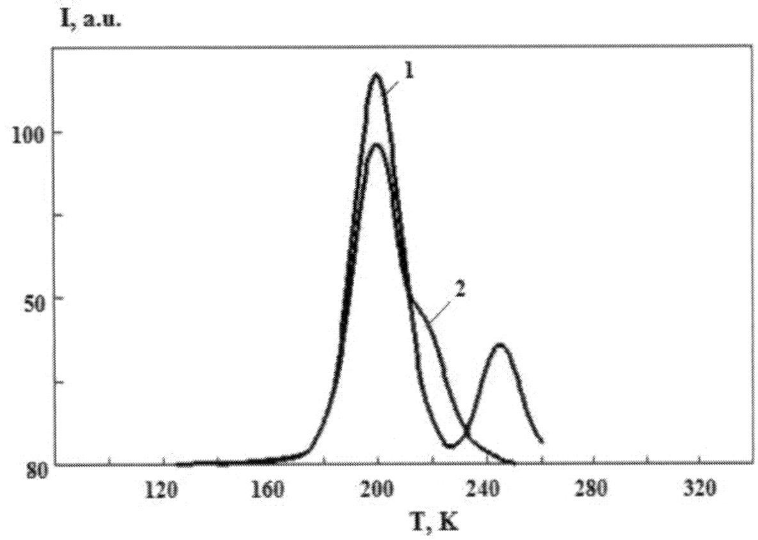

Figure 4. Curve for TSL of the potassium sulphate activated by 1 – tin, 2 – lead.

In the spectra of TSL of potassium sulphate crystals activated by ions of the transitional metals in addition to the relative decrease of the light sum of high-temperature peaks, its redistribution is observed as well. Similar redistribution of the light sum in crystals of potassium sulphate activated by ions of chlorine or mercury-like ions is not observed. The experimental data show that the influence of the studied ions of the transitional metals: Cu^{2+}, Co^{2+}, Ni^{2+} on high-temperature peaks of TSL is identical. It is known [1] that Cu^{2+} occupy cationic knot with larger coordination on oxygen in potassium sulphate lattice. It is known that the manganese ions which are transient, also advantageously occupy this cationic knot. Thus, it is possible to claim that Sn^{2+} ions replace base cations in nodal similar to copper ions.

4. Conclusion

According to the type of the emission spectrum, we can conclude that ions of divalent tin in the matrix of potassium sulphate form two types of luminescence centers. We assume that it is related to the fact that the tin ion is heterovalent concerning the replaced cation. Thus, the impurity ions of the transitional Sn^{2+} metals are the centers of electron capture in K_2SO_4. These impurity ions influence the emergence of additional cationic vacancies in K_2SO_4 crystal.

References

[1] Baltabekov A S, Koketaitegi T A and Kim L M 2010 *Eurasian Physical Technical J.* **7** 12
[2] Myrzakhmet M and Satayeva G, *et al* 2015 *J. of Luminescence* **169** B 804

Trace-transform invariants of tracks of high-velocity jets from the surface of tungsten droplets in the plasma flow

P Gulyaev[1], V Jordan[2,3], I Gulyaev[1,3], A Dolmatov[1]

[1] Physics and Chemistry Department, Ugra State University, 16 Chekhov street, Khanty-Mansiysk, 628012, Russian Federation
[2] Physics Department, Altai State University, 61 Lenin avenue, Barnaul, 656049, Russian Federation
[3] Laboratory for plasma dynamics of disperse systems, Khristianovich Institute of theoretical and applied mechanics Siberian branch of Russian Academy of Sciences, 4/1 Institutskaya street, Novosibirsk, 630090, Russian Federation

E-mail: p_gulyaev@ugrasu.ru

Abstract. The paper presents the analysis of the recorded tracks of high-velocity emission in the air-argon plasma flow during breaking up of tungsten microdroplets. This new physical effect of optical emission involves two stages. The first one includes thermionic emission of electrons from the surface of the melted tungsten droplet of 100–200 μm size and formation of the charged sphere of 3–5 mm diameter. After it reaches the breakdown electric potential, it collapses and produces a spherical shock wave and luminous radiation. The second stage includes previously unknown physical phenomenon of narrowly directed energy jet with velocity exceeding 4000 m/s from the surface of the tungsten droplet. The luminous spherical collapse and high-velocity jets were recorded using CMOS photo-array operating in a global shutter charge storage mode. Special features of the CMOS array scanning algorithm affect formation of distinctive signs of the recorded tracks, which stay invariant to trace transform (TT) with specific functional. The series of concentric circles were adopted as primitive object models (patterns) used in TT at the spherical collapse stage and linear segment of fixed thickness – at the high-velocity emission stage. The two invariants of the physical object, motion velocity and optical brightness distribution in the motion front, were adopted as desired identification features of tracks. The analytical expressions of the relation of 2D TT parameters and physical object motion invariants were obtained. The equations for spherical collapse stage correspond to Radon–Nikodym transform.

1. Introduction

This report focuses on the analysis of video frames capturing phenomenon of high-speed photoluminescence in a jet of low-temperature argon plasma [1,2]. We assume that this phenomenon can be attributed to guided ionization waves or "plasma bullets" (PB), which have been intensively investigated in atmospheric-pressure plasma jets in the past decade [3]. The peculiarity of our case is that high-speed emissions arose spontaneously and unpredictably in the process of fragmentation of liquid tungsten microdroplets. At present, the physical mechanism of initiation, evolution and formation of the propagation direction of such emissions is not clear and should be investigated in future work. As far as we know from the recorded video data, these high-speed emissions or PB do not interfere with each other. The main questions are: what is the nature of these high-speed objects, what factors determine their direction and what is the source of photoluminescence.

Photo-detector arrays (CMOS and CCD) operating in a charge accumulation mode produce specific distortions of images of fast-moving objects. These distortions appear as blurring of the image along the objects motion direction and complicate the analysis of its actual size and velocity. The goal of the present paper is determination of the basic invariants of motion: the velocity and the shape of the high-velocity objects which have to be immutable to variation of the exposure time and spatial resolution of the video camera.

2. Experiment and Equipment

The optical imaging system includes a digital camera HD1-1312 (PhotonFocus AG) with CMOS sensor of 1312*1082 size and photospectrometer LR-T1 (Aseq) with a spectral range of 300–1000 nm [4]. The bandpass filter with a central wavelength of 575 nm (FWHM 50 nm) was installed in the optical channel of the camera, and recording was conducted with a frequency of 55 frames per second. The applied camera lens provided a spatial resolution of 51.9 micrometer/pixel. The focal plane of the optical systems of the camera and photospectrometer were matched, the measurement region of the spectrometer was a circle 10 mm in diameter located in the center of the video camera frame.

Numerous video frames contain images of highly luminescent emission from melted tungsten particles moving in the plasma flow [2]. For example, figure 1 demonstrates the sources of intense radiation 1–2 mm in diameter around tungsten particles. In the same figure, ordinary tracks of tungsten droplets 100 microns in diameter and 0.5–1 mm in length can be seen. Notably, all high-velocity emissions are directed along the local direction of the motion a two-phase jet. The study of the spectrum of the total radiation of the jets (figure 2) showed no appearance of additional lines and bands in the spectral range of 555–595 nm when working with tungsten material compared to steel case. In addition, the spectral pyrometry method [5, 6] was applied to the continuous (thermal) spectrum in the 520–650 nm range which allowed us to find the temperature of the steel droplets in the cross section of 9 cm below the torch nozzle to be 2730–3090 K, and tungsten droplets – 3650–3890 K. The analysis of the video frames shows that plasma radiation in this area is not considerable, so we can assume that the measured temperature corresponds to the hottest particles of the dispersed phase. [7,8].

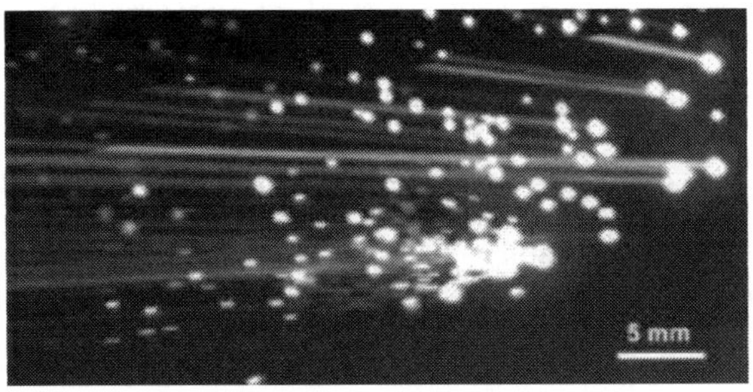

Figure 1. Group of tungsten droplets being a source of high-velocity emissions. Scale 51.9 microns/ pixel, frame exposure time is 10 μs.

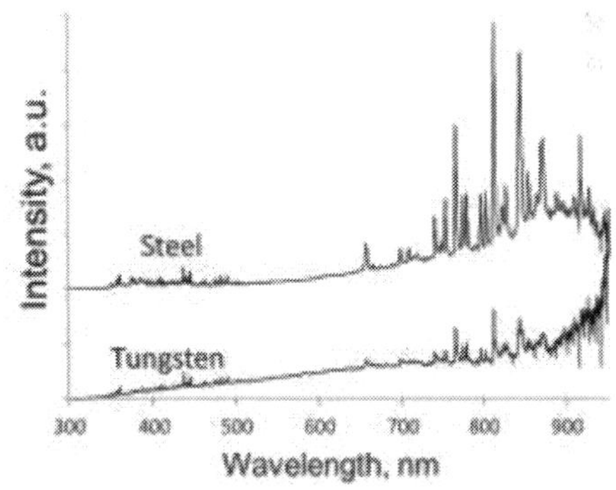

Figure 2. The spectra of emission of two-phase jets 9 cm downstream from the DC plasma torch nozzle when spraying steel (top) and tungsten (bottom) material.

Figure 3 (left) shows the images of several tracks of high-velocity emissions, in particular, the upper track corresponds to figure 4. As can be seen, the width of the tracks of various high-velocity emissions varies between 0.2–1 mm. Figure 3 (right) also shows the brightness distribution along the length of the respective tracks. It is important that the brightness of a track is not uniform along its length, and it has a wave shape. The length of such waves is different between the tracks and belongs to the 5–15 mm range. Simultaneously, an individual track wavelength increases with increasing distance from the source of emission (particles). In the upper graph of figure 3 one can see a tooth-shaped structure with a period of 1–1.2 mm. A close examination of the middle graph reveals similar structure with 4–5 mm wavelength.

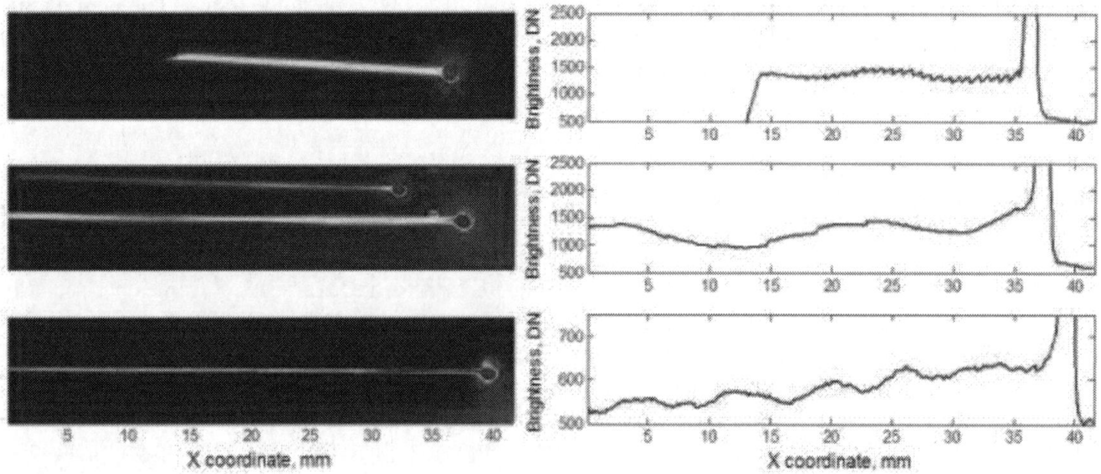

Figure 3. Images of individual tracks of high-speed emissions (left) and luminance distribution graphs along their length (right).

Let us consider a single particle with high-velocity emission in more detail (figure 4). The area in the immediate vicinity of the particle (about 1.5 mm in diameter) is occupied by radiation caused, probably, by the most intense stage of the gas discharge. A series of nearly spherical contours is observed around emitting particles, the diameters of spheres are about 2 mm, 4 mm, 6 mm and 9 mm, the brightness of spheres descends and the center is displaced from the particle to the outer region of the plasma stream (in figure 4 – upward) with diameter increase. The nature of these spheres cannot be attributed to light scattering by the particles of the surrounding atmosphere (such as halo or solar rainbow) since in the considered case the light wavefront is spherical rather than flat.

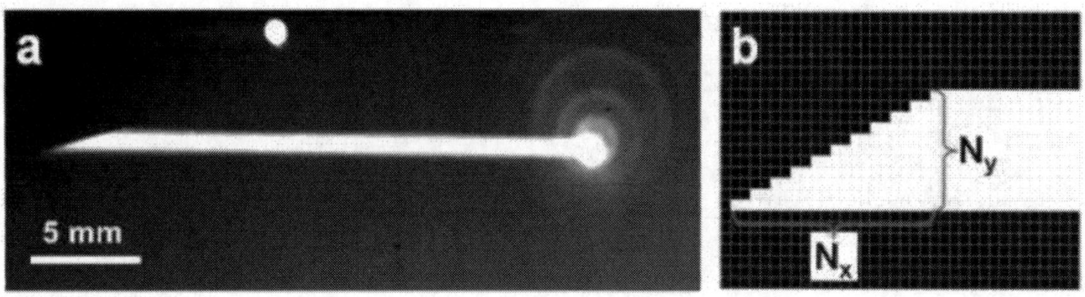

Figure 4. Typical image of a radiating particle and high-velocity emission. The frame exposure time 10 μs (a). Raster image of the tip of the track. For calculation of the propagation velocity of the emission (b).

All the video frames containing the entire (complete) tracks of high-velocity emissions demonstrate that their tips have a linearly cut shape (figure 4). A general analysis of the CMOS sensor operation algorithm provides explanation of this effect. Modern photo-arrays operate in a "global shutter" mode in which the start and end signal (charge) accumulation in the photosensitive region of each array cell takes place simultaneously, after which the charge is moved to a "floating diffusion" region (FD) and consequently is read out with a frequency of the onboard chip [6,7]. Depending on the architecture of CMOS array, accumulation of photo-electrons can occur even at the stage of storage in the FD region.

In fact, the exposure time of cells in adjacent rows will be different, which makes it possible to determine the velocity of the moving object from the slope of its image blurring. The clock frequency of the A1312 sensor (PhotonFocus AG) is 40 MHz, thus, the signal accumulation time differs in adjacent rows by 25 ns. Figure 4, b schematically illustrates a bitmap track tip and designates its length N_x and width N_y (in pixels). The velocity of the moving object can be calculated by the formula $V=(s \cdot N_x)/(\tau \cdot N_y)$ where s is the image scale. The calculation for the track in figure 4, using values $N_x=52$, $N_x=17$, s =51.9 microns, τ =25 ns, gives the value of the velocity V = 6350 m/s. Given that the total length of the track is 22 mm, one can suggest that the frame in figure 4 has been taken 3.5 μs after the emission has taken place. The measurements of other tracks velocity preferably lie in the range of 3–10 km/s, and the maximum value is 20 km/s. It may be noted that the obtained values are in good agreement with the above-obtained conservative estimate of 4.2 km/s.

Figure 5 provides description of three zones of interest for analysis of distortions of moving objects images recorded with CMOS photo-arrays operating in the charge storage mode with a global shutter.

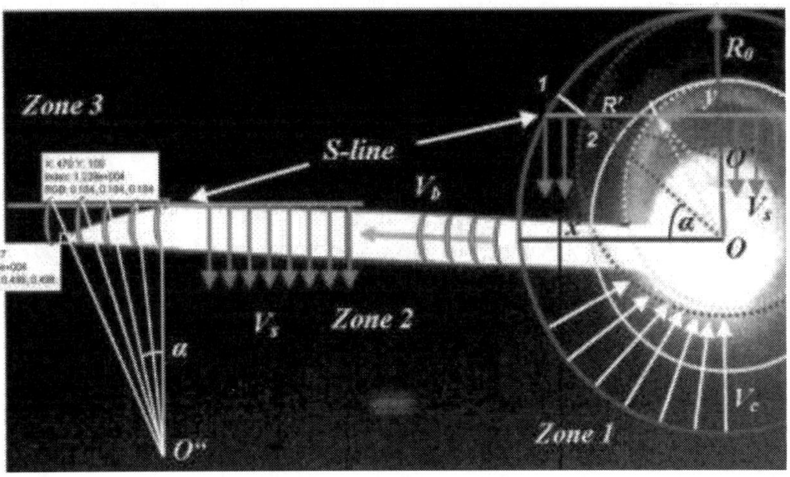

Figure 5. Three zones of interest in trace-transform during recording high-velocity discharge: zone 1 is the collapse and parallax of a spherical shock wave; zone 2 is the translator motion of the high-velocity emission and uniform "blur" of the image (Abel transform); zone 3 is the intersection of the photo-array read out lines and the image of the high-velocity object (Radon transform).

CMOS array scanning algorithm provides formation of distinctive signs of the recorded tracks, which stay invariant to trace transform (TT) with a specific functional. The series of concentric circles were adopted as primitive object models (patterns) used in TT at the spherical collapse stage (zone 1) and a linear segment of fixed thickness – at the high-velocity energy jet stage (zone 2). The two invariants of physical object – motion velocity and optical brightness distribution in the motion front, were adopted as desired identification features of tracks.

The following notation will be used: V_s is vertical speed of scanning lines (S-line) of the photo-array; V_c is radial velocity of a spherical shock-wave collapse; V_b is the velocity of the high-velocity object (plasma bullet); (x, y) are plane coordinates of the intersection of the scanning line (S-line) in the current time moment t. Given that in zone 1 all the luminescence objects are radially symmetric functions $f(x, y) = f(r)$, where r is the radius vector from the coordinate system center O, using the polar coordinates system, we have:

$$\begin{cases} x = r\cos\alpha \\ y = r\sin\alpha \end{cases}.$$

In accordance with figure 5, the coordinates of the image of the first spherical zone collapse can be written as follows:

$$\begin{cases} x = (R_0 - V_c t)\cos\alpha \\ y = R_0 - V_s t \end{cases}.$$

It is easy to calculate the coordinate of the center of a spherical wave image offset, subjected to the TT of Abel-Radon, using the Pythagoras' theorem:

$$y^{\|I} = R_0\left(1 - \frac{V_s}{V_s + V_c}\right).$$

Obviously, the subject of non-zero values of the variables t and difference $(V_c\text{-}V_s)$, one can determine the value V_c if V_s is known. To do this, one has to calculate the relative displacement of the centers of the two spherical shock waves (figure 4). After that, the expansion velocity is determined by the standard time-of-flight method [9,10]. It should be noted that the vertical displacement of the spherical shock waves centers is possible only if spherical front velocity has opposite signs in upper and lower parts of the zone, i.e. sphere is contracting to the center as shown in figure 5.

3. Basic physical phenomena and theoretical models

As mentioned above, the physical mechanism of the observed phenomenon is not completely clear, however, experimental results allow one to suggest that this is a form of gas discharge followed by directed high-velocity emission of charged particles – electrons. It is clear that realization of gas discharge requires the formation regions with high electric field. Under considered conditions of two-phase plasma flow this can be provided by thermionic emission of electrons from the surface of the metal droplets. As a result, the metal droplet itself acquires a positive charge and an electron cloud around it – negative charge. The reason why this effect was observed only with tungsten droplets is probably the high melting temperature of the material. As is known, thermionic emission current density obeys the Richardson-Dushman law: $j = A \cdot T^2 \cdot \exp(-W/kT)$, where T is the temperature of the material, W is electron work function of material, A is Richardson constant, and k is Boltzmann constant. Figure 6 shows the calculation of the thermionic current density for basic metal materials used in plasma spraying, based on the temperature range corresponding to their liquid state.

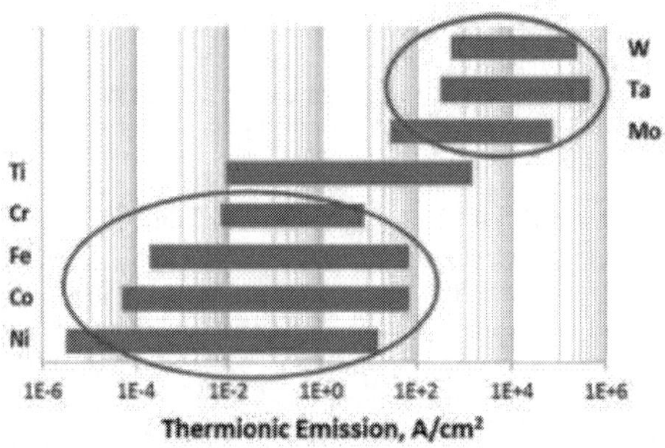

Figure 6. Thermionic current density for basic metals in the temperature range between material melting and boiling points.

The range of j weakly depends on the values of A and W ($A = 30 \div 60$ A·K^{-2}cm^{-2}, W =3.9-4.6 eV for these materials), and it is mainly determined by the material surface temperature. As can be seen, the intensity of thermionic emission of W, Ta and Mo metals group is 2–4 orders of magnitude higher than that of Cr, Fe, Co, Ni group. Thus, we believe that the appearance of the observed effect is due to the tungsten particle high thermionic emission of electrons at their melting temperature.

4. Conclusion

The analysis of the physical nature of high-velocity emissions should start from definition of the type of the observed moving objects. The drift velocity of the electrons in the plasma discharge depends on the value of the reduced electric field E/N (E is electric field, N is the density of charged particles) and it typically is 2–10 km/s. Obviously, the mobility and the drift velocity of heavy charged particles (ions) under the same conditions is several orders of magnitude lower due to the huge difference in their masses. For this reason, it can be assumed that the observed high-velocity emissions are caused by the motion of electrons.

Consider some of the possible causes of anisotropy of directed emission in a spherical discharge around metal droplets. As already noted, all high-velocity emissions propagate in the direction of gas stream. The gas pressure distribution on the surface of a spherical metal droplet is highly uneven: it has the maximum value in the front point, and the minimum value in the stern. Probably, the concentration of charged particles in the vicinity of the droplet follows uneven gas pressure distribution which leads to formation of the preferred direction of emission. Another explanation, more likely in our opinion, can be a displacement of the electron cloud (negatively charged) with respect to the metal droplet (positively charged) due to external gas flow. Thus, at the moment of a discharge, the center of the electron cloud is shifted downstream from the metal droplet. The result is a "cumulative emission" of electrons away from the surface of the metal droplet. Hydrodynamic analogy of such a scenario may be the collapse of cavitation bubbles near a solid surface or hollow droplets collision with a solid surface [11, 12].

The wave shape of the track brightness distribution (figure 3) requires close study as well. This could be explained by the formation of the "shock diamonds" (Mach disks) similarly to the supersonic gas jet. The sound velocity in the argon-air plasma under atmospheric pressure is about 1.5 km/s, i.e. the motion of the emission is supersonic. However, the formation of a "shock diamonds" requires quasi steady-state gas stream for a sufficient length of time, which is difficult to provide in our experimental conditions. Another possible explanation for the track images with a wave structure can be an effect similar to the wakefield acceleration, spatial separation of charges in the wake of a driver (dense cluster of charged particles or laser pulse) moving at nearly the light-speed. Assume that the temperature of the argon plasma is 3–5 thousand degrees, the ionization degree is approximately 10^{-4}, the density of charged particles is 10^{20} m^{-3}, and the plasma frequency $\omega_p = 10^{11}$ s^{-1}. Considering that in this case, the driver is a bunch of electrons at a speed of $V = 10^4$ m/s, we should expect antinodes in the wake of a size of $\lambda = V/\omega_p \sim 0.1$ μm. However, as we can see in figure 3, the typical wavelength values are 4 orders of magnitude higher – several millimeters. In such case, we have to suggest that the driver velocity should also be near the light-speed.

Acknowledgments

The reported study was funded by RFBR according to the research projects No. 15-42-00106 and No. 15-48-00100.

References

[1] Gulyaev I P, Dolmatov A V, Kharlamov M Yu, Gulyaev P Yu, Jordan V I, Krivtsun I V, Korzhyk V M and Demyanov O I 2015 *Journal of Thermal Spray Technology* **24** 1
[2] Gulyaev I P, Dolmatov A V, Gulyaev P Yu, Iordan V I, Kharlamov M Yu and Krivtsun I V 2016 *IOP Conference Series: Materials Science and Engineering* **110** 012057
[3] Lua X, Naidisb G V, Laroussic M and Ostrikov K 2014 *Physics Reports* **540** 123
[4] Dolmatov A V, Gulyaev I P and Jordan V I 2015 *IOP Conference Series: Materials Science and Engineering* **81** 012041

[5] Dolmatov A V, Gulyaev I P, Gulyaev P Yu and Iordan V I 2016 *IOP Conference Series: Materials Science and Engineering* **110** 012058
[6] Boronenko M P, Seregin A E, Gulyaev P Yu and Milyukova I V 2015 *Scientific Visualization* **7** 102
[7] Boronenko M P, Gulyaev P Yu, Seregin A E and Bebiya A G 2015 *IOP Conference Series: Materials Science and Engineering* 2015 **93** 012021
[8] Boronenko M P, Gulyaev P Yu, Seregin A E and Poluhina K G 2015 *Journal of Physics: Conference Series* 2015 **643** 012028
[9] Garkol' D A, Gulyaev P Y, Evstigneev V V and Mukhachev A B 1994 *Combustion, Explosion, and Shock Waves* **30** 72
[10] Gulyaev P Yu, Cui H Z, Gulyaev I P and Milyukova I V 2015 *High Temperatures - High Pressures* **44** 83
[11] Gulyaev I P, Solonenko O P, Gulyaev P Yu and Smirnov A V 2009 *Technical Physics Letters* **35** 885
[12] Gulyaev I P and Solonenko O P 2013 *Experiments in Fluids* **54** 1432

Proton-irradiation technology for high-frequency high-current silicon welding diode manufacturing

P B Lagov[1,2], A S Drenin[1] and M A Zinoviev[1]

[1] Department of Semiconductor Electronics and Semiconductor Physics, National University of Science and Technology "MISiS" (NUST "MISiS"), Moscow, 119049, Russian Federation
[2] Laboratory of Radiation Technologies, A.N. Frumkin Institute of Physical Chemistry and Electrochemistry Russian Academy of Sciences (IPCE RAS), Moscow, 119071, Russian Federation

E-mail: lagov2000@mail.ru

Abstract. Different proton irradiation regimes were tested to provide more than 20 kHz-frequency, soft reverse recovery "snap-less" behavior, low forward voltage drop and leakage current for 50 mm diameter 7 kA/400 V welding diode Al/Si/Mo structure. Silicon diode with such parameters is very suitable for high frequency resistance welding machines of new generation for robotic welding.

1. Introduction

Resistance welding is a technique used mainly for joining sheets of metal. In comparison with other welding methods, resistance welding is very efficient, as it causes little pollution and limited work piece deformation. It has high production rates, can easily be automated and requires no filler materials. Modern high power welding machinery requires diodes able to manage very high current levels in order to reduce the number of paralleled devices and to properly work with increased frequency. It is used extensively in the automotive industry since most cars have several thousand of spot welds made by industrial robots. Each welding cycle represents a load cycle for the diodes and the expected lifetime is generally more than ten million cycles. To keep the temperature swing as low as possible during the cycle, the diodes must be designed for very good trade-off between static and dynamic characteristics, lowest possible losses and thermal impedance [1, 2].

The main fundamental parameter that determines rectifier diode frequency except its geometry and doping levels is the minor charge carrier lifetime in active regions and its spatial distribution [3–6]. The lifetime value depends on characteristics of recombination centres in semiconductors, such as energy level in the forbidden zone, electrons and holes capture cross sections and concentration [7, 8].

Welding diode has the thinnest and lowest resistance base among deep-diffused high current silicon diodes, that is why accelerator-based [9–11] electron beam technology is efficient for frequency increasing from common 1–2 kHz to 10–12 kHz [12, 13]. However, further frequency increase by this method is limited by growth of the forward voltage drop, leakage current and not soft behavior during reverse recovery due to uniform distribution of recombination centres in the base region. Because of good previous results for high voltage diodes [14, 15] different proton treatments were tested in this work to achieve the desired 20–25 kHz frequency and high softness factor for welding diode. This aim is caused by the fact that the 20 kHz inverter (now transistor-based) technology shows numerous advantages. So the residual ripple of the welding current is negligible. The short current rise times enable extremely short welding operations of only several milliseconds without any flashes. The control cycle time of 25 μs ensures uniform and reproducible quality in the welded joint, even at these short welding times. The areas of application are projection welding and the welding of materials which possess good electrical conductivity, e.g. aluminum or copper. As a result of the high inverter frequency of 20 kHz, the welding transformers are working almost quietly. Welding diode should have reverse recovery time about 1.0–1.5 μs to meet these requirements surely.

2. Samples and Proton Irradiation Equipment

2.1. Welding diode samples

Welding diode structures of two types were involved into experimental investigations. First structures (Al/n^+np^+-Si/Al/Mo) were joined with Mo thermo-compensator via Al under high (about 800 °C) temperature and pressure. Second structures (Al/ p^+nn^+-Si/nano-Ag/Mo) were joined with Mo thermo-compensator via nano-Ag sintering paste using lower (about 250 °C) temperature. Both 50 mm diameter silicon diode structures (7 kA, 400 V) were formed identically using the same deep-diffusion process and electron-beam evaporation for 7 μm thick Al top contact making. Specific diode structure parameters are: 85–90 μm gradually Al- and B-doped p^+-anode; 50 μm uniformly P-doped (8–10 Ohm·cm initial substrate) n-base and 60 μm gradually P-doped n^+-cathode contact region.

2.2. Proton irradiation equipment

Two different accelerators were used for irradiation: pulsed 25 MeV proton linac I-2 in ITEP [16] and 3 MV tandem accelerator Tandetron 4130 in IPPE. The irradiation scheme is shown in figure 1.

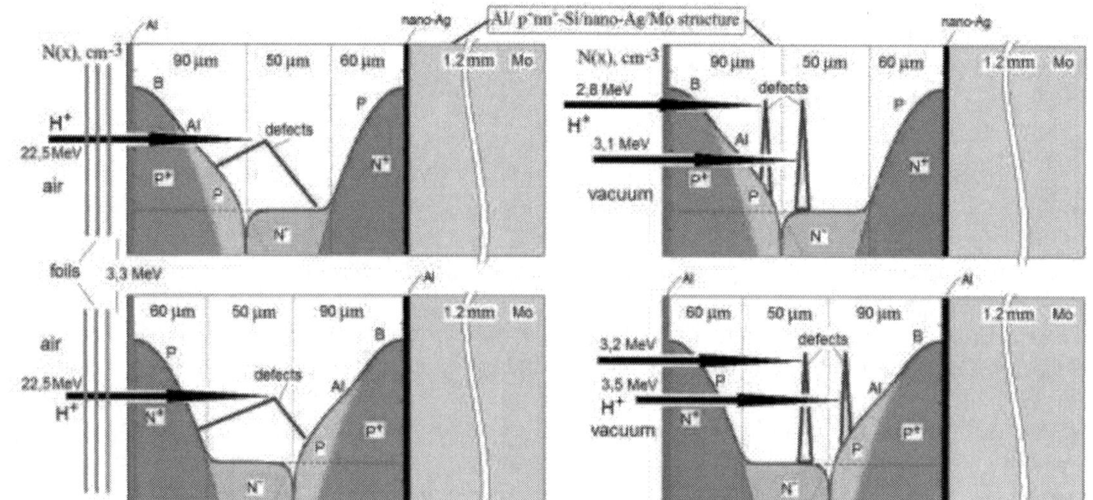

Figure 1. Schemes and variety of proton irradiation experiments.

Linac I-2 accelerates protons with fixed energy of 24.6 MeV and initial average current of 5 μA. Such high proton energy gives an opportunity to output the beam on air throw relatively thick and robust Al-window with low energy losses in it and without risks of its destruction or burning through during irradiation. After the window, protons have energy of 22.5 MeV and lower input energy needed on sample surface is achieved by using different stopping foils. Longitudinal straggle calculated using SRIM-2008 is about 55 μm in silicon for such irradiation conditions through Al-foils, and defects concentration decreases smoothly from the peak in the base near the junction towards the cathode region. In this case, welding diode structures were irradiated with energy of 3.3 MeV with fluencies $\Phi_{H+}=10^{11}...10^{12}$ cm^{-2}. Measurements of electrical parameters (Q_{rr} is reverse recovery charge, t_{rr} is reverse recovery time at dI/dt=100 A/µs and forward current of 1 kA, S is softness factor, I_{DRM} is leakage current at reverse voltage of 400 V and 150 °C, V_{FM} is forward voltage drop at forward current of 5 kA) using the DBC-226 measurement system were realized after thermal annealing (240 ºC, 3 hours) of irradiated diode structures. Oriented optimum proton fluence of $4\cdot10^{11}$ cm^{-2} was determined.

Tandetron 4130 accelerates protons with any energy in the range from 0.2 to 6.6 MeV with precise 0.1% adjustment and small proton and defects struggling (about some microns). In this case, irradiation experiments were carried out in a vacuum chamber with two different energies of 3.2 and 3.5 MeV (first type structures) and 2.8 and 3.1 MeV (second type structures) to form two defect peaks around and near the p^{+}n-junction [17] under fluence of $2\cdot10^{11}$ cm^{-2} for each energy. Irradiated diode structures also were annealed at the same conditions and then measured.

3. Experimental results

Essential experimental results are summarized in the table 1.

Table 1. Electrical parameters of proton irradiated welding diode structures

Diode type	E_{H+}, MeV	Straggle, мm	Q_{rr}, мC	t_{rr}, мs	S, a.u.	I_{DRM}, mA	V_{FM}, V
1	3.3	55	NA	1.2–1.3	0.4–0.6	15–20	1.2–1.3
1	3.2; 3.5	2–3	8–10	1.3–1.5	0.7–1.0	6–10	1.1–1.2
2	3.3	55	19–20	1.1–1.5	0.6–1.1	8–12	1.2–1.3
2	2.8; 3.1	2–3	7–9	1.0–1.4	1.5–4.6	0.2–2	1.1–1.2

Thus, all tested proton irradiation treatment could be applied for softness factor increasing from initial values of 0.2–0.4, but with different efficiency and influences on other characteristics. Silver sintered Al/p^{+}nn^{+}-Si/nano-Ag/Mo structures show better results for both irradiation techniques because protons do not cross the diode base region. Defects in the base body decrease the softness factor. Narrow straggle irradiation shows better results for both types of structures because of more local defect formation near the junction especially for Al/p^{+}nn^{+}-Si/nano-Ag/Mo structures that demonstrate the best electrical parameters. Nevertheless, cost-effective high volume proton treatment of bipolar device structures in vacuum needs energy efficient very reliable RFQ-accelerator similar to that described in [18] with wide treatment area up to 200x200 mm and equipped with high-performance high-volume vacuum automated chamber.

Acknowledgements

This work was carried out with the financial support of the Ministry of Education and Science of Russian Federation (the unique identifier of the project RFMEFI58415X0016). The authors also acknowledge ITEP (Moscow) and IPPE (Obninsk) for proton treatment.

References

[1] Cova P, Fasce F, Pampili P, Portesine M, Sozzi G and Zani P E 2004 High Reliable High Power Diode for Welding Applications *Microelectron. Reliab.* **44** pp 1437-1441

[2] Mei L F, Yan D B, Chen G Y, Xie D, Zhang M J and Ge X H 2015 Comparative Study on CO2 Laser Overlap Welding and Resistance Spot Welding for Automotive Body in White *Mater. Des.* **78** pp 107-117

[3] Vobecky J, Hazdra P, Humbel O and Galster N 2000 Crossing point current of electron and proton irradiated power P-i-N diodes *Microelectron. Reliab.* **40** pp 427-433

[4] Yamashita Y, Machida S and Sugiyama T 2014 Suppression of Reverse Recovery Surge Voltage of Silicon Power Diode by Adjusting Trap Energy Levels through Local Lifetime Control *Jpn. J. Appl. Phys.* **53**

[5] Wondrak W, Bethge K and Silber D 1987 Radiation Defect Distribution in Proton-Irradiated Silicon *J. Appl. Phys.* **62** pp 3464–66
[6] Kolesnikov N V, Lomasov V N and Malkhanov S E 1988 Spectrum and Spatial –Distribution of Radiation Defects in Proton-Irradiated Silicon *Sov. Phys. Semiconductors* **22** pp 329-330
[7] Khanna V K 2005 Physical understanding and technological control of carrier lifetimes in semiconductor materials and devices: A critique of conceptual development, state of the art and applications *Prog. Quantum Electron.* **29** pp 59-163
[8] Huhtinen M 2002 Simulation of Non-Ionizsing Energy Loss and Defect Formation in Silicon *Nucl. Instr. and Meth. A* 491 pp 194–215
[9] Pikaev A K, Glazunov P Y and Pavlov Yu S 1993 Radiation Center of Institute-of-Physical Chemistry in Moscow *Radiat. Phys. Chem.* **42** pp 887-890
[10] Pavlov Yu S and Lagov P B 2015 Magnetic Buncher Accelerator for Radiation Hardness Research and Pulse Detector Characterization *2015 15th European Conference on Radiation and Its Effects on Components and Systems (RADECS)* pp 336-338
[11] Pavlov Yu S, Solovev N G and Tomnikov A P 1985 Syncronization Circuit for Shaping Picosecond Accelerated-Electron Pulses *Instrum. Exp. Tech.* **28** pp 1335-1339
[12] ABB Application note 5SYA 2013-03. High current rectifier diodes for welding applications.
[13] Pavlov Y S, Surma A M, Lagov P B, Fomenko Y L and Geifman E M 2016 Accelerator-based electron beam technologies for modification of bipolar semiconductor devices *Journal of Physics: Conference Series* **747** 012085
[14] Cova P, Menozzi R, Portesine M, Bianconi M Gombia E and Mosca R 2005 Experimental and Numerical Study of H+ Irradiated p-i-n Diodes for Snubberless Applications *Solid-State Electron.* **49** pp 183-191
[15] Chernikov A A, Gubarev V N, Stavtsev A V, Surma A M and Vetrov I Y 2011 One more way to increase the recovery softness of fast highvoltage diodes *Proc. of the 2011-14th European Conf. on Power Electronics and Application* (Birmingham, ENGLAND)
[16] Lazarev N V, Andreev V A, Artemov V S, Batalin V A, Chuvilo I V, Edemsky V I, Kolomiets A A, Kondratiev B K, Kusmin Y N, Kouibida R P, Plotnikov V K, Porubai N I, Raskopin A M, Rybakov N I, Skachkov V S, Stasevich Y B, Stolbunov V S and Vengrov R M 1996 30 years operation of 25 MeV proton linac I-2 in ITEP at beam current of 200-230 mA *Proceedings of the XVIII International Linear Accelerator Conference* Ed. by Hill C, Vretenar M CERN Reports **96** pp 542-544
[17] Linder S 2006 *Power Semiconductors* (Lausanna: EPFL Press)
[18] Batalin V A, Volkov J N, Kulevoy T V and Petrenko S V 1994 Vacuum-Arc Ion-Source for ITEP RFQ Accelerator *Rev. Sci. Instrum.* 65 pp 3104-3108

Abnormal photothermal effect of laser radiation on highly defect oxide bronze nanoparticles under the sub-threshold excitation of absorption

P Gulyaev[1], **M Kotvanova**[1] **and A Omelchenko**[2]

[1] Physics and Chemistry Department, Ugra State University, 16 Chekhov St., Khanty-Mansiysk, 628012, Russian Federation
[2] Institute Photonic Technologies of Federal Scientific Research Centre "Crystallography and Photonics" of RAS, 2 Pionerskaya St., Moscow, 142190, Russian Federation

E-mail: p_gulyaev@ugrasu.ru

Abstract. The mechanism of abnormal photo-thermal effect of laser radiation on nanoparticles of oxide bronzes has been proposed in this paper. The basic features of the observed effect are: a) sub-threshold absorption of laser radiation by the excitation of donor-like levels formed in the energy gap due to superficial defects of the oxide bronze nano-crystals; b) an interband radiationless transition of energy of excitation on deep triplet levels and c) consequent recombination occurring at the plasmon absorption. K or Na atoms thermally intercalated to the octahedral crystal structure of TiO_2 in the wave SHS combustion generate acceptor levels in the gap. The prepared oxide bronzes of the non-stoichiometric composition Na_xTiO_2 and K_xTiO_2 were examined by high resolution TEM, and then grinded in a planetary mill with powerful dispersion energy density up to 4000 J/g. This made it possible to obtain nanoparticles about 50 nm with high surface defect density (10^{17}–10^{19} cm^{-2} at a depth of 10 nm). High photo-thermal effect of laser radiation on the defect nanocrystals observed after its impregnation into cartilaginous tissue exceeds 7 times in comparison with the intact ones.

1. Introduction

Photo-thermal effect of laser radiation on nanoparticles plays a key role in a number of medical technologies: laser hyperthermia, opto-acousic diagnosis of tissues and laser tissue engineering. Photo-physics of this effect is related to the complex nature of photon energy transition of laser radiation in condensed media with photo absorbing nanoparticles. An abnormal photo-thermal effect of near IR-laser radiation on nanoparticles of oxide bronzes of the transition metals impregnated in cartilage has been observed in [1].

It is known [2,3] that the oxide bronze nanoparticle dispersions, as well as the magnetite ones have similar optical properties and intensive absorption in the near IR spectral band. Laser heating of both nanoparticles in aqueous media should result in an amounted temperature response, but the observed photo-thermal effect in biotissue is different [1]. The origin of this difference is still unknown.

Note that the photo absorbing additives of the oxide bronze nanoparticles used for tissue impregnation have been produced by self-propagating high-temperature synthesis (SH-synthesis or SHS in brief) [4]. Solid products of the synthesis were grinded in a planetary mill into nano-sized particles [5].

Optical absorption of oxide bronze nanoparticles dispersions in water has shown a plasmon resonance peak of absorption [1,2] that was observed for magnetite [3] and "core-shell" iron oxide nanoparticles [6], as well. Optical absorption behavior of these nanoparticles dispersions in the near IR spectral band is variously discussed [2,3, 6-10]. The mechanism of photo-thermal effect of the near IR-laser radiation on biotissues impregnated with these nanoparticles is not fully understood.

To elucidate the mechanism of this effect we studied the influence of defect structure and electronic zone states of nano-crystals of oxide bronzes produced by SHS and serial high energy ball milling on absorption of nanoparticle dispersions in the near IR spectral band.

In this paper, the photo-physical mechanism of the near IR-absorption of laser radiation by the nano-sized crystals of oxide bronzes of the transition metals (W, Mo, Ti) accounting for their semiconductor properties and explaining the high photo-thermal effect by high density of electronic states in zones of highly defect nanocrystals has been proposed.

2. Experimental route and Theoretical approach

As starting materials for SH-synthesis, we used ultra disperse powder metal oxides TiO_2, MoO_3 and WO_3, and alkali metal salts of NaI and KI, which are mixed with the oxides in the stoichiometric ratio, and the synthesis was conducted in accordance with the previously developed method [4]. In the works on SHS of the transition metal oxides and their bronzes [1,4] grinding of solid materials in ball mills [5] used for preparing nanopowders from the synthesis products is widely spread. This method is simple, high energy effective, and has high efficiency, as well as comparatively not high temperature of grinding (~ 250–350^{0}C). Therefore, in this study it was used for fragmentation of Ti, Mo and W-bronze SH-synthesis, using methods [11] to determine the optimum modes of mechanical activation. The solid products of synthesis were crushed in an agate mortar before grinding. The final stage of grinding was carried out in planetary mills of the "Activator-2SL" and AGO-3, developing centrifugal acceleration up to 120 g.

The change in physical and chemical properties of the powder particles was monitored by measuring the particle size and specific surface area of particles by laser diffraction instrument LA-300 (Horiba, Japan) and permeability was measured by the instrument PSH-11M ("Company Hodakova", Russia) using Karman-Kozeny method. The rating of the process of grinding the oxide bronzes was estimated by differential calorimetry of the grinding bodies and the powder after each time interval of mechanical activation. Mechanical activation time changed by doubling, and ratio of grinding body volume loading in the reactor are discrete (in the vicinity of the optimal theoretical value of 50%) at a constant ratio of the mass of the crushed product to the weight of grinding bodies equal to 1:50. Ambient temperature influence on the measurement results and the deformation of materials were taken into account in accordance with the previously developed methods [11,12].

The absolute size of nanoparticles and crystal structure were determined by high-resolution TEM images at a magnification of 10^5–10^7 times obtained with an electron microscope JEM-2500SE (JEOL, Japan). The samples of Na_xMoO_3 bronze for TEM were prepared by manual machining of the powder. A sample was triturated in an agate mortar for 5 minutes, after which it stuck to the walls of the mortar in the form of a thin layer. A small amount was taken from the walls and deposited on the substrate by ultrasonic disperser. During grinding, the substance changed from dark blue (almost black) to blue (Na_xMoO_3). Preparation of the crystals for TEM included processing to obtain the desired shape and thickness, and etching. Electron microscopic images and digital electronograms obtained after processing were analyzed via "IMAGEG" program (FFT algorithm).

Cooked bronze powders were used for preparing aqueous dispersions of nanoparticles. For this purpose, an equal amount of powder was dissolved in water, purified by reverse osmosis. Using centrifugation and sedimentation of the prepared solution, it was carried out separation of aqueous dispersion in nanoparticle size. The separation process was monitored for optical control of nanoparticle size in aqueous dispersions. The average size of the nanoparticles was estimated using a scattering and absorption photometry of the solutions in the UV-Vis and near infrared regions. For this purpose, the spectrophotometers PE-5300B (EKROS, Russia) and U3600 (Shimadtsu, Japan) were used to obtain the absorbance spectra of solutions in the wavelength range from 300 nm to 1700 nm.

Monochromatic coefficients of laser radiation attenuation were measured for wavelengths λ = 1.45 μm and 1.56 μm in the concentration range from 10 mg/ml to 30 mg/ml of the dispersed phase in aqueous dispersions of tungsten oxide, molybdenum, and titanium bronzes. The samples of aqueous dispersions of the bronzes were confined in the volume between the two glass-plates of thin slide. The slide with bronze was irradiated by the near IR monochromatic light. We used the Er-fiber laser, λ = 1.56 μm (IRE-Polus, Russia), and a solid-state diode laser, λ = 1.45 μm (Lakhta Milon, Russia), as a light source. Measurement of optical transmittance of the colloidal solution of nanoparticles was performed by measuring the power of laser radiation passed through the solution. An optical path in the solution was about 0.5 mm. Power meter Field Master (Coherent, USA) was directly behind the slide onto which the laser beam was directed. In measurements, Fresnel reflection losses from the slide windows were taken into account. At normal incidence of a narrow beam of radiation on the cartilage tissue, the intensity of light I as it passes through the object varies according to the Bouguer-Lambert-Beer law:

$$I = I_0 \cdot e^{-\alpha l}, \qquad (1)$$

where I_0 is the intensity of the incident radiation, α is effective attenuation, l is the path traversed by the radiation.

Based on the results of measurements of power intensity of the laser radiation transmitted by (1), the reduced absorption coefficient was calculated (neglecting scattering due to small particle size). Note that the measurements of power of laser radiation transmitted through the slide with the probe were carried out to satisfy the condition of optically thin layer ($\alpha l \leq 1$).

A photothermal effect of laser radiation on the oxide bronze nanoparticle for different wavelengths in the UV-Vis-NIR-region of the spectrum was evaluated using two approaches: the classical electron theory of Drude-Lorentz and the quantum theory of solids. [13]. The energy diagram illustrating sub-threshold absorption in semiconductor oxides is shown in Figure 1.

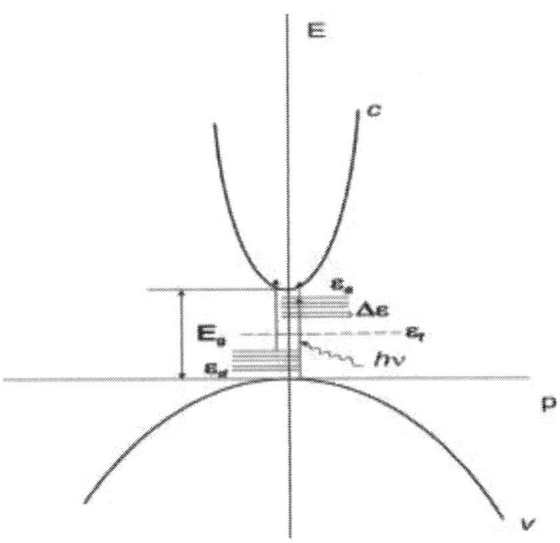

Figure 1. Energy diagram illustrating absorption in sub-threshold excitation of the oxide semiconductor defects (left); spherical nanocrystal surface defects (right).

The frequency of the plasmon resonance of nanoparticles in water is determined by Langmuir formula:

$$\nu^2 = \pi e^2 n / (m^* \varepsilon \varepsilon_0), \qquad (2)$$

where e is charge, n is density and m* is the effective mass of the conduction electrons. ε is relative permittivity of water, and ε_0 is dielectric constant of free space.

Nanoparticles with dimensions not exceeding 10–100 nm of complex oxides represented the metallic oxide bronzes of the transition metals: W, Mo, and Ti have a density of conduction electrons close to the metal: $n \sim 10^{23}$ cm^{-3}.

The effective mass is $m^* \approx m$, where m is the mass of free electrons $9.1 \cdot 10^{-31}$ kg. In this case, the frequency of the plasmon resonance of the nanoparticles is determined by the concentration of free charge carriers, i.e. $v \sim 1/\sqrt{n}$.

For nanoparticles with 10^3–10^5 atoms, it should be taken into account the quantum size effects [14], affecting the density of electron states in the band gap semiconducting oxide and oxide bronzes:

$$\Omega(\varepsilon_f) = 3N/2\varepsilon_f, \qquad (3)$$

where N is the number of states, ε_f is Fermi energy.

The average distance between the levels of the electrons according to [14]:

$$\Delta\varepsilon = 1/\Omega(\varepsilon_f) = 2\varepsilon_f/3N. \qquad (4)$$

3. Results and Discussions

Table 1 shows the measured values of the aqueous solutions of monochromatic absorption coefficient of potassium-hydrogen, and -molybdenum bronzes of concentrations varying from 10 mg/ml to 35 mg/ml. The extinction values are also calculated for these solutions.

We failed to obtain the stable colloids of titanium bronze for concentrations over 1 mg/ml in an aqueous medium. Nanoparticles of these bronzes stick together and settle to the bottom.

Table 1. The absorption coefficient (attenuation constant) of laser radiation for different wavelengths in aqueous solutions of potassium-hydrogen and -molybdenum bronzes of various concentrations.

Wave length, λ, nm	Concentration, C mg/ml	Tansmission, I_o/I_x		The absorption coefficient, α cm^{-1}		Extinction*, ml / (mg · cm)	
		K_xMoO_3	H_xMO_3	K_xMoO_3	H_xMO_3	K_xMoO_3	H_xMO_3
	30	540/90	475/60	35.8	39.6		
1560	20	480/145	500/130	23.9	25.4	1.2±0.02	1.3±0.015
	10	440/240	480/240	12.1	13.1		
	35	-	750/40	58.6	-		
1450	25	-	380/45	42.7	-	-	1.67±0.025

* Laser intensity (I) in mW/cm^2

Figure 2 shows the nanoparticle absorption spectra of aqueous dispersions of potassium- and hydrogen molybdenum oxide bronze in a wide wavelength range from UV-Vis to the near-IR. A different behavior of optical density (OD) is observed at long-wavelengths of absorption spectrum in the visible and the near-infrared region. For hydrogen molybdenum bronze, a visible intense absorption band is observed in the range from 700 nm to 900 nm. For potassium bronze, absorption decreases with increasing wavelength, it is asymptotically is closed for absorption of water.

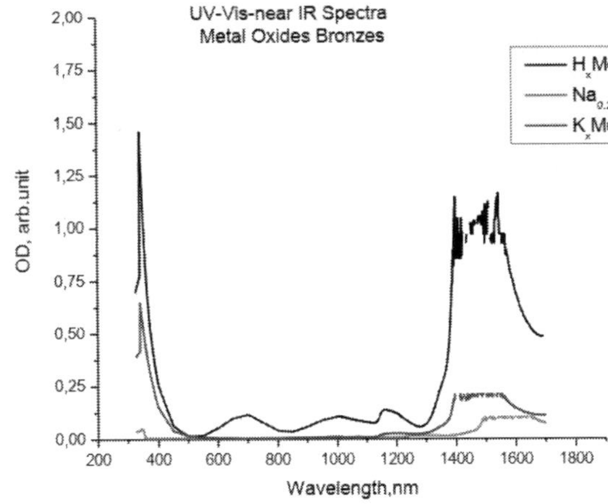

Figure 2. Absorption spectra of aqueous dispersions of nanoparticles $NaTiO_2$, and K, H molybdenum oxide bronzes.

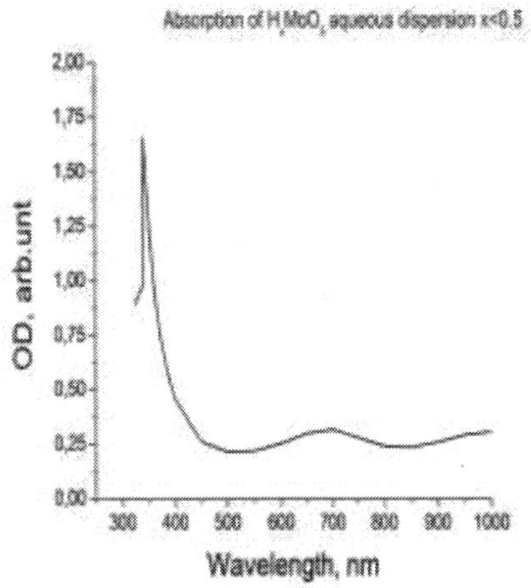

Figure 3. The absorption peak in the field of plasmon resonance observed for nanoparticles of hydrogen molybdenum oxide bronze

Figure 3 shows the absorption spectrum of an aqueous dispersion of nanoparticles of hydrogen molybdenum oxide bronze in the UV-Vis spectrum. There is a peak of the plasmon resonance at λ = 340 nm. Broad absorption bands are in the long-wavelength visible and near infrared regions of the spectrum.

Wide but less intense absorption bands in the infrared region are observed in aqueous solutions of tungsten bronzes K_xWO_3 [2]. In aqueous solutions of sodium and potassium titanium bronzes one can distinguish a plasmon resonance peak, but because of the low concentration of the colloids the marked absorption in the near-IR was not observed (Figure 2).

After machining of solid products of the synthesized Na and K titanium oxide bronzes the produced powders acquire magnetic properties. Thus, in the separation by sedimentation of the solutions with $Na_{0.2-0.25}$ TiO_2 nanoparticles it was found that the sediment at the bottom of the sedimentation column contains nanoparticles with magnetic properties. This was revealed by the attraction of the sediment particles to the permanent magnet with a magnetic flux density of ~0.2 T. To study the magnetic properties of oxide bronze nanoparticles, magnetic separation of the magnetic components from the solution of nanoparticles in the sedimentation was carried out [15]. We used a magnetic trap with axially symmetric alternating magnetic field with a constant gradient of 2 T/m and field amplitude of 0.05 T [15].

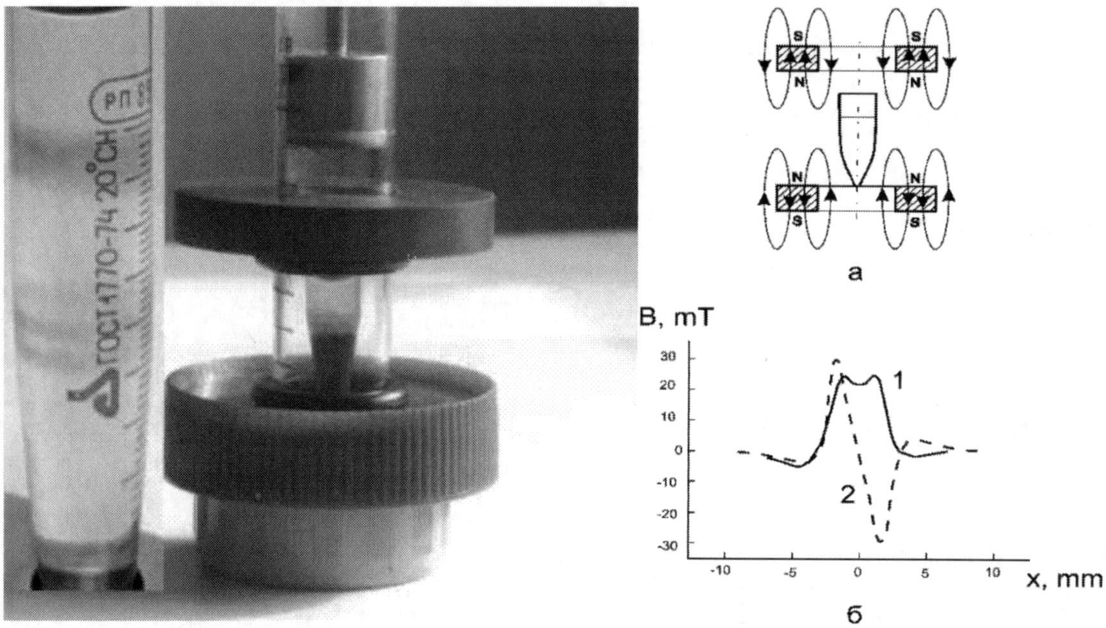

Figure 4. Magnetic nanoparticles (left) and a magnetic trap (right) used for the separation of magnetic nanoparticles: trap circuit (a) and the magnetic field distribution in the interaction of nanoparticles with a magnetic field (b) with unidirectional (1) and opposite (2) magnet switched on.

The magnetic component was separated from the colloidal solution of nanoparticles of the titanium oxide bronzes with the magnetic trap (Fig. 4), and its elemental analysis was conducted. The atomic absorption analysis of magnetic sludge showed that the spectrum has a line of alkali metals K and Na. The paramagnetic properties of nanoparticles sediment were studied by ESR. Hydrogen bronze of the metal oxides of Ti, W and Mo markedly differ from K and Na- titanium bronzes in magnetic properties. The magnetic field did not affect them. They were deposited on the bottom of the column by gravity.

The results of the study of paramagnetic properties of K_x TiO_2 powder bronzes obtained after grinding of SH-synthesis products revealed that paramagnetic powder is present in the signal samples corresponding to different times of grinding. Moreover, the signal amplitude increases with increasing grinding time (Figure 5).

ESR-spectra showed free iron absorption lines observed under magnetic field of 3400 gauss with g-factor closed to 2. The presence of magnetic nanoparticles in a grinded product is apparently due to the wear of the steel balls used as milling bodies and high hardness of crystals K_xTiO_2. In grinding softer W and Mo oxides, the magnetic component in bronze milling is absent.

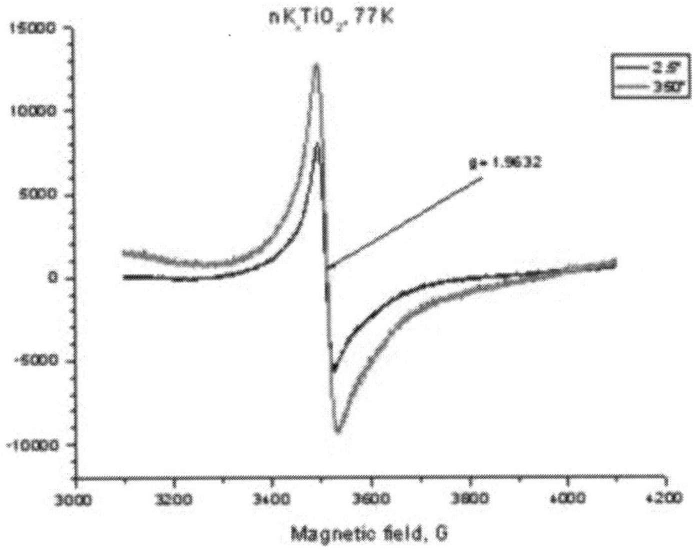

Figure 5. The ESR spectra of the K_xTiO_2 bronze powders obtained after SHS products grinding for different times of machining.

Structural studies a Na_xMoO_3 nanocrystals using a high-resolution TEM has shown the presence of high density of disturbed structures in the samples.

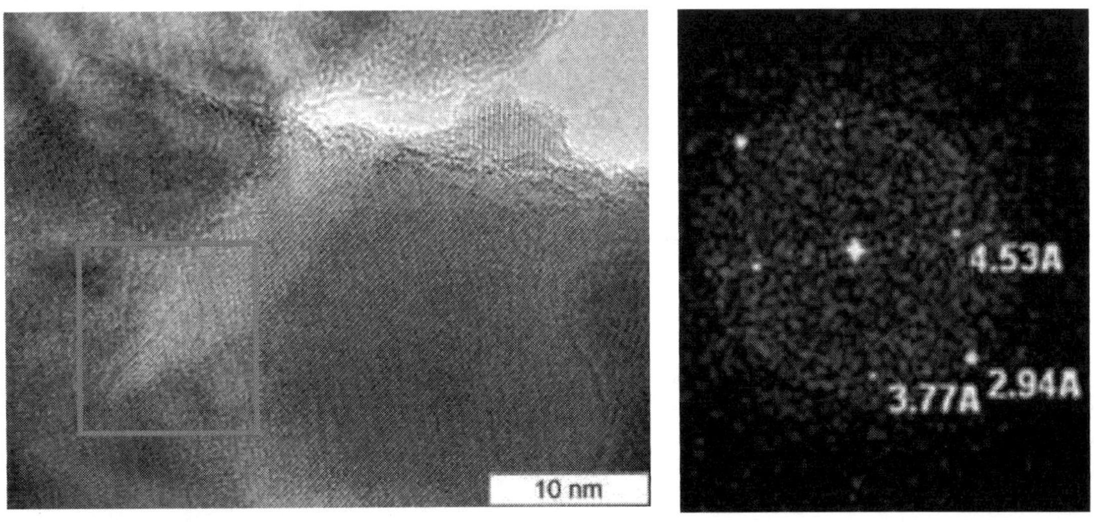

Figure 6. The images of nanocrystal (left) and electron diffraction (right) obtained for crystal structure of the Na_xMoO_3 -bronze with high resolution TEM.

The images in figure 6 present the structure of a highly defective bronze nano-crystal with varying degrees of layered single-crystal. The image of electron diffraction indicates the presence of the orthorhombic oxide bronze Na_xMoO_3 [16].The image on the left (Figure 6) indicates high density of the surface defects at the boundaries of the crystals. In the body of the crystal, there are sufficient ordering areas with layered organization, which occupy the range from 60 to 70% of the total sample. Regions with one-dimensional ordering are distinguished. These areas determine the properties of semiconductor crystallites, namely the properties of photonic crystals: the absorption peak of the plasmon resonance and broad bands in the near IR. The density of surface defects in the limit of a 10 nm x10 nm aria of the crystal, as indicated by square at the Fig.6, is ~ 10^{17}–10^{19} cm^{-2}. The high density of surface defects including non-stoichiometry defects, interstitial atoms and vacancies, as well as their number determines the intensity of the absorption bands, and the wavelengths of these bands are determined by the resonance properties of crystals caused by electron-phonon interaction. The broad absorption bands in the near infrared region of spectra (Figure 2) indicate a significant loss causing the photothermal effect.

Note that, the currently used metallic nanoparticles [17] (Au, Ag, etc.) have a lower photothermal effect of laser heating due to the narrow UV-Vis bands of plasmon absorption and weak electron-phonon interaction. As a result, to obtain photothermal effect of laser radiation on the nanoparticles based on plasmon resonance [17] it is needed to significantly increase the intensity of laser radiation (~10^5 W/cm^2 [17]). Using a shot pulse laser radiation, it is easy to reach high temperature T_e of electronic subsystem of nanocrystals. In this case, the equilibrium temperature of nanoparticle heating is not significant: $T \approx m_e c_e / mc \cdot T_e$, since $m_e << m$ and $c / c_e = 2$ [13], where m and m_e, c, c_e are the masses of the specific heat and the lattice and electron gas, respectively.

As shown in Figure 1, the average distance between electron levels $\Delta\varepsilon$, in accordance with (4), depends on the Fermi energy $\varepsilon_f \approx 1/2\ E_g$ and high density of electron states [14], which is characteristic of the defective semiconductor oxides and oxide bronzes. As a result, the energy of the quantum of infrared laser radiation ($h\nu$ ~ 0.5 eV) is sufficient to generate the high density of electronic states in the conduction band causing high absorption, in contrast to the defect-free semiconductors, where high energy of excitation of interband transitions is required.

In pure semiconductor crystals, higher energy of the photons ($h\nu > E_g$) is required for excitation of interband transitions of electrons to overcome the threshold photoabsorption. The impurity (defects) of semiconductor oxides and oxide bronzes, as shown, provides sub-threshold absorption mechanism of excitation.

4. Conclusions

When using laser heating of nanoparticles of metal oxides and oxide bronzes it is required a lower intensity laser radiation and photon energy ($h\nu$ ~ 0.5 eV) than that for metal nanoparticles based on plasmon resonance. Accounting a sub-threshold laser absorption character in the near infrared region of the spectrum allows us to explain abnormally high photothermal effect [1]. Photothermal effect of laser heating of cartilage impregnated with nanoparticles of hydrogen molybdenum oxide bronze [1] is more than 7.5 times greater than the heating effect of intact (without nanoparticles) cartilage [18].

Acknowledgments

The reported study was funded by the RFBR research project No. 15-42-00106.
Special thanks to prof. Radzig V.A. for useful discussion on ESR of the samples of metallic oxide bronzes.

References

[1] Gulyaev P, Kotvanova M, Pavlova S, Sobol E and Omelchenko A 2012 *Nanotechnologies in Russia* **7** 127
[2] Takeda H, Adachi K 2007 *J. Am. Ceram. Soc.* **90** 4059
[3] Tang J, Myer M, Bosnick K A and Brus L E 2003 *J. Phys. Chem.* **107** 7501
[4] Evstigneev V V, Guljaev P J, Miljukova I V, Goncharov V D, Vagner V A and Gladkih A A 2006 *Int. J. Self-Propag. High-Temp. Synth.* **15** 99
[5] Gusev A I, Kurlov A S 2008 Nanotechnology **19** 265302
[6] Omelchenko A I, Sobol E N, Simakin A V, Serkov A A, Sukhov I A and Shafeev G A 2015 *Laser Physics* **25** 025607
[7] Lynch D W, Rosei R, Weaver J H and Olson C J 1978 *J. Solid State Chem.* **8** 242
[8] Green M, Travlos A 1985 *Phil.Magn.* **51** 501
[9] Dickens P, Quilian R and Whittigham M 1968 *Mat. Res. Bull.* **3** 941

[10] Salje E, Guttler B 1984 *Phil.Magn.* **50** 607
[11] Song X J, Cui H Z,·Cao L L, Gulyaev·P Y 2016 *Transactions of Nonferrous Metals Society of China* **26** 1878
[12] Boronenko M P, Gulyaev P Yu, Seregin A E and Bebiya A G 2015 *IOP Conf. Ser.: Mater. Sci. Eng.* **93** 012021
[13] Davies D A 1978 *Waves, atoms and solids* (London - New York: Longman group Ltd) p 239
[14] Girifalco L A 1973 *Statistical Physics of Materials* (New York: John Wiley& Sons Ltd) p 362
[15] Omelchenko A I, Sobol E N 2010 *Int. Symposium on Laser Medical Applications, devoted to the 50th Anniversary of Laser Discovery* (Moscow July 5-6) 55
[16] Greenblatt M 1989 Transition metal oxide bronzes with quasi low-dimensional properties. **11** Ed. Schlenker C. Low-dimensional electronic properties of Molybdenum bronzes and oxides (Dordrecht Boston London: Kluwer Academic Publisher) p 450
[17] Govorov A, Zhang W, Skeini T, Richardson H, Lee J and Kotov N 2006 *Nanoscale Res. Lett.* **1** 84
[18] Youn J In and Milner T E 2008 *Laser Med. Sci.* **23** 229

IOP Conf. Series: Journal of Physics: Conf. Series **830** (2017) 012154 doi:10.1088/1742-6596/830/1/012154

UV excilamp inactivation of helminth eggs in wastewater

E I Lipatov, E A Sosnin, S M Avdeev and V F Tarasenko

Institute of High Current Electronics SB RAS, 2/3, Akademicheskii Ave., Tomsk, 634055, Russian Federation

E-mail: lipatov@loi.hcei.tsc.ru

Abstract. The inactivation of *Opisthorchis felineus* eggs in the wastewater was demonstrated. The wastewater samples were taken at the bio-filter outlet of district wastewater treatment plant of settlement "Airport" of rural settlement "Mirnenskoe" of Tomsk district of Tomsk region. The UV irradiation of wastewater samples was performed by the combined exposure of UV excilamps at 282 and 222 nm. There was less than 15% of the initial count of *Opisthorchis felineus* eggs in the wastewater after the UV treatment at the total surface dose of 25 mJ/cm^2. At the same time, 85% of the eggs lost the shell integrity and destroyed. It is proposed to use UV irradiation by excilamps to wastewater deworming on wastewater treatment plants of small capacity up to 200 m^3/day.

1. Introduction

District wastewater treatment plants designed and built in Soviet times to clean the drains of residential areas, settlements and neighborhoods located outside the city limits and having a population of 10 thousand people, as a rule, do not perform their design functions. The budgets of rural settlements and districts cannot cope with the timely reconstruction of treatment plants. Across the country, there are thousands of district wastewater treatment plants, which are pollution sources of water bodies by pathogens and helminthes eggs.

The main reason for poor performance of wastewater treatment plants is the presence in the drains of high concentrations of detergents and petroleum products that disrupt the estimated job of bio-filters. The solution may be the ultraviolet (UV) disinfection of wastewater at the outlet of the bio-filters.

The use of mercury UV lamps for wastewater purification may be dangerous to the environment in the event of bulb failure and requires special measures for the disposal of waste emitters. In addition, radiation in atomic line at 254 nm of mercury vapors (see. Figure 1) ineffectively destroying helminthes egg shell.

Alternative sources of spontaneous UV radiation are barrier discharge excilamps (see Figure 2), which emit in the molecular bands with maxima at wavelengths of 222 and 282 nm (see Figure 1). They do not contain mercury, effectively destroy the shell of helminthes eggs (KrCl, 222 nm), such as *Opisthorchis felineus* and *Diphyllobothrium latum* [1], provide disinfecting wastewater from thermotolerant coliform bacteria (XeBr, 282 nm), including bacteria of the *Proteus* genus [2].

The use of excilamps as light sources in the UV disinfection units at the output of bio-filters of district wastewater treatment plants producing the water release in the rivers of Ob basin can provide disinfection of wastewater from eggs of *O. felineus* and *D. latum*. In the long term of 20–30 years, it could reduce the intensity of the natural epidemiology centers of such diseases as opisthorchiasis and diphyllobothriasis.

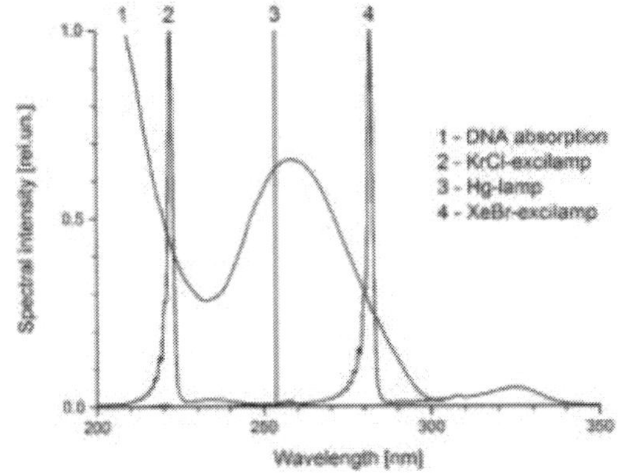

Figure 1. Absorption and emission spectra: the total absorption spectrum of DNA (1), the emission spectrum of barrier discharge KrCl-excilamp with a maximum at 222 nm (2); the atomic line of a mercury lamp at 253.7 nm (3); the emission spectrum of barrier discharge XeBr-excilamp with a maximum at 282 nm (4).

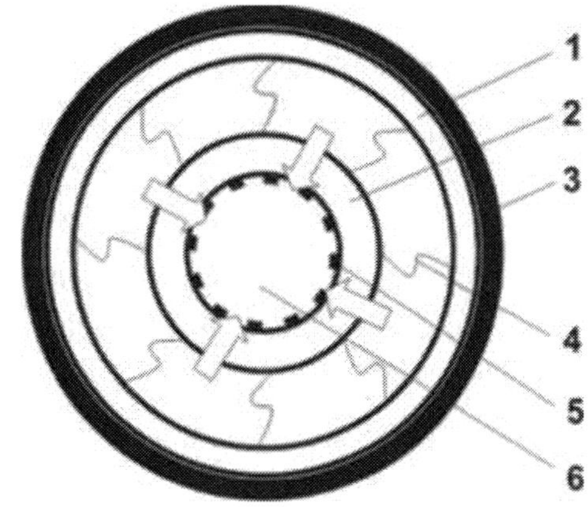

Figure 2. The cross section of the coaxial barrier discharge excilamp: quartz tubes, forming the outer and inner walls of excilamp (1, 2); the external reflective electrode (3); the discharge zone between quartz tubes (4); the inner translucent electrode (5); the inner irradiation cavity (6). White arrows indicate the concentrated UV radiation direction.

We report on the use of UV radiation of KrCl(222 nm)- and XeBr(282 nm)-excilamps for wastewater treatment in the laboratory. The water samples were taken at the bio-filter outlet of district wastewater treatment plant of settlement "Airport" of rural settlement "Mirnenskoe" of Tomsk district of Tomsk region. At the surface dose of 25 mJ/cm^2, it was achieved reduction in 6 times the count of viable eggs of *O. felineus*. Now the work is underway on the design and pre-production of prototype of UV wastewater disinfection unit based on KrCl- and XeBr-excilamps.

2. The action of UV excilamps on eggs of *O. felineus* and *D.latum*

Previously it has been shown [1] that the inactivation of *O. felineus* eggs by KrCl-excilamp radiation is more effective than the XeBr-excilamp radiation. The average power of KrCl-excilamp radiation was 2 times less than that of XeBr-excilamp. However, after irradiation by KrCl-excilamp, there were the viable eggs of *O. felineus* (survived after irradiation) less at 40–70% than in the case of XeBr-excilamp irradiation (see Figure 3). The spectral maximum of KrCl-excilamp band corresponds to photon energy of 5.6 eV, which is much higher than the photon energy of XeBr-excilamp (4.4 eV). We assume that the photons with higher energy destroy or embrittle efficiently the shell eggs of *O. felineus*. Under the conditions of experiments, the proportion of *O. felineus* eggs, surviving after irradiation at 222 nm, was 15–30% at surface radiation doses of 5.1–0.3 mJ/cm^2. Upon irradiation at 282 nm, the proportion of surviving *O. felineus* eggs was 70–90% at surface radiation doses of 116–4 mJ/cm^2.

For one of water samples containing the *O. felineus* eggs, after exposure to radiation at 222 nm at the surface dose of 8.5 mJ/cm^2, there was complete absence of eggs in the sample. The linear approximation (excluding zero experimental value at 8.5 mJ/cm^2) allows us to estimate the surface dose of 15.6 mJ/cm^2 required for complete inactivation of *O. felineus* eggs by radiation at 222 nm.

Irradiation of water samples containing *D. latum* eggs showed less effective disinfection by radiation at 222 nm than for samples containing *O. felineus* eggs. Under conditions of the experiments, the proportion of *D. latum* eggs surviving after irradiation at 222 nm was 44–68% at the surface radiation doses of 5.1–0.5 mJ/cm^2.

Differences in the inactivation levels of eggs of *O. felineus* and *D. latum*, achieved at 222 nm irradiation, may be associated with greater resistance of protective shells of *D. latum* eggs to incident photons.

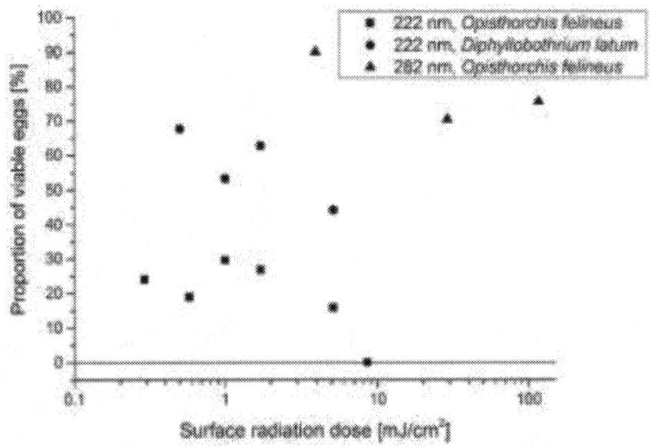

Figure3. Dependence of the proportion of eggs count after exposure to the eggs count before exposure from the surface dose of radiation of KrCl- and XeBr-excilamps.

3. The action of UV excilamp radiation on the wastewater from treatment plant

There were performed the comparative studies of the inactivation of microorganisms in the wastewater using the radiation of XeBr- and KrCl-excilamps [2]. It was shown that the simultaneously irradiation the wastewater by XeBr- and KrCl-excilamps had the greatest effect. The studies were conducted on the experimental setup, whose photo is shown in Figure 4.

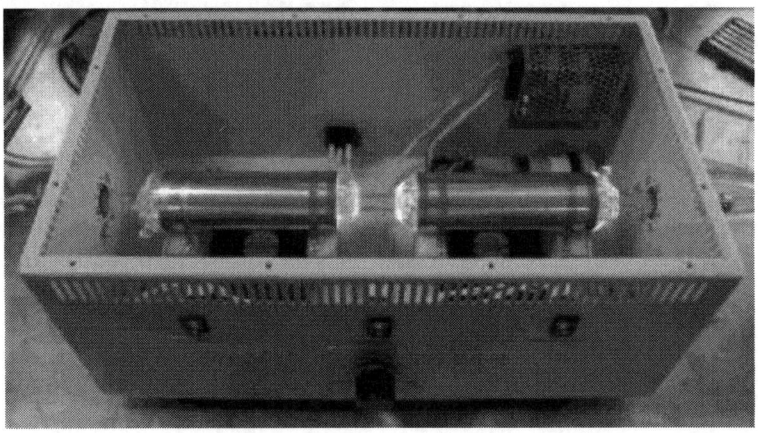

Figure 4. The photo of experimental setup.

The study of action of XeBr- and KrCl-excilamp irradiation on wastewater for the purpose of disinfection from *O. felineus* eggs was carried out on the same experimental setup. The setup provided the circulation of wastewater in the system: the original container, a quartz tube passing inside of coaxial KrCl- and XeBr-excilamps of barrier discharge, receiving container. The pumping rate through the system was 1.5 liter/min.

The wastewater for research was taken at the bio-filter outlet of district wastewater treatment plant in settlement "Airport" of rural settlement "Mirnenskoe" of Tomsk district of Tomsk region. For the studies there were taken three wastewater samples of 15 liters each. One of them was used for control. The wastewater from the two samples was subjected to simultaneous treatment by KrCl- and XeBr-excilamps. The total surface dose of irradiation at two wavelengths (222 and 282 nm) was about of 250 mJ/cm^2.

The treated water is tested for compliance with SanPiN 2.1.5.980-00 "Hygienic requirements for surface water" [3] using the method of investigation MUK 4.2.1884-04 "Sanitary-microbiological and sanitary-parasitological analysis of water from surface water bodies" [4] in the bacteriological laboratory of Center of Hygiene and Epidemiology in the Tomsk region.

There were found eggs of *O. felineus* in the wastewater samples. The UV treatment results for wastewater samples are shown in table 1.

Table 1. Counts of *O. felineus* eggs in samples #1 and #2 and proportions of count of surviving eggs after irradiation to count of eggs in the control sample

Sample	Count	Proportion (%)
Control	59	-
# 1	9	15
#2	5	8

Thus, a moderate surface dose of UV radiation by KrCl- and XeBr-excilamps (about 25 mJ/cm^2) can provide reduction of counts of *O. felineus* eggs in the wastewater to 6 or more times.

4. Conclusion

These encouraging results have allowed us to start the design and manufacture the prototype of industrial UV wastewater treatment setup for district wastewater treatment plants with productivity up to 200 m^3/day. Testing and endurance testing of this setup will be carried out at district wastewater treatment plant at settlement “Airport”, located in the rural settlement “Mirnenskoe” of Tomsk district of Tomsk region [5, 6]. In case of successful tests, it is supposed to patent and certificate the setup to achieve opportunities of industrial production of UV wastewater disinfection units.

Acknowledgments

The authors would like to appreciate the Administration of the Institute of High Current Electronics SB RAS, especially the Director of the institute N. A. Ratakhin and the Principal Director of Scientific Research I. Yu. Turchanovskiy for the organizing support of the work.

References

[1] Lipatov E I, Sosnin E A and Avdeev S M 2016 *Russian Journal of Parasitology* **35** 107

[2] Sosnin E A, Lipatov E I, Skakun V S, Panarin V S, Tarasenko V F, Zhdanova O S and Gol’tsova P A (in Russian) 2016 Modern scientific researches and innovations http://web.snauka.ru/issues/2016/03/65016

[3] SanPiN 2.1.5.980-00 2000 Hygienic requirements for surface water protection (Moscow: Federal Centre of Sanitary Inspection Ministry of Health of Russia) p 10

[4] MUK 4.2.1884-04 03.03.2004 Methodical instructions "Sanitary-microbiological and sanitary-parasitological analysis of water from surface water bodies" (app. Chief State Sanitary Doctor of the Russian Federation)

[5] Agreement on scientific and technical cooperation between IHCE SB RAS and Administration of rural settlement “Mirnenskoe” of Tomsk district from 01.12.2015 http://www.hcei.tsc.ru/ru/cat/documents/2016/2016_11_18.pdf

[6] Agreement on scientific and technical cooperation between Administration of Tomsk district and IHCE SB RAS from 19.05.2016 http://www.hcei.tsc.ru/ru/cat/documents/2016/2016_11_18_.pdf

About laser heat absorbing impurities in the transparence matrix of pentaerythritol tetranitrate

G Murastov, V Tsipilev, V Ovchinnikov and A Yakovlev

National Research Tomsk Polytechnic University, 30 Lenin Ave., 634050, Tomsk, Russia

E–mail: murastov.gennadiy@gmail.com

Abstract. Simulation of laser heat absorbing inclusions in a transparent matrix of explosive material is carried out. The inclusions are spherical particles with R_0 radius ranging from 10^{-5} to 10^{-2} cm. The duration of the laser pulse varies from 10^{-9} to 10^{-3} s and allows investigating process in both adiabatic and quasi–stationary heating modes. It has been found that the duration of laser pulse affects the maximum heating temperature for particles with a specific radius. A remarkable thing is that the temperature drastically decreases if the absorption cross section and small size of inclusions are taken into account. It leads to heat reduction in a thermal center, which is formed nearby a small–sized inclusion, and shows strongly decreased reactive capacity of explosive decomposition.

1. Introduction

Heat absorbing irregularities in a transparent matrix of explosive material (EM) have been studied previously to solve the problem of ignition of lead azide with different laser pulse duration [1,2]. It was assumed that lead particles are present in the matrix of explosives and laser pulse energy is stored in the particle volume. The critical heat storage in the center and temperature in its vicinity were calculated numerically in a wide range of laser pulse duration in terms of the thermal microcenter model of initiation from the adiabatic mode to the quasi-stationary mode.

The process of heating lead azide particles in the matrix was analyzed. Formation of heat centers ("hot spots") in unreactive medium was addressed in [3]. The features of heating inclusions with different size were shown. The maximum achievable values for surface temperatures of inclusions of different size were calculated numerically and analytically. The temperature profiles of heating in the vicinity of the particles were evaluated for a range of "short" and "long" laser pulses. And these features must be considered in both simulation of laser initiation and interpretation of the numerical calculation results.

Experimental works on laser initiation of PETN are considered to be of great interest nowadays [4–10]. A number of works report on PETN containing no inclusions [4–7] and provide experimental results for PETN containing inclusions of nano-sized carbon particles [9,11] or different metal particles[8,10]. It has been shown that nano-sized carbon additives and aluminum particles lead to 5–6 fold decrease in the initiation threshold. It seems to be interesting that the decrease for carbon impurities and aluminum impurities is similar despite strong differences in their optical performance and thermal performance, as well in chemical activity.

In-depth insight in the nature of laser initiation of composites based on PETN requires the study of the heating process of impurities of different nature exposed to laser beam. It is assumed that the matter is an unreactive medium. This question is the subject of this report.

2. Numerical simulation

Numerical simulation of laser heating of absorbing particle is conducted by the method described in [1,2], and it is assumed that the absorption cross section of the particles in the matrix of explosives does not correspond to geometric one. Illumination on the surface in numerical simulation is different from that produced on the sample surface by laser beam. Following [2], the temperature ΔT:

$$\Delta T \sim k \cdot F \cdot H_{\Pi}, \tag{1}$$

where $k = \dfrac{\sigma}{\pi \cdot R_0^2}$; $\sigma(R_0, \lambda_0, n_0, n_k)$ is absorption cross section [12] of the particle depending on radius R_0, laser wavelength λ_0, n_0 is refractive index of the medium and n_k is complex refractive index of the absorbing particles; $F = \dfrac{H_0}{H_{\Pi}}$; H_{π} is the energy density on the EM surface; $H_0(d_{\Pi}, \mu, \beta, \chi)$ is the energy density in the vicinity of the particles; for the general case, it depends on the size d_p of the laser beam on the sample surface, μ is absorption parameter and β scattering parameter, χ is scattering indicatrix of elementary volume.

Thus, k is the effective absorption coefficient of the particles. The relation between the spatial illuminance within the medium volume and the illuminance of the incident laser flux on the surface are expressed by F. Having normalized (1) to $k \cdot F$ and $H_{\Pi} = q_{\Pi} \cdot \tau$, we obtain:

$$\frac{\Delta T}{k \cdot F} \sim H_{\Pi} = \sigma \cdot \tau, \tag{2}$$

where q_{π} is laser flux density (W/cm^2).

The heating model takes the form from [2] in case of $\alpha_1 >> \alpha_2$

$$\begin{cases} \dfrac{\partial T_1}{\partial t} = \dfrac{3 \cdot \alpha_2 \cdot c_2 \rho_2}{R_0 \cdot c_1 \cdot \rho_1} \cdot \dfrac{\partial T_2}{\partial r} + \dfrac{3 \cdot q_{\Pi}}{4 \cdot R_0 \cdot c_1 \cdot \rho_1} \\ \dfrac{\partial T_2}{\partial t} = \alpha_2 \cdot \dfrac{1}{r^2} \dfrac{\partial}{\partial r}\left(r^2 \dfrac{\partial T_2}{\partial r} \right) \end{cases} \tag{3}$$

where T_1 is the temperature of the particle; T_2 is the EM temperature; c_1 is the specific heat capacity of the particle; ρ_1 is the density of one particle; c_2 is s the specific heat capacity of explosive; ρ_2 is the density of explosive; r is the coordinate of the heating point in a spherical coordinate system; R_0 is the radius of the particle; α_1, α_2 are thermal conductivity of particles and explosive material, respectively; q_{Π} is the laser flux density.

The solution of (3) was performed for different duration of the heating pulse and constant H_{Π} value equal to 100 mJ/cm^2, which is close to the experimental value obtained in [2]. The threshold power density was measured for a pure PETN sample (no impurities) with laser pulse duration of 10 ns. The pulse was Π–shaped in all the studied range of pulse widths. It is the best to suit the structure of long laser pulses in the region where the pulse waveform affects the result of heating [1,2]. The particle taken was spherical.

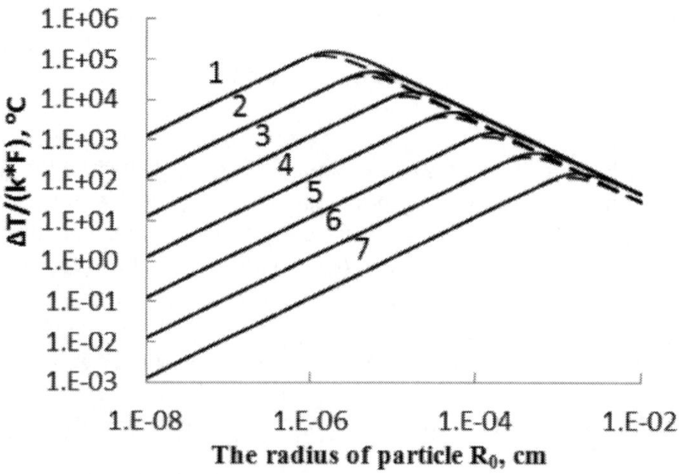

Figure 1. Dependence of the specified heating temperature of spherical carbon particle (solid curves) and aluminum particle (dotted line) on radius R_0. The energy density on the surface of the compacted PETN powder $H_п$=100mJ/cm^2. The duration of the laser pulse $\tau=10^{-9}$s(1); 10^{-8}s (2); 10^{-7}s (3); 10^{-6}s (4); 10^{-5}s (5); 10^{-4}s (6); 10^{-3}s (7).

The following constants were used for calculation c_1=760 J/kg·K; ρ_1=2250 kg/m^3; $\alpha_1=9.6\cdot10^{-7}$ m^2/s; c_2=1672 J/kg·K; ρ_2=1700 kg/m^3; $\alpha_2=6.67\cdot10^{-8}$ m^2/s.

The results of numerical calculations of ΔT particle heating for a wide range of laser pulses duration and sizes of carbon particles and aluminum particles are shown in figure 1. The set of curves on the scale $\frac{\Delta T}{k\cdot F}$ is of equal shape for carbon particles (solid curves) and for aluminum particles (dashed lines). The general patterns for particle heating are observed to be independent of the material used. First of all, each of the pulse durations corresponds to the particle size which exhibits best heating. Thus, the maximum heating point for carbon particles with R_0~100 nm, τ=1 μs – R_0~500 nm and τ=1 ms – R_0~20 μm in case of laser pulse τ=1ns. For aluminum particles, the pattern is similar, but the radius of the particles, wherein the maximum heating is realized, is somewhat less than that of carbon particles.

In the range of small particle size, the temperature of impurity increases linearly with increasing size R_0. This behavior corresponds to quasi-stationary heating ($\frac{\Delta T}{k\cdot F}\sim R_0$). The region becomes wider with increasing τ (figure 1, the left branch of the curves). At high values ($R_0>10^{-5}$cm), adiabatic mode is realized: $\frac{\Delta T}{k\cdot F}\sim\frac{1}{R_0}$, ΔT decreases with increasing R_0 (figure 1, the right branch of the curves).

The exact expressions for the heating temperature in asymptotics of these modes are shown in [3]. The formula for adiabatic heating mode is:

$$\frac{\Delta T}{k\cdot F}=\frac{3\cdot q_{\Pi}\cdot\tau}{4\cdot R_0\cdot c_1\cdot\rho_1}=\frac{3\cdot H_{\Pi}}{4\cdot c_1\rho_1\cdot R_0}, \quad (4)$$

for stationary heating mode, it is:

$$\frac{\Delta T}{k\cdot F}=\frac{q_{\Pi}\cdot R_0}{4\cdot\lambda_T}, \quad (5)$$

IOP Conf. Series: Journal of Physics: Conf. Series **830** (2017) 012155 doi:10.1088/1742-6596/830/1/012155

where λ_T is the thermal conductivity of the matrix.

It should be noted that expressions (4) and (5) can be used to check the accuracy of numerical calculations.

Thus, ΔT peak position for specific pulse duration τ conditionally divides the fields of stationary and adiabatic heating modes.

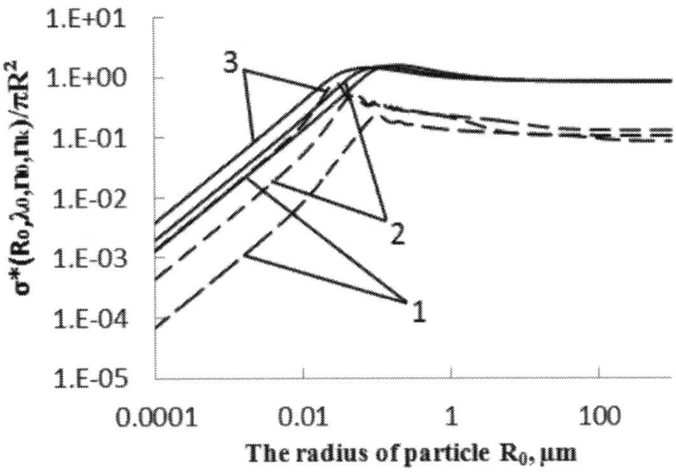

Figure 2. Dependence of the relative absorption cross section of carbon particles (solid curves) and aluminum particles (dashed lines) in the PETN matrix on radius R_0. The wavelength of laser is 1064 nm (1); 532 nm (2); 354.7 nm (3).

The issue of theoretical and practical interest is the heating curve taking into account the coefficients k and F. To determine the impact of the coefficient k, we calculated values using the expressions from Mie theory for aluminum and carbon particles in the range of R_0 sizes (wavelengths of laser radiation are 1064 nm, 532 nm and 354.7 nm that correspond to 1st, 2nd and 3rd harmonics of neodymium laser radiation and correspond to the PETN matrix transparency region[13]). The values of the complex refractive indices of carbon and aluminum particles were taken from [14,15] and accounted for the first harmonic 2.49–0.015i/1.028–9.25i; for the second harmonic 2.558–0.025i/0.727–5.663i and for the third harmonic 2.684–0.091i/0.291–3.717i, respectively. The calculation results are shown in figure 2. It can be seen that for both types of inclusions the relative absorption cross section decreases sharply with decreasing R_0, since $R_0 \sim 10^{-5}$ cm. For $R_0 > 10^{-5}$ cm (100 nm), the absorption cross section weakly depends on R_0, and for specified R_0, the k value may be greater than one. For each of the wavelengths, the maximum value k is generated (the first Mi extremum), and its position is shifted towards its reduction. For carbon particles, the maximum k attains the values R_0=200 nm, R_0=160 nm, R_0=70 nm for the 1st, 2nd and 3rd harmonics, respectively; for aluminum particles, the position of the maximum corresponds to the values R_0=100 nm, R_0=60 nm and R_0=40 nm. It can be assumed that the absorption cross section of the particles can have a significant influence on heating especially that of small-sized particles.

Figure 3 shows the results of calculations for the temperature of heating the particle surface with the values $k(R_0, \lambda_0, n_0, n_k)$ and $F(d_{II}, \lambda_0, n_0, P_{np}, \chi, \mu, \beta)$. The value of F≈10 was calculated for the case of wide laser beam using methods provided in [8]. It weakly depends on the considered wavelengths. The analysis of figure 3 shows that the basic patterns associated with maximum heating of a specified particle size for a given pulse duration remain unchanged. For carbon particles for τ=10 ns, the maximum temperature is observed in particles with R_0=100 nm, R_0=80 nm and R_0=50 nm upon irradiation with 1st, 2nd and 3rd harmonics, respectively. For the maximum heating point of aluminum particles, the size is R_0=80 nm, R_0=50 nm and R_0=65 nm, respectively. In the region of large values of

R_0, an adiabatic heating mode ($\Delta T \sim 1/R_0$) prevails; however at low values of R_0 ($R_0<100$ nm) accounting the absorption cross section the curve shape sharply changes so that in this region $\Delta T \sim R_0^2$. Thus, at constant energy of the laser pulse of different duration, we observe complex dependence of the heating temperature for absorption centers of different nature and, hence, the heat stored in the hot spot, on the size of these centers.

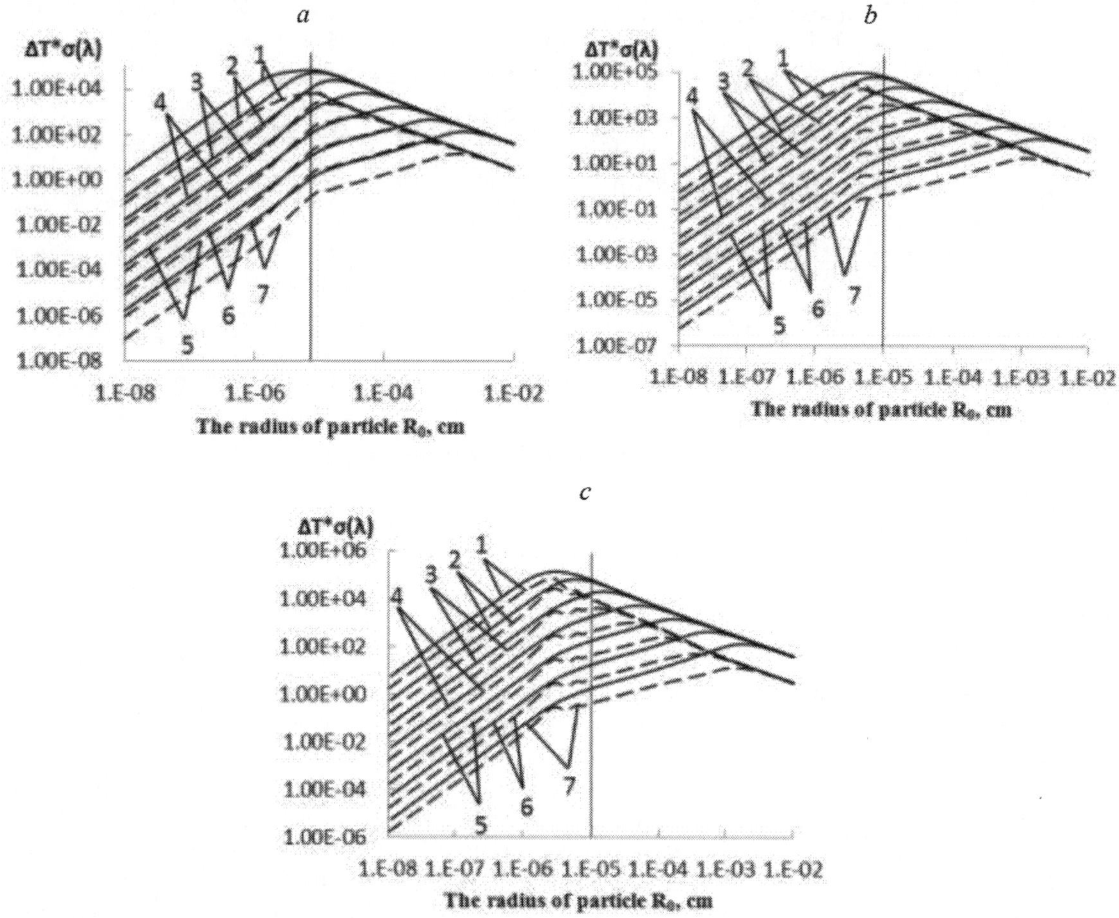

Figure 3. The dependence of the heating temperature on R_0 taking into account k (R_0) at λ_0=1064 nm for carbon particles (solid curves) and aluminum particles (dotted line) in the PETN matrix. *b* is the dependence for λ_0=532 nm; *c* the dependence λ_0=354.7 nm. The laser pulse duration τ: 10^{-9}s (1); 10^{-8}s (2); 10^{-7}s (3); 10^{-6}s (4); 10^{-5}s (5); 10^{-4}s (6); 10^{-3}s (7)

The heat storage Q in carbon particles and aluminum particles in the centers and in their neighborhood ($Q = k \cdot F \cdot H \cdot \pi \cdot R_0^2$) is shown in figure 4 for a wavelength of 1064 nm. Heat storage in the center does not depend on the heating mode, and is proportional to R_0^2; however for small R_0 ($R_0<10^{-5}$ cm), relative absorption cross section $k(R_0) \sim R_0$, which changes the dependence $Q(R_0)$, i.e. $Q(R_0) \sim R_0^3$, and it sharply decreases with R_0 reduced. This poses an ambiguous question about the reactivity of the centers. According to the adiabatic heating mode, as R_0 decreases, reactivity increases and then decreases sharply when $R_0<10^{-5}$ cm. In this mode, the criterion for initiation from data in [2] is to achieve a certain critical value of heat storage in the center. This value decreases monotonously with R_0 reducing. However, at first with reducing R_0, the reactivity increases but then at $R_0<10^{-5}$ cm, it sharply decreases [1].

In terms of the quasi-stationary mode, the heating initiation criterion is a specified critical temperature achieved in the center [2]. At the same time, the temperature of the center increases with R_0 growing (figure 1), the heating zone h ($h \approx R_0$) grows, and accordingly, their reactivity increases, which is confirmed by numerical calculation data [3] on ignition of lead azide. Note that in this mode, the heat storage in the absorbing particle is different and proportional to $k \cdot F \cdot H \cdot R_0^4$

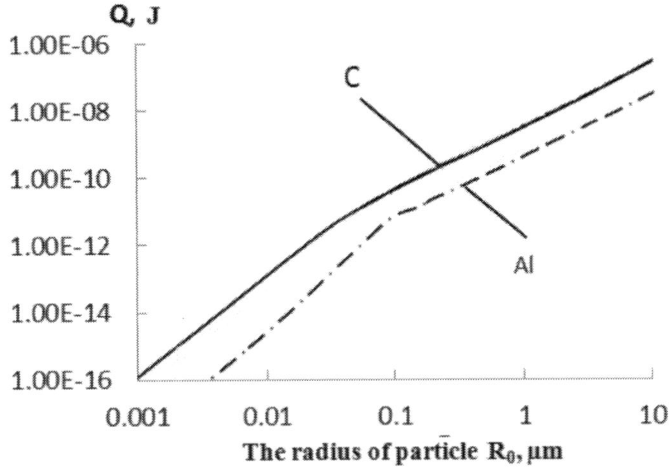

Figure 4. The heat storage near the aluminum particle (dotted line) and carbon particle (solid curve) depending on radius R_0 at energy density of the laser pulse up to 100 mJ/cm^2 for the wavelength λ_0=1064 nm.

3. Conclusion

The analysis of the results revealed a number of patterns of laser heating:

1. Micro- and nanoparticles of the carbon and aluminum impurity can be heated up to the values of tens of thousands degrees under laser pulse irradiation of mixed compositions based on PETN powders at near threshold levels of exposure (~100 mJ/cm^2). At the same time, the values of carbon particle heating can exceed those of aluminum heating by one order of magnitude.
2. For the impurity particles of different nature (carbon, aluminum) and for specified pulse duration, there is a certain size at which these are heated most. For experiments with laser pulse duration of about 10 ns, the maximum temperature for carbon particles is achieved for particles with size R_0=100, R_0=80 and R_0=50 nm at an irradiation wavelength of 1064, 532 and 354.7 nm, respectively. For aluminum particles, the maximum heating point is observed for R_0=80, R_0=50 and R_0=25 nm, respectively.
3. In case of R_0<100 nm, a quasi-stationary particle heating mode at $\Delta T \sim R_0$ can be observed in the entire investigated range of laser pulse durations. In case of R_0>1 μm, an adiabatic heating mode is generally observed, where $\Delta T \sim 1/R_0$. These characteristics determine the amount of heat in the microcenter formed in the vicinity of inclusions of different radius R_0, and their further development in the centers of explosive decomposition.

These patterns and features must be considered in theoretical and experimental studies of the sensitivity of mixed composition, as well as explosives containing intrinsic optical inhomogeneities, to the pulsed laser radiation.

It should be noted that the calculations were carried out through solid-phase approach without taking into account melting and evaporation of explosive matrix and impurity particles, which simplifies the calculation and analysis of the results but leads to overestimated values of the heating temperature. Furthermore, for the smallest values R_0<10^{-7}, the description of absorption and scattering processes may differ from that made under the Mie theory. However, not quantitative but qualitative patterns of the heating process are the issues of most interest in this paper.

Note that taking into account the chemical reaction in the vicinity of inclusions can provide a more accurate pattern of center heating and, in particular, specify the data on heat store in the centers and their reactivity. Further studies of the issue will be performed.

Acknowledgments
This work was financially supported by the Russian Foundation for Basic Research Grant 15-03-05385.

References

[1] Aleksandrov E, Sidonskii O and Tsipilev V 1991 Influence of combustion in the vicinity of absorbing inclusions on the laser ignition of a condensed medium *Combustion, Explosion, and Shock Waves* **27** pp 267–272
[2] Aleksandrov E and Tsipilev V 1984 Effect of the pulse length on the sensitivity of lead azide to laser radiation *Combustion, Explosion, and Shock Waves* **20** pp 690–694
[3] Morozova E, Tsipilev V and Burkina R 2011 Initiation of reactive substances radiation flux at its absorption optical inhomogeneities of matter *Combustion, Explosion, and Shock Waves* **47** pp 95–105
[4] Tsipilev V 2001 Laser initiation of PETN *Physico–chemical processes in inorganic materials* (Kemerovo) pp 113–114
[5] Zinchenko A and Tarzhanov V 1996 Laser initiation of PETN *Combustion, Explosion, and Shock Waves* **32** pp 113–119
[6] Karabanov Y and Bobolev V Burning of solid secondary explosives short pulse lasers *Combustion of Condensed Matter* (Chernogolovka) pp 5–8
[7] Skripin A, Morozova E and Tsipilev V 2010 Laser initiation of PETN powder in a volume compression *Bulletin of the Tomsk Polytechnic University.Energetic* (TPU, Tomsk) **317** pp 149–155
[8] Aleksandrov E, Voznyuk A and Tsipilev V 1989 Effect of absorbing impurities on explosive initiation by laser light *Combustion, Explosion, and Shock Waves* **25** pp 1–7
[9] Aduev B, Nurmukhametov D, Furega R, Zvekov A and Kalensky A 2013 The explosive expansion of PETN with nanoadditives aluminum when exposed to pulsed laser radiation of different wavelengths *Chemical Physics* **32** pp 39–44
[10] Aduev B, Nurmukhametov D, Furega R and Zvekov A 2014 Regulation of PETN sensitivity to laser exposure via nanoparticle additives metals nickel and aluminum *Chemical Physic* **33** pp 37–41
[11] Ovchinnikov V, Skripin A, Tsipilev V and Yakovlev A 2012 Dependence of explosion initiation threshold of PETN with absorptive additives on uniforn compression pressure of the sample *Russian Physics Journal* **55** pp 217–219
[12] Kriger V G, Kalensky A V, Zvekov A A, Zykov I Yu and Nikitin A P 2012 Effect of laser radiation absorption efficiency on the heating temperature of inclusions in transparent media *Combustion, Explosion, and Shock Waves* 48 pp 54–58
[13] Kalensky A V, Zvekov A A, Anan'eva M V, Zykov I Yu, Kriger V G and Aduev B P 2014 Influence of Laser Wavelength on the Critical Energy Density for Initiation of Energetic Materials *Combustion, Explosion, and Shock Waves* **50** pp 98–104
[14] McPeak K M, Jayanti S V, Kress S J, Meyer P S, Iotti S, Rossinelli A and Norris D J 2015 Plasmonic films can easily be better: Rules and recipes *ACS Photonics* **2** pp 326–333
[15] Larruquert J I, Rodríguez–de Marcos L V, Méndez J A, Martin P J and Bendavid A 2013 High reflectance ta–C coatings in the extreme ultraviolet *Opt. Exp.* **21** pp 27537–27549

Luminescence of LiF crystals doped with uranium

L A Lisitsyna[1], G S Denisov[2], A K Dauletbekova[3], Zh T Karipbayev[3], A A Markhabaeva[4], V A Vaganov[5], V M Lisitsyn[5] and A T Akilbekov[3]

[1]Tomsk State University of Architecture and Building, 2 Solyanaya sq., 634003 Tomsk
[2]Kyrgyz Russian Slavic University named after First President of Russia B.N.Yeltsyn
[3]L.N.Gumilyov Eurasian National University
[4]al-Farabi Kazakh National University
[5]National Research Tomsk Polytechnic University, 30 Lenin Ave., 634050, Tomsk

E-mail: zfl@mail.ru

Abstract. The work presents the results of the study of the luminescence of uranium doped LiF crystals. Introduction of the dopant (U) and the co-dopant (OH) shows the occurrence of additional absorption in the range of 260–320 nm. The IR spectrum in crystals containing OH has distinguished characteristic absorption bands at 3725 cm^{-1}. In crystals doped with (U), the bands 3550–3580 cm^{-1} are observed, which are responsible for OH ions. The doped crystals with co-dopant contain an additional band at 3342 cm^{-1}. The luminescence is observed in the 470–520 nm spectral range excited by ionizing and UV radiation.

1. Introduction

Lithium fluoride crystals activated by multivalent ions are effective converters of hard radiation flow energy into emission of visible range. The crystals have a potential for using as scintillation materials in medicine, since their density is close to the body tissue density. On the other hand, LiF crystals can be used as model crystals with complex defects [1]. The paper provides spectral and kinetic performance of lithium fluoride crystal luminescence doped with uranium and hydroxyl ions entered along with it to compensate for the difference between charge states and elastic stresses.

2. Research methodology

The luminescence and excitation spectra were measured using a spectrofluorimeter Agilent Cary Eclipse, and a xenon flash lamp was used as a light source. The transmittance spectra of samples in the area of 4000–2000 cm^{-1} were measured using a Jasco FT/IR-4700 spectrometer. Optical absorption spectra in a range of 195–1200 nm were measured with spectrophotometers SF-256 and SPECORD 250 PLUS. The luminescence decay kinetics of crystal after excitation by ns electron pulses was measured. Luminescence was recorded using a monochromator MDR-204 photoelectric multiplier Hamamatsu R928, GDS-2204 and LeCroy 6030A oscillographs.

The integrated luminescence spectra of crystals were also measured under excitation by UV radiation of the xenon lamp and pulsed electron flow with energy of 250 keV with λ exct=340 nm. Integrated luminescence spectra were recorded using a fiber-optic spectrometer AvaSpec-3648 in the spectral range of 200–1100 nm. Integration time was 100 ms.

3. Research samples

Pure and doped crystals of lithium fluoride were grown according to the improved Kyropoulos method in a platinum dish in the air from a chemical agent of extra pure grade. Dopants were added in the load before crystal growing in the form of uranyl nitrate and caustic lithium. Uranium content in samples was determined using spectral analysis. The presence of OH dopant was controlled within the absorption line of 3720 cm^{-1}. The amount of OH dopant added is indicated in the text by the concentration of LiOH in the load.

The following crystals were prepared for the study: LiF pure grown in the Research and Technology Institute of Optical Materials of Vavilov State Optical Institute: LiF+2.0 mol% OH, LiF+0.005 mol% U, LiF+0.05 mol% U, LiF+0.05 mol% U+0.5 mol% OH, and LiF+0.05 mol% U+2.0 mol% OH.

Doping gives rise to absorption bands in the spectral range with λ<400 nm (Fig.1). The pure LiF crystal is transparent in the examined spectral range from 195 to 1000 nm.

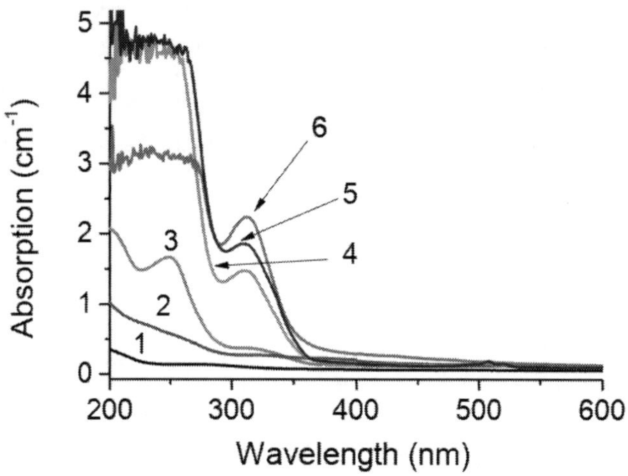

Figure 1. Absorption spectra of LiF crystals: pure (1); LiF + 2.0 mol% OH (2); LiF + 0.005 mol% U (3); LiF + 0.05% U + 0.5 mol% OH (4); LiF + 0.05 mol% U (5); LiF + 0.05 mol% U + 2.0 mol% OH (6).

Uranium doping gives rise to absorption in the spectral range with λ<400 nm, which grows up to 200 nm. When uranium concentration in the crystal is over 0.005 %, 260 and 320 nm, distinct bands appear in the spectrum. Doping the crystal with OH leads to absorption in the ultraviolet spectrum. Simultaneous adding uranium and OH groups results in crystal non-transparency in the region below 280 nm. Obviously, adding uranium in a forming crystal, when synthesizing, compensates for local elastic stresses of OH ions. These regularities of absorption spectra changes in the short-wave region, when adding multivalent dopants to lithium fluoride crystals, were described in [2-5].

Figure 2 shows an example of the results of the studied crystal transmission spectra measurements in the infrared region. Pure crystals are transparent in the spectral region from 4000 to 2000 cm^{-1}. Introduction of uranium and OH dopants into crystals gives rise to bands in the range of 3800–3200 cm^{-1}. The location of the revealed bands in the crystals spectra is shown in Table 1.

As presented in the table of research results, the spectra in the short-wave region are different for crystals containing OH and those free of OH. All OH doped crystals have a band with a peak at 3735 cm^{-1}, apparently, due to these radicals [6]. The spectra of crystals activated by uranium feature a band in the region of 3575 cm^{-1}, the position of the band shifts to 3562 cm^{-1} in the presence of the co-dopant (OH). The spectra of LiF crystals with U and OH co- dopants have a band in the region of 3342 cm^{-1}, the intensity of which increases at increased OH concentration. Apparently, the bands are due to complex defects incorporating OH in the area of U. [7, 8]. Moreover, there is an intensive band

IOP Conf. Series: Journal of Physics: Conf. Series **830** (2017) 012156 doi:10.1088/1742-6596/830/1/012156

for 3680 cm^{-1} in the spectrum of LiF crystal + 0.05% U + 2.0 mol% OH, the nature of which is not clear.

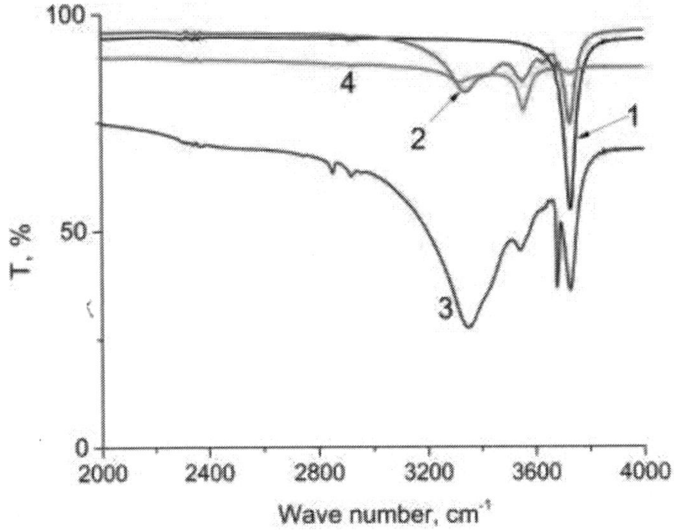

Figure 2. Absorption spectra of LiF crystals: pure (1); LiF + 2.0 mol%OH (2); LiF + 0.005% U (3); LiF + 0.05% U + 0.5 mol% OH (4); LiF + 0.05% U (5); LiF + 0.05% U + 2.0 mol % OH (6).

Table 1. Absorption band location in activated lithium fluoride crystals

Crystal	**Band location, cm^{-1}**				
LiF+2.0 mol% OH	3725				
LiF+0.005 mol% U			3575		
LiF+0.05 mol% U			3575		
LiF+0.05 mol% U + 0.5 mol% OH	3725			3562	3342
LiF+0.05 mol% U + 2.0 mol% OH	3725	3680		3562	3342

4. Experimental results

Fig. 3 shows the excitation spectra of the uranium doped LiF crystals in the crests of luminescence bands. The excitation spectrum depends on the uranium content in the crystal. There is a complex band at 250 nm (with distinct maxima at 230 and 270 nm) in the LiF crystal excitation spectrum with uranium concentration of 0.005 mol%. There are three bands with maxima at 220, 270 and 341 nm in the LiF crystal excitation spectrum with uranium concentration of 0.05 mol%. Apparently, the excitation band at 320 nm is related to the presence of OH in the dopant.

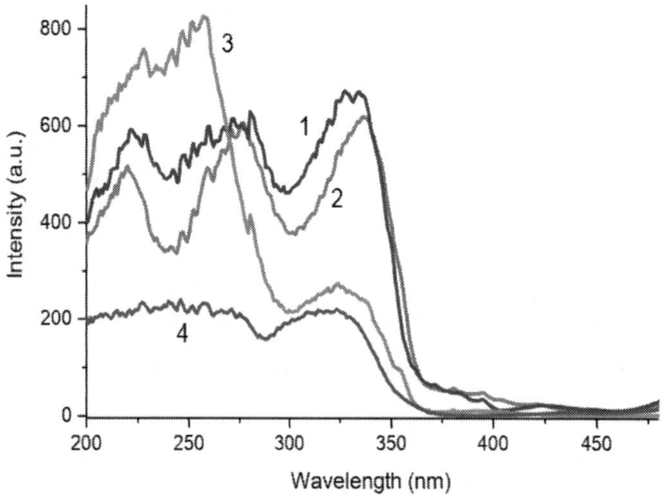

Figure 3. The excitation spectra of the uranium doped LiF crystals at the luminescence bands maxima: LiF + 0.05 mol% U + 0.5 mol% OH (λ=508 nm) (1); LiF + 0.05 mol% U (λ=520 nm) (2); LiF + 0.005 mol% U (λ=520 nm) (3); LiF + 0.05 mol% U + 2.0 mol% OH (λ=489 nm) (4).

Introduction of OH as a co-dopant leads to spectrum smoothing. Only the band at 330 nm is clearly observed in the excitation spectrum of LiF + 0.05% U + 2% OH crystal. Luminescence is not excited in the pure crystal.

Typical photoluminescence spectra of activated LiF crystals are shown in Figures 4 and 5. As shown in Figure 4, a group of spectra bands with maxima at 487, 508, 521 and 583 nm can be distinguished. The ratio between the bands depends on the concentration of uranium added and does not depend on the wavelength of exciting radiation. In crystals with a high concentration of U dopants, the base band falls at 508 nm, whereas in the crystal with a lower concentration, it falls at 521 nm.

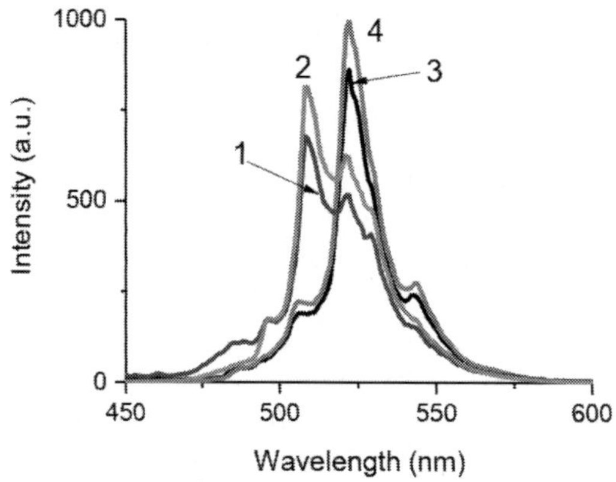

Figure 4. The luminescence spectra of crystals doped with uranium at different excitation wavelengths LiF + 0.05% U excited at λ=278 and 220 nm (1,2); LiF + 0.005% U excited at λ=259 and 220 nm (3,4)

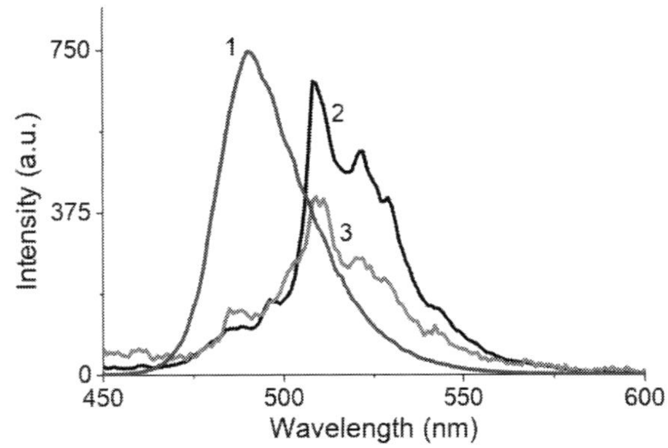

Figure 5. The luminescence spectra of LiF + 0.05% U + 2.0 mol% OH (λ_{em}=300 nm) (1), LiF + 0.05 mol% U (λ_{em}=222 nm) (2), LiF + 0.05% U + 0.5 mol% OH (λ_{em}=200 nm) crystals (3).

A broad band in the region from 320 to 600 nm with a maximum at about 425 nm can be observed in the luminescence spectrum of LiF + 0.5 mol% OH crystals. The spectrum does not depend on the excitation wavelength. In the spectral range of 480–550 nm, luminescence is not observed.

Figure 6 shows typical luminescence spectra of crystals doped with uranium (0.05 mol% U) and U + OH. Notes to Figure 6 show the wavelengths of monochromatic emission used for excitation. Luminescence spectrum type does not depend on the excitation wavelength. The type of all luminescence spectra with a dopant concentration of 0.05 mol% U does not depend on the presence of OH. This indicates that the crystal should be doped with uranium for OH entering for compensation. Introduction of uranium with a high concentration of OH dramatically changes the appearance of the luminescence spectrum. A band at 489 nm becomes the dominant band in the spectrum. It should be noted that there is low-intensity luminescence in crystals with a concentration of 0.5 mol% OH in the same area.

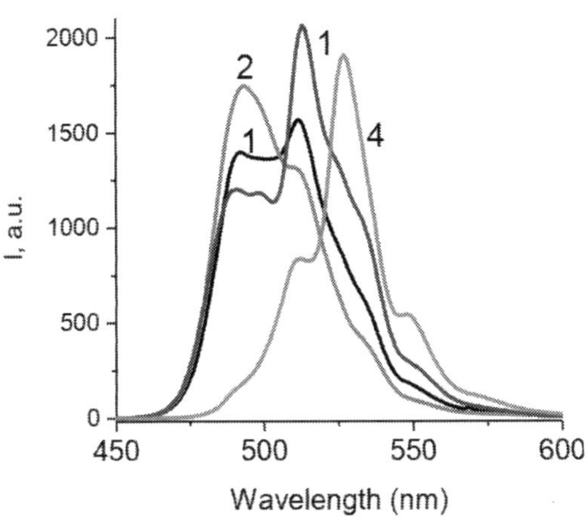

Figure 6. The luminescence spectra excited at λ_{em}=250 nm of LiF + 0.05 mol% U + 0.5 mol% OH (1), LiF + 0.05 mol% U + 2.0 mol% OH (2), LiF + 0.05 mol% U (3), LiF + 0.005 mol% U crystals (4).

The excitation of crystals with a pulse of high-energy electrons (250 keV) results in the excitation of luminescence of crystals with integral spectra, qualitatively similar to those measured under photoexcitation (Figure 7). A band at 425 nm is excited in pure and OH-doped crystals. Crystals containing U, or U and OH co-dopants quantitatively differ from the crystals measured under photoexcitation. A luminescence band at 490 nm is observed in the spectra of all crystals containing U and OH co-dopants. Its intensity is higher, the higher the concentration of introduced OH. It should be noted that the band also appears in the spectrum of the crystal with a high concentration of uranium. Obviously, this is a consequence of OH entering the crystal during synthesis.

Low-intensity luminescence in the region of 350–500 nm with a maximum at 425 nm is excited with a pulse of electron flow in pure crystals and crystals containing OH.

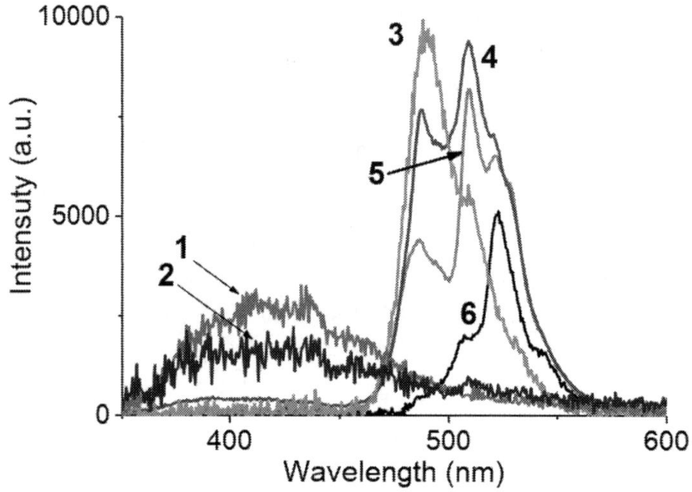

Figure 7. The cathodoluminescence spectra of LiF (1), LiF + 2.0 mol% OH (2), LiF + 0.05% U + 2.0 mol% OH (3), LiF + 0.05 mol% U + 0.5 mol% OH (4), LiF + 0.05 mol% U (5), LiF + 0.005 mol% U crystals (6).

Figure 8 shows the results of the study of the kinetic curves of cathodoluminescence decay for doped LiF crystals. From these results, it follows that there is a sharp contrast in the kinetic characteristics of the luminescence of LiF + 0.05 mol% U + 2.0 mol% OH crystal and those free of OH or containing a small concentration of uranium. Luminescence decays in LiF + 0.05 mol% U + 2.0 mol% OH crystal predominantly within a characteristic time of 230 μs. In LiF + 0.05 mol% U + 0.5 mol% OH, LiF + 0.05 mol% U and LiF + 0.005 mol% U crystals luminescence decays mainly 10–15 times with a characteristic time of 30–60 μs. All crystals has a component with a characteristic decay time of ~220 μs.

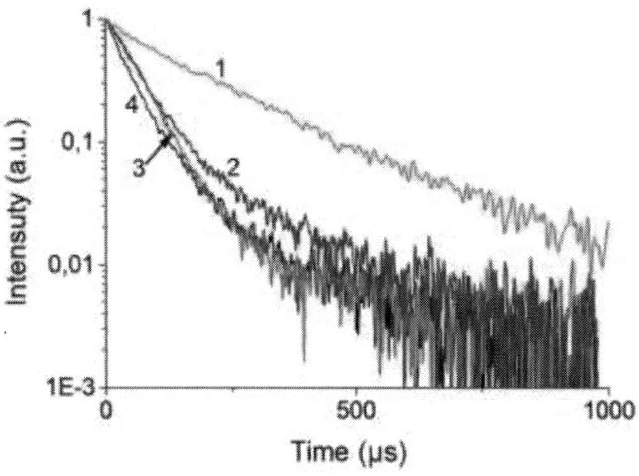

Figure 8. LiF + 0,05 mol% U + 2,0 mol% OH (1), LiF + 0,05 mol% U (2), LiF + 0,005 mol% U (3), LiF + 0,05 mol% U + 0,5 mol% OH (4) crystals cathodoluminescence decay kinetics

4. Conclusion

Introduction of OH as a co-dopant into the synthesized LiF crystal with uranium results in a complex defect in the crystal containing OH as a structural element. The presence of OH in the area of the introduced uranium ion is proved with the following results of the experimental studies.

1. The OH-containing uranium doped crystals have additional absorption bands in the spectral region of 3575, 3562 and 3342 cm^{-1} that can be apparently attributed to OH near the uranium ions.

2. The excitation band at 320 nm is observed only in crystals containing uranium and OH as a co-dopant.

3. The presence of OH co-dopant in uranium-doped crystals leads to occurrence of the luminescence spectrum at 490 nm. This band is also found in crystals, which were not intentionally doped with OH. Obviously, OH enters the crystal during growth to compensate for stresses in the dopant.

4. The type of all luminescence spectra with a dopant concentration of 0.05% U does not depend on the presence of OH. This indicates that the crystal should be doped with uranium for OH entering for compensation.

Acknowledgments

This work was performed with the financial support of the Ministry of Education and Science of Republic of Kazakhstan, contract #316 dated May 13, 2016

References

[1] Lisitsyna L A and Lisitsyn V M 2013 *Physics of the Solid State* **55** 11 pp 2297–2303
[2] Lisitsyna L A, Korepanov V I, Lisitsyn V M, Eliseev A E, Timoshenko N N and Dauletbekova A K 2011 *Optics and Spectroscopy* **110** 4 pp 568–573
[3] Nepomnyaschikh A I, Radzhabov E A and Egranov A V 1984 *LiF crystals color centers and luminescence* (Novosibirsk: Nauka)
[4] Jain V K 1986 *Journal of Physics D: Applied Physics* **19** 5 pp 1791–1796
[5] Rebane L A 1968 *Proceedings of the Institute for the Physics of the Atmosphere of the Estonian Soviet Socialist Republic* **37** pp 14–18
[6] Stoebe T G 1970 *J. Phys. Chem.* **31** 1291
[7] IKEZAWA M and ROLFE J 1973 *J. Chem. Phys.* **58** 2024
[8] Kamikawa T N *Phys. Stat.Sol. (B)* 1980 **99** 721

Research of a possibility of receiving sorbents for a sewage disposal from a wastage of coal preparation factory

S L Buyantuev[1], A S Kondratenko [2], S Y Shishulkin[2], Y Y Stebenkova[1], A B Khmelev[1]

[1] East Siberia State University of Technology and Management, Ulan-Ude, 670013, Republic of Buryatia, Russia
[2] Buryat State University, Ulan-Ude, 670000, Republic of Buryatia, Russia

E-mail: buyantuevsl@mail.ru

Abstract The paper presents the results of the studies of the structure and porosity of the coal cake processed by electric arc plasma. The main limiting factor in processing of coal cakes sorbents is their high water content. As a result of coal washing, the main share of water introduced into the cake falls on hard-hydrate and colloidal components. This makes impossible application of traditional processes of manufacturing from a cake of coal sorbents. Using the electric arc intensifies the processes of thermal activation of coal cakes associated with thermal shock, destruction and vapor-gas reactions occurring at the surfaces of the particles at an exposure temperature of up to 3000 ^{0}C, which increases the title product outlet (sorbent) and thereby reduces manufacturing costs and improves environmental performance. The investigation of the thermal activation zone is carried out in the plasma reactor chamber by thermal imaging method followed by mapping-and 3D-modeling of temperature fields. the most important physical and chemical properties of the sorbents from coal cake activated by plasma was studied. The obtained results showed the possibility of coal cake thermal activation by electric arc plasma to change its material composition, the appearance of porosity and associated sorption capacity applied for wastewater treatment.

1. Introduction

Now the coal sorbent is used in many processes of engineering chemistry. Besides, a sewage disposal is based mainly on adsorption by the absorbent carbon. Hardwood or high-grade coals briquettes-sorbents after carbonization are commonly used for production of sorbents [1].

In extraction of fossil coals, some of them in wet-process coal cleaning inevitably go to waste in the form of so-called highly watered coal cake. The major limiting factor of coal cake processing in sorbents is their high moisture content, and the main share of moisture entered in cake as a result of coal preparation is hardly removable that makes impossible application of such traditional processes of thermal activation as pyrolysis, devolatilization of coal, thermo-briquetting, etc. in production of coal sorbents. One of the perspective directions in this area is using of electric arc plasma allowing intensification of the processes of thermal activation of coal cake considerably related to thermal shock, the destruction and steam-gaseous reactions occur on the particle surfaces at exposure temperature up to 3000 ^{0}C. This increases the yield title product (sorbent) and thereby reduces manufacturing costs and improves the environmental performance associated with the use of waste product- coal cake [2].

The paper presents the results of the studies of thermal activation of the coal cake, processed by electric arc plasma treatment with change in its material composition, the appearance of porosity and associated sorption capacity applied for industrial wastewater treatment.

2. Measurements technique

Because of high moisture content it is impossible to use a coal cake in traditional production process of coal sorbent. Therefore, application of electric arc plasma for thermal activation of the coal cake reduces the cost of production of a sorbent and significantly improves environmental performance.

To study the process of the coal cake thermal activation, the combined plasma modular unit was created to produce coal sorbent by the electric plasma processing [3].

In the experiment, a reagent (coal cake) comes from above through the lid of the reactor into the interelectrode interval where the direct current arch is rotated by the external electromagnetic coil. Continuous plasma environment is formed with a uniform rotation of the anode arc spots between the annular cathode in the reactor with an average temperature of 2500–3000 K. Due to this, particles of the cake given from above to the reactor undergo thermochemical processing by plasma. The thermal activation zone in the camera of the electric plasma reactor was studied by the imaging infrared method for subsequent mapping with model operation of temperature profiles (figure 1) using the high-temperature infrared camera mcs-640 by Luma Spec. Particle residence time in the plasma zone depends on size fractions, and the arc rotation speed which is between 0.1 to 1 second (figure 2). The temperature and the rotational speed of the plasma are generally regulated by changing the power source and the coil current.

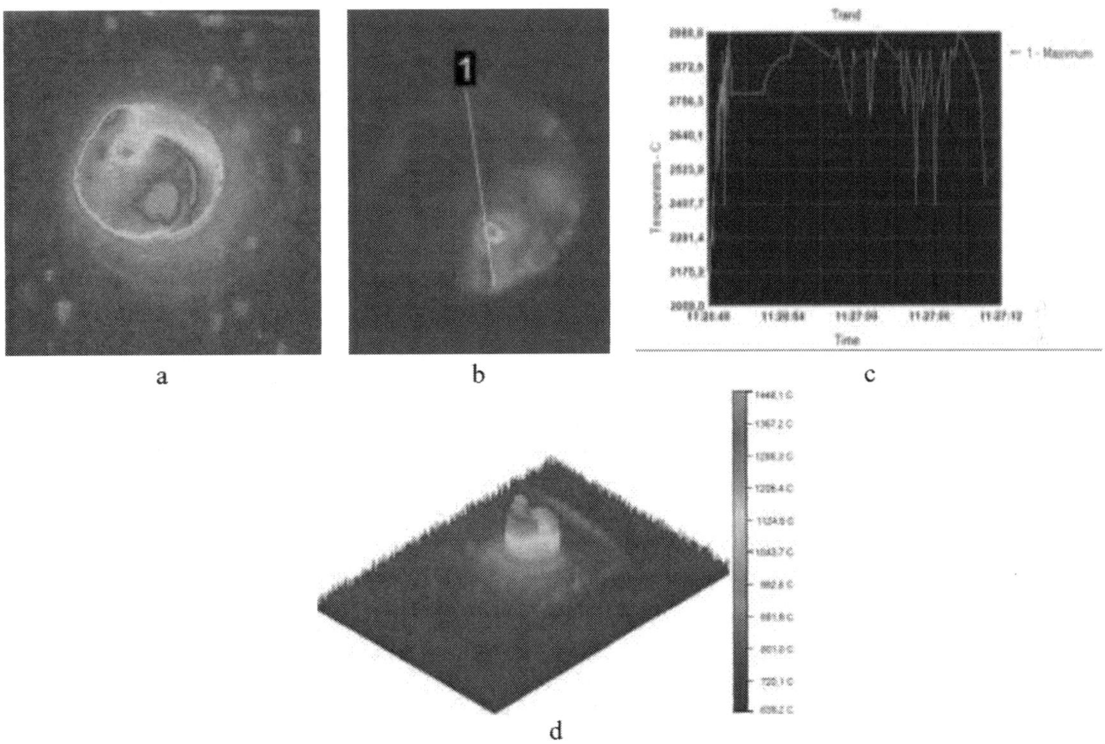

Figure 1. Temperature profiles of the reactor in the range of 1500–3000 ^{0}C (a, b) with the map of characteristic lines (c) and creation of a temperature profile (d)

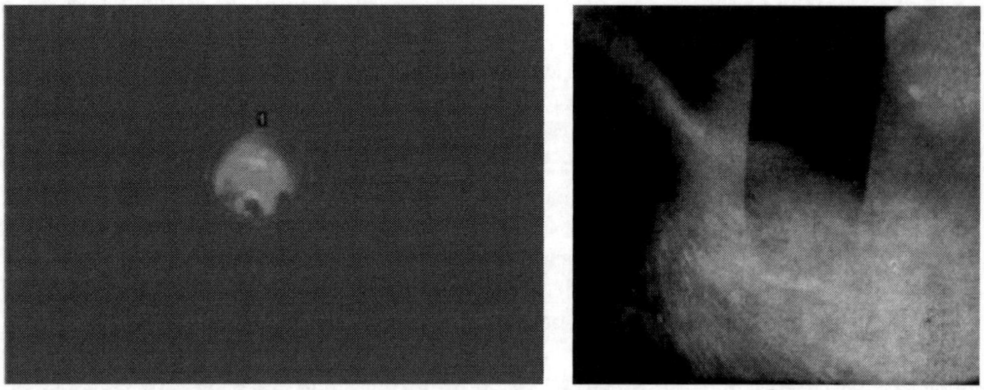

a b

Figure 2. The time of the coal cake in the reactor. Area of electric plasma zone (a) with capture of cake dust and its rotation (b) in a magnetic field

In electric plasma activation, in conditions of high temperature generated by rotating electric arc plasma, a thermal shock occurs that results in the release of the colloidal moisture present in the composition of the substance cake and thermal degradation accompanied by the release of volatile substances (flammable gases). Fiery steam and gas products interact with the newly incoming cake particles and the process becomes avalanche. This freed up excess moisture and combustible volatile substances, which loosen the cake particles, making them porous. A part of the mineral in the form of clay minerals softens and sticks together in the agglomerates under the influence of rotating electric arc plasma. It falls together with the heat-treated cake (sorbent) to the camera of the sorbent tank for further removal by sieving. There is an additional sorbent demineralization. This principle of creation of plasma installations will allow reaching the level of industrial production of carbon sorbents [4, 5].

The thermal energy is generated in the rotating arch zone under the influence of gas flow (plasma). This is obtained by thermal influence on the coal dust, with its distribution throughout the volume. Therefore, the power generation area was determined by condition of its uniform distribution. In the calculations of the reactor chamber diameter was taken as reference dimension, and all the basic laws of electric zone expressed through it. This makes it possible to extend the experiment results obtained in low-power reactors, reactors are geometrically similar to high power, and therefore, large diameter cells [6].

The current of the Ip reactor is expressed through the current density carried to a camera sectional area:

$$I_p = \Delta i_p \pi D_{к}^2/4 \quad (1)$$

Current density is $\Delta i_p = 4\, I_p / \pi D_{к}^2 = 1.132$ A/cm^2

Tension of the Up reactor is represented through a gradient of tension of an interelectrode interval:

$$Up = E_p D_{к}/2, \quad (2)$$

where $D_{к}/2$ is an interelectrode gap to which the arc voltage is supplied to determine a voltage gradient (in the reactor with a rod cathode).

Respectively tension gradient $E_p = 2\, U_p / D_{к} = 26.7$ V/cm.

Taking into account expressions for Ip and Up, the formula for reactor power is removed:

$$P_p = U_p * I_p = \pi/4 * 1/2\, E_p\, \Delta i_p D_{к}^3 \approx 0.39 E_p\, \Delta i_p D_{к}^3. \quad (3)$$

Reactor power is $P_p \approx 0.39 E_p\, \Delta i_p D_{к}^3 = 0.39 * 26{,}7$ V/cm $* 1.132$ A/cm$^{2}*$ $(15)^3$ cm $^3 = 40$ kW

As a result of transformations, it is concluded that allocated in the amount of thermal power generation area, which is equal to input electrical power, is proportional to the cube of the diameter of

IOP Conf. Series: Journal of Physics: Conf. Series **830** (2017) 012157 doi:10.1088/1742-6596/830/1/012157

the chamber, it is also proportional to Δi_p and voltage gradient density E_p, otherwise bulk power P_p, released in the area.

$\Delta i_p \equiv I_d/D_k2$ determine the size Δi_p by models [7] from conditions of technological indexes at high efficiency of installation. Then with the given diameter of the reactor, its size is provided with the choice of current.

3. Experimental results and discussion

At a tentative stage of studying of getter properties of the cake processed by plasma, the change in its element structure before thermal activation was investigated. The research data are presented in tables 1 and 2. The porosity of obtained sorbents was also determined, and analyzes were performed on the lightening capacity and sorption activity in wastewater treatment.

Table 1. Element structure of a coal cake before plasma processing

C	O	Mg	Al	Si	S	K	Ca	Ti	Fe	Zr	Result
42.98	39.69	0.16	4.76	7.77	0.33	0.41	0.62	0.25	1.66	1.37	100.00

Table 2. Element structure of a coal cake after plasma processing

C	O	Mg	Al	Si	S	K	Ca	Ti	Fe	Zr	Result
48.73	34.61	0.16	4.60	7.33	0.30	0.41	0.58	0.25	1.66	1.37	100.00

Then, the study of a cellular structure of a cake after electric plasma activation was conducted. For this purpose, the thermally activated exemplars were impregnated with luminescent dye to make the contrast with the aim of a clear distinction between particles and boundaries deep pores [8]. Light areas in the image correspond to the pores filled with a luminescent substance, and the dark ones indicate cake particles (fig. 3.). The study of the pore surface of the activated plasma cake was conducted by means of computer 3D-model operations [9].

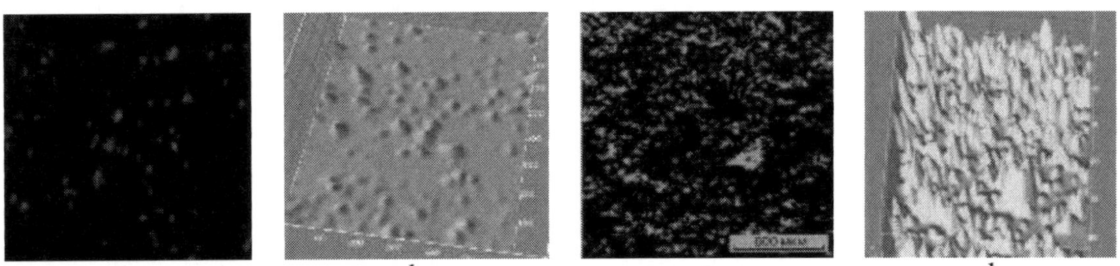

a b c d

Figure 3. Example of the coal cake 3D Model before electric plasma processing (a, b, increase in X200). Thermally processed coal (c), model of representation of the volume structure of the scan (d)

Further study of the possibility of technical application of the sorbents processed by plasma for a sewage disposal was carried out. In the course of the analyses, it was noticed that if coal cake processed in electric arc plasma pour into water, there is no washing away. The low leachability of fraction to 1 mm, being the main in fractional composition, is caused by rather high sedimentation speed of particles that is explained by a hydrophilic nature of their surface owing to an inexact dehydration of particles of a cake after thermal activation.

The samples of industrial waste water used for the experiments dealt with the urban CHP, the sampling point is the drain sump. To test the sorption activity was carried out to study the ability of lightening (absorbance) photocolorimetry samples both before and after passing through the sorbent cakes (fig. 4), and chemical analysis of water by capillary electrophoresis unit "Kapel 105M" by PNDF 14.1.2.3 and PNDF 14.1-95. methods of state environmental control [10].

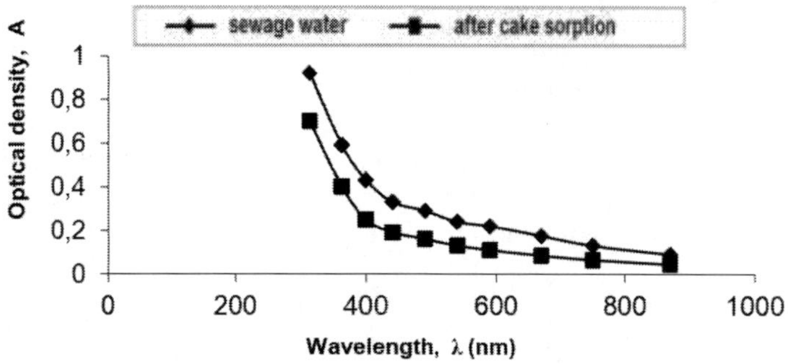

Figure 4. Electronic absorption spectra of exemplars of sewage before getter tests.

The results of studying the clarifying ability showed (fig. 4.) 2-fold decrease in an optical density (A) of the exemplars which passed through activated coal cake, in comparison with initial exemplars of sewage that indicates effectiveness of application of those sorbents.

The following stage of the research included the study of chemical composition of sewage. The samples were tested before and after passing through the column with activated cake. The results of the studies of sewage before cleaning are presented in tab. 3.

Table 3. Results of researches of sewage before and after cleaning

Chemical substance	Without sorbent, mg/l	After passing through a sorbent, mg/l	Cleaning, %
Sodium	5.87	1.45	75.30
Magnesium	10.6	7.54	28.87
Strontium	4.21	2.19	48.00
Barium	4.89	0.307	93.72
Calcium	78.5	17.1	78.21
Iron	0,08	0,04	50.00
Chloride	4.37	3.04	30.43
Sulphate	93.9	12.1	87.11
Fluoride	0.820	0.392	52.20
Ammonia	0,34	0,105	69.11

The carried-out tests showed the fitness of the coal cake processed by plasma as a sorbent for purification of the production sewage.

4. Conclusion

Thus, the coal cake, as the final stage of coal washing, is fundamentally suitable to extract the carbon sorbent by method of thermal processing in the electric plasma reactor. The tests are considered relevant since the presented methods improve an ecological situation in the region of coal mining by using technogenic wastage, coal cakes, and creation of carbon sorbents on their basis for purification of the production sewage.

It is also necessary to note that the coal sorbent can be considered as one of production components of coal plasma processing such as synthesis gas and carbon nanoparticle [2, 3, 4, 5, 10].

IOP Conf. Series: Journal of Physics: Conf. Series **830** (2017) 012157 doi:10.1088/1742-6596/830/1/012157

References

[1] Kienle H, Bader E. 1980 Active coals and their industrial applications (Stuttgart: 1984)
[2] Buyantuev S L, Kondratenko A S 2012 Structure and properties of carbon sorbents and carbon nanomaterials produced in the processing of plasma arc (Novosibirsk: Materials VIII All-Russian conference with international participation «The burning of solid fuel») pp 251-257
[3] Buyantuev S L, Kondratenko A S 2013 The sorption properties of coal after thermal degradation in an electric plasma (Tomsk: Proceedings of the higher educational institutions, ser. «Physics») **57** pp 65-70
[4] Buyantuev S L, Starinsky I V 2008 The process for producing activated carbon and installation for its implementation (RF patent for invention №2314996)
[5] Buyantuev, S L, Shishulkin S U, Kondratenko A S, Khmelev A B, Bazarsadaev E T 2016 The study of specific surface of coal treated by electric arc plasma according to the air permeability data of the layer and the coloring agent sorption *Research Journal of Pharmaceutical, Biological and Chemical Sciences* (London)
[6] Bron O B 1966 The movement of the electric arc in the magnetic field (Moscow: Electricity) pp 76-81
[7] Sergeev P V 1978 Arc in electric reactors (Alma-Ata. The science) p 140
[8] Konstantinov-Schlesinger M A 1961 Luminescent analysis (Moscow:Publishing house Nauka)
[9] Panteleev V G, Egorova O V, Klykova E I 2005 Computer microscopy (Moscow: Technosphere) p 304
[10] Buyantuev S L, Kondratenko A S, Starinsky I V, Khmelev A B 2015 Industrial testing and analysis of activated carbons derived from electric arc plasma *Journal of Physics: Conference Series* (London)

IOP Conf. Series: Journal of Physics: Conf. Series **830** (2017) 012158 doi:10.1088/1742-6596/830/1/012158

Self-trapped exciton configurations in Beryllium Oxide

M A Botov[1], A Yu Kuznetsov[2] and A B Sobolev[1]

[1]Dept. of High Mathematics, Ural Federal University, 19 Mira St., Yekaterinburg, Russia

[2]Dept. of Experimental Physics, Ural Federal University, 19 Mira St., Yekaterinburg, Russia

Abstract. The modern radiation technology, nuclear engineering, non-linear optics are associated with radiation-resistant optical material study. Evolution of electronic excitations in these materials is a complex multichannel process which currently has no integrated model. A special role belongs to the low-symmetry single crystals, such as beryllium oxide (BeO). We present theoretical results that advance our understanding of exciton-based channel of electronic excitations relaxation. The four possible self-trapped exciton (STE) configurations in beryllia single crystal have been investigated by using a quantum mechanical approach (Hartree-Fock and B3LYP HF-DFT hybrid functional, as implemented in the CRYSTAL09 code). B3LYP DFT functional with 30% of exact exchange was used (B3LYP30). All calculations were performed using periodic boundary conditions and full SC geometry relaxation. The lattice distortion and charge density distribution for considered defect configurations were obtained. STE-A1 luminescence energy was found to be 6.0 eV for HF and 6.5 eV for B3LYP30; STE-A2 luminescence energy was found to be 9.2 eV for HF and 7.8 eV for B3LYP30. STE-B1 luminescence energy was found to be 5.5 eV for HF, 6.2 eV for B3LYP30; STE-B2 luminescence energy was found to be 4.7 eV for HF.

1. Introduction

The exciton is well known as a bound state of an electron and a hole which are attracted to each other by the electrostatic Coulomb force [1]. It can be trapped at a lattice distortion and in this case become self-trapped exciton (STE). Such STEs recombine to the ground state, producing a characteristic luminescence. STE in oxides have a long history of study. Nevertheless, a generally accepted theory of excited state formation and STE models in wide-gap oxide crystals does not still exist. It is well established that exciton self-trapping takes place only in low symmetry oxide crystals, such as SiO_2, Al_2O_3, Y_2O_3 [2,3,4]. STE luminescence bands were also detected in BeO [5,6,7]. Beryllia has a wide range of applications, it is the only material apart from diamond which combines high thermal-shock resistance, high electrical resistivity, and high thermal conductivity at a similar level. It is very promising for personal dosimetry due to the proximity of its effective atomic number (Z=7.13) to biological tissue (Z=7.42) [8]. Theoretical investigation of intrinsic Schottky and Frenkel defects in beryllia by means of molecular statics method was performed in [9]. However, the absence of ab initio quantum-mechanical calculations of STE in BeO still occurs. In this paper, we shall present such ab initio modeling of STE in beryllium oxide.

2. Model and calculation details

Our study is based on ab initio calculations performed in CRYSTAL09 [10] package using Hartree-Fock (HF) approximation and density functional theory functional B3LYP[11,12] with 30% of exact exchange (B3LYP30). Atoms were described by their full-electron basis set (5-11G for Be and 8-411G for O)[13,14] with optimized valence shells. All calculations were performed using periodic boundary conditions and full supercell(SC) geometry relaxation. 108-atom supercell (symmetry group P63mc) was constructed as 3 × 3 × 3 expansions of BeO unit cell. A Monkhorst-Pack mesh of 13 k points in the irreducible part of the Brillouin zone was used for integration in the reciprocal space. Anderson's method [15] of Fock matrix mixing was used for better convergence. Our way to the stable STE model consists of two stages: the first is to model a hole and the second is to model STE itself. In the first stage, we place a hole at oxygen atom most remote from SC borders (by removing one electron from its outer shell). To facilitate initial hole localization, a small distortion at hole lattice site was created (such technique was inspired by [16,17,18]). An oxygen carrying hole was shifted up on 10% of lattice constant a (0.3 Å) along z axis. After that, SC geometry optimization procedure was performed. Shifting along x axis was also checked, but the result for z axis is energetically more favourable. The second stage was devoted to STE modelling. A peculiarity of STE structure is presence of two unpaired electrons; one of them belongs to an oxygen representing hole core, and another is localized in interatomic spacing near the hole (and we have to lock their spins to 1 during the calculation and use spin-unrestricted calculation schemes – UHF and corresponding DFT modification). Electron of STE is described by one diffuse Gauss-type s-orbital. This approach was earlier successfully applied for modelling of STE in alkali-halides [19] and corundum crystal [20]. Optimized values of exponent α-parameter in STE electron wave function are shown in Table 4. The final results after full SC geometry relaxation routine are discussed in the following section.

3. Results

The perfect BeO crystal parameters were calculated at the very first step in order to check basis sets correctness. The results are presented in Table 1. The results provided by B3LYP30 method are close to experimental data. HF typically overestimates energy gap.

Table 1. Calculated lattice constants *a*, *c* and internal parameter *z* (in Å), *a/c* ratio and energy gap Eg (in eV) of perfect BeO in comparison with experiment.

Parameter	HF	B3LYP30	Exp.
***a*, Å**	2.692	2.702	2.698[21]
***c*, Å**	4.336	4.369	4.380[21]
***c*/*a*, Å**	1.611	1.617	1.624[22]
z, Å	0.384	0.377	0.378[22]
Eg, eV	18.99	10.88	10.59[21]

3.1. Configurations of STE

Four off-centre STE configurations were studied. We denote them in terms of STE electron density center localization inside beryllium-oxygen tetrahedron: as "in(z)" (e inside tetrahedron, z-oriented) and "out(z)" (e outside tetrahedron, z-oriented), "in(x)" (e inside tetrahedron, initially x-oriented) and "out(x)" (e outside tetrahedron, initially x-oriented). We used 108 atoms supercell for both HF and B3LYP30 methods. Rather large SC is needed due to the size of the defect, delocalized character of its electronic component and defect-defect interactions between nearby SC. Figure 1 shows geometric structure and spin localization of mentioned configurations. It could be seen, that STE hole core is localized at one oxygen atom (O76 at Table.2 and Table.3). STE has shown itself as rather extensive formation – the distance between STE electron density center and O76 occupied by hole is about 50% of lattice constant a for axial orientation ("in(z)": 50.8% for HF and 47.6% for B3LYP30; "out(z)": 54.0% for HF and 61.5% for B3LYP30) and about 60% of a for non-axial orientation ("in(x)": 59% for HF and 58% for B3LYP30, "out(x)": 60% for HF). STE-"in" configurations are more compact than

STE-"out" ones. B3LYP30 give a 1% less size for STE_in(x) than HF. Unfortunately, B3LYP30 did not give stable solution corresponding to STE_out(x) configuration.

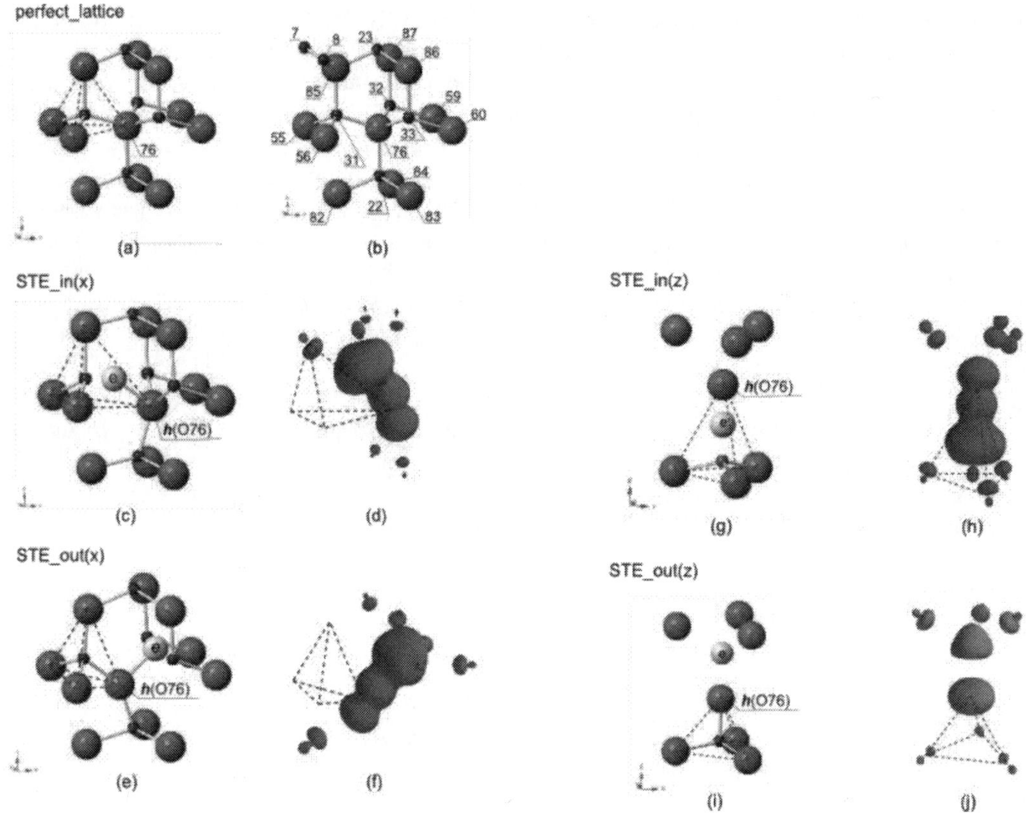

Figure 1. (Color online) Spin localization in the STE_in(x) and STE_out(x) configurations in BeO. Oxigens are red, Be-atoms are gray. Fictive atom "e" #109 (yellow) denotes location of STE electron density center. Fragments (a), (b), (c), (e), (g), (i) show the lattice around STE site and tetrahedron of interest. Fragments (d), (f), (h), (j) show the surface of the constant spin density with value S = 0.07.

It should be noted, that in spite of initial horizontal orientation of non-axial STE, after SC geometric relaxation STE become rotated around y-axis (sill laying in xz-plane). The final deviation from xy-plane reaches 35.5° for "in(x)" by HF, 35.9° for "in(x)" by B3LYP30 and 40.8° for "out(x)" by HF. STE appearance causes strong distortions of the surrounding lattice, especially at first coordination sphere.

3.2. Relaxation

Figure 1 (c), (e), (g) and (i) show STE surrounding atoms after geometry relaxation. Numbers of these atoms in SC are given at figure 1 (b). Shifting of STE nearest neighbors from their perfect positions exceeds 3% of lattice constant a. Oxygen O76 which carries a hole has the most noticeable displacement (excluding fictive atom "e" which represents the center of STE electron density. Table 2 and Table 3 demonstrate largest relaxations and spin charges in SC).

Table 2. Non-axial STE and surrounding atoms relaxations: top 10 of atoms with largest displacements. Displacements are denoted as Δ, measured in % of lattice constant a, spin charges are denoted as q. Correspondences between atomic numbers (at.#) and their positions in SC are shown in figure 1 (b)

STE_in(x)						STE_out(x) [a]		
HF			B3LYP30			HF		
at.#	Δ, %	q, |e|	at.#	Δ, %	q, |e|	at.#	Δ, %	q, |e|
109	36.05	1.356	109	35.47	1.499	109	36.34	1.339
76	22.98	0.753	76	22.47	0.595	76	24.12	0.765
31	8.580	0.098	31	7.893	0.098	22	8.756	0.009
22	7.599	0.009	22	7.365	0.006	32	6.896	0.051
85	7.159	0.100	85	7.079	0.137	33	6.896	0.051
83,84	3.720	0.002	23	3.406	0.002	31	6.891	0.008
7	3.222	0.001	83,84	2.241	0.002	82	4.669	0.044
23	3.193	0.000	7	2.769	0.001	86,87	3.632	0.067
32	3.156	0.008	82	2.269	0.002	55	2.738	0.005

[a] B3LYP30 did not give stable solution corresponding to STE_out(x) orientation.

Table 3. Axial STE and surrounding atoms relaxations: top 10 of atoms with largest displacements. Displacements are denoted as Δ, measured in % of lattice constant a, spin charges are denoted as q. Correspondences between atomic numbers (at.#) and their positions in SC are shown in figure 1 (b)

STE_in(z)						STE_out(z)					
HF			B3LYP30			HF			B3LYP30		
at.#	Δ,%	q,|e|	at.#	Δ,%	q,|e|	at.#	Δ,%	q,|e|	at.#	Δ,%	q,|e|
109	29.659	1.540	109	25.635	1.385	109	35.94	1.523	109	46.644	1.417
76	21.177	0.576	76	21.958	0.400	76	18.10	0.557	76	14.866	0.431
22	8.024	0.109	22	8.122	0.103	22	12.97	0.011	22	10.390	0.001
82	4.642	0.069	82	3.817	0.033	31	5.223	0.015	23	6.407	0.019
83	4.642	0.069	83	3.817	0.033	32	5.223	0.015	85	4.479	0.019
84	4.642	0.069	84	3.817	0.033	33	5.223	0.015	86	4.479	0.019
31	4.594	0.009	31	3.509	0.005	23	4.691	0.014	87	4.479	0.021
32	4.594	0.009	32	3.509	0.005	85	4.339	0.084	31	3.364	0.108
33	4.594	0.009	33	3.509	0.005	86	4.339	0.084	32	3.364	0.108
23	3.289	0.003	85	3.098	0.003	87	4.339	0.084	33	3.364	0.108

3.3. Electronic properties

Partial density of states (DOS) diagrams of STE components are shown in figure 2 and figure 3. Electron and hole levels are distinctly located in energy gap. Localized hole state splits from VB by 6.4 eV for HF and 1.0 eV for B3LYP30 as to STE_out(z) and by 13.7 eV for HF and by 3.7 eV for B3LYP30 as to STE_in(z). STE electron level is splitted off from CB bottom by 10.7 eV for HF and by 3.8 eV for B3LYP30 as to STE_out(z) and by 10.4 eV for HF and by 3.6 eV for B3LYP30 as to STE_in(z).

IOP Conf. Series: Journal of Physics: Conf. Series **830** (2017) 012158 doi:10.1088/1742-6596/830/1/012158

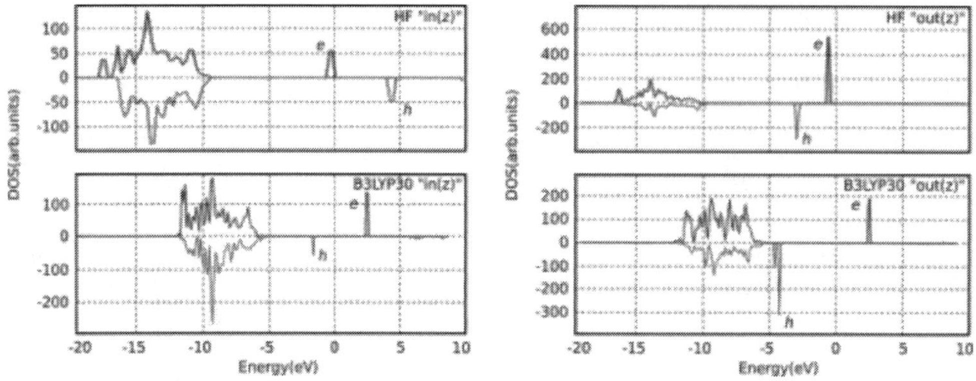

Figure 2. Density of states (DOS) diagrams for alpha (upper curves) and beta (lower curves) spin components of axial STE in beryllium oxide. Fermi level is at 0 eV.

Non axial STE is in conformable situation. Localized hole state splits off from a VB top by 13.8 eV for HF and 3.8 eV for B3LYP30 as to STE_in(x) and by 13.7 eV for HF as to STE_out(z). STE electron level is spitted off from CB bottom by 10.9 eV for HF and 4.0 eV for B3LYP30 as to STE_in(z) and by 11.7 eV for HF as to STE_out(z).

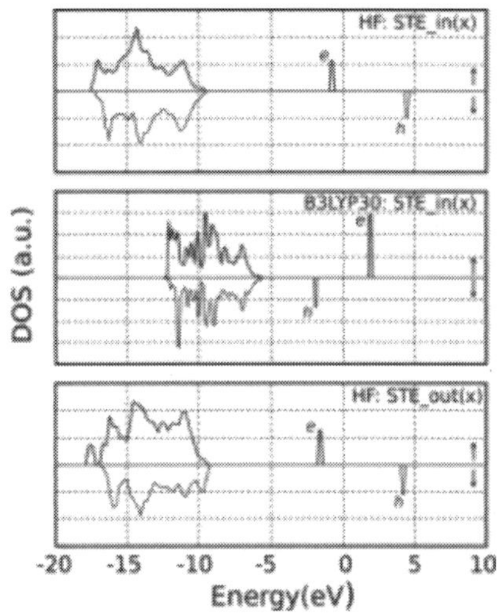

Figure 3. Density of states (DOS) diagrams for alpha (upper curves) and beta (lower curves) spin components of STE in beryllium oxide. Fermi level is at 0 eV.

Presence of two defect levels becomes apparent at luminescence spectrum also. STE triplet luminescence energies as a result of transition from low excited triplet state to ground state are given in Table 4 and Table 5. They were calculated using the ΔSCF method as the difference between the total energy of the fully relaxed triplet state and the ground singlet state at the triplet geometric structure. HF has better agreement with experimental data.

Table 4. STE(z) luminescence bands and α-parameter for exponent describing STE electronic component (SC 109 atoms)

Config.	Parameter	HF	B3LYP30	Exp.[5,6]
STE_in(z)	Elum	6.0	6.5	4.9
	α	0.11	0.12	-
STE_out(z)	Elum	9.2	7.8	6.7
	α	0.11	0.11	-

Table 5. STE(x) luminescence bands and α-parameter for exponent describing STE electronic component (SC 109 atoms)

Config.	Parameter	HF	B3LYP30	Exp.[5,6]
STE_in(x)	Elum	5.5	6.2	4.9
	α	0.12	0.10	-
STE_out(x)	Elum	4.7	-*	4.4
	α	0.12	-*	-

HF shows a reversed position of defect levels and overestimates Eg extent. Nevertheless, it gave reasonable results for all luminescence peaks under consideration (an average inaccuracy of 20%). Hybrid functional B3LYP30 have more noticeable deviation (25%) and did not give stable solution for one of the configurations (STE_out(x)). Divergence of STE_out(x) could be connected with its less tight site space for STE electron and excessive delocalization of electron over it.

4. Conclusions

In this paper, we have presented the results of STE computation modelling in BeO intended for better understanding of these type electronic excitations. It could be also interested from an aspect of modern methods application especially taking into consideration lack of theoretical investigation results relevant to STE in beryllia. Comparison of B3LYP functional with non-standard 30% of exact exchange and pure HF approach was done in the context of STE modelling in BeO crystal. B3LYP30 have advantages as to perfect crystal, but for some luminescence bands of STE it appears to be quantitatively less accurate than HF, in spite of correct relative position of defect levels. Calculated data for STE triplet luminescence reasonably agree with the experimental one.

5. References

[1] Knox R S 1963 *Theory of excitons, Solid state physics* (Ed. by Seitz and Turnbul, Academic, New York) **5**

[2] Lushchik Ch B 1985 *Excitons* (Moscow, Nauka) pp 362–385

[3] Song RT and. Williams R T 1993 *Self-Trapped Excitons* (Springer-Verlag, Berlin Heidelberg New York)

[4] Itoh N and Stoneham A M 2003 *Materials Modification by Electronic Excitation* (Cambridge University Press)

[5] Giniyatulin K N, Malysheva A F, Kruzhalov A V and Kyarner T N1982 *Tr.Inst.Fiz.Akad.Nauk Estonii* **53** 71

[6] Ivanov V Y, Pustovarov V A, Kruzhalov A V and Shulgin B V 1989 *Nucl. Instr. Meth. Phys. Res. A* **282** 559

[7] Gorbunov S V, Yakovlev V Yu, Ivanov V Yu and Kruzhalov A V 1990 *Soviet Physics – Solid State* **32** 10 1708

[8] Tausenev D S, Milman I I, Ivanov V Yu and. Kruzhalov A V 2008 *Radiation Measurements* **43** pp 349 – 352

[9] Kislov A N , Kruzhalov A V , Varaksin A N and Mazurenko V G 1991 *Soviet Physics – Solid State* **33** 10 1659

[10] Dovesi R, Sounders V R, Roetti C *et al* 2013 *CRYSTAL09 User's Manual* (University of Tourino)

[11] Becke A D 1993 *J.Chem.Phys.* **98** 7 5648-5652

[12] Stephens P J, Devlin F J, Chablowski C F and Frish M J 1994 *J.Phys.Chem.* **98** 45 11623-11627

[13] Lichanot A, Chaillet M, Larrieu C, Dovesi R and Pisani C 1992 *Chem. Phys.* **164** 383-394

[14] Towler M D, Allan N L, Harrison N M, Saunders V R, Mackrodt W C and Apra E 1994 *Phys. Rev. B* **50** 5041-5054

[15] Anderson D G 1964 *J. Assoc. Comput. Mach.* **12** 547
[16] Ramo D M, Sushko P V and Shluger A L 2012 *Phys.Rev.B* **85** 024120
[17] Shluger A L and Stefanovich E V 1990 *Phys.Rev.B* **42** 9664
[18] Gavartin J L and Shluger A L2001 *Phys.Rev.B* **64** 245111
[19] Kuznetsov A Y , Sobolev A B , Makarov A S and Botov M A 2009 *Russian Physics Journal*, **52** 8/2 83
[20] Kuznetsov A Yu, Botov M A, Makarov A S and Sobolev A B2014 *Russian Physics Journal*, **57** 12-3 36
[21] Martienssen W and Warlimont H (Eds.) 2005 *Springer Handbook of Condensed Matter and Materials Data* (ISBN 3-540-44376-2 Spinger Berlin Heidelberg New York)
[22] Hazen R M and Finger L W1986 *J.Appl.Phys.* **59** 3728

The portable detector of nitro-explosives in vapor phase with new sensing elements on a base of pyrimidine scaffold

A A Baranova[1], K O Khokhlov[1], R D Chuvashov[1], E V Verbitskiy[1,2], E M Cheprakova[1,2], G L Rusinov[1,2] and V N Charushin[1,2]

[1] Ural Federal University, Mira St., 19, Ekaterinburg, 620002, Russia.
[2] Postovsky Institute of Organic Synthesis, Ural Branch of the Russian Academy of Sciences,S. Kovalevskoy St., 22, Ekaterinburg, 620990, Russia.

E-mail: a.a.baranova@urfu.ru

Abstract. In this paper, the new sensors based on the C(4) or/and C(5) mono(heteroaryl) and di(heteroaryl) substituted pyrimidines are described. The effect of different substituents in pyrimidine ring on the detection limits and selectivity for prepared sensors are explained. The results of detection towards various nitro-explosives and volatile interferents at room temperature with use of the portable detector "Nitroscan" are shown.

1. Introduction

Development of an efficient and cost-effective method for detection of nitroaromatic explosives is highly desirable, since there are considerable environmental and security needs, especially with the surge of international terrorism and an enhanced use of explosives in terrorist attacks [1-3]. There are a number of handled sniffers (sensors working as an artificial dog's nose) using advantage of materials which are capable of searching explosives through smell of these compounds [4]. The most famous artificial nose for nitroaromatics has been launched successfully into the market, as "FIDO" by Nomadics Inc [5]. "Nitroscan", a Russian portable detector of nitro-explosives, has been designed and produced by "Promautomatika" Plant in collaboration with "UralProektAvtomatika" and "Security Intermolecular System" (Ekaterinburg, Russia) [6]. The main device functions are as follows: excitation of sensor element stationary luminescence and measuring its intensity in the absence of explosive vapors; registration of changes in the intensity of sensor element luminescence in time at the contact of its surface with explosive vapors; creation of external conditions for efficient interaction of the sensor element surface with explosive vapors. The device contains a ventilator for pumping air through the sensor element, an exciting luminescence light source, a photo element for luminescence intensity recording, an electronic tract, that converts the light flux into an electrical signal, the analog-digital converter and a microcontroller unit for control and display. Alarm on explosive detection is realized as a sound emitter and quantitative indicators are presented on the symbol display [7]. The development and investigation of new sensors for the device "Nitroscan" was the aim of our work.

2. Results and discussion

Recently, three types (A-C) of new push-pull dyes based on pyrimidines, bearing thiophene or/and phenyl linkers, have been studied as sensing fluorophores (figure 1). These compounds A-C have been obtained by using a combination of the Suzuki cross-coupling and nucleophilic aromatic substitution of hydrogen (S_N^H) reactions [8-10].

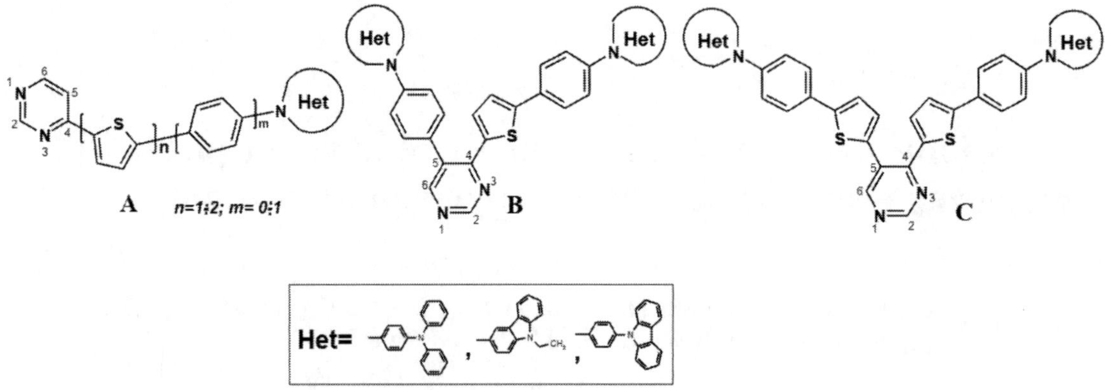

Figure 1. Molecular structures of the donor-π-acceptor organic dyes A-C on a pyrimidine scaffold.

To evaluate the utility of the fluorophores A-C for the detection of nitroaromatic explosives, the corresponding fluorescence measurements of the fluorophores A-C were carried out in acetonitrile solutions containing measured quantities of the compounds shown in figure 2, which serve as quenchers of the fluorescence.

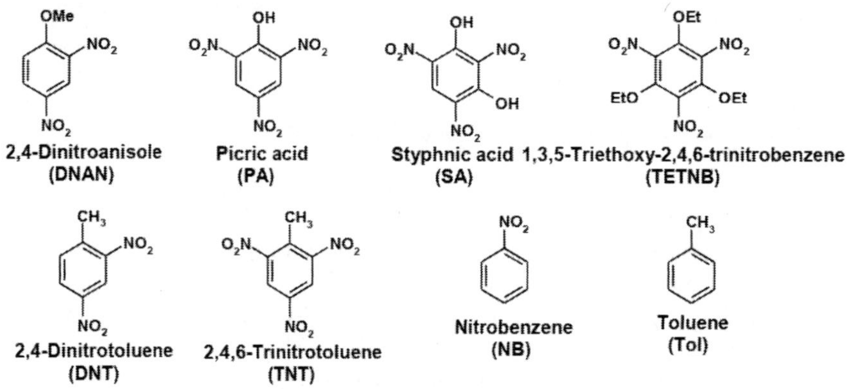

Figure 2. Structures of used quenchers and interferents.

Detection limits (DL) of the fluorophore to the analytes in acetonitrile solution were determined and calculated according to the following functions:

$$s_b = \sqrt{\frac{\sum_{i=1}^{n}(x_i - \bar{x})^2}{n-1}} \tag{1}$$

$$S = \frac{\Delta I}{\Delta c} \tag{2}$$

$$DL = \frac{3s_b}{S} \tag{3}$$

Firstly, the standard deviation (S_b) was calculated by measuring the intensity of the fluorophore in the blank solution for more than 10 times and then the average intensity ($\bar{x}$) was obtained. By fitting

IOP Conf. Series: Journal of Physics: Conf. Series **830** (2017) 012159 doi:10.1088/1742-6596/830/1/012159

the data into Function 1, the value of standard deviation (S_b) was obtained. Secondly, a certain amount of analyte such as DNAN, PA, SA, TETNB, DNT, TNT, TATB, NB or Tol was added into the blank solution and the resulting variation of the intensity (Δc) was recorded. By fitting the data into function 2, where ΔI is the variation of intensity, and Δc is the variation of quencher concentration, the value of precision S was calculated. Finally the detection limit, DL, was calculated according to function 3. The ranges of detection limits for different nitroaromatic compounds are given in table 1.

Table 1. Detection limits (DL) of DNAN, PA, SA, TETNB, DNT, TNT, TATB, NB and Tol towards fluorophores A-C in CH_3CN.

Dye	DL (mol×L^{-1})							
	DNAN	PA	SA	TETNB	DNT	TNT	NB	Tol
A	-	5.8×10^{-6} ÷2.4×10^{-7}	-	-	8.4×10^{-5} ÷1.2×10^{-6}	1.7×10^{-4} ÷1.4×10^{-6}	-	-
B	2.6×10^{-4} ÷1.2×10^{-5}	1.7×10^{-5} ÷1.8×10^{-7}	5.0×10^{-6} ÷2.1×10^{-6}	1.1×10^{-4} ÷3.5×10^{-6}	4.6×10^{-4} ÷3.6×10^{-5}	2.0×10^{-4} ÷6.5×10^{-5}	2.7×10^{-4} ÷4.3×10^{-5}	-
C	2.0×10^{-3} ÷2.9×10^{-5}	6.9×10^{-4} ÷4.0×10^{-5}	7.15×10^{-4} ÷3.7×10^{-5}	1.5×10^{-3} ÷4.7×10^{-5}	5.4×10^{-4} ÷2.9×10^{-5}	5.9×10^{-4} ÷3.0×10^{-5}	1.3×10^{-4} ÷3.8×10^{-3}	5.1×10^{-3} ÷1.1×10^{-4}

As can be seen in Table 1, detection limits are decreased for fluorophores from type A to type C. We suggest that an additional aryl substituent at C(5) causes a significant steric hindrance which has to be proceeding at the neighboring C(4) position of the pyrimidine ring. This steric hindrance deteriorates interaction of fluorophore with quencher and also interrupts π-electron conjugation in push-pull system.

On a base of fluorescence measurements in acetonitrile solutions, the two best fluorophores (see figure 3) have been proposed for fabrication of sensors prototypes and their application in detecting of explosive vapors [9,10].

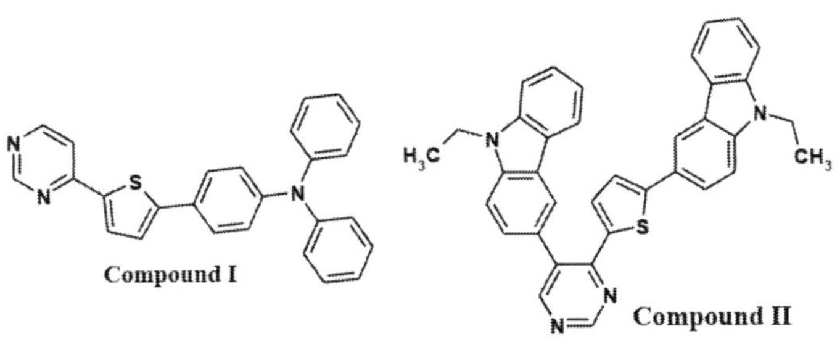

Figure 3. Structures of fluorophores I and II with the best data for detection limits in acetonitrile solutions.

For the fabrication of sensors, a non-woven dry wipe was used as a porous substrate (70% wood pulp, 30% polypropylene, Industrial Wipes purchased from Kunshan Yichen Clean Material Co., Ltd., China). A piece of the non-woven dry wipe (50 mm in diameter) was immerged in the fluorophore solution with a range of concentration: 1.0×10^{-3} (**I** or **II**), for 5 min. Then non-woven dry wipe or spunlace with immobilized fluorophore **I** or **II** was removed from the solution and dried at 70–80 °C for 30 min. To demonstrate its application as a fluorescence sensor for nitroaromatic compounds detection, the obtained non-woven dry wipe was placed in cartridges.

The resulted sensors have been used to detect nitrobenzene (NB) vapor in a hermetic glow-box (0.8 m×0.6 m×0.4 m) containing maximum permissible concentration (3 mg/m^3) and also 0.3 and 0.03 mg/m^3 of NB). It has been shown that all sensors are capable of fast reversible and reusable self-recovering after interruption of exposure to nitroaromatic explosive. Reversibility of the sensing process for sensors on the basis of fluorophores **I** and **II** in NB vapor was examined by using the following procedure: the device bearing a sensor was inserted into a glow box with NB vapor with various concentrations for 25 s, and then the fluorescence intensity was recorded. After that, the device was removed from the box for 25 s, and the ambient room air was pumped automatically through the sensor in the device. The recovered emission intensity of the sensor was measured. The sensing and pumped process was repeated 3 times, and the results obtained are shown in figure 4. As can be seen, the recovered fluorescence intensity is decreasing with each next cycle but the quenching efficiency does not falls down to a 0 level even up to cycle 30.

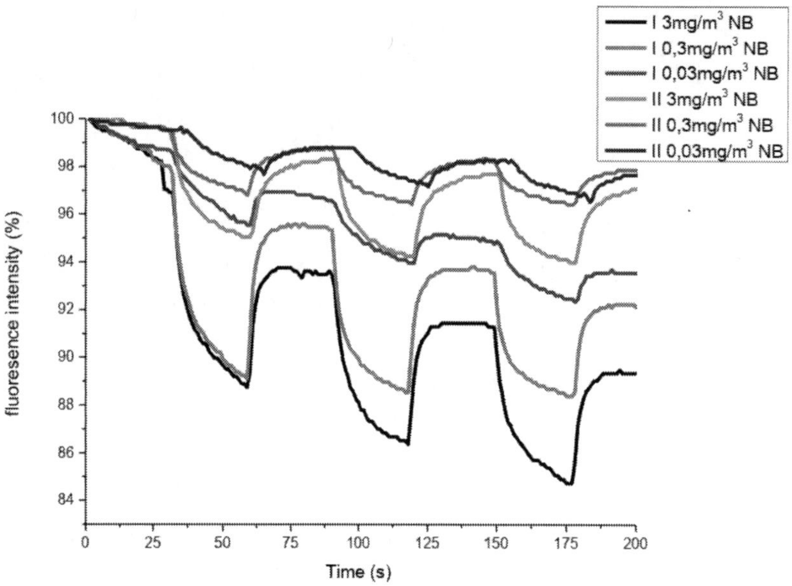

Figure 4. Fluorescent recovery cycles for the sensor on the basis of **I** and **II** in the device "Nitroscan" with exposure to nitrobenzene (NB) vapor with various concentrations.

In addition, both sensors **I** and **II** show a good selectivity to DNT in comparison with some other explosives [2,4,6-trinitrotoluene (TNT), 2,4,6-trinitrophenyl-N-methylnitramine (Tetryl), cyclotrimethylenetrinitramine (RDX), pentaerythritol tetranitrate (PETN)] and interferents, such as ethanol, ethylene glycol, acetone, acetic acid, toluene, nitrobenzene, phenol, and 1,2-dichlorobenzene (see figure 5), since their quenching sensitivities in these vapours are weaker than those in DNT.

3. Conclusion

In summary, fifteen dyes of the push-pull type on the basis of pyrimidines bearing various thiophene or/and phenyl linkers and various electron-donative fragments have been studied as sensing fluorophores. All dyes demonstrate high sensitivity towards nitroaromatic explosives. High fluorescence quenching response makes these fluorophores in vapour be promising fluorescence sensor materials for nitroaromatic explosives with the detecting limit of ~10−11 g/cm^3, as estimated in experiments with NB by utilizing the portable detector "Nitroscan".

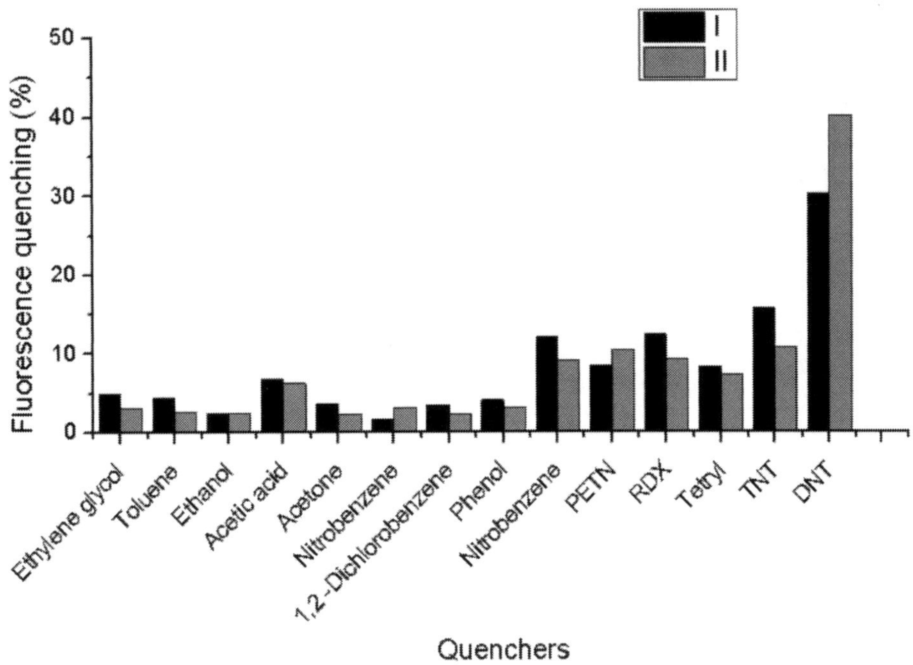

Figure 5. Sensor's **I** and **II** fluorescence quenching (%) in the first cycle towards various nitro-explosives and volatile interferents at room temperature.

Acknowledgements
The research was financially supported by the Program for the State Support of the Leading Scientific Schools of RF, Grant NSh-8922.2016 and the Russian Foundation for Basic Research (research projects No. 17-03-00011-A).

References
[1] Krausa M, Reznev A A 2004 *Vapor and trace detection of explosives for anti-terrorism purposes Kluwer* (Boston: Academic Publisher)
[2] Sunahara G I, Lotufo G, Kuperman R G, J Hawari 2009 *Ecotoxicology of explosives* (London, New York, Boca Raton: *CRC* Press Taylor & Francis)
[3] Yinon J 2007 *Counterterrorist detection techniques of explosives* (Amsterdam, Boston, Heidelberg, London, New York, Oxford, Paris, San Diego, San Francisco, Singapore, Sydney, Tokyo: Elsevier B V)
[4] Settles G S 2005 *Sniffers: fluid-dinamic sampling for olfactory trace detection in nature and homeland security* (Boston: Academic Publisher)
[5] Cumming C J, Aker C, Fisher M, Fok M, la Grone M J, Reust D, Rockley M G, Swager T M, Towers E, Williams V 2001 *Using novel fluorescent polymers as sensory materials for above-ground sensing of chemical signature compounds emanating from buried landmines* (*IEEE Trans. Geosci. Remote Sens.*39) pp 1119–28
[6] Website of the portable detector of nitro-explosives *«Nitroscan» http://nitroscan.pro*
[7] Baranova A A, Khokhlov K O 2014 Portable device for the detection of nitro-explosives based on optical properties of sensor's material *Journal of Physics: Conference Series* **552** 012034
[8] Verbitskiy E V, Cheprakova E M, Slepukhin P A, Kodess M I, Ezhikova M A, Pervova M G, Rusinov G L, Chupakhin O N, Charushin V N 2012 *Combination of the Suzuki-Miyaura cross-*

IOP Conf. Series: Journal of Physics: Conf. Series **830** (2017) 012159 doi:10.1088/1742-6596/830/1/012159

coupling and nucleophilic aromatic substitution of hydrogen (S_N^H) reactions as a versatile route to pyrimidines bearing thiophene fragments (*Tetrahedron vol 68*) pp 5445–52

[9] Verbitskiy E V, Baranova A A, Lugovik K I, Khokhlov K O, Cheprakova E M, Rusinov G L, Chupakhin O N, Charushin V N 2016 *New V-shaped push-pull systems based on 4,5-di(hetero)aryl substituted pyrimidines: their synthesis and application to the detection of nitroaromatic explosives* (ARKIVOC iii) pp 360–73

[10] Verbitskiy E V, Baranova A A, Lugovik K I, Shafikov M Z, Khokhlov K O, Cheprakova E M, Rusinov G L, Chupakhin O N, Charushin V N 2016 Detection of nitroaromatic explosives by new D–π–A sensing fluorophores on the basis of the pyrimidine scaffold *Analytical and Bioanalytical Chemistry* **408** pp 4093–101

IOP Conf. Series: Journal of Physics: Conf. Series **830** (2017) 012160 doi:10.1088/1742-6596/830/1/012160

Complex study on photoluminescence properties of YAG:Ce,Gd phosphors

V M Lisitsyn[1], Yangyang Ju[1], S A Stepanov[1] and N M Soschin[2]

[1] National Research Tomsk Polytechnic University, 30 Lenin Ave., Tomsk, 634050, Russia

[2] Scientific-Research Institute "Sycamore", Fryazino, 141195, Russia

E-mail: 1374586947@qq.com

Abstract: Luminescence characteristics of gadolinium co-doped yttrium aluminium garnet doped with cerium phosphors were studied. In this work, powder X-ray diffraction (XRD) spectra, elemental composition analyses, excitation and emission spectra, conversion efficiency of emission phosphor, corresponding (CIE) chromaticity colour coordinates and pulsed photoluminescence decay kinetic curves were investigated, all the measurements were performed at room temperature. The properties of the phosphors were studied by comparing the composition of the phosphors and their luminescent properties.

1. Introduction

Yellow colour phosphors based on yttrium aluminum garnet (YAG, $Y_3Al_5O_{12}$) are promising materials because of their good chemical and physical properties: excellent chemical stability at elevated temperature and under radiation stimulation, good emission characteristics under excitation.YAG phosphor is used to convert electrons to visible light in displays [1-5]. Ce^{3+}-activated phosphor $Y_{3-x}Al_5O_{12}$:$x$$Ce^{3+}$ (YAG:Ce) as a well-known efficient phosphor material is used to convert ultraviolet and blue LED radiation into yellow emission. It is assumed that the excitation of Ce^{3+} ion causes maximum absorbance in the blue region resulting in converting InGaN based blue LED radiation into a very broad intense yellow emission band and then combing the yellow emission with the non-absorbed blue emission from the blue LED to generate white light. However, it shows a low color rendering index (CRI). Thus in order to improve the quality of white light YAG:Ce can be doped with other rare earth ions, such as Gd^{3+} and Tb^{3+} [6]. Gd^{3+} (Gd^{3+} i.r. = 0.938 Å) has been added to substitute at the Y^{3+} (0.9 Å) site, which attributes to lattice distortion in the region of Ce^{3+} and shift of the emission band to a longer wavelength.

In the present study, industrial YAG phosphors with different concentration of Gd3+ were investigated.

2. Experimental details

Industrial YAG phosphors of SDL4000, SDL2700 and SDL3500 with different ratio of elemental composition were used for the present research. Information on the test samples is shown in Table 1. The elemental composition analyses of the samples were performed by a x-ray photoelectron spectroscopy (XPS) ESCALAB 250 with a PHI 5000Versaprobe-Scanning ESC Amicroprobe. The measurements were carried out by a monochromatic Al Ka X-Ray source (500 µm, 60 W, 10 kV) with photon energy of 1486.6 eV. Survey scans (binding energies ranging from 0 to 1400 eV) were performed using a 1 eV/step, while for higher resolution spectra of the Gd 4d, Ce 3d,O 1s, Al 2p,C ls, and Y 2d peaks, the hemispherical analyzer pass energy was maintained with 0.05 eV/step.

The phosphors of SDL2700 and SDL3500 contain Gd and Ce except for Y, Al and O. Gd and Ce in phosphor SDL4000 were not detected because of its small content. We emphasize that the used method allows us to determine the composition of the sample on its surface.

The crystal structure of the measured phosphors was identified by using the powder X-ray diffraction (XRD) Rigaku smart lab with Cu, Ka radiation generated at 40 kv/30 mA. The recorded XRD spectra were covered from 10° to 60° with increments of 0.02°.

The photoluminescence spectra of the samples were recorded by FLS980 spectrometer (Ediburgh instruments) with a xenon lamp (250–1000 nm) as an excitation source. CIE color coordinates in the yellow light region were determined by luminescence spectra of the investigated phosphors.

Table 1. Elemental composition of phosphors

No.	phosphor	Y	Al	O	Gd	Ce
1	SDL2700	4.48	24.13	64.49	6.22	0.68
2	SDL3500	13.47	26.8	57.18	1.26	1.29
3	SDL4000 (1)	16.6	25.83	57.57		
4	SDL4000 (2)	16.31	23.94	59.75		

3. Experimental results

3.1. Crystal structure

Figure 1 shows the XRD patterns of tested phosphors. For comparison, a standard YAG phase (PDF 09-1316 for $Y_3Al_5O_{12}$) was also inserted.

From the inset in figure 1, the diffraction peaks are in line with that of the standard YAG phase. It is clear that no impurity phases were formed, which are possible due to the dopant Ce^{3+} and Gd^{3+} ions. It can be found that the diffraction peak of (420) moves to the lower angle side with the increase of Gd^{3+} content relative to the position in the crystal. It is assumed [7] that the shift of the peak (420) is caused by lattice distortion and substitution of ions Y^{3+} (0.92Å) by ions Gd^{3+} (0.938Å).

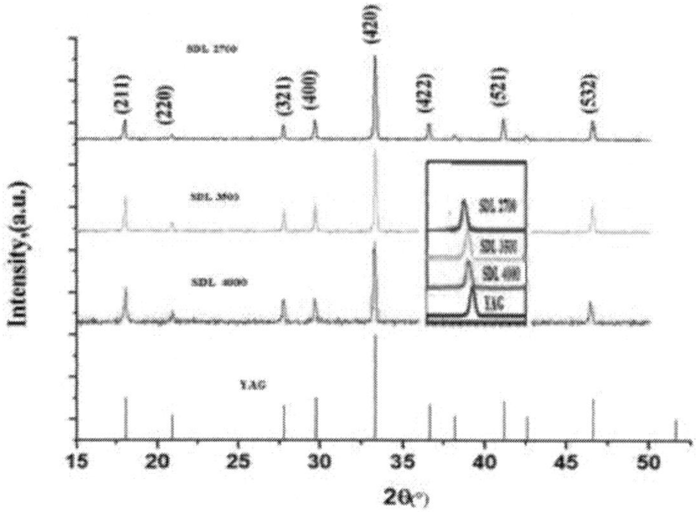

Figure 1. XRD patterns of SDL4000 ($Y_3Al_5O_{12}$), SDL3500 ($Y_{2.6}Gd_{0.25}Al5O12:Ce_{0.15}$), SDL2700 ($Y_{1.24}Gd_{1.56}Al5O12:Ce_{0.2}$) and standard $Y_3Al_5O_{12}$

3.2. Photoluminescence spectra and excitation spectra

Figure 2(a) shows excitation spectra of tested phosphors at the fixed emission wavelength of 560 nm, two bands can be seen in the region between 300 and 500 nm and these are peaked at 344 and 454 nm. The peak wavelength at 454 nm matches well with that of the light emitted by InGaN based blue LED; therefore, YAG phosphors can efficiently absorb blue light of the chip and convert it into a higher wavelength region.

Figure 2(b) (c) shows the photoluminescence (PL) emission spectra of YAG and YAG:Ce co-doped with Gd at a fixed excitation wavelength of 340 nm and 460 nm. The PL emission spectra show a broad emission band, which is supposed to be caused by Ce^{3+} ions with emission maxima occurring at 560 nm. It is known that Ce^{3+} is sensitive to the local crystalline field environment; therefore, lattice distortion by Gd ion substituting the Y^{3+} ion causes the shift in the emission band. It was observed from Figure 2 (b) and (c) that the substitution of Y ion results in red shift. However, the overall luminescence intensity decreases with the increase of Gd^{3+}.

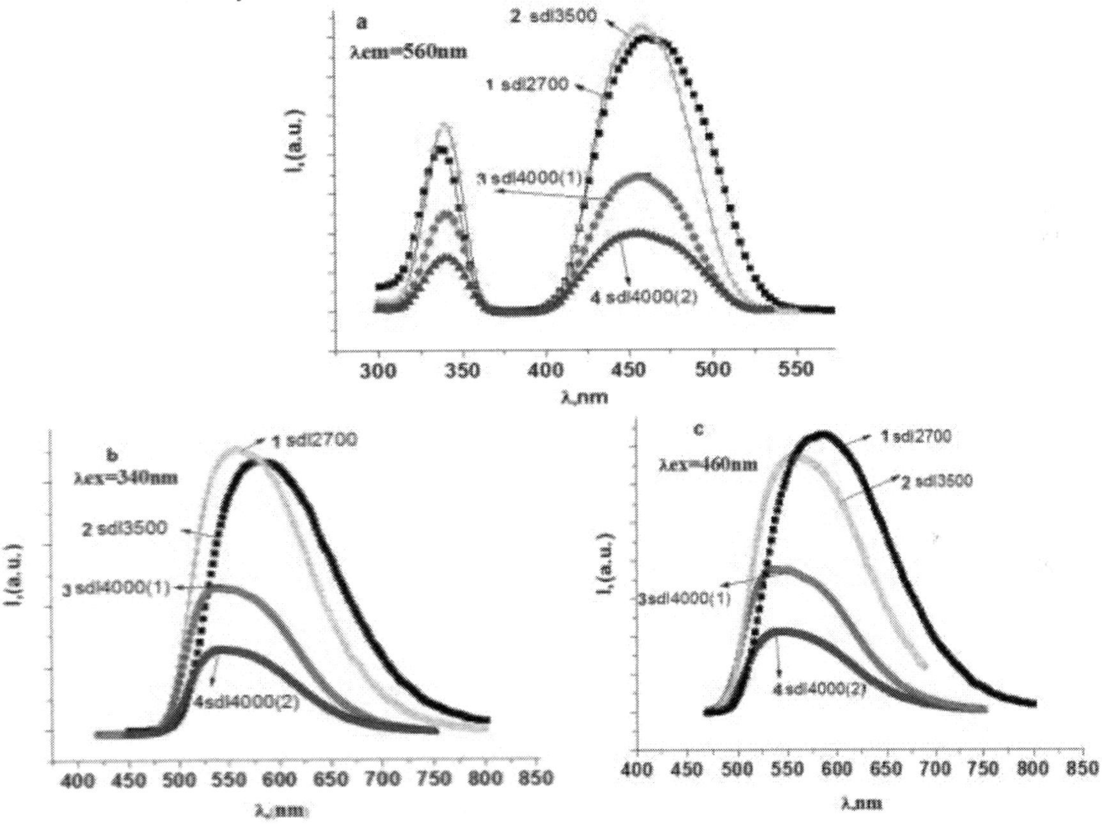

Figure 2. PL excitation spectra(a) and photoluminescent emission spectra (b, c) of phosphor samples SDL2700($Y_{1.24}Gd_{1.56}Al5O12:Ce_{0.2}$), SDL3500 ($Y_{2.6}Gd_{0.25}Al5O12:Ce_{0.15}$) and 2 SDL4000 ($Y_3Al_5O_{12}$)

The measurement results obtained for the spectral characteristics of the studied phosphors are summarized in Table 2. The positions of the luminescence bands do not depend on the area of the luminescence spectrum excited and the type of excitation. We can detect the following changes in the luminescence and excitation spectra.

The luminescence band of SDL4000 phosphor (~ 0.44 eV) under λ_{ex} = 454 nm is narrower than that of SDL2700 and SDL3500 (~ 0.485 eV) phosphors. The luminescence band of SDL4000 phosphor under λ_{ex} = 344nm (~ 0.495 eV) is wider than λ_{ex} = 454nm (~ 0.44 eV). The width of the excitation band at 340 nm of all studied phosphors varies from 0.264 to 0.324 eV. The width of the excitation band at 460 nm of all studied phosphors varies from 0.390 to 0.450 eV.

Table 2. Measurement results obtained for the luminescence spectra and excitation of phosphors: peak value position and half band-width

phosphor	luminescence				excitation			
	λ_{ex} = 460 nm		λ_{ex} = 340nm		In the region of 340nm		In the region of460nm	
	ΔE, eV	λ_{Max}, nm	ΔE, eV	λ_{Max},nm	ΔE, eV	λ_{Max}, nm	ΔE, eV	λ_{Max}, nm
SDL2700	0.487	584	0.489	582	0.264	336	0.450	459
SDL3500	0.484	560	0.487	558	0.294	339	0.390	459
SDL 4000(1)	0.435	539	0.486	540	0.313	339	0.396	456
SDL 4000(2)	0.447	541	0.504	542	0.324	339	0.426	456

3.3. *Evaluation limiting values of light output at conversion of spectra in LED*

The limiting values of the light output during spectrum conversion in phosphor were evaluated. The evaluation was carried out under the following conditions. We only take into account the energy losses in the process of converting chip radiation into luminescence in phosphor. In the spectrum of "white" light of LED is considered the proportion of "blue" radiation of chip that involved in the formation of the spectrum. The calculation method is described in [8], and the calculations were performed for two variants: the luminescence of phosphor is excited under λ_{ex} =344 nm in the UV radiation region and in the blue radiation region under λ_{ex} =454 nm. In this work, it was assumed that the value of emission intensity at peak wavelength λ=454 nm equals to the peak luminescence of phosphors at λ_{em}=560 nm under λ_{ex} =454 nm. The results are presented in Table 3.

Table 3. Energy losses during spectrum conversion in phosphor and limiting values of light output

Phosphor	λ_{ex} = 344 nm			λ_{ex} = 454 nm		
	η_e	η_t	η_l	η_e	η_t	η_l
SDL2700	0.566	0.434	229	0.73	0.27	267
SDL3500	0.58	0.42	261	0.76	0.24	327
SDL4000(1)	0.593	0.407	277	0.76	0.24	349
SDL4000(2)	0.59	0.41	278	0.76	0.24	350

As follows from the calculated results, the values of energy losses in the process of converting chip radiation under λ_{ex} =344 nm into luminescence of phosphor in the region of λ_{em}=560 nm are not less than 44%, the luminous efficiency of the LED cannot be more than 280 lm/W. Accordingly, under λ_{ex} = 454nm, the values of energy losses are more than 27%, the luminous efficiency of the LED cannot be more than 350 lm / W.

3.4. *CIE colour coordinates*

Table 4 shows the chromaticity coordinate of phosphor emission. With the increase in the content of Gd^{3+} emission spectrum is approaching the ideal range of white light with x = 0.33 and y = 0.33.

Table 4. CIE colour coordinates of each tested phosphors

Phosphors	CIE color coordinates	
	x	y
SDL2700($Y_{1.24}Gd_{1.56}Al5O12:Ce_{0.2}$)	0.34	0.49
SDL3500($Y_{2.6}Gd_{0.25}Al5O12:Ce_{0.15}$)	0.45	0.53
SDL4000 (1)($Y_3Al_5O_{12}$)	0.42	0.55
SDL4000(2) ($Y_3Al_5O_{12}$)	0.43	0.55

3.5. The light output of LED with phosphors

The integrated spectral efficiency was measured by using an integrating sphere and calibrated spectrophotometer AvaSpec-ULS3648. The energy output η is the radiation flux ratio of phosphor to absorb the excitation flux. Two types of chips were used for excitation, which generate radiation with a stream of 173 μW/ cm^2 in the band with a maximum at 460 nm and 16 730 μW/ cm^2 at 447 nm. Power of LED are almost 100 times different.

Table 5. The radiation energy efficiency of converting the phosphor emission into a blue band of the spectrum.

Phosphor	Radiation energy output η	
	λ_{ex} = 460 nm	λ_{ex} = 447nm
SDL2700	0.422	0.331
SDL3500	0.438	0.359
SDL4000(1)	0.377	0.373
SDL4000(2)	0.404	0.394

As follows from the results, radiation energy output η almost independent on the power of the exciting radiation during the conversion process of SDL4000 phosphor. The change does not exceed 2.5%. However, he energy output of SDL2700 phosphor by 22% and in the SDL3500 phosphor by 18% is reduced, with increase power of LED in 100 times. This difference in energy output can be attributed to the fact that phosphors are sensitive to temperature variation.

4. Conclusions

Complex research on a group of phosphors based on YAG with different content of Ce^{3+} activator and co-activator Gd^{3+} was conducted. These impurities are intentionally not introduced to the SDL4000 phosphor, in which Ce^{3+} and Gd^{3+} were not detected by using XPS method. All the studied phosphors have a crystal structure in accordance with a typical body centered cubic structure of YAG ($Y_3Al_5O_{12}$); however, lattice distortion causes the shift of the peak (420). It is assumed that the lattice distortion is caused by substitution of ions Y^{3+} (0.92Å) by ions Gd^{3+} (0.938Å). This explanation is well correlated with the measurement results obtained for XRD patterns of the studied phosphors. However, the peak shift in the SDL4000 phosphor relative to the peak in the crystal is evidently due to the other factors. Perhaps crystal lattice distorted in the SDL4000 phosphor is formed in the synthesis of intrinsic defects lattice.

Upon excitation in the region of 344 nm and 454 nm, the position of luminescence bands of all investigated phosphors respectively at 580 nm of SDL2700, 560 nm of SDL3500 and 540 nm of SDL4000 phosphor. With the increase of Gd^{3+} content, the emission wavelength shows a great red shift, which is due to the crystal lattice distortion formed in the region of the luminous center. It is assumed that the emission center of phosphors is Ce^{3+} ion. Apparently, luminous center of the SDL4000 phosphor has the same nature, though Ce^{3+} is not intentionally introduced during synthesis. It is interesting to observe that the luminescence band of SDL4000 phosphor under λ_{ex} = 344 nm is broader as compared to that under λ_{ex} = 454nm.

This difference can be explained either by the presence of two different crystal structures of luminous centers, or one type of luminous centers with different environments. One of the centers is dominant in SDL2700, 3500 phosphors.

It can be considered that the luminescence characteristics depend not only on the composition of the doping activators and co-activators, but also on the presence of high concentrations of native defects in the crystal which were included during synthesis. It is difficult to ensure compliance with stoichiometry during formation of the crystal with a complex structure. Therefore, native defects are introduced in the crystal during synthesis. Native defects together with activators and co-activators can form complex defects which are called nanodefects [9].

Acknowledgments
This work was supported by the *Key Laboratory of Supramolecular Structure, Chemical Institute of Jilin University*. The authors are greatly indebted to Prof. Yue Wangat Jilin University. Thanks to doctoral student Yue Wei Zhang for his assistance in experimental work.

Reference

[1] Jang H, BinIm W, Lee D, Jeon D and S Kim 2007 *J. Lumin.* **126** 371
[2] Jung K and W Lee 2007 *J. Lumin.* **126** 469.
[3] Zhou S, Fu Z, Zhang J and S Zhang 2006 *J. Lumin.* **118** 179.
[4] Zhou Y, Lin J, Yu M, Wang S and Zhang H 2002 *Mater. Lett.* **56** 628
[5] Blasse G and A Bril 1967 *Appl. Phys. Lett.* **11** 53
[6] Jun Wang, Tao Han,Tianchun Lang, Mingjing Tu and Lingling Penga 2015 *Optical Engineering* **54** 117106
[7] Lisitsyn V, Lukash V,Stepanov S and Y Ju 2016 *AIP Conference Proceedings* **1698** 060008
[8] Xu M, Zhang Z, Zhu J, Zhao J and X Chen 2014 *J. Phys. Chem. C.* **118** 27000
[9] Lisitsyn V, Stepanov S,Valiev D, Vishnyakova E, Abdullin H, Marhabaeva A and A Tulegenova 2016 *IOP Conf.Series: Materials Science and Engineering* **110** 012051

Dispersion characteristic of photoluminescence decay times of phosphor YAG: Ce, Gd

V M Lisitsyn[1], Yangyang Ju[1], S A Stepanov[1] and N M Soschin[2]

[1] National Research Tomsk Polytechnic University, 30 Lenin Ave., Tomsk, 634050, Russia
[2] Scientific-Research Institute "Sycamore", Fryazino, 141195, Russia

E-mail: 1374586947@qq.com

Abstract: The dispersion of the characteristic decay times of gadolinium co-doped yttrium aluminum garnet doped with cerium phosphors were studied. In the present work, an ultraviolet semiconductor laser (λ_{em}=375 nm, τ = 1 ns) was used as excitation source for measuring kinetics characteristics of phosphor groups based on YAG with different content of cerium.

1. Introduction

Phosphors based on yttrium aluminum garnet doped cerium ($Y_{3-x}Al_5O_{12}$: Ce_x^{3+}) have been found to be an efficient phosphor material, which are used for the conversion of ultraviolet or blue emission from LEDs into white light emitting LEDs. [1-6] On account of the 4f–5d transition of the Ce^{3+} ion, which shows a maximum absorbance in the blue region thus converting InGaN based blue LED radiation into a broad intense yellow emission band. White light is then a result of the combination of this yellow emission with the non-absorbed blue emission from the blue LED. In order to improve the color rendering index (CRI) of white light emission, co-activators are doped to change the composition of the matrix, and accordingly the luminescence spectrum also changes.

It is assumed that the change of luminescence spectrum is due to the lattice distortion formed in the region of luminous center of ion Ce^{3+}. In [7] suggested that the luminous center has a complex structure. Additional information about luminous centers can be obtained from kinetic characteristics of luminescence decay time. Luminous centers with different structures should have different kinetic characteristics.

2. Experimental details

Industrial phosphors SDL4000, SDL2700 and SDL3500 which are characterized by the different ratio of the elemental composition were used for this research. Among them, SDL2700, SDL3500 contain gadolinium and cerium in amounts of 6.22%, 1.26% and 0.68%, 1.29% of the total composition, respectively. SDL4000 does not contain gadolinium and cerium in amounts which are available for measurement by X-ray photoelectron spectroscopy (XPS).

The photoluminescence spectra of the phosphor samples were recorded by using FLS980 spectrometer with a xenon lamp (250-1000 nm) as the excitation source. Ultraviolet semiconductor laser (λ_{em} = 375 nm, τ = 1 ns) was used as the excitation source for the measurement of the kinetics of decay time

3. Experimental results

3.1. Photoluminescence properties and energy transfer

The excitation spectra (λ_{em} = 560 nm) of phosphor samples were measured in the region between 300 and 550 nm. The excitation spectrum allocated 2 peaks at 344 nm and 454 nm. The excitation peak at 454 nm agrees well with the peak of blue emission band of chip based on InGaN. Therefore, the studied YAG phosphors can efficiently absorb blue emission from chip and then convert it into a more long-wavelength region. Photoluminescence (PL) emission spectra of YAG and YAG:Ce were co-doped with Gd at a fixed excitation wavelength of 340 nm and 460 nm. The measurement results obtained from the spectral characteristics of the studied phosphors are summarized in Table 1. Spectra of SDL2700 and SDL3500 phosphor compared to SDL 4000 phosphor present a broader and shift to the red region.

Table1. Measurement results obtained for the luminescence spectra and excitation of phosphors: peak value position and half band-width

Phosphor	Luminescence				Excitation			
	λ_{ex} = 460 nm		λ_{ex} = 340nm		In the region of 340nm		In the region of460nm	
	ΔE, eV	λ_{Max}, nm	ΔE, eV	λ_{Max},nm	ΔE, eV	λ_{Max}, nm	ΔE, eV	λ_{Max}, nm
SDL2700	0.487	584	0.489	582	0.264	336	0.450	459
SDL3500	0.484	560	0.487	558	0.294	339	0.390	459
SDL 4000(1)	0.435	539	0.486	540	0.313	339	0.396	456
SDL 4000(2)	0.447	541	0.504	542	0.324	339	0.426	456

3.2. Kinetics curves of the luminescence decay

The kinetics curves of the luminescence decay were excited by pulsed laser radiation with λ = 375 nm. The results are shown in Figure 1.

The luminescence decay kinetic curves in nanosecond time domain are well described by a sum of two exponential functions.

$$I = A1 \exp(-t / \tau 1) + A2 \exp(-t / \tau 2) + I0 \quad [1]$$

Where A1 and A2 represent respective initial intensity; I0 represents the intensity of background, which is independent of the test range time; τ1 and τ2 - characteristic luminescence decay times in the nanosecond range.

Kinetic curves in the nanosecond time domain are well described by a sum of two exponential functions. The measurement results of the spectral dependence of the characteristic decay time phosphors are shown in Figure 1. From these results it can be concluded that investigated phosphors have different kinetic luminescence decay curves. The inset in Figure 3 shows the luminescence kinetic curves in semilogarithmic scale at different wavelengths. Results of the studied kinetic luminescence decay curves are summarized in Table 2.

Table2. Summarizes the values of the characteristic luminescence decay time τ for studied phosphors under pulsed laser excitation with λ = 375 nm

№	phosphor	τ_1 (ns)	Rel.%1	τ_2 (ns)	Rel.%2	τ (ns)
1	SDL2700	4.2	10.4	66.1	89.6	59.6
2	SDL3500	1.5	12.3	60.5	87.7	53.4
3	SDL4000(1)	3.0	12.7	61.8	87.3	54.4
4	SDL4000(2)	2.9	9.1	59.8	90.9	54.6

IOP Conf. Series: Journal of Physics: Conf. Series **830** (2017) 012161 doi:10.1088/1742-6596/830/1/012161

Short-time decay component in the nanosecond range for the investigated phosphors changes from 1.46 to 4.15 ns, long lifetime component is in the range of 59.8 - 61.8 ns. Long lifetime component accounts for 87.3% - 90.9% of total radiation. Averaged decay time τ is in the range of 53-59 ns, which corresponds to the measurement results with large duration excitation pulse about 60 ns in other working time.

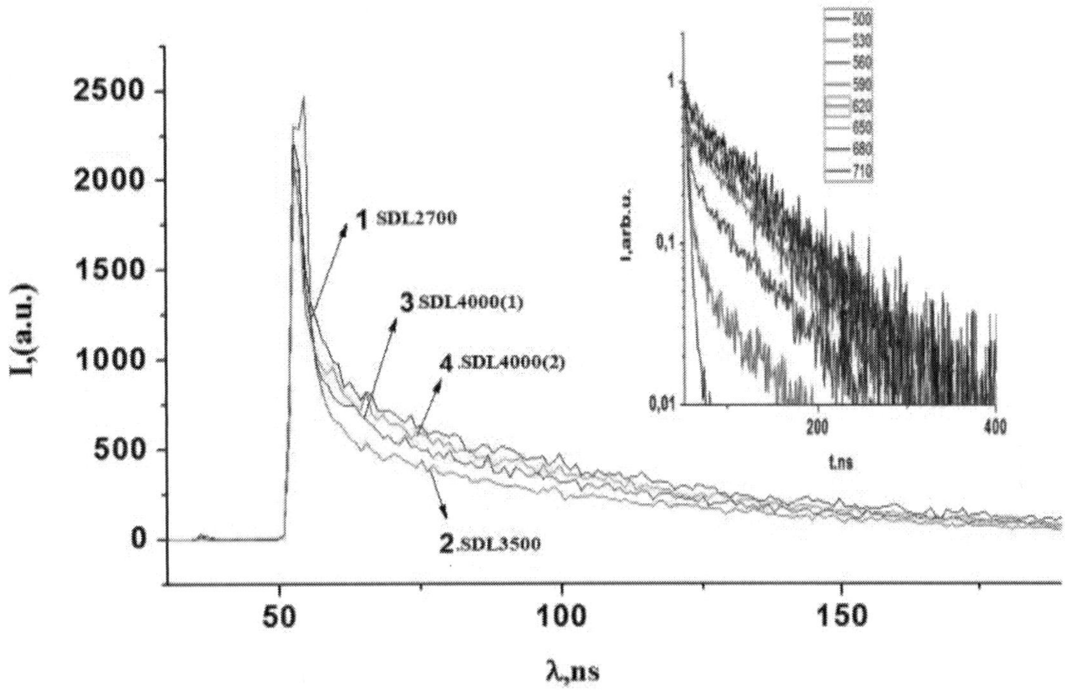

Figure1. Decay kinetic characteristics of the luminescence for investigated phosphors under pulsed laser excitation with λ = 375 nm

3.3. The kinetic curves dispersion of the luminescence decay

As shown in Figure 1, the kinetics of luminescence decay depends on the wavelength at which it is measured.

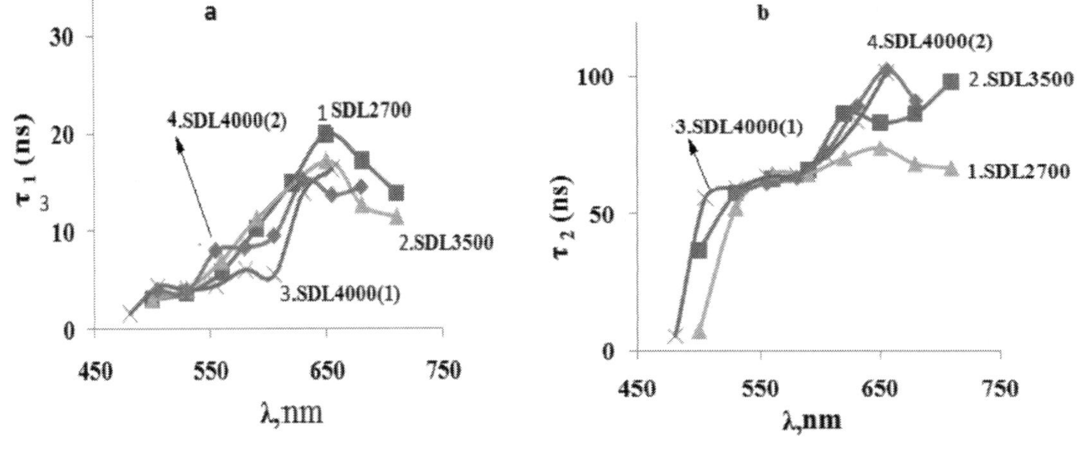

Figure 2.The dispersion of the characteristic decay times.

The dependence of the luminescence delay time with the excitation wavelength in the nanosecond range was studied. According to the measurement results, characteristic decay times were calculated to build a diagram of relationship between time and wavelength (dispersion characteristic times). The measurement results are shown in Figure 2: short –time component (2a) and long time component (2b).

As shown in Figure 2, the dispersion curve can be divided into three sections. Within a short time in the spectral region up to 500 nm is mainly recorded radiation of chip. Photoluminescence duration of two components is lower than at longer wavelengths in the region up to 610 nm. The emergence of this phenomenon may be caused by different luminous centers, or luminous centers are in different environments.

4. Conclusions

The investigation results make it possible to suggest the following conclusions: from the research of kinetic curves dispersion, it can be seen that characteristic decay times in the range of 480 - 610 are different from those in the range of 610 - 700 nm. This indicates that there are two kinds of luminous centers responsible for the luminescence of all investigated phosphors, luminescence independent on the content of cerium. The changes in the ratio of two types of luminous centers lead to a change in the luminescence spectrum of investigated phosphor. It is difficult to ensure compliance with stoichiometry during the crystal formation with a complex structure. Therefore, native defects are introduced in the crystal at the synthesis. Native defects together with activators and co-activators can form complex defects which are called nanodefects [7] Doped activators (Ce^{3+}) and co-activators (Gd^{3+}) result in changes in spatial and energy structure of nanodefects. As a result, the luminescence spectrum and kinetic characteristics change accordingly.

Acknowledgments

This work was supported by the *Key Laboratory of Supramolecular Structure, Chemical Institute of Jilin University*. The authors are greatly indebted to Prof. Yue Wangat Jilin University. Thanks to doctoral student Yue Wei Zhang for his assistance in experiment work.

Reference

[1] Yang H S and Kim Y S 2008 *J. Lumin.* **128** pp 1570-1576
[2] Kyeong Y J and Hyun W L 2007 *J. Lumin.* **127** pp 469-474
[3] Yang H S, Lee D K and Kim Y S 2009 *J. Materials Chemistry and Physics* **114** pp *665-669*
[4] Anupam S, Mahendra A, Manoj T, and Kapil D 2016 *AIP Conf. Proc.* **1728** 020263
[5] Mukherjee S, Sudarsan V, Vatsa R K and Tyagi A K 2009 *J. Lumin.* **129** pp 69-72
[6] Lisitsyn V, Lukash V, Stepanov S and Y Ju 2016 *AIP Conference Proceedings* **1698** 060008
[7] Lisitsyn V, Stepanov S, Valiev D, Vishnyakova E, Abdullin H, Marhabaeva A and A Tulegenova 2016 *IOP Conf.Series: Materials Science and Engineering* **110** 012051